Multiple Choice Quiz
(See related pages)

Results Reporter

Out of 25 questions, you answered 3 correctly, for a final grade of 12%.

3 correct (12%)
4 incorrect (16%)
18 unanswered (72%)

Your Results:

The correct answer for each question is indicated by a ☺.

1 CORRECT The nucleotide change _____ is an example of a transversion, while the nucleotide change _____ is an example of a transition.

- A) A→G; C→G
- ☺ ●B) C→G; A→G
- C) T→C; A→G
- D) C→T; G→A
- E) G→C; C→G

2 CORRECT Which type of mutation is most likely to revert?

- A) deletion
- B) translocation
- C) inversion
- ☺ ●D) transposition
- E) transition

3 CORRECT Which type of mutation is least likely to revert?

- ☺ ●A) deletion
- B) translocation
- C) inversion
- D) transposition
- E) transition

4 INCORRECT The hydrolysis of an -NH2 group from a base is called _____ causing _____:

- A) deamination; transversions

Routing Information

Date: Tue Sep 09 09:13:24 CDT 2003
My name: [   ]

Email these results to:

	Email address:	Format:
Me:	[ ]	Text ▾
My Instructor:	[ ]	Text ▾
My TA:	[ ]	Text ▾
Other:	[ ]	Text ▾

[E-Mail The Results]

Test Yourself

Take a quiz at the Genetics Online Learning Center to gauge your mastery of chapter content. Each chapter quiz is specifically constructed to test your comprehension of key concepts. Immediate feedback on your responses explains why an answer is correct or incorrect. You can even e-mail your quiz results to your professor!

Tweaking the Experiment

For the in-depth experiments that are found in each chapter, a self-help quiz is found. These questions ask you to predict how changes in the experimental parameters may affect the outcome of the experiment.

Interactive Activities

Fun and exciting learning experiences await you at the Genetics Online Learning Center! Each chapter offers a series of interactive crossword puzzles, vocabulary flashcards, and other engaging activities designed to reinforce learning.

Crossword Puzzle
(See related pages)

Across

6. Mendel's second law: the principle that segregation of alternative alleles at one locus into gametes is independent of the segregation of alleles at other loci. Only true for gene loci located on different chromosomes or those so far apart on one chromosome that crossing over is very frequent between the loci.

7. An essential element of meiosis occurring during prophase when nonsister chromatids exchange portions of DNA strands.

Down

2. The spread of cancerous cells to other parts of the body, forming new tumors at distant sites.

3. Unrestrained invasive cell growth. A tumor or cell mass resulting from uncontrollable cell division.

4. The vehicle by which hereditary information is physically transmitted from one generation to the next. In a eukaryotic cell, long threads of DNA that are associated with protein and that contain hereditary information.

5. The M phase of cell division in which the microtubular apparatus is

Crossword Compiler Software © 2001 x-word.com

TERM

Gene

DEFINITION
NEXT
REMOVE CARD

CLICK ON THE "TERM" OR "DEF...
PRESS "NEXT" TO MOVE ON TO...
PRESS "REMOVE CARD" TO RE...
PRESS "RANDOMIZE DECK" TO...

DEFINITION

basic unit of biological information; specific segments of DNA composed of distinctive sets of nucleotide pairs in a discrete region of a chromosome that encodes a particular protein; a coding locus.

TERM
NEXT
REMOVE CARD
RANDOMIZE DECK
CARDS REMAINING
34

CLICK ON THE "TERM" OR "DEFINITION" TO FLIP THE CARD OVER.
PRESS "NEXT" TO MOVE ON TO THE NEXT TERM.
PRESS "REMOVE CARD" TO REMOVE THE CURRENT CARD FROM THE DECK.
PRESS "RANDOMIZE DECK" TO PUT ALL THE CARDS BACK IN THE DECK AND RESTART.

:: ROBERT J. BROOKER

UNIVERSITY OF MINNESOTA–MINNEAPOLIS

third edition

GENETICS

ANALYSIS & PRINCIPLES

**McGraw-Hill
Higher Education**

Boston   Burr Ridge, IL   Dubuque, IA   New York   San Francisco   St. Louis
Bangkok   Bogotá   Caracas   Kuala Lumpur   Lisbon   London   Madrid   Mexico City
Milan   Montreal   New Delhi   Santiago   Seoul   Singapore   Sydney   Taipei   Toronto

McGraw-Hill
Irwin

GENETICS: ANALYSIS & PRINCIPLES, THIRD EDITION

Published by McGraw-Hill, a business unit of The McGraw-Hill Companies, Inc., 1221 Avenue of the Americas, New York, NY 10020. Copyright © 2009 by The McGraw-Hill Companies, Inc. All rights reserved. Previous edition © 2005. No part of this publication may be reproduced or distributed in any form or by any means, or stored in a database or retrieval system, without the prior written consent of The McGraw-Hill Companies, Inc., including, but not limited to, in any network or other electronic storage or transmission, or broadcast for distance learning.

Some ancillaries, including electronic and print components, may not be available to customers outside the United States.

 This book is printed on recycled, acid-free paper containing 10% postconsumer waste.

3 4 5 6 7 8 9 0 DOW/DOW 0 9

ISBN 978–0–07–299278–6
MHID 0–07–299278–6

Publisher: *Janice Roerig-Blong*
Executive Editor: *Patrick E. Reidy*
Senior Developmental Editor: *Lisa A. Bruflodt*
Marketing Manager: *Barbara Owca*
Senior Project Manager: *Jayne Klein*
Senior Production Supervisor: *Laura Fuller*
Senior Media Project Manager: *Jodi Banowetz*
Designer: *John Joran*
Interior Designer: *Rokusek Design, Inc.*
Cover Illustration: *Imagineering Media Services Inc.*
Senior Photo Research Coordinator: *John C. Leland*
Photo Research: *Pronk&Associates*
Supplement Producer: *Mary Jane Lampe*
Compositor: *Lachina Publishing Services*
Typeface: *10/12 Minion*
Printer: *R. R. Donnelley Williard, OH*

The credits section for this book begins on page 823 and is considered an extension of the copyright page.

Library of Congress Cataloging-in-Publication Data

Brooker, Robert J.
  Genetics : analysis & principles / Robert J. Brooker. — 3rd ed.
    p. cm.
  Includes bibliographical references and index.
  ISBN 978–0–07–299278–6 — ISBN 0–07–299278–6 (hard copy : alk. paper) 1. Genetics. I. Title.
QH430.B766 2009
576.5--dc22
                    2007038523

www.mhhe.com

ABOUT THE AUTHOR

Robert J. Brooker is a Professor in the Department of Genetics, Cell Biology, and Development at the University of Minnesota–Minneapolis. He received his B.A. in Biology from Wittenberg University in 1978 and his Ph.D. in Genetics from Yale University in 1983. At Harvard, he conducted post-doctoral studies on the lactose permease, which is the product of the *lacY* gene of the *lac* operon. He continues his work on transporters at the University of Minnesota. Dr. Brooker's laboratory primarily investigates the structure, function, and regulation of iron transporters found in bacteria and *C. elegans*. At the University of Minnesota, he teaches undergraduate courses in biology, genetics, and cell biology.

DEDICATION

To my wife, Deborah, and our children, Daniel, Nathan, and Sarah

BRIEF CONTENTS

::

TABLE OF CONTENTS

::

PREFACE

::

In the third edition of *Genetics: Analysis & Principles*, the content has been updated to reflect current trends in the field. In addition, the presentation of the content has been improved in a way that fosters active learning. As an author, researcher, and teacher, I want a textbook that gets students actively involved in learning genetics. To achieve this goal, I have worked with a talented team of editors, illustrators, and media specialists who have helped me to make the third edition of *Genetics: Analysis & Principles* a fun learning tool. The features that we feel are most appealing to students, and which have been added to or improved upon in the third edition, are the following.

- **Interactive exercises** Education specialists have crafted interactive exercises in which the students can make their own choices in problem-solving activities, and predict what the outcomes will be. These exercises often focus on inheritance patterns and human genetic diseases. (For example, see Chapters 4 and 22.)
- **Animations** Our media specialists have created over 40 animations for a variety of genetic processes. These animations were made specifically for this textbook, and use the art from the textbook. The animations literally make many of the figures in the textbook "come to life."
- **Experiments** As in the previous editions, each chapter (beginning with Chapter 2) incorporates one or two experiments that are presented according to the scientific method. These experiments are not "boxed off" from the rest of the chapter. Rather, they are integrated within the chapters and flow with the rest of the text. As you are reading the experiments, you will simultaneously explore the scientific method and the genetic principles that have been discovered using this approach. For students, I hope this textbook will help you to see the fundamental connection between scientific analysis and principles. For both students and instruc-

tors, I expect that this strategy will make genetics much more fun to explore.
- **Art** The art has been further refined for clarity and completeness. This makes it easier and more fun for students to study the illustrations without having to go back and forth between the art and the text.
- **Engaging text** A strong effort has been made in the third edition to pepper the text with questions. Sometimes these are questions that scientists considered when they were conducting their research. Sometimes they are questions that the students might ask themselves when they are learning about genetics.

Overall, an effective textbook needs to accomplish three goals. First, it needs to provide comprehensive, accurate, and up-to-date content in its field. Second, it needs to provide students with an exposure to the techniques and skills that are needed for them to become successful in that field. And finally, it should inspire students so they want to pursue that field as a career. The hard work that has gone into the third edition of *Genetics: Analysis & Principles* has been aimed at achieving all three of these goals.

HOW WE EVALUATED YOUR NEEDS

ORGANIZATION

In surveying many genetics instructors, it became apparent that most people fall into two camps: **Mendel first** versus **Molecular first.** I have taught genetics both ways. As a teaching tool, this textbook has been written with these different teaching strategies in mind. The organization and content lend themselves to various teaching formats.

Chapters 2 through 8 are largely inheritance chapters, while Chapters 24 through 26 examine population and quantitative genetics. The bulk of the molecular genetics is found in Chapters 9 through 23, although I have tried to weave a fair amount of molecular genetics into Chapters 2 through 8 as well. The information in

Chapters 9 through 23 *does not assume* that a student has already covered Chapters 2 through 8. Actually, each chapter is written with the perspective that instructors may want to vary the order of their chapters to fit their students' needs.

For those who like to discuss inheritance patterns first, a common strategy would be to cover Chapters 1 through 8 first, and then possibly 24 through 26. (However, many instructors like to cover quantitative and population genetics at the end. Either way works fine.) The more molecular and technical aspects of genetics would then be covered in Chapters 9 through 23. Alternatively, if you like the "Molecular first" approach, you would probably cover Chapter 1, then skip to Chapters 9 through 23, then return to Chapters 2 through 8, and then cover Chapters 24 through 26 at the end of the course. This textbook was written in such a way that either strategy works just fine.

ACCURACY

Both the publisher and I acknowledge the fact that inaccuracies can be a source of frustration for both the instructor and students. Therefore, throughout the writing and production of this textbook we have worked very hard to catch and correct errors during each phase of development and production.

Each chapter has been reviewed by a minimum of ten people. At least eight of these people were faculty members who teach the course and/or conduct research in genetics. In addition, two development editors have gone through the material, at least twice, to check for accuracy in art, and consistency between the text and art. We also had a team of students work through all the problem sets and one development editor also checked them. The author personally checked every question and answer when the chapters were completed.

PEDAGOGY

Based on our discussions with instructors from many institutions, some common goals have emerged. Instructors want a broad textbook that clearly explains concepts in a way that is interesting, accurate, concise, and up-to-date. Likewise, most instructors want students to understand the experimentation that revealed these genetic concepts. In this textbook, concepts and experimentation are woven together to provide a story that enables students to learn the important genetic concepts that they will need in their future careers, and also to be able to explain the types of experiments that allowed researchers to derive such concepts. The end-of-chapter problem sets are categorized according to their main focus, either conceptual or experimental, although some problems contain a little of both. The problems are meant to strengthen students' abilities in a wide variety of ways.

- By bolstering their understanding of genetic principles.
- By enabling students to apply genetic concepts to new situations.
- By analyzing scientific data.
- By organizing their thoughts regarding a genetic topic.
- By improving their writing skills.

Finally, since genetics is such a broad discipline ranging from the molecular to the populational levels, many instructors have told us that it is a challenge for students to see both "the forest and the trees." It is commonly mentioned that students often have trouble connecting the concepts they have learned in molecular genetics with the traits that occur at the level of a whole organism (i.e., What does transcription have to do with blue eyes?). To try to make this connection more meaningful, certain figure legends in each chapter, designated **Genes→ Traits,** remind students that molecular and cellular phenomena ultimately lead to the traits that are observed in each species (for example, see Figure 4.20).

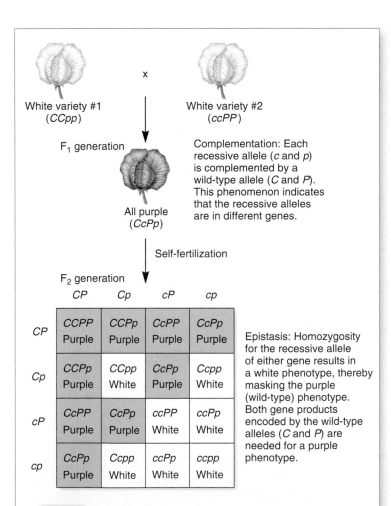

White variety #1
(*CCpp*)

X

White variety #2
(*ccPP*)

F_1 generation

Complementation: Each recessive allele (*c* and *p*) is complemented by a wild-type allele (*C* and *P*). This phenomenon indicates that the recessive alleles are in different genes.

All purple
(*CcPp*)

Self-fertilization

F_2 generation

	CP	Cp	cP	cp
CP	*CCPP* Purple	*CCPp* Purple	*CcPP* Purple	*CcPp* Purple
Cp	*CCPp* Purple	*CCpp* White	*CcPp* Purple	*Ccpp* White
cP	*CcPP* Purple	*CcPp* Purple	*ccPP* White	*ccPp* White
cp	*CcPp* Purple	*Ccpp* White	*ccPp* White	*ccpp* White

Epistasis: Homozygosity for the recessive allele of either gene results in a white phenotype, thereby masking the purple (wild-type) phenotype. Both gene products encoded by the wild-type alleles (*C* and *P*) are needed for a purple phenotype.

INTERACTIVE EXERCISE

FIGURE 4.20   A cross between two different white varieties of the sweet pea.

Genes →Traits The color of the sweet pea flower is controlled by two genes, which are epistatic to each other and show complementation. Each gene is necessary for the production of an enzyme required for pigment synthesis. The recessive allele of either gene encodes a defective enzyme. If an individual is homozygous recessive for either of the two genes, the purple pigment cannot be synthesized. This results in a white phenotype.

ILLUSTRATIONS

In surveying students whom I teach, I often hear it said that most of their learning comes from studying the figures. Likewise, instructors frequently use the illustrations from a textbook as a central teaching tool. For these reasons, the greatest amount of effort in improving the third edition has gone into the illustrations. The illustrations have been refined with four goals in mind:

1. **Completeness**   For most figures, it should be possible to understand an experiment or genetic concept by looking at the illustration alone. Students have complained that it is difficult to understand the content of an illustration if they have to keep switching back and forth between the figure and text. In cases where an illustration shows the steps in a scientific process, the steps are described in brief statements that allow the students to understand the whole process (for example, see Figure 12.10). Likewise, such illustrations should make it easier for instructors to explain these processes in the classroom.

2. **Clarity**   The figures have been extensively reviewed by students and instructors. This has helped us to avoid drawing things that may be confusing or unclear. I hope that no one looks at an element in any figure and wonders, "What is that thing?" Aside from being unmistakably drawn, all new elements within each figure are clearly labeled.

3. **Consistency**   Before we began to draw the figures for the second edition, we generated a style sheet that contained recurring elements that are found in many places in the textbook. Examples include the DNA double helix, DNA polymerase, and fruit flies. We agreed upon the best way(s) to draw these elements and also what colors they should be. Therefore, as students and instructors progress through this textbook, they become accustomed to the way things should look.

4. **Realism**   An important goal of this and previous editions is to make each figure as realistic as possible. When drawing macroscopic elements (for example, fruit flies, pea plants), the illustrations are based on real images, not on cartoonlike simplifications. Our most challenging goal, and one that we feel has been achieved most successfully, is the realism of our molecular drawings. Whenever possible, we have tried to depict molecular elements according to their actual structures, if such structures are known. For example, the ways we have drawn RNA polymerase, DNA polymerase, DNA helicase, and ribosomes are based on their crystal structures. When a student sees a figure in this textbook that illustrates an event in transcription, RNA polymerase is depicted in a way that is as realistic as possible (for example, see Figure 12.10).

WRITING STYLE

Motivation in learning often stems from enjoyment. If someone enjoys what they're reading, they are more likely to spend longer amounts of time with it and focus their attention more crisply.

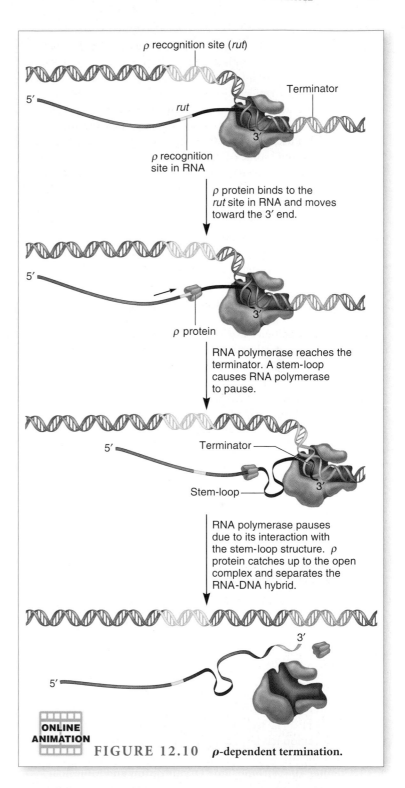

FIGURE 12.10   *ρ*-dependent termination.

The writing style of this book is meant to be interesting, down to earth, and easy to follow. Each section of every chapter begins with an overview of the contents of that section, usually with a table or figure that summarizes the broad points. The section then examines how those broad points were discovered experimentally, as well as explaining many of the finer scientific details. Important terms are introduced in a boldface font. These terms are also found in the glossary.

There are various ways to make a genetics book interesting and inspiring. The subject matter itself is pretty amazing, so it's not difficult to build on that. In addition to describing the concepts and experiments in ways that motivate students, it is important to draw upon examples that bring the concepts to life. In a genetics book, many of these examples come from the medical realm. This textbook contains lots of examples of human diseases that exemplify some of the underlying principles of genetics. Students often say that they remember certain genetic concepts because they remember how defects in certain genes can cause disease. For example, defects in DNA repair genes cause a higher predisposition to develop cancer. In addition, I have tried to be evenhanded in providing examples from the microbial and plant world. Finally, students are often interested in applications of genetics that impact their everyday lives. Because we hear about genetics in the news on a daily basis, it's inspiring for students to learn the underlying basis for such technologies. Chapters 18 to 21 are devoted to genetic technologies, and applications of these and other technologies are found throughout this textbook. By the end of their genetics course, students should come away with a greater appreciation for the impact of genetics in their lives.

EXAMPLES OF SIGNIFICANT CONTENT CHANGES IN THE THIRD EDITION

- **Interactive Exercises**   Many interactive exercises are available for students to understand inheritance patterns and human genetic diseases. These are indicated with an icon next to the relevant figures. They are largely found in Chapters 2, 4, 5, 7, and 22.
- **Animations**   Nearly every chapter has one or more custom animations to help students understand genetic processes. These are also indicated by an icon next to the relevant figure.
- **Chapter 3 (Reproduction and Chromosome Transmission)**   The illustrations and micrographs of mitosis and meiosis have been improved.
- **Chapter 4 (Extensions of Mendelian Inheritance)**   The topic of suppressor mutations has been expanded.
- **Chapter 8 (Variation in Chromosome Structure and Number)**   The technique of comparative genomic hybridization has been added, and included as a Feature Experiment.
- **Chapter 10 (Chromosome Organization and Molecular Structure)**   An illustration and expanded discussion has been added for SMC proteins, which are involved with chromosome condensation and sister chromatid cohesion. Also, the histone code hypothesis is discussed and illustrated.
- **Chapter 11 (DNA Replication)**   Illustrations in this chapter now depict the step-by-step formation of Okazaki fragments (for example, see Figure 11.10).

- **Chapter 13 (Translation of mRNA)**   This chapter has undergone some reorganization to bring the basic concepts of translation to the first section.
- **Chapter 15 (Gene Regulation in Eukaryotes)**   The topic of RNA interference has been expanded, including a Feature Experiment of the work of Fire and Mellow.
- **Chapter 16 (Gene Mutation and DNA Repair)**   To complement some new material in Chapter 4, several examples of suppressor mutations are described. In addition, homologous recombination repair and nonhomologous end joining repair are illustrated and discussed within the context of repairing double-stranded DNA breaks.
- **Chapter 19 (Biotechnology)**   Some new topics in biotechnology have been added including GloFish® and Bt corn.
- **Chapter 20 (Genomics I: Analysis of DNA)**   The approach of shotgun DNA sequencing is greatly expanded along with complementary illustrations.
- **Chapter 21 (Genomics II: Functional Genomics, Proteomics, and Bioinformatics)**   The method of chromatin immunoprecipitation (ChIP) has been added.
- **Chapter 22 (Medical Genetics and Cancer)**   Several new topics have been added including the use of molecular markers in the mapping of human genetic diseases, and the molecular profiling of cancer cells using DNA microarrays.
- **Chapter 24 (Population Genetics)**   This chapter contains an expanded discussion of natural selection with graphical representations of the various types. An entire new section on Sources of Genetic Variation has been added so students can appreciate how/why genetic variation is so prevalent.
- **Chapter 26 (Evolutionary Genetics)**   The cladistic approach has been greatly expanded. In addition, an entire section on Evolutionary Developmental Biology (Evo-Devo) has been added.

SUGGESTIONS WELCOME!

It seems very appropriate to use the word evolution to describe the continued development of this textbook. I welcome any and all comments. The refinement of any science textbook requires input from instructors and their students. These include comments regarding writing, illustrations, supplements, factual content, and topics that may need greater or less emphasis. You are invited to contact me at:

Dr. Rob Brooker
Dept. of Genetics, Cell Biology, and Development
University of Minnesota
6-160 Jackson Hall
321 Church St.
Minneapolis, MN 55455
brook005@umn.edu

TEACHING AND LEARNING SUPPLEMENTS

McGraw-Hill offers various tools and technology products to support the third edition of *Genetics: Analysis & Principles.*

FOR THE INSTRUCTOR:

Online Learning Center (www.mhhe.com/brookergenetics3e)

The text specific website offers an extensive array of learning and teaching tools. In addition to student resources, the site also includes an instructor's manual, test bank, and lecture outlines.

Presentation Center
Complete set of electronic book images and assets for instructors

Build instructional materials wherever, whenever, and however you want! Accessed from your textbook's Online Learning Center, Presentation Center is an online digital library containing photos, artwork, animations, and other media types that can be used to create customized lectures, visually enhanced tests and quizzes, compelling course websites, or attractive printed support materials. All assets are copyrighted by McGraw-Hill Higher Education, but can be used by instructors for classroom purposes. The visual resources in this collection include:

- **Art** Full-color digital files of all illustrations in the book can be readily incorporated into lecture presentations, exams, or custom-made classroom materials. In addition, all files are pre-inserted into PowerPoint slides for ease of lecture preparation.
- **Photos** The photo collection contains digital files of photographs from the text, which can be reproduced for multiple classroom uses.
- **Tables** Every table that appears in the text has been saved in electronic form for use in classroom presentations and/or quizzes.
- **Animations** Numerous full-color animations illustrating important processes are also provided. Harness the visual impact of concepts in motion by importing these files into classroom presentations or online course materials.

Also residing on your textbook's Online Learning Center are:

- **PowerPoint Lecture Outlines** Ready-made presentations that combine art and lecture notes are provided for each chapter of the text.
- **PowerPoint Slides** For instructors who prefer to create their lectures from scratch, all illustrations, photos, and tables are pre-inserted by chapter into blank PowerPoint slides.

FOR THE STUDENT:

Student Study Guide/Solutions Manual

The solutions to the end-of-chapter problems and questions will aid the students in developing their problem-solving skills by providing the steps for each solution. The Study Guide follows the order of sections and subsections in the textbook and summarizes the main points in the text, figures, and tables. It also contains concept-building exercises, self-help quizzes, and practice exams.

Online Learning Center (www.mhhe.com/brookergenetics3e)

The text specific website offers an extensive array of learning tools, including a variety of quizzes for each chapter, interactive genetics problems, animations, and more.

ACKNOWLEDGMENTS

The production of a textbook is truly a collaborative effort, and I am greatly indebted to a variety of people. The first, second, and third editions went through multiple rounds of rigorous revisions that involved the input of faculty, students, editors, educational specialists, and media specialists. Their collective contributions are reflected in the final outcome.

Let me begin by acknowledging the many people at McGraw-Hill whose efforts are amazing. My highest praise goes to Lisa Bruflodt (Senior Developmental Editor), who managed and scheduled nearly every aspect of this project. I simply don't understand how one person can be so well organized and manage to stay calm at the same time. I also would like to thank Patrick Reidy (Executive Editor) for his patience in overseeing this project. He has the unenviable job of managing the budget for the book and that is not an easy task. I'm sure that he had some lively discussions with Janice Roerig-Blong (Publisher) during the course of this project. Other people at McGraw-Hill have played key roles in producing an actual book and the supplements that go along with it. In particular, Jayne Klein (Project Manager) has done a superb job of managing the components that need to be assembled to produce a book, along with Laura Fuller (Production Supervisor). I would also like to thank John Leland (Photo Research Coordinator), who acted as an interface between me and the photo company. In addition, my gratitude goes to John Joran (Designer), who provided much input into the internal design of the book as well as creating an awesome cover. Finally, I would like to thank Barb Owca (Marketing Manager), whose major efforts begin when the third edition comes out!

With regard to the content of the book, two editors worked closely with me in developing a book that is clear, consistent, and easy for students to follow. Deborah Brooker (Art/Text Coordinating Editor) analyzed all of the chapters in the textbook and made improvements with regard to art and text coordination. With great care, she examined the content in the text and determined if it matched what is found in the illustrations; she also checked every illustration for clarity and completeness. The high quality and pedagogy of the art are largely due to her efforts. She also scrutinized the text for clarity and logic. In addition, Joni Fraser (Freelance Development Editor) analyzed each chapter for its overall content and made many valuable suggestions for improvements. She did a great job of sharpening the writing. I am particularly grateful for her meticulous analysis of the problem sets. In her spare time, I think she may have worked all of the problems. In addition to Deb and Joni, I would also like to thank Jane DeShaw (Freelance Copy Editor) for making grammatical improvements throughout the text and art.

I would also like to extend a special thanks to everyone at Imagineering, the art house that produced the illustrations in the second and third editions. Though we initially considered several art houses, when we saw how Imagineering could draw RNA polymerase, it was clear who the winner was. Their ability to draw three-dimensional molecular structures is outstanding. I am grateful to the many artists and contact people at Imagineering who have played important roles in developing the art for the second and third editions. Likewise, the people at Pronk & Associates have done a great job of locating many of the photographs that have been used in the third edition. Also, Lachina Publishing Services worked with great care in the paging of the book, making sure that the figures and relevant text are as close to each other as possible.

Finally, I want to thank the many scientists who reviewed the chapters of this textbook. Their broad insights and constructive suggestions were an important factor that shaped its final content and organization. I am truly grateful for their time and effort.

REVIEWERS

Shivanthi Anandan, *Drexel University*
Alan G. Atherly, *Iowa State University*
William Baird, *Clemson University*
Gail S. Begley, *Northeastern University*
Michael Benedik, *University of Houston*
Kelly Bidle, *Rider University*
Andrew J. Bohonak, *San Diego State University*
Mark A. Brick, *Colorado State University*
James J. Campanella, *Montclair State University*
Arden Campbell, *Iowa State University*
Gerard Campbell, *University of Pittsburgh*
Diane A. Caporale, *University of Wisconsin–Stevens Point*
J. Aaron Cassill, *University of Texas at San Antonio*
Richard W. Cheney, Jr., *Christopher Newport University*
Joseph P. Chinnici, *Virginia Commonwealth University*
Brian Condie, *University of Georgia*
Gregory Copenhaver, *University of North Carolina*
Stephen J. D'Surney, *University of Mississippi*
Sumana Datta, *Texas A&M University*
Jeff DeJong, *University of Texas at Dallas*
Robert S. Dotson, *Tulane University*
Johnny El-Rady, *University of South Florida*
Susan Elrod, *California Polytechnic State University*
Bert Ely, *University of South Carolina*
Les Erickson, *Salisbury University*
Michael L. Foster, *Eastern Kentucky University*
Robert G. Fowler, *San Jose State University*
Richard F. Gaber, *Northwestern University*
Sudhindra R. Gadagkar, *University of Dayton*
Matthew R. Gilg, *University of North Florida*
Jack R. Girton, *Iowa State University*
Michael A. Goldman, *San Francisco State University*
Elliott S. Goldstein, *Arizona State University*
Nels H. Granholm, *South Dakota State University*
Alison Hill, *Duke University*
Ralph Hillman, *Temple University*
David C. Hinkle, *University of Rochester*
J. Spencer Johnston, *Texas A&M University*
Gregg Jongeward, *University of the Pacific*
Christopher Kvaal, *St. Cloud State University*

Arlene T. Larson, *University of Colorado at Denver*
Elena Levine Keeling, *California Polytechnic State University*
Rebecca Lyczak, *Ursinus College*
Jocelyn Malamy, *University of Chicago*
Philip M. Mathis, *Middle Tennessee State University*
Terry R. McGuire, *Rutgers University*
Mark E. Meade, *Jacksonville State University*
Roderick M. Morgan, *Grand Valley State University*
Robert Moss, *Wofford College*
Jacqueline Peltier Horn, *Houston Baptist University*
David K. Peyton, *Morehead State University*
Chara J. Ragland, *Texas A&M University*
David H. Reed, *University of Mississippi*
Linda L. Restifo, *University of Arizona*
Inder Saxena, *University of Texas at Austin*
Deemah N. Schirf, *University of Texas at San Antonio*
Mark Schlueter, *Xavier University of Louisiana*
Ekaterina G. Sedia, *The Richard Stockton College of New Jersey*
Jeff Sekelsky, *University of North Carolina at Chapel Hill*
Richard N. Sherwin, *University of Pittsburgh*
Joan Smith Sonneborn, *University of Wyoming*
Adriana Stoica, *Georgetown University*
Martin Tracey, *Florida International University*
Carol Trent, *Western Washington University*
Sarah Ward, *Colorado State University*
Dan E. Wells, *University of Houston*
Douglas Wendell, *Oakland University*
Frederick Whipple, *California State University–Fullerton*
Mark Winey, *University of Colorado–Boulder*
David Wofford, *University of Florida*
Daniel Wang, *University of Miami*

ACCURACY CHECKERS

Ruth Ballard, *CSU–Sacramento*
Michel Baudry, *USC*
Michael Benedik, *University of Houston*
Carey Booth, *Portland State University*
Matt Bozovsky, *Loyola University*

Joel Chandlee, *University of Rhode Island*
Henry Chang, *Purdue University*
Karen Cichy, *Michigan State University*
Brian Condie, *University of Georgia*
Erin Cram, *Northeastern University*
Jeff DeJong, *University of Texas at Dallas*
Rodney Dyer, *Virginia Commonwealth University*
Kevin Edwards, *Illinois State University*
Aboubaker Elkharroubi, *Johns Hopkins University*
Cedric Feschotte, *UT–Arlington*
David Foltz, *Louisiana State University*
Michael Foster, *Eastern Kentucky University*
Thomas Fowler, *Southern Illinois University*
Anne Galbraith, *UW–Lacrosse*
Jack Girton, *Iowa State University*
Elliott Goldstein, *ASU*
Christine Gray, *University of Puget Sound*
Tara Harmer Luke, *The Richard Stockton College of NJ*
Marilyn Hart, *Mankato State University*
Daryl Henderson, *Stony Brooker University*
Adam Hrincevich, *LSU*
Michael Ibba, *Ohio State University*
Margaret Jefferson, *CSU–LA*
David Kass, *Eastern Michigan University*

Cindy Malone, *CSU–Northridge*
John Mishler, *Delaware Valley College*
James Morris, *Clemson University*
Sang-Chul Nam, *Baylor University*
Daniel Odom, *California State University–Northridge*
Michael Polymenis, *Texas A&M University*
David Reed, *University of Mississippi*
Laurie Russell, *St. Louis University*
Mark Seeger, *Ohio State University*
Theresa Spradling, *University of Northern Iowa*
JD Swanson, *University of Central Arkansas*
James Thompson, *University of Oklahoma*
John Tomkiel, *U of NC–Greensboro*
Carol Trent, *Western Washington University*
Fyodor Urnov, *University of California–Berkeley*
Harald Vaessin, *Ohio State University*
Alan Waldman, *University of South Carolina*
Danicl Wang, *University of Miami*
Sarah Ward, *Colorado State University*
Matt White, *Ohio University*
Dwayne Wise, *Mississippi State University*
Yang Yen, *South Dakota State University*
Malcolm Zellars, *Georgia State University*
Jianzhi Zhang, *University of Michigan*

A Visual Guide to
GENETICS: ANALYSIS & PRINCIPLES

::

ROBERT J. BROOKER
UNIVERSITY OF MINNESOTA–MINNEAPOLIS

Brooker's Genetics brings key concepts to life with its unique style of illustration.

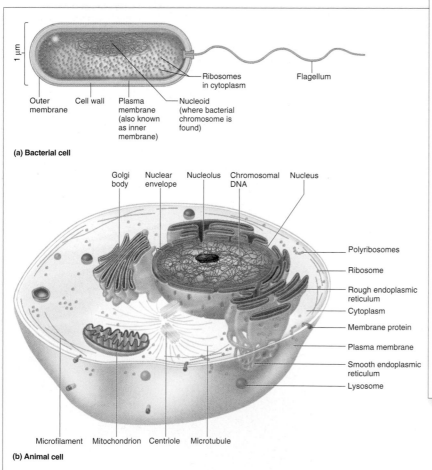

(a) Bacterial cell

1 μm

Outer membrane · Cell wall · Plasma membrane (also known as inner membrane) · Nucleoid (where bacterial chromosome is found) · Ribosomes in cytoplasm · Flagellum

(b) Animal cell

Golgi body · Nuclear envelope · Nucleolus · Chromosomal DNA · Nucleus

Polyribosomes · Ribosome · Rough endoplasmic reticulum · Cytoplasm · Membrane protein · Plasma membrane · Smooth endoplasmic reticulum · Lysosome

Microfilament · Mitochondrion · Centriole · Microtubule

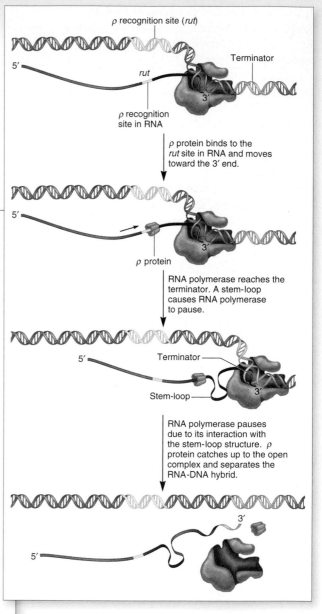

ρ recognition site (*rut*)

5′ · *rut* · Terminator · 3′

ρ recognition site in RNA

ρ protein binds to the *rut* site in RNA and moves toward the 3′ end.

5′ · ρ protein · 3′

RNA polymerase reaches the terminator. A stem-loop causes RNA polymerase to pause.

5′ · Terminator · Stem-loop · 3′

RNA polymerase pauses due to its interaction with the stem-loop structure. ρ protein catches up to the open complex and separates the RNA-DNA hybrid.

5′ · 3′

The digitally rendered images have a vivid three-dimensional look that will stimulate a student's interest and enthusiasm.

Each figure is carefully designed to follow closely with the text material.

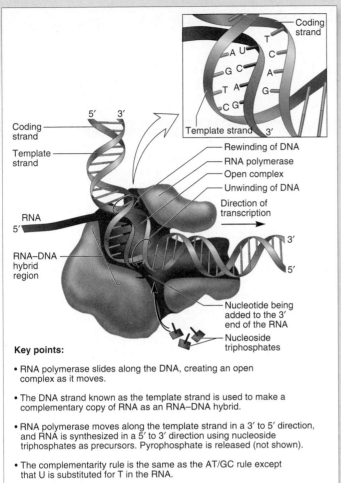

Coding strand

A–U
G–C
T–A
C–G
T C
C A
A G
G

Coding strand

Template strand

Coding strand

5′ 3′

Template strand

RNA 5′

RNA–DNA hybrid region

Rewinding of DNA
RNA polymerase
Open complex
Unwinding of DNA
Direction of transcription

3′
5′

Nucleotide being added to the 3′ end of the RNA

Nucleoside triphosphates

Key points:

• RNA polymerase slides along the DNA, creating an open complex as it moves.

• The DNA strand known as the template strand is used to make a complementary copy of RNA as an RNA–DNA hybrid.

• RNA polymerase moves along the template strand in a 3′ to 5′ direction, and RNA is synthesized in a 5′ to 3′ direction using nucleoside triphosphates as precursors. Pyrophosphate is released (not shown).

• The complementarity rule is the same as the AT/GC rule except that U is substituted for T in the RNA.

DNA strands separate at origin, creating 2 replication forks.

Origin of replication

Replication forks

Primers are needed to initiate DNA synthesis. The synthesis of the leading strand begins in the direction of the replication fork. The first Okazaki fragment of the lagging strand is made in the opposite direction.

Leading strand

Primer

3′
5′

3′

Direction of replication fork

5′
3′

Primer

First Okazaki fragment of the lagging strand

5′

The leading strand elongates, and a second Okazaki fragment is made.

5′
3′

3′
5′

Second Okazaki fragment

First Okazaki fragment

3′
5′

3′
5′

The leading strand continues to elongate. A third Okazaki fragment is made, and the first and second are connected together.

3′
5′

3′
5′

Third Okazaki fragment

First and second Okazaki fragments have been connected to each other.

3′
5′

Phage DNA — Bacteriophage

Bacterial chromosome

New phages can bind to bacterial cells.

Phage injects its DNA into cytoplasm.

On rare o... a prophage may be excised from host chromosome.

Lytic cycle

Lysogenic cycle

Cell lyses and releases the new phages.

Phage DNA directs the synthesis of many new phages. Host DNA is degraded.

Phage DNA integrates into host chromosome.

Prophage DNA is copied when cell divides.

Prophage

Every illustration was drawn with four goals in mind: completeness, clarity, consistency, and realism.

Each chapter (beginning with Chapter 2) incorporates one or two experiments that are presented according to the scientific method. These experiments are integrated within the chapters and flow with the rest of the textbook. As you read the experiments, you will simultaneously explore the scientific method and the genetic principles learned from this approach.

EXPERIMENT 5A

Creighton and McClintock Showed That Crossing Over Produced New Combinations of Alleles and Resulted in the Exchange of Segments Between Homologous Chromosomes

STEP 1: BACKGROUND OBSERVATIONS

Each experiment begins with a description of the information that led researchers to study an experimental problem. Detailed information about the researchers and the experimental challenges they faced help students to understand actual research.

As we have seen, Morgan's studies were consistent with the hypothesis that crossing over occurs between homologous chromosomes to produce new combinations of alleles. To obtain direct evidence that crossing over can result in genetic recombination, Harriet Creighton and Barbara McClintock used an interesting strategy involving parallel observations. In studies conducted in 1931, they first made crosses involving two linked genes to produce parental and recombinant offspring. Second, they used a microscope to view the structures of the chromosomes in the parents and in the offspring. Because the parental chromosomes had some unusual structural features, they could microscopically distinguish the two homologous chromosomes within a pair. As we will see, this enabled them to correlate the occurrence of recombinant offspring with microscopically observable exchanges in segments of homologous chromosomes.

Creighton and McClintock focused much of their attention on the pattern of inheritance of traits in corn. This species has 10 different chromosomes per set, which are named chromosome 1, chromosome 2, chromosome 3, and so on. In previous cytological examinations of corn chromosomes, some strains were found to have an unusual chromosome 9 with a darkly staining knob at one end. In addition, McClintock identified an abnormal version of chromosome 9 that also had an extra piece of chromosome 8 attached at the other end (**Figure 5.5a**). This chromosomal rearrangement is called a translocation.

Creighton and McClintock insightfully realized that this abnormal chromosome could be used to demonstrate that two homologous chromosomes physically exchange segments as a result of crossing over. They knew that a gene was located near the knobbed end of chromosome 9 that provided color to corn kernels. This gene existed in two alleles, the dominant allele *C* (colored) and the recessive allele *c* (colorless). A second gene, located near the translocated piece from chromosome 8, affected the texture of the kernel endosperm. The dominant allele *Wx* caused starchy endosperm, and the recessive *wx* allele caused waxy endosperm. Creighton and McClintock reasoned that a crossover involving a normal chromosome 9 and a knobbed/translocated chromosome 9 would produce a chromosome that had either a knob or a translocation, but not both. These two types of chromosomes would be distinctly different from either of the parental chromosomes (**Figure 5.5b**).

As shown in the experiment of **Figure 5.6**, Creighton and McClintock began with a corn strain that carried an abnormal chromosome that had a knob at one end and a translocation at the other. Genotypically, this chromosome was *C wx*. The cytologically normal chromosome in this strain was *c Wx*. This corn plant, termed parent A, had the genotype *Cc Wxwx*. It was

(a) Normal and abnormal chromosome 9

(b) Crossing over between normal and abnormal chromosome 9

FIGURE 5.5   **Crossing over between a normal and abnormal chromosome 9 in corn. (a)** A normal chromosome 9 in corn is compared to an abnormal chromosome 9 that contains a knob at one end and a translocation at the opposite end. **(b)** A crossover produces a chromosome that contains only a knob at one end and another chromosome that contains only a translocation at the other end.

crossed to a strain called parent B that carried two cytologically normal chromosomes and had the genotype *cc Wxwx*.

They then observed the kernels in two ways. First, they examined the phenotypes of the kernels to see if they were colored or colorless, and starchy or waxy. Second, the chromosomes in each kernel were examined under a microscope to determine their cytological appearance. Altogether, they observed a total of 25 kernels (see data of Figure 5.6).

■ THE HYPOTHESIS

Offspring with nonparental phenotypes are the product of a crossover. This crossover should create nonparental chromosomes via an exchange of chromosomal segments between homologous chromosomes.

STEP 2: HYPOTHESIS

The student is given a statement describing the possible explanation for the observed phenomenon that will be tested. The hypothesis section reinforces the scientific method and allows students to experience the process for themselves.

These problems are crafted to aid students in developing a wide range of skills. They also develop a student's cognitive, writing, analytical, computational, and collaborative abilities.

CONCEPTUAL QUESTIONS

Test the understanding of basic genetic principles. The student is given many questions with a wide range of difficulty. Some require critical thinking skills, and some require the student to write coherent essay questions.

EXPERIMENTAL QUESTIONS

Test the ability to analyze data, design experiments, or appreciate the relevance of experimental techniques.

STUDENT DISCUSSION/ COLLARBORATION QUESTIONS

Encourage students to consider broad concepts and practical problems. Some questions require a substantial amount of computational activities, which can be worked on as a group.

Conceptual Questions

C1. What is the meaning of the term genetic material?

C2. After the DNA from type IIIS bacteria is exposed to type IIR bacteria, list all of the steps that you think must occur for the bacteria to start making a type IIIS capsule.

C3. Look up the meaning of the word transformation in a dictionary and explain whether it is an appropriate word to describe the transfer of genetic material from one organism to another.

C4. What are the building blocks of a nucleotide? With regard to the 5' and 3' positions on a sugar molecule, how are nucleotides linked together to form a strand of DNA?

C5. Draw the structure of guanine, guanosine, and deoxyguanosine triphosphate.

C6. Draw the structure of a phosphodiester linkage.

C7. Describe how bases interact with each other in the double helix. This discussion should address the issues of complementarity, hydrogen bonding, and base stacking.

C8. If one DNA strand is 5'–GGCATTACACTAGGCCT–3' what is the sequence of the complementary strand?

C9. What is meant by the term DNA sequence?

C10. Make a side-by-side drawing of two DNA helices, one with 10 base pairs (bp) per 360° turn and the other with 15 bp per 360° turn.

C11. Discuss the differences in the structural features of A DNA, B DNA, and Z DNA.

C12. What parts of a nucleotide (namely, phosphate, sugar, and/or bases) occupy the major and minor grooves of double-stranded DNA, and what parts are found in the DNA backbone? If a DNA-binding protein does not recognize a specific nucleotide sequence, do you expect that it recognizes the major groove, the minor groove, or the DNA backbone? Explain.

C16. Compare the structural features of a double-stranded RNA structure with those of a DNA double helix.

C17. Which of the following DNA double helices would be more difficult to separate into single-stranded molecules by treatment with heat, which breaks hydrogen bonds?

A. GGCGTACCAGCGCAT

   CCGCATGGTCGCGTA

B. ATACGATTTACGAGA

   TATGCTAAATGCTCT

Explain your choice.

C18. What structural feature allows DNA to store information?

C19. Discuss the structural significance of complementarity in DNA and in RNA.

C20. An organism has a G + C content of 64% in its DNA. What are the percentages of A, T, G, and C?

C21. Let's suppose you have recently identified an organism that was scraped from an asteroid that hit the earth. (Fortunately, no one was injured.) When you analyze this organism, you discover that its DNA is a triple helix, composed of six different nucleotides: A, T, G, C, X, and Y. You measure the chemical composition of the bases and find the following amounts of these six bases: A = 24%, T = 23%, G = 11%, C = 12%, X = 21%, Y = 9%. What rules would you propose govern triplex DNA formation in this organism? Note: There is more than one possibility.

C22. Upon further analysis of the DNA described in conceptual question C21, you discover that the triplex DNA in this alien organism is composed of a double helix, with the third helix wound within the major groove (just like the DNA in Figure 9.20). How would you propose that this DNA is able to replicate itself? In your answer, be specific about the base pairing rules within

Experimental Questions

E1. Genetic material acts as a blueprint for an organism's traits. Explain how the experiments of Griffith indicated that genetic material was being transferred to the type IIR bacteria.

E2. With regard to the experiment described in Figure 9.3, answer the following:

A. List several possible reasons why only a small percentage of the type IIR bacteria was converted to type IIIS.

B. Explain why an antibody must be used to remove the bacteria that are not transformed. What would the results look like, in all five cases, if the antibody/centrifugation step had not been included in the experimental procedure?

C. The DNA extract was treated with DNase, RNase, or protease. Why was this done? (In other words, what were the researchers trying to demonstrate?)

E3. An interesting trait that some bacteria exhibit is resistance to killing by antibiotics. For example, certain strains of bacteria are resistant to tetracycline, whereas other strains are sensitive to tetracycline. Describe an experiment you would carry out to demonstrate that tetracycline resistance is an inherited trait encoded by the DNA of the resistant strain.

E4. With regard to the experiment of Figure 9.6, answer the following:

A. Provide possible explanations why some of the DNA is in the supernatant.

B. Plot the results if the radioactivity in the pellet, rather than in the supernatant, had been measured.

C. Why were ^{32}P and ^{35}S chosen as radioisotopes to label the phages?

D. List possible reasons why less than 100% of the phage protein was removed from the bacterial cells during the shearing process.

E5. Does the experiment of Figure 9.6 rule out the possibility that RNA is the genetic material of T2 phage? Explain your answer. If it does not, could you modify the approach of Hershey and Chase to show that it is DNA and not RNA that is the genetic material of T2 bacteriophage? Note: It is possible to specifically label DNA or RNA by providing bacteria with radiolabeled thymine or uracil, respectively.

E6. In Chapter 9, we considered two experiments—one by Avery, MacLeod, and McCarty and the second by Hershey and Chase— that indicated DNA is the genetic material. Discuss the strengths

answer the following:

A. What is the purpose of paper chromatography?

prior to its exposure to the plant tissue?

Questions for Student Discussion/Collaboration

1. Try to propose structures for a genetic material that are substantially different from the double helix. Remember that the genetic material must have a way to store information and a way to be faithfully replicated.

2. How might you provide evidence that DNA is the genetic material in mice?

Note: All answers appear at the website for this textbook; the answers to even-numbered questions are in the back of the textbook.

www.mhhe.com/brookergenetics3e

Visit the Online Learning Center for practice tests, answer keys, and other learning aids for this chapter. Enhance your understanding of genetics with our interactive exercises, quizzes, animations, and much more.

OVERVIEW OF GENETICS

1

Hardly a week goes by without a major news story involving a genetic breakthrough. The increasing pace of genetic discoveries has become staggering. The Human Genome Project is a case in point. This project began in the United States in 1990, when the National Institutes of Health and the Department of Energy joined forces with international partners to decipher the massive amount of information contained in our **genome**—the DNA found within all of our chromosomes (**Figure 1.1**). Working collectively, a large group of scientists from around the world has produced a detailed series of maps that help geneticists navigate through human DNA. Remarkably, in only a decade, they determined the DNA sequence (read in the bases of A, T, G, and C) covering over 90% of the human genome. The first draft of this sequence, published in 2001, is nearly 3 billion nucleotide base pairs in length. The completed sequence, published in 2003, has an accuracy greater than 99.99%; fewer than one mistake was made in every 10,000 base pairs!

Studying the human genome allows us to explore fundamental details about ourselves at the molecular level. The results of the project are expected to shed considerable light on basic questions, like how many genes we have, how genes direct the activities of living cells, how species evolve, how single cells develop into complex tissues, and how defective genes cause disease. Furthermore, such understanding may lend itself to improvements in modern medicine by leading to better diagnoses of diseases and the development of new medicines to treat them.

As scientists have attempted to unravel the mysteries within our genes, this journey has involved the invention of many new technologies. This textbook emphasizes a large number of these modern approaches. Paradoxically, one could argue that these advancements may have a greater impact on our lives than the original discoveries upon which they were based. For example, the technology of gene cloning originally arose from discoveries concerning bacterial enzymes that "cut and paste" DNA fragments together. These scientific discoveries have made it possible to create medicines that would otherwise be difficult or impossible to make. An example is human recombinant insulin, termed *Humulin,* which is synthesized in a strain of *Escherichia coli* bacteria that has been genetically altered by the addition of the gene for human insulin. The bacteria are grown in a laboratory and make large amounts of human insulin. As discussed in Chapter 19, this insulin is purified and administered to many people with insulin-dependent diabetes.

Copycat, the first cloned pet. In 2002, the cat shown here, called Copycat, was produced by cloning, a procedure described in Chapter 19.

PART I
Introduction

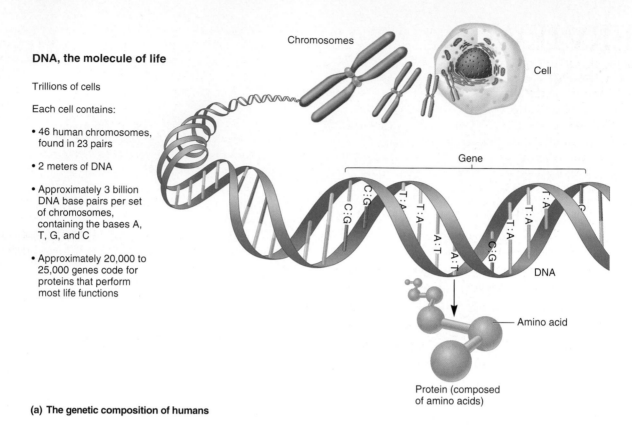

DNA, the molecule of life

Trillions of cells

Each cell contains:

- 46 human chromosomes, found in 23 pairs

- 2 meters of DNA

- Approximately 3 billion DNA base pairs per set of chromosomes, containing the bases A, T, G, and C

- Approximately 20,000 to 25,000 genes code for proteins that perform most life functions

(a) **The genetic composition of humans**

(b) **Genes on one human chromosome that are associated with disease when mutant**

FIGURE 1.1   **The Human Genome Project. (a)** The human genome is a complete set of human chromosomes. People have two sets of chromosomes, one from each parent. Collectively, each set of chromosomes is composed of a DNA sequence that is approximately 3 billion nucleotide base pairs long. Estimates suggest that each set contains about 20,000 to 25,000 different genes. This figure emphasizes the DNA found in the cell nucleus. Humans also have a small amount of DNA in their mitochondria, which has also been sequenced. **(b)** An important outcome of this work is the identification of genes that contribute to human diseases. This illustration depicts a map of a few genes that are located on human chromosome 4. When these genes carry certain rare mutations, they can cause the diseases designated in this figure.

New genetic technologies are often met with skepticism and sometimes even with disdain. An example would be DNA fingerprinting, a molecular method to identify an individual based on a DNA sample (see Chapter 24). Though this technology is now relatively common in the area of forensic science, it was not always universally accepted. High-profile crime cases in the news cause us to realize that not everyone believes in DNA fingerprinting, in spite of its extraordinary ability to uniquely identify individuals. A second controversial example would be mammalian cloning. In 1997, Ian Wilmut and his colleagues created clones of sheep, using mammary cells from an adult animal (**Figure 1.2**). More recently, such cloning has been achieved in several mammalian species, including sheep, cows, mice, goats, pigs, and cats. In 2002, the first pet was cloned, a cat named Copycat (see photo at the beginning of the chapter). The cloning of mammals provides the potential for many practical applications. With regard to livestock, cloning would enable farmers to use cells from their best individuals to create genetically homogeneous herds. This could be advantageous in terms of agricultural yield, although such a genetically homogeneous herd may be more susceptible to certain diseases. However, people have become greatly concerned with the possibility of human cloning. This prospect has raised serious ethical questions. Within the past few years, legislative bills have been introduced that involve bans on human cloning.

Finally, genetic technologies provide the means to modify the traits of animals and plants in ways that would have been unimaginable just a few decades ago. **Figure 1.3a** illustrates a bizarre example in which scientists introduced a gene from jellyfish into mice. As you may know, certain species of jellyfish emit a "green glow." These jellyfish have a gene that encodes a bioluminescent protein called green fluorescent protein (GFP). When exposed to blue or ultraviolet (UV) light, the protein emits a striking green-colored light. Scientists were able to clone the *GFP* gene from a sample of jellyfish cells and then introduce this gene into laboratory mice. The GFP protein is made throughout the cells of their bodies. As a result, their skin, eyes, and organs give off an eerie green glow when exposed to UV light. Only their fur does not glow.

(a) GFP expressed in mice

(b) GFP expressed in the gonads of a male mosquito

FIGURE 1.3   The introduction of a jellyfish gene into laboratory mice and mosquitoes. (a) A gene that naturally occurs in the jellyfish encodes a protein called green fluorescent protein (GFP). The GFP gene was cloned and introduced into mice. When these mice are exposed to UV light, GFP emits a bright green color. These mice glow green, just like jellyfish! **(b)** GFP was introduced next to a gene sequence that causes the expression of GFP only in the gonads of male mosquitoes. This allows researchers to identify and sort males from females.

FIGURE 1.2   The cloning of a mammal. The lamb on the left is Dolly, the first mammal to be cloned. She was cloned from the cells of a Finn Dorset (a white-faced sheep). The sheep on the right is Dolly's surrogate mother, a Blackface ewe. A description of how Dolly was created is presented in Chapter 19.

GFP allows researchers to identify particular proteins in cells or specific body parts. For example, Andrea Crisanti and colleagues have altered mosquitoes to express a green fluorescent protein only in the gonads of males (**Figure 1.3b**). This enables the researchers to identify and sort males and females. Why is this useful? The ability to rapidly sort mosquitoes makes it possible to produce populations of sterile males and then release the sterile males without the risk of releasing additional females. The release of sterile males may be effective at controlling mosquito populations because females only breed once before they die. Mating with a sterile male will prevent a female from producing offspring.

Overall, as we move forward in the twenty-first century, the excitement level in the field of genetics is high, perhaps higher than it has ever been. Nevertheless, the excitement generated by new genetic knowledge and technologies will also create many ethical and societal challenges. In this chapter, we will begin with an overview of genetics and then explore the various fields of genetics and their experimental approaches.

1.1 THE RELATIONSHIP BETWEEN GENES AND TRAITS

Genetics is the branch of biology that deals with heredity and variation. It stands as the unifying discipline in biology by allowing us to understand how life can exist at all levels of complexity, ranging from the molecular to the population level. Genetic variation is the root of the natural diversity that we observe among members of the same species as well as among different species.

Genetics is centered on the study of genes. A gene is classically defined as a unit of heredity, but such a vague definition does not do justice to the exciting characteristics of genes as intricate molecular units that manifest themselves as potent contributors to cell structure and function. At the molecular level, a **gene** is a segment of DNA that produces a functional product. The functional product of most genes is a polypeptide, which is a linear sequence of amino acids that folds into units that constitute proteins. In addition, genes are commonly described according to the way they affect **traits,** which are the characteristics of an organism. In humans, for example, we speak of traits such as eye color, hair texture, and height. The ongoing theme of this textbook is the relationship between genes and traits. As an organism grows and develops, its collection of genes provides a blueprint that determines its characteristics.

In this section of Chapter 1, we will examine the general features of life, beginning with the molecular level and ending with populations of organisms. As will become apparent, genetics is the common thread that explains the existence of life and its continuity from generation to generation. For most students, this chapter should serve as a cohesive review of topics they learned in other introductory courses such as General Biology. Even so, it is usually helpful to see the "big picture" of genetics before delving into the finer details that are covered in Chapters 2 through 26.

Living Cells Are Composed of Biochemicals

To fully understand the relationship between genes and traits, we need to begin with an examination of the composition of living organisms. Every cell is constructed from intricately organized chemical substances. Small organic molecules such as glucose and amino acids are produced from the linkage of atoms via chemical bonds. The chemical properties of organic molecules are essential for cell vitality in two key ways. First, the breakage of chemical bonds during the degradation of small molecules provides energy to drive cellular processes. A second important function of these small organic molecules is their role as the building blocks for the synthesis of larger molecules. Four important categories of larger cellular molecules are **nucleic acids** (i.e., DNA and RNA), **proteins, carbohydrates,** and **lipids.** Three of these—nucleic acids, proteins, and carbohydrates—form **macromolecules** that are composed of many repeating units of smaller building blocks. Proteins, RNA, and carbohydrates can be made from hundreds or even thousands of repeating building blocks. DNA is the largest macromolecule found in living cells. A single DNA molecule can be composed of a linear sequence of hundreds of millions of nucleotides!

The formation of cellular structures relies on the interactions of molecules and macromolecules. For example, nucleotides are the building blocks of DNA, which is a constituent of cellular chromosomes (**Figure 1.4**). In addition, the DNA is associated with a myriad of proteins that provide organization to the structure of chromosomes. Within a eukaryotic cell, the chromosomes are contained in a compartment called the cell nucleus. The nucleus is bounded by a membrane composed of lipids and proteins that shields the chromosomes from the rest of the cell. The organization of chromosomes within a cell nucleus protects the chromosomes from mechanical breakage and provides a single compartment for genetic activities such as gene transcription. As a general theme, the formation of large cellular structures arises from interactions among different molecules and macromolecules. These cellular structures, in turn, are organized to make a complete living cell.

Each Cell Contains Many Different Proteins That Determine Cellular Structure and Function

To a great extent, the characteristics of a cell depend on the types of proteins that it makes. As we will learn throughout this textbook, proteins are the "workhorses" of all living cells. The range of functions among different types of proteins is truly remarkable. Some proteins help determine the shape and structure of a given cell. For example, the protein known as tubulin can assemble into large structures known as microtubules, which provide the cell with internal structure and organization. Other proteins are inserted into cell membranes and aid in the transport of ions and small molecules across the membrane. Proteins may also function as biological motors. An interesting case is the protein known as myosin, which is involved in the contractile properties of muscle cells. Within multicellular organisms, certain proteins also function in cell-to-cell recognition and signaling. For example, hormones such as insulin are secreted by endocrine cells and bind to the insulin receptor protein found within the plasma membrane of target cells.

Enzymes, which accelerate chemical reactions, are a particularly important category of proteins. Some enzymes play a role in the breakdown of molecules or macromolecules into smaller units. These are known as catabolic enzymes and are important in the utilization of energy. Alternatively, anabolic enzymes and accessory proteins function in the synthesis of molecules and macromolecules throughout the cell. The construction of a cell greatly depends on its proteins involved in anabolism because these are required to synthesize all cellular macromolecules.

Molecular biologists have come to realize that the functions of proteins underlie the cellular characteristics of every organism. At the molecular level, proteins can be viewed as the active participants in the enterprise of life.

DNA Stores the Information for Protein Synthesis

The genetic material is composed of a substance called **deoxyribonucleic acid,** abbreviated **DNA.** The DNA stores the information needed for the synthesis of all cellular proteins. In other words, the main function of the genetic blueprint is to code for the production of cellular proteins in the correct cell, at the proper time, and in suitable amounts. This is an extremely complicated task because living cells make thousands of different proteins. Genetic analyses have shown that a typical bacterium can make a few thousand different proteins, and estimates among higher eukaryotes range in the tens of thousands.

DNA's ability to store information is based on its molecular structure. DNA is composed of a linear sequence of **nucleotides.** Each nucleotide contains one nitrogen-containing base, either adenine (A), thymine (T), guanine (G), or cytosine (C). The linear order of these bases along a DNA molecule contains information similar to the way that groups of letters of the alphabet represent words. For example, the "meaning" of the sequence of bases ATGGGCCTTAGC differs from that of TTTAAGCTTGCC. DNA sequences within most genes contain the information to direct the order of amino acids within polypeptides according to the **genetic code.** In the code, a three-base sequence specifies one particular **amino acid** among the 20 possible choices. One or more polypeptides form a functional protein. In this way, the DNA can store the information to specify the proteins made by an organism.

DNA Sequence	Amino Acid Sequence
ATG GGC CTT AGC	METHIONINE GLYCINE LEUCINE SERINE
TTT AAG CTT GCC	PHENYLALANINE LYSINE LEUCINE ALANINE

In living cells, the DNA is found within large structures known as **chromosomes. Figure 1.5** is a photograph of the 46 chromosomes contained in a cell from a human male. The DNA of an average human chromosome is an extraordinarily long, linear, double-stranded structure that contains well over a hundred

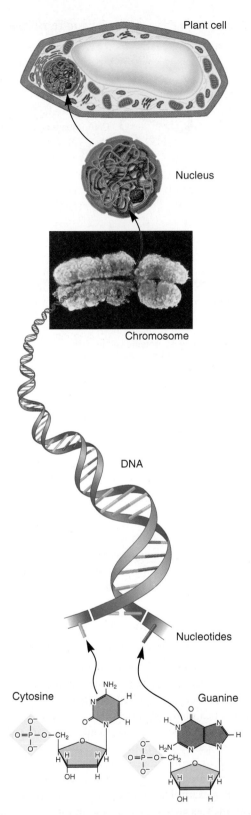

FIGURE 1.4   **Chemical composition of living cells.** Cellular structures are constructed from smaller building blocks. In this example, DNA is formed from the linkage of nucleotides to produce a very long macromolecule. The DNA associates with proteins to form a chromosome. The chromosomes are located within a membrane-bounded organelle called the nucleus, which, along with many different types of organelles, is found within a complete cell.

FIGURE 1.5   A micrograph of the 46 chromosomes found in a cell from a human male.

Functioning of proteins within living cells influences an organism's traits.

FIGURE 1.6   Gene expression at the molecular level. The expression of a gene is a multistep process. During transcription, one of the DNA strands is used as a template to make an RNA strand. During translation, the RNA strand is used to specify the sequence of amino acids within a polypeptide. One or more polypeptides produce a protein that functions within the cell, thereby influencing an organism's traits.

ONLINE ANIMATION

million nucleotides. Along the immense length of a chromosome, the genetic information is parceled into functional units known as genes. An average-sized human chromosome is expected to contain about 1,000 different genes.

The Information in DNA Is Accessed During the Process of Gene Expression

To synthesize its proteins, a cell must be able to access the information that is stored within its DNA. The process of using a gene sequence to affect the characteristics of cells and organisms is referred to as **gene expression.** At the molecular level, the information within genes is accessed in a stepwise process. In the first step, known as **transcription,** the DNA sequence within a gene is copied into a nucleotide sequence of **ribonucleic acid (RNA).** Most RNAs contain the information for the synthesis of a particular polypeptide. This type of RNA is called **messenger RNA (mRNA).** For polypeptide synthesis to occur, the sequence of nucleotides transcribed in an mRNA must be **translated** (using the genetic code) into the amino acid sequence of a polypeptide (**Figure 1.6**). After a polypeptide is made, it folds into a three-dimensional structure. As mentioned, a protein is a functional unit. Some proteins are composed of a single polypeptide, and other proteins consist of two or more polypeptides.

Gene expression results in the production of proteins with specific structures and functions. The unique relationship between gene sequences and protein structures is of paramount importance because the distinctive structure of each protein determines its function within a living cell or organism. Mediated by the process of gene expression, therefore, the sequence of nucleotides in DNA stores the information required for synthesizing proteins with specific structures and functions.

The Molecular Expression of Genes Within Cells Leads to an Organism's Traits

A trait is any characteristic that an organism displays. In genetics, we often focus our attention on **morphological traits** that affect the appearance of an organism. The color of a flower and the height of a pea plant are morphological traits. Geneticists frequently study these types of traits because they are easy to evaluate. For example, an experimenter can simply look at a plant and tell if it has red or white flowers. However, not all traits are morphological. **Physiological traits** affect the ability of an organism to function. For example, the rate at which a bacterium metabolizes a sugar such as lactose is a physiological trait. Like morphological traits, physiological traits are controlled, in part, by the expression of genes. **Behavioral traits** also affect the ways that an organism responds to its environment. An example would be the mating calls of bird species. In animals, the nervous system plays a key role in governing such traits.

A difficult, yet very exciting, aspect of genetics is that our observations and theories span four levels of biological organization: molecules, cells, organisms, and populations. This can make it difficult to appreciate the relationship between genes and traits. To understand this connection, we need to relate the following phenomena:

1. Genes are expressed at the **molecular level.** In other words, gene transcription and translation lead to the production of a particular protein, which is a molecular process.
2. Proteins often function at the **cellular level.** The function of a protein within a cell will affect the structure and workings of that cell.
3. An organism's traits are determined by the characteristics of its cells. We do not have microscopic vision, yet when we view morphological traits, we are really observing the properties of an individual's cells. For example, a red flower has its color because the flower cells make a red pigment. The trait of red flower color is an observation at the **organism level.** Yet the trait is rooted in the molecular characteristics of the organism's cells.
4. A **species** is a group of organisms that maintains a distinctive set of attributes in nature. The occurrence of a trait within a species is an observation at the **population level.** Along with learning how a trait occurs, we also want to understand why a trait becomes prevalent in a particular species. In many cases, researchers discover that a trait predominates within a population because it promotes the reproductive success of the members of the population. This leads to the evolution of beneficial traits.

As a schematic example to illustrate the four levels of genetics, **Figure 1.7** shows the trait of pigmentation in butterflies. One is light-colored and the other is very dark. Let's consider how we can explain this trait at the molecular, cellular, organism, and population levels.

At the molecular level, we need to understand the nature of the gene or genes that govern this trait. As shown in Figure 1.7a, a gene, which we will call the pigmentation gene, is responsible for the amount of pigment that is produced. The pigmentation gene can exist in two different forms called **alleles.** In this example, one allele confers a dark pigmentation and one causes a light pigmentation. Each of these alleles encodes a protein that functions as a pigment-synthesizing enzyme. However, the DNA sequences of the two alleles differ slightly from each other. This difference in the DNA sequence leads to a variation in the structure and function of the respective pigmentation enzymes.

At the cellular level (Figure 1.7b), the functional differences between the pigmentation enzymes affect the amount of pigment that is produced. The allele causing dark pigmentation, which is shown on the left, encodes a protein that functions very well. Therefore, when this gene is expressed in the cells of the wings, a large amount of pigment is made. By comparison, the allele causing light pigmentation encodes an enzyme that func-

(a) Molecular level

(b) Cellular level

(c) Organism level

(d) Population level

FIGURE 1.7   **The relationship between genes and traits at the (a) molecular, (b) cellular, (c) organism, and (d) population levels.**

tions poorly. Therefore, when this allele is the only pigmentation gene that is expressed, little pigment is made.

At the organism level (Figure 1.7c), the amount of pigment in the wing cells governs the color of the wings. If the pigment cells produce high amounts of pigment, the wings are dark-colored; if the pigment cells produce little pigment, the wings are light.

Finally, at the population level (Figure 1.7d), geneticists would like to know why a species of butterfly would contain some members with dark wings and other members with light wings. One possible explanation is differential predation. The butterflies with dark wings might avoid being eaten by birds if they happen to live within the dim light of a forest. The dark wings would help to camouflage the butterfly if it were perched on a dark surface such as a tree trunk. In contrast, the lightly colored wings would be an advantage if the butterfly inhabited a brightly lit meadow. Under these conditions, a bird may be less likely to notice a light-colored butterfly that is perched on a sunlit surface. A population geneticist might study this species of butterfly and find that the dark-colored members usually live in forested areas and the light-colored members reside in unforested regions.

Inherited Differences in Traits Are Due to Genetic Variation

In Figure 1.7, we considered how gene expression could lead to variation in a trait of an organism, such as dark- versus light-colored butterflies. Variation in traits among members of the same species is very common. For example, some people have brown hair, while others have blond hair; some petunias have white flowers, while others have purple flowers. These are examples of **genetic variation.** This term describes the differences in inherited traits among individuals within a population.

In large populations that occupy a wide geographic range, genetic variation can be quite striking. In fact, morphological differences have often led geneticists to misidentify two members of the same species as belonging to separate species. As an example, **Figure 1.8** compares four garter snakes that are members of the same species, *Thamnophis ordinoides*. They display dramatic differences in their markings. Such contrasting forms within a single species are termed **morphs.** You can easily imagine how someone might mistakenly conclude that these four snakes are not members of the same species.

Changes in the nucleotide sequence of DNA underlie the genetic variation that we see among individuals. Throughout this textbook, we will routinely examine how variation in the genetic material results in changes in the outcome of traits. At the molecular level, genetic variation can be attributed to different types of modifications.

1. Small or large differences can occur within gene sequences. These are called **gene mutations.** This type of variation, which produces two or more alleles of the same gene, was previously described in Figure 1.7. In many cases, gene mutations will alter the expression or function of the protein that the gene specifies.

FIGURE 1.8   **Four garter snakes showing different morphs within a single species.**

2. Major alterations can also occur in the structure of a chromosome. A large segment of a chromosome can be lost, rearranged, or reattached to another chromosome.
3. Variation may also occur in the total number of chromosomes. In some cases, an organism may inherit one too many or one too few chromosomes. In other cases, it may inherit an extra set of chromosomes.

Variations within the sequences of genes are a common source of genetic variation among members of the same species. In humans, familiar examples of variation involve genes for eye color, hair texture, and skin pigmentation. Chromosome variation—a change in chromosome structure and/or number—is also found, but this type of change is often detrimental. Many human genetic disorders are the result of chromosomal alterations. The most common example is Down syndrome, which is due to the presence of an extra chromosome (**Figure 1.9a**). By comparison, chromosome variation in plants is common and often can lead to strains of plants with superior characteristics, such as increased resistance to disease. Plant breeders have frequently exploited this observation. Cultivated varieties of wheat, for example, have many more chromosomes than the wild species (**Figure 1.9b**).

Traits Are Governed by Genes and by the Environment

In our discussion thus far, we have considered the role that genes play in the outcome of traits. Another critical factor is the **environment**—the surroundings in which an organism exists. A variety of factors in an organism's environment profoundly affect its morphological and physiological features. For example, a person's diet greatly influences many traits such as height, weight, and even intelligence. Likewise, the amount of sunlight a plant receives affects its growth rate and the color of its flowers. The

(a)   **(b)**

FIGURE 1.9 **Examples of chromosome variation.** (a) A person with Down syndrome competing in the Special Olympics. This person has 47 chromosomes rather than the common number of 46, because she has an extra copy of chromosome 21. (b) A wheat plant. Bread wheat is derived from the contributions of three related species with two sets of chromosomes each, producing an organism with six sets of chromosomes.

FIGURE 1.10 **Environmental influence on the outcome of PKU within a single family.** All three children pictured here have inherited the alleles that cause PKU. The child in the middle was raised on a phenylalanine-free diet and developed normally. The other two children were born before the benefits of a phenylalanine-free diet were known and were raised on diets that contained phenylalanine. Therefore, they manifest a variety of symptoms, including mental retardation. People born today with this disorder are usually diagnosed when infants. (*Photo from the March of Dimes Birth Defects Foundation.*)

term **norm of reaction** refers to the effects of environmental variation on an individual's traits.

External influences may dictate the way that genetic variation is manifested in an individual. An interesting example is the human genetic disease **phenylketonuria (PKU).** Humans possess a gene that encodes an enzyme known as phenylalanine hydroxylase. Most people have two functional copies of this gene. People with one or two functional copies of the gene can eat foods containing the amino acid phenylalanine and metabolize it properly.

A rare variation in the sequence of the phenylalanine hydroxylase gene results in a nonfunctional version of this protein. Individuals with two copies of this rare, inactive allele cannot metabolize phenylalanine. This occurs in about 1 in 8,000 births among Caucasians in the United States. When given a standard diet containing phenylalanine, individuals with this disorder are unable to break down this amino acid. Phenylalanine accumulates and is converted into phenylketones, which are detected in the urine. PKU individuals manifest a variety of detrimental traits, including mental retardation, underdeveloped

teeth, and foul-smelling urine. In contrast, when PKU individuals are identified at birth and raised on a restricted diet that is low in phenylalanine, they develop normally (**Figure 1.10**). Fortunately, through routine newborn screening, most affected babies in the United States are now diagnosed and treated early. PKU provides a dramatic example of how the environment and an individual's genes can interact to influence the traits of the organism.

During Reproduction, Genes Are Passed from Parent to Offspring

Now that we have considered how genes and the environment govern the outcome of traits, we can turn to the issue of inheritance. A centrally important matter in genetics is the manner in which traits are passed from parents to offspring. The foundation for our understanding of inheritance came from the studies of Gregor Mendel in the nineteenth century. His work revealed that genetic determinants, which we now call genes, are passed from parent to offspring as discrete units. We can predict the outcome of genetic crosses based on Mendel's laws of inheritance.

The inheritance patterns identified by Mendel can be explained by the existence of chromosomes and their behavior during cell division. As in Mendel's pea plants, sexually reproducing species are commonly **diploid.** This means they contain two copies of each chromosome, one from each parent. The two

(a) Chromosomal composition found in most female human cells (46 chromosomes)

(b) Chromosomal composition found in a human gamete (23 chromosomes)

FIGURE 1.11   **The complement of human chromosomes in somatic cells and gametes. (a)** A schematic drawing of the 46 chromosomes of a human. With the exception of the sex chromosomes, these are always found in homologous pairs. **(b)** The chromosomal composition of a gamete, which contains only 23 chromosomes, one from each pair. This gamete contains an X chromosome. Half of the gametes from human males would contain a Y chromosome instead of the X chromosome.

copies are called **homologues** of each other. Because genes are located within chromosomes, diploid organisms have two copies of most genes. Humans, for example, have 46 chromosomes, which are found in homologous pairs (**Figure 1.11a**). With the exception of the sex chromosomes (namely, X and Y), each homologous pair contains the same kinds of genes. For example, both copies of human chromosome 12 carry the gene that encodes phenylalanine hydroxylase, which was discussed previously. Therefore, an individual has two copies of this gene. The two copies may or may not be identical alleles.

Most cells of the human body that are not directly involved in sexual reproduction contain 46 chromosomes. These cells are called **somatic cells.** In contrast, the **gametes**—sperm and egg cells—contain half that number and are termed **haploid** (**Figure 1.11b**). The union of gametes during fertilization restores the diploid number of chromosomes. The primary advantage of sexual reproduction is that it enhances genetic variation. For example, a tall person with blue eyes and a short person with brown eyes may have short offspring with blue eyes or tall offspring with brown eyes. Therefore, sexual reproduction can result in new combinations of two or more traits that differ from those of either parent.

The Genetic Composition of a Species Evolves over the Course of Many Generations

As we have just seen, sexual reproduction has the potential to enhance genetic variation. This can be an advantage for a population of individuals as they struggle to survive and compete within their natural environment. The term **biological evolution**

refers to the phenomenon that the genetic makeup of a population can change over the course of many generations.

As suggested by Charles Darwin, the members of a species are in competition with each other for essential resources. Random genetic changes (i.e., mutations) occasionally occur within an individual's genes, and sometimes these changes lead to a modification of traits that promote reproductive success. For example, over the course of many generations, random gene mutations have lengthened the neck of the giraffe, enabling it to feed on leaves that are high in the trees. When a mutation creates a new allele that is beneficial, the allele may become prevalent in future generations because the individuals carrying the allele are more likely to reproduce and pass the beneficial allele to their offspring. This process is known as **natural selection.** In this way, a species becomes better adapted to its environment.

Over a long period of time, the accumulation of many genetic changes leads to rather striking modifications in a species' characteristics. As an example, **Figure 1.12** depicts the evolution of the modern-day horse. A variety of morphological changes occurred, including an increase in size, fewer toes, and modified jaw structure.

1.2 FIELDS OF GENETICS

Genetics is a broad discipline encompassing molecular, cellular, organism, and population biology. Many scientists who are interested in genetics have been trained in supporting disciplines such as biochemistry, biophysics, cell biology, mathematics, microbi-

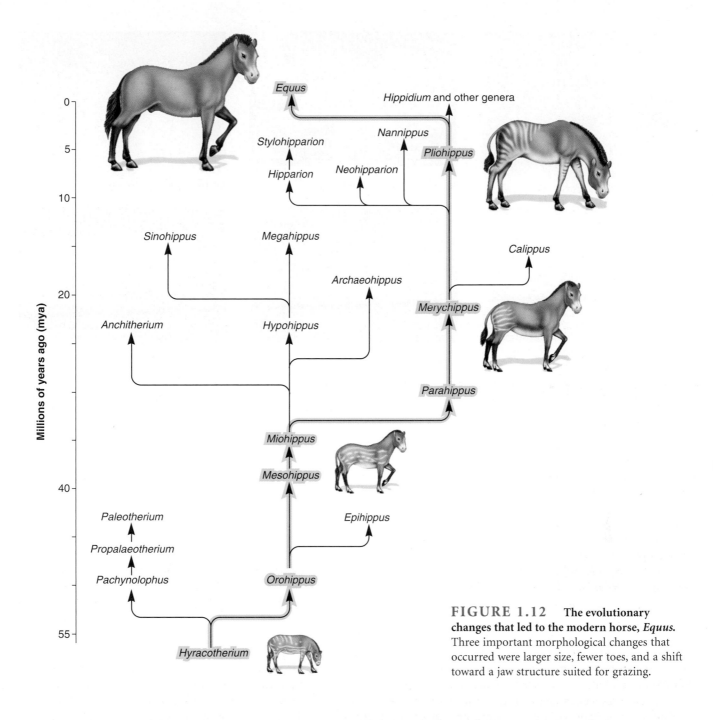

FIGURE 1.12   **The evolutionary changes that led to the modern horse, *Equus*.** Three important morphological changes that occurred were larger size, fewer toes, and a shift toward a jaw structure suited for grazing.

ology, population biology, ecology, agriculture, and medicine. Experimentally, geneticists often focus their efforts on **model organisms**—organisms studied by many different researchers so they can compare their results and determine scientific principles that apply more broadly to other species. **Figure 1.13** shows some common examples, including *Escherichia coli* (a bacterium), *Saccharomyces cerevisiae* (a yeast), *Drosophila melanogaster* (fruit fly), *Caenorhabditis elegans* (a nematode worm), *Danio rerio* (zebrafish), *Mus musculus* (mouse), and *Arabidopsis thaliana* (a flowering plant). By limiting their work to a few such model organisms, researchers can more easily unravel the genetic composition of a given species. Furthermore, the genes found in model organisms often function in a similar way to those found in humans.

The study of genetics has been traditionally divided into three areas: transmission, molecular, and population genetics, although overlap is found among these three fields. In this section, we will examine the general questions that scientists in these areas are attempting to answer.

Transmission Genetics Explores the Inheritance Patterns of Traits as They Are Passed from Parents to Offspring

A scientist working in the field of transmission genetics examines the relationship between the transmission of genes from parent to offspring and the outcome of the offspring's traits. For example,

(a) *Escherichia coli*

(b) *Saccharomyces cerevisiae*

(c) *Drosophila melanogaster*

(d) *Caenorhabditis elegans*

(e) *Danio rerio*

(f) *Mus musculus*

(g) *Arabidopsis thaliana*

FIGURE 1.13   **Examples of model organisms studied by geneticists.** (**a**) *Escherichia coli* (a bacterium), (**b**) *Saccharomyces cerevisiae* (a yeast), (**c**) *Drosophila melanogaster* (fruit fly), (**d**) *Caenorhabditis elegans* (a nematode worm), (**e**) *Danio rerio* (zebrafish), (**f**) *Mus musculus* (mouse), and (**g**) *Arabidopsis thaliana* (a flowering plant).

how can two brown-eyed parents produce a blue-eyed child? Or why do tall parents tend to produce tall children, but not always? Our modern understanding of transmission genetics began with the studies of Gregor Mendel. His work provided the conceptual framework for transmission genetics. In particular, he originated the idea that genetic determinants, which we now call genes, are passed as discrete units from parents to offspring via sperm and egg cells. Since these pioneering studies of the 1860s, our knowledge of genetic transmission has greatly increased. Many patterns of genetic transmission are more complex than the simple Mendelian patterns that are described in Chapter 2. The additional complexities of transmission genetics are examined in Chapters 3 through 8.

Experimentally, the fundamental approach of a transmission geneticist is the **genetic cross**. A genetic cross involves breeding two selected individuals and the subsequent analysis of their offspring in an attempt to understand how traits are passed from parents to offspring. In the case of experimental organisms, the researcher chooses two parents with particular traits and then categorizes the offspring according to the traits they possess. In many cases, this analysis is quantitative in nature. For example, an experimenter may cross two tall pea plants and obtain 100 offspring that fall into two categories: 75 tall and 25 dwarf. As we will see in Chapter 2, the ratio of tall and dwarf offspring provides important information concerning the inheritance pattern of this trait.

Throughout Chapters 2 to 8, we will learn how researchers seek to answer many fundamental questions concerning the passage of traits from parents to offspring. Some of these questions are as follows:

What are the common patterns of inheritance for genes?
Chapters 2–4

When two or more genes are located on the same chromosome, how does this affect the pattern of inheritance? **Chapters 5, 6**

Are there unusual patterns of inheritance that cannot be explained by the simple transmission of genes located on chromosomes in the cell nucleus? **Chapter 7**

How do variations in chromosome structure or chromosome number occur, and how are they transmitted from parents to offspring? **Chapter 8**

Molecular Genetics Is Focused on a Biochemical Understanding of the Hereditary Material

The goal of molecular genetics, as the name of the field implies, is to understand how the genetic material works at the molecular level. In other words, molecular geneticists want to understand the molecular features of DNA and how these features underlie the expression of genes. The experiments of molecular geneticists are usually conducted within the confines of a laboratory. Their efforts frequently progress to a detailed analysis of DNA, RNA, and/or protein, using a variety of techniques that are described throughout Parts III, IV, and V of this book.

Molecular geneticists often study mutant genes that have abnormal function. This is called a **genetic approach** to the study of a research question. In many cases, researchers analyze the effects of gene mutations that eliminate the function of a gene. This type of mutation is called a **loss-of-function mutation,** and the resulting gene is called a **loss-of-function allele.** By studying the effects of such mutations, the role of the functional, nonmutant gene is often revealed. For example, let's suppose that a particular plant species produces purple flowers. If a loss-of-function mutation within a given gene causes a plant to produce white flowers, one would suspect the role of the functional gene involves the production of purple pigmentation.

Studies within molecular genetics interface with other disciplines such as biochemistry, biophysics, and cell biology. In addition, advances within molecular genetics have shed considerable light on the areas of transmission and population genetics. Our quest to understand molecular genetics has spawned a variety of modern molecular technologies and computer-based approaches. Furthurmore, discoveries within molecular genetics have had widespread applications in agriculture, medicine, and biotechnology.

Some general questions within the field of molecular genetics are the following:

What are the molecular structures of DNA and RNA? **Chapters 9, 18**

What is the composition and conformation of chromosomes? **Chapters 10, 20**

How is the genetic material copied? **Chapter 11**

How are genes expressed at the molecular level? **Chapters 12, 13, 18, 19, 21**

How is gene expression regulated so that it occurs under the appropriate conditions and in the appropriate cell type? **Chapters 14, 15, 18, 23**

What is the molecular nature of mutations? How are mutations repaired? **Chapter 16**

How does the genetic material become rearranged at the molecular level? **Chapter 17**

What is the underlying relationship between genes and genetic diseases? **Chapter 22**

How do genes govern the development of multicellular organisms? **Chapter 23**

Population Genetics Is Concerned with Genetic Variation and Its Role in Evolution

The foundations of population genetics arose during the first few decades of the twentieth century. Although many scientists of this era did not accept the findings of Mendel and/or Darwin, the theories of population genetics provided a compelling way to connect the two viewpoints. Mendel's work and that of many succeeding geneticists gave insight into the nature of genes and how they are transmitted from parents to offspring. The work of Darwin provided a natural explanation for the various types of characteristics observed among the members of a species. To relate these two phenomena, population geneticists have developed mathematical theories to explain the prevalence of certain forms of genes within populations of individuals. The work of population geneticists helps us understand how the forces of nature have produced and favored the existence of individuals that carry particular genes.

Population geneticists are particularly interested in genetic variation and how that variation is related to an organism's environment. In this field, the prevalence of alleles within a population is of central importance. Some general questions in population genetics are the following:

Why are two or more different alleles of a gene maintained in a population? **Chapter 24**

What factors alter the prevalence of alleles within a population? **Chapter 24**

What are the contributions of genetics and environment in the outcome of a trait? **Chapter 25**

How do genetics and the environment influence quantitative traits, such as size and weight? **Chapter 25**

What factors have the most impact on the process of evolution? **Chapter 26**

How does evolution occur at the molecular level? **Chapter 26**

Genetics Is an Experimental Science

Science is a way of knowing about our natural world. The science of genetics allows us to understand how the expression of our genes produces the traits that we possess. Researchers typically follow two general types of scientific approaches—hypothesis testing and discovery-based science. In **hypothesis testing,** also called the **scientific method,** scientists follow a series of steps to reach verifiable conclusions about the world in which we live. Although scientists arrive at their theories in different ways, the scientific method provides a way to validate (or invalidate) a particular hypothesis. Alternatively, research may also involve the collection of data without a preconceived hypothesis. For example, researchers might analyze the genes found in cancer cells to identify those genes that have become mutant. In this case, the scientists may not have a hypothesis about which particular genes may be involved. The collection and analysis of data without the need for a preconceived hypothesis is called **discovery-based science** or, simply, discovery science.

In traditional science textbooks, the emphasis often lies on the product of science. Namely, many textbooks are aimed primarily at teaching the student about the observations that scientists have made and the theories that they have proposed to explain these observations. Along the way, the student is provided with many bits and pieces of experimental techniques and data. Likewise, this textbook also provides you with many observations and theories. However, it attempts to go one step further. Each of the following chapters contains one or two experiments that have been "dissected" into five individual components to help you to understand the entire scientific process:

1. Background information is provided so that you may appreciate what previous observations were known prior to conducting the experiment.
2. Most experiments involve hypothesis testing. In those cases, the figure states the hypothesis that the scientists were trying to test. In other words, what scientific question was the researcher trying to answer?
3. Next, the figure follows the experimental steps the scientist took to test the hypothesis. The steps necessary to carry out the experiment are listed in the order they were conducted. The figure contains two parallel illustrations

labeled Experimental Level and Conceptual Level. The illustration shown in the Experimental Level helps you to understand the techniques that were followed. The Conceptual Level helps you to understand what is actually happening at each step in the procedure.

4. The raw data for each experiment are then presented.
5. Last, an interpretation of the data is offered within the text.

The rationale behind this approach is that it enables you to see the experimental process from beginning to end. Hopefully, you will find this a more interesting and rewarding way to learn about genetics. As you read through the chapters, the experiments will help you to see the relationship between science and scientific theories.

As a student of genetics, you will be given the opportunity to involve your mind in the experimental process. As you are reading an experiment, you may find yourself thinking about different approaches and alternative hypotheses. Different people can view the same data and arrive at very different conclusions. As you progress through the experiments in this book, you will enjoy genetics far more if you try to develop your own skills at formulating hypotheses, designing experiments, and interpreting data. Also, some of the questions in the problem sets are aimed at refining these skills.

Finally, it is worthwhile to point out that science is a social discipline. As you develop your skills at scrutinizing experiments, it is fun to discuss your ideas with other people, including fellow students and faculty members. Keep in mind that you do not need to "know all the answers" before you enter into a scientific discussion. Instead, it is more rewarding to view science as an ongoing and never-ending dialogue.

PROBLEM SETS & INSIGHTS

Solved Problems

S1. A human gene called the *CF* gene (for cystic fibrosis) encodes a protein that functions in the transport of chloride ions across the cell membrane. Most people have two copies of a functional *CF* gene and do not have cystic fibrosis. However, a mutant version of the cystic fibrosis gene is found in some people. If a person has two mutant copies of the gene, he or she develops the disease known as cystic fibrosis. Are the following examples a description of genetics at the molecular, cellular, organism, or population level?

A. People with cystic fibrosis have lung problems due to a buildup of mucus in their lungs.

B. The mutant *CF* gene encodes a defective chloride transporter.

C. A defect in the chloride transporter causes a salt imbalance in lung cells.

D. Scientists have wondered why the mutant cystic fibrosis gene is relatively common. In fact, it is the most common mutant gene that causes a severe disease in Caucasians. Usually, mutant genes that cause severe diseases are relatively rare. One possible explanation why the *CF* gene is so common is that people who have one copy of the functional *CF* gene and one copy of the mutant gene may be more resistant to diarrheal diseases such

as cholera. Therefore, even though individuals with two mutant copies are very sick, people with one mutant copy and one functional copy might have a survival advantage over people with two functional copies of the gene.

Answer:

A. Organism. This is a description of a trait at the level of an entire individual.

B. Molecular. This is a description of a gene and the protein it encodes.

C. Cellular. This is a description of how protein function affects the cell.

D. Population. This is a possible explanation why two versions of the gene occur within a population.

S2. Explain the relationship between the following pairs of terms:

A. RNA and DNA

B. RNA and transcription

C. Gene expression and trait

D. Mutation and allele

Answer:

A. DNA is the genetic material. DNA is used to make RNA. RNA is then used to specify a sequence of amino acids within a polypeptide.

B. Transcription is a process in which RNA is made using DNA as a template.

C. Genes are expressed at the molecular level to produce functional proteins. The functioning of proteins within living cells ultimately affects an organism's traits.

D. Alleles are alternative forms of the same gene. For example, a particular human gene affects eye color. The gene can exist as a blue allele or a brown allele. The difference between these two alleles is caused by a mutation. Perhaps the brown allele was the first eye color allele in the human population. Within some ancestral person, however, a mutation may have occurred in the eye color gene that converted the brown allele to the blue allele. Now the human population has both the brown allele and the blue allele.

S3. How are genes passed from generation to generation?

Answer: When a diploid individual makes haploid cells for sexual reproduction, the cells contain half the number of chromosomes. When two haploid cells (e.g., sperm and egg) combine with each other, a zygote is formed that begins the life of a new individual. This zygote has inherited half of its chromosomes and, therefore, half of its genes from each parent. This is how genes are passed from parents to offspring.

Conceptual Questions

C1. Pick any example of a genetic technology and describe how it has directly impacted your life.

C2. At the molecular level, what is a gene? Where are genes located?

C3. Most genes encode proteins. Explain how the structure and function of proteins produce an organism's traits.

C4. Briefly explain how gene expression occurs at the molecular level.

C5. A human gene called the β-globin gene encodes a polypeptide that functions as a subunit of the protein known as hemoglobin. Hemoglobin is found within red blood cells; it carries oxygen. In human populations, the β-globin gene can be found as the common allele called the Hb^A allele, but it can also be found as the rare Hb^S allele. Individuals who have two copies of the Hb^S allele have the disease called sickle-cell disease. Are the following examples a description of genetics at the molecular, cellular, organism, or population level?

A. The Hb^S allele encodes a polypeptide that functions slightly differently from the polypeptide encoded by the Hb^A allele.

B. If an individual has two copies of the Hb^S allele, that person's red blood cells form a sickle shape.

C. Individuals who have two copies of the Hb^A allele do not have sickle-cell disease, but they are not resistant to malaria. People who have one Hb^A allele and one Hb^S allele do not have sickle-cell disease, and they are resistant to malaria. People who have two copies of the Hb^S allele have sickle-cell anemia, and this disease may significantly shorten their lives.

D. Individuals with sickle-cell disease have anemia because their red blood cells are easily destroyed by the body.

C6. What is meant by the term genetic variation? Give two examples of genetic variation not discussed in Chapter 1. What causes genetic variation at the molecular level?

C7. What is the cause of Down syndrome?

C8. Your textbook describes how the trait of phenylketonuria (PKU) is greatly influenced by the environment. Pick a trait in your favorite plant and explain how genetics and environment may play important roles.

C9. What is meant by the term diploid? Which cells of the human body are diploid, and which cells are not?

C10. What is a DNA sequence?

C11. What is the genetic code?

C12. Explain the relationships between the following pairs of genetic terms:

A. Gene and trait

B. Gene and chromosome

C. Allele and gene

D. DNA sequence and amino acid sequence

C13. With regard to biological evolution, which of the following statements is not correct? Explain why.

A. During its lifetime, an animal evolves to become better adapted to its environment.

B. The process of biological evolution has produced species that are better adapted to their environments.

C. When an animal is better adapted to its environment, the process of natural selection makes it more likely for that animal to reproduce.

C14. What are the primary interests of researchers working in the following fields of genetics?

A. Transmission genetics

B. Molecular genetics

C. Population genetics

Experimental Questions

E1. What is a genetic cross?

E2. The technique known as DNA sequencing (described in Chapter 18) enables researchers to determine the DNA sequence of genes. Would this technique be used primarily by transmission geneticists, molecular geneticists, or population geneticists?

E3. Figure 1.5 shows a micrograph of chromosomes from a normal human cell. If you performed this type of experiment using cells from a person with Down syndrome, what would you expect to see?

E4. Many organisms are studied by geneticists. Of the following species, do you think it would be more likely for them to be studied by a transmission geneticist, a molecular geneticist, or a population geneticist? Explain your answer. Note: More than one answer may be possible.

A. Dogs

B. *E. coli*

C. Fruit flies

D. Leopards

E. Corn

E5. Pick any trait you like in any species of wild plant or animal. The trait must somehow vary among different members of the species. For example, some butterflies have dark wings and others have light wings (see Figure 1.7).

A. Discuss all of the background information that you already have (from personal observations) regarding this trait.

B. Propose a hypothesis that would explain the genetic variation within the species. For example, in the case of the butterflies, your hypothesis might be that the dark butterflies survive better in dark forests, while the light butterflies survive better in lighter fields.

C. Describe the experimental steps you would follow to test your hypothesis.

D. Describe the possible data you might collect.

E. Interpret your data.

Note: When picking a trait to answer this question, do not pick the trait of wing color in butterflies.

Note: All answers appear at the website for this textbook; the answers to even-numbered questions are in the back of the textbook.

www.mhhe.com/brookergenetics3e

Visit the Online Learning Center for practice tests, answer keys, and other learning aids for this chapter. Enhance your understanding of genetics with our interactive exercises, quizzes, animations, and much more.

MENDELIAN INHERITANCE

2

An appreciation for the concept of heredity can be traced far back in human history. Hippocrates, a famous Greek physician, was the first person to provide an explanation for hereditary traits (ca. 400 B.C.E.). He suggested that "seeds" are produced by all parts of the body, which are then collected and transmitted to the offspring at the time of conception. Furthermore, he hypothesized that these seeds cause certain traits of the offspring to resemble those of the parents. This idea, known as **pangenesis,** was the first attempt to explain the transmission of hereditary traits from generation to generation.

For the next 2,000 years, the ideas of Hippocrates were accepted by some and rejected by many. After the invention of the microscope in the late seventeenth century, some people observed sperm and thought they could see a tiny creature inside, which they termed a homunculus (little man). This homunculus was hypothesized to be a miniature human waiting to develop within the womb of its mother. Those who held that thought, known as spermists, suggested that only the father was responsible for creating future generations and that any resemblance between mother and offspring was due to influences "within the womb." During the same time, an opposite school of thought also developed. According to the ovists, the egg was solely responsible for human characteristics. The only role of the sperm was to stimulate the egg onto its path of development. Of course, neither of these ideas was correct.

The first systematic studies of genetic crosses were carried out by Joseph Kölreuter from 1761 to 1766. In crosses between different strains of tobacco plants, he found that the offspring were usually intermediate in appearance between the two parents. This led Kölreuter to conclude that both parents make equal genetic contributions to their offspring. Furthermore, his observations were consistent with **blending inheritance.** According to this view, the factors that dictate hereditary traits can blend together from generation to generation. The blended traits would then be passed to the next generation. The popular view before the 1860s, which combined the notions of pangenesis and blending inheritance, was that hereditary traits were rather malleable and could change and blend over the course of one or two generations. However, the pioneering work of Gregor Mendel would prove instrumental in refuting this viewpoint.

In Chapter 2, we will first examine the outcome of Mendel's crosses in pea plants. We begin our inquiry into genetics here because the inheritance patterns observed in peas are fundamentally related to inheritance patterns found in other eukaryotic species such as humans, mice, fruit flies, and corn. We will

The garden pea, studied by Mendel.

PART II
Patterns of Inheritance

discover how Mendel's insights into the patterns of inheritance in pea plants revealed some simple rules that govern the process of inheritance. In Chapters 3 through 8, we will explore more complex patterns of inheritance and also consider the role that the chromosomes play as the carriers of the genetic material.

In the second part of this chapter, we will become familiar with general concepts in probability and statistics. How are statistical methods useful? First, probability calculations allow us to predict the outcomes of simple genetic crosses, as well as the outcomes of more complicated crosses described in later chapters. In addition, we will learn how to use statistics to test the validity of genetic hypotheses that attempt to explain the inheritance patterns of traits.

2.1 MENDEL'S LAWS OF INHERITANCE

Gregor Johann Mendel, born in 1822, is now remembered as the father of genetics (**Figure 2.1**). He grew up on a small farm in Hynčice (formerly Heinzendorf) in northern Moravia, which was then a part of Austria and is now a part of the Czech Republic. As a young boy, he worked with his father grafting trees to improve the family orchard. Undoubtedly, his success at grafting taught him that precision and attention to detail are important elements of success. These qualities would later be important in his experiments as an adult scientist. Instead of farming, however, Mendel was accepted into the Augustinian monastery of St. Thomas, completed his studies for the priesthood, and was ordained in 1847. Soon after becoming a priest, Mendel worked for a short time as a substitute teacher. To continue that role, he needed to obtain a teaching license from the government. Surprisingly, he failed the licensing exam due to poor answers in the areas of physics and natural history. Therefore, Mendel then enrolled at the University of Vienna to expand his knowledge in these two areas. Mendel's training in physics and mathematics taught him to perceive the world as an orderly place, governed by natural laws. In his studies, Mendel learned that these natural laws could be stated as simple mathematical relationships.

In 1856, Mendel began his historic studies on pea plants. For eight years, he grew and crossed thousands of pea plants on a small 115- by 23-foot plot. He kept meticulously accurate records that included quantitative data concerning the outcome of his crosses. He published his work, entitled "Experiments on Plant Hybrids," in 1866. This paper was largely ignored by scientists at that time, possibly because of its title or because it was published in a rather obscure journal (*The Proceedings of the Brünn Society of Natural History*). Another reason his work went unrecognized could be tied to a lack of understanding of chromosomes and their transmission, a topic we will discuss in Chapter 3. Nevertheless, Mendel's groundbreaking work allowed him to propose the natural laws that now provide a framework for our understanding of genetics.

Prior to his death in 1884, Mendel reflected, "My scientific work has brought me a great deal of satisfaction and I am convinced that it will be appreciated before long by the whole

FIGURE 2.1   **Gregor Johann Mendel, the father of genetics.**

world." Sixteen years later, in 1900, the work of Mendel was independently rediscovered by three biologists with an interest in plant genetics: Hugo de Vries of Holland, Carl Correns of Germany, and Erich von Tschermak of Austria. Within a few years, the impact of Mendel's studies was felt around the world. In this section, we will examine Mendel's experiments and consider their monumental significance in the field of genetics.

Mendel Chose Pea Plants as His Experimental Organism

Mendel's study of genetics grew out of his interest in ornamental flowers. Prior to his work with pea plants, many plant breeders had conducted experiments aimed at obtaining flowers with new varieties of colors. When two distinct individuals with different characteristics are mated or **crossed** to each other, this is called a **hybridization** experiment, and the offspring are referred to as **hybrids.** For example, a hybridization experiment could involve a cross between a purple-flowered plant and a white-flowered plant. Mendel was particularly intrigued, in such experiments, by the consistency with which offspring of subsequent generations showed characteristics of one or the other parent. His intellectual foundation in physics and the natural sciences led him

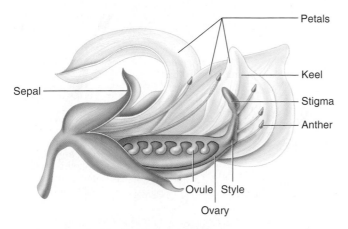

Petals

Keel

Stigma

Anther

Sepal

Ovule | Style

Ovary

(a) Structure of a pea flower

(b) A flowering pea plant

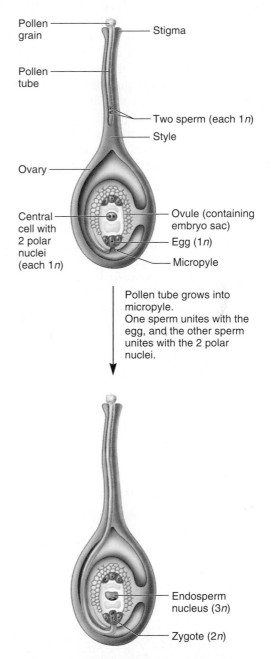

Pollen grain — Stigma

Pollen tube

Two sperm (each 1*n*)

Style

Ovary

Central cell with 2 polar nuclei (each 1*n*)

Ovule (containing embryo sac)

Egg (1*n*)

Micropyle

Pollen tube grows into micropyle.
One sperm unites with the egg, and the other sperm unites with the 2 polar nuclei.

Endosperm nucleus (3*n*)

Zygote (2*n*)

(c) Pollination and fertilization in angiosperms

FIGURE 2.2   **Flower structure and pollination in pea plants. (a)** The pea flower can produce both pollen and egg cells. The pollen grains are produced within the anthers, and the egg cells are produced within the ovules that are contained within the ovary. A modified petal called a keel encloses the anthers and ovaries. **(b)** Photograph of a flowering pea plant. **(c)** A pollen grain must first land on the stigma. After this occurs, the pollen sends out a long tube through which two sperm cells travel toward an ovule to reach an egg cell. The fusion between a sperm and an egg cell results in fertilization and creates a zygote. A second sperm fuses with a central cell containing two polar nuclei to create the endosperm. The endosperm provides a nutritive material for the developing embryo.

to consider that this regularity might be rooted in natural laws that could be expressed mathematically. To uncover these laws, he realized that he would need to carry out quantitative experiments in which the numbers of offspring carrying certain traits were carefully recorded and analyzed.

Mendel chose the garden pea, *Pisum sativum*, to investigate the natural laws that govern plant hybrids. The morphological features of this plant are shown in **Figure 2.2a** and **b.** Several properties of this species were particularly advantageous for studying plant hybridization. First, the species

was available in several varieties that had decisively different physical characteristics. Many strains of the garden pea were available that varied in the appearance of their height, flowers, seeds, and pods.

A second important issue is the ease of making crosses. In flowering plants, reproduction occurs by a pollination event (**Figure 2.2c**). Male gametes (**sperm**) are produced within **pollen grains** formed in the **anthers,** while the female gametes (**eggs**) are contained within **ovules** that form in the **ovaries.** For fertilization to occur, a pollen grain lands on the **stigma,** which stimulates the growth of a pollen tube. This enables sperm cells to enter the stigma and migrate toward an ovule. Fertilization occurs when a sperm enters the micropyle, an opening in the ovule wall, and fuses with an egg cell. The term **gamete** is used to describe haploid reproductive cells that can unite to form a zygote. It should be emphasized, however, that the process that produces gametes in animals is quite different from the way that gametes are produced in plants and fungi. These processes are described in greater detail in Chapter 3.

In some experiments, Mendel wanted to carry out **self-fertilization,** which means that the pollen and egg are derived from the same plant. In peas, a modified petal known as the keel covers the reproductive structures of the plant. Because of this covering, pea plants naturally reproduce by self-fertilization. In fact, pollination occurs even before the flower opens. In other experiments, however, Mendel wanted to make crosses between different plants. How did he accomplish this goal? Fortunately, pea plants contain relatively large flowers that are easy to manipulate, making it possible to make crosses between two particular plants and study their outcomes. This process, known as **cross-fertilization,** requires that the pollen from one plant be placed on the stigma of another plant. This procedure is shown in **Figure 2.3**. Mendel was able to pry open immature flowers and remove the anthers before they produced pollen. Therefore, these flowers could not self-fertilize. He would then obtain pollen from another plant by gently touching its mature anthers with a paintbrush. Mendel applied this pollen to the stigma of the flower that already had its anthers removed. In this way, he was able to cross-fertilize his pea plants and thereby obtain any type of hybrid he wanted.

Mendel Studied Seven Traits That Bred True

When he initiated his studies, Mendel obtained several varieties of peas that were considered to be distinct. These plants were different with regard to many morphological characteristics. Such characteristics of an organism are called characters, or **traits.** Over the course of two years, Mendel tested the strains to determine if their characteristics bred true. This means that a trait did not vary in appearance from generation to generation. For example, if the seeds from a pea plant were yellow, the next generation would also produce yellow seeds. Likewise, if these offspring were allowed to self-fertilize, all of their offspring would also produce yellow seeds, and so on. A variety that continues to produce the same characteristic after several generations of self-fertilization is called a **true-breeding line,** or **strain.**

Mendel next concentrated his efforts on the analysis of characteristics that were clearly distinguishable between different true-breeding lines. **Figure 2.4** illustrates the seven traits that

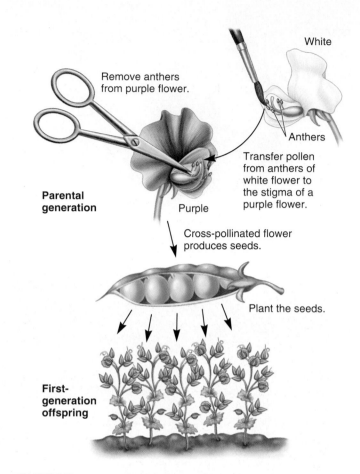

FIGURE 2.3   **How Mendel cross-fertilized two different pea plants.** This illustration depicts a cross between a plant with purple flowers and another plant with white flowers. The offspring from this cross are the result of pollination of the purple flower using pollen from a white flower.

Mendel eventually chose to follow in his breeding experiments. All seven were found in two variants. The term **variant** refers to a particular trait that may be found in two or more versions within a single species. For example, one trait he followed was height, which was found in two variants—tall and dwarf plants. Mendel studied this trait by crossing the variants to each other. A cross in which an experimenter is observing only one trait is called a **single-factor cross,** also called a **monohybrid cross.** When the two parents are different variants for a given trait, this type of cross produces single-trait hybrids, also known as **monohybrids.**

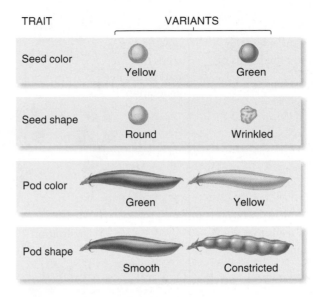

TRAIT   VARIANTS

Height   Tall   Dwarf

Flower color   Purple   White

Flower position   Axial   Terminal

TRAIT   VARIANTS

Seed color   Yellow   Green

Seed shape   Round   Wrinkled

Pod color   Green   Yellow

Pod shape   Smooth   Constricted

FIGURE 2.4   An illustration of the seven traits that Mendel studied. Each trait was found as two variants that were decisively different from each other.

EXPERIMENT 2A

Mendel Followed the Outcome of a Single Trait for Two Generations

Prior to conducting his studies, Mendel did not already have a hypothesis to explain the formation of hybrids. However, his educational background caused him to realize that a quantitative analysis of crosses may uncover mathematical relationships that would otherwise be mysterious. His experiments were designed to determine the relationships that govern hereditary traits. This rationale is called an **empirical approach.** Laws that are deduced from an empirical approach are known as empirical laws.

Mendel's experimental procedure is shown in **Figure 2.5.** He began with true-breeding plants that differed with regard to a single trait. These are termed the **parental generation,** or **P** gen-

eration. When the true-breeding parents were crossed to each other, this is called a P cross, and the offspring constitute the **F$_1$ generation,** for first filial generation. As seen in the data, all plants of the F$_1$ generation showed the phenotype of one parent but not the other. This prompted Mendel to follow the transmission of this trait for one additional generation. To do so, the plants of the F$_1$ generation were allowed to self-fertilize to produce a second generation called the **F$_2$ generation,** for second filial generation.

■ THE GOAL

Mendel speculated that the inheritance pattern for a single trait may follow quantitative natural laws. The goal of this experiment was to uncover such laws.

■ ACHIEVING THE GOAL — FIGURE 2.5   Mendel's analysis of single-factor crosses.

Starting material: Mendel began his experiments with true-breeding pea plants that varied with regard to only one of seven different traits (see Figure 2.4).

1. For each of seven traits, Mendel cross-fertilized two different true-breeding lines. Keep in mind that each cross involved two plants that differed in regard to only one of the seven traits studied. The illustration at the right shows one cross between a tall and dwarf plant. This is called a P (parental) cross.

Experimental level

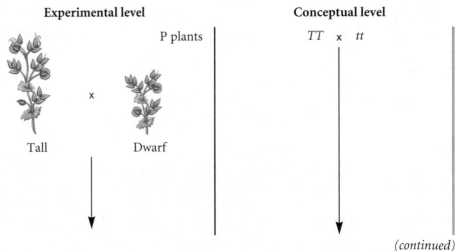

P plants

Tall   x   Dwarf

Conceptual level

TT   x   *tt*

(continued)

2. Collect many seeds. The following spring, plant the seeds and allow the plants to grow. These are the plants of the F_1 generation.

3. Allow the F_1 generation plants to self-fertilize. This produces seeds that are part of the F_2 generation.

4. Collect the seeds and plant them the following spring to obtain the F_2 generation plants.

5. Analyze the characteristics found in each generation.

Note: The P cross produces seeds that are part of the F_1 generation.

F_1 seeds

F_1 plants

All tall

Self-fertilization

F_2 seeds

F_2 plants

Tall   Tall   Dwarf   Tall

All Tt

Tt

Self-fertilization

$TT + 2\ Tt + tt$

■ THE DATA

P Cross	F_1 Generation	F_2 Generation	Ratio
Tall × dwarf stem	All tall	787 tall, 277 dwarf	2.84:1
Purple × white flowers	All purple	705 purple, 224 white	3.15:1
Axial × terminal flowers	All axial	651 axial, 207 terminal	3.14:1
Yellow × green seeds	All yellow	6,022 yellow, 2,001 green	3.01:1
Round × wrinkled seeds	All round	5,474 round, 1,850 wrinkled	2.96:1
Green × yellow pods	All green	428 green, 152 yellow	2.82:1
Smooth × constricted pods	All smooth	882 smooth, 299 constricted	2.95:1
Total	All dominant	14,949 dominant, 5,010 recessive	2.98:1

■ INTERPRETING THE DATA

The data shown in Figure 2.5 are the results of producing an F_1 generation via cross-fertilization, and an F_2 generation via self-fertilization of the F_1 monohybrids. A quantitative analysis of these data allowed Mendel to propose three important ideas:

1. Mendel's data argued strongly against a blending mechanism of heredity. In all seven cases, the F_1 generation displayed characteristics that were distinctly like one of the two parents rather than traits intermediate in character. Using genetic terms that Mendel originated and are still used today, his first proposal was that the variant for one trait is **dominant** over another variant. For example, the variant of green pods is dominant to that of yellow pods. The term **recessive** is used to describe a variant that is masked by the presence of a dominant trait but reappears in subsequent generations. Yellow pods and dwarf stems are examples of recessive variants. They can also be referred to as recessive traits.

2. When a true-breeding plant with a dominant trait was crossed to a true-breeding plant with a recessive trait, the dominant trait was always observed in the F_1 generation. In the F_2 generation, some offspring displayed the dominant characteristic, while a smaller proportion showed the recessive trait. However, none of the offspring exhibited intermediate traits. How did Mendel explain this observation? Because the recessive trait appeared in the F_2 generation, he formed a second proposal—the genetic determinants of traits are passed along as "unit factors" from generation to generation. His data were consistent with a **particulate theory of inheritance,** in which the genes that govern traits are inherited as discrete units that remain

unchanged as they are passed from parent to offspring. Mendel called them unit factors, but we now call them genes.

3. A third important interpretation of Mendel's data is related to the proportions of offspring. When Mendel compared the numbers of dominant and recessive offspring in the F_2 generation, he noticed a recurring pattern. Within experimental variation, he always observed approximately a 3:1 ratio between the dominant trait and the recessive trait. Mendel was the first scientist to apply this type of quantitative analysis in a biological experiment. As described next, this quantitative approach allowed him to make a third proposal—genes **segregate** from each other during the process that gives rise to gametes.

A self-help quiz involving this experiment can be found at the Online Learning Center.

The 3:1 Phenotypic Ratio That Mendel Observed Is Consistent with the Segregation of Alleles, Now Known as Mendel's Law of Segregation

Mendel's research was aimed at understanding the laws that govern the inheritance of traits. At that time, scientists did not understand the molecular composition of the genetic material or its mode of transmission during gamete formation and fertilization. We now know that the genetic material is composed of deoxyribonucleic acid (DNA), a component of chromosomes. Each chromosome contains hundreds or thousands of shorter segments that function as genes—a term that was originally coined by the Danish botanist Wilhelm Johannsen in 1909. A **gene** is defined as a "unit of heredity" that may influence the outcome of an organism's traits. Each of the seven traits that Mendel studied is influenced by a different gene.

Most eukaryotic species, such as pea plants and humans, have their genetic material organized into pairs of chromosomes. For this reason, eukaryotes have two copies of most genes. These copies may be the same or they may differ. The term **allele** refers to different versions of the same gene. With this modern knowledge, the results shown in Figure 2.5 are consistent with the idea that each parent transmits only one copy of each gene (i.e., one allele) to each offspring. **Mendel's law of segregation** states that:

The two copies of a gene segregate (or separate) from each other during transmission from parent to offspring.

Therefore, only one copy of each gene is found in a gamete. At fertilization, two gametes combine randomly, potentially producing different allelic combinations.

Let's use Mendel's cross of tall and dwarf pea plants to illustrate how alleles are passed from parents to offspring (**Figure 2.6**). The letters T and t are used to represent the alleles of the gene that determines plant height. By convention, the uppercase letter represents the dominant allele (T for tall height, in this case), and the recessive allele is represented by the same letter in lowercase (t, for dwarf height). For the P cross, both parents are true-breeding plants. Therefore, we know each has identical copies of the height gene. When an individual possesses two identical copies of a gene, the individual is said to be **homozygous** with respect to that gene. (The prefix *homo* means like, and the suffix *zygo* means pair.) In the P cross, the tall plant is homozygous for the tall allele T, while the dwarf plant is homozygous for the dwarf allele t. The term **genotype** refers to the genetic composition of an individual. TT and tt are the genotypes of the P generation in this experiment. The term **phenotype** refers to an observable characteristic of an organism. In the P generation, half of the plants are phenotypically tall and half are dwarf.

In contrast, the F_1 generation is **heterozygous,** with the genotype Tt, because every individual carries one copy of the tall allele and one copy of the dwarf allele. A heterozygous individual carries different alleles of a gene. (The prefix *hetero* means different.) Although these plants are heterozygous, their phenotypes are tall because they have a copy of the dominant tall allele.

The law of segregation predicts that the phenotypes of the F_2 generation will be tall and dwarf in a ratio of 3:1 (see Figure 2.6). Because the parents of the F_2 generation are heterozygous, each parent can transmit either a T allele or a t allele to a particular offspring, but not both, because each gamete carries only one of the two alleles. Therefore, TT, Tt, and tt are the possible genotypes of the F_2 generation (note that the genotype Tt is the same as tT). By randomly combining these alleles, the genotypes are produced in a 1:2:1 ratio. Because TT and Tt both produce tall phenotypes, a 3:1 phenotypic ratio is observed in the F_2 generation.

A Punnett Square Can Be Used to Predict the Outcome of Crosses

An easy way to predict the outcome of simple genetic crosses is to use a **Punnett square,** a method originally proposed by Reginald Punnett. To construct a Punnett square, you must know the genotypes of the parents. With this information, the Punnett square enables you to predict the types of offspring the parents are expected to produce and in what proportions. We will follow a step-by-step description of the Punnett square approach using a cross of heterozygous tall plants as an example.

Step 1. Write down the genotypes of both parents. In this example, a heterozygous tall plant is crossed to another heterozygous tall plant. The plant providing the pollen is considered the male parent and the plant providing the eggs, the female parent.

Male parent: Tt

Female parent: Tt

Step 2. *Write down the possible gametes that each parent can make.* Remember that the law of segregation tells us that a gamete can contain only one copy of each gene.

Male gametes: T or t

Female gametes: T or t

PHENOTYPE   GENOTYPE

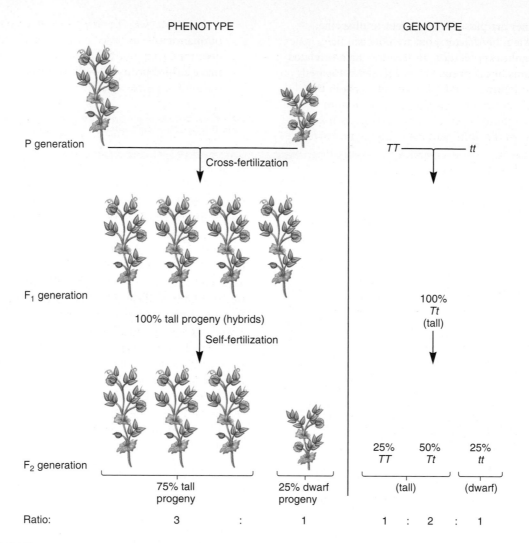

P generation — Cross-fertilization

TT ——— tt

F₁ generation

100% tall progeny (hybrids)

Self-fertilization

100%
Tt
(tall)

F₂ generation

75% tall progeny   25% dwarf progeny

25% 50% 25%
TT Tt tt

(tall)   (dwarf)

Ratio: 3   :   1   1 : 2 : 1

FIGURE 2.6   Mendel's law of segregation. This illustration shows a cross between a true-breeding tall plant and a true-breeding dwarf plant and the subsequent segregation of the tall (T) and dwarf (t) alleles in the F₁ and F₂ generations.

Step 3. *Create an empty Punnett square.* In the examples shown in this textbook, the number of columns equals the number of male gametes, and the number of rows equals the number of female gametes. Our example has two rows and two columns. Place the male gametes across the top of the Punnett square and the female gametes along the side.

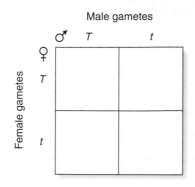

Step 4. *Fill in the possible genotypes of the offspring by combining the alleles of the gametes in the empty boxes.*

Male gametes

♂ T t

♀

T TT Tt

t Tt tt

Female gametes

Step 5. *Determine the relative proportions of genotypes and phenotypes of the offspring.* The genotypes are obtained directly from the Punnett square. They are contained within the boxes that have been filled in. In this example, the genotypes are *TT*, *Tt*, and *tt* in a 1:2:1 ratio. To determine the phenotypes, you must know the dominant/recessive relationship between the alleles. For plant height, we know that *T* (tall) is dominant to *t* (dwarf).

The genotypes *TT* and *Tt* are tall, whereas the genotype *tt* is dwarf. Therefore, our Punnett square shows us that the ratio of phenotypes is 3:1, or 3 tall plants : 1 dwarf plant. Additional problems of this type are provided in the Solved Problems at the end of this chapter.

EXPERIMENT 2B

Mendel Also Analyzed Crosses Involving Two Different Traits

Though his experiments described in Figure 2.5 revealed important ideas regarding hereditary laws, Mendel realized that additional insights might be uncovered if he conducted more complicated experiments. In particular, he conducted crosses in which he simultaneously investigated the pattern of inheritance for two different traits. In other words, he carried out **two-factor crosses,** also called **dihybrid crosses,** in which he followed the inheritance of two different traits within the same groups of individuals. For example, let's consider an experiment in which one of the traits was seed shape, found in round or wrinkled variants; the second trait was seed color, which existed as yellow and green variants. In this dihybrid cross, Mendel followed the inheritance pattern for both traits simultaneously.

What results are possible from a dihybrid cross? One possibility is that the genetic determinants for two different traits are always linked to each other and inherited as a single unit (**Figure 2.7a**). If this were the case, the F₁ offspring could produce only two types of gametes, *RY* and *ry*. A second possibility is that they are not linked and can assort themselves independently into hap-

loid gametes (**Figure 2.7b**). According to independent assortment, an F₁ offspring could produce four types of gametes, *RY*, *Ry*, *rY*, and *ry*. Keep in mind that the results of Figure 2.5 have already shown us that a gamete carries only one allele for each gene.

The experimental protocol of one of Mendel's two-factor crosses is shown in **Figure 2.8**. He began with two different strains of true-breeding pea plants that were different with regard to two traits. In this example, one plant was produced from seeds that were round and yellow; the other plant from seeds that were wrinkled and green. When these plants were crossed, the seeds, which contain the plant embryo, are considered part of the F₁ generation. As expected, the data revealed that the F₁ seeds displayed a phenotype of round and yellow. This was observed because round and yellow are dominant traits. It is the F₂ generation that supports the independent assortment model and refutes the linkage model.

■ THE HYPOTHESES

The inheritance pattern for two different traits follows one or more quantitative natural laws. Two possible hypotheses are described in Figure 2.7.

(a) HYPOTHESIS: Linked assortment

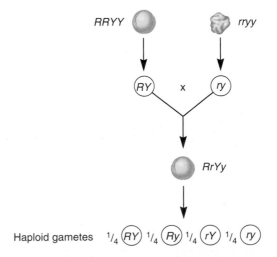

(b) HYPOTHESIS: Independent assortment

FIGURE 2.7 **Two hypotheses to explain how two different genes assort during gamete formation.** (**a**) According to the linked hypothesis, the two genes always stay associated with each other. (**b**) In contrast, the independent assortment hypothesis proposes that the two different genes randomly segregate into haploid cells.

■ **TESTING THE HYPOTHESES — FIGURE 2.8** Mendel's analysis of two-factor crosses.

Starting material: In this experiment, Mendel began with two types of true-breeding pea plants that were different with regard to two traits. One plant had round, yellow seeds (*RRYY*); the other plant had wrinkled, green seeds (*rryy*).

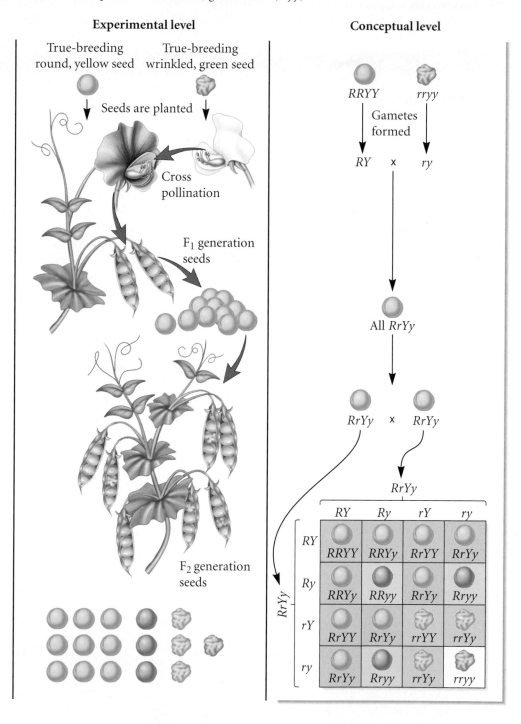

	Experimental level	**Conceptual level**

1. Cross the two true-breeding plants to each other. This produces F$_1$ generation seeds.

2. Collect many seeds and record their phenotype.

3. F$_1$ seeds are planted and grown, and the F$_1$ plants are allowed to self-fertilize. This produces seeds that are part of the F$_2$ generation.

4. Analyze the characteristics found in the F$_2$ generation seeds.

■ **THE DATA**

P cross	F$_1$ generation	F$_2$ generation
Round, yellow × wrinkled, green seeds	All round, yellow	315 round, yellow seeds
		108 round, green seeds
		101 wrinkled, yellow seeds
		32 wrinkled, green seeds

■ **INTERPRETING THE DATA**

The F$_2$ generation had seeds that were round and green and seeds that were wrinkled and yellow. These two categories of F$_2$ seeds are called **nonparentals** because these combinations of traits were not found in the true-breeding plants of the parental generation. The occurrence of nonparental variants contradicts the linkage model. According to the linkage model, the *R* and *Y* variants should be linked together and so should the *r* and *y* variants.

If this were the case, the F₁ plants could produce gametes that are only *RY* or *ry*. These would combine to produce *RRYY* (round, yellow), *RrYy* (round, yellow), or *rryy* (wrinkled, green) in a 1:2:1 ratio. Nonparental seeds could not be produced. However, Mendel did not obtain this result. Instead, he observed a phenotypic ratio of 9:3:3:1 in the F₂ generation.

Mendel's results from many dihybrid experiments rejected the hypothesis of linked assortment and, instead, supported the hypothesis that different traits assort themselves independently during reproduction. Using the modern notion of genes, **Mendel's law of independent assortment** states:

Two different genes will randomly assort their alleles during the formation of haploid cells.

In other words, the allele for one gene will be found within a resulting gamete independently of whether the allele for a different gene is found in the same gamete. Using the example given in Figure 2.8, the round and wrinkled alleles will be assorted into haploid gametes independently of the yellow and green alleles. Therefore, a heterozygous *RrYy* parent can produce four different gametes—*RY*, *Ry*, *rY*, and *ry*—in equal proportions.

In an F₁ self-fertilization experiment, any two gametes can combine randomly during fertilization. This allows for 4², or 16, possible offspring, although some offspring will be genetically identical to each other. As shown in **Figure 2.9**, these 16 possible combinations result in seeds with the following phenotypes: 9 round, yellow; 3 round, green; 3 wrinkled, yellow; and 1 wrinkled, green. This 9:3:3:1 ratio is the expected outcome when a dihybrid is allowed to self-fertilize. Mendel was clever enough

to realize that the data for his dihybrid experiments were close to a 9:3:3:1 ratio. In Figure 2.8, for example, his F₁ generation produced F₂ seeds with the following characteristics: 315 round, yellow seeds; 108 round, green seeds; 101 wrinkled, yellow seeds; and 32 wrinkled, green seeds. If we divide each of these numbers by 32 (the number of plants with wrinkled, green seeds), the phenotypic ratio of the F₂ generation is 9.8:3.2:3.4:1.0. Within experimental error, Mendel's data approximated the predicted 9:3:3:1 ratio for the F₂ generation.

The law of independent assortment held true for all seven traits that Mendel studied in pea plants. However, in other cases, the inheritance pattern of two different genes is consistent with the linkage model described earlier in Figure 2.7a. In Chapter 5, we will examine the inheritance of genes that are linked to each other because they are physically within the same chromosome. As we will see, linked genes do not assort independently.

An important consequence of the law of independent assortment is that a single individual can produce a vast array of genetically different gametes. As mentioned in Chapter 1, diploid species have pairs of homologous chromosomes, which may differ with respect to the alleles they carry. When an offspring receives a combination of alleles that differs from those in the parental generation, this phenomenon is termed **genetic recombination.** One mechanism that accounts for genetic recombination is independent assortment. A second mechanism, discussed in Chapter 5, is crossing over, which can reassort alleles that happen to be linked along the same chromosome.

The phenomenon of independent assortment is rooted in the random pattern by which the homologues assort themselves

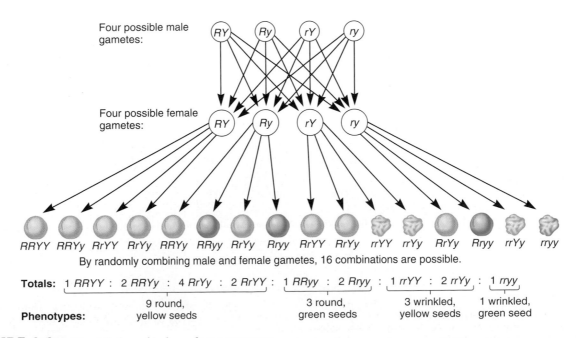

FIGURE 2.9   Mendel's law of independent assortment.

Genes→Traits The cross is between two parents that are heterozygous for seed shape and seed color (*RrYy* × *RrYy*). Four types of male gametes are possible: *RY*, *Ry*, *rY*, and *ry*. Likewise, four types of female gametes are possible: *RY*, *Ry*, *rY*, and *ry*. These four types of gametes are the result of the independent assortment of the seed shape and seed color alleles relative to each other. During fertilization, any one of the four types of male gametes can combine with any one of the four types of female gametes. This results in 16 types of offspring, each one containing two copies of the seed shape gene and two copies of the seed color gene.

during the process of meiosis, a topic addressed in Chapter 3. If a species contains a large number of homologous chromosomes, this creates the potential for an enormous amount of genetic diversity. For example, human cells contain 23 pairs of homologous chromosomes. These pairs can randomly assort into gametes during meiosis. The number of different gametes an individual can make equals 2^n, where n is the number of pairs of chromosomes. Therefore, humans can make 2^{23}, or over 8 million, possible gametes, due to independent assortment. The capacity to make so many genetically different gametes enables a species to produce individuals with many different combinations of traits. This allows environmental forces to select for those combinations of traits that favor reproductive success.

A self-help quiz involving this experiment can be found at the Online Learning Center.

A Punnett Square Can Also Be Used to Solve Independent Assortment Problems

As already depicted in Figure 2.8, we can use a Punnett square to predict the outcome of crosses involving two or more genes that assort independently. Let's see how such a Punnett square is made by considering a cross between two plants that are heterozygous for height and seed color (**Figure 2.10**). This cross is *TtYy* × *TtYy*. When we construct a Punnett square for this cross, we must keep in mind that each gamete has a single allele for each of two genes. In this example, the four possible gametes from each parent are

TY, Ty, tY, and *ty*

In this dihybrid experiment, we need to make a Punnett square containing 16 boxes. The phenotypes of the resulting offspring are predicted to occur in a ratio of 9:3:3:1.

In crosses involving three or more genes, the construction of a single large Punnett square to predict the outcome of crosses becomes very unwieldy. For example, in a trihybrid cross between two pea plants that are *Tt Rr Yy*, each parent can make 2^3, or 8, possible gametes. Therefore, the Punnett square must contain $8 \times 8 = 64$ boxes. As a more reasonable alternative, we can consider each gene separately and then algebraically combine them by multiplying together the expected outcomes for each gene. Two such methods, termed the **multiplication method** and the **forked-line method**, are shown in solved problem S3 at the end of this chapter.

Independent assortment is also revealed by a **dihybrid testcross.** In this type of experiment, dihybrid individuals are mated to individuals that are doubly homozygous recessive for the two traits. For example, individuals with a *TtYy* genotype could be crossed to *ttyy* plants. As shown here, independent assortment would predict a 1:1:1:1 ratio among the resulting offspring.

	TY	Ty	tY	ty
ty	*TtYy* Tall, yellow	*Ttyy* Tall, green	*ttYy* Dwarf, yellow	*ttyy* Dwarf, green

Cross: *TtYy* x *TtYy*

♀ \ ♂	TY	Ty	tY	ty
TY	*TTYY* Tall, yellow	*TTYy* Tall, yellow	*TtYY* Tall, yellow	*TtYy* Tall, yellow
Ty	*TTYy* Tall, yellow	*TTyy* Tall, green	*TtYy* Tall, yellow	*Ttyy* Tall, green
tY	*TtYY* Tall, yellow	*TtYy* Tall, yellow	*ttYY* Dwarf, yellow	*ttYy* Dwarf, yellow
ty	*TtYy* Tall, yellow	*Ttyy* Tall, green	*ttYy* Dwarf, yellow	*ttyy* Dwarf, green

Genotypes: 1 *TTYY* : 2 *TTYy* : 4 *TtYy* : 2 *TtYY* : 1 *TTyy* : 2 *Ttyy* : 1 *ttYY* : 2 *ttYy* : 1 *ttyy*

Phenotypes: 9 tall plants with yellow seeds | 3 tall plants with green seeds | 3 dwarf plants with yellow seeds | 1 dwarf plant with green seeds

FIGURE 2.10   A Punnett square for a dihybrid cross. The Punnett square shown here involves a cross between two pea plants that are heterozygous for height and seed color. The cross is *TtYy* × *TtYy*.

Modern Geneticists Are Often Interested in the Relationship Between the Molecular Expression of Genes and the Outcome of Traits

Mendel's work with pea plants was critically important because his laws of inheritance pertain to all eukaryotic organisms, such as fruit flies, corn, roundworms, mice, and humans, that transmit their genes through sexual reproduction. During the past several decades, many researchers have focused their attention on the relationship between the phenotypic appearance of traits and the molecular expression of genes. This theme will recur throughout the textbook (and we will draw attention to it by designating certain figure legends with a "Genes → Traits" label). As mentioned in Chapter 1, most genes encode proteins that function within living cells. The specific function of individual proteins affects the outcome of an individual's traits. A genetic approach can help us understand the relationship between a protein's function and its effect on phenotype. Most commonly, a geneticist will try to identify an individual that has a defective copy of a gene to see how that will affect the phenotype of the organism. These defective genes are called **loss-of-function alleles,** and they provide geneticists with a great amount of information. Unknowingly, Gregor Mendel had studied seven loss-of-function alleles among his strains of plants. The recessive characteristics in his pea plants were due to genes that had been rendered defective by a mutation. Such alleles are often inherited in a recessive manner, though this is not always the case.

How are loss-of-function alleles informative? In many cases, such alleles provide critical clues concerning the purpose of the protein's function within the organism. For example, we expect the gene affecting flower color (purple versus white) to encode a protein that is necessary for pigment production. This protein may function as an enzyme that is necessary for the synthesis of purple pigment. Furthermore, a reasonable guess is that the white allele is a loss-of-function allele that is unable to express this protein and therefore cannot make the purple pigment. To confirm this idea, a biochemist could analyze the petals from purple and white flowers and try to identify the protein that is defective or missing in the white petals but functionally active in the purple ones. The identification and characterization of this protein would provide a molecular explanation for this phenotypic characteristic.

Pedigree Analysis Can Be Used to Follow the Mendelian Inheritance of Traits in Humans

Before we end our discussion of simple Mendelian traits, let's address the question of how we can analyze inheritance patterns among humans. In his experiments, Mendel selectively made crosses and then analyzed a large number of offspring. When studying human traits, however, researchers cannot control parental crosses. Instead, they must rely on the information that is contained within family trees. This type of approach, known as a **pedigree analysis,** is aimed at determining the type of inheritance pattern that a gene will follow. Although this method may be less definitive than the results described in Mendel's experiments, a pedigree analysis can often provide important clues concerning the pattern of inheritance of traits within human families. An expanded discussion of human pedigrees is provided

in Chapter 22, which concerns the inheritance patterns of many different human diseases.

In order to discuss the applications of pedigree analyses, we need to understand the organization and symbols of a pedigree (**Figure 2.11**). The oldest generation is at the top of the pedigree, and the most recent generation is at the bottom. Vertical

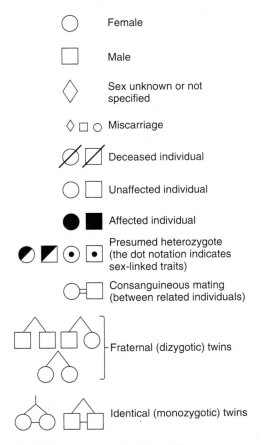

(a) Human pedigree showing cystic fibrosis

○ Female

□ Male

◇ Sex unknown or not specified

◇ □ ○ Miscarriage

⊘ ⊘ Deceased individual

○ □ Unaffected individual

● ■ Affected individual

◐ ◨ ⊙ ⊡ Presumed heterozygote (the dot notation indicates sex-linked traits)

○—□ Consanguineous mating (between related individuals)

Fraternal (dizygotic) twins

Identical (monozygotic) twins

(b) Symbols used in a human pedigree

FIGURE 2.11   Pedigree analysis. (a) A family pedigree in which some of the members are affected with cystic fibrosis. Individuals I-1, I-2, II-4, and II-5 are depicted as presumed heterozygotes because they produce affected offspring. **(b)** The symbols used in a pedigree analysis. Note: In pedigrees shown in this textbook, such as those found in the problem sets, the heterozygotes are not shown as half-filled symbols. Most pedigrees throughout the book show individuals' phenotypes—open symbols are unaffected individuals and filled (closed) symbols are affected individuals.

lines connect each succeeding generation. A man (square) and woman (circle) who produce one or more offspring are directly connected by a horizontal line. A vertical line connects parents with their offspring. If parents produce two or more offspring, the group of siblings (brothers and sisters) is denoted by two or more individuals projecting from the same horizontal line.

When a pedigree involves the transmission of a human trait or disease, affected individuals are depicted by filled symbols (in this case, black) that distinguish them from unaffected individuals. Each generation is given a roman numeral designation, and individuals within the same generation are numbered from left to right. A few examples of the genetic relationships in Figure 2.11a are described here:

> Individuals I-1 and I-2 are the grandparents of III-1, III-2, III-3, III-4, III-5, III-6, and III-7
>
> Individuals III-1, III-2, and III-3 are brother and sisters
>
> Individual III-4 is affected by a genetic disease

The symbols shown in Figure 2.11 depict certain individuals, such as I-1, I-2, II-4, and II-5, as presumed heterozygotes because they are unaffected with a disease but produce homozygous offspring that are affected with a recessive genetic disease. However, in many pedigrees, such as those found in the problem sets at the end of the chapter, the inheritance pattern may not be known, so the symbols reflect only phenotypes. In most pedigrees, affected individuals are shown with closed symbols, and unaffected individuals, including those that might be heterozygous for a recessive disease, are depicted with open symbols.

Pedigree analysis is commonly used to determine the inheritance pattern of human genetic diseases. Human geneticists are routinely interested in knowing whether a genetic disease is inherited as a recessive or dominant trait. One way to discern the dominant/recessive relationship between two alleles is by a pedigree analysis. Genes that play a role in disease may exist as a normal allele or a mutant allele that causes disease symptoms. If the disease follows a simple Mendelian pattern of inheritance and is caused by a recessive allele, an individual must inherit two copies of the mutant allele to exhibit the disease. Therefore, a recessive pattern of inheritance makes two important predictions. First, two heterozygous normal individuals will, on average, have 1/4 of their offspring affected. Second, all offspring of two affected individuals will be affected. Alternatively, a dominant trait predicts that affected individuals will have inherited the gene from at least one affected parent (unless a new mutation has occurred during gamete formation).

The pedigree in Figure 2.11a concerns a human genetic disease known as cystic fibrosis (CF). Among Caucasians, approximately 3% of the population are heterozygous carriers of this recessive allele. In homozygotes, the disease symptoms include abnormalities of the pancreas, intestine, sweat glands, and lungs. These abnormalities are caused by an imbalance of ions across the plasma membrane. In the lungs, this leads to a buildup of thick, sticky mucus. Respiratory problems may lead to early death, although modern treatments have greatly increased the life span of CF patients. In the late 1980s, the gene for CF was identified. The CF gene encodes a protein called the cystic fibrosis transmembrane conductance regulator (CFTR). This protein regulates the ion balance across the cell membrane in tissues of the pancreas, intestine, sweat glands, and lungs. The mutant allele causing CF alters the encoded CFTR protein. The altered CFTR protein is not correctly inserted into the plasma membrane, resulting in a decreased function that causes the ionic imbalance. As seen in the pedigree, the pattern of affected and unaffected individuals is consistent with a recessive mode of inheritance. Two unaffected individuals can produce an affected offspring. Although not shown in this pedigree, a recessive mode of inheritance is also characterized by the observation that two affected individuals will produce 100% affected offspring. However, for human genetic diseases that limit survival and/or fertility, there may never be cases where two affected individuals produce offspring.

2.2 PROBABILITY AND STATISTICS

A powerful application of Mendel's work is that the laws of inheritance can be used to predict the outcome of genetic crosses. In agriculture, for example, plant and animal breeders are concerned with the types of offspring their crosses will produce. This information is used to produce commercially important crops and livestock. In addition, people are often interested in predicting the characteristics of the children they may have. This may be particularly important to individuals who carry alleles that cause inherited diseases. Of course, we cannot see into the future and definitively predict what will happen. Nevertheless, genetic counselors can help couples to predict the likelihood of having an affected child. This probability is one factor that may influence a couple's decision whether to have children.

In this section, we will see how probability calculations are used in genetic problems to predict the outcome of crosses. To compute probability, we will use three mathematical operations known as the sum rule, the product rule, and the binomial expansion equation. These methods allow us to determine the probability that a cross between two individuals will produce a particular outcome. To apply these operations, we must have some knowledge regarding the genotypes of the parents and the pattern of inheritance of a given trait.

Probability calculations can also be used in hypothesis testing. In many situations, a researcher would like to discern the genotypes and patterns of inheritance for traits that are not yet understood. A traditional approach to this problem is to conduct crosses and then analyze their outcomes. The proportions of offspring may provide important clues that allow the experimenter to propose a hypothesis, based on the quantitative laws of inheritance, that explains the transmission of the trait from parent to offspring. Statistical methods, such as the chi square test, can then be used to evaluate how well the observed data from crosses fit the expected data. We will end this chapter with an example that applies the chi square test to a genetic cross.

Probability Is the Likelihood That an Event Will Occur

The chance that an event will occur in the future is called the event's **probability.** For example, if you flip a coin, the probability is 0.50, or 50%, that the head side will be showing when the coin lands. Probability depends on the number of possible outcomes. In this case, two possible outcomes (heads or tails) are equally likely. This allows us to predict a 50% chance that a coin flip will produce heads. The general formula for probability (P) is

$$\text{Probability} = \frac{\text{Number of times an event occurs}}{\text{Total number of events}}$$

$$P_{\text{heads}} = 1 \text{ heads} / (1 \text{ heads} + 1 \text{ tails}) = 1/2 = 50\%$$

In genetic problems, we are often interested in the probability that a particular type of offspring will be produced. Recall that when two heterozygous tall pea plants (Tt) are crossed, the phenotypic ratio of the offspring is 3 tall:1 dwarf. This information can be used to calculate the probability for either type of offspring.

$$\text{Probability} = \frac{\text{Number of individuals with a given phenotype}}{\text{Total number of individuals}}$$

$$P_{\text{tall}} = 3 \text{ tall} / (3 \text{ tall} + 1 \text{ dwarf}) = 3/4 = 75\%$$

$$P_{\text{dwarf}} = 1 \text{ dwarf} / (3 \text{ tall} + 1 \text{ dwarf}) = 1/4 = 25\%$$

The probability is 75% of obtaining a tall plant and 25% of obtaining a dwarf plant. When we add together the probabilities of all the possible outcomes (tall and dwarf), we should get a sum of 100% (here, 75% + 25% = 100%).

A probability calculation allows us to predict the likelihood that an event will occur in the future. The accuracy of this prediction, however, depends to a great extent on the size of the sample. For example, if we toss a coin six times, our probability prediction would suggest that 50% of the time we should get heads (i.e., three heads and three tails). In this small sample size, however, we would not be too surprised if we came up with four heads and two tails. Each time we toss a coin, there is a random chance that it will be heads or tails. The deviation between the observed and expected outcomes is called the **random sampling error**. In a small sample, the error between the predicted percentage of heads and the actual percentage observed may be quite large. By comparison, if we flipped a coin 1,000 times, the percentage of heads would be fairly close to the predicted 50% value. In a larger sample, we expect the random sampling error to be a much smaller percentage.

The Sum Rule Can Be Used to Predict the Occurrence of Mutually Exclusive Events

Now that we have an understanding of probability, we can see how mathematical operations using probability values allow us to predict the outcome of genetic crosses. Our first genetic problem involves the use of the **sum rule,** which states that *the probability that one of two or more mutually exclusive events will occur is equal to the sum of the individual probabilities of the events.* As an example, let's consider a cross between two mice that are both heterozygous for genes affecting the ears and tail. One gene can be found as an allele designated *de,* which is a recessive allele that

causes droopy ears; the normal allele is *De.* An allele of a second gene causes a crinkly tail. This crinkly tail allele (*ct*) is recessive to the normal allele (*Ct*). If a cross is made between two heterozygous mice (*Dede Ctct*), the predicted ratio of offspring is 9 with normal ears and normal tails, 3 with normal ears and crinkly tails, 3 with droopy ears and normal tails, and 1 with droopy ears and a crinkly tail. These four phenotypes are mutually exclusive. For example, a mouse with droopy ears and a normal tail cannot have normal ears and a crinkly tail.

The sum rule allows us to determine the probability that we will obtain any one of two or more different types of offspring. For example, in a cross between two heterozygotes (*Dede Ctct × Dede Ctct*), we can ask the following question: What is the probability that an offspring will have normal ears and a normal tail or have droopy ears and a crinkly tail? In other words, if we closed our eyes and picked an offspring out of a litter from this cross, what are the chances that we would be holding a mouse that has normal ears and a normal tail or a mouse with droopy ears and a crinkly tail? In this case, the investigator wants to predict whether one of two mutually exclusive events will occur. A strategy for solving such genetic problems using the sum rule is described here.

The Cross: *Dede Ctct × Dede Ctct*

The Question: What is the probability that an offspring will have normal ears and a normal tail or have droopy ears and a crinkly tail?

Step 1. *Calculate the individual probabilities of each phenotype.* This can be accomplished using a Punnett square.

The probability of normal ears and a normal tail is $9/(9 + 3 + 3 + 1) = 9/16$

The probability of droopy ears and a crinkly tail is $1/(9 + 3 + 3 + 1) = 1/16$

Step 2. *Add together the individual probabilities.*

$$9/16 + 1/16 = 10/16$$

This means that 10/16 is the probability that an offspring will have either normal ears and a normal tail or droopy ears and a crinkly tail. We can convert 10/16 to 0.625, which means that 62.5% of the offspring are predicted to have normal ears and a normal tail or droopy ears and a crinkly tail.

The Product Rule Can Be Used to Predict the Probability of Independent Events

We can use probability to make predictions regarding the likelihood of two or more independent outcomes from a genetic cross. When we say that events are independent, we mean that the occurrence of one event does not affect the probability of another event. As an example, let's consider a rare, recessive human trait known as congenital analgesia. Persons with this trait can distinguish between sharp and dull, and hot and cold, but do not perceive extremes of sensation as being painful. The first case of congenital analgesia, described in 1932, was a man who made his living entertaining the public as a "human pincushion."

For a phenotypically unaffected couple, each being heterozygous for the recessive allele causing congenital analgesia, we can ask the question, What is the probability that the couple's first three offspring will have congenital analgesia? To answer this question, the **product rule** is used. According to this rule, *the probability that two or more independent events will occur is equal to the product of their individual probabilities.* A strategy for solving this type of problem is shown here.

The Cross: $Pp \times Pp$ (where P is the common allele and p is the recessive congenital analgesia allele)

The Question: What is the probability that the couple's first three offspring will have congenital analgesia?

Step 1. *Calculate the individual probability of this phenotype.* As described previously, this is accomplished using a Punnett square.

The probability of an affected offspring is 1/4 (25%).

Step 2. *Multiply the individual probabilities.* In this case, we are asking about the first three offspring, and so we multiply 1/4 three times.

$$1/4 \times 1/4 \times 1/4 = 1/64 = 0.016$$

Thus, the probability that the first three offspring will have this trait is 0.016. In other words, we predict that 1.6% of the time the first three offspring of a couple, each heterozygous for the recessive allele, will all have congenital analgesia. In this example, the phenotypes of the first, second, and third offspring are independent events. In this case, the phenotype of the first offspring does not have an effect on the phenotype of the second or third offspring.

In the problem described here, we have used the product rule to determine the probability that the first three offspring will all have the same phenotype (congenital analgesia). We can also apply the rule to predict the probability of a sequence of events that involves combinations of different offspring. For example, consider the question, What is the probability that the first offspring will be unaffected, the second offspring will have congenital analgesia, and the third offspring will be unaffected? Again, to solve this problem, begin by calculating the individual probability of each phenotype.

Unaffected = 3/4

Congenital analgesia = 1/4

The probability that these three phenotypes will occur in this specified order is

$$3/4 \times 1/4 \times 3/4 = 9/64 = 0.14, \text{ or } 14\%$$

In other words, this sequence of events is expected to occur only 14% of the time.

The product rule can also be used to predict the outcome of a cross involving two or more genes. Let's suppose an individual with the genotype *Aa Bb CC* was crossed to an individual with the genotype *Aa bb Cc*. We could ask the question, What is the probability that an offspring will have the genotype *AA bb Cc*? If the three genes independently assort, the probability of

inheriting alleles for each gene is independent of the other two genes. Therefore, we can separately calculate the probability of the desired outcome for each gene.

Cross: *Aa Bb CC* × *Aa bb Cc*

Probability that an offspring will be *AA* = 1/4, or 0.25

Probability that an offspring will be *bb* = 1/2, or 0.5

Probability that an offspring will be *Cc* = 1/2, or 0.5

We can use the product rule to determine the probability that an offspring will be *AA bb Cc*;

$$P = (0.25)(0.5)(0.5) = 0.0625, \text{ or } 6.25\%$$

The Binomial Expansion Equation Can Be Used to Predict the Probability of an Unordered Combination of Events

A third predictive problem in genetics is to determine the probability that a certain proportion of offspring will be produced with particular characteristics; here they can be produced in an unspecified order. For example, we can consider a group of children produced by two heterozygous brown-eyed (*Bb*) individuals. We can ask the question, What is the probability that two out of five children will have blue eyes?

In this case, we are not concerned with the order in which the offspring are born. Instead, we are only concerned with the final numbers of blue-eyed and brown-eyed offspring. One possible outcome would be the following: firstborn child with blue eyes, second child with blue eyes, and then the next three with brown eyes. Another possible outcome could be firstborn child with brown eyes, second with blue eyes, third with brown eyes, fourth with blue eyes, and fifth with brown eyes. Both of these scenarios would result in two offspring with blue eyes and three with brown eyes. In fact, several other ways to have such a family could occur.

To solve this type of question, the **binomial expansion equation** can be used. This equation represents all of the possibilities for a given set of unordered events.

$$P = \frac{n!}{x!(n-x)!} p^x q^{n-x}$$

where

P = the probability that the unordered outcome will occur

n = total number of events

x = number of events in one category (e.g., blue eyes)

p = individual probability of x

q = individual probability of the other category (e.g., brown eyes)

Note: In this case, $p + q = 1$.

The symbol ! denotes a factorial. $n!$ is the product of all integers from n down to 1. For example, $4! = 4 \times 3 \times 2 \times 1 = 24$. An exception is 0!, which equals 1.

The use of the binomial expansion equation is described next.

The Cross: $Bb \times Bb$

The Question: What is the probability that two out of five offspring will have blue eyes?

Step 1. *Calculate the individual probabilities of the blue-eye and brown-eye phenotypes.* If we constructed a Punnett square, we would find the probability of blue eyes is 1/4 and the probability of brown eyes is 3/4:

$$p = 1/4$$

$$q = 3/4$$

Step 2. *Determine the number of events in category* x *(in this case, blue eyes) versus the total number of events.* In this example, the number of events in category x is two blue-eyed children among a total number of five.

$$x = 2$$

$$n = 5$$

Step 3. *Substitute the values* for p, q, x, *and* n *in the binomial expansion equation.*

$$P = \frac{n!}{x!(n-x)!} p^x q^{n-x}$$

$$P = \frac{5!}{2!(5-2)!} (1/4)^2 (3/4)^{5-2}$$

$$P = \frac{5 \times 4 \times 3 \times 2 \times 1}{(2 \times 1)(3 \times 2 \times 1)} (1/16)(27/64)$$

$$P = 0.26 = 26\%$$

Thus, the probability is 0.26 that two out of five offspring will have blue eyes. In other words, 26% of the time we expect a $Bb \times Bb$ cross yielding five offspring to contain two blue-eyed children and three brown-eyed children.

In solved problem S7 at the end of this chapter, we consider an expanded version of this approach that uses a **multinomial expansion equation.** This equation is needed to solve unordered genetic problems that involve three or more phenotypic categories.

The Chi Square Test Can Be Used to Test the Validity of a Genetic Hypothesis

We now look at a different issue in genetic problems, namely **hypothesis testing.** Our goal here is to determine if the data from genetic crosses are consistent with a particular pattern of inheritance. For example, a geneticist may study the inheritance of body color and wing shape in fruit flies over the course of two generations. The following question may be asked about the F_2 generation: Do the observed numbers of offspring agree with the predicted numbers based on Mendel's laws of segregation and independent assortment? As we will see in Chapters 3 through 8, not all traits follow a simple Mendelian pattern of inheritance. Some genes do not segregate and independently assort themselves the same way that Mendel's seven traits did in pea plants.

To distinguish between inheritance patterns that obey Mendel's laws versus those that do not, a conventional strategy is to make crosses and then quantitatively analyze the offspring. Based on the observed outcome, an experimenter may make a tentative hypothesis. For example, it may seem that the data are obeying Mendel's laws. Hypothesis testing provides an objective, statistical method to evaluate whether or not the observed data really agree with the hypothesis. In other words, we use statistical methods to determine whether the data that have been gathered from crosses are consistent with predictions based on quantitative laws of inheritance.

The rationale behind a statistical approach is to evaluate the **goodness of fit** between the observed data and the data that are predicted from a hypothesis. This is sometimes called a **null hypothesis** because it assumes there is no real difference between the observed and expected values. Any actual differences that occur are presumed to be due to random sampling error. If the observed and predicted data are very similar, we can conclude that the hypothesis is consistent with the observed outcome. In this case, it is reasonable to accept the hypothesis. However, it should be emphasized that this does not prove a hypothesis is correct. Statistical methods can never prove a hypothesis is correct. They can provide insight as to whether or not the observed data seem reasonably consistent with the hypothesis. Alternative hypotheses, perhaps even ones that the experimenter has failed to realize, may also be consistent with the data. In some cases, statistical methods may reveal a poor fit between hypothesis and data. In other words, a high deviation would be found between the observed and expected values. If this occurs, the hypothesis is rejected. Hopefully, the experimenter can subsequently propose an alternative hypothesis that has a reasonable fit with the data.

One commonly used statistical method to determine goodness of fit is the **chi square test** (often written χ^2). We can use the chi square test to analyze population data in which the members of the population fall into different categories. This is the kind of data we have when we evaluate the outcome of genetic crosses, because these usually produce a population of offspring that differ with regard to phenotypes. The general formula for the chi square test is

$$\chi^2 = \sum \frac{(O - E)^2}{E}$$

where

O = observed data in each category

E = expected data in each category based on the experimenter's hypothesis

Σ means to sum this calculation for each category. For example, if the population data fell into two categories, the chi square calculation would be

$$\chi^2 = \frac{(O_1 - E_1)^2}{E_1} + \frac{(O_2 - E_2)^2}{E_2}$$

We can use the chi square test to determine if a genetic hypothesis is consistent with the observed outcome of a genetic cross. The strategy described next provides a step-by-step outline for applying the chi square testing method. In this problem, the experimenter wants to determine if a dihybrid cross is obeying the laws of Mendel. The experimental organism is *Drosophila melanogaster* (the common fruit fly), and the two traits affect wing shape and body color. Straight wing shape and curved wing shape are designated by c^+ and c, respectively; gray body color and ebony body color are designated by e^+ and e, respectively. Note: In certain species, such as *Drosophila melanogaster*, the convention is to designate the common (wild-type) allele with a plus sign. Recessive mutant alleles are designated with lowercase letters and dominant mutant alleles with capital letters.

The Cross: A true-breeding fly with straight wings and a gray body ($c^+c^+e^+e^+$) is crossed to a true-breeding fly with curved wings and an ebony body ($ccee$). The flies of the F_1 generation are then allowed to mate with each other to produce an F_2 generation.

The Outcome:

F_1 generation:	All offspring have straight wings and gray bodies
F_2 generation:	193 straight wings, gray bodies
	69 straight wings, ebony bodies
	64 curved wings, gray bodies
	26 curved wings, ebony bodies
Total:	352

Step 1. *Propose a hypothesis that allows us to calculate the expected values based on Mendel's laws.* The F_1 generation suggests that the trait of straight wings is dominant to curved wings and gray body coloration is dominant to ebony. Looking at the F_2 generation, it appears that offspring are following a 9:3:3:1 ratio. If so, this is consistent with an independent assortment of the two traits.

Based on these observations, the hypothesis is:

Straight (c^+) is dominant to curved (c), and gray (e^+) is dominant to ebony (e). The two traits segregate and assort independently from generation to generation.

Step 2. *Based on the hypothesis, calculate the expected values of the four phenotypes.* We first need to calculate the individual probabilities of the four phenotypes. According to our hypothesis, there should be a 9:3:3:1 ratio in the F_2 generation. Therefore, the expected probabilities are:

9/16 = straight wings, gray bodies
3/16 = straight wings, ebony bodies
3/16 = curved wings, gray bodies
1/16 = curved wings, ebony bodies

The observed F_2 generation contained a total of 352 individuals. Our next step is to calculate the expected numbers of each type of offspring when the total equals 352. This can be accomplished by multiplying each individual probability by 352.

9/16 × 352 = 198 (expected number with straight wings, gray bodies)

3/16 × 352 = 66 (expected number with straight wings, ebony bodies)

3/16 × 352 = 66 (expected number with curved wings, gray bodies)

1/16 × 352 = 22 (expected number with curved wings, ebony bodies)

Step 3. *Apply the chi square formula, using the data for the expected values that have been calculated in step 2.* In this case, the data include four categories, and thus the sum has four terms.

$$\chi^2 = \frac{(O_1 - E_1)^2}{E_1} + \frac{(O_2 - E_2)^2}{E_2} + \frac{(O_3 - E_3)^2}{E_3} + \frac{(O_4 - E_4)^2}{E_4}$$

$$\chi^2 = \frac{(193 - 198)^2}{198} + \frac{(69 - 66)^2}{66} + \frac{(64 - 66)^2}{66} + \frac{(26 - 22)^2}{22}$$

$$\chi^2 = 0.13 + 0.14 + 0.06 + 0.73 = 1.06$$

Step 4. *Interpret the calculated chi square value. This is done using a chi square table.*

Before interpreting the chi square value we have obtained, we must understand how to use **Table 2.1**. The probabilities, called **P values,** listed in the chi square table allow us to determine the likelihood that the amount of variation indicated by a given chi square value is due to random chance alone, based on a particular hypothesis. For example, let's consider a value—0.00393—listed in row 1. (The meaning of the rows will be explained shortly.) Chi square values that are equal to or greater than 0.00393 are expected to occur 95% of the time when a hypothesis is correct. In other words, 95 out of 100 times we would expect that random chance alone would produce a deviation between the experimental data and hypothesized model that is equal to or greater than 0.00393. A low chi square value indicates a high probability that the observed deviations could be due to random chance alone. By comparison, chi square values that are equal to or greater than 3.841 are expected to occur less than 5% of the time due to random sampling error. If a high chi square value is obtained, an experimenter becomes suspicious that the high deviations have occurred because the hypothesis is incorrect. A common convention is to reject the null hypothesis if the chi square value results in a probability that is less than 0.05—less than 5%.

In our problem involving flies with straight or curved wings and gray or ebony bodies, we have calculated a chi square value of 1.06. Before we can determine the probability that this deviation would have occurred as a matter of random chance, we must first determine the degrees of freedom (*df*) in this experiment. The **degrees of freedom** is a measure of the number of categories that are independent of each other. When phenotype categories are derived from a Punnett square, it is typically $n - 1$, where n equals the total number of categories. In the preceding problem, $n = 4$ (the categories are the phenotypes: straight wings and gray body; straight wings and ebony body; curved wings and gray body; and curved wings and

TABLE 2.1

Chi Square Values and Probability

Degrees of Freedom	$P = 0.99$	0.95	0.80	0.50	0.20	Null Hypothesis rejected	
						0.05	0.01
1.	0.000157	0.00393	0.0642	0.455	1.642	3.841	6.635
2.	0.020	0.103	0.446	1.386	3.219	5.991	9.210
3.	0.115	0.352	1.005	2.366	4.642	7.815	11.345
4.	0.297	0.711	1.649	3.357	5.989	9.488	13.277
5.	0.554	1.145	2.343	4.351	7.289	11.070	15.086
6.	0.872	1.635	3.070	5.348	8.558	12.592	16.812
7.	1.239	2.167	3.822	6.346	9.803	14.067	18.475
8.	1.646	2.733	4.594	7.344	11.030	15.507	20.090
9.	2.088	3.325	5.380	8.343	12.242	16.919	21.666
10.	2.558	3.940	6.179	9.342	13.442	18.307	23.209
15.	5.229	7.261	10.307	14.339	19.311	24.996	30.578
20.	8.260	10.851	14.578	19.337	25.038	31.410	37.566
25.	11.524	14.611	18.940	24.337	30.675	37.652	44.314
30.	14.953	18.493	23.364	29.336	36.250	43.773	50.892

From Fisher, R. A., and Yates, F. (1943) *Statistical Tables for Biological, Agricultural, and Medical Research.* Oliver and Boyd, London.

ebony body); thus, the degrees of freedom equals 3.* We now have sufficient information to interpret our chi square value of 1.06.

With $df = 3$, the chi square value of 1.06 we have obtained is slightly greater than 1.005, which gives a P value of 0.80, or 80%. What does this P value mean? If the hypothesis is correct, chi square values equal to or greater than 1.005 are expected to

occur 80% of the time based on random chance alone. To reject the hypothesis, the chi square would have to be greater than 7.815. Because it was actually far less than this value, we are inclined to accept that the null hypothesis is correct.

We must keep in mind that the chi square test does not prove a hypothesis is correct. It is a statistical method for evaluating whether or not the data and hypothesis have a good fit.

CONCEPTUAL SUMMARY

The work of Gregor Mendel has led to great insight into the transmission of traits from parent to offspring. First, his work showed us that the determinants of traits are discrete units that are passed via gametes from generation to generation. This particulate theory of inheritance refuted previous ideas such as blending inheritance and pangenesis. In addition, Mendel deduced two fundamental laws that pertain to the transmission of many traits in eukaryotic organisms. The law of segregation tells us that two variants for a single trait will segregate during their passage to offspring. This means that the two alleles for a particular gene will segregate from each other during the formation of haploid cells, so a gamete will contain only one copy of an allele. Mendel's law of independent assortment says that two different genes assort independently of each other during the formation of haploid cells.

An understanding of genetic inheritance allows us to make predictions about future offspring and also to test the validity of genetic hypotheses. Probability is the likelihood that an event will occur in the future. By knowing the genotypes of parents, we can determine the probability of obtaining an offspring with a particular trait. Furthermore, mathematical operations can be used to predict combinations of offspring that may, or may not, occur in a specified order. The sum rule is used in genetic problems that involve mutually exclusive events; the product rule is used to analyze independent events; and the binomial expansion equation is used to predict unordered events.

In other genetic problems, the goal is to deduce the pattern of inheritance from parents to offspring. This requires an analysis of genetic crosses in order to propose a genetic hypothesis that is consistent with the observed data. A chi square test is used to evaluate whether or not the genetic hypothesis predicts values that have a reasonable agreement with the observed data. In this approach, the amount of deviation between the observed and

* If our hypothesis already assumed that the law of segregation is obeyed, the degrees of freedom would be one (see Chapter 5).

expected values is used to determine the **goodness of fit** between the data and hypothesis. If the deviation is so large that it would be expected to occur less than 5% of the time as a result of random sampling error, then it is appropriate to reject the hypothesis. If, on the other hand, the chi square value is low, the null hypothesis is accepted.

In Chapter 2, a foundation has been provided on which to base our understanding of simple inheritance patterns. The two laws that we examined, segregation and independent assortment, provide us with insight into the transmission of traits from parent to offspring. While they form a cornerstone for our understanding of genetics, we will see in later chapters that not all genes obey these laws. In Chapter 3, we will take a close look at the composition and behavior of chromosomes, particularly during meiosis. As we will see, the segregation and assortment of chromosomes provides a cellular explanation for Mendel's laws.

EXPERIMENTAL SUMMARY

In Chapter 2, we examined the research studies of Gregor Mendel. By making single-factor crosses and examining the traits of offspring for two generations, Mendel found that the determinants of traits do not blend. Rather, they are discrete units that are passed along unaltered from generation to generation. He quantitatively analyzed the outcome of his self-fertilization experiments among monohybrids and obtained a 3:1 phenotypic ratio for the F_2 generation. This observation allowed him to deduce the law of segregation.

In dihybrid experiments, Mendel followed the inheritance pattern of two genes relative to each other. His data supported the idea that two different genes assort independently of each other during the formation of haploid cells, producing a 9:3:3:1 phenotypic ratio in the F_2 generation. The law of independent assortment held true for all seven traits that Mendel studied. However, as we will see in Chapter 5, some genes follow a linked pattern of inheritance.

PROBLEM SETS & INSIGHTS

Solved Problems

S1. A heterozygous pea plant that is tall with yellow seeds, $TtYy$, is allowed to self-fertilize. What is the probability that an offspring will be either tall with yellow seeds, tall with green seeds, or dwarf with yellow seeds?

Answer: This problem involves three mutually exclusive events, and so we use the sum rule to solve it. First, we must calculate the individual probabilities for the three phenotypes. The outcome of the cross can be determined using a Punnett square.

Cross: $TtYy$ x $TtYy$

♂ TY	TY	Ty	tY	ty
TY	TTYY Tall, yellow	TTYy Tall, yellow	TtYY Tall, yellow	TtYy Tall, yellow
Ty	TTYy Tall, yellow	TTyy Tall, green	TtYy Tall, yellow	Ttyy Tall, green
tY	TtYY Tall, yellow	TtYy Tall, yellow	ttYY Dwarf, yellow	ttYy Dwarf, yellow
ty	TtYy Tall, yellow	Ttyy Tall, green	ttYy Dwarf, yellow	ttyy Dwarf, green

$P_{\text{tall with yellow seeds}} = 9/(9 + 3 + 3 + 1) = 9/16$

$P_{\text{tall with green seeds}} = 3/(9 + 3 + 3 + 1) = 3/16$

$P_{\text{dwarf with yellow seeds}} = 3/(9 + 3 + 3 + 1) = 3/16$

Sum rule: $9/16 + 3/16 + 3/16 = 15/16 = 0.94 = 94\%$

We expect to get one of these three phenotypes 15/16, or 94%, of the time.

S2. As described in Chapter 2, a human disease known as cystic fibrosis is inherited as a recessive trait. Two unaffected individuals have a first child with the disease. What is the probability that their next two children will not have the disease?

Answer: An unaffected couple has already produced an affected child. To be affected, the child must be homozygous for the disease allele and thus has inherited one copy from each parent. Therefore, because the parents are unaffected with the disease, we know that both of them must be heterozygous carriers for the recessive disease-causing allele. With this information, we can calculate the probability that they will produce an unaffected offspring. Using a Punnett square, this couple should produce a ratio of 3 unaffected : 1 affected offspring.

♂	N	n
N	NN	Nn
n	Nn	nn

N = common allele
n = cystic fibrosis allele

The probability of a single unaffected offspring is

$$P_{\text{unaffected}} = 3/(3 + 1) = 3/4$$

To obtain the probability of getting two unaffected offspring in a row (i.e., in a specified order), we must apply the product rule.

$$3/4 \times 3/4 = 9/16 = 0.56 = 56\%$$

The chance that their next two children will be unaffected is 56%.

S3. A cross was made between two heterozygous pea plants, $TtYy \times TtYy$. The following Punnett square was constructed:

Phenotypic ratio:

9 tall, yellow seeds : 3 tall, green seeds : 3 dwarf, yellow seeds : 1 dwarf, green seed

What is wrong with this Punnett square?

Answer: The outside of the Punnett square is supposed to contain the possible types of gametes. A gamete should contain one copy of each type of gene. Instead, the outside of this Punnett square contains two copies of one gene and zero copies of the other gene. The outcome happens to be correct (i.e., it yields a 9:3:3:1 ratio), but this is only a coincidence. The outside of the Punnett square must contain one copy of each type of gene. In this example, the correct possible types of gametes are TY, Ty, tY, and ty for each parent.

S4. A pea plant is heterozygous for three genes ($Tt\ Rr\ Yy$), where $T =$ tall, $t =$ dwarf, $R =$ round seeds, $r =$ wrinkled seeds, $Y =$ yellow seeds, and $y =$ green seeds. If this plant is self-fertilized, what are the predicted phenotypes of the offspring, and what fraction of the offspring will occur in each category?

Answer: You could solve this problem by constructing a large Punnett square and filling in the boxes. However, in this case, eight different male gametes and eight different female gametes are possible: TRY, TRy, TrY, tRY, trY, Try, tRy, and try. It would become rather tiresome to construct and fill in this Punnett square, which would contain 64 boxes. As an alternative, we can consider each gene separately and then algebraically combine them by multiplying together the expected phenotypic outcomes for each gene. In the cross $Tt\ Rr\ Yy \times Tt\ Rr\ Yy$, the following Punnett squares can be made for each gene:

3 tall : 1 dwarf

3 round : 1 wrinkled

3 yellow : 1 green

Instead of constructing a large, 64-box Punnett square, we can use two similar ways to determine the phenotypic outcome of this trihybrid cross. In the **multiplication method,** we can simply multiply these three combinations together:

(3 tall + 1 dwarf)(3 round + 1 wrinkled)(3 yellow + 1 green)

This multiplication operation can be done in a stepwise manner. First, multiply (3 tall + 1 dwarf) by (3 round + 1 wrinkled).

(3 tall + 1 dwarf)(3 round + 1 wrinkled) = 9 tall, round + 3 tall, wrinkled + 3 dwarf, round, + 1 dwarf, wrinkled

Next, multiply this product by (3 yellow + 1 green).

(9 tall, round + 3 tall, wrinkled + 3 dwarf, round + 1 dwarf, wrinkled)(3 yellow + 1 green) = 27 tall, round, yellow + 9 tall, round, green + 9 tall, wrinkled, yellow + 3 tall, wrinkled, green + 9 dwarf, round, yellow + 3 dwarf, round, green + 3 dwarf, wrinkled, yellow + 1 dwarf, wrinkled, green

Even though the multiplication steps are also somewhat tedious, this approach is much easier than making a Punnett square with 64 boxes, filling them in, deducing each phenotype, and then adding them up!

A second approach that is analogous to the multiplication method is the **forked-line method.** In this case, the genetic proportions are determined by multiplying together the probabilities of each phenotype.

Tall or dwarf	Round or wrinkled	Yellow or green	Observed product	Phenotype

$3/4$ tall
- $3/4$ round
  - $3/4$ yellow → $(3/4)(3/4)(3/4) = 27/64$ tall, round, yellow
  - $1/4$ green → $(3/4)(3/4)(1/4) = 9/64$ tall, round, green
- $1/4$ wrinkled
  - $3/4$ yellow → $(3/4)(1/4)(3/4) = 9/64$ tall, wrinkled, yellow
  - $1/4$ green → $(3/4)(1/4)(1/4) = 3/64$ tall, wrinkled, green

$1/4$ dwarf
- $3/4$ round
  - $3/4$ yellow → $(1/4)(3/4)(3/4) = 9/64$ dwarf, round, yellow
  - $1/4$ green → $(1/4)(3/4)(1/4) = 3/64$ dwarf, round, green
- $1/4$ wrinkled
  - $3/4$ yellow → $(1/4)(1/4)(3/4) = 3/64$ dwarf, wrinkled, yellow
  - $1/4$ green → $(1/4)(1/4)(1/4) = 1/64$ dwarf, wrinkled, green

S5. For an individual expressing a dominant trait, how can you tell if it is a heterozygote or a homozygote?

Answer: One way is to conduct a testcross with an individual that expresses the recessive version of the same trait. If the individual is heterozygous, half of the offspring will show the recessive trait, whereas if the individual is homozygous, none of the offspring will express the recessive trait.

$Dd \times dd$ or $DD \times dd$
↓   ↓
$1 Dd$ (dominant trait)   All Dd (dominant trait)

$1\ dd$ (recessive trait)

Another way to determine heterozygosity involves a more careful examination of the individual at the cellular or molecular level. At the cellular level, the heterozygote may not look exactly like the homozygote. This phenomenon is described in Chapter 4. Also, gene cloning methods described in Chapter 18 can be used to distinguish between heterozygotes and homozygotes.

S6. In dogs, black fur color is dominant to white. Two heterozygous black dogs are mated. What would be the probability of the following combinations of offspring?

A. A litter of six pups, four with black fur and two with white fur.

B. A litter of six pups, the firstborn with white fur, and among the remaining five pups, two with white fur and three with black fur.

C. A first litter of six pups, four with black fur and two with white fur, and then a second litter of seven pups, five with black fur and two with white fur.

D. A first litter of five pups, four with black fur and one with white fur, and then a second litter of seven pups in which the firstborn is homozygous, the second born is black, and the remaining five pups are three black and two white.

Answer:

A. Because this is an unordered combination of events, we use the binomial expansion equation, where $n = 6$, $x = 4$, $p = 0.75$ (probability of black), and $q = 0.25$ (probability of white).

The answer is 0.297, or 29.7%, of the time.

B. We use the product rule because the order is specified. The first pup is white and then the remaining five are born later. We also need to use the binomial expansion equation to determine the probability of the remaining five pups.

(probability of a white pup)(binomial expansion for the remaining five pups)

The probability of the white pup is 0.25. In the binomial expansion equation, $n = 5$, $x = 2$, $p = 0.25$, and $q = 0.75$.

The answer is 0.066, or 6.6%, of the time.

C. The order of the two litters is specified, so we need to use the product rule. We multiply the probability of the first litter times the probability of the second litter. We need to use the binomial expansion equation for each litter.

(binomial expansion of the first litter)(binomial expansion of the second litter)

For the first litter, $n = 6$, $x = 4$, $p = 0.75$, $q = 0.25$. For the second litter, $n = 7$, $x = 5$, $p = 0.75$, $q = 0.25$.

The answer is 0.092, or 9.2%, of the time.

D. The order of the litters is specified, so we need to use the product rule to multiply the probability of the first litter times the probability of the second litter. We use the binomial expansion equation to determine the probability of the first litter. The probability of the second litter is a little more complicated. The firstborn is homozygous. There are two mutually exclusive ways to be homozygous, BB and bb. We use the sum rule to determine the probability of the first pup, which equals $0.25 + 0.25 = 0.5$. The probability of the second pup is 0.75, and we use the binomial expansion equation to determine the probability of the remaining pups.

(binomial expansion of first litter)([0.5][0.75][binomial expansion of second litter])

For the first litter, $n = 5$, $x = 4$, $p = 0.75$, $q = 0.25$. For the last five pups in the second litter, $n = 5$, $x = 3$, $p = 0.75$, $q = 0.25$.

The answer is 0.039, or 3.9%, of the time.

S7. In Chapter 2, the binomial expansion equation was used in situations where only two phenotypic outcomes are possible. When more than two outcomes are possible, we use a **multinomial**

expansion equation to solve a problem involving an unordered number of events. A general expression for this equation is

$$P = \frac{n!}{a!b!c!\ldots} p^a q^b r^c \ldots$$

where P = the probability that the unordered number of events will occur.

n = total number of events

$$a + b + c + \ldots = n$$

$$p + q + r + \ldots = 1$$

(p is the likelihood of a, q is the likelihood of b, r is the likelihood of c, and so on)

The multinomial expansion equation can be useful in many genetic problems where more than two combinations of offspring are possible. For example, this formula can be used to solve problems involving an unordered sequence of events in a dihybrid experiment. This approach is illustrated next.

A cross is made between two heterozygous tall plants with axial flowers (*TtAa*), where tall is dominant to dwarf and axial is dominant to terminal flowers. What is the probability that a group of five offspring will be composed of two tall plants with axial flowers, one tall plant with terminal flowers, one dwarf plant with axial flowers, and one dwarf plant with terminal flowers?

Answer:

Step 1. *Calculate the individual probabilities of each phenotype.* This can be accomplished using a Punnett square.

The phenotypic ratios are 9 tall with axial flowers, 3 tall with terminal flowers, 3 dwarf with axial flowers, and 1 dwarf with terminal flowers.

The probability of a tall plant with axial flowers is $9/(9 + 3 + 3 + 1) = 9/16$.

The probability of a tall plant with terminal flowers is $3/(9 + 3 + 3 + 1) = 3/16$.

The probability of a dwarf plant with axial flowers is $3/(9 + 3 + 3 + 1) = 3/16$.

The probability of a dwarf plant with terminal flowers is $1/(9 + 3 + 3 + 1) = 1/16$.

$$p = 9/16$$
$$q = 3/16$$
$$r = 3/16$$
$$s = 1/16$$

Step 2. *Determine the number of each type of event versus the total number of events.*

$$n = 5$$
$$a = 2$$
$$b = 1$$
$$c = 1$$
$$d = 1$$

Step 3. *Substitute the values in the multinomial expansion equation.*

$$P = \frac{n!}{a!b!c!d!} p^a q^b r^c s^d$$

$$P = \frac{5!}{2!1!1!1!} (9/16)^2 (3/16)^1 (3/16)^1 (1/16)^1$$

$$P = 0.04 = 4\%$$

This means that 4% of the time we would expect to obtain five offspring with the phenotypes described in the question.

Conceptual Questions

C1. Why did Mendel's work refute the idea of blending inheritance?

C2. What is the difference between cross-fertilization and self-fertilization?

C3. Describe the difference between genotype and phenotype. Give three examples. Is it possible for two individuals to have the same phenotype but different genotypes?

C4. With regard to genotypes, what is a true-breeding organism?

C5. How can you determine whether an organism is heterozygous or homozygous for a dominant trait?

C6. In your own words, describe what Mendel's law of segregation means. Do not use the word segregation in your answer.

C7. Based on genes in pea plants that we have considered in this chapter, which statement(s) is not correct?

A. The gene causing tall plants is an allele of the gene causing dwarf plants.

B. The gene causing tall plants is an allele of the gene causing purple flowers.

C. The alleles causing tall plants and purple flowers are dominant.

C8. In a cross between a heterozygous tall pea plant and a dwarf plant, predict the ratios of the offspring's genotypes and phenotypes.

C9. Do you know the genotype of an individual with a recessive trait and/or a dominant trait? Explain your answer.

C10. A cross is made between a pea plant that has constricted pods (a recessive trait; smooth is dominant) and is heterozygous for seed color (yellow is dominant to green) and a plant that is heterozygous for both pod texture and seed color. Construct a Punnett square that depicts this cross. What are the predicted outcomes of genotypes and phenotypes of the offspring?

C11. A pea plant that is heterozygous with regard to seed color (yellow is dominant to green) is allowed to self-fertilize. What are the predicted outcomes of genotypes and phenotypes of the offspring?

C12. Describe the significance of nonparentals with regard to the law of independent assortment. In other words, explain how the appearance of nonparentals refutes a linkage hypothesis.

C13. For the following pedigrees, describe what you think is the most likely inheritance pattern (dominant versus recessive). Explain your reasoning. Filled (black) symbols indicate affected individuals.

(a)

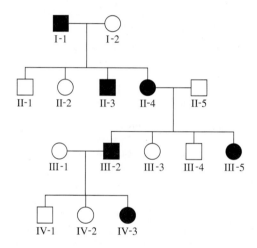

(b)

C14. Ectrodactyly, also known as "lobster claw syndrome," is a recessive disorder in humans. If a phenotypically unaffected couple produces an affected offspring, what are the following probabilities?

A. Both parents are heterozygotes.

B. An offspring is a heterozygote.

C. The next three offspring will be phenotypically unaffected.

D. Any two out of the next three offspring will be phenotypically unaffected.

C15. Identical twins are produced from the same sperm and egg (which splits after the first mitotic division), whereas fraternal twins are produced from separate sperm and separate egg cells. If two parents with brown eyes (a dominant trait) produce one twin boy with blue eyes, what are the following probabilities?

A. If the other twin is identical, he will have blue eyes.

B. If the other twin is fraternal, he/she will have blue eyes.

C. If the other twin is fraternal, he/she will transmit the blue eye allele to his/her offspring.

D. The parents are both heterozygotes.

C16. In cocker spaniels, solid coat color is dominant over spotted coat color. If two heterozygous dogs were crossed to each other, what would be the probability of the following combinations of offspring?

A. A litter of five pups, four with solid fur and one with spotted fur.

B. A first litter of six pups, four with solid fur and two with spotted fur, and then a second litter of five pups, all with solid fur.

C. A first litter of five pups, the firstborn with solid fur, and then among the next four, three with solid fur and one with spotted fur, and then a second litter of seven pups in which the firstborn is spotted, the second born is spotted, and the remaining five are composed of four solid and one spotted animal.

D. A litter of six pups, the firstborn with solid fur, the second born spotted, and among the remaining four pups, two with spotted fur and two with solid fur.

C17. A cross was made between a white male dog and two different black females. The first female gave birth to eight black pups, while the second female gave birth to four white and three black pups. What are the likely genotypes of the male parent and the two female parents? Explain whether you are uncertain about any of the genotypes.

C18. In humans, the allele for brown eye color (*B*) is dominant to blue eye color (*b*). If two heterozygous parents produce children, what are the following probabilities?

A. The first two children have blue eyes.

B. A total of four children, two with blue eyes and the other two with brown eyes.

C. The first child has blue eyes, and the next two have brown eyes.

C19. Albinism, a condition characterized by a partial or total lack of skin pigment, is a recessive human trait. If a phenotypically unaffected couple produced an albino child, what is the probability that their next child will be albino?

C20. A true-breeding tall plant was crossed to a dwarf plant. Tallness is a dominant trait. The F_1 individuals were allowed to self-fertilize. What are the following probabilities for the F_2 generation?

A. The first plant is dwarf.

B. The first plant is dwarf or tall.

C. The first three plants are tall.

D. For any seven plants, three are tall and four are dwarf.

E. The first plant is tall, and then among the next four, two are tall and the other two are dwarf.

C21. For pea plants with the following genotypes, list the possible gametes that the plant can make:

A. *TT Yy Rr*

B. *Tt YY rr*

C. *Tt Yy Rr*

D. *tt Yy rr*

C22. An individual has the genotype *Aa Bb Cc* and makes an abnormal gamete with the genotype *Aa Bc*. Does this gamete violate the law of independent assortment and/or the law of segregation? Explain your answer.

C23. Maple syrup urine disease is a disease found in humans in which the body is unable to metabolize the amino acids leucine, isoleucine, and valine. One of the symptoms is that the urine smells like maple syrup. An unaffected couple produced six children in the following order: unaffected daughter, affected daughter, unaffected son, unaffected son, affected son, and unaffected son. The youngest unaffected son marries an unaffected woman and has three children in the following order: affected daughter, unaffected daughter, and unaffected son. Draw a pedigree that describes this family. What type of inheritance (dominant or recessive) would you propose to explain maple syrup urine disease?

C24. Marfan syndrome is a rare inherited human disorder characterized by unusually long limbs and digits plus defects in the heart (especially the aorta) and the eyes. Following is a pedigree for this disorder. Affected individuals are shown with filled (black) symbols. What type of inheritance pattern do you think is the most likely?

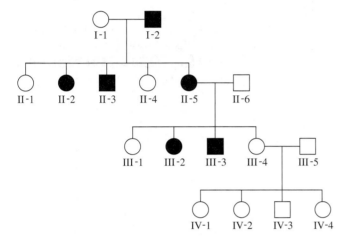

C25. A true-breeding pea plant with round and green seeds was crossed to a true-breeding plant with wrinkled and yellow seeds. Round and yellow seeds are the dominant traits. The F_1 plants were allowed to self-fertilize. What are the following probabilities for the F_2 generation?

A. An F_2 plant with wrinkled, yellow seeds.

B. Three out of three F_2 plants with round, yellow seeds.

C. Five F_2 plants in the following order: two have round, yellow seeds; one has round, green seeds; and two have wrinkled, green seeds.

D. An F_2 plant will not have round, yellow seeds.

C26. A true-breeding tall pea plant was crossed to a true-breeding dwarf plant. What is the probability that an F_1 individual will be true-breeding? What is the probability that an F_1 individual will be a true-breeding tall plant?

C27. What are the expected phenotypic ratios from the following cross: $Tt\ Rr\ yy\ Aa \times Tt\ rr\ YY\ Aa$, where T = tall, t = dwarf, R = round, r = wrinkled, Y = yellow, y = green, A = axial, a = terminal; T, R, Y, and A are dominant alleles. Note: See solved problem S3 for help in answering this problem.

C28. When an abnormal organism contains three copies of a gene (instead of the normal number of two copies), the alleles for the gene usually segregate so that a gamete will contain one or two copies of the gene. Let's suppose that an abnormal pea plant has three copies of the height gene. Its genotype is TTt. The plant is also heterozygous for the seed color gene, Yy. How many types of gametes can this plant make, and in what proportions? (Assume that it is equally likely that a gamete will contain one or two copies of the height gene.)

C29. Honeybees are unusual in that male bees (drones) have only one copy of each gene, while female bees have two copies of their genes. That is because drones develop from eggs that have not been fertilized by sperm cells. In bees, the trait of long wings is dominant over short wings, and the trait of black eyes is dominant over white eyes. If a drone with short wings and black eyes was mated to a queen bee that is heterozygous for both genes, what are the predicted genotypes and phenotypes of male and female offspring? What are the phenotypic ratios if we assume an equal number of male and female offspring?

C30. A pea plant that is dwarf with green, wrinkled seeds was crossed to a true-breeding plant that is tall with yellow, round seeds. The F_1 generation was allowed to self-fertilize. What types of gametes, and in what proportions, would the F_1 generation make? What would be the ratios of genotypes and phenotypes of the F_2 generation?

C31. A true-breeding plant with round and green seeds was crossed to a true-breeding plant with wrinkled and yellow seeds. The F_1 plants were allowed to self-fertilize. What is the probability of obtaining the following plants in the F_2 generation: two that have round, yellow seeds; one with round, green seeds; and two with wrinkled, green seeds? (Note: See solved problem S7 for help.)

C32. Wooly hair is a rare dominant trait found in people of Scandinavian descent in which the hair resembles the wool of a sheep. A male with wooly hair, who has a mother with straight hair, moves to an island that is inhabited by people who are not of Scandinavian descent. Assuming that he never leaves the island and that no other Scandinavians immigrate to the island, what is the probability that a great-grandchild of this male will have wooly hair? (Hint: You may want to draw a pedigree to help you figure this out.) If this wooly-haired male has eight great-grandchildren, what is the probability that one out of eight will have wooly hair?

C33. Huntington disease is a rare dominant trait that causes neurodegeneration later in life. A man in his thirties, who already has three children, discovers that his mother has Huntington disease though his father is unaffected. What are the following probabilities?

A. That the man in his thirties will develop Huntington disease.

B. That his first child will develop Huntington disease.

C. That one out of three of his children will develop Huntington disease.

C34. A woman with achondroplasia (a dominant form of dwarfism) and a phenotypically unaffected man have seven children, all of whom have achondroplasia. What is the probability of producing such a family if this woman is a heterozygote? What is the probability that the woman is a heterozygote if her eighth child does not have this disorder?

Experimental Questions

E1. Describe three advantages of using pea plants as an experimental organism.

E2. Explain the technical differences between a cross-fertilization experiment versus a self-fertilization experiment.

E3. How long did it take Mendel to complete the experiment in Figure 2.5?

E4. For all seven traits described in the data of Figure 2.5, Mendel allowed the F_2 plants to self-fertilize. He found that when F_2 plants with recessive traits were crossed to each other, they always bred true. However, when F_2 plants with dominant traits were crossed, some bred true while others did not. A summary of Mendel's results is shown here.

The Ratio of True-Breeding and Non-True-Breeding Parents of the F_2 Generation

F_2 Parents	True-Breeding	Non-True-Breeding	Ratio
Round	193	372	1:1.93
Yellow	166	353	1:2.13
Gray	36	64	1:1.78
Smooth	29	71	1:2.45
Green	40	60	1:1.5
Axial	33	67	1:2.08
Tall	28	72	1:2.57
TOTAL:	525	1,059	1:2.02

When considering the data in this table, keep in mind that it describes the characteristics of the F_2 generation parents that had displayed a dominant phenotype. These data were deduced by analyzing the outcome of the F_3 generation. Based on Mendel's laws, explain the 1:2 ratio obtained in these data.

E5. From the point of view of crosses and data collection, what are the experimental differences between a monohybrid and a dihybrid experiment?

E6. As in many animals, albino coat color is a recessive trait in guinea pigs. Researchers removed the ovaries from an albino female guinea pig and then transplanted ovaries from a true-breeding black guinea pig. They then mated this albino female (with the transplanted ovaries) to an albino male. The albino female produced three offspring. What were their coat colors? Explain the results.

E7. The fungus *Melampsora lini* causes a disease known as flax rust. Different strains of *M. lini* cause varying degrees of the rust disease. Conversely, different strains of flax are resistant or sensitive to the various varieties of rust. The Bombay variety of flax is resistant to *M. lini*-strain 22 but sensitive to *M. lini*-strain 24. A strain of flax called 770B is just the opposite; it is resistant to strain 24 but sensitive to strain 22. When 770B was crossed to Bombay, all the F_1 individuals were resistant to both strain 22 and strain 24. When F_1 individuals were self-fertilized, the following data were obtained:

43 resistant to strain 22 but sensitive to strain 24

9 sensitive to strain 22 and strain 24

32 sensitive to strain 22 but resistant to strain 24

110 resistant to strain 22 and strain 24

Explain the inheritance pattern for flax resistance and sensitivity to *M. lini* strains.

E8. For Mendel's data shown in Figure 2.8, conduct a chi square analysis to determine if the data agree with Mendel's law of independent assortment.

E9. Would it be possible to deduce the law of independent assortment from a monohybrid experiment? Explain your answer.

E10. In fruit flies, curved wings are recessive to straight wings, and ebony body is recessive to gray body. A cross was made between true-breeding flies with curved wings and gray bodies to flies with straight wings and ebony bodies. The F_1 offspring were then mated to flies with curved wings and ebony bodies to produce an F_2 generation.

A. Diagram the genotypes of this cross, starting with the parental generation and ending with the F_2 generation.

B. What are the predicted phenotypic ratios of the F_2 generation?

C. Let's suppose the following data were obtained for the F_2 generation:

114 curved wings, ebony body

105 curved wings, gray body

111 straight wings, gray body

114 straight wings, ebony body

Conduct a chi square analysis to determine if the experimental data are consistent with the expected outcome based on Mendel's laws.

E11. A recessive allele in mice results in an abnormally long neck. Sometimes, during early embryonic development, the abnormal neck causes the embryo to die. An experimenter began with a population of true-breeding normal mice and true-breeding mice with long necks. Crosses were made between these two populations to produce an F_1 generation of mice with normal necks. The F_1 mice were then mated to each other to obtain an F_2 generation. For the mice that were born alive, the following data were obtained:

522 mice with normal necks

62 mice with long necks

What percentage of homozygous mice (that would have had long necks if they had survived) died during embryonic development?

E12. The data in Figure 2.5 show the results of the F_2 generation for seven of Mendel's crosses. Conduct a chi square analysis to determine if these data are consistent with the law of segregation.

E13. Let's suppose you conducted an experiment involving genetic crosses and calculated a chi square value of 1.005. There were four categories of offspring (i.e., the degrees of freedom equaled 3). Explain what the 1.005 value means. Your answer should include the phrase "80% of the time."

E14. A tall pea plant with axial flowers was crossed to a dwarf plant with terminal flowers. Tall plants and axial flowers are dominant traits. The following offspring were obtained: 27 tall, axial flowers; 23 tall, terminal flowers; 28 dwarf, axial flowers; and 25 dwarf, terminal flowers. What are the genotypes of the parents?

E15. A cross was made between two strains of plants that are agriculturally important. One strain was disease resistant but herbicide sensitive; the other strain was disease sensitive but herbicide resistant. A plant breeder crossed the two plants and then allowed the F_1 generation to self-fertilize. The following data were obtained:

F_1 generation:	All offspring are disease sensitive and herbicide resistant
F_2 generation:	157 disease sensitive, herbicide resistant
	57 disease sensitive, herbicide sensitive
	54 disease resistant, herbicide resistant
	20 disease resistant, herbicide sensitive
Total:	288

Formulate a hypothesis that you think is consistent with the observed data. Test the goodness of fit between the data and your hypothesis using a chi square test. Explain what the chi square results mean.

E16. A cross was made between a plant that has blue flowers and purple seeds to a plant with white flowers and green seeds. The following data were obtained:

F_1 generation:	All offspring have blue flowers with purple seeds
F_2 generation:	103 blue flowers, purple seeds
	49 blue flowers, green seeds
	44 white flowers, purple seeds
	104 white flowers, green seeds
Total:	300

Start with the hypothesis that blue flowers and purple seeds are dominant traits and that the two genes assort independently. Calculate a chi square value. What does this value mean with regard to your hypothesis? If you decide to reject your hypothesis, which aspect of the hypothesis do you think is incorrect (i.e., blue flowers and purple seeds are dominant traits, or the idea that the two genes assort independently)?

Questions for Student Discussion/Collaboration

1. Consider the following cross in pea plants: $Tt\ Rr\ yy\ Aa \times Tt\ rr\ Yy\ Aa$, where T = tall, t = dwarf, R = round, r = wrinkled, Y = yellow, y = green, A = axial, a = terminal. What is the expected phenotypic outcome of this cross? Have one group of students solve this problem by making one big Punnett square, and have another group solve it by making four single-gene Punnett squares and using the product rule. Time each other to see who gets done first.

2. A cross was made between two pea plants, $TtAa$ and $Ttaa$, where T = tall, t = dwarf, A = axial, and a = terminal. What is the probability that the first three offspring will be tall with axial flowers or dwarf with terminal flowers and the fourth offspring will be tall with axial flowers. Discuss what operation(s) (e.g., sum rule, product rule, and/or binomial expansion equation) you used to solve them and in what order they were used.

3. Consider the following tetrahybrid cross: $Tt\ Rr\ yy\ Aa \times Tt\ RR\ Yy\ aa$, where T = tall, t = dwarf, R = round, r = wrinkled, Y = yellow, y = green, A = axial, a = terminal. What is the probability that the first three plants will have round seeds? What is the easiest way to solve this problem?

Note: All answers appear at the website for this textbook; the answers to even-numbered questions are in the back of the textbook.

www.mhhe.com/brookergenetics3e

Visit the Online Learning Center for practice tests, answer keys, and other learning aids for this chapter. Enhance your understanding of genetics with our interactive exercises, quizzes, animations, and much more.

REPRODUCTION AND CHROMOSOME TRANSMISSION

In Chapter 2, we considered some patterns of inheritance that explain the passage of traits from parent to offspring. In this chapter, we will survey reproduction at the cellular level and pay close attention to the inheritance of chromosomes. An examination of chromosomes at the microscopic level provides us with insights into understanding the inheritance patterns of traits. To appreciate this relationship, we will first consider how cells distribute their chromosomes during the process of cell division. We will see that in bacteria and most unicellular eukaryotes, simple cell division provides a way to reproduce asexually. Then we will explore a form of cell division called meiosis, which produces cells with half the number of chromosomes. By closely examining this process, we will see how the transmission of chromosomes accounts for the inheritance patterns that were observed by Mendel.

3.1 GENERAL FEATURES OF CHROMOSOMES

The **chromosomes** are structures within living cells that contain the genetic material. Genes are physically located within chromosomes. Biochemically, chromosomes contain a very long segment of DNA, which is the genetic material, and proteins, which are bound to the DNA and provide it with an organized structure. In eukaryotic cells, this complex between DNA and proteins is called **chromatin.** In this chapter, we will focus on the cellular mechanics of chromosome transmission to better understand the patterns of gene transmission that we considered in Chapter 2. In particular, we will examine how chromosomes are copied and sorted into newly made cells. In later chapters, particularly Chapters 8, 10, and 11, we will examine the molecular features of chromosomes in greater detail.

Before we begin a description of chromosome transmission, we need to consider the distinctive cellular differences between bacterial and eukaryotic species. Bacteria and archaea are referred to as **prokaryotes,** from the Greek meaning prenucleus, because their chromosomes are not contained within a separate nucleus of the cell. Prokaryotes usually have a single type of circular chromosome in a region of the cytoplasm called the **nucleoid** (**Figure 3.1a**). The cytoplasm is enclosed by a plasma membrane that regulates the uptake of nutrients and the excretion of waste products. Outside the plasma membrane is found a rigid cell wall that protects the cell from breakage. Certain species of bacteria also have an outer membrane located beyond the cell wall.

Eukaryotes, from the Greek meaning true nucleus, include some simple species, such as single-celled protists and some fungi (such as yeast), and more complex multicellular species, such as plants, animals, and other fungi. The cells of eukaryotic species have internal membranes that enclose highly specialized compartments (**Figure 3.1b**). These compartments form membrane-bounded

Chromosome sorting during cell division. When eukaryotic cells divide, they replicate and sort their chromosomes so that each cell receives the correct amount.

(a) Bacterial cell

(b) Animal cell

FIGURE 3.1   **The basic organization of cells.** (**a**) A bacterial cell. The example shown here is typical of a bacterium such as *Escherichia coli*, which has an outer membrane. (**b**) A eukaryotic cell. The example shown here is typical of an animal cell.

organelles with specific functions. For example, the lysosomes play a role in the degradation of macromolecules. The endoplasmic reticulum and Golgi body play a role in protein modification and trafficking. A particularly conspicuous organelle is the **nucleus,** which is bounded by two membranes that constitute the nuclear envelope. Most of the genetic material is found within chromosomes that are located in the nucleus. In addition to the nucleus, certain organelles in eukaryotic cells contain a small amount of their own DNA. These include the mitochondrion, which plays a role in ATP synthesis, and, in plant cells, the chloroplast, which plays a role in photosynthesis. The DNA found in these organelles is referred to as extranuclear or extrachromosomal DNA to distinguish it from the DNA that is found in the cell nucleus. We will examine the role of mitochondrial and chloroplast DNA in Chapter 7.

In this section, we will focus on the composition of chromosomes found in the nucleus of eukaryotic cells. As you will learn, eukaryotic species contain genetic material that comes in sets of linear chromosomes.

Eukaryotic Chromosomes Are Examined Cytologically to Yield a Karyotype

Insights into inheritance patterns have been gained by observing the behavior of chromosomes under the microscope. **Cytogenetics** is the field of genetics that involves the microscopic examination of chromosomes. The most basic observation that a **cytogeneticist** can make is to examine the chromosomal composition of a particular cell. For eukaryotic species, this is usually accomplished by observing the chromosomes as they are found in actively dividing cells. When a cell is preparing to divide, the chromosomes become more tightly coiled, which shortens them and thereby increases their diameter. The consequence of this shortening is that distinctive shapes and numbers of chromosomes become visible with a light microscope. Each species has a particular chromosome composition. For example, most human cells contain 23 pairs of chromosomes, for a total of 46.

On rare occasions, some individuals may inherit an abnormal number of chromosomes or a chromosome with an abnormal structure. Such abnormalities can often be detected by a microscopic examination of the chromosomes within actively dividing cells. In addition, a cytogeneticist may examine chromosomes as a way to distinguish between two closely related species.

(b) The slide is viewed by a light microscope; the sample is seen on a video screen. The chromosomes can be arranged electronically on the screen.

A sample of blood is collected, treated with drugs that stimulate the cells to divide, and subjected to centrifugation.

Supernatant

Blood cells — Pellet

The supernatant is discarded, and the cell pellet is suspended in a hypotonic solution. This causes the cells to swell.

— Hypotonic solution

The sample is subjected to centrifugation a second time to concentrate the cells. The cells are suspended in a fixative, stained, and placed on a slide.

Fix
Stain

Blood cells —

(a) Preparing cells for a karyotype

11 μm

(c) For a diploid human cell, two complete sets of chromosomes from a single cell constitute a karyotype of that cell.

FIGURE 3.2   **The procedure for making a human karyotype.**

Figure 3.2a shows the general procedure for preparing human chromosomes to be viewed by microscopy. In this example, the cells were obtained from a sample of human blood; more specifically, the chromosomes within lymphocytes (a type of white blood cell) were examined. Blood cells are a type of **somatic cell.** This term refers to any cell of the body that is not a gamete or a precursor to a gamete. The **gametes** (sperm and egg cells or their precursors) are also called **germ cells.**

After the blood cells have been removed from the body, they are treated with drugs that stimulate them to divide. As shown in Figure 3.2a, these actively dividing cells are centrifuged to concentrate them. The concentrated preparation is then mixed with a hypotonic solution that makes the cells swell. This swelling causes the chromosomes to spread out within the cell and thereby makes it easier to see each individual chromosome. Next, the cells are treated with a fixative that chemically freezes them so that the chromosomes will no longer move around. The cells are then treated with a chemical dye that binds to the chromosomes and stains them. As discussed in greater detail in Chapter 8, this gives chromosomes a distinctive banding pattern that greatly enhances their visualization and ability to be uniquely identified (also see Figure 8.1c, d). The cells are then placed on a slide and viewed with a light microscope.

In a cytogenetics laboratory, the microscopes are equipped with a camera that can photograph the chromosomes. In recent years, advances in technology have allowed cytogeneticists to view microscopic images on a computer screen (**Figure 3.2b**). On the computer screen, the chromosomes can be organized in a standard way, usually from largest to smallest. As seen in **Figure 3.2c**, the human chromosomes have been lined up, and a number is given to designate each type of chromosome. An exception would be the sex chromosomes, which are designated with the letters X and Y. A photographic representation of the chromosomes within a cell is called a **karyotype.** A karyotype reveals how many chromosomes are found within an actively dividing somatic cell.

Eukaryotic Chromosomes Are Inherited in Sets

Most eukaryotic species are **diploid,** which means that each type of chromosome is a member of a pair. In diploid species, the somatic cells have two sets of chromosomes. For example, most human cells have 46 chromosomes—two sets of 23 each. Other diploid species, however, can have different numbers of chromosomes. For example, the dog has 39 chromosomes per set (78 total), the fruit fly has 4 chromosomes per set (8 total), and the tomato has 12 per set (24 total).

When a species is diploid, the members of a pair of chromosomes are called **homologues;** each type of chromosome is found in a homologous pair. As shown in Figure 3.2c, for example, a somatic cell has two copies of chromosome 1, two copies of chromosome 2, and so forth. Within each pair, the chromosome on the left is a homologue to the one on the right, and vice versa. The two chromosomes in a homologous pair are nearly identical in size, have the same banding pattern, and contain a similar composition of genetic material. If a par-

ticular gene is found on one copy of a chromosome, it is also found on the other homologue. However, the two homologues may carry different alleles of a given gene. As an example, let's consider a gene in humans, called *OCA2*, which is one of a few different genes that affect eye color. The *OCA2* gene is located on chromosome 15 and comes in variants that result in brown, green, or blue eyes. In a person with brown eyes, one copy of chromosome 15 might carry a dominant brown allele, while its homologue could carry a recessive blue allele.

At the molecular level, how similar are homologous chromosomes? The answer is that the sequence of bases of one homologue would usually differ by less than 1% compared to the sequence of the other homologue. For example, the DNA sequence of chromosome 1 that you inherited from your mother would be greater than 99% identical to the sequence of chromosome 1 that you inherited from your father. Nevertheless, it should be emphasized that the sequences are not identical. The slight differences in DNA sequences provide the allelic differences in genes. Again, if we use the eye color gene as an example, a slight difference in DNA sequence distinguishes the brown, green, and blue alleles. It should also be noted that the striking similarities between homologous chromosomes do not apply to the pair of sex chromosomes—X and Y. These chromosomes differ in size and genetic composition. Certain genes that are found on the X chromosome are not found on the Y chromosome, and vice versa. The X and Y chromosomes are not considered homologous chromosomes though they do have short regions of homology.

Figure 3.3 considers two homologous chromosomes that are labeled with three different genes. An individual carrying these two chromosomes would be homozygous for the dominant allele of gene *A*. The individual would be heterozygous, *Bb*, for the second gene. For the third gene, the individual is homozygous for a recessive allele, *c*. The physical location of a gene is called its **locus** (plural: **loci**). As seen in Figure 3.3, for example, the locus of gene *C* is toward one end of this chromosome, while the locus of gene *B* is more in the middle.

FIGURE 3.3   A comparison of homologous chromosomes. Each pair of homologous chromosomes carries the same types of genes, but, as shown here, the alleles may or may not be different.

3.2 CELL DIVISION

Now that we have an appreciation for the chromosomal composition of living cells, we can consider how chromosomes are copied and transmitted when cells divide. One purpose of cell division is **asexual reproduction.** In this process, a preexisting cell divides to produce two new cells. By convention, the original cell is usually called the mother cell, and the new cells are the two daughter cells. When species are unicellular, the mother cell is judged to be one individual, and the two daughter cells are two new separate organisms. Asexual reproduction is how bacterial cells proliferate. In addition, certain unicellular eukaryotes, such as the amoeba and baker's yeast (*Saccharomyces cerevisiae*), can reproduce asexually.

A second important reason for cell division is multicellularity. Species such as plants, animals, most fungi, and certain protists are derived from a single cell that has undergone repeated cellular divisions. Humans, for example, begin as a single fertilized egg; repeated cellular divisions produce an adult with several trillion cells. The precise transmission of chromosomes during every cell division is critical in order for all the cells of the body to receive the correct amount of genetic material.

In this section, we will consider how the process of cell division requires the duplication, organization, and distribution of the chromosomes. In bacteria, which have a single circular chromosome, the division process is relatively simple. Prior to cell division, bacteria duplicate their circular chromosome; they then distribute a copy into each of the two daughter cells. This process, known as binary fission, is described first. Eukaryotes have multiple numbers of chromosomes that occur as sets. Compared to bacteria, this added complexity requires a more complicated sorting process to ensure that each newly made cell receives the correct number and types of chromosomes. As described later in this section, a mechanism known as mitosis entails the organization and distribution of eukaryotic chromosomes during cell division.

Bacteria Reproduce Asexually by Binary Fission

As discussed earlier (see Figure 3.1a), bacterial species are typically unicellular, although individual bacteria may associate with each other to form pairs, chains, or clumps. Unlike eukaryotes, which have their chromosomes in a separate nucleus, the circular chromosomes of bacteria are in direct contact with the cytoplasm. In Chapter 10, we will consider the molecular structure of bacterial chromosomes in greater detail.

The capacity of bacteria to divide is really quite astounding. Some species, such as *Escherichia coli*, a common bacterium of the intestine, can divide every 20 to 30 minutes. Prior to cell division, bacterial cells copy, or replicate, their chromosomal DNA. This produces two identical copies of the genetic material, as shown at the top of **Figure 3.4.** Following DNA replication, a bacterial cell divides into two daughter cells by a process known as **binary fission.** During this event, the two daughter cells become separated from each other by the formation of a septum. As seen in the figure, each cell receives a copy of the chromosomal genetic

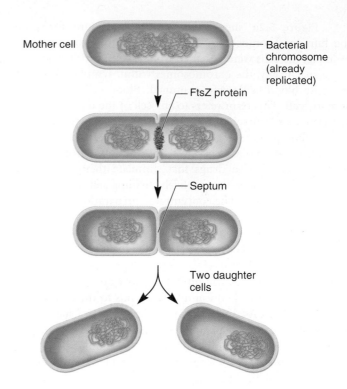

FIGURE 3.4   **Binary fission: the process by which bacterial cells divide.** Prior to division, the chromosome replicates to produce two identical copies. These two copies segregate from each other, with one copy going to each daughter cell.

material. Except when rare mutations occur, the daughter cells are usually genetically identical because they contain exact copies of the genetic material from the mother cell.

Recent evidence has shown that bacterial species produce a protein called FtsZ, which is important in cell division. This protein assembles into a ring at the future site of the septum. FtsZ is thought to be the first protein to move to this division site, and recruits other proteins that produce a new cell wall between the daughter cells. FtsZ is evolutionarily related to a eukaryotic protein called tubulin. As discussed later in this chapter, tubulin is the main component of microtubules, which play a key role in chromosome sorting in eukaryotes. Both FtsZ and tubulin form structures that provide cells with organization and play key roles in cell division.

Binary fission is an asexual form of reproduction because it does not involve genetic contributions from two different gametes. On occasion, bacteria can exchange small pieces of genetic material with each other. We will consider some interesting mechanisms of genetic exchange in Chapter 6.

Eukaryotic Cells Progress Through a Cell Cycle to Produce Genetically Identical Daughter Cells

The common outcome of eukaryotic cell division is to produce two daughter cells that have the same number and types of chromosomes as the original mother cell. This requires a replica-

FIGURE 3.5   The eukaryotic cell cycle. Dividing cells progress through a series of phases, denoted G_1, S, G_2, and M phases (mitosis). This diagram shows the progression of a cell through mitosis to produce two daughter cells. The original diploid cell had three pairs of chromosomes, for a total of six individual chromosomes. During S phase, these have replicated to yield 12 chromatids found in six pairs of sister chromatids. After mitosis is complete, each of the two daughter cells contains six individual chromosomes, just like the mother cell. Note: The chromosomes in G_0, G_1, S, and G_2 phases are not condensed. In this drawing, they are shown partially condensed so they can be easily counted.

tion and division process that is more complicated than simple binary fission. Eukaryotic cells that are destined to divide progress through a series of phases known as the **cell cycle** (**Figure 3.5**). These phases are G for gap, S for synthesis (of the genetic material), and M for mitosis. There are two G phases, G_1 and G_2. The term "gap" originally described the gaps between S phase and mitosis in which it was not microscopically apparent that significant changes were occurring in the cell. However, we now know that both gap phases are critical periods in the cell cycle that involve many molecular changes. In actively dividing cells, the G_1, S, and G_2 phases are collectively known as **interphase.** In addition, cells may remain permanently, or for long periods of time, in a phase of the cell cycle called G_0. A cell in the G_0 phase is either temporarily not progressing through the cell cycle or, in the case of terminally differentiated cells, such as most nerve cells in an adult mammal, will never divide again.

During the G_1 phase, a cell may prepare to divide. Depending on the cell type and the conditions that it encounters, a cell in the G_1 phase may accumulate molecular changes (e.g., syn-

thesis of proteins) that cause it to progress through the rest of the cell cycle. When this occurs, cell biologists say that a cell has reached a **restriction point** and is committed on a pathway that leads to cell division. Once past the restriction point, the cell will then advance to the S phase, during which the chromosomes are replicated. After replication, the two copies are called **chromatids.** They are joined to each other at a region of DNA called the **centromere** to form a unit known as a pair of **sister chromatids** (**Figure 3.6**). The **kinetochore** is a group of proteins that are bound to the centromere. These proteins help to hold the sister chromatids together and also play a role in chromosome sorting, as discussed later. When S phase is completed, a cell actually has twice as many chromatids compared to the number of chromosomes in the G_1 phase. For example, a human cell in the G_1 phase has 46 distinct chromosomes, whereas in G_2, it would have 46 pairs of sister chromatids, for a total of 92 chromatids. The term chromosome—meaning colored body—can be a bit confusing because it originally meant a distinct structure that is observable with the microscope. Therefore, the term

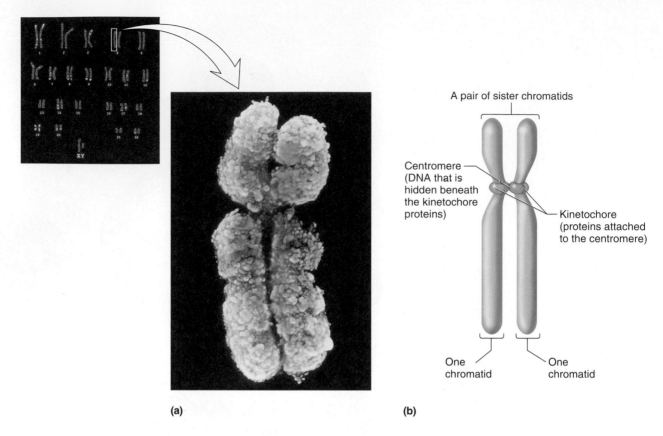

(a)   (b)

FIGURE 3.6   Chromosomes following DNA replication. (a) A photomicrograph of a chromosome in a form called a pair of sister chromatids. This chromosome is in the metaphase stage of mitosis, which is described later in the chapter. Note: Each of the 46 chromosomes that are viewed in a human karyotype is actually a pair of sister chromatids. See the white rectangular box in the inset. (b) A schematic drawing of sister chromatids. This structure has two chromatids that lie side by side. As seen here, each chromatid is a distinct unit. The two chromatids are held together by kinetochore proteins that bind to each other and to the centromeres of each chromatid.

chromosome can refer to either a pair of sister chromatids during the G_2 and early stages of M phase or to the structures that are observed at the end of M phase and those during G_1 that contain the equivalent of one chromatid (refer back to Figure 3.5).

During the G_2 phase, the cell accumulates the materials that are necessary for nuclear and cell division. It then progresses into the M phase of the cell cycle, when **mitosis** occurs. The primary purpose of mitosis is to distribute the replicated chromosomes, dividing one cell nucleus into two nuclei, so that each daughter cell receives the same complement of chromosomes. For example, a human cell in the G_2 phase has 92 chromatids, which are found in 46 pairs. During mitosis, these pairs of chromatids are separated and sorted so that each daughter cell receives 46 chromosomes.

Mitosis was first observed microscopically in the 1870s by the German biologist Walter Flemming, who coined the term mitosis (from the Greek *mitos*, thread). He studied the dividing epithelial cells of salamander larvae and noticed that chromosomes were constructed of two parallel "threads." These threads separated and moved apart, one going to each of the two daughter nuclei. By this mechanism, Flemming pointed out that the two daughter cells received an identical group of threads, of quantity comparable to the number of threads in the parent cell.

The Mitotic Spindle Apparatus Organizes and Sorts Eukaryotic Chromosomes

Before we discuss the events of mitosis, let's first consider the structure of the **mitotic spindle apparatus** (also known simply as the **mitotic spindle**), which is involved in the organization and sorting of chromosomes (**Figure 3.7**). In animal and plant cells, the mitotic spindle is formed from two structures called the **centrosomes.** Each centrosome is located at a **spindle pole**. In animal cells, a pair of **centrioles** at right angles to each other is found within each centrosome. However, centrioles are not found in many other eukaryotic species, including plants, and are not required for spindle formation. Each centrosome organizes the construction of structures called microtubules. These microtubules are produced from the rapid polymerization of tubulin proteins. The mitotic spindle has three types of microtubules. The aster microtubules emanate outward from the centrosome toward the plasma membrane. They are important for the positioning of the spindle apparatus within the cell and later in the process of cell division. The polar microtubules project toward the region where the chromosomes will be found during mitosis—the region between the two spindle poles. Polar microtubules

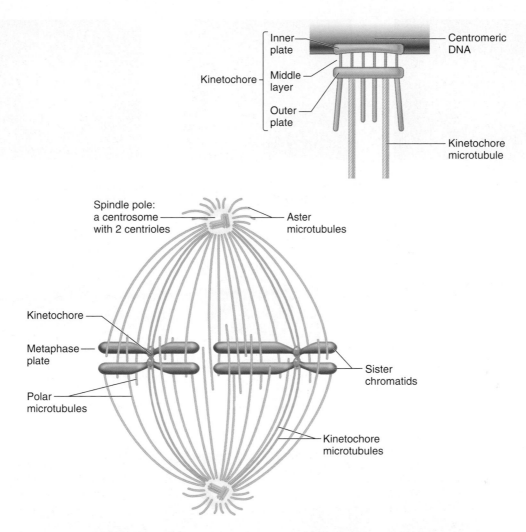

FIGURE 3.7   The structure of the mitotic spindle. A single centrosome duplicates during S phase and the two centrosomes separate at the beginning of M phase. The mitotic spindle is formed from microtubules that are rooted in the centrosomes. Each centrosome is located at a spindle pole. The aster microtubules emanate away from the region between the poles. They help to position the spindle within the cell and are used as reference points for cell division. The polar microtubules project into the region between the two poles; they play a role in pole separation. The kinetochore microtubules are attached to the kinetochore of sister chromatids. As seen in the inset, the kinetochore is composed of a group of proteins that form three layers: the inner plate, which recognizes the centromere; the outer plate, which recognizes a kinetochore microtubule; and the middle layer, which connects the inner and outer plates.

that overlap with each other play a role in the separation of the two poles. They help to "push" the poles away from each other. Finally, the kinetochore microtubules have attachments to the kinetochore, which is a complex of proteins that is bound to the centromere of individual chromosomes. As seen in the inset to Figure 3.7, the kinetochore proteins form three layers. The proteins of the inner plate make direct contact with the centromeric DNA, while the outer plate contacts the kinetochore microtubules. The role of the middle layer is to connect these two regions.

The mitotic spindle allows cells to organize and separate chromosomes so that each daughter cell receives the same complement of chromosomes. This sorting process, known as mitosis, is described next.

The Transmission of Chromosomes During the Division of Eukaryotic Cells Requires a Process Known as Mitosis

In **Figure 3.8**, the process of mitosis is shown for a diploid animal cell. In the simplified diagrams shown along the bottom of this figure, the original mother cell contains six chromosomes; it is diploid ($2n$) and contains three chromosomes per set ($n = 3$). One set is shown in blue, and the homologous set is shown in red. As discussed next, mitosis is subdivided into phases known as prophase, prometaphase, metaphase, anaphase, and telophase.

Prophase Prior to mitosis, the cells are in interphase, during which the chromosomes are **decondensed**—less tightly compacted—and

FIGURE 3.8   **The process of mitosis in an animal cell.** The top panels illustrate cells of a fish embryo progressing through mitosis. The bottom panels are schematic drawings that emphasize the sorting and separation of the chromosomes. In this case, the original diploid cell had six chromosomes (three in each set). At the start of mitosis, these have already replicated into 12 chromatids. The final result is two daughter cells each containing six chromosomes.

found in the nucleus (**Figure 3.8a**). At the start of mitosis, in **prophase,** the chromosomes have already replicated to produce 12 chromatids, joined as six pairs of sister chromatids (**Figure 3.8b**). As prophase proceeds, the nuclear membrane begins to dissociate into small vesicles. At the same time, the chromatids **condense** into more compact structures that are readily visible by light microscopy. The mitotic spindle also begins to form and the nucleolus disappears.

Prometaphase   As mitosis progresses from prophase to prometaphase, the centrosomes move to opposite ends of the cell and demarcate two spindle poles, one within each of the future daughter cells. Once the nuclear membrane has dissociated into vesicles, the spindle fibers can interact with the sister chromatids. This interaction occurs in a phase of mitosis called **prometaphase** (**Figure 3.8c**). How do sister chromatids become attached to the spindle? Initially, kinetochore microtubules are rapidly formed and can be seen growing out from the two poles. As it grows, if the end of a kinetochore microtubule happens to make contact with a kinetochore, its end is said to be "captured" and remains firmly attached to the kinetochore. This random process is how sister chromatids become attached to kinetochore microtubules. Alternatively, if the end of a kinetochore microtubule does not collide with a kinetochore, the microtubule eventually depolymerizes and retracts to the centrosome. As the end of prometaphase nears, the kinetochore on a pair of sister chromatids is attached to kinetochore microtubules from opposite poles. As these events are occurring, the sister chromatids are seen to undergo jerky movements as they are tugged, back and forth, between the two poles. By the end of prometaphase, the mitotic spindle is completely formed.

Metaphase   Eventually, the pairs of sister chromatids align themselves along a plane called the **metaphase plate.** As shown in **Figure 3.8d**, when this alignment is complete, the cell is in **metaphase** of mitosis. At this point, each pair of chromatids is attached to both poles by kinetochore microtubules. The pairs of sister chromatids have become organized into a single row along the metaphase plate. When this organizational process is finished, the chromatids can be equally distributed into two daughter cells.

Anaphase   The next step in the division process occurs during **anaphase** (**Figure 3.8e**). At this stage, the connection that is responsible for holding the pairs of chromatids together is broken. (We will examine the process of sister chromatid cohesion and separation in more detail in Chapter 10.) Each chromatid, now an individual chromosome, is linked to only one of the two poles. As anaphase proceeds, the chromosomes move toward the pole to which they are attached. This involves a shortening of the kinetochore microtubules. In addition, the two poles themselves move farther apart due to the elongation of the polar microtubules, which slide in opposite directions due to the actions of motor proteins.

Telophase   During **telophase,** the chromosomes reach their respective poles and decondense. The nuclear membrane now re-forms to produce two separate nuclei. In **Figure 3.8f,** this has

produced two nuclei that contain six chromosomes each. The nucleoli will also reappear.

Cytokinesis   In most cases, mitosis is quickly followed by **cytokinesis,** in which the two nuclei are segregated into separate daughter cells. Likewise, cytokinesis also segregates cell organelles such as mitochondria and chloroplasts into daughter cells. In animal cells, cytokinesis begins shortly after anaphase. A contractile ring, comprised of myosin motor proteins and actin filaments, assembles adjacent to the plasma membrane. Myosin hydrolyzes ATP, which shortens the ring and thereby constricts the plasma membrane to form a **cleavage furrow** that ingresses, or moves inwards (**Figure 3.9a**). Ingression continues until a midbody structure is formed that physically pinches one cell into two.

In plants, the two daughter cells are separated by the formation of a **cell plate** (**Figure 3.9b**). At the end of anaphase, Golgi-derived vesicles carrying cell wall materials are transported

(a) Cleavage of an animal cell

(b) Formation of a cell plate in a plant cell

FIGURE 3.9   **Cytokinesis in an animal and plant cell. (a)** In an animal cell, cytokinesis involves the formation of a cleavage furrow. **(b)** In a plant cell, cytokinesis occurs via the formation of a cell plate between the two daughter cells.

to the equator of a dividing cell. These vesicles are directed to their location via the phragmoplast, which is composed of parallel aligned microtubules and actin filaments that serve as tracks for vesicle movement. The fusion of these vesicles gives rise to the cell plate, which is a membrane-bound compartment. The cell plate begins in the middle of the cell and expands until it attaches to the mother cell wall. Once this attachment has taken place, the cell plate undergoes a process of maturation and eventually separates the mother cell into two daughter cells.

Outcome of mitotic cell division Mitosis and cytokinesis ultimately produce two daughter cells having the same number of chromosomes as the mother cell. Barring rare mutations, the two daughter cells are genetically identical to each other and to the mother cell from which they were derived. Thus, the critical consequence of this sorting process is to ensure genetic consistency from one somatic cell to the next. The development of multicellularity relies on the repeated process of mitosis and cytokinesis. For diploid organisms that are multicellular, most of the somatic cells are diploid and genetically identical to each other.

3.3 SEXUAL REPRODUCTION

In Section 3.2, we considered how a cell divides to produce two new cells with identical complements of genetic material. Now we will turn our attention to sexual reproduction, a common way for eukaryotic organisms to produce offspring. During sexual reproduction, gametes are made that contain half the amount of genetic material. These gametes fuse with each other in the process of fertilization to begin the life of a new organism. Gametes are highly specialized cells that are designed to locate each other and provide nutrients to the developing embryo. The process whereby gametes form is called **gametogenesis.**

Some simple eukaryotic species are **isogamous,** which means that the gametes are morphologically similar. Examples of isogamous organisms include many species of fungi and algae. Most eukaryotic species, however, are **heterogamous**—they produce two morphologically different types of gametes. Male gametes, or **sperm cells,** are relatively small and usually travel far distances to reach the female gamete. The mobility of the male gamete is an important characteristic, making it likely that it will come in close proximity to the female gamete. The sperm of most animal species contain a single flagellum that enables them to swim. The sperm of ferns and nonvascular plants such as bryophytes may have multiple flagella. In flowering plants, however, the sperm are contained within pollen grains. Pollen is a small mobile structure that can be carried by the wind or on the feet or hairs of insects. In flowering plants, sperm are delivered to egg cells via pollen tubes. Compared to sperm cells, the female gamete, known as the **egg cell,** or **ovum,** is usually very large and nonmotile. In animal species, the egg stores a large amount of nutrients that will be available to the growing embryo.

Gametes are typically **haploid,** which means they contain half the number of chromosomes as diploid cells. Compared to a diploid cell, a haploid gamete contains a single set of chromo-

somes. In this case, the haploid gametes are $1n$, while the diploid cells are $2n$. For example, a diploid human cell contains 46 chromosomes, but a gamete (sperm or egg cell) contains only 23 chromosomes.

During the process known as **meiosis** (from the Greek meaning less), haploid cells are produced from a cell that was originally diploid. For this to occur, the chromosomes must be correctly sorted and distributed in a way that reduces the chromosome number to half its original value. In the case of humans, for example, each gamete must receive half the total number of chromosomes, but not just any 23 chromosomes will do. A gamete must receive one chromosome from each of the 23 pairs. In this section, we will examine the cellular events of gamete development in animal and plant species and how the stages of meiosis lead to the formation of cells with a haploid complement of chromosomes.

Meiosis Produces Cells That Are Haploid

The process of meiosis bears striking similarities to mitosis. Like mitosis, meiosis begins after a cell has progressed through the G_1, S, and G_2 phases of the cell cycle. However, meiosis involves two successive divisions rather than one (as in mitosis). Prior to meiosis, the chromosomes are replicated in S phase to produce pairs of sister chromatids. This single replication event is then followed by two sequential cell divisions called meiosis I and II. As in mitosis, each of these is subdivided into prophase, prometaphase, metaphase, anaphase, and telophase.

Figure 3.10 emphasizes some of the important events that occur during prophase of meiosis I, which is further subdivided into periods known as leptotena, zygotena, pachytena, diplotena, and diakinesis. During the **leptotene** stage, the replicated chromosomes begin to condense and become visible with a light microscope. In Figure 3.10 (leptotene and zygotene stages), each structure is actually a pair of chromatids. Unlike mitosis, the **zygotene** stage of prophase of meiosis I involves a recognition process known as **synapsis,** in which the homologous chromosomes recognize each other and then align themselves along their entire lengths. The associated chromatids are known as **bivalents.** Each bivalent contains two pairs of sister chromatids, or a total of four chromatids.

In most eukaryotic species, a **synaptonemal complex** is formed between the homologous chromosomes. As shown in **Figure 3.11**, this complex is composed of parallel lateral elements, which are bound to the chromosomal DNA, and a central element, which promotes the binding of the lateral elements to each other via transverse filaments. The synaptonemal complex may not be required for the pairing of homologous chromosomes, because some species, such as *Aspergillus nidulans* and *Schizosaccharomyces pombe,* completely lack such a complex, yet their chromosomes synapse correctly. At present, the precise role of the synaptonemal complex is not clearly understood, and it remains the subject of intense research. It may play more than one role. First, although it may not be required for synapsis, the synaptonemal complex could help to maintain homologous pairing in situations where the normal process has failed. Second, the complex may play a role in meiotic chromosome structure. And

FIGURE 3.10   The events that occur during prophase of meiosis I.

FIGURE 3.11   **The synaptonemal complex formed during prophase of meiosis I. (a)** Micrograph of a synaptonemal complex. **(b)** Lateral elements are bound to the chromosomal DNA of homologous chromatids. A central element provides a link between the lateral elements via transverse filaments.

third, the synaptonemal complex may serve to regulate the process of crossing over, which is described next.

Prior to the **pachytene** stage, when synapsis is complete, an event known as **crossing over** usually occurs. Crossing over involves a physical exchange of chromosome pieces. Depending on the size of the chromosome and the species, an average eukaryotic chromosome incurs a couple to a couple dozen crossovers. During spermatogenesis in humans, for example, an average chromosome undergoes slightly more than two crossovers, whereas chromosomes in certain plant species may undergo 20 or more crossovers. Recent research has shown that crossing over is critical for the proper segregation of chromosomes. In fact, abnormalities in chromosome segregation may be related to a defect in crossing over. In a high percentage of people with Down syndrome, in which an individual has three copies of chromosome 21 instead of two, research has shown that the presence of the extra chromosome is associated with a lack of crossing over between homologous chromosomes.

In Figure 3.10, crossing over has occurred at a single site between two of the larger chromatids. The connection that results from crossing over is called a **chiasma** (plural: **chiasmata**), because it physically resembles the Greek letter chi, χ. We will consider the genetic consequences of crossing over in Chapter 5 and the molecular process of crossing over in Chapter 17. By the end of the **diplotene** stage, the synaptonemal complex has largely disappeared. The bivalent pulls apart slightly, and microscopically it becomes easier to see that it is actually composed of four chromatids. A bivalent is also called a **tetrad** (from the prefix *tetra-*, meaning four) because it is composed of four chromatids. In the last stage of prophase of meiosis I, **diakinesis,** the synaptonemal complex completely disappears.

Figure 3.10 emphasizes the pairing and crossing over that occurs during prophase of meiosis I. In **Figure 3.12**, we turn our attention to the remaining events in meiosis. In prometaphase of meiosis I, the spindle apparatus is complete, and the chromatids are attached via kinetochore microtubules. At metaphase of meiosis I, the bivalents are organized along the metaphase plate. However, their pattern of alignment is strikingly different from that observed during mitosis (refer back to Figure 3.8d). Before we consider the rest of meiosis I, a particularly critical feature for you to appreciate is how the bivalents are aligned along the metaphase plate. In particular, the alignment of pairs of sister chromatids is in a double row rather than a single row as in mitosis. Furthermore, the arrangement of sister chromatids within this double row is random with regard to the blue and red homologues. In Figure 3.12, one of the blue homologues is above the metaphase plate and the other two are below, while one of the red homologues is below the metaphase plate and other two are above.

In an organism that produces many gametes, meiosis in other cells could produce a different arrangement of homologues—three blues above and none below, or none above and three below, and so on. As discussed later in this chapter, the random arrangement of homologues is consistent with Mendel's law of independent assortment. Because most eukaryotic species have several chromosomes per set, the sister chromatids can be randomly aligned along the metaphase plate in many possible ways. For example, consider humans that have 23 chromosomes per set. The possible number of different, random alignments equals 2^n, where n equals the number of chromosomes per set. Thus, in humans, this would equal 2^{23}, or over 8 million, possibilities! Because the homologues are genetically similar but not identical, we see from this calculation that the random alignment of homologous chromosomes provides a mechanism to promote a vast amount of genetic diversity.

In addition to the random arrangement of homologues within a double row, a second distinctive feature of metaphase of meiosis I is the attachment of kinetochore microtubules to the sister chromatids (**Figure 3.13**). One pair of sister chromatids is linked to one of the poles, and the homologous pair is linked to the opposite pole. This arrangement is quite different from the kinetochore attachment sites during mitosis in which a pair of sister chromatids is linked to both poles (see Figure 3.8).

During anaphase of meiosis I, the two pairs of sister chromatids within a bivalent separate from each other (see Figure 3.12). However, the connection that holds sister chromatids together does not break. Instead, each joined pair of chromatids migrates to one pole, and the homologous pair of chromatids moves to the opposite pole. Finally, at telophase of meiosis I, the sister chromatids have reached their respective poles, and decondensation occurs in many, but not all, species. The nuclear membrane may re-form to produce two separate nuclei. The end result of meiosis I is two cells, each with three pairs of sister chromatids. It is thus a reduction division. The original diploid cell had its chromosomes in homologous pairs, while the two cells produced at the end of meiosis I are considered to be haploid; they do not have pairs of homologous chromosomes.

The sorting events that occur during meiosis II are similar to those that occur during mitosis, but the starting point is different. For a diploid organism with six chromosomes, mitosis begins with 12 chromatids that are joined as six pairs of sister chromatids (refer back to Figure 3.8). By comparison, the two cells that begin meiosis II each have six chromatids that are joined as three pairs of sister chromatids. Otherwise, the steps that occur during prophase, prometaphase, metaphase, anaphase, and telophase of meiosis II are analogous to a mitotic division.

If we compare the outcome of meiosis (see Figure 3.12) to that of mitosis (see Figure 3.8), the results are quite different. (A comparison is also made in solved problem S3 at the end of this chapter.) In these examples, mitosis produced two diploid daughter cells with six chromosomes each, whereas meiosis produced four haploid daughter cells with three chromosomes each. In other words, meiosis has halved the number of chromosomes per cell. With regard to alleles, the results of mitosis and meiosis are also different. The daughter cells produced by mitosis are genetically identical. However, the haploid cells produced by meiosis are not genetically identical to each other because they contain only one homologous chromosome from each pair. Later, we will consider how the gametes may differ in the alleles that they carry on their homologous chromosomes.

MEIOSIS I

Centrosomes with centrioles

EARLY PROPHASE

Mitotic spindle

Sister chromatids

Synapsis of homologous chromatids and crossing over

LATE PROPHASE

Bivalent

Nuclear membrane fragmenting

PROMETAPHASE

Metaphase plate

METAPHASE

ANAPHASE

Cleavage furrow

TELOPHASE AND CYTOKINESIS

MEIOSIS II

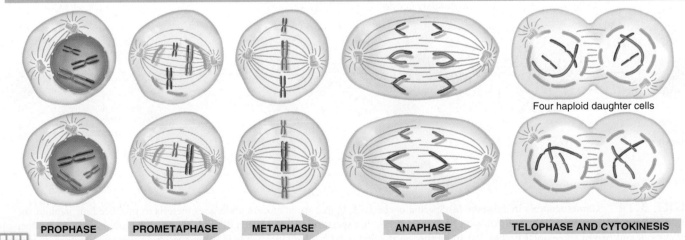

Four haploid daughter cells

PROPHASE   PROMETAPHASE   METAPHASE   ANAPHASE   TELOPHASE AND CYTOKINESIS

ONLINE ANIMATION

FIGURE 3.12   The stages of meiosis in an animal cell. See text for details.

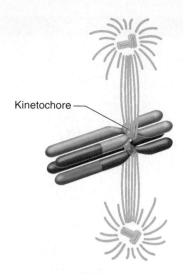

Kinetochore

FIGURE 3.13 **Attachment of the kinetochore microtubules to replicated chromosomes during meiosis.** The kinetochore microtubules from a given pole are attached to one pair of chromatids in a bivalent, but not both. Therefore, each pair of sister chromatids is attached to only one pole.

In Animals, Spermatogenesis Produces Four Haploid Sperm Cells and Oogenesis Produces a Single Haploid Egg Cell

In male animals, **spermatogenesis,** the production of sperm, occurs within glands known as the testes. The testes contain spermatogonial cells that divide by mitosis to produce two cells. One of these remains a spermatogonial cell, and the other cell becomes a primary spermatocyte. As shown in **Figure 3.14a,** the spermatocyte progresses through meiosis I and meiosis II to produce four haploid cells, which are known as spermatids. These cells then mature into sperm cells. The structure of a sperm cell includes a long flagellum and a head. The head of the sperm contains little more than a haploid nucleus and an organelle at its tip, known as an acrosome. The acrosome contains digestive enzymes that are released when a sperm meets an egg cell. These enzymes enable the sperm to penetrate the outer protective layers of the egg and gain entry into the egg cell's cytosol. In animal species without a mating season, sperm production is a continuous process in mature males. A mature human male, for example, produces several hundred million sperm each day.

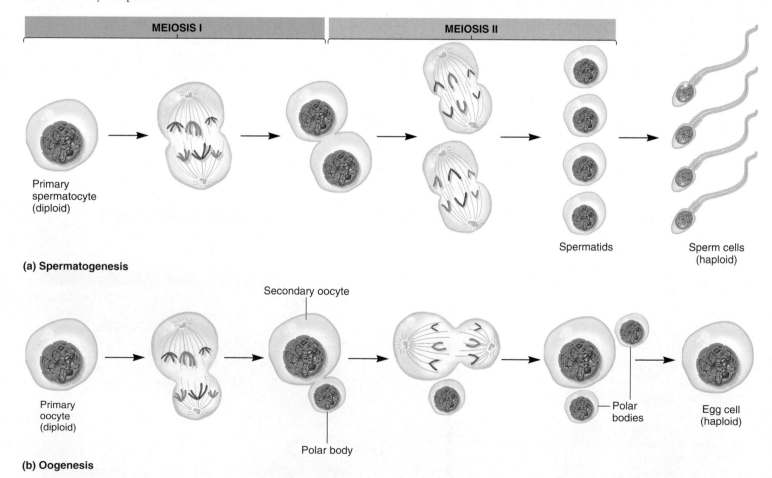

FIGURE 3.14 **Gametogenesis in animals.** (**a**) Spermatogenesis. A diploid spermatocyte undergoes meiosis to produce four haploid (*n*) spermatids. These differentiate during spermatogenesis to become mature sperm. (**b**) Oogenesis. A diploid oocyte undergoes meiosis to produce one haploid egg cell and two or three polar bodies. For some species, the first polar body divides, while in other species, it does not. Because of asymmetric cytokinesis, the amount of cytoplasm the egg receives is maximized. The polar bodies degenerate.

In female animals, **oogenesis,** the production of egg cells, occurs within specialized diploid cells of the ovary known as oogonia. Quite early in the development of the ovary, the oogonia initiate meiosis to produce primary oocytes. For example, in humans, approximately 1 million primary oocytes per ovary are produced before birth. These primary oocytes are arrested—enter a dormant phase—at prophase of meiosis I, remaining at this stage until the female becomes sexually mature. Beginning at this stage, primary oocytes are periodically activated to progress through the remaining stages of oocyte development.

During oocyte maturation, meiosis produces only one cell that is destined to become an egg, as opposed to the four gametes produced from each primary spermatocyte during spermatogenesis. How does this occur? As shown in **Figure 3.14b,** the first meiotic division is asymmetric and produces a secondary oocyte and a much smaller cell, known as a polar body. Most of the cytoplasm is retained by the secondary oocyte and very little by the polar body, allowing the oocyte to become a larger cell with more stored nutrients. The secondary oocyte then begins meiosis II. In mammals, the secondary oocyte is released from the ovary—an event called ovulation—and travels down the oviduct toward the uterus. During this journey, if a sperm cell penetrates the secondary oocyte, it is stimulated to complete meiosis II; the secondary oocyte produces a haploid egg and a second polar body. The haploid egg and sperm nuclei then unite to create the diploid nucleus of a new individual.

Plant Species Alternate Between Haploid (Gametophyte) and Diploid (Sporophyte) Generations

Most species of animals are diploid, and their haploid gametes are considered to be a specialized type of cell. By comparison, the life cycles of plant species alternate between haploid and diploid generations. The haploid generation is called the **gametophyte,** whereas the diploid generation is called the **sporophyte.** Meiosis produces haploid cells called spores, which divide by mitosis to produce the gametophyte. In simpler plants, such as mosses, a haploid spore can produce a large multicellular gametophyte by repeated mitoses and cellular divisions. In flowering plants, however, spores develop into gametophytes that contain only a few cells. In this case, the organism that we think of as a "plant" is the sporophyte, while the gametophyte is very inconspicuous. In fact, the gametophytes of many plant species are small structures produced within the much larger sporophyte. Certain cells within the haploid gametophytes then become specialized as haploid gametes.

Figure 3.15 provides an overview of gametophyte development and gametogenesis in flowering plants. Meiosis occurs within two different structures of the sporophyte: the anthers and the ovaries, which produce male and female gametophytes, respectively. This diagram depicts a flower from an angiosperm, which is a plant that produces seeds within an ovary.

In the anther, diploid cells called microsporocytes undergo meiosis to produce four haploid microspores. These separate into

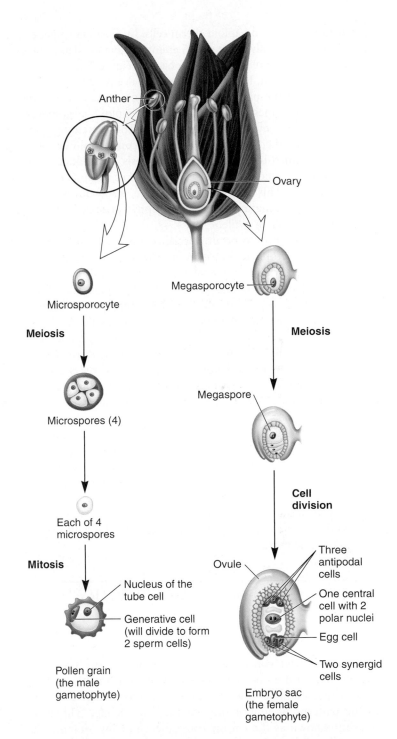

FIGURE 3.15 **The formation of male and female gametes by the gametophytes of angiosperms.**

individual microspores. In many angiosperms, each microspore undergoes mitosis to produce a two-celled structure containing one tube cell and one generative cell, both of which are haploid. This structure differentiates into a **pollen grain,** which is the male gametophyte with a thick cell wall. Later, the generative cell undergoes mitosis to produce two haploid sperm cells. In most plant species, this mitosis occurs only if the pollen grain germinates—lands on a stigma and forms a pollen tube (refer back to Figure 2.2c).

By comparison, female gametophytes are produced within ovules found in the plant ovaries. A cell known as a megasporocyte undergoes meiosis to produce four haploid megaspores. Three of the four megaspores degenerate. The remaining haploid megaspore then undergoes three successive mitotic divisions accompanied by asymmetric cytokinesis to produce seven individual cells—one egg, two synergids, three antipodals, and one central cell. This seven-celled structure, also known as the **embryo sac,** is the mature female gametophyte. Each embryo sac is contained within an ovule.

For fertilization to occur, specialized cells within the male and female gametophytes must meet. The steps of plant fertilization were described in Chapter 2. To begin this process, a pollen grain lands on a stigma (refer back to Figure 2.2c). This stimulates the tube cell to sprout a tube that grows through the style and eventually makes contact with an ovule. As this is occurring, the generative cell undergoes mitosis to produce two haploid sperm cells. The sperm cells migrate through the pollen tube and eventually reach the ovule. One of the sperm enters the central cell, which contains the two polar nuclei. This creates a cell that is triploid (3n). This cell divides mitotically to produce **endosperm,** which acts as a food-storing tissue. The other sperm enters the egg cell. The egg and sperm nuclei fuse to create a diploid cell, the zygote, which becomes a plant embryo. Therefore, fertilization in flowering plants is actually a double fertilization. The result is that the endosperm, which uses a large amount of plant resources, will develop only when an egg cell has been fertilized. After fertilization is complete, the ovule develops into a seed, and the surrounding ovary develops into the fruit, which encloses one or more seeds.

3.4 THE CHROMOSOME THEORY OF INHERITANCE AND SEX CHROMOSOMES

Thus far, we have considered how chromosomes are transmitted during cell division and gamete formation. In this section, we will first examine how chromosomal transmission is related to the patterns of inheritance observed by Mendel. This relationship, known as the chromosome theory of inheritance, was a major breakthrough in our understanding of genetics because it established the framework for understanding how chromosomes carry and transmit the genetic determinants that govern the outcome of traits. This theory dramatically unfolded as a result of three lines of scientific inquiry (**Table 3.1**). One avenue concerned Mendel's breeding studies, in which he analyzed the transmission of traits from parent to offspring. A second line of inquiry involved the material basis for heredity. A Swiss botanist, Carl Nägeli, and a German biologist, August Weismann, championed the idea that a substance found in living cells is responsible for the transmission of traits from parents to offspring. Nägeli also suggested that both parents contribute equal amounts of this substance to their offspring. Several scientists, including Oscar

TABLE 3.1

Chronology for the Development and Proof of the Chromosome Theory of Inheritance

1866	Gregor Mendel: analyzed the transmission of traits from parents to offspring and showed that it follows a pattern of segregation and independent assortment.
1876–77	Oscar Hertwig and Hermann Fol: observed that the nucleus of the sperm enters the egg during animal cell fertilization.
1877	Eduard Strasburger: observed that the sperm nucleus of plants (and no detectable cytoplasm) enters the egg during plant fertilization.
1878	Walter Flemming: described mitosis in careful detail.
1883	Carl Nägeli and August Weismann: proposed the existence of a genetic material, which Nägeli called idioplasm and Weismann called germ plasm.
1883	Wilhelm Roux: proposed that the most important event of mitosis is the equal partitioning of "nuclear qualities" to the daughter cells.
1883	Edouard van Beneden: showed that gametes contain half the number of chromosomes and that fertilization restores the normal diploid number.
1884–85	Hertwig, Strasburger, and Weismann: proposed that chromosomes are carriers of the genetic material.
1889	Theodore Boveri: showed that enucleated sea urchin eggs that are fertilized by sperm from a different species develop into larva that have characteristics that coincide with the sperm's species.
1900	Hugo de Vries, Carl Correns, and Erich von Tschermak: rediscovered Mendel's work.
1901	Thomas Montgomery: determined that maternal and paternal chromosomes pair with each other during meiosis.
1901	C. E. McClung: discovered that sex determination in insects is related to differences in chromosome composition.
1902	Boveri: showed that when sea urchin eggs were fertilized by two sperm, the abnormal development of the embryo was related to an abnormal number of chromosomes.
1903	Walter Sutton: showed that even though the chromosomes seem to disappear during interphase, they do not actually disintegrate. Instead, he argued that chromosomes must retain their continuity and individuality from one cell division to the next.
1902–3	Boveri and Sutton: independently proposed tenets of the chromosome theory of inheritance. Some historians primarily credit this theory to Sutton.
1910	Thomas Hunt Morgan: showed that a genetic trait (i.e., white-eyed phenotype in *Drosophila*) was linked to a particular chromosome.
1913	E. Eleanor Carothers: demonstrated that homologous pairs of chromosomes show independent assortment.
1916	Calvin Bridges: studied chromosomal abnormalities as a way to confirm the chromosome theory of inheritance.

For a description of these experiments, the student is encouraged to read Voeller, B. R. (1968), *The Chromosome Theory of Inheritance. Classic Papers in Development and Heredity.* New York: Appleton-Century-Crofts.

Hertwig, Eduard Strasburger, and Walter Flemming, conducted studies suggesting that the chromosomes are the carriers of the genetic material. We now know the DNA within the chromosomes is the genetic material.

Finally, the third line of evidence involved the microscopic examination of the processes of fertilization, mitosis, and meiosis. Researchers became increasingly aware that the characteristics of organisms are rooted in the continuity of cells during the life of an organism and from one generation to the next. When the work of Mendel was rediscovered, several scientists noted striking parallels between the segregation and assortment of traits noted by Mendel and the behavior of chromosomes during meiosis. Among them were Theodore Boveri, a German biologist, and Walter Sutton at Columbia University. They independently proposed the chromosome theory of inheritance, which was a milestone in our understanding of genetics. The principles of this theory are described at the beginning of this section.

The remainder of this section focuses on sex chromosomes. The experimental connection between the chromosome theory of inheritance and sex chromosomes is profound. Even though an examination of meiosis provided compelling evidence that Mendel's laws could be explained by chromosome sorting, researchers still needed to correlate chromosome behavior with the inheritance of particular traits. Because sex chromosomes, such as the X and Y chromosome, look very different under the microscope, and because many genes found on the X chromosome are not found on the Y chromosome, geneticists were able to correlate the inheritance of certain traits with the transmission of specific sex chromosomes. In particular, early studies identified genes on the X chromosome that govern eye color in fruit flies. The phenomenon researchers discovered is called **X-linked inheritance**. This inheritance pattern confirmed the idea that genes are found on chromosomes. In addition, X-linked inheritance showed us that not all traits follow simple Mendelian rules. In later chapters, we will examine a variety of traits that are governed by chromosomal genes yet follow inheritance patterns that are more complex than those observed by Mendel.

The Chromosome Theory of Inheritance Relates the Behavior of Chromosomes to the Mendelian Inheritance of Traits

According to the **chromosome theory of inheritance,** the inheritance patterns of traits can be explained by the transmission patterns of chromosomes during gametogenesis and fertilization. This theory is based on a few fundamental principles.

1. Chromosomes contain the genetic material that is transmitted from parent to offspring and from cell to cell.
2. Chromosomes are replicated and passed along, generation after generation, from parent to offspring. They are also passed from cell to cell during the multicellular development of an organism. Each type of chromosome retains its individuality during cell division and gamete formation.
3. The nuclei of most eukaryotic cells contain chromosomes that are found in homologous pairs—they are diploid. One member of each pair is inherited from the mother, the other from the father. At meiosis, one of the two members of each pair segregates into one daughter nucleus, and the homologue segregates into the other daughter nucleus. Gametes contain one set of chromosomes—they are haploid.

4. During the formation of haploid cells, different types of (nonhomologous) chromosomes segregate independently of each other.
5. Each parent contributes one set of chromosomes to its offspring. The maternal and paternal sets of homologous chromosomes are functionally equivalent; each set carries a full complement of genes.

The chromosome theory of inheritance allows us to see the relationship between Mendel's laws and chromosomal transmission. As shown in **Figure 3.16**, Mendel's law of segregation can be

Heterozygous (*Yy*) cell from a plant with yellow seeds

Meiosis I
Prophase

Metaphase

Anaphase
Telophase

Meiosis II

Haploid cells

FIGURE 3.16   Mendel's law of segregation can be explained by the segregation of homologues during meiosis. The two copies of a gene are contained on homologous chromosomes. In this example using pea seed color, the two alleles are *Y* (yellow) and *y* (green). During meiosis, the homologous chromosomes segregate from each other, leading to segregation of the two alleles into separate gametes.

Genes→Traits The gene for seed color exists in two alleles, *Y* (yellow) and *y* (green). During meiosis, the homologous chromosomes that carry these alleles segregate from each other. The resulting cells receive the *Y* or *y* allele but not both. When two gametes unite during fertilization, the alleles that they carry determine the traits of the resulting offspring.

explained by the homologous pairing and segregation of chromosomes during meiosis. This figure depicts the behavior of a pair of homologous chromosomes that carry a gene for seed color. One of the chromosomes carries a dominant allele that confers yellow seed color, while the homologous chromosome carries a recessive allele that confers green color. A heterozygous individual would pass only one of these alleles to each offspring. In other words, a gamete may contain the yellow allele or the green allele but not both. Because homologous chromosomes segregate from each other, a gamete will contain only one copy of each type of chromosome.

How is the law of independent assortment explained by the behavior of chromosomes? **Figure 3.17** considers the segregation of two types of chromosomes, each carrying a different gene. One pair of chromosomes carries the gene for seed color: the yellow (Y) allele is on one chromosome, and the green (y) allele is on the homologue. The other pair of (smaller) chromosomes carries the gene for seed shape: one copy has the round (R) allele,

and the homologue carries the wrinkled (r) allele. At metaphase of meiosis I, the different types of chromosomes have randomly aligned along the metaphase plate. As shown in Figure 3.17, this can occur in more than one way. On the left, the R allele has sorted with the y allele, while the r allele has sorted with the Y allele. On the right, the opposite situation has occurred. Therefore, the random alignment of chromatid pairs during meiosis I can lead to an independent assortment of genes that are found on nonhomologous chromosomes. As we will see in Chapter 5, this law is violated if two different genes are located close to one another on the same chromosome.

Sex Differences Often Correlate with the Presence of Sex Chromosomes

According to the chromosome theory of inheritance, chromosomes carry the genes that determine an organism's traits and are

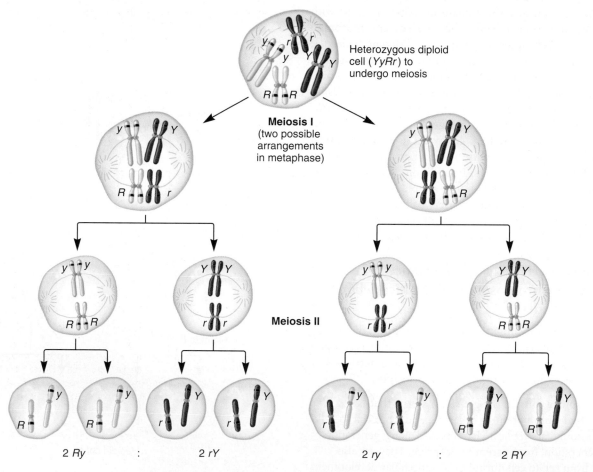

FIGURE 3.17   **Mendel's law of independent assortment can be explained by the random alignment of bivalents during metaphase of meiosis I.** This figure shows the assortment of two genes located on two different chromosomes, using pea seed color and shape as an example ($YyRr$). During metaphase of meiosis I, different possible arrangements of the homologues within bivalents can lead to different combinations of the alleles in the resulting gametes. For example, on the left, the dominant R allele has sorted with the recessive y allele; on the right, the dominant R allele has sorted with the dominant Y allele.

Genes→Traits Most species have several different chromosomes that carry many different genes. In this example, the gene for seed color exists in two alleles, Y (yellow) and y (green), and the gene for seed shape is found as R (round) and r (wrinkled) alleles. The two genes are found on different (nonhomologous) chromosomes. During meiosis, the homologous chromosomes that carry these alleles segregate from each other. In addition, the chromosomes carrying the Y or y alleles will independently assort from the chromosomes carrying the R or r alleles. As shown here, this provides a reassortment of alleles, potentially creating combinations of alleles that are different from the parental combinations. When two gametes unite during fertilization, the alleles they carry affect the traits of the resulting offspring.

the basis of Mendel's law of segregation and independent assortment. Some early evidence supporting this theory involved the determination of sex. Many species are divided into male and female sexes. In 1901, C. E. McClung, who studied fruit flies, was the first to suggest that male and female sexes are due to the inheritance of particular chromosomes. Since McClung's initial observations, we now know that a pair of chromosomes, called the **sex chromosomes,** determines sex in many different species. Some examples are described in **Figure 3.18.**

In the X-Y system of sex determination, which operates in mammals, the male contains one X chromosome and one Y chromosome, whereas the female contains two X chromosomes (Figure 3.18a). In this case, the male is called the **heterogametic sex.** Two types of sperm are produced: one that carries only the X chromosome, and another type that carries the Y. In contrast, the female is the **homogametic sex** because all eggs carry a single X chromosome. The 46 chromosomes carried by humans consist of 1 pair of sex chromosomes and 22 pairs of **autosomes**—chromosomes that are not sex chromosomes. In the human male, each of the four sperm produced during gametogenesis contains 23 chromosomes. Two sperm contain an X chromosome, and the other two have a Y chromosome. The sex of the offspring is determined by whether the sperm that fertilizes the egg carries an X or a Y chromosome.

What causes an offspring to develop into a male or female? One possibility is that two X chromosomes are required for female development. A second possibility is that the Y chromosome promotes male development. In the case of mammals, the second possibility is correct. This is known from the analysis of rare individuals who carry chromosomal abnormalities. For example, mistakes that occasionally occur during meiosis may produce an individual who carries two X chromosomes and one Y chromosome. Such an individual develops into a male.

Other mechanisms of sex determination include the X-0, Z-W, and haplo-diploid systems. The X-0 system of sex determination operates in many insects (Figure 3.18b). In such species, the male has one sex chromosome (the X) and is designated X0, while the female has a pair (two Xs). In other insect species, such as *Drosophila melanogaster*, the male is XY. For both types of insect species (i.e., X0 or XY males, and XX females), the ratio between X chromosomes and the number of autosomal sets determines sex. If a fly has one X chromosome and is diploid for the autosomes (2*n*), the ratio is 1/2, or 0.5. This fly will become a male even if it does not receive a Y chromosome. In contrast to mammals, the Y chromosome in the X-0 system does not determine maleness. If a fly receives two X chromosomes and is diploid, the ratio is 2/2, or 1.0, and the fly becomes a female.

For the Z-W system, which determines sex in birds and some fish, the male is ZZ and the female is ZW (Figure 3.18c). The letters Z and W are used to distinguish these types of sex chromosomes from those found in the X-Y pattern of sex determination of other species. In the Z-W system, the male is the homogametic sex, and the female is heterogametic.

Another interesting mechanism of sex determination, known as the haplo-diploid system, is found in bees (Figure 3.18d). The male bee, called the drone, is produced from unfer-

(a) X–Y system in mammals

(b) The X–0 system in certain insects

(c) The Z–W system in birds

(d) The haplo-diploid system in bees

FIGURE 3.18   Different mechanisms of sex determination in animals. See text for a description.

Genes→Traits Certain genes that are found on the sex chromosomes play a key role in the development of sex (male versus female). For example, in mammals, genes on the Y chromosome initiate male development. In the X-0 system, the ratio of X chromosomes to the sets of autosomes plays a key role in governing the pathway of development toward male or female.

tilized haploid eggs. Female bees, both worker bees and queen bees, are produced from fertilized eggs and therefore are diploid.

The chromosomal basis for sex determination is rooted in the location of particular genes on the sex chromosomes. The molecular basis for sex determination is described in Chapter 23.

Although sex in many species of animals is determined by chromosomes, other mechanisms are also known. In certain reptiles and fish, sex is controlled by environmental factors such as temperature. For example, in the American alligator (*Alligator mississippiensis*), temperature controls sex development. When fertilized eggs of this alligator are incubated at 33°C, nearly 100%

of them produce male individuals. When the eggs are incubated at a temperature a few degrees below 33°C, they produce nearly all females, while at a temperature a few degrees above 33°C, they produce 95% females.

EXPERIMENT 3A

Morgan's Experiments Showed a Connection Between a Genetic Trait and the Inheritance of a Sex Chromosome in *Drosophila*

In the early 1900s, Thomas Hunt Morgan carried out a particularly influential study that confirmed the chromosome theory of inheritance. Morgan was trained as an embryologist, and much of his early research involved descriptive and experimental work in that field. He was particularly interested in ways that organisms change. He wrote, "The most distinctive problem of zoological work is the change in form that animals undergo, both in the course of their development from the egg (embryology) and in their development in time (evolution)." Throughout his life, he usually had dozens of different experiments going on simultaneously, many of them unrelated to each other. He jokingly said there are three kinds of experiments—those that are foolish, those that are damn foolish, and those that are worse than that!

In one of his most famous studies, Morgan engaged one of his graduate students to rear the fruit fly *Drosophila melanogaster* in the dark, hoping to produce flies whose eyes would atrophy from disuse and disappear in future generations. Even after many consecutive generations, however, the flies appeared to have no noticeable changes despite repeated attempts at inducing muta-

tions by treatments with agents such as X-rays and radium. After two years, Morgan finally obtained an interesting result when a true-breeding line of *Drosophila* produced a male fruit fly with white eyes rather than the normal red eyes. Because this had been a true-breeding line of flies, this white-eyed male must have arisen from a new mutation that converted a red-eye allele (denoted w^+) into a white-eye allele (denoted w). Morgan is said to have carried this fly home with him in a jar, put it by his bedside at night while he slept, and then taken it back to the laboratory during the day.

Much like Mendel, Morgan studied the inheritance of this white-eye trait by making crosses and quantitatively analyzing their outcome. In the experiment described in **Figure 3.19**, he began with his white-eyed male and crossed it to a true-breeding red-eyed female. All of the F_1 offspring had red eyes, indicating that red is dominant to white. The F_1 offspring were then mated to each other to obtain an F_2 generation.

■ THE GOAL

This is an example of discovery-based science rather than hypothesis testing. In this case, a quantitative analysis of genetic crosses may reveal the inheritance pattern for the white-eye allele.

■ ACHIEVING THE GOAL — FIGURE 3.19   Inheritance pattern of an X-linked trait in fruit flies.

Starting material: A true-breeding line of red-eyed fruit flies plus one white-eyed male fly that was discovered in the culture.

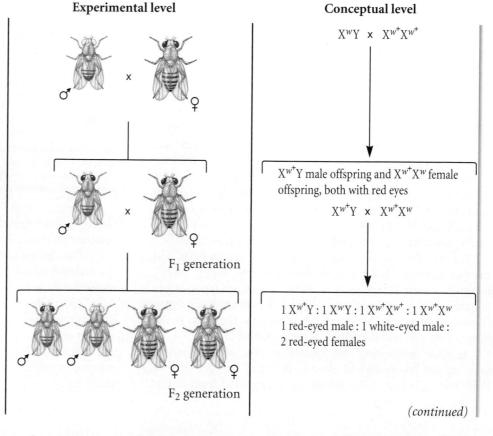

	Experimental level	Conceptual level
1. Cross the white-eyed male to a true-breeding red-eyed female.	♂ × ♀	$X^wY \times X^{w^+}X^{w^+}$
2. Record the results of the F_1 generation. This involves noting the eye color and sexes of many offspring.	♂ × ♀ F_1 generation	$X^{w^+}Y$ male offspring and $X^{w^+}X^w$ female offspring, both with red eyes $X^{w^+}Y \times X^{w^+}X^w$
3. Cross F_1 offspring with each other to obtain F_2 offspring. Also record the eye color and sex of the F_2 offspring.	♂ ♂ ♀ ♀ F_2 generation	$1\ X^{w^+}Y : 1\ X^wY : 1\ X^{w^+}X^{w^+} : 1\ X^{w^+}X^w$ 1 red-eyed male : 1 white-eyed male : 2 red-eyed females

(continued)

4. In a separate experiment, perform a testcross between a white-eyed male and a red-eyed female from the F_1 generation. Record the results.

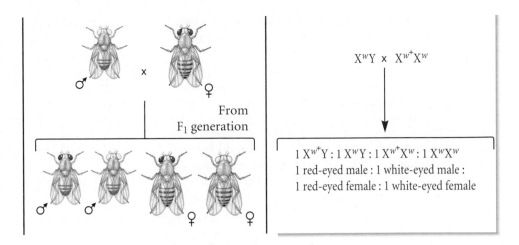

$X^wY \times X^{w^+}X^w$

From F_1 generation

$1\ X^{w^+}Y : 1\ X^wY : 1\ X^{w^+}X^w : 1\ X^wX^w$
1 red-eyed male : 1 white-eyed male :
1 red-eyed female : 1 white-eyed female

THE DATA

Cross	Results	
Original white-eyed male to red-eyed female	F_1 generation:	All red-eyed flies
F_1 male to F_1 female	F_2 generation:	2,459 red-eyed females
		1,011 red-eyed males
		0 white-eyed females
		782 white-eyed males
White-eyed male to F_1 female	Testcross:	129 red-eyed females
		132 red-eyed males
		88 white-eyed females
		86 white-eyed males

INTERPRETING THE DATA

As seen in the data table, the F_2 generation consisted of 2,459 red-eyed females, 1,011 red-eyed males, and 782 white-eyed males. Most notably, no white-eyed female offspring were observed in the F_2 generation. These results suggested that the pattern of transmission from parent to offspring depends on the sex of the offspring and on the alleles that they carry. As shown in the Punnett square here, the data are consistent with the idea that the eye color alleles are located on the X chromosome:

F_1 male is $X^{w^+}Y$
F_1 female is $X^{w^+}X^w$

Male gametes

	♂ X^{w^+}	Y
♀ X^{w^+}	$X^{w^+}X^{w^+}$ Red, female	$X^{w^+}Y$ Red, male
X^w	$X^{w^+}X^w$ Red, female	X^wY White, male

Female gametes

The Punnett square predicts that the F_2 generation will not have any white-eyed females. This prediction was confirmed experimentally. These results indicated that the eye color alleles are located on the X chromosome. Genes that are physically located within the X chromosome are called **X-linked genes** or **X-linked alleles.** However, it should also be pointed out that the experimental ratio in the F_2 generation of red eyes to white eyes is $(2,459 + 1,011){:}782$, which equals 4.4:1. This ratio deviates significantly from the predicted ratio of 3:1. How can this discrepancy be explained? Later work revealed that the lower-than-expected number of white-eyed flies is due to their decreased survival rate.

Morgan also conducted a **testcross** (step 4, Figure 3.19) in which an individual with a dominant phenotype and unknown genotype is crossed to an individual with a recessive phenotype. In this case, he mated an F_1 red-eyed female to a white-eyed male. This cross produced red-eyed males and females, and white-eyed males and females, in approximately equal numbers. The testcross data are also consistent with an X-linked pattern of inheritance. As shown in the following Punnett square, the testcross predicts a 1:1:1:1 ratio:

Testcross:
Male is X^wY
F_1 female is $X^{w^+}X^w$

Male gametes

	♂ X^w	Y
♀ X^{w^+}	$X^{w^+}X^w$ Red, female	$X^{w^+}Y$ Red, male
X^w	X^wX^w White, female	X^wY White, male

Female gametes

The observed data are 129:132:88:86, which is a ratio of 1.5:1.5:1:1. Again, the lower-than-expected numbers of white-eyed males and females can be explained by a lower survival rate for white-eyed flies. In his own interpretation, Morgan concluded that R (red eye color) and X (a sex factor that is present in two copies in the female) are combined and have never existed apart. In other words, this gene for eye color is on the X chromosome. Morgan was the first geneticist to receive the Nobel Prize.

Calvin Bridges, a graduate student in the laboratory of Morgan, also examined the transmission of X-linked traits. Bridges conducted hundreds of crosses involving several different types of X-linked alleles. In his crosses, he occasionally obtained offspring that had unexpected phenotypes and abnormalities in sex chromosome composition. In all cases, the parallel between the cytological presence of sex chromosome abnormalities and the occurrence of unexpected traits confirmed the idea that the sex chromosomes carry X-linked genes. Together, the work of

Morgan and Bridges provided an impressive body of evidence confirming the idea that traits following an X-linked pattern of inheritance are governed by genes that are physically located on the X chromosome. Bridges wrote, "There can be no doubt that the complete parallelism between the unique behavior of chromosomes and the behavior of sex-linked genes and sex in this case means that the sex-linked genes are located in and borne by the X chromosomes." An example of Bridges's work is described in solved problem S5 at the end of this chapter.

A self-help quiz involving this experiment can be found at the Online Learning Center.

CONCEPTUAL SUMMARY

In Chapter 3, we have considered the process of reproduction at the cellular level. During binary fission in bacteria and mitotic cell division in eukaryotes, a mother cell divides to produce two daughter cells that are genetically identical to each other. In bacteria, which contain a single circular chromosome, binary fission provides a way to reproduce asexually. Eukaryotic species, which contain many linear chromosomes, must sort their chromosomes during mitotic cell division. This ensures that the chromosomes are correctly distributed to the daughter cells. In simple eukaryotes, mitotic divisions can produce new unicellular offspring; it is a form of asexual reproduction. Otherwise, the primary reason that eukaryotic species undergo mitotic cell divisions is to produce organisms that are multicellular. Diploid eukaryotes develop from a single fertilized egg cell that undergoes repeated mitotic cell divisions to produce an organism containing many cells. Most of the cells of the body are genetically identical to each other.

Sexual reproduction requires the formation of gametes, such as sperm cells or egg cells, that contain half the genetic material. For diploid organisms that have two sets of chromosomes ($2n$), the process of meiosis results in cells that are haploid, which have one set ($1n$). Meiosis requires two successive divisions. Meiosis I involves a pairing between homologous chromosomes that does not occur during mitosis. This pairing is necessary to ensure that the resultant cells will contain one complete set of chromosomes. Meiosis II is very similar to a mitotic division, except that the cells entering meiosis II begin with half the number of chromatids as compared to cells entering mitosis.

In animals, meiosis occurs in conjunction with the cellular specialization of gametes. During spermatogenesis, meiosis results

in the production of four haploid sperm cells from one primary spermatocyte. In comparison, oogenesis produces a single haploid egg from each primary oocyte. Fertilization in animals involves the union between a single sperm and an egg cell. In plants, sexual reproduction involves an alternation between diploid organisms (sporophytes) and haploid organisms (gametophytes). However, the gametophytes of more complex plants contain only a few cells, which are primarily involved in producing gametes. In flowering plants, sexual reproduction is a double fertilization event in which one sperm fertilizes the egg and a second sperm unites with the central cell to produce the endosperm. The proliferation of endosperm provides nutrients to the plant embryo.

A cellular understanding of mitosis, meiosis, and fertilization allows us to appreciate how the transmission of traits is due to the passage of chromosomes. Genes, which govern the outcome of traits, are physically located within the structure of chromosomes. The chromosome theory of inheritance, proposed by Boveri and Sutton, describes how the transmission of chromosomes accounts for Mendel's laws of segregation and independent assortment. Later, researchers were able to correlate the inheritance of particular traits with the transmission of specific sex chromosomes. In particular, early studies identified genes on the X chromosome that govern eye color in fruit flies. This phenomenon, called X-linked inheritance, confirmed the idea that chromosomes carry genes that govern the outcome of traits. In some species, chromosomes also play a role in sex determination. Particular chromosomes, termed sex chromosomes, determine sex in many species.

EXPERIMENTAL SUMMARY

A central method that has been used to understand chromosomal transmission is light microscopy. Cytogeneticists can view the chromosomes in actively dividing cells under a microscope. This has allowed researchers to study the behavior of chromosomes during mitosis and meiosis. In the late nineteenth and early twentieth century, microscopic examinations of chromosomes yielded observations consistent with the idea that the chromosomes are the carriers of genes. This idea, along with Mendel's experiments in pea plants, led to the formulation of the chromosome theory

of inheritance. Many subsequent experiments have confirmed the validity of this theory. In particular, investigations involving the inheritance of the X chromosome were instrumental in establishing that the inheritance of certain traits paralleled the inheritance of the X chromosome. Morgan was the first scientist to show that the inheritance of a recessive X-linked trait was consistent with the respective gene's location on the X chromosome. Since that time, the chromosomal locations of many genes have been firmly established.

PROBLEM SETS & INSIGHTS

Solved Problems

S1. A diploid cell has eight chromosomes, four per set. For the following diagram, in what phase of mitosis, meiosis I or meiosis II, is this cell?

Answer: The cell is in metaphase of meiosis II. You can tell because the chromosomes are lined up in a single row along the metaphase plate, and the cell has only four pairs of sister chromatids. If it were mitosis, the cell would have eight pairs of sister chromatids.

S2. An unaffected woman (i.e., without disease symptoms) who is heterozygous for the X-linked allele causing Duchenne muscular dystrophy has children with a man with a normal allele. What are the probabilities of the following combinations of offspring?

A. An unaffected son

B. An unaffected son or daughter

C. A family of three children, all of whom are affected

Answer: The first thing we must do is construct a Punnett square to determine the outcome of the cross. N represents the normal allele, n the recessive allele causing Duchenne muscular dystrophy. The mother is heterozygous, and the father has the normal allele.

Male gametes

	♂ X^N	Y
♀ X^N	$X^N X^N$	$X^N Y$
X^n	$X^N X^n$	$X^n Y$

Female gametes

Phenotype ratio is

2 normal daughters :
1 normal son :
1 affected son

A. There are four possible children, one of whom is an unaffected son. Therefore, the probability of an unaffected son is 1/4.

B. Use the sum rule: 1/4 + 1/2 = 3/4.

C. You could use the product rule because there would be three offspring in a row with the disorder: $(1/4)(1/4)(1/4) = 1/64 = 0.016 = 1.6\%$.

S3. What are the major differences between prophase, metaphase, and anaphase when comparing mitosis, meiosis I, and meiosis II?

Answer: The table summarizes key differences.

A Comparison of Mitosis, Meiosis I, and Meiosis II

Phase	Event	Mitosis	Meiosis I	Meiosis II
Prophase	Synapsis:	No	Yes	No
Prophase	Crossing over:	Rarely	Commonly	Rarely
Metaphase	Alignment along the metaphase plate:	Sister chromatids	Bivalents	Sister chromatids
Anaphase	Separation of:	Sister chromatids	Bivalents	Sister chromatids

S4. Among different plant species, both male and female gametophytes can be produced by single individuals or by separate sexes. In some species, such as the garden pea, a single individual can produce both male and female gametophytes. Fertilization takes place via self-fertilization or cross-fertilization. A plant species that has a single type of flower producing both pollen and eggs is termed a monoclinous plant. In other plant species, two different types of flowers produce either pollen or eggs. When both flower types are on a single individual, such a species is termed monoecious. It is most common for the "male flowers" to be produced near the top of the plant and the "female flowers" toward the bottom. Though less common, some species of plants are dioecious. For dioecious species, one individual makes either male flowers or female flowers, but not both.

Based on your personal observations of plants, try to give examples of monoclinous, monoecious, and dioecious plants. What would be the advantages and disadvantages of each?

Answer: Monoclinous plants—pea plant, tulip, and roses. The same flower produces pollen on the anthers and egg cells within the ovary.

Monoecious plants—corn and pine trees. In corn, the tassels are the male flowers and the ears result from fertilization within the female flowers. In pine trees, pollen is produced in cones near the top of the tree, and eggs cells are found in larger cones nearer the bottom.

Dioecious plants—holly and ginkgo trees. Certain individuals produce only pollen, while others produce only eggs.

An advantage of being monoclinous or monoecious is that fertilization is relatively easy because the pollen and egg cells are produced on the same individual. This is particularly true for monoclinous plants. The proximity of the pollen to the egg cells makes it more likely for self-fertilization to occur. This is advantageous if the plant population is relatively sparse. On the other hand, a dioecious species can reproduce only via cross-fertilization. The advantage of cross-fertilization is that it enhances genetic variation. Over the long run, this can be an advantage because cross-fertilization is more likely to produce a varied population of individuals, some of which may possess combinations of traits that promote survival.

S5. To test the chromosome theory of inheritance, Calvin Bridges made crosses involving the inheritance of X-linked traits. One of his experiments concerned two different X-linked genes affecting eye color and wing size. For the eye color gene, the red-eye allele

(w^+) is dominant to the white-eye allele (w). A second X-linked trait is wing size; the allele called miniature is recessive to the normal allele. In this case, m represents the miniature allele and m^+ the normal allele. A male fly carrying a miniature allele on its single X chromosome has small (miniature) wings. A female must be homozygous, mm, in order to have miniature wings.

Bridges made a cross between $X^{w,m^+} X^{w,m^+}$ female flies (white eyes and normal wings) to $X^{w^+,m}$ Y male flies (red eyes and miniature wings). He then examined the eyes, wings, and sexes of thousands of offspring. Most of the offspring were females with red eyes and normal wings, and males with white eyes and normal wings. On rare occasions (approximately 1 out of 1,700 flies), however, he also obtained female offspring with white eyes or males with red eyes. He also noted the wing shape in these flies and then cytologically examined their chromosome composition using a microscope. The following results were obtained:

Offspring	Eye Color	Wing Size	Sex Chromosomes
Expected females	Red	Normal	XX
Expected males	White	Normal	XY
Unexpected females (rare)	White	Normal	XXY
Unexpected males (rare)	Red	Miniature	X0

Data from: Bridges, C. B. "Non-Disjunction as Proof of the Chromosome Theory of Heredity," *Genetics 1*, 1–52, 107–63.

Explain these data.

Answer: Remember that in fruit flies, the number of X chromosomes (not the presence of the Y chromosome) determines sex. As seen in the data, the flies with unexpected phenotypes were abnormal in their sex chromosome composition. The white-eyed female flies were due to the union between an abnormal XX female gamete and a normal Y male gamete. Likewise, the unexpected male offspring contained only one X chromosome and no Y. These male offspring were due to the union between an abnormal egg without any X chromosome and a normal sperm containing one X chromosome. The wing size of the unexpected males was a particularly significant result. The red-eyed males showed a miniature wing size. As noted by Bridges, this means they inherited their X chromosome from their father rather than their mother. This observation provided compelling evidence that the inheritance of the X chromosome correlates with the inheritance of particular traits.

At the time of his work, Bridges's results were particularly striking because chromosomal abnormalities had been rarely observed in *Drosophila*. Nevertheless, Bridges first predicted how chromosomal abnormalities would cause certain unexpected phenotypes, and then he actually observed the abnormal number of chromosomes using a microscope. Together, his work provided evidence confirming the idea that traits that follow an X-linked pattern of inheritance are governed by genes physically located on the X chromosome.

Conceptual Questions

C1. The process of binary fission begins with a single mother cell and ends with two daughter cells. Would you expect the mother and daughter cells to be genetically identical? Explain why or why not.

C2. What is a homologue? With regard to genes and alleles, how are homologues similar to and different from each other?

C3. What is a sister chromatid? Are sister chromatids genetically similar or identical? Explain.

C4. With regard to sister chromatids, which phase of mitosis is the organization phase, and which is the separation phase?

C5. A species is diploid containing three chromosomes per set. Draw what the chromosomes would look like in the G_1 and G_2 phases of the cell cycle.

C6. How does the attachment of kinetochore microtubules to the kinetochore differ in metaphase of meiosis I compared to metaphase of mitosis? Discuss what you think would happen if a sister chromatid was not attached to a kinetochore microtubule.

C7. For the following events, specify whether they occur during mitosis, meiosis I, and/or meiosis II:

A. Separation of conjoined chromatids within a pair of sister chromatids

B. Pairing of homologous chromosomes

C. Alignment of chromatids along the metaphase plate

D. Attachment of sister chromatids to both poles

C8. Describe the key events during meiosis that result in a 50% reduction in the amount of genetic material per cell.

C9. A cell is diploid and contains three chromosomes per set. Draw the arrangement of chromosomes during metaphase of mitosis and metaphase of meiosis I and II. In your drawing, make one set dark and the other lighter.

C10. The arrangement of homologues during metaphase of meiosis I is a random process. In your own words, explain what this means.

C11. A eukaryotic cell is diploid containing 10 chromosomes (5 in each set). For mitosis and meiosis, how many daughter cells would be produced, and how many chromosomes would each one contain?

C12. If a diploid cell contains six chromosomes (i.e., three per set), how many possible random arrangements of homologues could occur during metaphase of meiosis I?

C13. A cell has four pairs of chromosomes. Assuming that crossing over does not occur, what is the probability that a gamete will contain all of the paternal chromosomes? If n equals the number of chromosomes in a set, which of the following expressions can be used to calculate the probability that a gamete will receive all of the paternal chromosomes: $(1/2)^n$, $(1/2)^{n-1}$, or $n^{1/2}$?

C14. With regard to question C13, how would the phenomenon of crossing over affect the results? In other words, would the probability of a gamete inheriting only paternal chromosomes be higher or lower? Explain your answer.

C15. Eukaryotic cells must sort their chromosomes during mitosis so that each daughter cell receives the correct number of chromosomes. Why don't bacteria need to sort their chromosomes?

C16. Why is it necessary that the chromosomes condense during mitosis and meiosis? What do you think might happen if the chromosomes were not condensed?

C17. Nine-banded armadillos almost always give birth to four offspring that are genetically identical quadruplets. Explain how you think this happens.

C18. A diploid species contains four chromosomes per set for a total of eight chromosomes in its somatic cells. Draw the cell as it would look in late prophase of meiosis II and prophase of mitosis. Discuss how prophase of meiosis II and prophase of mitosis differ from each other, and explain how the difference originates.

C19. Explain why the products of meiosis may not be genetically identical while the products of mitosis are.

C20. The period between meiosis I and meiosis II is called interphase II. Does DNA replication take place during interphase II? Explain your answer.

C21. List several ways in which telophase appears to be the reverse of prophase and prometaphase.

C22. In corn, there are 10 chromosomes per set and the sporophyte of the species is diploid. If you performed a karyotype, what is the total number of chromosomes that you would expect to see in the following types of cells?

A. A leaf cell

B. The sperm nucleus of a pollen grain

C. An endosperm cell after fertilization

D. A root cell

C23. The arctic fox has 50 chromosomes (25 per set), and the common red fox has 38 chromosomes (19 per set). These species can interbreed to produce viable but infertile offspring. How many chromosomes would the offspring have? What problems do you think may occur during meiosis that would explain the offspring's infertility?

C24. Let's suppose that a gene affecting pigmentation is found on the X chromosome (in mammals or insects) or the Z chromosome (in birds) but not on the Y or W chromosome. It is found on an autosome in bees. This gene is found in two alleles, *D* (dark), which is dominant to *d* (light). What would be the phenotypic results of crosses between a true-breeding dark female and true-breeding light male, and the reciprocal crosses involving a true-breeding light female and true-breeding dark male, in the following species? Refer back to Figure 3.18 for the mechanism of sex determination in these species.

A. Birds

B. *Drosophila*

C. Bees

D. Humans

C25. Describe the cellular differences between male and female gametes.

C26. At puberty, the testes contain a finite number of cells and produce an enormous number of sperm cells during the life span of a male. Explain why testes do not run out of spermatogonial cells.

C27. Describe the timing of meiosis I and II during human oogenesis.

C28. Three genes (*A*, *B*, and *C*) are found on three different chromosomes. For the following diploid genotypes, describe all the possible gamete combinations.

A. *Aa Bb Cc*

B. *AA Bb CC*

C. *Aa BB Cc*

D. *Aa bb cc*

C29. A phenotypically normal woman with an abnormally long chromosome 13 (and a normal homologue of chromosome 13) marries a phenotypically normal man with an abnormally short chromosome 11 (and a normal homologue of chromosome 11). What is the probability of producing an offspring that will have both a long chromosome 13 and a short chromosome 11? If such a child is produced, what is the probability that this child would eventually pass both abnormal chromosomes to one of his/her offspring?

C30. Assuming that such a fly would be viable, what would be the sex of a fruit fly with the following chromosomal composition?

A. One X chromosome and two sets of autosomes

B. Two X chromosomes, one Y chromosome, and two sets of autosomes

C. Two X chromosomes and four sets of autosomes

D. Four X chromosomes, two Y chromosomes, and four sets of autosomes

C31. What would be the sex of a human with the following numbers of sex chromosomes?

A. XXX

B. X (also described as X0)

C. XYY

D. XXY

Experimental Questions

E1. When studying living cells in a laboratory, researchers sometimes use drugs as a way to make cells remain at a particular stage of the cell cycle. For example, aphidicolin inhibits DNA synthesis in eukaryotic cells and causes them to remain in the G_1 phase because they cannot replicate their DNA. In what phase of the cell cycle—G_1, S, G_2, prophase, metaphase, anaphase, or telophase—would you expect somatic cells to stay if the following types of drug were added?

A. A drug that inhibits microtubule formation

B. A drug that allows microtubules to form but prevents them from shortening

C. A drug that inhibits cytokinesis

D. A drug that prevents chromosomal condensation

E2. In Morgan's experiments, which result do you think is the most convincing piece of evidence pointing to X-linkage of the eye color gene? Explain your answer.

E3. In his original studies of Figure 3.19, Morgan first suggested that the original white-eyed male had two copies of the white-eye allele. In this problem, let's assume that he meant the fly was X^wY^w instead of X^wY. Are his data in Figure 3.19 consistent with this hypothesis? What crosses would need to be made to rule out the possibility that the Y chromosome carries a copy of the eye color gene?

E4. How would you set up crosses to determine if a gene was Y linked versus X linked?

E5. Occasionally during meiosis, a mistake can happen whereby a gamete may receive zero or two sex chromosomes rather than one. Calvin Bridges made a cross between white-eyed female flies and red-eyed male flies. As you would expect, most of the offspring were red-eyed females and white-eyed males. On rare occasions, however, he found a white-eyed female or a red-eyed male. These rare flies were not due to new gene mutations but instead were due to mistakes during meiosis in the parent flies. Consider the mechanism of sex determination in fruit flies and propose how this could happen. In your answer, describe the sex chromosome composition of these rare flies.

E6. Let's suppose that you have karyotyped a female fruit fly with red eyes and found that it has three X chromosomes instead of the normal two. Although you do not know its parents, you do know that this fly came from a mixed culture of flies in which some had red eyes, some had white eyes, and some had eosin eyes. Eosin is an allele of the same gene that has white and red alleles. Eosin is a pale orange color. The red allele is dominant and the white allele is recessive. The expression of the eosin allele, however, depends on the number of copies of the allele. When females have two copies of this allele, they have eosin eyes. When females are heterozygous for the eosin allele and white allele, they have light-eosin eyes. When females are heterozygous for the red allele and the eosin allele, they have red eyes. Males that have a single copy of eosin allele have light-eosin eyes.

You cross this female with a white-eyed male and count the number of offspring. You may assume that this unusual female makes half of its gametes with one X chromosome and half of its gametes with two X chromosomes. The following results were obtained:

	Females*	Males
Red eyes	50	11
White eyes	0	0
Eosin	20	0
Light-eosin	21	20

*A female offspring can be XXX, XX, or XXY.

Explain the 3:1 ratio between female and male offspring. What was the genotype of the original mother, which had red eyes and three X chromosomes? Construct a Punnett square that is consistent with these data.

E7. With regard to thickness and length, what do you think the chromosomes would look like if you microscopically examined them during interphase? How would that compare to their appearance during metaphase?

E8. White-eyed flies have a lower survival rate compared to red-eyed flies. Based on the data in Figure 3.19, what percentage of white-eyed flies survived compared to red-eyed flies, assuming 100% survival of red-eyed flies?

E9. A rare form of dwarfism that also included hearing loss was found to run in a particular family. It is inherited in a dominant manner. It was discovered that an affected individual had one normal copy of chromosome 15 and one abnormal copy of chromosome 15 that was unusually long. How would you determine if the unusually long chromosome 15 was causing this disorder?

E10. Discuss why crosses (i.e., the experiments of Mendel) and the microscopic observations of chromosomes during mitosis and meiosis were both needed to deduce the chromosome theory of inheritance.

E11. A cross was made between female flies with white eyes and miniature wings (both X-linked recessive traits) to male flies with red eyes and normal wings. On rare occasions, female offspring were produced with white eyes. If we assume these females are due to errors in meiosis, what would be the most likely chromosomal composition of such flies? What would be their wing shape?

E12. Experimentally, how do you think researchers were able to determine that the Y chromosome causes maleness in mammals, whereas the ratio of X chromosomes to the sets of autosomes causes sex determination in fruit flies?

Questions for Student Discussion/Collaboration

1. In Figure 3.19, Morgan obtained a white-eyed male fly in a population containing many red-eyed flies that he thought were true-breeding. As mentioned in the experiment, he crossed this fly with several red-eyed sisters, and all the offspring had red eyes. But actually this is not quite true. Morgan observed 1,237 red-eyed flies and 3 white-eyed males. Provide two or more explanations why he obtained 3 white-eyed males in the F_1 generation.

2. A diploid eukaryotic cell has 10 chromosomes (5 per set). As a group, take turns having one student draw the cell as it would look during a phase of mitosis, meiosis I, or meiosis II; then have the other students guess which phase it is.

3. Discuss the principles of the chromosome theory of inheritance. Which principles were deduced via light microscopy, and which were deduced from crosses? What modern techniques could be used to support the chromosome theory of inheritance?

Note: All answers appear at the website for this textbook; the answers to even-numbered questions are in the back of the textbook.

Visit the Online Learning Center for practice tests, answer keys, and other learning aids for this chapter. Enhance your understanding of genetics with our interactive exercises, quizzes, animations, and much more.

EXTENSIONS OF MENDELIAN INHERITANCE

The term **Mendelian inheritance** describes inheritance patterns that obey two laws: the law of segregation and the law of independent assortment. Until now, we have mainly considered traits that are affected by a single gene that is found in two different alleles. In these cases, one allele is dominant over the other. This type of inheritance is sometimes called **simple Mendelian inheritance** because the observed ratios in the offspring readily obey Mendel's laws. For example, when two different true-breeding pea plants are crossed (e.g., tall and dwarf) and the F_1 generation is allowed to self-fertilize, the F_2 generation shows a 3:1 phenotypic ratio of tall and dwarf offspring.

In Chapter 4, we will extend our understanding of Mendelian inheritance by examining the transmission patterns for several traits that do not display a simple dominant/recessive relationship. Geneticists have discovered an amazing diversity of mechanisms by which alleles affect the outcome of traits. Many alleles don't produce the ratios of offspring that are expected from a simple Mendelian relationship. This does not mean that Mendel was wrong. Rather, the inheritance patterns of many traits are more complex and interesting than he had realized. In this chapter, we will examine how the outcome of a trait may be influenced by a variety of factors such as the level of protein expression, the sex of the individual, the presence of multiple alleles of a given gene, and environmental effects. We will also explore how two different genes can contribute to the outcome of a single trait. Later, in Chapters 5 and 7, we will examine eukaryotic inheritance patterns that actually violate the laws of segregation and independent assortment.

4.1 INHERITANCE PATTERNS OF SINGLE GENES

We begin Chapter 4 with the further exploration of traits that are influenced by a single gene. **Table 4.1** describes the general features of several types of Mendelian inheritance patterns that have been observed by researchers. These various patterns occur because the outcome of a trait may be governed by two or more alleles in many different ways. In this section, we will examine these patterns with two goals in mind. First, we want to understand how the molecular expression of genes can account for an individual's phenotype. In other words, we will explore the underlying relationship between molecular genetics—the expression of genes to produce functional proteins—and the traits of individuals that inherit the genes. Our second goal concerns the outcome of crosses. Many of the

Inheritance patterns and alleles. In the petunia, multiple alleles can result in flowers with several different colors, such as the three shown here.

TABLE 4.1

Types of Mendelian Inheritance Patterns Involving Single Genes

Type	Description
Simple Mendelian	**Inheritance:** This term is commonly applied to the inheritance of alleles that obey Mendel's laws and follow a strict dominant/recessive relationship. In Chapter 4, we will see that some genes can be found in three or more alleles, making the relationship more complex. **Molecular:** 50% of the protein encoded by two copies of the dominant (functional) allele is sufficient to produce the dominant trait.
Incomplete dominance	**Inheritance:** This pattern occurs when the heterozygote has a phenotype that is intermediate between either corresponding homozygote. For example, a cross between homozygous red-flowered and homozygous white-flowered parents will have heterozygous offspring with pink flowers. **Molecular:** 50% of the protein encoded by two copies of the functional allele is not sufficient to produce the same trait as the homozygote making 100%.
Incomplete penetrance	**Inheritance:** This pattern occurs when a dominant phenotype is not expressed even though an individual carries a dominant allele. An example is an individual who carries the polydactyly allele but has a normal number of fingers and toes. **Molecular:** Even though a dominant gene may be present, the protein encoded by the gene may not exert its effects. This can be due to environmental influences or due to other genes that may encode proteins that counteract the effects of the protein encoded by the dominant allele.
Overdominance	**Inheritance:** This pattern occurs when the heterozygote has a trait that is more beneficial than either homozygote. **Molecular:** Three common ways that heterozygotes gain benefits: (1) Their cells may have increased resistance to infection by microorganisms; (2) they may produce more forms of protein dimers, with enhanced function; or (3) they may produce proteins that function under a wider range of conditions.
Codominance	**Inheritance:** This pattern occurs when the heterozygote expresses both alleles simultaneously. For example, in blood typing, an individual carrying the *A* and *B* alleles will have an AB blood type. **Molecular:** The codominant alleles encode proteins that function slightly differently from each other, and the function of each protein in the heterozygote affects the phenotype uniquely.
X linked	**Inheritance:** This pattern involves the inheritance of genes that are located on the X chromosome. In mammals and fruit flies, males are hemizygous for X-linked genes, while females have two copies. **Molecular:** If a pair of X-linked alleles shows a simple dominant/recessive relationship, 50% of the protein encoded by two copies of the dominant allele is sufficient to produce the dominant trait (in the female).
Sex-influenced inheritance	**Inheritance:** This pattern refers to the impact of sex on the phenotype of the individual. Some alleles are recessive in one sex and dominant in the opposite sex. An example would be pattern baldness in humans. **Molecular:** Sex hormones may regulate the molecular expression of genes. This can have an impact on the phenotypic effects of alleles.
Sex-limited inheritance	**Inheritance:** This refers to traits that occur in only one of the two sexes. An example would be breast development in mammals. **Molecular:** Sex hormones may regulate the molecular expression of genes. This can have an impact on the phenotypic effects of alleles. In this case, sex hormones that are primarily produced in only one sex are essential to produce a particular phenotype.
Lethal alleles	**Inheritance:** An allele that has the potential of causing the death of an organism. **Molecular:** Lethal alleles are most commonly loss-of-function alleles that encode proteins that are necessary for survival. In rare cases, the allele may be due to a mutation in a nonessential gene that changes a protein to function with abnormal and detrimental consequences.

inheritance patterns described in Table 4.1 do not produce a 3:1 phenotypic ratio when two heterozygotes produce offspring. In this section, we consider how allelic interactions produce ratios that differ from a simple Mendelian pattern. However, as our starting point, we will begin by reconsidering a simple dominant/recessive relationship from a molecular perspective.

Recessive Alleles Often Cause a Reduction in the Amount or Function of the Encoded Proteins

For any given gene, geneticists refer to prevalent alleles in a natural population as **wild-type alleles.** In large populations, more than one wild-type allele may occur—a phenomenon known as **genetic polymorphism.** For example, **Figure 4.1** illustrates a striking example of polymorphism in the elderflower orchid, *Dactylorhiza sambucina.* Throughout the range of this species in Europe, both yellow- and red-flowered individuals are prevalent.

Both colors are considered wild type. At the molecular level, a wild-type allele typically encodes a protein that is made in the proper amount and functions normally. As discussed in Chapter 24, wild-type alleles tend to promote the reproductive success of organisms in their native environments.

In addition, random mutations occur in populations and alter preexisting alleles. Geneticists sometimes refer to these kinds of alleles as **mutant alleles** to distinguish them from the more common wild-type alleles. Because random mutations are more likely to disrupt gene function, mutant alleles are often defective in their ability to express a functional protein. Such mutant alleles tend to be rare in natural populations. They are typically, but not always, inherited in a recessive fashion.

Among Mendel's seven traits discussed in Chapter 2, the wild-type alleles are tall plants, purple flowers, axial flowers, yellow seeds, round seeds, green pods, and smooth pods (refer back to Figure 2.4). The mutant alleles are dwarf plants, white flowers, terminal flowers, green seeds, wrinkled seeds, yellow pods, and constricted pods. You may have already noticed that the seven wild-type alleles are dominant over the seven mutant alleles. Likewise, red eyes and normal wings are examples of wild-type alleles in *Drosophila*, and white eyes and miniature wings are recessive mutant alleles.

The idea that recessive alleles usually cause a substantial decrease in the expression of a functional protein is supported by the analysis of many human genetic diseases. Keep in mind that a genetic disease is usually caused by a mutant allele. **Table 4.2** lists several examples of human genetic diseases in which the recessive allele fails to produce a specific cellular protein in its active form. In many cases, molecular techniques have enabled researchers to clone these genes and determine the differences between the wild-type and mutant alleles. They have found that the recessive

allele usually contains a mutation that causes a defect in the synthesis of a fully functional protein.

To understand why many defective mutant alleles are inherited recessively, we need to take a quantitative look at protein function. With the exception of sex-linked genes, diploid individuals have two copies of every gene. In a simple dominant/recessive relationship, the recessive allele does not affect the phenotype of the heterozygote. In other words, a single copy of the dominant allele is sufficient to mask the effects of the recessive allele. If the recessive allele cannot produce a functional protein, how do we explain the wild-type phenotype of the heterozygote? As described in **Figure 4.2**, a common explanation is that 50% of the functional protein is adequate to provide the wild-type phenotype. In this example, the *PP* homozygote and *Pp* heterozygote

FIGURE 4.1   An example of genetic polymorphism. Both yellow and red flowers are common in natural populations of the elderflower orchid, *Dactylorhiza sambucina*, and both are considered wild type.

Dominant (functional) allele: *P* (purple)
Recessive (defective) allele: *p* (white)

Genotype	*PP*	*Pp*	*pp*
Amount of functional protein P	100%	50%	0%

Phenotype	Purple	Purple	White
Simple dominant/ recessive relationship			

FIGURE 4.2   A comparison of protein levels among homozygous and heterozygous genotypes *PP*, *Pp*, and *pp*.

Genes → Traits In a simple dominant/recessive relationship, 50% of the protein encoded by two copies of the dominant allele is sufficient to produce the wild-type phenotype, in this case, purple flowers. A complete lack of the functional protein results in white flowers.

TABLE 4.2

Examples of Recessive Human Diseases

Disease	Protein That Is Produced by the Normal Gene*	Description
Phenylketonuria	Phenylalanine hydroxylase	Inability to metabolize phenylalanine. The disease can be prevented by following a phenylalanine-free diet. If the diet is not followed early in life, it can lead to severe mental retardation and physical degeneration.
Albinism	Tyrosinase	Lack of pigmentation in the skin, eyes, and hair.
Tay-Sachs disease	Hexosaminidase A	Defect in lipid metabolism. Leads to paralysis, blindness, and early death.
Sandhoff disease	Hexosaminidase B	Defect in lipid metabolism. Muscle weakness in infancy, early blindness, and progressive mental and motor deterioration.
Cystic fibrosis	Chloride transporter	Inability to regulate ion balance across epithelial cells. Leads to production of thick lung mucus and chronic lung infections.
Lesch-Nyhan syndrome	Hypoxanthine-guanine phosphoribosyl transferase	Inability to metabolize purines, which are bases found in DNA and RNA. Leads to self-mutilation behavior, poor motor skills, and usually mental impairment and kidney failure.

*Individuals who exhibit the disease are either homozygous for a recessive allele or hemizygous (for X-linked genes in human males). The disease symptoms result from a defect in the amount or function of the normal protein.

each make sufficient functional protein to yield purple flowers. This means that the homozygous individual makes twice as much of the wild-type protein than it really needs to produce purple flowers. Therefore, if the amount is reduced to 50%, as in the heterozygote, the individual still has plenty of this protein to accomplish whatever cellular function it performs. The phenomenon that "50% of the normal protein is enough" is fairly common among many genes.

A second possible explanation for other genes is that the heterozygote actually produces more than 50% of the functional protein. Due to the phenomenon of gene regulation, the expression of the normal gene may be increased or "up regulated" in the heterozygote to compensate for the lack of function of the defective allele. The topic of gene regulation is discussed in Chapters 14 and 15.

Incomplete Dominance Occurs When Two Alleles Produce an Intermediate Phenotype

Although many alleles display a simple dominant/recessive relationship, geneticists have also identified some cases in which a heterozygote exhibits **incomplete dominance**—a condition in which the phenotype is intermediate between the corresponding homozygous individuals. In 1905, the German botanist Carl Correns first observed this phenomenon in the four-o'clock (*Mirabilis jalapa*), involving flower color. **Figure 4.3** describes Correns's experiment, in which a homozygous red-flowered four-o'clock plant was crossed to a homozygous white-flowered plant. The wild-type allele for red flower color is designated C^R and the white allele is C^W. As shown here, the offspring had pink flowers. If these F_1 offspring were allowed to self-fertilize, the F_2 generation consisted of 1/4 red-flowered plants, 1/2 pink-flowered plants, and 1/4 white-flowered plants. The pink plants in the F_2 generation were heterozygotes with an intermediate phenotype. As noted in the Punnett square in Figure 4.3, the F_2 generation displayed a 1:2:1 phenotypic ratio, which is different from the 3:1 ratio observed for simple Mendelian inheritance.

In Figure 4.3, incomplete dominance resulted in a heterozygote with an intermediate phenotype. At the molecular level, the allele that causes a white phenotype is expected to result in a lack of a functional protein required for pigmentation. Depending on the effects of gene regulation, the heterozygotes may produce only 50% of the normal protein, but this amount is not sufficient to create the same phenotype as the C^RC^R homozygote, which may make twice as much of this protein. In this example, a reasonable explanation is that 50% of the functional protein cannot accomplish the same level of pigment synthesis that 100% of the protein can.

Finally, our opinion of whether a trait is dominant or incompletely dominant may depend on how closely we examine the trait in the individual. The more closely we look, the more likely we are to discover that the heterozygote is not quite the same as the wild-type homozygote. For example, Mendel studied the characteristic of pea seed shape and visually concluded that the *RR* and *Rr* genotypes produced round seeds and the *rr*

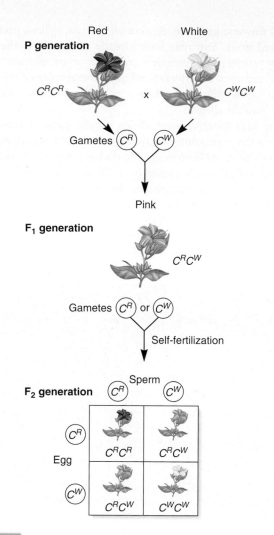

FIGURE 4.3   **Incomplete dominance in the four-o'clock plant, *Mirabilis jalapa*.**

INTERACTIVE EXERCISE

Genes → Traits When two different homozygotes (C^RC^R and C^WC^W) are crossed, the resulting heterozygote, C^RC^W, has an intermediate phenotype of pink flowers. In this case, 50% of the functional protein encoded by the C^R allele is not sufficient to produce a red phenotype.

genotype produced wrinkled seeds. The peculiar morphology of the wrinkled seed is caused by a large decrease in the amount of starch deposition in the seed due to a defective *r* allele. Since the time of Mendel's work, other scientists have dissected round and wrinkled seeds and examined their contents under the microscope. They have found that round seeds from heterozygotes actually contain an intermediate number of starch grains compared to seeds from the corresponding homozygotes (**Figure 4.4**). Within the seed, an intermediate amount of the functional protein is not enough to produce as many starch grains as in the homozygote carrying two copies of the *R* allele. Nevertheless, at the level of our unaided eyes, heterozygotes produce seeds that appear to be round. With regard to phenotypes, the *R* allele is dominant to the *r* allele at the level of visual examination, but the *R* and *r* alleles show incomplete dominance at the level of starch biosynthesis.

Dominant (functional) allele: *R* (round)
Recessive (defective) allele: *r* (wrinkled)

Genotype	*RR*	*Rr*	*rr*
Amount of functional (starch-producing) protein	100%	50%	0%

Phenotype	Round	Round	Wrinkled
With unaided eye (simple dominant/recessive relationship)			
With microscope (incomplete dominance)			

FIGURE 4.4   **A comparison of phenotype at the macroscopic and microscopic levels.**

Genes → Traits This illustration shows the effects of a heterozygote having only 50% of the functional protein needed for starch production. This seed appears to be as round as those of the homozygote carrying the *R* allele, but when examined microscopically, it has produced only half the amount of starch.

Traits May Skip a Generation Due to Incomplete Penetrance and Vary in Their Expressivity

As we have seen, dominant alleles are expected to influence the outcome of a trait when they are present in heterozygotes. Occasionally, however, this may not occur. **Figure 4.5a** illustrates a human pedigree for a dominant trait known as polydactyly. This trait causes the affected individual to have additional fingers and/or toes (**Figure 4.5b**). Polydactyly is due to an autosomal dominant allele—the allele is found in a gene located on an autosome (not a sex chromosome) and a single copy of this allele is sufficient to cause this condition. Sometimes, however, individuals carry the dominant allele but do not exhibit the trait. In Figure 4.5a, individual III-2 has inherited the polydactyly allele from his mother and passed the allele to a daughter and son. However, individual III-2 does not actually exhibit the trait himself, even though he is a heterozygote. This phenomenon, called **incomplete penetrance,** indicates that a dominant allele does not always "penetrate" into the phenotype of the individual. The measure of penetrance is described at the populational level. For example, if 60% of the heterozygotes carrying a dominant allele exhibit the trait, we would say that this trait is 60% penetrant. At the individual level, the trait is either present or not.

Another term used to describe the outcome of traits is the degree to which the trait is expressed, or its **expressivity.** In the case of polydactyly, the number of extra digits can vary. For example, one individual may have an extra toe on only one foot, whereas a second individual may have extra digits on both the hands and feet. Using genetic terminology, a person with several extra digits would have high expressivity of this trait, while a

(a)

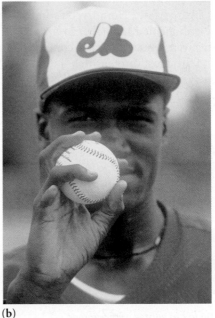

(b)

FIGURE 4.5   **Polydactyly, a dominant trait that shows incomplete penetrance. (a)** A family pedigree. Affected individuals are shown in black. Notice that offspring IV-1 and IV-3 have inherited the trait from a parent, III-2, who is heterozygous but does not exhibit polydactyly. **(b)** Antonio Alfonseca, a baseball player with polydactyly. His extra finger does not give him an advantage when pitching because it is small and does not touch the ball.

person with a single extra digit would have low expressivity. The extreme case, when a heterozygote does not express the trait at all, is incomplete penetrance.

How do we explain incomplete penetrance and variable expressivity? While the answer may not always be understood, the range of phenotypes is often due to environmental influences and/or due to effects of a modifier gene in which one gene modifies the phenotypic effects of another gene. We will consider the issue of the environment next. The effects of modifier genes will be discussed later in the chapter.

The Outcome of Traits Is Influenced by the Environment

Throughout this book, our study of genetics tends to focus on the roles of genes in the outcome of traits. In addition to genetics, environmental conditions have a great impact on the phenotype of the individual. For example, the arctic fox (*Alopex lagopus*) goes through two color phases. During the cold winter, the arctic fox is primarily white, while in the warmer summer, it is mostly brown (**Figure 4.6a**). As discussed later, such temperature-sensitive alleles affecting fur color are found among many species of mammals.

A dramatic example of the relationship between environment and phenotype can be seen in the human genetic disease known as phenylketonuria (PKU). This autosomal recessive disease is caused by a defect in a gene that encodes the enzyme phenylalanine hydroxylase. Homozygous individuals with this defective allele are unable to metabolize the amino acid phenylalanine. When given a standard diet containing phenylalanine, which is found in most protein-rich foods, PKU individuals manifest a variety of detrimental traits including mental retarda-

tion, underdeveloped teeth, and foul-smelling urine. In contrast, when PKU individuals are diagnosed early and follow a restricted diet free of phenylalanine, they develop normally (**Figure 4.6b**). Since the 1960s, testing methods have been developed that can determine if an individual is lacking the phenylalanine hydroxylase enzyme. These tests permit the identification of infants who have PKU. Their diets can then be modified before the harmful effects of phenylalanine ingestion have occurred. As a result of government legislation, more than 90% of infants born in the United States are now tested for PKU. This test both prevents a great deal of human suffering and is cost-effective. In the United States, the annual cost of PKU testing is estimated to be a few million dollars, whereas the cost of treating severely retarded individuals with the disease would be hundreds of millions of dollars.

The examples of the arctic fox and PKU represent dramatic effects of very different environmental conditions. When considering the environment, geneticists often examine a range of conditions, rather than simply observing phenotypes under two different conditions. The term **norm of reaction** refers to the effects of environmental variation on a phenotype. Specifically, it is the

(a) Arctic fox in winter and summer

(b) Healthy person with PKU

Facet

(c) Norm of reaction

FIGURE 4.6   Variation in the expression of traits due to environmental effects. (**a**) The arctic fox in the winter and summer. (**b**) A person with PKU who has followed a restricted diet and developed normally. (**c**) Norm of reaction. In this experiment, fertilized eggs from a population of genetically identical *Drosophila melanogaster* were allowed to develop into adult flies at different environmental temperatures. This graph shows the relationship between temperature (an environmental factor) and facet number in the eyes of the resulting adult flies. The micrograph shows an eye of *D. melanogaster*.

phenotypic range seen in individuals with a particular genotype. To evaluate the norm of reaction, researchers study members of true-breeding strains that have the same genotypes and subject them to different environmental conditions. As an example, let's consider facet number in the eyes of fruit flies, *Drosophila melanogaster*. This species has compound eyes composed of many individual facets. **Figure 4.6c** shows the norm of reaction for facet number in genetically identical fruit flies that developed at different temperatures. As shown in the figure, the facet number varies with changes in temperature. At a higher temperature (30°C), the facet number is approximately 750, while at a lower temperature (15°C), it is over 1,000.

Overdominance Occurs When Heterozygotes Have Superior Traits

As we have just seen, the environment plays a key role in the outcome of traits. For certain genes, heterozygotes may display characteristics that are more beneficial for their survival in a particular environment. Such heterozygotes may be more likely to survive and reproduce. For example, a heterozygote may be larger, disease resistant, or better able to withstand harsh environmental conditions. The phenomenon in which a heterozygote has greater reproductive success compared to either of the corresponding homozygotes is called **overdominance** or **heterozygote advantage.**

A well-documented example involves a human allele that causes sickle-cell disease in homozygous individuals. This disease is an autosomal recessive disorder in which the affected individual produces an altered form of the protein hemoglobin, which carries oxygen within red blood cells. Most people carry the Hb^A allele and make hemoglobin A. Individuals affected with sickle-cell anemia are homozygous for the Hb^S allele and produce only hemoglobin S. This causes their red blood cells to deform into a sickle shape under conditions of low oxygen concentration (**Figure 4.7a, b**). The sickling phenomenon causes the life span of these cells to be greatly shortened to only a few weeks compared with a normal span of four months, and therefore, anemia results. In addition, abnormal sickled cells can become clogged in the capillaries throughout the body, leading to localized areas of oxygen depletion. Such an event, called a crisis, causes pain and sometimes tissue and organ damage. For these reasons, the homozygous Hb^SHb^S individual usually has a shortened life span compared to an individual producing hemoglobin A.

In spite of the harmful consequences to homozygotes, the sickle-cell allele has been found at a fairly high frequency among human populations that are exposed to malaria. The protozoan genus that causes malaria, *Plasmodium*, spends part of its life cycle within the *Anopheles* mosquito and another part within the red blood cells of humans who have been bitten by an infected mosquito. However, red blood cells of heterozygotes, Hb^AHb^S, are likely to rupture when infected by this parasite, thereby preventing the parasite from propagating. People who are heterozygous have better resistance to malaria compared to Hb^AHb^A homozygotes, while not incurring the ill effects of sickle-cell disease. Therefore, even though the homozygous Hb^SHb^S condition is detrimental, the greater survival of the heterozygote has selected for the presence of the Hb^S allele within populations where malaria is prevalent. When viewing survival in such a region, overdominance explains the prevalence of the sickle-cell allele. In Chapter 24, we will consider the role that natural selection plays in maintaining alleles that are beneficial to the heterozygote but harmful to the homozygote.

Figure 4.7c illustrates the predicted outcome when two heterozygotes have children. In this example, 1/4 of the offspring are Hb^AHb^A (unaffected, not malaria resistant), 1/2 are Hb^AHb^S

(a) **Normal red blood cell**   (b) **Sickled red blood cell**   (c) **Example of sickle-cell inheritance pattern**

FIGURE 4.7   **Inheritance of sickle-cell disease.** A comparison of (**a**) normal red blood cells and (**b**) those from a person with sickle-cell disease. (**c**) The outcome of a cross between two heterozygous individuals.

(unaffected, malaria resistant) and 1/4 are $Hb^S Hb^S$ (sickle-cell disease). This 1:2:1 ratio deviates from a simple Mendelian 3:1 phenotypic ratio.

Overdominance is usually due to two alleles that produce proteins with slightly different amino acid sequences. How can we explain the observation that two protein variants in the heterozygote produce a more favorable phenotype? There are three common explanations. In the case of sickle-cell disease, the phenotype is related to the infectivity of *Plasmodium* (**Figure**

(a) Disease resistance

(b) Homodimer formation

E1
27°–32°C
(optimum
temperature
range)

E2
30°–37°C
(optimum
temperature
range)

(c) Variation in functional activity

FIGURE 4.8   **Three possible explanations for overdominance at the molecular level. (a)** The successful infection of cells by certain microorganisms depends on the function of particular cellular proteins. In this example, functional differences between A1A1 and A1A2 proteins affect the ability of a pathogen to propagate in the cells. **(b)** Some proteins function as homodimers. In this example, a gene exists in two alleles designated *A1* and *A2*, which encode polypeptides also designated A1 and A2. The homozygotes that are *A1A1* or *A2A2* will make homodimers that are A1A1 and A2A2, respectively. The *A1A2* heterozygote can make A1A1 and A2A2 and can also make A1A2 homodimers, which may have better functional activity. **(c)** In this example, a gene exists in two alleles designated *E1* and *E2*. The *E1* allele encodes an enzyme that functions well in the temperature range of 27° to 32°C. *E2* encodes an enzyme that functions in the range of 30° to 37°C. A heterozygote, *E1E2*, would produce both enzymes and have a broader temperature range (i.e., 27° to 37°C) in which the enzyme would function.

4.8a). In the heterozygote, the infectious agent is less likely to propagate within red blood cells. Interestingly, researchers have speculated that other alleles in humans may confer disease resistance in the heterozygous condition but are detrimental in the homozygous state. These include PKU, in which the heterozygous fetus may be resistant to miscarriage caused by a fungal toxin, and Tay-Sachs disease, in which the heterozygote may be resistant to tuberculosis.

A second way to explain overdominance is related to the subunit composition of proteins. In some cases, proteins function as a complex of multiple subunits; each subunit is composed of one polypeptide. A protein composed of two subunits is called a dimer. When both subunits are encoded by the same gene, the protein is a homodimer. The prefix homo- means that the subunits come from the same type of gene although the gene may exist in different alleles. **Figure 4.8b** considers a situation in which a gene exists in two alleles that encode polypeptides designated A1 and A2. Homozygous individuals can produce only A1A1 or A2A2 homodimers, whereas a heterozygote can also produce an A1A2 homodimer. Thus, heterozygotes can produce three forms of the homodimer, homozygotes only one. For some proteins, A1A2 homodimers may have better functional activity because they are more stable·or able to function under a wider range of conditions. The greater activity of the homodimer protein may be the underlying reason why a heterozygote has characteristics superior to either homozygote.

A third molecular explanation of overdominance is that the proteins encoded by each allele exhibit differences in their functional activity. For example, suppose that a gene encodes a metabolic enzyme that can be found in two forms (corresponding to the two alleles), one that functions better at a lower temperature and the other that functions optimally at a higher temperature (**Figure 4.8c**). The heterozygote, which makes a mixture of both enzymes, may be at an advantage under a wider temperature range than either of the corresponding homozygotes.

Before ending this topic, let's also compare overdominance with a related phenomenon. Among plant and animal breeders, a common mating strategy is to begin with two different highly inbred strains and cross them together to produce hybrids. When the hybrids display traits that are superior to both corresponding parental strains, the outcome is known as **heterosis,** or **hybrid vigor.** Within the field of agriculture, heterosis has been particularly valuable in improving quantitative traits such as size, weight, growth rate, and disease resistance. It was first described by George Shull in crosses involving different strains of corn. Heterosis differs from overdominance, because the hybrid may be heterozygous for many genes, not just a single gene. Some of the beneficial effects of heterosis may be caused by the occurrence of overdominance in one or more heterozygous genes. However, as we will see in Chapter 25, heterosis can also result from the masking of deleterious recessive alleles that tend to accumulate in highly inbred domesticated strains.

Many Genes Exist as Three or More Different Alleles

Thus far, we have considered examples in which a gene exists in two different alleles. As researchers have probed genes at the molecular level within natural populations of organisms, they have discovered that most genes exist in **multiple alleles.** Within a population, genes are typically found in three or more alleles.

An interesting example of multiple alleles involves coat color in rabbits. **Figure 4.9** illustrates the relationship between genotype and phenotype for a combination of four different alleles, which are designated C (full coat color), c^{ch} (chinchilla pattern of coat color), c^h (himalayan pattern of coat color), and c (albino). In this case, the gene encodes an enzyme called tyrosinase, which is the first enzyme in a metabolic pathway that leads to the synthesis of melanin from the amino acid tyrosine. This pathway results in the formation of two forms of melanin. Eumelanin, a black pigment, is made first, and then phaeomelanin, an orange/yellow pigment, is made from eumelanin. Alleles of other genes can also influence the relative amounts of eumelanin and phaeomelanin.

Differences in the various alleles are related to the function of tyrosinase. The C allele encodes a fully functional tyrosinase that allows the synthesis of both eumelanin and phaeomelanin, resulting in a full brown coat color. The C allele is dominant to the other three alleles. The chinchilla allele (c^{ch}) is a partial defect in tyrosinase that leads to a slight reduction in black pigment and a greatly diminished amount of orange/yellow pigment, which makes the animal look more gray. The albino allele, designated c, is a complete loss of tyrosinase, resulting in white color. The himalayan pattern of coat color, determined by the c^h allele, is an example of a **temperature-sensitive allele.** The mutation in this gene has caused a change in the structure of tyrosinase, so it only works enzymatically at low temperature. Because of this property, the enzyme functions only in cooler regions of the body, primarily the tail, the paws, and the tips of the nose and ears. As shown in **Figure 4.10,** similar types of temperature-sensitive alleles have been found in other species of domestic animals, such as the Siamese cat.

Alleles of the ABO Blood Group Can Be Dominant, Recessive, or Codominant

The ABO group of antigens, which determine blood type in humans, is another example of multiple alleles and illustrates yet another allelic relationship called codominance. To understand this concept, we first need to examine the molecular characteristics of human blood types. The plasma membranes of red blood cells have groups of interconnected sugars—carbohydrate

(a) Full coat color CC, Cc^h, Cc^h, or Cc.

(b) Chinchilla coat color $c^{ch}c^{ch}$, $c^{ch}c^h$, or $c^{ch}c$.

(c) Himalayan coat color c^hc^h or c^hc.

(d) Albino coat color cc.

INTERACTIVE EXERCISE   FIGURE 4.9   **The relationship between genotype and phenotype in rabbit coat color.**

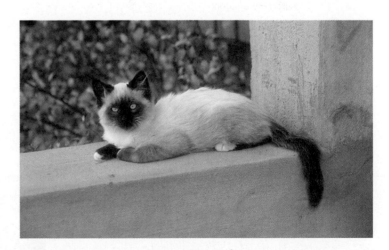

FIGURE 4.10   **The expression of a temperature-sensitive conditional allele produces a Siamese pattern of coat color.**

Genes → Traits The allele affecting fur pigmentation encodes a pigment-producing protein that functions only at lower temperatures. For this reason, the dark fur is produced only in the cooler parts of the animal, including the tips of the ears, nose, paws, and tail.

trees—that act as surface antigens (**Figure 4.11a**). Antigens are molecular structures that are recognized by antibodies produced by the immune system. On red blood cells, three different types of surface antigens, known as A, B, and O, may be found.

The synthesis of these surface antigens is controlled by three alleles, designated I^A, I^B, and i, respectively. The i allele is recessive to both I^A and I^B. A person who is homozygous ii will have type O blood. A homozygous I^AI^A or heterozygous I^Ai individual will have type A blood. The red blood cells of this individual will contain the surface antigen known as A. Similarly, a homozygous I^BI^B or heterozygous I^Bi individual will produce the surface antigen B. As Figure 4.11a indicates, surface antigens A and B have significantly different molecular structures. A person who is I^AI^B will have the blood type AB and express both surface antigens A and B. The phenomenon in which two alleles are both expressed in the heterozygous individual is called **codominance.** In this case, the I^A and I^B alleles are codominant to each other.

As an example of the inheritance of blood type, let's consider the possible offspring between two parents who are I^Ai and I^Bi (**Figure 4.11b**). The I^Ai parent makes I^A and i gametes, while the I^Bi parent makes I^B and i gametes. These combine to produce I^AI^B, I^Ai, I^Bi, and ii offspring in a 1:1:1:1 ratio. The resulting blood types are AB, A, B, and O, respectively.

Biochemists have analyzed the carbohydrate trees produced on the surfaces of cells of differing blood types. In type O, the tree is smaller than type A or type B because a sugar has not been attached to a specific site on the tree. This idea is schematically shown in Figure 4.11a. How do we explain this difference at the molecular level? The gene that determines ABO blood type encodes an enzyme called glycosyl transferase that attaches a sugar to the carbohydrate tree. The i allele carries a mutation that renders this enzyme inactive, which prevents the attachment of an additional sugar. By comparison, the two types of glycosyl transferase encoded by the I^A and I^B alleles have different structures in their active sites. The active site is the part of the protein that recognizes the sugar molecule that will be attached to the carbohydrate tree. The glycosyl transferase encoded by the I^A allele recognizes UDP-GalNAc and attaches GalNAc

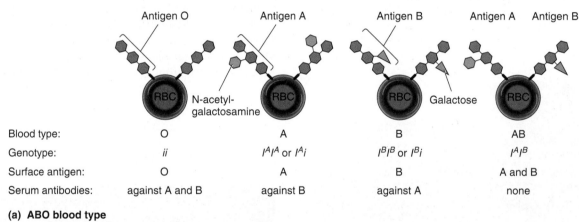

Blood type: O   A   B   AB
Genotype: ii I^AI^A or I^Ai I^BI^B or I^Bi I^AI^B
Surface antigen: O   A   B   A and B
Serum antibodies: against A and B   against B   against A   none

(a) ABO blood type

(b) Example of the ABO inheritance pattern

(c) Formation of A and B antigen by glycosyl transferase

INTERACTIVE EXERCISE

FIGURE 4.11   **ABO blood type. (a)** A schematic representation of blood type at the cellular level. **(b)** The predicted offspring from parents who are I^Ai and I^Bi. **(c)** The glycosyl transferase encoded by the I^A and I^B alleles recognizes different sugars due to changes in its active site. The i allele results in a nonfunctional enzyme.

(i.e., N-acetylgalactosamine) to the carbohydrate tree (**Figure 4.11c**). N-acetylgalactosamine is symbolized as a green hexagon. This creates the structure of surface antigen A. In contrast, the glycosyl transferase encoded by the I^B allele recognizes UDP-galactose and attaches galactose to the carbohydrate tree. Galactose is symbolized as an orange triangle. This creates the molecular structure of surface antigen B. A person with type AB blood makes both types of enzymes and thereby has a tree with both types of sugar attached.

A small difference in the structure of the carbohydrate tree, namely, an N-acetylgalactosamine in antigen A versus galactose in antigen B, explains why the two antigens are different from each other at the molecular level. These differences enable them to be recognized by different antibodies. A person who has blood type A makes antibodies to blood type B (refer back to Figure 4.11a). The antibodies against blood type B require a galactose in the carbohydrate tree for their proper recognition. Their antibodies will not recognize and destroy their own blood cells, but they will recognize and destroy the blood cells from a type B person.

With this in mind, let's consider why blood typing is essential for safe blood transfusions. The donor's blood must be an appropriate match with the recipient's blood. A person with type O blood naturally produces antibodies against both A and B antigens. If a person with type O blood is given type A, type B, or type AB blood, the antibodies in the recipient will react with the donated blood cells and cause them to agglutinate (clump together). This is a life-threatening situation that causes the blood vessels to clog. Other incompatible combinations include a type A person receiving type B or type AB blood, and a type B person receiving type A or type AB blood. Because individuals with type AB blood do not produce antibodies to either A or B antigens, they can receive any type of blood and are known as universal recipients. By comparison, type O persons are universal donors because their blood can be given to type O, A, B, and AB people.

The Inheritance Pattern of X-Linked Genes Can Be Revealed by Reciprocal Crosses

Let's now turn our attention to inheritance patterns of single genes in which the sexes of the parents and offspring play a critical role. As discussed in Chapter 3, many species have males and females that differ in their sex chromosome composition. In mammals, for example, females are XX and males are XY. In such species, certain traits are governed by genes that are located on a sex chromosome. For these traits, the outcome of crosses depends on the genotypes and sexes of the parents and offspring.

As an example, let's consider a human disease known as Duchenne muscular dystrophy (DMD), which was first described by the French neurologist Guillaume Duchenne in the 1860s. Affected individuals show signs of muscle weakness as early as age three. The disease gradually weakens the skeletal muscles and eventually affects the heart and breathing muscles. Survival is rare beyond the early 30s. The gene for DMD, found on the X chromosome, encodes a protein called dystrophin that is required inside muscle cells for structural support. Dystrophin is thought to strengthen muscle cells by anchoring elements of the internal cytoskeleton to the plasma membrane. Without it, the plasma membrane becomes permeable and may rupture.

Duchenne muscular dystrophy is inherited in an **X-linked recessive** pattern—the allele causing the disease is recessive and located on the X chromosome. In the pedigree shown in **Figure 4.12**, several males are affected by this disorder, as indicated by filled squares. The mothers of these males are presumed heterozygotes for this X-linked recessive allele. This recessive disorder is very rare among females because daughters would have to inherit a copy of the mutant allele from their mother and a copy from an affected father.

X-linked muscular dystrophy has also been found in certain breeds of dogs such as golden retrievers (**Figure 4.13a**). Like humans, the mutation occurs in the dystrophin gene, and the symptoms include severe weakness and muscle atrophy that begin at about six to eight weeks of age. Many dogs that inherit this disorder die within the first year of life, though some can live three to five years and reproduce.

Figure 4.13b (left side) considers a cross between an unaffected female dog with two copies of the wild-type gene and a male dog with muscular dystrophy that carries the mutant allele. When setting up a Punnett square involving X-linked traits, we must consider the alleles on the X chromosome as well as the fact that males may transmit a Y chromosome instead of the X chromosome. The male makes two types of gametes, one that carries the X chromosome and one that carries the Y. The Punnett square must also include the Y chromosome even though this chromosome does not carry any X-linked genes. The X chromosomes from the female and male are designated with their corresponding alleles. When the Punnett square is filled in, it predicts the X-linked genotypes and sexes of the offspring. As seen on the left side of Figure 4.13b, none of the offspring from this cross are affected with the disorder, although all female offspring are carriers.

The right side of Figure 4.13b shows a **reciprocal cross**—a second cross in which the sexes and phenotypes are reversed. In this case, an affected female animal is crossed to an unaffected

FIGURE 4.12   A human pedigree for Duchenne muscular dystrophy, an X-linked recessive trait. Affected individuals are shown with filled symbols. Females who are unaffected with the disease, but are heterozygous carriers, are shown with half-filled symbols.

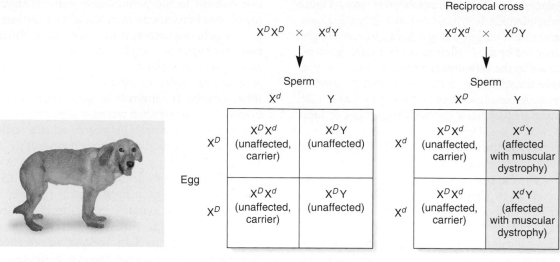

$X^D X^D \times X^d Y$ $X^d X^d \times X^D Y$

Reciprocal cross

Sperm

	X^d	Y
X^D	$X^D X^d$ (unaffected, carrier)	$X^D Y$ (unaffected)
X^D	$X^D X^d$ (unaffected, carrier)	$X^D Y$ (unaffected)

Egg

Sperm

	X^D	Y
X^d	$X^D X^d$ (unaffected, carrier)	$X^d Y$ (affected with muscular dystrophy)
X^d	$X^D X^d$ (unaffected, carrier)	$X^d Y$ (affected with muscular dystrophy)

(a) Male golden retriever with X-linked muscular dystrophy

(b) Examples of X-linked muscular dystrophy inheritance patterns

INTERACTIVE EXERCISE

FIGURE 4.13   **X-linked muscular dystrophy in dogs. (a)** The golden retriever shown here is a male that has the disease. **(b)** The left side shows a cross between an unaffected female and an affected male. The right shows a reciprocal cross between an affected female and an unaffected male. *D* represents the normal allele for the dystrophin gene, and *d* is the mutant allele that causes a defect in dystrophin function.

male. This cross produces female offspring that are carriers and all male offspring will be affected with muscular dystrophy.

When comparing the two Punnett squares, the outcome of the reciprocal cross yielded different results. This is expected of X-linked genes, because the male transmits the gene only to female offspring, while the female transmits an X chromosome to both male and female offspring. Because the male parent does not transmit the X chromosome to his sons, he does not contribute to their X-linked phenotypes. This explains why X-linked traits do not behave equally in reciprocal crosses. Experimentally, the observation that reciprocal crosses do not yield the same results is an important clue that a trait may be X linked.

Genes Located on Mammalian Sex Chromosomes Can Be Transmitted in an X-Linked, a Y-Linked, or a Pseudoautosomal Pattern

Our discussion of sex chromosomes has concentrated on genes that are located on the X chromosome but not on the Y chromosome. The term **sex-linked gene** refers to a gene that is found on one of the two types of sex chromosomes but not on both. Hundreds of X-linked genes have been identified in humans and other mammals.

The inheritance pattern of X-linked genes shows certain distinctive features, such as males transmit X-linked genes only to their daughters, and sons receive their X-linked genes from their mothers. The term **hemizygous** is used to describe the single copy of an X-linked gene in the male. A male mammal is said to be hemizygous for X-linked genes. Because males of certain species, such as humans, have a single copy of the X chromosome, another distinctive feature of X-linked inheritance is that

males are more likely to be affected by rare, recessive X-linked disorders.

By comparison, relatively few genes are located only on the Y chromosome. These few genes are called **holandric genes.** An example of a holandric gene is the *Sry* gene found in mammals. Its expression is necessary for proper male development. A Y-linked inheritance pattern is very distinctive—the gene is transmitted only from fathers to sons.

Besides sex-linked genes, the X and Y chromosomes also contain short regions of homology where the X and Y chromosomes carry the same genes. In addition to several smaller regions, the human sex chromosomes have three homologous regions (**Figure 4.14**). These regions, which are evolutionarily related, promote the necessary pairing of the X and Y chromosomes that occurs during meiosis I of spermatogenesis. Relatively

Mic2 gene

X

Y

Mic2 gene

FIGURE 4.14   **A comparison of the homologous and nonhomologous regions of the X and Y chromosome in humans.** The brackets show three regions of homology between the X and Y chromosome. A few pseudoautosomal genes, such as *Mic2*, are found on both the X and Y chromosomes in these small regions of homology. Many X-linked genes are found in the nonhomologous regions of the X chromosome. A few Y-linked genes are found in the nonhomologous regions of the Y chromosome.

few genes are located in these homologous regions. One example is a human gene called *Mic2*, which encodes a cell surface antigen. The *Mic2* gene is found on both the X and Y chromosomes. It follows a pattern of inheritance called **pseudoautosomal inheritance.** The term pseudoautosomal refers to the idea that the inheritance pattern of the *Mic2* gene is the same as the inheritance pattern of a gene located on an autosome even though the *Mic2* gene is actually located on the sex chromosomes. As in autosomal inheritance, males have two copies of pseudoautosomally inherited genes, and they can transmit the genes to both daughters and sons.

Some Traits Are Influenced by the Sex of the Individual

As we have just seen, the transmission pattern of sex-linked genes depends on the sex of the parents and offspring. Sex can influence traits in other ways as well. The term **sex-influenced inheritance** refers to the phenomenon in which an allele is dominant in one sex but recessive in the opposite sex. Therefore, sex influence is a phenomenon of heterozygotes. Sex-influenced inheritance should not be confused with sex-linked inheritance. The genes that govern sex-influenced traits are usually autosomal, not on the X or Y chromosome.

In humans, the common form of pattern baldness provides an example of sex-influenced inheritance. As shown in **Figure 4.15**, the balding pattern is characterized by hair loss on the front and top of the head but not on the sides. This type of pattern baldness is inherited as an autosomal trait. (A common misconception is that this gene is X linked.) When a male is heterozygous for the baldness allele, he will become bald.

Genotype	Phenotype	
	Males	**Females**
BB	Bald	Bald
Bb	Bald	Nonbald
bb	Nonbald	Nonbald

In contrast, a heterozygous female will not be bald. Women who are homozygous for the baldness allele will develop the trait, but it is usually characterized by a significant thinning of the hair that occurs relatively late in life.

The sex-influenced nature of pattern baldness is related to the production of the male sex hormone testosterone. The gene that affects pattern baldness encodes an enzyme called 5-α-reductase, which converts testosterone to 5-α-dihydrotestosterone (DHT). DHT binds to cellular receptors and affects the expression of many genes, including those in the cells of the scalp. The allele that causes pattern baldness results in an overexpression of this enzyme. Because mature males normally make more testosterone than females, this allele has a greater phenotypic impact in males. However, a rare tumor of the adrenal gland can cause the secretion of abnormally large amounts of testosterone in females. If this occurs in a woman who is heterozygous *Bb*, she will become bald. If the tumor is removed surgically, her hair will return to its normal condition.

The autosomal nature of pattern baldness has been revealed by the analysis of many human pedigrees. An example is shown in **Figure 4.16a.** A bald male may inherit the bald allele from either parent, and thus a striking observation is that bald fathers can pass this trait to their sons. This could not occur if the trait was X linked, because fathers do not transmit an X chromosome to their sons. The analyses of many human pedigrees have shown that bald fathers, on average, have at least 50% bald sons. They are expected to produce an even higher percentage of bald male offspring if they are homozygous for the bald allele and/or the mother also carries one or two copies of the bald allele. For example, a heterozygous bald male and heterozygous (nonbald) female will produce 75% bald sons, while a homozygous bald male or homozygous bald female will produce all bald sons.

Figure 4.16b shows the predicted offspring if two heterozygotes produce offspring. In this Punnett square, the phenotypes are designated for both sons and daughters. *BB* offspring are bald, while *bb* offspring are nonbald. *Bb* offspring are bald if they are sons and nonbald if they are daughters. The predicted

(a) John Adams (father)   **(b) John Quincy Adams (son)**   **(c) Charles Francis Adams (grandson)**   **(d) Henry Adams (great-grandson)**

FIGURE 4.15   **Pattern baldness in the Adams family line.**

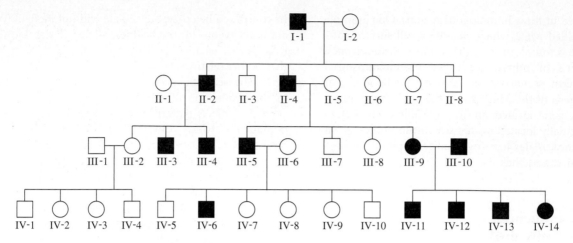

(a) A pedigree for human pattern baldness

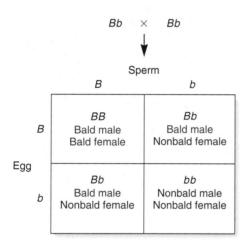

(b) Example of an inheritance pattern involving baldness

FIGURE 4.16 **Inheritance of pattern baldness, a sex-influenced trait involving an autosomal gene.** (**a**) A family pedigree. Bald individuals are shown in black. (**b**) The predicted offspring from two heterozygous parents.

genotypic ratios from this cross would be 1 *BB* bald son : 1 *BB* bald daughter : 2 *Bb* bald sons : 2 *Bb* nonbald daughters : 1 *bb* nonbald son : 1 *bb* nonbald daughter. The predicted phenotypic ratios would be 3 bald sons : 1 bald daughter : 3 nonbald daughters : 1 nonbald son. The ratio of bald to nonbald offspring is 4:4, which is the same as 1:1.

Another example in which sex affects an organism's phenotype is provided by **sex-limited inheritance,** in which the trait occurs in only one of the two sexes. In humans, for example, breast development is a trait that is normally limited to females, whereas beard growth is limited to males. Among many types of birds, the male of the species has more ornate plumage than the female. As shown in **Figure 4.17,** roosters have a larger comb and wattles and longer neck and tail feathers than do hens. These are sex-limited features that may be found in roosters but never in normal hens. However, some varieties of chickens have males that are hen-feathered. In these species, hen-feathering is controlled by a dominant allele (*H*) that is expressed both in males and females, while cock-feathering is controlled by a recessive allele (*h*) that is expressed only in males. These alleles are located on an autosome; they are not on a sex chromosome.

Genotype	Phenotype	
	Females	**Males**
hh	Hen-feathered	Cock-feathered
Hh	Hen-feathered	Hen-feathered
HH	Hen-feathered	Hen-feathered

Like baldness in humans, hen-feathering depends on the production of sex hormones. If a newly hatched female with an *hh* genotype has her single ovary removed surgically, she will develop cock-feathering and look indistinguishable from a male.

Mutations That Cause a Loss of Function in an Essential Gene Result in a Lethal Phenotype

Let's now turn our attention to alleles that have the most detrimental effect on phenotype—those that result in death. An allele that has the potential to cause the death of an organism is called a **lethal allele.** These are usually inherited in a recessive manner. When the absence of a specific protein results in a lethal phenotype, the gene that encodes the protein is considered an **essential**

(a) Hen

(b) Rooster

INTERACTIVE EXERCISE

FIGURE 4.17 **Differences in the feathering pattern in male and female chickens, an example of sex-limited inheritance.**

gene for survival. Though it varies according to species, researchers estimate that approximately 1/3 of all genes are essential genes. By comparison, **nonessential genes** are not absolutely required for survival, although they are likely to be beneficial to the organism. A loss-of-function mutation in a nonessential gene will not usually cause death. On rare occasions, however, a nonessential gene may acquire a mutation that causes the gene product to be abnormally expressed in a way that may interfere with normal cell function and lead to a lethal phenotype. Therefore, not all lethal mutations occur in essential genes, although the great majority do.

Many lethal alleles prevent cell division and thereby kill an organism at a very early stage. Others, however, may only exert their effects later in life, or under certain environmental conditions. For example, a human genetic disease known as Huntington disease is caused by a dominant allele. The disease is characterized by a progressive degeneration of the nervous system, dementia, and early death. The age when these symptoms appear, or the **age of onset,** is usually between 30 and 50.

Other lethal alleles may kill an organism only when certain environmental conditions prevail. Such **conditional lethal alleles** have been extensively studied in experimental organisms. For example, some conditional lethals will kill an organism only in a particular temperature range. These alleles, called **temperature-sensitive (ts) lethal alleles,** have been observed in many organisms, including *Drosophila*. A ts lethal allele may be fatal for a developing larva at a high temperature (30°C), but the larva will survive if grown at a lower temperature (22°C). Temperature-sensitive lethal alleles are typically caused by mutations that alter the structure of the encoded protein so it does not function correctly at the nonpermissive temperature or becomes unfolded and is rapidly degraded. Conditional lethal alleles may also be identified when an individual is exposed to a particular agent in the environment. For example, people with a defect in the gene that encodes the enzyme glucose-6-phosphate dehydrogenase (G6PD) have a negative reaction to the ingestion of fava beans. This can lead to an acute hemolytic syndrome with 10% mortality if not treated properly.

Finally, it is surprising that certain lethal alleles act only in some individuals. These are called **semilethal alleles.** Of course, any particular individual cannot be semidead. However, within a population, a semilethal allele will kill some individuals but not all of them. The reasons for semilethality are not always understood, but environmental conditions and the actions of other genes within the organism may help to prevent the detrimental effects of certain semilethal alleles.

In some cases, a lethal allele may produce ratios that seemingly deviate from Mendelian ratios. An example is an allele in a breed of cats known as Manx, which originated on the Isle of Man. (**Figure 4.18a**). The Manx cat carries a dominant mutation that affects the spine. This mutation shortens the tail, resulting in a range of tail lengths from normal to tailless. When two Manx cats are crossed to each other, the ratio of offspring is 1 normal : 2 Manx. How do we explain the 1:2 ratio? The answer is that about 1/4 of the offspring die during early embryonic development (**Figure 4.18b**). Therefore, in the homozygous condition, this dominant allele is lethal.

(a) A Manx cat

Mm (Manx) $\times$ Mm (Manx)

Sperm

	M	m
M	MM (early embryonic death)	Mm (Manx)
m	Mm (Manx)	mm (non-Manx)

Egg

1:2 ratio of kittens that are born

(b) Example of a Manx inheritance pattern

INTERACTIVE EXERCISE

FIGURE 4.18 **The Manx cat, which carries a lethal allele.** (a) Photo of a Manx cat, which typically has a shortened tail. (b) Outcome of a cross between two Manx cats. Animals that are homozygous for the dominant Manx allele (*M*) die during early embryonic development.

Single Genes Have Pleiotropic Effects

Before ending our discussion of single-gene inheritance patterns, let's take a broader look at how a single gene may affect phenotype. Although we tend to discuss genes within the context of how they influence a single trait, most genes actually have multiple effects throughout a cell and/or throughout a multicellular organism. The multiple effects of a single gene on the phenotype of an organism is called **pleiotropy**. Pleiotropy occurs for several reasons, including the following:

1. The expression of a single gene can affect cell function in more than one way. For example, a defect in a microtubule protein may affect cell division and cell movement.
2. A gene may be expressed in different cell types in a multicellular organism.
3. A gene may be expressed at different stages of development.

In all or nearly all cases, the expression of a gene is pleiotropic with regard to the characteristics of an organism. The expression of any given gene influences the expression of many other genes in the genome, and vice versa. Pleiotropy is revealed when researchers study the effects of gene mutations. As an example of a pleiotropic mutation, let's consider cystic fibrosis, which is a recessive human disorder. In the late 1980s, the gene for cystic fibrosis was identified. It encodes a protein called the cystic fibrosis transmembrane conductance regulator (CFTR), which regulates ionic balance by allowing the transport of chloride ions (Cl^-) across epithelial-cell membranes. The mutation that causes cystic fibrosis diminishes the function of this Cl^- transporter, affecting several parts of the body in different ways. Because the movement of Cl^- affects water transport across membranes, the most severe symptom of cystic fibrosis is thick mucus in the lungs that occurs because of a water imbalance. In sweat glands, the normal Cl^- transporter has the function of recycling salt out of the glands and back into the skin before it can be lost to the outside world. Persons with cystic fibrosis have excessively salty sweat due to their inability to recycle salt back into their skin cells—a common test for cystic fibrosis is measurement of salt on the skin. Another effect is seen in the reproductive system of males who are homozygous for the cystic fibrosis allele. Males with cystic fibrosis may be infertile because the vas deferens, the tubules that transport sperm from the testes, may be absent or undeveloped. Presumably, a normally functioning Cl^- transporter is needed for the proper development of the vas deferens in the embryo. Taken together, we can see that a defect in CFTR has multiple effects throughout the body.

4.2 GENE INTERACTIONS

In Section 4.1, we considered the effects of a single gene on the outcome of a trait. This approach helps us to understand the various ways that alleles can influence traits. Researchers often examine the effects of a single gene on the outcome of a single trait as a way to simplify the genetic analysis. For example, Mendel studied one gene that affected the height of pea plants—tall versus dwarf alleles. Actually, many other genes in pea plants also affect height, but Mendel did not happen to study variants in those other height genes. How then did Mendel study the affects of a single gene? The answer lies in the genotypes of his strains. Although many genes affect the height of pea plants, Mendel chose true-breeding strains that differed with regard to only one of those genes. As a hypothetical example, let's suppose that pea plants have 10 genes affecting height, which we will call K, L, M, N, O, P, Q, R, S, and T. The genotypes of two hypothetical strains of pea plants may be

| Tall strain: | *KK LL MM NN OO PP QQ RR SS TT* |
| Dwarf strain: | *KK LL MM NN OO PP QQ RR SS tt* |

In this example, the alleles affecting height may differ at only a single gene. One strain is *TT* and the other is *tt*, and this accounts for the difference in their height. If we make crosses between these tall and dwarf strains, the genotypes of the F_2 offspring will differ with regard to only one gene; the other nine genes will be identical in all of them. This approach allows a researcher to study the effects of a single gene even though many genes may affect a single trait.

Researchers now appreciate that essentially all traits are affected by the contributions of many genes. Morphological features such as height, weight, growth rate, and pigmentation are all affected by the expression of many different genes in combination with environmental factors. In this section, we will further our understanding of genetics by considering how the allelic variants of two different genes affect a single trait. This phenomenon is known as a **gene interaction**. **Table 4.3** considers several

TABLE **4.3**

Types of Mendelian Inheritance Patterns Involving Two Genes

Type	Description
Epistasis	An inheritance pattern in which the alleles of one gene mask the phenotypic effects of the alleles of a different gene.
Complementation	A phenomenon in which two different parents that express the same or similar recessive phenotypes produce offspring with a wild-type phenotype.
Modifying genes	A phenomenon in which the allele of one gene modifies the phenotypic outcome of the alleles of a different gene.
Gene redundancy	A pattern in which the loss of function in a single gene has no phenotypic effect, but the loss of function of two genes has an effect. Functionality of only one of the two genes is necessary for a normal phenotype; the genes are functionally redundant.
Intergenic suppressors	An inheritance pattern in which the phenotypic effects of one mutation are reversed by a suppressor mutation in another gene.

examples in which two different genes interact to influence the outcome of particular traits. In this section, we will examine these examples in greater detail.

A Cross Involving a Two-Gene Interaction Can Produce Four Distinct Phenotypes

The first case of two different genes interacting to affect a single trait was discovered by William Bateson and Reginald Punnett in 1906 while they were investigating the inheritance of comb morphology in chickens. Several common varieties of chicken possess combs with different morphologies, as illustrated in **Figure 4.19a**. In their studies, Bateson and Punnett crossed a Wyandotte breed having a rose comb to a Brahma having a pea comb. All F_1 offspring had a walnut comb.

When these F_1 offspring were mated to each other, the F_2 generation consisted of chickens with four types of combs in the following phenotypic ratio: 9 walnut : 3 rose : 3 pea : 1 single comb. As we have seen in Chapter 2, a 9:3:3:1 ratio is obtained in the F_2 generation when the F_1 generation is heterozygous for two different genes and these genes assort independently. However, an important difference here is that we have four distinct categories of a single trait. Based on the 9:3:3:1 ratio, Bateson and Punnett reasoned that a single trait (comb morphology) was determined by two different genes.

R (rose comb) is dominant to r

P (pea comb) is dominant to p

R and P (walnut comb) are codominant

$rrpp$ produces a single comb

As shown in the Punnett square of **Figure 4.19b**, each of the genes can exist in two alleles, and the two genes show independent assortment.

When the alleles of one gene mask the phenotypic effects of the alleles of another gene, the phenomenon is called **epistasis.** Geneticists consider epistasis relative to a particular phenotype. For example, let's consider walnut comb as our reference phenotype. Relative to a walnut comb, some geneticists may view rr and pp to be epistatic to this phenotype. In other words, both rr and pp mask a walnut comb. This is an example of **recessive epistasis** because an individual must be homozygous for either recessive allele to mask the walnut comb phenotype. As shown in the Punnett square of Figure 4.19b, four distinct phenotypes are produced.

A Cross Involving a Two-Gene Interaction Can Produce Two Distinct Phenotypes Due to Epistasis

Bateson and Punnett also discovered an unexpected gene interaction when studying crosses involving the sweet pea, *Lathyrus odoratus*. The wild sweet pea has purple flowers. However, they obtained several true-breeding mutant varieties with white flowers. Not surprisingly, when they crossed a true-breeding purple-flowered plant to a true-breeding white-flowered plant, the F_1 generation contained all purple-flowered plants and the F_2 generation (produced by self-fertilization of the F_1 generation) consisted of purple- and white-flowered plants in a 3:1 ratio.

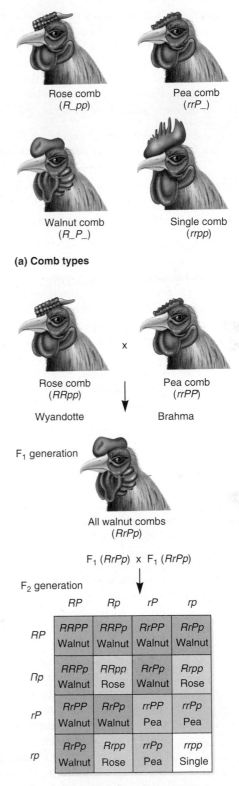

(a) Comb types

(b) The crosses of Bateson and Punnett

FIGURE 4.19 **Inheritance of comb morphology in chickens.** This trait is influenced by two different genes, which can each exist in two alleles. (**a**) Four phenotypic outcomes are possible. The underline symbol indicates the allele could be either dominant or recessive. (**b**) The crosses of Bateson and Punnett examined the interaction of the two genes.

INTERACTIVE EXERCISE

A surprising result came in an experiment where they crossed two different varieties of white-flowered plants (**Figure 4.20**). All of the F₁ generation plants had purple flowers! Bateson and Punnett then allowed the F₁ offspring to self-fertilize. The F₂ generation resulted in purple and white flowers in a ratio of 9 purple : 7 white. From this result, Bateson and Punnett deduced that two different genes were involved, with the following relationship:

C (one purple-color-producing) allele is dominant to c (white)

P (another purple-color-producing) allele is dominant to p (white)

cc or pp masks the P or C alleles, producing white color

If possible, geneticists use the wild-type phenotype as their reference phenotype when describing an epistatic interaction. In this case, purple flowers are wild type. Homozygosity for the white allele of one gene masks the expression of the purple-producing allele of another gene. In other words, the cc genotype is epistatic to a purple phenotype, and the pp genotype is also epistatic to a purple phenotype. At the level of genotypes, cc is epistatic to PP or Pp, and pp is epistatic to CC or Cc. This is another example of recessive epistasis. As seen in Figure 4.20, this epistatic interaction produces only two phenotypes—purple or white flowers—in a 9:7 ratio.

Epistatis often occurs because two (or more) different proteins participate in a common function. For example, two or more proteins may be part of an enzymatic pathway leading to the formation of a single product. To illustrate this idea, let's consider the formation of a purple pigment in the sweet pea.

$$\text{Colorless precursor} \xrightarrow{\text{Enzyme C}} \text{Colorless intermediate} \xrightarrow{\text{Enzyme P}} \textbf{Purple pigment}$$

In this example, a colorless precursor molecule must be acted on by two different enzymes to produce the purple pigment. Gene C encodes a functional protein called enzyme C, which converts the colorless precursor into a colorless intermediate. Two copies of the recessive allele (cc) result in a lack of production of this enzyme in the homozygote. Gene P encodes a functional enzyme P, which converts the colorless intermediate into the purple pigment. Like the c allele, the recessive p allele encodes a defective enzyme P. If an individual is homozygous for either recessive allele (cc or pp), it will not make any functional enzyme C or enzyme P, respectively. When one of these enzymes is missing, purple pigment cannot be made, and the flowers remain white.

The parental cross shown in Figure 4.20 illustrates another genetic phenomenon called **complementation.** This term refers to the production of offspring with a wild-type phenotype from parents that both display the same or similar recessive phenotype. In this case, purple-flowered F₁ offspring were obtained from two white-flowered parents. Complementation typically occurs because the recessive phenotype in the parents is due to homozgyosity at two different genes. In our sweet pea example, one parent is CCpp and the other is ccPP. In the F₁ offspring, the C and P alleles, which are wild-type and dominant, complement

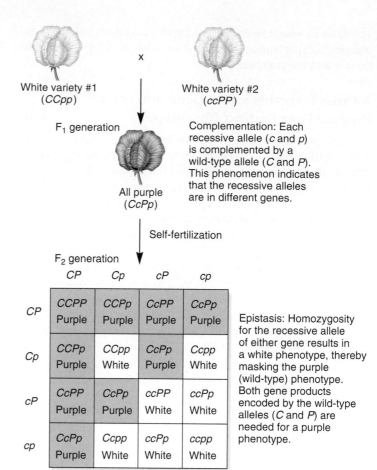

FIGURE 4.20   A cross between two different white varieties of the sweet pea.

Genes → Traits The color of the sweet pea flower is controlled by two genes, which are epistatic to each other and show complementation. Each gene is necessary for the production of an enzyme required for pigment synthesis. The recessive allele of either gene encodes a defective enzyme. If an individual is homozygous recessive for either of the two genes, the purple pigment cannot be synthesized. This results in a white phenotype.

the c and p alleles, which are recessive. The offspring needs one wild-type allele of both genes to display the wild-type phenotype. Why is complementation an important experimental observation? When geneticists observe complementation in a genetic cross, the results suggest that the recessive phenotype in the two parent strains is caused by mutant alleles in two different genes.

A Cross Involving a Two-Gene Interaction Can Produce Three Distinct Phenotypes Due to Epistasis

Thus far, we have observed two epistatic interactions, one producing four phenotypes and the other producing only two. Coat color in rodents provides an example that produces three phenotypes. If a true-breeding black rat is crossed to a true-breeding albino rat, the result is a rat with agouti coat color. Animals with agouti coat color have black pigmentation at the tips of each hair that changes to orange pigmentation near the root. If two agouti animals of the F₁ generation are crossed to each other, they produce agouti, black, and albino offspring in a 9:3:4 ratio (**Figure 4.21**).

How do we explain this ratio? This cross involves two genes that are called *A* (for agouti) and *C* (for colored). The dominant *A* allele of the agouti gene encodes a protein that regulates hair color such that the pigmentation shifts from black (eumelanin) at the tips to orange (phaeomelanin) near the roots. The recessive

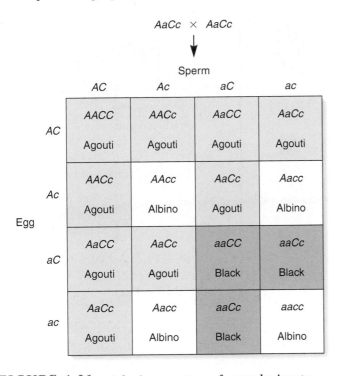

$AaCc \times AaCc$

FIGURE 4.21   Inheritance pattern of coat color in rats involving a gene interaction between the agouti gene (*A* or *a*) and the colored gene (*C* or *c*).

allele, *a*, inhibits the shift to orange pigmentation and thereby results in black pigment production throughout the entire hair, when an animal is *aa*. As with rabbits, the colored gene encodes tyrosinase, which is needed for the first step in melanin synthesis. The *C* allele allows pigmentation to occur, while the *c* allele causes the loss of tyrosinase function. The *C* allele is dominant to the *c* allele; *cc* homozygotes are albino and have white coat color.

As shown in Figure 4.21, the F₁ offspring are heterozygous for the two genes. In this case, *C* is dominant to *c*, and *A* is dominant to *a*. If an animal has at least one copy of both dominant alleles, the result is agouti coat color. Let's consider agouti as our reference phenotype. In the F₂ generation, if an animal has a dominant *C* allele but is *aa* homozygous, it will develop a black coat. The four cases of albino animals are all *cc* homozygous. This occurs even when an animal carries the dominant *A* allele. The *a* allele is epistatic to *C* and causes a black coat color in the *aa* homozygote, while *c* is epistatic to *A* and results in white coat color in the *cc* homozygote. This is a third example of recessive epistasis. At the molecular level, *aa* is masking the orange pigmentation, and *cc* is masking any pigmentation. Unlike the sweet pea, in which loss-of-function alleles had the same effect (white flowers), the effects of these loss-of-function alleles are not the same; *aa* yields black fur, and *cc* produces white fur.

Though the molecular effects of the *aa* genotype can be viewed as epistasis because the orange pigmentation is actually masked, some geneticists may not view this effect as epistasis but instead would call it a **gene modifier effect**—the alleles of one gene modify the phenotypic effect of the alleles of a different gene. From this alternative viewpoint, the pigmentation is not totally masked, but instead the agouti color is modified to black. As described next, geneticists may classify some alleles as gene modifiers.

E X P E R I M E N T   4 A

Bridges Observed an 8:4:3:1 Ratio Because the Cream-Eye Gene Can Modify the X-Linked Eosin Allele but Not the Red or White Alleles

As we have seen, geneticists view epistasis as a situation in which the alleles of a given gene mask the phenotypic effects of the alleles of another gene. In some cases, however, two genes may interact to influence a particular phenotype, but the interaction of particular alleles seems to modify the phenotype, not mask it.

Calvin Bridges discovered an early example in which one gene modifies the phenotypic effects of an X-linked eye color gene in *Drosophila*. As discussed in Chapter 3, the X-linked red allele (*w+*) is dominant to the white allele (*w*). Besides these two alleles, Thomas Hunt Morgan and Calvin Bridges found another allele of this gene that they called eosin (*w-e*), which results in eyes that are a pale orange color. The red allele is dominant to the eosin allele. In addition, the expression of the eosin allele depends on the number of copies of the allele. When females have two copies of this allele, they have eosin eyes. When females are heterozygous for the eosin allele and white allele, they have light-eosin eyes. Within true-breeding cultures

of flies with eosin eyes, he occasionally found a fly that had a noticeably different eye color. In particular, he identified a rare fly with cream-colored eyes. Bridges reasoned that this new eye color could be explained in two different ways. One possibility is that the cream-colored phenotype could be the result of a new mutation that changed the eosin allele into a cream allele. A second possibility is that a different gene may have incurred a mutation that modified the phenotypic expression of the eosin allele. This second possibility is an example of a gene interaction. To distinguish between these two possibilities, he carried out the crosses described in **Figure 4.22**. He crossed males with cream-colored eyes to wild-type females and then allowed the F₁ generation flies, which all had red eyes, to mate with each other. As shown in the data, all F₂ females had red eyes, while males had red eyes, eosin eyes, or cream eyes.

■ THE HYPOTHESES

Cream-colored eyes in fruit flies are due to the effect of an allele that is in the same gene as the eosin allele or in a second gene that modifies the expression of the eosin allele.

■ TESTING THE HYPOTHESES — FIGURE 4.22   A gene interaction between the cream allele and eosin allele.

Starting material: From a culture of flies with eosin eyes, Bridges obtained a fly with cream-colored eyes and used it to produce a true-breeding culture of flies with cream-colored eyes. The allele was called *cream a* (c^a).

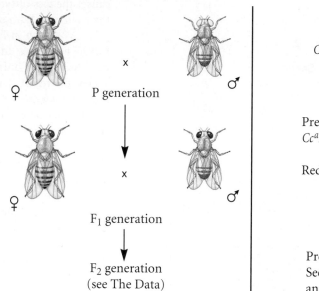

Experimental level

1. Cross males with cream-colored eyes to wild-type females.

2. Observe the F$_1$ offspring and then allow the offspring to mate with each other.

3. Observe and record the eye color and sex of the F$_2$ generation.

♀   ×   ♂

P generation

♀   ×   ♂

F$_1$ generation

F$_2$ generation
(see The Data)

Conceptual level

$CCX^{w^+}X^{w^+} \times c^a c^a X^{w-e}Y$

Predicted outcome:
$Cc^a X^{w^+}X^{w-e}$ and $Cc^a X^{w^+}Y$

Red-eyed females and males

Allow F$_1$
offspring
to mate.

Predicted outcome:
See Punnett square
and The Data.

■ THE DATA

Cross	Outcome
P cross:	
Cream-eyed male × wild-type female	F$_1$: All red eyes
F$_1$ cross:	
F$_1$ brother × F$_1$ sister	F$_2$: 104 females with red eyes
	47 males with red eyes
	44 males with eosin eyes
	14 males with cream eyes

■ INTERPRETING THE DATA

To interpret these data, keep in mind that Bridges already knew that the eosin allele is X linked. However, he did not know whether the cream allele was in the same gene as the eosin allele, in a different gene on the X chromosome, or on an autosome. The F$_2$ generation indicates that the cream allele is not in the same gene as the eosin allele. If the cream allele was in the same gene as the eosin allele, none of the F$_2$ males would have had eosin eyes; there would have been a 1:1 ratio of red-eyed males and cream-eyed males in the F$_2$ generation. This result was not obtained. Instead, the actual results are consistent with the idea that the male flies of the parental generation possessed both the eosin and cream alleles. Therefore, Bridges concluded that the cream allele was an allele of a different gene.

One possibility is that the cream allele is an autosomal recessive allele. If so, we can let C represent the dominant allele (which does not modify the eosin phenotype) and c^a represent the cream allele that modifies the eosin color to cream. We already know that the eosin allele is X linked and recessive to the red allele. The parental cross is expected to produce all red-eyed F$_1$ flies in which

the males are $Cc^a X^{w^+}Y$ and the females are $Cc^a X^{w^+}X^{w-e}$. When these F$_1$ offspring are allowed to mate with each other, the Punnett square shown here would predict the following outcome:

$Cc^a X^{w^+}X^{w-e} \times Cc^a X^{w^+}Y$

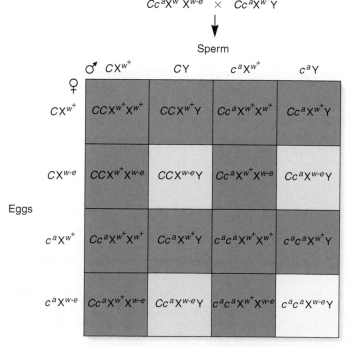

Outcome:
1 $CCX^{w^+}X^{w^+}$: 1 $CCX^{w^+}X^{w-e}$: 2 $Cc^a X^{w^+}X^{w^+}$: 2 $Cc^a X^{w^+}X^{w-e}$:
1 $c^a c^a X^{w^+}X^{w^+}$: 1 $c^a c^a X^{w^+}X^{w-e}$ = **8 red-eyed females**

1 $CCX^{w^+}Y$: 2 $Cc^a X^{w^+}Y$: 1 $c^a c^a X^{w^+}Y$ = **4 red-eyed males**

1 $CCX^{w-e}Y$: 2 $Cc^a X^{w-e}Y$ = **3 light eosin-eyed males**

1 $c^a c^a X^{w-e}Y$ = **1 cream-eyed male**

This phenotypic outcome proposes that the specific modifier allele, c^a, can modify the phenotype of the eosin allele but not the red-eye allele. The eosin allele can be modified only when the c^a allele is homozygous. The predicted 8:4:3:1 ratio agrees reasonably well with Bridges's data.

A self-help quiz involving this experiment can be found at the Online Learning Center.

Due to Gene Redundancy, Loss-of-Function Alleles May Have No Effect on Phenotype

During the past several decades, researchers have discovered new kinds of gene interactions by studying model organisms such as *Escherichia coli* (a bacterium), *Saccharomyces cerevisiae* (baker's yeast), *Arabidopsis thaliana* (a model plant), *Drosophila melanogaster* (fruit fly), *Caenorhabditis elegans* (a nematode worm), and *Mus musculus* (the laboratory mouse). The isolation of mutants that alter the phenotypes of these organisms has become a powerful way to investigate gene function and has provided ways for researchers to identify new kinds of gene interactions. With the advent of modern molecular techniques (described in Chapters 16, 18, and 19), a common approach for investigating gene function is to intentionally produce loss-of-function alleles in a gene of interest. When a geneticist abolishes gene function by creating an organism that is homozygous for a loss-of-function allele, the resulting organism is said to have undergone a **gene knockout.**

Why are gene knockouts useful? The primary reason for making a gene knockout is to understand how a gene affects the structure and function of cells or the phenotypes of organisms. For example, if a researcher knocked out a particular gene in a mouse and the resulting animal was unable to hear, the researcher would suspect that the role of the functional gene is to promote the formation of ear structures that are vital for hearing.

Interestingly, by studying many gene knockouts in a variety of experimental organisms, geneticists have discovered that many knockouts have no obvious effect on phenotype at the cellular level or the level of discernable traits. To explore gene function further, researchers may make two or more gene knockouts in the same organism. In some cases, gene knockouts in two different genes produce a phenotypic change even though the single knockouts have no effect (**Figure 4.23**). Geneticists may attribute this change to **gene redundancy**—the phenomenon that one gene can compensate for the loss of function of another gene.

Gene redundancy may be due to different underlying causes. One common reason is gene duplication. Certain genes have been duplicated during evolution, so a species may contain two or more copies of similar genes. These copies, which are not identical due to the accumulation of random changes during evolution, are called **paralogs.** When one gene is missing, a paralog may be able to carry out the missing function. For example, genes *A* and *B* in Figure 4.23 could be paralogs of each other. Alternatively, gene redundancy may involve proteins that are involved in a common cellular function. When one of the proteins is missing due to a gene knockout, the function of another protein may be increased to compensate for the missing protein and thereby overcome the defect.

Let's explore the consequences of gene redundancy in a genetic cross. George Shull conducted one of the first studies

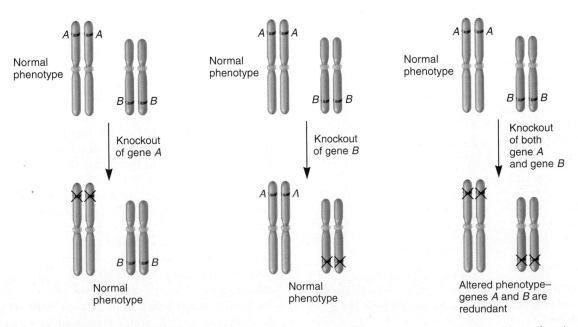

FIGURE 4.23 **A molecular explanation for gene redundancy.** To have a normal phenotype, an organism must have a functional copy of gene *A* or gene *B*, but not both. If both gene *A* and gene *B* are knocked out, an altered phenotype occurs.

that illustrated the phenomenon of gene redundancy. His work involved a weed known as shepherd's purse, a member of the mustard family. The trait he followed was the shape of the seed capsule, which is commonly triangular (**Figure 4.24**). Strains producing smaller ovate capsules are due to loss-of-function alleles in two different genes (*ttvv*). The ovate strain is an example of a double gene knockout. When Shull crossed a true-breeding plant with triangular capsules to a plant having ovate capsules, the F₁ generation all had triangular capsules. When the F₁ plants were self-fertilized, a surprising result came in the F₂ generation. Shull observed a 15:1 ratio of plants having triangular to ovate capsules. The result can be explained by gene redundancy. Having one functional copy of either gene (*T* or *V*) is sufficient to produce the triangular phenotype. *T* and *V* are redundant because only one of them is necessary for a triangular shape. When the functions of both genes are knocked out, as in the *ttvv* homozygote, the capsule becomes smaller and ovate.

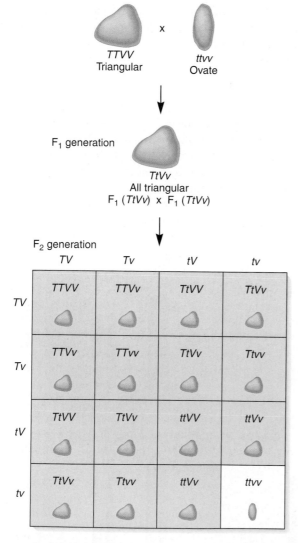

FIGURE 4.24   Inheritance of capsule shape in shepherd's purse, an example of gene redundancy. In this case, triangular shape requires a dominant allele in one of two genes, but not both. The *T* and *V* alleles are redundant.

The Phenotypic Effects of a Mutation Can Be Reversed by a Suppressor Mutation

When studying an experimental organism, a common approach to gain a deeper understanding of gene interaction is the isolation of a **suppressor mutation**—a second mutation that reverses the phenotypic effects of a first mutation. When a suppressor mutation is in a different gene than the first mutation, it is called an **intergenic suppressor** or **extragenic suppressor**.

What type of information might a researcher gain from the analysis of intergenic suppressor mutants? Usually, the primary goal is to identify proteins that participate in a common cellular process that ultimately affects the traits of an organism. In *Drosophila*, several different proteins work together in a signaling pathway that determines whether or not certain parts of the body contain sensory cells, such as those that make up mechanosensory bristles. Researchers have isolated dominant mutants that result in flies with fewer bristles. The mutated gene was named *Hairless* to reflect this phenotype. In this case, the wild-type allele is designated *h*, and the dominant mutant is *H*. After the *Hairless* mutant was obtained, researchers then isolated mutants that suppressed the hairless phenotype. Such suppressor mutants, which are in a different gene, produced flies that have a wild-type number of bristles. These mutants, which are also dominant, are in a gene that was named *Suppressor of Hairless*. The wild-type allele is designated *soh*, and the dominant mutant allele is *SoH*.

How do we explain the effects of these mutations at the molecular level? Let's first consider the functions of the proteins encoded by the normal (wild-type) genes (**Figure 4.25**). The role of the SoH protein, encoded by the *soh* allele of the *Suppressor of Hairless* gene, is to prevent the formation of sensory structures such as bristles in regions of the body where they should not be made. The Hairless protein is made in regions of the body where bristles should form, and binds to the SoH protein and inhibits its function. When the Hairless protein is properly expressed on the surface of the fly, as in an *hh* homozygote, bristles will form there.

Now let's consider the effects of a single mutation in the *Hairless* gene. In a heterozygote carrying the dominant allele (*H*), only half the amount of functional Hairless protein is made. This is not enough to inhibit all of the SoH proteins that are made. Therefore, the uninhibited SoH proteins prevent bristle formation and result in a hairless (bristleless) phenotype.

What happens in the double mutant? The suppressor mutation eliminates one of the two functional *soh* alleles. The double mutant expresses only one functional *h* allele and one functional *soh* allele. In the double mutant, the reduced amount of Hairless protein is able to inhibit the reduced amount of the SoH protein. Therefore, the ability of the SoH proteins to prevent bristle formation is stopped. Bristles form in the double heterozygote.

The analysis of a mutant and its suppressor often provides key information that two proteins participate in a common function. In some cases, the analysis reveals that two proteins physically interact with each other. As we have just seen, this type of interaction occurs between the Hairless and SoH proteins. Alternatively, two distinct proteins encoded by different genes may

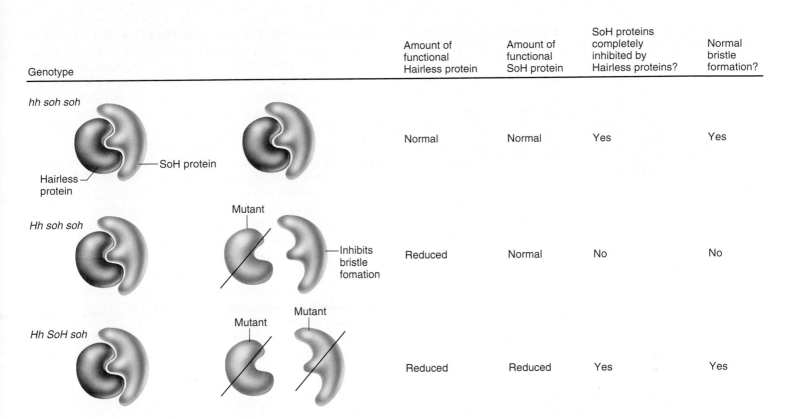

Genotype	Amount of functional Hairless protein	Amount of functional SoH protein	SoH proteins completely inhibited by Hairless proteins?	Normal bristle formation?
hh soh soh	Normal	Normal	Yes	Yes
Hh soh soh	Reduced	Normal	No	No
Hh SoH soh	Reduced	Reduced	Yes	Yes

FIGURE 4.25   An example of a gene interaction involving an intergenic suppressor. The *Hairless* mutation, which produces the dominant *H* allele, results in flies with fewer bristles. A dominant suppressor mutation in a second gene restores bristle formation. This dominant allele is designated *SoH*. Examination of the interactions between the mutant and its suppressor reveals that the Hairless and SoH proteins physically interact with each other to determine whether bristles are formed.

participate in a common function, but do not directly interact with each other. For example, two enzymes may be involved in a biochemical pathway that leads to the synthesis of an amino acid. A mutation that greatly decreases the amount of one enzyme may limit the ability of an organism to make the amino acid. If this occurs, the amino acid would have to be supplied to the organism for it to survive. A suppressor mutation could increase the function of another enzyme in the pathway and thereby restore the ability of the organism to make adequate amounts of the amino acid. Such a suppressor would alleviate the need for the organism to have the amino acid supplemented in its diet.

Other suppressors exert their effects by altering the amount of protein encoded by a mutant gene. For example, a mutation may decrease the functional activity of a protein that is needed for sugar metabolism. An organism harboring such a mutation may not be able to metabolize the sugar at a sufficient rate for growth or survival. A suppressor mutation in a different gene could alter genetic regulatory proteins and thereby increase the amount of the protein encoded by the mutant gene. (The proteins involved in gene regulation are described in Chapters 14 and 15.) This suppressor mutation would increase the amount of the defective protein and thereby result in a faster rate of sugar metabolism.

CONCEPTUAL SUMMARY

In Chapter 4, we have considered many examples that extend our understanding of Mendelian inheritance. We began with a consideration of the inheritance patterns involving single genes. Recessive alleles are often due to mutations that abolish gene function. In many simple dominant/recessive relationships, 50% of the normal protein is sufficient to yield a dominant phenotype. By comparison, incomplete dominance occurs when the heterozygote has an intermediate phenotype. In this case, 50% of the encoded pro-

tein does not produce a wild-type phenotype. In the pattern called incomplete penetrance, even though the dominant allele is present, the expected phenotype does not result. In other cases, the outcome depends on the degree to which the trait is expressed, or its expressivity. The outcome of traits is also due to environmental influences and/or modifier genes. The term norm of reaction refers to the effects of environmental variation on a phenotype. For certain genes, heterozygotes may display characteristics that are more beneficial for their survival in a particular environment, a phenomenon called overdominance. Overdominance may result in the prevalence of alleles, such as the sickle-cell allele, that are detrimental in the homozygous condition.

With regard to single genes, we also considered that, in natural populations, multiple alleles exist for most genes. Examples include coat color in rabbits and human blood types. The ABO blood types illustrate an inheritance pattern in which two alleles are distinctly expressed in the heterozygote, a phenomenon called codominance. We also examined how sex impacts the expression and inheritance patterns of single genes. Some genes are sex linked, meaning they are found on a sex chromosome. In mammals, X-linked genes are fairly common and produce inheritance patterns in which a reciprocal cross does not produce the same result. In sex-influenced inheritance, the phenotype of the heterozygote depends on its sex. A more extreme form of sex influence involves sex-limited traits, which are found in only one sex. Finally, we considered lethal alleles that cause the death of an organism. These are often loss-of-function alleles in essential genes. Such alleles can produce unusual ratios in a genetic cross.

We have also explored cases in which two different genes affect the outcome of a single trait, a concept known as a gene interaction. Epistasis is an inheritance pattern in which the alleles of one gene mask the phenotypic effects of the alleles of a different gene. Examples include comb shape in chickens, flower color in sweet peas, and coat color in rodents. Complementation is a phenomenon in which two different parents, both expressing the same or similar recessive phenotypes, produce offspring with a wild-type phenotype. In the example we explored, two different white strains of sweet peas, when crossed to each other, produced purple-flowered offspring. Modifying genes, such as the cream allele in *Drosophila*, alter the phenotypic outcome of the alleles of a different gene.

We also considered gene interactions that are often observed in experimental organisms. When the loss of function in a single gene has no phenotypic effect, but the loss of function of two or more genes has an effect, the genes are termed redundant with each other. Intergenic suppressors are mutations that reverse the effects of a first mutation. From the analysis of mutants and their intergenic suppressors, researchers often can identify proteins that participate in a common cellular process. The proteins may, but do not always, interact directly. Other intergenic suppressors exert their effects by altering the amount of protein encoded by a mutant gene.

EXPERIMENTAL SUMMARY

To understand how inheritance may deviate from a simple Mendelian pattern, either a genetic or molecular approach can be taken. In a genetic approach, a researcher makes crosses and then analyzes the phenotypes of the offspring. In Chapter 4, we have seen that the experimenter must weigh many factors when designing a cross and analyzing its outcome. These include the dominant/recessive relationships of alleles (namely, dominant, recessive, incompletely dominant, overdominant, or codominant), the presence of multiple alleles in a population, the relationship between the sex of the offspring and their phenotype, and the possibility of lethal alleles. In addition, complicating factors such as incomplete penetrance, the influence of the environment, and gene interactions must be considered. Based on these factors, a researcher can construct a Punnett square to see if a pattern of inheritance fits the available data obtained from crosses.

Researchers may also seek to isolate particular types of mutants such as gene knockouts and suppressor mutants.

Another level of understanding in Mendelian inheritance is to appreciate how the molecular expression of genes underlies their impact on the phenotype of an organism. Using modern molecular tools that are described later in this textbook, researchers can quantitatively compare the level of gene expression between wild-type and mutant alleles. This helps us to understand how the amount of gene expression is correlated with the phenotype of the organism. Also, when gene interactions occur, an investigator may want to understand, at the molecular level, how the gene products participate in a common cellular function such as an enzymatic pathway. Therefore, the molecular investigation of gene interactions frequently involves research collaborations between geneticists, cell biologists, and biochemists.

PROBLEM SETS & INSIGHTS

Solved Problems

S1. In humans, why are X-linked recessive traits more likely to occur in males compared to females?

Answer: Because a male is hemizygous for X-linked traits, the phenotypic expression of X-linked traits depends on only a single copy of the gene. When a male inherits a recessive X-linked allele, he will automatically exhibit the trait because he does not have another copy of the gene on the corresponding Y chromosome. This phenomenon is particularly relevant to the inheritance of recessive X-linked alleles that cause human disease. (Some examples will be described in Chapter 22.)

S2. In Ayrshire cattle, the spotting pattern of the animals can be either red and white or mahogany and white. The mahogany and white pattern is caused by the allele *M*. The red and white phenotype is controlled by the allele *m*. When mahogany and white animals are mated to red and white animals, the following results are obtained:

Genotype	Phenotype	
	Females	**Males**
MM	Mahogany and white	Mahogany and white
Mm	Red and white	Mahogany and white
mm	Red and white	Red and white

Explain the pattern of inheritance.

Answer: The inheritance pattern for this trait is sex-influenced inheritance. The *M* allele is dominant in males but recessive in females, while the *m* allele is dominant in females but recessive in males.

S3. For the following pedigree involving a single gene causing an inherited disease, indicate which modes of inheritance are not possible. (Affected individuals are shown as filled symbols.)

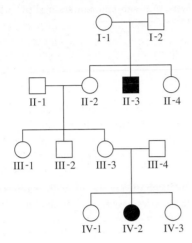

A. Recessive

B. Dominant

C. X linked, recessive

D. Sex influenced, dominant in females

E. Sex limited, recessive in females

Answer:

A. It could be recessive.

B. It is probably not dominant unless it is incompletely penetrant.

C. It could not be X-linked recessive because individual IV-2 does not have an affected father.

D. It could not be sex influenced, dominant in females because individual II-3 (who would have to be homozygous) has an unaffected mother (who would have to be heterozygous and affected).

E. It is not sex limited because individual II-3 is an affected male and IV-2 is an affected female.

S4. Red-green color blindness is inherited as a recessive X-linked trait. What are the following probabilities?

A. A woman with phenotypically normal parents and a color-blind brother will have a color-blind son. Assume that she has no previous children.

B. The next child of a phenotypically normal woman, who has already had one color-blind son, will be a color-blind son.

C. The next child of a phenotypically normal woman, who has already had one color-blind son, and who is married to a color-blind man, will have a color-blind daughter.

Answer:

A. The woman's mother must have been a heterozygote. So there is a 50% chance that the woman is a carrier. If she has children, 1/4

(i.e., 25%) will be affected sons if she is a carrier. However, there is only a 50% chance that she is a carrier. We multiply 50% times 25%, which equals $0.5 \times 0.25 = 0.125$, or a 12.5% chance.

B. If she already had a color-blind son, then we know she must be a carrier, so the chance is 25%.

C. The woman is heterozygous and her husband is hemizygous for the color-blind allele. This couple will produce 1/4 offspring that are color-blind daughters. The rest are 1/4 carrier daughters, 1/4 normal sons, and 1/4 color-blind sons. Answer is 25%.

S5. Pattern baldness is an example of a sex-influenced trait that is dominant in males and recessive in females. A couple, neither of whom is bald, produced a bald son. What are the genotypes of the parents?

Answer: Because the father is not bald, we know he must be homozygous, *bb*. Otherwise, he would be bald. A female who is not bald can be either *Bb* or *bb*. Because she has produced a bald son, we know that she must be *Bb* in order to pass the *B* allele to her son.

S6. Two pink-flowered four-o'clocks were crossed to each other. What are the following probabilities for the offspring?

A. A red-flowered plant.

B. The first three plants examined will be white.

C. A plant will be either white or pink.

D. A group of six plants contain one pink, two whites, and three reds.

Answer: The first thing we need to do is construct a Punnett square to determine the individual probabilities for each type of offspring.

Because flower color is incompletely dominant, the cross is *Rr* × *Rr*.

	♂ *R*	*r*
♀ *R*	*RR* Red	*Rr* Pink
r	*Rr* Pink	*rr* White

The phenotypic ratio is 1 red : 2 pink : 1 white. In other words, 1/4 are expected to be red, 1/2 pink, and 1/4 white.

A. The probability of a red-flowered plant is 1/4, which equals 25%.

B. Use the product rule.

$1/4 \times 1/4 \times 1/4 = 1/64 = 1.6\%$

C. Use the sum rule because these are mutually exclusive events. A given plant cannot be both white and pink.

$1/4 + 1/2 = 3/4 = 75\%$

D. Use the multinomial expansion equation. See solved problem S7 in Chapter 2 for an explanation of the multinomial expansion equation. In this case, three phenotypes are possible.

$$P = \frac{n!}{a!b!c!} p^a q^b r^c$$

where

    n = total number of offspring = 6
    a = number of reds = 3
    p = probability of reds = (1/4)
    b = number of pinks = 1
    q = probability of pink = (1/2)
    c = number of whites = 2
    r = probability of whites = (1/4)

If we substitute these values into the equation,

$$P = \frac{6!}{3!1!2!} (1/4)^3 (1/2)^1 (1/4)^2$$
$$P = 0.029 = 2.9\%$$

This means that 2.9% of the time we would expect to obtain six plants, three with red flowers, one with pink flowers, and two with white flowers.

Conceptual Questions

C1. Describe the differences among dominance, incomplete dominance, codominance, and overdominance.

C2. Discuss the differences among sex-influenced, sex-limited, and sex-linked inheritance. Describe examples.

C3. What is meant by a gene interaction? How can a gene interaction be explained at the molecular level?

C4. Let's suppose a recessive allele encodes a completely defective protein. If the functional allele is dominant, what does that tell us about the amount of the functional protein that is sufficient to cause the phenotype? What if the allele shows incomplete dominance?

C5. A nectarine is a peach without the fuzz. The difference is controlled by a single gene that is found in two alleles, D and d. At the molecular level, would it make more sense to you that the nectarine is homozygous for a recessive allele or that the peach is homozygous for the recessive allele? Explain your reasoning.

C6. An allele in *Drosophila* produces a "star-eye" trait in the heterozygous individual. However, the star-eye allele is lethal in homozygotes. What would be the ratio and phenotypes of surviving flies if star-eyed flies were crossed to each other?

C7. A seed dealer wants to sell four-o'clock seeds that will produce only red, white, or pink flowers. Explain how this should be done.

C8. The serum from one individual (let's call this person individual 1) is known to agglutinate the red blood cells from a second individual (individual 2). List the pairwise combinations of possible genotypes that individuals 1 and 2 could be. If individual 1 is the parent of individual 2, what are his/her possible genotypes?

C9. Which blood phenotypes (A, B, AB, and/or O) provide an unambiguous genotype? Is it possible for a couple to produce a family of children with all four blood types? If so, what would the genotypes of the parents have to be?

C10. A woman with type B blood has a child with type O blood. What are the possible genotypes and blood types of the father?

C11. A type A woman is the daughter of a type O father and type A mother. If she has children with a type AB man, what are the following probabilities?

A. A type AB child

B. A type O child

C. The first three children with type AB blood

D. A family containing two children with type B blood and one child with type AB

C12. In Shorthorn cattle, coat color is controlled by a single gene that can exist as a red allele (R) or white allele (r). The heterozygotes

(Rr) have a color called roan that looks less red than the RR homozygotes. However, when examined carefully, the roan phenotype in cattle is actually due to a mixture of completely red hairs and completely white hairs. Should this be called incomplete dominance, codominance, or something else? Explain your reasoning.

C13. In chickens, the Leghorn variety has white feathers due to a dominant allele. Silkies have white feathers due to a recessive allele in a second (different) gene. If a true-breeding white Leghorn is crossed to a true-breeding white Silkie, what is the expected phenotype of the F_1 generation? If members of the F_1 generation are mated to each other, what is the expected phenotypic outcome of the F_2 generation? Assume the chickens in the parental generation are homozygous for the white allele at one gene and homozygous for the brown allele at the other gene. In subsequent generations, nonwhite birds will be brown.

C14. Propose the most likely mode of inheritance (autosomal dominant, autosomal recessive, or X-linked recessive) for the following pedigrees. Affected individuals are shown with filled (black) symbols.

(a)

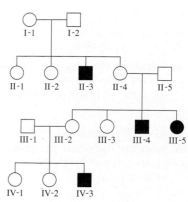

(b)

C15. A human disease known as vitamin D−resistant rickets is inherited as an X-linked dominant trait. If a male with the disease produces children with a female who does not have the disease, what is the expected ratio of affected and unaffected offspring?

C16. Hemophilia is a X-linked recessive trait in humans. If a heterozygous woman has children with a an unaffected man, what are the odds of the following combinations of children?

A. An affected son

B. Four unaffected offspring in a row

C. An unaffected daughter or son

D. Two out of five offspring that are affected

C17. Explain whether the following events could or could not happen.

A. A male has the same X chromosome as his paternal grandfather.

B. A male has the same X chromosome as his maternal grandfather.

C. A female has the same X chromosome as her maternal grandmother.

D. A female has the same X chromosome as her maternal great-great-grandmother.

C18. Incontinentia pigmenti is a rare, X-linked dominant disorder in humans characterized by swirls of pigment in the skin. If an affected female, who had an unaffected father, has children with an unaffected male, what would be the predicted ratios of affected and unaffected sons and daughters?

C19. With regard to pattern baldness in humans (a sex-influenced trait), a woman who is not bald and whose mother is bald has children with a bald man whose father is not bald. What are their probabilities of having the following types of families?

A. Their first child will not become bald.

B. Their first child will be a male who will not become bald.

C. Their first three children will be females who are not bald.

C20. In rabbits, the color of body fat is controlled by a single gene with two alleles, designated Y and y. The outcome of this trait is affected by the diet of the rabbit. When raised on a standard vegetarian diet, the dominant Y allele confers white body fat, and the y allele confers yellow body fat. However, when raised on a xanthophyll-free diet, the homozygote yy animal has white body fat. If a heterozygous animal is crossed to a rabbit with yellow body fat, what are the proportions of offspring with white and yellow body fat when raised on a standard vegetarian diet? How do the proportions change if the offspring are raised on a xanthophyll-free diet?

C21. A Siamese cat that spends most of its time outside was accidentally injured in a trap and required several stitches in its right front paw. The veterinarian had to shave the fur from the paw and leg, which originally had rather dark fur. Later, when the fur grew back, it was much lighter than the fur on the other three legs. Do you think this injury occurred in the hot summer or cold winter? Explain your answer.

C22. A true-breeding male fly with eosin eyes is crossed to a white-eyed female that is heterozygous for the wild-type (C) and cream alleles (c^a). What are the expected proportions of their offspring?

C23. In Chapter 4, we considered the trait of hen- versus cock-feathering. Starting with two heterozygous fowl that are hen-feathered, explain how you would obtain a true-breeding line that always produced cock-feathered males.

C24. In the pedigree shown here for a trait determined by a single gene (affected individuals are shown in black), state whether it would be possible for the trait to be inherited in each of the following ways:

A. Recessive

B. X-linked recessive

C. Dominant, complete penetrance

D. Sex influenced, dominant in males

E. Sex limited

F. Dominant, incomplete penetrance

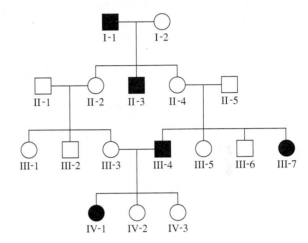

C25. The pedigree shown here also concerns a trait determined by a single gene (affected individuals are shown in black). Which of the following patterns of inheritance are possible?

A. Recessive

B. X-linked recessive

C. Dominant

D. Sex influenced, recessive in males

E. Sex limited

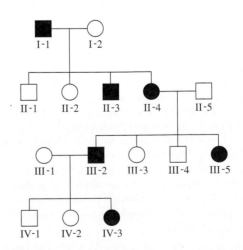

C26. Let's suppose you have pedigree data from thousands of different families involving a particular genetic disease. How would you decide whether the disease is inherited as a recessive trait as opposed to one that is dominant with incomplete penetrance?

C27. Compare phenotypes at the molecular, cellular, and organism levels for individuals who are homozygous for the hemoglobin allele, $Hb^A Hb^A$, and the sickle-cell allele, $Hb^S Hb^S$.

C28. A very rare dominant allele that causes the little finger to be crooked has a penetrance value of 80%. In other words, 80% of heterozygotes carrying the allele will have a crooked little finger. If a homozygous unaffected person has children with a heterozygote carrying this mutant allele, what is the probability that an offspring will have little fingers that are crooked?

C29. A sex-influenced trait in humans is one that affects the length of the index finger. A "short" allele is dominant in males and recessive in females. Heterozygous males have an index finger that is significantly shorter than the ring finger. The gene affecting index finger length is located on an autosome. A woman with short index fingers has children with a man who has normal index fingers. They produce five children in the following order: female, male, male, female, male. The oldest female offspring marries a man with normal fingers and then has one daughter. The youngest male among the five children marries a woman with short index fingers, and then they have two sons. Draw the pedigree for this family. Indicate the phenotypes of every individual (filled symbols for individuals with short index fingers and open symbols for individuals with normal index fingers).

C30. In horses, there are three coat-color patterns termed cremello (beige), chestnut (brown), and palomino (golden with light mane and tail). If two palomino horses are mated, they produce about 1/4 cremello, 1/4 chestnut, and 1/2 palomino offspring. In contrast, cremello horses and chestnut horses breed true. (In other words, two cremello horses will produce only cremello offspring and two chestnut horses will produce only chestnut offspring.) Explain this pattern of inheritance.

C31. Briefly describe three explanations for how a suppressor mutation exerts its effects at the molecular level.

Experimental Questions

E1. Mexican hairless dogs have little hair and few teeth. When a Mexican hairless is mated to another breed of dog, about half the puppies are hairless. When two Mexican hairless dogs are mated to each other, about 1/3 of the surviving puppies have hair, and about 2/3 of the surviving puppies are hairless. However, about two out of eight puppies from this type of cross are born grossly deformed and do not survive. Explain this pattern of inheritance.

E2. In chickens, some varieties have feathered shanks (legs), while others do not. In a cross between a Black Langhans (feathered shanks) and Buff Rocks (unfeathered shanks), the shanks of the F_1 generation are all feathered. When the F_1 generation is crossed, the F_2 generation contains chickens with feathered shanks to unfeathered shanks in a ratio of 15:1. Suggest an explanation for this result.

E3. In sheep, the formation of horns is a sex-influenced trait; the allele that results in horns is dominant in males and recessive in females. Females must be homozygous for the horned allele to have horns. A horned ram was crossed to a polled (unhorned) ewe, and the first offspring they produced was a horned ewe. What are the genotypes of the parents?

E4. A particular breed of dog can have long hair or short hair. When true-breeding long-haired animals were crossed to true-breeding short-haired animals, the offspring all had long hair. The F_2 generation produced a 3:1 ratio of long- to short-haired offspring. A second trait involves the texture of the hair. The two variants are wiry hair and straight hair. F_1 offspring from a cross of these two varieties all had wiry hair, and F_2 offspring showed a 3:1 ratio of wiry- to straight-haired puppies. Recently, a breeder of the short-, wiry-haired dogs found a female puppy that was albino. Similarly, another breeder of the long-, straight-haired dogs found a male puppy that was albino. Because the albino trait is always due to a recessive allele, the two breeders got together and mated the two dogs. Surprisingly, all the puppies in the litter had black hair. How would you explain this result?

E5. In the clover butterfly, males are always yellow, but females can be yellow or white. In females, white is a dominant allele. Two yellow butterflies were crossed to yield an F_1 generation consisting of 50% yellow males, 25% yellow females, and 25% white females. Describe how this trait is inherited and the genotypes of the parents.

E6. The *Mic2* gene in humans is present on both the X and Y chromosome. Let's suppose the *Mic2* gene exists in a dominant *Mic2* allele, which results in normal surface antigen, and a recessive *mic2* allele, which results in defective surface antigen production. Using molecular techniques, it is possible to identify homozygous and heterozygous individuals. By following the transmission of the *Mic2* and *mic2* alleles in a large human pedigree, would it be possible to distinguish between pseudoautosomal inheritance and autosomal inheritance? Explain your answer.

E7. Duroc Jersey pigs are typically red, but a sandy variation is also seen. When two different varieties of true-breeding sandy pigs were crossed to each other, they produced F_1 offspring that were red. When these F_1 offspring were crossed to each other, they produced red, sandy, and white pigs in a 9:6:1 ratio. Explain this pattern of inheritance.

E8. As discussed in Chapter 4, comb morphology in chickens is governed by a gene interaction. Two walnut comb chickens were crossed to each other. They produced only walnut comb and rose comb offspring, in a ratio of 3:1. What are the genotypes of the parents?

E9. In certain species of summer squash, fruit color is determined by two interacting genes. A dominant allele, *W*, determines white color, and a recessive allele allows the fruit to be colored. In a homozygous *ww* individual, a second gene determines fruit color: *G* (green) is dominant to *g* (yellow). A white squash and a yellow squash were crossed, and the F_1 generation yielded approximately 50% white fruit and 50% green fruit. What are the genotypes of the parents?

E10. Certain species of summer squash can exist in long, spherical, or disk shapes. When a true-breeding long-shaped strain was crossed to a true-breeding disk-shaped strain, all of the F_1 offspring were disk-shaped. When the F_1 offspring were allowed to self-fertilize, the F_2 generation consisted of a ratio of 9 disk-shaped : 6 round-shaped : 1 long-shaped. Assuming the shape of summer squash is governed by two different genes, with each gene existing in two alleles, propose a mechanism to account for this 9:6:1 ratio.

E11. In a species of plant, two genes control flower color. The red allele (*R*) is dominant to the white allele (*r*); the color-producing allele (*C*) is dominant to the non-color-producing allele (*c*). You suspect that either an *rr* homozygote or a *cc* homozygote will produce white flowers. In other words, *rr* is epistatic to *C*, and *cc* is epistatic to *R*. To test your hypothesis, you allowed heterozygous plants (*RrCc*) to self-fertilize and counted the offspring. You obtained the following data: 201 plants with red flowers and 144 with white flowers. Conduct a chi square analysis to see if your observed data are consistent with your hypothesis.

E12. In *Drosophila*, red eyes is the wild-type phenotype. There are several different genes (with each gene existing in two or more alleles) that affect eye color. One allele causes purple eyes, and a different allele causes sepia eyes. Both of these alleles are recessive compared to red eye color. When flies with purple eyes were crossed to flies with sepia eyes, all of the F_1 offspring had red eyes. When the F_1 offspring were allowed to mate with each other, the following data were obtained:

146 purple eyes

151 sepia eyes

50 purplish sepia eyes

444 red eyes

Explain this pattern of inheritance. Conduct a chi square analysis to see if the experimental data fit with your hypothesis.

E13. As mentioned in Experimental question E12, red eyes is the wild-type phenotype in *Drosophila*, and there are several different genes (with each gene existing in two or more alleles) that affect eye

color. One allele causes purple eyes, and a different allele causes vermilion eyes. The purple and vermilion alleles are recessive compared to red eye color. The following crosses were made, and the following data were obtained:

Cross 1: Males with vermillion eyes x females with purple eyes

354 offspring, all with red eyes

Cross 2: Males with purple eyes x females with vermillion eyes

212 male offspring with vermillion eyes

221 female offspring with red eyes

Explain the pattern of inheritance based on these results. What additional crosses might you make to confirm your hypothesis?

E14. Let's suppose that you were looking through a vial of fruit flies in your laboratory and noticed a male fly that has pink eyes. What crosses would you make to determine if the pink allele is an X-linked gene? What crosses would you make to determine if the pink allele is an allele of the same X-linked gene that has white and eosin alleles? Note: The white and eosin alleles are discussed in Figure 4.22.

E15. When examining a human pedigree, what features do you look for to distinguish between X-linked recessive inheritance versus autosomal recessive inheritance? How would you distinguish X-linked dominant inheritance from autosomal dominant inheritance in a human pedigree?

Questions for Student Discussion/Collaboration

1. Let's suppose a gene exists as a functional wild-type allele and a nonfunctional mutant allele. At the organism level, the wild-type allele is dominant. In a heterozygote, discuss whether dominance occurs at the cellular or molecular level. Discuss examples in which the issue of dominance depends on the level of examination.

2. A true-breeding rooster with a rose comb, feathered shanks, and cock-feathering was crossed to a hen that is true-breeding for pea comb and unfeathered shanks but is heterozygous for hen-feathering. If you assume these genes can assort independently, what is the expected outcome of the F_1 generation?

3. In oats, the color of the chaff is determined by a two-gene interaction. When a true-breeding black plant was crossed to a true-breeding white plant, the F_1 generation was composed of all black plants. When the F_1 offspring were crossed to each other, the ratio produced was 12 black : 3 gray : 1 white. First, construct a Punnett square that accounts for this pattern of inheritance. Which genotypes produce the gray phenotype? Second, at the level of protein function, how would you explain this type of inheritance?

Note: All answers appear at the website for this textbook; the answers to even-numbered questions are in the back of the textbook.

www.mhhe.com/brookergenetics3e

Visit the Online Learning Center for practice tests, answer keys, and other learning aids for this chapter. Enhance your understanding of genetics with our interactive exercises, quizzes, animations, and much more.

LINKAGE AND GENETIC MAPPING IN EUKARYOTES

In Chapter 2, we were introduced to Mendel's laws of inheritance. According to these principles, we expect that two different genes will segregate and independently assort themselves during the process that creates gametes. After Mendel's work was rediscovered at the turn of the twentieth century, chromosomes were identified as the cellular structures that carry genes. The chromosome theory of inheritance explained how the transmission of chromosomes is responsible for the passage of genes from parents to offspring.

When geneticists first realized that chromosomes contain the genetic material, they began to suspect that a conflict might sometimes occur between the law of independent assortment of genes and the behavior of chromosomes during meiosis. In particular, geneticists assumed that each species of organism must contain thousands of different genes, yet cytological studies revealed that most species have at most a few dozen chromosomes. Therefore, it seemed likely, and turned out to be true, that each chromosome would carry many hundreds or even thousands of different genes. The transmission of genes located close to each other on the same chromosome violates the law of independent assortment.

In this chapter, we will consider the pattern of inheritance that occurs when different genes are situated on the same chromosome. In addition, we will briefly explore how the data from genetic crosses are used to construct a **genetic map**—a diagram that describes the order of genes along a chromosome. Newer strategies for gene mapping are described in Chapter 20. However, an understanding of traditional mapping studies, as described in this chapter, will strengthen our appreciation for these newer molecular approaches. More importantly, traditional mapping studies further illustrate how the location of two or more genes on the same chromosome can affect the transmission patterns from parents to offspring.

5.1 LINKAGE AND CROSSING OVER

In eukaryotic species, each linear chromosome contains a very long segment of DNA. A chromosome contains many individual functional units—called genes—that influence an organism's traits. A typical chromosome is expected to contain many hundreds or perhaps a few thousand different genes. The term **linkage** has two related meanings. First, linkage refers to the phenomenon that two or more genes may be located on the same chromosome. The genes are physically linked to each other, because each eukaryotic chromosome contains a single, continuous, linear molecule of DNA. Second, genes that are close

Crossing over during meiosis. This event provides a way to reassort the alleles of genes that are located on the same chromosome.

together on the same chromosome tend to be transmitted as a unit. This second meaning indicates that linkage has an influence on inheritance patterns.

Chromosomes are sometimes called **linkage groups,** because a chromosome contains a group of genes that are physically linked together. In each species, the number of linkage groups equals the number of chromosome types. For example, human somatic cells have 46 chromosomes, which are composed of 22 types of autosomes that come in pairs plus one pair of sex chromosomes, the X and Y. Therefore, humans have 22 autosomal linkage groups, an X chromosome linkage group, and a Y chromosome linkage group.

Geneticists are often interested in the transmission of two or more traits in a genetic cross. When a geneticist follows the variants of two different traits in a cross, this is called a **dihybrid cross;** when three traits are followed, it is a **trihybrid cross;** and so on. The outcome of a dihybrid or trihybrid cross depends on whether or not the genes are linked to each other along the same chromosome. In this section, we will examine how linkage affects the transmission patterns of two or more traits.

Crossing Over May Produce Recombinant Genotypes

Even though the alleles for different genes may be linked along the same chromosome, the linkage can be altered during meiosis. In diploid eukaryotic species, homologous chromosomes can exchange pieces with each other, a phenomenon called **crossing over.** This event occurs during prophase of meiosis I. As discussed in Chapter 3, the replicated chromosomes, known as sister chromatids, associate with the homologous sister chromatids to form a structure known as a **bivalent.** A bivalent is composed of two pairs of sister chromatids. In prophase of meiosis I, a sister chromatid of one pair commonly crosses over with a sister chromatid from the homologous pair.

Figure 5.1 considers meiosis when two genes are linked on the same chromosome. One of the parental chromosomes carries the *A* and *B* alleles, while the homologue carries the *a* and *b* alleles. In Figure 5.1a, no crossing over has occurred. Therefore, the resulting haploid cells contain the same combination of alleles as the original chromosomes. In this case, two haploid cells carry the dominant *A* and *B* alleles, and the other two carry the recessive *a* and *b* alleles. The arrangement of linked alleles has not been altered.

In contrast, Figure 5.1b illustrates what can happen when crossing over occurs. Two of the haploid cells contain combinations of alleles, namely *A* and *b* or *a* and *B*, which differ from those in the original chromosomes. In these two cells, the grouping of linked alleles has been changed. An event such as this, leading to a new combination of alleles, is known as **genetic recombination.** The haploid cells carrying the *A* and *b*, or the *a* and *B*, alleles are called **nonparental** or **recombinant cells.** Likewise, if such haploid cells were gametes that participated in fertilization, the resulting offspring are called nonparental or recombinant offspring. These offspring can display combinations of traits that are different from those of either parent. In contrast, offspring

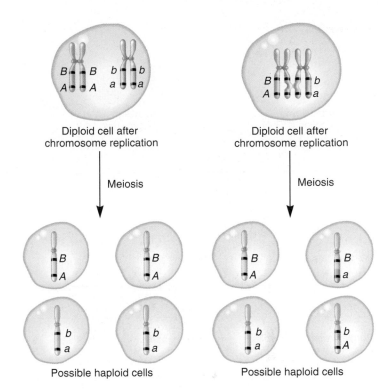

(a) Without crossing over, linked alleles segregate together.

(b) Crossing over can reassort linked alleles.

 FIGURE 5.1 **Consequences of crossing over during meiosis. (a)** In the absence of crossing over, the *A* and *B* alleles and the *a* and *b* alleles are maintained in the same arrangement found in the parental chromosomes. **(b)** Crossing over has occurred in the region between the two genes, creating two nonparental haploid cells with a new combination of alleles.

that have inherited the same combination of alleles that are found in the chromosomes of their parents are known as **parental** or **nonrecombinant** offspring.

In this section, we will consider how crossing over affects the pattern of inheritance for genes linked on the same chromosome. In Chapter 17, we will consider the molecular events that cause crossing over to occur.

Bateson and Punnett Discovered Two Traits That Did Not Assort Independently

An early study indicating that some traits may not assort independently was carried out by William Bateson and Reginald Punnett in 1905. According to Mendel's law of independent assortment, a dihybrid cross between two individuals, heterozygous for two genes, should yield a 9:3:3:1 phenotypic ratio among the offspring. However, a surprising result occurred when Bateson and Punnett conducted a cross in the sweet pea involving two different traits—flower color and pollen shape.

As seen in **Figure 5.2,** they began by crossing a truebreeding strain with purple flowers (*PP*) and long pollen (*LL*) to

P generation

F₁ offspring

Purple flowers,
long pollen (*PpLl*)

Self-fertilization

F₂ offspring	Observed number	Ratio	Expected number	Ratio
Purple flowers, long pollen	296	15.6	240	9
Purple flowers, round pollen	19	1.0	80	3
Red flowers, long pollen	27	1.4	80	3
Red flowers, round pollen	85	4.5	27	1

FIGURE 5.2   **An experiment of Bateson and Punnett with sweet peas, showing that independent assortment does not always occur. Note: The expected numbers are rounded to the nearest whole number.**

Genes →Traits Two genes that govern flower color and pollen shape are found on the same chromosome. Therefore, the offspring tend to inherit the parental combinations of alleles (*PL* or *pl*). Due to occasional crossing over, a lower percentage of offspring inherit nonparental combinations of alleles (*Pl* or *pL*).

P generation

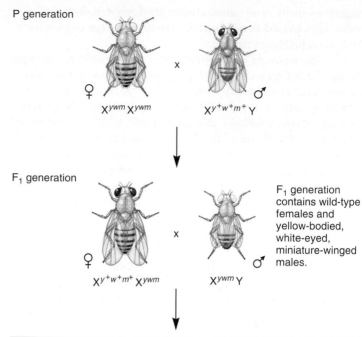

F₁ generation

F₁ generation contains wild-type females and yellow-bodied, white-eyed, miniature-winged males.

F₂ generation	Females	Males	Total
Gray body, red eyes, long wings	439	319	758
Gray body, red eyes, miniature wings	208	193	401
Gray body, white eyes, long wings	1	0	1
Gray body, white eyes, miniature wings	5	11	16
Yellow body, red eyes, long wings	7	5	12
Yellow body, red eyes, miniature wings	0	0	0
Yellow body, white eyes, long wings	178	139	317
Yellow body, white eyes, miniature wings	365	335	700

FIGURE 5.3   **Morgan's trihybrid cross involving three X-linked traits in *Drosophila*.**

Genes →Traits Three genes that govern body color, eye color, and wing length are all found on the X chromosome. Therefore, the offspring tend to inherit the parental combinations of alleles (*y⁺ w⁺ m⁺* or *y w m*). Figure 5.4 explains how single and double crossovers can create nonparental combinations of alleles.

a strain with red flowers (*pp*) and round pollen (*ll*). This yielded an F₁ generation of plants that all had purple flowers and long pollen (*PpLl*). An unexpected result came from the F₂ generation. Even though the F₂ generation had four different phenotypic categories, the observed numbers of offspring did not conform to a 9:3:3:1 ratio. Bateson and Punnett found that the F₂ generation had a much greater proportion of the two phenotypes found in the parental generation—purple flowers with long pollen and red flowers with round pollen. Therefore, they suggested that the transmission of these two traits from the parental generation to the F₂ generation was somehow coupled and not easily assorted in an independent manner. However, Bateson and Punnett did not realize that this coupling was due to the linkage of the flower color gene and the pollen shape gene on the same chromosome.

Morgan Provided Evidence for the Linkage of X-Linked Genes and Proposed That Crossing Over Between X Chromosomes Can Occur

The first direct evidence that different genes are physically located on the same chromosome came from the studies of Thomas Hunt Morgan in 1911, who investigated the inheritance pattern of different traits that had been shown to follow an X-linked pat-

tern of inheritance. **Figure 5.3** illustrates an experiment involving three traits that Morgan studied. His parental crosses were wild-type male fruit flies mated to females that had yellow bodies (*yy*), white eyes (*ww*), and miniature wings (*mm*). The wild-type alleles for these three genes are designated *y⁺* (gray body), *w⁺* (red eyes), and *m⁺* (long wings). As expected, the phenotypes of the F₁ generation were wild-type females, and males with yellow bodies, white eyes, and miniature wings. The linkage of these genes was revealed when the F₁ flies were mated to each other and the F₂ generation examined.

Instead of equal proportions of the eight possible phenotypes, Morgan observed a much higher proportion of the combinations of traits found in the parental generation. He observed 758 flies with gray bodies, red eyes, and long wings, and 700 flies with yellow bodies, white eyes, and miniature wings. The combination of gray body, red eyes, and long wings was found in the males of the parental generation, while the combination of yellow body, white eyes, and miniature wings was the same as the females of the parental generation. Morgan's explanation for this higher proportion of parental combinations was that all three genes are located on the X chromosome and, therefore, tend to be transmitted together as a unit.

However, to fully account for the data shown in Figure 5.3, Morgan needed to interpret two other key observations. First, he needed to explain why a significant proportion of the F_2 generation had nonparental combinations of alleles. Along with the two parental phenotypes, five other phenotypic combinations appeared that were not found in the parental generation. Second, he needed to explain why a quantitative difference was observed between nonparental combinations involving body color and eye color versus eye color and wing length. This quantitative difference is revealed by reorganizing the data from Morgan's cross by pairs of genes.

Gray body, red eyes	1,159	
Yellow body, white eyes	1,017	
Gray body, white eyes	17	} Nonparental
Yellow body, red eyes	12	} offspring
Total	2,205	

Red eyes, long wings	770	
White eyes, miniature wings	716	
Red eyes, miniature wings	401	} Nonparental
White eyes, long wings	318	} offspring
Total	2,205	

Morgan found a substantial difference between the numbers of nonparental offspring when pairs of genes were considered separately. Nonparental combinations involving only eye color and wing length were fairly common—401 + 318 nonparental offspring. In sharp contrast, nonparental combinations for body color and eye color were quite rare—17 + 12 nonparental offspring.

How did Morgan explain these data? He considered the studies conducted in 1909 of the Belgian cytologist F. A. Janssens, who observed chiasmata under the microscope and proposed that crossing over involves a physical exchange between homologous chromosomes. Morgan shrewdly realized that crossing over between homologous X chromosomes was consistent with his data. He assumed that crossing over did not occur between the X and Y chromosome and that these three genes are not found on the Y chromosome. With these ideas in mind, he made three important hypotheses to explain his results:

1. The genes for body color, eye color, and wing length are all located on the same chromosome, namely, the X chromosome. Therefore, the alleles for all three traits are most likely to be inherited together.
2. Due to crossing over, the homologous X chromosomes (in the female) can exchange pieces of chromosomes and create new (nonparental) combinations of alleles.
3. The likelihood of crossing over depends on the distance between two genes. If two genes are far apart from each other, crossing over is more likely to occur between them.

With these ideas in mind, **Figure 5.4** illustrates the possible events that occurred in the F_1 female flies of Morgan's experiment. One of the X chromosomes carried all three dominant alleles, while the other had all three recessive alleles. During oogenesis in the F_1 female flies, crossing over may or may not have occurred in this region of the X chromosome. If no crossing over occurred, the parental phenotypes were produced in the F_2 offspring. Alternatively, a crossover sometimes occurred between the eye color gene and the wing length gene to produce nonparental offspring with gray bodies, red eyes, and miniature wings or yellow bodies, white eyes, and long wings. According to Morgan's proposal, such an event is fairly likely because these two genes are far apart from each other on the X chromosome. In contrast, he proposed that the body color and eye color genes are very close together, which makes crossing over between them an unlikely event. Nevertheless, it occasionally occurred, yielding offspring with gray bodies, white eyes, and miniature wings, or with yellow bodies, red eyes, and long wings. Finally, it was also possible for two homologous chromosomes to cross over twice. This double crossover is very unlikely. Among the 2,205 offspring Morgan examined, he found only one fly with a gray body, white eyes, and long wings that could be explained by this phenomenon.

A Chi Square Analysis Can Be Used to Distinguish Between Linkage and Independent Assortment

Now that we have an appreciation for linkage and the production of recombinant offspring, let's consider how an experimenter can objectively decide whether two genes are linked or assort independently. In Chapter 2, we used chi square analysis to evaluate the goodness of fit between a genetic hypothesis and observed experimental data. This method can similarly be employed to determine if the outcome of a dihybrid cross is consistent with linkage or independent assortment.

To conduct a chi square analysis, we must first propose a hypothesis. In a dihybrid cross, the standard hypothesis is that the two genes are not linked. This hypothesis is chosen even if the observed data suggest linkage, because an independent assortment hypothesis allows us to calculate the expected number of offspring based on the genotypes of the parents and the law of independent assortment. In contrast, for two linked genes that have not been previously mapped, we cannot calculate the expected number of offspring from a genetic cross because we do not know how likely it is for a crossover to occur between the two genes. Without expected numbers of recombinant and parental offspring, we cannot conduct a chi square test. Therefore, we begin with the hypothesis that the genes are not linked. Recall from Chapter 2 that the hypothesis we are testing is called a **null hypothesis,** because it assumes there is no real difference between the observed and expected values. The goal is to determine whether or not the data fit the hypothesis. If the chi square value is low and we cannot reject the null hypothesis, we infer that the genes assort independently. On the other hand, if the chi square value is so high that our hypothesis is rejected, we will accept the alternative hypothesis, namely, that the genes are linked.

Sex chromosomes in

X chromosomes in F$_1$ female before oogenesis	F$_1$ female gametes	F$_1$ male gametes	F$_2$ offspring phenotypes

(a) No crossing over between y^+ and m^+ (most likely)

Gray body, red eyes, long wings; total = 758

Yellow body, white eyes, miniature wings; total = 700

FIGURE 5.4   **The likelihood of crossing over provides an explanation of Morgan's trihybrid cross. (a)** In the absence of crossing over, the parental chromosomes carry the $y^+w^+m^+$ or $y\,w\,m$ alleles. Alternatively, a crossover may occur between homologous X chromosomes in the female parent during oogenesis. As described in Chapter 3, crossing over actually occurs at the bivalent stage, but for simplicity, this figure shows only two X chromosomes (one of each homologue) rather than four chromatids, which would occur during the bivalent stage of meiosis. In part **(b)**, a crossover has occurred between the eye color and wing length genes to produce recombinant chromosomes that carry the y^+w^+m or $y\,w\,m^+$ alleles. Because the eye color and wing length genes are relatively far apart, such recombinant offspring are common. **(c)** A crossover occurs between the body color and eye color genes to produce recombinant chromosomes that contain the $y^+w\,m$ or $y\,w^+\,m^+$ alleles. Because these two genes are very close together, these two types of recombinant offspring are fairly uncommon. **(d)** Finally, a double crossover could produce chromosomes carrying the $y^+w\,m^+$ or $y\,w^+\,m$ alleles. This occurs very rarely. Note: In this experiment, crossing over occurred between homologous X chromosomes in the female during oogenesis. A peculiar characteristic of *Drosophila* is that crossing over does not occur during spermatogenesis. In many other animal species, homologous recombination commonly occurs during spermatogenesis. Also note that this figure shows only a portion of the X chromosome. A map of the entire X chromosome is shown in Figure 5.8.

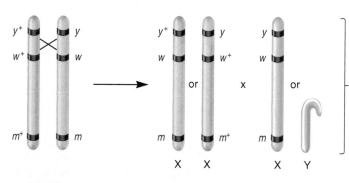

(b) Crossing over between w^+ and m^+ (fairly likely)

Gray body, red eyes, miniature wings; total = 401

Yellow body, white eyes, long wings; total = 317

(c) Crossing over between y^+ and w^+ (unlikely)

Gray body, white eyes, miniature wings; total = 16

Yellow body, red eyes, long wings; total = 12

(d) Double crossing over (very unlikely)

Gray body, white eyes, long wings; total = 1

Yellow body, red eyes, miniature wings; total = 0

Of course, a statistical analysis cannot prove that a hypothesis is true. If the chi square value is high, we accept the linkage hypothesis because we are assuming that only two explanations for a genetic outcome are possible: the genes are either linked or not linked. However, if other factors affect the outcome of the cross, such as a decreased viability of particular phenotypes, these may result in large deviations between the observed and expected values and cause us to reject the independent assortment hypothesis even though it may be correct.

Let's consider Morgan's data concerning body color and eye color (refer back to page 103). This cross produced the following offspring: 1,159 gray body, red eyes; 1,017 yellow body, white eyes; 17 gray body, white eyes; and 12 yellow body, red eyes. However, when a heterozygous female ($X^{y^+w^+} X^{yw}$) is crossed to a hemizygous male ($X^{yw}Y$), the laws of segregation and independent assortment predict the following outcome:

F$_1$ male gametes

Mendel's laws predict a 1:1:1:1 ratio among the four phenotypes. The observed data obviously seem to conflict with this expected outcome. Nevertheless, we stick to the strategy just discussed. We begin with the hypothesis that the two genes are not linked, and then we conduct a chi square analysis to see if the data fit this hypothesis. If the data do not fit, we will reject the idea that the genes assort independently and conclude the genes are linked.

A step-by-step outline for applying the chi square test to distinguish between linkage and independent assortment is described next.

Step 1. *Propose a hypothesis.* Even though the observed data appear inconsistent with this hypothesis, we propose that the two genes for eye color and body color obey Mendel's law of independent assortment. This hypothesis allows us to calculate expected values. Because the data seem to conflict with this hypothesis, we actually anticipate that the chi square analysis will allow us to reject the independent assortment hypothesis in favor of a linkage hypothesis. We are also assuming the alleles follow the law of segregation, and the four phenotypes are equally viable.

Step 2. *Based on the hypothesis, calculate the expected values of each of the four phenotypes.* Each phenotype has an equal probability of occurring (see the Punnett square given previously). Therefore, the probability of each phenotype is 1/4. The observed F$_2$ generation had a total of 2,205 individuals. Our next step is to calculate the expected numbers of offspring with each phenotype when the total equals 2,205; 1/4 of the offspring should be each of the four phenotypes:

$$1/4 \times 2{,}205 = 551 \text{ (expected number of each phenotype, rounded to the nearest whole number)}$$

Step 3. *Apply the chi square formula, using the data for the observed values (O) and the expected values (E) that have been calculated in step 2.* In this case, the data consist of four phenotypes.

$$\chi^2 = \frac{(O_1 - E_1)^2}{E_1} + \frac{(O_2 - E_2)^2}{E_2} + \frac{(O_3 - E_3)^2}{E_3} + \frac{(O_4 - E_4)^2}{E_4}$$

$$\chi^2 = \frac{(1{,}159 - 551)^2}{551} + \frac{(17 - 551)^2}{551}$$

$$+ \frac{(12 - 551)^2}{551} + \frac{(1{,}017 - 551)^2}{551}$$

$$\chi^2 = 670.9 + 517.5 + 527.3 + 394.1 = 2{,}109.8$$

Step 4. *Interpret the calculated chi square value.* This is done with a chi square table, as discussed in Chapter 2. The four phenotypes are based on the law of segregation and the law of independent assortment. By itself, the law of independent assortment predicts only two categories, recombinant and nonrecombinant. Therefore, based on a hypothesis of independent assortment, the degrees of freedom equals $n-1$, which is $2-1$, or 1.

The calculated chi square value is enormous! This means that the deviation between observed and expected values is very large. With one degree of freedom, such a large deviation is expected to occur by chance alone less than 1% of the time (see Table 2.1). Therefore, we reject the hypothesis that the two genes assort independently. As an alternative, we accept the hypothesis that the genes are linked.

Creighton and McClintock Showed That Crossing Over Produced New Combinations of Alleles and Resulted in the Exchange of Segments Between Homologous Chromosomes

As we have seen, Morgan's studies were consistent with the hypothesis that crossing over occurs between homologous chromosomes to produce new combinations of alleles. To obtain direct evidence that crossing over can result in genetic recombination, Harriet Creighton and Barbara McClintock used an interesting strategy involving parallel observations. In studies conducted in 1931, they first made crosses involving two linked genes to produce parental and recombinant offspring. Second, they used a microscope to view the structures of the chromosomes in the parents and in the offspring. Because the parental chromosomes had some unusual structural features, they could microscopically distinguish the two homologous chromosomes within a pair. As we will see, this enabled them to correlate the occurrence of recombinant offspring with microscopically observable exchanges in segments of homologous chromosomes.

Creighton and McClintock focused much of their attention on the pattern of inheritance of traits in corn. This species has 10 different chromosomes per set, which are named chromosome 1, chromosome 2, chromosome 3, and so on. In previous cytological examinations of corn chromosomes, some strains were found to have an unusual chromosome 9 with a darkly staining knob at one end. In addition, McClintock identified an abnormal version of chromosome 9 that also had an extra piece of chromosome 8 attached at the other end (**Figure 5.5a**). This chromosomal rearrangement is called a translocation.

Creighton and McClintock insightfully realized that this abnormal chromosome could be used to demonstrate that two homologous chromosomes physically exchange segments as a result of crossing over. They knew that a gene was located near the knobbed end of chromosome 9 that provided color to corn kernels. This gene existed in two alleles, the dominant allele *C* (colored) and the recessive allele *c* (colorless). A second gene, located near the translocated piece from chromosome 8, affected the texture of the kernel endosperm. The dominant allele *Wx* caused starchy endosperm, and the recessive *wx* allele caused waxy endosperm. Creighton and McClintock reasoned that a crossover involving a normal chromosome 9 and a knobbed/translocated chromosome 9 would produce a chromosome that had either a knob or a translocation, but not both. These two types of chromosomes would be distinctly different from either of the parental chromosomes (**Figure 5.5b**).

As shown in the experiment of **Figure 5.6**, Creighton and McClintock began with a corn strain that carried an abnormal chromosome that had a knob at one end and a translocation at the other. Genotypically, this chromosome was *C wx*. The cytologically normal chromosome in this strain was *c Wx*. This corn plant, termed parent A, had the genotype *Cc Wxwx*. It was

(a) Normal and abnormal chromosome 9

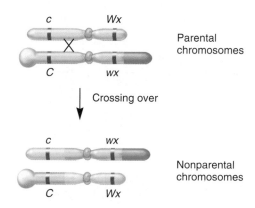

(b) Crossing over between normal and abnormal chromosome 9

FIGURE 5.5   Crossing over between a normal and abnormal chromosome 9 in corn. (**a**) A normal chromosome 9 in corn is compared to an abnormal chromosome 9 that contains a knob at one end and a translocation at the opposite end. (**b**) A crossover produces a chromosome that contains only a knob at one end and another chromosome that contains only a translocation at the other end.

crossed to a strain called parent B that carried two cytologically normal chromosomes and had the genotype *cc Wxwx*.

They then observed the kernels in two ways. First, they examined the phenotypes of the kernels to see if they were colored or colorless, and starchy or waxy. Second, the chromosomes in each kernel were examined under a microscope to determine their cytological appearance. Altogether, they observed a total of 25 kernels (see data of Figure 5.6).

■ THE HYPOTHESIS

Offspring with nonparental phenotypes are the product of a crossover. This crossover should create nonparental chromosomes via an exchange of chromosomal segments between homologous chromosomes.

■ **TESTING THE HYPOTHESIS — FIGURE 5.6**   **Experimental correlation between genetic recombination and crossing over.**

Starting materials: Two different strains of corn. One strain, referred to as parent A, had an abnormal chromosome 9 (knobbed/translocation) with a dominant *C* allele and a recessive *wx* allele. It also had a cytologically normal copy of chromosome 9 that carried the recessive *c* allele and the dominant *Wx* allele. Its genotype was *Cc Wxwx*. The other strain (referred to as parent B) had two normal versions of chromosome 9. The genotype of this strain was *cc Wxwx*.

Experimental level

1. Cross the two strains described. The tassel is the pollen-bearing structure, and the silk (equivalent to the stigma and style) is connected to the ovary. After fertilization, the ovary will develop into an ear of corn.

Tassel

Silk

Parent A
Cc Wxwx

x

Parent B
cc Wxwx

2. Observe the kernels from this cross.

F₁ ear of corn

F₁ kernels

3. Microscopically examine chromosome 9 in the kernels.

Microscope

Conceptual level

C   *c*   *c*   *c*

x

wx   *Wx*   *Wx*   *wx*

Each kernel is a separate seed that has inherited a set of chromosomes from each parent.

Colored/waxy   Colorless/waxy

From parent B

C   *c*   *c*   *c*

A recombinant chromosome

wx   *wx*   *wx*   *wx*

From parent A

This illustrates only 2 possible outcomes in the F₁ kernels. The recombinant chromosome on the right is due to crossing over during meiosis in parent A. As shown in The Data, there are several possible outcomes.

■ THE DATA

Phenotype of F$_1$ Kernel	Number of Kernels Analyzed	Cytological Appearance of a Chromosome in F$_1$ Offspring*		Did a Crossover Occur During Gamete Formation in Parent A?
Colored/waxy	3	Knobbed/translocation C wx	Normal c wx	No
Colorless/starchy	11	Knobless/normal c Wx	Normal c or Wx c wx	No
Colorless/starchy	4	Knobless/translocation c wx	Normal c Wx	Yes
Colorless/waxy	2	Knobless/translocation c wx	Normal c wx	Yes
Colored/starchy	5	Knobbed/normal C Wx	Normal c or Wx c wx	Yes
Total	25			

*In this table, the chromosome on the left was inherited from parent A, and the chromosome on the right was inherited from parent B.

■ INTERPRETING THE DATA

By combining the gametes in a Punnett square, the following types of offspring can be produced:

	Parent A	Parent B
	C wx (nonrecombinant)	c Wx
	c Wx (nonrecombinant)	c wx
	C Wx (recombinant)	
	c wx (recombinant)	

Parent B

♀ \ ♂	c Wx	c wx	
C wx	Cc Wxwx Colored, starchy	Cc wxwx Colored, waxy	Nonrecombinant
c Wx	cc WxWx Colorless, starchy	cc Wxwx Colorless, starchy	Nonrecombinant
C Wx	Cc WxWx Colored, starchy	Cc Wxwx Colored, starchy	Recombinant
c wx	cc Wxwx Colorless, starchy	cc wxwx Colorless, waxy	Recombinant

In this experiment, the researchers were interested in whether or not crossing over had occurred in parent A, which was heterozygous for both genes. This parent could produce four types of gametes, while parent B could produce only two types.

As seen in the Punnett square, two of the phenotypic categories, colored, starchy (Cc Wxwx or Cc WxWx) and colorless, starchy (cc WxWx or cc Wxwx), were ambiguous because they could arise from a nonrecombinant and from a recombinant gamete. In other words, these phenotypes could be produced whether or not recombination occurred in parent A. Therefore, let's focus on the two unambiguous phenotypic categories: colored, waxy (Cc wxwx) and colorless, waxy (cc wxwx). The colored, waxy phenotype could happen only if recombination did not occur in parent A and if parent A passed the knobbed, translocated chromosome to its offspring. As shown in the data, three kernels were obtained with this phenotype, and all of them had the knobbed, translocated chromosome. By comparison, the colorless, waxy phenotype could be obtained only if genetic recombination occurred in parent A and this parent passed a chromosome 9 that had a translocation but was knobless. Two kernels were obtained with this phenotype, and both of them had the expected chromosome that had a translocation but was knobless. Taken together, these results showed a perfect correlation between genetic recombination of alleles and the cytological presence of a chromosome displaying a genetic exchange of chromosomal pieces from parent A.

Overall, the observations described in this experiment were consistent with the idea that a crossover occurred in the region between the *C* and *wx* genes that involved an exchange of segments between two homologous chromosomes. As stated by Creighton and McClintock, "Pairing chromosomes, heteromorphic in two regions, have been shown to exchange parts at the same time they exchange genes assigned to these regions." These results supported the view that genetic recombination involves a physical exchange between homologous chromosomes. This microscopic evidence helped to convince geneticists that recombinant offspring arise from the physical exchange of segments of homologous chromosomes. As shown in solved problem S4 at the end of this chapter, an experiment by Curt Stern was also consistent with the conclusion that crossing over between homologous chromosomes accounts for the formation of offspring with recombinant phenotypes.

A self-help quiz involving this experiment can be found at the Online Learning Center.

Crossing Over Occasionally Occurs During Mitosis

In multicellular organisms, the union of egg and sperm is followed by many cellular divisions, which occur in conjunction with mitotic divisions of the cell nuclei. As discussed in Chapter 3, mitosis normally does not involve the homologous pairing of chromosomes to form a bivalent. Therefore, crossing over during mitosis is expected to occur much less frequently than during meiosis. Nevertheless, it does happen on rare occasions. Mitotic crossing over may produce a pair of recombinant chromosomes that have a new combination of alleles, an event known as **mitotic recombination.** If it occurs during an early stage of embryonic development, the daughter cells containing the recombinant chromosomes continue to divide many times to produce a patch of tissue in the adult. This may result in a portion of tissue with characteristics different from those of the rest of the organism.

In 1936, Curt Stern identified unusual patches on the bodies of certain *Drosophila* strains. He was working with strains carrying X-linked alleles affecting body color and bristle morphology (**Figure 5.7**). A recessive allele confers yellow body color (*y*), and another recessive allele causes shorter body bristles that look singed (*sn*). The corresponding wild-type alleles result in gray body color (y^+) and long bristles (sn^+). Females that are $y^+y\ sn^+sn$ are expected to have gray body color and long bristles. This was generally the case. However, when Stern carefully observed the bodies of these female flies under a low-power microscope, he occasionally noticed places in which two adjacent regions were different from the rest of the body—a twin spot. He concluded that twin spotting was too frequent to be explained by the random positioning of two independent single spots that happened to occur close together. How then did Stern explain the phenomenon of twin spotting? He proposed that twin spots are due to a single mitotic recombination within one cell during embryonic development.

As shown in Figure 5.7, the X chromosomes of the fertilized egg are $y^+\ sn$ and $y\ sn^+$. During development, a rare crossover can occur during mitosis to produce two adjacent daughter cells that are $y^+y^+\ snsn$ and $yy\ sn^+sn^+$. As embryonic development proceeds, the cell on the left will continue to divide to produce many cells, eventually producing a patch on the body that has gray color with singed bristles. The daughter cell next to it will produce a patch of yellow body color with long bristles. These two adjacent patches—a twin spot—will be surrounded by cells that are $y^+y\ sn^+sn$ and have gray color and long bristles. Twin spots provide evidence that mitotic recombination occasionally occurs.

5.2 GENETIC MAPPING IN PLANTS AND ANIMALS

The purpose of **genetic mapping,** also known as gene mapping or chromosome mapping, is to determine the linear order and distance of separation among genes that are linked to each other along the same chromosome. **Figure 5.8** illustrates a simplified genetic map of *Drosophila melanogaster* depicting the locations of many different genes along the individual chromosomes. As shown here, each gene has its own unique **locus**—the site where the gene is found within a particular chromosome. For example, the gene designated *brown eyes* (*bw*), which affects eye color, is located near one end of chromosome 2. The gene designated *black body* (*b*), which affects body color, is found near the middle of the same chromosome.

Why is genetic mapping useful? First, it allows geneticists to understand the overall complexity and genetic organization of a particular species. The genetic map of a species portrays the underlying basis for the inherited traits that an organism displays. In some cases, the known locus of a gene within a genetic map can help molecular geneticists to clone that gene and thereby obtain greater information about its molecular features. In addition, genetic maps are useful from an evolutionary point of view. A comparison of the genetic maps for different species can improve our understanding of the evolutionary relationships among those species.

Along with these scientific uses, genetic maps have many practical benefits. For example, many human genes that play a role in human disease have been genetically mapped. This information can be used to diagnose and perhaps someday treat inherited human diseases. It can also help genetic counselors predict the likelihood that a couple will produce children with certain inherited diseases. In addition, genetic maps are gaining increasing importance in agriculture. A genetic map can provide plant and animal breeders with helpful information for improving agriculturally important strains through selective breeding programs.

In this section, we will examine traditional genetic mapping techniques that involve an analysis of crosses of individuals that are

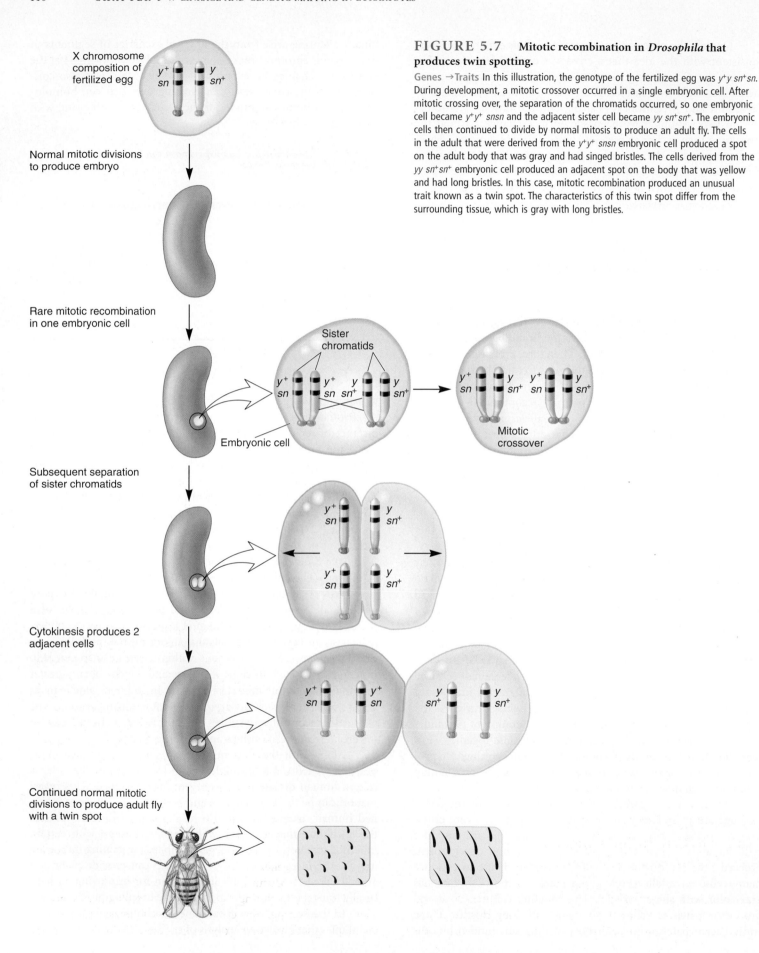

X chromosome composition of fertilized egg

Normal mitotic divisions to produce embryo

Rare mitotic recombination in one embryonic cell

Sister chromatids

Embryonic cell

Mitotic crossover

Subsequent separation of sister chromatids

Cytokinesis produces 2 adjacent cells

Continued normal mitotic divisions to produce adult fly with a twin spot

FIGURE 5.7 **Mitotic recombination in *Drosophila* that produces twin spotting.**

Genes → Traits In this illustration, the genotype of the fertilized egg was $y^+y\ sn^+sn$. During development, a mitotic crossover occurred in a single embryonic cell. After mitotic crossing over, the separation of the chromatids occurred, so one embryonic cell became $y^+y^+\ snsn$ and the adjacent sister cell became $yy\ sn^+sn^+$. The embryonic cells then continued to divide by normal mitosis to produce an adult fly. The cells in the adult that were derived from the $y^+y^+\ snsn$ embryonic cell produced a spot on the adult body that was gray and had singed bristles. The cells derived from the $yy\ sn^+sn^+$ embryonic cell produced an adjacent spot on the body that was yellow and had long bristles. In this case, mitotic recombination produced an unusual trait known as a twin spot. The characteristics of this twin spot differ from the surrounding tissue, which is gray with long bristles.

FIGURE 5.8   **A simplified genetic linkage map of *Drosophila melanogaster*.** This simplified map illustrates a few of the many thousands of genes that have been identified in this organism.

heterozygous for two or more genes. The frequency of nonparental offspring due to crossing over provides a way to deduce the linear order of genes along a chromosome. As depicted in Figure 5.8, this linear arrangement of genes is known as a **genetic linkage map.** This approach has been useful for analyzing organisms that are easily crossed and produce a large number of offspring in a short period of time. Genetic linkage maps have been constructed for several plant species and certain species of animals, such as *Drosophila*. For many organisms, however, traditional mapping approaches are difficult due to long generation times or the inability to carry out experimental crosses (as in humans). Fortunately, many alternative methods of gene mapping have been developed to replace the need to carry out crosses. As described in Chapter 20, molecular approaches are increasingly used to map genes.

The Frequency of Recombination Between Two Genes Can Be Correlated with Their Map Distance Along a Chromosome

Genetic mapping allows us to estimate the relative distances between linked genes based on the likelihood that a crossover will occur between them. If two genes are very close together on the same chromosome, a crossover is unlikely to begin in the

region between them. However, if two genes are very far apart, a crossover is more likely to be initiated in this region and thereby recombine the alleles of the two genes. Experimentally, the basis for genetic mapping is that the percentage of recombinant offspring is correlated with the distance between two genes. If two genes are far apart, many recombinant offspring will be produced. However, if two genes are close together, very few recombinant offspring will be observed.

To interpret a genetic mapping experiment, the experimenter must know if the characteristics of an offspring are due to crossing over during meiosis in a parent. This is accomplished by conducting a **testcross.** Most testcrosses are between an individual that is heterozygous for two or more genes and an individual that is recessive and homozygous for the same genes. The goal of the testcross is to determine if recombination has occurred during meiosis in the heterozygous parent. Thus, genetic mapping is based on the level of recombination that occurs in just one parent—the heterozygote. In a testcross, new combinations of alleles cannot occur in the gametes of the other parent, which is homozygous for these genes.

Figure 5.9 illustrates how a testcross provides an experimental strategy to distinguish between recombinant and nonrecombinant offspring. This cross concerns two linked genes affecting bristle length and body color in fruit flies. The recessive alleles

$s^+ s\, e^+e$ $s s\, e e$

s^+ s s s
e^+ e e e

Parent   x   Parent
♀   ♂

s^+ s s s s^+ s s s
e^+ e e e e e e^+ e

$s^+ s\, e^+e$ $s s\, e e$ $s^+ s\, e e$ $s s\, e^+e$

Long bristles	Short bristles	Long bristles	Short bristles
Gray body	Ebony body	Ebony body	Gray body
Nonrecombinant	**Nonrecombinant**	**Recombinant**	**Recombinant**

Total: 537   542   76   75

FIGURE 5.9   **Use of a testcross to distinguish between recombinant and nonrecombinant offspring.** The cross involves one *Drosophila* parent that is homozygous recessive for short bristles (*ss*) and ebony body (*ee*), and one parent heterozygous for both genes ($s^+s\ e^+e$). (Note: *Drosophila* geneticists normally designate the short allele as *ss* and a homozygous fly with short bristles as *ssss*. In the text, the allele causing short bristles is designated with a single *s* to avoid confusion between the allele designation and the genotype of the fly.)

INTERACTIVE EXERCISE

are *s* (short bristles) and *e* (ebony body), and the dominant (wild-type) alleles are s^+ (long bristles) and e^+ (gray body). One parent displays both recessive traits. Therefore, we know this parent is homozygous for the recessive alleles of the two genes (*ss ee*). The other parent is heterozygous for the linked genes affecting bristle length and body color. This parent was produced from a cross involving a true-breeding wild-type fly and a true-breeding fly with short bristles and an ebony body. Therefore, in this heterozygous parent, we know that the *s* and *e* alleles are located on one chromosome and the corresponding s^+ and e^+ alleles are located on the homologous chromosome.

Now let's take a look at the four possible types of offspring these parents can produce. The offspring's phenotypes are long

bristles, gray body; short bristles, ebony body; long bristles, ebony body; and short bristles, gray body. All four types of offspring have inherited a chromosome carrying the *s* and *e* alleles from their homozygous parent (shown on the right in each pair). Focus your attention on the other chromosome. The offspring with long bristles and gray bodies have inherited a chromosome carrying the s^+ and e^+ alleles from the heterozygous parent. This chromosome is not the product of a crossover. The offspring with short bristles and ebony bodies have also inherited a chromosome carrying the *s* and *e* alleles from the heterozygous parent. Again, this chromosome is not the product of a crossover.

The other two types of offspring, however, can be produced only if crossing over has occurred in the region between these two genes. Those with long bristles and ebony bodies or short bristles and gray bodies have inherited a chromosome that is the product of a crossover during meiosis in the heterozygous parent. As noted in Figure 5.9, the recombinant offspring are fewer in number than are the nonrecombinant offspring.

The frequency of recombination can be used as an estimate of the physical distance between two genes on the same chromosome. The **map distance** is defined as the number of recombinant offspring divided by the total number of offspring, multiplied by 100. We can calculate the map distance between these two genes using this formula:

$$\text{Map distance} = \frac{\text{Number of recombinant offspring}}{\text{Total number of offspring}} \times 100$$

$$= \frac{76 + 75}{537 + 542 + 76 + 75} \times 100$$

$$= 12.3 \text{ map units}$$

The units of distance are called **map units (mu)** or sometimes **centiMorgans (cM)** in honor of Thomas Hunt Morgan. One map unit is equivalent to a 1% frequency of recombination. In this example, we would conclude that the *s* and *e* alleles are 12.3 map units apart from each other along the same chromosome.

EXPERIMENT 5B

Alfred Sturtevant Used the Frequency of Crossing Over in Dihybrid Crosses to Produce the First Genetic Map

In 1911, the first individual to construct a (very small) genetic map was Alfred Sturtevant, an undergraduate who spent time in the laboratory of Thomas Hunt Morgan. Sturtevant wrote: "In conversation with Morgan . . . I suddenly realized that the variations in the strength of linkage, already attributed by Morgan to differences in the spatial separation of the genes, offered the possibility of determining sequences [of different genes] in the linear dimension of a chromosome. I went home and spent most of the night (to the neglect of my undergraduate homework) in producing the first chromosome map, which included the sex-linked genes, *y*, *w*, *v*, *m*, and *r*, in the order and approximately the relative spacing that they still appear on the standard maps."

In the experiment of **Figure 5.10**, Sturtevant considered the outcome of crosses involving six different mutant alleles that altered the phenotype of flies. All of these alleles were known to be recessive and X linked. They are *y* (yellow body color), *w* (white eye color), *w-e* (eosin eye color), *v* (vermilion eye color), *m* (miniature wings), and *r* (rudimentary wings). The *w* and *w-e* alleles are alleles of the same gene. In contrast, the *v* allele (vermilion eye color) is an allele of a different gene that also affects eye color. The two alleles that affect wing length, *m* and *r*, are also in different genes. Therefore, Sturtevant studied the inheritance of six recessive alleles, but since *w* and *w-e* are alleles of the same gene, his genetic map contained only five genes. The corresponding wild-type alleles are *y⁺* (gray body), *w⁺* (red eyes), *v⁺* (red eyes), *m⁺* (long wings), and *r⁺* (long wings).

■ THE HYPOTHESIS

When genes are located on the same chromosome, the distance between the genes can be estimated from the proportion of recombinant offspring. This provides a way to map the order of genes along a chromosome.

■ **TESTING THE HYPOTHESIS — FIGURE 5.10   The first genetic mapping experiment.**

Starting materials: Sturtevant began with several different strains of *Drosophila* that contained the six alleles already described.

Experimental level	Conceptual level

1. Cross a female that is heterozygous for two different genes to a male that is hemizygous recessive for the same two genes. In this example, cross a female that is $X^{y^+w^+}X^{yw}$ to a male that is $X^{yw}Y$.

This strategy was employed for many dihybrid combinations of the six alleles already described.

2. Observe the outcome of the crosses.

3. Calculate the percentages of offspring that are the result of crossing over (number of nonparental/total).

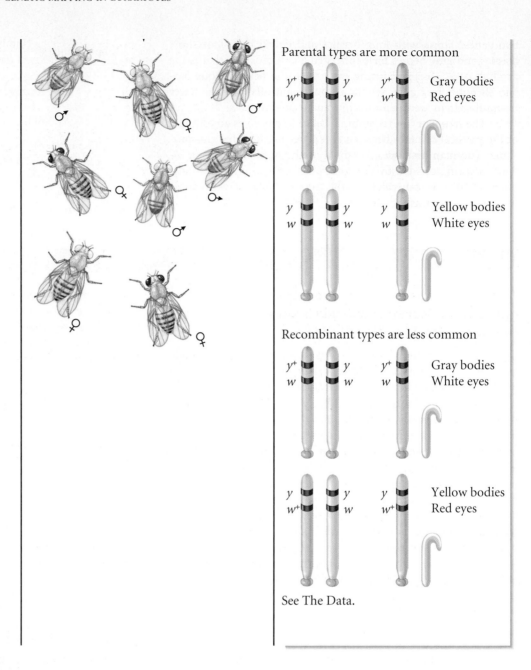

Parental types are more common

y^+ y y^+ Gray bodies
w^+ w w^+ Red eyes

y y y Yellow bodies
w w w White eyes

Recombinant types are less common

y^+ y y^+ Gray bodies
w w w White eyes

y y y Yellow bodies
w^+ w w^+ Red eyes

See The Data.

THE DATA

Alleles Concerned	Number Recombinant/Total Number	Percent Recombinant Offspring
y and w/w-e	214/21,736	1.0
y and v	1,464/4,551	32.2
y and m	115/324	35.5
y and r	260/693	37.5
w/w-e and v	471/1,584	29.7
w/w-e and m	2,062/6,116	33.7
w/w-e and r	406/898	45.2
v and m	17/573	3.0
v and r	109/405	26.9

INTERPRETING THE DATA

As shown in Figure 5.10, Sturtevant made pairwise testcrosses and then counted the number of offspring in the four phenotypic categories. Two of the categories were nonrecombinant and two were recombinant, requiring a crossover between the X chromosomes in the female heterozygote. Let's begin by contrasting the results between particular pairs of genes, shown in the data. In some dihybrid crosses, the percentage of nonparental offspring was rather low. For example, dihybrid crosses involving the y allele and the w or w-e allele yielded 1% recombinant offspring. This result suggested that these two genes are very close together. By comparison, other dihybrid crosses showed a higher percentage of nonparental offspring. For example, crosses involving the *v* and *r* alleles produced 26.9% recombinant offspring. These two genes are expected to be farther apart.

To construct his map, Sturtevant began with the assumption that the map distances would be more accurate between genes that are closely linked. Therefore, his map is based on the distance between *y* and *w* (1.0), *w* and *v* (29.7), *v* and *m* (3.0), and *v* and *r* (26.9). He also considered other features of the data to deduce the order of the genes. For example, the percentage of crossovers between *w* and *m* was 33.7. The percentage of crossovers between *w* and *v* was 29.7, suggesting that *v* is between *w* and *m*, but closer to *m*. The proximity of *v* and *m* is confirmed by the low percentage of crossovers between *v* and *m* (3.0). Sturtevant collectively considered the data and proposed the genetic map shown here.

In this genetic map, Sturtevant began at the *y* allele and mapped the genes from left to right. For example, the *y* and *v* alleles are 30.7 mu apart, and the *v* and *m* alleles are 3.0 mu apart. This study by Sturtevant was a major breakthrough, because it showed how to map the locations of genes along chromosomes by making the appropriate crosses.

If you look carefully at Sturtevant's data, you will notice a few observations that do not agree very well with his genetic map. For example, the percentage of recombinant offspring for the *y* and *r* dihybrid cross was 37.5 (but the map distance is 57.6), and the crossover percentage between *w* and *r* was 45.2 (but the map distance is 56.6). As the percentage of recombinant offspring approaches a value of 50%, this value becomes a progressively more inaccurate measure of actual map distance (**Figure 5.11**). What is the basis for this inaccuracy? When the distance between

two genes is large, the likelihood of multiple crossovers in the region between them causes the observed number of recombinant offspring to underestimate this distance. In addition, multiple crossovers set a quantitative limit on the relationship between map distance and the percentage of recombinant offspring. Even though two different genes can be on the same chromosome and more than 50 mu apart, a testcross is expected to yield a maximum of only 50% recombinant offspring. (This idea is also discussed in solved problem S5 at the end of this chapter.) When two different genes are more than 50 mu apart, they follow the law of independent assortment in a testcross.

Actual map distance along the chromosome
(computed from the analysis of many closely linked genes)

FIGURE 5.11 **Relationship between the percentage of recombinant offspring in a testcross and the actual map distance between genes.** The y-axis depicts the percentage of recombinant offspring that would be observed in a dihybrid testcross. The actual map distance, shown on the x-axis, is calculated by analyzing the percentages of recombinant offspring from a series of many dihybrid crosses involving closely linked genes. Even though two genes may be more than 50 map units apart, the percentage of recombinant offspring will not exceed 50%.

A self-help quiz involving this experiment can be found at the Online Learning Center.

Trihybrid Crosses Can Be Used to Determine the Order and Distance Between Linked Genes

Thus far, we have considered the construction of genetic maps using dihybrid testcrosses to compute map distance. The data from trihybrid crosses can yield additional information about map distance and gene order. In a trihybrid cross, the experimenter crosses two individuals that differ in three traits. The following experiment outlines a common strategy for using trihybrid crosses to map genes. In this experiment, the parental generation consists of fruit flies that differ in body color, eye color, and wing shape. We must begin with true-breeding lines so that we know which alleles are initially linked to each other on the same chromosome. In this example, all the dominant alleles are linked on the same chromosome.

Step 1. *Cross two true-breeding strains that differ with regard to three alleles.* In this example, we will cross a fly that has a black body (*bb*), purple eyes (*prpr*), and vestigial wings

(*vgvg*) to a homozygous wild-type fly with a gray body (*b⁺b⁺*), red eyes (*pr⁺pr⁺*), and long wings (*vg⁺vg⁺*):

Parental flies

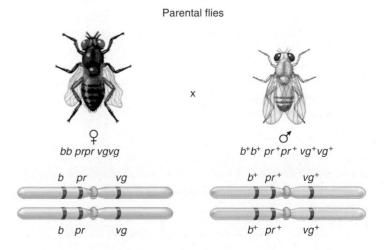

The goal in this step is to obtain F_1 individuals that are heterozygous for all three genes. In the F_1 heterozygotes, all dominant alleles are located on one chromosome, and all recessive alleles are on the other homologous chromosome.

Step 2. *Perform a testcross by mating F_1 female heterozygotes to male flies that are homozygous recessive for all three alleles* (bb prpr vgvg).

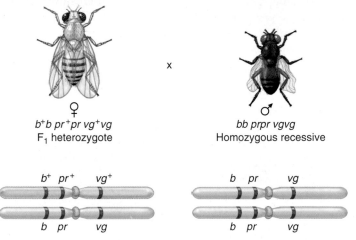

$b^+b\ pr^+pr\ vg^+vg$
F$_1$ heterozygote

x

$bb\ prpr\ vgvg$
Homozygous recessive

$b^+\ pr^+\ \ \ \ vg^+$
$b\ \ pr\ \ \ \ vg$

$b\ \ pr\ \ \ vg$
$b\ \ pr\ \ \ vg$

During gametogenesis in the heterozygous female F_1 flies, crossovers may produce new combinations of the three alleles.

Step 3. *Collect data for the F_2 generation.* As shown in **Table 5.1**, eight phenotypic combinations are possible. An analysis of the F_2 generation flies allows us to map these three genes. Because the three genes exist as two alleles each, we have $2^3 = 8$ possible combinations of offspring. If these alleles assorted independently, all eight combinations would occur in equal proportions. However, we see that the proportions of the eight phenotypes are far from equal.

The genotypes of the parental generation correspond to the phenotypes gray body, red eyes, and long wings, and black body, purple eyes, and vestigial wings. In crosses involving linked genes, the parental phenotypes occur most frequently in the offspring. The remaining six phenotypes are due to crossing over.

The double crossover is always expected to be the least frequent category of offspring. Two of the phenotypes—gray body, purple eyes, and long wings, and black body, red eyes, and vestigial wings—arose from a double crossover between two pairs of genes. Also, the combination of traits in the double crossover tells us which gene is in the middle. When a chromatid undergoes a double crossover, the gene in the middle becomes separated from the other two genes at either end.

$b^+\ pr^+\ \ \ vg^+$

$b\ \ pr\ \ \ vg$

→

$b^+\ pr\ \ \ vg^+$

$b\ \ pr^+\ \ vg$

TABLE 5.1

Data from a Trihybrid Cross (see step 2)

Phenotype	Number of Observed Offspring (males and females)	Chromosome Inherited from F$_1$ Female
Gray body, red eyes, long wings	411	$b^+\ \ pr^+\ \ \ vg^+$
Gray body, red eyes, vestigial wings	61	$b^+\ \ pr^+\ \ \ vg$
Gray body, purple eyes, long wings	2	$b^+\ \ pr\ \ \ vg^+$
Gray body, purple eyes, vestigial wings	30	$b^+\ \ pr\ \ \ vg$
Black body, red eyes, long wings	28	$b\ \ pr^+\ \ \ vg^+$
Black body, red eyes, vestigial wings	1	$b\ \ pr^+\ \ \ vg$
Black body, purple eyes, long wings	60	$b\ \ pr\ \ \ vg^+$
Black body, purple eyes, vestigial wings	412	$b\ \ pr\ \ \ vg$

In the double crossover categories, the recessive purple-eye allele is separated from the other two recessive alleles. When mated to a homozygous recessive fly in the testcross, this yields flies with gray bodies, purple eyes, and long wings, or with black bodies, red eyes, and vestigial wings. This observation indicates that the gene for eye color lies between the genes for body color and wing shape.

Step 4. *Calculate the map distance between pairs of genes.* To do this, we need to understand which gene combinations are recombinant and which are nonrecombinant. The recombinant offspring are due to crossing over in the heterozygous female parent. If you look back at step 2, you can see the arrangement of alleles in the heterozygous female parent in the absence of crossing over. Let's consider this arrangement with regard to gene pairs:

b^+ is linked to pr^+, and b is linked to pr
pr^+ is linked to vg^+, and pr is linked to vg
b^+ is linked to vg^+, and b is linked to vg

With regard to body color and eye color, the recombinant offspring have gray bodies and purple eyes $(2 + 30)$ or black bodies and red eyes $(28 + 1)$. As shown along the right side of Table 5.1, these offspring were produced by crossovers in the female parents. The total number of these recombinant offspring is 61. The map distance between the body color and eye color genes is thus

$$\text{Map distance} = \frac{61}{944 + 61} \times 100 = 6.1\ \text{mu}$$

With regard to eye color and wing shape, the recombinant offspring have red eyes and vestigial wings (61 + 1) or purple eyes and long wings (2 + 60). The total number is 124. The map distance between the eye color and wing shape genes is

$$\text{Map distance} = \frac{124}{881 + 124} \times 100 = 12.3 \text{ mu}$$

With regard to body color and wing shape, the recombinant offspring have gray bodies and vestigial wings (61 + 30) or black bodies and long wings (28 + 60). The total number is 179. The map distance between the body color and wing shape genes is

$$\text{Map distance} = \frac{179}{826 + 179} \times 100 = 17.8 \text{ mu}$$

Step 5. *Construct the map.* Based on the map unit calculation, the body color (*b*) and wing shape (*vg*) genes are farthest apart. The eye color gene (*pr*) must lie in the middle. As mentioned earlier, this order of genes is also confirmed by the pattern of traits found in the double crossovers. To construct the map, we use the distances between the genes that are closest together.

In our example, we have placed the body color gene first and the wing shape gene last. The data also are consistent with a map in which the wing shape gene comes first and the body color gene comes last. In detailed genetic maps, the locations of genes are mapped relative to the centromere.

You may have noticed that our calculations underestimate the distance between the body color and wing shape genes. We obtained a value of 17.8 map units even though the distance seems to be 18.4 map units when we add together the distance between body color and eye color genes (6.1 mu) and the distance between eye color and wing shape genes (12.3 mu). What accounts for this discrepancy? The answer is double crossovers. If you look at the data in Table 5.1, the offspring with gray bodies, purple eyes, and long wings or with black bodies, red eyes, and vestigial wings are due to a double crossover. From a phenotypic perspective, these offspring are not recombinant with regard to the body color and wing shape alleles. Even so, we know that they arose from a double crossover between these two genes. Therefore, we should consider these crossovers when calculating the distance between the body color and wing shape genes. In this case, three offspring (2 + 1) were due to double crossovers. Because they are double crossovers, we multiply two times the number of double crossovers (2 + 1) and add this number to our previous value of recombinant offspring:

$$\text{Map distance} = \frac{179 + 2(2 + 1)}{826 + 179} \times 100 = 18.4 \text{ mu}$$

Interference Can Influence the Number of Double Crossovers That Occur in a Short Region

In Chapter 2, we considered the product rule to determine the probability that two independent events will both occur. The product rule allows us to predict the expected likelihood of a double crossover provided we know the individual probabilities of each single crossover. Let's reconsider the data of the trihybrid testcross just described to see if the frequency of double crossovers is what we would expect based on the product rule. If each crossover is an independent event, we can multiply the likelihood of a single crossover between *b* and *pr* (0.061) times the likelihood of a single crossover between *pr* and *vg* (0.123). The product rule predicts

Expected likelihood of a double crossover = 0.061 × 0.123 = 0.0075 = 0.75%

Based on a total of 1,005 offspring produced:

Expected number of offspring due to a double crossover = 1,005 × 0.0075 = 7.5

In other words, we would expect about seven or eight offspring to be produced as a result of a double crossover. The observed number of offspring was only three (namely, two with gray bodies, purple eyes, and long wings, and one with a black body, red eyes, and vestigial wings). What accounts for the lower number? This lower-than-expected value is probably not due to random sampling error. Instead, the likely cause is a common genetic phenomenon known as **positive interference,** in which the occurrence of a crossover in one region of a chromosome decreases the probability that a second crossover will occur nearby. In other words, the first crossover interferes with the ability to form a second crossover in the immediate vicinity. To provide interference with a quantitative value, we first calculate the coefficient of coincidence (*C*), which is the ratio of the observed number of double crossovers to the expected number.

$$C = \frac{\text{Observed number of double crossovers}}{\text{Expected number of double crossovers}}$$

Interference (*I*) is expressed as

$$I = 1 - C$$

For the data of the trihybrid testcross, the observed number of crossovers is 3 and the expected number is 7.5, so the coefficient of coincidence equals 3/7.5 = 0.40. In other words, only 40% of the expected number of double crossovers were actually observed. The value for interference equals 1 − 0.4 = 0.60, or 60%. This means that 60% of the expected number of crossovers did not occur. Because *I* has a positive value, this is called positive interference. Rarely, the outcome of a testcross yields a negative value for interference. A negative interference value suggests that a first crossover enhanced the rate of a second crossover in a nearby region. Although the molecular mechanisms that cause interference are not entirely understood, in most organisms the number of crossovers is regulated so that very few occur per chromosome. The reasons for positive and negative interference will require further research.

5.3 GENETIC MAPPING IN HAPLOID EUKARYOTES

Before ending our discussion of genetic mapping, let's consider some pioneering studies that involved the genetic mapping of haploid organisms. You may find it surprising that certain species of simple eukaryotes, particularly unicellular algae and fungi, which spend part of their life cycle in the haploid state, have also been used in genetic mapping studies. The sac fungi, called ascomycetes, have been particularly useful to geneticists because of their unique style of sexual reproduction. In fact, much of our earliest understanding of genetic recombination came from the genetic analyses of fungi.

Fungi may be unicellular or multicellular organisms. Fungal cells are typically haploid ($1n$) and can reproduce asexually. In addition, fungi can also reproduce sexually by the fusion of two haploid cells to create a diploid zygote ($2n$) (**Figure 5.12**). The diploid zygote can then proceed through meiosis to produce four haploid cells, which are called **spores.** This group of four spores is known as a **tetrad** (not to be confused with a tetrad of four sister chromatids). In some species, meiosis is followed by a mitotic division to produce eight spores, known as an **octad.** In *Ascomycete* fungi, the cells of a tetrad or octad are contained within a sac known as an **ascus** (plural: *asci*). In other words, the products of a single meiotic division are contained within one sac. This key feature is useful to geneticists, and it dramatically differs from sexual reproduction in animals and plants. For example, in animals, oogenesis produces a single functional egg, and spermatogenesis occurs in the testes, where the resulting sperm become mixed with millions of other sperm.

Using a microscope, researchers can dissect asci and study the traits of each haploid spore. In this way, these organisms offer a unique opportunity for geneticists to identify and study all of the cells that are derived from a single meiotic division. In this section, we will consider how the analysis of asci can be used to map genes in fungi.

Ordered Tetrad Analysis Can Be Used to Map the Distance Between a Gene and the Centromere

The arrangement of spores within an ascus varies from species to species (**Figure 5.13a**). In some cases, the ascus provides enough space for the tetrads or octads of spores to randomly mix together. This creates an **unordered tetrad** or **octad.** These occur in fungal species such as *Saccharomyces cerevisiae* and *Aspergillus nidulans* and also in certain unicellular algae (*Chlamydomonas reinhardtii*). By comparison, other species of fungi produce a very tight ascus that prevents spores from randomly moving around, which results in an **ordered tetrad** or **octad. Figure 5.13b** illustrates how an ordered octad is formed in *Neurospora crassa.* In this example, spores that carry the *A* allele have orange pigmentation, while spores having the *a* (albino) allele are white.

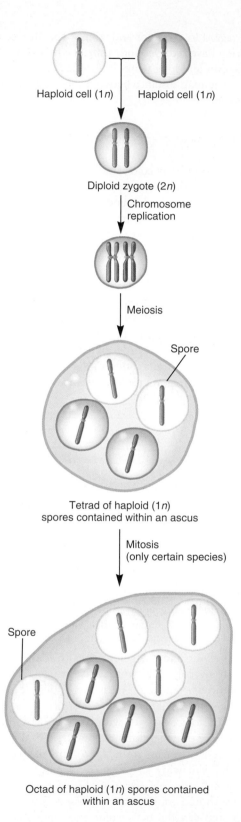

FIGURE 5.12 **Sexual reproduction in ascomycetes.** For simplicity, this diagram shows each haploid cell as having only one chromosome per haploid set. However, fungal species actually contain several chromosomes per haploid set.

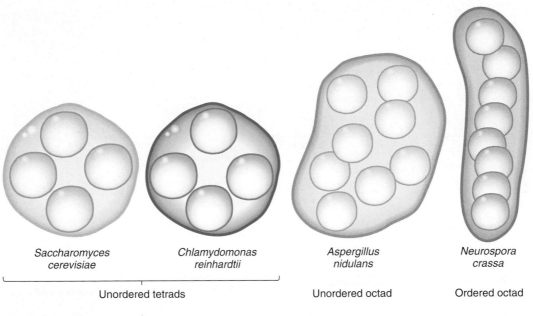

(a) Different arrangements of spores

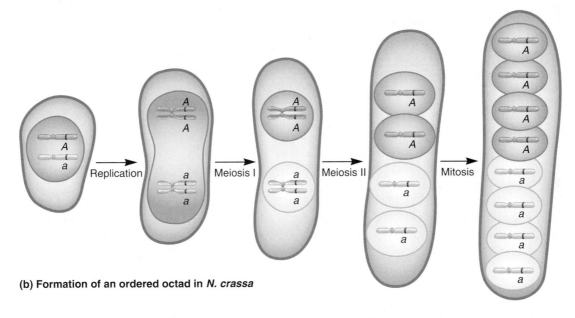

(b) Formation of an ordered octad in *N. crassa*

FIGURE 5.13   **Arrangement of spores within asci of different species. (a)** *Saccharomyces cerevisiae* and *Chlamydomonas reinhardtii* (an alga) produce unordered tetrads, *Aspergillus nidulans* produces an unordered octad, and *Neurospora crassa* produces an ordered octad. **(b)** Ordered octads are produced in *N. crassa* by meiosis and mitosis in such a way that the eight resulting cells are arranged linearly.

A key feature of ordered tetrads or octads is that the position and order of spores within the ascus reflects their relationship to each other as they were produced by meiosis and mitosis. This idea is schematically shown in Figure 5.13b. After the original diploid cell has undergone chromosome replication, the first meiotic division produces two cells that are arranged next to each other within the sac. The second meiotic division then produces four cells that are also arranged in a row. Due to the tight enclosure of the sac around the cells, each pair of daughter cells is forced to lie next to each other in a linear fashion. Likewise, when these four cells divide by mitosis, each pair of daughter cells is located next to each other.

In species that make ordered tetrads or octads, experimenters can determine the genotypes of the spores within the asci and map the distance between a single gene and the centromere. Because the location of the centromere can be seen under the microscope, the mapping of a gene relative to the centromere provides a way to correlate a gene's location with the cytological characteristics of a chromosome. This approach has been extensively exploited in *N. crassa*.

Figure 5.14 compares the arrangement of cells within a *Neurospora* ascus depending on whether or not a crossover has occurred between two homologues that differ at a gene with

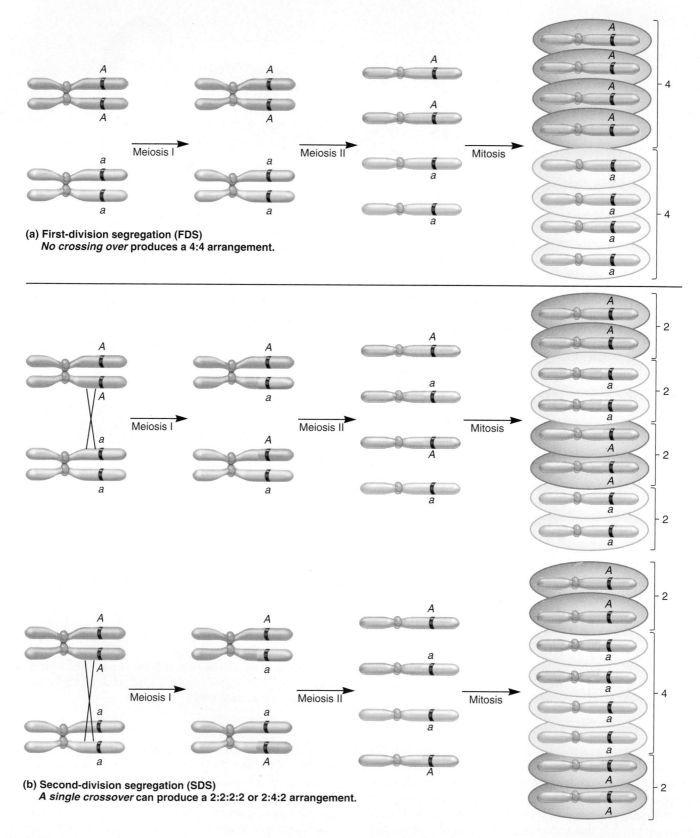

(a) First-division segregation (FDS)
No crossing over produces a 4:4 arrangement.

(b) Second-division segregation (SDS)
A single crossover can produce a 2:2:2:2 or 2:4:2 arrangement.

FIGURE 5.14   **A comparison of the arrangement of cells within an ordered octad, depending on whether or not crossing over has occurred.** (a) If no crossing over has occurred, the octad will have a 4:4 arrangement of spores known as an FDS or M1 pattern. (b) If a crossover has occurred between the centromere and the gene of interest, a 2:2:2:2 or 2:4:2 pattern, known as an SDS or M2 pattern, is observed.

alleles *A* (orange pigmentation) and *a* (albino, which results in a white phenotype). In Figure 5.14a, a crossover has not occurred, so the octad contains a linear arrangement of four haploid cells carrying the *A* allele, which are adjacent to four haploid cells that contain the *a* allele. This 4:4 arrangement of spores within the ascus is called **first-division segregation** (**FDS**) or an M1 pattern. It is called a first-division segregation pattern because the *A* and *a* alleles have segregated from each other after the first meiotic division.

In contrast, as shown in Figure 5.14b, if a crossover occurs between the centromere and the gene of interest, the ordered octad will deviate from the 4:4 pattern. Depending on the relative locations of the two chromatids that participated in the crossover, the ascus will contain a 2:2:2:2 or 2:4:2 pattern. These patterns are called **second-division segregation** (**SDS**) or M2 patterns. In this case, the *A* and *a* alleles do not segregate until the second meiotic division is completed.

Because a pattern of second-division segregation is a result of crossing over, the percentage of SDS asci can be used to calculate the map distance between the centromere and the gene of interest. To understand why this is possible, let's consider the relationship between a crossover site and the centromere. As shown in **Figure 5.15**, a crossover will separate a gene from its original centromere only if it begins in the region between the centromere and that gene. Therefore, the chances of getting a 2:2:2:2 or 2:4:2 pattern depend on the distance between the gene of interest and the centromere.

To determine the map distance between the centromere and a gene, the experimenter must count the number of SDS asci and the total number of asci. In SDS asci, only half of the spores are actually the product of a crossover. Therefore, the map distance is calculated as

$$\text{Map distance} = \frac{(1/2)\,(\text{Number of SDS asci})}{\text{Total number of asci}} \times 100$$

Unordered Tetrad Analysis Can Be Used to Map Genes in Dihybrid Crosses

Unordered tetrads contain a group of spores that are the product of meiosis and randomly arranged in an ascus. An experimenter can conduct a dihybrid cross, remove the spores from each ascus, and determine the phenotypes of the spores. This analysis can determine if two genes are linked or assort independently. If two genes are linked, a tetrad analysis can also be used to compute map distance.

Figure 5.16 illustrates the possible outcomes starting with two haploid yeast strains. One strain carries the wild-type alleles *ura*+ and *arg*+, which are required for uracil and arginine biosynthesis, respectively. The other strain has defective alleles *ura-2* and *arg-3*; these result in yeast strains that require uracil and arginine in the growth medium. A diploid zygote with the genotype *ura*+*ura-2 arg*+ *arg-3* was produced from the fusion of haploid cells from these two strains. The diploid cell then proceeds through meiosis to produce four haploid cells. After the completion of meiosis, three distinct types of tetrads could be produced. One possibility is that the tetrad will contain four spores with the parental combinations of alleles. This ascus is said to have the **parental ditype** (**PD**). Alternatively, an ascus may have two parental cells and two nonparental cells, which is called a **tetratype** (**T**). Finally, an ascus with a **nonparental ditype** (**NPD**) contains four cells with nonparental genotypes.

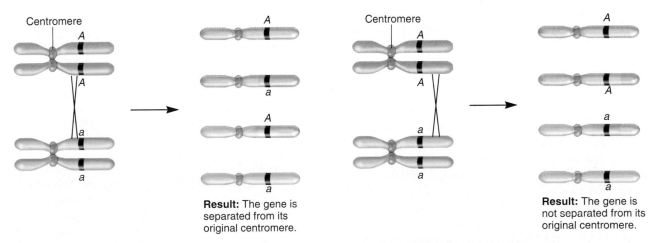

Result: The gene is separated from its original centromere.

(a) Crossover begins between centromere and gene of interest.

Result: The gene is not separated from its original centromere.

(b) Crossover does not begin between centromere and gene of interest.

FIGURE 5.15   **The relationship between a crossover site and the separation of an allele from its original centromere.** (**a**) If a crossover initially forms between the centromere and the gene of interest, the gene will be separated from its original centromere. (**b**) If a crossover initiates outside this region, the gene remains attached to its original centromere.

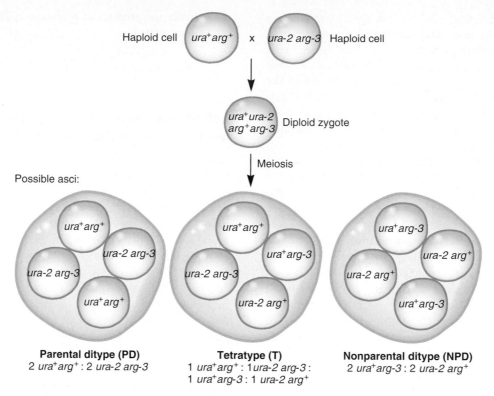

Haploid cell $ura^+ arg^+$ x ura-2 arg-3 Haploid cell

$ura^+ ura$-2
$arg^+ arg$-3 Diploid zygote

Meiosis

Possible asci:

$ura^+ arg^+$
ura-2 arg-3
ura-2 arg-3
$ura^+ arg^+$

$ura^+ arg^+$
ura-2 arg-3
$ura^+ arg$-3
ura-2 arg^+

$ura^+ arg$-3
ura-2 arg^+
ura-2 arg^+
$ura^+ arg$-3

Parental ditype (PD)
2 $ura^+ arg^+$: 2 ura-2 arg-3

Tetratype (T)
1 $ura^+ arg^+$: 1 ura-2 arg-3 :
1 $ura^+ arg$-3 : 1 ura-2 arg^+

Nonparental ditype (NPD)
2 $ura^+ arg$-3 : 2 ura-2 arg^+

INTERACTIVE EXERCISE

FIGURE 5.16   **The assortment of two genes in an unordered tetrad.** If the tetrad contains 100% parental cells, this ascus has the parental ditype (PD). If it contains 50% parental and 50% recombinant cells, it is a tetratype (T). Finally, an ascus with 100% recombinant cells is called a nonparental ditype (NPD). This figure does not illustrate the chromosomal locations of the alleles. In this type of experiment, the goal is to determine whether the two genes are linked on the same chromosome.

When two genes assort independently, the number of asci having a parental ditype is expected to equal the number having a nonparental ditype, thus yielding 50% recombinant spores. For linked genes, **Figure 5.17** illustrates the relationship between crossing over and the type of ascus that will result. If no crossing over occurs in the region between the two genes, the parental ditype will be created (Figure 5.17a). A single crossover event will produce a tetratype (Figure 5.17b). Double crossovers can yield a parental ditype, tetratype, or nonparental ditype, depending on the combination of chromatids that are involved (Figure 5.17c). A nonparental ditype is produced when a double crossover involves all four chromatids. A tetratype will result from a three-chromatid crossover. Finally, a double crossover between the same two chromatids will produce the parental ditype.

The data from a tetrad analysis can be used to calculate the map distance between two linked genes. As in conventional mapping, the map distance is calculated as the percentage of offspring that carry recombinant chromosomes. As mentioned, a tetratype contains 50% recombinant chromosomes; a nonparental ditype, 100%. Therefore, the map distance is computed as

$$\text{Map distance} = \frac{\text{NPD} + (1/2)(\text{T})}{\text{total number of asci}} \times 100$$

Over short map distances, this calculation provides a fairly reliable measure of distance. However, it does not adequately

account for double crossovers. When two genes are far apart on the same chromosome, the calculated map distance using this equation underestimates the actual map distance due to double crossovers. Fortunately, a particular strength of tetrad analysis is that we can derive another equation that accounts for double crossovers and thereby provides a more accurate value for map distance. To begin this derivation, let's consider a more precise way to calculate map distance.

$$\text{Map distance} = \frac{\text{Single crossover tetrads} + (2)(\text{Double crossover tetrads})}{\text{Total number of asci}} \times 0.5 \times 100$$

This equation includes the number of single and double crossovers in the computation of map distance. The total number of crossovers equals the number of single crossovers plus two times the number of double crossovers. Overall, the tetrads that contain single and double crossovers also contain 50% nonrecombinant chromosomes. To calculate map distance, therefore, we divide the total number of crossovers by the total number of asci and multiply by 0.5 and 100.

To be useful, we need to relate this equation to the number of parental ditypes, nonparental ditypes, and tetratypes that are obtained by experimentation. To derive this relationship, we must consider the types of tetrads that are produced from no crossing over, a single crossover, and double crossovers. To do so, let's take

FIGURE 5.17   Relationship between crossing over and the production of the parental ditype, tetratype, and nonparental ditype for two linked genes. In the case of double crossovers, this figure shows the outcome in which the crossover on the left occurs first.

another look at Figure 5.17. As shown there, the parental ditype and tetratype are ambiguous. The parental ditype can be derived from no crossovers or a double crossover; the tetratype can be derived from a single crossover or a double crossover. However, the nonparental ditype is unambiguous, because it can be produced only from a double crossover. We can use this observation

as a way to determine the actual number of single and double crossovers. As seen in Figure 5.17, 1/4 of all the double crossovers are nonparental ditypes. Therefore, the total number of double crossovers equals four times the number of nonparental ditypes.

Next, we need to know the number of single crossovers. A single crossover will yield a tetratype, but double crossovers can also yield a tetratype. Therefore, the total number of tetratypes overestimates the true number of single crossovers. Fortunately, we can compensate for this overestimation. Because two types of tetratypes are due to a double crossover, the actual number of tetratypes arising from a double crossover should equal 2NPD. Therefore, the true number of single crossovers is calculated as T−2NPD.

Now we have accurate measures of both single and double crossovers. The number of single crossovers equals T−2NPD, and the number of double crossovers equals 4NPD. We can substitute these values into our previous equation.

$$\text{Map distance} = \frac{(T - 2NPD) + (2)(4NPD)}{\text{Total number of asci}} \times 0.5 \times 100$$

$$= \frac{T + 6NPD}{\text{Total number of asci}} \times 0.5 \times 100$$

This equation provides a more accurate measure of map distance because it considers both single and double crossovers.

CONCEPTUAL SUMMARY

Linkage refers to the phenomenon that many different genes may be located on the same chromosome. Chromosomes are sometimes called linkage groups because they contain a group of linked genes. Linkage affects the pattern of inheritance, because closely linked genes do not assort independently during meiosis. This produces a greater percentage of offspring that display parental phenotypes. Nevertheless, nonparental offspring can be produced as a result of crossing over.

The likelihood of crossing over depends on the distance between two genes. If two genes are far apart from each other on the same chromosome, it is more likely that crossing over will occur between them. Therefore, when two genes are widely separated, a substantial percentage of recombinant offspring will be obtained from a testcross. However, the percentage of recombinant offspring cannot exceed a value of 50%, even when two genes are more than 50 map units apart on the same chromosome. The relationship between the percentage of recombinant offspring and the linear distance between genes is the basis for genetic mapping.

EXPERIMENTAL SUMMARY

Experimentally, the phenomenon of linkage was deduced from genetic crosses. Bateson and Punnett were the first scientists to notice that certain genes do not assort independently. Morgan conducted crosses involving X-linked traits in fruit flies and correctly proposed that linkage is due to the location of particular genes on the same chromosome. He also hypothesized that recombinant phenotypes occur because of crossing over during meiosis. Morgan realized that the likelihood of crossing over depends on the distance between two genes. The hypothesis that crossing over can result in genetic recombination was confirmed cytologically by the studies of Creighton and McClintock, which showed that the production of recombinant offspring correlates with the physical exchange of material between chromosomes.

Genetic mapping is the determination of gene order and distance along chromosomes. In this chapter, we have considered how testcrosses are conducted as a method to map genes. Sturtevant was the first person to understand that the percentage of recombinant offspring in a testcross could be used as a measure of the relative distance between two genes. Map distance is computed as the number of recombinant offspring divided by the total number of offspring multiplied by 100. This approach can be readily applied to map genes using dihybrid and trihybrid testcrosses. Genetic mapping is most accurate when map distances are calculated between closely linked genes. As the map distance approaches 50 map units (mu) and above, the percentage of observed recombinant offspring is not a reliable measure of map distance. In Chapter 20, several molecular methods of genetic mapping are described.

The chapter ended with a discussion of gene mapping methods in fungi. A group of fungi known as the ascomycetes have been extensively used in genetic studies, because all products of a single meiosis are enclosed within an ascus. For fungi such as *Neurospora* that make an ordered octad, the spores are arranged in a manner that reflects their relationship to each other during meiosis and mitosis. Ordered octads can be analyzed to map the location of a single gene relative to the centromere. Fungal species and unicellular algae that produce unordered asci have also been used in mapping studies. In yeast, for example, dihybrid crosses are made, and the distance between the two genes can be computed by determining the proportions of parental ditypes, tetratypes, and nonparental ditypes.

In Chapter 6, we will consider the linkage of genes within bacterial chromosomes and bacteriophages. Although bacteria normally reproduce asexually, they still can transfer genetic material by various mechanisms. As we will see, these mechanisms also provide a way to map genes along the bacterial chromosome.

PROBLEM SETS & INSIGHTS

Solved Problems

S1. In the garden pea, orange pods (*orp*) are recessive to green pods (*Orp*), and sensitivity to pea mosaic virus (*mo*) is recessive to resistance to the virus (*Mo*). A plant with orange pods and sensitivity to the virus was crossed to a true-breeding plant with green pods and resistance to the virus. The F_1 plants were then testcrossed to plants with orange pods and sensitivity to the virus. The following results were obtained:

> 160 orange pods, virus sensitive
>
> 165 green pods, virus resistant
>
> 36 orange pods, virus resistant
>
> 39 green pods, virus sensitive
> _____
> 400 total

A. Conduct a chi square analysis to see if these genes are linked.

B. If they are linked, calculate the map distance between the two genes.

Answer:

A. Chi square analysis.

1. Our hypothesis is that the genes are not linked.

2. Calculate the predicted number of offspring based on the hypothesis. The testcross is

The predicted outcome of this cross under our hypothesis is a 1:1:1:1 ratio of plants with the four possible phenotypes. In other words, 1/4 should have the phenotype orange pods, virus sensitive; 1/4 should have green pods, virus resistant; 1/4 should have orange pods, virus resistant; and 1/4 should have green pods, virus sensitive. Because a total of 400 offspring were produced, our hypothesis predicts 100 offspring in each category.

3. Calculate the chi square.

$$\chi^2 = \frac{(O_1 - E_1)^2}{E_1} + \frac{(O_2 - E_2)^2}{E_2} + \frac{(O_3 - E_3)^2}{E_3} + \frac{(O_4 - E_4)^2}{E_4}$$

$$\chi^2 = \frac{(160 - 100)^2}{100} + \frac{(165 - 100)^2}{100} + \frac{(36 - 100)^2}{100} + \frac{(39 - 100)^2}{100}$$

$$\chi^2 = 36 + 42.3 + 41 + 37.2 = 156.5$$

4. Interpret the chi square value. The calculated chi square value is quite large. This indicates that the deviation between observed and expected values is very high. For one degree of freedom in Table 2.1, such a large deviation is expected to

occur by chance alone less than 1% of the time. Therefore, we reject the hypothesis that the genes assort independently. As an alternative, we may infer that the two genes are linked.

B. Calculate the map distance.

$$\text{Map distance} = \frac{\text{Number of nonparental offspring}}{\text{Total number of offspring}} \times 100$$

$$= \frac{36 + 39}{36 + 39 + 160 + 165} \times 100$$

$$= 18.8 \text{ mu}$$

The genes are approximately 18.8 mu apart.

S2. Two recessive disorders in mice—droopy ears and flaky tail—are caused by genes that are located 6 mu apart on chromosome 3. A true-breeding mouse with normal ears (*De*) and a flaky tail (*ft*) was crossed to a true-breeding mouse with droopy ears (*de*) and a normal tail (*Ft*). The F_1 offspring were then crossed to mice with droopy ears and flaky tails. If this testcross produced 100 offspring, what is the expected outcome?

Answer: The testcross is

The parental offspring are

Dede ftft   Normal ears, flaky tail

dede Ftft   Droopy ears, normal tail

The recombinant offspring are

dede ftft   Droopy ears, flaky tail

Dede Ftft   Normal ears, normal tail

Because the two genes are located 6 mu apart on the same chromosome, 6% of the offspring will be recombinants. Therefore, the expected outcome for 100 offspring is

> 3 droopy ears, flaky tail
>
> 3 normal ears, normal tail
>
> 47 normal ears, flaky tail
>
> 47 droopy ears, normal tail

S3. The following X-linked recessive traits are found in fruit flies: vermilion eyes are recessive to red eyes, miniature wings are recessive to long wings, and sable body is recessive to gray body. A cross was made between wild-type males with red eyes, long wings, and gray bodies to females with vermilion eyes, miniature wings, and sable bodies. The heterozygous females from this cross, which had red eyes, long wings, and gray bodies, were then crossed to males with vermilion eyes, miniature wings, and sable bodies. The following outcome was obtained:

Males and Females

1,320 vermilion eyes, miniature wings, sable body

1,346 red eyes, long wings, gray body

102 vermilion eyes, miniature wings, gray body

90 red eyes, long wings, sable body

42 vermilion eyes, long wings, gray body

48 red eyes, miniature wings, sable body

2 vermilion eyes, long wings, sable body

1 red eyes, miniature wings, gray body

A. Calculate the map distance between the three genes.

B. Is positive interference occurring?

Answer:

A. The first step is to determine the order of the three genes. We can do this by evaluating the pattern of inheritance in the double crossovers. The double crossover group occurs with the lowest frequency. Thus, the double crossovers are vermilion eyes, long wings, and sable body, and red eyes, miniature wings, and gray body. Compared to the parental combinations of alleles (vermilion eyes, miniature wings, sable body and red eyes, long wings, gray body), the gene for wing length has been reassorted. Two flies have long wings associated with vermilion eyes and sable body, and one fly has miniature wings associated with red eyes and gray body. Taken together, these results indicate that the wing length gene is found in between the eye color and body color genes.

Eye color——wing length——body color

We now calculate the distance between eye color and wing length, and between wing length and body color. To do this, we consider the data according to gene pairs:

vermilion eyes, miniature wings = 1,320 + 102 = 1,422

red eyes, long wings = 1,346 + 90 = 1,436

vermilion eyes, long wings = 42 + 2 = 44

red eyes, miniature wings = 48 + 1 = 49

The recombinants are vermilion eyes, long wings and red eyes, miniature wings. The map distance between these two genes is

$(44 + 49)/(1,422 + 1,436 + 44 + 49) \times 100 = 3.2$ mu

Likewise, the other gene pair is wing length and body color.

miniature wings, sable body = 1,320 + 48 = 1,368

long wings, gray body = 1,346 + 42 = 1,388

miniature wings, gray body = 102 + 1 = 103

long wings, sable body = 90 + 2 = 92

The recombinants are miniature wings, gray body and long wings, sable body. The map distance between these two genes is

$(103 + 92)/(1,368 + 1,388 + 103 + 92) \times 100 = 6.6$ mu

With these data, we can produce the following genetic map:

B. To calculate the interference value, we must first calculate the coefficient of coincidence.

$$C = \frac{\text{Observed number of double crossovers}}{\text{Expected number of double crossovers}}$$

Based on our calculation of map distances in part A, the percentage of single crossovers equals 3.2% (0.032) and 6.6% (0.066). The expected number of double crossovers equals 0.032×0.066, which is 0.002, or 0.2%. A total of 2,951 offspring were produced. If we multiply 2,951 × 0.002, we get 5.9, which is the expected number of double crossovers. The observed number was 3. Therefore,

$C = 3/5.9 = 0.51$

$I = 1 - C = 1 - 0.51 = 0.49$

In other words, approximately 49% of the expected double crossovers did not occur due to interference.

S4. Around the same time as the study of Creighton and McClintock, described in Figure 5.6, Curt Stern conducted similar experiments with *Drosophila*. He had strains of flies with microscopically detectable abnormalities in the X chromosome. In one case, the X chromosome was shorter than normal due to a deletion at one end. In another case, the X chromosome was longer than normal because an extra piece of the Y chromosome was attached at the other end of the X chromosome, where the centromere is located. He had female flies that had both abnormal chromosomes. On the short X chromosome, a recessive allele (*car*) was located that results in carnation-colored eyes, and a dominant allele (*B*) that causes bar-shaped eyes was also found on this chromosome. On the long X chromosome were located the wild-type alleles for these two genes (designated *car+* and *B+*), which confer red eyes and round eyes, respectively. Stern realized that a crossover between the two X chromosomes in such female flies would result in recombinant chromosomes that would be cytologically distinguishable from the parental chromosomes. If a crossover occurred between the *B* and *car* genes on the X chromosome, this is expected to produce a normal-sized X chromosome and an abnormal chromosome with a deletion at one end and an extra piece of the Y chromosome at the other end.

Stern crossed these female flies to male flies that had a normal-length X chromosome with the *car* allele and the allele for round eyes (*car B+*). Using a microscope, he could discriminate between the morphologies of parental chromosomes—like those contained within the original parental flies—and recombinant chromosomes that may be found in the offspring. What would be the predicted phenotypes and chromosome characteristics in the offspring if crossing over did or did not occur between the X chromosomes in the female flies of this cross?

Answer: To demonstrate that genetic recombination is due to crossing over, Stern needed to correlate recombinant phenotypes (due to genetic recombination) with the inheritance of recombinant chromosomes (due to crossing over). Because he knew the arrangement of alleles in the female flies, he could predict the phenotypes of parental and nonparental offspring. The male flies could contribute the *car* and *B*⁺ alleles (on a cytologically normal X chromosome) or contribute a Y chromosome. In the absence of crossing over, the female flies could contribute a short X chromosome with the *car* and *B* alleles or a long X chromosome with the *car*⁺ and *B*⁺ alleles. If crossing over occurred in the region between these two genes, the female flies would contribute recombinant X chromosomes. One possible recombinant X chromosome would be normal-sized and carry the *car* and *B*⁺ alleles, while the other recombinant X chromosome would be deleted at one end with a piece of the Y chromosome at the other end and carry the *car*⁺ and *B* alleles. When combined with an X or Y chromosome from the males, the parental offspring would have carnation, bar eyes or wild-type eyes; the nonparental offspring would have carnation, round eyes or red, bar eyes.

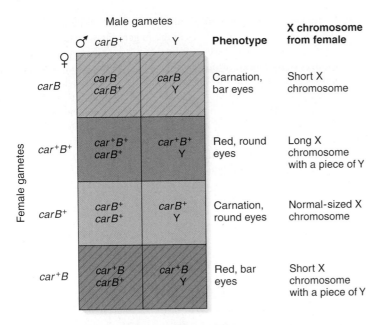

The results shown in the Punnett square are the actual results that Stern observed. His interpretation was that crossing over between homologous chromosomes—in this case, the X chromosome—accounts for the formation of offspring with recombinant phenotypes.

S5. Researchers have discovered a limit to the relationship between map distance and the percentage of recombinant offspring. Even though two genes on the same chromosome may be much more than 50 mu apart, we do not expect to obtain greater than 50% recombinant offspring in a testcross. You may be wondering why this is so. The answer lies in the pattern of multiple crossovers. At the pachytene stage of meiosis, a single crossover in the region between two genes will produce only 50% recombinant chromosomes (see Figure 5.1b). Therefore, to exceed a 50% recombinant level, it would seem necessary to have multiple crossovers within the tetrad.

Let's suppose that two genes are far apart on the same chromosome. A testcross is made between a heterozygous individual, *AaBb*, and a homozygous individual, *aabb*. In the heterozygous individual, the dominant alleles (*A* and *B*) are linked on the same chromosome, and the recessive alleles (*a* and *b*) are linked on the same chromosome. Draw out all of the possible double crossovers (between two, three, or four chromatids) and determine the average number of recombinant offspring, assuming an equal probability of all of the double crossover possibilities.

Answer: A double crossover between the two genes could involve two chromatids, three chromatids, or four chromatids. The possibilities for all types of double crossovers are shown here:

Double crossover (involving 4 chromatids)

Double crossover (involving 3 chromatids)

Double crossover (involving 3 chromatids)

Double crossover (involving 2 chromatids)

This drawing considers the situation where two crossovers are expected to occur in the region between the two genes. Because the tetrad is composed of two pairs of homologues, there are several possible ways that a double crossover could occur between homologues. In this illustration, the crossover on the right has occurred first. Because all of these double crossing over events are equally probable, we take the average of them to determine the maximum recombination frequency. This average equals 50%.

Conceptual Questions

C1. What is the difference in meaning between the terms genetic recombination and crossing over?

C2. When applying a chi square approach in a linkage problem, explain why an independent assortment hypothesis is used.

C3. What is mitotic recombination? A heterozygous individual (*Bb*) with brown eyes has one eye with a small patch of blue. Provide two or more explanations for how the blue patch may have occurred.

C4. Mitotic recombination can occasionally produce a twin spot. Let's suppose an animal species can be heterozygous for two genes that govern fur color and length: One gene affects pigmentation, with dark pigmentation (*A*) dominant to albino (*a*); the other gene affects hair length, with long hair (*L*) dominant to short hair (*l*). The two genes are linked on the same chromosome. Let's assume an animal is *AaLl*; *A* is linked to *l*, and *a* is linked to *L*. Draw the chromosomes labeled with these alleles, and explain how mitotic recombination could produce a twin spot with one spot having albino pigmentation and long fur, the other having dark pigmentation and short fur.

C5. A crossover has occurred in the bivalent shown here.

If a second crossover occurs in the same region between these two genes, which two chromatids would be involved in order to produce the following outcomes?

A. 100% recombinants

B. 0% recombinants

C. 50% recombinants

C6. A crossover has occurred in the bivalent shown here.

What would be the outcome of this single crossover event? If a second crossover occurs somewhere between *A* and *C*, explain which two chromatids it would involve and where it would occur (i.e., between which two genes) to produce the types of chromosomes shown here:

A. *A B C*, *A b C*, *a B c*, and *a b c*

B. *A b c*, *A b c*, *a B C*, and *a B C*

C. *A B c*, *A b c*, *a B C*, and *a b C*

D. *A B C*, *A B C*, *a b c*, and *a b c*

C7. A diploid organism has a total of 14 chromosomes and about 20,000 genes per haploid genome. Approximately how many genes are in each linkage group?

C8. If you try to throw a basketball into a basket, the likelihood of succeeding depends on the size of the basket. It is more likely that you will get the ball into the basket if the basket is bigger. In your own words, explain how this analogy also applies to the idea that the likelihood of crossing over is greater when two genes are far apart compared to when they are close together.

C9. By conducting testcrosses, researchers have found that the sweet pea has seven linkage groups. How many chromosomes would you expect to find in leaf cells?

C10. In humans, a rare dominant disorder known as nail-patella syndrome causes abnormalities in the fingernails, toenails, and kneecaps. Researchers have examined family pedigrees with regard to this disorder and, within the same pedigree, also examined the individuals with regard to their blood types. (A description of blood genotypes is found in Chapter 4.) In the following pedigree, individuals affected with nail-patella disorder are shown with filled symbols. The genotype of each individual with regard to their ABO blood type is also shown. Does this pedigree suggest any linkage between the gene that causes nail-patella syndrome and the gene that causes blood type?

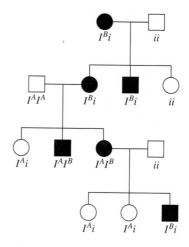

C11. When true-breeding mice with brown fur and short tails (*BBtt*) were crossed to true-breeding mice with white fur and long tails (*bbTT*), all F_1 offspring had brown fur and long tails. The F_1 offspring were crossed to mice with white fur and short tails. What are the possible phenotypes of the F_2 offspring? Which F_2 offspring are recombinant, and which are nonrecombinant? What will be the ratios of the F_2 offspring if independent assortment is taking place? How will the ratios be affected by linkage?

C12. Though we often think of genes in terms of the phenotypes they produce (e.g., curly leaves, flaky tail, brown eyes), the molecular function of most genes is to encode proteins. Many cellular proteins function as enzymes. The table that follows describes the map distances between six different genes that encode six different enzymes: *Ada*, adenosine deaminase; *Hao-1*, hydroxyacid oxidase-1; *Hdc*, histidine decarboxylase; *Odc-2*, ornithine decarboxylase-2; *Sdh-1*, sorbitol dehydrogenase-1; and *Ass-1*, arginosuccinate synthetase-1.

Map distances between two genes:

	Ada	Hao-1	Hdc	Odc-2	Sdh-1	Ass-1
Ada		14		8	28	
Hao-1	14		9		14	
Hdc		9		15	5	
Odc-2	8		15			63
Sdh-1	28	14	5			43
Ass-1				63	43	

Construct a genetic map that describes the locations of all six genes.

C13. If the likelihood of a single crossover in a particular chromosomal region is 10%, what is the theoretical likelihood of a double or triple crossover in that same region? How would positive interference affect these theoretical values?

C14. Except for fungi that form asci, in most dihybrid crosses involving linked genes, we cannot tell if a double crossover between the two genes has occurred because the offspring will inherit the parental combination of alleles. How does the inability to detect double crossovers affect the calculation of map distance? Is map distance underestimated or overestimated because of our inability to detect double crossovers? Explain your answer.

C15. Researchers have discovered that certain regions of chromosomes are much more likely to cross over compared to other regions. We might call such a region a "hot spot" for crossing over. Let's suppose that two genes, gene A and gene B, are 5,000,000 base pairs apart on the same chromosome. Genes A and B are in a hot spot for crossing over. Two other genes, let's call them gene C and gene D, are also 5,000,000 base pairs apart but are not in a hot spot for recombination. If we conducted dihybrid crosses to compute the map distance between genes A and B, and other dihybrid crosses to compute the map distance between genes C and D, would the map distances be the same between A and B compared to C and D? Explain.

C16. Describe the unique features of ascomycetes that lend themselves to genetic analysis.

C17. In fungi, what is the difference between a tetrad and an octad? What cellular process occurs in an octad that does not occur in a tetrad?

C18. Explain the difference between an unordered versus an ordered octad.

C19. In *Neurospora*, a cross is made between a wild-type and an albino mutant strain, which produce orange and white spores, respectively. Draw two different ways that an octad might look if it was displaying second-division segregation.

C20. One gene in *Neurospora*, let's call it gene A, is located close to a centromere, while a second gene, let's call it gene B, is located more toward the end of the chromosome. Would the percentage of octads exhibiting first-division segregation be higher with respect to gene A or gene B? Explain your answer.

Experimental Questions (Includes Most Mapping Questions)

E1. Figure 5.2 shows the first experimental results that indicated linkage between two different genes. Conduct a chi square analysis to confirm that the genes are really linked and the data could not be explained by independent assortment.

E2. In the experiment of Figure 5.6, the researchers followed the inheritance pattern of chromosomes that were abnormal at both ends to correlate genetic recombination with the physical exchange of chromosome pieces. Is it necessary to use a chromosome that is abnormal at both ends, or could the researchers have used a parental strain with two abnormal versions of chromosome 9, one with a knob at one end and its homologue with a translocation at the other end?

E3. The experiment of Figure 5.6 is not like a standard testcross, because neither parent is homozygous recessive for both genes. If you were going to carry out this same kind of experiment to verify that crossing over can explain the recombination of alleles of different genes, how would you modify this experiment to make it a standard testcross? For both parents, you should designate which alleles are found on an abnormal chromosome (i.e., knobbed, translocation chromosome 9) and which alleles are found on normal chromosomes.

E4. How would you determine that genes in mammals are located on the Y chromosome linkage group? Is it possible to conduct crosses (let's say in mice) to map the distances between genes along the Y chromosome? Explain.

E5. Explain the rationale behind a testcross. Is it necessary for one of the parents to be homozygous recessive for the genes of interest?

In the heterozygous parent of a testcross, must all of the dominant alleles be linked on the same chromosome and all of the recessive alleles be linked on the homologue?

E6. In your own words, explain why a testcross cannot produce more than 50% recombinant offspring. When a testcross does produce 50% recombinant offspring, what do these results mean?

E7. Explain why the percentage of recombinant offspring in a testcross is a more accurate measure of map distance when two genes are close together. When two genes are far apart, is the percentage of recombinant offspring an underestimate or overestimate of the actual map distance?

E8. If two genes are more than 50 mu apart, how would you ever be able to show experimentally that they are located on the same chromosome?

E9. In Morgan's trihybrid testcross of Figure 5.3, he realized that crossing over was more frequent between the eye color and wing length genes than between the body color and eye color genes. Explain how he determined this.

E10. In the experiment of Figure 5.10, list the gene pairs from the particular dihybrid crosses that Sturtevant used to construct his genetic map.

E11. In the tomato, red fruit (R) is dominant over yellow fruit (r), and yellow flowers (Wf) are dominant over white flowers (wf). A cross was made between true-breeding plants with red fruit and yellow flowers, and plants with yellow fruit and white flowers. The F_1 generation plants were then crossed to plants with yellow fruit and white flowers. The following results were obtained:

333 red fruit, yellow flowers

64 red fruit, white flowers

58 yellow fruit, yellow flowers

350 yellow fruit, white flowers

Calculate the map distance between the two genes.

E12. Two genes are located on the same chromosome and are known to be 12 mu apart. An *AABB* individual was crossed to an *aabb* individual to produce *AaBb* offspring. The *AaBb* offspring were then crossed to *aabb* individuals.

A. If this cross produces 1,000 offspring, what are the predicted numbers of offspring with each of the four genotypes: *AaBb*, *Aabb*, *aaBb*, and *aabb*?

B. What would be the predicted numbers of offspring with these four genotypes if the parental generation had been *AAbb* and *aaBB* instead of *AABB* and *aabb*?

E13. Two genes, designated *A* and *B*, are located 10 mu from each other. A third gene, designated *C*, is located 15 mu from *B* and 5 mu from *A*. A parental generation consisting of *AA bb CC* and *aa BB cc* individuals were crossed to each other. The F_1 heterozygotes were then testcrossed to *aa bb cc* individuals. If we assume no double crossovers occur in this region, what percentage of offspring would you expect with the following genotypes?

A. *Aa Bb Cc*

B. *aa Bb Cc*

C. *Aa bb cc*

E14. Two genes in tomatoes are 61 mu apart; normal fruit (*F*) is dominant to fasciated fruit (*f*), and normal numbers of leaves (*Lf*) is dominant to leafy (*lf*). A true-breeding plant with normal leaves and fruit was crossed to a leafy plant with fasciated fruit. The F_1 offspring were then crossed to leafy plants with fasciated fruit. If this cross produced 600 offspring, what are the expected numbers of plants in each of the four possible categories: normal leaves, normal fruit; normal leaves, fasciated fruit; leafy, normal fruit; and leafy, fasciated fruit?

E15. In the tomato, three genes are linked on the same chromosome. Tall is dominant to dwarf, skin that is smooth is dominant to skin that is peachy, and fruit with a normal tomato shape is dominant to oblate shape. A plant that is true-breeding for the dominant traits was crossed to a dwarf plant with peachy skin and oblate fruit. The F_1 plants were then testcrossed to dwarf plants with peachy skin and oblate fruit. The following results were obtained:

151 tall, smooth, normal

33 tall, smooth, oblate

11 tall, peach, oblate

2 tall, peach, normal

155 dwarf, peach, oblate

29 dwarf, peach, normal

12 dwarf, smooth, normal

0 dwarf, smooth, oblate

Construct a genetic map that describes the order of these three genes and the distances between them.

E16. A trait in garden peas involves the curling of leaves. A dihybrid cross was made involving a plant with yellow pods and curling leaves to a wild-type plant with green pods and normal leaves. All F_1 offspring had green pods and normal leaves. The F_1 plants were then crossed to plants with yellow pods and curling leaves. The following results were obtained:

117 green pods, normal leaves

115 yellow pods, curling leaves

78 green pods, curling leaves

80 yellow pods, normal leaves

A. Conduct a chi square analysis to determine if these two genes are linked.

B. If they are linked, calculate the map distance between the two genes. How accurate do you think this distance is?

E17. In mice, the gene that encodes the enzyme inosine triphosphatase is 12 mu from the gene that encodes the enzyme ornithine decarboxylase. Let's suppose you have identified a strain of mice homozygous for a defective inosine triphosphatase gene that does not produce any of this enzyme and is also homozygous for a defective ornithine decarboxylase gene. In other words, this strain of mice cannot make either enzyme. You crossed this homozygous recessive strain to a normal strain of mice to produce heterozygotes. The heterozygotes were then backcrossed to the strain that cannot produce either enzyme. What is the probability of obtaining a mouse that cannot make either enzyme?

E18. In the garden pea, several different genes affect pod characteristics. A gene affecting pod color (green is dominant to yellow) is approximately 7 mu away from a gene affecting pod width (wide is dominant to narrow). Both genes are located on chromosome 5. A third gene, located on chromosome 4, affects pod length (long is dominant to short). A true-breeding wild-type plant (green, wide, long pods) was crossed to a plant with yellow, narrow, short pods. The F_1 offspring were then testcrossed to plants with yellow, narrow, short pods. If the testcross produced 800 offspring, what are the expected numbers of the eight possible phenotypic combinations?

E19. A sex-influenced trait is dominant in males and causes bushy tails. The same trait is recessive in females and results in a normal tail. Fur color is not sex influenced. Yellow fur is dominant to white fur. A true-breeding female with a bushy tail and yellow fur was crossed to a white male without a bushy tail. The F_1 females were then crossed to white males without bushy tails. The following results were obtained:

Males	Females
28 normal tails, yellow	102 normal tails, yellow
72 normal tails, white	96 normal tails, white
68 bushy tails, yellow	0 bushy tails, yellow
29 bushy tails, white	0 bushy tails, white

A. Conduct a chi square analysis to determine if these two genes are linked.

B. If the genes are linked, calculate the map distance between them. Explain which data you used in your calculation.

E20. Three recessive traits in garden pea plants are as follows: yellow pods are recessive to green pods, bluish green seedlings are recessive to green seedlings, creeper (a plant that cannot stand up) is recessive to normal. A true-breeding normal plant with green pods and green seedlings was crossed to a creeper with yellow pods and bluish green seedlings. The F_1 plants were then crossed to creepers with yellow pods and bluish green seedlings. The following results were obtained:

2,059 green pods, green seedlings, normal

151 green pods, green seedlings, creeper

281 green pods, bluish green seedlings, normal

15 green pods, bluish green seedlings, creeper

2,041 yellow pods, bluish green seedlings, creeper

157 yellow pods, bluish green seedlings, normal

282 yellow pods, green seedlings, creeper

11 yellow pods, green seedlings, normal

Construct a genetic map that describes the map distance between these three genes.

E21. In mice, a trait called snubnose is recessive to a wild-type nose, a trait called pintail is dominant to a normal tail, and a trait called jerker (a defect in motor skills) is recessive to a normal gait. Jerker mice with a snubnose and pintail were crossed to normal mice, and then the F_1 mice were crossed to jerker mice that have a snubnose and normal tail. The outcome of this cross was as follows:

560 jerker, snubnose, pintail

548 normal gait, normal nose, normal tail

102 jerker, snubnose, normal tail

104 normal gait, normal nose, pintail

77 jerker, normal nose, normal tail

71 normal gait, snubnose, pintail

11 jerker, normal nose, pintail

9 normal gait, snubnose, normal tail

Construct a genetic map that describes the order and distance between these genes.

E22. In *Drosophila*, an allele causing vestigial wings is 12.5 mu away from another gene that causes purple eyes. A third gene that affects body color has an allele that causes black body color. This third gene is 18.5 mu away from the vestigial wings gene and 6 mu away from the gene causing purple eyes. The alleles causing vestigial wings, purple eyes, and black body are all recessive. The dominant (wild-type) traits are long wings, red eyes, and gray body. A researcher crossed wild-type flies to flies with vestigial wings, purple eyes, and black bodies. All F_1 flies were wild type. F_1 female flies were then crossed to male flies with vestigial wings, purple eyes, and black bodies. If 1,000 offspring were observed, what are the expected numbers of the following types of flies?

Long wings, red eyes, gray body

Long wings, purple eyes, gray body

Long wings, red eyes, black body

Long wings, purple eyes, black body

Short wings, red eyes, gray body

Short wings, purple eyes, gray body

Short wings, red eyes, black body

Short wings, purple eyes, black body

Which kinds of flies can be produced only by a double crossover event?

E23. Three autosomal genes are linked along the same chromosome. The distance between gene *A* and *B* is 7 mu, the distance between *B* and *C* is 11 mu, and the distance between *A* and *C* is 4 mu. An individual who is *AA bb CC* was crossed to an individual who is *aa BB cc* to produce heterozygous F_1 offspring. The F_1 offspring were then crossed to homozygous *aa bb cc* individuals to produce F_2 offspring.

A. Draw the arrangement of alleles on the chromosomes in the parents and in the F_1 offspring.

B. Where would a crossover have to occur to produce an F_2 offspring that was heterozygous for all three genes?

C. If we assume that no double crossovers occur in this region, what percentage of F_2 offspring is likely to be homozygous for all three genes?

E24. Let's suppose that two different X-linked genes exist in mice, designated with the letters *N* and *L*. Gene *N* exists in a dominant, normal allele and in a recessive allele, *n*, that is lethal. Similarly, gene *L* exists in a dominant, normal allele and in a recessive allele, *l*, that is lethal. Heterozygous females are normal, while males that carry either recessive allele are born dead. Explain whether or not it would be possible to map the distance between these two genes by making crosses and analyzing the number of living and dead offspring. You may assume that you have strains of mice in which females are heterozygous for one or both genes.

E25. The alleles *his-5* and *lys-1*, found in baker's yeast, result in cells that require histidine and lysine for growth, respectively. A cross was made between two haploid yeast strains that are *his-5 lys-1* and *his+ lys+*. From the analysis of 818 individual tetrads, the following numbers of tetrads were obtained:

2 spores: *his-5 lys+* + 2 spores: *his+ lys-1* = 4

2 spores: *his-5 lys-1* + 2 spores: *his+ lys+* = 502

1 spore: *his-5 lys-1* + 1 spore: *his-5 lys+* + 1 spore: *his+ lys-1* + 1 spore: *his+ lys+* = 312

A. Compute the map distance between these two genes using the method of calculation that considers double crossovers and the one that does not. Which method gives a higher value? Explain why.

B. What is the frequency of single crossovers between these two genes?

C. Based on your answer to part B, how many NPDs are expected from this cross? Explain your answer. Is positive interference occurring?

E26. On chromosome 4 in *Neurospora*, the allele *pyr-1* results in a pyrimidine requirement for growth. A cross was made between a *pyr-1* and a *pyr⁺* (wild-type) strain, and the following results were obtained:

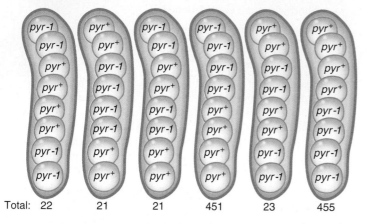

Total: 22   21   21   451   23   455

What is the distance between the *pyr-1* gene and the centromere?

E27. On chromosome 3 in *Neurospora*, the *pro-1* gene is located approximately 9.8 mu from the centromere. Let's suppose a cross was made between a *pro-1* and a *pro⁺* strain and 1,000 asci were analyzed.

A. What are the six types of asci that can be produced?

B. What are the expected numbers of each type of ascus?

Questions for Student Discussion/Collaboration

1. In mice, a dominant gene that causes a short tail is located on chromosome 2. On chromosome 3, a recessive gene causing droopy ears is 6 mu away from another recessive gene that causes a flaky tail. A recessive gene that causes a jerker (uncoordinated) phenotype is located on chromosome 4. A jerker mouse with droopy ears and a short, flaky tail was crossed to a normal mouse. All F₁ generation mice were phenotypically normal, except they had short tails. These F₁ mice were then testcrossed to jerker mice with droopy ears and long, flaky tails. If this cross produced 400 offspring, what would be the proportions of the 16 possible phenotypic categories?

2. In Chapter 3, we discussed the idea that the X and Y chromosomes have a few genes in common. These genes are inherited in a pseudoautosomal pattern. With this phenomenon in mind, discuss whether or not the X and Y chromosomes are really distinct linkage groups.

3. Mendel studied seven traits in pea plants, and the garden pea happens to have seven different chromosomes. It has been pointed out that Mendel was very lucky not to have conducted crosses involving two traits governed by genes that are closely linked on the same chromosome because the results would have confounded his theory of independent assortment. It has even been suggested that Mendel may not have published data involving traits that were linked! An article by Stig Blixt ("Why Didn't Gregor Mendel Find Linkage?" *Nature* 256:206, 1975) considers this issue. Look up this article and discuss why Mendel did not find linkage.

Note: All answers appear at the website for this textbook; the answers to even-numbered questions are in the back of the textbook.

---

www.mhhe.com/brookergenetics3e

Visit the Online Learning Center for practice tests, answer keys, and other learning aids for this chapter. Enhance your understanding of genetics with our interactive exercises, quizzes, animations, and much more.

GENETIC TRANSFER AND MAPPING IN BACTERIA AND BACTERIOPHAGES

One reason researchers are so interested in bacteria and viruses is related to their impact on health. Infectious diseases caused by these agents are a leading cause of human death, accounting for a quarter to a third of deaths worldwide. The spread of infectious diseases results from human behavior, and in recent times has been accelerated by increased trade and travel, and the inappropriate use of antibiotic drugs. Although the incidence of fatal infectious diseases in the United States is relatively low compared to the worldwide average, an alarming increase in more deadly strains of bacteria and viruses has occurred over the past few decades. Since 1980, the number of deaths in the United States due to infectious diseases has approximately doubled.

Thus far, our attention in Part II of this textbook has focused on genetic analyses of eukaryotic species such as fungi, plants, and animals. As we have seen, these organisms are amenable to genetic studies for two reasons. First, allelic differences, such as white versus red eyes in *Drosophila* and tall versus dwarf pea plants, provide readily discernible traits among different individuals. Second, because most eukaryotic species reproduce sexually, crosses can be made, and the pattern of transmission of traits from parent to offspring can be analyzed. The ability to follow allelic differences in a genetic cross is a basic tool in the genetic examination of eukaryotic species.

In Chapter 6, we turn our attention to the genetic analysis of bacteria. Like their eukaryotic counterparts, bacteria often possess allelic differences that affect their cellular traits. Common allelic variations among bacteria that are readily discernible involve traits such as sensitivity to antibiotics and differences in their nutrient requirements for growth. In these cases, the allelic differences are between different strains of bacteria, because any given bacterium is usually haploid for a particular gene. In fact, the haploid nature of bacteria is one advantage that makes it easier to identify mutations that produce phenotypes such as altered nutritional requirements. Loss-of-function mutations, which are often recessive in diploid eukaryotes, are not masked by dominant alleles in haploid species. Throughout this chapter, we will consider interesting experiments that examine bacterial strains with allelic differences.

Compared to eukaryotes, another striking difference in prokaryotic species is their mode of reproduction. Because bacteria reproduce asexually, researchers do not use crosses in the genetic analysis of bacterial species. Instead, they rely on a similar mechanism, called genetic transfer, in which a segment of bacterial DNA is transferred from one bacterium to another. In the first part of this chapter,

Conjugating bacteria. The bacteria shown here are transferring genetic material by a process called conjugation.

we will explore the different routes of genetic transfer. We will see how researchers have used genetic transfer to map the locations of genes along the chromosome of many bacterial species.

In the second part of this chapter, we will examine **bacteriophages** (also known as **phages**), which are viruses that infect bacteria. Bacteriophages contain their own genetic material that governs the traits of the phage. As we will see, the genetic analysis of phages can yield a highly detailed genetic map of a short region of DNA. These types of analyses have provided researchers with insights regarding the structure and function of genes.

6.1 GENETIC TRANSFER AND MAPPING IN BACTERIA

Genetic transfer is a process by which one bacterium transfers genetic material to another bacterium. Why is genetic transfer an advantage? Like sexual reproduction in eukaryotes, genetic transfer in bacteria is thought to enhance the genetic diversity of bacterial species. For example, a bacterial cell carrying a gene that provides antibiotic resistance may transfer this gene to another bacterial cell, allowing that bacterial cell to survive exposure to the antibiotic.

Bacteria can naturally transfer genetic material in three ways (**Table 6.1**). The first route, known as **conjugation,** involves a direct physical interaction between two bacterial cells. One bacte-

TABLE 6.1

Three Mechanisms of Genetic Transfer Found in Bacteria

Mechanism	Description
Conjugation	Requires direct contact between a donor and a recipient cell. The donor cell transfers a strand of DNA to the recipient. In the example shown here, DNA known as a plasmid is transferred to the recipient cell.
Transduction	When a virus infects a donor cell, it incorporates a fragment of bacterial chromosomal DNA into a newly made virus particle. The virus then transfers this fragment of DNA to a recipient cell.
Transformation	When a bacterial cell dies, it releases a fragment of its DNA into the environment. This DNA fragment is taken up by a recipient cell, which incorporates the DNA into its chromosome.

rium acts as a donor and transfers genetic material to a recipient cell. A second means of transfer is called **transduction.** This occurs when a virus infects a bacterium and then transfers bacterial genetic material from that bacterium to another. The last mode of genetic transfer is **transformation.** In this case, genetic material is released into the environment when a bacterial cell dies. This material then binds to a living bacterial cell, which can take it up. These three mechanisms of genetic transfer have been extensively investigated in research laboratories, and their molecular mechanisms continue to be studied with great interest. In this section, we will examine these three systems of genetic transfer in greater detail.

We will also learn how genetic transfer between bacterial cells has provided unique ways to accurately map bacterial genes. The mapping methods described in this chapter have been largely replaced by molecular approaches described in Chapter 20. Even so, the mapping of bacterial genes serves to illuminate the mechanisms by which genes are transferred between bacterial cells and also helps us to appreciate the strategies of newer mapping approaches.

Bacteria Can Transfer Genetic Material During Conjugation

The natural ability of some bacteria to transfer genetic material between each other was first recognized by Joshua Lederberg and Edward Tatum in 1946. They were studying strains of *Escherichia coli* that had different nutritional requirements for growth. A strain that cannot synthesize a particular nutrient and needs that nutrient supplemented in its growth medium is called an **auxotroph.** For example, a strain that cannot make the amino acid methionine would need this amino acid in its growth medium and would be called a methionine auxotroph. By comparison, a strain that could make this amino acid would be termed a methionine prototroph. A **prototroph** does not need the nutrient in its growth medium.

The experiment in **Figure 6.1** considers one strain, designated *met⁻ bio⁻ thr⁺ leu⁺ thi⁺*, which required one amino acid, methionine (met), and one vitamin, biotin (bio), in order to grow. This strain did not require the amino acids threonine (thr) or leucine (leu), or the vitamin thiamine (thi) for growth. Another strain, designated *met⁺ bio⁺ thr⁻ leu⁻ thi⁻*, had just the opposite requirements. It was an auxotroph for threonine, leucine, and thiamine, but a prototroph for methionine and biotin. These differences in nutritional requirements correspond to variations in the genetic material of the two strains. The first strain had two defective genes encoding enzymes necessary for methionine and biotin synthesis. The second strain contained three defective genes required to make threonine, leucine, and thiamine.

Figure 6.1 compares the results when the two strains were mixed together and when they were not mixed. Without mixing, about 100 million (10^8) *met⁻ bio⁻ thr⁺ leu⁺ thi⁺* cells were applied to plates on a growth medium lacking amino acids, biotin, and thiamine; no colonies were observed to grow. This result is expected because the media did not contain methionine or biotin. Likewise, when 10^8 *met⁺ bio⁺ thr⁻ leu⁻ thi⁻* cells were plated, no colonies were observed because threonine, leucine, and thiamine were missing from this growth medium. However, when the two strains were mixed together and then 10^8 cells plated, approximately 10 bacterial colonies formed. Because growth occurred,

FIGURE 6.1   **Experiment of Lederberg and Tatum demonstrating genetic transfer during conjugation in *E. coli*.** When plated on growth media lacking amino acids, biotin, and thiamine, the *met⁻ bio⁻ thr⁺ leu⁺ thi⁺* or *met⁺ bio⁺ thr⁻ leu⁻ thi⁻* strains were unable to grow. However, if they were mixed together and then plated, some colonies were observed. These colonies were due to the transfer of genetic material between these two strains by conjugation. Note: In bacteria, it is common to give genes a three-letter name (shown in italics) that is related to the function of the gene. A plus superscript (⁺) indicates a functional gene, and a minus superscript (⁻) indicates a mutation that has caused the gene or gene product to be inactive. In some cases, several genes have related functions. These may have the same three-letter name followed by different capital letters. For example, different genes involved with leucine biosynthesis may be called *leuA, leuB, leuC,* and so on. In the experiment described in Figure 6.1, the genes involved in leucine biosynthesis had not been distinguished, so the gene involved is simply referred to as *leu⁺* (for a functional gene) and *leu⁻* (for a nonfunctional gene).

the genotype of the cells within these colonies must have been *met⁺ bio⁺ thr⁺ leu⁺ thi⁺*. How could this genotype occur? Lederberg and Tatum hypothesized that some genetic material was transferred between the two strains. One possibility is that the genetic material providing the ability to synthesize methionine and biotin (*met⁺ bio⁺*) was transferred to the *met⁻ bio⁻ thr⁺ leu⁺ thi⁺* strain. Alternatively, the ability to synthesize threonine, leucine, and thiamine (*thr⁺ leu⁺ thi⁺*) may have been transferred to the *met⁺ bio⁺ thr⁻ leu⁻ thi⁻* cells. The results of this experiment did not distinguish between these two possibilities.

In 1950, Bernard Davis conducted experiments showing that two strains of bacteria must make physical contact with each other to transfer genetic material. The apparatus he used, known as a U-tube, is shown in **Figure 6.2.** At the bottom of the U-tube is a filter with pores small enough to allow the passage of genetic material (i.e., DNA molecules) but too small to permit the passage of bacterial cells. On one side of the filter, Davis added a bacterial strain with a certain combination of nutritional requirements (the *met⁻ bio⁻ thr⁺ leu⁺ thi⁺* strain). On the other side, he added a different bacterial strain (the *met⁺ bio⁺ thr⁻ leu⁻ thi⁻* strain). The application of alternating pressure and suction

FIGURE 6.2   A U-tube apparatus like that used by Bernard Davis. The fluid in the tube is forced through the filter by alternating suction and pressure. However, the pores in the filter are too small for the passage of bacteria.

promoted the movement of liquid through the filter. Because the bacteria were too large to pass through the pores, the movement of liquid did not allow the two types of bacterial strains to mix with each other. However, any genetic material that was released from a bacterium could pass through the filter.

After incubation in a U-tube, bacteria from either side of the tube were placed on media that could select for the growth of cells that were *met⁺ bio⁺ thr⁺ leu⁺ thi⁺*. These selective media lacked methionine, biotin, threonine, leucine, and thiamine but contained all other nutrients essential for growth. In this case, no bacterial colonies grew on the plates. The experiment showed that, without physical contact, the two bacterial strains did not transfer genetic material to one another.

The term conjugation is now used to describe the natural process of genetic transfer between bacterial cells that requires

direct cell-to-cell contact. Many, but not all, species of bacteria can conjugate. Working independently, Joshua and Esther Lederberg, William Hayes, and Luca Cavalli-Sforza discovered in the early 1950s that only certain bacterial strains can act as donors of genetic material. For example, only about 5% of natural isolates of *E. coli* can act as donor strains. Research studies showed that a strain incapable of acting as a donor could subsequently be converted to a donor strain after being mixed with another donor strain. Hayes correctly proposed that donor strains contain a fertility factor that can be transferred to conjugation-defective strains to make them conjugation proficient.

We now know that certain donor strains of *E. coli* contain a small circular segment of genetic material known as an **F factor** (for fertility factor) in addition to their circular chromosome. Strains of *E. coli* that contain an F factor are designated *F⁺*, while strains without F factors are termed *F⁻*. In recent years, the molecular details of the conjugation process have been extensively studied. Though the mechanisms vary somewhat from one bacterial species to another, some general themes have emerged. F factors carry several genes that are required for conjugation to occur. For example, **Figure 6.3** shows the arrangement of genes on the F factor found in certain strains of *E. coli*. The functions of the proteins encoded by these genes are needed to transfer a strand of DNA from the donor cell to a recipient cell.

Figure 6.4a describes the molecular events that occur during conjugation in *E. coli*. Contact between donor and recipient cells is a key step that initiates the conjugation process. **Sex pili** (singular: **pilus**) are made by *F⁺* strains (**Figure 6.4b**). The gene encoding the pilin protein (*traA*) is located on the F factor. The pili act as attachment sites that promote the binding of bacteria to each other. In this way, an *F⁺* strain makes physical contact with an *F⁻* strain. In certain species, such as *E. coli*, long pili project from F⁺ cells and attempt to make contact with nearby F⁻ cells.

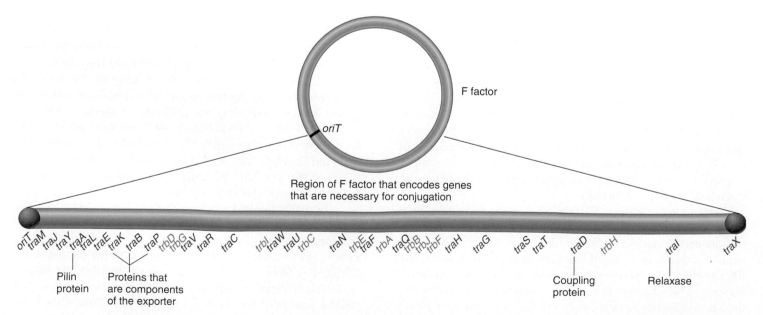

FIGURE 6.3   Genes on the F factor that play a role during conjugation. A region of the F factor contains a few dozen genes that play a role in the conjugative process. Because they play a role in the transfer of DNA from donor to recipient cell, the genes are designated with the three-letter names of *tra* or *trb*, followed by a capital letter. The *tra* genes are shown in red, and the *trb* genes are shown in blue. The functions of a few examples are indicated. The origin of transfer is designated *oriT*.

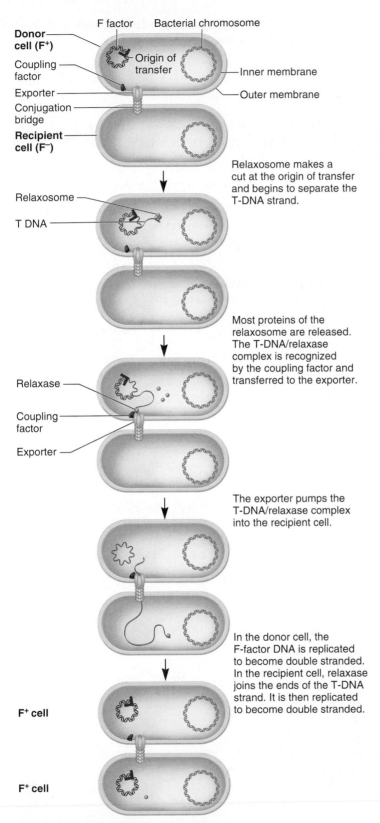

(a) Transfer of an F factor via conjugation

Once contact is made, the pili shorten and thereby draw the donor and recipient cells closer together. A **conjugation bridge** is then formed between the two cells, which provides a passageway for DNA transfer.

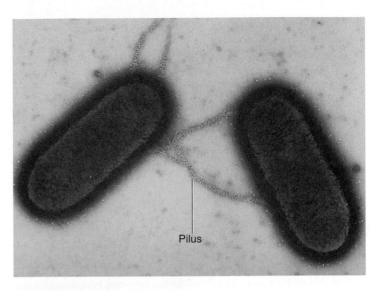

(b) Conjugating *E. coli*

FIGURE 6.4 **The transfer of an F factor during bacterial conjugation.** (a) The mechanism of transfer. The end result is that both cells have an F factor. (b) Two *E. coli* cells in the act of conjugation. The cell on the left is F⁺, and the one on the right is F⁻. The two cells make contact with each other via sex pili that are made by the F⁺ cell.

The successful contact between a donor and recipient cell stimulates the donor cell to begin the transfer process. Genes within the F factor encode a protein complex called the **relaxosome**. This complex first recognizes a DNA sequence in the F factor known as the **origin of transfer**. Upon recognition, the site is cut in one DNA strand. This cut strand of DNA is called **T DNA** because it is the strand that will be transferred to the recipient cell. The relaxosome also catalyzes the separation of the T-DNA strand from its complementary strand. As the DNA strands separate, most of the proteins within the relaxosome are released, but one protein, called relaxase, remains bound to the end of the T DNA. The complex between the T DNA and relaxase is called a **nucleoprotein** because it contains both nucleic acid (DNA) and protein (relaxase).

The next phase of conjugation involves the export of the nucleoprotein complex from the donor cell to the recipient cell. To begin this process, the T-DNA/relaxase complex is recognized by a coupling factor that promotes the entry of the nucleoprotein into the exporter, a complex of proteins that spans both inner and outer membranes of the donor cell. In bacterial species, this complex is formed from 10 to 15 different proteins that are encoded by genes within the F factor.

Once the T-DNA/relaxase complex is pumped out of the donor cell, it travels through the conjugation bridge and then into the recipient cell. As shown in Figure 6.4a, the other strand of the F factor DNA remains in the donor cell, where DNA replication restores the F factor DNA to its original double-stranded condition. After the recipient cell receives a single strand of the F factor DNA, relaxase catalyzes the joining of the ends of the linear DNA molecule to form a circular molecule. This single-stranded DNA is replicated in the recipient cell to become double stranded.

The result of conjugation is that the recipient cell has acquired an F factor, converting it from an F⁻ to an F⁺ cell. The genetic composition of the donor strain has not changed. In some cases, the F⁺ factor that is transferred to the recipient strain may carry genes that were once found in the bacterial chromosome. These types of F factors are called **F' factors.** For example, in the experiment of Lederberg and Tatum described in Figure 6.1, an F' factor carrying the *bio⁺* and *met⁺* genes or an F' factor carrying the *thr⁺*, *leu⁺*, and *thi⁺* genes may have been transferred to the recipient strain. Therefore, conjugation may introduce new genes into the recipient strain and thereby alter its genotype.

Hfr Strains Contain an F Factor Integrated into the Bacterial Chromosome

Luca Cavalli-Sforza discovered a strain of *E. coli* that was very efficient at transferring many chromosomal genes to recipient F⁻ strains. Cavalli-Sforza designated this bacterial strain an **Hfr strain** (for high frequency of recombination). In an *Hfr* strain, an F factor has become integrated into the bacterial chromosome (**Figure 6.5a**). William Hayes, who independently isolated

FIGURE 6.5   Formation of an Hfr cell and its ability to transfer portions of the *E. coli* chromosome. (a) An Hfr cell is created when an F factor integrates into the bacterial chromosome. **(b)** The transfer of the bacterial chromosome begins at the origin of transfer and then proceeds around the circular chromosome. After a segment of chromosome has been transferred to the F⁻ recipient cell, it recombines with the recipient cell's chromosome. If mating occurs for a brief period, only a short segment of the chromosome is transferred. If mating is prolonged, a longer segment of the bacterial chromosome is transferred.

Genes → Traits The F⁻ recipient cell was originally *lac⁻* (unable to metabolize lactose) and *pro⁻* (unable to synthesize proline). If mating occurred for a short period of time, the recipient cell acquired *lac⁺*, allowing it to metabolize lactose. If mating occurred for a longer period of time, the recipient cell also acquired *pro⁺*, enabling it to synthesize proline.

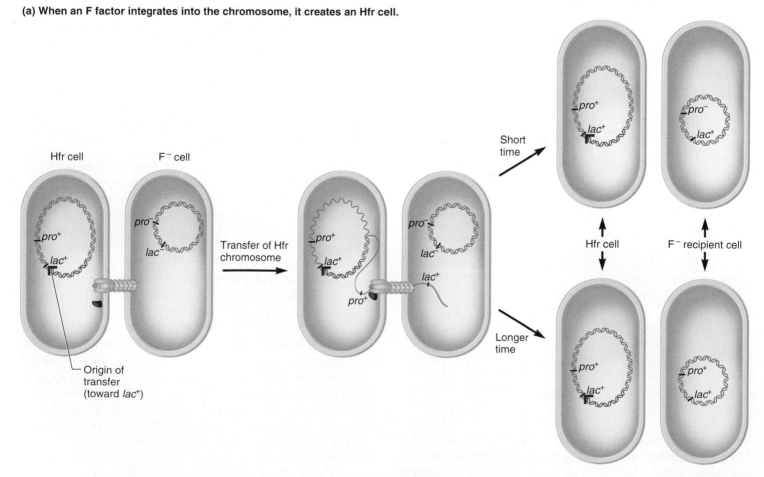

(a) When an F factor integrates into the chromosome, it creates an Hfr cell.

(b) An Hfr donor cell can pass a portion of its chromosome to an F⁻ recipient cell.

another *Hfr* strain, demonstrated that conjugation between an *Hfr* strain and an *F⁻* strain involves the transfer of a portion of the bacterial chromosome from the *Hfr* strain to the F⁻ cell (**Figure 6.5b**). The origin of transfer within the integrated F factor determines the starting point and direction of this transfer process. One of the DNA strands is cut at the origin of transfer. This cut, or nicked, site is the starting point at which the Hfr chromosome will enter the F⁻ recipient cell. From this starting point, a strand of the DNA of the Hfr chromosome begins to enter the F⁻ cell in a linear manner. The transfer process occurs in conjunction with chromosomal replication, so the Hfr cell retains its original chromosomal composition. About 1.5 to 2 hours are required for the entire Hfr chromosome to pass into the F⁻ cell. Because most matings do not last that long, usually only a portion of the Hfr chromosome is transmitted to the F⁻ cell.

Once inside the F⁻ cell, the chromosomal material from the Hfr cell can swap, or recombine, with the homologous region of the recipient cell's chromosome. (Chapter 17 describes the process of homologous recombination.) How does this process affect the recipient cell? As illustrated in Figure 6.5b, this recombination may provide the recipient cell with a new combination of alleles. In this example, the recipient strain was originally *lac⁻* (unable to metabolize lactose) and *pro⁻* (unable to synthesize proline). If mating occurred for a short time, the recipient cell received a short segment of chromosomal DNA from the donor. In this case, the recipient cell has become *lac⁺* but remains *pro⁻*. If the mating is prolonged, the recipient cell will receive a longer segment of chromosomal DNA from the donor. After a longer mating, the recipient becomes *lac⁺* and *pro⁺*. As shown in Figure 6.5b, an important feature of Hfr mating is that the bacterial chromosome is transferred linearly to the recipient strain. In this example, *lac⁺* is always transferred first, and *pro⁺* is transferred later.

In any particular *Hfr* strain, the origin of transfer has a specific orientation that promotes either a counterclockwise or clockwise transfer of genes. Also, among different *Hfr* strains, the origin of transfer may be located in different regions of the chromosome. Therefore, the order of gene transfer depends on the location and orientation of the origin of transfer. For example, another *Hfr* strain could have its origin of transfer next to *pro⁺* and transfer *pro⁺* first and then *lac⁺*.

EXPERIMENT 6A

Conjugation Experiments Can Map Genes Along the *E. coli* Chromosome

The first genetic mapping experiments in bacteria were carried out by Elie Wollman and François Jacob in the 1950s. At the time of their studies, not much information was known about the organization of bacterial genes along the chromosome. A few key advances made Wollman and Jacob realize that the process of genetic transfer could be used to map the order of genes in *E. coli*. First, the discovery of conjugation by Lederberg and Tatum and the identification of *Hfr* strains by Cavalli-Sforza and Hayes made it clear that bacteria can transfer genes from donor to recipient cells in a linear fashion. In addition, Wollman and Jacob were aware of previous microbiological studies concerning bacteriophages—viruses that bind to *E. coli* cells and subsequently infect them. These studies showed that bacteriophages can be sheared from the surface of *E. coli* cells if they are spun in a blender. In this treatment, the bacteriophages are detached from the surface of the bacterial cells, but the bacteria themselves remain healthy and viable. Wollman and Jacob reasoned that a blender treatment could also be used to separate bacterial cells that were in the act of conjugation without killing them. This technique is known as an **interrupted mating.**

The rationale behind Wollman and Jacob's mapping strategy is that the time it takes for genes to enter a donor cell is directly related to their order along the bacterial chromosome. They hypothesized that the chromosome of the donor strain in an Hfr mating is transferred in a linear manner to the recipient strain. If so, the order of genes along the chromosome can be deduced by determining the time it takes various genes to enter the recipient strain. Assuming the Hfr chromosome is transferred linearly, they realized that interruptions of mating at different times would lead to various lengths of the Hfr chromosome being transferred to the F⁻ recipient cell. If two bacterial cells had mated for a short period of time, only a small segment of the Hfr chromosome would be transferred to the recipient bacterium. However, if the bacterial cells were allowed to mate for a longer period of time before being interrupted, a longer segment of the Hfr chromosome could be transferred (see Figure 6.5b). By determining which genes were transferred during short matings and which required longer mating times, Wollman and Jacob were able to deduce the order of particular genes along the *E. coli* chromosome.

As shown in the experiment of **Figure 6.6**, Wollman and Jacob began with two *E. coli* strains. The donor (*Hfr*) strain had the following genetic composition:

thr⁺: able to synthesize threonine, an essential amino acid for growth

leu⁺: able to synthesize leucine, an essential amino acid for growth

aziˢ: sensitive to killing by azide (a toxic chemical)

tonˢ: sensitive to infection by bacteriophage T1. (As discussed later, when bacteriophages infect bacteria, they may cause lysis, which results in plaque formation.)

lac⁺: able to metabolize lactose and use it for growth

gal⁺: able to metabolize galactose and use it for growth

strˢ: sensitive to killing by streptomycin (an antibiotic)

The recipient (*F⁻*) strain had the opposite genotype: *thr⁻ leu⁻ aziʳ tonʳ lac⁻ gal⁻ strʳ* (*r* = resistant). Before the experiment, Wollman and Jacob already knew the *thr⁺* gene was transferred first, followed by the *leu⁺* gene, and both were transferred relatively soon (5 to 10 minutes) after mating. Their main goal in this experiment was to determine the times at which the other genes (*aziˢ tonˢ lac⁺ gal⁺*) were transferred to the recipient strain. The transfer of the *strˢ* gene was not examined because streptomycin was used to kill the donor strain following conjugation.

Before discussing the conclusions of this experiment, let's consider how Wollman and Jacob monitored gene transfer. To determine if particular genes had been transferred after mating, they took the mated cells and first plated them on growth media that lacked threonine (thr) and leucine (leu) but contained streptomycin (str). On these plates, the original donor and recipient strains could not grow because the donor strain was streptomycin sensitive and the recipient strain required threonine and leucine. However, mated cells in which the donor had transferred chromosomal DNA carrying the *thr*$^+$ and *leu*$^+$ genes to the recipient cell would be able to grow.

To determine the order of gene transfer of the *azi*^{s}, *ton*^{s}, *lac*$^+$, and *gal*$^+$ genes, Wollman and Jacob picked colonies from the first plates and restreaked them on media that contained azide or bacteriophage T1 or on media that contained lactose or galactose as the sole source of energy for growth. The plates were incubated overnight to observe the formation of visible bacterial growth. Whether or not the bacteria could grow depended on their genotypes. For example, a cell that is *azi*^{s} cannot grow on media containing azide, and a cell that is *lac*$^-$ cannot grow on media containing lactose as the carbon source for growth. By comparison, a cell that is *azi*^{r} and *lac*$^+$ can grow on both types of media.

■ THE GOAL
(DISCOVERY-BASED SCIENCE)

The chromosome of the donor strain in an Hfr mating is transferred in a linear manner to the recipient strain. The order of genes along the chromosome can be deduced by determining the time various genes take to enter the recipient strain.

■ ACHIEVING THE GOAL — FIGURE 6.6   The use of conjugation to map the order of genes along the *E. coli* chromosome.

Starting materials: The two *E. coli* strains already described, one *Hfr* strain (*thr*$^+$ *leu*$^+$ *azi*^{s} *ton*^{s} *lac*$^+$ *gal*$^+$ *str*^{s}) and one F$^-$ (*thr*$^-$ *leu*$^-$ *azi*^{r} *ton*^{r} *lac*$^-$ *gal*$^-$ *str*^{r}).

	Experimental level	**Conceptual level**

1. Mix together a large number of Hfr donor and F$^-$ recipient cells.

Flask with bacteria

2. After different periods of time, take a sample of cells and interrupt conjugation in a blender.

Hfr   F$^-$

Separate by blending; donor DNA recombines with recipient cell chromosome.

3. Plate the cells on growth media lacking threonine and leucine but containing streptomycin. Note: The general methods for growing bacteria in a laboratory are described in the Appendix.

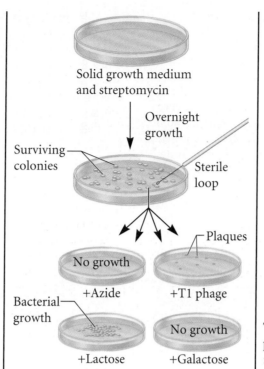

Solid growth medium and streptomycin

Overnight growth

Surviving colonies

Sterile loop

Plaques

No growth

+Azide

+T1 phage

Bacterial growth

No growth

+Lactose

+Galactose

4. Pick each surviving colony, which would have to be *thr⁺ leu⁺ strʳ*, and test to see if it is sensitive to killing by azide, sensitive to infection by T1 bacteriophage, and able to metabolize lactose or galactose.

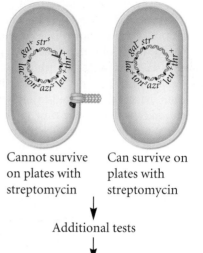

In this conceptual example, the cells have been incubated about 20 minutes.

Cannot survive on plates with streptomycin

Can survive on plates with streptomycin

Additional tests

The conclusion is that the colony that was picked contained cells with a genotype of *thr⁺ leu⁺ aziˢ tonˢ lac⁺ gal⁻ strʳ*.

■ THE DATA

Minutes That Bacterial Cells Were Allowed to Mate Before Blender Treatment	*Percent of Surviving Bacterial Colonies with the Following Genotypes:*				
	thr⁺ leu⁺	*aziˢ*	*tonˢ*	*lac⁺*	*gal⁺*
5	—*	—	—	—	—
10	100	12	3	0	0
15	100	70	31	0	0
20	100	88	71	12	0
25	100	92	80	28	0.6
30	100	90	75	36	5
40	100	90	75	38	20
50	100	91	78	42	27
60	100	91	78	42	27

*There were no surviving colonies within the first 5 minutes of mating.

■ INTERPRETING THE DATA

Now let's discuss the data shown in Figure 6.6. After the first plating, all survivors would be cells in which the *thr⁺* and *leu⁺* alleles had been transferred to the F⁻ recipient strain, which was already streptomycin resistant. As seen in the data, 5 minutes was not sufficient time to transfer the *thr⁺* and *leu⁺* alleles because no surviving colonies were observed. After 10 minutes or longer, however, surviving bacterial colonies with the *thr⁺ leu⁺* genotype were obtained. To determine the order of the remaining genes (*aziˢ, tonˢ, lac⁺,* and *gal⁺*), each surviving colony was tested to see if it was sensitive to killing by azide, sensitive to infection by T1

bacteriophage, able to use lactose for growth, or able to use galactose for growth. The likelihood of surviving colonies depended on whether the *aziˢ, tonˢ, lac⁺,* and *gal⁺* genes were close to the origin of transfer or farther away. For example, when cells were allowed to mate for 25 minutes, 80% carried the *tonˢ* gene, whereas only 0.6% carried the *gal⁺* gene. These results indicate that the *tonˢ* gene is closer to the origin of transfer compared to the *gal⁺* gene. When comparing the data in Figure 6.6, a consistent pattern emerged. The gene that conferred sensitivity to azide (*aziˢ*) was transferred first, followed by *tonˢ, lac⁺,* and finally, *gal⁺*. From these data, as well as those from other experiments, Wollman and Jacob constructed a genetic map that described the order of these genes along the *E. coli* chromosome.

| thr | leu | azi | ton | lac | gal |

This work provided the first method for bacterial geneticists to map the order of genes along the bacterial chromosome. Throughout the course of their studies, Wollman and Jacob identified several different *Hfr* strains in which the origin of transfer had been integrated at different places along the bacterial chromosome. When they compared the order of genes among different *Hfr* strains, their results were consistent with the idea that the *E. coli* chromosome is circular (see solved problem S2).

A self-help quiz involving this experiment can be found at the Online Learning Center.

A Genetic Map of the *E. coli* Chromosome Has Been Obtained from Many Conjugation Studies

Conjugation experiments have been used to map more than 1,000 genes along the circular *E. coli* chromosome. A map of the *E. coli* chromosome is shown in **Figure 6.7.** This simplified map shows the locations of only a few dozen genes. Because the chromosome is circular, we must arbitrarily assign a starting point on the map, in this case the gene *thrA*. Researchers scale genetic maps from bacterial conjugation studies in units of **minutes.** This unit refers to the relative time it takes for genes to first enter an F⁻ recipient strain during a conjugation experiment. The *E. coli* genetic map shown in Figure 6.7 is 100 minutes long, which is approximately the time that it takes to transfer the complete chromosome during an Hfr mating.

The distance between two genes is determined by comparing their times of entry during a conjugation experiment. As shown in **Figure 6.8,** the time of entry is found by conducting mating experiments at different time intervals before interruption. We compute the time of entry by extrapolating the data back to the x-axis. In this experiment, the time of entry of the *lacZ* gene was approximately 16 minutes, and that of the *galE* gene was 25 minutes. Therefore, these two genes are approximately 9 minutes apart from each other along the *E. coli* chromosome.

Let's look back at Figure 6.7 and consider where the origin of transfer must have been located in the donor strain that was used in the experiment of Figure 6.8. The *lacZ* gene is located at 7 minutes on the chromosome map, and the *galE* is found at 16 minutes. For the donor strain used in Figure 6.8, we can deduce that the origin of transfer was located at approximately 91 minutes on the chro-

FIGURE 6.8   Time course of an interrupted *E. coli* conjugation experiment. By extrapolating the data back to the origin, the approximate time of entry of the *lacZ* gene is found to be 16 minutes; that of the *galE* gene, 25 minutes. Therefore, the distance between these two genes is 9 minutes.

mosome map and transferred DNA in the clockwise direction. If we assume the origin was located here, it would take about 16 minutes to transfer *lacZ* and about 25 minutes to transfer *galE*.

Bacteria May Contain Different Types of Plasmids

Thus far, we have considered F factors, which are one type of DNA that can exist independently of the chromosomal DNA. The more general term for this structure is **plasmid.** Most known plasmids are circular, although some are linear. Plasmids occur naturally in many strains of bacteria and in a few types of eukaryotic cells such as yeast. The smallest plasmids consist of just a few thousand base pairs (bp) and carry only a gene or two; the largest are in the range of 100,000 to 500,000 bp and carry several dozen or even hundreds of genes. Some plasmids, such as F factors, can integrate into the chromosome. These are also called **episomes.**

A plasmid has its own origin of replication that allows it to be replicated independently of the bacterial chromosome. The DNA sequence of the origin of replication influences how many copies of the plasmid are found within a cell. Some origins are said to be very strong because they result in many copies of the plasmid, perhaps as many as 100 per cell. Other origins of replication have sequences that are described as much weaker, in that the number of copies created is relatively low, such as one or two per cell.

Why do bacteria have plasmids? Plasmids are not usually necessary for bacterial survival. However, in many cases, certain genes within a plasmid provide some type of growth advantage to the cell. By studying plasmids in many different species, researchers have discovered that most plasmids fall into five different categories:

1. Fertility plasmids, also known as F factors, allow bacteria to mate with each other.
2. Resistance plasmids, also known as R factors, contain genes that confer resistance against antibiotics and other types of toxins.
3. Degradative plasmids carry genes that enable the bacterium to digest and utilize an unusual substance. For example, a degradative plasmid may carry genes that allow a bacterium to digest an organic solvent such as toluene.

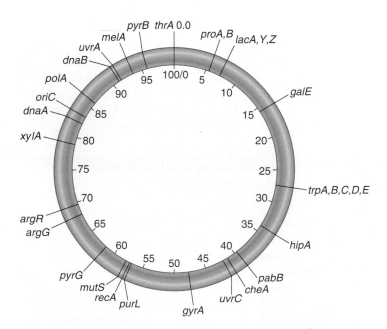

FIGURE 6.7   A simplified genetic map of the *E. coli* chromosome indicating the positions of several genes. *E. coli* has a circular chromosome with about 4,200 different genes. This map shows the locations of several of them. The map is scaled in units of minutes, and it proceeds in a clockwise direction. The starting point on the map is the gene *thrA*.

4. Col-plasmids contain genes that encode colicines, which are proteins that kill other bacteria.
5. Virulence plasmids carry genes that turn a bacterium into a pathogenic strain.

Bacteriophages Transfer Genetic Material from One Bacterial Cell to Another Via Transduction

We now turn to a second method of genetic transfer, one that involves bacteriophages. Before we discuss the ability of bacteriophages to transfer genetic material between bacterial cells, let's consider some general features of a phage's life cycle. Bacteriophages are composed of genetic material that is surrounded by a protein coat. As shown in **Figure 6.9,** bacteriophages bind to the surface of a bacterium and inject their genetic material into the bacterial cytoplasm. At this point, depending on the specific type of virus and its growth conditions, a phage may follow a lytic cycle or a lysogenic cycle. During the **lytic cycle,** the bacteriophage directs the synthesis of many copies of the phage genetic material and coat proteins (see Figure 6.9, left side). These components then assemble to make new phages. When synthesis and assembly are completed, the bacterial host cell is lysed (broken apart), releasing the newly made phages into the environment. **Virulent phages** follow only a lytic cycle and thus infection results in the death of the host cell.

In other cases, a bacteriophage infects a bacterium and follows the lysogenic cycle (see Figure 6.9, right side). During the **lysogenic cycle,** most types of phages integrate their genetic material into the chromosome of the bacterium. This integrated phage DNA is known as a **prophage.** A prophage can exist in a dormant state for a long time during which no new bacteriophages are made. When a bacterium containing a lysogenic prophage divides to produce two daughter cells, the prophage's genetic material is copied along with the bacterial chromosome. Therefore, both daughter cells will inherit the prophage. At some later time, a prophage may become activated to excise itself from the bacterial chromosome and enter the lytic cycle. When this happens, it will promote the synthesis of new phages and eventually lyse the host cell. A bacteriophage that usually exists in the lysogenic life cycle is called a **temperate phage.** Under most conditions, temperate phages do not produce new phages and will not kill the host bacterial cell.

With a general understanding of bacteriophage life cycles, we may now examine the ability of phages to transfer genetic material between bacteria. This process is called transduction. Examples of phages that can transfer bacterial chromosomal DNA from one bacterium to another are the P22 and P1 phages, which infect the bacterial species *Salmonella typhimurium* and *E. coli*, respectively. The P22 and P1 phages can follow either the lytic or lysogenic cycle. In the lytic cycle, when the phage infects the bacterial cell, the bacterial chromosome becomes fragmented into small pieces

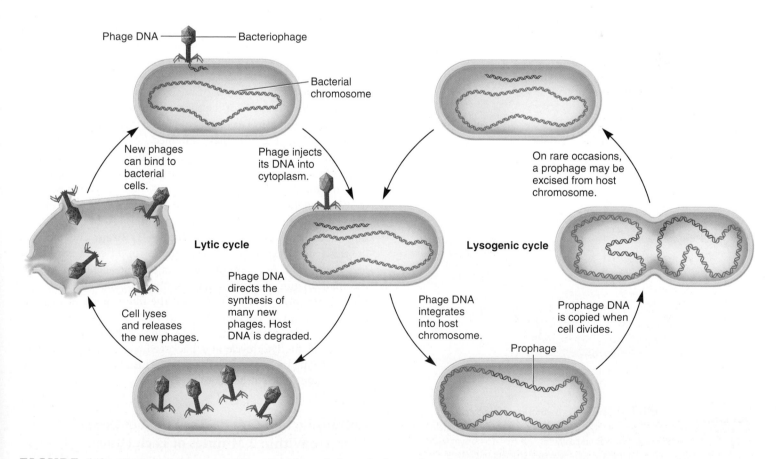

FIGURE 6.9   **The lytic and lysogenic life cycles of certain bacteriophages.** Some bacteriophages, such as temperate phages, can follow both cycles. Other phages, known as virulent phages, can follow only a lytic cycle.

of DNA. The phage DNA directs the synthesis of more phage DNA and proteins, which then assemble to make new phages (**Figure 6.10**).

How does a bacteriophage transfer bacterial chromosomal genes from one cell to another? Occasionally, a mistake can happen in which a piece of bacterial DNA assembles with phage proteins. This creates a phage that contains bacterial chromosomal DNA. When phage synthesis is completed, the bacterial cell is lysed and releases the newly made phage into the environment.

Phage DNA

his+   *lys+*

Phage infects bacterial cell.

his+   *lys+*

Host DNA is hydrolyzed into pieces, and phage DNA and proteins are made.

Phages assemble; occasionally a phage carries a piece of the host cell chromosome.

lys+

Transducing phage with host DNA — *his+*

Transducing phage injects its DNA into a new recipient cell.

Recombination

Recipient cell (*his−lys−*)

his+

his−   *lys−*

The transduced DNA is recombined into the chromosome of the recipient cell.

Recombinant bacterium

his+   *lys−*

The recombinant bacterium's genotype has changed from *his−lys−* to *his+lys−*.

FIGURE 6.10   Transduction in bacteria.

ONLINE ANIMATION

Genes → Traits The transducing phage introduced DNA into a new recipient cell that was originally *his− lys−* (unable to synthesize histidine and lysine). During transduction, it received a segment of bacterial chromosomal DNA that carried *his+*. Following recombination, the recipient cell's genotype was changed to *his+*, so it now could synthesize histidine.

Following release, this abnormal phage can bind to a living bacterial cell and inject its genetic material into the bacterium. The DNA fragment, which was derived from the chromosomal DNA of the first bacterium, can then recombine with the recipient cell's bacterial chromosome. In this case, the recipient bacterium has been changed from a cell that was *his− lys−* (unable to synthesize histidine and lysine) to a cell that is *his+ lys−* (able to synthesize histidine but unable to synthesize lysine). In the example shown in Figure 6.10, any piece of the bacterial chromosomal DNA can be incorporated into the phage. This type of transduction is called **generalized transduction.** By comparison, some phages carry out specialized transduction in which only particular bacterial genes are transferred to recipient cells (see solved problem S5 at the end of the chapter).

Transduction was first discovered in 1952 by Joshua Lederberg and Norton Zinder, using an experimental strategy similar to that depicted in Figure 6.1. They mixed together two strains of the bacterium *S. typhimurium*. One strain, designated LA-22, was *phe− trp− met+ his+*. This strain was unable to synthesize phenylalanine or tryptophan but was able to synthesize methionine and histidine. The other strain, LA-2, was *phe+ trp+ met− his−*. It was able to synthesize phenylalanine and tryptophan but not methionine or histidine. When a mixture of these cells was placed on plates with growth media lacking these four amino acids, approximately one cell in 100,000 was observed to grow. The genotype of the surviving bacterial cells must have been *phe+ trp+ met+ his+*. Therefore, Lederberg and Zinder concluded that genetic material had been transferred between the two strains.

A novel result occurred when Lederberg and Zinder repeated this experiment using a U-tube apparatus, as previously shown in Figure 6.2. They placed the LA-22 strain (*phe− trp− met+ his+*) on one side of the filter and LA-2 (*phe+ trp+ met− his−*) on the other. After an incubation period of several minutes, they removed samples from either side of the tube and plated the cells on media lacking the four amino acids. Surprisingly, they obtained colonies from the side of the tube that contained LA-22 but not from the side that contained LA-2. From these results, they concluded that some filterable agent was being transferred from LA-2 to LA-22 that converted LA-22 to a *phe+ trp+ met+ his+* genotype. In other words, a filterable agent carrying the *phe+* and *trp+* genes was produced on the side of the tube containing LA-2, traversed the filter, and then was taken up by the LA-22 strain. By conducting this type of experiment using filters with different pore sizes, Lederberg and Zinder found that the filterable agent was slightly less than 0.1 μm in diameter, a size much smaller than a bacterium. They correctly concluded that the filterable agent in these experiments was a bacteriophage. In this case, the LA-2 strain contained a prophage, such as P22. On occasion, the prophage switched to the lytic cycle and packaged a segment of bacterial DNA carrying the *phe+* and *trp+* genes. This phage then traversed the filter and injected the genes into LA-22.

Cotransduction Can Be Used to Map Genes That Are Within 2 Minutes of Each Other

Can transduction be used to map the distance between bacterial genes? The answer is yes, but only if the genes are relatively close together. During transduction, P1 phages cannot package pieces that are greater than 2 to 2.5% of the entire *E. coli* chromosome, and P22 phages cannot package pieces that are greater than 1% of the length of the *S. typhimurium* chromosome. If two genes are close together along the chromosome, a bacteriophage may package a single piece of the chromosome that carries both genes and transfer that piece to another bacterium. This phenomenon is called **cotransduction.** Thus, the likelihood that two genes will be cotransduced depends on how close together they lie. If two genes are far apart along a bacterial chromosome, they will never be cotransduced because the bacteriophage cannot physically package a DNA fragment that is larger than 1 to 2.5% of the bacterial chromosome. In genetic mapping studies, cotransduction is used to determine the order and distance between genes that lie fairly close to each other.

To map genes using cotransduction, a researcher selects for the transduction of one gene and then monitors whether or not a second gene is cotransduced along with it. As an example, let's consider a donor strain of *E. coli* that is *arg⁺ met⁺ strˢ* (able to synthesize arginine and methionine but sensitive to killing by streptomycin) and a recipient strain that is *arg⁻ met⁻ strʳ* (**Figure 6.11**). The donor strain is infected with phage P1. Some of the *E. coli* cells are lysed by P1, and this P1 lysate is mixed with the recipient cells. After allowing sufficient time for transduction, the recipient cells are plated on a growth medium that contains arginine and streptomycin but not methionine. Therefore, these plates select for the growth of cells in which the *met⁺* gene has been transferred to the recipient strain, but they do not select for the growth of cells in which the *arg⁺* gene has been transferred, because the growth media are supplied with arginine.

Nevertheless, the *arg⁺* gene may have been cotransduced with the *met⁺* gene if the two genes are close together. To determine this, a sample of cells from each bacterial colony can be picked up with a wire loop and restreaked on media that lack both amino acids. If the colony can grow, the cells must have also obtained the *arg⁺* gene during transduction. In other words, cotransduction of both the *arg⁺* and *met⁺* genes has occurred. Alternatively, if the restreaked colony does not grow, it must have received only the *met⁺* gene during transduction. Data from this type of experiment are shown at the bottom of Figure 6.11. These data indicate a cotransduction frequency of 21/50 = 0.42 or 42%.

FIGURE 6.11 **The steps in a cotransduction experiment.** The donor strain, which is *arg⁺ met⁺ strˢ* (able to synthesize arginine and methionine but sensitive to streptomycin), is infected with phage P1. Some of the cells are lysed by P1, and this P1 lysate is mixed with cells of the recipient strain, which are *arg⁻ met⁻ strʳ*. P1 phages in this lysate may carry fragments of the donor cell's chromosome, and the P1 phage may inject that DNA into the recipient cells. To identify recipient cells that have received the *met⁺* gene from the donor strain, the recipient cells are plated on a growth medium that contains arginine and streptomycin but not methionine. To determine if the *arg⁺* gene has also been cotransduced, cells from each bacterial colony (on the plates containing arginine and streptomycin) are lifted with a sterile wire loop and streaked on media that lack both amino acids. If the cells can grow, cotransduction of the *arg⁺* and *met⁺* genes has occurred. Alternatively, if the restreaked colony does not grow, it must have received only the *met⁺* gene during transduction.

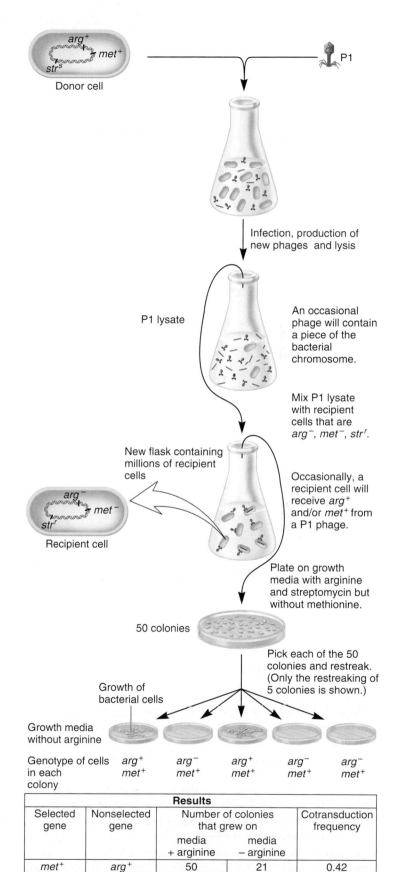

Results				
Selected gene	Nonselected gene	Number of colonies that grew on		Cotransduction frequency
		media + arginine	media – arginine	
met⁺	*arg⁺*	50	21	0.42

In 1966, Tai Te Wu derived a mathematical expression that relates cotransduction frequency with map distance obtained from conjugation experiments. This equation is

$$\text{Cotransduction frequency} = (1 - d/L)^3$$

where

d = distance between two genes in minutes
L = the size of the chromosomal pieces (in minutes) that the phage carries during transduction (For P1 transduction, this size is approximately 2% of the bacterial chromosome, which equals about 2 minutes.)

This equation assumes the bacteriophage randomly packages pieces of the bacterial chromosome that are similar in size. Depending on the type of phage used in a transduction experiment, this assumption may not always be valid. Nevertheless, this equation has been fairly reliable in computing map distance for P1 transduction experiments in *E. coli*. We can use this equation to estimate the distance between the two genes described in Figure 6.11.

$$0.42 = (1 - d/2)^3$$

$$(1 - d/2) = \sqrt[3]{0.42}$$

$$1 - d/2 = 0.75$$

$$d/2 = 0.25$$

$$d = 0.5 \text{ minutes}$$

This equation tells us that the distance between the *met*⁺ and *arg*⁺ genes is approximately 0.5 minutes.

Historically, genetic mapping strategies in bacteria often involved data from both conjugation and transduction experiments. Conjugation has been used to determine the relative order and distance of genes, particularly those that are far apart along the chromosome. In comparison, transduction experiments can provide very accurate mapping data for genes that are fairly close together.

Bacteria Can Also Transfer Genetic Material by Transformation

A third mechanism for the transfer of genetic material from one bacterium to another is known as transformation. This process was first discovered by Frederick Griffith in 1928 while working with strains of *Streptococcus pneumoniae* (formerly known as *Diplococcus pneumoniae* or pneumococcus). During transformation, a living bacterial cell will take up DNA that is released from a dead bacterium. This DNA may then recombine into the living bacterium's chromosome, producing a bacterium with genetic material that it has received from the dead bacterium. (This experiment is discussed in detail in Chapter 9.) Transformation may be either a natural process that has evolved in certain bacteria, in which case it is called **natural transformation,** or an artificial process in which the bacterial cells are forced to take up DNA, an experimental approach termed **artificial transformation.** For example, a technique known as electroporation, in which an electric current causes the uptake of DNA, is used by researchers to promote the transport of DNA into a bacterial cell.

Since the initial studies of Griffith, we have learned a great deal about the events that occur in natural transformation. This form of genetic transfer has been reported in a wide variety of bacterial species. Bacterial cells that are able to take up DNA are known as **competent cells.** Those that can take up DNA naturally carry genes that encode proteins called **competence factors.** These proteins facilitate the binding of DNA fragments to the cell surface, the uptake of DNA into the cytoplasm, and its subsequent incorporation into the bacterial chromosome. Temperature, ionic conditions, and nutrient availability can affect whether or not a bacterium will be competent to take up genetic material from its environment. These conditions influence the expression of competence genes.

In recent years, geneticists have unraveled some of the steps that occur when competent bacterial cells are transformed by genetic material in their environment. **Figure 6.12** describes the steps of transformation. First, a large fragment of genetic material binds to the surface of the bacterial cell. Competent cells express DNA receptors that promote such binding. Before entering the cell, however, this large piece of chromosomal DNA must be cut into smaller fragments. This cutting is accomplished by an extracellular bacterial enzyme known as an endonuclease, which makes occasional random cuts in the long piece of chromosomal DNA. At this stage, the DNA fragments are composed of double-stranded DNA.

In the next step, the DNA fragment begins its entry into the bacterial cytoplasm. For this to occur, the double-stranded DNA interacts with proteins in the bacterial membrane. One of the DNA strands is degraded, and the other strand enters the bacterial cytoplasm via an uptake system, which is structurally similar to the one described for conjugation (as shown earlier in Figure 6.4a) but is involved with DNA uptake rather than export.

To be stably inherited, the DNA strand must be incorporated into the bacterial chromosome. If the DNA strand has a sequence that is similar to a region of DNA in the bacterial chromosome, the DNA may be incorporated into the chromosome by a process known as **homologous recombination,** discussed in detail in Chapter 17. For this to occur, the single-stranded DNA aligns itself with the homologous location on the bacterial chromosome. In the example shown in Figure 6.12, the foreign DNA carries a functional *lys*⁺ gene that aligns itself with a nonfunctional (mutant) *lys*⁻ gene already present within the bacterial chromosome. The foreign DNA then recombines with one of the strands in the bacterial chromosome of the competent cell. In other words, the foreign DNA replaces one of the chromosomal strands of DNA, which is subsequently degraded. During homologous recombination, alignment of the *lys*⁻ and the *lys*⁺ alleles results in a region of DNA called a **heteroduplex** that contains one or more base sequence mismatches. However, the heteroduplex exists only temporarily. DNA repair enzymes in the recipient cell recognize the heteroduplex and repair it. In this case, the heteroduplex has been repaired by eliminating the mutation that caused the *lys*⁻ genotype, thereby creating a *lys*⁺ gene. In this example, the recipient cell has been transformed from a *lys*⁻ strain to a *lys*⁺ strain. Alternatively, a DNA fragment that has entered a

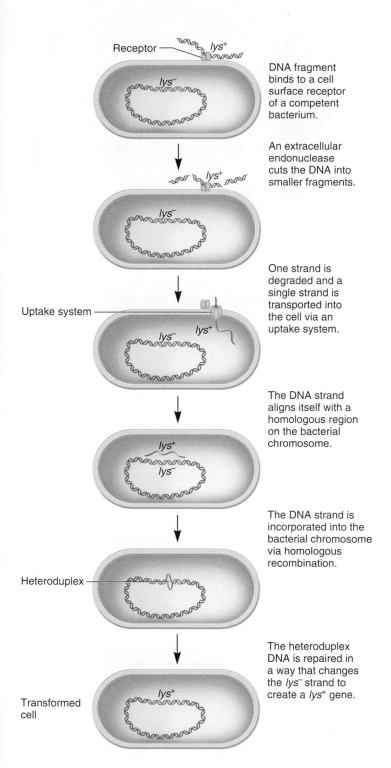

FIGURE 6.12 **The steps of bacterial transformation.** In this example, a fragment of DNA carrying a *lys*⁺ gene enters the competent cell and recombines with the chromosome, transforming the bacterium from *lys*⁻ to *lys*⁺.

Genes → Traits Bacterial transformation can also lead to new traits for the recipient cell. The recipient cell was *lys*⁻ (unable to synthesize the amino acid lysine). Following transformation, it became *lys*⁺. This result would transform the recipient bacterial cell into a cell that could synthesize lysine and grow on media that lacked this amino acid. Before transformation, the recipient *lys*⁻ cell would not have been able to grow on media lacking lysine.

- Receptor — *lys*⁺
- *lys*⁻ — DNA fragment binds to a cell surface receptor of a competent bacterium.
- *lys*⁺ — An extracellular endonuclease cuts the DNA into smaller fragments.
- Uptake system / *lys*⁺ / *lys*⁻ — One strand is degraded and a single strand is transported into the cell via an uptake system.
- *lys*⁺ / *lys*⁻ — The DNA strand aligns itself with a homologous region on the bacterial chromosome.
- Heteroduplex — The DNA strand is incorporated into the bacterial chromosome via homologous recombination.
- Transformed cell / *lys*⁺ — The heteroduplex DNA is repaired in a way that changes the *lys*⁻ strand to create a *lys*⁺ gene.

cell may not be homologous to any genes that are already found in the bacterial chromosome. In this case, the DNA strand may be incorporated at a random site in the chromosome. This process is known as **nonhomologous** or **illegitimate recombination.**

Some bacteria preferentially take up DNA fragments from other bacteria of the same species or closely related species. How does this occur? Recent research has shown that the mechanism can vary among different species. In *Streptococcus pneumoniae*, the cells secrete a short peptide called the **competence-stimulating peptide (CSP).** When many *S. pneumoniae* cells are in the vicinity of one another, the concentration of CSP becomes high, and this stimulates the cells, via a cell-signaling pathway, to express the competence proteins that are needed for the uptake of DNA and its incorporation in the chromosome. Because competence requires a high external concentration of CSP, *S. pneuomoniae* cells are more likely to take up DNA from nearby *S. pneumoniae* cells that have died and released their DNA into the environment.

Other bacterial species promote the uptake of DNA among members of their own species via **DNA uptake signal sequences,** which are 9 or 10 bp long. In the human pathogens *Neisseria meningitidis* (a causative agent of meningitis), *N. gonorrhoeae* (a causative agent of gonorrhea), and *Haemophilus influenzae* (a causative agent of ear, sinus, and respiratory infections), these sequences are found at many locations within their respective genomes. For example, *H. influenzae* contains approximately 1,500 copies of the sequence 5'-AAGTGCGGT-3' in its genome, while *N. meningitidis* contains about 1,900 copies of the sequence 5'-GCCGTCTGAA-3'. DNA fragments that contain their own uptake signal sequence are preferentially taken up by these species compared to other DNA fragments. For example, *H. influenzae* is much more likely to take up a DNA fragment with the sequence 5'-AAGTGCGGT-3'. For this reason, transformation is more likely to involve DNA uptake between members of the same species.

Transformation has also been used to map many bacterial genes, using methods similar to the cotransduction experiments described earlier. If two genes are close together, the **cotransformation** frequency is expected to be high, whereas genes that are far apart will have a cotransformation frequency that is very low or even zero. Like cotransduction, genetic mapping via cotransformation is used only to map genes that are relatively close together.

Horizontal Gene Transfer Is the Transfer of Genes Between Different Species

The transmission of genes from mother cell to daughter cell or from parent to offspring is called **vertical gene transfer.** In comparison, a gene transferred between two different species is termed **horizontal gene transfer.** The three mechanisms of genetic transfer that we have considered—conjugation, transduction, and transformation—are important mechanisms for horizontal gene transfer among bacterial species. When analyzing the genomes of bacterial species, researchers have discovered that a sizable fraction of their genes are derived from horizontal gene transfer. For example, over the past 100 million years, *E. coli* and *Salmonella typhimurium* have acquired roughly 17% of their genes via horizontal gene transfer.

The types of genes that bacteria acquire via horizontal gene transfer are quite varied, though they commonly involve functions that are readily acted upon by natural selection. These include genes that confer antibiotic resistance, the ability to degrade toxic compounds, and pathogenicity. Geneticists have suggested that much of the speciation that has occurred in prokaryotic species is the result of horizontal gene transfer. In many cases, the acquisition of new genes allows a novel survival strategy that has led to the formation of a new species. These processes are considered in detail in Chapters 24 and 26.

The medical relevance of horizontal gene transfer is quite profound. Antibiotics are commonly prescribed to treat many bacterial illnesses, including infections of the respiratory tract, urinary tract, skin, ears, and eyes. In addition, antibiotics are used in agriculture as a supplement in animal feed and to control certain bacterial diseases of high-value fruits and vegetables. Unfortunately, however, the widespread and uncontrolled use of antibiotics has promoted the prevalence of antibiotic-resistant strains of bacteria. This phenomenon, termed **acquired antibiotic resistance,** may occur via genetic alterations in the bacteria's own genome or by the horizontal transfer of resistance genes from a resistant strain to a sensitive strain. Resistant strains carry genes that counteract the effects of antibiotics. Such resistance genes encode proteins that either break down the drug, pump the drug out of the cell, or prevent the drug from inhibiting cellular processes.

Bacterial resistance to antibiotics in community-acquired respiratory tract infections, such as pneumonia, as well as other medical illnesses, is a serious problem, and it is increasing in prevalence worldwide at an alarming rate. As often mentioned in the news media, antibiotic resistance has increased dramatically over the past few decades, and resistance has been reported in almost all species of bacteria. In many countries, for example, penicillin resistance in *Streptococcus pneumoniae* is found in over 50% of all strains, with resistance to other drugs rising as well. Likewise, the antibiotic-resistance problem in hospitals continues to worsen. Resistant strains of *Klebsiella pneumoniae* and *Enterococcus* are significant causes of morbidity and mortality among critically ill patients in intensive care units. Treating infections caused by these pathogens poses increasingly difficult therapeutic dilemmas.

6.2 INTRAGENIC MAPPING IN BACTERIOPHAGES

Let's now turn our attention to **viruses**—small particles that have genetic material and can propagate with the aid of a host cell. Biologists do not consider viruses to be living entities because they rely on a host cell for their existence and proliferation. Nevertheless, we can think of viruses as having traits because they have unique biological structures and functions. Each type of virus has its own genetic material, which contains many genes. In this section, we will focus our attention on a bacteriophage called T4. Its genetic material contains several dozen different genes encoding proteins that carry out a variety of functions. For example, some of the genes encode proteins needed for the synthesis of new viruses and the lysis of the host cell. Other genes encode the viral coat proteins that are found in the head, shaft,

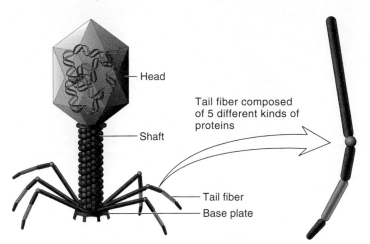

Head

Shaft

Tail fiber composed of 5 different kinds of proteins

Tail fiber

Base plate

FIGURE 6.13   **Structure of the T4 virus.**

Genes → Traits The inset to this figure shows the five different proteins, encoded by five different genes, that make up a tail fiber. The expression of these genes provides the bacteriophage with the trait of having tail fibers, enabling it to attach itself to the surface of a bacterium.

base plate, and tail fibers. **Figure 6.13** illustrates the structure of T4. As seen here, five different proteins bind to each other to form a tail fiber. The expression of T4 genes to make these proteins provides the bacteriophage with the trait of having tail fibers, enabling it to attach to the surface of a bacterium.

The study of viral genes has been instrumental in our basic understanding of how the genetic material works. During the 1950s, Seymour Benzer embarked on a 10-year study that focused on the function of viral genes in the T4 bacteriophage. In this section, we will examine some of his pivotal results. We will also explore how he conducted a detailed type of genetic mapping known as **intragenic mapping** or **fine structure mapping.**

Earlier in this chapter and in Chapter 5, we explored intergenic mapping, the goal of which is to determine the distance between two different genes. The determination of the distance between genes *A* and *B* is an example of intergenic mapping:

gene *A*   gene *B*

DNA

Distance between gene *A* and gene *B*

Intergenic mapping

In comparison, intragenic mapping seeks to establish distances between two or more mutations within the same gene. For example, in a population of viruses, gene *C* may exist as two different mutant alleles: One allele may be due to a mutation near the beginning of the gene; the second, to a mutation near the end:

gene *C* (allele 1)

gene *C* (allele 2)

Distance between 2 mutations in gene *C*

Intragenic mapping (also known as fine structure mapping)

In this section, we will explore the pioneering studies that led to the development of intragenic mapping and advanced our knowledge of gene function. Benzer's results showed that, rather than being an indivisible particle, a gene must be composed of a large structure that can be subdivided during intragenic crossing over.

Mutations in Viral Genes Can Alter Plaque Morphology

As they progress through the lytic cycle, bacteriophages ultimately produce new phages, which are released when the bacterial cell lyses (refer back to Figure 6.9, left side). In the laboratory, researchers can visually observe the consequences of bacterial cell lysis in the following way (**Figure 6.14**). A sample of bacterial cells and lytic bacteriophages are mixed together and then poured onto petri plates containing nutrient agar for bacterial cell growth. Bacterial cells that are not infected by a bacteriophage will rapidly grow and divide to produce a "lawn" of bacteria. This lawn of bacteria is opaque—you cannot see through it to the underlying agar. In the experiment shown in Figure 6.14, 11 bacterial cells have been infected by bacteriophages and these infected cells are found at random locations in the lawn of uninfected bacteria. The infected cells will lyse and release newly made bacteriophages. These bacteriophages will then infect the nearby bacteria within the lawn. These cells eventually lyse and also release newly made phages. Over time, these repeated cycles of infection and lysis will produce an observable clear area, or **plaque,** where the bacteria have been lysed around the original site where a phage infected a bacterial cell.

Now that we have an appreciation for the composition of a viral plaque, let's consider how the characteristics of a plaque can be viewed as a trait of a bacteriophage. As we have seen, the genetic analysis of any organism requires strains with allelic differences. Because bacteriophages can be visualized only with an electron microscope, it would be rather difficult for geneticists to analyze mutations that affect phage morphology. However, some mutations in the bacteriophage's genetic material can alter the ability of the phage to cause plaque formation. Therefore, we can view the morphology of plaques as a trait of the bacteriophage. Because plaques are visible with the unaided eye, mutations affecting this trait lend themselves to a much easier genetic analysis. An example is a rapid-lysis mutant of bacteriophage T4, which tends to form unusually large plaques (**Figure 6.15**). The plaques are large because the mutant phages lyse the bacterial cells more rapidly than do the wild-type phages. Rapid-lysis mutants form large, clearly defined plaques, as opposed to wild-type bacteriophages, which produce smaller, fuzzy-edged plaques. Mutations in different bacteriophage genes can produce a rapid-lysis phenotype.

Benzer studied one category of T4 phage mutants, designated *rII* (*r* stands for rapid lysis). In the bacterial strain called *E. coli* B, *rII* phages produce abnormally large plaques. Nevertheless, *E. coli* B strains produce low yields of *rII* phages because the *rII* phages lyse the bacterial cells so quickly they do not have sufficient time to produce many new phages. To help study this phage, Benzer wanted to obtain large quantities of it. Therefore, to improve his yield of *rII* phage, he decided to test its yield in other bacterial strains.

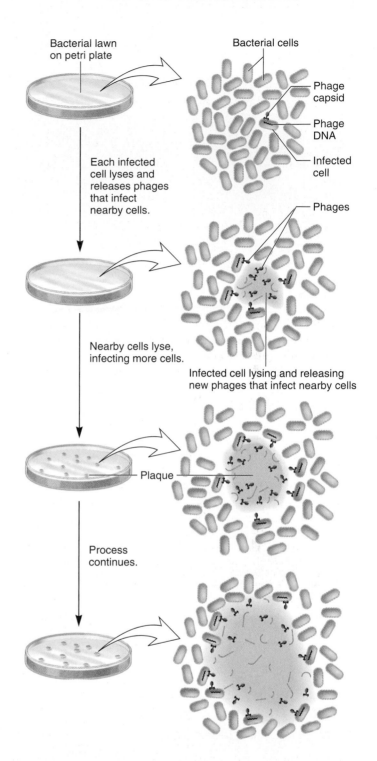

FIGURE 6.14   The formation of phage plaques on a lawn of bacteria in the laboratory. In this experiment, bacterial cells were mixed with a small number of lytic bacteriophages. In this particular example, eleven bacterial cells were initially infected by phages. The cells were poured onto petri plates containing nutrient agar for bacterial cell growth. Bacterial cells will rapidly grow and divide to produce an opaque lawn of densely packed bacteria. The 11 infected cells will lyse and release newly made bacteriophages. These bacteriophages will then infect the nearby bacteria within the lawn. Likewise, these newly infected cells will lyse and release new phages. By this repeated process, the area of cell lysis creates a clear zone known as a viral plaque.

Wild-type plaque   Plaque caused by a rapid-lysis mutant

Plaques caused by wild-type
bacteriophages

Plaques caused by rapid-lysis
bacteriophage strains

FIGURE 6.15   A comparison of plaques produced by the wild-type T4 bacteriophage and a rapid-lysis mutant.
Genes → Traits **(a)** These plaques were caused by the infection and lysis of *E. coli* cells by the wild-type T4 phage. **(b)** A mutation in a phage gene, called a rapid-lysis mutation, caused the phage to lyse *E. coli* cells more quickly. A phage carrying a rapid-lysis mutant allele yields much larger plaques with clearly defined edges.

On the day Benzer decided to do this, he happened to be teaching a phage genetics class. For that class, he was growing two *E. coli* strains designated *E. coli* K12S and *E. coli* K12(λ). He was growing these two strains to teach his class about the lysogenic cycle.

E. coli K12(λ) has DNA from another phage, called lambda, integrated into its chromosome, whereas *E. coli* K12S does not. To see if the use of these strains might improve phage yield, *E. coli* B, *E. coli* K12S, and *E. coli* K12(λ) were infected with the *rII* and wild-type T4 phage strains. As expected, the wild-type phage could infect all three bacterial strains. However, the *rII* mutant strains behaved quite differently. In *E. coli* B, the *rII* strains produced large plaques that had poor yields of bacteriophage. In *E. coli* K12S, the *rII* mutants produced normal plaques that gave good yields of phage. Surprisingly, in *E. coli* K12(λ), the *rII* mutants were unable to produce plaques at all, for reasons that were not understood. Nevertheless, as we will see later, this fortuitous observation was a critical feature that allowed intragenic mapping in this bacteriophage.

A Complementation Test Can Reveal If Mutations Are in the Same Gene or in Different Genes

In his experiments, Benzer was interested in a single trait, namely, the ability to form plaques. He had isolated many *rII* mutant strains that could form large plaques in *E. coli* B but could not produce plaques in *E. coli* K12(λ). To attempt gene mapping, he needed to know if the various *rII* mutations were in the same gene or if they involved mutations in different genes. To accomplish this, he conducted a **complementation test.** In this type of approach, the goal is to determine if two different mutations that affect the same trait are in the same gene or in two different genes (also see Figure 4.20).

The possible outcomes of complementation tests involving mutations that affect plaque formation are shown in **Figure 6.16.** This example involves four different *rII* mutations in T4

rII strain 1
(gene *A* is defective,
gene *B* is normal)

gene *A*   gene *B*

↓ Coinfect *E. coli* K12 (λ)

↓ Plate and observe if plaques are formed.

No plaques

No complementation occurs, because the coinfected cell is unable to make the normal product of gene *A*. The coinfected cell will not produce viral particles, thus no bacterial cell lysis and no plaque formation.

(a) Noncomplementation: The phage mutations are in the same gene.

rII strain 3
(gene *A* is defective,
gene *B* is normal)

gene *A*   gene *B*

rII strain 4
(gene *A* is normal,
gene *B* is defective)

gene *A*   gene *B*

↓ Coinfect *E. coli* K12 (λ)

↓ Plate and observe if plaques are formed.

Viral plaques

Complementation occurs, because the coinfected cell is able to make normal products of gene *A* and gene *B*. The coinfected bacterial cell will produce viral particles that lyse the cell, resulting in the appearance of clear plaques.

(b) Complementation: The phage mutations are in different genes.

FIGURE 6.16   A comparison of noncomplementation and complementation. Four different T4 phage strains (designated 1 through 4) that carry *rII* mutations were coinfected into *E. coli* K12(λ). **(a)** If two *rII* phage strains possess mutations in the same gene, noncomplementation will occur. **(b)** If the *rII* mutations are in different genes (such as gene *A* and gene *B*), a coinfected cell will have two mutant genes but also two wild-type genes. Doubly infected cells with a wild-type copy of each gene can produce new phages and form plaques. This result is called complementation because the defective genes in each *rII* strain are complemented by the corresponding wild-type genes.

bacteriophage, designated strains 1 through 4, that prevent plaque formation in *E. coli* K12(λ). To conduct this complementation experiment, bacterial cells were coinfected with two different strains of T4 phage. Two distinct outcomes are possible. In Figure 6.16a, the two *rII* phage strains possess deleterious mutations in the same gene (gene *A*). Because they cannot make a wild-type gene *A* product when coinfected into an *E. coli* K12(λ) cell, plaques will not form. This phenomenon is called **noncomplementation.**

Alternatively, if each *rII* mutation is in a different phage gene (e.g., gene *A* and gene *B*), a bacterial cell that is coinfected by both types of phages will have two mutant genes as well as two wild-type genes (Figure 6.16b). If the mutant phage genes behave in a recessive fashion, the doubly infected cell will have a wild-type phenotype. Why does this phenotype occur? The coinfected cells will produce normal proteins that are encoded by the wild-type versions of both genes *A* and *B*. For this reason, coinfected cells will be lysed in the same manner as if infected by the wild-type strain. Therefore, this coinfection should be able to produce plaques in *E. coli* K12(λ). This result is called **complementation** because the defective genes in each *rII* strain are complemented by the corresponding wild-type genes. It should be noted that, for a variety of reasons, intergenic complementation may not always work. One possibility is that a mutation may behave in a dominant fashion. In addition, mutations that affect regulatory genetic regions rather than the protein-coding region may not show complementation.

By carefully considering the pattern of complementation and noncomplementation, Benzer found that the *rII* mutations occurred in two different genes, which were termed *rIIA* and *rIIB*. The identification of two distinct genes affecting plaque formation was a necessary step that preceded his intragenic mapping analysis, which is described next. Benzer coined the term **cistron** to refer to the smallest genetic unit that gives a negative complementation test. In other words, if two mutations occur within the same cistron, they cannot complement each other. Since these studies, researchers have learned that a cistron is equivalent to a gene. In recent decades, the term gene has gained wide popularity while the term cistron is not commonly used. However, the term polycistronic is still used to describe bacterial mRNAs that carry two or more gene sequences, as described in Chapter 14.

Intragenic Maps Were Constructed Using Data from a Recombinational Analysis of Mutants Within the *rII* Region

As we have seen in Figure 6.16, the ability of strains with mutations in two different genes to produce viral plaques upon coinfection is due to complementation. Noncomplementation occurs when two different strains have mutations in the same gene. However, at an extremely low rate, two noncomplementing strains of viruses can produce an occasional viral plaque if intragenic recombination has taken place. For example, **Figure 6.17** shows a coinfection experiment between two phage strains that both contain *rII* mutations in gene *A*. These mutations are located at different places within the same gene. On rare occasions, a crossover may occur in the very short region between

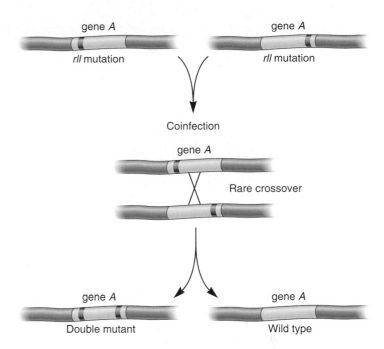

FIGURE 6.17 **Intragenic recombination.** Following coinfection, a rare crossover has occurred between the sites of the two mutations. This produces a wild-type phage with no mutations and a double mutant phage with both mutations.

each mutation. This crossover produces a double mutant gene *A* and a wild-type gene *A*. Because this event has produced a wild-type gene *A*, the function of the protein encoded by gene *A* will be restored. Therefore, new phages can be made in *E. coli* K12(λ), resulting in the formation of viral plaques.

Figure 6.18 describes the general strategy for intragenic mapping of *rII* phage mutations. Bacteriophages from two different noncomplementing *rII* phage mutants (here, *r103* and *r104*) were mixed together in equal numbers and then infected into *E. coli* B. In this strain, the *rII* mutants grew and propagated. Recall from Figure 6.17, when two different mutants coinfect the same cell, intragenic recombination can occur, producing wild-type phages and double mutant phages. However, these intragenic recombinants were produced at a very low rate. Following coinfection and lysis of *E. coli* B, a new population of phages was isolated. This population was expected to contain predominantly nonrecombinant phages. However, due to intragenic recombination, it should also contain a very low percentage of wild-type phages and double mutant phages (refer back to Figure 6.17).

How could Benzer determine the number of rare phages that were produced by intragenic recombination? The key approach is that *rII* mutant phages cannot grow in *E. coli* K12(λ). Following coinfection, he took this new population of phages and used some of them to infect *E. coli* B and some to infect *E. coli* K12(λ). After plating, the *E. coli* B infection was used to determine the total number of phages, because *rII* mutants as well as wild-type phages can produce plaques in this strain. The overwhelming majority of these phages were expected to be nonrecombinant phages. The *E. coli* K12(λ) infection was used to determine the number of rare intragenic recombinants that produce wild-type phages.

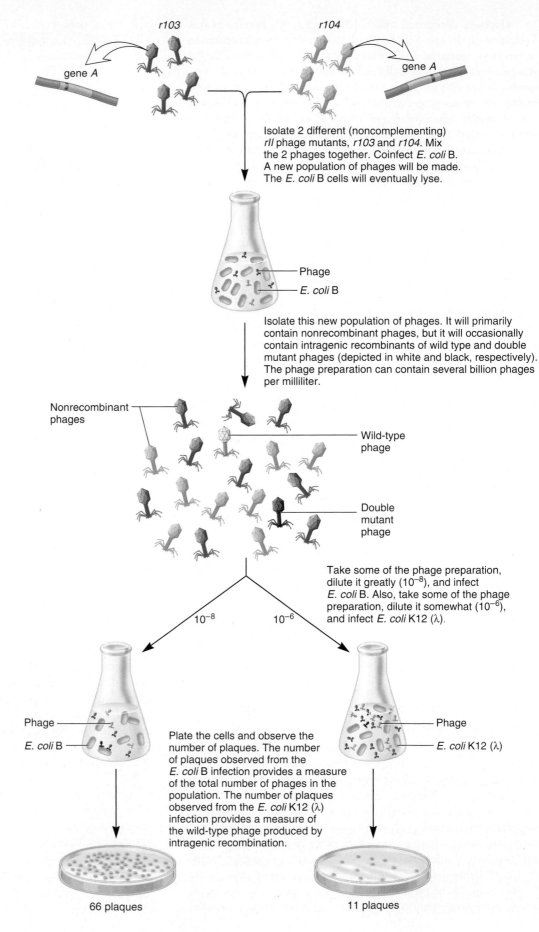

r103

r104

gene *A*

gene *A*

Isolate 2 different (noncomplementing) *rII* phage mutants, *r103* and *r104*. Mix the 2 phages together. Coinfect *E. coli* B. A new population of phages will be made. The *E. coli* B cells will eventually lyse.

Phage

E. coli B

Isolate this new population of phages. It will primarily contain nonrecombinant phages, but it will occasionally contain intragenic recombinants of wild type and double mutant phages (depicted in white and black, respectively). The phage preparation can contain several billion phages per milliliter.

Nonrecombinant phages

Wild-type phage

Double mutant phage

Take some of the phage preparation, dilute it greatly (10^{-8}), and infect *E. coli* B. Also, take some of the phage preparation, dilute it somewhat (10^{-6}), and infect *E. coli* K12 (λ).

10^{-8}

10^{-6}

Phage

E. coli B

Plate the cells and observe the number of plaques. The number of plaques observed from the *E. coli* B infection provides a measure of the total number of phages in the population. The number of plaques observed from the *E. coli* K12 (λ) infection provides a measure of the wild-type phage produced by intragenic recombination.

Phage

E. coli K12 (λ)

FIGURE 6.18   **Benzer's method of intragenic mapping in the *rII* region.**

66 plaques

11 plaques

Figure 6.18 illustrates the great advantage of this experimental system in detecting a low percentage of recombinants. In the laboratory, phage preparations containing several billion phages per milliliter are readily made. Among billions of phages, a low percentage (e.g., 1 in every 1,000) may be wild-type phages arising from intragenic recombination. The wild-type recombinants can produce plaques in *E. coli* K12(λ), whereas the *rII* mutant strains cannot. In other words, only the tiny fraction of wild-type recombinants would produce plaques in *E. coli* K12(λ).

The frequency of recombinant phages can be determined by comparing the number of wild-type phages, produced by intragenic recombination, and the total number of phages. As shown in Figure 6.18, the total number of phages can be deduced from the number of plaques obtained from the infection of *E. coli* B. In this experiment, the phage preparation was diluted by 10^8 (1:100,000,000), and 1 ml was used to infect *E. coli* B. Because this plate produced 66 plaques, the total number of phages in the original preparation was $66 \times 10^8 = 6.6 \times 10^9$, or 6.6 billion phages per milliliter. By comparison, the phage preparation used to infect *E. coli* K12(λ) was diluted by only 10^6 (1 : 1,000,000). This plate produced 11 plaques. Therefore, the number of wild-type phages was 11×10^6, which equals 11 million wild-type phages per milliliter.

As we have already seen in Chapter 5, genetic mapping distance is computed by dividing the number of recombinants by the total population (nonrecombinants and recombinants) times 100. In this experiment, intragenic recombination produced an equal number of two types of recombinants: wild-type phages and double mutant phages. Only the wild-type phages are detected in the infection of *E. coli* K12(λ). Therefore, to obtain the total number of recombinants, the number of wild-type phages must be multiplied by two. With all this information, we can compute the frequency of recombinants using the experimental approach described in Figure 6.18.

$$\text{Frequency of recombinants} = \frac{2\,[\text{Wild-type plaques obtained in } E.\ coli\ \text{K12}(\lambda)]}{\text{Total number of plaques obtained in } E.\ coli\ \text{B}}$$

We can use this equation to calculate the frequency of recombinants obtained in the experiment described in Figure 6.18.

$$\text{Frequency of recombinants} = \frac{2(11 \times 10^6)}{6.6 \times 10^9}$$

$$= 3.3 \times 10^{-3} = 0.0033$$

In this example, approximately 3.3 recombinants were produced per 1,000 phages.

The frequency of recombinants provides a measure of map distance. In eukaryotic mapping studies, we compute the map distance by multiplying the frequency of recombinants by 100 to give a value in map units (also known as centiMorgans). Similarly, in these experiments, the frequency of recombinants can provide a measure of map distance along the bacteriophage DNA. In this case, the map distance is between two mutations

within the same gene. Like intergenic mapping, the frequency of intragenic recombinants is correlated with the distance between the two mutations; the farther apart they are, the higher the frequency of recombinants. If two mutations happen to be located at exactly the same site within a gene, coinfection would not be able to produce any wild-type recombinants, and so the map distance would be zero. These are known as **homoallelic** mutations.

Deletion Mapping Can Be Used to Localize Many *rII* Mutations to Specific Regions in the *rIIA* or *rIIB* Genes

Now that we have seen the general approach to intragenic mapping, let's consider a method to efficiently map hundreds of *rII* mutations within the two genes designated *rIIA* and *rIIB*. As you may have realized, the coinfection experiments described in Figure 6.18 are quite similar to Sturtevant's strategy of making dihybrid crosses to map genes along the X chromosome of *Drosophila* (refer back to Figure 5.10). Similarly, Benzer wanted to coinfect different *rII* mutants in order to map the sites of the mutations within the *rIIA* and *rIIB* genes. During the course of his work, he obtained hundreds of different *rII* mutant strains that he wanted to map. However, making all the pairwise combinations would have been an overwhelming task. Instead, Benzer used an approach known as **deletion mapping** as a first step in localizing his *rII* mutations to a fairly short region within gene *A* or gene *B*.

Figure 6.19 describes the general strategy used in deletion mapping. This approach is easier to understand if we use an example. Let's suppose that the goal is to know the approximate location of an *rII* mutation, such as *r103*. To do so, *E. coli* K12(λ) would be coinfected with *r103* and a deletion strain. Each deletion strain is a T4 bacteriophage that is missing a known segment of the *rIIA* and/or *rIIB* gene. If the deleted region includes the same region that contains the *r103* mutation, a coinfection cannot produce intragenic wild-type recombinants. Therefore, plaques will not be formed. However, if a deletion strain recombines with *r103* to produce a wild-type phage, the deleted region does not overlap with the *r103* mutation.

In the example shown in Figure 6.19, the *r103* strain produced wild-type recombinants when coinfected with deletion strains *PB242*, *A105*, and *638*. However, coinfection of *r103* with *PT1*, *J3*, *1241*, and *1272* did not produce intragenic wild-type recombinants. Because coinfection with *PB242* produced recombinants and *PT1* did not, the *r103* mutation must be located in the region that is missing in *PT1* but not missing in *PB242*. As shown at the bottom of Figure 6.19, this region is called A4 (the A refers to the *rIIA* gene). In other words, the *r103* mutation is located somewhere within the A4 region, but not in the other six regions (A1, A2, A3, A5, A6, and B).

As described in Figure 6.19, this first step in the deletion mapping strategy localized an *rII* mutation to one of seven regions; six of these were in *rIIA* and one was in *rIIB*. Other deletion strains were used to eventually localize each *rII* mutation to one of 47 short regions; 36 were in *rIIA*, 11 in *rIIB*. At this point, pairwise coinfections were made between mutant strains that had been localized to the same region by deletion mapping. For

FIGURE 6.19   **The use of deletion strains to localize rII mutants to short regions within the rIIA or rIIB gene.** The deleted regions are shown in gray.

example, 24 mutations were deletion-mapped to a region called A5d. Pairwise coinfection experiments were conducted among this group of 24 mutants to precisely map their locations relative to each other in the A5d region. Similarly, all mutants in each of the 46 other groups were mapped by pairwise coinfections. In this way, a fine structure map was constructed depicting the locations of hundreds of different rII mutations (**Figure 6.20**). As seen in this figure, certain locations contained a relatively high number of mutations compared to other sites. These were termed **hot spots** for mutation.

Intragenic Mapping Experiments Provided Insight into the Relationship Between Traits and Molecular Genetics

Intragenic mapping studies were a pivotal achievement in our early understanding of gene structure. Since the time of Mendel, geneticists had considered a gene to be the smallest unit of heredity, which provided an organism with its inherited traits. In the late 1950s, however, the molecular nature of the gene was not understood. Because it is a unit of heredity, some scientists envisioned a gene as being a particle-like entity that could not be further subdivided into additional parts. However, intragenic mapping studies revealed, convincingly, that this is not the case. These studies

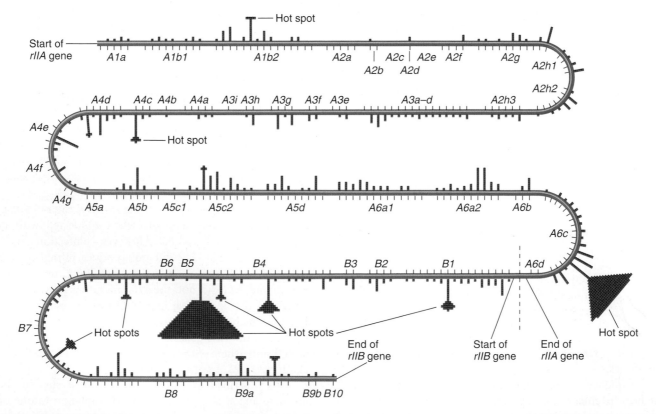

FIGURE 6.20   **The outcome of intragenic mapping of many rII mutations.** The blue line represents the linear sequence of the rIIA and rIIB genes, which are found within the T4 phage's genetic material. Each small purple box attached to the blue line symbolizes a mutation that was mapped by intragenic mapping. Among hundreds of independent mutant phages, several mutations sometimes mapped to the same site. In this figure, mutations at the same site form columns of boxes. Hot spots contain a large number of mutations and are represented as a group of boxes attached to a column of boxes. A hot spot contains many mutations at the same site within the rIIA or rIIB gene.

showed that mutations can occur at many different sites within a single gene. Furthermore, intragenic crossing over could recombine these mutations, resulting in wild-type genes. Therefore, rather than being an indivisible particle, a gene must be composed of a large structure that can be subdivided during crossing over.

Benzer's results were published in the late 1950s and early 1960s, not long after the physical structure of DNA had been elu-cidated by Watson and Crick. We now know that a gene is a segment of DNA that is composed of smaller building blocks called nucleotides. A typical gene is a linear sequence of several hundred to many thousand base pairs. As the genetic map of Figure 6.20 indicates, mutations can occur at many sites along the linear structure of a gene; intragenic crossing over can recombine mutations that are located at different sites within the same gene.

CONCEPTUAL SUMMARY

This chapter has been concerned with genetic transfer and mapping in bacteria. Three mechanisms are known for the transfer of genetic material from one bacterial cell to another: conjugation, transduction, and transformation. During conjugation, bacteria make direct physical contact with each other. Donor cells can be either F⁺, F', or Hfr cells. An F⁺ cell contains a circular piece of genetic material, a plasmid called an F factor that is transferred to the recipient (F⁻) cell during mating. An F' factor is an F factor that also carries genes derived from the bacterial chromosome. An Hfr cell (for high frequency of recombination) has an integrated F factor and can transfer chromosomal DNA to a recipient cell. A second route of genetic transfer is transduction. In this case, a bacteriophage accidentally packages a portion of the bacterial chromosome, which is then transferred to another bacterium upon infection. Finally, a third mechanism of genetic transfer is transformation. For this to occur, a dead bacterial cell must release its genetic material into the environment. A living bacterial cell in a competent state subsequently imports the DNA, and then recombination causes the imported DNA to replace segments of genetic material along the bacterial chromosome. Overall, these three mechanisms promote gene transfer within bacterial species and also allow horizontal gene transfer among different bacterial species. From a medical perspective, this phenomenon has led to a dramatic rise in the prevalence of strains that are resistant to antibiotics.

Chapter 6 concluded with a description of intragenic mapping studies in T4 bacteriophages. Benzer constructed a fine structure genetic map of two bacteriophage genes, *rIIA* and *rIIB*. In the early 1960s, his fine structure map provided important insights into the molecular nature of the gene. It revealed that a gene is not an indivisible unit. Rather, his results indicated that mutations can occur at many sites along the linear structure of a gene and that intragenic crossing over can recombine mutations located at different sites within the same gene. These results are consistent with our current knowledge of the molecular structure of genes.

EXPERIMENTAL SUMMARY

Research studies of genetic transfer in bacteria have provided a unique experimental strategy for mapping the linear order and relative locations of bacterial genes. Conjugation has been used to map the locations of many genes along the bacterial chromosome. In this approach, map distances are determined by the number of minutes it takes for a gene to enter a recipient cell during conjugation. In addition, cotransduction and cotransformation experiments have been commonly used to accurately map bacterial genes that are relatively close to each other on the chromosome.

As mentioned in the Conceptual Summary, intragenic mapping studies of T4 bacteriophages have provided great insight into the molecular nature of genes. Benzer constructed a fine structure genetic map of two bacteriophage genes, *rIIA* and *rIIB*. This was accomplished using a coinfection strategy, in which the very low percentage of intragenic recombinants yielding wild-type phages could be identified by their unique ability to infect a strain of bacteria known as *E. coli* K12(λ). This approach illustrates the power of phage genetics compared to the analysis of offspring in eukaryotic species. The selective inability of *rII* mutants to lyse *E. coli* K12(λ) allowed detection of the very few intragenic recombinants that produced wild-type phages that could lyse *E. coli* K12(λ). In this way, the very short distance that occurs between mutations within a single gene could be determined.

PROBLEM SETS & INSIGHTS

Solved Problems

S1. In *E. coli*, the gene *bioD*⁺ encodes an enzyme involved in biotin synthesis, and *galK*⁺ encodes an enzyme involved in galactose utilization. An *E. coli* strain that contained wild-type versions of both genes was infected with P1, and then a P1 lysate was obtained. This lysate was used to transduce (infect) a strain that was *bioD*⁻ and *galK*⁻. The cells were plated on media containing galactose as the sole carbon source for growth to select for transduction of the *galK*⁺ gene. These media also were supplemented with biotin. The colonies were then restreaked on media that lacked biotin to see if the *bioD*⁺ gene had been cotransduced. The following results were obtained:

| | | Number of Colonies That Grew On: | | |
Selected Gene	Non-selected Gene	Galactose + Biotin	Galactose − Biotin	Cotrans-duction Frequency
galK⁺	bioD⁺	80	10	0.125

How far apart are these two genes?

Answer: We can use the cotransduction frequency to calculate the distance between the two genes (in minutes) using the equation

$$\text{Cotransduction frequency} = (1 - d/2)^3$$

$$0.125 = (1 - d/2)^3$$

$$1 - d/2 = \sqrt[3]{0.125}$$

$$1 - d/2 = 0.5$$

$$d/2 = 1 - 0.5$$

$$d = 1.0 \text{ minute}$$

The two genes are approximately 1 minute apart on the *E. coli* chromosome.

S2. By conducting mating experiments between a single *Hfr* strain and a recipient strain, Wollman and Jacob mapped the order of many bacterial genes. Throughout the course of their studies, they identified several different *Hfr* strains in which the F factor DNA had been integrated at different places along the bacterial chromosome. A sample of their experimental results is shown in the following table:

Order of Transfer of Several Different Bacterial Genes

Hfr strain	Origin	First								Last
H	O	thr	leu	azi	ton	pro	lac	gal	str	met
1	O	leu	thr	met	str	gal	lac	pro	ton	azi
2	O	pro	ton	azi	leu	thr	met	str	gal	lac
3	O	lac	pro	ton	azi	leu	thr	met	str	gal
4	O	met	str	gal	lac	pro	ton	azi	leu	thr
5	O	met	thr	leu	azi	ton	pro	lac	gal	str
6	O	met	thr	leu	azi	ton	pro	lac	gal	str
7	O	ton	azi	leu	thr	met	str	gal	lac	pro

A. Explain how these results are consistent with the idea that the bacterial chromosome is circular.

B. Draw a map that shows the order of genes and the locations of the origins of transfer among these different *Hfr* strains.

Answer:

    A. In comparing the data among different *Hfr* strains, the order of the nine genes was always the same or the reverse of the same order. For example, *HfrH* and *Hfr1* transfer the same genes but their orders are reversed relative to each other. In addition, the *Hfr* strains showed an overlapping pattern of transfer with regard to the origin. For example, *Hfr1* and *Hfr2* had the same order of genes, but *Hfr1* began with *leu* and ended with *azi*, while *Hfr2* began with *pro* and ended with *lac*. From these findings, Wollman and Jacob concluded that the segment of DNA that was the origin of transfer had been inserted at different points within a circular *E. coli* chromosome in different *Hfr* strains. They also concluded that the origin can be inserted in either orientation, so the direction of gene transfer can be clockwise or counterclockwise around the circular bacterial chromosome.

B. A genetic map consistent with these results is shown here.

S3. An *Hfr* strain that is *leuA⁺* and *thiL⁺* was mated to a strain that is *leuA⁻* and *thiL⁻*. In the data points shown here, the mating was interrupted and the percentage of recombinants for each gene was determined by streaking on media that lacked either leucine or thiamine. The results are shown.

What is the map distance (in minutes) between these two genes?

Answer: This problem is solved by extrapolating the data points to the x-axis to determine the time of entry. For *leuA⁺*, they extrapolate back to 10 minutes. For *thiL⁺*, they extrapolate back to 20 minutes. Therefore, the distance between the two genes is approximately 10 minutes.

S4. Genetic transfer via transformation can also be used to map genes along the bacterial chromosome. In this approach, fragments of chromosomal DNA are isolated from one bacterial strain and used to transform another strain. The experimenter examines the transformed bacteria to see if they have incorporated two or more different genes. For example, the DNA may be isolated from a donor *E. coli* bacterium that has functional copies of the *araB* and *leuD* genes. Let's call these genes *araB⁺* and *leuD⁺* to indicate the genes are functional. These two genes are required for arabinose metabolism and leucine synthesis, respectively. To map the distance between these two genes via transformation, a recipient bacterium would be used that is *araB⁻* and *leuD⁻*. Following transformation, the recipient bacterium may become *araB⁺* and *leuD⁺*. This phenomenon is called cotransformation because two genes from the donor bacterium have been transferred to the recipient via transformation. In this type of experiment, the recipient cell is exposed to a fairly low concentration of donor DNA, making it unlikely that the recipient bacterium will take up more than one fragment of DNA. Therefore, under these conditions, cotransformation is likely only when two genes are fairly close together and are found on one fragment of DNA.

In a cotransformation experiment, a researcher has isolated DNA from an *araB*+ and *leuD*+ donor strain. This DNA was transformed into a recipient strain that was *araB*− and *leuD*−. Following transformation, the cells were plated on media containing arabinose and leucine. On this media, only bacteria that are *araB*+ can grow. The bacteria can be either *leuD*+ or *leuD*− because leucine is provided in the media. Colonies, which grew on this media, were then restreaked on media that contained arabinose but lacked leucine. Only *araB*+ and *leuD*+ cells could grow on these secondary plates. Following this protocol, a researcher obtained the following results:

Number of colonies growing on arabinose plus leucine media: 57

Number of colonies that grew when restreaked on arabinose media without leucine: 42

What is the map distance between these two genes? Note: This problem can be solved using the strategy of a cotransduction experiment except that the researcher must determine the average size of DNA fragments that are taken up by the bacterial cells. This would correspond to the value of *L* in a cotransduction experiment.

Answer: As mentioned, the basic principle of gene mapping via cotransformation is identical to the method of gene mapping via cotransduction described in this chapter. One way to calculate the map distance is to use the same equation that we used for cotransduction data, except that we substitute cotransformation frequency for cotransduction frequency.

$$\text{Cotransformation frequency} = (1 - d/L)^3$$

(Note: Cotransformation is not quite as accurate as cotransduction because the sizes of chromosomal pieces tend to vary significantly from experiment to experiment, so the value of *L* is not quite as reliable. Nevertheless, cotransformation has been used extensively to map the order and distance between closely linked genes along the bacterial chromosome.)

The researcher needs to experimentally determine the value of *L* by running the DNA on a gel and estimating the average size of the DNA fragments. Let's assume they are about 2% of the bacterial chromosome, which, for *E. coli*, would be about 80,000 base pairs in length. So *L* equals 2 minutes, which is the same as 2%.

$$\text{Cotransformation frequency} = (1 - d/L)^3$$

$$42/57 = (1 - d/2)^3$$

$$d = 0.2 \text{ minutes}$$

The distance between *araB* and *leuD* is approximately 0.2 minutes.

S5. In our discussion of transduction via P1 or P22, the life cycle of the bacteriophage sometimes resulted in the packaging of many different pieces of the bacterial chromosome. For other bacteriophages, however, transduction may involve the transfer of only a few specific genes from the donor cell to the recipient. This phenomenon is known as specialized transduction. The key event that causes specialized transduction to occur is that the lysogenic phase of the phage life cycle involves the integration of the viral DNA at a single specific site within the bacterial chromosome. The transduction of particular bacterial genes involves an abnormal excision of the phage DNA from this site within the chromosome that would carry adjacent bacterial genes. For example, a bacteriophage called lambda (λ) that infects *E. coli* specifically integrates between two genes designated *gal*+ and *bio*+ (required for galactose utilization and biotin synthesis, respectively). Either of these genes could be packaged into the phage if an abnormal excision event occurred. How would specialized transduction be different from generalized transduction?

Answer: Generalized transduction can involve the transfer of any bacterial gene, while specialized transduction can transfer only genes that are adjacent to the site where the phage integrates. As mentioned, a bacteriophage that infects *E. coli* cells, known as lambda (λ), provides a well-studied example of specialized transduction. In the case of phage lambda, the lysogenic life cycle results in the integration of the phage DNA at a site that is called the attachment site (described further in Chapter 17). The attachment site is located between two bacterial genes, *gal*+ and *bio*+. An *E. coli* strain that is lysogenic for phage lambda will have the lambda DNA integrated between these two bacterial genes. On occasion, the phage may enter the lytic cycle and excise its DNA from the bacterial chromosome. When this occurs normally, the phage excises its entire viral DNA from the bacterial chromosome. The excised phage DNA is then replicated and becomes packaged into newly made phages. However, an abnormal excision does occur at a low rate (i.e., about one in a million). In this abnormal event, the phage DNA is excised in such a way that an adjacent bacterial gene is included and some of the phage DNA is not included in the final product. For example, the abnormal excision may yield a fragment of DNA that includes the *gal*+ gene and some of the lambda DNA but is missing part of the lambda DNA. If this DNA fragment is packaged into a virus, it is called a defective phage because it is missing some of the phage DNA. If it carries the *gal*+ gene, it is designated λ*dgal* (the letter *d* designates a defective phage). Alternatively, an abnormal excision may carry the *bio*+ gene. This phage is designated λ*dbio*. Defective lambda phages can then transduce the *gal*+ or *bio*+ genes to other *E. coli* cells.

Conceptual Questions

C1. The terms conjugation, transduction, and transformation are used to describe three different natural forms of genetic transfer between bacterial cells. Briefly discuss the similarities and differences between these processes.

C2. Conjugation is sometimes called "bacterial mating." Is it a form of sexual reproduction? Explain.

C3. If you mix together an equal number of F+ and F− cells, how would you expect the proportions to change over time? In other words, do you expect an increase in the relative proportions of F+ or of F− cells? Explain your answer.

C4. What is the difference between an F+ and an *Hfr* strain? Which type of strain do you expect to transfer many bacterial genes to recipient cells?

C5. What is the role of the origin of transfer during F+- and Hfr-mediated conjugation? What is the significance of the direction of transfer in Hfr-mediated conjugation?

C6. What is the role of sex pili during conjugation?

C7. Think about the structure and transmission of F factors and discuss how you think F factors may have originated.

C8. Each species of bacteria has its own distinctive cell surface. The characteristics of the cell surface play an important role in processes such as conjugation and transduction. For example, certain strains of *E. coli* have pili on their cell surface. These pili enable *E. coli* to mate with other *E. coli*, and the pili also enable certain bacteriophages (such as M13) to bind to the surface of *E. coli* and gain entry into the cytoplasm. With these ideas in mind, explain which forms of genetic transfer (i.e., conjugation, transduction, and transformation) are more likely to occur between different species of bacteria. Discuss some of the potential consequences of interspecies genetic transfer.

C9. Briefly describe the lytic and lysogenic cycles of bacteriophages. In your answer, explain what a prophage is.

C10. What is cotransduction? What determines the likelihood that two genes will be cotransduced?

C11. When bacteriophage P1 causes *E. coli* to lyse, the resulting material is called a P1 lysate. What type of genetic material would be found in most of the P1 phages in the lysate? What kind of genetic material would occasionally be found within a P1 phage?

C12. As described in Figure 6.10, host DNA is hydrolyzed into small pieces, which are occasionally assembled with phage proteins, creating a phage with bacterial chromosomal DNA. If the breakage of the chromosomal DNA is not random (i.e., it is more likely to break at certain spots compared to other spots), how might nonrandom breakage affect cotransduction frequency?

C13. Describe the steps that occur during bacterial transformation. What is a competent cell? What factors may determine whether a cell will be competent?

C14. Which bacterial genetic transfer process does not require recombination with the bacterial chromosome?

C15. Researchers who study the molecular mechanism of transformation have identified many proteins in bacteria that function in the uptake of DNA from the environment and its recombination into the host cell's chromosome. This means that bacteria have evolved molecular mechanisms for the purpose of transformation by extracellular DNA. Of what advantage(s) would it be for a bacterium to import DNA from the environment and/or incorporate it into its chromosome?

C16. Antibiotics such as tetracycline, streptomycin, and bacitracin are small organic molecules that are synthesized by particular species of bacteria. Microbiologists have hypothesized that the reason why certain bacteria make antibiotics is to kill other species that occupy the same environment. Bacteria that produce an antibiotic may be able to kill competing species. This provides more resources for the antibiotic-producing bacteria. In addition, bacteria that have the genes necessary for antibiotic biosynthesis contain genes that confer resistance to the same antibiotic. For example, tetracycline is made by the soil bacterium *Streptomyces aureofaciens*. Besides the genes that are needed to make tetracycline, *S. aureofaciens* also contains genes that confer tetracycline resistance; otherwise it would kill itself when it makes tetracycline. In recent years, however, many other species of bacteria that do not synthesize tetracycline have acquired the genes that confer tetracycline resistance. For example, certain strains of *E. coli* carry tetracycline-resistance genes, even though *E. coli* does not synthesize tetracycline. When these genes are analyzed at the molecular level, it has been found that they are evolutionarily related to the genes in *S. aureofaciens*. This observation indicates that the genes from *S. aureofaciens* have been transferred to *E. coli*.

A. What form of genetic transfer (i.e., conjugation, transduction, or transformation) would be the most likely mechanism of interspecies gene transfer?

B. Because *S. aureofaciens* is a nonpathogenic soil bacterium and *E. coli* is an enteric bacterium, do you think it was direct gene transfer, or do you think it may have occurred in multiple steps (i.e., from *S. aureofaciens* to other bacterial species and then to *E. coli*)?

C. How could the widespread use of antibiotics to treat diseases have contributed to the proliferation of many bacterial species that are resistant to antibiotics?

C17. What does the term complementation mean? If two different mutations that produce the same phenotype can complement each other, what can you conclude about the locations of each mutation?

C18. Intragenic mapping is sometimes called interallelic mapping. Explain why the two terms mean the same thing. In your own words, explain what an intragenic map is.

C19. As discussed in Chapter 12, genes are composed of a sequence of nucleotides. A typical gene in a bacteriophage is a few hundred or a few thousand nucleotides in length. If two different strains of bacteriophage T4 have a mutation in the *rIIA* gene that gives a rapid-lysis phenotype, yet they never produce wild-type phages by intragenic recombination when they are coinfected into *E. coli* B, what would you conclude about the locations of the mutations in the two different T4 strains?

Experimental Questions

E1. In the experiment of Figure 6.1, a *met⁻ bio⁻ thr⁺ leu⁺ thi⁺* cell could become *met⁺ bio⁺ thr⁺ leu⁺ thi⁺* by a (rare) double mutation that converts the *met⁻ bio⁻* genetic material into *met⁺ bio⁺*. Likewise, a *met⁺ bio⁺ thr⁻ leu⁻ thi⁻* cell could become *met⁺ bio⁺ thr⁺ leu⁺ thi⁺* by three mutations that convert the *thr⁻ leu⁻ thi⁻* genetic material into *thr⁺ leu⁺ thi⁺*. From the results of Figure 6.1, how do you know that the occurrence of 10 *met⁺ bio⁺ thr⁺ leu⁺ thi⁺* colonies is not due to these types of rare double or triple mutations?

E2. In the experiment of Figure 6.1, Lederberg and Tatum could not discern whether *met⁺ bio⁺* genetic material was transferred to the *met⁻ bio⁻ thr⁺ leu⁺ thi⁺* strain or if *thr⁺ leu⁺ thi⁺* genetic material was transferred to the *met⁺ bio⁺ thr⁻ leu⁻ thi⁻* strain. Let's suppose that one strain is streptomycin resistant (say, *met⁺ bio⁺ thr⁻ leu⁻ thi⁻*) while the other strain is sensitive to streptomycin. Describe an experiment that could determine whether the *met⁺ bio⁺* genetic material was transferred to the *met⁻ bio⁻ thr⁺ leu⁺ thi⁺* strain or if the *thr⁺ leu⁺ thi⁺* genetic material was transferred to the *met⁺ bio⁺ thr⁻ leu⁻ thi⁻* strain.

E3. Explain how a U-tube apparatus can distinguish between genetic transfer involving conjugation and genetic transfer involving transduction. Do you think a U-tube could be used to distinguish between transduction and transformation?

E4. What is an interrupted mating experiment? What type of experimental information can be obtained from this type of study? Why is it necessary to interrupt mating?

E5. In a conjugation experiment, what is meant by the time of entry? How is the time of entry determined experimentally?

E6. In your laboratory, you have an F^- strain of E. coli that is resistant to streptomycin and is unable to metabolize lactose, but it can metabolize glucose. Therefore, this strain can grow on media that contain glucose and streptomycin, but it cannot grow on media containing lactose. A researcher has sent you two E. coli strains in two separate tubes. One strain, let's call it strain A, has an F factor that carries the genes that are required for lactose metabolism. On its chromosome, it also has the genes that are required for glucose metabolism. However, it is sensitive to streptomycin. This strain can grow on media containing lactose or glucose, but it cannot grow if streptomycin is added to the media. The second strain, let's call it strain B, is an F^- strain. On its chromosome, it has the genes that are required for lactose and glucose metabolism. Strain B is also sensitive to streptomycin. Unfortunately, when strains A and B were sent to you, the labels had fallen off the tubes. Describe how you could determine which tubes contain strain A and strain B.

E7. As mentioned in solved problem S2, origins of transfer can be located in many different locations, and their direction of transfer can be clockwise or counterclockwise. Let's suppose a researcher mated six different Hfr strains that were thr^+ leu^+ ton^s str^r azi^s lac^+ gal^+ pro^+ met^+ to an F^- strain that was thr^- leu^- ton^r str^s azi^r lac^- gal^- pro^- met^-, and obtained the following results:

Strain	Order of Gene Transfer
1	ton^s azi^s leu^+ thr^+ met^+ str^r gal^+ lac^+ pro^+
2	leu^+ azi^s ton^s pro^+ lac^+ gal^+ str^r met^+ thr^+
3	lac^+ gal^+ str^r met^+ thr^+ leu^+ azi^s ton^s pro^+
4	leu^+ thr^+ met^+ str^r gal^+ lac^+ pro^+ ton^s azi^s
5	ton^s pro^+ lac^+ gal^+ str^r met^+ thr^+ leu^+ azi^s
6	met^+ str^r gal^+ lac^+ pro^+ ton^s azi^s leu^+ thr^+

Draw a circular map of the E. coli chromosome and describe the locations and orientations of the origins of transfer in these six Hfr strains.

E8. An Hfr strain that is $hisE^+$ and $pheA^+$ was mated to a strain that is $hisE^-$ and $pheA^-$. The mating was interrupted and the percentage of recombinants for each gene was determined by streaking on media that lacked either histidine or phenylalanine. The following results were obtained:

A. Determine the map distance (in minutes) between these two genes.

B. In a previous experiment, it was found that $hisE$ is 4 minutes away from the gene $pabB$. PheA was shown to be 17 minutes from this gene. Draw a genetic map describing the locations of all three genes.

E9. Acridine orange is a chemical that inhibits the replication of F factor DNA but does not affect the replication of chromosomal DNA, even if the chromosomal DNA contains an Hfr. Let's suppose that you have an E. coli strain that is unable to metabolize lactose and has an F factor that carries a streptomycin-resistant gene. You also have an F^- strain of E. coli that is sensitive to streptomycin and has the genes that allow the bacterium to metabolize lactose. This second strain can grow on lactose-containing media. How would you generate an Hfr strain that is resistant to streptomycin and can metabolize lactose? (Hint: F factors occasionally integrate into the chromosome to become Hfr strains, and occasionally Hfr strains excise their DNA from the chromosome to become F^+ strains that carry an F' factor.)

E10. In a P1 transduction experiment, the P1 lysate contains phages that carry pieces of the host chromosomal DNA, but the lysate also contains broken pieces of chromosomal DNA (see Figure 6.10). If a P1 lysate is used to transfer chromosomal DNA to another bacterium, how could you show experimentally that the recombinant bacterium has been transduced (i.e., taken up a P1 phage with a piece of chromosomal DNA inside) versus transformed (i.e., taken up a piece of chromosomal DNA that is not within a P1 phage coat)?

E11. Could you devise an experimental strategy to get P1 phage to transduce the entire lambda genome from one strain of bacterium to another strain? (Note: The general features of phage lambda's life cycle are described in Chapter 14.) Phage lambda has a genome size of 48,502 nucleotides (about 1% of the size of the E. coli chromosome) and can follow the lytic or lysogenic life cycle. Growth of E. coli on minimal growth media favors the lysogenic life cycle, whereas growth on rich media and/or UV light promotes the lytic cycle.

E12. Let's suppose a new strain of P1 has been identified that packages larger pieces of the E. coli chromosome. This new P1 strain packages pieces of the E. coli chromosome that are 5 minutes long. If two genes are 0.7 minutes apart along the E. coli chromosome, what would be the cotransduction frequency using a normal strain of P1 and using this new strain of P1 that packages large pieces? What would be the experimental advantage of using this new P1 strain?

E13. If two bacterial genes are 0.6 minutes apart on the bacterial chromosome, what frequency of cotransductants would you expect to observe in a P1 transduction experiment?

E14. In an experiment involving P1 transduction, the cotransduction frequency was 0.53. How far apart are the two genes?

E15. In a cotransduction experiment, the transfer of one gene is selected for and the presence of the second gene is then determined. If 0 out of 1,000 P1 transductants that carry the first gene also carry the second gene, what would you conclude about the minimum distance between the two genes?

E16. In a cotransformation experiment (see solved problem S4), DNA was isolated from a donor strain that was $proA^+$ and $strC^+$ and sensitive to tetracycline. (The $proA$ and $strC$ genes confer the ability to synthesize proline and confer streptomycin resistance, respectively.) A recipient strain is *proA⁻* and *strC⁻* and is resistant to tetracycline. After transformation, the bacteria were first streaked on media containing proline, streptomycin, and tetracycline. Colonies were then restreaked on media containing streptomycin and tetracycline. (Note: Both types of media had carbon and nitrogen sources for growth.) The following results were obtained:

> 70 colonies grew on media containing proline, streptomycin, and tetracycline, while only 2 of these 70 colonies grew when restreaked on media containing streptomycin and tetracycline but lacking proline.

A. If we assume the average size of the DNA fragments is 2 minutes, how far apart are these two genes?

B. What would you expect the cotransformation frequency to be if the average size of the DNA fragments was 4 minutes and the two genes are 1.4 minutes apart?

E17. If you took a pipette tip and removed a phage plaque from a petri plate, what would it contain?

E18. As shown in Figure 6.16, phages with *rII* mutations cannot produce plaques in *E. coli* K12(λ), but wild-type phages can. From an experimental point of view, explain why this observation is so significant.

E19. In the experimental strategy described in Figure 6.18, explain why it was necessary to dilute the phage preparation used to infect *E. coli* B so much more than the phage preparation used to infect *E. coli* K12(λ).

E20. Here are data from several complementation experiments, involving rapid-lysis mutations in genes *rIIA* and *rIIB*. The strain designated *L51* is known to have a mutation in *rIIB*.

Phage Mixture	Complementation
L91 and L65	No
L65 and L62	No
L33 and L47	Yes
L40 and L51	No
L47 and L92	No
L51 and L47	Yes
L51 and L92	Yes
L33 and L40	No
L91 and L92	Yes
L91 and L33	No

List which groups of mutations are in the *rIIA* gene and which groups are in the *rIIB* gene.

E21. A researcher has several different strains of T4 phage with single mutations in the same gene. In these strains, the mutations render the phage temperature sensitive. This means that temperature-sensitive phages can propagate when the bacterium (*E. coli*) is grown at 32°C but cannot propagate themselves when *E. coli* is grown at 37°C. Think about Benzer's strategy for intragenic mapping and propose an experimental strategy to map the temperature-sensitive mutations.

E22. Explain how Benzer's results indicated that a gene is not an indivisible unit.

E23. Explain why deletion mapping was used as a step in the intragenic mapping of *rII* mutations.

Questions for Student Discussion/Collaboration

1. Discuss the advantages of the genetic analysis of bacteria and bacteriophages. Make a list of the types of allelic differences among bacteria and phages that are suitable for genetic analyses.

2. Complementation occurs when two defective alleles in two different genes are found within the same organism and produce a normal phenotype. What other examples of complementation have we encountered in previous chapters of this textbook?

Note: All answers appear at the website for this textbook; the answers to even-numbered questions are in the back of the textbook.

Visit the Online Learning Center for practice tests, answer keys, and other learning aids for this chapter. Enhance your understanding of genetics with our interactive exercises, quizzes, animations, and much more.

7

NON-MENDELIAN INHERITANCE

Mendelian inheritance patterns involve genes that directly influence the outcome of an offspring's traits and obey Mendel's laws. To predict phenotype, we must consider several factors. These include the dominant/recessive relationship of alleles, gene interactions that may affect the expression of a single trait, and the roles that sex and the environment play in influencing the individual's phenotype. Once these factors are understood, we can predict the phenotypes of offspring from their genotypes.

Most genes in eukaryotic species follow a Mendelian pattern of inheritance. However, many genes do not. For example, in Chapter 5 we saw how genes that are closely linked did not obey Mendel's law of independent assortment. One could argue that such inheritance patterns are also non-Mendelian. In this chapter, we will examine several additional and even bizarre types of inheritance patterns that deviate from a Mendelian pattern. In the first two sections of this chapter, we will consider two important examples of non-Mendelian inheritance called the maternal effect and epigenetic inheritance. Even though these inheritance patterns involve genes on chromosomes within the cell nucleus, the genotype of the offspring does not directly govern their phenotype in ways predicted by Mendel. We will see how the timing of gene expression and gene inactivation can cause a non-Mendelian pattern of inheritance.

In the third section, we will examine deviations from Mendelian inheritance that arise because some genetic material is not located in the cell nucleus. Certain cellular organelles, such as mitochondria and chloroplasts, contain their own genetic material. We will survey the inheritance of organellar genes and a few other examples in which traits are influenced by genetic material that exists outside of the cell nucleus.

7.1 MATERNAL EFFECT

We will begin by considering genes that have a **maternal effect.** This term refers to an inheritance pattern for certain **nuclear genes**—genes located on chromosomes that are found in the cell nucleus—in which the genotype of the mother directly determines the phenotype of her offspring. Surprisingly, for maternal effect genes, the genotypes of the father and offspring themselves do not affect the phenotype of the offspring. We will see that this phenomenon is explained by the accumulation of gene products that the mother provides to her developing eggs.

Shell coiling in the water snail, *Lymnaea peregra*. In this species, some snails coil to the left, while others coil to the right. This is due to an inheritance pattern called the maternal effect.

The Genotype of the Mother Determines the Phenotype of the Offspring for Maternal Effect Genes

The first example of a maternal effect gene was studied in the 1920s by A. E. Boycott and involved morphological features of the water snail, *Lymnaea peregra*. In this species, the shell and internal organs can be arranged in either a right-handed (dextral) or left-handed (sinistral) direction. The dextral orientation is more common and is dominant to the sinistral orientation. **Figure 7.1** describes the results of a genetic analysis carried out by Boycott. In this experiment, he began with two different true-breeding strains of snails with either a dextral or sinistral morphology. Many combinations of crosses produced results that could not be explained by a Mendelian pattern of inheritance. When a dextral female (*DD*) was crossed to a sinistral male (*dd*), all F$_1$ offspring were dextral. However, in the **reciprocal cross,** where a sinistral female (*dd*) was crossed to a dextral male (*DD*), all F$_1$ offspring were sinistral. Taken together, these results contradict a Mendelian pattern of inheritance.

Parental generation

♀ *DD* x *dd* ♂   ♀ *dd* x *DD* ♂

F$_1$ generation

♀ *Dd* x ♂ *Dd*   ♀ *Dd* x ♂ *Dd*
All dextral   All sinistral

F$_2$ generation

1 *DD* : 2 *Dd* : 1 *dd*   Males and females
All dextral

Cross to each other

F$_3$ generation

3 dextral : 1 sinistral   Males and females

FIGURE 7.1   Experiment showing the inheritance pattern of snail coiling. In this experiment, *D* (dextral) is dominant to *d* (sinistral). The genotype of the mother determines the phenotype of the offspring. This phenomenon is known as the maternal effect. In this case, a *DD* or *Dd* mother produces dextral offspring, and a *dd* mother produces sinistral offspring. The genotypes of the father and offspring do not affect the offspring's phenotype.

How can we explain the unusual results obtained in Figure 7.1? Alfred Sturtevant proposed the idea that snail coiling is due to a maternal effect gene that exists as a dextral (*D*) or sinistral (*d*) allele. His conclusions were drawn from the inheritance patterns of the F$_2$ and F$_3$ generations. In this experiment, the genotype of the F$_1$ generation is expected to be heterozygous (*Dd*). When these F$_1$ individuals were crossed to each other, a genotypic ratio of 1 *DD* : 2 *Dd* : 1 *dd* is predicted for the F$_2$ generation. Because the *D* allele is dominant to the *d* allele, a 3:1 phenotypic ratio of dextral to sinistral snails should be produced according to a Mendelian pattern of inheritance. Instead of this predicted phenotypic ratio, however, the F$_2$ generation was composed of all dextral snails. This incongruity with Mendelian inheritance is due to the maternal effect. The phenotype of the offspring depended solely on the genotype of the mother. The F$_1$ mothers were *Dd*. The *D* allele in the mothers is dominant to the *d* allele and caused the offspring to be dextral, even if the offspring's genotype was *dd*. When the members of the F$_2$ generation were crossed, the F$_3$ generation exhibited a 3:1 ratio of dextral to sinistral snails. This ratio corresponds to the genotypes of the F$_2$ females, which were the mothers of the F$_3$ generation. The ratio of F$_2$ females was 1 *DD* : 2 *Dd* : 1 *dd*. The *DD* and *Dd* females produced dextral offspring, whereas the *dd* females produced sinistral offspring. This explains the 3:1 ratio of dextral and sinistral offspring in the F$_3$ generation.

Female Gametes Receive Gene Products from the Mother That Affect Early Developmental Stages of the Embryo

At the molecular and cellular level, the non-Mendelian inheritance pattern of maternal effect genes can be explained by the process of oogenesis in female animals (**Figure 7.2a**). As an animal oocyte (egg) matures, many surrounding maternal cells called nurse cells provide the egg with nutrients and other materials. In Figure 7.2a, a female is heterozygous for the snail-coiling maternal effect gene, with the alleles designated *D* and *d*. Depending on the outcome of meiosis, the haploid egg may receive the *D* allele or the *d* allele, but not both. The surrounding nurse cells, however, produce both *D* and *d* gene products (mRNA and proteins). These gene products are then transported into the egg. As shown here, the egg has received both the *D* allele gene products and the *d* allele gene products. These gene products persist for a significant time after the egg has been fertilized and begins its embryonic development. In this way, the gene products of the nurse cells, which reflect the genotype of the mother, influence the early developmental stages of the embryo.

Now that we have an understanding of the relationship between oogenesis and maternal effect genes, let's reconsider the topic of snail coiling. As shown in **Figure 7.2b**, a female snail that is *DD* will transmit only the *D* gene products to the egg. During the early stages of embryonic development, these gene products will cause the egg cleavage to occur in a way that promotes a right-handed body plan. A heterozygous female will transmit both *D* and *d* gene products. Because the *D* allele is dominant, the maternal effect will also cause a right-handed body plan.

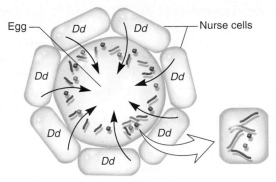

Egg — Nurse cells

The nurse cells express mRNA and/or protein from genes of the *d* allele (red) and the *D* allele (green) and transfer those products to the egg.

(a) Transfer of gene products from nurse cells to egg

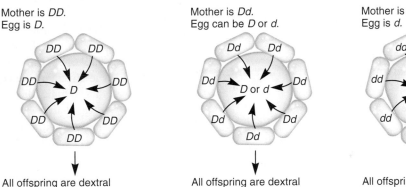

Mother is *DD*.
Egg is *D*.

Mother is *Dd*.
Egg can be *D* or *d*.

Mother is *dd*.
Egg is *d*.

All offspring are dextral because the egg received the gene products of the *D* allele.

All offspring are dextral because the egg received the gene products of the dominant *D* allele.

All offspring are sinistral because the egg only received the gene products of the *d* allele.

(b) Maternal effect in snail coiling

FIGURE 7.2 **The mechanism of maternal effect in snail coiling.** **(a)** Transfer of gene products from nurse cells to an egg. The nurse cells are heterozygous (*Dd*). Both the *D* and *d* alleles are activated in the nurse cells to produce *D* and *d* gene products (mRNA and/or proteins). These products are transported into the cytoplasm of the egg, where they accumulate to significant amounts. **(b)** Explanation of the maternal effect in snail coiling. **(c)** The direction of snail coiling is determined by differences in the cleavage planes during early embryonic development.

Genes → Traits If the nurse cells are *DD* or *Dd*, they will transfer the *D* gene product to the egg and thereby cause the resulting offspring to be dextral. If the nurse cells are *dd*, only the *d* gene product will be transferred to the egg, so the resulting offspring will be sinistral.

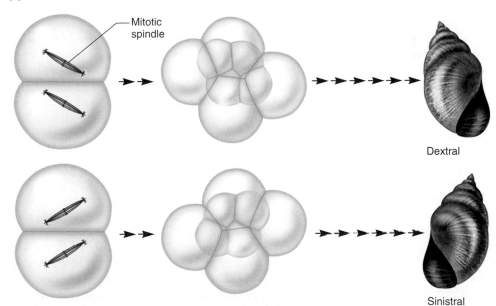

Mitotic spindle

Dextral

Sinistral

(c) An explanation of coiling direction at the cellular level

Finally, a *dd* mother will contribute only *d* gene products that promote a left-handed body plan, even if the egg is fertilized by a sperm carrying a *D* allele. The sperm's genotype is irrelevant, because the expression of the sperm's gene would occur too late. The origin of dextral and sinistral coiling can be traced to the orientation of the mitotic spindle at the two- to four-cell stage of embryonic development. The dextral and sinistral snails develop as mirror images of each other (**Figure 7.2c**).

Since these initial studies, researchers have found that maternal effect genes encode proteins that are important in the early steps of embryogenesis. The accumulation of maternal gene products in the egg allows embryogenesis to proceed quickly after fertilization. Maternal effect genes often play a role in cell division, cleavage pattern, and body axis orientation. Therefore, defective alleles in maternal effect genes tend to have a dramatic effect on the phenotype of the individual, altering major features of morphology, often with dire consequences.

Our understanding of maternal effect genes has been greatly aided by their identification in experimental organisms such as *Drosophila melanogaster*. In such organisms with a short generation time, geneticists have successfully searched for mutant alleles that prevent the normal process of embryonic development. In *Drosophila*, geneticists have identified several maternal effect genes with profound effects on the early stages of development. The pattern of development of a *Drosophila* embryo occurs along axes, such as the anteroposterior axis and the dorsoventral axis. The proper development of each axis requires a distinct set of maternal gene products. For example, the maternal effect gene called *bicoid* produces a gene product that accumulates in a region of the egg that will eventually become anterior structures in the developing embryo. Mutant alleles of maternal effect genes often lead to abnormalities in the anteroposterior or the dorsoventral pattern of development. More recently, several maternal effect genes have been identified in mice and humans that are required for proper embryonic development. Chapter 23 examines the relationships among the actions of several maternal effect genes during embryonic development.

7.2 EPIGENETIC INHERITANCE

As we have just seen, events during oogenesis can cause the inheritance pattern of traits to deviate from a Mendelian pattern. Likewise, **epigenetic inheritance** is a pattern in which a modification occurs to a nuclear gene or chromosome that alters gene expression, but is not permanent over the course of many generations. As we will see, epigenetic inheritance patterns are the result of DNA and chromosomal modifications that occur during oogenesis, spermatogenesis, or early stages of embryogenesis. Once they are initiated during these early stages, epigenetic changes alter the expression of particular genes in a way that may be fixed during an individual's lifetime. Therefore, epigenetic changes can permanently affect the phenotype of the individual. However, epigenetic modifications are not permanent over the course of many generations, and they do not change the actual DNA sequence. For example, a gene may undergo an epigenetic change that inactivates it for the lifetime of an individual. However, when this individual makes gametes, the gene may become activated and remain operative during the lifetime of an offspring who inherits the active gene.

In this section, we will examine two examples of epigenetic inheritance called dosage compensation and genomic imprinting. The effect of dosage compensation is to offset differences in the number of sex chromosomes. One of the sex chromosomes is altered, with the result that males and females have similar levels of gene expression, even though they do not possess the same complement of sex chromosomes. In mammals, dosage compensation is initiated during the early stages of embryonic development. By comparison, genomic imprinting happens prior to fertilization; it involves a change in a single gene or chromosome during gamete formation. Depending on whether the modification occurs during spermatogenesis or oogenesis, imprinting governs whether an offspring expresses a gene that has been inherited from its mother or father.

Dosage Compensation Is Necessary to Ensure Genetic Equality Between the Sexes

Dosage compensation refers to the phenomenon that the level of expression of many genes on the sex chromosomes (such as the X chromosome) is similar in both sexes even though males and females have a different complement of sex chromosomes. This term was coined in 1932 by Hermann Muller to explain the effects of eye color mutations in *Drosophila*. Muller observed that female flies homozygous for certain X-linked eye color alleles had a similar phenotype to hemizygous males. He noted that an X-linked gene conferring an apricot eye color produces a very similar phenotype in homozygous females and hemizygous males. In contrast, a female that has one copy of the apricot allele and a deletion of the apricot gene on the other X chromosome has eyes of paler color. Therefore, one copy of the allele in the female is not equivalent to one copy of the allele in the male. Instead, two copies of the allele in the female produce a phenotype that is similar to that produced by one copy in the male. In other words, the difference in gene dosage—two copies in females versus one copy in males—is being compensated at the level of gene expression.

Since these initial studies, dosage compensation has been studied extensively in mammals, *Drosophila*, and *Caenorhabditis elegans* (a nematode). Depending on the species, dosage compensation occurs via different mechanisms (**Table 7.1**). Female mammals equalize the expression of X-linked genes by turning off one of their two X chromosomes. This process is known as **X inactivation.** In *Drosophila*, the male accomplishes dosage compensation by doubling the expression of most X-linked genes. In *C. elegans*, the XX animal is a hermaphrodite that produces both sperm and egg cells, while an animal carrying a single X chromosome is a male that produces only sperm. The XX hermaphrodite diminishes the expression of X-linked genes on both X chromosomes to approximately 50% of that in the male.

In birds, the Z chromosome is a large chromosome, usually the fourth or fifth largest, which contains almost all of the known sex-linked genes. The W chromosome is generally a much smaller microchromosome containing a high proportion of repeat sequence DNA that does not encode genes. Males are ZZ, and females are ZW. Several years ago, researchers studied the level of expression of a Z-linked gene that encodes an enzyme called aconitase. They discovered that males express twice as much aconitase as females do. These results suggested that dosage compensation does not occur in birds. However, more recently the expression of nine Z-linked genes was examined in chickens, and at least six of the genes showed expression levels that were

TABLE **7.1**

Mechanisms of Dosage Compensation Among Different Species

| Species | Sex Chromosomes in: | | Mechanism of Compensation |
	Females	Males	
Placental mammals	XX	XY	One of the X chromosomes in the somatic cells of females is inactivated. In certain species, the paternal X chromosome is inactivated, while in other species, such as humans, either the maternal or paternal X chromosome is randomly inactivated throughout the somatic cells of females.
Marsupial mammals	XX	XY	The paternally derived X chromosome is inactivated in the somatic cells of females.
Drosophila melanogaster	XX	XY	The level of expression of genes on the X chromosome in males is increased 2-fold.
Caenorhabditis elegans	XX*	X0	The level of expression of genes on both X chromosomes in hermaphrodites is decreased to 50% levels compared to males.

*In *C. elegans*, an XX individual is a hermaphrodite, not a female.

similar in males and females. The interpretation of these recent results is that dosage compensation occurs in birds, but perhaps not for every gene. The molecular mechanism of dosage compensation is not well understood in birds. Recent evidence by Heather McQueen and colleagues suggests that dosage compensation in birds may be achieved by the modification of histone proteins along the single Z chromosome in the female. Such a modification could provide dosage compensation by increasing the expression of Z-linked genes twofold.

Dosage Compensation Occurs in Female Mammals by the Inactivation of One X Chromosome

In 1961, Mary Lyon proposed that dosage compensation in mammals occurs by the inactivation of a single X chromosome in females. Liane Russell also proposed the same theory around the same time. This proposal brought together two lines of study. The first type of evidence came from cytological studies. In 1949, Murray Barr and Ewart Bertram identified a highly condensed structure in the interphase nuclei of somatic cells in female cats that was not found in male cats. This structure became known as the **Barr body** (**Figure 7.3a**). In 1960, Susumu Ohno correctly proposed that the Barr body is a highly condensed X chromosome.

In addition to this cytological evidence, Lyon was also familiar with mammalian examples in which the coat color had a variegated pattern. **Figure 7.3b** illustrates a calico cat, which is a female that is heterozygous for an X-linked gene that can occur as an orange or a black allele. (The white underside is due to a dominant allele in a different gene.) The orange and black patches are randomly distributed in different female individuals. The calico pattern does not occur in male cats, but similar kinds of mosaic patterns have been identified in the female mouse. Lyon suggested that both the Barr body and the calico pattern are the result of X inactivation in the cells of female mammals.

The mechanism of X inactivation, also known as the **Lyon hypothesis,** is schematically illustrated in **Figure 7.4**. This example involves a white and black variegated coat color found in certain strains of mice. As shown here, a female mouse has inherited an X chromosome from its mother that carries an allele conferring white coat color (X^b). The X chromosome from its father carries a black coat color allele (X^B). How can X inactivation explain a variegated coat pattern? Initially, both X chromosomes are active. However, at an early stage of embryonic development, one of the two X chromosomes is randomly inactivated in each somatic cell and becomes a Barr body. For example, one embryonic cell may have the X^B chromosome inactivated. As

(a) Nucleus with a Barr body

(b) A calico cat

FIGURE 7.3   X chromosome inactivation in female mammals. (a) The left micrograph shows the Barr body on the periphery of a human nucleus after staining with a DNA-specific dye. Because it is compact, the Barr body is the most brightly staining. The white scale bar is 5 μm. The right micrograph shows the same nucleus using a yellow fluorescent probe that recognizes the X chromosome. The Barr body (inactive X) is more compact compared to the active X chromosome, which is to the left of the Barr body. **(b)** The fur pattern of a calico cat. Genes → Traits The pattern of black and orange fur on this cat is due to random X inactivation during embryonic development. The orange patches of fur are due to the inactivation of the X chromosome that carries a black allele; the black patches are due to the inactivation of the X chromosome that carries the orange allele. In general, only heterozygous female cats can be calico. A rare exception would be a male cat (XXY) that has an abnormal composition of sex chromosomes.

INTERACTIVE EXERCISE

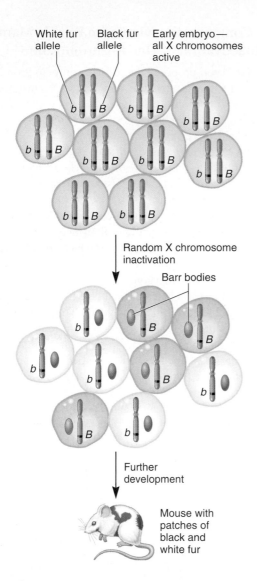

White fur allele   Black fur allele   Early embryo— all X chromosomes active

Random X chromosome inactivation

Barr bodies

Further development

Mouse with patches of black and white fur

the embryo continues to grow and mature, this embryonic cell will divide and may eventually give rise to billions of cells in the adult animal. The epithelial (skin) cells that are derived from this embryonic cell will produce a patch of white fur because the X^B chromosome has been permanently inactivated. Alternatively, another embryonic cell may have the other X chromosome inactivated (i.e., X^b). The epithelial cells derived from this embryonic cell will produce a patch of black fur. Because the primary event of X inactivation is a random process that occurs at an early stage of development, the result is an animal with some patches of white fur and other patches of black fur. This is the basis of the variegated phenotype.

During inactivation, the chromosomal DNA becomes highly compacted in a Barr body, so most of the genes on the inactivated X chromosome cannot be expressed. When cell division occurs and the inactivated X chromosome is replicated, both copies remain highly compacted and inactive. Likewise, during subsequent cell divisions, X inactivation is passed along to all future somatic cells.

FIGURE 7.4   The mechanism of X chromosome inactivation.

Genes → Traits The top of this figure represents a mass of several cells that compose the early embryo. Initially, both X chromosomes are active. At an early stage of embryonic development, random inactivation of one X chromosome occurs in each cell. This inactivation pattern is maintained as the embryo matures into an adult.

EXPERIMENT 7A

In Adult Female Mammals, One X Chromosome Has Been Permanently Inactivated

According to the Lyon hypothesis, each somatic cell of female mammals will express the genes on one of the X chromosomes, but not both. If an adult female is heterozygous for an X-linked gene, only one of two alleles will be expressed in any given cell. In 1963, Ronald Davidson, Harold Nitowsky, and Barton Childs set out to test the Lyon hypothesis at the cellular level. To do so, they analyzed the expression of a human X-linked gene that encodes an enzyme involved with sugar metabolism known as glucose-6-phosphate dehydrogenase (G-6-PD).

Prior to the Lyon hypothesis, biochemists had found that individuals vary with regard to the G-6-PD enzyme. This variation can be detected when the enzyme is subjected to gel electrophoresis (see the Appendix for a description of gel electrophore-

sis). One *G-6-PD* allele encodes a G-6-PD enzyme that migrates very quickly during gel electrophoresis (the "fast" enzyme), whereas another *G-6-PD* allele produces an enzyme that migrates more slowly (the "slow" enzyme). As shown in **Figure 7.5**, a sample of cells from heterozygous adult females produces both types of enzymes, whereas hemizygous males produce either the fast or slow type. The difference in migration between the fast and slow G-6-PD enzymes is due to minor differences in the structures of these enzymes. These minor differences do not significantly affect G-6-PD function, but they do enable geneticists to distinguish the proteins encoded by the two X-linked alleles.

As shown in **Figure 7.6**, Davidson, Nitowsky, and Childs tested the Lyon hypothesis using cell culturing techniques. They removed small samples of epithelial cells from a heterozygous female and grew them in the laboratory. When combined together, these samples contained a mixture of both types of

FIGURE 7.5 **Mobility of G-6-PD protein on gels.** *G-6-PD* can exist as a fast allele that encodes a protein that migrates more quickly to the bottom of the gel and a slow allele that migrates more slowly. The protein encoded by the fast allele is seen more toward the bottom of the gel.

enzymes because the adult cells were derived from many different embryonic cells, some that had the slow allele inactivated and some that had the fast allele inactivated. In the experiment

of Figure 7.6, these cells were sparsely plated onto solid growth media. After several days, each cell grew and divided to produce a colony, also called a **clone** of cells. All cells within a colony were derived from a single cell. The researchers reasoned that all cells within a single clone would express only one of the two *G-6-PD* alleles if the Lyon hypothesis was correct. Nine colonies were grown in liquid cultures, and then the cells were lysed to release the G-6-PD proteins inside of them. The proteins were then subjected to SDS gel electrophoresis.

THE HYPOTHESIS

According to the Lyon hypothesis, an adult female who is heterozygous for the fast and slow *G-6-PD* alleles should express only one of the two alleles in any particular somatic cell and its descendants, but not both.

TESTING THE HYPOTHESIS – FIGURE 7.6   Evidence that adult female mammals contain one X chromosome that has been permanently inactivated.

Starting material: Small skin samples taken from a woman who was heterozygous for the fast and slow alleles of *G-6-PD*.

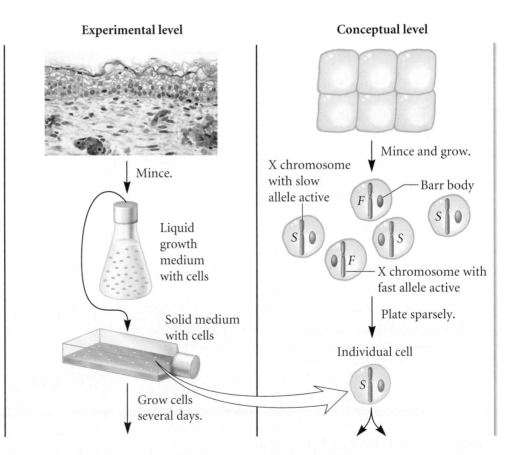

1. Mince the tissue to separate the individual cells.

2. Grow the cells in a liquid growth medium and then plate (sparsely) onto solid growth medium. The cells then divide to form a clone of many cells.

(continued)

3. Take nine isolated clones and grow in liquid cultures. (Only three are shown here.)

Clone of cells

Grow each clone in a separate flask.

Many cell divisions

Produces a clone of cells

4. Take cells from the liquid cultures, lyse cells to obtain proteins, and subject to gel electrophoresis. (This technique is described in the Appendix.)

Note: As a control, lyse cells from step 1, and subject the proteins to gel electrophoresis. This control sample is not from a clone. It is a mixture of cells derived from a woman's skin sample.

Sample

Control

Origin

Slow
Fast

⊖

⊕

From a clone with the fast allele active

From a clone with the slow allele active

■ THE DATA

All cells   Clones
1   2 3 4 5 6 7 8 9 10

← Slow G-6-PD
← Fast G-6-PD

■ INTERPRETING THE DATA

In the data shown in Figure 7.6, the control (lane 1) was a protein sample obtained from a mixture of epithelial cells from a heterozygous woman who produced both types of G-6-PD enzymes. Bands corresponding to the fast and slow enzymes were observed in this lane. As described in steps 2 to 4, this mixture of epithelial cells was also used to generate nine clones. The proteins obtained from these clones are shown in lanes 2 to 10. Each clone was a population of cells independently derived from a single epithelial cell. Because the epithelial cells were obtained from an adult female, the Lyon hypothesis predicts that each epithelial cell would already have one of its X chromosomes permanently inactivated and would pass this trait to its progeny cells. For example, suppose that an epithelial cell had inactivated the X chromosome that encoded the fast G-6-PD. If this cell was allowed to form a clone of cells on a plate, all cells in this clonal population would be expected to have the same X chromosome inactivated—the X chromosome encoding the fast G-6-PD. Therefore, this clone of cells should express only the slow G-6-PD. As shown in the data, all nine clones expressed either the fast or slow G-6-PD protein, but not both. These results are consistent with the hypothesis that X inactivation has already occurred in any given epithelial cell and that this pattern of inactivation is passed to all of its progeny cells.

A self-help quiz involving this experiment can be found at the Online Learning Center.

X Inactivation in Mammals Depends on the X-Inactivation Center and the *Xist* Gene

Since the Lyon hypothesis was confirmed, the genetic control of X inactivation has been investigated further by several laboratories. Research has shown that mammalian cells possess the ability to count their X chromosomes and allow only one of them to remain active. How was this determined? A key observation came from comparisons of the chromosome composition of people who have been born with normal or abnormal numbers of sex chromosomes.

Phenotype	Chromosome Composition	Number of X Chromosomes	Number of Barr Bodies
Normal female	XX	2	1
Normal male	XY	1	0
Turner syndrome (female)	X0	1	0
Triple X syndrome (female)	XXX	3	2
Klinefelter syndrome (male)	XXY	2	1

In normal females, two X chromosomes are counted and one is inactivated, while in males, one X chromosome is counted and none inactivated. If the number of X chromosomes exceeds two, as in triple X syndrome, additional X chromosomes are converted to Barr bodies.

Although the genetic control of inactivation is not entirely understood at the molecular level, a short region on the X chromosome called the **X-inactivation center (Xic)** is known to play a critical role (**Figure 7.7**). Eeva Therman and Klaus Patau identified the Xic from its key role in X inactivation. The counting of human X chromosomes is accomplished by counting the number of Xics. A Xic must be found on an X chromosome for inactivation to occur. Therman and Patau discovered that if one of the two X chromosomes in a female is missing its Xic due to a chromosome mutation, a cell counts only one Xic and X inactivation does not occur. Having two active X chromosomes is a lethal condition for a human female embryo.

Let's consider how the molecular expression of certain genes controls X inactivation. The expression of a specific gene within the Xic is required for the compaction of the X chromosome into a Barr body. This gene, discovered in 1991, is named *Xist* (for X-inactive specific transcript). The *Xist* gene on the inactivated X chromosome is active, which is unusual because most other genes on the inactivated X chromosome are silenced. The *Xist* gene product is an RNA molecule that does not encode a protein. Instead, the role of the *Xist* RNA is to coat the X chromosome and inactivate it. After coating, other proteins associate with the *Xist* RNA and promote chromosomal compaction into a Barr body.

Another gene found within the Xic, designated *Tsix*, also plays a role in X inactivation. As shown in Figure 7.7, the *Xist* and *Tsix* genes are overlapping and transcribed in opposite directions. (The name *Tsix* is *Xist* spelled backwards). The role of the *Tsix* gene is not completely understood, but it appears to decrease the expression of the *Xist* gene. On the inactive X chromosome, the *Xist* gene is expressed, and the *Tsix* gene is not. The opposite situation occurs on the active X chromosome—*Tsix* is expressed, and *Xist* is not. The choice of which X chromosome to inactivate involves a complex interplay between the expression of the *Xist* and *Tsix* genes. Researchers have studied heterozygous females that carry a normal *Tsix* gene on one X chromosome and a defective, mutant *Tsix* gene on the other. The X chromosome carrying the mutant *Tsix* gene is preferentially inactivated.

Another region termed the **X chromosomal controlling element (Xce)** also affects the choice of the X chromosome to be inactivated. Genetic variation occurs in the Xce. An X chromosome that carries a strong Xce is more likely to remain active than an X chromosome that carries a weak Xce, thereby leading to skewed (nonrandom) X inactivation. As shown in Figure 7.7, the Xce is very close to the end of the Xic and may even lie within the Xic. Although the mechanism by which the Xce exerts its effects are not well understood, some researchers speculate that Xce serves as a binding site for proteins that regulate the expression of genes in the Xic, such as *Xist* or *Tsix*. Genetic variation in Xce that enhances *Xist* expression would tend to promote Barr body formation, whereas Xce variation that enhances *Tsix* expression would tend to prevent X inactivation.

The process of X inactivation can be divided into three phases: initiation, spreading, and maintenance (**Figure 7.8**). During initiation, one of the X chromosomes is targeted to remain active, and the other is chosen to be inactivated. During the spreading phase, the chosen X chromosome is inactivated. This spreading requires the expression of the *Xist* gene. The *Xist* RNA coats the inactivated X chromosome and recruits proteins that promote compaction. This compaction occurs via the modification of histone proteins, which are described in Chapter 10. The spreading phase is so named because inactivation begins near the X-inactivation center and spreads in both directions along the X chromosome. Both initiation and spreading occur during embryonic development. Subsequently, the inactivated X chromosome is maintained as such during future cell divisions. When a cell divides, the Barr body is replicated, and both copies remain compacted.

Some genes on the inactivated X chromosome are expressed in the somatic cells of adult female mammals. These genes are said to escape the effects of X inactivation. As mentioned, *Xist* is an example of a gene that is expressed from the highly condensed Barr body. In humans, up to a quarter of X-linked genes may escape inactivation to some degree. Many of these genes occur in clusters. Among these are the pseudoautosomal genes found on the X and Y chromosomes in the regions of homology described in Chapter 4. Dosage compensation is not necessary for X-linked pseudoautosomal genes because they are located on both the X

Portion of the X chromosome

FIGURE 7.7 **The X-inactivation center (Xic) of the X chromosome.** The *Xist* gene is transcribed into RNA from the inactive X chromosome but not from the active X chromosome. This *Xist* RNA binds to the inactive X chromosome and recruits proteins that promote its compaction. The precise location of the Xce is not known, but it may be within the Xic or adjacent to it. Xce may regulate the transcription of the *Xist* and/or *Tsix* genes and thereby influence the choice of the X chromosome that remains active.

Initiation: Occurs during embryonic development. The number of X-inactivation centers (Xics) are counted and one of the X chromosomes remains active and the other is targeted for inactivation.

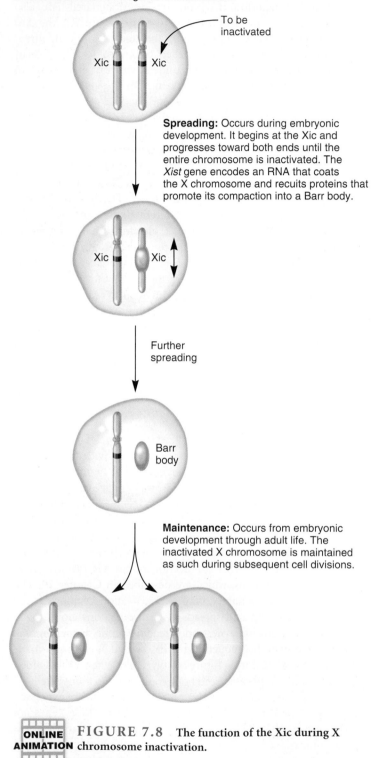

To be inactivated

Xic   Xic

Spreading: Occurs during embryonic development. It begins at the Xic and progresses toward both ends until the entire chromosome is inactivated. The *Xist* gene encodes an RNA that coats the X chromosome and recuits proteins that promote its compaction into a Barr body.

Xic   Xic

Further spreading

Barr body

Maintenance: Occurs from embryonic development through adult life. The inactivated X chromosome is maintained as such during subsequent cell divisions.

ONLINE ANIMATION   **FIGURE 7.8**   **The function of the Xic during X chromosome inactivation.**

and Y chromosomes. How are genes on the Barr body expressed? While the mechanism is not understood, perhaps these genes are found in localized regions where the chromatin is less tightly packed and able to be transcribed.

The Expression of an Imprinted Gene Depends on the Sex of the Parent from Which the Gene Was Inherited

As we have just seen, dosage compensation changes the level of expression of many genes located on the X chromosome. We now turn to another epigenetic phenomenon known as imprinting. The term imprinting implies a type of marking process that has a memory. For example, newly hatched birds identify marks on their parents, which allows them to distinguish their parents from other individuals. The term **genomic imprinting** refers to an analogous situation in which a segment of DNA is marked, and that mark is retained and recognized throughout the life of the organism inheriting the marked DNA. The phenotypes caused by imprinted genes follow a non-Mendelian pattern of inheritance because the marking process causes the offspring to distinguish between maternally and paternally inherited alleles. Depending on how the genes are marked, the offspring expresses only one of the two alleles. This phenomenon is termed **monoallelic expression.**

To understand genomic imprinting, let's consider a specific example. In the mouse, a gene designated *Igf2* encodes a protein growth hormone called insulin-like growth factor 2. Imprinting occurs in a way that results in the expression of the paternal *Igf2* allele but not the maternal allele. The paternal allele is transcribed into RNA, while the maternal allele is transcriptionally silent. With regard to phenotype, a functional *Igf2* gene is necessary for normal size. A loss-of-function allele of this gene, designated *Igf2*[−], is defective in the synthesis of a functional Igf2 protein. This may cause a mouse to be a dwarf, but the dwarfism depends on whether the mutant allele is inherited from the male or female parent, as shown in **Figure 7.9**. On the left side, an offspring has inherited the *Igf2* allele from its father and the *Igf2*[−] allele from its mother. Due to imprinting, only the *Igf2* allele is expressed in the offspring. Therefore, this mouse grows to a normal size. Alternatively, in the reciprocal cross on the right side, an individual has inherited the *Igf2*[−] allele from its father and the *Igf2* allele from its mother. In this case, the *Igf2* allele is not expressed. In this mouse, the *Igf2*[−] allele would be transcribed into mRNA, but the mutation renders the Igf2 protein defective. Therefore, the offspring on the right has a dwarf phenotype. As shown here, both offspring have the same genotype; they are heterozygous for the *Igf2* alleles (i.e., *Igf2 Igf2*[−]). They are phenotypically different, however, because only the paternally inherited allele is expressed.

At the cellular level, imprinting is an epigenetic process that can be divided into three stages: (1) the establishment of the imprint during gametogenesis, (2) the maintenance of the imprint during embryogenesis and in adult somatic cells, and (3) the erasure and reestablishment of the imprint in the germ cells. These stages are described in **Figure 7.10**, which shows the imprinting of the *Igf2* gene. The two mice shown here have inherited the *Igf2* allele from their father and the *Igf2*[−] allele from their mother. Due to imprinting, both mice express the *Igf2* allele in their somatic cells, and the pattern of imprinting is maintained in the somatic cells throughout development. In the germ cells (i.e., sperm and eggs), the imprint is erased; it will be reestablished according to the sex of the animal. The female mouse on the left will transmit only transcriptionally inactive

Igf2⁻ Igf2⁻   Igf2 Igf2   Igf2 Igf2   Igf2⁻ Igf2⁻
(mother's   x   (father's   (mother's   x   (father's
genotype)   genotype)   genotype)   genotype)

Igf2● Igf2⁻▲   Igf2▲ Igf2⁻●
Normal offspring   Dwarf offspring

(Only the Igf2 allele is   (Only the Igf2⁻ allele is
expressed in somatic cells   expressed in somatic cells
of this heterozygous offspring.)   of this heterozygous offspring.)

▲ Denotes an allele that is silent in the offspring
● Denotes an allele that is expressed in the offspring

INTERACTIVE EXERCISE

FIGURE 7.9   An example of genomic imprinting in the mouse. In the cross on the left, a homozygous male with the normal *Igf2* allele is crossed to a homozygous female carrying a defective allele, designated *Igf2⁻*. An offspring is heterozygous and normal because the paternal allele is active. In the reciprocal cross on the right, a homozygous male carrying the defective allele is crossed to a homozygous normal female. In this case, the offspring is heterozygous and dwarf. This is because the paternal allele is defective due to mutation and the maternal allele is not expressed. The photograph shows normal-size (left) and dwarf littermates (right) derived from a cross between a wild-type female and a heterozygous male carrying a loss-of-function *Igf2* allele (courtesy of A. Efstratiadis). The loss-of-function allele was created using gene knockout methods described in Chapter 19 (see Figure 19.7).

alleles to her offspring. The male mouse on the right will transmit transcriptionally active alleles. However, because this male is a heterozygote, it will transmit either a functionally active *Igf2* allele or a functionally defective mutant allele (*Igf2⁻*). An *Igf2⁻* allele, which is inherited from a male mouse, can be expressed into mRNA (i.e., it is transcriptionally active), but it will not produce a functional Igf2 protein due to the deleterious mutation that created the *Igf2⁻* allele; a dwarf phenotype will result.

As seen in Figure 7.10, genomic imprinting is permanent in the somatic cells of an animal, but the marking of alleles can be altered from generation to generation. For example, the female mouse on the left possesses an active copy of the *Igf2* allele, but any allele this female transmits to its offspring will be transcriptionally inactive.

Genomic imprinting occurs in several species, including numerous insects, plants, and mammals. Imprinting may involve

Establishment of the imprint
In this example, imprinting occurs during gametogenesis in the *Igf2* gene, which exists in the *Igf2* allele from the male and the *Igf2⁻* allele from the female. This imprinting occurs so that only the paternal allele is expressed.

Maintenance of the imprint
After fertilization, the imprint pattern is maintained throughout development. In this example, the maternal *Igf2⁻* allele will not be expressed in the somatic cells. Note that the offspring on the left is a female and the one on the right is a male; both are normal in size.

Erasure and reestablishment
During gametogenesis, the imprint is erased. The female mouse produces eggs in which the gene is silenced. The male produces sperm in which the gene can be transcribed into mRNA.

Eggs carry silenced alleles   Sperm carry expressed alleles

▲ Silenced allele
● Transcribed allele

FIGURE 7.10   Genomic imprinting during gametogenesis. This example involves a mouse gene *Igf2*, which is found in two alleles designated *Igf2* and *Igf2⁻*. The left side shows a female mouse that was produced from a sperm carrying the *Igf2* allele and an egg carrying the *Igf2⁻* allele. In the somatic cells of this female animal, the *Igf2* allele is active. However, when this female produces eggs, both alleles are transcriptionally inactive when they are transmitted to offspring. The right side of this figure shows a male mouse that was also produced from a sperm carrying the *Igf2* allele and an egg carrying the *Igf2⁻* allele. In the somatic cells of this male animal, the *Igf2* allele is active. However, the sperm from this male will contain either a functionally active *Igf2* allele or a functionally defective *Igf2⁻* allele. Note: The *Igf2⁻* allele transmitted by the sperm can be expressed into mRNA in the somatic cells of offspring, but the expression will not produce a functional Igf2 protein due to the mutation that inhibits the protein's function.

a single gene, a part of a chromosome, an entire chromosome, or even all of the chromosomes from one parent. Helen Crouse discovered the first example of imprinting, which involved an entire chromosome in the housefly, *Sciara coprophila*. In this species, the fly normally inherits three sex chromosomes, rather than two as in most other species. One X chromosome is inherited from the female, and two are inherited from the male. In male flies, both paternal X chromosomes are lost from somatic cells during embryogenesis. In female flies, only one of the paternal X chromosomes is lost. In both sexes, the maternally inherited X chromosome is never lost. These results indicate that the maternal X chromosome is marked to promote its retention or paternal X chromosomes are marked to promote their loss.

Genomic imprinting can also be involved in the process of X inactivation, described previously. In certain species, imprinting plays a role in the choice of the X chromosome that will be inactivated. For example, in marsupials, the paternal X chromosome is marked so that it is the X chromosome that is always inactivated in the somatic cells of females. In marsupials, X inactivation is not random; the maternal X chromosome is always active.

The Imprinting of Genes and Chromosomes Is a Molecular Marking Process That Involves DNA Methylation

As we have seen, genomic imprinting must involve a marking process. A particular gene or chromosome must be marked dissimilarly during spermatogenesis versus oogenesis. After fertilization takes place, this differential marking affects the expression of particular genes. What is the molecular explanation for genomic imprinting? As discussed in Chapter 15, **DNA methylation**—the attachment of methyl groups onto a cytosine base—is a common way that eukaryotic genes may be regulated. Research indicates that genomic imprinting involves a **differentially methylated region (DMR)** that is located near the imprinted gene. Depending on the particular gene, the DMR is methylated in the egg or the sperm, but not both. The DMR contains binding sites for one or more proteins that regulate the transcription of a nearby gene.

For most imprinted genes, methylation at a DMR results in an inhibition of gene expression. Methylation could enhance the binding of proteins that inhibit transcription and/or inhibit the binding of proteins that enhance transcription. (The relationship between methylation and gene expression is described in Chapter 15.) For this reason, imprinting is usually described as a marking process that silences gene expression by preventing transcription. However, this is not always the case. Two imprinted human genes, *H19* and *Igf2*, provide an interesting example. These two genes lie close to each other on human chromosome 11 and appear to be controlled by the same DMR, which is a 2,000-base pair region that lies upstream of the *H19* gene (**Figure 7.11a**). It contains binding sites for proteins that regulate the transcription of the *H19* or *Igf2* genes. This DMR is highly methylated on the paternally inherited chromosome. Methylation of the DMR silences the *H19* gene because a specific protein or group of proteins bind to the DMR only after it has been methylated, and protein binding inhibits the transcription of the *H19* gene.

The same DMR that controls *H19* also controls *Igf2*, but in a different way. The unmethylated DMR inherited from a female binds a protein that causes an alteration in the structure of chromatin over a fairly long distance that includes the *Igf2* gene. This alteration in structure blocks the ability of transcriptional activator proteins to bind to the *Igf2* gene. If these activator proteins are unable to bind to the *Igf2* gene, transcription is silenced. In contrast, the protein that changes chromatin structure does not bind to a methylated DMR. Therefore, the *Igf2* gene that is inherited from a male is recognized by transcriptional activator proteins and transcribed into mRNA.

Now that we have an understanding of how methylation may affect gene transcription, let's consider the methylation process during the life of an animal. A key issue is that certain imprinted genes are differentially methylated during oogenesis compared to spermatogenesis (**Figure 7.11b**). In this example, a female and male offspring have inherited a methylated DMR from their father and a nonmethylated DMR from their mother. If this was the DMR next to the *H19* gene, the somatic cells of the offspring would express only the maternal *H19* allele. This pattern of imprinting is maintained in the somatic cells of both offspring. However, when the female offspring makes gametes, the imprinting is erased during early oogenesis, so the female will pass an unmethylated DMR to its offspring. In the male, the imprinting is also erased during early spermatogenesis, but then *de novo* (new) methylation occurs in both DMRs. In the case of the *H19* gene, the result of DMR methylation is that the male transmits inactive alleles of this gene.

Genomic imprinting is a fairly new and exciting area of research. Imprinting has been identified in many mammalian genes (**Table 7.2**). In some cases, the female alleles are transcriptionally active in the somatic cells of offspring, whereas in other cases, the male alleles are active. The biological significance of imprinting is still a matter of much speculation. Several hypotheses have been proposed to explain the potential benefits of genomic imprinting. One example, described by David Haig, involves differences in female versus male reproductive patterns in

TABLE **7.2**

Examples of Mammalian Genes and Inherited Human Diseases That Involve Imprinted Genes*

Gene	Allele Expressed	Function
WT1	Maternal	Wilms tumor–suppressor gene; suppresses cell growth
INS	Paternal	Insulin; hormone involved in cell growth and metabolism
Igf2	Paternal	Insulin-like growth factor II; similar to insulin
Igf2R	Maternal	Receptor for insulin-like growth factor II
H19	Maternal	Unknown
SNRPN	Paternal	Splicing factor
Gabrb	Maternal	Neurotransmitter receptor

*Researchers estimate that approximately 1–2% of human genes are subjected to genomic imprinting, but fewer than 100 have actually been demonstrated to be imprinted.

(a) Actions of regulatory proteins that bind to an unmethylated DMR or to a methylated DMR

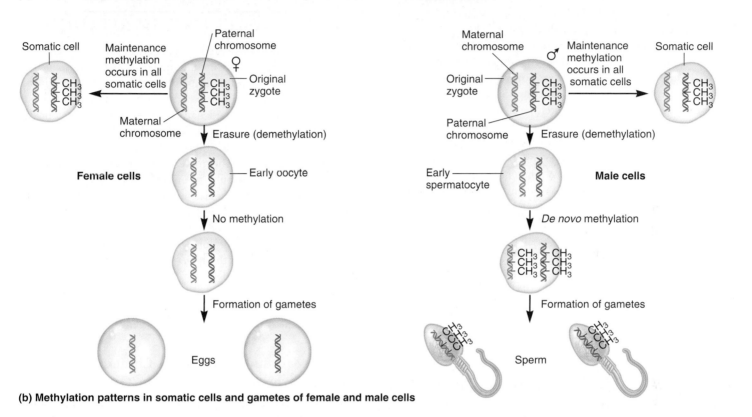

(b) Methylation patterns in somatic cells and gametes of female and male cells

FIGURE 7.11 **DNA methylation in the imprinting process.** (a) The effects of methylation on gene transcription. A differentially methylated region (DMR) is located next to two genes called *H19* and *Igf2*. This DMR is methylated on the paternally inherited chromosome. When it is methylated, it binds a protein that inhibits the *H19* gene. Therefore, the paternally inherited *H19* allele is silenced. In contrast, when the DMR is not methylated, it binds a different protein that inhibits the *Igf2* gene. In this way, the maternally inherited *Igf2* allele is silenced. (b) The pattern of DMR methylation during the life of an animal. In this example, a male and a female offspring have inherited a methylated DMR and nonmethylated DMR from their father and mother, respectively. Maintenance methylation retains the imprinting in somatic cells during embryogenesis and in adulthood. Demethylation occurs in cells that are destined to become gametes. In this example, *de novo* methylation occurs only in cells that are destined to become sperm. Haploid male gametes transmit a methylated DMR, whereas haploid female gametes transmit an unmethylated DMR.

mammals. As discussed in Chapter 24, natural selection will favor types of genetic variation that confer a survival advantage. The likelihood that favorable variation will be passed to offspring may differ between the sexes. In many mammalian species, females may mate with multiple males, perhaps generating embryos in the same uterus fathered by different males. For males, silencing genes that inhibit embryonic growth would be an advantage. The embryos of males that silence such genes would grow faster than other embryos in the same uterus, making it more likely for the males to pass their genes to future generations. For females, however, rapid growth of embryos might be a disadvantage because it could drain too many resources from the mother. According to this scenario, the mother would silence genes that cause rapid embryonic growth. From the mother's perspective, she would give

all of her offspring an equal chance of survival without sapping her own strength. This would make it more likely for the female to pass her genes to future generations. The Haig hypothesis seems to be consistent with the imprinting of several mammalian genes that are involved with growth, such as *Igf2*. Females silence this growth-enhancing gene, whereas males do not. However, several imprinted genes do not seem to play a role in embryonic development. Therefore, an understanding of the biological role(s) of genomic imprinting will require further investigation.

Imprinting plays a role in the inheritance of certain human diseases such as Prader-Willi syndrome (PWS) and Angelman syndrome (AS). PWS is characterized by reduced motor function, obesity, and mental deficiencies. AS patients are thin and hyperactive, have unusual seizures and repetitive symmetrical muscle movements, and exhibit mental deficiencies. Most commonly, both PWS and AS involve a small deletion in human chromosome 15. If this deletion is inherited from the maternal parent, it leads to Angelman syndrome; if inherited from the father, it leads to Prader-Willi syndrome (**Figure 7.12**).

Researchers have discovered that this region contains closely linked but distinct genes that are maternally or paternally imprinted. AS results from the lack of expression of a single gene (*UBE3A*) that codes for a protein called E6-AP, which functions to transfer small ubiquitin molecules to certain proteins to target their degradation. Both copies of this gene are active in many of the body's tissues. In the brain, however, only the copy inherited from a person's mother (the maternal copy) is active. The paternal allele of *UBE3A* is silenced. Therefore, if the maternal allele is deleted, as in the left side of Figure 7.12, the individual will develop AS because he or she will not have an active copy of the *UBE3A* gene.

The gene(s) responsible for PWS has not been definitively determined, although five imprinted genes in this region of chro-

mosome 15 are known. One possible candidate involved in PWS is a gene designated *SNRPN*. The gene product is part of a <u>s</u>mall <u>n</u>uclear <u>r</u>ibonucleoprotein <u>p</u>olypeptide <u>N</u>, which is a complex that controls RNA splicing and is necessary for the synthesis of critical proteins in the brain. The maternal allele of *SNRPN* is silenced, and only the paternal copy is active.

7.3 EXTRANUCLEAR INHERITANCE

Thus far, we have considered several types of non-Mendelian inheritance patterns. These include maternal effect genes, dosage compensation, and genomic imprinting. All of these inheritance patterns involve genes found on chromosomes in the cell nucleus. Another cause of non-Mendelian inheritance patterns concerns genes that are not located in the cell nucleus. In eukaryotic species, the most biologically important example of extranuclear inheritance involves genetic material in cellular organelles. In addition to the cell nucleus, the mitochondria and chloroplasts contain their own genetic material. Because these organelles are found within the cytoplasm of the cells, the inheritance of organellar genetic material is called **extranuclear inheritance** (the prefix *extra-* means outside of) or **cytoplasmic inheritance.** In this section, we will examine the genetic composition of mitochondria and chloroplasts and explore the pattern of transmission of these organelles from parent to offspring. We will also consider a few other examples of inheritance patterns that cannot be explained by the transmission of nuclear genes.

Mitochondria and Chloroplasts Contain Circular Chromosomes with Many Genes

In 1951, Y. Chiba was the first to suggest that chloroplasts contain their own DNA. He based his conclusion on the staining properties of a DNA-specific dye known as Feulgen. Researchers later developed techniques to purify organellar DNA. In addition, electron microscopy studies provided interesting insights into the organization and composition of mitochondrial and chloroplast chromosomes. More recently, the advent of molecular genetic techniques in the 1970s and 1980s has allowed researchers to determine the genome sequences of organellar DNAs. From these types of studies, the chromosomes of mitochondria and chloroplasts were found to resemble smaller versions of bacterial chromosomes.

The genetic material of mitochondria and chloroplasts is located inside the organelle in a region known as the **nucleoid** (**Figure 7.13**). The genome is a single circular chromosome (composed of double-stranded DNA), although a nucleoid contains several copies of this chromosome. In addition, a mitochondrion or chloroplast often has more than one nucleoid. In mice, for example, each mitochondrion has one to three nucleoids, with each nucleoid containing two to six copies of the circular mitochondrial genome. However, this number is variable and depends on the type of cell and the stage of development. In comparison, the chloroplasts of algae and higher plants tend to have more nucleoids per organelle. **Table 7.3** describes the genetic composition of mitochondria and chloroplasts for a few selected species.

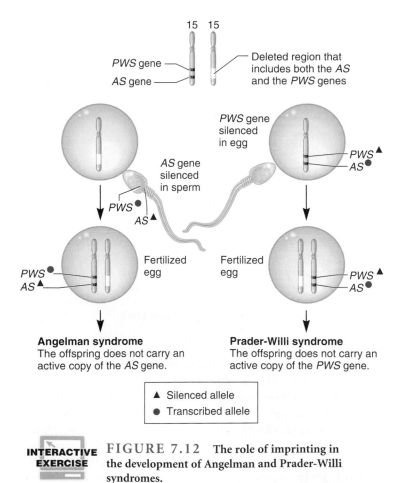

INTERACTIVE EXERCISE

FIGURE 7.12 **The role of imprinting in the development of Angelman and Prader-Willi syndromes.**

Genes → Traits A small region on chromosome 15 contains two different genes designated the *AS* gene and *PWS* gene in this figure. If a chromosome 15 deletion is inherited from the maternal parent, Angelman syndrome occurs because the offspring does not inherit an active copy of the *AS* gene (left). Alternatively, the chromosome 15 deletion may be inherited from the male parent, leading to Prader-Willi syndrome. The phenotype of this syndrome occurs because the offspring does not inherit an active copy of the *PWS* gene (right).

Nucleoid

(a)

nucleoid   nucleoid

(b)

FIGURE 7.13   Nucleoids within (a) mitochondria and (b) chloroplasts. The mitochondrial and chloroplast chromosomes are found within the nucleoid region of the organelle.

tRNA^Phe   tRNA^Pro   tRNA^Thr
tRNA^Val   Small rRNA   tRNA^Glu
Large rRNA
tRNA^Leu
tRNA^Ile
tRNA^Gly   tRNA^Leu
tRNA^Met   tRNA^Ser
tRNA^Trp   tRNA^His
tRNA^Ala
tRNA^Asp
tRNA^Cys   tRNA^Arg
tRNA^Tyr
tRNA^Ser   tRNA^Gly
tRNA^Leu   tRNA^Lys

- Ribosomal RNA genes
- Transfer RNA genes
- NADH dehydrogenase genes
- Cytochrome *b* gene
- Cytochrome *c* oxidase genes
- ATP synthase genes
- Noncoding DNA

FIGURE 7.14   A genetic map of human mitochondrial DNA (mtDNA). This diagram illustrates the locations of many genes along the circular mitochondrial chromosome. The genes shown in red encode transfer RNAs. For example, tRNA^Arg encodes the tRNA that carries arginine. The genes that encode ribosomal RNA are shown in light brown. The remaining genes encode proteins that function within the mitochondrion. The mitochondrial genomes from numerous species have been determined.

TABLE **7.3**

Genetic Composition of Mitochondria and Chloroplasts

Species	Organelle	Nucleoids per Organelle	Total Number of Chromosomes per Organelle
Tetrahymena	Mitochondrion	1	6–8
Mouse	Mitochondrion	1–3	5–6
Chlamydomonas	Chloroplast	5–6	~80
Euglena	Chloroplast	20–34	100–300
Higher plants	Chloroplast	12–25	~60

Data from: Gillham, N. W. (1994). *Organelle Genes and Genomes.* Oxford University Press, New York.

Besides variation in copy number, the sizes of mitochondrial and chloroplast genomes also vary greatly among different species. For example, a 400-fold variation is found in the sizes of mitochondrial chromosomes. In general, the mitochondrial genomes of animal species tend to be fairly small; those of fungi and protists are intermediate in size; and those of plant cells tend to be fairly large. Among algae and plants, substantial variation is also found in the sizes of chloroplast chromosomes.

Figure 7.14 illustrates a map of human **mitochondrial DNA (mtDNA).** Each copy of the mitochondrial chromosome consists of a circular DNA molecule that is only 17,000 base pairs (bp) in length. This size is less than 1% of a typical bacterial chromosome. The human mtDNA carries relatively few genes. Thirteen genes encode proteins that function within the mitochondrion. In addition, mtDNA carries genes that encode ribosomal RNA and transfer RNA. These rRNAs and tRNAs are necessary for the synthesis of the 13 proteins that are encoded by the mtDNA. The primary role of mitochondria is to provide cells with the bulk of their adenosine triphosphate (ATP), which is used as an energy source to drive cellular reactions. These 13 proteins function in a process known as oxidative phosphorylation, which enables the mitochondria to synthesize ATP. However, mitochondria require many additional proteins to carry out oxidative phosphorylation and other mitochondrial functions. Most mitochondrial proteins are encoded by genes within the cell nucleus. When these nuclear genes are expressed, the mitochondrial proteins are first synthesized outside the mitochondria in the cytosol of the cell and then transported into the mitochondria.

Chloroplast genomes tend to be larger than mitochondrial genomes, and they have a correspondingly greater number of genes. A typical chloroplast genome is approximately 100,000 to 200,000 bp in length, which is about 10 times larger than the mitochondrial genome of animal cells. **Figure 7.15** shows the **chloroplast DNA (cpDNA)** of the tobacco plant, which is a circular DNA molecule that contains 156,000 bp of DNA and car-

ries between 110 and 120 different genes. These genes encode ribosomal RNAs, transfer RNAs, and many proteins required for photosynthesis. As with mitochondria, many chloroplast proteins are encoded by genes found in the plant cell nucleus. These proteins contain chloroplast-targeting signals that direct them into the chloroplasts.

Extranuclear Inheritance Produces Non-Mendelian Results in Reciprocal Crosses

In diploid eukaryotic species, most genes within the nucleus obey a Mendelian pattern of inheritance because the homologous pairs of chromosomes segregate during gamete formation. Except for sex-linked traits, offspring inherit one copy of each gene from both the maternal and paternal parents. The sorting of chromosomes during meiosis explains the inheritance patterns of nuclear genes. By comparison, the inheritance of extranuclear genetic material does not display a Mendelian pattern. Mitochondria and chloroplasts are not sorted during meiosis and therefore do not segregate into gametes in the same way as nuclear chromosomes.

In 1909, Carl Correns discovered a trait that showed a non-Mendelian pattern of inheritance involving pigmentation

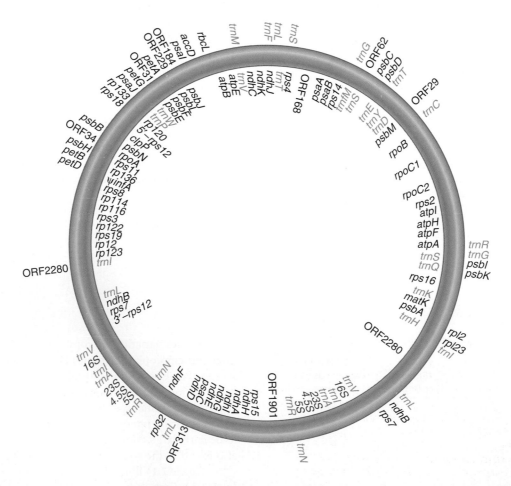

FIGURE 7.15 **A genetic map of tobacco chloroplast DNA (cpDNA).** This diagram illustrates the locations of many genes along the circular chloroplast chromosome. The gene names shown in blue encode transfer RNAs. The genes that encode ribosomal RNA are shown in red. The remaining genes shown in black encode polypeptides that function within the chloroplast. The genes designated ORF (open reading frame) encode polypeptides with unknown functions.

in *Mirabilis jalapa* (the four-o'clock plant). Leaves can be green, white, or variegated with both green and white sectors. Correns demonstrated that the pigmentation of the offspring depended solely on the maternal parent (**Figure 7.16**). If the female parent had white pigmentation, all offspring had white leaves. Similarly, if the female was green, all offspring were green. When the female was variegated, the offspring could be green, white, or variegated.

The pattern of inheritance observed by Correns is a type of extrachromosomal inheritance called **maternal inheritance** (not to be confused with maternal effect). Chloroplasts are a type of plastid that makes chlorophyll, a green photosynthetic pigment. Maternal inheritance occurs because the chloroplasts are inherited only through the cytoplasm of the egg. The pollen grains of *M. jalapa* do not transmit chloroplasts to the offspring.

The phenotypes of leaves can be explained by the types of chloroplasts within the leaf cells. The green phenotype, which is the wild-type condition, is due to the presence of normal chloroplasts that make green pigment. By comparison, the white phenotype is due to a mutation in a gene within the chloroplast DNA that diminishes the synthesis of green pigment. A cell may contain both types of chloroplasts, a condition known as **heteroplasmy.** A leaf cell containing both types of chloroplasts is green because the normal chloroplasts produce green pigment.

How does a variegated phenotype occur? **Figure 7.17** considers the leaf of a plant that began from a fertilized egg that contained both types of chloroplasts (i.e., a heteroplasmic cell). As a plant grows, the two types of chloroplasts are irregularly distributed to daughter cells. On occasion, a cell may receive only the chloroplasts that have a defect in making green pigment. Such a cell continues to divide and produce a sector of the plant that is entirely white. In this way, the variegated phenotype is created. Similarly, if we consider the results of Figure 7.16, a female parent that is variegated may transmit green, white, or a mixture of these types of chloroplasts to the egg cell and thereby produce green, white, or variegated offspring, respectively.

Studies in Yeast and *Chlamydomonas* Provided Genetic Evidence for Extranuclear Inheritance of Mitochondria and Chloroplasts

The research of Correns and others indicated that some traits, such as leaf pigmentation, are inherited in a non-Mendelian manner. However, such studies did not definitively determine that

Cross 1

× ♀ ♂

All white offspring

Reciprocal cross of cross 1

× ♀ ♂

All green offspring

Cross 2

× ♀ ♂

Green, white, or variegated offspring

Reciprocal cross of cross 2

× ♀ ♂

All green offspring

INTERACTIVE EXERCISE

FIGURE 7.16   Maternal inheritance in the four-o'clock plant, *Mirabilis jalapa*. The reciprocal crosses of four-o'clock plants by Carl Correns consisted of a pair of crosses between white-leaved and green-leaved plants, and a pair of crosses between variegated-leaved and green-leaved plants.

Genes → Traits In this example, the white phenotype is due to chloroplasts that carry a mutant allele that diminishes green pigmentation. The variegated phenotype is due to a mixture of chloroplasts, some of which carry the normal (green) allele and some of which carry the white allele. In the crosses shown here, the parent providing the eggs determines the phenotypes of the offspring. This is due to maternal inheritance. The egg contains the chloroplasts that are inherited by the offspring. (Note: The defective chloroplasts that give rise to white sectors are not completely defective in chlorophyll synthesis. Therefore, entirely white plants can survive, though they are smaller than green or variegated plants.)

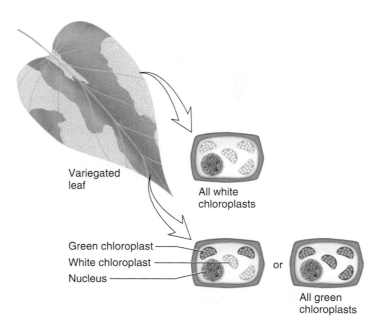

Variegated leaf

All white chloroplasts

Green chloroplast
White chloroplast
Nucleus

or

All green chloroplasts

FIGURE 7.17   A cellular explanation of the variegated phenotype in *Mirabilis jalapa*. This plant inherited two types of chloroplasts—those that can produce green pigment and those that are defective. As the plant grows, the two types of chloroplasts are irregularly distributed to daughter cells. On occasion, a cell may receive only the chloroplasts that are defective at making green pigment. Such a cell continues to divide and produces a sector of the plant that is entirely white. Cells that contain both types of chloroplasts or cells that contain only green chloroplasts would produce green tissue, which may be adjacent to a sector of white tissue. This is the basis for the variegated phenotype of the leaves.

maternal inheritance is due to genetic material within organelles. Further progress in the investigation of extranuclear inheritance was provided by detailed genetic analyses of eukaryotic microorganisms, such as yeast and algae, by isolating and characterizing mutant phenotypes that specifically affected the chloroplasts or mitochondria.

During the 1940s and 1950s, yeasts and molds became model eukaryotic organisms for investigating the inheritance of mitochondria. Because mitochondria produce energy for cells in the form of ATP, mutations that yield defective mitochondria are expected to make cells grow much more slowly. Boris Ephrussi and his colleagues identified mutations in *Saccharomyces cerevisiae* that had such a phenotype. These mutants were called **petites** to describe their formation of small colonies on agar plates as opposed to wild-type strains that formed larger colonies. Biochemical and physiological evidence indicated that petite mutants had defective mitochondria. The researchers found that petite mutants could not grow when the cells only had an energy source requiring the metabolic activity of mitochondria, but could form small colonies when grown on sugars metabolized by the glycolytic pathway, which occurs outside the mitochondria.

Because yeast exist in two mating types, designated a and α, Ephrussi was able to mate a wild-type strain to his petite mutants. Genetic analyses showed that petite mutants were inherited in different ways. When a wild-type strain was crossed to a segregational petite mutant, he obtained a ratio of 2 wild-type cells to 2 petite cells (**Figure 7.18a**). This result is consistent with a Mendelian pattern of inheritance (see the discussion of tetrad analysis in Chapter 5). Therefore, segregational petite mutations cause defects in genes located in the cell nucleus. These genes encode proteins necessary for mitochondrial function. Such proteins are synthesized in the cytosol and are then taken up by the mitochondria, where they perform their functions. Segregational petites get their name because they segregate in a Mendelian manner during meiosis.

By comparison, the second category of petite mutants, known as vegetative petite mutants, did not segregate in a Mendelian manner (**Figure 7.18b**). Ephrussi identified two types of vegetative petites, called neutral petites and suppressive petites. In a cross between a wild-type strain and a neutral petite, all four haploid daughter cells were wild type. This type of inheritance contradicts the normal 2:2 ratio expected for the segregation of Mendelian traits. In comparison, a cross between a wild-type strain and a suppressive petite usually yielded all petite colonies. Thus, both types of vegetative petites are defective in mitochondrial function and show a non-Mendelian pattern of inheritance. These results occurred because vegetative petites carry mutations in the mitochondrial genome itself.

Since these initial studies, researchers have found that neutral petites lack most of their mitochondrial DNA, whereas suppressive petites usually lack small segments of the mitochondrial genetic material. When two yeast cells are mated, the daughter cells inherit mitochondria from both parents. For example, in a cross between a wild-type and a neutral petite strain, the daughter cells inherit both types of mitochondria. Because wild-type mitochondria are inherited, the cells display a normal phenotype. The inheritance pattern of suppressive petites is more difficult to explain because the daughter cells inherit both normal and suppressive petite mitochondria. One possibility is that the suppressive petite mitochondria replicate more rapidly so that the wild-type mitochondria are not maintained in the cytoplasm for many doublings. Alternatively, some experimental evidence suggests that genetic exchanges between the mitochondrial genomes of wild-type and suppressive petites may ultimately produce a defective population of mitochondria.

Let's now turn our attention to the inheritance of chloroplasts that are found in eukaryotic species capable of photosynthesis (namely, algae and plants). The unicellular alga *Chlamydomonas reinhardtii* has been used as a model organism to investigate the inheritance of chloroplasts. This organism contains a single chloroplast that occupies approximately 40% of the cell volume. Genetic studies of chloroplast inheritance began when Ruth Sager identified a mutant strain of *Chlamydomonas* with resistance to the antibiotic streptomycin (sm^r). By comparison, most strains are sensitive to killing by streptomycin (sm^s).

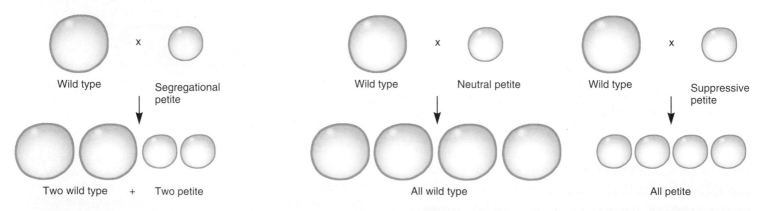

Wild type   x   Segregational petite

Two wild type   +   Two petite

Wild type   x   Neutral petite

All wild type

Wild type   x   Suppressive petite

All petite

(a) A cross between a wild type and a segregational petite

(b) A cross between a wild type and a neutral or suppressive vegetative petite

INTERACTIVE EXERCISE   FIGURE 7.18   **Transmission of the petite trait in *Saccharomyces cerevisiae*.** (a) A wild-type strain crossed to a segregational petite. (b) A wild-type strain crossed to a neutral vegetative petite and to a suppressive vegetative petite.

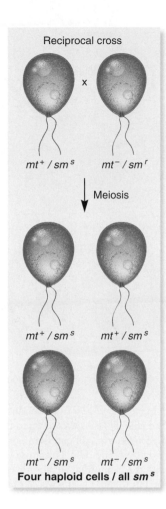

FIGURE 7.19   Chloroplast inheritance in *Chlamydomonas.* *Mt*⁺ and *mt*⁻ indicate the two mating types of the organism. *Sm*ʳ indicates streptomycin resistance, whereas *sm*ˢ indicates sensitivity to this antibiotic.

Sager conducted crosses to determine the inheritance pattern of the *sm*ʳ gene. During mating, two haploid cells unite to form a diploid cell, which then undergoes meiosis to form four haploid cells. Like yeast, *Chlamydomonas* is an organism that can be found in two mating types, in this case, designated *mt*⁺ and *mt*⁻. Mating type is due to nuclear inheritance and segregates in a 1:1 manner. By comparison, Sager and her colleagues discovered that the *sm*ʳ gene was inherited from the *mt*⁺ parent but not from the *mt*⁻ parent (**Figure 7.19**). Therefore, this *sm*ʳ gene was not inherited in a Mendelian manner. This pattern occurred because only the *mt*⁺ parent transmits chloroplasts to daughter cells and the *sm*ʳ gene is found in the chloroplast genome.

The Pattern of Inheritance of Mitochondria and Chloroplasts Varies Among Different Species

The inheritance of traits via genetic material within mitochondria and chloroplasts is now a well-established phenomenon that geneticists have investigated in many different species. In **heterogamous** species, two kinds of gametes are made. The female gamete tends to be large and provides most of the cytoplasm

to the zygote, whereas the male gamete is small and often provides little more than a nucleus. Therefore, mitochondria and chloroplasts are most often inherited from the maternal parent. However, this is not always the case. **Table 7.4** describes the inheritance patterns of mitochondria and chloroplasts in several selected species.

In species where maternal inheritance is generally observed, the paternal parent may occasionally provide mitochondria via the sperm. This phenomenon, called **paternal leakage,** occurs in many species that primarily exhibit maternal inheritance of their organelles. In the mouse, for example, approximately one to four paternal mitochondria are inherited for every 100,000 maternal mitochondria per generation of offspring. Most offspring do not inherit any paternal mitochondria, but a rare individual may inherit a mitochondrion from the sperm.

A Few Rare Human Diseases Are Caused by Mitochondrial Mutations

As noted previously, the human mitochondrial genome has 13 genes that encode polypeptides necessary for the synthesis of ATP. In addition, the mtDNA has genes that encode ribosomal RNA and transfer RNA molecules. Human mtDNA is maternally inherited because it is transmitted from mother to offspring via the cytoplasm of the egg. Therefore, the transmission of human mitochondrial diseases follows a strict maternal inheritance pattern.

Table 7.5 describes several mitochondrial diseases that have been discovered in humans and are caused by mutations in mitochondrial genes. These are usually chronic degenerative disorders that affect cells requiring a high level of ATP, such as nerve and muscle cells. For example, Leber's hereditary optic neuropathy (LHON) affects the optic nerve and may lead to the progressive

TABLE 7.4

Transmission of Organelles Among Different Species

Species	Organelle	Transmission
Mammals	Mitochondria	Maternal inheritance
S. cerevisiae	Mitochondria	Biparental inheritance
Molds	Mitochondria	Usually maternal inheritance; paternal inheritance has been found in the genus *Allomyces*
Chlamydomonas	Mitochondria	Inherited from the parent with the *mt*⁻ mating type
Chlamydomonas	Chloroplasts	Inherited from the parent with the *mt*⁺ mating type
Plants		
Angiosperms	Mitochondria and chloroplasts	Often maternal inheritance, although biparental inheritance is found among some species
Gymnosperms	Mitochondria and chloroplasts	Usually paternal inheritance

TABLE 7.5

Examples of Human Mitochondrial Diseases

Disease	Mitochondrial Gene Mutated
Leber's hereditary optic neuropathy	A mutation in one of several mitochondrial genes that encode respiratory chain proteins: *ND1, ND2, CO1, ND4, ND5, ND6,* and *cytb*
Neurogenic muscle weakness	A mutation in the *ATPase6* gene that encodes a subunit of the mitochondrial ATP-synthetase, which is required for ATP synthesis
Mitochondrial encephalomyopathy, lactic acidosis, and strokelike episodes	A mutation in genes that encode tRNAs for leucine and lysine
Mitochondrial myopathy	A mutation in a gene that encodes a tRNA for leucine
Maternal myopathy and cardiomyopathy	A mutation in a gene that encodes a tRNA for leucine
Myoclonic epilepsy with ragged-red muscle fibers	A mutation in a gene that encodes a tRNA for lysine

Data from: Wallace, D. C. (1993). Mitochondrial diseases: genotype versus phenotype. *Trends Genet. 9,* 128–33.

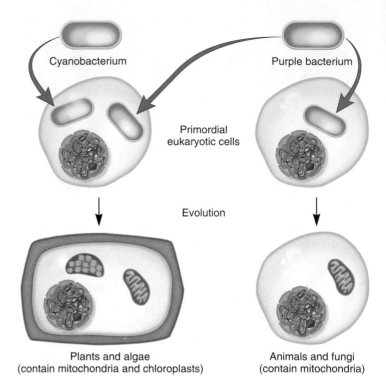

FIGURE 7.20   The endosymbiotic origin of mitochondria and chloroplasts. According to the endosymbiotic theory, chloroplasts descended from an endosymbiotic relationship between cyanobacteria and eukaryotic cells. This arose when a bacterium took up residence within a primordial eukaryotic cell. Over the course of evolution, the intracellular bacterial cell gradually changed its characteristics, eventually becoming a chloroplast. Similarly, mitochondria are derived from an endosymbiotic relationship between purple bacteria and eukaryotic cells.

loss of vision in one or both eyes. LHON can be caused by a defective mutation in one of several different mitochondrial genes. Researchers are still investigating how a defect in these mitochondrial genes produces the symptoms of this disease.

Extranuclear Genomes of Mitochondria and Chloroplasts Evolved from an Endosymbiotic Relationship

The idea that the nucleus, mitochondria, and chloroplasts contain their own separate genetic material may at first seem puzzling. Wouldn't it be simpler to have all of the genetic material in one place in the cell? The underlying reason for distinct genomes of mitochondria and chloroplasts can be traced back to their evolutionary origin, which is thought to involve a symbiotic association.

A symbiotic relationship occurs when two different species live together in a close association. The symbiont is the smaller of the two species; the host is the larger. The term **endosymbiosis** describes a symbiotic relationship in which the symbiont actually lives inside (*endo-*, inside) the host. In 1883, Andreas Schimper proposed that chloroplasts were descended from an endosymbiotic relationship between cyanobacteria and eukaryotic cells. This idea, now known as the **endosymbiosis theory,** suggested that the ancient origin of chloroplasts was initiated when a cyanobacterium took up residence within a primordial eukaryotic cell (**Figure 7.20**). Over the course of evolution, the characteristics of the intracellular bacterial cell gradually changed to those of a chloroplast. In 1922, Ivan Wallin also proposed an endosymbiotic origin for mitochondria.

In spite of these hypotheses, the question of endosymbiosis was largely ignored until researchers in the 1950s discovered that chloroplasts and mitochondria contain their own genetic material. The issue of endosymbiosis was hotly debated after Lynn Margulis published a book entitled *Origin of Eukaryotic Cells* (1970). During the 1970s and 1980s, the advent of molecular genetic techniques allowed researchers to analyze genes from chloroplasts, mitochondria, bacteria, and eukaryotic nuclear genomes. They found that genes in chloroplasts and mitochondria are very similar to bacterial genes but not as similar to those found within the nucleus of eukaryotic cells. This observation provided strong support for the endosymbiotic origin of mitochondria and chloroplasts, which is now widely accepted.

The endosymbiosis theory proposes that the relationship provided eukaryotic cells with useful cellular characteristics. Chloroplasts were derived from cyanobacteria, a bacterial species that is capable of photosynthesis. The ability to carry out photosynthesis provided algal and plant cells with the ability to use the energy from sunlight. By comparison, mitochondria are thought to have been derived from a different type of bacteria known as Gram-negative nonsulfur purple bacteria. In this case, the endosymbiotic relationship enabled eukaryotic cells to synthesize greater amounts of ATP. It is less clear how the relationship would have been beneficial to cyanobacteria or purple bacteria, though the cytosol of a eukaryotic cell may have provided a stable environment with an adequate supply of nutrients.

During the evolution of eukaryotic species, most genes that were originally found in the genome of the primordial cyanobacteria and purple bacteria have been lost or transferred from the organelles to the nucleus. The sequences of certain genes within the nucleus are consistent with their origin within an organelle. Such genes are more similar in their DNA sequence to known bacterial genes than to their eukaryotic counterparts. Therefore, researchers have concluded that these genes have been removed from the mitochondrial and chloroplast chromosomes, and relocated to the nuclear chromosomes. This has occurred many times throughout evolution, so modern mitochondria and chloroplasts have lost most of the genes that are still found in present-day purple bacteria and cyanobacteria.

Most of this gene transfer occurred early in mitochondrial and chloroplast evolution. The functional transfer of mitochondrial genes seems to have ceased in animals, but gene transfer from mitochondria and chloroplasts to the nucleus continues to occur in plants at a low rate. The molecular mechanism of gene transfer is not entirely understood, but the direction of transfer is well established. During evolution, gene transfer has occurred primarily from the organelles to the nucleus. Transfer of genes from the nucleus to the organelles has almost never occurred, although one example is known of a nuclear gene in plants that has been transferred to the mitochondrial genome. This unidirectional gene transfer from organelles to the nucleus partly explains why the organellar genomes now contain relatively few genes. In addition, gene transfer can occur between organelles. It can happen between two mitochondria, between two chloroplasts, and between a chloroplast and mitochondrion. Overall, the transfer of genetic material between the nucleus, chloroplasts, and mitochondria is an established phenomenon, although its biological benefits remain unclear.

Eukaryotic Cells Occasionally Contain Symbiotic Infective Particles

Other unusual endosymbiotic relationships have been identified in eukaryotic organisms. Several examples are known in which infectious particles establish a symbiotic relationship with their host. In some cases, research indicates that symbiotic infectious particles are bacteria that exist within the cytoplasm of eukaryotic cells. While symbiotic infectious particles are relatively uncommon, they have provided interesting and even bizarre examples of the extranuclear inheritance of traits.

In the 1940s, Tracy Sonneborn studied a trait in the protozoan *Paramecia aurelia* known as the killer trait. Killer paramecia secrete a substance called paramecin, which kills some but not all strains of paramecia. Sonneborn found that killer strains contain particles in their cytoplasm known as kappa particles. Each kappa particle is 0.4 μm long and has its own DNA. Genes within the kappa particle encode the paramecin toxin. In addition, other kappa particle genes provide the killer paramecia with resistance to paramecin.

Nonkiller paramecia are killed when mixed with killer paramecia. However, Sonnenborn found that when nonkiller paramecia were mixed with a cell extract derived from killer paramecia, the kappa particles within the extract are taken up by the nonkiller strains and convert them into killer strains. In other words, the extranuclear particle that determines the killer trait is infectious.

Infectious particles have also been identified in fruit flies. P. l'Heritier identified strains of *Drosophila melanogaster* that are highly sensitive to killing by CO_2. Reciprocal crosses between CO_2-sensitive and normal flies revealed that the trait is inherited in a non-Mendelian manner. Furthermore, cell extracts from a sensitive fly can infect a normal fly and make it sensitive to CO_2.

Another example of an infectious particle in fruit flies involves a trait known as sex ratio in which affected flies produce progenies with a large excess of females. Chana Malogolowkin and Donald Poulson discovered one strain of *Drosophila willistoni* where most of the offspring of female flies were daughters; nearly all the male offspring died. The sex ratio trait is transmitted from mother to offspring. The rare surviving males do not transmit this trait to their male or female offspring. This result indicates a maternal inheritance pattern for the sex ratio trait. The agent in the cytoplasm of female flies responsible for the sex ratio trait was later found to be a symbiotic bacterium, which was named *Spiroplasma poulsonii*. Its presence is usually lethal to males but not to females. This infective agent can be extracted from the tissues of adult females and used to infect the females of a normal strain of flies.

CONCEPTUAL SUMMARY

In this chapter, we have considered inheritance patterns that differ from Mendelian inheritance. In some cases, genes that are physically located on nuclear chromosomes fail to follow a Mendelian pattern of inheritance. For example, genes that display a maternal effect are expressed in the mother's nurse cells during oogenesis and affect the phenotype of the offspring. Epigenetic inheritance describes another class of examples in which nuclear genes do not follow a Mendelian pattern of inheritance. During dosage compensation, the level of expression of genes on the sex chromosomes is altered. In mammals, dosage compensation occurs via X inactivation. According to the Lyon hypothesis, one of the two X chromosomes in females is randomly inactivated during embryonic development to produce a transcriptionally inactive Barr body. The pattern of active and inactive X chromosomes is then inherited during later stages of development to create a female animal in which a single X chromosome is active in each cell. A second example of epigenetic inheritance is known as genomic imprinting. In this case, a gene or even an entire chromosome is marked in such a way as to be inherited in an inactive form. Imprinting can be divided into three stages: (1) the establishment of the imprint during gametogenesis, (2) the maintenance of the imprint during embryogenesis and in adult somatic cells, and (3) the erasure and reestablishment of the imprint in the germ cells. Differential DNA methylation of regions near the imprinted genes is the key event in this marking process.

Extranuclear or cytoplasmic inheritance is due to the existence of genetic material in organelles outside the nucleus. The two most important examples of extranuclear inheritance are due to genetic material within mitochondria and chloroplasts. In many cases, extranuclear genetic material follows a maternal inheritance pattern because the egg is large and, in most species, is more likely than the sperm to transmit organelles to the offspring. However, some species display biparental inheritance in which the organelle is inherited from both parents, and a few species show paternal inheritance of organelles. Some human diseases are caused by mutations in mitochondrial genes.

The extranuclear genome of mitochondria and chloroplasts is believed to be the result of the endosymbiotic origin of these organelles. According to the endosymbiosis theory, mitochondria are derived from purple bacteria, and chloroplasts are derived from cyanobacteria that were engulfed by eukaryotic cells. During evolution, many genes have been transferred from the organellar genomes to the nuclear genomes, so modern mitochondria and chloroplast genomes are small and contain relatively few genes. Other examples of extranuclear inheritance include symbiotic infective particles identified in *Paramecium aurelia* and fruit flies.

EXPERIMENTAL SUMMARY

The primary way that researchers have identified non-Mendelian patterns of inheritance is by making a series of genetic crosses and then analyzing the phenotypes of the offspring. In many cases, reciprocal crosses have yielded results that deviate from Mendelian inheritance. For genes with a maternal effect, reciprocal crosses reveal that the genotype of the mother governs the phenotype of the offspring. In genomic imprinting, reciprocal crosses indicate that one parent transmits an active form of a gene while the other parent transmits an allele that is inactive.

Similarly, the outcomes of reciprocal crosses are usually different for extrachromosomal inheritance. In most species, organelles are inherited from the maternal parent, although many exceptions are known. The inheritance of the genetic material within mitochondria and chloroplasts, first identified in studies of yeast and algae, is non-Mendelian.

Researchers have also investigated non-Mendelian inheritance at the cellular and molecular levels. The cellular explanation of maternal effect genes is that the nurse cells transfer their gene products to the developing oocyte. In Chapter 23, we will examine methods that demonstrate the accumulation of maternal effect gene products within the oocyte. Genomic imprinting is also being investigated at the molecular level. Researchers exploring the biochemistry of imprinted genes have identified DNA methylation as a key to marking. Further experimentation is needed to elucidate the mechanisms of erasure and *de novo* methylation that occur during gametogenesis.

Likewise, researchers have explored the cellular and molecular mechanisms of X inactivation. Cytological examination of female mammalian cells revealed the presence of a highly condensed X chromosome, called the Barr body. A study of the expression of an enzyme known as glucose-6-phosphate dehydrogenase (G-6-PD) at the cellular level confirmed that one X chromosome is permanently inactivated in the somatic cell lineages of females. A region on the X chromosome called Xic is required for the proper counting and inactivation of X chromosomes. Expression of the *Xist* gene located in that region is required for the X-inactivation mechanism. The embryonic expression of a gene termed *Tsix* may play a role in choosing the active X chromosome. Another region termed Xce is also important in the choice of the inactivated chromosome.

Finally, molecular tools have been used to dissect the genetic composition of organellar chromosomes. Some of these methods, which include DNA cloning and sequencing, are described in Chapter 18. These techniques have enabled researchers to identify the genes located in mitochondrial DNA (mtDNA) and chloroplast DNA (cpDNA).

PROBLEM SETS & INSIGHTS

Solved Problems

S1. Our understanding of maternal effect genes has been greatly aided by their identification in experimental organisms such as *Drosophila melanogaster* and *Caenorhabditis elegans*. In experimental organisms with a short generation time, geneticists have successfully searched for mutant alleles that prevent the normal process of embryonic development. In many cases, the offspring will die at early embryonic or larval stages. These are called maternal effect lethal alleles. How would a researcher identify a mutation that produced a recessive maternal effect lethal allele?

Answer: A maternal effect lethal allele can be identified when a phenotypically normal mother produces only offspring with gross developmental abnormalities. For example, let's call the normal allele *N* and the maternal effect lethal allele *n*. A cross between two flies that are heterozygous for a maternal effect lethal allele would produce 1/4 of the offspring with a homozygous genotype, *nn*. These flies are viable because of the maternal effect. Their mother would be *Nn* and provide the *n* egg with a sufficient amount of *N* gene product so that the *nn* flies would develop properly. However, homozygous *nn* females cannot provide their eggs with any normal gene product. Therefore, all of their offspring are abnormal and die during early stages.

S2. A maternal effect gene in *Drosophila*, called *torso*, is found as a recessive allele that prevents the correct development of anterior- and posterior-most structures. A wild-type male is crossed to a female of unknown genotype. This mating produces 100% larva that are missing their anterior- and posterior-most structures and therefore die during early development. What is the genotype and phenotype of the female fly in this cross? What are the genotypes and phenotypes of the female fly's parents?

Answer: Because this cross produces 100% abnormal offspring, the female fly must be homozygous for the abnormal *torso* allele. Even so, the female fly must be phenotypically normal in order to reproduce. This female fly had a mother that was heterozygous for a normal and abnormal *torso* allele and a father that was either heterozygous or homozygous for the abnormal *torso* allele.

$torso^+ torso^-$ × $torso^+ torso^-$ or $torso^- torso^-$

(grandmother)   (grandfather)

↓

$torso^- torso^-$ (mother of 100% abnormal offspring)

This female fly is phenotypically normal because its mother was heterozygous and provided the gene products of the $torso^+$ allele from the nurse cells. However, this homozygous female will produce only abnormal offspring because it cannot provide them with the normal $torso^+$ gene products.

S3. An individual with Angelman syndrome produced an offspring with Prader-Willi syndrome. Why does this occur? What are the sexes of the parent with Angelman syndrome and the offspring with Prader-Willi syndrome?

Answer: These two different syndromes are most commonly caused by a small deletion in chromosome 15. In addition, genomic imprinting plays a role because genes in this deleted region are differentially imprinted, depending on sex. If this deletion is inherited from the paternal parent, the offspring develops Prader-Willi syndrome. Therefore, in this problem, the individual with Angelman syndrome must have been a male because he produced a child with Prader-Willi syndrome. The child could be either a male or female.

S4. In yeast, a haploid petite mutant also carries a mutant gene that requires the amino acid histidine for growth. The petite his^- strain is crossed to a wild-type his^+ strain to yield the following tetrad:

2 cells: petite his^-

2 cells: petite his^+

Explain the inheritance of the petite and his^- mutations.

Answer: The his^- and his^+ alleles are segregating in a 2:2 ratio. This result indicates a nuclear pattern of inheritance. By comparison, all four cells in this tetrad have a petite phenotype. This is a suppressive petite that arises from a mitochondrial mutation.

S5. Let's suppose that you are a horticulturist who has recently identified an interesting plant with variegated leaves. How would you determine if this trait is nuclearly or cytoplasmically inherited?

Answer: Make crosses and reciprocal crosses involving normal and variegated strains. In many species, chloroplast genomes are inherited maternally, although this is not always the case. In addition, a significant percentage of paternal leakage may occur. Nevertheless, when reciprocal crosses yield different outcomes, an organellar mode of inheritance is possibly at work.

S6. A phenotype that is similar to a yeast suppressive petite was also identified in the mold *Neurospora crassa*. Mary and Herschel Mitchell identified a slow-growing mutant that they called *poky*. Unlike yeast, which are isogamous (i.e., produce one type of gamete), *Neurospora* is sexually dimorphic and produces male and female reproductive structures. When a *poky* strain of *Neurospora* was crossed to a wild-type strain, the results were different between reciprocal crosses. If a *poky* mutant was the female parent, all spores exhibited the *poky* phenotype. By comparison, if the wild-type strain was the female parent, all spores were wild type. Explain these results.

Answer: These genetic studies indicate that the *poky* mutation is maternally inherited. The cytoplasm of the female reproductive cells provides the offspring with their mitochondria. Besides these genetic studies, the Mitchells and their collaborators showed that *poky* mutants are defective in certain cytochromes, which are iron-containing proteins that are known to be located in the mitochondria.

Conceptual Questions

C1. Define the term epigenetic inheritance, and describe two examples.

C2. Describe the inheritance pattern of maternal effect genes. Explain how the maternal effect occurs at the cellular level. What are the expected functional roles of the proteins that are encoded by maternal effect genes?

C3. A maternal effect gene exists in a dominant *N* (normal) allele and a recessive *n* (abnormal) allele. What would be the ratios of genotypes and phenotypes for the offspring of the following crosses?

A. *nn* female × *NN* male

B. *NN* female × *nn* male

C. *Nn* female × *Nn* male

C4. A *Drosophila* embryo dies during early embryogenesis due to a recessive maternal effect allele called *bicoid*. The wild-type allele is designated *bicoid*$^+$. What are the genotypes and phenotypes of the embryo's mother and maternal grandparents?

C5. For Mendelian traits, the nuclear genotype (i.e., the alleles found on chromosomes in the cell nucleus) directly influences an offspring's traits. In contrast, for non-Mendelian inheritance patterns, the offspring's phenotype cannot be reliably predicted solely from its genotype. For the following traits, what do you need to know to predict the phenotypic outcome?

A. Dwarfism due to a mutant *Igf2* allele

B. Snail coiling direction

C. Leber's hereditary optic neuropathy

C6. Let's suppose a maternal effect gene exists as a normal dominant allele and an abnormal recessive allele. A mother who is phenotypically abnormal produces all normal offspring. Explain the genotype of the mother.

C7. Let's suppose that a gene affects the anterior morphology in house flies and is inherited as a maternal effect gene. The gene exists in a normal allele, *H*, and a recessive allele, *h*, which causes a small head. A female fly with a normal head is mated to a true-breeding male with a small head. All of the offspring have small heads. What are the genotypes of the mother and offspring? Explain your answer.

C8. Explain why maternal effect genes exert their effects during the early stages of development.

C9. As described in Chapter 19, researchers have been able to "clone" mammals by fusing a cell having a diploid nucleus (i.e., a somatic cell) with an egg that has had its (haploid) nucleus removed.

A. With regard to maternal effect genes, would the phenotype of such a cloned animal be determined by the animal that donated the egg or by the animal that donated the somatic cell? Explain.

B. Would the cloned animal inherit extranuclear traits from the animal that donated the egg or by the animal that donated the somatic cell? Explain.

C. In what ways would you expect this cloned animal to be similar to or different from the animal that donated the somatic cell? Is it fair to call such an animal a "clone" of the animal that donated the diploid nucleus?

C10. With regard to the numbers of sex chromosomes, explain why dosage compensation is necessary.

C11. What is a Barr body? How is its structure different from that of other chromosomes in the cell? How does the structure of a Barr body affect the level of X-linked gene expression?

C12. Among different species, describe three distinct strategies for accomplishing dosage compensation.

C13. Describe when X inactivation occurs and how this leads to phenotypic results at the organism level. In your answer, you should explain why X inactivation causes results such as variegated coat patterns in mammals. Why do two different calico cats have their patches of orange and black fur in different places? Explain whether or not a variegated coat pattern due to X inactivation could occur in marsupials.

C14. Describe the molecular process of X inactivation. This description should include the three phases of inactivation and the role of the Xic. Explain what happens to X chromosomes during embryogenesis, in adult somatic cells, and during oogenesis.

C15. On rare occasions, an abnormal human male is born who is somewhat feminized compared to normal males. Microscopic examination of the cells of one such individual revealed that he has a single Barr body in each cell. What is the chromosomal composition of this individual?

C16. How many Barr bodies would you expect to find in humans with the following abnormal compositions of sex chromosomes?

A. XXY

B. XYY

C. XXX

D. X0 (a person with just a single X chromosome)

C17. Certain forms of human color blindness are inherited as X-linked recessive traits. Hemizygous males are color-blind, while heterozygous females are not. However, heterozygous females sometimes have partial color blindness.

A. Discuss why heterozygous females may have partial color blindness.

B. Doctors identified an unusual case in which a heterozygous female was color-blind in her right eye but had normal color vision in her left eye. Explain how this might have occurred.

C18. A black female cat ($X^B X^B$) and an orange male cat ($X^O Y$) were mated to each other and produced a male cat that was calico. Which sex chromosomes did this male offspring inherit from its mother and father? Remember that the presence of the Y chromosome determines maleness in mammals.

C19. What is the spreading stage of X inactivation? Why do you think it is called a spreading stage? Discuss the role of the *Xist* gene in the spreading stage of X inactivation.

C20. When does the erasure and reestablishment phase of genomic imprinting occur? Explain why it is necessary to erase an imprint and then reestablish it in order to always maintain imprinting from the same sex of parent.

C21. In what types of cells would you expect *de novo* methylation to occur? In what cell types would it not occur?

C22. On rare occasions, people are born with a condition known as uniparental disomy. It happens when an individual inherits both copies of a chromosome from one parent and no copies from the other parent. This occurs when two abnormal gametes happen to complement each other to produce a diploid zygote. For example, an abnormal sperm that lacks chromosome 15 could fertilize an egg that contains two copies of chromosome 15. In this situation, the individual would be said to have maternal uniparental disomy 15 because both copies of chromosome 15 were inherited from the mother. Alternatively, an abnormal sperm with two copies of chromosome 15 could fertilize an egg with no copies. This is known as paternal uniparental disomy 15. If a female is born with paternal uniparental disomy 15, would you expect her to be phenotypically normal, have Angelman syndrome (AS), or have Prader-Willi syndrome (PWS)? Explain. Would you expect her to produce normal offspring or offspring affected with AS or PWS?

C23. Genes that cause Prader-Willi syndrome and Angelman syndrome are closely linked along chromosome 15. Although people with these syndromes do not usually reproduce, let's suppose that a couple produces two children with Angelman syndrome. The oldest child (named Pat) grows up and has two children with Prader-Willi syndrome. The second child (named Robin) grows up and has one child with Angelman syndrome.

A. Are Pat and Robin's parents both normal or does one of them have Angelman or Prader-Willi syndrome? If one of them has a disorder, explain why it is the mother or the father.

B. What are the sexes of Pat and Robin? Explain.

C24. How is the process of X inactivation similar to genomic imprinting? How is it different?

C25. What is extranuclear inheritance? Describe three examples.

C26. What is a reciprocal cross? Let's suppose that a gene is found as a wild-type allele and a recessive mutant allele. What would be the expected outcomes of reciprocal crosses if a true-breeding normal individual were crossed to a true-breeding individual carrying the mutant allele? What would be the results if the gene were maternally inherited?

C27. Among different species, does extranuclear inheritance always follow a maternal inheritance pattern? Why or why not?

C28. What is the phenotype of a petite mutant? Where can a petite mutation occur: in nuclear genes, extranuclear genetic material, or both? What is the difference between a neutral and suppressive petite?

C29. Extranuclear inheritance often correlates with maternal inheritance. Even so, paternal leakage is not uncommon. What is paternal leakage? If a cross produced 200 offspring and the rate of mitochondrial paternal leakage was 3%, how many offspring would be expected to contain paternal mitochondria?

C30. Discuss the structure and organization of the mitochondrial and chloroplast genomes. How large are they, how many genes do they contain, and how many copies of the genome are found in each organelle?

C31. Explain the likely evolutionary origin of mitochondrial and chloroplast genomes. How have the sizes of the mitochondrial and chloroplast genomes changed since their origin? How has this occurred?

C32. Which of the following traits or diseases are determined by nuclear genes?

A. Snail coiling pattern

B. Prader-Willi syndrome

C. Streptomycin resistance in *Chlamydomonas*

D. Leber's hereditary optic neuropathy

C33. Acute murine leukemia virus (AMLV) causes leukemia in mice. This virus is easily passed from mother to offspring through the mother's milk. (Note: Even though newborn offspring acquire the virus, they may not develop leukemia until much later in life. Testing can determine if an animal carries the virus.) Describe how the formation of leukemia via AMLV resembles a maternal inheritance pattern. How could you determine that this form of leukemia is not caused by extranuclear inheritance?

C34. Describe how a biparental pattern of extranuclear inheritance would resemble a Mendelian pattern of inheritance for a particular gene. How would they differ?

C35. According to the endosymbiosis theory, mitochondria and chloroplasts are derived from bacteria that took up residence within eukaryotic cells. However, at one time, prior to being taken up by eukaryotic cells, these bacteria were free-living organisms. However, we cannot take a mitochondrion or chloroplast out of a living eukaryotic cell and get it to survive and replicate on its own. Why not?

Experimental Questions

E1. Figure 7.1 describes an example of a maternal effect gene. Explain how Sturtevant deduced a maternal effect gene based on the F_2 and F_3 generations.

E2. Discuss the types of experimental observations that Mary Lyon brought together in proposing her hypothesis concerning X inactivation. In your own words, explain how these observations were consistent with her hypothesis.

E3. Chapter 18 describes three blotting methods (i.e., Southern blotting, Northern blotting, and Western blotting) that are used to detect specific genes and gene products. Southern blotting detects DNA, Northern blotting detects RNA, and Western blotting detects proteins. Let's suppose that a female fruit fly is heterozygous for a maternal effect gene, which we will call gene *B*. The female is *Bb*. The normal allele, *B*, encodes a functional mRNA that is 550 nucleotides long. A recessive allele, *b*, encodes a shorter mRNA that is 375 nucleotides long. (Allele *b* is due to a deletion within this gene.) How could you use one or more of these techniques to show that nurse cells transfer gene products from gene *B* to developing eggs? You may assume that you can dissect the ovaries of fruit flies and isolate eggs separately from nurse cells. In your answer, describe your expected results.

E4. As a hypothetical example, a trait in mice results in mice with very long tails. You initially have a true-breeding strain with normal tails and a true-breeding strain with long tails. You then make the following types of crosses:

Cross 1: When true-breeding females with normal tails are crossed to true-breeding males with long tails, all F_1 offspring have long tails.

Cross 2: When true-breeding females with long tails are crossed to true-breeding males with normal tails, all F_1 offspring have normal tails.

Cross 3: When F_1 females from cross 1 are crossed to true-breeding males with normal tails, all offspring have normal tails.

Cross 4: When F_1 males from cross 1 are crossed to true-breeding females with long tails, half of the offspring have normal tails and half have long tails.

Explain the pattern of inheritance of this trait.

E5. You have a female snail that coils to the right, but you do not know its genotype. You may assume that right coiling (*D*) is dominant to left coiling (*d*). You also have male snails at your disposal of known genotype. How would you determine the genotype of this female snail? In your answer, describe your expected results depending on whether the female is *DD*, *Dd*, or *dd*.

E6. On a recent camping trip, you find one male snail on a deserted island that coils to the right. However, in this same area, you find several shells (not containing living snails) that coil to the left. Therefore, you conclude that you are not certain of the genotype of this male snail. On a different island, you find a large colony of snails of the same species. All of these snails coil to the right, and every snail shell that you find on this second island coils to the right. With regard to the maternal effect gene that determines coiling pattern, how would you determine the genotype of the male snail that you found on the deserted island? In your answer, describe your expected results.

E7. Figure 7.6 describes the results of X inactivation in mammals. If fast and slow alleles of glucose-6-phosphate dehydrogenase (*G-6-PD*) exist in other species, what would be the expected results of gel electrophoresis for a heterozygous female of the following species?

A. Marsupial

B. *Drosophila melanogaster*

C. *Caenorhabditis elegans* (Note: We are considering the hermaphrodite in *C. elegans* to be equivalent to a female.)

E8. Two male mice, which we will call male A and male B, are both phenotypically normal. Male A was from a litter that contained half phenotypically normal mice and half dwarf mice. The mother of male A was known to be homozygous for the normal *Igf2* allele. Male B was from a litter of eight mice that were all phenotypically normal. The parents of male B were a phenotypically normal male and a dwarf female. Male A and male B were put into a cage with two female mice that we will call female A and female B. Female A is dwarf, and female B is phenotypically normal. The parents of these two females were unknown, although it was known that they were from the same litter. The mice were allowed to mate with each other, and the following data were obtained:

Female A gave birth to three dwarf babies and four normal babies.

Female B gave birth to four normal babies and two dwarf babies.

Which male(s) mated with female A and female B? Explain.

E9. In the experiment of Figure 7.6, why does a clone of cells produce only one type of G-6-PD enzyme? What would you expect to happen if a clone was derived from an early embryonic cell? Why does the initial sample of tissue produce both forms of G-6-PD?

E10. Chapter 18 describes a blotting method known as Northern blotting that can be used to determine the amount of mRNA produced by a particular gene. In this method, the amount of a specific mRNA produced by cells is detected as a band on a gel. If one type of cell produces twice as much of a particular mRNA compared to another cell, the band will appear twice as dark. Also, sometimes mutations affect the length of mRNA that is transcribed from a gene. For example, a small deletion within a gene may shorten an mRNA. Northern blotting also can discern the sizes of mRNAs.

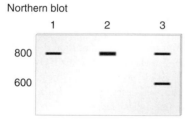

Lane 1 is a Northern blot of mRNA from cell type A that is 800 nucleotides long.

Lane 2 is a Northern blot of the same mRNA from cell type B. (Cell type B produces twice as much of this RNA compared to cell type A.)

Lane 3 shows a heterozygote in which one of the two genes has a deletion, which shortens the mRNA by 200 nucleotides.

Here is the question. Let's suppose an X-linked gene exists as two alleles, which we will call *B* and *b*. Allele *B* encodes an mRNA that is 750 nucleotides long, while allele *b* encodes a shorter mRNA that is 675 nucleotides long. Draw the expected results of a Northern blot using mRNA isolated from the same type of somatic cells taken from the following individuals:

A. First lane is mRNA from an X^bY male fruit fly.

   Second lane is mRNA from an X^bX^b female fruit fly.

   Third lane is mRNA from an X^BX^b female fruit fly.

B. First lane is mRNA from an X^BY male mouse.

   Second lane is mRNA from an X^BX^b female mouse.[*]

   Third lane is mRNA from an X^BX^B female mouse.[*]

   [*]The sample is taken from an adult female mouse. It is not a clone of cells. It is a tissue sample, like the one described in the experiment of Figure 7.6.

C. First lane is mRNA from an X^B0 male *C. elegans*.

   Second lane is mRNA from an X^BX^b hermaphrodite *C. elegans*.

   Third lane is mRNA from an X^BX^B hermaphrodite *C. elegans*.

E11. A variegated trait in plants is analyzed using reciprocal crosses. The following results are obtained:

Variegated female × Normal male	Normal female × Variegated male
↓	↓
1,024 variegated + 52 normal	1,113 normal + 61 variegated

Explain this pattern of inheritance.

E12. Ruth Sager and her colleagues discovered that the mode of inheritance of streptomycin resistance in *Chlamydomonas* could be altered if the mt^+ cells were exposed to UV irradiation prior to mating. This exposure dramatically increased the frequency of biparental inheritance. What would be the expected outcome of a cross between an mt^+ sm^r and an mt^- sm^s strain in the absence of UV irradiation? How would the result differ if the mt^+ strain was exposed to UV light?

E13. Take a look at Figure 7.18 and describe how you could experimentally distinguish between yeast strains that are neutral petites versus suppressive petites.

Questions for Student Discussion/Collaboration

1. Recessive maternal effect genes are identified in flies (for example) when a phenotypically normal mother cannot produce any normal offspring. Because all the offspring are dead, this female fly cannot be used to produce a strain of heterozygous flies that could be used in future studies. How would you identify heterozygous individuals that are carrying a recessive maternal effect allele? How would you maintain this strain of flies in a laboratory over many generations?

2. What is an infective particle? Discuss the similarities and differences between infective particles and organelles such as mitochondria and chloroplasts. Do you think the existence of infective particles supports the endosymbiosis theory of the origin of mitochondria and chloroplasts?

Note: All answers appear at the website for this textbook; the answers to even-numbered questions are in the back of the textbook.

Visit the Online Learning Center for practice tests, answer keys, and other learning aids for this chapter. Enhance your understanding of genetics with our interactive exercises, quizzes, animations, and much more.

VARIATION IN CHROMOSOME STRUCTURE AND NUMBER

8

The term **genetic variation** refers to genetic differences among members of the same species or those between different species. Throughout Chapters 2 to 7, we have focused primarily on variation in specific genes, which is called **allelic variation.** In Chapter 8, our emphasis will shift to larger types of genetic changes that affect the structure or number of eukaryotic chromosomes. You might think small mutations that produce allelic differences would happen much more frequently than such larger changes. Surprisingly, changes in chromosome structure and number are actually fairly common. These larger alterations may affect the expression of many genes and thereby affect phenotypes. Variation in chromosome structure and number are of great importance in the field of genetics because they are critical in the evolution of new species and have widespread medical relevance. In addition, agricultural geneticists have discovered that such variation can lead to the development of new strains of crops, which may be quite profitable.

In the first section of Chapter 8, we will begin by exploring how the structure of a eukaryotic chromosome can be modified, either by altering the total amount of genetic material or by rearranging the order of genes along a chromosome. Such changes can usually be detected microscopically. The rest of the chapter is concerned with changes in the total number of chromosomes. Variation in chromosome number occurs in two subtypes: changes in the number of sets of chromosomes and changes in the numbers of individual chromosomes within a set. Natural variation in the number of sets of chromosomes is relatively common among different species, particularly in the plant kingdom. Within the same species, changes in chromosome number within a set can also occur, but such alterations are usually detrimental to the organism. We will explore how variation in chromosome number occurs and consider examples where it has significant phenotypic consequences. We will conclude by examining how changes in chromosome number can be induced through experimental treatments and how these approaches have applications in research and in agriculture.

The chromosome composition of humans. Somatic cells in humans contain 46 chromosomes, which come in 23 pairs.

8.1 VARIATION IN CHROMOSOME STRUCTURE

Chromosomes in the nuclei of eukaryotic cells contain long, linear DNA molecules that carry hundreds or even thousands of genes. In this section, we will explore how the composition of a chromosome can be changed. As you will see, segments of a chromosome can be lost, duplicated, or rearranged in a new way.

We will also examine the cellular mechanisms that underlie these changes in chromosome structure. Unusual events during meiosis may affect how altered chromosomes are transmitted from parents to offspring. Also, we will consider many examples in which chromosomal alterations affect an organism's phenotypic characteristics.

Natural Variation Exists in Chromosome Structure

To appreciate changes in chromosome structure, researchers need to have a reference point for a normal set of chromosomes. To determine what the normal chromosomes of a species look like, a **cytogeneticist**—a scientist who studies chromosomes microscopically—examines the chromosomes from several members of a given species. In most cases, two phenotypically normal individuals of the same species will have the same number and types of chromosomes.

To determine the chromosomal composition of a species, the chromosomes in actively dividing cells are examined microscopically. **Figure 8.1a** shows micrographs of chromosomes from three species: a human, a fruit fly, and a corn plant. As seen here, a human has 46 chromosomes (23 pairs), a fruit fly has 8 chromosomes (4 pairs), and corn has 20 chromosomes (10 pairs). Except for the sex chromosomes, which differ between males and females, most members of the same species have very similar chromosomes. For example, the overwhelming majority of people have 46 chromosomes in their somatic cells. By comparison, the chromosomal compositions of distantly related species, such as humans and fruit flies, may be very different. A total of 46 chromosomes is normal for humans, while 8 chromosomes is the norm for fruit flies.

Cytogeneticists have various ways to classify and identify chromosomes. The three most commonly used features are location of the centromere, size, and banding patterns that are revealed when the chromosomes are treated with stains. As shown in **Figure 8.1b**, chromosomes are classified as **metacentric** (in which the centromere is near the middle), **submetacentric** (in which the centromere is slightly off center), **acrocentric** (in which the centromere is significantly off center but not at the end), and **telocentric** (in which the centromere is at one end). Because the centromere is never exactly in the center of a chromosome, each chromosome has a short arm and a long arm. The short arm is designated with the letter *p* (for the French, petite), and the long arm is designated with the letter *q*. In the case of telocentric chromosomes, the short arm may be nearly nonexistent.

Figure 8.1c shows a human karyotype. The procedure for making a karyotype was described in Chapter 3 (see Figure 3.2). A **karyotype** is a micrograph in which all of the chromosomes within a single cell have been arranged in a standard fashion. When preparing a karyotype, the chromosomes are aligned with the short arms on top and the long arms on the bottom. By convention, the chromosomes are numbered roughly according to their size, with the largest chromosomes having the smallest numbers. For example, human chromosomes 1, 2, and 3 are relatively large, whereas 21 and 22 are the two smallest. An exception to the numbering system involves the sex chromosomes, which are designated with letters (for humans, X and Y).

Because different chromosomes often have similar sizes and centromeric locations (e.g., compare human chromosomes 8, 9, and 10), geneticists must use additional methods to accurately identify each type of chromosome within a karyotype. For detailed identification, chromosomes are treated with stains to produce characteristic banding patterns. Several different staining procedures are used by cytogeneticists to identify specific chromosomes. An example is **G banding,** which is shown in Figure 8.1c. In this procedure, chromosomes are treated with mild heat or with proteolytic enzymes that partially digest chromosomal proteins. When exposed to the dye called Giemsa, named after its inventor Gustav Giemsa, some chromosomal regions bind the dye heavily and produce a dark band. In other regions, the stain hardly binds at all and a light band results. Though the mechanism of staining is not completely understood, the dark bands are thought to represent regions that are more tightly compacted. As shown in Figure 8.1c and d, the alternating pattern of G bands is a unique feature for each chromosome.

In the case of human chromosomes, approximately 300 G bands can usually be distinguished during metaphase. A larger number of G bands (in the range of 800) can be observed in prometaphase chromosomes because they are more extended than metaphase chromosomes. **Figure 8.1d** shows the conventional numbering system that is used to designate G bands along a set of human chromosomes. The left chromatid in each pair of sister chromatids shows the expected banding pattern during metaphase, while the right chromatid shows the banding pattern as it would appear during prometaphase.

Why is the banding pattern of eukaryotic chromosomes useful? First, when stained, individual chromosomes can be distinguished from each other, even if they have similar sizes and centromeric locations. For example, compare the differences in banding patterns between human chromosomes 8 and 9 (Figure 8.1d). These differences permit us to distinguish these two chromosomes even though their sizes and centromeric locations are very similar. Banding patterns are also used to detect changes in chromosome structure. As discussed next, chromosomal rearrangements or changes in the total amount of genetic material are more easily detected in banded chromosomes. Also, chromosome banding can be used to assess evolutionary relationships between species. Research studies have shown that the similarity of chromosome banding patterns is a good measure of genetic relatedness.

Changes in Chromosome Structure Include Deletions, Duplications, Inversions, and Translocations

With an understanding that chromosomes typically come in a variety of shapes and sizes, let's consider how the structures of normal chromosomes can be modified. In some cases, the total amount of genetic material within a single chromosome can be increased or decreased significantly. Alternatively, the genetic material in one or more chromosomes may be rearranged without affecting the total amount of material. As shown in **Figure 8.2**, these mutations are categorized as deletions, duplications, inversions, and translocations.

Deletions and duplications are changes in the total amount of genetic material within a single chromosome. In Figure 8.2,

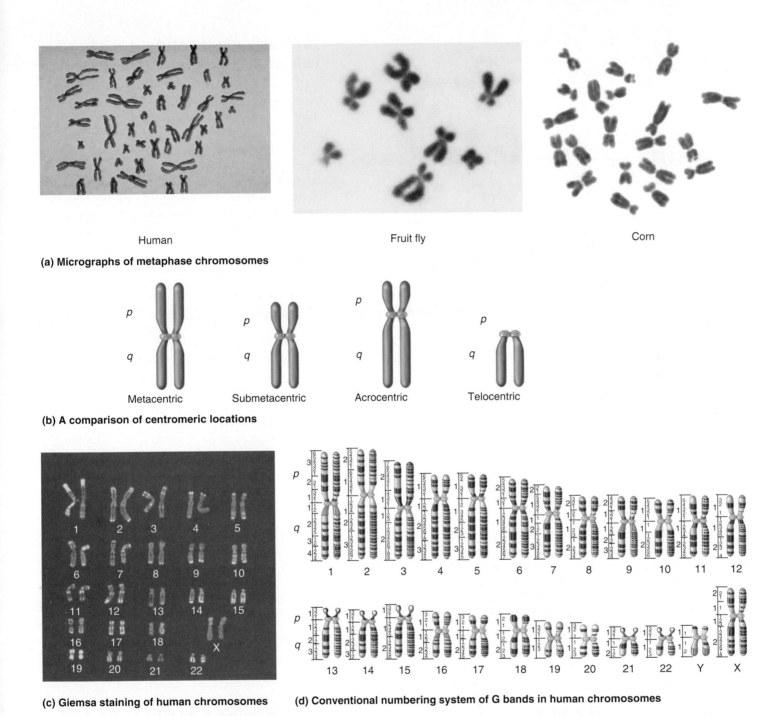

(a) Micrographs of metaphase chromosomes

Human

Fruit fly

Corn

(b) A comparison of centromeric locations

Metacentric   Submetacentric   Acrocentric   Telocentric

(c) Giemsa staining of human chromosomes

(d) Conventional numbering system of G bands in human chromosomes

FIGURE 8.1   Features of normal chromosomes. (a) Micrographs of chromosomes from a human, a fruit fly, and corn. **(b)** A comparison of centromeric locations. Centromeres can be metacentric, submetacentric, acrocentric (near one end), or telocentric (at the end). **(c)** Human chromosomes that have been stained with Giemsa. **(d)** The conventional numbering of bands in Giemsa-stained human chromosomes. The numbering is divided into broad regions, which then are subdivided into smaller regions. The numbers increase as the region gets farther away from the centromere. For example, if you take a look at the left chromatid of chromosome 1, the uppermost dark band would be at a location designated p35. The banding patterns of chromatids change as the chromatids condense. The left chromatid of each pair of sister chromatids shows the banding pattern of a chromatid in metaphase, and the right side shows the banding pattern as it would appear in prometaphase. Note: In prometaphase, the chromatids are more extended than in metaphase.

human chromosomes are labeled according to their normal G banding patterns. When a **deletion** occurs, a segment of chromosomal material is missing. In other words, the affected chromosome is deficient in a significant amount of genetic material. The term **deficiency** is also used to describe a missing region of a chromosome. In contrast, a **duplication** occurs when a section

of a chromosome is repeated compared to the normal parent chromosome.

Inversions and translocations are chromosomal rearrangements. An **inversion** involves a change in the direction of the genetic material along a single chromosome. For example, in Figure 8.2c, a segment of one chromosome has been inverted,

FIGURE 8.2   Types of changes in chromosome structure. The large chromosome shown throughout is human chromosome 1. The smaller chromosome seen in (d) and (e) is human chromosome 21. (**a**) A deletion occurs that removes a large portion of the q2 region, indicated by the red arrows. (**b**) A duplication occurs that doubles the q2-q3 region. (**c**) An inversion occurs that inverts the q2-q3 region. (**d**) The q2-q4 region of chromosome 1 is translocated to chromosome 21. A region of a chromosome cannot be inserted directly to the tip of another chromosome because telomeres at the tips of chromosomes would prevent such an event. In this example, a small piece at the end of chromosome 21 would have to be removed in order for the q2-q4 region of chromosome 1 to be attached to chromosome 21. (**e**) The q2-q4 region of chromosome 1 is exchanged with most of the q1-q2 region of chromosome 21.

so the order of four G bands is opposite to that of the parent chromosome. A **translocation** occurs when one segment of a chromosome becomes attached to a different chromosome or to a different part of the same chromosome. A **simple translocation** occurs when a single piece of chromosome is attached to another chromosome. In a **reciprocal translocation,** two different types of chromosomes exchange pieces, thereby producing two abnormal chromosomes carrying translocations.

Figure 8.2 illustrates the common ways that the structure of chromosomes can be altered. Throughout the rest of this section, we will consider how these changes occur, how the changes are detected experimentally, and how they affect the phenotypes of the individuals who inherit them.

The Loss of Genetic Material in a Deletion Tends to Be Detrimental to an Organism

A chromosomal deletion occurs when a chromosome breaks in one or more places and a fragment of the chromosome is lost. In **Figure 8.3a**, a normal chromosome has broken into two separate pieces. The piece without the centromere will be lost and degraded. This event produces a chromosome with a **terminal deletion.** In **Figure 8.3b**, a chromosome has broken in two places to produce three chromosomal fragments. The central fragment is lost, and the two outer pieces reattach to each other. This process has created a chromosome with an **interstitial deletion.** Deletions can also be created when recombination takes place at incorrect locations between two homologous chromosomes. The products of this type of aberrant recombination event are one chromosome with a deletion and another chromosome with a duplication. This process is examined later in this chapter.

The phenotypic consequences of a chromosomal deletion depend on the size of the deletion and whether it includes genes or portions of genes that are vital to the development of the organism. When deletions have a phenotypic effect, they are usually detrimental. Larger deletions tend to be more harmful because more genes

FIGURE 8.3   Production of terminal and interstitial deletions. This illustration shows the production of deletions in human chromosome 1.

are missing. Many examples are known in which deletions have significant phenotypic influences. For example, a human genetic disease known as cri-du-chat syndrome is caused by a deletion in a segment of the short arm of human chromosome 5 (**Figure 8.4a**). Individuals who carry a single copy of this abnormal chromosome along with a normal chromosome 5 display an array of abnormalities including mental deficiencies, unique facial anomalies, and an unusual catlike cry in infancy, which is the meaning of the French name for the syndrome (**Figure 8.4b**). Some other human genetic diseases, such as Angelman syndrome and Prader-Willi syndrome, described in Chapter 7, are due to a deletion in chromosome 15.

Duplications Tend to Be Less Harmful Than Deletions

Duplications result in extra genetic material. They are usually caused by abnormal events during recombination. Under normal circumstances, crossing over occurs at analogous sites between homologous chromosomes. On rare occasions, a crossover may occur at misaligned sites on the homologues (**Figure 8.5**). This results in one chromatid with an internal duplication and another chromatid with a deletion. In Figure 8.5, the chromosome with the extra genetic material carries a **gene duplication,** because the number of copies of gene *C* has been increased from one to two. In most cases, gene duplications happen as rare, sporadic events during the evolution of species. Later in this section,

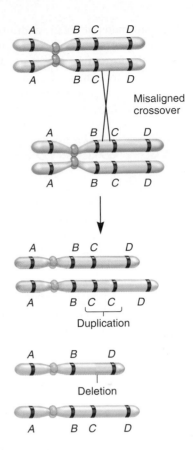

FIGURE 8.5   **Abnormal crossing over, leading to a duplication and a deletion.** A crossover has occurred at sites between genes *C* and *D* in one chromatid and between genes *B* and *C* in another chromatid. After crossing over is completed, one chromatid contains a duplication, and the other contains a deletion.

(a) Chromosome 5   **(b) A child with cri-du-chat syndrome**

FIGURE 8.4   **Cri-du-chat syndrome. (a)** Chromosome 5 from the karyotype of an individual with this disorder. A section of the short arm of one chromosome is missing. **(b)** An affected individual.
Genes → Traits Compared to an individual who has two copies of each gene on chromosome 5, an individual with cri-du-chat syndrome has only one copy of the genes that are located within the missing segment. This genetic imbalance (one versus two copies of many genes on chromosome 5) causes the phenotypic characteristics of this disorder, which include a catlike cry in infancy, short stature, characteristic facial anomalies (e.g., a triangular face, almond-shaped eyes, broad nasal bridge, and low-set ears), and microencephaly (a smaller than normal brain).

we will consider how multiple copies of genes can evolve into a family of genes with specialized functions.

Like deletions, the phenotypic consequences of duplications tend to be correlated with size. Duplications are more likely to have phenotypic effects if they involve a large piece of the chromosome. In general, small duplications are less likely to have harmful effects than are deletions of comparable size. This observation suggests that having only one copy of a gene is more harmful than having three copies. In humans, relatively few well-defined syndromes are caused by small chromosomal duplications. An example is Charcot-Marie-Tooth disease (type 1A), a peripheral neuropathy characterized by numbness in the hands and feet that is caused by a small duplication on the short arm of chromosome 17.

Duplications Provide Additional Material for Gene Evolution, Sometimes Leading to the Formation of Gene Families

In contrast to the gene duplication that causes Charcot-Marie-Tooth disease, the majority of small chromosomal duplications have no phenotypic effect. Nevertheless, they are vitally important because they provide raw material for the addition of more genes into a species' chromosomes. Over the course of many generations, this can lead to the formation of a **gene family** consisting of two or more genes that are similar to each other. As shown in **Figure 8.6**, the members of a gene family are derived from the same ancestral gene. Over time, two copies of an ancestral gene

can accumulate different mutations. Therefore, after many generations, the two genes will be similar but not identical. During evolution, this type of event can occur several times, creating a family of many similar genes.

When two or more genes are derived from a single ancestral gene, the genes are said to be **homologous.** Homologous genes within a single species are called **paralogs** and constitute a gene family. A well-studied example of a gene family is shown in **Figure 8.7**, which illustrates the evolution of the globin gene family found in humans. The globin genes encode polypeptides that are subunits of proteins that function in oxygen binding. For example, hemoglobin is a protein found in red blood cells; its function is to carry oxygen throughout the body. The globin gene family is composed of 14 paralogs that were originally derived from a single ancestral globin gene. According to an evolutionary analysis, the ancestral globin gene first duplicated about 500 million years ago and became separate genes encoding myoglobin and the hemoglobin group of genes. The primordial hemoglobin gene duplicated into an α-chain gene and a β-chain gene, which subsequently duplicated to produce several genes located on chromosomes 16 and 11, respectively. Currently, 14 globin genes are found on three different human chromosomes.

Why is it advantageous to have a family of globin genes? Although all globin polypeptides are subunits of proteins that play a role in oxygen binding, the accumulation of different mutations in the various family members has produced globins that are more specialized in their function. For example, myoglobin is better at binding and storing oxygen in muscle cells, and the hemoglobins are better at binding and transporting oxygen

ONLINE ANIMATION

FIGURE 8.6   Gene duplication and the evolution of paralogs. An abnormal crossover event like the one described in Figure 8.5 leads to a gene duplication. Over time, each gene accumulates different mutations.

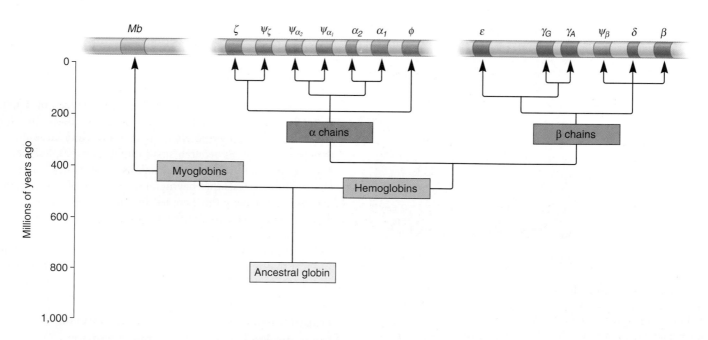

FIGURE 8.7   The evolution of the globin gene family in humans. The globin gene family evolved from a single ancestral globin gene. The first gene duplication produced two genes that accumulated mutations and became the genes encoding myoglobin (on chromosome 22) and the group of hemoglobins. The primordial hemoglobin gene then duplicated to produce several α-chain and β-chain genes, which are found on chromosomes 16 and 11, respectively. The four genes shown in gray are nonfunctional pseudogenes.

via the red blood cells. Also, different globin genes are expressed during different stages of human development. The ε- and ζ-globin genes are expressed very early in embryonic life, whereas the α-globin and γ-globin genes are expressed during the second and third trimesters of gestation. Following birth, the α-globin gene remains turned on, while the γ-globin genes are turned off and the β-globin gene is turned on. These differences in the expression of the globin genes reflect the differences in the oxygen transport needs of humans during the embryonic, fetal, and postpartum stages of life.

EXPERIMENT 8A

Comparative Genomic Hybridization Is Used to Detect Chromosome Deletions and Duplications

As we have seen, chromosome deletions and duplications can occur by a variety of mechanisms and may have an impact on the phenotypes of individuals who inherit them. One very important reason why researchers have become interested in these types of chromosomal changes is related to cancer. As discussed in Chapter 22, chromosomal deletions and duplications have been associated with many types of human cancers. Though such changes may be detectable by traditional chromosomal staining and karyotyping methods, small deletions and duplications may be difficult to detect in this manner. Fortunately, researchers have been able to develop more sensitive methods to identify changes in chromosome structure.

In 1992, Anne Kallioniemi, Daniel Pinkel, and colleagues devised a method called **comparative genomic hybridization (CGH).** This technique is largely used to determine if cancer cells have changes in chromosome structure, such as deletions or duplications. To begin this procedure, DNA is isolated from a test sample, which in this case was a sample of breast cancer cells, and also from a normal reference sample (**Figure 8.8**). The DNA from the breast cancer cells was used as a template to make green fluorescent DNA, and the DNA from normal cells was used to make red fluorescent DNA. These DNA molecules were 600 bp to 1,000 bp in length. Both types of DNA were then dena-tured by heat treatment. Equal amounts of the two fluorescently labeled DNA samples were mixed together and applied to normal metaphase chromosomes in which the DNA had also been denatured. Because the fluorescently labeled DNA fragments and the metaphase chromosomes had both been denatured, the fluorescently labeled DNA strands could bind to complementary regions on the metaphase chromosomes. This process is called **hybridization** because the DNA from one sample (a green or red DNA strand) forms a double-stranded region with a DNA strand from another sample (an unlabeled metaphase chromosome). Following hybridization, the metaphase chromosomes were visualized using a fluorescence microscope, and the images were analyzed by a computer that could determine the relative intensities of green and red fluorescence. What are the expected results? If a chromosomal region in the breast cancer cells and the normal cells are present in the same amount, the ratio between green and red fluorescence should be 1. If a chromosomal region is deleted in the breast cancer cell line, the ratio will be less than one, or if a region is duplicated, it will be greater than one.

■ THE GOAL

Deletions or duplications in cancer cells can be detected by comparing the ability of fluorescently labeled DNA from cancer cells and normal cells to bind (hybridize) to normal metaphase chromosomes.

■ ACHIEVING THE GOAL — FIGURE 8.8   **The use of comparative genomic hybridization to detect deletions and duplications in cancer cells.**

Starting materials: Breast cancer cells and normal cells.

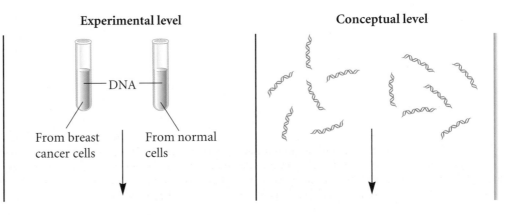

1. Isolate DNA from human breast cancer cells and normal cells. This involved breaking open the cells and isolating the DNA by chromatography. (See Appendix for description of chromatography.)

Experimental level

DNA

From breast cancer cells   From normal cells

Conceptual level

2. Label the breast cancer DNA with a green fluorescent molecule and the normal DNA with a red fluorescent molecule. This was done by using the DNA from step 1 as a template, and incorporating fluorescently labeled nucleotides into newly made DNA strands.

3. The DNA strands were then denatured by heat treatment. Mix together equal amounts of fluorescently labeled DNA and add it to a preparation of metaphase chromosomes from white blood cells. The procedure for preparing metaphase chromosomes is described in Figure 3.2. The metaphase chromosomes were also denatured.

4. Allow the fluorescently labeled DNA to hybridize to the metaphase chromosomes.

5. Visualize the chromosomes with a fluorescence microscope. Analyze the amount of green and red fluorescence along each chromosome with a computer.

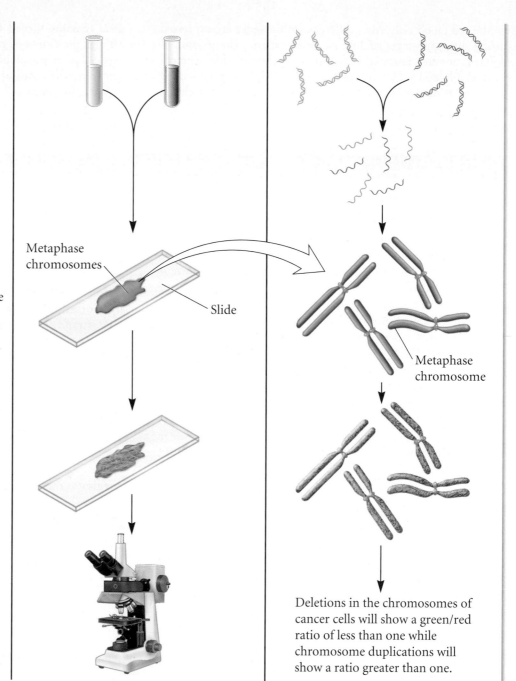

Metaphase chromosomes

Slide

Metaphase chromosome

Deletions in the chromosomes of cancer cells will show a green/red ratio of less than one while chromosome duplications will show a ratio greater than one.

THE DATA

Note: Unlabeled repetitive DNA was also included in this experiment to decrease the level of nonspecific, background labeling. This repetitive DNA also prevents labeling near the centromere. As seen in the data, regions in the chromosomes where the curves are missing are due to the presense of highly repetitive sequences near the centromere.

INTERPRETING THE DATA

The data of Figure 8.8 show the ratio of green (cancer DNA) to red (normal DNA) fluorescence along five different metaphase chromosomes. Chromosome 1 shows a large duplication, as indicated by the ratio of 2. One interpretation of this observation is that both copies of chromosome 1 carry a duplication. In comparison, chromosomes 9, 11, 16, and 17 have regions with a value of 0.5. This value indicates one of the two chromosomes of these four types in the cancer cells carries a deletion, while the other chromosome of these four types does not. (If the value was zero, this would indicate both copies of a chromosome had deleted the same region.) Overall, these results illustrate how this technique can be used to map chromosomal duplications and deletions in cancer cells.

This method is named comparative genomic hybridization because a comparison is made between the ability of two DNA samples (cancer versus normal cells) to hybridize to an entire genome. In this case, the entire genome is in the form of metaphase chromosomes. As discussed in Chapter 20, the fluorescently labeled DNAs can be hybridized to a DNA microarray instead of metaphase chromosomes. This newer method, called array comparative genomic hybridization (aCGH), is gaining widespread use in the analysis of cancer cells.

A self-help quiz involving this experiment can be found at the Online Learning Center.

Inversions Often Occur Without Phenotypic Consequences

We now turn our attention to inversions, changes in chromosome structure that involve a rearrangement in the genetic material. A chromosome with an inversion contains a segment that has been flipped to the opposite direction. Geneticists classify inversions according to the location of the centromere. If the centromere lies within the inverted region of the chromosome, the inverted region is known as a **pericentric inversion** (**Figure 8.9b**). Alternatively, if the centromere is found outside the inverted region, the inverted region is called a **paracentric inversion** (**Figure 8.9c**).

When a chromosome contains an inversion, the total amount of genetic material remains the same as in a normal chromosome. Therefore, the great majority of inversions do not have any phenotypic consequences. In rare cases, however, an inversion can alter the phenotype of an individual. Whether or not this occurs is related to the boundaries of the inverted segment. When an inversion occurs, the chromosome is broken in two places, and the center piece flips around to produce the inversion. If either breakpoint occurs within a vital gene, the function of the gene is expected to be disrupted, possibly producing a phenotypic effect. For example, some people with hemophilia (type A) have inherited an X-linked inversion in which the breakpoint has inactivated the gene for factor VIII, which is a blood-clotting protein. In other cases, an inversion (or translocation) may reposition a gene on a chromosome in a way that alters its normal level of expression. This is a type of **position effect**—a change in

phenotype that occurs when the position of a gene changes from one chromosomal site to a different location. This topic is also discussed in Chapter 16 (see Figures 16.2 and 16.3).

Because inversions seem like an unusual genetic phenomenon, it is perhaps surprising that they are found in human populations in significant numbers. About 2% of the human population carries inversions that are detectable with a light microscope. In most cases, these individuals are phenotypically normal and live their lives without knowing they carry an inversion. In a few cases, however, an individual with an inversion chromosome may produce offspring with phenotypic abnormalities. This event

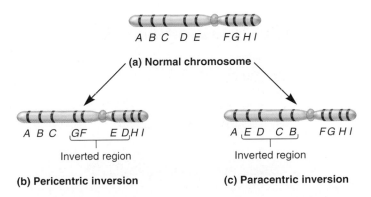

FIGURE 8.9 **Types of inversions.** (a) Depicts a normal chromosome with the genes ordered from *A* through *I.* A pericentric inversion (b) includes the centromere, whereas a paracentric inversion (c) does not.

may prompt a physician to request a microscopic examination of the individual's chromosomes. In this way, phenotypically normal individuals may discover they have a chromosome with an inversion. Next, we will examine why an individual carrying an inversion may produce offspring with phenotypic abnormalities.

Inversion Heterozygotes May Produce Abnormal Chromosomes Due to Crossing Over

An individual that carries one copy of a normal chromosome and one copy of an inverted chromosome is known as an **inversion heterozygote.** Such an individual, though possibly phenotypically normal, may have a high probability of producing haploid cells that are abnormal in their total genetic content. This likelihood

depends on the size of the inverted segment. During meiosis, an inversion heterozygote with a fairly large inverted segment may produce a sizable fraction of abnormal haploid cells, perhaps 1/3 or even higher.

The underlying cause of abnormality is the phenomenon of crossing over within the inverted region. During meiosis I, pairs of homologous sister chromatids synapse with each other. **Figure 8.10** illustrates how this occurs in an inversion heterozygote. For the normal chromosome and inversion chromosome to synapse properly, an **inversion loop** must form to permit the homologous genes on both chromosomes to align next to each other despite the inverted sequence. If a crossover occurs within the inversion loop, highly abnormal chromosomes are produced.

(a) Pericentric inversion

(b) Paracentric inversion

FIGURE 8.10   **The consequences of crossing over in the inversion loop. (a)** Crossover within a pericentric inversion. **(b)** Crossover within a paracentric inversion.

The consequences of this type of crossover depend on whether the inversion is pericentric or paracentric. Figure 8.10a describes a crossover in the inversion loop when one of the homologues has a pericentric inversion in which the centromere lies within the inverted region of the chromosome. This event consists of a single crossover that involves only two of the four sister chromatids. Following the completion of meiosis, this single crossover yields two abnormal chromosomes. Both of these abnormal chromosomes have a segment that is deleted and a different segment that is duplicated. In this example, one of the abnormal chromosomes is missing genes *H* and *I* and has an extra copy of genes *A*, *B*, and *C*. The other abnormal chromosome has the opposite situation; it is missing genes *A*, *B*, and *C* and has an extra copy of genes *H* and *I*. These abnormal chromosomes may result in gametes that are inviable. Alternatively, if these abnormal chromosomes are passed to offspring, they are likely to produce phenotypic abnormalities, depending on the amount and nature of the duplicated/deleted genetic material. A large deletion is likely to be lethal.

Figure 8.10b shows the outcome of a crossover involving a paracentric inversion in which the centromere lies outside the inverted region. This single crossover event produces a very strange outcome. One chromosome, called a **dicentric** chromosome, contains two centromeres. The region of the chromosome connecting the two centromeres is a **dicentric bridge.** The crossover also produces a piece of chromosome without any centromere—an **acentric fragment,** which will be lost and degraded in subsequent cell divisions. The dicentric chromosome is a temporary condition. If the two centromeres try to move toward opposite poles during anaphase, the dicentric bridge will be forced to break at some random location. Therefore, the net result of this crossover is to produce one normal chromosome, one chromosome with an inversion, and two chromosomes that contain deletions. These two chromosomes with deletions result from the breakage of the dicentric chromosome. They are missing the genes that were located on the acentric fragment.

Translocations Involve Exchanges Between Different Chromosomes

Another type of chromosomal rearrangement is a translocation in which a piece from one chromosome is attached to another chromosome. Eukaryotic cells have evolved telomeres, which tend to prevent translocations from occurring. As described in Chapters 10 and 11, **telomeres**—specialized repeated sequences of DNA—are found at the ends of normal chromosomes. Telomeres allow cells to distinguish where a chromosome ends and prevent the attachment of chromosomal DNA to the natural ends of a chromosome.

If cells are exposed to agents that cause chromosomes to break, the broken ends lack telomeres and are said to be reactive—a reactive end readily binds to another reactive end. If a single chromosome break occurs, DNA repair enzymes will usually recognize the two reactive ends and join them back together; the chromosome is repaired properly. However, if multiple chromosomes are broken, the reactive ends may be joined incorrectly to produce abnormal chromosomes (**Figure 8.11a**). This is one mechanism that causes reciprocal translocations to occur.

(b) Nonhomologous crossover

(a) Chromosomal breakage and DNA repair

FIGURE 8.11   Two mechanisms that cause a reciprocal translocation. **(a)** When two different chromosomes break, the broken ends are recognized by DNA repair enzymes, which attempt to reattach broken ends. If two different chromosomes are broken at the same time, the incorrect ends may become attached to each other. **(b)** A nonhomologous crossover has occurred between chromosome 1 and chromosome 7. This crossover yields two chromosomes that carry translocations.

A second mechanism that can cause a translocation is an abnormal crossover. As shown in **Figure 8.11b**, a reciprocal translocation can be produced when two nonhomologous chromosomes cross over. This type of rare aberrant event results in a rearrangement of the genetic material, though not a change in the total amount of genetic material.

The reciprocal translocations we have considered thus far are also called **balanced translocations** because the total amount of genetic material is not altered. Like inversions, balanced translocations usually occur without any phenotypic consequences because the individual has a normal amount of genetic material. In a few cases, balanced translocations can result in position effects similar to those that can occur in inversions. In addition, carriers of a reciprocal translocation are at risk of having offspring with an **unbalanced translocation,** in which significant portions of genetic material are duplicated and/or deleted. Unbalanced translocations are generally associated with phenotypic abnormalities or even lethality.

Let's consider how a person with a balanced translocation may produce gametes and offspring with an unbalanced translocation. An inherited human syndrome known as familial Down syndrome provides an example. A person with a normal phenotype may have one copy of chromosome 14, one copy of chromosome 21, and one copy of a chromosome that is a fusion between chromosome 14 and 21 (**Figure 8.12a**). The individual has a normal phenotype because the total amount of genetic material is present (with the exception of the short arms of these chromosomes that do not carry vital genetic material). During meiosis, these three types of chromosomes will replicate and segregate from each other. However, because the three chromosomes cannot segregate evenly, six possible types of gametes may be produced. One gamete is normal, and one has a balanced amount of genetic material. The four gametes to the right, however, are unbalanced, either containing too much or too little material from chromosome 14 or 21. The unbalanced gametes may be inviable, of they could combine with a normal gamete. The three offspring on the right will not survive.

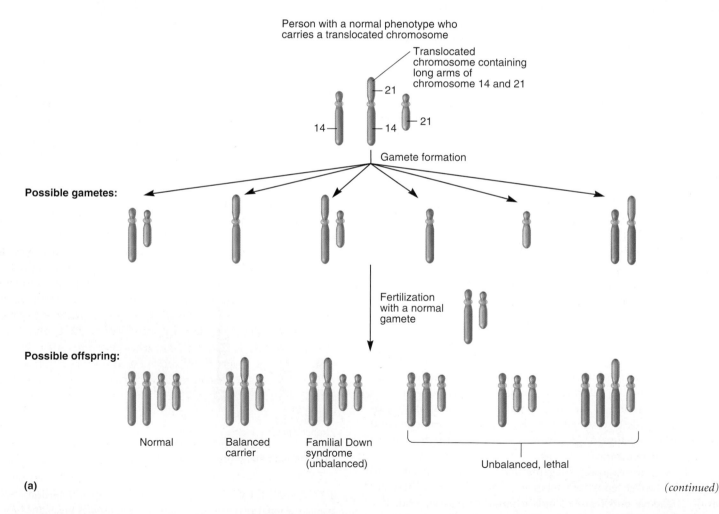

Person with a normal phenotype who carries a translocated chromosome

Translocated chromosome containing long arms of chromosome 14 and 21

14 ─   ─ 14   ─ 21

Gamete formation

Possible gametes:

Fertilization with a normal gamete

Possible offspring:

Normal   Balanced   Familial Down   Unbalanced, lethal
              carrier   syndrome
                              (unbalanced)

(a)   *(continued)*

FIGURE 8.12   **Transmission of familial Down syndrome. (a)** Potential transmission of familial Down syndrome. The individual with the chromosome composition shown at the top of this figure may produce a gamete carrying chromosome 21 and a fused chromosome containing the long arms of chromosomes 14 and 21. Such a gamete can give rise to an offspring with familial Down syndrome. **(b)** The karyotype of an individual with familial Down syndrome. This karyotype shows that the long arm of chromosome 21 has been translocated to chromosome 14 (see arrow). In addition, the individual also carries two normal copies of chromosome 21. **(c)** An individual with this disorder.

46,XY,-14,+t(14q21q)

(b)

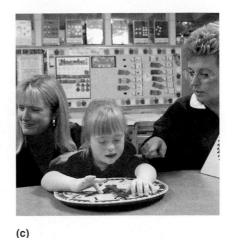

(c)

FIGURE 8.12 *(continued)*

In comparison, the unbalanced gamete that carries chromosome 21 and the fused chromosome will result in an offspring with familial Down syndrome (also see karyotype in **Figure 8.12b**). Such an offspring has three copies of the genes that are found on the long arm of chromosome 21. **Figure 8.12c** shows a person with this disorder. She has characteristics similar to those of an individual who has the more prevalent form of Down syndrome, which is due to three entire copies of chromosome 21. We will examine this common form of Down syndrome later in this chapter.

The abnormal chromosome that occurs in familial Down syndrome is an example of a **Robertsonian translocation,** named after William Robertson, who first described this type of fusion in grasshoppers. This type of translocation arises from breaks near the centromeres of two nonhomologous acrocentric chromosomes. In the example shown in Figure 8.12, the long arms of chromosomes 14 and 21 had fused, creating one large single chromosome; the two short arms are lost. This type of fusion between two nonhomologous acrocentric chromosomes is the most common type of chromosome rearrangement in humans, occurring at a frequency of approximately one in 900 live births. In humans, Robertsonian translocations involve only the acrocentric chromosomes 13, 14, 15, 21, and 22.

Individuals with Reciprocal Translocations May Produce Abnormal Gametes Due to the Segregation of Chromosomes

As we have seen, individuals who carry balanced translocations have a greater risk of producing gametes with unbalanced combinations of chromosomes. Whether or not this occurs depends on the segregation pattern during meiosis I (**Figure 8.13**). In this example, the parent carries a reciprocal translocation and is likely to be phenotypically normal. During meiosis, the homologous chromosomes attempt to synapse with each other. Because of the translocations, the pairing of homologous regions leads to the formation of an unusual structure that contains four pairs of sister chromatids (i.e., eight chromatids), termed a **translocation cross.**

To understand the segregation of translocated chromosomes, pay close attention to the centromeres, which are numbered in Figure 8.13. For these translocated chromosomes, the expected segregation pattern is governed by the centromeres. Each haploid gamete should receive one centromere located on chromosome 1 and one centromere located on chromosome 2. This can occur in two ways. One possibility is alternate segregation. As shown in Figure 8.13a, this occurs when the chromosomes diagonal to each other within the translocation cross sort into the same cell. One daughter cell receives two normal chromosomes, and the other cell gets two translocated chromosomes. Following meiosis II, four haploid cells are produced: two have normal chromosomes, and two have reciprocal (balanced) translocations.

Another possible segregation pattern is called adjacent-1 segregation (Figure 8.13b). This occurs when adjacent chromosomes (one of each type of centromere) segregate into the same cell. Following anaphase of meiosis I, each daughter cell receives one normal chromosome and one translocated chromosome. After meiosis II is completed, this produces four haploid cells, all of which are genetically unbalanced because part of one chromosome has been deleted and part of another has been duplicated. If these haploid cells give rise to gametes that unite with a normal gamete, the zygote is expected to be abnormal genetically and possibly phenotypically.

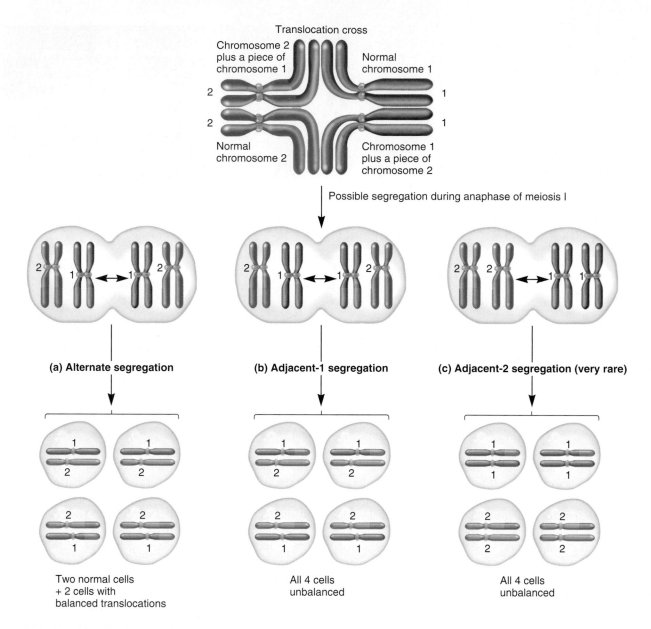

FIGURE 8.13   **Meiotic segregation of a reciprocal translocation.** Follow the numbered centromeres through each process. **(a)** Alternate segregation gives rise to balanced haploid cells, whereas **(b)** adjacent-1 and **(c)** adjacent-2 produce haploid cells with an unbalanced amount of genetic material.

On very rare occasions, adjacent-2 segregation can occur (Figure 8.13c). In this case, the centromeres do not segregate as they should. One daughter cell has received both copies of the centromere on chromosome 1; the other, both copies of the centromere on chromosome 2. This rare segregation pattern also yields four abnormal haploid cells that contain an unbalanced combination of chromosomes.

Alternate and adjacent-1 segregation patterns are the likely outcomes when an individual carries a reciprocal translocation. Depending on the sizes of the translocated segments, both types may be equally likely to occur. In many cases, the haploid cells from adjacent-1 segregation are not viable, thereby lowering the fertility of the parent. This condition is called **semisterility.**

8.2 VARIATION IN CHROMOSOME NUMBER

As we have seen in Section 8.1, chromosome structure can be altered in a variety of ways. Likewise, the total number of chromosomes can vary. Eukaryotic species typically contain several chromosomes that are inherited as one or more sets. Variations in chromosome number can be categorized in two ways: variation in the number of sets of chromosomes and variation in the number of particular chromosomes within a set.

Organisms that are **euploid** have a chromosome number that is an exact multiple of a chromosome set. In *Drosophila melanogaster*, for example, a normal individual has 8 chromosomes.

The species is diploid, having two sets of 4 chromosomes each (**Figure 8.14a**). A normal fruit fly is euploid because 8 chromosomes divided by 4 chromosomes per set equals two exact sets. On rare occasions, an abnormal fruit fly can be produced with 12 chromosomes, containing three sets of 4 chromosomes each. This alteration in euploidy produces a **triploid** fruit fly with 12 chromosomes. Such a fly is also euploid because it has exactly three sets of chromosomes. Organisms with three or more sets of chromosomes are also called **polyploid** (**Figure 8.14b**). Geneticists use the letter n to represent a set of chromosomes. A diploid organism is referred to as $2n$, a triploid organism as $3n$, a **tetraploid** organism as $4n$, and so on.

A second way in which chromosome number can vary is by **aneuploidy.** Such variation involves an alteration in the number of particular chromosomes, so the total number of chromosomes is not an exact multiple of a set. For example, an abnormal fruit fly could contain nine chromosomes instead of eight because it has three copies of chromosome 2 instead of the normal two copies (**Figure 8.14c**). Such an animal is said to have trisomy 2 or to be **trisomic.** Instead of being perfectly diploid (2n), a trisomic animal is $2n + 1$. By comparison, a fruit fly could be lacking a single chromosome, such as chromosome 1, and contain a total of seven chromosomes ($2n - 1$). This animal would be **monosomic** and be described as having monosomy 1.

In this section, we will begin by considering several examples of aneuploidy. This is generally regarded as an abnormal condition that usually has a negative impact on phenotype. We will then examine euploid variation that occurs occasionally in animals and quite frequently in plants, and consider how it affects phenotypic variation.

Aneuploidy Causes an Imbalance in Gene Expression That Is Often Detrimental to the Phenotype of the Individual

The phenotype of every eukaryotic species is influenced by thousands of different genes. In humans, for example, a single set of chromosomes contains approximately 20,000 to 25,000 different genes. To produce a phenotypically normal individual, intricate coordination has to occur in the expression of thousands of genes. In the case of humans and other diploid species, evolution has resulted in a developmental process that works correctly when somatic cells have two copies of each chromosome. In other words, when a human is diploid, the balance of gene expression among many different genes usually produces a person with a normal phenotype.

Aneuploidy commonly causes an abnormal phenotype. To understand why, let's consider the relationship between gene

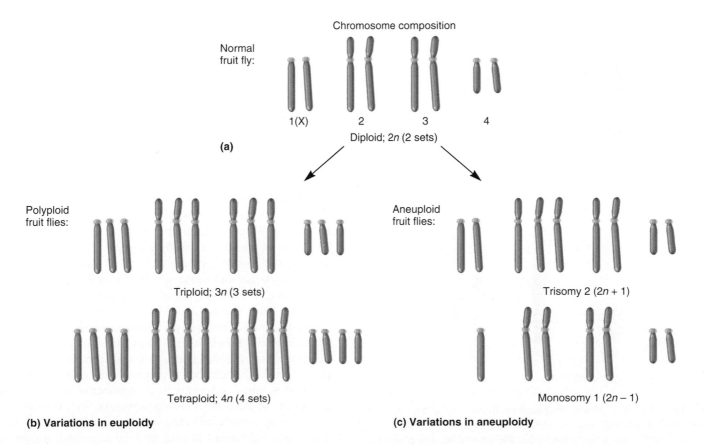

FIGURE 8.14   Types of variation in chromosome number. (a) Depicts the normal diploid number of chromosomes in *Drosophila*. **(b)** Examples of polyploidy. **(c)** Examples of aneuploidy.

expression and chromosome number in a species that has three pairs of chromosomes (**Figure 8.15**). The level of gene expression is influenced by the number of genes per cell. Compared to a diploid cell, if a gene is carried on a chromosome that is present in three copies instead of two, more of the gene product is typically made. For example, a gene present in three copies instead of two may produce 150% of the gene product, though that number may vary due to effects of gene regulation. Alternatively, if only one copy of that gene is present due to a missing chromosome, less of the gene product will usually be made, perhaps only 50%. Therefore, in trisomic and monosomic individuals, an imbalance occurs between the level of gene expression on the chromosomes found in pairs versus the one type that is not.

FIGURE 8.15   Imbalance of gene products in trisomic and monosomic individuals. Aneuploidy of chromosome 2 (i.e., trisomy and monosomy) leads to an imbalance in the amount of gene products from chromosome 2 compared to the amounts from chromosomes 1 and 3.

At first glance, the difference in gene expression between euploid and aneuploid individuals may not seem terribly dramatic. Keep in mind, however, that a eukaryotic chromosome carries hundreds or even thousands of different genes. Therefore, when an organism is trisomic or monosomic, many gene products will occur in excessive or deficient amounts. This imbalance among many genes appears to underlie the abnormal phenotypic effects that aneuploidy frequently causes. In most cases, these effects are detrimental and produce an individual that is less likely to survive than a euploid individual.

In the 1920s, Albert Blakeslee and his colleagues were the first to recognize the harmful effects of aneuploidy by studying Jimson weed (*Datura stramonium*). **Figure 8.16** compares normal and trisomic strains with regard to morphology of the capsule, which is the dry fruit. All 12 trisomies had capsules that were morphologically different from those of the normal diploid strain. These aneuploid plants also had many other morphologically distinguishable traits, including changes in leaf shape, size, and so forth. In many cases, the observed changes in chromosome number produced detrimental traits. For example, Blakeslee noted that the cocklebur plant (trisomy 6) is "weak and lopping with the leaves narrow and twisted."

Aneuploidy in Humans Causes Abnormal Phenotypes

A key reason why geneticists are so interested in aneuploidy is its relationship to certain inherited disorders in humans. Even though most people are born with a normal number of chromosomes (i.e., 46), alterations in chromosome number occur fairly frequently during gamete formation. About 5 to 10% of all fertilized human eggs result in an embryo with an abnormality in chromosome number! In most cases, these abnormal embryos do not develop properly and result in a spontaneous abortion very early in pregnancy. Approximately 50% of all spontaneous abortions are due to alterations in chromosome number.

In some cases, an abnormality in chromosome number produces an offspring that survives to birth or longer. Several human disorders involve abnormalities in chromosome number. The most common are trisomies of chromosomes 13, 18, or 21, and abnormalities in the number of the sex chromosomes (**Table 8.1**). Most of the known trisomies involve chromosomes that are relatively small—chromosome 13, 18, or 21—and may carry fewer genes than larger chromosomes. Trisomies of the other human autosomes and monosomies of all autosomes are presumed to produce a lethal phenotype, and many have been found in spontaneously aborted embryos and fetuses. For example, all possible human trisomies have been found in spontaneously aborted embryos except trisomy 1. It is believed that trisomy 1 is lethal at such an early stage that it prevents the successful implantation of the embryo. Variation in the number of X chromosomes, unlike that of other large chromosomes, is often nonlethal. The survival of trisomy X individuals may be explained by X inactivation, which is described in Chapter 7. In an individual with more than one X chromosome, all additional X chromosomes are converted to Barr bodies in the somatic cells of adult tissues. In an

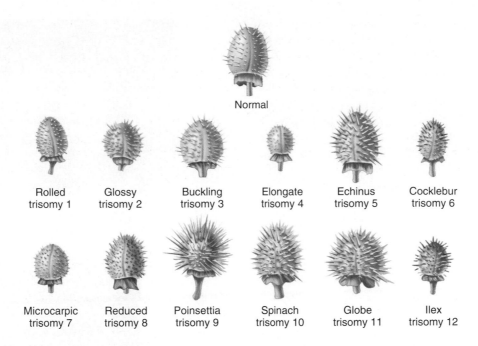

FIGURE 8.16 The effects of trisomy on the phenotype of the capsules of Jimson weed, *Datura stramonium*.

Genes → Traits As described in Figure 8.15, trisomy leads to an imbalance in the copy number of genes on the trisomic chromosomes (here, 150% of normal) versus those on the other chromosomes (100%). This drawing compares the morphology of the capsule of the *Datura* plant in a normal individual and in those carrying a trisomy of each of the 12 types of chromosomes in this species. As seen here, trisomy of each chromosome causes changes in the phenotype of the capsule that are due to changes in the balance of gene expression.

TABLE 8.1			
Aneuploid Conditions in Humans			
Condition	**Frequency**	**Syndrome**	**Characteristics**
Autosomal			
Trisomy 13	1/15,000	Patau	Mental and physical deficiencies, wide variety of defects in organs, large triangular nose, early death
Trisomy 18	1/6,000	Edward	Mental and physical deficiencies, facial abnormalities, extreme muscle tone, early death
Trisomy 21	1/800	Down	Mental deficiencies, abnormal pattern of palm creases, slanted eyes, flattened face, short stature
Sex Chromosomal			
XXY	1/1,000 (males)	Klinefelter	Sexual immaturity (no sperm), breast swelling
XYY	1/1,000 (males)	Jacobs	Tall and thin
XXX	1/1,500 (females)	Triple X	Tall and thin, menstrual irregularity
X0	1/5,000 (females)	Turner	Short stature, webbed neck, sexually undeveloped

individual with trisomy X, for example, two out of three X chromosomes are converted to inactive Barr bodies. Unlike the level of expression for autosomal genes, the normal level of expression for X-linked genes is from a single X chromosome. In other words, the correct level of mammalian gene expression results from two copies of each autosomal gene and one copy of each X-linked gene. This explains how the expression of X-linked genes in males (XY) can be maintained at the same levels as in females (XX). It may also explain why trisomy X is not a lethal condition. The phenotypic effects noted in Table 8.1 involving sex chromosomal abnormalities may be due to the expression of X-linked genes prior to embryonic X inactivation or to the expression of genes on the inactivated X chromosome. As described in Chapter 7, pseudoautosomal genes and some other genes on the inactivated X chromosome are expressed in humans. Having one or three copies of the sex chromosomes would result in an under- or overexpression of these X-linked genes, respectively.

Some human abnormalities in chromosome number are influenced by the age of the parents. Older parents are more likely to produce children with abnormalities in chromosome number. Down syndrome provides an example. The common form of this disorder is caused by the inheritance of three copies of chromosome 21. The incidence of Down syndrome rises with the age of either parent. In males, however, the rise occurs relatively late in life, usually past the age when most men have children. By comparison, the likelihood of having a child with Down syndrome

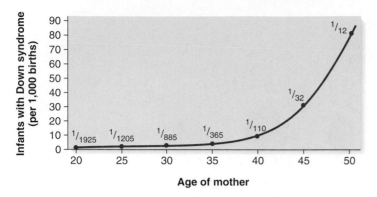

FIGURE 8.17   The incidence of Down syndrome births according to the age of the mother. The y-axis shows the number of infants born with Down syndrome per 1,000 live births, and the x-axis plots the age of the mother at the time of birth. The data points indicate the fraction of live offspring born with Down syndrome.

rises dramatically during a woman's reproductive age (**Figure 8.17**). This syndrome was first described by the English physician John Langdon Down in 1866. The association between maternal age and Down syndrome was later discovered by L. S. Penrose in 1933, even before the chromosomal basis for the disorder was identified by the French scientist Jérôme Lejeune in 1959. Down syndrome is most commonly caused by **nondisjunction,** which means that the chromosomes do not segregate properly. In this case, nondisjunction of chromosome 21 most commonly occurs during meiosis I in the oocyte.

Different hypotheses have been proposed to explain the relationship between maternal age and Down syndrome. One popular idea suggests that it may be due to the age of the oocytes. Human primary oocytes are produced within the ovary of the female fetus prior to birth and are arrested at prophase of meiosis I and remain in this stage until the time of ovulation. Therefore, as a woman ages, her primary oocytes have been in prophase I for a progressively longer period of time. This added length of time may contribute to an increased frequency of nondisjunction. About 5% of the time, Down syndrome is due to an extra paternal chromosome. Prenatal tests can determine if a fetus has Down syndrome and some other genetic abnormalities. The topic of genetic testing is discussed in Chapter 22.

Variations in Euploidy Occur Naturally in a Few Animal Species

We now turn our attention to changes in the number of sets of chromosomes, referred to as variations in euploidy. Most species of animals are diploid. In some cases, changes in euploidy are not well tolerated. For example, polyploidy in mammals is generally a lethal condition. However, many examples of naturally occurring variations in euploidy occur. In **haplodiploid** species, which includes many species of bees, wasps, and ants, one of the sexes is haploid, usually the male, and the other is diploid. For example, male bees, which are called drones, contain a single set of chromosomes. They are produced from unfertilized eggs. By

(a) *Hyla chrysoscelis*

(b) *Hyla versicolor*

FIGURE 8.18   Differences in euploidy in two closely related frog species. The frog in (**a**) is diploid, whereas the frog in (**b**) is tetraploid. Both species grow to the same size.

Genes → Traits Though similar in appearance, these two species differ in their number of chromosome sets. At the level of gene expression, this observation suggests that the number of copies of each gene (two versus four) does not critically affect the phenotype of these two species.

comparison, female bees are produced from fertilized eggs and are diploid.

Many examples of vertebrate polyploid animals have been discovered. Interestingly, on several occasions, animals that are morphologically very similar to each other can be found as a diploid species as well as a separate polyploid species. This situation occurs among certain amphibians and reptiles. **Figure 8.18** shows photographs of a diploid and a tetraploid ($4n$) frog. As you see, they look indistinguishable from each other. Their difference can be revealed only by an examination of the chromosome number in the somatic cells of the animals and by mating calls—*H. chrysoscelis* has a faster trill rate than *H. versicolor*.

Variations in Euploidy Can Occur in Certain Tissues Within an Animal

Thus far, we have considered variations in chromosome number that occur at fertilization, so all the somatic cells of an individual contain this variation. In many animals, certain tissues of the body display normal variations in the number of sets of chromosomes. Diploid animals sometimes produce tissues that are polyploid. For example, the cells of the human liver can vary to a great degree in their ploidy. Liver cells contain nuclei that can be triploid, tetraploid, and even octaploid ($8n$). The occurrence of polyploid tissues or cells in organisms that are otherwise diploid is known as **endopolyploidy.** What is the biological significance of endopolyploidy? One possibility is that the increase in chromosome number in certain cells may enhance their ability to produce specific gene products that are needed in great abundance.

An unusual example of natural variation in the ploidy of somatic cells occurs in *Drosophila* and some other insects. Within certain tissues, such as the salivary glands, the chromosomes undergo repeated rounds of chromosome replication without cellular division. For example, in the salivary gland cells of *Drosophila,* the pairs of chromosomes double approximately nine times ($2^9 = 512$). **Figure 8.19a** illustrates how repeated rounds of chromosomal replication produce a bundle of chromosomes that lie together in a parallel fashion. This bundle, termed a **polytene chromosome,** was first observed by E. G. Balbiani in 1881. Later,

in the 1930s, Theophilus Painter and colleagues recognized that the size and morphology of polytene chromosomes provided geneticists with unique opportunities to study chromosome structure and gene organization.

Figure 8.19b shows a micrograph of a polytene chromosome. The structure of polytene chromosomes is different from other forms of endopolyploidy because the replicated chromosomes remain attached to each other. Prior to the formation of polytene chromosomes, *Drosophila* cells contain eight chromosomes (two sets of four chromosomes each; see Figure 8.14a). In the salivary gland cells, the homologous chromosomes synapse with each other and replicate to form a polytene structure. During this process, the four types of chromosomes aggregate to form a single structure with several polytene arms. The central point where the chromosomes aggregate is known as the **chromocenter.** Each of the four types of chromosome is attached to the chromocenter near its centromere. The X and Y and chromosome 4 are telocentric, and chromosomes 2 and 3 are metacentric. Therefore, chromosomes 2 and 3 have two arms that radiate from the chromocenter, while the X and Y and chromosome 4 have a single arm projecting from the chromocenter (**Figure 8.19c**).

Because of their considerable size, polytene chromosomes lend themselves to an easy microscopic examination. Ordinarily, we use light microscopy to visualize the highly condensed metaphase chromosomes seen during mitosis or meiosis, as in Figure 8.1a. Because polytene chromosomes are so large, we

(a) Repeated chromosome replication produces polytene chromosome.

(b) A polytene chromosome

Each polytene arm is composed of hundreds of chromosomes aligned side by side.

Chromocenter

(c) Relationship between a polytene chromosome and regular *Drosophila* chromosomes

FIGURE 8.19   Polytene chromosomes in *Drosophila*. (a) A schematic illustration of the formation of polytene chromosomes. Several rounds of repeated replication without cellular division result in a bundle of sister chromatids that lie side by side. Both homologues also lie parallel to each other. This replication does not occur in highly condensed, heterochromatic DNA near the centromere. **(b)** A photograph of a polytene chromosome. **(c)** This drawing shows the relationship between the four pairs of chromosomes and the formation of a polytene chromosome in the salivary gland. The heterochromatic regions of the chromosomes aggregate at the chromocenter, and the arms of the chromosomes project outward.

can see them during interphase, when normal chromosomes are not readily visible. Remarkably, a polytene chromosome during interphase is 100 to 200 times larger than the average metaphase chromosome. As shown in Figure 8.19b, polytene chromosomes exhibit a characteristic banding pattern. Each dark band is known as a **chromomere.** The structure of the genetic material within a dark band is more compact than the interband region. More than 95% of the DNA is found within these dark bands. The banding patterns of polytene chromosomes are much more detailed than those observed in metaphase chromosomes. Cytogeneticists have identified approximately 5,000 bands along polytene chromosomes. At one time, each chromomere was thought to correspond to one gene. However, this idea was found to be incorrect because the sequencing of the entire *Drosophila* genome has revealed an approximate gene number of 14,000.

Polytene chromosomes have allowed geneticists to study the organization and functioning of interphase chromosomes in great detail. When a gene is deleted or duplicated via mutation, researchers can map the change if it results in a microscopically visible alteration in the structure of a polytene chromosome. In addition, because polytene chromosomes can be observed during interphase, the expression of particular genes in salivary cells can be correlated with changes in the compaction of certain bands in the polytene chromosome.

Variations in Euploidy Are Common in Plants

We now turn our attention to variations of euploidy that occur in plants. Compared to animals, plants more commonly exhibit polyploidy. Among ferns and flowering plants, at least 30 to 35% of species are polyploid. Polyploidy is also important in agriculture. Many of the fruits and grains we eat are produced from polyploid plants. For example, the species of wheat that we use to make bread, *Triticum aestivum*, is a hexaploid (6n) that arose from the union of diploid genomes from three closely related species (**Figure 8.20a**).

In many instances, polyploid strains of plants display outstanding agricultural characteristics. They are often larger in size and more robust. These traits are clearly advantageous in the production of food. In addition, polyploid plants tend to exhibit a greater adaptability, which allows them to withstand harsher environmental conditions. Also, polyploid ornamental plants often produce larger flowers than their diploid counterparts (**Figure 8.20b**).

Polyploid plants having an odd number of chromosome sets, such as triploids (3n) or pentaploids (5n), usually cannot reproduce. Why are they sterile? The sterility arises because they produce highly aneuploid gametes. During prophase of meiosis I, homologous pairs of sister chromatids will form bivalents. However,

Tetraploid

Diploid

(a) Cultivated wheat, a hexaploid species

(b) A comparison of diploid and tetraploid petunias

FIGURE 8.20   **Examples of polyploid plants. (a)** Cultivated wheat, *Triticum aestivum*, is a hexaploid. It was derived from three different diploid species of grasses that originally were found in the Middle East and were cultivated by ancient farmers in that region. **(b)** Differences in euploidy may exist in two closely related petunia species. The flower at the bottom is diploid, whereas the large one at the top is tetraploid.

Genes → Traits An increase in chromosome number from diploid to tetraploid or hexaploid affects the phenotype of the individual. In the case of many plant species, a polyploid individual is larger and more robust than its diploid counterpart. This suggests that having additional copies of each gene is somewhat better than having two copies of each gene. This phenomenon in plants is rather different from the situation in animals. Tetraploidy in animals may have little effect (as in Figure 8.18b), and it is also common for polyploidy in animals to be detrimental.

organisms with an odd number of chromosomes, such as three, will display an unequal separation of homologous chromosomes during anaphase of meiosis I (**Figure 8.21**). An odd number cannot be divided equally between two daughter cells. For each type of chromosome, a daughter cell randomly gets one or two copies. For example, one daughter cell might receive one copy of chromosome 1, two copies of chromosome 2, two copies of chromosome 3, one copy of chromosome 4, and so forth. For a triploid species containing many different chromosomes in a set, meiosis is very unlikely to produce a daughter cell that is euploid. If we assume that a daughter cell will receive either one copy or two copies of each kind of chromosome, the probability that meiosis will produce a cell that will be perfectly haploid or diploid is $(1/2)^{n-1}$, where n is the number of chromosomes in a set. As an example, in a triploid organism containing 20 chromosomes per set, the probability of producing a haploid or diploid cell is 0.000001907, or 1 in 524,288. Thus, meiosis is almost certain to produce cells that contain one copy of some chromosomes and two copies of the other chromosomes. This high probability of aneuploidy underlies the reason for triploid sterility.

Though sterility is generally a detrimental trait, it can be desirable agriculturally because it may result in a seedless fruit. For example, domestic bananas and seedless watermelons are triploid varieties. The domestic banana was originally derived from a seed-producing diploid species and has been asexually propagated by humans via cuttings. The small black spots in the center of a domestic banana are degenerate seeds. In the case of flowers, the seedless phenotype can also be beneficial. Seed producers such as Burpee have developed triploid varieties of

flowering plants such as marigolds. Because the triploid marigolds are sterile and unable to set seed, more of their energy goes into flower production. According to Burpee, "They bloom and bloom, unweakened by seed bearing."

8.3 NATURAL AND EXPERIMENTAL WAYS TO PRODUCE VARIATIONS IN CHROMOSOME NUMBER

As we have seen, variations in chromosome number are fairly widespread and usually have a significant impact on the phenotypes of plants and animals. For these reasons, researchers have wanted to understand the cellular mechanisms that cause variations in chromosome number. In some cases, a change in chromosome number is the result of nondisjunction. The term nondisjunction refers to an event in which the chromosomes do not segregate properly. As we will see, it may be caused by an improper separation of homologous pairs in a bivalent in meiosis or a failure of the centromeres to disconnect during mitosis.

Meiotic nondisjunction can produce haploid cells that have too many or too few chromosomes. If such a cell gives rise to a gamete that fuses with a normal gamete during fertilization, the resulting offspring will have an abnormal chromosome number in all of its cells. An abnormal nondisjunction event also may occur after fertilization in one of the somatic cells of the body. This second mechanism is known as **mitotic nondisjunction.** When this occurs during embryonic stages of development, it may lead to a patch of tissue in the organism that has an altered chromosome number. A third common way in which the chromosome number of an organism can vary is by interspecies crosses. An **alloploid** organism contains sets of chromosomes from two or more different species. This term refers to the occurrence of chromosome sets (ploidy) from the genomes of different (allo) species.

In this section, we will examine these three mechanisms in greater detail. Also, in the past few decades, researchers have devised several methods to manipulate chromosome number in experimentally and agriculturally important species. We will conclude this section by exploring how the experimental manipulation of chromosome number has had an important impact on genetic research and agriculture.

Meiotic Nondisjunction Can Produce Aneuploidy or Polyploidy

Nondisjunction during meiosis can occur during anaphase of meiosis I or meiosis II. If it happens during meiosis I, an entire bivalent migrates to one pole (**Figure 8.22a**). Following the completion of meiosis, the four resulting haploid cells produced from this event are abnormal. If nondisjunction occurs during anaphase of meiosis II (**Figure 8.22b**), the net result is two abnormal

FIGURE 8.21 Schematic representation of anaphase of meiosis I in a triploid organism containing three sets of four chromosomes. In this example, the homologous chromosomes (three each) do not evenly separate during anaphase. Each cell receives one copy of some chromosomes and two copies of other chromosomes. This produces aneuploid gametes.

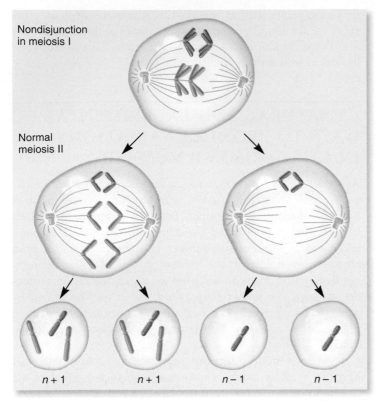

(a) Nondisjunction in meiosis I

(b) Nondisjunction in meiosis II

FIGURE 8.22   **Nondisjunction during meiosis I and II.** The chromosomes shown in purple are behaving properly during meiosis I and II, so each cell receives one copy of this chromosome. The chromosomes shown in blue are not disjoining correctly. In (**a**), nondisjunction occurred in meiosis I, so the resulting four cells receive either two copies of the blue chromosome or zero copies. In (**b**), nondisjunction occurred during meiosis II, so one cell has two blue chromosomes and another cell has zero. The remaining two cells are normal.

and two normal haploid cells. If a gamete that is missing a chromosome is viable and participates in fertilization, the resulting offspring is monosomic for the missing chromosomes. Alternatively, if a gamete carrying an extra chromosome unites with a normal gamete, the offspring will be trisomic.

In rare cases, all of the chromosomes can undergo nondisjunction and migrate to one of the daughter cells. The net result of **complete nondisjunction** is a diploid cell and a cell without any chromosomes. While the cell without chromosomes is nonviable, the diploid cell might participate in fertilization with a normal haploid gamete to produce a triploid individual. Therefore, complete nondisjunction can produce individuals that are polyploid.

Mitotic Nondisjunction or Chromosome Loss Can Produce a Patch of Tissue with an Altered Chromosome Number

Abnormalities in chromosome number occasionally occur after fertilization takes place. In this case, the abnormal event happens during mitosis rather than meiosis. One possibility is that the sister chromatids separate improperly, so one daughter cell has three copies of that chromosome while the other daughter cell has only one (**Figure 8.23a**). Alternatively, the sister chromatids could sep-

arate during anaphase of mitosis, but one of the chromosomes could be improperly attached to the spindle, so that it would not migrate to a pole (**Figure 8.23b**). A chromosome will be degraded if it is left outside the nucleus when the nuclear membrane reforms. In this case, one of the daughter cells would have two copies of that chromosome, while the other would have only one.

When genetic abnormalities occur after fertilization, the organism will contain a subset of cells that are genetically different from those of the rest of the organism. This condition is referred to as **mosaicism.** The size and location of the mosaic region depend on the timing and location of the original abnormal event. If a genetic alteration happens very early in the embryonic development of an organism, the abnormal cell will be the precursor for a large section of the organism. In the most extreme case, an abnormality could take place at the first mitotic division. As a bizarre example, consider a fertilized *Drosophila* egg that is XX. One of the X chromosomes may be lost during the first mitotic division, producing one daughter cell that is XX and one that is X0. Flies that are XX develop into females, and X0 flies develop into males. Therefore, in this example, one-half of the organism will become female and one-half will become male! This peculiar and rare individual is referred to as a **bilateral gynandromorph** (**Figure 8.24**).

(a) Mitotic nondisjunction

Not attached
to spindle

(b) Chromosome loss

FIGURE 8.23   **Nondisjunction and chromosome loss during mitosis in somatic cells.** (**a**) Mitotic nondisjunction produces a trisomic and a monosomic daughter cell. (**b**) Chromosome loss produces a normal and a monosomic daughter cell.

FIGURE 8.24   **A bilateral gynandromorph of *Drosophila melanogaster*.**

Genes → Traits   In *Drosophila*, the ratio between genes on the X chromosome and genes on the autosomes determines sex. This fly began as an XX female. One X chromosome carried the recessive white-eye and miniature wing alleles, while the other X chromosome carried the wild-type alleles. The X chromosome carrying the wild-type alleles was lost from one of the cells during the first mitotic division, producing one XX cell and one X0 cell. The XX cell became the precursor for the left side of the fly, which developed as female. The X0 cell became the precursor for the other side of the fly, which developed as male with a white eye and a miniature wing.

Changes in Euploidy Can Occur by Autopolyploidy, Alloploidy, and Allopolyploidy

Different mechanisms account for changes in the number of chromosome sets among natural populations of plants and animals (**Figure 8.25**). As previously mentioned, complete nondisjunction, due to a general defect in the spindle apparatus, can produce an individual with one or more extra sets of chromosomes. This individual is known as an **autopolyploid** (Figure 8.25a). The prefix auto- (meaning self) and term polyploid (meaning many sets of chromosomes) refer to an increase in the number of chromosome sets within a single species.

A much more common mechanism for change in chromosome number, called **alloploidy,** is a result of interspecies crosses (Figure 8.25b). An alloploid that has one set of chromosomes from two different species is called an **allodiploid.** This event is most likely to occur between species that are close evolutionary relatives. For example, closely related species of grasses may interbreed to produce allodiploids. As shown in Figure 8.25c, an **allopolyploid** contains two (or more) sets of chromosomes from two (or more) species. In this case, the **allotetraploid** contains two complete sets of chromosomes from two different species, for a total of four sets. In nature, allotetraploids usually arise from allodiploids. This can occur when a somatic cell in an allodiploid undergoes complete nondisjunction to create an allotetraploid cell. In plants, such a cell can continue to grow and produce a section of the plant that is allotetraploid. If this part of the plant produced seeds, the seeds would give rise to allotetraploid offspring. Cultivated wheat (refer back to Figure 8.20a) is a plant in which two species must have interbred to create an allotetraploid, and then a third species interbred with the allotetraploid to create an allohexaploid.

Allodiploids Are Often Sterile, but Allotetraploids Are More Likely to Be Fertile

Geneticists are interested in the production of alloploids and allopolyploids as ways to generate interspecies hybrids with desirable traits. For example, if one species of grass can withstand hot temperatures and a closely related species is adapted to survive cold winters, a plant breeder may attempt to produce an interspecies hybrid that combines both qualities—good growth in the heat and survival through the winter. Such an alloploid may be desirable in climates with both hot summers and cold winters.

An important determinant of success in producing a fertile allodiploid is the degree of similarity of the different species' chromosomes. In two very closely related species, the number and types of chromosomes might be very similar. **Figure 8.26** shows a karyotype of an interspecies hybrid between the roan antelope (*Hippotragus equinus*) and the sable antelope (*Hippotragus niger*). As seen here, these two closely related species have the same number of chromosomes. The sizes and banding patterns of the chromosomes show that they correspond to one another. For example, chromosome 1 from both species is fairly large, has very similar banding patterns, and carries many of the same genes. Evolutionarily related chromosomes from two different species are called **homeologous** chromosomes (not to be confused with

(a) Autopolyploidy (tetraploid)

(b) Alloploidy (allodiploid)

(c) Allopolyploidy (allotetraploid)

FIGURE 8.25   A comparison of autopolyploidy, alloploidy, and allopolyploidy.

homologous). This allodiploid is fertile because the homeologous chromosomes can properly synapse during meiosis to produce haploid gametes.

The critical relationship between chromosome pairing and fertility was first recognized by the Russian cytogeneticist Georgi Karpechenko in 1928. He crossed a radish (*Raphanus*) and a cabbage (*Brassica*), both of which are diploid and contain 18 chromosomes. Each of these organisms produces haploid cells containing 9 chromosomes. Therefore, the allodiploid produced from this interspecies cross contains 18 chromosomes. However, because the radish and cabbage are not closely related species, the nine *Raphanus* chromosomes are distinctly different from the nine *Brassica* chromosomes. During meiosis I, the radish

and cabbage chromosomes cannot synapse with each other. This prevents the proper chromosome pairing and results in a high degree of aneuploidy (**Figure 8.27a**). Therefore, the radish/cabbage hybrid is sterile.

Among his strains of sterile alloploids, Karpechenko discovered that on rare occasions a plant would produce a viable seed. When such seeds were planted and subjected to karyotyping, the plants were found to be allotetraploids with two sets of chromosomes from each of the two species. In the example shown in **Figure 8.27b**, the radish/cabbage allotetraploid contains 36 chromosomes instead of 18. The homologous chromosomes from each of the two species can synapse properly. When anaphase of meiosis I occurs, the pairs of synapsed chromosomes

FIGURE 8.26 **The karyotype of a hybrid animal produced from two closely related antelope species.** In each chromosome pair in this karyotype, one chromosome was inherited from the roan antelope (*Hippotragus equinus*) and the other from the sable antelope (*Hippotragus niger*). When it is possible to distinguish the chromosomes from the two species, the roan chromosomes are shown on the left side of each pair.

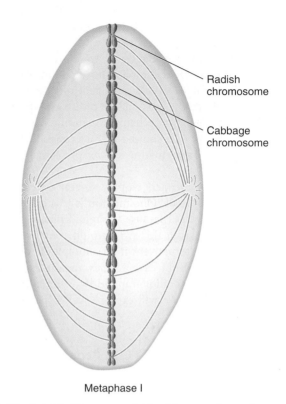

Metaphase I

(a) Allodiploid with a monoploid set from each species

Metaphase I

(b) Allotetraploid with a diploid set from each species

FIGURE 8.27 **A comparison of metaphase I in an allodiploid and an allotetraploid.** The purple chromosomes are from the radish (*Raphanus*), and the blue are from the cabbage (*Brassica*). **(a)** In the allodiploid, the radish and cabbage chromosomes have not synapsed and are randomly aligned during metaphase. **(b)** In the allotetraploid, the homologous radish and homologous cabbage chromosomes are properly aligned along the metaphase plate.

can disjoin equally to produce cells with 18 chromosomes each (a haploid set from the radish plus a haploid set from the cabbage). These cells can give rise to gametes with 18 chromosomes that can combine with each other to produce an allotetraploid containing 36 chromosomes. In this way, the allotetraploid is a fertile organism. Karpechenko's goal was to create a "vegetable for the masses" that would combine the nutritious roots of a radish with the flavorful leaves of a cabbage. Unfortunately, however, the allotetraploid had the leaves of the radish and the roots of the cabbage! Nevertheless, Karpechenko's scientific contribution was still important because he showed that it is possible to artificially produce a new self-perpetuating species of plant by creating an allotetraploid.

Modern plant breeders employ several different strategies to produce allotetraploids. One method is to start with two different tetraploid species. Because tetraploid plants make diploid gametes, the hybrid from this cross would contain a diploid set from each species. Alternatively, if an allodiploid containing one set of chromosomes from each species already exists, a second approach is to create an allotetraploid by using agents that alter chromosome number. We will explore such experimental treatments next.

Experimental Treatments Can Promote Polyploidy

Because polyploid and allopolyploid plants often exhibit desirable traits, the development of polyploids is of considerable interest among plant breeders. Experimental studies on the ability of environmental agents to promote polyploidy began in the early 1900s. Since that time, various agents have been shown to promote nondisjunction and thereby lead to polyploidy. These include abrupt temperature changes during the initial stages of seedling growth and the treatment of plants with chemical agents that interfere with the formation of the spindle apparatus.

The drug colchicine is commonly used to promote polyploidy. Once inside the cell, colchicine binds to tubulin (a protein found in the spindle apparatus) and thereby interferes with normal chromosome segregation during mitosis or meiosis. In 1937, Alfred Blakeslee and Amos Avery applied colchicine to plant tissue and, at high doses, were able to cause complete mitotic nondisjunction and produce polyploidy in plant cells. Colchicine can be applied to seeds, young embryos, or rapidly growing regions of a plant (**Figure 8.28**). This application may produce aneuploidy, which is usually an undesirable outcome, but it often produces polyploid cells, which may grow faster than the surrounding diploid tissue. In a diploid plant, colchicine may cause complete mitotic nondisjunction, yielding tetraploid ($4n$) cells. As the tetraploid cells continue to divide, they generate a portion of the plant that is often morphologically distinguishable from the remainder. For example, a tetraploid stem may have a larger diameter and produce larger leaves and flowers. Because individual plants can be propagated asexually from pieces of plant tissue (i.e., cuttings), the polyploid portion of the plant can be removed, treated with the proper growth hormones, and

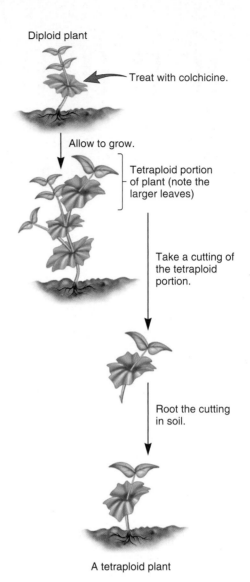

FIGURE 8.28   Use of colchicine to promote polyploidy in plants. Colchicine interferes with the mitotic spindle apparatus and promotes nondisjunction. If complete nondisjunction occurs in a diploid cell, a daughter cell will be formed that is tetraploid. Such a tetraploid cell may continue to divide and produce a segment of the plant with more robust characteristics. This segment may be cut from the rest of the plant and rooted. In this way, a tetraploid plant can be propagated.

grown as a separate plant. Alternatively, the tetraploid region of a plant may have flowers that produce seeds. A tetraploid flower will produce diploid pollen and eggs, which combine to produce tetraploid offspring. In this way, the use of colchicine provides a straightforward method to produce polyploid strains of plants.

Cell Fusion Techniques Can Be Used to Make Hybrid Plants

Thus far, we have examined several mechanisms that produce variations in chromosome number. Some of these processes occur naturally and have figured prominently in speciation and evolution. In addition, agricultural geneticists can administer treatments such as colchicine to promote nondisjunction and thereby obtain useful strains of organisms. More recently, researchers have developed cellular approaches to produce hybrids with altered chromosomal composition. As described here, these cellular approaches have important applications in research and agriculture.

In the technique known as **cell fusion,** individual cells are mixed together and made to fuse. In agriculture, cell fusion can create new strains of plants. An advantage of this approach is that researchers can artificially "cross" two species that cannot interbreed naturally. As an example, **Figure 8.29** illustrates the use of cell fusion to produce a hybrid grass. The parent cells were derived from tall fescue grass (*Festuca arundinacea*) and Italian ryegrass (*Lolium multiflorum*). Prior to fusion, the cells from these two species were treated with agents that gently digest the cell wall without rupturing the plasma membrane. A plant cell without a cell wall is called a **protoplast.** The protoplasts were mixed together and treated with agents that promote cellular fusion. Immediately after this takes place, a cell containing two separate nuclei is formed. This cell, known as a **heterokaryon,** will spontaneously go through a nuclear fusion process to produce a **hybrid cell** with a single nucleus. After nuclear fusion, the hybrid cells can be grown on laboratory media and eventually regenerate an entire plant. The allotetraploid shown in Figure 8.29 has phenotypic characteristics that are intermediate between tall fescue grass and Italian ryegrass.

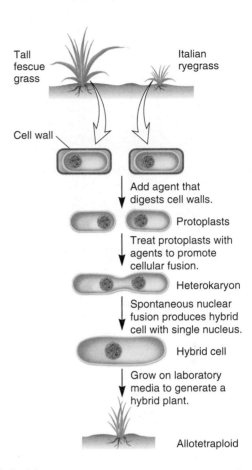

FIGURE 8.29   The technique of cell fusion. This technique is shown here with cells from tall fescue grass (*Festuca arundinacea*) and Italian ryegrass (*Lolium multiflorum*).The resulting hybrid is an allotetraploid.

Genes → Traits The allotetraploid contains two copies of genes from each parent. In this case, the allotetraploid displays characteristics that are intermediate between the tall fescue grass and the Italian ryegrass.

Monoploids Produced in Agricultural and Genetic Research Can Be Used to Make Homozygous and Hybrid Strains

A goal of some plant breeders is to have diploid strains of crop plants that are homozygous for all of their genes. One true-breeding strain can then be crossed to a different true-breeding strain to produce an F_1 hybrid that is heterozygous for many genes. Such hybrids are often more vigorous than the corresponding homozygous strains. This phenomenon, known as **hybrid vigor** or **heterosis,** is described in greater detail in Chapter 25. Seed companies often use this strategy to produce hybrid seed for many crops, such as corn and alfalfa. To achieve this goal, the companies must have homozygous parental strains that can be crossed to each other to produce the hybrid seed. One way to obtain these homozygous strains involves inbreeding over many generations. This may be accomplished after several rounds of self-fertilization. As you might imagine, this can be a rather time-consuming endeavor.

As an alternative, the production of **monoploids**—organisms that have a single set of chromosomes in their somatic cells—can be used as part of an experimental strategy to develop homozygous diploid strains of plants. Monoploids have been used to improve agricultural crops such as wheat, rice, corn, barley, and potato. In 1964, Sipra Guha and Satish Maheshwari developed a method to produce monoploid plants directly from pollen grains. **Figure 8.30** describes the experimental technique called **anther culture,** which has been extensively used to produce diploid strains of crop plants that are homozygous for all of their genes. The method involves alternation between monoploid and diploid generations. The parental plant is diploid but not homozygous for all of its genes. The anthers from this diploid plant are collected, and the haploid pollen grains are induced to begin development by a cold shock—an abrupt exposure to cold temperature. After several weeks, monoploid plantlets emerge, and these can be grown on agar media in a laboratory. However, due to the presence of deleterious alleles that are recessive, many of the pollen grains may fail to produce viable plantlets. Therefore, anther culture has been described as a "monoploid sieve" that weeds out individuals that carry deleterious recessive alleles.

Eventually, plantlets that are healthy can be transferred to small pots. After the plants grow to a reasonable size, a section of the monoploid plant can be treated with colchicine to convert it to diploid tissue. A cutting from this diploid section can then be used to generate a separate plant. This diploid plant is homozygous for all of its genes because it was produced by the chromosomal doubling of a monoploid strain.

In certain animal species, monoploids can be produced by experimental treatments that induce the eggs to begin development without fertilization by sperm. This process is known as **parthenogenesis.** In many cases, however, the haploid zygote will develop only for a short period of time before it dies. Nevertheless, a short phase of development can be useful to research scientists. For example, the zebrafish (*Danio rerio*), a common aquarium fish, has recently gained popularity among researchers interested in vertebrate development. The haploid egg can be induced to begin development by exposure to sperm rendered biologically inactive by UV irradiation.

FIGURE 8.30 **The experimental production of monoploids with anther culture.** In the technique of anther culture, the anthers are collected, and the haploid pollen within them is induced to begin development by a cold shock treatment. After several weeks, monoploid plantlets emerge, and these can be grown on an agar medium. Eventually, the plantlets can be transferred to small pots. After the plantlets have grown to a reasonable size, a section of the monoploid plantlet can be treated with colchicine to convert it to diploid tissue. A cutting from this diploid section can then be used to generate a diploid plant.

CONCEPTUAL SUMMARY

We classify variations in chromosome structure as deletions (or deficiencies), duplications, inversions, and translocations. Deletions and duplications involve changes in the total amount of genetic material within a chromosome; inversions and translocations are chromosomal rearrangements. Deletions tend to be detrimental to an organism, although this depends on the size and location of the deletion. By comparison, duplications tend to be less harmful. From an evolutionary perspective, duplications provide the raw material for the addition of genes into a species' genome. This can lead to the evolution of gene families in which a group of genes encodes proteins that carry out similar yet specialized functions.

Inversions often do not affect the phenotype of the individual that carries them. They are classified according to the location of the centromere, as either pericentric (if the centromere lies within the inverted region) or paracentric (if the centromere is found outside the inverted region). A translocation occurs when one segment of a chromosome becomes attached to a different chromosome or to a different part of the same chromosome. Translocations are categorized as simple, in which a single piece of chromosome is attached to another chromosome, or reciprocal, in which two different types of chromosomes exchange pieces. In a Robertsonian translocation, the long arms of two acrocentric chromosomes fuse, creating one large single chromosome, and the two short arms are lost. In balanced translocations, there is no resulting change in the total amount of genetic material, and the resulting phenotype is usually normal.

In a few cases, inversions and balanced translocations can affect the phenotype because a breakpoint is within a vital gene or the rearrangement results in a position effect that alters gene expression or regulation. Inversions and translocations are often associated with fertility problems and a higher probability of producing abnormal offspring. Even though an individual carrying an inversion or balanced translocation may be phenotypically normal, crossing over and chromosomal segregation may lead to the production of gametes that do not have a balanced amount of genetic material. Familial Down syndrome provides an example of this phenomenon.

Chromosome number is another critical factor that determines the phenotype of an organism. Aneuploidy involves an alteration in the number of chromosomes, so the total number is not an exact multiple of a set. In aneuploidy, an individual may have an extra chromosome (trisomy) or may be missing a chromosome (monosomy); both of these are usually detrimental to the individual's phenotype. For example, various human genetic diseases, such as Down syndrome (trisomy 21), are due to irregularities in chromosome number. Likewise, in plants, aneuploidy also significantly affects the phenotype of an individual.

Organisms that are euploids have a chromosome number that is an exact multiple of a chromosome set. Variations in the number of sets of chromosomes are also common, particularly in the plant kingdom. Many of our modern crops are polyploid plants that exhibit characteristics superior to those of their diploid counterparts. Polyploids with an odd number of chromosome sets are usually sterile, and these are useful in producing seedless varieties of plants. In animals, some cells of the body may be endopolyploid—that is, they have more sets of chromosomes than other somatic cells. The polytene chromosomes found in *Drosophila* are a dramatic example of this phenomenon.

Changes in chromosome number can arise via several different mechanisms. In natural populations, the most important of these are meiotic or mitotic nondisjunction and interspecies crosses. Meiotic nondisjunction can produce cells with alterations in chromosome number, which can lead to aneuploidy or polyploidy. Either mitotic nondisjunction or chromosome loss can occur after fertilization and can lead to patches of tissue with an altered chromosomal composition, a condition termed mosaicism. Interspecies crosses can produce allodiploids, which have one set of chromosomes from each of two different species, or allopolyploids, which have two or more sets from each species. Allotetraploids, which have two complete sets of chromosomes from two different species, are more likely to be fertile because the chromosomes are able to form homologous pairs during meiosis.

EXPERIMENTAL SUMMARY

The primary experimental strategy for studying variation in chromosome structure and number is the microscopic examination of chromosomes. A karyotype is a micrograph that shows the composition of chromosomes from an actively dividing eukaryotic cell. When preparing a karyotype, the chromosomes may be stained with Giemsa dye, which gives individual chromosomes a unique G banding pattern. By observing the banding patterns and number of chromosomes under a microscope, a cytogeneticist can determine if an individual carries a change in chromosome structure or number. The technique of comparative genomic hybridization can also be used to detect changes in chromosome structure, such as deletions and duplications. This

method has found wide use in the analysis of cancer cells, which often exhibit such changes.

Experimental methods that cause changes in chromosome number are also available. The drug colchicine can increase chromosome number by promoting mitotic nondisjunction. Using cell fusion techniques, researchers have artificially crossed two species that cannot interbreed naturally, creating fertile interspecies hybrids. The production of monoploid plants can be used as part of a strategy to develop homozygous diploid strains. In certain animal species, experimental treatments have induced eggs to begin development without fertilization by sperm.

PROBLEM SETS & INSIGHTS

Solved Problems

S1. Describe how a gene family is produced. Discuss the common and unique features of the family members in the globin gene family.

Answer: A gene family is produced when a single gene is copied one or more times by a gene duplication event. This duplication may occur by an abnormal (misaligned) crossover, which produces a chromosome with a deletion and another chromosome with a gene duplication.

Over time, this type of duplication may occur several times to produce many copies of a particular gene. In addition, translocations may move the duplicated genes to other chromosomes, so the members of the gene family may be dispersed among several different chromosomes. Eventually, each member of a gene family will accumulate mutations, which may subtly alter their function.

All the members of the globin gene family bind oxygen. Myoglobin tends to bind it more tightly; therefore, it is good at storing oxygen. Hemoglobin binds it more loosely, so it can transport oxygen throughout the body (via red blood cells) and release it to the tissues that need oxygen. The polypeptides that form hemoglobins are predominantly expressed in red blood cells, whereas myoglobin genes are expressed in many different cell types. The expression pattern of the globin genes changes during different stages of development. The ε- and ζ-globin genes are expressed in the early embryo. They are turned off near the end of the first trimester, and then the γ-globin genes exhibit their maximal expression during the second and third

trimesters of gestation. Following birth, the γ-globin genes are silenced, and the β-globin gene is expressed for the rest of a person's life. These differences in the expression of the globin genes reflect the differences in the oxygen transport needs of humans during the different stages of life. Overall, the evolution of gene families has resulted in gene products that are better suited to a particular tissue or stage of development. This has allowed a better "fine-tuning" of human traits.

S2. An inversion heterozygote has the following inverted chromosome:

What is the result if a crossover occurs between genes *F* and *G* on one inversion and one normal chromosome?

Answer: The resulting product is four chromosomes. One chromosome is normal, one is an inversion chromosome, and two chromosomes have duplications and deletions. The two duplicated/deficient chromosomes are shown here:

S3. In humans, the number of chromosomes per set equals 23. Even though the following conditions are lethal, what would be the total number of chromosomes for the following individuals?

   A. Trisomy 22

   B. Monosomy 11

   C. Triploid individual

Answer:

   A. 47 (the diploid number, 46, plus 1)

   B. 45 (the diploid number, 46, minus 1)

   C. 69 (3 times 23)

S4. A diploid species with 44 chromosomes (i.e., 22/set) is crossed to another diploid species with 38 chromosomes (i.e., 19/set). What would be the number of chromosomes in an allodiploid or allotetraploid produced from this cross? Would you expect the offspring to be sterile or fertile?

Answer: An allodiploid would have 22 + 19 = 41 chromosomes. This individual would likely be sterile, because all the chromosomes would not have homeologous partners to pair with during meiosis. This yields aneuploidy, which usually causes sterility. An allotetraploid would have 44 + 38 = 82 chromosomes. Because each chromosome would have a homologous partner, the allotetraploid would likely be fertile.

S5. **Pseudodominance** occurs when a single copy of a recessive allele is phenotypically expressed because the second copy of the gene has been deleted from the homologous chromosome; the individual is hemizygous for the recessive allele. As an example, we can consider the notch phenotype in *Drosophila*, which is an X-linked trait. Fruit flies with this condition have wings with a notched appearance at their edges. Female flies that are heterozygous for this mutation have notched wings; homozygous females and hemizygous males are unable to survive. The notched phenotype is due to a defect in a single gene called *notch* (*N*). Geneticists studying fruit flies with this phenotype have discovered that some of the mutant flies are due to a small deletion that includes the *notch* gene as well as a few genes on either side of it. Other notch mutations are due to small mutations confined within the *notch* gene itself. A genetic analysis can distinguish between notched fruit flies carrying a deletion versus those that carry a single-gene mutation. This is possible because the *notch* gene happens to be located next to the red/white eye color gene on the X chromosome. How would you distinguish between a notched phenotype due to a deletion that included the *notch* gene and the adjacent eye color gene versus a notch phenotype due to a small mutation only within the *notch* gene itself?

Answer: To determine if the notch mutation is due to a deletion, red-eyed females with the notched phenotype can be crossed to white-eyed males. Only the daughters with notched wings need to be analyzed. If they have red eyes, this means the notch mutation has not deleted the red-eye allele from the X chromosome. Alternatively, if they have white eyes, this indicates the red-eye allele has been deleted from the X chromosome that carries the notch mutation. In this case, the white-eye allele is expressed because it is present in a single copy in a female fly with two X chromosomes. This phenomenon is pseudodominance.

S6. Albert Blakeslee began using the Jimson weed (*Datura stramonium*) as an experimental organism to teach his students the laws of Mendelian inheritance. Although this plant has not gained widespread use in genetic studies, Blakeslee's work provided a convincing demonstration that changes in chromosome number have an impact on the phenotype of organisms (see Figure 8.16). Blakeslee's assistant, B. T. Avery, identified a Jimson weed mutant that he called "globe" because the capsule is more rounded than normal. In genetic crosses, he found that the globe mutant had a peculiar pattern of inheritance. The globe trait was passed to about 25% of the offspring when the globe plants were allowed to self-fertilize. Unexpectedly, about 25% of the offspring also had the globe phenotype when globe plants were pollinated by a normal plant. In contrast, when pollen from a globe plant was used to pollinate a normal plant, less than 2% of the offspring had the globe phenotype. This non-Mendelian pattern of inheritance caused Blakeslee and his colleagues to investigate the nature of this trait further. We now know the globe phenotype is due to trisomy 11. Can you explain this unusual pattern of inheritance knowing it is due to trisomy 11?

Answer: This unusual pattern of inheritance of these aneuploid strains can be explained by the viability of euploid versus aneuploid gametes and/or gametophytes. When an individual is trisomic, there is a 50% chance that an egg or sperm will inherit an extra chromosome and a 50% chance that a gamete will be normal. Blakeslee's results indicate that when a pollen grain inherited an extra copy of chromosome 11, it was almost always nonviable and unable to produce an aneuploid offspring. However, a significant percentage of aneuploid eggs were viable, so some (25%) of the offspring were aneuploid. Because an aneuploid plant should produce a 1:1 ratio between euploid and aneuploid eggs, the observation that only 25% of the offspring were aneuploid also indicates that about half of the female gametophytes or aneuploid eggs from such gametophytes were also nonviable. Overall, these results provide compelling evidence that imbalances in chromosome number can alter reproductive viability and also cause significant phenotypic consequences.

Conceptual Questions

C1. Which changes in chromosome structure cause a change in the total amount of genetic material, and which do not?

C2. Explain why small deletions and duplications are less likely to have a detrimental effect on an individual's phenotype than large ones. If a small deletion within a single chromosome happens to have a phenotypic effect, what would you conclude about the genes in this region?

C3. How does a chromosomal duplication occur?

C4. What is a gene family? How are gene families produced over time? With regard to gene function, what is the biological significance of a gene family?

C5. Following a gene duplication, two genes will accumulate different mutations, causing them to have slightly different sequences. In Figure 8.7, which pair of genes would you expect to have more similar sequences, α_1 and α_2 or ψ_{α_1} and α_2? Explain your answer.

C6. Two chromosomes have the following order of genes:

Normal: *A B C* centromere *D E F G H I*

Abnormal: *A B G F E D* centromere *C H I*

Does the abnormal chromosome have a pericentric or paracentric inversion? Draw a sketch showing how these two chromosomes would pair during prophase of meiosis I.

C7. An inversion heterozygote has the following inverted chromosome:

Centromere

A B J I HGF ED   C K L M

Inverted region

What would be the products if a crossover occurred between genes *H* and *I* on one inverted and one normal chromosome?

C8. An inversion heterozygote has the following inverted chromosome:

Centromere

A   B CD J I HGF E KL M

Inverted region

What would be the products if a crossover occurred between genes *H* and *I* on one inverted and one normal chromosome?

C9. Explain why inversions and reciprocal translocations do not usually cause a phenotypic effect. In a few cases, however, they do. Explain how.

C10. An individual has the following reciprocal translocation:

What would be the outcome of alternate and adjacent-1 segregation?

C11. A phenotypically normal individual has the following combinations of abnormal chromosomes:

The normal chromosomes are shown on the left of each pair. Suggest a series of events (breaks, translocations, crossovers, etc.) that may have produced this combination of chromosomes.

C12. Two phenotypically normal parents produce a phenotypically abnormal child in which chromosome 5 is missing part of its long arm but has a piece of chromosome 7 attached to it. The child also has one normal copy of chromosome 5 and two normal copies of chromosome 7. With regard to chromosomes 5 and 7, what do you think are the chromosomal compositions of the parents?

C13. In the segregation of centromeres, why is adjacent-2 segregation less frequent than alternate or adjacent-1 segregation?

C14. Which of the following types of chromosomal changes would you expect to have phenotypic consequences? Explain your choices.

A. Pericentric inversion

B. Reciprocal translocation

C. Deletion

D. Unbalanced translocation

C15. Explain why a translocation cross occurs during metaphase of meiosis I when a cell contains a reciprocal translocation.

C16. A phenotypically abnormal individual has a phenotypically normal father with an inversion on one copy of chromosome 7 and a normal mother without any changes in chromosome structure. The order of genes along chromosome 7 in the father is as follows:

R T D M centromere *P U X Z C*   (normal chromosome 7)

R T D U P centromere *M X Z C*   (inverted chromosome 7)

The phenotypically abnormal offspring has a chromosome 7 with the following order of genes:

R T D M centromere *P U D T R*

With a sketch, explain how this chromosome was formed. In your answer, explain where the crossover occurred (i.e., between which two genes).

C17. A diploid fruit fly has eight chromosomes. How many total chromosomes would be found in the following flies?

A. Tetraploid

B. Trisomy 2

C. Monosomy 3

D. $3n$

E. $4n + 1$

C18. A person is born with one X chromosome, zero Y chromosomes, trisomy 21, and two copies of the other chromosomes. How many chromosomes does this person have altogether? Explain whether this person is euploid or aneuploid.

C19. Two phenotypically unaffected parents produce two children with familial Down syndrome. With regard to chromosomes 14 and 21, what are the chromosomal compositions of the parents?

C20. Aneuploidy is typically detrimental, whereas polyploidy is sometimes beneficial, particularly in plants. Discuss why you think this is the case.

C21. Explain how aneuploidy, deletions, and duplications cause genetic imbalances. Why do you think that deletions and monosomies are more detrimental than duplications and trisomies?

C22. Female fruit flies homozygous for the X-linked white-eye allele are crossed to males with red eyes. On very rare occasions, an offspring is a male with red eyes. Assuming these rare offspring are not due to a new mutation in one of the mother's X chromosomes that converted the white-eye allele into a red-eye allele, explain how this red-eyed male arose.

C23. A cytogeneticist has collected tissue samples from members of the same butterfly species. Some of the butterflies were located in Canada, and others were found in Mexico. Upon karyotyping, the cytogeneticist discovered that chromosome 5 of the Canadian butterflies had a large inversion compared to the Mexican butterflies. The Canadian butterflies were inversion homozygotes, whereas the Mexican butterflies had two normal copies of chromosome 5.

A. Explain whether a mating between the Canadian and Mexican butterflies would produce phenotypically normal offspring.

B. Explain whether the offspring of a cross between Canadian and Mexican butterflies would be fertile.

C24. Why do you think that human trisomies 13, 18, and 21 can survive but the other trisomies are lethal? Even though X chromosomes are large, aneuploidies of this chromosome are also tolerated. Explain why.

C25. A zookeeper has collected a male and female lizard that look like they belong to the same species. They mate with each other and produce phenotypically normal offspring. However, the offspring are sterile. Suggest one or more explanations for their sterility.

C26. What is endopolyploidy? What is its biological significance?

C27. What is mosaicism? How is it produced?

C28. Explain how polytene chromosomes of *Drosophila melanogaster* are produced and how they form a six-armed structure.

C29. Describe some of the advantages of polyploid plants. What are the consequences of having an odd number of chromosome sets?

C30. While conducting field studies on a chain of islands, you decide to karyotype two phenotypically identical groups of turtles, which are found on different islands. The turtles on one island have 24 chromosomes, while the turtles on another island have 48 chromosomes. How would you explain this observation? How do you think the turtles with 48 chromosomes came into being? If you mated the two types of turtles together, would you expect their offspring to be phenotypically normal? Would you expect them to be fertile? Explain.

C31. A diploid fruit fly has eight chromosomes. Which of the following terms should not be used to describe a fruit fly with four sets of chromosomes?

    A. Polyploid

    B. Aneuploid

    C. Euploid

    D. Tetraploid

    E. $4n$

C32. Which of the following terms should not be used to describe a human with three copies of chromosome 12?

    A. Polyploid

    B. Triploid

    C. Aneuploid

    D. Euploid

    E. $2n + 1$

    F. Trisomy 12

C33. The kidney bean, *Phaseolus vulgaris*, is a diploid species containing a total of 22 chromosomes in somatic cells. How many possible types of trisomic individuals could be produced in this species?

C34. The karyotype of a young girl who is affected with familial Down syndrome revealed a total of 46 chromosomes. Her older brother, however, who is phenotypically unaffected, actually had 45 chromosomes. Explain how this could happen. What would you expect to be the chromosomal number in the parents of these two children?

C35. A triploid plant has 18 chromosomes (i.e., 6 chromosomes per set). If we assume a gamete has an equal probability of receiving one or two copies of each of the six types of chromosome, what are the odds of this plant producing a monoploid or a diploid gamete? What are the odds of producing an aneuploid gamete? If the plant is allowed to self-fertilize, what are the odds of producing a euploid offspring?

C36. Describe three naturally occurring ways that the chromosome number can change.

C37. Meiotic nondisjunction is much more likely than mitotic nondisjunction. Based on this observation, would you conclude that meiotic nondisjunction is usually due to nondisjunction during meiosis I or meiosis II? Explain your reasoning.

C38. A woman who is heterozygous, *Bb*, has brown eyes. *B* (brown) is a dominant allele, and *b* (blue) is recessive. In one of her eyes, however, there is a patch of blue color. Give three different explanations for how this might have occurred.

C39. What is an allodiploid? What factor determines the fertility of an allodiploid? Why are allotetraploids more likely to be fertile?

C40. What are homeologous chromosomes?

C41. Meiotic nondisjunction usually occurs during meiosis I. What is not separating properly: bivalents or sister chromatids? What is not separating properly during mitotic nondisjunction?

C42. Table 8.1 shows that Turner syndrome occurs when an individual inherits one X chromosome but lacks a second sex chromosome. Can Turner syndrome be due to nondisjunction during oogenesis, spermatogenesis, or both? If a phenotypically normal couple has a color-blind child (due to a recessive X-linked allele) with Turner syndrome, did nondisjunction occur during oogenesis or spermatogenesis in this child's parents? Explain your answer.

C43. Male honeybees, which are monoploid, produce sperm by meiosis. Explain what unusual event (compared to other animals) must occur during spermatogenesis in honeybees to produce sperm? Does this unusual event occur during meiosis I or meiosis II?

Experimental Questions

E1. What is the main goal of comparative genome hybridization? Explain how the ratio of green/red fluorescence provides information regarding chromosome structure.

E2. Let's suppose a researcher conducted comparative genomic hybridization (see Figure 8.8) and accidentally added two-fold too much (red) DNA from normal cells. What green/red ratio would you expect in a region from a chromosome from a cancer cell that carried a duplication on both chromosomal copies? What ratio would be observed for a region that was deleted on just one of the chromosomes from cancer cells?

E3. With regard to the analysis of chromosome structure, explain the experimental advantage that polytene chromosomes offer. Discuss why changes in chromosome structure are more easily detected in polytene chromosomes compared to ordinary (nonpolytene) chromosomes.

E4. Describe how colchicine can be used to alter chromosome number.

E5. Describe the steps you would take to produce a tetraploid plant that is homozygous for all of its genes.

E6. In agriculture, what is the primary purpose of anther culture?

E7. What are some experimental advantages of cell fusion techniques as opposed to interbreeding approaches?

E8. It is an exciting time to be a plant breeder because so many options are available for the development of new types of agriculturally useful plants. Let's suppose you wish to develop a seedless tomato that could grow in a very hot climate and is resistant to a viral pathogen that commonly infects tomato plants. At your disposal, you have a seed-bearing tomato strain that is heat resistant and produces great-tasting tomatoes. You also have a wild strain of tomato plants (which have lousy-tasting tomatoes) that is resistant to the viral pathogen. Suggest a series of steps you might follow to produce a great-tasting, seedless tomato that is resistant to heat and the viral pathogen.

E9. What is a G band? Discuss how G bands are useful in the analysis of chromosome structure.

E10. A female fruit fly contains one normal X chromosome and one X chromosome with a deletion. The deletion is in the middle of the X chromosome and is about 10% of the entire length of the X chromosome. If you stained and observed the chromosomes of this female fly in salivary gland cells, draw what the polytene arm of the X chromosome would look like. Explain your drawing.

E11. Describe two different experimental strategies to create an allotetraploid from two different diploid species of plants.

E12. In the procedure of anther culture (see Figure 8.30), an experimenter may begin with a diploid plant and then cold shock

the pollen to get them to grow as haploid plantlets. In some cases, the pollen may come from a phenotypically vigorous plant that is heterozygous for many genes. Even so, many of the haploid plantlets appear rather weak and nonvigorous. In fact, many of them fail to grow at all. In contrast, some of the plantlets are fairly healthy. Explain why some plantlets would be weak, while others could be quite healthy.

Questions for Student Discussion/Collaboration

1. A chromosome involved in a reciprocal translocation also has an inversion. In addition, the cell also contains two normal chromosomes.

Make a drawing that shows how these chromosomes will pair during metaphase of meiosis I.

2. Besides the ones mentioned in this textbook, look for other examples of variations in euploidy. Perhaps you might look in more advanced textbooks concerning population genetics, ecology, or the like. Discuss the phenotypic consequences of these changes.

3. Cell biology textbooks often discuss cellular proteins encoded by genes that are members of a gene family. Examples of such proteins include myosins and glucose transporters. Take a look through a cell biology textbook and identify some proteins encoded by members of gene families. Discuss the importance of gene families at the cellular level.

4. Discuss how variation in chromosome number has been useful in agriculture.

Note: All answers appear at the website for this textbook; the answers to even-numbered questions are in the back of the textbook.

www.mhhe.com/brookergenetics3e

Visit the Online Learning Center for practice tests, answer keys, and other learning aids for this chapter. Enhance your understanding of genetics with our interactive exercises, quizzes, animations, and much more.

9

MOLECULAR STRUCTURE OF DNA AND RNA

In Chapters 2 through 8, we focused on the relationship between the inheritance of genes and chromosomes, and the outcome of an organism's traits. In Chapter 9, we will shift our attention to **molecular genetics**—the study of DNA structure and function at the molecular level. An exciting goal of molecular genetics is to use our knowledge of DNA structure to understand how DNA functions as the genetic material. Using molecular techniques, researchers have determined the organization of many genes. This information, in turn, has helped us understand how the expression of such genes governs the outcome of an individual's inherited traits.

The past several decades have seen dramatic advances in techniques and approaches to investigate and even to alter the genetic material. These advances have greatly expanded our understanding of molecular genetics and also have provided key insights into the mechanisms underlying transmission and population genetics. Molecular genetic technology is also widely used in supporting disciplines such as biochemistry, cell biology, and microbiology.

To a large extent, our understanding of genetics comes from our knowledge of the molecular structure of DNA (deoxyribonucleic acid) and RNA (ribonucleic acid). In this chapter, we will begin by considering classic experiments that showed DNA is the genetic material. We will then survey the molecular features of DNA and RNA that underlie their function.

9.1 IDENTIFICATION OF DNA AS THE GENETIC MATERIAL

In his pioneering experiments, Gregor Mendel studied several different traits in pea plants. By conducting the appropriate crosses, he showed that traits are inherited as discrete units as they pass from parent to offspring. This implies that living organisms contain a genetic material that governs an individual's traits, a substance that is transferred during the process of reproduction.

To fulfill its role, the genetic material must meet several criteria.

1. **Information:** The genetic material must contain the information necessary to construct an entire organism. In other words, it must provide the blueprint to determine the inherited traits of an organism.
2. **Transmission:** During reproduction, the genetic material must be passed from parents to offspring.

A molecular model showing the structure of the DNA double helix.

PART III

Molecular Structure and Replication of the Genetic Material

3. **Replication:** Because the genetic material is passed from parents to offspring, and from mother cell to daughter cells during cell division, it must be copied.

4. **Variation:** Within any species, a significant amount of phenotypic variability occurs. For example, Mendel studied several traits in pea plants that were variable among different plants. These included height (tall versus dwarf) and seed color (yellow versus green). Therefore, the genetic material must also have variation that can account for the known phenotypic differences within each species.

Along with Mendel's work, the data of many other geneticists in the early 1900s were consistent with these four properties: information, transmission, replication, and variation. However, the experimental study of genetic crosses cannot, by itself, identify the chemical nature of the genetic material.

In the 1880s, August Weismann and Carl Nägeli championed the idea that a chemical substance within living cells is responsible for the transmission of traits from parents to offspring. The chromosome theory of inheritance was developed, and experimentation demonstrated that the chromosomes are the carriers of the genetic material (see Chapter 3). Nevertheless, the story was not complete because chromosomes contain both DNA and proteins. Also, RNA is found in the vicinity of chromosomes. Therefore, further research was needed to precisely identify the genetic material. In this section, we will examine the first experimental approaches to achieve this goal.

Experiments with Pneumococcus Suggested That DNA Is the Genetic Material

Some early work in microbiology was important in developing an experimental strategy to identify the genetic material. Frederick Griffith studied a type of bacterium known then as pneumococci and now classified as *Streptococcus pneumoniae*. Certain strains of *S. pneumoniae* secrete a polysaccharide capsule, whereas other strains do not. When streaked on petri plates containing solid growth media, capsule-secreting strains have a smooth colony morphology, whereas those strains unable to secrete a capsule have a rough appearance.

When comparing two different smooth strains of *S. pneumoniae*, researchers found that the chemical composition of their capsules can differ significantly. Using biochemical techniques, these different types of smooth strains were characterized. For example, a type II smooth strain and a type III smooth strain both make a capsule, but their capsules differ from each other biochemically. This idea is schematically shown in **Figure 9.1**. Rare mutations can occasionally convert a smooth bacterium into a rough bacterium, and vice versa. These infrequent interconversions are type-specific. For example, a type III smooth strain can mutate to become a type III rough strain. On rare occasions, a bacterium from this rough strain may mutate back to become a smooth strain. In this case, it can mutate to become only a type III smooth strain, never a type II smooth strain.

The different forms of *S. pneumoniae* also affect their virulence, or ability to cause disease. When smooth strains of *S. pneumonia* infect a mouse, the capsule allows the bacteria to escape attack by the mouse's immune system. As a result, the bacteria can grow and eventually kill the mouse. In contrast, the nonencapsulated (rough) bacteria are destroyed by the animal's immune system.

In 1928, Griffith conducted experiments that involved the injection of live and/or heat-killed bacteria into mice. He then observed whether or not the bacteria caused a lethal infection. Griffith was working with two strains of *S. pneumoniae*, a type IIIS (S for smooth) and a type IIR (R for rough). When injected into a live mouse, the type IIIS bacteria proliferated within the mouse's bloodstream and ultimately killed the mouse (**Figure 9.2a**). Following the death of the mouse, Griffith found many type IIIS bacteria within the mouse's blood. In contrast, when type IIR bacteria were injected into a mouse, the mouse lived (**Figure 9.2b**). To verify that the proliferation of the smooth bacteria was causing the death of the mouse, Griffith killed the smooth bacteria with heat treatment before injecting them into the mouse. In this case, the mouse also survived (**Figure 9.2c**).

The critical and unexpected result was obtained in the experiment outlined in **Figure 9.2d**. In this experiment, live type IIR bacteria were mixed with heat-killed type IIIS bacteria. As shown here, the mouse died. Furthermore, extracts from tissues of the dead mouse were found to contain living type IIIS bacteria! What can account for these results? Because type IIR bacteria cannot mutate to type IIIS, the interpretation of these data is that something from the dead type IIIS bacteria was transforming the

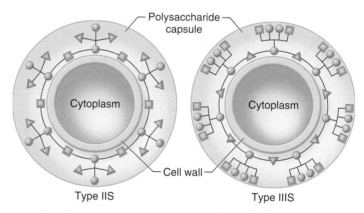

FIGURE 9.1   **Schematic illustration of biochemical differences between the capsules of two strains of pneumococci.** The capsule is composed of sugar units linked to form a polysaccharide material. This figure illustrates how the types of sugars (represented by circles, squares, and triangles) and their arrangement to form a polysaccharide capsule can differ among different strains. This figure is not meant to represent the actual polysaccharide capsule, which is much more complex than shown here. Instead, it emphasizes that the structures of the polysaccharide capsules are different in type IIS and type IIIS bacteria.

Genes → Traits The chemical composition of the polysaccharide capsule is governed by genes within the bacterial cell. These genes encode enzymes that synthesize the polysaccharide capsule. The genes within type IIS bacteria encode enzymes that synthesize the polysaccharide capsule depicted on the left. The genes within type IIIS bacteria encode enzymes that function somewhat differently to synthesize the capsule shown on the right.

Living type IIIS bacteria were injected into a mouse.

Living type IIR bacteria were injected into a mouse.

Heat-killed type IIIS bacteria were injected into a mouse.

Living type IIR and heat-killed type IIIS bacteria were injected into a mouse.

After several days

After several days

After several days

After several days

Mouse died

Mouse survived

Mouse survived

Mouse died

Type IIIS bacteria were isolated from the dead mouse.

No living bacteria were isolated from the mouse.

No living bacteria were isolated from the mouse.

Type IIIS bacteria were isolated from the dead mouse.

(a) Live IIIS

(b) Live IIR

(c) Dead IIIS

(d) Live IIR + dead IIIS

FIGURE 9.2   Griffith's experiments on genetic transformation in pneumococcus.

type IIR bacteria into type IIIS. Griffith called this process **transformation,** and the unidentified substance causing this to occur was termed the transformation principle. The steps of bacterial transformation are described in Chapter 6 (see Figure 6.12).

At this point, let's look at what Griffith's observations mean in genetic terms. The transformed bacteria acquired the *information* to make a type III capsule. Among different strains, *variation* exists both in the ability to create a capsule and the type that is made. The genetic material for the type III trait must be *replicated* so that it can be *transmitted* from mother to daughter cells during cell division. Taken together, these observations are consistent with the idea that the formation of a capsule is governed by the bacteria's genetic material, meeting the four criteria described previously. Griffith's experiments showed that some genetic material from the dead bacteria had been transferred to the living bacteria and provided them with a new trait. However, Griffith did not know what the transforming substance was.

Important scientific discoveries often take place when researchers recognize that someone else's experimental observations can be used to address a particular scientific question. Oswald Avery, Colin MacLeod, and Maclyn McCarty realized that Griffith's observations could be used as part of an experimental strategy to identify the genetic material. They asked the question, What substance is being transferred from the dead type IIIS bacteria to the live type IIR? To answer this question, they incorporated additional biochemical techniques into their experimental methods.

At the time of these experiments in the 1940s, researchers already knew that DNA, RNA, proteins, and carbohydrates are major constituents of living cells. To separate these components and to determine if any of them was the genetic material, Avery, MacLeod, and McCarty used established biochemical purification procedures and prepared bacterial extracts from type IIIS strains containing each type of these molecules. After many repeated attempts with different types of extracts, they discovered that only one of the extracts, namely, the one that contained purified DNA, was able to convert the type IIR bacteria into type IIIS. As shown in **Figure 9.3**, when this extract was mixed with type IIR bacteria, some of the bacteria were converted to type IIIS. However, if no DNA extract was added, no type IIIS bacterial colonies were observed on the Petri plates.

A biochemist might point out that a DNA extract may not be 100% pure. In fact, any purified extract might contain small traces of some other substances. Therefore, one can argue that a small amount of contaminating material in the DNA extract might actually be the genetic material. The most likely contaminating substances in this case would be RNA or protein. To further verify that the DNA in the extract was indeed responsible for the transformation, Avery, MacLeod, and McCarty treated samples of

FIGURE 9.3 **Experimental protocol used by Avery, MacLeod, and McCarty to identify the transforming principle.** Samples of *S. pneumonia* cells were either not exposed to a type IIIS DNA extract (tube 1) or exposed to a type IIIS DNA extract (tubes 2–5). Tubes 3, 4, and 5 also contained DNase, RNase, or protease, respectively. After incubation, the cells were exposed to antibodies, which are molecules that can specifically recognize the molecular structure of macromolecules. In this experiment, the antibodies recognized the cell surface of type IIR bacteria and caused them to clump together. The clumped bacteria were removed by a gentle centrifugation step. Only the bacteria that were not recognized by the antibody (namely, the type IIIS bacteria) remained in the supernatant. The cells in the supernatant were plated on solid growth media. After overnight incubation, visible colonies may be observed.

the DNA extract with enzymes that digest DNA (called **DNase**), RNA (**RNase**), or protein (**protease**)(see Figure 9.3). When the DNA extracts were treated with RNase or protease, they still converted type IIR bacteria into type IIIS. These results indicated that any remaining RNA or protein in the extract was not acting as the genetic material. However, when the extract was treated with DNase, it lost its ability to convert type IIR into type IIIS bacteria. These results indicated that the degradation of the DNA in the extract by DNase prevented conversion of type IIR to type IIIS. This interpretation is consistent with the hypothesis that DNA is the genetic material. A more elegant way of saying this is that the transforming principle is DNA.

Hershey and Chase Provided Evidence That the Genetic Material Injected into the Bacterial Cytoplasm Is T2 Phage DNA

A second experimental approach indicating that DNA is the genetic material came from the studies of Alfred Hershey and Martha Chase in 1952. Their research centered on the study of a virus known as T2. This virus infects *Escherichia coli* bacterial cells and is therefore known as a **bacteriophage** or simply a **phage.** As shown in **Figure 9.4,** the external structure of the T2 phage, known as the capsid or phage coat, consists of a head, sheath, tail fibers, and base plate. Biochemically, the phage coat is composed entirely of protein, which includes several different polypeptides. DNA is found inside the head of the T2 capsid. From a molecular point of view, this virus is rather simple, because it is composed of only two types of macromolecules: DNA and proteins.

Although the viral genetic material contains the blueprint to make new viruses, a virus itself cannot synthesize new viruses. Instead, a virus must introduce its genetic material into the

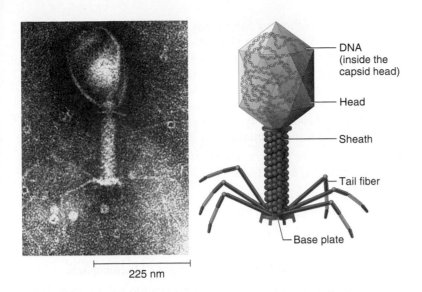

225 nm

FIGURE 9.4 **Structure of the T2 bacteriophage.** The T2 bacteriophage is composed of a phage coat, or capsid, with genetic material inside the head of the capsid. The capsid is divided into regions called the head, sheath, tail fibers, and base plate. These components are composed of proteins. The genetic material is composed of DNA.

Genes → Traits The genetic material of a bacteriophage contains many genes, which provide the blueprint for making new viruses. When the bacteriophage injects its genetic material into a bacterium, these genes are activated and direct the host cell to make new bacteriophages, as described in Figure 9.5.

Phage binds to host cell.

Phage injects its DNA into host cell.

Phage DNA directs the synthesis of viral components.

Phage components are assembled.

Host cell lyses and new phages are released.

ONLINE ANIMATION **FIGURE 9.5**   **Life cycle of the T2 bacteriophage.**

cytoplasm of a living cell. In the case of T2, this first involves the attachment of its tail fibers to the bacterial cell wall and the subsequent injection of its genetic material into the cytoplasm of the cell (**Figure 9.5**). The phage coat remains attached on the outside of the bacterium and does not enter the cell. After the entry of the viral genetic material, the bacterial cytoplasm provides all the synthetic machinery necessary to make viral proteins and DNA. The viral proteins and DNA assemble to make new viruses that are subsequently released from the cell by **lysis** (i.e., cell breakage).

To verify that DNA is the genetic material of T2, Hershey and Chase devised a method to separate the phage coat, which is attached to the outside of the bacterium, from the genetic material, which is injected into the cytoplasm. They were aware of microscopy experiments by Thomas Anderson showing that the T2 phage attaches itself to the outside of a bacterium by its tail fibers. Hershey and Chase reasoned that this is a fairly precarious attachment that could be disrupted by subjecting the bacteria to high shear forces, such as those produced in a kitchen blender. Their method was to expose bacteria to T2 phage, allowing sufficient time for the viruses to attach to bacteria and inject their genetic material. They then sheared the phage coats from the surface of the bacteria by a blender treatment. In this way, the phages' genetic material, which had been injected into the cytoplasm of the bacterial cells, could be separated from the phage coats that were sheared away.

Hershey and Chase used radioisotopes to distinguish proteins from DNA. Sulfur atoms are found in proteins but not in

DNA, whereas phosphorus atoms are found in DNA but not in phage proteins. Therefore, ^{35}S (a radioisotope of sulfur) and ^{32}P (a radioisotope of phosphorus) were used to specifically label phage proteins and DNA, respectively. Researchers can grow *E. coli* cells in media that contain ^{35}S or ^{32}P and then infect the *E. coli* cells with T2 phages. When new phages are produced, these will be labeled with ^{35}S or ^{32}P. In the experiment described in **Figure 9.6**, they began with *E. coli* cells and two preparations of T2 phage that were obtained in this manner. One preparation was labeled with ^{35}S to label the phage proteins, and the other preparation was labeled with ^{32}P to label the phage DNA. In separate flasks, each type of phage was mixed with a new sample of *E. coli* cells. The phages were given sufficient time to inject their genetic material into the bacterial cells, and then the sample was

subjected to shearing force using a blender. This treatment was expected to remove the phage coat from the surface of the bacterial cell. The sample was then subjected to centrifugation at a speed that would cause the heavier bacterial cells to form a pellet at the bottom of the tube, while the light phage coats would remain in the supernatant, the liquid found above the pellet. The amount of radioactivity in the supernatant (emitted from either ^{35}S or ^{32}P) was determined using a scintillation counter.

■ THE HYPOTHESIS

Only the genetic material of the phage is injected into the bacterium. Isotope labeling will reveal if it is DNA or protein.

■ TESTING THE HYPOTHESIS — FIGURE 9.6   **Evidence that DNA is the genetic material of T2 bacteriophage.**

Starting materials: The starting materials were *E. coli* cells and two preparations of T2 phage. One phage preparation had phage proteins labeled with ^{35}S, and the other preparation had phage DNA labeled with ^{32}P.

Experimental level   Conceptual level

1. Grow bacterial cells. Divide into two flasks.

2. Into one flask, add ^{35}S-labeled phage; in the second flask, add ^{32}P-labeled phage.

3. Allow infection to occur.

4. Agitate solutions in blenders for different lengths of time to shear the empty phages off the bacterial cells.

5. Centrifuge at 10,000 rpm.

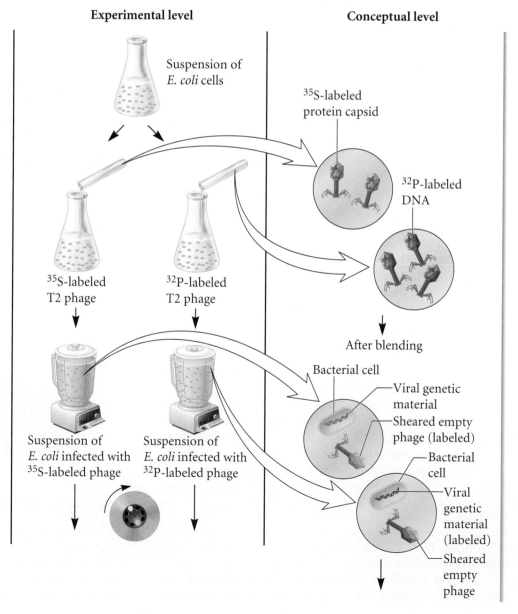

Suspension of *E. coli* cells

^{35}S-labeled protein capsid

^{32}P-labeled DNA

^{35}S-labeled T2 phage

^{32}P-labeled T2 phage

After blending

Bacterial cell

Viral genetic material

Sheared empty phage (labeled)

Suspension of *E. coli* infected with ^{35}S-labeled phage

Suspension of *E. coli* infected with ^{32}P-labeled phage

Bacterial cell

Viral genetic material (labeled)

Sheared empty phage

(continued)

6. The heavy bacterial cells sediment to the pellet, while the lighter phages remain in the supernatant. (See Appendix for explanation of centrifugation.)

7. Count the amount of radioisotope in the supernatant with a scintillation counter (see Appendix). Compare it with the starting amount.

■ THE DATA

■ INTERPRETING THE DATA

As seen in the data, most of the ^{35}S isotope was found in the supernatant. Because the shearing force was expected to remove the phage coat, this result indicates that the empty phages contain primarily protein. By comparison, only about 35% of the ^{32}P was found in the supernatant following shearing. Therefore, most of the DNA was located within the bacterial cells in the pellet. These results are consistent with the idea that the DNA is injected into the bacterial cytoplasm during infection, which would be the expected result if DNA is the genetic material.

By themselves, the results described in Figure 9.6 were not conclusive evidence that DNA is the genetic material. For exam-

ple, you may have noticed that less than 100% of the phage protein was found in the supernatant. Therefore, some of the phage protein could have been introduced into the bacterial cells (and could function as the genetic material). Nevertheless, the results of Hershey and Chase were consistent with the conclusion that the genetic material is DNA rather than protein. Overall, their studies of the T2 phage were quite influential in convincing the scientific community that DNA is the genetic material.

A self-help quiz involving this experiment can be found at the Online Learning Center.

RNA Functions as the Genetic Material in Some Viruses

We now know that bacteria, archaea, protists, fungi, plants, and animals all use DNA as their genetic material. As mentioned, viruses also have their own genetic material. Hershey and Chase concluded from their experiments that this genetic material is DNA. In the case of T2 bacteriophage, that is the correct conclusion. However, many viruses use RNA, rather than DNA, as their genetic material. In 1956, Alfred Gierer and Gerhard Schramm

TABLE 9.1

Examples of DNA- and RNA-Containing Viruses

Virus	Host	Nucleic Acid
Tomato bushy stunt virus	Tomato	RNA
Tobacco mosaic virus	Tobacco	RNA
Influenza virus	Humans	RNA
HIV	Humans	RNA
f2	*E. coli*	RNA
Qß	*E. coli*	RNA
Cauliflower mosaic virus	Cauliflower	DNA
Herpes virus	Humans	DNA
SV40	Primates	DNA
Epstein-Barr virus	Humans	DNA
T2	*E. coli*	DNA
M13	*E. coli*	DNA

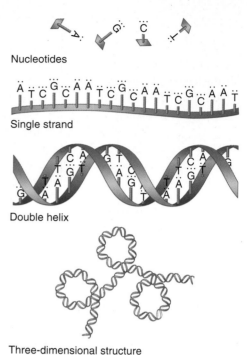

FIGURE 9.7 **Levels of nucleic acid structure.**

isolated RNA from the tobacco mosaic virus (TMV), which infects plant cells. When this purified RNA was applied to plant tissue, the plants developed the same types of lesions that occurred when they were exposed to intact tobacco mosaic viruses. Gierer and Schramm correctly concluded that the viral genome of tobacco mosaic virus is composed of RNA. Since that time, many other viruses have been found to contain RNA as their genetic material. **Table 9.1** compares the genetic compositions of several different types of viruses.

9.2 NUCLEIC ACID STRUCTURE

Geneticists, biochemists, and biophysicists have been interested in the molecular structure of nucleic acids for many decades. Both DNA and RNA are macromolecules composed of smaller building blocks. To fully appreciate their structures, we need to consider four levels of complexity (**Figure 9.7**):

1. **Nucleotides** form the repeating structural unit of nucleic acids.
2. Nucleotides are linked together in a linear manner to form a **strand** of DNA or RNA.
3. Two strands of DNA (and sometimes RNA) can interact with each other to form a **double helix.**
4. The three-dimensional structure of DNA results from the folding and bending of the double helix. Within living cells, DNA is associated with a wide variety of proteins that influence its structure. Chapter 10 examines the roles of these proteins in creating the three-dimensional structure of DNA found within chromosomes.

In this section, we will first examine the structure of individual nucleotides and then progress through the structural features of DNA and RNA. Along the way, we will also review some of the pivotal experiments that led to the discovery of the double helix.

Nucleotides Are the Building Blocks of Nucleic Acids

The **nucleotide** is the repeating structural unit of DNA and RNA. A nucleotide has three components: at least one phosphate group, a pentose sugar, and a nitrogenous base. As shown in **Figure 9.8**, nucleotides can vary with regard to the sugar and the nitrogenous base. The two types of sugars are **deoxyribose** and **ribose,** which are found in DNA and RNA, respectively. The five different bases are subdivided into two categories: the **purines** and the **pyrimidines.** The purine bases, **adenine (A)** and **guanine (G),** contain a double-ring structure; the pyrimidine bases, **thymine (T), cytosine (C),** and **uracil (U),** contain a single-ring structure. The sugar in DNA is always deoxyribose; in RNA, ribose. Also, the base thymine is not found in RNA. Rather, uracil is found in RNA instead of thymine. Adenine, guanine, and cytosine occur in both DNA and RNA. As noted in Figure 9.8, the bases and sugars have a standard numbering system. The nitrogen and carbon atoms found in the ring structure of the bases are given numbers 1 through 9 for the purines and 1 through 6 for the pyrimidines. In comparison, the five carbons found in the sugars are designated with primes, such as 1', to distinguish them from the numbers found in the bases.

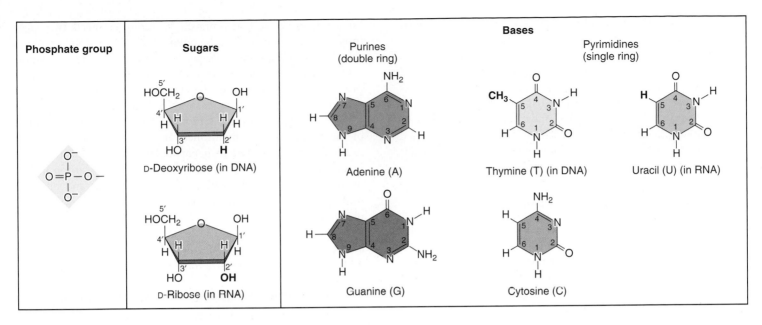

FIGURE 9.8 **The components of nucleotides.** The three building blocks of a nucleotide are one or more phosphate groups, a sugar, and a base. The bases are categorized as purines (adenine and guanine) and pyrimidines (thymine, cytosine, and uracil).

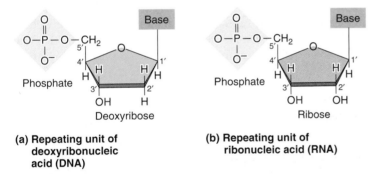

(a) Repeating unit of deoxyribonucleic acid (DNA)

(b) Repeating unit of ribonucleic acid (RNA)

FIGURE 9.9 **The structure of nucleotides found in (a) DNA and (b) RNA.** DNA contains deoxyribose as its sugar and the bases A, T, G, and C. RNA contains ribose as its sugar and the bases A, U, G, and C. In a DNA or RNA strand, the oxygen on the 3' carbon is linked to the phosphorus atom of phosphate in the adjacent nucleotide. The two atoms (O and H) shown in red would be found within individual nucleotides but not when nucleotides are joined together to make strands of DNA and RNA.

Figure 9.9 shows the repeating unit of nucleotides found in DNA and RNA. The locations of the attachment sites of the base and phosphate to the sugar molecule are important to the nucleotide's function. In the sugar ring, carbon atoms are numbered in a clockwise direction, beginning with a carbon atom adjacent to the ring oxygen atom. The fifth carbon is outside the ring structure. In a single nucleotide, the base is always attached to the 1' carbon atom and one or more phosphate groups are attached at the 5' position. As discussed later, the –OH group attached to the 3' carbon is important in allowing nucleotides to form covalent linkages with each other.

The terminology used to describe nucleic acid units is based on three structural features: the type of base, the type of sugar, and the number of phosphate groups. When a base is attached to only a sugar, we call this pair a **nucleoside.** If adenine is attached to ribose, this nucleoside is called adenosine (**Figure 9.10**). Nucleosides containing guanine, thymine, cytosine, or uracil are called guanosine, thymidine, cytidine, and uridine, respectively. When only the bases are attached to deoxyribose, they are called deoxyadenosine,

FIGURE 9.10 **A comparison between the structures of an adenine-containing nucleoside and nucleotides.**

deoxyguanosine, deoxythymidine, and deoxycytidine. The covalent attachment of one or more phosphate molecules to a nucleoside creates a nucleotide. If a nucleotide contains adenine, ribose, and one phosphate, it is adenosine monophosphate, abbreviated AMP. If a nucleotide contains adenine, ribose, and three phosphate groups, it is called adenosine triphosphate, or ATP. If it contains guanine, ribose, and three phosphate groups, it is guanosine triphosphate, or GTP. A nucleotide can also be composed of adenine, deoxyribose, and three phosphate groups. This nucleotide is deoxyadenosine triphosphate (dATP).

Nucleotides Are Linked Together to Form a Strand

A strand of DNA or RNA has nucleotides that are covalently attached to each other in a linear fashion. **Figure 9.11** depicts a short strand of DNA with four nucleotides. A few structural features are worth noting. First, the linkage involves an ester bond between a phosphate group on one nucleotide and the sugar molecule on the adjacent nucleotide. Another way of viewing this

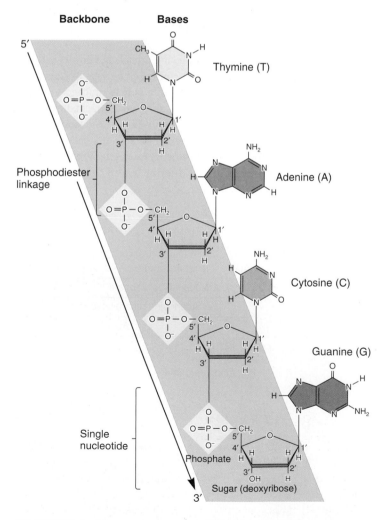

FIGURE 9.11   A short strand of DNA containing four nucleotides. Nucleotides are covalently linked together to form a strand of DNA.

linkage is to notice that a phosphate group connects two sugar molecules. For this reason, the linkage in DNA or RNA strands is called a **phosphodiester linkage.** The phosphates and sugar molecules form the **backbone** of a DNA or RNA strand. The bases project from the backbone. The backbone is negatively charged due to a negative charge on each phosphate.

A second important structural feature is the orientation of the nucleotides. As mentioned, the carbon atoms in a sugar molecule are numbered in a particular way. A phosphodiester linkage involves a phosphate attachment to the 5' carbon in one nucleotide and to the 3' carbon in the other. In a strand, all sugar molecules are oriented in the same direction. As shown in Figure 9.11, the 5' carbons in every sugar molecule are above the 3' carbons. Therefore, a strand has a **directionality** based on the orientation of the sugar molecules within that strand. In Figure 9.11, the direction of the strand is 5' to 3' when going from top to bottom.

A critical aspect regarding DNA and RNA structure is that a strand contains a specific sequence of bases. In Figure 9.11, the sequence of bases is thymine–adenine–cytosine–guanine, abbreviated TACG. Furthermore, to show the directionality, the strand should be abbreviated 5'–TACG–3'. The nucleotides within a strand are covalently attached to each other, so the sequence of bases cannot shuffle around and become rearranged. Therefore, the sequence of bases in a DNA strand will remain the same over time, except in rare cases when mutations occur. As we will see throughout this textbook, the sequence of bases within DNA and RNA is the defining feature that allows them to carry information.

A Few Key Events Led to the Discovery of the Double-Helix Structure

A major discovery in molecular genetics was made in 1953 by James Watson and Francis Crick. At that time, DNA was already known to be composed of nucleotides. However, it was not understood how the nucleotides are bonded together to form the structure of DNA. Watson and Crick committed themselves to determine the structure of DNA because they felt this knowledge was needed to understand the functioning of genes. Other researchers, such as Rosalind Franklin and Maurice Wilkins, shared this view. Before we examine the characteristics of the double helix, let's consider the events that provided the scientific framework for Watson and Crick's breakthrough.

In the early 1950s, Linus Pauling proposed that regions of proteins can fold into a secondary structure known as an α helix (**Figure 9.12a**). To elucidate this structure, Pauling built large models by linking together simple ball-and-stick units (**Figure 9.12b**). By carefully scaling the objects in his models, he could visualize if atoms fit together properly in a complicated three-dimensional structure. Is this approach still used today? The answer is "Yes," except that today researchers construct their three-dimensional models on computers. As we will see, Watson and Crick also used a ball-and-stick approach to solve the structure of the DNA double helix. Interestingly, they were well aware that Pauling might figure out the structure of DNA before they did. This provided a stimulating rivalry between the researchers.

(a) An α helix in a protein   **(b) Linus Pauling**

FIGURE 9.12   **Linus Pauling and the α-helix protein structure.**
(**a**) An α helix is a secondary structure found in proteins. This structure emphasizes the polypeptide backbone (shown as a tan ribbon), which is composed of amino acids linked together in a linear fashion. Hydrogen bonding between hydrogen and oxygen atoms stabilizes the helical conformation. (**b**) Linus Pauling with a ball-and-stick model.

(a) Rosalind Franklin

(b) X-ray diffraction of wet DNA fibers

FIGURE 9.13   **X-ray diffraction of DNA.**

A second important development that led to the elucidation of the double helix was X-ray diffraction data. When a purified substance, such as DNA, is subjected to X-rays, it will produce a well-defined diffraction pattern if the molecule is organized into a regular structural pattern. An interpretation of the diffraction pattern (using mathematical theory) can ultimately provide information concerning the structure of the molecule. Rosalind Franklin (**Figure 9.13a**), working in the same laboratory as Maurice Wilkins, used X-ray diffraction to study wet DNA fibers. Franklin made marked advances in X-ray diffraction techniques while working with DNA. She adjusted her equipment to produce an extremely fine beam of X-rays. She extracted finer DNA fibers than ever before and arranged them in parallel bundles. Franklin also studied the fibers' reactions to humid conditions. The

diffraction pattern of Franklin's DNA fibers is shown in **Figure 9.13b**. This pattern suggested several structural features of DNA. First, it was consistent with a helical structure. Second, the diameter of the helical structure was too wide to be only a single-stranded helix. Finally, the diffraction pattern indicated that the helix contains about 10 base pairs per complete turn. These observations were instrumental in solving the structure of DNA.

EXPERIMENT 9B

Chargaff Found That DNA Has a Biochemical Composition in Which the Amount of A Equals T and the Amount of G Equals C

Another piece of information that led to the discovery of the double-helix structure came from the studies of Erwin Chargaff. In the 1940s and 50s, he pioneered many of the biochemical techniques for the isolation, purification, and measurement of

nucleic acids from living cells. This was not a trivial undertaking, because the biochemical composition of living cells is complex. At the time of Chargaff's work, researchers already knew that the building blocks of DNA are nucleotides containing the bases adenine, thymine, guanine, or cytosine. Chargaff analyzed the base composition of DNA, which was isolated from many different species. He expected that the results might provide important clues concerning the structure of DNA.

The experimental protocol of Chargaff is described in **Figure 9.14**. He began with various types of cells as starting material. The chromosomes were extracted from cells and then treated with protease to separate the DNA from chromosomal proteins. The DNA was then subjected to a strong acid treatment that cleaved the bonds between the sugars and bases. Therefore, the strong acid treatment would release the individual bases from the DNA strands. This mixture of bases was then subjected to paper chromatography to separate the four types. The amounts of the four bases were determined spectroscopically.

■ THE GOAL

An analysis of the base composition of DNA in different organisms may reveal important features about the structure of DNA.

■ ACHIEVING THE GOAL — FIGURE 9.14   An analysis of base composition among different DNA samples.

Starting material: The following types of cells were obtained: *Escherichia coli*, *Streptococcus pneumoniae* (type III), yeast, turtle red blood cells, salmon sperm cells, chicken red blood cells, and human liver cells.

Experimental level   **Conceptual level**

1. For each type of cell, extract the chromosomal material. This can be done in a variety of ways, including the use of high salt, detergent, or mild alkali treatment. Note: The chromosomes contain both DNA and protein.

2. Remove the protein. This can be done in several ways, including treament with protease.

3. Hydrolize the DNA to release the bases from the DNA strands. A common way to do this is by strong acid treatment.

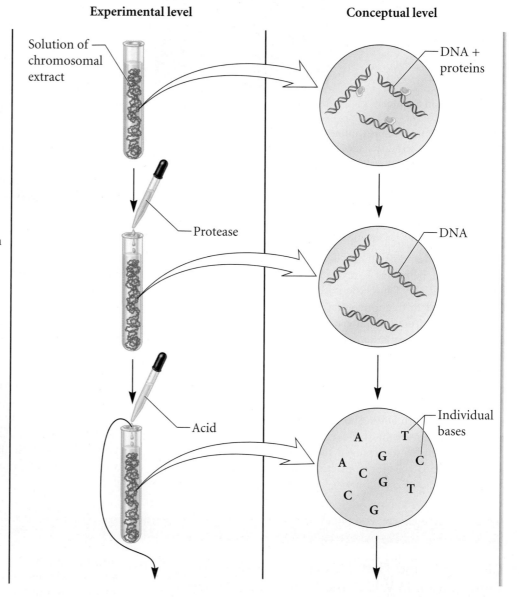

Solution of chromosomal extract

DNA + proteins

Protease

DNA

Acid

Individual bases

4. Separate the bases by chromatography. Paper chromatography provides an easy way to separate the four types of bases. (The technique of chromatography is described in the Appendix.)

5. Extract bands from paper into solutions and determine the amounts of each base by spectroscopy. Each base will absorb light at a particular wavelength. By examining the absorption profile of a sample of base, it is then possible to calculate the amount of the base. (Spectroscopy is described in the Appendix.)

6. Compare the base content in the DNA from different organisms.

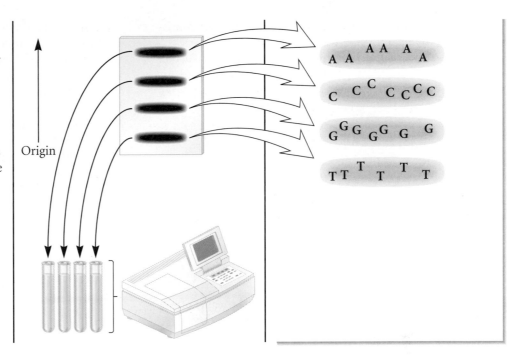

THE DATA

Base Content in the DNA from a Variety of Organisms*

Organism	% of Bases (based on molarity)			
	Adenine	Thymine	Guanine	Cytosine
Escherichia coli	26.0	23.9	24.9	25.2
Streptococcus pneumoniae (type III)	29.8	31.6	20.5	18.0
Yeast	31.7	32.6	18.3	17.4
Turtle red blood cells	28.7	27.9	22.0	21.3
Salmon sperm	29.7	29.1	20.8	20.4
Chicken red blood cells	28.0	28.4	22.0	21.6
Human liver cells	30.3	30.3	19.5	19.9

*When the base contents from different tissues within the same species were measured, similar results were obtained.

INTERPRETING THE DATA

The data shown in Figure 9.14 are only a small sampling of Chargaff's results. During the late 1940s and early 1950s, Chargaff published many papers concerned with the chemical composition of DNA from biological sources. Hundreds of measurements were made. The compelling observation was that the amount of adenine was similar to thymine, and the amount of guanine was similar to cytosine. These results were not sufficient to propose a model for the structure of DNA. However, they provided the important clue that DNA is structured so that each molecule of adenine interacts with thymine, and each molecule of guanine interacts with cytosine. A DNA structure in which A binds to T, and G to C, would explain the equal amounts of A and T, and G and C observed in Chargaff's experiments. As we will see, this observation became crucial evidence that Watson and Crick used to elucidate the structure of the double helix.

A self-help quiz involving this experiment can be found at the Online Learning Center.

Watson and Crick Deduced the Double-Helical Structure of DNA

Thus far, we have examined key pieces of information used to determine the structure of DNA. In particular, the X-ray diffraction work of Franklin suggested a helical structure composed of two or more strands with 10 bases per turn. In addition, the work of Chargaff indicated that the amount of A equals T, and the amount of G equals C. Furthermore, Watson and Crick were familiar with Pauling's success in using ball-and-stick models to deduce the secondary structure of proteins. With these key observations, they set out to solve the structure of DNA.

Watson and Crick assumed DNA is composed of nucleotides that are linked together in a linear fashion. They also assumed the chemical linkage between two nucleotides is always the same. With these ideas in mind, they tried to build ball-and-stick models that incorporated the known experimental observations. Because the diffraction pattern suggested the helix must have two (or more) strands, a critical question was, How could two strands interact? As discussed in his book *The Double Helix*, James Watson noted that in an early attempt at model building, they considered the possibility that the negatively charged phosphate groups, together with magnesium ions, were promoting an interaction between the backbones of DNA strands (**Figure**

9.15). However, more detailed diffraction data were not consistent with this model.

Because the magnesium hypothesis for DNA structure appeared to be incorrect, it was back to the drawing board (or back to the ball-and-stick units) for Watson and Crick. During

FIGURE 9.15   **An incorrect hypothesis for the structure of the DNA double helix.** This illustration shows an early hypothesis of Watson and Crick's, suggesting that two DNA strands interact by a cross-link between the negatively charged phosphate groups in the backbone and divalent Mg^{2+} cations.

this time, Rosalind Franklin had produced even clearer X-ray diffraction patterns, which provided greater detail concerning the relative locations of the bases and backbone of DNA. This was a major breakthrough that suggested a two-strand interaction that was helical. In their model building, the emphasis shifted to models containing the two backbones on the outside of the model, with the bases projecting toward each other. At first, a structure was considered in which the bases form hydrogen bonds with the identical base in the opposite strand (A to A, T to T, G to G, and C to C). However, the model building revealed that the bases could not fit together this way. The final hurdle was overcome when it was realized that the hydrogen bonding of adenine to thymine was structurally similar to that of guanine to cytosine. With an interaction between A and T and between G and C, the ball-and-stick models showed that the two strands would fit together properly. This ball-and-stick model, shown in **Figure 9.16**, was consistent with all the known data regarding DNA structure.

For their work, Watson, Crick, and Maurice Wilkins were awarded the Nobel Prize in 1962. The contribution of Rosalind Franklin to the discovery of the double helix was also critical and has been acknowledged in several books and articles. Franklin

(a) Watson and Crick

FIGURE 9.16   **Watson and Crick and their model of the DNA double helix.** (a) James Watson is shown here on the left and Francis Crick on the right. (b) The molecular model they originally proposed for the double helix. Each strand contains a sugar-phosphate backbone. In opposite strands, A hydrogen bonds to T, and G hydrogen bonds with C.

(b) Original model of the DNA double helix

was independently trying to solve the structure of DNA. However, Wilkins, who worked in the same laboratory, shared Franklin's X-ray data with Watson and Crick, presumably without her knowledge. This provided important information that helped them solve the structure of DNA, which was published in the journal *Nature* in April 1953. Though she was not given credit in the original publication of the double-helix structure, Franklin's key contribution became known in later years. Unfortunately, however, Rosalind Franklin died in 1958, and the Nobel Prize is not awarded posthumously.

The Molecular Structure of the DNA Double Helix Has Several Key Features

The general structural features of the double helix are shown in **Figure 9.17**. In a DNA double helix, two DNA strands are twisted together around a common axis to form a structure that resem-

bles a circular staircase. This double-stranded structure is stabilized by **base pairs (bp)**—pairs of bases in opposite strands that are hydrogen bonded to each other. Counting the bases, if you move past 10 base pairs, you have gone 360° around the backbone. The linear distance of a complete turn is 3.4 nm; each base pair traverses 0.34 nm.

A distinguishing feature of the hydrogen bonding between base pairs is its specificity. An adenine base in one strand hydrogen bonds with a thymine base in the opposite strand, or a guanine base hydrogen bonds with a cytosine. This **AT/GC rule,** also known as Chargaff's rule, explained the earlier data of Chargaff showing that the DNA from many organisms contains equal amounts of A and T, and equal amounts of G and C (see Figure 9.14). The AT/GC rule indicates that purines (A and G) always bond with pyrimidines (T and C). This keeps the width of the double helix relatively constant. As noted in Figure 9.17, three hydrogen bonds occur between G and C but only two between A and T. For this

Key Features

- Two strands of DNA form a right-handed double helix.

- The bases in opposite strands hydrogen bond according to the AT/GC rule.

- The 2 strands are antiparallel with regard to their 5′ to 3′ directionality.

- There are ~10.0 nucleotides in each strand per complete 360° turn of the helix.

FIGURE 9.17   Key features of the structure of the double helix.

reason, DNA sequences that have a high proportion of G and C tend to form more stable double-stranded structures.

The AT/GC rule implies that we can predict the sequence in one DNA strand if the sequence in the opposite strand is known. For example, if one strand has the sequence of 5'–ATG-GCGGATTT–3', then the opposite strand must be 3'–TACC-GCCTAAA–5'. In genetic terms, we would say that these two sequences are **complementary** to each other or that the two sequences exhibit complementarity. In addition, you may have noticed that the sequences are labeled with 5' and 3' ends. These numbers designate the direction of the DNA backbones. The direction of DNA strands is depicted in the inset to Figure 9.17. When going from the top of this figure to the bottom, one strand is running in the 5' to 3' direction, while the other strand is 3' to 5'. This opposite orientation of the two DNA strands is referred to as an **antiparallel** arrangement. An antiparallel structure was initially proposed in the models of Watson and Crick.

Figure 9.18a is a schematic model that emphasizes certain molecular features of DNA structure. The helical structure is formed by the DNA backbone. The bases in this model are depicted as flat rectangular structures that hydrogen bond in pairs. (The hydrogen bonds are the dotted lines.) Although the bases are not actually rectangular, they do form flattened planar structures. Within DNA, the bases are oriented so that the flattened regions are facing each other, an arrangement referred to as base stacking. In other words, if you think of the bases as flat plates, these plates are stacked on top of each other in the double-stranded DNA structure. Along with hydrogen bonding, base stacking is a structural feature that stabilizes the double helix.

By convention, the direction of the DNA double helix shown in Figure 9.18a spirals in a direction that is called "right-handed." To understand this terminology, imagine that a double helix is laid on your desk; one end of the helix is close to you, and the other end is at the opposite side of the desk. As it spirals away from you, a right-handed helix turns in a clockwise direction. By comparison, a left-handed helix would spiral in a counterclockwise manner. Both strands in Figure 9.18a spiral in a right-handed direction.

Figure 9.18b is a space-filling model for DNA in which the atoms are represented by spheres. This model emphasizes the surface features of DNA. Note that the backbone—composed of sugar and phosphate groups—is on the outermost surface. In a living cell, the backbone has the most direct contact with water. In contrast, the bases are more internally located within the double-stranded structure. Biochemists use the term **grooves** to describe the indentations where the atoms of the bases are in contact with the surrounding water. As you travel around the DNA helix, the structure of DNA has two grooves: the **major groove** and the **minor groove.** As will be discussed in later chapters, certain proteins can bind within these grooves and interact with a particular sequence of bases, thereby altering DNA structure and regulating gene expression.

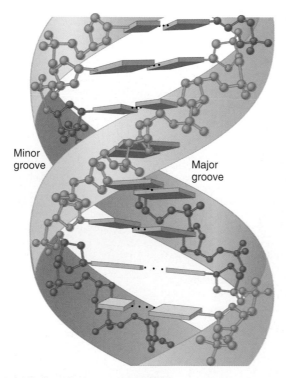

Minor groove

Major groove

(a) Ball-and-stick model of DNA

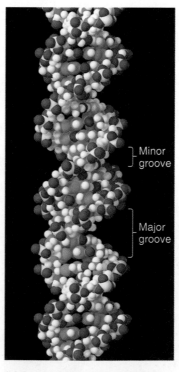

Minor groove

Major groove

(b) Space-filling model of DNA

ONLINE ANIMATION

FIGURE 9.18 **Two models of the double helix.** (a) Ball-and-stick model of the double helix. The ribose–phosphate backbone is shown in detail, while the bases are depicted as flattened rectangles. (b) Space-filling model of the double helix.

DNA Can Form Alternative Types of Double Helices

The DNA double helix can form different types of structures. **Figure 9.19** compares the structures of **A DNA, B DNA,** and **Z DNA.** The highly detailed structures shown here were deduced by X-ray crystallography on short segments of DNA. B DNA is the predominant form of DNA found in living cells. However, under certain in vitro conditions, the two strands of DNA can twist into A DNA or Z DNA, which differ significantly from B DNA. A and B DNA are right-handed helices; Z DNA has a left-handed conformation. In addition, the helical backbone in Z

(a) **Molecular structures**

A DNA   B DNA   Z DNA

Minor groove

Major groove

Backbone

Bases

B DNA   Z DNA

(b) **Space-filling models**

FIGURE 9.19 **Comparison of the structures of A DNA, B DNA, and Z DNA.** (**a**) The highly detailed structures shown here were deduced by X-ray crystallography performed on short segments of DNA. In contrast to the less detailed structures obtained from DNA wet fibers, the diffraction pattern obtained from the crystallization of short segments of DNA provides much greater detail concerning the exact placement of atoms within a double-helical structure. Alexander Rich, Richard Dickerson, and their colleagues were the first researchers to crystallize a short piece of DNA. (**b**) Space-filling models of the B-DNA and Z-DNA structures. In the case of Z DNA, the black lines connect the phosphate groups in the DNA backbone. As seen here, they travel along the backbone in a zigzag pattern.

DNA appears to zigzag slightly as it winds itself around the double-helical structure. The numbers of base pairs per 360° turn are 11.0, 10.0, and 12.0 in A, B, and Z DNA, respectively. In B DNA, the bases tend to be centrally located, and the hydrogen bonds between base pairs occur relatively perpendicular to the central axis. In contrast, the bases in A DNA and Z DNA are substantially tilted relative to the central axis.

The ability of the predominant B DNA to adopt A-DNA and Z-DNA conformations depends on certain conditions. In X-ray diffraction studies, A DNA occurs under conditions of low humidity. The ability of a double helix to adopt a Z-DNA conformation depends on various factors. At high ionic strength (i.e., high salt concentration), formation of a Z-DNA conformation is favored by a sequence of bases that alternates between purines and pyrimidines. One such sequence is

```
5'-GCGCGCGCGCG-3'

3'-CGCGCGCGCGC-5'
```

At lower ionic strength, the methylation of cytosine bases can favor Z-DNA formation. Cytosine **methylation** occurs when a cellular enzyme attaches a methyl group ($-CH_3$) to the cytosine base. In addition, negative supercoiling (a topic discussed in Chapter 10) favors the Z-DNA conformation.

What is the biological significance of A and Z DNA? Research has not found any biological role for A DNA. However, accumulating evidence suggests a possible biological role for Z DNA in the process of transcription. Recent research has identified cellular proteins that specifically recognize Z DNA. In 2005, Alexander Rich and colleagues reported that the Z-DNA binding region of one such protein played a role in regulating the transcription of particular genes. In addition, other research has suggested that Z DNA may play a role in chromosome structure by affecting the level of compaction.

DNA Can Form a Triple Helix, Called Triplex DNA

A surprising discovery made in 1957 by Alexander Rich, David Davies, and Gary Felsenfeld was that DNA can form a triple-helical structure called **triplex DNA.** This triplex was formed in vitro using pieces of DNA that were made synthetically. Although this result was interesting, it seemed to have little, if any, biological relevance.

About 30 years later, interest in triplex DNA was renewed by the observation that triplex DNA can form in vitro by mixing natural double-stranded DNA and a third short strand that is synthetically made. The synthetic strand binds into the major groove of the naturally occurring double-stranded DNA (**Figure 9.20**). As shown here, an interesting feature of triplex DNA formation is that it is sequence specific. In other words, the synthetic third strand incorporates itself into a triple helix due to specific interactions between the synthetic DNA and the biological DNA. The pairing rules are that a thymine in the synthetic DNA will hydrogen bond at an AT pair in the biological DNA and that a cytosine in the synthetic DNA will hydrogen bond at a GC pair.

The formation of triplex DNA has been implicated in several cellular processes, including recombination, which is described in Chapter 17. In addition, researchers are interested

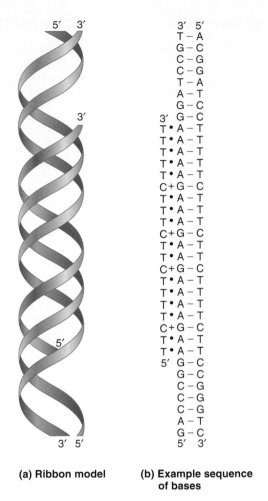

(a) Ribbon model   **(b) Example sequence of bases**

FIGURE 9.20   The structure of triplex DNA. (a) As seen in the ribbon model, the third, synthetic strand binds within the major groove of the double-stranded structure. **(b)** Within triplex DNA, the third strand hydrogen bonds according to the rule T to AT, and C to GC. The cytosine bases in the third strand are protonated (i.e., positively charged).

in triplex DNA due to its potential as a tool to specifically inhibit particular genes. As shown in Figure 9.20, the synthetic DNA strand binds into the major groove according to specific base-pairing rules. Therefore, researchers can design a synthetic DNA to recognize the base sequence found in a particular gene. When the synthetic DNA binds to a gene, it inhibits transcription. In addition, the synthetic DNA can cause mutations in a gene that inactivate its function. Researchers are excited about the possibility of using such synthetic DNA to silence the expression of particular genes. For example, this approach could be used to silence genes that become overactive in cancer cells. However, future research will be needed to develop effective ways to promote the uptake of synthetic DNAs into the appropriate target cells.

The Three-Dimensional Structure of DNA Within Chromosomes Requires Additional Folding and the Association with Proteins

To fit within a living cell, the long double-helical structure of chromosomal DNA must be extensively compacted into a three-dimensional conformation. With the aid of DNA-binding proteins, such as histone proteins, the double helix becomes greatly twisted and folded. **Figure 9.21** depicts the relationship between the DNA double helix and the compaction that occurs within a eukaryotic chromosome. Chapter 10 is devoted to the topic of chromosome organization and the molecular mechanisms responsible for the packaging of genetic material in cells.

RNA Molecules Are Composed of Strands That Fold into Specific Structures

Let's now turn our attention to RNA structure, which bears many similarities to DNA structure. The structure of an RNA strand is much like a DNA strand (**Figure 9.22**). Strands of RNA are typically several hundred or several thousand nucleotides in length, which is much shorter than chromosomal DNA. When RNA is made during transcription, the DNA is used as a template to make a copy of single-stranded RNA. In most cases, only one of the two DNA strands is used as a template for RNA synthesis. Therefore, only one complementary strand of RNA is usually made. Nevertheless, relatively short sequences within one RNA molecule or between two separate RNA molecules can form double-stranded regions.

The helical structure of RNA molecules is due to the ability of complementary regions to form base pairs between A and U and between G and C. This base pairing allows short segments to form a double-stranded region. As shown in **Figure 9.23**, different types of structural patterns are possible. These include bulge loops, internal loops, multibranched junctions, and stem-loops (also called hairpins). These structures contain regions of complementarity punctuated by regions of noncomplementarity. As shown in Figure 9.23, the complementary regions are held together by connecting hydrogen bonds, while the noncomplementary regions have their bases projecting away from the double-stranded region.

Many factors contribute to the structure of RNA molecules. These include the base-paired double-stranded helices, stacking between bases, and hydrogen bonding between bases and backbone regions. In addition, interactions with ions, small molecules, and large proteins may influence RNA structure. **Figure 9.24** depicts the structure of a transfer RNA molecule known as tRNA^{phe}, which is a tRNA molecule that carries the amino acid phenylalanine. It was the first naturally occurring RNA to have its structure elucidated. This RNA molecule has several double-stranded and single-stranded regions. RNA double helices are antiparallel and right-handed with 11 to 12 base pairs per turn. In a living cell, the various regions of an RNA molecule fold and interact with each other to produce the three-dimensional structure.

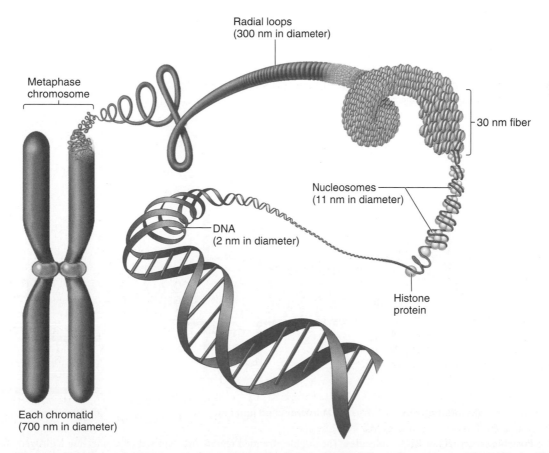

FIGURE 9.21 **The steps in eukaryotic chromosomal compaction leading to the metaphase chromosome.** The DNA double helix is wound around histone proteins and then is further compacted to form a highly condensed metaphase chromosome. The levels of DNA compaction will be described in greater detail in Chapter 10.

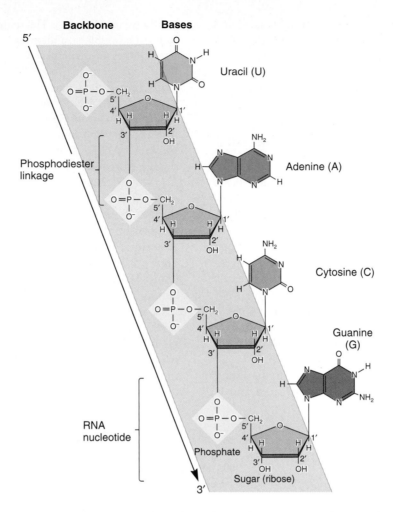

FIGURE 9.22   **A strand of RNA.** This structure is very similar to a DNA strand (see Figure 9.11), except that the sugar is ribose instead of deoxyribose, and uracil is substituted for thymine.

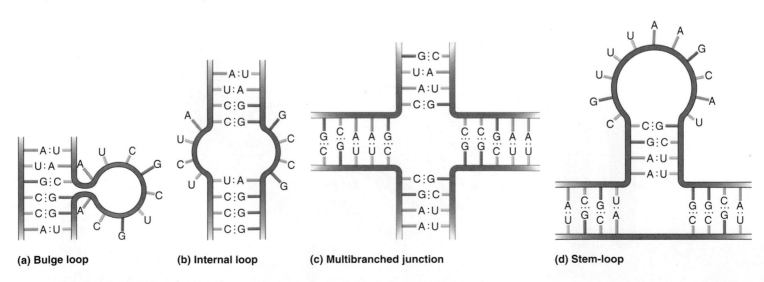

(a) Bulge loop   **(b) Internal loop**   **(c) Multibranched junction**   **(d) Stem-loop**

FIGURE 9.23   **Possible structures of RNA molecules.** The double-stranded regions are depicted by connecting hydrogen bonds. Loops are noncomplementary regions that are not hydrogen bonded with complementary bases. Double-stranded RNA structures can form within a single RNA molecule or between two separate RNA molecules.

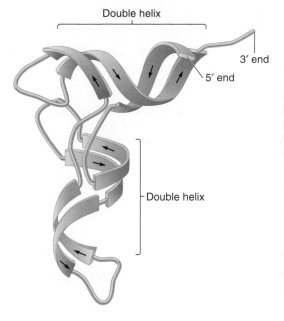

Double helix

3' end
5' end

Double helix

(a) Ribbon model

(b) Space-filling model

FIGURE 9.24 **The structure of tRNA^phe, the transfer RNA molecule that carries phenylalanine.** (a) The double-stranded regions of the molecule are shown as antiparallel ribbons. (b) A space-filling model of tRNA^phe.

CONCEPTUAL SUMMARY

The molecular structure of nucleic acids underlies their function. Nucleotides, which are composed of a phosphate group, pentose sugar, and nitrogenous base, form the repeating structural unit of nucleic acids. A strand of DNA or RNA contains a linear sequence of nucleotides. Strands of DNA (and occasionally RNA) can form a double helix where complementary strands interact due to hydrogen bonding between adenine and thymine (or uracil in RNA) and between guanine and cytosine. Base stacking also stabilizes a double-stranded structure. The most common form of DNA is a right-handed double helix; the backbones in the double helix are composed of sugar–phosphate linkages, with the bases projecting inward from the backbones and hydrogen bonding with each other. The two strands are antiparallel, with one strand running in the 5' to 3' direction, and the other running in the 3' to 5' direction. Three different helical conformations have been identified: A DNA, B DNA, and Z DNA. B DNA is the predominant form found in living cells, but short regions of Z DNA may play an important functional role in transcription. Within chromosomes, DNA is folded into a three-dimensional conformation with the aid of proteins. Short segments within RNA may also form double-helical structures such as bulge loops, internal loops, multibranched junctions, and stem-loops. The structure of RNA is dictated by several factors including double-helical regions, base stacking, hydrogen bonding between bases and backbone, and interactions with other molecules.

EXPERIMENTAL SUMMARY

Two different experimental approaches were used to show that DNA is the genetic material. Griffith's experiments with *S. pneumonia* showed that a "transforming principle," later identified as DNA, was capable of transferring genetic information. Avery, MacLeod, and McCarty purified DNA from type IIIS strains of *S. pneumonia* and used it to transform type IIR strains into type IIIS. Further testing showed that DNase prevented the transformation, whereas RNase and protease did not. These observations indicated that DNA is the transforming substance that provides type IIR bacteria with a new trait. Hershey and Chase showed that T2 bacteriophage injects DNA into bacterial cells, indicating that DNA is the genetic material of this virus. Taken together, the latter two studies were instrumental in convincing scientists that the genetic material is composed of DNA.

Several experimental observations enabled Watson and Crick to determine the structure of DNA. The X-ray diffraction work of Franklin indicated that DNA had a helical structure with two strands and 10 base pairs per 360° turn. Furthermore, the work of Chargaff indicated that the amount of adenine (A) equals that of thymine (T), and the amount of guanine (G) equals that of cytosine (C). With these observations, and employing Pauling's strategy of using ball-and-stick models to visualize the structure of macromolecules, Watson and Crick deduced the structure of DNA, which is a double helix in which the outer backbone is composed of antiparallel strands of alternating deoxyribose and phosphate molecules. The four types of bases are within the interior of the double helix; bases in opposite strands hydrogen bond with each other according to the AT/GC rule.

PROBLEM SETS & INSIGHTS

Solved Problems

S1. A hypothetical sequence at the beginning of an mRNA molecule is

5'–A<u>UUUGCCCUA</u>G<u>CAAAC</u>GUAG<u>CAAAC</u>G...rest of the coding sequence

Using two out of the three underlined sequences, draw two possible models for potential stem-loop structures at the 5' end of this mRNA.

Answer:

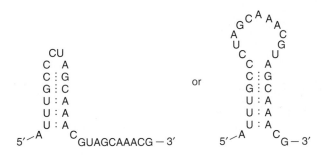

S2. Describe the previous experimental evidence that led Watson and Crick to the discovery of the DNA double helix.

Answer:

1. The chemical structure of single nucleotides was understood by the 1950s.

2. Watson and Crick assumed DNA is composed of nucleotides linked together in a linear fashion to form a strand. They also assumed the chemical linkage between two nucleotides is always the same.

3. Franklin's diffraction patterns suggested several structural features. First, it was consistent with a helical structure. Second, the diameter of the helical structure was too wide to be only a single-stranded helix. Finally, the line spacing on the diffraction pattern indicated the helix contains about 10 base pairs per complete turn.

4. In the chemical analysis of the DNA from different species, the work of Chargaff indicated the amount of adenine equaled the amount of thymine, and the amount of cytosine equaled that of guanine.

5. In the early 1950s, Linus Pauling proposed that regions of proteins can fold into a secondary structure known as an α helix. To discover this, Pauling built large models by linking together simple ball-and-stick units. In this way, he could determine if atoms fit together properly in a complicated three-dimensional structure. A similar approach was used by Watson and Crick to solve the structure of the DNA double helix.

S3. Within living cells, a myriad of different proteins play important functional roles by binding to DNA and RNA. As described throughout your textbook, the dynamic interactions between nucleic acids and proteins lie at the heart of molecular genetics. Some proteins bind to DNA (or RNA) but not in a sequence-specific manner. For example, histones are proteins important in the formation of chromosome structure. In this case, the positively charged histone proteins actually bind to the negatively charged phosphate groups in DNA. In addition, several other proteins interact with DNA but do not require a specific nucleotide sequence to carry out their function. For example, DNA polymerase, which catalyzes the synthesis of new DNA strands, does not bind to DNA in a sequence-dependent manner. By comparison, many other proteins do interact with nucleic acids in a sequence-dependent fashion. This means that a specific sequence of bases can provide a structure that will be recognized by a particular protein. Throughout the textbook, the functions of many of these proteins will be described. Some examples include transcription factors that affect the rate of transcription, proteins that bind to centromeres, and proteins that bind to origins of replication. With regard to the three-dimensional structure of DNA, where would you expect DNA-binding proteins to bind if they recognize a specific base sequence? What about DNA-binding proteins that do not recognize a base sequence?

Answer: DNA-binding proteins that recognize a base sequence must bind into a major or minor groove of the DNA, which is where the bases would be accessible to a DNA-binding protein. Most DNA-binding proteins, which recognize a base sequence, fit into the major groove. By comparison, other DNA-binding proteins, such as histones, which do not recognize a base sequence, bind to the DNA backbone.

S4. The formation of a double-stranded structure must obey the rule that adenine hydrogen bonds to thymine (or uracil) and cytosine hydrogen bonds to guanine. Based on your previous understanding of genetics (from this course or a general biology course), discuss reasons why complementarity is an important feature of DNA and RNA structure and function.

Answer: Note: Many of the topics described below are discussed in Chapters 10 through 13. One way that complementarity underlies function is that it provides the basis for the synthesis of new strands of DNA and RNA. During replication, the synthesis of the new DNA strands occurs in such a way that adenine hydrogen bonds to thymine, and cytosine hydrogen bonds to guanine. In other words, the molecular feature of a complementary double-stranded structure makes it possible to produce exact copies of DNA. Likewise, the ability to transcribe DNA into RNA is based on complementarity. During transcription, one strand of DNA is used as a template to make a complementary strand of RNA.

In addition to the synthesis of new strands of DNA and RNA, complementarity is important in other ways. As mentioned in this chapter, the folding of RNA into a particular structure is driven by the hydrogen bonding of complementary regions. This event is necessary to produce functionally active tRNA molecules. Likewise, stem-loop structures also occur in other types of RNA. For example, the rapid formation of stem-loop structures is known to occur as RNA is being transcribed and to affect the termination of transcription.

A third way that complementarity can be functionally important is that it can promote the interaction of two separate RNA molecules. During translation, codons in mRNA bind to the anticodons in tRNA (see Chapter 13). This binding is due to complementarity. For example, if a codon is 5'–AGG–3', the anticodon is 3'–UCC–5'. This type of specific interaction between codons and anticodons is an important step that enables the nucleotide sequence in mRNA to code for an amino acid sequence within a protein. In addition, many other examples of RNA-RNA interactions are known and will be described throughout this textbook.

S5. An important feature of triplex DNA formation is that it is sequence specific. The synthetic third strand incorporates itself into a triple helix, so a thymine in the synthetic DNA will bind near an AT pair in the biological DNA, and a cytosine in the synthetic DNA will bind near a GC pair. From a practical point of view, this opens the possibility of synthesizing a short strand of DNA that will form a triple helix at a particular target site. For example, if the sequence of a particular gene is known, researchers can make a synthetic piece of DNA that will form a triple helix somewhere within that gene according to the T to AT, and C to GC rule. Triplex DNA formation is known to inhibit gene transcription. In other words, when the synthetic DNA binds within the DNA of a gene, the formation of triplex DNA prevents that gene from being transcribed into RNA. Discuss how this observation might be used to combat diseases.

Answer: Triplex DNA formation opens the exciting possibility of designing synthetic pieces of DNA to inhibit the expression of particular genes. Theoretically, such a tool could be used to combat viral diseases or to inhibit the growth of cancer cells. To combat a viral disease, a synthetic DNA could be made that specifically binds to an essential viral gene and thereby prevents viral proliferation. To inhibit cancer, a synthetic DNA could be made to bind to an oncogene. (Note: As described in Chapter 22, an oncogene is a gene that promotes cancerous growth.) Inhibition of an oncogene could prevent cancer. At this point, a primary obstacle in applying this approach is devising a method of getting the synthetic DNA into living cells.

Conceptual Questions

C1. What is the meaning of the term genetic material?

C2. After the DNA from type IIIS bacteria is exposed to type IIR bacteria, list all of the steps that you think must occur for the bacteria to start making a type IIIS capsule.

C3. Look up the meaning of the word transformation in a dictionary and explain whether it is an appropriate word to describe the transfer of genetic material from one organism to another.

C4. What are the building blocks of a nucleotide? With regard to the 5' and 3' positions on a sugar molecule, how are nucleotides linked together to form a strand of DNA?

C5. Draw the structure of guanine, guanosine, and deoxyguanosine triphosphate.

C6. Draw the structure of a phosphodiester linkage.

C7. Describe how bases interact with each other in the double helix. This discussion should address the issues of complementarity, hydrogen bonding, and base stacking.

C8. If one DNA strand is 5'–GGCATTACACTAGGCCT–3' what is the sequence of the complementary strand?

C9. What is meant by the term DNA sequence?

C10. Make a side-by-side drawing of two DNA helices, one with 10 base pairs (bp) per 360° turn and the other with 15 bp per 360° turn.

C11. Discuss the differences in the structural features of A DNA, B DNA, and Z DNA.

C12. What parts of a nucleotide (namely, phosphate, sugar, and/or bases) occupy the major and minor grooves of double-stranded DNA, and what parts are found in the DNA backbone? If a DNA-binding protein does not recognize a specific nucleotide sequence, do you expect that it recognizes the major groove, the minor groove, or the DNA backbone? Explain.

C13. List the structural differences between DNA and RNA.

C14. Draw the structure of deoxyribose and number the carbon atoms. Describe the numbering of the carbon atoms in deoxyribose with regard to the directionality of a DNA strand. In a DNA double helix, what does the term antiparallel mean?

C15. Write out a sequence of an RNA molecule that could form a stem-loop with 24 nucleotides in the stem and 16 nucleotides in the loop.

C16. Compare the structural features of a double-stranded RNA structure with those of a DNA double helix.

C17. Which of the following DNA double helices would be more difficult to separate into single-stranded molecules by treatment with heat, which breaks hydrogen bonds?

A. GGCGTACCAGCGCAT

   CCGCATGGTCGCGTA

B. ATACGATTTACGAGA

   TATGCTAAATGCTCT

Explain your choice.

C18. What structural feature allows DNA to store information?

C19. Discuss the structural significance of complementarity in DNA and in RNA.

C20. An organism has a G + C content of 64% in its DNA. What are the percentages of A, T, G, and C?

C21. Let's suppose you have recently identified an organism that was scraped from an asteroid that hit the earth. (Fortunately, no one was injured.) When you analyze this organism, you discover that its DNA is a triple helix, composed of six different nucleotides: A, T, G, C, X, and Y. You measure the chemical composition of the bases and find the following amounts of these six bases: A = 24%, T = 23%, G = 11%, C = 12%, X = 21%, Y = 9%. What rules would you propose govern triplex DNA formation in this organism? Note: There is more than one possibility.

C22. Upon further analysis of the DNA described in conceptual question C21, you discover that the triplex DNA in this alien organism is composed of a double helix, with the third helix wound within the major groove (just like the DNA in Figure 9.20). How would you propose that this DNA is able to replicate itself? In your answer, be specific about the base pairing rules within the double helix and which part of the triplex DNA would be replicated first.

C23. A DNA-binding protein recognizes the following double-stranded sequence:

   5'–GCCCGGGC–3'

   3'–CGGGCCCG–5'

This type of double-stranded structure could also occur within the stem region of an RNA stem-loop molecule. Discuss the structural differences between RNA and DNA that might prevent this DNA-binding protein from recognizing a double-stranded RNA molecule.

C24. Within a protein, certain amino acids are positively charged (e.g., lysine and arginine), some are negatively charged (e.g., glutamate and aspartate), some are polar but uncharged, and some are nonpolar. If you knew that a DNA-binding protein was recognizing the DNA backbone rather than base sequences, which amino acids in the protein would be good candidates for interacting with the DNA?

C25. In what ways are the structures of an α helix in proteins and the DNA double helix similar, and in what ways are they different?

C26. A double-stranded DNA molecule contains 560 nucleotides. How many complete turns would be found in this double helix?

C27. As the minor and major grooves of the DNA wind around a DNA double helix, do they ever intersect each other, or do they always run parallel to each other?

C28. What chemical group (phosphate group, hydroxyl group, or a nitrogenous base) is found at the 3' end of a DNA strand? What group is found at the 5' end?

C29. The base composition of an RNA virus was analyzed and found to be 14.1% A, 14.0% U, 36.2% G, and 35.7% C. Would you conclude that the viral genetic material is single-stranded RNA or double-stranded RNA?

C30. The genetic material found within some viruses is single-stranded DNA. Would this genetic material contain equal amounts of A and T and equal amounts of G and C?

C31. A medium-sized human chromosome contains about 100 million bp. If the DNA were stretched out in a linear manner, how long would it be?

C32. A double-stranded DNA molecule is 1 cm long, and the percentage of adenine is 15%. How many cytosines would be found in this DNA molecule?

C33. Could single-stranded DNA form a stem-loop structure? Why or why not?

C34. As described in Chapter 15, the methylation of cytosine bases can have an important impact on gene expression. For example, the methylation of cytosines may inhibit the transcription of genes. A methylated cytosine base has the following structure:

Would you expect the methylation of cytosine to affect the hydrogen bonding between cytosine and guanine in a DNA double helix? Why or why not? (Hint: See Figure 9.17 for help.) Take a look at solved problem S3 and speculate as to how methylation could affect gene expression.

C35. An RNA molecule has the following sequence:

Region 1   Region 2   Region 3
5'-CAUCC AUCCAUUCCCC AUCCGAUAA GGGGAAUGG AUCC GAAUGGAU AAC-3'

Parts of region 1 can form a stem-loop with region 2 and with region 3. Can region 1 form a stem-loop with region 2 and region 3 at the same time? Why or why not? Which stem-loop would you predict to be more stable: a region 1/region 2 interaction or a region 1/region 3 interaction? Explain your choice.

Experimental Questions

E1. Genetic material acts as a blueprint for an organism's traits. Explain how the experiments of Griffith indicated that genetic material was being transferred to the type IIR bacteria.

E2. With regard to the experiment described in Figure 9.3, answer the following:

A. List several possible reasons why only a small percentage of the type IIR bacteria was converted to type IIIS.

B. Explain why an antibody must be used to remove the bacteria that are not transformed. What would the results look like, in all five cases, if the antibody/centrifugation step had not been included in the experimental procedure?

C. The DNA extract was treated with DNase, RNase, or protease. Why was this done? (In other words, what were the researchers trying to demonstrate?)

E3. An interesting trait that some bacteria exhibit is resistance to killing by antibiotics. For example, certain strains of bacteria are resistant to tetracycline, whereas other strains are sensitive to tetracycline. Describe an experiment you would carry out to demonstrate that tetracycline resistance is an inherited trait encoded by the DNA of the resistant strain.

E4. With regard to the experiment of Figure 9.6, answer the following:

A. Provide possible explanations why some of the DNA is in the supernatant.

B. Plot the results if the radioactivity in the pellet, rather than in the supernatant, had been measured.

C. Why were ^{32}P and ^{35}S chosen as radioisotopes to label the phages?

D. List possible reasons why less than 100% of the phage protein was removed from the bacterial cells during the shearing process.

E5. Does the experiment of Figure 9.6 rule out the possibility that RNA is the genetic material of T2 phage? Explain your answer. If it does not, could you modify the approach of Hershey and Chase to show that it is DNA and not RNA that is the genetic material of T2 bacteriophage? Note: It is possible to specifically label DNA or RNA by providing bacteria with radiolabeled thymine or uracil, respectively.

E6. In Chapter 9, we considered two experiments—one by Avery, MacLeod, and McCarty and the second by Hershey and Chase—that indicated DNA is the genetic material. Discuss the strengths and weaknesses of the two approaches. Which experimental approach did you find the most convincing? Why?

E7. The type of model building used by Pauling and Watson and Crick involved the use of ball-and-stick units. Now we can do model building on a computer screen. Even though you may not be familiar with this approach, discuss potential advantages that computers might provide in molecular model building.

E8. With regard to Chargaff's experiment described in Figure 9.14, answer the following:

A. What is the purpose of paper chromatography?

B. Explain why it is necessary to remove the bases in order to determine the base composition of DNA.

C. Would Chargaff's experiments have been convincing if they had been done on only one species? Discuss.

E9. Gierer and Schramm exposed plant tissue to purified RNA from tobacco mosaic virus, and the plants developed the same types of lesions as if they were exposed to the virus itself. What would be the results if the RNA was treated with DNase, RNase, or protease prior to its exposure to the plant tissue?

Questions for Student Discussion/Collaboration

1. Try to propose structures for a genetic material that are substantially different from the double helix. Remember that the genetic material must have a way to store information and a way to be faithfully replicated.

2. How might you provide evidence that DNA is the genetic material in mice?

Note: All answers appear at the website for this textbook; the answers to even-numbered questions are in the back of the textbook.

www.mhhe.com/brookergenetics3e

Visit the Online Learning Center for practice tests, answer keys, and other learning aids for this chapter. Enhance your understanding of genetics with our interactive exercises, quizzes, animations, and much more.

10

CHROMOSOME ORGANIZATION AND MOLECULAR STRUCTURE

Chromosomes are the structures within cells that contain the genetic material. The term **genome** describes all the types of genetic material that an organism has. For bacteria, the genome is typically a single circular chromosome. For eukaryotes, the nuclear genome refers to one complete set of chromosomes that resides in the cell nucleus. In other words, the haploid complement of chromosomes is considered a nuclear genome. Eukaryotes have a mitochondrial genome, and plants also have a chloroplast genome. Unless otherwise noted, the term eukaryotic genome refers to the nuclear genome.

The primary function of the genetic material is to store the information needed to produce the characteristics of an organism. As we have seen in Chapter 9, the sequence of bases in a DNA molecule can store information. To fulfill their role at the molecular level, chromosomal sequences facilitate four important processes: (1) the synthesis of RNA and cellular proteins, (2) the replication of chromosomes, (3) the proper segregation of chromosomes, and (4) the compaction of chromosomes so they can fit within living cells. In this chapter, we will examine the general organization of the genetic material within viral, bacterial, and eukaryotic chromosomes. In addition, the molecular mechanisms that account for the packaging of the genetic material in viruses, bacteria, and eukaryotic cells will be described. We will begin by considering the comparatively simple genomes of viruses.

10.1 VIRAL GENOMES

Viruses are small infectious particles that contain nucleic acid as their genetic material, surrounded by a protein coat, or capsid (**Figure 10.1a**). The capsid of **bacteriophages,** which are viruses that infect bacteria, may also contain a sheath, base plate, and tail fibers (see Figure 9.4). Certain eukaryotic viruses also have an envelope consisting of a membrane embedded with spike proteins (**Figure 10.1b**). By themselves, viruses are not cellular organisms. They do not contain energy-producing enzymes, ribosomes, or cellular organelles. Instead, viruses rely on their **host cells**—the cells they infect—for replication. In general, most viruses exhibit a limited **host range,** the spectrum of host species that a virus can infect. Many viruses can infect only specific types of cells of one host species. Depending on the life cycle of the virus, the host cell may or may not be destroyed during the process of viral replication and release. In this section, we will consider the genetic composition of viruses and examine how viral genomes are packaged into virus particles.

Structure of a bacterial chromosome. This is an electron micrograph of a bacterial chromosome, which has been released from a bacterial cell.

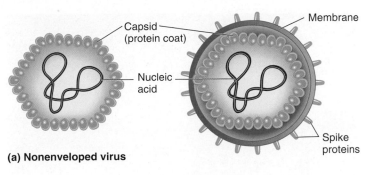

(a) Nonenveloped virus

(b) Enveloped virus with spikes

FIGURE 10.1 **General structure of viruses.** (a) The simplest viruses contain a nucleic acid molecule (DNA or RNA) surrounded by a capsid, or protein coat. (b) Other viruses also contain an envelope composed of a membrane and spike proteins. The membrane is obtained from the host cell when the virus buds through the plasma membrane.

Viral Genomes Are Relatively Small and Are Composed of DNA or RNA

A **viral genome** is the genetic material that a virus contains. The term viral chromosome is also used to describe the viral genome. Surprisingly, the nucleic acid composition of viral genomes varies markedly among different types of viruses. **Table 10.1** describes the genome characteristics of a few selected viruses. The genome can be DNA or RNA, but not both. In some cases, it is single stranded, whereas in others, it is double stranded. Depending on the type of virus, the genome can be linear or circular.

As shown in Table 10.1, viral genomes vary in size from several thousand to more than a hundred thousand nucleotides in length. For example, the genomes of some simple viruses, such as Qβ virus, are only a few thousand nucleotides in length and contain only a few genes. Other viruses, particularly those with a complex structure, have many more genes. The T even phages (T2, T4, etc.), discussed in Chapters 6 and 9, are examples of more complex viruses.

Viral Genomes Are Packaged into the Capsid in an Assembly Process

In an infected cell, the life cycle of the virus eventually leads to the synthesis of viral nucleic acids and proteins. Newly synthesized viral chromosomes and capsid proteins must then come together and assemble to make mature virus particles. Viruses with a simple structure may self-assemble, which means that the nucleic acid and capsid proteins spontaneously bind to each other to form a mature virus. The structure of one self-assembling virus, the tobacco mosaic virus, is shown in **Figure 10.2**. As shown here, the proteins assemble around the RNA genome, which becomes trapped inside the hollow capsid. This assembly process can occur in vitro if purified capsid proteins and RNA are mixed together.

Some viruses, such as T2 bacteriophage, have more complicated structures that do not self-assemble. The correct assembly of this virus requires the help of proteins not found within the mature virus particle itself. When virus assembly requires the participation of noncapsid proteins, the process is called directed assembly, because the noncapsid proteins direct the proper assembly of the virus. What are the functions of these noncapsid proteins? Some proteins, called scaffolding proteins, catalyze the assembly process and are transiently associated with the capsid. However, as viral assembly nears completion, the scaffolding proteins are expelled from the mature virus. In addition, other noncapsid proteins act as proteases that specifically cleave viral

TABLE 10.1

Characteristics of Selected Viral Genomes

Virus	Host	Type of Nucleic Acid*	Size**	Number of Genes***
Parvovirus	Mammals	ssDNA	5.0	5
Fd	*E. coli*	ssDNA	6.4	10
Lambda	*E. coli*	dsDNA	48.5	71
T4	*E. coli*	dsDNA	169.0	288
Qβ	*E. coli*	ssRNA	4.2	4
TMV	Many plants	ssRNA	6.4	6
Influenza virus	Mammals	ssRNA	13.5	11
Human immuno-deficiency virus (HIV)	Primates	ssRNA	9.7	9
Herpes simplex virus, type 2 (genital herpes)	Humans	dsDNA	158.4	77

*ss refers to single stranded, and ds refers to double stranded.
**Number of thousands of nucleotides or nucleotide base pairs
***This number refers to the number of protein-encoding units. In some cases, two or more proteins are made from a single gene due to events such as protein processing.

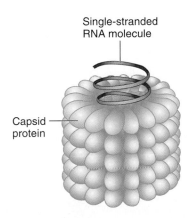

FIGURE 10.2 **Structure of the tobacco mosaic virus.** This self-assembling virus is composed of a coiled RNA molecule surrounded by 2,130 identical protein subunits. Only a portion of the tobacco mosaic virus is shown here. Several layers of proteins have been omitted from this illustration to reveal the RNA genome, which is trapped inside the protein coat.

capsid proteins. This cleavage produces a capsid protein that is somewhat smaller and able to assemble correctly. For many viruses, the cleavage of capsid proteins into smaller units is an important event that precedes viral assembly.

10.2 BACTERIAL CHROMOSOMES

Let's now turn our attention to the organization of chromosomes found in bacterial species. Inside a bacterial cell, the chromosome is highly compacted and found within a region of the cell known as the **nucleoid.** Although bacteria usually contain a single type of chromosome, more than one copy of that chromosome may be found within one bacterial cell. Depending on the growth conditions and phase of the cell cycle, bacteria may have one to four identical chromosomes per cell. In addition, the number of copies varies depending on the bacterial species. As shown in **Figure 10.3,** each chromosome occupies its own distinct nucleoid region within the cell. Unlike the eukaryotic nucleus, the bacterial nucleoid is not a separate cellular compartment bounded by a membrane. Rather, the DNA in a nucleoid is in direct contact with the cytoplasm of the cell.

In this section, we will explore two important features of bacterial chromosomes. First, the organization of DNA sequences along the chromosome will be examined. Second, we will consider the mechanisms that cause the chromosome to become a compacted structure within a nucleoid of the bacterium.

Bacterial Chromosomes Contain a Few Thousand Gene Sequences Interspersed with Other Functionally Important Sequences

Bacterial chromosomal DNA is usually a circular molecule, though some bacteria have linear chromosomes. A typical chro-

FIGURE 10.3 **The localization of nucleoids within** *Bacillus subtilis* **bacteria.** The nucleoids are fluorescently labeled and seen as bright, oval-shaped regions within the bacterial cytoplasm. Note that two or more nucleoids are found within each cell. Some of the cells seen here are in the process of dividing.

mosome is a few million base pairs (bp) in length. For example, the chromosome of one strain of *Escherichia coli* has approximately 4.6 million bp, and the *Haemophilus influenzae* chromosome has roughly 1.8 million bp. A bacterial chromosome commonly has a few thousand different genes. These genes are interspersed throughout the entire chromosome (**Figure 10.4**). **Structural genes**—nucleotide sequences that encode proteins—account for the majority of bacterial DNA. The nontranscribed regions of DNA located between adjacent genes are termed **intergenic regions.**

Other sequences in chromosomal DNA influence DNA replication, gene transcription, and chromosome structure. For

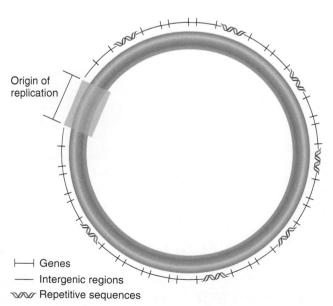

Key features:

- Most, but not all, bacterial species contain circular chromosomal DNA.

- A typical chromosome is a few million base pairs in length.

- Most bacterial species contain a single type of chromosome, but it may be present in multiple copies.

- Several thousand different genes are interspersed throughout the chromosome. The short regions between adjacent genes are called intergenic regions.

- One origin of replication is required to initiate DNA replication.

- Repetitive sequences may be interspersed throughout the chromosome.

FIGURE 10.4   Organization of sequences in bacterial chromosomal DNA.

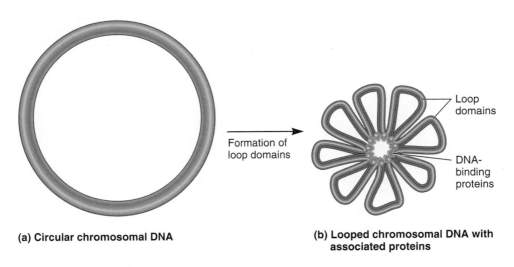

(a) Circular chromosomal DNA

(b) Looped chromosomal DNA with associated proteins

FIGURE 10.5   **The formation of loop domains within the bacterial chromosome.** To promote compaction, (**a**) the large, circular chromosomal DNA is organized into (**b**) smaller, looped chromosomal DNA with loop domains and associated proteins.

example, bacterial chromosomes have one **origin of replication,** a sequence that is a few hundred nucleotides in length. This nucleotide sequence functions as an initiation site for the assembly of several proteins required for DNA replication. Also, a variety of **repetitive sequences** have been identified in many bacterial species. These sequences are found in multiple copies and are usually interspersed within the intergenic regions throughout the bacterial chromosome. Repetitive sequences may play a role in a variety of genetic processes, including DNA folding, DNA replication, gene regulation, and genetic recombination. As discussed in Chapter 17, some repetitive sequences are transposable elements that can move throughout the genome. Figure 10.4 summarizes the key features of sequence organization within bacterial chromosomes.

The Formation of Chromosomal Loops Helps Make the Bacterial Chromosome More Compact

To fit within the bacterial cell, the chromosomal DNA must be compacted about 1,000-fold. Part of this compaction process involves the formation of **loop domains** within the bacterial chromosome (**Figure 10.5**). As its name suggests, a loop domain is a segment of chromosomal DNA folded into a structure that resembles a loop. DNA-binding proteins anchor the base of the loops in place. The number of loops varies according to the size of the bacterial chromosome and the species. In *E. coli*, a chromosome has 50 to 100 loop domains with about 40,000 to 80,000 bp of DNA in each loop. This looped structure compacts the circular chromosome about 10-fold.

DNA Supercoiling Further Compacts the Bacterial Chromosome

Because DNA is a long thin molecule, twisting forces can dramatically change its conformation. This effect is similar to twisting a rubber band. If twisted in one direction, a rubber band will even-

tually coil itself into a compact structure as it absorbs the energy applied by the twisting motion. Because the two strands within DNA already coil around each other, the formation of additional coils due to twisting forces is referred to as **DNA supercoiling** (**Figure 10.6**).

How do twisting forces affect DNA structure? **Figure 10.7** illustrates four possibilities. In Figure 10.7a, a double-stranded DNA molecule with five complete turns is anchored between two plates. In this hypothetical example, the ends of the DNA molecule cannot rotate freely. Both underwinding and overwinding of the DNA double helix can induce supercoiling of the helix. Because B DNA is a right-handed helix, underwinding is a left-handed twisting motion, and overwinding is a right-handed twist. Along the left side of Figure 10.7, one of the plates has been given a turn in the direction that tends to unwind the helix. As the helix absorbs this force, two things can happen. The underwinding motion can cause fewer turns (Figure 10.7b) or cause a negative supercoil to form (Figure 10.7c). On the right side of

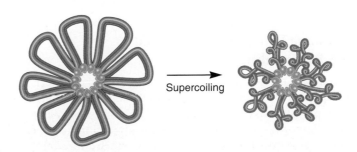

(a) Looped chromosomal DNA

(b) Looped and supercoiled DNA

FIGURE 10.6   **DNA supercoiling leads to further compaction of the looped chromosomal DNA.** (**a**) The looped chromosomal DNA becomes much more compacted due to (**b**) supercoiling within the loops.

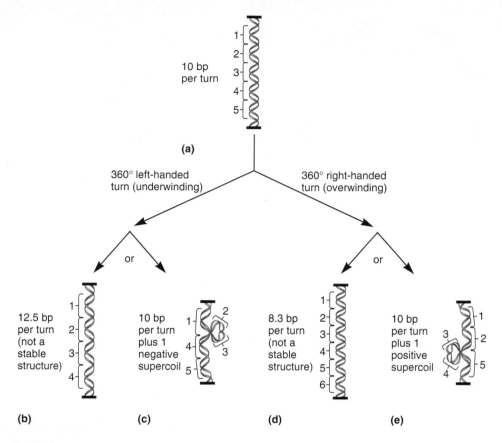

FIGURE 10.7   **Schematic representation of DNA supercoiling.** In this example, the DNA in (**a**) is anchored between two plates and given a twist as noted by the arrows. A left-handed twist (underwinding) could produce either (**b**) fewer turns or (**c**) a negative supercoil. A right-handed twist (overwinding) produces (**d**) more turns or (**e**) a positive supercoil. The structures shown in (b) and (d) are unstable.

Figure 10.7, one of the plates has been given a right-handed turn, which overwinds the double helix. This can lead to either more turns (Figure 10.7d) or the formation of a positive supercoil (Figure 10.7e). The DNA conformations shown in Figure 10.7a, c, and e differ only with regard to supercoiling. These three DNA conformations are referred to as **topoisomers** of each other. The DNA conformations shown in Figure 10.7b and d are not structurally favorable and would not occur in living cells.

Chromosome Function Is Influenced by DNA Supercoiling

The chromosomal DNA in living bacteria is negatively supercoiled. In the chromosome of *E. coli*, about one negative supercoil occurs per 40 turns of the double helix. Negative supercoiling has several important consequences. As already mentioned, the supercoiling of chromosomal DNA makes it much more compact (see Figure 10.6). Therefore, supercoiling helps to greatly decrease the size of the bacterial chromosome. In addition, negative supercoiling also affects DNA function. To understand how, remember that negative supercoiling is due to an underwinding force on the DNA. Therefore, negative supercoiling creates tension on the DNA strands that may be released by DNA strand

separation (**Figure 10.8**). Although most of the chromosomal DNA is negatively supercoiled and compact, the force of negative supercoiling may promote DNA strand separation in small regions. This enhances genetic activities such as replication and transcription that require the DNA strands to be separated.

How does bacterial DNA become supercoiled? Two bacterial enzymes are primarily responsible for the level of supercoiling within living bacteria. In 1976, Martin Gellert and collaborators discovered the enzyme **DNA gyrase,** also known as topoisomerase II. This enzyme, which contains four subunits, introduces negative supercoils into DNA using energy from ATP (**Figure 10.9**) or relaxes positive supercoils when they occur. A second enzyme, **topoisomerase I,** can relax negative supercoils. The competing actions of DNA gyrase and topoisomerase I govern the overall supercoiling of the bacterial DNA.

The ability of gyrase to introduce negative supercoils into DNA is critical for bacteria to survive. For this reason, much research has been aimed at identifying drugs that will specifically block bacterial gyrase function as a way to cure or alleviate diseases caused by bacteria. Two main classes, quinolones and coumarins, inhibit gyrase and other bacterial topoisomerases and thereby block bacterial cell growth. These drugs do not inhibit eukaryotic topoisomerases, which are structurally different from

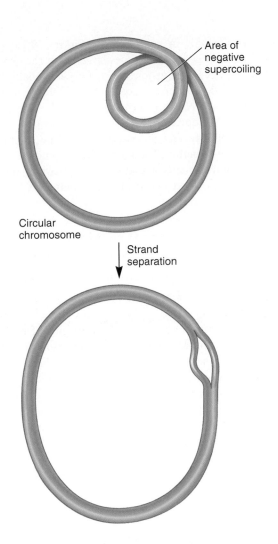

FIGURE 10.8   **Negative supercoiling promotes strand separation.**

their bacterial counterparts. This finding has been the basis for the production of many drugs with important antibacterial applications. An example is ciprofloxacin (e.g., Cipro®), which is used to treat a wide spectrum of bacterial diseases, including anthrax.

10.3 EUKARYOTIC CHROMOSOMES

Eukaryotic species have one or more sets of chromosomes; each set is composed of several different linear chromosomes (refer back to Figure 8.1). Humans, for example, have two sets of 23

FIGURE 10.9   The action of DNA gyrase. DNA gyrase, also known as topoisomerase II, is composed of two A and two B subunits. The A subunits cleave the DNA. The unbroken segment of DNA then passes through the break, and the break is repaired. The B subunits capture the energy from ATP hydrolysis to catalyze this process. The result is that two negative turns have been introduced into the DNA molecule. In this example, the DNA originally contained one positive supercoil, and now it has one net negative supercoil.

chromosomes each, for a total of 46. The total amount of DNA in cells of eukaryotic species is usually much greater than that in bacterial cells. This enables eukaryotic genomes to contain many more genes than their bacterial counterparts. A distinguishing feature of eukaryotic cells is that their chromosomes are located within a separate cellular compartment known as the **nucleus.** To fit within the nucleus, the length of DNA must be compacted by a remarkable amount. As in bacterial chromosomes, this is accomplished by the binding of the DNA to many different cellular proteins. The term **chromatin** is used to describe the DNA-protein complex found within eukaryotic chromosomes. Chromatin is a dynamic structure that can change its shape and composition during the life of a cell.

In this section, we will examine the sizes of eukaryotic genomes and the organization of DNA sequences along the length of eukaryotic chromosomes. We will consider techniques to analyze the complexity of DNA sequences found in chromosomes. We will conclude by examining the mechanisms that account for the compaction of eukaryotic chromosomes during different stages of the cell cycle.

The Sizes of Eukaryotic Genomes Vary Substantially

When comparing different eukaryotic species, dramatic variation in genome size is often observed (**Figure 10.10a**; note that this is a log scale). In many cases, this variation is not related to the complexity of the species. For example, two closely related species of salamander, *Plethodon richmondi* and *Plethodon lar-*

selli, differ considerably in genome size (**Figure 10.10b,c**). The genome of *P. larselli* is over two times larger than the genome of *P. richmondi.* However, the genome of *P. larselli* probably doesn't contain more genes. How do we explain the difference in genome size? The additional DNA in *P. larselli* is due to the accumulation of repetitive DNA sequences present in many copies. In some species, these repetitive sequences can accumulate to enormous levels. Such highly repetitive sequences do not encode proteins, and their function remains a matter of controversy and great interest. The structure and significance of repetitive DNA will be discussed later in this chapter.

Eukaryotic Chromosomes Have Many Functionally Important Regions Including Genes, Origins of Replication, Centromeres, and Telomeres

Each eukaryotic chromosome contains a long, linear DNA molecule (**Figure 10.11**). Three types of regions are required for chromosomal replication and segregation: origins of replication, centromeres, and telomeres. **Origins of replication** are chromosomal sites that are necessary to initiate DNA replication. Unlike most bacterial chromosomes, which contain only one origin of replication, eukaryotic chromosomes contain many origins, interspersed approximately every 100,000 bp apart. The function of origins of replication is discussed in greater detail in Chapter 11. **Centromeres** are DNA regions that play a role in the proper segregation of chromosomes during mitosis and meiosis. For most species, each eukaryotic chromosome contains a single centro-

(b) *Plethodon richmondi*

(c) *Plethodon larselli*

(a) Genome size (nucleotide base pairs per haploid genome)

FIGURE 10.10   **Haploid genome sizes among groups of eukaryotic species.** (**a**) Ranges of genome sizes among different groups of eukaryotes. (**b**) A species of salamander, *Plethodon richmondi*, and (**c**) a close relative, *Plethodon larselli*. The genome of *P. larselli* is over two times as large as that of *P. richmondi.*

Genes→Traits The two species of salamander shown here have very similar traits, even though the genome of *P. larselli* is over twice as large as that of *P. richmondi*. However, the genome of *P. larselli* is not likely to contain more genes. Rather, the additional DNA is due to the accumulation of short repetitive DNA sequences that do not code for genes and are present in many copies.

Key features:

- Eukaryotic chromosomes are usually linear.

- A typical chromosome is tens of millions to hundreds of millions of base pairs in length.

- Eukaryotic chromosomes occur in sets. Many species are diploid, which means that somatic cells contain 2 sets of chromosomes.

- Genes are interspersed throughout the chromosome. A typical chromosome contains between a few hundred and several thousand different genes.

- Each chromosome contains many origins of replication that are interspersed about every 100,000 base pairs.

- Each chromosome contains a centromere that forms a recognition site for the kinetochore proteins.

- Telomeres contain specialized sequences located at both ends of the linear chromosome.

- Repetitive sequences are commonly found near centromeric and telomeric regions, but they may also be interspersed throughout the chromosome.

⊢—⊣ Genes

〜〜 Repetitive sequences

FIGURE 10.11   **Organization of eukaryotic chromosomes.**

mere. The centromere serves as an attachment site for the **kinetochore,** a group of cellular proteins that link the centromere to the spindle apparatus during mitosis and meiosis, ensuring the proper segregation of the chromosomes to each daughter cell. Finally, at the ends of linear chromosomes are found specialized regions known as **telomeres.** Telomeres serve several important functions in the replication and stability of the chromosome. As discussed in Chapter 8, telomeres prevent chromosomes from having "sticky ends" and thereby inhibit chromosomal rearrangements such as translocations. In addition, they prevent chromosome shortening in two ways. First, the telomeres protect chromosomes from digestion via enzymes called exonucleases that recognize the ends of DNA. Secondly, as discussed in Chapter 11, an unusual form of DNA replication occurs at the telo-

mere to ensure that eukaryotic chromosomes do not become shortened with each round of DNA replication.

Genes are located between the centromeric and telomeric regions along the entire eukaryotic chromosome. A single chromosome usually has a few hundred to several thousand different genes. The sequence of a typical eukaryotic gene is a few thousand to tens of thousands of base pairs in length. In less complex eukaryotes such as yeast, genes are relatively small and primarily contain nucleotide sequences that encode the amino acid sequences within proteins. In more complex eukaryotes such as mammals and higher plants, structural genes tend to be much longer due to the presence of **introns**—noncoding intervening sequences. Introns range in size from less than 100 bp to more than 10,000 bp. Therefore, the presence of large introns can greatly increase the lengths of eukaryotic genes.

The Genomes of Eukaryotes Contain Sequences That Are Unique, Moderately Repetitive, or Highly Repetitive

The term **sequence complexity** refers to the number of times a particular base sequence appears throughout the genome. Unique or nonrepetitive sequences are those found once or a few times within the genome. Structural genes are typically unique sequences of DNA. The vast majority of proteins in eukaryotic cells are encoded by genes present in one or a few copies. In the case of humans, unique sequences make up roughly 40% of the entire genome.

Moderately repetitive sequences are found a few hundred to several thousand times in the genome. In a few cases, moderately repetitive sequences are multiple copies of the same gene. For example, the genes that encode ribosomal RNA (rRNA) are found in many copies. Ribosomal RNA is necessary for the functioning of ribosomes. Cells need a large amount of rRNA for making ribosomes, and this is accomplished by having multiple copies of the genes that encode rRNA. Likewise, the histone genes are also found in multiple copies because a large number of histone proteins are needed for the structure of chromatin. In addition, other types of functionally important sequences can be moderately repetitive. For example, moderately repetitive sequences may play a role in the regulation of gene transcription and translation. By comparison, some moderately repetitive sequences do not play a functional role and are derived from **transposable elements**—segments of DNA that have the ability to move within a genome. This category of repetitive sequences is discussed in greater detail in Chapter 17.

Highly repetitive sequences are found tens of thousands or even millions of times throughout the genome. Each copy of a highly repetitive sequence is relatively short, ranging from a few nucleotides to several hundred in length. A widely studied example is the *Alu* family of sequences found in humans and other primates. The *Alu* sequence is approximately 300 bp long. This sequence derives its name from the observation that it contains a site for cleavage by a restriction enzyme known as *Alu*I. (The function of restriction enzymes is described in Chapter 18.) The *Alu* sequence is present in about 1,000,000 copies in the human genome. It represents about 10% of the total human DNA and occurs approximately every 5,000 to 6,000 base pairs! Evolutionary studies suggest that the *Alu* sequence arose 65 million years ago from a section of a single ancestral gene known as the *7SL RNA* gene. Since that time, this gene has become a type of transposable element called a **retroelement,** which can be transcribed into RNA, copied into DNA, and then inserted into the genome. Remarkably, over the course of 65 million years, the *Alu* sequence has been copied and inserted into the human genome to achieve the modern number of about 1,000,000 copies.

Some highly repetitive sequences, like the *Alu* family, are interspersed throughout the genome. However, other highly repetitive sequences are clustered together in a **tandem array,** also known as tandem repeats. In a tandem array, a very short nucleotide sequence is repeated many times in a row. In *Drosophila*, for example, 19% of the chromosomal DNA is highly repetitive DNA found in tandem arrays. An example is shown here.

```
AATATAATATAATATAATATAATATATAATAT

TTATATTATATTATATTATATTATATATTATA
```

In this particular tandem array, two related sequences, AATAT and AATATAT, are repeated. Tandem arrays of short sequences are commonly found in centromeric regions of chromosomes and can be quite long, sometimes more than 1,000,000 bp in length!

What is the functional significance of highly repetitive sequences? Whether highly repetitive sequences play any significant functional role is controversial. Some experiments in *Drosophila* indicate that highly repetitive sequences may be important in the proper segregation of chromosomes during meiosis. It is not yet clear if highly repetitive DNA plays the same role in other species. The sequences within highly repetitive DNA vary greatly from species to species. In fact, the amount of highly repetitive DNA can vary a great deal even among closely related species (as noted earlier in Figure 10.10).

Sequence Complexity Can Be Evaluated in a Renaturation Experiment

One approach that has proven useful in understanding genome complexity has come from renaturation studies. These kinds of experiments were first carried out by Roy Britten and David Kohne in 1968. In a renaturation study, the DNA is broken up into pieces containing several hundred base pairs. The double-stranded DNA is then denatured (separated) into single-stranded pieces by heat treatment (**Figure 10.12a**). When the temperature is lowered, the pieces of DNA that are complementary can reassociate, or renature, with each other to form double-stranded molecules.

The rate of renaturation of complementary DNA strands provides a way to distinguish between unique, moderately repetitive, and highly repetitive sequences. For a given category of DNA sequences, the renaturation rate will depend on the concentration of its complementary partner. Highly repetitive DNA sequences renature much faster because there are many copies of the complementary sequences. In contrast, unique sequences, such as those

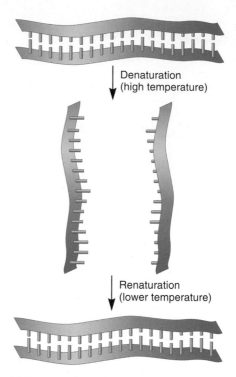

(a) Renaturation of DNA strands

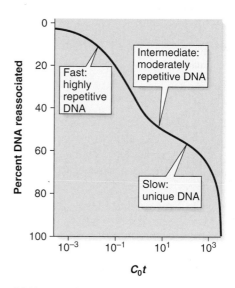

(b) Human chromosomal DNA $C_0 t$ curve

FIGURE 10.12   Renaturation and DNA sequence complexity. (**a**) Denaturation and renaturation (or reassociation) of DNA strands. (**b**) A $C_0 t$ curve for human chromosomal DNA.

found within most genes, take longer to renature because of the added time it takes for the unique sequences to find each other.

The renaturation of two DNA strands is a bimolecular reaction that involves the collision of two complementary DNA strands. Its rate is proportional to the product of the concentrations of both strands. If C is the concentration of a single-stranded

DNA, then for any DNA derived from a double-stranded fragment, the concentration of one DNA strand (denoted C_1) equals the concentration of its complementary partner (denoted C_2). Letting $C = C_1 = C_2$, we see the rate of renaturation is represented by the second-order equation

$$- dC/dt = kC^2$$

This is called a second-order equation because the rate depends on the concentration of both reactants—C_1 and C_2. In this case, this product is simplified to C^2 because $C_1 = C_2$.

This equation says that a change in concentration of a single strand ($- dC$) with respect to time (dt) equals a rate constant (k) times the concentration of the single-stranded molecule squared (C^2). This equation can then be integrated to determine how the concentration of the single-stranded DNA will change from time zero to a later time.

$$\frac{C}{C_0} = \frac{1}{1 + k_2 C_0 t}$$

Where

> C = the concentration of single-stranded DNA at a later time, t
>
> C_0 = the concentration of single-stranded DNA at time zero
>
> k_2 = the second-order rate constant for renaturation

In this equation, C/C_0 is the fraction of DNA still in single-stranded form after a given length of time. For example, if C/C_0 equals 0.4 after a certain period of time, 40% of the DNA is still in the single-stranded form, while 60% has renatured into the double-stranded form. A renaturation experiment can provide quantitative information about the complexity of DNA sequences within chromosomal DNA. In the experiment shown in **Figure 10.12b**, human DNA was sheared into small pieces (each about 600 bp in length), subjected to heat, and then allowed to renature at a lower temperature. The rates of renaturation for the DNA pieces can be represented in a plot of C/C_0 versus $C_0 t$, which is referred to as a $C_0 t$ **curve** (called a Cot curve). The term Cot refers to the DNA concentration (C_0) multiplied by the incubation time (t). A fair amount of the DNA renatures very rapidly. This is the highly repetitive DNA. Some of the DNA reassociates at a moderate rate, and the remaining DNA renatures fairly slowly. From these data, the relative amounts of highly repetitive, moderately repetitive, and unique DNA sequences can be approximated. As seen in Figure 10.12b, about 40% of human DNA fragments are unique DNA sequences that renature slowly.

Eukaryotic Chromatin Must Be Compacted to Fit Within the Cell

We now turn our attention to ways that eukaryotic chromosomes are folded to fit within a living cell. A typical eukaryotic chromosome contains a single, linear double-stranded DNA molecule that may be hundreds of millions of base pairs in length. If the DNA from a single set of human chromosomes was stretched from end to end, the length would be over 1 meter! By comparison, most eukaryotic cells are only 10 to 100 μm in diameter, and the cell nucleus is only about 2 to 4 μm in diameter. Therefore,

the DNA in a eukaryotic cell must be folded and packaged by a staggering amount to fit inside the nucleus.

The compaction of linear DNA within eukaryotic chromosomes is accomplished through mechanisms that involve interactions between DNA and several different proteins. In recent years, it has become increasingly clear that the proteins bound to chromosomal DNA are subject to change during the life of the cell. These changes in protein composition, in turn, affect the degree of compaction of the chromatin. Chromosomes are very dynamic structures that alternate between tight and loose compaction states in response to changes in protein composition. In the remaining parts of Chapter 10, we will focus our attention on two issues of chromosome structure. First, we will consider how chromosomes are compacted and organized during interphase—the period of the cell cycle that includes the G_1, S, and G_2 phases. Later, we will examine the additional compaction that is necessary to produce the highly condensed chromosomes found in M phase.

Linear DNA Wraps Around Histone Proteins to Form Nucleosomes, the Repeating Structural Unit of Chromatin

The repeating structural unit within eukaryotic chromatin is the **nucleosome**—a double-stranded segment of DNA wrapped around an octamer of **histone proteins** (**Figure 10.13a**). Each octamer contains eight histone subunits, two copies each of four different histone proteins. The DNA lies on the surface and makes 1.65 negative superhelical turns around the histone octamer. The amount of DNA that is required to wrap around the histone octamer is 146 or 147 bp. At its widest point, a single nucleosome is about 11 nm in diameter.

The chromatin of eukaryotic cells contains a repeating pattern in which the nucleosomes are connected by linker regions of DNA that vary in length from 20 to 100 bp, depending on the species and cell type. It has been suggested that the overall structure of connected nucleosomes resembles beads on a string. This structure shortens the length of the DNA molecule about sevenfold.

Each of the histone proteins consists of a globular domain and a flexible, charged amino terminus called an amino terminal tail. Histone proteins are very basic proteins because they contain a large number of positively charged lysine and arginine amino acids. The arginines, in particular, play a major role in binding to the DNA. Arginines within the histone proteins form electrostatic and hydrogen-bonding interactions with the phosphate groups along the DNA backbone. The octamer of histones contains two molecules each of four different histone proteins: H2A, H2B, H3, and H4. These are called the core histones. In 1997, Timothy Richmond and colleagues determined the structure of a nucleosome by X-ray crystallography (**Figure 10.13b**).

Another histone, H1, is found in most eukaryotic cells and is called the linker histone. It binds to the DNA in the linker region between nucleosomes and may help to compact adjacent nucleosomes (**Figure 10.13c**). The linker histones are less tightly bound to the DNA than are the core histones. In addition, nonhistone proteins bound to the linker region play a role in the organization and compaction of chromosomes, and their presence may affect the expression of nearby genes.

(a) Nucleosomes showing core histone proteins

Nucleosome Core Particle

(b) Molecular model for nucleosome structure

(Image courtesy of Timothy J. Richmond. Reprinted by permission from Macmillan Publishers Ltd. *Nature*. Crystal structure of the nucleosome core particle at 2.8 A resolution. Luger K., Mader, AW, Richmond, RK, Sargent, DF, Richmond, TJ. 389:6648, 251–260, 1997.)

(c) Nucleosomes showing linker histones and nonhistone proteins

FIGURE 10.13   Nucleosome structure. (a) A nucleosome consists of 146 or 147 bp of DNA wrapped around an octamer of core histone proteins. **(b)** A model for the structure of a nucleosome as determined by X-ray crystallography. **(c)** The linker region of DNA connects adjacent nucleosomes. The linker histone H1 and nonhistone proteins also bind to this linker region.

EXPERIMENT 10A

The Repeating Nucleosome Structure Is Revealed by Digestion of the Linker Region

The model of nucleosome structure was originally proposed by Roger Kornberg in 1974. He based his proposal on several observations. Biochemical experiments had shown that chromatin contains a ratio of one molecule of each of the four core histones (namely, H2A, H2B, H3, and H4) per 100 bp of DNA. Approximately one H1 protein was found per 200 bp of DNA. In addition, purified core histone proteins were observed to bind to each other via specific pairwise interactions. Subsequent X-ray diffraction studies showed that chromatin is composed of a repeating pattern of smaller units. Finally, electron microscopy of chromatin fibers revealed a diameter of approximately 11 nm. Taken together, these observations led Kornberg to propose a model in which the DNA double helix is wrapped around an octamer of core histone proteins. Including the linker region, this would involve about 200 bp of DNA.

Markus Noll decided to test Kornberg's model by digesting chromatin with DNase I, an enzyme that cuts the DNA backbone, and then accurately measuring the molecular mass of the DNA fragments by gel electrophoresis. Noll assumed that the linker region of DNA will be more accessible to DNase I and, therefore,

DNase I is more likely to make cuts in the linker region than in the 146-bp region that is tightly bound to the core histones. If this is correct, incubation with DNase I is expected to make cuts in the linker region and thereby produce DNA pieces approximately 200 bp in length. The size of the DNA fragments may vary somewhat because the linker region is not of constant length and because the cut within the linker region may occur at different sites.

Figure 10.14 describes the experimental protocol of Noll. He began with nuclei from rat liver cells and incubated them with low, medium, or high concentrations of DNase I. The DNA was extracted into an aqueous phase and then loaded onto an agarose gel that separated the fragments according to their molecular mass. The DNA fragments within the gel were stained with a UV-sensitive dye, ethidium bromide, which made it possible to view the DNA fragments under UV illumination.

■ THE HYPOTHESIS

This experiment seeks to test the beads-on-a-string model for chromatin structure. According to this model, DNase I should preferentially cut the DNA in the linker region and thereby produce DNA pieces that are about 200 bp in length.

■ TESTING THE HYPOTHESIS — FIGURE 10.14   DNase I cuts chromatin into repeating units containing 200 bp of DNA.

Starting material: Nuclei from rat liver cells.

1. Incubate the nuclei with low, medium, and high concentrations of DNase I. The conceptual level illustrates a low DNase I concentration.

2. Extract the DNA. This involves dissolving the nuclear membrane with detergent and extracting with the organic solvent phenol.

3. Load the DNA into a well of an agarose gel and run the gel to separate the DNA pieces according to size. On this gel, also load DNA fragments of known molecular mass (marker lane).

4. Visualize the DNA fragments by staining the DNA with ethidium bromide, a dye that binds to DNA and is fluorescent when excited by UV light.

Experimental level

Conceptual level

■ THE DATA

Low   Medium   High

600bp →
400bp →
200bp →

DNase concentration: 30 units ml⁻¹   150 units ml⁻¹   600 units ml⁻¹

(Reprinted by permission from Macmillan Publishers Ltd. *Nature*. Subunit structure of chromatin. Markus Noll. 251:5472, 249–251. 1974.)

■ INTERPRETING THE DATA

As shown in the data of Figure 10.14, at high DNase I concentrations, the entire sample of chromosomal DNA was digested into fragments of approximately 200 bp in length. This result is predicted by the beads-on-a-string model. Furthermore, at lower DNase I concentrations, longer pieces were observed, and these were in multiples of 200 bp (400, 600, etc.). How do we explain these longer pieces? They occurred because occasional linker regions remained uncut at lower DNase I concentrations. For example, if one linker region was not cut, a DNA piece would contain two nucleosomes and be 400 bp in length. If two consecutive linker regions were not cut, this would produce a piece with three nucleosomes containing about 600 bp of DNA. Taken together, these results strongly supported the nucleosome model for chromatin structure.

A self-help quiz involving this experiment can be found at the Online Learning Center.

Nucleosomes Become Closely Associated to Form a 30 nm Fiber

In eukaryotic chromatin, nucleosomes associate with each other to form a more compact structure that is 30 nm in diameter. Evidence for the packaging of nucleosomes was obtained in the microscopy studies of Fritz Thoma in 1977. Chromatin samples were treated with a resin that removed histone H1, but the removal depended on the salt concentration. A moderate salt solution (100 mM NaCl) removed H1, but a solution with no added NaCl did not remove H1. These samples were then observed with an electron microscope. At moderate salt concentrations (**Figure 10.15a**), the chromatin exhibited the classic beads-on-a-string morphology. Without added NaCl (when H1 is expected to remain bound to the DNA), these "beads" associated with each other into a more compact conformation (**Figure 10.15b**). These results suggest that the nucleosomes are packaged to create a more compact unit and that H1 has a role in the packaging and compaction of nucleosomes. However, the precise role of H1 in chromatin compaction remains unclear. Recent data suggest that the core histones also play a key role in the compaction and relaxation of chromatin.

The experiment of Figure 10.15 and other experiments have established that nucleosome units are organized into a more compact structure that is 30 nm in diameter, known as the **30 nm fiber** (**Figure 10.16a**). The 30 nm fiber shortens the total length of DNA another sevenfold. The structure of the 30 nm fiber has proven difficult to determine, because the conformation of the DNA may be substantially altered when it is extracted from living cells. Most models for the 30 nm fiber fall into two main classes. The solenoid model suggests a helical structure in which contact between nucleosomes produces a symmetrically compact structure within the 30 nm fiber (**Figure 10.16b**). This type of model is still favored by some researchers in the field. However, experimental data also suggest that the 30 nm fiber may not form such a regular structure. Instead, an alternative zigzag model, advocated

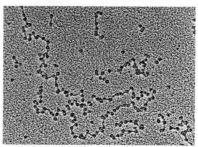

(a) H1 histone not bound— beads on a string

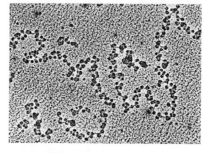

(b) H1 histone bound to linker region—nucleosomes more compact

FIGURE 10.15   The nucleosome structure of eukaryotic chromatin as viewed by electron microscopy. The chromatin in (**a**) has been treated with moderate salt concentrations to remove the linker histone H1. It exhibits the classic beads-on-a-string morphology. The chromatin in (**b**) has been incubated without added NaCl and shows a more compact morphology.

by Rachel Horowitz, Christopher Woodcock, and others, is based on techniques such as cryoelectron microscopy (electron microscopy at low temperature). According to the zigzag model, linker regions within the 30 nm structure are variably bent and twisted, and little face-to-face contact occurs between nucleosomes (**Figure 10.16c**). At this level of compaction, the overall picture of chromatin that emerges is an irregular, fluctuating, three-dimensional zigzag structure with stable nucleosome units

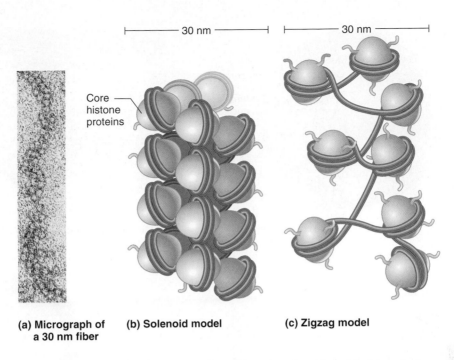

├─── 30 nm ───┤   ├─── 30 nm ───┤

Core
histone
proteins

(a) Micrograph of   **(b) Solenoid model**   **(c) Zigzag model**
a 30 nm fiber

FIGURE 10.16   **The 30 nm fiber. (a)** A photomicrograph of the 30 nm fiber. **(b)** In the solenoid model, the nucleosomes are packed in a spiral configuration. **(c)** In the zigzag model, the linker DNA forms a more irregular structure, and less contact occurs between adjacent nucleosomes. The zigzag model is consistent with more recent data regarding chromatin conformation.

connected by deformable linker regions. In 2005, Timothy Richmond and colleagues were the first to solve the crystal structure of a segment of DNA containing multiple nucleosomes, in this case four. The structure with four nucleosomes revealed that the linker DNA zigzags back and forth between each nucleosome, a feature consistent with the zigzag model.

Chromosomes Are Further Compacted by Anchoring the 30 nm Fiber into Radial Loop Domains Along the Nuclear Matrix

Thus far, we have examined two mechanisms that compact eukaryotic DNA. These involve the wrapping of DNA within nucleosomes and the arrangement of nucleosomes to form a 30 nm fiber. Taken together, these two events shorten the DNA nearly 50-fold. A third level of compaction involves interactions between the 30 nm fibers and a filamentous network of proteins in the nucleus called the **nuclear matrix.** As shown in **Figure 10.17a,** the nuclear matrix consists of two parts. The **nuclear lamina** is a collection of fibers that line the inner nuclear membrane. These fibers are composed of intermediate filament proteins. The second part is an **internal nuclear matrix,** which is connected to the nuclear lamina and fills the interior of the nucleus. The internal nuclear matrix, whose structure and functional role remain controversial, is hypothesized to be an intricate fine network of irregular protein fibers plus many other proteins that bind to these fibers. Even when the chromatin is extracted from the

nucleus, the internal nuclear matrix may remain intact (**Figure 10.17b** and **c**). However, the matrix should not be considered a static structure. Research indicates that the protein composition of the internal nuclear matrix is very dynamic and complex, consisting of dozens or perhaps hundreds of different proteins. The protein composition varies depending on species, cell type, and environmental conditions. This complexity has made it difficult to propose models regarding its overall organization. Further research will be necessary to understand the structure and dynamic nature of the internal nuclear matrix.

The proteins of the nuclear matrix are involved in compacting the DNA into **radial loop domains,** similar to those described for the bacterial chromosome. During interphase, chromatin is organized into loops, often 25,000 to 200,000 bp in size, which are anchored to the nuclear matrix. The chromosomal DNA of eukaryotic species contains sequences called **matrix-attachment regions (MARs)** or **scaffold-attachment regions (SARs),** which are interspersed at regular intervals throughout the genome. The MARs bind to specific proteins in the nuclear matrix, thus forming chromosomal loops (**Figure 10.17d**).

Why is the attachment of radial loops to the nuclear matrix important? In addition to compaction, the nuclear matrix serves to organize the chromosomes within the nucleus. Each chromosome in the cell nucleus is located in a discrete **chromosome territory.** As shown in studies by Thomas Cremer, Christoph Cremer, and others, these territories can be viewed when interphase cells are exposed to multiple fluorescent molecules that recognize

(a) Proteins that form the nuclear matrix

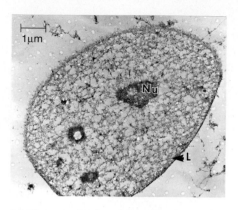

(b) Micrograph of nucleus with chromatin removed

(c) Micrograph showing a close-up of nuclear matrix

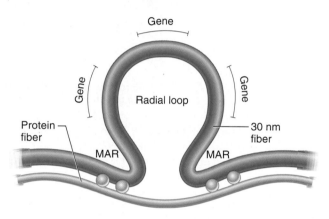

(d) Radial loop bound to a nuclear matrix fiber

FIGURE 10.17   **Structure of the nuclear matrix. (a)** This schematic drawing shows the arrangement of the matrix within a cell nucleus. The nuclear lamina (depicted in red) is a collection of fibrous proteins that line the inner nuclear membrane. The internal nuclear matrix is composed of protein filaments (depicted in green) that are interconnected. These fibers also have many other proteins associated with them (depicted in orange). **(b)** An electron micrograph of the nuclear matrix during interphase after the chromatin has been removed. The nucleolus is labeled Nu, and the lamina is labeled L. **(c)** At higher magnification, the protein fibers are more easily seen (arrowheads point at fibers). **(d)** The matrix-attachment regions (MARs), which contain a high percentage of A and T bases, bind to the nuclear matrix and create radial loops. This causes a greater compaction of eukaryotic chromosomal DNA.

specific sequences on particular chromosomes. **Figure 10.18** illustrates an experiment in which chicken cells were exposed to a mixture of probes that recognize multiple sites along several of the larger chromosomes found in this species (*Gallus gallus*). Figure 10.18a shows the chromosomes in metaphase. The probes label each type of metaphase chromosome with a different color. Figure 10.18b shows the use of the same probes during interphase, when the chromosomes are less condensed and found in the cell nucleus. As seen here, each chromosome occupies its own distinct territory.

Before ending the topic of interphase chromosome compaction, let's consider how the compaction level of interphase chromosomes may vary. This variability can be seen with a light microscope and was first observed by the German cytologist E. Heitz in 1928. He coined the term **heterochromatin** to describe the tightly compacted regions of chromosomes. In general, these regions of the chromosome are transcriptionally inactive. By

comparison, the less condensed regions, known as **euchromatin,** reflect areas that are capable of gene transcription. In euchromatin, the 30 nm fiber forms radial loop domains. In heterochromatin, these radial loop domains become compacted even further.

Figure 10.19 illustrates the distribution of euchromatin and heterochromatin in a typical eukaryotic chromosome during interphase. The chromosome contains regions of both heterochromatin and euchromatin. Heterochromatin is most abundant in the centromeric regions of the chromosomes and, to a lesser extent, in the telomeric regions. The term **constitutive heterochromatin** refers to chromosomal regions that are always heterochromatic and permanently inactive with regard to transcription. Constitutive heterochromatin usually contains highly repetitive DNA sequences, such as tandem repeats, rather than gene sequences. **Facultative heterochromatin** refers to chromatin that can occasionally interconvert between heterochromatin and euchromatin. An example of facultative heterochromatin occurs in female mammals when

(a) Metaphase chromosomes   **(b) Chromosomes in the cell nucleus during interphase**

FIGURE 10.18   Chromosome territories in the cell nucleus. (a) Several metaphase chromosomes from the chicken were labeled with chromosome-specific probes. Each of seven types of chicken chromosomes (i.e., 1, 2, 3, 4, 5, 6, and Z) is labeled a different color. **(b)** The same probes were used to label interphase chromosomes in the cell nucleus. Each of these chromosomes occupies its own distinct, nonoverlapping territory within the cell nucleus. (Note: Chicken cells are diploid, with two copies of each chromosome.)

(Reprinted by permission from Macmillan Publishers Ltd. *Nature Reviews/Genetics.* Chromosome territories, nuclear architecture and gene regulation in mammalian cells. Cremer, T. & Cremer, C. 2:4, 292–301, 2001.)

FIGURE 10.19   Chromatin structure during interphase. Heterochromatic regions are more highly condensed and tend to be localized in centromeric and telomeric regions.

one of the two X chromosomes is converted to a heterochromatic Barr body. As discussed in Chapter 7, most of the genes on the Barr body are transcriptionally inactive. The conversion of one X chromosome to heterochromatin occurs during embryonic development in the somatic cells of the body.

The Histone Code Controls Chromatin Compaction

Thus far, we have learned that the genomes of eukaryotic species are greatly compacted to fit inside the cell nucleus. Even euchromatin, which is looser than heterochromatin, is still compacted to such a degree that it is difficult for transcription factors and RNA polymerase to access and transcribe genes. As discussed in Chapters 12 and 15, euchromatin must be loosened up so that genes can be transcribed into RNA. Gene transcription is greatly influenced by a process known as **chromatin remodeling.** This term refers to changes in chromatin structure that regulate the ability of transcription factors to gain access to genes so they may be transcribed into RNA.

As described earlier, each of the core histone proteins consists of a globular domain and a flexible, charged amino terminus

called an amino terminal tail (refer back to Figure 10.13a). The DNA wraps around the globular domains, and the amino terminal tails protrude from the chromatin. In recent years, researchers have discovered that particular amino acids in the amino terminal tails are subject to several types of covalent modifications, including acetylation, methylation, and phosphorylation. Over 50 different enzymes have been identified in mammals that selectively modify amino terminal tails. **Figure 10.20** shows examples of sites in the tails of H2A, H2B, H3, and H4 that can be modified.

How do histone modifications affect the level of chromatin compaction? First, they may directly influence interactions between nucleosomes. Second, histone modifications provide binding sites that are recognized by proteins. According to the **histone code hypothesis,** proposed by Brian Strahl, C. David Allis, and Bryan Turner in 2000, the pattern of histone modification acts much like a language or code in specifying alterations in chromatin structure. For example, one pattern might involve phosphorylation of the serine at the first position in H2A and acetylation of the fifth and eighth amino acids in H4, which are lysines. A different pattern could involve acetylation of the fifth

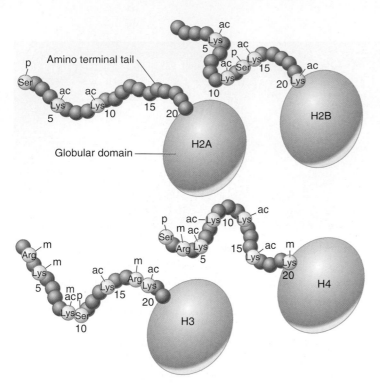

FIGURE 10.20   **Examples of covalent modifications that may occur in the amino terminal tails of core histone proteins.** The amino acids are numbered from the amino terminus. The modifications shown here are m for methylation, p for phosphorylation, and ac for acetylation. Many more modifications can occur to the amino terminal tails; the ones shown here represent common examples.

amino acid, a lysine, in H2B and methylation of the third amino acid in H4, which is an arginine.

The pattern of covalent modifications to the amino terminal tails provides binding sites for proteins that subsequently affect the degree of chromatin compaction. One pattern of histone modification may attract proteins that cause the chromatin to become even more compact, which would silence the transcription of genes in the region. A different combination of histone modifications may attract proteins, such as the chromatin remodeling enzymes discussed in Chapter 15, which serve to loosen the chromatin and thereby promote gene transcription. In this way, the histone code plays a key role in making the information within the genomes of eukaryotic species accessible. Researchers are trying to unravel which patterns of histone modifications promote compaction and which promote a loosening of chromatin structure. In other words, they are trying to decipher the effects of the covalent modifications that comprise the histone code.

Condensin and Cohesin Promote the Formation of Metaphase Chromosomes

As we have seen, several mechanisms can alter the level of chromosomal compaction. Furthermore, the degree of compaction

can vary along a single eukaryotic chromosome. When cells prepare to divide, the chromosomes become even more compacted or condensed. This aids in their proper sorting during metaphase. **Figure 10.21** illustrates the levels of compaction that lead to a metaphase chromosome. During interphase, most of the chromosomal DNA is found in euchromatin, in which the 30 nm fibers form radial loop domains that are attached to a protein scaffold. The average distance that loops radiate from the protein scaffold is approximately 300 nm. This structure can be further compacted via additional folding of the radial loop domains and protein scaffold. This additional level of compaction greatly shortens the overall length of a chromosome and produces a diameter of approximately 700 nm, which is the compaction level found in heterochromatin. During interphase, most chromosomal regions are euchromatic, and some localized regions, such as those near centromeres, are heterochromatic.

As cells enter M phase, the level of compaction changes dramatically. By the end of prophase, sister chromatids are entirely heterochromatic. Two parallel chromatids have a larger diameter of approximately 1,400 nm but a much shorter length compared to interphase chromosomes. These highly condensed metaphase chromosomes undergo little gene transcription because it is difficult for transcription proteins to gain access to the compacted DNA. Therefore, most transcriptional activity ceases during M phase, although a few specific genes may be transcribed. M phase is usually a short period of the cell cycle.

In highly condensed chromosomes, such as those found in metaphase, the radial loops are highly compacted and remain anchored to a **scaffold,** which is formed from nonhistone proteins of the nuclear matrix. Experimentally, researchers can delineate the nonhistone proteins of the scaffold that hold the loops in place. **Figure 10.22a** shows a human metaphase chromosome. In this condition, the radial loops of DNA are in a very compact configuration. If this chromosome is treated with a high concentration of salt to remove both the core and linker histones, the highly compact configuration is lost, but the bottoms of the elongated loops remain attached to the scaffold composed of nonhistone proteins. In **Figure 10.22b**, an arrow points to an elongated DNA strand emanating from the darkly staining scaffold. Remarkably, the scaffold retains the shape of the original metaphase chromosome even though the DNA strands have become greatly elongated. These results illustrate that the structure of metaphase chromosomes is determined by the nuclear matrix proteins, which form the scaffold, and by the histones, which are needed to compact the radial loops.

Researchers are trying to understand the steps that lead to the formation and organization of metaphase chromosomes. During the past several years, studies in yeast and frog oocytes have been aimed at the identification of proteins that promote the conversion of interphase chromosomes into metaphase chromosomes. In yeast, mutants have been characterized that have alterations in the condensation or the segregation of chromosomes. Similarly, biochemical studies using frog oocytes resulted in the purification of protein complexes that promote chromosomal condensation or sister chromatid alignment. These two lines of independent research produced the same results. Researchers

2 nm

DNA double helix

60 nm

Wrapping of DNA around a histone octamer

Histone H1

11 nm

Histone octamer

Nucleosome

33 nm

(a) Nucleosomes ("beads on a string")

Formation of a three-dimensional zigzag structure via histone H1 and other DNA-binding proteins

30 nm

200 nm

(b) 30 nm fiber

Nucleosome

Anchoring of radial loops to the nuclear matrix

300 nm

90 nm

(c) Radial loop domains

Protein scaffold

Further compaction of radial loops

700 nm

180 nm

Formation of a scaffold from the nuclear matrix and further compaction of all radial loops

1,400 nm

(d) Metaphase chromosome

640 nm

FIGURE 10.21 The steps in eukaryotic chromosomal compaction leading to the metaphase chromosome.

(a) Metaphase chromosome   **(b) Metaphase chromosome treated with high salt to remove histone proteins**

FIGURE 10.22   **The importance of histone proteins and scaffolding proteins in the compaction of eukaryotic chromosomes.** (a) A metaphase chromosome. (b) A metaphase chromosome following treatment with high salt concentration to remove the histone proteins. The arrow on the left points to the scaffold (composed of nonhistone proteins), which anchors the bases of the radial loops. The arrow on the right points to an elongated strand of DNA.

found that cells contain two multiprotein complexes called **condensin** and **cohesin,** which play a critical role in chromosomal condensation and sister chromatid alignment, respectively.

Condensin and cohesin are two completely distinct complexes, but both contain a category of proteins called **SMC proteins.** SMC stands for structural maintenance of chromosomes. These proteins use energy from ATP to catalyze changes in chromosome structure. Together with topoisomerases, SMC proteins have been shown to promote major changes in DNA structure. An emerging theme is that SMC proteins actively fold, tether, and manipulate DNA strands. They are dimers that have a V-shaped structure. The monomers, which are connected at a hinge region, have two long coiled arms with a head region that binds ATP (**Figure 10.23**). The length of each monomer is about 50 nm, which is equivalent to approximately 150 bp of DNA.

As their names suggest, condensin and cohesin play different roles in metaphase chromosome structure. Prior to M phase, condensin is found outside the nucleus (**Figure 10.24**). However, as M phase begins, condensin is observed to coat the individual chromatids as euchromatin is converted into heterochromatin. The role of condensin in the compaction process is not well understood. Although condensin is often implicated

in the process of chromosomal condensation, researchers have been able to deplete condensin from actively dividing cells, and the chromosomes are still able to condense. However, such condensed chromosomes show abnormalities in their ability to separate from each other during cell division. These results suggest that condensin is important in the proper organization of highly condensed chromosomes, such as those found during metaphase.

In comparison, the function of cohesin is to promote the binding (i.e., cohesion) between sister chromatids. After S phase and until the middle of prophase, sister chromatids remain attached to each other along their length. As shown in **Figure 10.25**, this attachment is promoted by cohesin, which is found along the entire length of each chromatid. In certain species, such as mammals, cohesins located along the chromosome arms are released during prophase. This allows the chromosome arms to separate. However, some cohesins remain attached, primarily to the centromeric regions, leaving the centromeric region as the main linkage before anaphase. At anaphase, the cohesins bound to the centromere are rapidly degraded by a protease aptly named separase, thereby allowing sister chromatid separation.

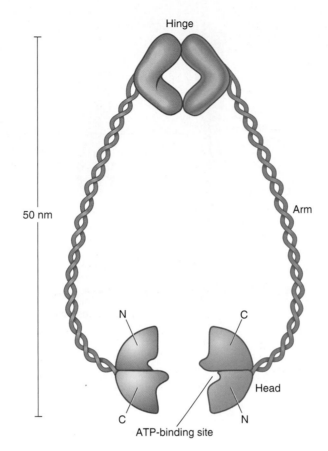

FIGURE 10.23   The structure of SMC proteins. This figure shows the generalized structure of SMC proteins, which are dimers consisting of hinge, arm, and head regions. The head regions bind and hydrolyze ATP. Condensin and cohesin have additional protein subunits not shown here.

FIGURE 10.24   The localization of condensin during interphase and the start of M phase. During interphase (G_1, S, and G_2), most of the condensin protein is found outside the nucleus. The interphase chromosomes are largely euchromatic. At the start of M phase, condensin travels into the nucleus and binds to the chromosomes, which become heterochromatic due to a greater compaction of the radial loop domains.

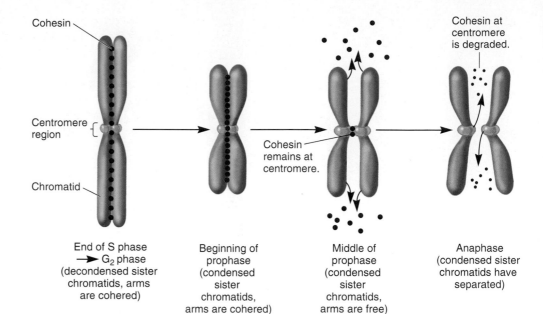

ONLINE ANIMATION

FIGURE 10.25
The alignment of sister chromatids via cohesin.
After S phase is completed, many cohesin complexes bind along each chromatid, thereby facilitating their attachment to each other. During the middle of prophase, cohesin is released from the chromosome arms, but some cohesin remains in the centromeric regions. At anaphase, the remaining cohesin complexes are rapidly degraded by separase, which promotes sister chromatid separation.

Cohesin

Centromere region

Chromatid

Cohesin remains at centromere.

Cohesin at centromere is degraded.

End of S phase → G_2 phase (decondensed sister chromatids, arms are cohered)

Beginning of prophase (condensed sister chromatids, arms are cohered)

Middle of prophase (condensed sister chromatids, arms are free)

Anaphase (condensed sister chromatids have separated)

CONCEPTUAL SUMMARY

The chromosomal location and function of DNA sequences are central to our understanding of genetics. Viruses contain relatively small genomes that are composed of DNA or RNA. The viral genome is packaged into mature virus particles via self-assembly or by directed assembly. In bacteria, a few thousand different genes are interspersed throughout a circular chromosome; a single origin of replication is also present. Short repetitive sequences, which may play a role in chromosome structure and gene function, are found throughout the chromosome. By comparison, eukaryotes have many linear chromosomes. A single centromere is found on each chromosome and functions as a recognition site for the group of proteins that form the kinetochore. At the ends of the linear chromosomes are telomeres. Hundreds or even thousands of different genes are located between the centromeres and telomeres. An unusual feature of many eukaryotic species is that their DNA contains an abundance of highly repetitive sequences. The amount of highly repetitive DNA varies a great deal, even among closely related species, and the functional importance of these sequences remains unclear.

A second important issue that underlies chromosomal structure and function is the amount of compaction that occurs within chromosomal DNA. In bacteria, two main events, the formation of loop domains within the chromosome and DNA supercoiling, are responsible for the folding of the circular chromosome into a compact structure within a nucleoid. Certain DNA-binding proteins are involved in this process. In eukaryotic chromatin, the DNA is first folded into nucleosomes, which contain the DNA double helix wrapped around histone proteins. The nucleosomes are then compacted into a 30 nm fiber. The 30 nm fiber is further organized into radial loop domains, which are anchored to the nuclear matrix. This is termed euchromatin. In heterochromatin, the radial loop domains are compacted even further. During interphase, eukaryotic chromosomes are found within the cell nucleus. They consist of short, highly compacted regions that are heterochromatic and less condensed euchromatic regions. Histone modifications play a key role in controlling the level of chromatin compaction. During M phase, the chromosomes become highly condensed and entirely heterochromatic. Condensin functions in the proper formation of highly condensed chromosomes. Cohesin promotes the binding between sister chromatids.

EXPERIMENTAL SUMMARY

Our understanding of genome complexity and organization has been aided by various cytological, genetic, biochemical, and molecular techniques. In this chapter, we considered renaturation experiments as a technique to evaluate the sequence complexity of DNA. In Chapters 5 and 6, we examined ways to genetically map the locations of genes along eukaryotic and bacterial chromosomes. In Chapter 20, we will explore the use of cytological

techniques such as in situ hybridization to identify particular sequences within an intact chromosome. In addition, Chapter 20 will describe molecular methods in which segments of a genome are cloned and analyzed molecularly.

The second topic in this chapter was the compaction of chromosomes to fit within living cells. This phenomenon can also be examined via several different approaches. The microscopic analysis of chromosomes provides key insight into their structure. In addition, chromatin can be analyzed biochemically by many different methods. For example, the model of repeating nucleosome structure was verified by the studies of Noll, which involved digestion of chromatin by the enzyme DNase I. More recently, the detailed structure of nucleosomes has been solved by crystallization studies. Currently, much research is aimed at identifying the proteins bound to chromosomal DNA and elucidating the roles these proteins play in determining chromatin structure.

PROBLEM SETS & INSIGHTS

Solved Problems

S1. Here is a C_0t curve for a hypothetical eukaryotic species:

Estimate the amount of highly repetitive DNA, moderately repetitive DNA, and unique DNA.

Answer: About 20% is highly repetitive and renatures quickly, about 50% is moderately repetitive, and about 30% is unique and renatures very slowly.

S2. Let's suppose a bacterial DNA molecule is given a left-handed twist. How does this affect the structure and function of the DNA?

Answer: A left-handed twist is negative supercoiling. Negative supercoiling makes the bacterial chromosome more compact. It also promotes DNA functions that involve strand separation, including gene transcription and DNA replication.

S3. To hold bacterial DNA in a more compact configuration, specific proteins must bind to the DNA and stabilize its conformation (as shown in Figure 10.5). Several different proteins are involved in this process. These proteins have been collectively referred to as "histonelike" due to their possible functional similarity to the histone proteins found in eukaryotes. Based on your knowledge of eukaryotic histone proteins, what biochemical properties would you expect from bacterial histonelike proteins?

Answer: The histonelike proteins have the properties expected for proteins involved in DNA folding. They are all small proteins found in relative abundance within the bacterial cell. In some cases, the histonelike proteins are biochemically similar to eukaryotic histones. For example, they tend to be basic (positively charged) and bind to DNA in a non-sequence-dependent fashion. However, other proteins appear to bind to bacterial DNA at specific sites in order to promote DNA bending.

Conceptual Questions

C1. In viral replication, what is the difference between self-assembly and directed assembly?

C2. Bacterial chromosomes have one origin of replication, whereas eukaryotic chromosomes have several. Would you expect viral chromosomes to have an origin of replication? Why or why not?

C3. What is a bacterial nucleoid? With regard to cellular membranes, what is the difference between a bacterial nucleoid and a eukaryotic nucleus?

C4. In Part II of this textbook, we considered inheritance patterns for diploid eukaryotic species. Bacteria frequently contain two or more nucleoids. With regard to genes and alleles, how is a bacterium that contains two nucleoids similar to a diploid eukaryotic cell, and how is it different?

C5. Describe the two main mechanisms by which the bacterial DNA becomes compacted.

C6. As described in Chapter 9, 1 bp of DNA is approximately 0.34 nm in length. A bacterial chromosome is about 4 million bp in length and is organized into about 100 loops that are about 40,000 bp in length.

A. If it was stretched out linearly, how long (in micrometers) would one loop be?

B. If a bacterial chromosomal loop is circular, what would be its diameter? (Note: Circumference = πD, where D is the diameter of the circle.)

C. Is the diameter of the circular loop calculated in part B small enough to fit inside a bacterium? The dimensions of the bacterial cytoplasm, such as *E. coli*, are roughly 0.5 μm wide and 1.0 μm long.

C7. Why is DNA supercoiling called supercoiling rather than just coiling? Why is positive supercoiling called overwinding and negative supercoiling called underwinding? How would you define the terms positive and negative supercoiling for Z DNA (described in Chapter 9)?

C8. Coumarins and quinolones are two classes of drugs that inhibit bacterial growth by directly inhibiting DNA gyrase. Discuss two reasons why inhibiting DNA gyrase might inhibit bacterial growth.

C9. Take two pieces of string that are approximately 10 inches each, and create a double helix by wrapping the two strings around each other to make 10 complete turns. Tape one end of the strings to a table, and now twist the strings three times (360° each time) in a right-handed direction. Note: As you are looking down at the strings from above, a right-handed twist is in the clockwise direction.

A. Did the three turns create more or fewer turns in your double helix? How many turns are now in your double helix?

B. Is your double helix right-handed or left-handed? Explain your answer.

C. Did the three turns create any supercoils?

D. If you had coated your double helix with rubber cement and allowed the cement to dry before making the three additional right-handed turns, would the rubber cement make it more or less likely for the three turns to create supercoiling? Would a pair of cemented strings be more or less like a real DNA double helix compared to an uncemented pair of strings? Explain your answer.

C10. Try to explain the function of DNA gyrase with a drawing.

C11. How are two topoisomers different from each other? How are they the same?

C12. On rare occasions, a chromosome can suffer a small deletion that removes the centromere. When this occurs, the chromosome usually is not found within subsequent daughter cells. Explain why a chromosome without a centromere is not transmitted very efficiently from mother to daughter cells. (Note: If a chromosome is located outside the nucleus after telophase, it is degraded.)

C13. What is the function of a centromere? At what stage of the cell cycle would you expect the centromere to be the most important?

C14. Describe the characteristics of highly repetitive DNA.

C15. Describe the structures of a nucleosome and a 30 nm fiber.

C16. Beginning with the G_1 phase of the cell cycle, describe the level of compaction of the eukaryotic chromosome. How does the level of compaction change as the cell progresses through the cell cycle? Why is it necessary to further compact the chromatin during mitosis?

C17. If you assume the average length of linker DNA is 50 bp, approximately how many nucleosomes are found in the haploid human genome, which contains 3 billion bp?

C18. Draw the binding between the nuclear matrix and MARs.

C19. Compare heterochromatin and euchromatin. What are the differences between them?

C20. Compare the structure and cell localization of chromosomes during interphase and M phase.

C21. What types of genetic activities occur during interphase? Explain why these activities cannot occur during M phase.

C22. Let's assume the linker DNA averages 54 bp in length. How many molecules of H2A would you expect to find in a DNA sample that is 46,000 bp in length?

C23. In Figure 10.15, what are we looking at in part b? Is this an 11 nm fiber, a 30 nm fiber, or a 300 nm fiber? Does this DNA come from a cell during M phase or interphase?

C24. What are the roles of the core histone proteins compared to the role of histone H1 in the compaction of eukaryotic DNA?

C25. A typical eukaryotic chromosome found in humans contains about 100 million bp of DNA. As described in Chapter 9, 1 bp of DNA has a linear length of 0.34 nm.

A. What is the linear length of the DNA for a typical human chromosome in micrometers?

B. What is the linear length of a 30 nm fiber of a typical human chromosome?

C. Based on your calculation of part B, would a typical human chromosome fit inside the nucleus (with a diameter of 5 μm) if the 30 nm fiber were stretched out in a linear manner? If not, explain how a typical human chromosome fits inside the nucleus during interphase.

C26. Which of the following terms should not be used to describe a Barr body?

A. Chromatin

B. Euchromatin

C. Heterochromatin

D. Chromosome

E. Genome

C27. Discuss the differences in the compaction levels of metaphase chromosomes compared to interphase chromosomes. When would you expect gene transcription and DNA replication to take place, during M phase or interphase? Explain why.

C28. Explain two ways that histone modifications may affect the level of chromatin compaction.

C29. What is an SMC protein? Describe two examples.

Experimental Questions

E1. Two circular DNA molecules, which we will call molecule A and molecule B, are topoisomers of each other. When viewed under the electron microscope, molecule A appears more compact compared to molecule B. The level of gene transcription is much lower for molecule A. Which of the following three possibilities could account for these observations?

First possibility: Molecule A has 3 positive supercoils, and molecule B has 3 negative supercoils.

Second possibility: Molecule A has 4 positive supercoils, and molecule B has 1 negative supercoil.

Third possibility: Molecule A has 0 supercoils, and molecule B has 3 negative supercoils.

E2. Explain how a renaturation experiment can provide quantitative information about genome sequence complexity.

E3. In a renaturation experiment, does the copy number affect only the rate of renaturation, or does it also affect the rate of denaturation? Explain your answer.

E4. Let's suppose that you have isolated DNA from a cell and have viewed it under a microscope. It looks supercoiled. What experiment would you perform to determine if it is positively or negatively supercoiled? In your answer, describe your expected results. You may assume that you have purified topoisomerases at your disposal.

E5. We seem to know more about the structure of eukaryotic chromosomal DNA than bacterial DNA. Discuss why you think this is so, and list several experimental procedures that have yielded important information concerning the compaction of eukaryotic chromatin.

E6. An organism contains 20% highly repetitive DNA, 10% moderately repetitive DNA, and 70% unique sequences. Draw the expected C_0t curve that would be obtained from this organism.

E7. When chromatin is treated with a moderate salt concentration, the linker histone H1 is removed (see Figure 10.15a). Higher salt concentration removes the rest of the histone proteins (see Figure 10.22b). If the experiment of Figure 10.14 were carried out after the DNA was treated with moderate or high salt, what would be the expected results?

E8. Let's suppose you have isolated chromatin from some bizarre eukaryote with a linker region that is usually 300 to 350 bp in length. The nucleosome structure is the same as in other eukaryotes. If you digested this eukaryotic organism's chromatin with a high concentration of DNase I, what would be your expected results?

E9. If you were given a sample of chromosomal DNA and asked to determine if it is bacterial or eukaryotic, what experiment would you perform, and what would be your expected results?

E10. Consider how histone proteins bind to DNA and then explain why a high salt concentration can remove histones from DNA (as shown in Figure 10.22b).

E11. In Chapter 20, the technique of fluorescence *in situ* hybridization (FISH) is described. This is another method used to examine sequence complexity within a genome. In this method, a particular DNA sequence, such as a particular gene sequence, can be detected within an intact chromosome by using a DNA probe that is complementary to the sequence. For example, let's consider the β-globin gene, which is found on human chromosome 11. A probe complementary to the β-globin gene will bind to the β-globin gene and show up as a brightly colored spot on human chromosome 11. In this way, researchers can detect where the β-globin gene is located within a set of chromosomes. Because the β-globin gene is unique and because human cells are diploid (i.e., have two copies of each chromosome), a FISH experiment would show two bright spots per cell; the probe would bind to each copy of chromosome 11. What would you expect to see if you used the following types of probes?

A. A probe complementary to the *Alu*I sequence

B. A probe complementary to a tandemly repeated sequence near the centromere of the X chromosome

Questions for Student Discussion/Collaboration

1. Bacterial and eukaryotic chromosomes are very compact. Discuss the advantages and disadvantages of having a compact structure.

2. The prevalence of highly repetitive sequences seems rather strange to many geneticists. Do they seem strange to you? Why or why not? Discuss whether or not you think they have an important function.

3. Discuss and make a list of the similarities and differences between bacterial and eukaryotic chromosomes.

Note: All answers appear at the website for this textbook; the answers to even-numbered questions are in the back of the textbook.

www.mhhe.com/brookergenetics3e

Visit the Online Learning Center for practice tests, answer keys, and other learning aids for this chapter. Enhance your understanding of genetics with our interactive exercises, quizzes, animations, and much more.

11

DNA REPLICATION

As discussed throughout Chapters 2 to 8, genetic material is transmitted from parent to offspring and from cell to cell. For transmission to occur, the genetic material must be copied. During this process, known as **DNA replication,** the original DNA strands are used as templates for the synthesis of new DNA strands. We will begin Chapter 11 with a consideration of the structural features of the double helix that underlie the replication process. Then we will examine how chromosomes are replicated within living cells, addressing the following questions: where does DNA replication begin, how does it proceed, and where does it end? We first consider bacterial DNA replication and examine how DNA replication occurs within living cells, and then turn our attention to the unique features of the replication of eukaryotic DNA. At the molecular level, it is rather remarkable that the replication of chromosomal DNA occurs very quickly, very accurately, and at the appropriate time in the life of the cell. For this to happen, many cellular proteins play vital roles. In this chapter, we will examine the mechanism of DNA replication and consider the functions of several proteins involved in the process.

A model for DNA replication. This molecular model shows a DNA replication fork, the site where new DNA strands are made. In this model, the original DNA is yellow and blue. The newly made strands are purple.

11.1 STRUCTURAL OVERVIEW OF DNA REPLICATION

Let's begin by recalling a few important structural features of the double helix from Chapter 9 because they bear directly on the replication process. The double helix is composed of two DNA strands, and the individual building blocks of each strand are nucleotides. The nucleotides contain one of four bases: adenine, thymine, guanine, or cytosine. The double-stranded structure is held together by base stacking and by hydrogen bonding between the bases in opposite strands. A critical feature of the double-helix structure is that adenine hydrogen bonds with thymine, and guanine hydrogen bonds with cytosine. This rule, known as the AT/GC rule or Chargaff's rule, is the basis for the complementarity of the base sequences in double-stranded DNA.

Another feature worth noting is that the strands within a double helix have an antiparallel alignment. This directionality is determined by the orientation of sugar molecules within the sugar–phosphate backbone. If one strand is running in the 5' to 3' direction, the complementary strand is running in the 3' to 5' direction. The issue of directionality will be important when we consider the function of the enzymes that synthesize new DNA strands. In this section, we will consider how the structure of the DNA double helix provides the basis for DNA replication.

Existing DNA Strands Act as Templates for the Synthesis of New Strands

As shown in **Figure 11.1a**, DNA replication relies on the complementarity of DNA strands according to the AT/GC rule. During the replication process, the two complementary strands of DNA come apart and serve as **template strands,** or **parental strands,** for the synthesis of two new strands of DNA. After the double helix has separated, individual nucleotides have access to the template strands. Hydrogen bonding between individual nucleotides and the template strands must obey the AT/GC rule. To complete the replication process, a covalent bond is formed between the phosphate of one nucleotide and the sugar of the previous nucleotide. The two newly made strands are referred to as the **daughter strands.** Note that the base sequences are identical in both double-stranded molecules after replication (**Figure 11.1b**). Therefore, DNA is replicated so that both copies retain the same information—the same base sequence—as the original molecule.

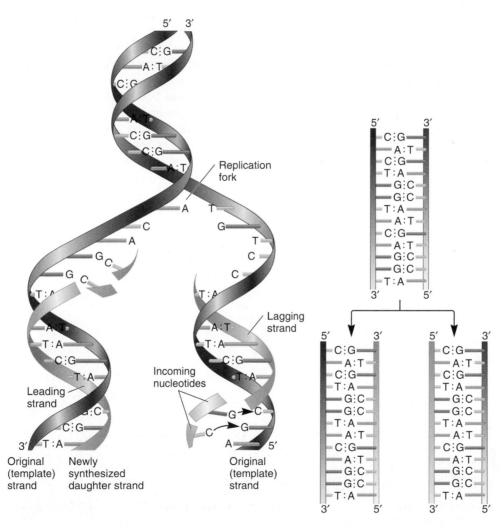

(a) The mechanism of DNA replication

(b) The products of replication

FIGURE 11.1   **The structural basis for DNA replication. (a)** The mechanism of DNA replication as originally proposed by Watson and Crick. As we will see, the synthesis of one newly made strand (the leading strand) occurs in the direction toward the replication fork, whereas the synthesis of the other newly made strand (the lagging strand) occurs in small segments away from the replication fork. **(b)** DNA replication produces two copies of DNA with the same sequence as the original DNA molecule.

Three Different Models Were Proposed That Described the Net Result of DNA Replication

Scientists in the late 1950s had considered three different mechanisms to explain the net result of DNA replication. These mechanisms are shown in **Figure 11.2.** The first is referred to as a **conservative model.** According to this hypothesis, both strands of parental DNA remain together following DNA replication. In this model, the original arrangement of parental strands is completely conserved, while the two newly made daughter strands also remain together following replication. The second is called a **semiconservative model.** In this mechanism, the double-stranded DNA is half conserved following the replication process. In other words, the newly made double-stranded DNA contains one parental strand and one daughter strand. The third, called the **dispersive model,** proposes that segments of parental DNA and newly made DNA are interspersed in both strands following the replication process. Only the semiconservative model shown in Figure 11.2b is actually correct.

In 1958, Matthew Meselson and Franklin Stahl devised a method to experimentally distinguish newly made daughter strands from the original parental strands. The technique they used involved heavy isotope labeling. Nitrogen, which is found within the bases of DNA, occurs in a light (^{14}N) form and a heavy (^{15}N) form. Prior to their experiment, they grew *E. coli* cells in the presence of ^{15}N for many generations. This produced a population of cells in which all of the DNA was heavy labeled. At the start of their experiment, shown in **Figure 11.3** (generation 0), they switched the bacteria to a medium that contained only ^{14}N and then collected samples of cells after various time points. Under the growth conditions they employed, 30 minutes is the time required for one doubling, or one generation time. Because the bacteria were doubling in a medium that contained only ^{14}N, all of the newly made DNA strands would be labeled with light nitrogen, while the original strands would remain in the heavy form.

Meselson and Stahl then analyzed the density of the DNA by centrifugation, using a cesium chloride (CsCl) gradient. (The procedure of gradient centrifugation is described in the Appendix.) If both DNA strands contained ^{14}N, the DNA would have a light density and sediment near the top of the tube. If one strand contained ^{14}N and the other strand contained ^{15}N, the DNA would be half-heavy and have an intermediate density. Finally, if both strands contained ^{15}N, the DNA would be heavy and would sediment closer to the bottom of the centrifuge tube.

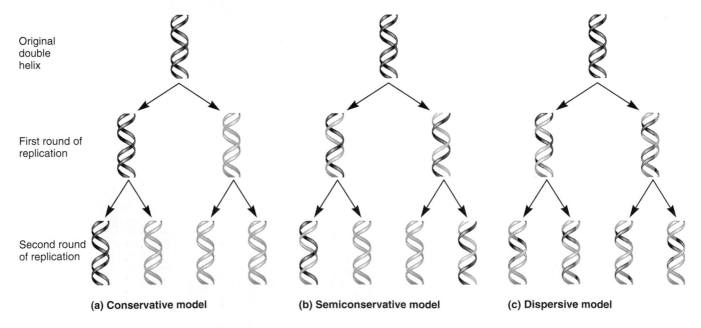

Original double helix

First round of replication

Second round of replication

(a) Conservative model   **(b) Semiconservative model**   **(c) Dispersive model**

FIGURE 11.2   **Three possible models for DNA replication.** The two original parental DNA strands are shown in purple, and the newly made strands after one and two generations are shown in light blue.

■ THE HYPOTHESIS

Based on Watson's and Crick's ideas, the hypothesis was that DNA replication is semiconservative. Figure 11.2 also shows two alternative models.

TESTING THE HYPOTHESIS — FIGURE 11.3   **Evidence that DNA replication is semiconservative.**

Starting material: A strain of *E. coli* that has been grown for many generations in the presence of ^{15}N. All of the nitrogen in the DNA is labeled with ^{15}N.

Experimental level

Conceptual level

1. Add an excess of ^{14}N-containing compounds to the bacterial cells so that all the newly made DNA will contain ^{14}N.

2. Incubate the cells for various lengths of time. Note: The ^{15}N-labeled DNA is shown in purple and the ^{14}N-labeled DNA is shown in blue.

3. Lyse the cells by the addition of lysozyme and detergent, which disrupt the bacterial cell wall and cell membrane, respectively.

4. Load a sample of the lysate onto a CsCl gradient. (Note: The average density of DNA is around 1.7 g/cm^3, which is well isolated from other cellular macromolecules.)

5. Centrifuge the gradients until the DNA molecules reach their equilibrium densities.

6. DNA within the gradient can be observed under a UV light.

^{14}N solution

Suspension of bacterial cells labeled with ^{15}N

37°C   Lyse cells

Lysate

CsCl gradient

UV light

Generation 0

Add ^{14}N

1

2

Up to 4 generations

DNA

Cell wall

Cell membrane

Density centrifugation

Light DNA

Half-heavy DNA

Heavy DNA

(Result shown here is after 2 generations.)

■ **THE DATA**

Generations After ^{14}N Addition

4.1   3.0   2.5   1.9   1.5   1.1   1.0   0.7   0.3

← Light
← Half-heavy
← Heavy

■ **INTERPRETING THE DATA**

As seen in the data of Figure 11.3, after one round of DNA replication (i.e., one generation), all of the DNA sedimented at a density that was half-heavy. Which of the three models is consistent with this result? Both the semiconservative and dispersive models are consistent. In contrast, the conservative model would predict two separate DNA types: a light type and a heavy type. Because all of the DNA had sedimented as a single band, this model was disproved. According to the semiconservative model, the replicated DNA would contain one original strand (a heavy strand) and a newly made daughter strand (a light strand). Likewise, in a dispersive model, all of the DNA should have been half-heavy after one generation as well. To determine which of these two remaining models is correct, therefore, Meselson and Stahl had to investigate future generations.

After approximately two rounds of DNA replication (i.e., 1.9 generations), a mixture of light DNA and half-heavy DNA was observed. This result was consistent with the semiconservative model of DNA replication, because some DNA molecules should contain all light DNA, while other molecules should be half-heavy (see Figure 11.2b). The dispersive model predicts that after two generations, the heavy nitrogen would be evenly dispersed among four strands, each strand containing 1/4 heavy nitrogen and 3/4 light nitrogen (see Figure 11.2c). However, this result was not obtained. Instead, the results of the Meselson and Stahl experiment provided compelling evidence in favor of only the semiconservative model for DNA replication.

A self-help quiz involving this experiment can be found at the Online Learning Center.

11.2 BACTERIAL DNA REPLICATION

Thus far, we have considered how a complementary, double-stranded structure underlies the ability of DNA to be copied. In addition, the experiments of Meselson and Stahl showed that DNA replication results in two double helices, each one containing an original parental strand and a newly made daughter strand. We will now turn our attention to how DNA replication actually occurs within living cells. Much research has focused on the bacterium *Escherichia coli*. The results of these studies have provided the foundation for our current molecular understanding of DNA replication. The replication of the bacterial chromosome is a stepwise process in which many cellular proteins participate. In this section, we will follow this process from beginning to end.

Bacterial Chromosomes Contain a Single Origin of Replication

Figure 11.4 presents an overview of the process of bacterial chromosomal replication. The site on the bacterial chromosome where DNA synthesis begins is known as the **origin of replication.** Bacterial chromosomes have a single origin of replication. The synthesis of new daughter strands is initiated within the origin and proceeds in both directions, or **bidirectionally,** around the bacterial chromosome. This means that two **replication forks** move in opposite directions outward from the origin. A replication fork is the site where the parental DNA strands have separated and new daughter strands are being made. Eventually, these replication forks meet each other on the opposite side of the bacterial chromosome to complete the replication process.

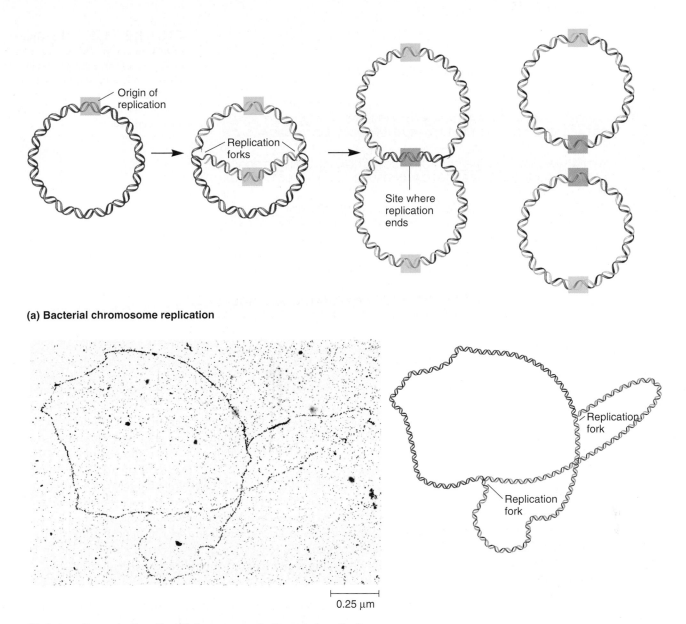

(a) Bacterial chromosome replication

(b) Autoradiograph of an *E. coli* chromosome in the act of replication

FIGURE 11.4   **The process of bacterial chromosome replication.** (a) An overview of the process of bacterial chromosome replication. (b) A replicating *E. coli* chromosome visualized by autoradiography. This chromosome was radiolabeled by growing bacterial cells in media containing radiolabeled thymidine. The diagram at the right shows the locations of the two replication forks. The chromosome is about one-third replicated. New strands are shown in blue.

Replication Is Initiated by the Binding of DnaA Protein to the Origin of Replication

Considerable research has focused on the origin of replication in *E. coli*. This origin is named *oriC* for origin of Chromosomal replication (**Figure 11.5**). Three types of DNA sequences are found within *oriC*: an AT-rich region, DnaA box sequences, and GATC methylation sites. The GATC methylation sites will be discussed later in this chapter when we consider the regulation of replication.

DNA replication is initiated by the binding of **DnaA proteins** to sequences within the origin known as **DnaA box sequences.** The DnaA box sequences serve as recognition sites for the binding of the DnaA proteins. When DnaA proteins are in their ATP-bound form, they bind to the five DnaA boxes in *oriC* to initiate DNA replication. DnaA proteins also bind to each other to form a complex (**Figure 11.6**). With the aid of other DNA-binding proteins, such as HU and IHF, this causes the DNA to bend around the complex of DnaA proteins and results in the separation of the AT-rich region. Because only two hydrogen bonds form between AT base pairs, while three hydrogen bonds occur between G and C, the DNA strands are more easily separated at an AT-rich region.

Following separation of the AT-rich region, the DnaA proteins, with the help of the DnaC protein, recruit **DNA helicase** enzymes to this site. DNA helicase is also known as DnaB protein. When a DNA helicase encounters a double-stranded region,

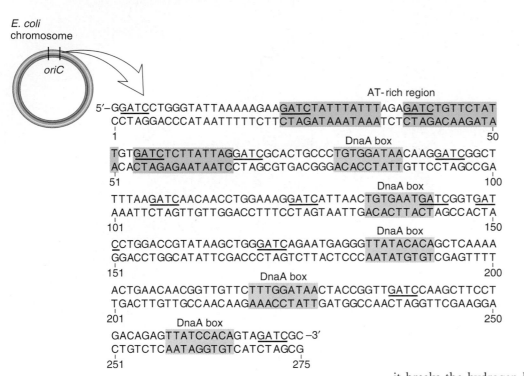

FIGURE 11.5 **The sequence of *oriC* in *E. coli*.** The AT-rich region is composed of three tandem repeats that are 13 base pairs long and highlighted in blue. The five DnaA boxes are highlighted in orange. The GATC methylation sites are underlined.

DnaA proteins bind to DnaA boxes and to each other. Additional proteins that cause the DNA to bend also bind (not shown). This causes the region to wrap around the DnaA proteins and separates the AT-rich region.

DNA helicase (DnaB protein) binds to the origin. DnaC protein (not shown) assists this process.

DNA helicase separates the DNA in both directions, creating 2 replication forks.

it breaks the hydrogen bonds between the two strands, thereby generating two single strands. Two DNA helicases begin strand separation within the *oriC* region and continue to separate the DNA strands beyond the origin. These enzymes use the energy from ATP hydrolysis to catalyze the separation of the double-stranded parental DNA. In *E. coli*, DNA helicases bind to single-stranded DNA and travel along the DNA in a 5' to 3' direction to keep the replication fork moving. As shown in Figure 11.6, the action of DNA helicases promotes the movement of two replication forks outward from *oriC* in opposite directions. This initiates the replication of the bacterial chromosome in both directions, an event termed **bidirectional replication.**

Several Proteins Are Required for DNA Replication at the Replication Fork

Figure 11.7 provides an overview of the molecular events that occur as one of the two replication forks moves around the bacterial chromosome, and **Table 11.1** summarizes the functions of the major proteins involved in *E. coli* DNA replication. Let's begin with strand separation. To act as a template for DNA replication, the strands of a double helix must separate. As mentioned previously, the function of DNA helicase is to break the hydrogen bonds between base pairs and thereby unwind the strands; this action generates positive supercoiling ahead of each replication

ONLINE ANIMATION

FIGURE 11.6 **The events that occur at *oriC* to initiate the DNA replication process.** To initiate DNA replication, DnaA proteins bind to the five DnaA boxes, which causes the DNA strands to separate at the AT-rich region. DnaA and DnaC proteins then recruit DNA helicase (DnaB) into this region. Each DNA helicase is composed of six subunits, which form a ring around one DNA strand and migrates in the 5' to 3' direction. As shown here, the movement of two DNA helicase proteins serves to separate the DNA strands beyond the *oriC* region.

Functions of key proteins involved with DNA replication

- DNA helicase breaks the hydrogen bonds between the DNA strands.

- Topoisomerase alleviates positive supercoiling.

- Single-strand binding proteins keep the parental strands apart.

- Primase synthesizes an RNA primer.

- DNA polymerase III synthesizes a daughter strand of DNA.

- DNA polymerase I excises the RNA primers and fills in with DNA (not shown).

- DNA ligase covalently links the Okazaki fragments together.

ONLINE ANIMATION

FIGURE 11.7   The proteins involved with DNA replication.

Note: The drawing of DNA polymerase III depicts the catalytic subunit that synthesizes DNA.

fork. As shown in Figure 11.7, an enzyme known as a **topoisomerase (type II),** also called **DNA gyrase,** travels in front of DNA helicase and alleviates positive supercoiling.

After the two parental DNA strands have been separated and the supercoiling relaxed, they must be kept that way until the complementary daughter strands have been made. What prevents the DNA strands from coming back together? DNA replication requires **single-strand binding proteins** that bind to the strands of parental DNA and prevent them from re-forming a double helix. In this way, the bases within the parental strands are kept in an exposed condition that enables them to hydrogen bond with individual nucleotides.

The next event in DNA replication involves the synthesis of short strands of RNA (rather than DNA) called **RNA primers.** These strands of RNA are synthesized by the linkage of ribonucleotides via an enzyme known as **DNA primase,** or simply **primase.** This enzyme synthesizes short strands of RNA, typically 10 to 12 nucleotides in length. These short RNA strands start, or prime, the process of DNA replication. In the **leading strand,** a single primer is made at the origin of replication. In the **lagging strand,** multiple primers are made. As discussed later, the RNA primers are eventually removed.

A type of enzyme known as **DNA polymerase** is responsible for synthesizing the DNA of the leading and lagging strands. This enzyme catalyzes the formation of covalent bonds between adjacent nucleotides and thereby makes the new daughter strands. In *E. coli,* five distinct proteins function as DNA polymerases and are designated polymerase I, II, III, IV, and V. DNA polymerases I and III are involved in normal DNA replication, while DNA polymerases II, IV, and V play a role in DNA repair and the replication of damaged DNA.

TABLE 11.1

Proteins Involved in *E. coli* DNA Replication

Common Name	Function
DnaA protein	Binds to DnaA boxes within the origin to initiate DNA replication
DnaC protein	Aids DnaA in the recruitment of DNA helicase to the origin
DNA helicase (DnaB)	Separates double-stranded DNA
Topoisomerase	Removes positive supercoiling ahead of the replication fork
Single-strand binding protein	Binds to single-stranded DNA and prevents it from re-forming a double-stranded structure
Primase	Synthesizes short RNA primers
DNA polymerase III	Synthesizes DNA in the leading and lagging strands
DNA polymerase I	Removes RNA primers, fills in gaps with DNA
DNA ligase	Covalently attaches adjacent Okazaki fragments
Tus	Binds to ter sequences and prevents the advancement of the replication fork

DNA polymerase III is responsible for most of the DNA replication. It is a large enzyme consisting of 10 different subunits that play various roles in the DNA replication process (**Table 11.2**). The α subunit actually catalyzes the bond formation between adjacent nucleotides, while the remaining nine subunits fulfill other functions. The complex of all 10 subunits together is called DNA polymerase III holoenzyme. By comparison, DNA polymerase I is composed of a single subunit. Its role during DNA replication is to remove the RNA primers and fill in the vacant regions with DNA.

Though the various DNA polymerases in *E. coli* and other bacterial species vary in their subunit composition, several common structural features have emerged. The catalytic subunit of all DNA polymerases has a structure that resembles a human hand. As shown in **Figure 11.8**, the template DNA is threaded through the palm of the hand; the thumb and fingers are wrapped around the DNA. The incoming dNTPs enter the catalytic site, bind to the template strand according to the AT/GC rule, and then are covalently attached to the 3' end of the growing strand. DNA polymerase also contains a 3' exonuclease site that removes mismatched bases, as described later.

As researchers began to unravel the function of DNA polymerase, two features seemed unusual (**Figure 11.9**). DNA polymerase cannot begin DNA synthesis by linking together the first two individual nucleotides. Rather, this type of enzyme can elongate only a preexisting strand starting with an RNA primer or existing DNA strand (Figure 11.9a). A second unusual feature is the directionality of strand synthesis. DNA polymerase can attach nucleotides only in the 5' to 3' direction, not in the 3' to 5' direction (Figure 11.9b).

Due to these two unusual features, the synthesis of the leading and lagging strands shows distinctive differences (**Figure 11.10**). The synthesis of RNA primers by DNA primase allows DNA polymerase III to begin the synthesis of complementary daughter strands of DNA. DNA polymerase III catalyzes the attachment of nucleotides to the 3' end of each primer, in a 5' to 3' direction. In the leading strand, one RNA primer is made at the origin, and then DNA polymerase III can attach nucleotides in a 5' to 3' direction as it slides toward the opening of the replication fork. The synthesis of the leading strand is therefore continuous.

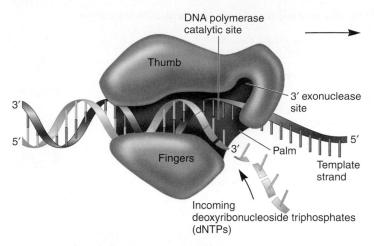

(a) Schematic side view of DNA polymerase III

(b) Molecular model for DNA polymerase bound to DNA

(Reprinted by permission from Macmillan Publishers Ltd. *The Embo Journal*. Crystal structures of open and closed forms of binary and ternary complexes of the large fragment of *Thermus aquaticus* DNA polymerase I: structural basis for nucleotide incorporation. Ying Li et al. 17:24, 7514–7525, 1998.)

FIGURE 11.8   The action of DNA polymerase. (a) DNA polymerase slides along the template strand as it synthesizes a new strand by connecting deoxyribonucleoside triphosphates (dNTPs) in a 5' to 3' direction. The catalytic subunit of DNA polymerase resembles a hand that is wrapped around the template strand. In this regard, the movement of DNA polymerase along the template strand is similar to a hand that is sliding along a rope. **(b)** The molecular structure of DNA polymerase I from the bacterium *Thermus aquaticus*. This model shows a portion of DNA polymerase I that is bound to DNA. This molecular structure depicts a front view of DNA polymerase; part (a) is a schematic side view.

In the lagging strand, the synthesis of DNA also elongates in a 5' to 3' manner, but it does so in the direction away from the replication fork. In the lagging strand, RNA primers must repeatedly initiate the synthesis of short segments of DNA; thus, the synthesis has to be discontinuous. The length of these fragments in bacteria is typically 1,000 to 2,000 nucleotides. In eukaryotes, the fragments are shorter—100 to 200 nucleotides. Each fragment contains a short RNA primer at the 5' end, which is made by primase. The remainder of the fragment is a strand of DNA

TABLE 11.2

Subunit Composition of DNA Polymerase III Holoenzyme from *E. coli*

Subunit(s)	Function
α	Synthesizes DNA
ε	3' to 5' proofreading (removes mismatched nucleotides)
θ	Accessory protein that stimulates the proofreading function
β	Clamp protein, which allows DNA polymerase to slide along the DNA without falling off
τ, γ, δ, δ', ψ, and χ	Clamp loader complex, involved with helping the clamp protein bind to the DNA

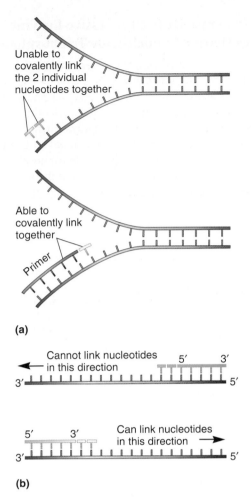

(a)

(b)

FIGURE 11.9 **Unusual features of DNA polymerase function.**
(a) DNA polymerase can elongate a strand only from an RNA primer or existing DNA strand. (b) DNA polymerase can attach nucleotides only in a 5′ to 3′ direction. Note the template strand is in the opposite, 3′ to 5′, direction.

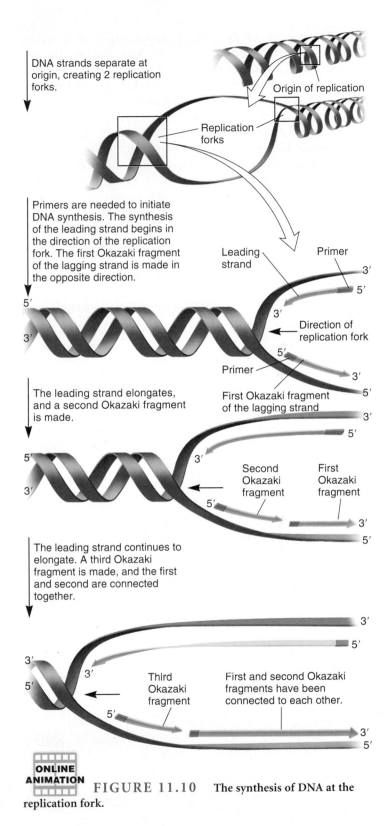

FIGURE 11.10 **The synthesis of DNA at the replication fork.**

made by DNA polymerase III. The DNA fragments made in this manner are known as **Okazaki fragments,** after Reiji and Tuneko Okazaki, who initially discovered them in the late 1960s.

To complete the synthesis of Okazaki fragments within the lagging strand, three additional events must occur: removal of the RNA primers, synthesis of DNA in the area where the primers have been removed, and the covalent attachment of adjacent fragments of DNA (see Figure 11.10 and refer back to Figure 11.7). In *E. coli*, the RNA primers are removed by the action of DNA polymerase I. This enzyme has a 5′ to 3′ exonuclease activity, which means that DNA polymerase I digests away the RNA primers in a 5′ to 3′ direction, leaving a vacant area. DNA polymerase I then fills in this region by synthesizing DNA there. It uses the 3′ end of an adjacent Okazaki fragment as a primer. For example, in Figure 11.10, DNA polymerase I would remove the RNA primer from the first Okazaki fragment and then synthesize DNA in the vacant region by attaching nucleotides to the 3′ end of the second Okazaki fragment. After the gap has been completely filled in, a covalent bond is still missing between the last nucleotide added by DNA polymerase I and the adjacent

DNA strand that had been previously made by DNA polymerase III. An enzyme known as **DNA ligase** catalyzes a covalent bond between adjacent fragments to complete the replication process in the lagging strand (refer back to Figure 11.7). In *E. coli*, DNA ligase requires NAD$^+$ to carry out this reaction, whereas the DNA ligases found in archaea and eukaryotes require ATP.

The synthesis of the lagging strand was studied by the Okazakis using radiolabeled nucleotides. They incubated *E. coli* cells with radiolabeled thymidine for 15 seconds and then added an excess of nonlabeled thymidine. This is termed a **pulse/chase experiment** because the cells were given the radiolabeled compound for a brief period of time—a pulse—followed by an excess amount of unlabeled compound—a chase. They then isolated DNA from samples of cells at timed intervals after the pulse/chase. The DNA was denatured into single-stranded molecules, and the sizes of the radiolabeled DNA strands were determined by centrifugation. At quick time intervals, such as only a few seconds following the thymidine incubation, the fragments were found to be short, in the range of 1,000 to 2,000 nucleotides in length. At extended time intervals, the radiolabeled strands became much longer. At these later time points, the adjacent Okazaki fragments would have had enough time to link together.

DNA Polymerase III Is a Processive Enzyme That Uses Deoxyribonucleoside Triphosphates

Let's now turn our attention to other enzymatic features of DNA polymerase. As shown in **Figure 11.11**, DNA polymerases catalyze the covalent attachment between the phosphate in one nucleotide and the sugar in the previous nucleotide. The formation of this covalent (ester) bond requires an input of energy. Prior to bond formation, the nucleotide about to be attached to the growing strand is a deoxyribonucleoside triphosphate (dNTP). It contains three phosphate groups attached at the 5'–carbon atom of deoxyribose. The deoxyribonucleoside triphosphate first enters the catalytic site of DNA polymerase and binds to the template strand according to the AT/GC rule. Next, the 3'–OH group on the previous nucleotide reacts with the phosphate group adjacent to the sugar on the incoming nucleotide. The breakage of a covalent bond between two phosphates in deoxyribonucleoside triphosphate is a highly exergonic reaction that provides the energy to

FIGURE 11.11 **The enzymatic action of DNA polymerase.** An incoming deoxyriboucleoside triphosphate (dNTP) is cleaved to form a nucleoside monophosphate and pyrophosphate. The energy released from this exergonic reaction allows the nucleoside monophosphate to form a covalent (ester) bond at the 3' end of the growing strand. This reaction is catalyzed by DNA polymerase. Pyrophosphate (PP$_i$) is released.

form a covalent (ester) bond between the sugar at the 3' end of the DNA strand and the phosphate of the incoming nucleotide. The formation of this covalent bond causes the newly made strand to grow in the 5' to 3' direction. As shown in Figure 11.11, pyrophosphate (PP_i) is released.

As noted in Chapter 9 (Figure 9.11), the term phosphodiester linkage (also called a phosphodiester bond) is used to describe the linkage between a phosphate and two sugar molecules. As its name implies, a phosphodiester linkage involves two ester bonds. In comparison, as a DNA strand grows, a single covalent (ester) bond is formed between adjacent nucleotides (see Figure 11.11). The other ester bond in the phosphodiester linkage—the bond between the 5'-oxygen and phosphorus—is already present in the incoming nucleotide.

DNA polymerase catalyzes the covalent attachment of nucleotides with great speed. In *E. coli*, DNA polymerase III attaches approximately 750 nucleotides per second! DNA polymerase III can catalyze the synthesis of the daughter strands so quickly because it is a **processive enzyme.** This means it does not dissociate from the growing strand after it has catalyzed the covalent joining of two nucleotides. Rather, as depicted in Figure 11.8a, it remains clamped to the DNA template strand and slides along the template as it catalyzes the synthesis of the daughter strand. The β (beta) subunit of the holoenzyme, also known as the clamp protein, promotes the association of the holoenzyme with the DNA as it glides along the template strand (refer back to Table 11.2). The β subunit forms a dimer in the shape of a ring; the hole of the ring is large enough to accommodate a double-stranded DNA molecule, and its width is about one turn of DNA. A complex of several subunits functions as a clamp loader that allows the DNA polymerase holoenzyme to initially clamp onto the DNA.

The effects of processivity are really quite remarkable. In the absence of the β subunit, DNA polymerase can synthesize DNA at a rate of approximately only 20 nucleotides per second. On average, it falls off the DNA template after about ten nucleotides have been linked together. By comparison, when the β subunit is present, as in the holoenzyme, the synthesis rate is approximately 750 nucleotides per second. In the leading strand, DNA polymerase III has been estimated to synthesize a segment of DNA that is over 500,000 nucleotides in length before it inadvertently falls off.

Replication Is Terminated When the Replication Forks Meet at the Termination Sequences

On the opposite side of the *E. coli* chromosome from *oriC* is a pair of **termination sequences** called ter sequences. A protein known as the termination utilization substance (Tus) binds to the ter sequences and stops the movement of the replication forks. As shown in **Figure 11.12**, one of the ter sequences designated T1 prevents the advancement of the fork moving left to right, but allows the movement of the other fork (see the inset to Figure 11.12). Alternatively, T2 prevents the advancement of the fork moving right to left, but allows the advancement of the other fork. In any given cell, only one ter sequence is required to stop the advancement of one replication fork, and then the other fork ends its synthesis of DNA when it reaches the halted replica-

FIGURE 11.12   The termination of DNA replication. Two sites in the bacterial chromosome, shown with rectangles, are ter sequences designated T1 and T2. The T1 site prevents the further advancement of the fork moving left to right, while T2 prevents the advancement of the fork moving right to left. As shown in the inset, the binding of Tus prevents the replication forks from proceeding past the ter sequences in a particular direction.

tion fork. In other words, DNA replication ends when oppositely advancing forks meet, usually at T1 or T2. Finally, DNA ligase covalently links the two daughter strands, creating two circular, double-stranded molecules.

After DNA replication is completed, one last problem may exist. DNA replication often results in two intertwined DNA molecules known as **catenanes** (**Figure 11.13**). Fortunately, catenanes are only transient structures in DNA replication. In *E. coli*, topoisomerase introduces a temporary break into the DNA strands and then rejoins them after the strands have become unlocked. This allows the catenanes to be separated into individual circular molecules.

Certain Enzymes of DNA Replication Bind to Each Other to Form a Complex

Figure 11.14 provides a more three-dimensional view of the DNA replication process. DNA helicase and primase are physically bound to each other to form a complex known as a **primosome.** This complex leads the way at the replication fork. The primosome tracks along the DNA, separating the parental strands and synthesizing RNA primers at regular intervals along the lagging strand. By acting within a complex, the actions of DNA helicase and primase can be better coordinated.

The primosome is physically associated with two DNA polymerase holoenzymes to form a **replisome.** As shown in Figure 11.14, two DNA polymerase III proteins act in concert to replicate the leading and lagging strands. The term **dimeric DNA polymerase** is used to describe two DNA polymerase holoenzymes that move as a unit toward the replication fork. For this to occur, the lagging strand is looped out with respect to the DNA polymerase that synthesizes the lagging strand. This loop allows

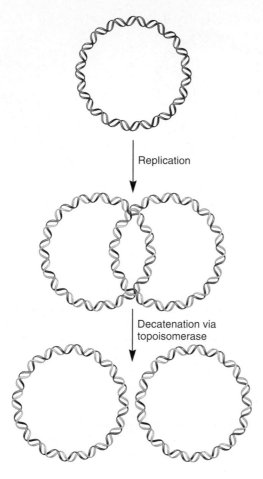

FIGURE 11.13 **Separation of catenanes.** DNA replication can result in two intertwined chromosomes called catenanes. These catenanes can be separated by the action of topoisomerase.

the lagging-strand polymerase to make DNA in a 5' to 3' direction yet move toward the opening of the replication fork. Interestingly, when this DNA polymerase reaches the end of an Okazaki fragment, it must be released from the template DNA and "hop" to the RNA primer that is closest to the fork. The clamp loader complex (see Table 11.2), which is part of DNA polymerase holoenzyme, then reloads the enzyme at the site where the next RNA primer has been made. Similarly, after primase synthesizes an RNA primer in the 5' to 3' direction, it must hop over the primer and synthesize the next primer closer to the replication fork.

The Fidelity of DNA Replication Is Ensured by Proofreading Mechanisms

With replication occurring so rapidly, one might imagine that mistakes could happen in which the wrong nucleotide could be incorporated into the growing daughter strand. Although mistakes can happen during DNA replication, they are extraordinarily rare. In the case of DNA synthesis via DNA polymerase III, only one mistake per 100 million nucleotides is made. Therefore, DNA synthesis occurs with a high degree of accuracy or **fidelity.**

Why is the fidelity so high? First, the hydrogen bonding between G and C or A and T is much more stable than between mismatched pairs. However, this stability accounts for only part of the fidelity, because mismatching due to stability considerations would account for one mistake per 1,000 nucleotides. Two additional characteristics of DNA polymerases contribute to the fidelity of DNA replication. The active site of DNA polymerase is such that it will preferentially catalyze the attachment of nucleotides when the correct bases are located in opposite strands. In other words, DNA polymerase is unlikely to catalyze bond formation between adjacent nucleotides if a mismatched base pair is formed.

FIGURE 11.14 **A three-dimensional view of DNA replication.** DNA helicase and primase associate together to form a primosome. The primosome associates with two DNA polymerase enzymes to form a replisome.

Mismatch causes DNA polymerase to pause, leaving mismatched nucleotide near the 3′ end.

3′ exonuclease site

Template strand

3′

5′

T

C

3′

Base pair mismatch near the 3′ end

The 3′ end enters the exonuclease site.

3′

5′

3′

5′

At the 3′ exonuclease site, the strand is digested in the 3′ to 5′ direction until the incorrect nucleotide is removed.

Incorrect nucleotide removed

3′

3′

5′

5′

ONLINE ANIMATION

FIGURE 11.15   The proofreading function of DNA polymerase. When a base pair mismatch is found, the end of the newly made strand is shifted into the 3′ exonuclease site. The DNA is digested in the 3′ to 5′ direction to release the incorrect nucleotide.

This phenomenon, called induced fit, decreases the error rate to a range of 1 in 100,000 to 1 million.

Another way that DNA polymerase decreases the error rate is by the enzymatic removal of mismatched nucleotides. As shown in **Figure 11.15**, DNA polymerase can identify a mismatched nucleotide and remove it from the daughter strand. This occurs by exonuclease cleavage of the bonds between adjacent nucleotides at the 3′ end of the newly made strand. The ability to remove mismatched bases by this mechanism is called the **proofreading function** of DNA polymerase. Proofreading occurs by the removal of nucleotides in the 3′ to 5′ direction at the 3′ exonuclease site. After the mismatched nucleotide is removed, DNA polymerase resumes DNA synthesis in the 5′ to 3′ direction.

Bacterial DNA Replication Is Coordinated with Cell Division

Bacterial cells can divide into two daughter cells at an amazing rate. Under optimal conditions, certain bacteria such as *E. coli* can divide every 20 to 30 minutes. DNA replication should take place only when a cell is about to divide. If DNA replication occurs too frequently, too many copies of the bacterial chromosome will be found in each cell. Alternatively, if DNA replication does not occur frequently enough, a daughter cell will be left without a chromosome. Therefore, cell division in bacterial cells must be coordinated with DNA replication.

Bacterial cells regulate the DNA replication process by controlling the initiation of replication at *oriC*. This control has been extensively studied in *E. coli*. In this bacterium, several different mechanisms may control DNA replication. In general, the regulation prevents the premature initiation of DNA replication at *oriC*.

After the initiation of DNA replication, DnaA protein hydrolyzes its ATP and therefore switches to an ADP-bound form. DnaA-ADP has a lower affinity for DnaA boxes and does not readily form a complex. This prevents premature initiation. In addition, the initiation of replication is controlled by the amount of the DnaA protein (**Figure 11.16**). As discussed previously, the DnaA protein binds to DnaA boxes within the origin of replication, causing the DNA helix to separate at the AT-rich region within the origin. To initiate DNA replication, the concentration of the DnaA protein must be high enough so that it can bind to all of the DnaA boxes and form a complex. Immediately following DNA replication, the number of DnaA boxes is double, so an insufficient amount of DnaA protein is available to initiate a second round of replication. Also, some of the DnaA protein may be rapidly degraded and some of it may be inactive because it becomes attached to other regions of chromosomal DNA and to the cell membrane during cell division. Because it takes time to accumulate newly made DnaA protein, DNA replication cannot occur until the daughter cells have had time to grow.

FIGURE 11.16 **The amount of DnaA protein provides a way to regulate DNA replication.** To begin replication, enough DnaA protein must be present to bind to all of the DnaA boxes. Immediately after DNA replication, insufficient DnaA protein is available to reinitiate a second (premature) round of DNA replication at the two origins of replication. This is because twice as many DnaA boxes are found after DNA replication and because some DnaA proteins may be degraded or stuck to other chromosomal sites and to the cell membrane.

(a) Structure of adenine and methyladenine

(b) Results immediately after DNA replication

FIGURE 11.17 **Methylation of GATC sites in *oriC*.** (**a**) The action of Dam, which covalently attaches a methyl group to adenine to form methyladenine. (**b**) Prior to DNA replication, the action of Dam causes both adenines within the GATC sites to be methylated. After DNA replication, only the adenines in the original strands are methylated. Several minutes will pass before Dam will methylate these unmethylated adenines.

Another way to regulate DNA replication involves the GATC methylation sites within *oriC*. These sites can be methylated by an enzyme known as <u>D</u>NA <u>a</u>denine <u>m</u>ethyltransferase (Dam). The Dam enzyme recognizes the 5'–GATC–3' sequence, binds there, and attaches a methyl group onto the adenine base, forming methyladenine (**Figure 11.17a**). DNA methylation within *oriC* helps regulate the replication process. Prior to DNA replication, these sites are methylated in both strands. This full methylation of the 5'–GATC–3' sites facilitates the initiation of DNA replication at the origin. Following DNA replication, the newly made strands are not methylated, because adenine rather than methyladenine is incorporated into the daughter strands (**Figure 11.17b**). The initiation of DNA replication at the origin does not readily occur until after it has become fully methylated. Because it takes several minutes for Dam to methylate the 5'–GATC–3' sequences within this region, DNA replication will not occur again too quickly.

EXPERIMENT 11B

DNA Replication Can Be Studied In Vitro

Much of our understanding of bacterial DNA replication has come from thousands of experiments in which DNA replication has been studied in vitro. This approach was pioneered by Arthur Kornberg in the 1950s, who received a Nobel Prize for his efforts in 1959.

Figure 11.18 describes Kornberg's approach to monitor DNA replication in vitro. In this experiment, an extract of proteins from *E. coli* was used. Although we will not consider the procedures for purifying replication proteins, an alternative approach is to purify individual proteins from the extract and study their functions individually. In either case, the proteins would be mixed with template DNA and radiolabeled nucleotides. Kornberg correctly hypothesized that deoxyribonucleoside triphosphates (dNTPs) are the precursors for DNA synthesis. Also, he knew that deoxyribonucleoside triphosphates are soluble

in an acidic solution, whereas long strands of DNA will precipitate out of solution at an acidic pH. This precipitation event provides a method to separate nucleotides—in this case, deoxyribonucleoside triphosphates—from strands of DNA. Therefore, after the proteins, template DNA, and nucleotides were incubated for a sufficient time to allow the synthesis of new strands, step 3 of this procedure involved the addition of perchloric acid. This step precipitated strands of DNA, which were then separated from the radiolabeled nucleotides via centrifugation. Newly made strands of DNA, which were radiolabeled, sediment to the pellet, while radiolabeled nucleotides that had not been incorporated into new strands remained in the supernatant.

■ THE HYPOTHESIS

DNA synthesis can occur in vitro if all the necessary components are present.

■ TESTING THE HYPOTHESIS — FIGURE 11.18   In vitro synthesis of DNA strands.

Starting material: An extract of proteins from *E. coli*.

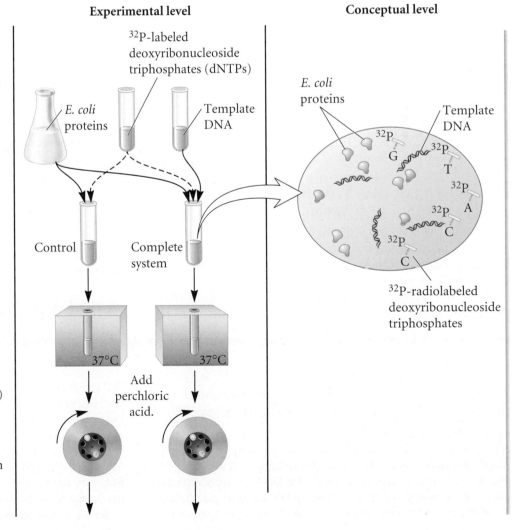

Experimental level

Conceptual level

1. Mix together the extract of *E. coli* proteins, template DNA that is not radiolabeled, and ^{32}P-radiolabeled deoxyribonucleoside triphosphates. This is expected to be a complete system that contains everything necessary for DNA synthesis. As a control, a second sample is made in which the template DNA was omitted from the mixture.

2. Incubate the mixture for 30 minutes at 37°C.

3. Add perchloric acid to precipitate DNA. (It does not precipitate free nucleotides.)

4. Centrifuge the tube.
   Note: The radiolabeled deoxyribonucleoside triphosphates that have not been incorporated into DNA will remain in the supernatant.

^{32}P-labeled deoxyribonucleoside triphosphates (dNTPs)

E. coli proteins

Template DNA

Control

Complete system

37°C   37°C

Add perchloric acid.

E. coli proteins

Template DNA

^{32}P G

^{32}P T

^{32}P A

^{32}P C

^{32}P C

^{32}P-radiolabeled deoxyribonucleoside triphosphates

Free labeled nucleotides

Supernatant

Pellets

5. Collect the pellet, which contains precipitated DNA and proteins. (The control pellet is not expected to contain DNA.)

6. Count the amount of radioactivity in the pellet using a scintillation counter. (See the Appendix.)

DNA with ^{32}P-labeled nucleotides in new strand

■ THE DATA

Conditions	Amount of Radiolabeled DNA*
Complete system	3,300
Control (template DNA omitted)	0

*Calculated in picomoles of ^{32}P-labeled DNA.

■ INTERPRETING THE DATA

As shown in the data of Figure 11.18, when the *E. coli* proteins were mixed with nonlabeled template DNA and radiolabeled deoxyribonucleoside triphosphates, an acid-precipitable, radiolabeled product was formed. This product was newly synthesized DNA strands. As a control, if nonradiolabeled template DNA was omitted from the assay, no radiolabeled DNA was made. This is the expected result, because the template DNA is necessary to make new daughter strands. Taken together, these results indicate that this technique can be used to measure the synthesis of DNA in vitro.

The in vitro approach has provided the foundation to study the replication process at the molecular level. A common experimental strategy is to purify proteins from cell extracts and to determine their roles in the replication process. In other words, purified proteins, such as those described in Table 11.1, can be mixed with nucleotides, template DNA, and other substances in a test tube to determine if the synthesis of new DNA strands occurs. This approach still continues, particularly as we try to understand the added complexities of eukaryotic DNA replication.

A self-help quiz involving this experiment can be found at the Online Learning Center.

The Isolation of Mutants Has Been Instrumental to Our Understanding of DNA Replication

In the previous experiment, we considered an experimental strategy to study DNA synthesis in vitro. In his early experiments, Arthur Kornberg used crude extracts containing *E. coli* proteins and monitored their ability to synthesize DNA. In such extracts, the predominant polymerase enzyme is DNA polymerase I. Surprisingly, its activity is so high that it is nearly impossible to detect the activities of the other DNA polymerases. For this reason, researchers in the 1950s and 1960s thought that DNA polymerase I was the only enzyme responsible for DNA replication. This situation dramatically changed as a result of mutant isolation.

In 1969, Paula DeLucia and John Cairns isolated a mutant in which DNA polymerase I lacked its 5' to 3' polymerase function but retained its 5' to 3' exonuclease function, which is needed to remove RNA primers. This mutant was identified by screening thousands of bacterial colonies that had been subjected to mutagens—agents that cause mutations. This result indicated that the DNA-synthesizing function of DNA polymerase I is not absolutely required for bacteria to replicate their DNA, because the strain harboring this mutation could grow normally. How is this possible? The researchers concluded that *E. coli* must have other DNA polymerases. Therefore, DeLucia and Cairns set out to find these seemingly elusive enzymes.

The isolation of mutants was one way that helped researchers identify additional DNA polymerase enzymes, namely DNA polymerase II and III. In addition, mutant isolation played a key role in the identification of other proteins needed to replicate the leading and lagging strands, as well as proteins that recognize the origin of replication and the ter sites. Because DNA replication is vital for cell division, most mutations that block DNA replication would be lethal to a growing population of bacterial cells. For this reason, if researchers want to identify loss-of-function mutations in vital genes, they must screen for **conditional mutants.** A type of conditional mutant is a **temperature-sensitive (ts) mutant.** In the case of a vital gene, an organism harboring a temperature-sensitive mutation can survive at the permissive temperature but not at the nonpermissive temperature. For example, a ts mutant might survive and grow at 30°C (the permissive temperature) but fail to grow at 42°C (the nonpermissive temperature). The higher temperature inactivates the function of the protein encoded by the mutant gene.

Figure 11.19 shows a general strategy for the isolation of ts mutants. Researchers expose bacterial cells to a mutagen that increases the likelihood of mutations. The mutagenized cells are plated on growth media and incubated at the permissive temperature. The colonies are then replica plated onto two plates, one incubated at the permissive temperature and one at the nonpermissive temperature. As seen here, this enables researchers to identify ts mutations that are lethal at the nonpermissive temperature.

With regard to the study of DNA replication, researchers analyzed a large number of ts mutants to discover if any of them had a defect in DNA replication. For example, one could expose a ts mutant to radiolabeled thymine (a base that is incorporated into DNA), shift to the nonpermissive temperature, and determine if a mutant strain could make radiolabeled DNA, using procedures that are similar to those described in Figure 11.18. Because *E. coli* has many vital genes not involved with DNA replication, only a small subset of ts mutants would be expected to have mutations in genes that encode proteins that are critical to the replication process. Therefore, researchers had to screen many thousands of ts mutants to identify the few involved in DNA replication. This approach is sometimes called a "brute force" genetic screen.

Table 11.3 summarizes some of the genes that were identified using this type of strategy. The genes were originally designated with the name dna, followed by a capital letter that generally refers to the order in which they were discovered. When shifted to the nonpermissive temperature, certain mutants showed a rapid arrest of DNA synthesis. These so-called rapid-stop mutations inactivated genes that encode enzymes needed for DNA replication. By comparison, other mutants were able to

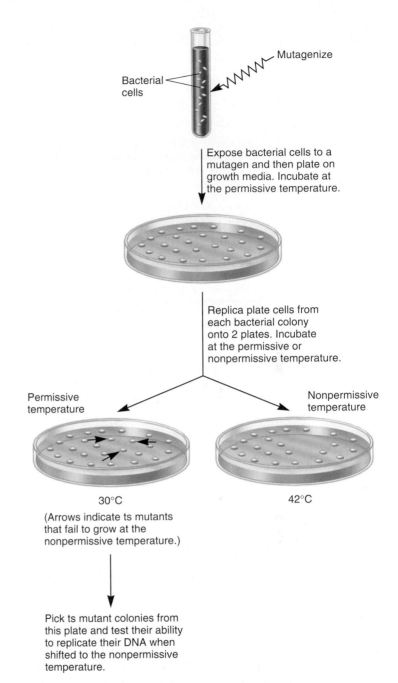

FIGURE 11.19   A strategy to identify ts mutations in vital genes. In this approach, bacteria are mutagenized, which increases the likelihood of mutation, and then grown at the permissive temperature. Colonies are then replica plated and grown at both the permissive and nonpermissive temperatures. (Note: The procedure of replica plating is shown in Chapter 16, Figure 16.7.) Ts mutants fail to grow at the nonpermissive temperature. The appropriate colonies can be picked from the plates, grown at the permissive temperature, and analyzed to see if DNA replication is altered at the nonpermissive temperature.

complete their current round of replication but could not start another round of replication. These slow-stop mutants involved genes that encode proteins needed for the initiation of replication at the origin. In later studies, the proteins encoded by these

TABLE 11.3

Examples of ts Mutants Involved in DNA Replication in *E. coli*

Gene Name	Protein Function
Rapid-Stop Mutants	
dnaE	α subunit of DNA polymerase III, synthesizes DNA
dnaX	τ subunit of DNA polymerase III, promotes the dimerization of two DNA polymerase III proteins together at the replication fork and stimulates DNA helicase
dnaN	β subunit of DNA polymerase III, functions as a clamp protein that makes DNA polymerase a processive enzyme
dnaZ	γ subunit of DNA polymerase III, helps the β subunit bind to the DNA
dnaG	Primase, needed to make RNA primers
dnaB	Helicase, needed to unwind the DNA strands during replication
Slow-Stop Mutants	
dnaA	DnaA protein that recognizes the DnaA boxes at the origin
dnaC	DnaC protein that recruits DNA helicase to the origin

genes were purified, and their functions were studied in vitro. This work contributed greatly to our modern understanding of DNA replication at the molecular level.

11.3 EUKARYOTIC DNA REPLICATION

Eukaryotic DNA replication is not as well understood as bacterial replication. Much research has been carried out on a variety of experimental organisms, particularly yeast and mammalian cells. Many of these studies have found extensive similarities between the general features of DNA replication in prokaryotes and eukaryotes. For example, DNA helicases, topoisomerases, single-strand binding proteins, primases, DNA polymerases, and DNA ligases—the types of bacterial enzymes described in Table 11.1—have also been identified in eukaryotes. Nevertheless, at the molecular level, eukaryotic DNA replication appears to be substantially more complex. These additional intricacies of eukaryotic DNA replication are related to several features of eukaryotic cells. In particular, eukaryotic cells have larger, linear chromosomes, the chromatin is tightly packed within nucleosomes, and cell cycle regulation is much more complicated. This section will emphasize some of the unique features of eukaryotic DNA replication.

Initiation Occurs at Multiple Origins of Replication on Linear Eukaryotic Chromosomes

Because eukaryotes have long, linear chromosomes, the chromosomes require multiple origins of replication so that the DNA can be replicated in a reasonable length of time. In 1968, Joel Huberman and Arthur Riggs provided evidence for multiple origins of replication by adding a radiolabeled nucleoside (^{3}H-thymidine) to a culture of actively dividing cells, followed by a chase with nonlabeled thymidine. The radiolabeled thymidine was taken up by the cells and incorporated into their newly made DNA strands for a brief period. The chromosomes were then isolated from the cells and subjected to autoradiography. As seen in **Figure 11.20**, radiolabeled segments were interspersed among nonlabeled segments. This result is consistent with the hypothesis that eukaryotic chromosomes contain multiple origins of replication.

As shown schematically in **Figure 11.21a**, DNA replication proceeds bidirectionally from many origins of replication during S phase of the cell cycle. The multiple replication forks eventually make contact with each other to complete the replication process. **Figure 11.21b** shows a region of eukaryotic DNA in the process of replication. As DNA replication radiates bidirectionally from each origin, regions are formed that contain two double helices. These regions are punctuated by other regions that have not yet replicated and consist of one double helix.

The molecular features of eukaryotic origins of replication may have some similarities to the origins found in bacteria. At the molecular level, eukaryotic origins of replication have been extensively studied in the yeast *Saccharomyces cerevisiae*. In this organism, several replication origins have been identified and sequenced. They have been named **ARS elements** (for Autonomously Replicating Sequence). ARS elements, which are about 50 bp in length, are necessary to initiate chromosome replication. ARS elements have unique features of their DNA sequences. First, they contain a higher percentage of A and T bases than the rest of the chromosomal DNA. In addition, they contain a copy of the ARS consensus sequence (ACS), ATTTAT(A or G)TTTA, along with additional elements that enhance origin function. This arrangement is similar to bacterial origins. In *S. cerevisiae*, origins of replication are determined primarily by their DNA sequences.

FIGURE 11.20   Evidence for multiple origins of replication in eukaryotic chromosomes. In this experiment, cells were given a pulse/chase of ^{3}H-thymidine and unlabeled thymidine. The chromosomes were isolated and subjected to autoradiography. In this micrograph, radiolabeled segments were interspersed among nonlabeled segments, indicating that eukaryotic chromosomes contain multiple origins of replication.

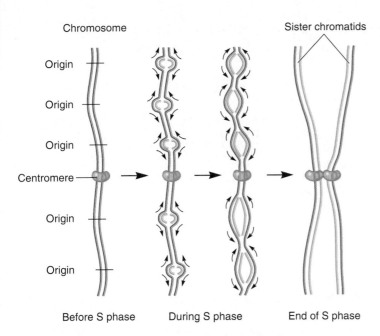

Chromosome   Sister chromatids

Origin

Origin

Origin

Centromere

Origin

Origin

Before S phase   During S phase   End of S phase

(a) DNA replication from multiple origins of replication

5 kb

(b) A micrograph of a replicating eukaryotic chromosome

FIGURE 11.21   **The replication of eukaryotic chromosomes.**
(a) At the beginning of synthesis (S) phase of the cell cycle, eukaryotic chromosome replication begins from multiple origins of replication. As S phase continues, the replication forks move bidirectionally to replicate the DNA. By the end of S phase, all of the replication forks have merged. The net result is two sister chromatids attached to each other at the centromere. **(b)** A micrograph of a replicating chromosome in a eukaryotic species, showing DNA replication radiating bidirectionally from each origin.

In animals, the critical features that define origins of replication are not completely understood. In some cases, origins are not determined by particular DNA sequences but instead occur at specific sites along a chromosome due to chromatin structure and protein modifications.

DNA replication in eukaryotes begins with the assembly of a **prereplication complex (preRC)** consisting of at least 14 different proteins. Part of the preRC is a group of six proteins

called the **origin recognition complex (ORC)** that acts as the initiator of eukaryotic DNA replication. ORC was originally identified in yeast as a protein complex that binds directly to ARS elements. DNA replication at the origin begins with the binding of ORC, which usually occurs during G_1 phase. Other proteins of the preRC then bind, including a group of proteins called **MCM helicase**.[1] The binding of MCM helicase at the origin completes a process called **DNA replication licensing**; only those origins with MCM helicase can initiate DNA synthesis. During S phase, DNA synthesis begins when preRCs are acted upon by at least 22 additional proteins that activate MCM helicase and assemble two divergent replication forks at each replication origin. An important role of these additional proteins is to carefully regulate the initiation of DNA replication so that it happens at the correct time during the cell cycle and occurs only once during the cell cycle. The precise roles of these proteins are under active research investigation.

Eukaryotes Contain Several Different DNA Polymerases

Eukaryotes have many types of DNA polymerases. For example, mammalian cells have well over a dozen different DNA polymerases (**Table 11.4**). Four of these, designated α (alpha), δ (delta), ε (epsilon), and γ (gamma), have the primary function of replicating DNA. DNA polymerase γ functions in the mitochondria to replicate mitochondrial DNA, whereas α, δ, and ε are involved with DNA replication in the cell nucleus during S phase. In the nucleus, DNA polymerase α is the only eukaryotic polymerase

TABLE 11.4

Eukaryotic DNA Polymerases

Polymerase Types*	Function
$\alpha, \delta, \varepsilon$	Replication of nondamaged DNA in the cell nucleus during S phase
γ	Replication of mitochondrial DNA
η, κ, ι, ξ (lesion-replicating polymerases)	Replication of damaged DNA
$\alpha, \beta, \delta, \varepsilon, \sigma, \lambda, \mu, \varphi, \theta, \eta$	DNA repair or other functions†

*The designations are those of mammalian enzymes.
†Many DNA polymerases have dual functions. For example, DNA polymerases α, δ, and ε are involved in the replication of normal DNA and also play a role in DNA repair. In cells of the immune system, certain genes that encode antibodies (i.e., immunoglobulin genes) undergo a phenomenon known as hypermutation. This increases the variation in the kinds of antibodies the cells can make. Certain polymerases in this list, such as η, may play a role in hypermutation of immunoglobulin genes. DNA polymerase σ may play a role in sister chromatid cohesion, a topic discussed in Chapter 10.

---

[1] MCM is an acronym for minichromosome maintenance. The genes encoding MCM proteins were originally identified in mutant yeast strains that are defective in the maintenance of minichromosomes in the cell. MCM proteins have since been shown to play a role in DNA replication.

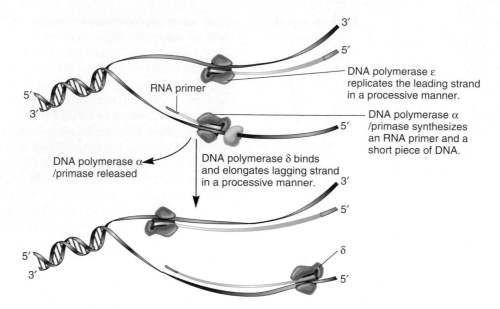

FIGURE 11.22   An example of a DNA polymerase switch. After DNA polymerase α/primase has made an RNA primer and short segment of DNA, this complex is released, and a processive polymerase, such as DNA polymerase δ, attaches to the template strand and rapidly synthesizes DNA. A similar mechanism could switch the α/primase complex for ε in the leading strand.

that associates with primase. The functional role of the DNA polymerase α/primase complex is to synthesize a short RNA-DNA primer of approximately 10 RNA nucleotides followed by 20 to 30 DNA nucleotides. This short RNA-DNA strand is then used by DNA polymerase δ or ε for the processive elongation of the leading and lagging strands (**Figure 11.22**). The exchange of DNA polymerase α for δ or ε is called a **polymerase switch.** It occurs only after the RNA-DNA primer has been made by the complex of DNA polymerase α and primase. The relative importance of DNA polymerase δ versus ε is not completely understood. Some studies have suggested that DNA polymerase ε may be involved with leading-strand synthesis, while DNA polymerase δ has a greater role in lagging-strand synthesis. However, this hypothesis will require further experimentation.

What are the functions of the other DNA polymerases? Several of them also play an important role in DNA repair, a topic that will be examined in Chapter 16. DNA polymerase β, which has been studied for several decades, is not involved in the replication of normal DNA, but plays an important role in removing incorrect bases from damaged DNA. More recently, several addi-

tional DNA polymerases have been identified. While their precise roles have not been elucidated, many of these are in a category called **lesion-replicating polymerases.** When DNA polymerase α, δ, and ε encounter abnormalities in DNA structure, such as abnormal bases or cross-links, they may be unable to replicate over the aberration. When this occurs, lesion-replicating polymerases are attracted to the damaged DNA and have special properties that enable them to synthesize a complementary strand over the abnormal region. Each type of lesion-replicating polymerase may be able to replicate over a different kind of DNA damage.

The Ends of Eukaryotic Chromosomes Are Replicated by Telomerase

Linear eukaryotic chromosomes contain **telomeres** at both ends. The term telomere refers to the complex of telomeric sequences within the DNA and the special proteins that are bound to these sequences. Telomeric sequences consist of a moderately repetitive tandem array and a 3' overhang region that is 12 to 16 nucleotides in length (**Figure 11.23**).

FIGURE 11.23   General structure of telomeric sequences. The telomere DNA consists of a tandemly repeated sequence and a 12- to 16-nucleotide overhang.

The tandem array that occurs within the telomere has been studied in a wide variety of eukaryotic organisms. A common feature is that the telomeric sequence contains several guanine nucleotides and often many thymine nucleotides (Table 11.5). Depending on the species and the cell type, this sequence can be tandemly repeated up to several hundred times in the telomere region.

One reason why telomeric repeat sequences are needed is because DNA polymerase is unable to replicate the 3' ends of DNA strands. Why is DNA polymerase unable to replicate this region? The answer lies in the two unusual enzymatic features of this enzyme. As discussed previously, DNA polymerase synthesizes DNA only in a 5' to 3' direction, and it cannot link together the first two individual nucleotides; it can elongate only preexisting strands. These two features of DNA polymerase function pose a problem at the 3' ends of linear chromosomes. As shown in **Figure 11.24**, the 3' end of a DNA strand cannot be replicated by DNA polymerase because a primer cannot be made upstream from this point. Therefore, if this problem were not solved, the chromosome would become progressively shorter with each round of DNA replication.

To prevent the loss of genetic information due to chromosome shortening, additional DNA sequences are attached to the ends of telomeres. In 1984, Carol Greider and Elizabeth Blackburn discovered an enxyme called **telomerase** that prevents chromosome shortening. It recognizes the sequences at the ends of eukaryotic chromosomes and synthesizes additional repeats of telomeric sequences. **Figure 11.25** shows the interesting mechanism

TABLE 11.5

Telomeric Repeat Sequences Within Selected Organisms

Group	Example	Telomeric Repeat Sequence
Mammals	Humans	TTAGGG
Slime molds	*Physarum, Didymium*	TTAGGG
	Dictyostelium	$AG_{(1-8)}$
Filamentous fungi	*Neurospora*	TTAGGG
Budding yeast	*Saccharomyces cerevisiae*	$TG_{(1-3)}$
Ciliates	*Tetrahymena*	TTGGGG
	Paramecium	TTGGG(T/G)
	Euplotes	TTTTGGGG
Higher plants	*Arabidopsis*	TTTAGGG

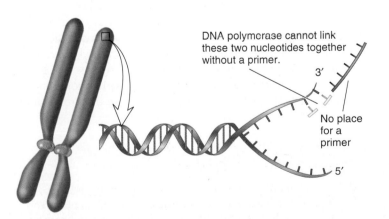

FIGURE 11.24   The replication problem at the ends of linear chromosomes. DNA polymerase cannot synthesize a DNA strand that is complementary to the 3' end because a primer cannot be made upstream from this site.

FIGURE 11.25   The enzymatic action of telomerase. A short, three-nucleotide segment of RNA within telomerase causes it to bind to the 3' overhang. The adjacent part of the RNA is used as a template to make a short, six-nucleotide repeat of DNA. After the repeat is made, telomerase moves six nucleotides to the right and then synthesizes another repeat. This process is repeated many times to lengthen the top strand shown in this figure. The bottom strand is made by DNA polymerase, using an RNA primer at the end of the chromosome that is complementary to the telomeric repeat sequence in the top strand. DNA polymerase fills in the region, which is sealed by ligase.

by which telomerase works. The telomerase enzyme contains both protein subunits and RNA. The RNA part of telomerase contains a sequence complementary to the DNA sequence found in the telomeric repeat. This allows telomerase to bind to the 3' overhang region of the telomere. Following binding, the RNA sequence beyond the binding site functions as a template allowing the synthesis of a six-nucleotide sequence at the end of the DNA strand. This is called polymerization, because it is analogous to the function of DNA polymerase. It is catalyzed by two identical protein subunits called telomerase reverse transcriptase (TERT). TERT's name indicates that is uses an RNA template to synthesize DNA. Following polymerization, the telomerase can then move—a process called translocation—to the new end of this DNA strand and attach another six nucleotides to the end. This binding-polymerization-translocation cycle occurs many times in a row and thereby greatly lengthens the 3' end of the DNA strand in the telomeric region. The complementary strand is then synthesized by DNA polymerase.

CONCEPTUAL SUMMARY

The structural basis for DNA replication is the double-stranded helix of DNA, in which the AT/GC rule is obeyed. This complementarity between strands allows DNA to be replicated in a semiconservative fashion. DNA replication begins at a specific site within the DNA, known as the origin of replication. Circular bacterial chromosomes have a single origin of replication. Specific proteins, such as the DnaA protein found in *E. coli*, recognize the origin and initiate the DNA replication process. The synthesis of DNA strands proceeds bidirectionally from the origin. This synthesis requires various proteins. DNA helicase breaks the hydrogen bonding between the parental strands, topoisomerase removes positive supercoils, and single-strand binding proteins hold the parental strands in a single-stranded state. Primase synthesizes RNA primers, which are necessary for DNA polymerase to elongate new daughter strands in a 5' to 3' direction. In the leading strand, synthesis of the daughter strand is continuous. In the lagging strand, the synthesis of DNA occurs in short Okazaki fragments. To complete the synthesis of Okazaki fragments within the lagging strand, the RNA primers are removed, DNA polymerase fills in the gaps, and DNA ligase covalently attaches the fragments together. DNA replication is terminated when both replication forks reach the ter sequences. If replication has resulted in two intertwined DNA molecules known as catenanes, the catenanes are separated into individual circular molecules.

Coordination of cell division and DNA replication is accomplished via the cellular regulation of replication. In bacteria, the control of DNA replication occurs at the origin. Several mechanisms can prevent premature DNA replication. Two examples are the availability of the DnaA protein and the methylation of GATC sites.

DNA replication in eukaryotes has several unique features. Eukaryotic chromosomes contain multiple origins of replication. Eukaryotes have multiple types of DNA polymerases involved in synthesizing the leading and lagging strand, DNA repair, and replicating of DNA lesions. Also, because eukaryotic chromosomes are linear, a specialized mechanism exists for the replication at the telomeres. An enzyme known as telomerase recognizes the sequences at the ends of eukaryotic chromosomes and synthesizes additional repeats of telomeric sequences. This prevents chromosome shortening with each round of DNA replication.

EXPERIMENTAL SUMMARY

The semiconservative mechanism of DNA replication was initially proposed based on the double helix structure determined by Watson and Crick. Nevertheless, researchers needed to experimentally demonstrate that this mechanism was correct. The results of the isotope labeling experiments of Meselson and Stahl were consistent with a semiconservative mode of replication in which DNA contains one parental strand and one newly synthesized or daughter strand.

Once the semiconservative model was established, researchers focused their attention on the details of DNA replication within living cells. Arthur Kornberg and his colleagues developed methods to synthesize bacterial DNA in vitro. In conjunction with the isolation of ts mutants, this allowed researchers to identify the role of individual proteins in DNA replication. Many of these components were found to be enzymes, such as DNA polymerase, primase, and DNA helicase. In eukaryotes, similar findings have been obtained, but more enzymes are involved and the sequences of the origins of replication are more complex. These added complexities have made it more difficult to analyze DNA replication in eukaryotes, although much progress has been made. As we will see in Chapter 22, the regulation of DNA replication is an important topic and an area of intense research because it may underlie the proliferation of cancer cells.

PROBLEM SETS & INSIGHTS

Solved Problems

S1. Describe three ways to account for the high fidelity of DNA replication. Discuss the quantitative contributions of each of the three ways.

Answer:

First: AT and GC pairs are preferred in the double helix structure. This provides fidelity to around one mistake per 1,000.

Second: Induced fit by DNA polymerase prevents covalent bond formation unless the proper nucleotides are in place. This increases fidelity another 100- to 1,000-fold, to about one error in 100,000 to 1 million.

Third: Exonuclease proofreading increases fidelity another 100- to 1,000-fold, to about one error per 100 million nucleotides added.

S2. What do you think would happen if the ter sequences were deleted from the bacterial DNA?

Answer: Instead of meeting at the ter sequences, the two replication forks would meet somewhere else. This would depend on how fast they were moving. For example, if one fork was advancing faster than the other fork, they would meet closer to where the slower-moving fork started. In fact, researchers have actually conducted this experiment. Interestingly, *E. coli* without the ter sequences seemed to survive just fine.

S3. Summarize the steps that occur in the process of chromosomal DNA replication in *E. coli*.

Answer:

Step 1. DnaA proteins bind to the origin of replication, resulting in the separation of the AT-rich region.

Step 2. DNA helicase breaks the hydrogen bonds between the DNA strands, topoisomerases alleviate positive supercoiling, and single-strand binding proteins hold the parental strands apart.

Step 3. Primase synthesizes one RNA primer in the leading strand and many RNA primers in the lagging strand. DNA polymerase III then synthesizes the daughter strands of DNA. In the lagging strand, many short segments of DNA (Okazaki

fragments) are made. DNA polymerase I removes the RNA primers and fills in with DNA, and DNA ligase covalently links the Okazaki fragments together.

Step 4. The processes described in steps 2 and 3 continue until the two replication forks reach the ter sequences on the other side of the circular bacterial chromosome.

Step 5. Topoisomerases unravel the intertwined chromosomes, if necessary.

S4. If a strain of *E. coli* overproduced the Dam enzyme, how would that affect the DNA replication process? Would you expect such a strain to have more or less chromosomes per cell compared to a normal strain of *E. coli*? Explain why.

Answer: If a strain overproduced the Dam enzyme, DNA would replicate more rapidly. The GATC methylation sites in the origin of replication have to be fully methylated for DNA replication to occur. Immediately after DNA replication, a delay occurs before the next round of DNA replication because the two copies of newly replicated DNA are hemimethylated. A strain that overproduces Dam would rapidly convert the hemimethylated DNA into fully methylated DNA and more quickly allow the next round of DNA replication to occur. For this reason, the overproducing strain might have more copies of the *E. coli* chromosome because it would not have a long delay in DNA replication.

Conceptual Questions

C1. What are the key structural features of the DNA molecule that underlie its ability to be faithfully replicated?

C2. With regard to DNA replication, define the term bidirectional replication.

C3. Which of the following statements is not true? Explain why.

A. A DNA strand can serve as a template strand on many occasions.

B. Following semiconservative DNA replication, one strand is a newly made daughter strand and the other strand is a parental strand.

C. A DNA double helix may contain two strands of DNA that were made at the same time.

D. A DNA double helix obeys the AT/GC rule.

E. A DNA double helix could contain one strand that is 10 generations older than its complementary strand.

C4. The compound known as nitrous acid is a reactive chemical that replaces amino groups ($-NH_2$) with keto groups ($=O$). When nitrous acid reacts with the bases in DNA, it can change cytosine to uracil and change adenine to hypoxanthine. A DNA double helix has the following sequence:

```
TTGGATGCTGG
AACCTACGACC
```

A. What would be the sequence of this double helix immediately after reaction with nitrous acid? Let the letter H represent hypoxanthine and U represent uracil.

B. Let's suppose this DNA was reacted with nitrous acid. The nitrous acid was then removed, and the DNA was replicated for two generations. What would be the sequences of the DNA products after the DNA had replicated two times? Your answer should contain the sequences of four double helices. Note: During DNA replication, uracil hydrogen bonds with adenine, and hypoxanthine hydrogen bonds with cytosine.

C5. One way that bacterial cells regulate DNA replication is by GATC methylation sites within the origin of replication. Would this mechanism work if the DNA was conservatively (rather than semiconservatively) replicated?

C6. The chromosome of *E. coli* contains 4.6 million bp. How long will it take to replicate its DNA? Assuming DNA polymerase III is the primary enzyme involved and this enzyme can actively proofread during DNA synthesis, how many base pair mistakes will be made in one round of DNA replication in a bacterial population containing 1,000 bacteria?

C7. Here are two strands of DNA.

——————————————————DNA polymerase–>

————————————————————————

The one on the bottom is a template strand, and the one on the top is being synthesized by DNA polymerase in the direction shown by the arrow. Label the 5' and 3' ends of the top and bottom strands.

C8. A DNA strand has the following sequence:

```
5'-GATCCCGATCCGCATACATTTACCAGATCACCACC-3'
```

In which direction would DNA polymerase slide along this strand (from left to right or from right to left)? If this strand was used as a template by DNA polymerase, what would be the sequence of the newly made strand? Indicate the 5' and 3' ends of the newly made strand.

C9. List and briefly describe the three types of sequences within bacterial origins of replication that are functionally important.

C10. As shown in Figure 11.5, five DnaA boxes are found within the origin of replication in *E. coli*. Take a look at these five sequences carefully.

   A. Are the sequences of the five DnaA boxes very similar to each other? (Hint: Remember that DNA is double stranded; think about these sequences in the forward and reverse direction.)

   B. What is the most common sequence for the DnaA box? In other words, what is the most common base in the first position, second position, and so on until the ninth position? The most common sequence is called the consensus sequence.

   C. The *E. coli* chromosome is about 4.6 million bp long. Based on random chance, is it likely that the consensus sequence for a DnaA box occurs elsewhere in the *E. coli* chromosome? If so, why aren't there multiple origins of replication in *E. coli*?

C11. Obtain two strings of different colors (e.g., black and white) that are the same length. A length of 20 inches is sufficient. Tie a knot at one end of the black string, and tie a knot at one end of the white string. Each knot designates the 5' end of your strings. Make a double helix with your two strings. Now tape one end of the double helix to a table so that the tape is covering the knot on the black string.

   A. Pretend your hand is DNA helicase and use your hand to unravel the double helix, beginning at the end that is not taped to the table. Should your hand be sliding along the white string or the black string?

   B. As in Figure 11.14, imagine that your two hands together form a dimeric replicative DNA polymerase. Unravel your two strings halfway to create a replication fork. Grasp the black string with your left hand and the white string with your right hand. Your thumbs should point toward the 5' end of each string. You need to loop one of the strings so that one of the DNA polymerases can synthesize the lagging strand. With such a loop, the dimeric replicative DNA polymerase can move toward the replication fork and synthesize both DNA strands in the 5' to 3' direction. In other words, with such a loop, your two hands can touch each other with both of your thumbs pointing toward the fork. Should the black string be looped, or should the white string be looped?

C12. Sometimes DNA polymerase makes a mistake, and the wrong nucleotide is added to the growing DNA strand. With regard to pyrimidines and purines, two general types of mistakes are possible. The addition of an incorrect pyrimidine instead of the correct pyrimidine (e.g., adding cytosine where thymine should be added) is called a transition. If a pyrimidine is incorrectly added to the growing strand instead of purine (e.g., adding cytosine where an adenine should be added), this type of mistake is called a transversion. If a transition or transversion is not detected by DNA polymerase, this will create a mutation that permanently changes the DNA sequence. Though both types of mutations are rare, transition mutations are more frequent than transversion mutations. Based on your understanding of DNA replication and DNA polymerase, offer three explanations why transition mutations are more common.

C13. A short genetic sequence, which may be recognized by DNA primase, is repeated many times throughout the *E. coli* chromosome. Researchers have hypothesized that DNA primase may recognize this sequence as a site to begin the synthesis of an RNA primer for DNA replication. The *E. coli* chromosome is roughly 4.6 million bp in length. How many copies of the DNA primase recognition sequence would be necessary to replicate the entire *E. coli* chromosome?

C14. Single-strand binding proteins keep the two parental strands of DNA separated from each other until DNA polymerase has an opportunity to replicate the strands. Suggest how single-strand binding proteins keep the strands separated and yet do not impede the ability of DNA polymerase to replicate the strands.

C15. The ability of DNA polymerase to digest a DNA strand from one end is called its exonuclease activity. Exonuclease activity is used to digest RNA primers and also to proofread a newly made DNA strand. Note: DNA polymerase I does not change direction while it is removing an RNA primer and synthesizing new DNA. It does change direction during proofreading.

   A. In which direction, 5' to 3' or 3' to 5' is the exonuclease activity occurring during the removal of RNA primers and during the proofreading and removal of mistakes following DNA replication?

   B. Figure 11.15 shows a drawing of the 3' exonuclease site. Do you think this site would be used by DNA polymerase I to remove RNA primers? Why or why not?

C16. In the following drawing, the top strand is the template DNA, and the bottom strand shows the lagging strand prior to the action of DNA polymerase I. The lagging strand contains three Okazaki fragments. The RNA primers have not yet been removed.

The top strand is the template DNA

```
3'—————————————————————————————5'
5'*************_____*************_____************_____3'
   RNA primer      ↑ RNA primer ↑      RNA primer
|————————————||————————————||————————————|
   Left Okazaki    Middle Okazaki    Right Okazaki
   fragment        fragment          fragment
```

   A. Which Okazaki fragment was made first, the one on the left or the one on the right?

   B. Which RNA primer would be the first one to be removed by DNA polymerase I, the primer on the left or the primer on the right? For this primer to be removed by DNA polymerase I and for the gap to be filled in, is it necessary for the Okazaki fragment in the middle to have already been synthesized? Explain why.

   C. Let's consider how DNA ligase connects the left Okazaki fragment with the middle Okazaki fragment. After DNA polymerase I removes the middle RNA primer and fills in the gap with DNA, where does DNA ligase function? See the arrows on either side of the middle RNA primer. Is ligase needed at the left arrow, at the right arrow, or both?

   D. When connecting two Okazaki fragments, DNA ligase needs to use NAD+ or ATP as a source of energy to catalyze this reaction. Explain why DNA ligase needs another source of energy to connect two nucleotides, but DNA polymerase needs nothing more than the incoming nucleotide and the existing DNA strand. Note: You may want to refer to Figure 11.11 to answer this question.

C17. What is DNA methylation? Why is DNA in a hemimethylated condition immediately after DNA replication? What are the functional consequences of methylation in the regulation of DNA replication?

C18. Describe the three important functions of the DnaA protein.

C19. If a strain of bacteria was making too much DnaA protein, how would you expect this to affect its ability to regulate DNA replication? With regard to the number of chromosomes per cell, how might this strain differ from a normal bacterial strain?

C20. Draw a picture that illustrates how DNA helicase works.

C21. What is an Okazaki fragment? In which strand of DNA are Okazaki fragments found? Based on the properties of DNA polymerase, why is it necessary to make these fragments?

C22. Discuss the similarities and differences in the synthesis of DNA in the lagging and leading strands. What is the advantage of a primosome and a replisome as opposed to having all replication enzymes functioning independently of each other?

C23. Explain the proofreading function of DNA polymerase.

C24. What is a processive enzyme? Explain why this is an important feature of DNA polymerase.

C25. Why is it important for living organisms to regulate DNA replication?

C26. What enzymatic features of DNA polymerase prevent it from replicating one of the DNA strands at the ends of linear chromosomes? Compared to DNA polymerase, how is telomerase different in its ability to synthesize a DNA strand? What does telomerase use as its template for the synthesis of a DNA strand? How does the use of this template result in a telomere sequence that is tandemly repetitive?

C27. As shown in Figure 11.25, telomerase attaches additional DNA, six nucleotides at a time, to the ends of eukaryotic chromosomes. However, it works in only one DNA strand. Describe how the opposite strand is replicated.

C28. If a eukaryotic chromosome has 25 origins of replication, how many replication forks does it have at the beginning of DNA replication?

C29. A diagram of a linear chromosome is shown here. The end of each strand is labeled with an A, B, C, or D. Which ends could not be replicated by DNA polymerase? Why not?

5'-A————————————————————————————B-3'

3'-C————————————————————————————D-5'

C30. As discussed in Chapter 10, some viruses contain RNA as their genetic material. Certain RNA viruses can exist as a provirus in which the viral genetic material has been inserted into the chromosomal DNA of the host cell. For this to happen, the viral RNA must be copied into a strand of DNA. An enzyme called reverse transcriptase, encoded by the viral genome, copies the viral RNA into a complementary strand of DNA. The strand of DNA is then used as a template to make a double-stranded DNA molecule. This double-stranded DNA molecule is then inserted into the chromosomal DNA, where it may exist as a provirus for a long period of time.

A. How is the function of reverse transcriptase similar to the function of telomerase?

B. Unlike DNA polymerase, reverse transcriptase does not have a proofreading function. How might this affect the proliferation of the virus?

C31. Telomeres contain a 3' overhang region, as shown in Figure 11.23. Does telomerase require a 3' overhang to replicate the telomere region? Explain.

Experimental Questions

E1. Answer the following questions that pertain to the experiment of Figure 11.3.

A. What would be the expected results if the Meselson and Stahl experiment were carried out for four or five generations?

B. What would be the expected results of the Meselson and Stahl experiment after three generations if the mechanism of DNA replication was dispersive?

C. As shown in the data, explain why three different bands (i.e., light, half-heavy, and heavy) can be observed in the CsCl gradient.

E2. An absentminded researcher follows the steps of Figure 11.3, and when the gradient is viewed under UV light, the researcher does not see any bands at all. Which of the following mistakes could account for this observation? Explain how.

A. The researcher forgot to add ^{14}N-containing compounds.

B. The researcher forgot to add lysozyme.

C. The researcher forgot to turn on the UV lamp.

E3. Figure 11.4b shows an autoradiograph of a replicating bacterial chromosome. If you analyzed many replicating chromosomes,

what types of information could you learn about the mechanism of DNA replication?

E4. The experiment of Figure 11.18 described a method for determining the amount of DNA made during replication. Let's suppose that you can purify all of the proteins required for DNA replication. You then want to "reconstitute" DNA synthesis by mixing together all of the purified components necessary to synthesize a complementary strand of DNA. If you started with single-stranded DNA as a template, what additional proteins and molecules would you have to add for DNA polymerization to occur? What additional proteins would be necessary if you started with a double-stranded DNA molecule?

E5. Using the reconstitution strategy described in experimental question E4, what components would you have to add to measure the ability of telomerase to synthesize DNA? Be specific about the type of template DNA that you would add to your mixture.

E6. As described in Figure 11.18, perchloric acid precipitates strands of DNA, but it does not precipitate free nucleotides. (Note: The term free nucleotide means nucleotides that are not connected covalently to other nucleotides.) Explain why this is a critical step in the experimental procedure. If a researcher used a different reagent that precipitated DNA strands and free nucleotides instead of using perchloric acid (which precipitates only DNA strands), how would that affect the results?

E7. Would the experiment of Figure 11.18 work if the ^{32}P-labeled nucleotides were deoxyribonucleoside monophosphates instead of deoxyribonucleoside triphosphates? Explain why or why not.

E8. To synthesize DNA in vitro, single-stranded DNA can be used as a template. As described in Figure 11.18, you also need to add DNA polymerase, dNTPs, and a primer in order to synthesize a complementary strand of DNA. The primer can be a short sequence of DNA or RNA. The primer must be complementary to the template DNA. Let's suppose a single-stranded DNA molecule is 46 nucleotides long and has the following sequence:

GCCCCGGTACCCCGTAATATACGGGACTAGGCCGGAGGTCCGGGCG

This template DNA is mixed with a primer with the sequence 5'–CGCCCGGACC–3', DNA polymerase, and dNTPs. In this case, a double-stranded DNA molecule is made. However, if the researcher substitutes a primer with the sequence 5'–CCAG-GCCCGC–3', a double-stranded DNA molecule is not made.

A. Which is the 5' end of the DNA molecule shown, the left end or the right end?

B. If you added a primer that was 10 nucleotides long and complementary to the left end of the single-stranded DNA, what would be the sequence of the primer? You should designate the 5' and 3' ends of the primer. Could this primer be used to replicate the single-stranded DNA?

E9. The technique of dideoxy sequencing of DNA is described in Chapter 18. The technique relies on the use of dideoxyribonucleotides (shown in Figures 18.14 and 18.15). A dideoxyribonucleotide has a hydrogen atom attached to the 3' carbon atom instead of an –OH group. When a dideoxyribonucleotide is incorporated into a newly made strand, the strand cannot grow any longer. Explain why.

E10. Another technique described in Chapter 18 is the polymerase chain reaction (PCR) (see Figure 18.6). This method is based on our understanding of DNA replication. In this method, a small amount of double-stranded template DNA is mixed with a high concentration of primers. Nucleotides and DNA polymerase are also added. The template DNA strands are separated by heat treatment, and when the temperature is lowered, the primers can bind to the single-stranded DNA, and then DNA polymerase replicates the DNA. This increases the amount of DNA made from the primers. This cycle of steps (i.e., heat treatment, lower temperature, allow DNA replication to occur) is repeated again and again and again. Because the cycles are repeated many times, this method is called a chain reaction. It is called a polymerase chain reaction because DNA polymerase is the enzyme needed to increase the amount of DNA with each cycle. In a PCR experiment, the template DNA is placed in a tube, and the primers, nucleotides, and DNA polymerase are added to the tube. The tube is then placed in a machine called a thermocycler, which raises and lowers the temperature. During one cycle, the temperature is raised (e.g., to 95°C) for a brief period and then lowered (e.g., to 60°C) to allow the primers to bind. The sample is incubated at the lower temperature for a few minutes to allow DNA replication to proceed. In a typical PCR experiment, the tube may be left in the thermocycler for 25 to 30 cycles. The total time for a PCR experiment is a few hours.

A. Why is DNA helicase not needed in a PCR experiment?

B. How is the sequence of each primer important in a PCR experiment? Do the two primers recognize the same strand or opposite strands?

C. The DNA polymerase used in PCR experiments is a DNA polymerase isolated from thermophilic bacteria. Why is this kind of polymerase used?

D. If a tube initially contained 10 copies of double-stranded DNA, how many copies of double-stranded DNA (in the region flanked by the two primers) would be obtained after 27 cycles?

Questions for Student Discussion/Collaboration

1. The complementarity of double-stranded DNA is the underlying reason that DNA can be faithfully copied. Propose alternative chemical structures that could be faithfully copied.

2. The technique described in Figure 11.18 makes it possible to measure DNA synthesis in vitro. Let's suppose you have purified the following enzymes: DNA polymerase, DNA helicase, ligase, primase, single-strand binding protein, and topoisomerase. You also have the following reagents available:

A. Radiolabeled nucleotides (labeled with ^{32}P, a radioisotope of phosphorus)

B. Nonlabeled double-stranded DNA

C. Nonlabeled single-stranded DNA

D. An RNA primer that binds to one end of the nonlabeled single-stranded DNA

With these reagents, how could you show that DNA helicase is necessary for strand separation and primase is necessary for the synthesis of an RNA primer? Note: In this question, think about conditions where DNA helicase or primase would be necessary to allow DNA replication and other conditions where they would be unnecessary.

3. DNA replication is fast, virtually error-free, and coordinated with cell division. Discuss which of these three features you think is the most important.

Note: All answers appear at the website for this textbook; the answers to even-numbered questions are in the back of the textbook.

Visit the Online Learning Center for practice tests, answer keys, and other learning aids for this chapter. Enhance your understanding of genetics with our interactive exercises, quizzes, animations, and much more.

GENE TRANSCRIPTION AND RNA MODIFICATION

12

T he function of the genetic material is that of a blueprint. It stores the information necessary to create a living organism. The information is contained within units called genes. At the molecular level, a definition of a **gene** is a segment of DNA that is used to make a functional product, either an RNA molecule or a polypeptide. How is the information within a gene accessed? The first step in this process is called **transcription,** which literally means the act or process of making a copy. In genetics, this term refers to the process of synthesizing RNA from a DNA sequence (**Figure 12.1**). The structure of DNA is not altered as a result of transcription. Rather, the DNA base sequence has only been accessed to make a copy in the form of RNA. Therefore, the same DNA can continue to store information. DNA replication, which was discussed in Chapter 11, provides a mechanism for copying that information so it can be transmitted from cell to cell and from parent to offspring.

Structural genes encode the amino acid sequence of a polypeptide. When a structural gene is transcribed, the first product is an RNA molecule known as

A molecular model showing the enzyme RNA polymerase in the act of sliding along the DNA and synthesizing a copy of RNA.

DNA replication: makes DNA copies that are transmitted from cell to cell and from parent to offspring.

Gene

Chromosomal DNA: stores information in units called genes.

Transcription: produces an RNA copy of a gene.

Messenger RNA: a temporary copy of a gene that contains information to make a polypeptide.

Translation: produces a polypeptide using the information in mRNA.

Polypeptide: becomes part of a functional protein that contributes to an organism's traits.

FIGURE 12.1   **The central dogma of genetics.** The usual flow of genetic information is from DNA to mRNA to polypeptide. Note: The direction of informational flow shown in this figure is the most common direction found in living organisms, but exceptions occur. For example, RNA viruses and certain transposable elements use an enzyme called reverse transcriptase to make a copy of DNA from RNA.

PART IV
Molecular Properties of Genes

messenger RNA (mRNA). During polypeptide synthesis—a process called **translation**—the sequence of nucleotides within the mRNA determines the sequence of amino acids in a polypeptide. One or more polypeptides then assemble into a functional protein. The synthesis of functional proteins ultimately determines an organism's traits. The model depicted in Figure 12.1, which is called the **central dogma of genetics** (also called the central dogma of molecular biology), was first enunciated by Francis Crick in 1958. It forms a cornerstone for our understanding of genetics at the molecular level. The flow of genetic information occurs from DNA to mRNA to polypeptide.

In this chapter, we begin to study the molecular steps in gene expression, with an emphasis on transcription and the modifications that may occur to an RNA transcript after it has been made. In Chapter 13, we will examine the process of translation, and Chapters 14 and 15 will focus on how the level of gene expression is regulated at the molecular level.

12.1 OVERVIEW OF TRANSCRIPTION

One key concept important in the process of transcription is that short base sequences define the beginning and ending of a gene and also play a role in regulating the level of RNA synthesis. In this section, we will begin by examining the sequences that determine where transcription starts and ends, and also briefly consider DNA sequences, called regulatory sites, that influence whether a gene is turned on or off. The functions of regulatory

sites will be examined in greater detail in Chapters 14 and 15. A second important concept is the role of proteins in transcription. DNA sequences, in and of themselves, just exist. For genes to be actively transcribed, proteins must recognize particular DNA sequences and act upon them in a way that affects the transcription process. In the later part of this section, we will consider how proteins participate in the general steps of transcription, and the types of RNA transcripts that can be made.

Gene Expression Requires Base Sequences That Perform Different Functional Roles

At the molecular level, **gene expression** is the overall process by which the information within a gene is used to produce a functional product, such as a polypeptide. Along with environmental factors, the molecular expression of genes determines an organism's traits. For a gene to be expressed, a few different types of base sequences perform specific roles. **Figure 12.2** shows a common organization of base sequences needed to create a structural gene that functions in a bacterium such as *E. coli*. Each type of base sequence performs its role during a particular stage of gene expression. For example, the promoter and terminator are base sequences used during gene transcription. Specifically, the **promoter** provides a site to begin transcription, and the **terminator** specifies the end of transcription. These two sequences cause RNA synthesis to occur within a defined location. As shown in Figure 12.2, the DNA is transcribed into RNA from the end of the promoter to the termina-

DNA:

- **Regulatory sequences:** site for the binding of regulatory proteins; the role of regulatory proteins is to influence the rate of transcription. Regulatory sequences can be found in a variety of locations.

- **Promoter:** site for RNA polymerase binding; signals the beginning of transcription.

- **Terminator:** signals the end of transcription.

mRNA:

- **Ribosomal binding site:** site for ribosome binding; translation begins near this site in the mRNA. In eukaryotes, the ribosome scans the mRNA for a start codon.

- **Start codon:** specifies the first amino acid in a polypeptide sequence, usually a formylmethionine (in bacteria) or a methionine (in eukaryotes).

- **Codons:** 3-nucleotide sequences within the mRNA that specify particular amino acids. The sequence of codons within mRNA determines the sequence of amino acids within a polypeptide.

- **Stop codon:** specifies the end of polypeptide synthesis.

- Bacterial mRNA may be polycistronic, which means it encodes two or more polypeptides.

FIGURE 12.2   **Organization of sequences of a bacterial gene and its mRNA transcript.** This figure depicts the general organization of sequences that are needed to create a functional gene that encodes an mRNA.

tor. As described later, the base sequence in the RNA transcript is complementary to the **template strand** of DNA, also called the **antisense strand.** For genes that encode proteins, the non-template DNA strand is called the **coding strand** or the **sense strand.** The sequence of nucleotides in the transcribed mRNA is the same as that found in the coding strand, except that the T's are replaced by U's.

A category of proteins called **transcription factors** recognizes base sequences in the DNA and controls transcription. Some transcription factors bind directly to the promoter and facilitate transcription. Other transcription factors recognize **regulatory sequences**—short stretches of DNA involved in the regulation of transcription. Certain transcription factors bind to such regulatory sequences and increase the rate of transcription while others inhibit transcription.

Base sequences within an mRNA are used during the translation process. In bacteria, a short sequence within the mRNA, the **ribosomal-binding site,** provides a location for the ribosome to bind and begin translation. The bacterial ribosome recognizes this site because it is complementary to a sequence in ribosomal RNA. In addition, mRNA contains a series of **codons,** read as

groups of three nucleotides, which contain the information for a polypeptide's sequence. The first codon, which is very close to the ribosomal-binding site, is the **start codon.** This is followed by many more codons that dictate the sequence of amino acids within the synthesized polypeptide. Finally, a **stop codon** signals the end of translation. Chapter 13 will examine the process of translation in greater detail.

The Three Stages of Transcription Are Initiation, Elongation, and Termination

Transcription occurs in three stages: **initiation; elongation,** or synthesis of the RNA transcript; and **termination (Figure 12.3).** These steps involve protein-DNA interactions in which proteins such as **RNA polymerase,** the enzyme that synthesizes RNA, interact with DNA sequences. What causes transcription to begin? The initiation stage in the transcription process is a recognition step. The sequence of bases within the promoter region is recognized by transcription factors. The specific binding of transcription factors to the promoter sequence identifies the starting site for transcription.

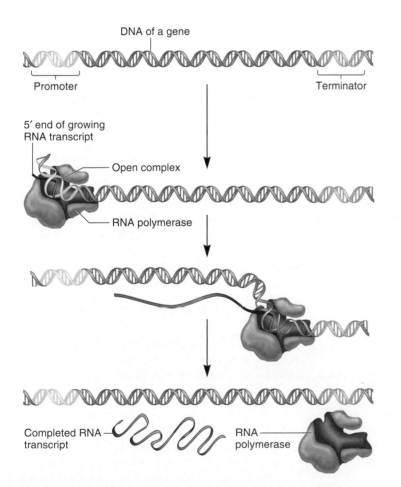

DNA of a gene

Promoter   Terminator

5′ end of growing RNA transcript

Open complex

RNA polymerase

Initiation: The promoter functions as a recognition site for transcription factors (not shown). The transcription factor(s) enables RNA polymerase to bind to the promoter. Following binding, the DNA is denatured into a bubble known as the open complex.

Elongation/synthesis of the RNA transcript: RNA polymerase slides along the DNA in an open complex to synthesize RNA.

Termination: A terminator is reached that causes RNA polymerase and the RNA transcript to dissociate from the DNA.

Completed RNA transcript

RNA polymerase

ONLINE ANIMATION

FIGURE 12.3   **Stages of transcription.**

Genes→Traits The ability of genes to produce an organism's traits relies on the molecular process of gene expression. Transcription is the first step in gene expression. During transcription, the gene's sequence within the DNA is used as a template to make a complementary copy of RNA. In Chapter 13, we will examine how the sequence in mRNA is translated into a polypeptide chain. After polypeptides are made within a living cell, they fold into functional proteins that govern an organism's traits.

Transcription factors and RNA polymerase first bind to the promoter region when the DNA is in the form of a double helix. For transcription to occur, the DNA strands must be separated. This allows one of the two strands to be used as a template for the synthesis of a complementary strand of RNA. This synthesis occurs as RNA polymerase slides along the DNA, creating a small bubblelike structure known as the open promoter complex, or simply as the **open complex.** Eventually, RNA polymerase reaches a terminator, which causes both RNA polymerase and the newly made RNA transcript to dissociate from the DNA.

RNA Transcripts Have Different Functions

Once they are made, RNA transcripts play different functional roles (Table 12.1). Well over 90% of all genes are structural genes, which are transcribed into mRNA. For structural genes, mRNAs are made first, but the final, functional products are polypeptides that operate within proteins. The remaining types of RNAs described in Table 12.1 are never translated. The RNA transcripts from such nonstructural genes have various important cellular functions. For nonstructural genes, the functional product is the RNA. In some cases, the RNA transcript becomes part of a complex that contains both protein subunits and one or more RNA molecules. Examples of protein-RNA complexes include ribosomes, signal recognition particles, RNaseP, spliceosomes and telomerase.

12.2 TRANSCRIPTION IN BACTERIA

Our molecular understanding of gene transcription initially came from studies involving bacteria and bacteriophages. Several early investigations focused on the production of viral RNA after bacteriophage infection. The first suggestion that RNA is derived from the transcription of DNA was made by Elliot Volkin and Lazarus Astrachan in 1956. When the researchers exposed *E. coli* cells to T2 bacteriophage, they observed that the RNA made immediately after infection had a base composition substantially different from the base composition of RNA prior to infection. Furthermore, the base composition after infection was very similar to the base composition in the T2 DNA, except that the RNA contained uracil instead of thymine. These results were consistent with the idea that the bacteriophage DNA is used as a template for the synthesis of bacteriophage RNA.

In 1960, Matthew Meselson and François Jacob found that proteins are synthesized on ribosomes. One year later, Jacob and his colleague Jacques Monod proposed that a certain type of RNA acts as a genetic messenger (from the DNA to the ribosome) to provide the information for protein synthesis. They hypothesized that this RNA, which they called messenger RNA, is transcribed from the sequence within DNA and then directs the synthesis of particular polypeptides. In the early 1960s, this proposal was remarkable, considering that it was made before the actual isolation and characterization of the mRNA molecules in vitro. In 1961, the hypothesis was confirmed by Sydney Brenner in collaboration with Jacob and Meselson. They found that when

a virus infects a bacterial cell, a virus-specific RNA is made that rapidly associates with preexisting ribosomes in the cell.

Since these pioneering studies, a great deal has been learned about the molecular features of bacterial gene transcription. Much of our knowledge comes from studies in *E. coli*. In this section, we will examine the three steps in the gene transcription process as they occur in bacteria.

TABLE 12.1

Functions of RNA Molecules

Type of RNA	Description
mRNA	Messenger RNA (mRNA) encodes the sequence of amino acids within a polypeptide. In bacteria, some mRNAs encode a single polypeptide. Other mRNAs are polycistronic—a single mRNA encodes two or more polypeptides. In most species of eukaryotes, each mRNA usually encodes a single polypeptide. However, in some species, such as *Caenorhabditis elegans* (a nematode worm), polycistronic mRNAs are relatively common.
tRNA	Transfer RNA (tRNA) is necessary for the translation of mRNA. The structure and function of transfer RNA are outlined in Chapter 13.
rRNA	Ribosomal RNA (rRNA) is necessary for the translation of mRNA. Ribosomes are composed of both rRNAs and protein subunits. The structure and function of ribosomes are examined in Chapter 13.
MicroRNA	MicroRNAs (miRNAs) are short RNA molecules that are involved in gene regulation in eukaryotes (see Chapter 15).
scRNA	Small cytoplasmic RNA (scRNA) is found in the cytoplasm of bacteria and eukaryotes. In bacteria, scRNA is needed for protein secretion. An example in eukaryotes is 7S RNA, which is necessary in the targeting of proteins to the endoplasmic reticulum. It is a component of a complex known as signal recognition particle (SRP), which is composed of 7S RNA and six different protein subunits.
RNA of RNaseP	RNaseP is an enzyme necessary in the processing of all bacterial tRNA molecules. The RNA is the catalytic component of this enzyme. RNaseP is composed of a 350- to 410-nucleotide RNA and one protein subunit.
snRNA	Small nuclear RNA (snRNA) is necessary in the splicing of eukaryotic pre-mRNA. snRNAs are components of a spliceosome, which is composed of both snRNAs and protein subunits. The structure and function of spliceosomes are examined later in this chapter.
Telomerase RNA	The enzyme telomerase, which is involved in the replication of eukaryotic telomeres, is composed of an RNA molecule and protein subunits.
snoRNA	Small nucleolar RNA (snoRNA) is necessary in the processing of eukaryotic rRNA transcripts. snoRNAs are also associated with protein subunits. In eukaryotes, snoRNAs are found in the nucleolus, where rRNA processing and ribosome assembly occur.
Viral RNAs	Some types of viruses use RNA as their genome, which is packaged within the viral capsid.

A Promoter Is a Short Sequence of DNA That Is Necessary to Initiate Transcription

The type of DNA sequence known as the promoter gets its name from the idea that it "promotes" gene expression. More precisely, this sequence of bases directs the exact location for the initiation of RNA transcription. Most of the promoter region is located just ahead of or upstream from the site where transcription of a gene actually begins. By convention, the bases in a promoter sequence are numbered in relation to the **transcriptional start site** (**Figure 12.4**). This site is the first base used as a template for RNA transcription and is denoted +1. The bases preceding this site are numbered in a negative direction. No base is numbered zero. Therefore, most of the promoter region is labeled with negative numbers that describe the number of bases preceding the beginning of transcription.

Although the promoter may encompass a region several dozen nucleotides in length, short **sequence elements** are particularly critical for promoter recognition. By comparing the sequence of DNA bases within many promoters, researchers have learned that certain sequences of bases are necessary to create a functional promoter. In many promoters found in *E. coli* and similar species, two sequence elements are important. These are located at approximately the −35 and −10 sites in the promoter region (Figure 12.4). The sequence at the −35 region is 5′–TTGACA–3′, and the one at the −10 region is 5′–TATAAT–3′. The TATAAT sequence is sometimes called the **Pribnow box** after David Pribnow, who initially discovered it in 1975.

The sequences at the −35 and −10 sites can vary among different genes. For example, **Figure 12.5** illustrates the sequences found in several different *E. coli* promoters. The most commonly occurring bases within a sequence element form the **consensus sequence.** This sequence is efficiently recognized by proteins that

FIGURE 12.5 **Examples of −35 and −10 sequences within a variety of bacterial promoters.** This figure shows the −35 and −10 sequences for seven different bacterial and bacteriophage promoters. The consensus sequence is shown at the bottom. The spacer regions contain the designated number of nucleotides between the −35 and −10 region or between the −10 region and the transcriptional start site. For example, N_{17} means there are 17 nucleotides between the end of the −35 region and the beginning of the −10 region.

initiate transcription. For many bacterial genes, a strong correlation is found between the maximal rate of RNA transcription and the degree to which the −35 and −10 regions agree with their consensus sequences.

Bacterial Transcription Is Initiated When RNA Polymerase Holoenzyme Binds at a Promoter Sequence

Thus far, we have considered the DNA sequences that constitute a functional promoter. Let's now turn our attention to the proteins that recognize those sequences and carry out the transcription process. The enzyme that catalyzes the synthesis of RNA is **RNA polymerase.** In *E. coli*, the **core enzyme** is composed of five subunits, $\alpha_2\beta\beta'\omega$. The association of a sixth subunit, **sigma (σ) factor,** with the core enzyme is referred to as RNA polymerase **holoenzyme.** The different subunits within the holoenzyme play distinct functional roles. The two α subunits are important in the proper assembly of the holoenzyme and in the process of binding to DNA. The β and β' subunits are also needed for binding to the DNA and carry out the catalytic synthesis of RNA. The ω (omega) subunit is important for the proper assembly of the core enzyme. The holoenzyme is required to initiate transcription; the primary role of σ factor is to recognize the promoter. Proteins, such as σ factor, that influence the function of RNA polymerase are types of transcription factors.

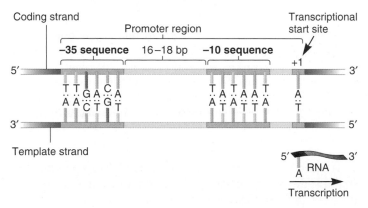

FIGURE 12.4 **The conventional numbering system of promoters.** The first nucleotide that acts as a template for transcription is designated +1. The numbering of nucleotides to the left of this spot is in a negative direction, while the numbering to the right is in a positive direction. For example, the nucleotide that is immediately to the left of the +1 nucleotide is numbered −1, and the nucleotide to the right of the +1 nucleotide is numbered +2. There is no zero nucleotide in this numbering system. In many bacterial promoters, sequence elements at the −35 and −10 regions play a key role in promoting transcription.

FIGURE 12.6 **The binding of σ factor protein to the DNA double helix.** In this example, the protein contains two α helices connected by a turn, termed a helix-turn-helix motif. Two α helices of the protein can fit within the major groove of the DNA. Amino acids within the α helices form hydrogen bonds with the bases in the DNA.

After RNA polymerase holoenzyme is assembled into its six subunits, it binds loosely to the DNA and then slides along the DNA, much as a train rolls down the tracks. How is a promoter identified? When the holoenzyme encounters a promoter sequence, σ factor recognizes the bases at both the –35 and –10 regions. σ factor protein contains a structure called a **helix-turn-helix motif** that can bind tightly to these regions. Alpha helices within the protein fit into the major groove of the DNA double helix and form hydrogen bonds with the bases. This phenomenon of molecular recognition is shown in **Figure 12.6**. Hydrogen bonding occurs between nucleotides in the –35 and –10 regions of the promoter and amino acid side chains in the helix-turn-helix structure of σ factor.

As shown in **Figure 12.7**, the process of transcription is initiated when σ factor within the holoenzyme has bound to the promoter region to form the **closed complex.** For transcription to begin, the double-stranded DNA must then be unwound into an open complex. This unwinding first occurs at the TATAAT sequence in the –10 region, which contains only AT base pairs, as shown in Figure 12.4. AT base pairs form only two hydrogen bonds, whereas GC pairs form three. Therefore, DNA in an AT-rich region is more easily separated because fewer hydrogen bonds must be broken. A short strand of RNA is made within the open complex, and then σ factor is released from the core enzyme. The release of σ factor marks the transition to the elongation phase of transcription. The core enzyme may now slide down the DNA to synthesize a strand of RNA.

The RNA Transcript Is Synthesized During the Elongation Stage

After the initiation stage of transcription is completed, the RNA transcript is made in the elongation stage of transcription. During the synthesis of the RNA transcript, RNA polymerase moves along the DNA, causing it to unwind (**Figure 12.8**). As previously mentioned, the DNA strand used as a template for RNA synthe-

FIGURE 12.7 **The initiation stage of transcription in bacteria.** The σ factor subunit of the RNA polymerase holoenzyme recognizes the –35 and –10 regions of the promoter. The DNA unwinds in the –10 region to form an open complex, and a short RNA is made. σ factor then dissociates from the holoenzyme, and the RNA polymerase core enzyme can proceed down the DNA to transcribe RNA, forming an open complex as it goes.

sis is called the template or antisense strand. The opposite DNA strand is the coding or sense strand; it has the same sequence as the RNA transcript except that T in the DNA corresponds to U in the RNA. Within a given gene, only the template strand is used for RNA synthesis, while the coding strand is never used. As it moves along the DNA, the open complex formed by the action of RNA polymerase is approximately 17 bp long. On average, the rate of RNA synthesis is about 43 nucleotides per second! Behind the open complex, the DNA rewinds back into a double helix.

As described in Figure 12.8, the chemistry of transcription by RNA polymerase is similar to the synthesis of DNA via DNA polymerase, which is discussed in Chapter 11. RNA polymerase always

Key points:

- RNA polymerase slides along the DNA, creating an open complex as it moves.

- The DNA strand known as the template strand is used to make a complementary copy of RNA as an RNA–DNA hybrid.

- RNA polymerase moves along the template strand in a 3′ to 5′ direction, and RNA is synthesized in a 5′ to 3′ direction using nucleoside triphosphates as precursors. Pyrophosphate is released (not shown).

- The complementarity rule is the same as the AT/GC rule except that U is substituted for T in the RNA.

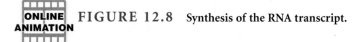

FIGURE 12.8   Synthesis of the RNA transcript.

connects nucleotides in the 5′ to 3′ direction. During this process, RNA polymerase catalyzes the formation of a bond between the 5′ phosphate group on one nucleotide and the 3′–OH group on the previous nucleotide. The complementarity rule is similar to the AT/GC rule, except that uracil substitutes for thymine in the RNA. In other words, RNA synthesis obeys an $A_{DNA}U_{RNA}$/$T_{DNA}A_{RNA}$/$G_{DNA}C_{RNA}$/$C_{DNA}G_{RNA}$ rule.

When considering the transcription of multiple genes within a chromosome, the direction of transcription and the DNA strand used as a template can vary among different genes. **Figure 12.9** shows three genes adjacent to each other within a chromosome. Genes *A* and *B* are transcribed from left to right, using the bottom DNA strand as a template. By comparison, gene *C* is transcribed from right to left and uses the top DNA strand as a template. Note that in all three cases, the template strand is read in the 3′ to 5′ direction, and the synthesis of the RNA transcript occurs in a 5′ to 3′ direction.

Transcription Is Terminated by Either an RNA-Binding Protein or an Intrinsic Terminator

The end of RNA synthesis is referred to as termination. Prior to termination, the hydrogen bonding between the DNA and RNA within the open complex is of central importance in preventing dissociation of RNA polymerase from the template strand. Termination occurs when this short RNA-DNA hybrid region is forced to separate, thereby releasing RNA polymerase as well as the newly made RNA transcript. In *E. coli*, two different mechanisms for termination have been identified. For certain genes, an RNA-binding protein known as *ρ* (**rho**) is responsible for terminating transcription, in a mechanism called *ρ*-dependent termination. For other genes, termination does not require the involvement of the *ρ* protein. This is referred to as *ρ*-independent termination.

In **ρ-dependent termination,** the termination process requires two components. First, a sequence near the 3′ end of the newly made RNA, called the *rut* site for <u>r</u>ho <u>ut</u>ilization site,

FIGURE 12.9   The transcription of three different genes found in the same chromosome. RNA polymerase synthesizes each RNA transcript in a 5′ to 3′ direction, sliding along a DNA template strand in a 3′ to 5′ direction. However, the use of the template strand can vary from gene to gene. For example, genes *A* and *B* use the bottom strand, while gene *C* uses the top strand.

acts as a recognition site for the binding of the ρ protein (**Figure 12.10**). How does ρ protein facilitate termination? The ρ protein functions as a helicase, an enzyme that can separate RNA-DNA hybrid regions. After the *rut* site is synthesized in the RNA, ρ protein binds to the RNA and moves in the direction of RNA polymerase. The second component of ρ-dependent termination is the site where termination actually takes place. At this terminator site, the DNA encodes an RNA sequence containing several GC base pairs that form a stem-loop structure. RNA synthesis termi-

nates several nucleotides beyond this stem-loop structure. As discussed in Chapter 9, a stem-loop structure, also called a hairpin, can form due to complementary sequences within the RNA (refer back to Figure 9.23). This stem-loop forms almost immediately after the RNA sequence is synthesized and quickly binds to RNA polymerase. This binding results in a conformational change that causes RNA polymerase to pause in its synthesis of RNA. The pause allows ρ protein to catch up to the stem-loop, pass through it, and break the hydrogen bonds between the DNA and RNA within the open complex. When this occurs, the completed RNA strand is separated from the DNA along with RNA polymerase.

Let's now turn our attention to **ρ-independent termination,** a process that does not require the ρ protein. In this case, the terminator is composed of two adjacent nucleotide sequences that function within the RNA (**Figure 12.11**). One is a uracil-rich sequence located at the 3' end of the RNA. The second sequence is adjacent to the uracil-rich sequence and promotes the formation of a stem-loop structure. As shown in Figure 12.11, the formation of the stem-loop causes RNA polymerase to pause in its synthesis of RNA. This pausing is stabilized by other proteins that bind to RNA polymerase. For example, a protein called NusA, which is bound to RNA polymerase, promotes pausing at stem-

ρ recognition site (*rut*)

Terminator

5′

rut

3′

ρ recognition site in RNA

ρ protein binds to the *rut* site in RNA and moves toward the 3′ end.

5′

3′

ρ protein

RNA polymerase reaches the terminator. A stem-loop causes RNA polymerase to pause.

5′

Terminator

Stem-loop

3′

RNA polymerase pauses due to its interaction with the stem-loop structure. ρ protein catches up to the open complex and separates the RNA-DNA hybrid.

3′

5′

ONLINE ANIMATION

FIGURE 12.10 ρ-dependent termination.

U-rich RNA in the RNA-DNA hybrid

5′

Stem-loop that causes RNA polymerase to pause

NusA

While RNA polymerase pauses, the U-rich sequence is not able to hold the RNA-DNA hybrid together. Termination occurs.

Terminator

5′

U U U U 3′

FIGURE 12.11   **ρ-independent or intrinsic termination.** When RNA polymerase reaches the end of the gene, it transcribes a uracil-rich sequence. As this uracil-rich sequence is transcribed, a stem-loop forms just upstream from the open complex. The formation of this stem-loop causes RNA polymerase to pause in its synthesis of the transcript. This pausing is stabilized by NusA, which binds near the region where RNA exits the open complex. While it is pausing, the RNA in the RNA-DNA hybrid is a uracil-rich sequence. Because hydrogen bonds between U and A are relatively weak interactions, the transcript and RNA polymerase dissociate from the DNA.

loop sequences. At the precise time RNA polymerase pauses, the uracil-rich sequence in the RNA transcript is bound to the DNA template strand. As previously mentioned, the hydrogen bonding of RNA to DNA keeps RNA polymerase clamped onto the DNA. However, the binding of this uracil-rich sequence to the DNA template strand is relatively weak, causing the RNA transcript to spontaneously dissociate from the DNA and cease further transcription. Because this process does not require a protein (the ρ protein) to physically remove the RNA transcript from the DNA, it is also referred to as **intrinsic termination.** In *E. coli*, about half of the genes show intrinsic termination, and the other half are terminated by ρ protein.

12.3 TRANSCRIPTION IN EUKARYOTES

Many of the basic features of gene transcription are very similar in bacterial and eukaryotic species. Much of our understanding of transcription has come from studies in *Saccharomyces cerevisiae* (baker's yeast) and other eukaryotic species, including mammals. In general, gene transcription in eukaryotes is more complex than that of their bacterial counterparts. Eukaryotic cells are larger and contain a variety of compartments known as organelles. This added level of cellular complexity dictates that eukaryotes contain many more genes encoding cellular proteins. In addition, most eukaryotic species are multicellular, being composed of many different cell types. Multicellularity adds the requirement that genes be transcribed in the correct type of cell and during the proper stage of development. Therefore, in any given species, the transcription of the thousands of different genes that an organism possesses requires appropriate timing and coordination. In this section, we will examine features of gene transcription unique to eukaryotes. The regulation of eukaryotic gene transcription is covered in Chapter 15.

Eukaryotes Have Multiple RNA Polymerases That Are Structurally Similar to the Bacterial Enzyme

The genetic material within the nucleus of a eukaryotic cell is transcribed by three different RNA polymerase enzymes, designated RNA polymerase I, II, and III. What are the roles of these enzymes? Each of the three RNA polymerases transcribes different categories of genes. RNA polymerase I transcribes all of the genes that encode ribosomal RNA (rRNA) except for the 5S rRNA. RNA polymerase II plays a major role in cellular transcription because it transcribes all of the structural genes. It is responsible for the synthesis of all mRNA and also transcribes certain snRNA genes, which are needed for pre-mRNA splicing. RNA polymerase III transcribes all tRNA genes and the 5S rRNA gene.

All three RNA polymerases are structurally very similar and are composed of many subunits. They contain two large catalytic subunits similar to the β and β' subunits of bacterial RNA polymerase. The structures of RNA polymerase from a few different species have been determined by X-ray crystallography. A remarkable similarity exists between the bacterial enzyme and its

(a) Structure of a bacterial RNA polymerase

Structure of a eukaryotic RNA polymerase II (yeast)

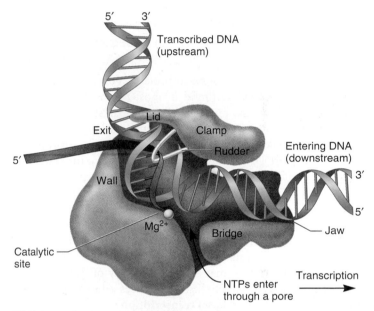

(b) Schematic structure of RNA polymerase

FIGURE 12.12   Structure and molecular function of RNA polymerase. (a) A comparison of the crystal structures of a bacterial RNA polymerase (left) to a eukaryotic RNA polymerase II (right). The bacterial enzyme is from *Thermus aquaticus*. The eukaryotic enzyme is from *Saccharomyces cerevisiae*. **(b)** A mechanism for transcription based on the crystal structure. In this diagram, the direction of transcription is from left to right. The double-stranded DNA enters the polymerase along a bridge surface that is between the jaw and clamp. At a region termed the wall, the RNA-DNA hybrid is forced to make a right-angle turn, which enables nucleotides to bind to the template strand. Mg^{2+} is located at the catalytic site. Nucleoside triphosphates (NTPs) enter the catalytic site via a pore region and bind to the template DNA. At the catalytic site, the nucleotides are covalently attached to the 3' end of the RNA. As RNA polymerase slides down the template, a small region of the protein termed the rudder separates the RNA-DNA hybrid. The single-stranded RNA then exits under a small lid.

eukaryotic counterparts. **Figure 12.12a** compares the structures of a bacterial RNA polymerase with RNA polymerase II from yeast. As seen here, both enzymes have a very similar structure. Also, it is very exciting that this structure provides a way to envision how the transcription process works. As seen in **Figure 12.12b,** DNA enters the enzyme through the jaw and lies on a surface within RNA polymerase termed the bridge. The part of the enzyme called the clamp is thought to control the movement of the DNA through RNA polymerase. A wall in the enzyme forces the RNA-DNA hybrid to make a right-angle turn. This bend facilitates the ability of nucleotides to bind to the template strand. Mg^{2+} is located at the catalytic site, which is precisely at the 3' end of the growing RNA strand. Nucleoside triphosphates (NTPs) enter the catalytic site via a pore region. The correct nucleotide binds to the template DNA and is covalently attached to the 3' end. As RNA polymerase slides down the template, a rudder, which is about 9 bp away from the 3' end of the RNA, forces the RNA-DNA hybrid apart. The single-stranded RNA then exits under a small lid.

Eukaryotic Structural Genes Have a Core Promoter and Regulatory Elements

In eukaryotes, the promoter sequence is more variable and often more complex than that found in bacteria. For structural genes, at least three features are found in most promoters: regulatory elements, a TATA box, and a transcriptional start site. **Figure 12.13** shows a common pattern of sequences found within the promoters of eukaryotic structural genes. The **core promoter** is a relatively short DNA sequence that is necessary for transcription to take place. It consists of a TATAAA sequence called the **TATA box** and the **transcriptional start site,** where transcription begins. The TATA box, which is usually about 25 bp upstream from a transcriptional start site, is important in determining the precise starting point for transcription. If it is missing from the core promoter, the transcription start site point becomes undefined, and transcription may start at a variety of different locations. The core promoter, by itself, produces a low level of transcription. This is termed **basal transcription.**

Regulatory elements affect the ability of RNA polymerase to recognize the core promoter and begin the process of transcription. There are two categories of regulatory elements. Activating sequences, known as **enhancers,** are needed to stimulate transcription. In the absence of enhancer sequences, most eukaryotic genes have very low levels of basal transcription. Under certain conditions, it may also be necessary to prevent transcription of a given gene. This occurs via **silencers**—DNA sequences that inhibit transcription. As seen in Figure 12.13, a common location for regulatory elements is the −50 to −100 region. However, the locations of regulatory elements are quite variable among different eukaryotic genes. These elements can be far away from the core promoter yet exert strong effects on the ability of RNA polymerase to initiate transcription. Eukaryotic regulatory elements are recognized by transcription factors—proteins that bind to regulatory elements and influence the rate of transcription.

DNA sequences such as the TATA box, enhancers, and silencers exert their effects only over a particular gene. They are called *cis*-acting elements. The term *cis* comes from chemistry nomenclature meaning "next to." *Cis*-acting elements, though possibly far away from the core promoter, are always found within the same chromosome as the genes they regulate. By comparison, the regulatory transcription factors that bind to such elements are called ***trans*-acting factors** (the term *trans* means "across from"). The transcription factors that control the expression of a gene are themselves encoded by genes; regulatory genes that encode transcription factors may be far away from the genes they control. When a gene encoding a *trans*-acting factor is expressed, the transcription factor protein that is made can diffuse throughout the cell and bind to its appropriate *cis*-acting element. Let's now turn our attention to the function of such proteins.

Transcription of Eukaryotic Structural Genes Is Initiated When RNA Polymerase II and General Transcription Factors Bind to a Promoter Sequence

Thus far, we have considered the DNA sequences that play a role in the promoter region of eukaryotic structural genes. By studying transcription in a variety of eukaryotic species, researchers

FIGURE 12.13   **A common pattern found for the promoter of structural genes recognized by RNA polymerase II.** The start site usually occurs at adenine; two pyrimidines (Py) and a cytosine precede this adenine, and five pyrimidines (Py; cytosine or thymine) follow it. A TATA box is approximately 25 bp upstream. However, the sequences that constitute eukaryotic promoters are quite diverse, and not all structural genes have a TATA box. Regulatory elements, such as GC or CAAT boxes, are variable in their locations but often are found in the −50 to −100 region. The core promoters for RNA polymerase I and III are quite different. A single upstream regulatory element is involved in the binding of RNA polymerase I to its promoter, while two regulatory elements called A and B boxes facilitate the binding of RNA polymerase III.

have discovered that three categories of proteins are needed for basal transcription at the core promoter: RNA polymerase II, general transcription factors, and mediator (**Table 12.2**).

Five different proteins called **general transcription factors (GTFs)** are always needed for RNA polymerase II to initiate transcription of structural genes. **Figure 12.14** describes the assembly of GTFs and RNA polymerase II at the TATA box. As shown here, a series of interactions leads to the formation of the open complex. Transcription factor IID (TFIID) first binds to the TATA box and thereby plays a critical role in the recognition of the core promoter. TFIID is composed of several subunits, including TATA-binding protein (TBP), which directly binds to the TATA box, and several other proteins called TBP-associated factors (TAFs). After TFIID binds to the TATA box, it associates with TFIIB. TFIIB promotes the binding of RNA polymerase II and TFIIF to the core promoter. Lastly, TFIIE and TFIIH bind to the complex. This completes the assembly of proteins to form a closed complex, also known as a **preinitiation complex.**

TFIIH plays a major role in the formation of the open complex. TFIIH has several subunits that perform different func-

tions. Certain subunits act as helicases, which break the hydrogen bonding between the double-stranded DNA and thereby promote the formation of the open complex. Another subunit hydrolyzes ATP and phosphorylates a domain in RNA polymerase II known as the carboxyl terminal domain (CTD). Phosphorylation of the CTD releases the contact between RNA polymerase II and TFIIB.

TABLE **12.2**

Proteins Needed for Transcription via the Core Promoter of Eukaryotic Structural Genes

RNA polymerase II: The enzyme that catalyzes the linkage of ribonucleotides in the 5' to 3' direction, using DNA as a template. Essentially all eukaryotic RNA polymerase II proteins are composed of 12 subunits. The two largest subunits are structurally similar to the β and β' subunits found in *E. coli* RNA polymerase.

General transcription factors:

TFIID: Composed of TATA-binding protein (TBP) and other TBP-associated factors (TAFs). Recognizes the TATA box of eukaryotic structural gene promoters.

TFIIB: Binds to TFIID and then enables RNA polymerase II to bind to the core promoter. Also promotes TFIIF binding.

TFIIF: Binds to RNA polymerase II and plays a role in its ability to bind to TFIIB and the core promoter. Also plays a role in the ability of TFIIE and TFIIH to bind to RNA polymerase II.

TFIIE: Plays a role in the formation and/or the maintenance of the open complex. It may exert its effects by facilitating the binding of TFIIH to RNA polymerase II and regulating the activity of TFIIH.

TFIIH: A multisubunit protein that has multiple roles. First, certain subunits act as helicases and promote the formation of the open complex. Other subunits phosphorylate the carboxyl terminal domain (CTD) of RNA polymerase II, which releases its interaction with TFIIB and thereby allows RNA polymerase II to proceed to the elongation phase.

Mediator: A multisubunit complex that mediates the effects of regulatory transcription factors on the function of RNA polymerase II. Though mediator typically has certain core subunits, many of its subunits vary, depending on the cell type and environmental conditions. The ability of mediator to affect RNA polymerase II function is thought to occur via the CTD of RNA polymerase II. Mediator can influence the ability of TFIIH to phosphorylate CTD, and subunits within mediator itself have the ability to phosphorylate CTD. Because CTD phosphorylation is needed to release RNA polymerase II from TFIIB, mediator plays a key role in the ability of RNA polymerase II to switch from the initiation to the elongation stage of transcription.

TFIID binds to the TATA box. TFIID is a complex of proteins that includes the TATA-binding protein (TBP) and several TBP-associated factors (TAFs).

TATA box

TFIIB binds to TFIID.

TFIIB acts as a bridge to bind RNA polymerase II and TFIIF.

RNA polymerase II

TFIIE and TFIIH bind to RNA polymerase II to form a preinitiation or closed complex.

Preinitiation complex

TFIIH acts as a helicase to form an open complex. TFIIH also phosphorylates the CTD domain of RNA polymerase II. CTD phosphorylation breaks the contact between TFIIB and RNA polymerase II. TFIIB, TFIIE, and TFIIH are released.

Open complex

PO_4

PO_4

CTD domain of RNA polymerase II

FIGURE 12.14 **Steps leading to the formation of the open complex.**

Next, TFIIB, TFIIE, and TFIIH dissociate, and RNA polymerase II is free to proceed to the elongation stage of transcription.

In vitro, when researchers mix together TFIID, TFIIB, TFIIF, TFIIE, TFIIH, RNA polymerase II, and a DNA sequence containing a TATA box and transcriptional start site, the DNA is transcribed into RNA. Therefore, these components are referred to as the **basal transcription apparatus.** In a living cell, however, additional components regulate transcription and allow it to proceed at a reasonable rate.

In addition to GTFs and RNA polymerase II, another component required for transcription is a large protein complex termed mediator. This complex was discovered by Roger Kornberg and colleagues in 1990. In 2006, Kornberg was awarded the Nobel Prize for his studies regarding the molecular basis of eukaryotic transcription. **Mediator** derives its name from the observation that it mediates interactions between RNA polymerase II and regulatory transcription factors that bind to enhancers or silencers. It serves as an interface between RNA polymerase II and many, diverse regulatory signals. The subunit composition of mediator is quite complex and variable. The core subunits form an elliptical-shaped complex that partially wraps around RNA polymerase II. Mediator itself may phosphorylate the CTD of RNA polymerase II and it may regulate the ability of TFIIH to phosphorylate the CTD. Therefore, it can play a pivotal role in the switch between transcriptional initiation and elongation. The function of mediator during eukaryotic gene regulation is explored in greater detail in Chapter 15.

Chromatin Structure Plays a Key Role in Gene Transcription

As we have learned, eukaryotic transcription involves the binding of general transcription factors and RNA polymerase to the promoter region and the subsequent movement of RNA polymerase along the DNA double helix, allowing one strand to function as a template for transcription. The compaction of DNA to form chromatin can be an obstacle to the transcription process. During interphase, when most transcription occurs, the chromatin of eukaryotes is found in 30 nm fibers organized into radial loop domains. Within the 30 nm fiber, the DNA is wound around histone octamers to form nucleosomes (refer back to Chapter 10, Figure 10.16). The size of a histone octamer is roughly five times smaller than the complex of RNA polymerase II and GTFs. Therefore, because RNA polymerase is a very large enzyme compared to a nucleosome, the tight wrapping of DNA within a nucleosome is expected to inhibit the ability of RNA polymerase to transcribe the DNA.

How is chromatin structure loosened so that transcription can occur? Two common mechanisms alter chromatin structure. First, the amino terminal tails of the core histone proteins are covalently modified in a variety of ways, including the acetylation of lysines, methylation of lysines, and phosphorylation of serines (refer back to Chapter 10, Figure 10.20). These covalent modifications play a key role in the level of chromatin compaction. For example, positively charged lysine residues within the amino terminal tails of the core histone proteins can be acetylated by

(a) Histone acetylation

(b) ATP-dependent chromatin remodeling

FIGURE 12.15   Mechanisms that disrupt the tight binding of DNA and histones within nucleosomes. (**a**) The covalent modification of the amino terminal tails of histones by acetylation, for example, alters their ability to bind to the DNA backbone. Histone acetyltransferase catalyzes the attachment of acetyl groups to lysine residues, thereby loosening the interactions between histones and DNA. Histone deacetylase reverses this process. (**b**) In ATP-dependent chromatin remodeling, proteins of the SWI/SNF family catalyze an ATP-dependent change in the locations of nucleosomes. This may involve a shift in nucleosomes to a new location (left) or a change in the spacing of nucleosomes over a long stretch of DNA (right). These effects may significantly alter gene transcription.

enzymes called **histone acetyltransferases (Figure 12.15a).** The attachment of the acetyl group ($-COCH_3$) eliminates the positive charge on the lysine side chain and thereby disrupts the favorable interaction between the histone protein and the negatively charged DNA backbone. This may loosen the structure of nucleosomes and the 30 nm fiber and thereby facilitate the ability of RNA polymerase to transcribe a gene. Some studies suggest that histones are completely displaced, whereas others suggest they are loosened but remain attached to the DNA. In addition, covalently modified histones can be recognized by proteins that alter the compaction level of the chromatin. Other enzymes, called histone deacetylases, can remove the acetyl groups and thereby restore a tighter interaction. As discussed further in Chapter 15, the competing actions of acetyltransferases and deacetylases play an important role in the transcriptional regulation of genes.

A second way to alter chromatin structure is through **ATP-dependent chromatin remodeling** (**Figure 12.15b**). In this process, the energy of ATP hydrolysis is used to drive a change in the structure of nucleosomes and thereby makes the DNA more or less accessible for transcription. Therefore, chromatin remodeling is important for both activation and repression of transcription. The remodeling process is carried out by a protein complex that recognizes nucleosomes and uses ATP to alter their configuration. The complexes that catalyze ATP-dependent chromatin remodeling are members of the **SWI/SNF family,** a large group of proteins found in all eukaryotic species. They were first identified in yeast as proteins that play a role in the activation of gene transcription. The abbreviations SWI and SNF refer to the effects that occur in yeast when these remodeling enzymes are defective. SWI mutants are defective in mating type <u>swi</u>tching, and SNF mutations create a <u>s</u>ucrose <u>n</u>on-fermenting phenotype.

As shown in Figure 12.15b, one result of ATP-dependent chromatin remodeling is a change in the locations of nucleosomes, involving either a shift in nucleosomes to a new location or an alteration in the spacing of nucleosomes over a long stretch of DNA. These effects on nucleosome location may substantially alter the ability of general transcription factors to recognize the promoter and also influence the ability of RNA polymerase to transcribe a given gene. More recently, research has shown that chromatin remodeling may also affect the configuration of the 30 nm fiber and/or the radial loop domains found in euchromatin.

12.4 RNA MODIFICATION

During the 1960s and 1970s, studies in bacteria established the physical structure of the gene. The analysis of bacterial genes showed that the sequence of DNA within the coding strand corresponds to the sequence of nucleotides in the mRNA, except that T is replaced with U. During translation, the sequence of codons in the mRNA is then read, providing the instructions for the correct amino acid sequence in a polypeptide. The one-to-one correspondence between the sequence of codons in the DNA coding strand and the amino acid sequence of the polypeptide has been termed the **colinearity** of gene expression.

The situation dramatically changed in the late 1970s, when the tools became available to study eukaryotic genes at the molecular level. The scientific community was astonished by the discovery that eukaryotic structural genes are not always colinear with their functional mRNAs. Instead, the coding sequences within many eukaryotic genes are separated by DNA sequences that are not translated into protein. The coding sequences are called **exons,** and the sequences that interrupt them are called **intervening sequences** or **introns.** During transcription, an RNA is made corresponding to the entire gene sequence. Subsequently, the sequences in the RNA that correspond to the introns are removed, while the RNA sequences derived from the exons are connected or spliced together. This process is called **RNA splicing.** Since the 1970s, research has revealed that splicing is a common genetic phenomenon in eukaryotic species. Splicing occurs occasionally in bacteria as well.

Aside from splicing, research has also shown that RNA transcripts can be modified in several other ways. **Table 12.3** describes the general types of RNA modifications. For example, rRNAs and tRNAs are synthesized as long transcripts that are processed into smaller functional pieces. In addition, most eukaryotic mRNAs have a cap attached to their 5' end and a tail attached at their 3' end. In this section, we will examine the molecular mechanisms that account for several types of RNA modifications and consider why they are functionally important.

Some Large RNA Transcripts Are Processed into Smaller Functional Transcripts by Enzymatic Cleavage

For many nonstructural genes, the RNA transcript initially made during gene transcription is processed or cleaved into smaller pieces. As an example, **Figure 12.16** shows the processing of mammalian ribosomal RNA. The ribosomal RNA gene is transcribed by RNA polymerase I to make a long primary transcript, known as 45S rRNA. The term 45S refers to the sedimentation characteristics of this transcript in Svedberg units. Following the synthesis of the 45S rRNA, cleavage occurs at several points to produce three fragments, termed 18S, 5.8S, and 28S rRNA. These are functional rRNA molecules that play a key role in creating the structure of the ribosome. In eukaryotes, the processing of 45S rRNA into smaller rRNAs and the assembly of ribosomal subunits occur in a structure within the cell nucleus known as the **nucleolus.**

The production of tRNA molecules requires processing via exonucleases and endonucleases. An **exonuclease** is a type of enzyme that cleaves a covalent bond between two nucleotides at one end of a strand. Starting at one end, an exonuclease can digest a strand, one nucleotide at a time. Some exonucleases can begin this digestion only from the 3' end, traveling in the 3' to 5' direction, while others can begin only at the 5' end and digest in the 5' to 3' direction. By comparison, an **endonuclease** can cleave the bond between two adjacent nucleotides within a strand.

Like ribosomal RNA, tRNAs are synthesized as large precursor tRNAs that must be cleaved to produce mature, functional tRNAs that bind to amino acids. This processing has been studied extensively in *E. coli*. **Figure 12.17** shows the processing of a precursor tRNA, which involves the action of two endonucleases and one exonuclease. The precursor tRNA is recognized by an enzyme known as RNaseP. This enzyme is an endonuclease that cuts the precursor tRNA. The action of RNaseP produces the correct 5' end of the mature tRNA. A different endonuclease cleaves the precursor tRNA to remove a 170-nucleotide segment from the 3' end. Next, an exonuclease, called RNaseD, binds to the 3' end and digests the RNA in the 3' to 5' direction. When it reaches an ACC sequence, the exonuclease stops digesting the precursor tRNA molecule. Therefore, all tRNAs in *E. coli* have

TABLE 12.3

Modifications That May Occur to RNAs

Modification	Description
Processing	The cleavage of a large RNA transcript into smaller pieces. One or more of the smaller pieces becomes a functional RNA molecule. Processing occurs for rRNA and tRNA transcripts.
Splicing	Splicing involves both cleavage and joining of RNA molecules. The RNA is cleaved at two sites, which allows an internal segment of RNA, known as an intron, to be removed. After the intron is removed, the two ends of the RNA molecules are joined together. Splicing is common among eukaryotic pre-mRNAs, and it also occurs occasionally in rRNAs, tRNAs, and a few bacterial RNAs.
5' capping	The attachment of a 7-methylguanosine cap (m^7G) to the 5' end of mRNA. This occurs in eukaryotic mRNAs. The cap plays a role in the splicing of introns, the exit of mRNA from the nucleus, and the binding of mRNA to the ribosome.
3' polyA tailing	The attachment of a string of adenine-containing nucleotides to the 3' end of mRNA at a site where the mRNA is cleaved (see upward arrow). This occurs in eukaryotic mRNAs. It is important for RNA stability and translation.
RNA editing	The change of the base sequence of an RNA after it has been transcribed (described in Chapter 15).
Base modification	The covalent modification of a base within an RNA molecule. As described in Chapter 13, base modification commonly occurs in tRNA molecules.

an ACC sequence at their 3' ends. Finally, certain bases in tRNA molecules may be covalently modified to alter their structure. The functional importance of modified bases in tRNAs is discussed in Chapter 13.

As researchers studied tRNA processing, they discovered certain enzymatic features that were very unusual and exciting, changing the way biologists view the actions of enzymes. RNaseP has been found to be an enzyme that contains both RNA and protein subunits. In 1983, Sidney Altman and colleagues made the surprising discovery that the RNA portion of this enzyme, not the protein subunit, contains the catalytic ability to cleave the precursor tRNA. RNaseP is an example of a **ribozyme,** an RNA molecule with enzymatic activity. Prior to the study of RNaseP and the identification of self-splicing RNAs (discussed later), biochemists had staunchly believed that only protein molecules could function as enzymes.

FIGURE 12.16   **The processing of ribosomal RNA in eukaryotes.** The large ribosomal RNA gene is transcribed into a long 45S rRNA primary transcript. This transcript is cleaved to produce 18S, 5.8S, and 28S rRNA molecules, which become associated with protein subunits in the ribosome. This processing occurs within the nucleolus of the cell.

^mG = Methylguanosine
P = Pseudouridine
T = 4-Thiouridine
IP = 2-Isopentenyladenosine

FIGURE 12.17   **The processing of precursor tRNA molecules.** RNaseP is an endonuclease that makes a cut that creates the 5' end of the mature tRNA. To produce the 3' end of mature tRNA, an endonuclease makes a cut, and then the exonuclease RNaseD removes nine nucleotides at the 3' end. In addition to these cleavage steps, several bases within the tRNA molecule are modified to other bases.

EXPERIMENT 12A

Introns Were Experimentally Identified Via Microscopy

Although the discovery of ribozymes was very surprising, the observation that tRNA and rRNA transcripts are processed to a smaller form did not seem unusual to geneticists and biochemists, because the enzymatic cleavage of RNA was similar to the cleavage that can occur for other macromolecules such as DNA and proteins. In sharp contrast, when splicing was detected in the 1970s, it was a novel concept. Splicing involves cleavage at two sites. An intron is removed, and—in a unique step—the remaining fragments are hooked back together again.

Eukaryotic introns were first detected by comparing the base sequence of viral genes and their mRNA transcripts during viral infection of mammalian cells by adenovirus. This research was carried out in 1977 by two groups headed by Philip Sharp and Richard Roberts. This pioneering observation led to the next question: Are introns a peculiar phenomenon that occurs only in viral genes, or are they found in eukaryotic genes as well?

In the late 1970s, several research groups, including those of Pierre Chambon, Bert O'Malley, and Phillip Leder, investigated the presence of introns in eukaryotic structural genes. The experiments of Leder used electron microscopy to identify introns in the β-globin gene. β globin is a polypeptide that is a subunit of hemoglobin, the protein that carries oxygen in red blood cells. To detect introns within the gene, Leder considered the possible effects of mRNA binding to a gene. **Figure 12.18a** considers the situation in which a gene does not contain an intron. In this experiment, a segment of double-stranded chromosomal DNA containing a gene was first denatured and mixed with mature mRNA encoded by that gene. Because the mRNA is complementary to the template strand of the DNA, the template strand and the mRNA will bind to each other to form a hybrid molecule. This event is called **hybridization.** Later, when the DNA is allowed to renature, the binding of the mRNA to the template strand of DNA prevents the two strands of DNA from forming a double helix. In the absence of any introns, the single-stranded DNA will form a loop. Because the RNA has displaced one of the DNA strands, this structure is known as an RNA displacement loop, or **R loop,** as shown in Figure 12.18a.

In contrast, Leder and colleagues realized that a different type of R loop structure would be formed if the gene contained an intron (**Figure 12.18b**). When mRNA is hybridized to a region of a gene containing one intron and then the other DNA strand is allowed to renature, two single-stranded R loops will form that are separated by a double-stranded DNA region. The intervening double-stranded region occurs because an intron has been spliced out of the mature mRNA, so the mRNA cannot hybridize to this segment of the gene.

(a) No introns in the DNA

(b) One intron in the DNA. The intron in the pre-mRNA is spliced out.

FIGURE 12.18 **Hybridization of mRNA to double-stranded DNA.** In this experiment, the DNA is denatured and then allowed to renature under conditions that favor an RNA-DNA hybrid. (**a**) If the DNA does not contain an intron, the binding of the mRNA to the template strand of DNA prevents the two strands of DNA from forming a double helix. The single-stranded region of DNA will form an R loop. (**b**) When mRNA hybridizes to a gene containing one intron, two single-stranded R loops will form that are separated by a double-stranded DNA region. The intervening double-stranded region occurs because an intron has been spliced out of the mRNA and the mRNA cannot hybridize to this segment of the gene.

As shown in steps 1 through 4 of **Figure 12.19**, this hybridization approach was used to identify introns within the β-globin gene. Following hybridization, the samples were placed on a microscopy grid, shadowed with heavy metal, and then observed by electron microscopy.

■ **THE HYPOTHESIS**

The mouse β-globin gene contains one or more introns.

TESTING THE HYPOTHESIS — FIGURE 12.19   RNA hybridization to the β-globin gene reveals an intron.

Starting material: A cloned fragment of chromosomal DNA that contains the mouse β-globin gene.

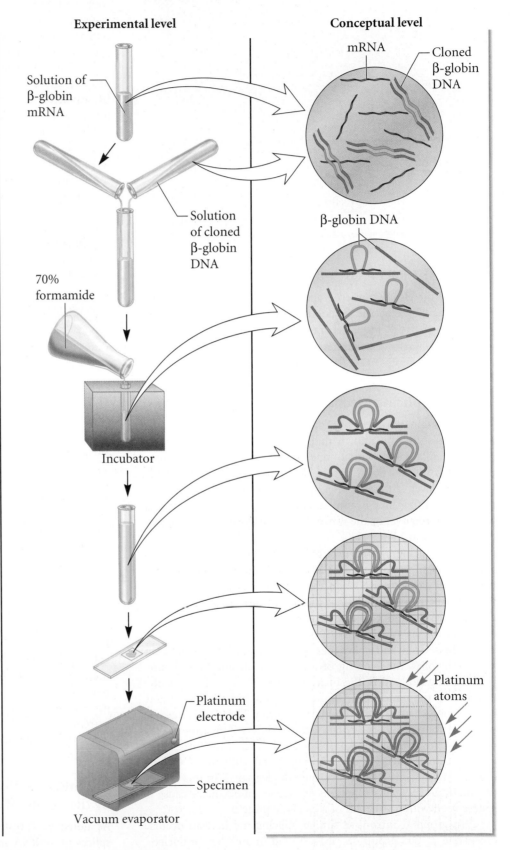

Experimental level

Conceptual level

1. Isolate mature mRNA for the mouse β-globin gene. Note: Globin mRNA is abundant in reticulocytes, which are immature red blood cells.

2. Mix together the β-globin mRNA and cloned DNA of the β-globin gene.

3. Separate the double-stranded DNA and allow the mRNA to hybridize. This is done using 70% formamide, at 52°C, for 16 hours.

4. Dilute the sample to decrease the formamide concentration. This allows the DNA to re-form a double-stranded structure. Note: The DNA cannot form a double-stranded structure in regions where the mRNA has already hybridized.

5. Spread the sample onto a microscopy grid.

6. Stain with uranyl acetate and shadow with heavy metal. Note: The technique of electron microscopy is described in the Appendix.

7. View the sample under the electron microscope.

THE DATA

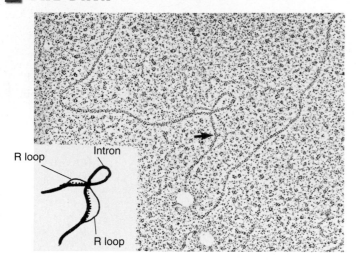

R loop   Intron

R loop

INTERPRETING THE DATA

As seen in the electron micrograph, the β-globin mRNA hybrid-ized to the DNA of the β-globin gene, which resulted in the formation of two R loops separated by a double-stranded DNA

region. These data were consistent with the idea that the DNA of the β-globin gene contains an intron. Similar results were obtained by Chambon and O'Malley for other structural genes. Since these initial discoveries, introns have been found in many eukaryotic genes. The prevalence and biological significance of introns are discussed later in this chapter.

Since the late 1970s, DNA sequencing methods have per-mitted an easier and more precise way of detecting introns. Researchers can clone a fragment of chromosomal DNA that contains a particular gene. This is called a **genomic clone.** In addition, mRNA can be used as a starting material to make a copy of DNA known as **complementary DNA (cDNA).** The cDNA will not contain introns, because the introns have been previously removed during RNA splicing. In contrast, if a gene contains introns, a genomic clone for a eukaryotic gene will also contain introns. Therefore, a comparison of the DNA sequences from genomic and cDNA clones can provide direct evidence that a particular gene contains introns. Compared to genomic DNA, the cDNA will be missing base sequences that were removed dur-ing splicing.

> *A self-help quiz involving this experiment can be found at the Online Learning Center.*

Different Splicing Mechanisms Can Remove Introns

Since the original discovery of introns, the investigations of many research groups have shown that most structural genes among complex eukaryotes contain one or more introns. Less com-monly, introns can occur within tRNA and rRNA genes. At the molecular level, different RNA splicing mechanisms have been identified. In the three examples shown in **Figure 12.20,** splicing leads to the removal of the intron RNA and the covalent connec-tion of the exon RNA.

The splicing among **group I** and **group II introns** occurs via **self-splicing**—splicing that does not require the aid of other enzymes. Instead, the RNA functions as its own ribozyme. Group I and II differ in the ways that the introns are removed and the exons are connected. Group I introns that occur within the rRNA of *Tetrahymena* (a protozoan) have been studied extensively by Thomas Cech and colleagues. In this organism, the splicing process involves the binding of a single guanosine to a guano-sine binding site within the intron (Figure 12.20a). This guano-sine breaks the bond between the first exon and the intron and becomes attached to the 5' end of the intron. The 3'—OH group of exon 1 then breaks the bond next to a different guanine nucle-otide that lies at the boundary between the end of the intron and exon 2; exon 1 forms a phosphoester bond with the 5' end of exon 2. The intron RNA is subsequently degraded. In this exam-ple, the RNA molecule functions as its own ribozyme, because it splices itself without the aid of a catalytic protein.

In group II introns, a similar splicing mechanism occurs, except the 2'—OH group on ribose found in an adenine nucleo-tide already within the intron strand begins the catalytic process

(Figure 12.20b). Experimentally, group I and II self-splicing can occur in vitro without the addition of any proteins. However, in a living cell, proteins known as **maturases** often enhance the rate of splicing of group I and II introns.

In eukaryotes, the transcription of structural genes produces a long transcript known as **pre-mRNA,** which is located within the nucleus. These large RNA transcripts are also known as **heteroge-neous nuclear RNA (hnRNA).** This pre-mRNA is usually altered by splicing and other modifications before it exits the nucleus. Unlike group I and II introns, which may undergo self-splicing, pre-mRNA splicing requires the aid of a multicomponent struc-ture known as the **spliceosome.** As discussed shortly, this is needed to recognize the intron boundaries and to properly remove it.

Table 12.4 describes the occurrence of introns among the genes of different species. The biological significance of group I and II introns is not understood. By comparison, pre-mRNA splicing is a widespread phenomenon among complex eukary-otes. In mammals and flowering plants, most structural genes have at least one intron that can be located anywhere within the gene. In some cases, a single gene can have many introns. As an extreme example, the human dystrophin gene that, when mutant, causes Duchenne muscular dystrophy has 79 exons punctuated by 78 introns.

Pre-mRNA Splicing Occurs by the Action of a Spliceosome

As noted previously, the spliceosome is a large complex that splices pre-mRNA in eukaryotes. It is composed of several sub-units known as **snRNPs** (pronounced "snurps"). Each snRNP

FIGURE 12.20 **Mechanisms of RNA splicing.** Group I and II introns are self-splicing. **(a)** The splicing of group I introns involves the binding of a free guanosine to a site within the intron, leading to the cleavage of RNA at the 3' end of exon 1. The bond between a different guanine nucleotide (in the intron strand) and the 5' end of exon 2 is cleaved. The 3' end of exon 1 then forms a covalent bond with the 5' end of exon 2. **(b)** In group II introns, a similar splicing mechanism occurs, except that the 2' –OH group on an adenine nucleotide (already within the intron) begins the catalytic process. **(c)** Pre-mRNA splicing requires the aid of a multicomponent structure known as the spliceosome.

TABLE **12.4**		
Occurrence of Introns		
Type of Intron	**Mechanism of Removal**	**Occurrence**
Group I	Self-splicing	Found in rRNA genes within the nucleus of *Tetrahymena* and other simple eukaryotes. Found in a few structural, tRNA, and rRNA genes within the mitochondrial DNA (fungi and plants) and in chloroplast DNA. Found very rarely in tRNA genes within bacteria.
Group II	Self-splicing	Found in a few structural, tRNA, and rRNA genes within the mitochondrial DNA (fungi and plants) and in chloroplast DNA. Also found rarely in bacterial genes.
Pre-mRNA	Spliceosome	Very commonly found in structural genes within the nucleus of eukaryotes.

contains small nuclear RNA and a set of proteins. During splicing, the subunits of a spliceosome carry out several functions. First, spliceosome subunits bind to an intron sequence and precisely recognize the intron-exon boundaries. In addition, the spliceosome must hold the pre-mRNA in the correct configuration to ensure the splicing together of the exons. And finally, the spliceosome catalyzes the chemical reactions that cause the introns to be removed and the exons to be covalently linked.

Intron RNA is defined by particular sequences within the intron and at the intron-exon boundaries. The consensus sequences for the splicing of mammalian pre-mRNA are shown in **Figure 12.21**. These sequences serve as recognition sites for the binding of the spliceosome. The bases most commonly found at these sites—those that are highly conserved evolutionarily—are shown in bold. The 5' and 3' splice sites occur at the ends of the intron, while the branch site is somewhere in the middle. These sites are recognized by components of the spliceosome.

FIGURE 12.21   **Consensus sequences for pre-mRNA splicing in complex eukaryotes.** Consensus sequences exist at the intron-exon boundaries and at a branch site found within the intron itself. The adenine nucleotide shown in blue in this figure corresponds to the adenine nucleotide at the branch site in Figure 12.22. The nucleotides shown in bold are highly conserved. Designations: A/C = A or C, Pu = purine, Py = pyrimidine, N = any of the four bases. The 5' splice site is also called the splice donor site, and the 3' splice site is the splice acceptor site.

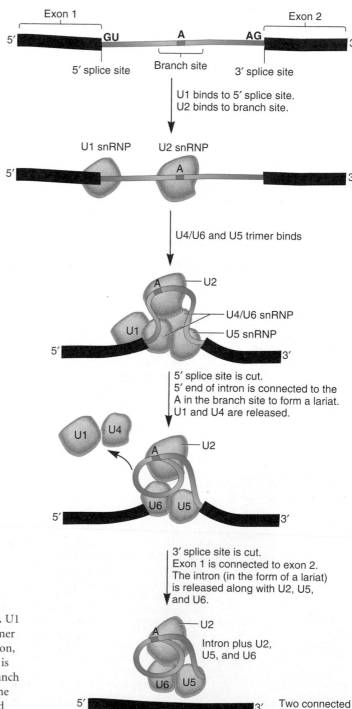

The molecular mechanism of pre-mRNA splicing is depicted in **Figure 12.22**. The snRNP designated U1 binds to the 5' splice site, while U2 binds to the branch site. This is followed by the binding of a trimer of three snRNPs: a U4/U6 dimer plus U5. The intron loops outward, and the two exons are brought closer together. The 5' splice site is then cut, and the 5' end of the intron becomes covalently attached to the 2'–OH group of a specific adenine nucleotide in the branch site. U1 and U4 are released. In the final step, the 3' splice site is cut, and then the exons are covalently attached to each other. The three snRNPs, U2, U5, and U6, remain attached to the intron, which is in a lariat configuration. Eventually, the intron will be degraded, and the snRNPs will be used again to splice other pre-mRNAs.

The chemical reactions that occur during pre-mRNA splicing are not completely understood. Though further research is needed, evidence is accumulating that certain snRNA molecules within the spliceosome may play an enzymatic role in the removal of introns and the connection of exons. In other words, snRNAs may function as ribozymes that cleave the RNA at the exon-intron boundaries and connect the remaining exons. Researchers have speculated that RNA molecules within U2 and U6 may have this catalytic function.

Before ending our discussion of pre-mRNA splicing, let's consider why it may be an advantage for a species to have genes that contain introns. One benefit is a phenomenon called **alternative splicing.** When a pre-mRNA has multiple introns, variation may occur in the pattern of splicing, so the resulting mRNAs contain alternative combinations of exons. The variation in splicing may happen in different cell types or during different stages of development. Such alternative splicing allows mature mRNAs to contain different patterns of exons.

FIGURE 12.22   **Splicing of pre-mRNA via a spliceosome.** U1 binds to the 5' splice site, while U2 binds to the branch site. A trimer of three snRNPs, a U4/U6 dimer plus U5, binds to the intron region, causing it to form a loop. This brings the exons close together. The 5' splice site is cut, and the 5' end of the intron is attached to the adenine nucleotide in the branch site. U1 and U4 are released. In the final step, the 3' splice site is cut, and then the exons are connected together. The three snRNPs, U2, U5, and U6, remain bound to the intron, which is found in a lariat configuration. Later, the intron will be degraded, and the snRNPs will be used over again.

As an example, let's suppose a mammalian pre-mRNA contained seven exons. In one cell type (e.g., muscle cells), it may be spliced to produce a mature mRNA with the following pattern of exons: 1-2-4-5-6-7. In a different cell type (e.g., nerve cells), it might be spliced in an alternative pattern: 1-2-3-5-6-7. In this example, the mRNA from muscle cells contains exon 4, while the mRNA from nerve cells contains exon 3. When alternative splicing occurs, proteins with significant differences in their amino acid sequences are produced. In most cases, the alternative versions of a protein will have similar functions, because much of their amino acid sequences will be identical to each other. Nevertheless, alternative splicing produces differences in amino acid sequences that provide each protein with its own unique characteristics. The biological advantage of alternative splicing is that two or more different proteins can be derived from a single gene. This allows an organism to carry fewer genes in its genome. The molecular mechanism of alternative splicing is examined in greater detail in Chapter 15. It involves the actions of proteins (not shown in Figure 12.22) that influence whether or not U1 and U2 can begin the splicing process.

The Ends of Eukaryotic Pre-mRNAs Have a 5' Cap and a 3' Tail

In addition to splicing, pre-mRNAs in eukaryotes are also subjected to modifications at their 5' and 3' ends. At their 5' end, most mature mRNAs have a 7-methylguanosine covalently attached, an event known as **capping.** Capping occurs while the pre-mRNA is being made by RNA polymerase II, usually when the transcript is only 20 to 25 nucleotides in length. As shown in **Figure 12.23**, it is a three-step process. The nucleotide at the 5' end of the transcript has three phosphate groups. First, an enzyme called RNA 5'-triphosphatase removes one of the phosphates, and then a second enzyme, guanylyltransferase, uses GTP to attach a guanosine monophosphate (GMP) to the 5' end. Finally, a methyltransferase attaches a methyl group to the guanine base.

What are the functions of the 7-methylguanosine cap? The cap structure is recognized by cap-binding proteins, which perform various roles. For example, cap-binding proteins are required for the proper exit of most mRNAs from the nucleus. Also, the cap structure is recognized by initiation factors that are needed during the early stages of translation. Finally, the cap structure may be important in the efficient splicing of introns, particularly the first intron located nearest the 5' end.

Let's now turn our attention to the 3' end of the RNA molecule. Most mature mRNAs have a string of adenine nucleotides, referred to as a **polyA tail,** which is important for mRNA stability and in the synthesis of polypeptides. The polyA tail is not encoded in the gene sequence. Instead, it is added enzymatically after the pre-mRNA has been completely transcribed. This process is termed **polyadenylation.**

The steps required to synthesize a polyA tail are shown in **Figure 12.24**. To acquire a polyA tail, the pre-mRNA contains a polyadenylation sequence near its 3' end. In higher eukaryotes, the consensus sequence is AAUAAA. This sequence is downstream

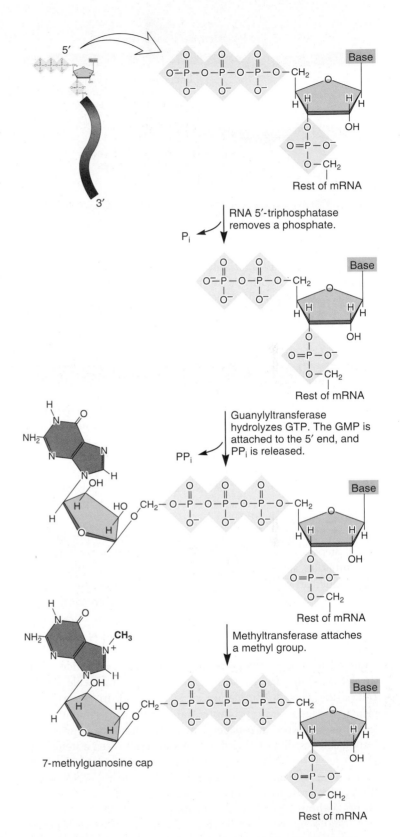

FIGURE 12.23   **Attachment of a 7-methylguanosine cap to the 5' end of mRNA.** When the transcript is about 20 to 25 nucleotides in length, RNA 5'-triphosphatase removes one of the three phosphates, and then a second enzyme, guanylyltransferase, attaches GMP to the 5' end. Finally, a methyltransferase attaches a methyl group to the guanine base.

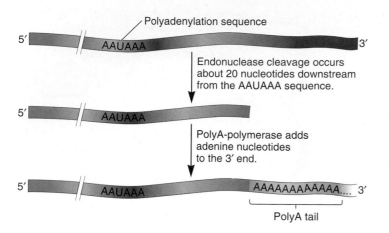

FIGURE 12.24 **Attachment of a polyA tail.** First, an endonuclease cuts the RNA at a location that is 11 to 30 nucleotides after the AAUAAA polyadenylation sequence, making the RNA shorter at its 3' end. Adenine-containing nucleotides are then attached, one at a time, to the 3' end by the enzyme polyA-polymerase.

(toward the 3' end) from the stop codon in the pre-mRNA. An endonuclease recognizes the polyadenylation sequence and cleaves the pre-mRNA at a location that is about 20 nucleotides beyond the 3' end of the AAUAAA sequence. The fragment beyond the 3' cut is degraded. Next, an enzyme known as polyA-polymerase attaches many adenine-containing nucleotides. The length of the polyA tail varies among different mRNAs, from a few dozen to several hundred adenine nucleotides. As discussed in Chapter 15, a long polyA tail increases the stability of mRNA and plays a role during translation.

CONCEPTUAL SUMMARY

Transcription is the process of RNA synthesis from a DNA template. Different types of RNA transcripts can be made. These include mRNA, which specifies the sequence of amino acids within a polypeptide, and tRNA and rRNA, which are necessary to translate mRNA. In addition, several other small RNA molecules, such as microRNA, 7S RNA, scRNA, the RNA component of RNaseP, snRNA, telomerase RNA, and snoRNA have various functions in the cell.

The synthesis of RNA during gene transcription involves three steps: initiation, elongation, and termination. During the initiation stage, proteins that bind to DNA recognize specific sequences within the promoter region. In the case of many bacterial promoters, the −35 and −10 sequences are recognized by σ factor, which is part of RNA polymerase holoenzyme. To begin the synthesis or elongation of a transcript, the DNA must be converted to an open complex by denaturing a small double-stranded region. At this point, RNA polymerase can slide down the DNA, synthesizing a complementary RNA molecule. At the end of the gene, a termination signal is found. In bacteria, termination signals can promote transcriptional termination in two different ways. In ρ-dependent termination, the ρ protein is responsible for dissociating the RNA transcript when a stem-loop structure causes RNA polymerase to pause. In ρ-independent termination, the formation of a stem-loop within the RNA causes RNA polymerase to stall and eventually fall off the DNA, but this occurs without the help of ρ protein.

Transcription in eukaryotes is similar but more complex compared to the process in bacteria. In eukaryotes, three RNA polymerases designated I, II, and III transcribe different types of genes. Transcriptional initiation of structural genes requires five general transcription factors that enable RNA polymerase II to recognize the TATA box in the −25 region. Transcription is initiated by an assembly process that leads to the formation of the open complex. A large protein complex termed mediator plays a role in the switch between initiation and elongation. For transcription to occur, the chromatin structure must be loosened so that RNA polymerase can transcribe the template DNA. This involves two mechanisms: covalent modifications of histone proteins and ATP-dependent chromatin remodeling.

Following transcription, a newly made RNA transcript may be modified in various ways. For example, tRNA transcripts are processed to smaller forms by endo- and exonucleases. Some genes contain introns that must be removed from the RNA via splicing. Group I and Group II introns are self-splicing, which means they catalyze their own removal. These types of introns are found in the rRNA genes of *Tetrahymena* and other simpler eukaryotes, in organelle DNA, and occasionally in bacterial genes. In contrast, introns are commonly found in eukaryotic pre-mRNA, where they are removed by a multicomponent structure known as a spliceosome. Many eukaryotic pre-mRNAs undergo alternative splicing, with the result that a single gene can produce more than one type of polypeptide. Besides splicing, most eukaryotic pre-mRNA is also capped with 7-methylguanosine at its 5' end, and a polyA tail is attached to its 3' end.

EXPERIMENTAL SUMMARY

Experimentally, many methods are used to study the transcription of genes at the molecular level. Most of these are described in Chapter 18. Several of these techniques are considered in the experimental questions at the end of this chapter. For example, the quantity of an RNA transcript can be determined by Northern blotting, in which RNA transcripts are identified through a DNA-RNA hybridization technique. This method can also be used to detect alternative splicing of mRNAs.

The binding and assembly of proteins during transcription can be examined in a variety of ways. In Chapter 11, we considered how Arthur Kornberg developed an in vitro system to study DNA replication. Similarly, researchers have succeeded in developing in vitro systems to study transcription. In this way,

they have identified the basal transcriptional apparatus needed for transcription to occur.

In this chapter, we have also considered how hybridization and electron microscopy can be used to detect introns. In this approach, the hybridization of mRNA to a DNA sequence blocks the DNA's ability to re-form double-stranded regions. Using an electron microscope to examine the pattern of single-stranded DNA and double-stranded DNA regions, the hybridization of RNA to DNA allowed researchers to identify introns within the β-globin gene. Since the late 1970s, DNA sequencing methods that compare the DNA sequences of genomic clones and complementary DNA (cDNA) clones have provided an easier and more precise way of detecting whether a particular gene contains introns.

PROBLEM SETS & INSIGHTS

Solved Problems

S1. Describe the important events that occur during the three stages of gene transcription in bacteria. What proteins play critical roles in the three stages?

Answer: The three stages are initiation, elongation, and termination.

Initiation: RNA polymerase holoenzyme slides along the DNA until σ factor recognizes a promoter. σ factor binds tightly to this sequence, forming a closed complex. The DNA is then denatured to form a bubblelike structure known as the open complex.

Elongation: RNA polymerase core enzyme slides along the DNA, synthesizing RNA as it goes. The α subunits of RNA polymerase keep the enzyme bound to the DNA, while the β subunits are responsible for binding and for the catalytic synthesis of RNA. The ω (omega) subunit is also important for the proper assembly of the core enzyme. During elongation, RNA is made according to the AU/GC rule, with nucleotides being added in the 5' to 3' direction.

Termination: RNA polymerase eventually reaches a sequence at the end of the gene that signals the cessation of transcription. In ρ-independent termination, the properties of the termination sequences in the DNA are sufficient to cause termination. In ρ-dependent termination, the ρ (rho) protein recognizes a sequence within the RNA, binds there, and travels toward RNA polymerase. When the formation of a stem-loop structure causes RNA polymerase to pause, ρ catches up and separates the RNA-DNA hybrid region, releasing the RNA polymerase.

S2. What is the difference between a structural gene and a nonstructural gene?

Answer: Structural genes encode mRNA that is translated into a polypeptide sequence. Nonstructural genes encode RNAs that are never translated. Products of nonstructural genes include tRNA and rRNA, which function during translation; microRNA, which is involved in gene regulation; 7S RNA, which is part of a complex known as SRP; scRNA, small cytoplasmic RNA found in bacteria; the RNA of RNaseP; telomerase RNA, which is involved in telomere replication;

snoRNA, which is involved in rRNA trimming; and snRNA, which is a component of spliceosome. In many cases, the RNA from nonstructural genes becomes part of a complex composed of RNA molecules and protein subunits.

S3. When RNA polymerase transcribes DNA, only one of the two DNA strands is used as a template. Take a look at Figure 12.4 and explain how RNA polymerase determines which DNA strand is the template strand.

Answer: The binding of σ factor and RNA polymerase depends on the sequence of the promoter. RNA polymerase binds to the promoter in such a way that the −35 sequence TTGACA and the −10 sequence TATAAT are within the coding strand, while the −35 sequence AACTGT and the −10 sequence ATATTA are within the template strand.

S4. The process of transcriptional termination is not as well understood in eukaryotes as it is in bacteria. Nevertheless, current evidence suggests several different mechanisms exist for termination. Like bacteria, the termination of certain genes appears to occur via intrinsic terminators (as in ρ-independent termination), while the termination of other genes may involve RNA-binding proteins (as in ρ-dependent termination). In eukaryotes, a third type of mechanism is found for the termination of rRNA genes by RNA polymerase I. In this case, a protein known as TTFI (transcription termination factor I) binds to the DNA downstream from the termination site. Discuss how the binding of a protein downstream from the termination site could promote transcriptional termination.

Answer: First, the binding of TTFI could act as a roadblock to the movement of RNA polymerase I. Second, TTFI could promote the dissociation of the RNA transcript and RNA polymerase I from the DNA; it may act like a helicase. Third, it could cause a change in the structure of the DNA that prevents RNA polymerase from moving past the termination site. Though multiple effects are possible, the third effect seems the most likely because TTFI is known to cause a bend in the DNA when it binds to the termination sequence.

Conceptual Questions

C1. Genes may be structural genes that encode polypeptides, or they may be nonstructural genes.

A. Describe three examples of genes that are not structural genes.

B. For structural genes, one DNA strand is called the template strand, and the complementary strand is called the coding strand. Are these two terms appropriate for nonstructural genes? Explain.

C. Do nonstructural genes have a promoter and terminator?

C2. In bacteria, what event marks the end of the initiation stage of transcription?

C3. What is the meaning of the term consensus sequence? Give an example. Describe the locations of consensus sequences within bacterial promoters. What are their functions?

C4. What is the consensus sequence of the following six DNA molecules?

GGCATTGACT

GCCATTGTCA

CGCATAGTCA

GGAAATGGGA

GGCTTTGTCA

GGCATAGTCA

C5. Mutations in bacterial promoters may increase or decrease the level of gene transcription. Promoter mutations that increase transcription are termed up promoter mutations, and those that decrease transcription are termed down promoter mutations. As shown in Figure 12.5, the sequence of the −10 region of the promoter for the *lac* operon is TATGTT. Would you expect the following mutations to be up promoter or down promoter mutations?

A. TATGTT to TATATT

B. TATGTT to TTTGTT

C. TATGTT to TATGAT

C6. According to the examples shown in Figure 12.5, which positions of the −35 sequence (i.e., first, second, third, fourth, fifth, and/or sixth) are more tolerant of changes? Do you think that these positions play a more or less important role in the binding of σ factor? Explain why.

C7. In Chapter 9, we considered the dimensions of the double helix (see Figure 9.17). In an α helix of a protein, there are 3.6 amino acids per complete turn. Each amino acid advances the α helix by 0.15 nm; a complete turn of an α helix is 0.54 nm in length. As shown in Figure 12.6, two α helices of a transcription factor occupy the major groove of the DNA. According to Figure 12.6, estimate the number of amino acids that bind to this region. How many complete turns of the α helices occupy the major groove of DNA?

C8. A mutation within a gene sequence changes the start codon to a stop codon. How will this mutation affect the transcription of this gene?

C9. What is the subunit composition of bacterial RNA polymerase holoenzyme? What are the functional roles of the different subunits?

C10. At the molecular level, describe how σ factor recognizes bacterial promoters. Be specific about the structure of σ factor and the type of chemical bonding.

C11. Let's suppose a DNA mutation changes the consensus sequence at the −35 location so that σ factor is no longer able to bind there. Explain how a mutation would prevent σ factor from binding to the DNA. Look at Figure 12.5 and describe two specific base substitutions you think would inhibit the binding of σ factor. Explain why you think your base substitutions would have this effect.

C12. What is the complementarity rule that governs the synthesis of an RNA molecule during transcription? An RNA transcript has the following sequence:

5'-GGCAUGCAUUACGGCAUCACACUAGGGAUC-3'

What is the sequence of the template and coding strands of the DNA that encodes this RNA? On which side (5' or 3') of the template strand is the promoter located?

C13. Describe the movement of the open complex along the DNA.

C14. Describe what happens to the chemical bonding interactions when transcriptional termination occurs. Be specific about the type of chemical bonding.

C15. Discuss the differences between ρ-dependent and ρ-independent termination.

C16. In Chapter 11, we discussed the function of DNA helicase, which is involved in DNA replication. The structure and function of DNA helicase and ρ protein are rather similar to each other. Explain how the function of these two proteins is similar and how it is different.

C17. Discuss the similarities and differences between RNA polymerase (described in this chapter) and DNA polymerase (described in Chapter 11).

C18. Mutations that occur at the end of a gene may alter the sequence of the gene and prevent transcriptional termination.

A. What types of mutations would prevent ρ-independent termination?

B. What types of mutations would prevent ρ-dependent termination?

C. If a mutation prevented transcriptional termination at the end of a gene, where would gene transcription end? Or would it end?

C19. If the following RNA polymerases were missing from a eukaryotic cell, what types of genes would not be transcribed?

A. RNA polymerase I

B. RNA polymerase II

C. RNA polymerase III

C20. What sequence elements are found within the core promoter of structural genes in eukaryotes? Describe their locations and specific functions.

C21. For each of the following transcription factors, how would eukaryotic transcriptional initiation be affected if it were missing?

A. TFIIB

B. TFIID

C. TFIIH

C22. Discuss how the binding of DNA to histones in a nucleosome structure may affect the process of transcription. Do you think that the nucleosome structure would affect the ability of transcription factors to recognize the promoter sequence in eukaryotic genes? Explain.

C23. Which eukaryotic transcription factor(s) shown in Figure 12.14 plays an equivalent role to σ factor found in bacterial cells?

C24. The initiation phase of eukaryotic transcription via RNA polymerase II is considered an assembly and disassembly process. Which types of biochemical interactions—hydrogen bonding, ionic bonding, covalent bonding, and/or hydrophobic interactions—would you expect to drive the assembly and disassembly process? How would temperature and salt concentration affect assembly and disassembly?

C25. A eukaryotic structural gene contains two introns and three exons: exon 1–intron 1–exon 2–intron 2–exon 3. The 5' splice site at the boundary between exon 2 and intron 2 has been eliminated by a small deletion in the gene. Describe how the pre-mRNA encoded by this mutant gene would be spliced. Indicate which introns and exons would be found in the mRNA after splicing occurs.

C26. Describe the processing events that occur during the production of tRNA in *E. coli*.

C27. Describe the structure and function of a spliceosome. Speculate why the spliceosome subunits contain snRNA. In other words, what do you think is/are the functional role(s) of snRNA during splicing?

C28. What is the unique feature of ribozyme function? Give two examples described in this chapter.

C29. What does it mean to say that gene expression is colinear?

C30. What is meant by the term self-splicing? What types of introns are self-splicing?

C31. In eukaryotes, what types of modification occur to pre-mRNA?

C32. What is alternative splicing? What is its biological significance?

C33. The processing of ribosomal RNA in eukaryotes is shown in Figure 12.16. Why is this called cleavage or processing but not splicing?

C34. In the splicing of group I introns shown in Figure 12.20, does the 5' end of the intron have a phosphate group? Explain.

C35. According to the mechanism shown in Figure 12.22, several snRNPs play different roles in the splicing of pre-mRNA. Identify the snRNP that recognizes the following sites:

A. 5' splice site

B. 3' splice site

C. Branch site

C36. After the intron (which is in a lariat configuration) is released during pre-mRNA splicing, a brief moment occurs before the two exons are connected to each other. Which snRNP(s) holds the exons in place so they can be covalently connected to each other?

C37. A lariat contains a closed loop and a linear end. An intron has the following sequence: 5'–GUPuAGUA–60 nucleotides–UACUUAUCC–100 nucleotides–Py$_{12}$NPyAG–3'. Which sequence would be found within the closed loop of the lariat, the 60-nucleotide sequence or the 100-nucleotide sequence?

Experimental Questions

E1. A research group has sequenced the cDNA and genomic DNA from a particular gene. The cDNA is derived from mRNA, so it does not contain introns. Here are the DNA sequences.

cDNA:

```
5'-ATTGCATCCAGCGTATACTATCTCGGGCCCAATTAAT-
GCCAGCGGCCAGACTATCACCCAACTCGGTTACCTACTAG-
TATATCCCATATACTAGCATATATTTTACCCATAATTTGTGTGT-
GGGTATACAGTATAATCATATA-3'
```

Genomic DNA (contains one intron):

```
5'-ATTGCATCCAGCGTATACTATCTCGGGCCCAAT-
TAATGCCAGCGGCCAGACTATCACCCAACTCG-
GCCCACCCCCCAGGTTTACACAGTCATACCATACA
TACAAAAATCGCAGTTACTTATCCCAAAAAAACCTAG-
ATACCCCACATACTATTAACTCTTTCTTTCTAGGTTACCTAC-
TAGTATATCCCATATACTAGCATATATTTTACCCATAATTTGT-
GTGTGGGTATACAGTATAATCATATA-3'
```

Indicate where the intron is located. Does the intron contain the normal consensus splice site sequences based on those described in Figure 12.21? Underline the splice site sequences, and indicate whether or not they fit the consensus sequence.

E2. What is an R loop? In an R loop experiment, to which strand of DNA does the mRNA bind, the coding strand or the template strand?

E3. If a gene contains three introns, draw what it would look like in an R loop experiment.

E4. Chapter 18 describes a technique known as Northern blotting that can be used to detect RNA transcribed from a particular gene. In this method, a specific RNA is detected using a short segment of cloned DNA as a probe. The DNA probe, which is radioactive, is complementary to the RNA that the researcher wishes to detect. After the radioactive probe DNA binds to the RNA, the RNA is visualized as a dark (radioactive) band on an X-ray film. As shown here, the method of Northern blotting can be used to determine the amount of a particular RNA transcribed in a given cell type. If one type of cell produces twice as much of a particular mRNA compared to another cell, the band will appear twice as intense. Also, the method can distinguish if alternative RNA splicing has occurred to produce an RNA that has a different molecular mass.

Northern blot

Lane 1   is a sample of RNA isolated from nerve cells.

Lane 2   is a sample of RNA isolated from kidney cells. Nerve cells produce twice as much of this RNA compared to kidney cells.

Lane 3   is a sample of RNA isolated from spleen cells. Spleen cells produce an alternatively spliced version of this RNA that is about 200 nucleotides longer than the RNA produced in nerve and kidney cells.

Let's suppose a researcher was interested in the effects of mutations on the expression of a particular structural gene in eukaryotes. The gene has one intron that is 450 nucleotides long. After this intron is removed from the pre-mRNA, the mRNA transcript is 1,100 nucleotides in length. Diploid somatic cells have two copies of this gene. Make a drawing that shows the expected results of a Northern blot using mRNA from the cytosol of somatic cells, which were obtained from the following individuals:

Lane 1: A normal individual

Lane 2: An individual homozygous for a deletion that removes the −50 to −100 region of the gene that encodes this mRNA

Lane 3: An individual heterozygous in which one gene is normal and the other gene had a deletion that removes the −50 to −100 region

Lane 4: An individual homozygous for a mutation that introduces an early stop codon into the middle of the coding sequence of the gene

Lane 5: An individual homozygous for a three-nucleotide deletion that removes the AG sequence at the 3' splice site

E5. A gel retardation assay can be used to study the binding of proteins to a segment of DNA. This method is described in Chapter 18. When a protein binds to a segment of DNA, it retards the movement of the DNA through a gel, so the DNA appears at a higher point in the gel (see the following).

Lane 1: 900 bp fragment alone

Lane 2: 900 bp fragment plus a protein that binds to the 900 bp fragment

In this example, the segment of DNA is 900 bp in length, and the binding of a protein causes the DNA to appear at a higher point in the gel. If this 900 bp fragment of DNA contains a eukaryotic promoter for a structural gene, draw a gel that shows the relative locations of the 900 bp fragment under the following conditions:

Lane 1: 900 bp plus TFIID

Lane 2: 900 bp plus TFIIB

Lane 3: 900 bp plus TFIID and TFIIB

Lane 4: 900 bp plus TFIIB and RNA polymerase II

Lane 5: 900 bp plus TFIID, TFIIB, and RNA polymerase II/TFIIF

E6. As described in Chapter 18 and in experimental question E5, a gel retardation assay can be used to determine if a protein binds to DNA. This method can also determine if a protein binds to RNA. In the combinations described here, would you expect the migration of the RNA to be retarded due to the binding of a protein?

A. mRNA from a gene that is terminated in a ρ-independent manner plus ρ-protein

B. mRNA from a gene that is terminated in a ρ-dependent manner plus ρ-protein

C. pre-mRNA from a structural gene that contains two introns plus the snRNP called U1

D. Mature mRNA from a structural gene that contains two introns plus the snRNP called U1

E7. The technique of DNA footprinting is described in Chapter 18. If a protein binds over a region of DNA, it will protect chromatin in that region from digestion by DNase I. To carry out a DNA footprinting experiment, a researcher has a sample of a cloned DNA fragment. The fragments are exposed to DNase I in the presence and absence of a DNA-binding protein. Regions of the DNA fragment not covered by the DNA-binding protein will be digested by DNase I, and this will produce a series of bands on a gel. Regions of the DNA fragment not digested by DNase I (because a DNA-binding protein is preventing DNase I from gaining access to the DNA) will be revealed, because a region of the gel will not contain any bands.

In the DNA footprinting experiment shown here, a researcher began with a sample of cloned DNA 300 bp in length. This DNA contained a eukaryotic promoter for RNA polymerase II. For the sample loaded in lane 1, no proteins were added. For the sample loaded in lane 2, the 300 bp fragment was mixed with RNA polymerase II plus TFIID and TFIIB.

A. How long of a region of DNA is "covered up" by the binding of RNA polymerase II and the transcription factors?

B. Describe how this binding would occur if the DNA was within a nucleosome structure. (Note: The structure of nucleosomes is described in Chapter 10.) Do you think that the DNA is in a nucleosome structure when RNA polymerase and transcription factors are bound to the promoter? Explain why or why not.

E8. As described in Table 12.1, several different types of RNA are made, especially in eukaryotic cells. Researchers are sometimes interested in focusing their attention on the transcription of structural genes in eukaryotes. Such researchers want to study mRNA. One method that is used to isolate mRNA is column chromatography. (Note: See the Appendix for a general description of chromatography.) Researchers can covalently attach short pieces of DNA that contain stretches of thymine (i.e., TTTTTTTTTTT) to the column matrix. This is called a poly-dT column. When a cell extract is poured over the column, mRNA binds to the column, while other types of RNA do not.

A. Explain how you would use a poly-dT column to obtain a purified preparation of mRNA from eukaryotic cells. In your description, explain why mRNA binds to this column and what you would do to release the mRNA from the column.

B. Can you think of ways to purify other types of RNA, such as tRNA or rRNA?

Questions for Student Discussion/Collaboration

1. Based on your knowledge of introns and pre-mRNA splicing, discuss whether or not you think alternative splicing fully explains the existence of introns. Can you think of other possible reasons to explain the existence of introns?

2. Discuss the types of RNA transcripts and the functional roles they play. Why do you think some RNAs form complexes with protein subunits?

Note: All answers appear at the website for this textbook; the answers to even-numbered questions are in the back of the textbook.

www.mhhe.com/brookergenetics3e

Visit the Online Learning Center for practice tests, answer keys, and other learning aids for this chapter. Enhance your understanding of genetics with our interactive exercises, quizzes, animations, and much more.

TRANSLATION OF mRNA

T he synthesis of cellular proteins occurs via the translation of the sequence of codons within mRNA into a sequence of amino acids of a polypeptide. The general steps that occur in this process have already been outlined in Chapter 1. In this chapter, we will explore the current state of knowledge regarding translation, with an eye toward the specific molecular interactions responsible for this process. During the past few decades, the concerted efforts of geneticists, cell biologists, and biochemists have profoundly advanced our understanding of translation. Even so, many questions remain unanswered, and this topic continues to be an exciting area of research investigation.

We will begin by considering the classic experiments that revealed the purpose of some genes is to encode proteins that function as enzymes. Next, we will examine how the genetic code is used to decipher the information within mRNA to produce a polypeptide with a specific amino acid sequence. The rest of this chapter is devoted to a molecular understanding of translation as it occurs in living cells. This will involve an examination of the cellular components—including many different proteins, RNAs, and small molecules—needed for the translation process. We will consider the structure and function of tRNA molecules, which act as the translators of the genetic information within mRNA, and then examine the composition of ribosomes. Finally, we will explore the differences between translation in bacterial cells and eukaryotic cells.

A molecular model for the structure of a ribosome. This is a model of ribosome structure based on X-ray crystallography. Ribosomes are needed to synthesize polypeptides, using mRNA as a template.

13.1 THE GENETIC BASIS FOR PROTEIN SYNTHESIS

Proteins are critically important as active participants in cell structure and function. The primary role of DNA is to store the information needed for the synthesis of all the proteins that an organism makes. As we discussed in Chapter 12, genes that encode an amino acid sequence are known as **structural genes.** The RNA transcribed from structural genes is called **messenger RNA (mRNA).** The main function of the genetic material is to encode the production of cellular proteins in the correct cell, at the proper time, and in suitable amounts. This is an extremely complicated task because living cells make thousands of different proteins. Genetic analyses have shown that a typical bacterium can make a few thousand different proteins, and estimates for eukaryotes range from several thousand in simple eukaryote organisms, such as yeast, to tens of thousands in plants and animals.

In this section, we will begin by considering early experiments that showed the role of genes is to encode proteins. We will examine the general features of the genetic code—the sequence of bases in a codon that specifies an amino acid—and explore the experiments through which the code was deciphered or "cracked." Finally, we will look at the biochemistry of polypeptide synthesis to see how this determines the structure and function of proteins, which are ultimately responsible for the characteristics of living cells and an organism's traits.

Archibald Garrod Proposed That Some Genes Code for the Production of a Single Enzyme

The idea that a relationship exists between genes and the production of proteins was first suggested at the beginning of the twentieth century by Archibald Garrod, a British physician. Prior to Garrod's studies, biochemists had studied many metabolic pathways within living cells. These pathways consist of a series of metabolic conversions of one molecule to another, each step catalyzed by a specific enzyme. Each enzyme is a distinctly different protein that catalyzes a particular chemical reaction. **Figure 13.1** illustrates part of the metabolic pathway for the degradation of phenylalanine, an amino acid commonly found in human diets. The enzyme phenylalanine hydroxylase catalyzes the conversion of phenylalanine to tyrosine, and a different enzyme, tyrosine aminotransferase, converts tyrosine into p-hydroxyphenylpyruvic acid, and so on. In all of the steps shown in Figure 13.1, a specific enzyme catalyzes a single type of chemical reaction.

Garrod studied patients who had defects in their ability to metabolize certain compounds. He was particularly interested in the inherited disease known as **alkaptonuria.** In this disorder, the patient's body accumulates abnormal levels of homogentisic acid (also called alkapton), which is excreted in the urine, causing it to appear black on exposure to air. In addition, the disease is characterized by bluish black discoloration of cartilage and skin (ochronosis). Garrod proposed that the accumulation of homogentisic acid in these patients is due to a missing enzyme, namely, homogentisic acid oxidase (see Figure 13.1).

How did Garrod realize that certain genes encode enzymes? He already knew that alkaptonuria is an inherited trait that follows a recessive pattern of inheritance. Therefore, an individual with alkaptonuria must have inherited the mutant gene that causes this disorder from both parents. From these observations, Garrod proposed that a relationship exists between the inheritance of the trait and the inheritance of a defective enzyme. Namely, if an individual inherited the mutant gene (which causes a loss of enzyme function), he or she would not produce any normal enzyme and would be unable to metabolize homogentisic acid. Garrod described alkaptonuria as an **inborn error of metabolism.** This hypothesis was the first suggestion that a connection exists between the function of genes and the production of enzymes. At the turn of the century, this idea was particularly insightful, because the structure and function of the genetic material were completely unknown.

FIGURE 13.1   The metabolic pathway of phenylalanine breakdown. This diagram shows part of the pathway of phenylalanine metabolism, which consists of enzymes that successively convert one molecule to another. Certain human genetic diseases (shown in red boxes) are caused when enzymes in this pathway are missing or defective.

Genes→Traits When a person inherits two defective copies of the gene that encodes homogentisic acid oxidase, he or she cannot convert homogentisic acid into maleylacetoacetic acid. Such a person accumulates large amounts of homogentisic acid in the urine and has other symptoms of the disease known as alkaptonuria. Similarly, if a person has two mutant alleles of the gene encoding phenylalanine hydroxylase, he or she is unable to synthesize the enzyme phenylalanine hydroxylase and has the disease called phenylketonuria (PKU).

Beadle and Tatum's Experiments with *Neurospora* Led Them to Propose the One Gene–One Enzyme Hypothesis

In the early 1940s, George Beadle and Edward Tatum were also interested in the relationship among genes, enzymes, and traits. They developed an experimental system for investigating the relationship between genes and the production of particular enzymes. Consistent with the ideas of Garrod, the underlying assumption behind their approach was that a relationship exists between genes and the production of enzymes. However, the quantitative nature of this relationship was unclear. In particular, they asked the question, Does one gene control the production of one enzyme, or does one gene control the synthesis of many enzymes involved in a complex biochemical pathway?

At the time of their studies, many geneticists were trying to understand the nature of the gene by studying morphological traits. However, Beadle and Tatum realized that morphological traits are likely to be based on systems of biochemical reactions so complex as to make analysis exceedingly difficult. Therefore, they turned their genetic studies to the analysis of simple nutritional requirements in *Neurospora crassa*, a common bread mold. *Neurospora* can be easily grown in the laboratory and has few nutritional requirements: a carbon source (sugar), inorganic salts, and the vitamin biotin. Normal *Neurospora* cells produce many different enzymes that can synthesize the organic molecules, such as amino acids and other vitamins, which are essential for growth.

Beadle and Tatum wanted to understand how enzymes are controlled by genes. They reasoned that a mutation in a gene, causing a defect in an enzyme needed for the cellular synthesis of an essential molecule, would prevent that mutant strain from growing on minimal medium, which contains only a carbon source, inorganic salts, and biotin. For example, if a *Neurospora* mutant strain could not make the vitamin pantothenic acid, it would be unable to grow on minimal medium. However, it would grow if the growth medium was supplemented with pantothenic acid.

Beadle and Tatum analyzed more than 2,000 strains that had been irradiated to induce mutations. They found three strains that were unable to grow on minimal medium (**Table 13.1**). In each of these cases, the mutant strain required only a single vitamin to restore its growth on minimal medium. One of these strains required pyridoxine for growth, the second strain required thiamine, and the third required *p*-aminobenzoic acid. In the normal strain, these vitamins are synthesized by cellu-

lar enzymes. In the mutant strains, a genetic defect in one gene prevented the synthesis of a functional enzyme required for the synthesis of one particular vitamin. Therefore, Beadle and Tatum concluded that a single gene controlled the synthesis of a single enzyme. This was referred to as the **one gene–one enzyme hypothesis.**

In later decades, this idea had to be modified in two ways. First, enzymes are only one category of cellular proteins. All proteins are encoded by genes, and many of them do not function as enzymes. Second, some proteins are composed of two or more different polypeptides. Therefore, it is more accurate to say that a structural gene encodes a polypeptide. The term **polypeptide** refers to a structure; it is a linear sequence of amino acids. A structural gene encodes a polypeptide. By comparison, the term **protein** denotes function. Some proteins are composed of one polypeptide. In such cases, a single gene does encode a single protein. In other cases, however, a functional protein is composed of two or more different polypeptides. An example would be hemoglobin, which is composed of two α-globin and two β-globin polypeptides. In this case, the expression of two genes—the α-globin and β-globin genes—is needed to create one functional protein.

During Translation, the Genetic Code Within mRNA Is Used to Make a Polypeptide with a Specific Amino Acid Sequence

Let's now turn to a general description of translation. Why have researchers named this process translation? At the molecular level, **translation** involves an interpretation of one language—the language of mRNA, a nucleotide sequence—into the language of proteins—an amino acid sequence. The ability of mRNA to be translated into a specific sequence of amino acids relies on the **genetic code.** The sequence of bases within an mRNA molecule provides coded information that is read in groups of three nucleotides known as codons (**Figure 13.2**). The sequence of three bases in most codons specifies a particular amino acid. These codons are termed **sense codons.** For example, the codon AGC specifies the amino acid serine, while the codon GGG encodes the amino acid glycine. The codon AUG, which specifies methionine, is used as a **start codon;** it is usually the first codon that begins a polypeptide sequence. The AUG codon can also be used to specify additional methionines within the coding sequence. Finally, three codons are used to end the process of translation. These are UAA, UAG, and UGA, which are known as **stop codons.** They are also known as **termination** and **nonsense codons.**

The codons in mRNA are recognized by the anticodons in transfer RNA (tRNA) molecules (Figure 13.2). **Anticodons** are three-nucleotide sequences that are complementary to codons in mRNA. The tRNA molecules carry the amino acids that correspond to the codons in the mRNA. In this way, the order of codons in mRNA dictates the order of amino acids within a polypeptide.

The details of the genetic code are shown in **Table 13.2.** Because polypeptides are composed of 20 different kinds of amino acids, a minimum of 20 codons is needed in order to

TABLE 13.1

Identification of Mutant Strains of *Neurospora crassa* That Were Unable to Synthesize Particular Vitamins

Single Vitamin Added to Minimal Media	Observed Growth:		
	Strain 1	Strain 2	Strain 3
None	No	No	No
Riboflavin	No	No	No
Thiamine	No	Yes	No
Pantothenic acid	No	No	No
Pyridoxine	Yes	No	No
Niacin	No	No	No
Folic acid	No	No	No
p-aminobenzoic acid	No	No	Yes

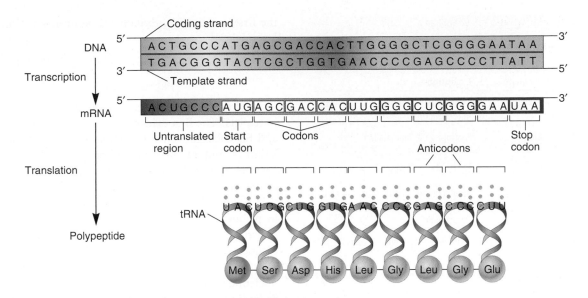

FIGURE 13.2 **The relationships among the DNA coding sequence, mRNA codons, tRNA anticodons, and amino acids in a polypeptide.** The sequence of nucleotides within DNA is transcribed to make a complementary sequence of nucleotides within mRNA. This sequence of nucleotides in mRNA is translated into a sequence of amino acids of a polypeptide. tRNA molecules act as intermediates in this translation process.

TABLE 13.2

The Genetic Code

		Second base			
First base		U	C	A	G	Third base
U	UUU UUC Phenylalanine (Phe) UUA UUG Leucine (Leu)	UCU UCC UCA UCG Serine (Ser)	UAU UAC Tyrosine (Tyr) UAA Stop codon UAG Stop codon	UGU UGC Cysteine (Cys) UGA Stop codon UGG Tryptophan (Trp)	U C A G
C	CUU CUC CUA CUG Leucine (Leu)	CCU CCC CCA CCG Proline (Pro)	CAU CAC Histidine (His) CAA CAG Glutamine (Gln)	CGU CGC CGA CGG Arginine (Arg)	U C A G
A	AUU AUC AUA Isoleucine (Ile) AUG Methionine (Met); start codon	ACU ACC ACA ACG Threonine (Thr)	AAU AAC Asparagine (Asn) AAA AAG Lysine (Lys)	AGU AGC Serine (Ser) AGA AGG Arginine (Arg)	U C A G
G	GUU GUC GUA GUG Valine (Val)	GCU GCC GCA GCG Alanine (Ala)	GAU GAC Aspartic acid (Asp) GAA GAG Glutamic acid (Glu)	GGU GGC GGA GGG Glycine (Gly)	U C A G

specify each type. With four types of bases in mRNA (A, U, G, and C), a genetic code containing two bases in a codon would not be sufficient because it would only have 4^2, or 16, possible types. By comparison, a three-base codon system can specify 4^3, or 64, different codons. Because the number of possible codons exceeds 20—which is the number of different types of amino acids—the genetic code is termed **degenerate.** This means that more than one codon can specify the same amino acid. For example, the codons GGU, GGC, GGA, and GGG all specify the amino acid

glycine. Such codons are termed **synonymous codons.** In most instances, the third base in the codon is the base that varies. The third base is sometimes referred to as the **wobble base.** This term is derived from the idea that the complementary base in the tRNA can "wobble" a bit during the recognition of the third base of the codon in mRNA. The significance of the wobble base will be discussed later in this chapter.

From the analysis of many different species, including bacteria, protists, fungi, plants, and animals, researchers have

Examples of Exceptions to the Genetic Code*

Codon	Universal Meaning	Exception
AUA	Isoleucine	Methionine in yeast and mammalian mitochondria
UGA	Stop	Tryptophan in mammalian mitochondria
CUU, CUA, CUC, CUG	Leucine	Threonine in yeast mitochondria
AGA, AGG	Arginine	Stop codon in ciliated protozoa and in yeast and mammalian mitochondria
UAA, UAG	Stop	Glutamine in ciliated protozoa

*Several other exceptions, sporadically found among various species, are also known.

found that the genetic code is nearly universal. Relatively few exceptions to the genetic code have been noted (**Table 13.3**). The eukaryotic organelles known as mitochondria have their own DNA, which encodes a few structural genes. In mammals, the mitochondrial genetic code contains differences such as AUA = methionine and UGA = tryptophan. Also, in mitochondria and certain ciliated protists, AGA and AGG specify stop codons instead of arginine. Except for rare exceptions such as the ones listed in Table 13.3, the genetic code shown in Table 13.2 has been found to be universal in all species studied thus far.

The start codon (AUG) defines the **reading frame** of an mRNA—a sequence of codons determined by reading bases in groups of three, beginning with the start codon. This concept is best understood with a few examples. The mRNA sequence shown below encodes a short polypeptide with 7 amino acids:

5'-AUGCCCGGAGGCACCGUCCAAU-3'

Met-Pro-Gly-Gly-Thr-Val-Gln

If we remove one base (C) adjacent to the start codon, this changes the reading frame to produce a different polypeptide sequence:

5'-AUGCCGGAGGCACCGUCCAAU-3'

Met-Pro-Glu-Ala-Pro-Ser-Asn

Alternatively, if we remove three bases (CCC) next to the start codon, the resulting polypeptide has the same reading frame as the first polypeptide, though one amino acid (Pro, proline) has been deleted:

5'-AUGGGAGGCACCGUCCAAU-3'

Met-Gly-Gly-Thr-Val-Gln

How did researchers discover that the genetic code is read in triplets? The first evidence came from studies of Francis Crick and his colleagues in 1961. These experiments involved the isolation of mutants in a bacteriophage called T4. As described in Chapter 6, mutations in T4 genes that affect plaque morphology are easily identified (see Figure 6.15). In particular, loss-of-function mutations within certain T4 genes, designated *rII*, resulted in plaques that were larger and had a clear boundary. In comparison, wild-type phages, designated *r*⁺, produced smaller plaques with a fuzzy boundary. Crick and colleagues exposed T4 phages to a chemical called proflavin, which causes single-nucleotide additions or deletions in gene sequences. The mutagenized phages were plated to identify large (*rII*) plaques. Though proflavin can cause either single-nucleotide additions or deletions, the first mutant strain that the researchers identified was arbitrarily called a (+) mutation. Many years later, when methods of DNA sequencing became available, it was determined that the (+) mutation is a single-nucleotide addition. **Table 13.4** shows a hypothetical wild-type sequence in a phage gene (first line) and considers how nucleotide additions and/or deletions could affect the resulting amino acid sequence. A single-nucleotide addition (+) would alter the reading frame beyond the point of insertion and thereby abolish the proper function of the encoded protein. This is called a frameshift mutation, because it has changed the reading frame. This mutation resulted in a loss-of-function for the protein encoded by this viral gene and thereby produced an *rII* plaque phenotype.

The (+) mutant strain was then subjected to a second round of mutagenesis via proflavin. Several plaques were identified that had reverted to a wild-type (*r*⁺) phenotype. By analyzing these strains using methods described in Chapter 6, it was determined that each one contained a second mutation that was close to the original (+) mutation. These second mutations were designated (−) mutations. Three different (−) mutations, designated a, b, and c, were identified. Each of these (−) mutations was a single-nucleotide deletion that was close to the original (+) mutation. Therefore, it restored the reading frame and produced a protein with a nearly normal amino acid sequence.

The critical experiment that suggested the genetic code is read in triplets came by combining different (−) mutations together. Mutations in different phages can be brought together into the same phage via crossing over, as described in Chapter 6. Using such an approach, the researchers constructed strains containing one, two, or three (−) mutations. The results showed that a wild-type plaque morphology was obtained only when three (−) mutations were combined in the same phage (Table 13.4). The three (−) mutations in the same phage restored the normal reading frame. These results were consistent with the hypothesis that the genetic code is read in multiples of three nucleotides.

TABLE 13.4

Evidence That the Genetic Code Is Read in Triplets*

Strain	Plaque Phenotype	DNA Coding Sequence/Polypeptide Sequence§	Downstream Sequence‡
Wild type	r+	ATG GGG CCC GTC CAT CCG TAC GCC GGA ATT ATA Met Gly Pro Val His Pro Tyr Ala Gly Ile Ile----------	In frame
		↓A	
(+)	rII	ATG GGG ACC CGT CCA TCC GTA CGC CGG AAT TAT A Met Gly Thr Arg Pro Ser Val Arg Arg Asn Tyr---------	Out of frame
		↓A ↑C	
(+)(−)a	r+	ATG GGG ACC GTC CAT CCG TAC GCC GGA ATT ATA Met Gly Thr Val His Pro Tyr Ala Gly Ile Ile----------	In frame
		↓A ↑T	
(+)(−)b	r+	ATG GGG ACC CGC CAT CCG TAC GCC GGA ATT ATA Met Gly Thr Arg His Pro Tyr Ala Gly Ile Ile----------	In frame
		↓A ↑G	
(+)(−)c	r+	ATG GGG ACC CTC CAT CCG TAC GCC GGA ATT ATA Met Gly Thr Leu His Pro Tyr Ala Gly Ile Ile----------	In frame
		↑C	
(−)a	rII	ATG GGG CCG TCC ATC CGT ACG CCG GAA TTA TA Met Gly Pro Ser Ile Arg Thr Pro Glu Leu----------------	Out of frame
		↑C ↑T	
(−)a(−)b	rII	ATG GGG CCG CCA TCC GTA CGC CGG AAT TAT A Met Gly Pro Pro Ser Val Arg Arg Asn Tyr----------------	Out of frame
		↑↑↑GTC	
(−)a(−)b(−)c	r+	ATG GGG CCC CAT CCG TAC GCC GGA ATT ATA Met Gly Pro His Pro Tyr Ala Gly Ile Ile--------------	In frame [Only Val is missing]

*This table shows only a small portion of a hypothetical coding sequence.

§A down arrow (↓) indicates the location of a single nucleotide addition; an up arrow (↑) indicates the location of a single nucleotide deletion.

‡The term downstream sequence refers to the remaining part of the sequence that is not shown in this figure. It could include hundreds of codons. An "in-frame" sequence would be wild type, whereas an "out-of-frame" sequence (caused by the addition or deletion of one or two base pairs) would not.

EXPERIMENT 13A

Synthetic RNA Helped to Decipher the Genetic Code

Having determined that the genetic code is read in triplets, how did scientists determine the functions of the 64 codons of the genetic code? During the early 1960s, three research groups headed by Marshall Nirenberg, Severo Ochoa, and H. Gobind Khorana set out to decipher the genetic code. Though they used different methods, all of these groups used synthetic mRNA in their experimental approaches to "crack the code." We will first consider the work of Nirenberg and his colleagues. Prior to their studies, several laboratories had already determined that extracts from bacterial cells, containing a mixture of components including ribosomes, tRNAs, and other factors required for translation, are able to synthesize polypeptides if mRNA and amino acids are added. This mixture is termed an in vitro or **cell-free translation system.** If radiolabeled amino acids are added to a cell-free translation system, the synthesized polypeptides are radiolabeled and easy to detect.

To decipher the genetic code, Nirenberg and colleagues needed to gather information regarding the relationship between mRNA composition and polypeptide composition. To accomplish this goal, they made mRNA molecules of a known base composition, added them to a cell-free translation system, and then analyzed the amino acid composition of the resultant polypeptides. For example, if an mRNA molecule consisted of a string of adenine-containing nucleotides (e.g., 5'–AAAAAAAAAAAAAAAAA–3'), researchers could add this polyA mRNA to a cell-free translation system and ask the question, Which amino acid is specified by a codon that contains only adenine nucleotides? (As Table 13.2 shows, it is lysine.)

Before discussing the details of this type of experiment, let's consider how the synthetic mRNA molecules were made. To synthesize mRNA, an enzyme known as polynucleotide phosphorylase was used. In the presence of excess ribonucleoside diphosphates, also called nucleoside diphosphates (NDPs), this enzyme catalyzes the covalent linkage of nucleotides to make a polymer of RNA. Because it does not use a template, the order of the nucleotides is random. For example, if only uracil-containing diphosphates (UDPs) are added, then a polyU mRNA (5'–UUUUUUUUUUUUUUUU–3') is made. If nucleotides containing two different bases, such as uracil and guanine, are added, then the phosphorylase makes a random polymer containing both nucleotides (5'–GGGUGUGUGGUGGGUG–3'). An experimenter can control the amounts of the nucleotides that are added. For example, if 70% G and 30% U are mixed together with polynucleotide phosphorylase, the predicted amounts of the codons within the random polymer are as follows:

Codon Possibilities	Percentage in the Random Polymer
GGG	$0.7 \times 0.7 \times 0.7 = 0.34 =$ 34%
GGU	$0.7 \times 0.7 \times 0.3 = 0.15 =$ 15%
GUU	$0.7 \times 0.3 \times 0.3 = 0.06 =$ 6%
UUU	$0.3 \times 0.3 \times 0.3 = 0.03 =$ 3%
UUG	$0.3 \times 0.3 \times 0.7 = 0.06 =$ 6%
UGG	$0.3 \times 0.7 \times 0.7 = 0.15 =$ 15%
UGU	$0.3 \times 0.7 \times 0.3 = 0.06 =$ 6%
GUG	$0.7 \times 0.3 \times 0.7 = 0.15 =$ <u>15%</u>
	100%

By controlling the amounts of the NDPs in the phosphorylase reaction, the relative amounts of the possible codons can be predicted.

The first experiment that demonstrated the ability to synthesize polypeptides from synthetic mRNA was performed by Marshall Nirenberg and J. Heinrich Matthaei in 1961. As shown in **Figure 13.3**, a cell-free translation system was added to 20 different tubes. An mRNA template made via polynucleotide phosphorylase was then added to each tube. In this example, the mRNA was made using 70% G and 30% U. Next, the 20 amino acids were added to each tube, but each tube differed with regard to the type of radiolabeled amino acid. For example, radiolabeled glycine would be found in only 1 of the 20 tubes. The tubes were incubated a sufficient length of time to allow translation to occur. The newly made polypeptides were then precipitated onto a filter by treatment with trichloroacetic acid. This precipitates polypeptides but not amino acids. A washing step caused amino acids that had not been incorporated into polypeptides to pass through the filter. Finally, the amount of radioactivity trapped on the filter was determined by liquid scintillation counting.

■ THE GOAL

The researchers assumed that the sequence of bases in mRNA determines the incorporation of specific amino acids into a polypeptide. The purpose of this experiment was to provide information that would help to decipher the relationship between base composition and particular amino acids.

■ ACHIEVING THE GOAL — FIGURE 13.3   Elucidation of the genetic code.

Starting material: A cell-free translation system that can synthesize polypeptides if mRNA and amino acids are added.

	Experimental level	Conceptual level
1. Add the cell-free translation system to each of 20 tubes.	Cell-free translation system	

2. To each tube, add random mRNA polymers of G and U made via polynucleotide phosphorylase using 70% G and 30% U.

3. Add a different radiolabeled amino acid to each tube, and add the other 19 non-radiolabeled amino acids.

4. Incubate for 60 minutes to allow translation to occur.

5. Add 15% trichloroacetic acid (TCA), which precipitates polypeptides but not amino acids.

6. Place the precipitate onto a filter and wash to remove unused amino acids.

7. Count the radioactivity on the filter in a scintillation counter (see the Appendix for a description).

8. Calculate the amount of radiolabeled amino acids in the precipitated polypeptides.

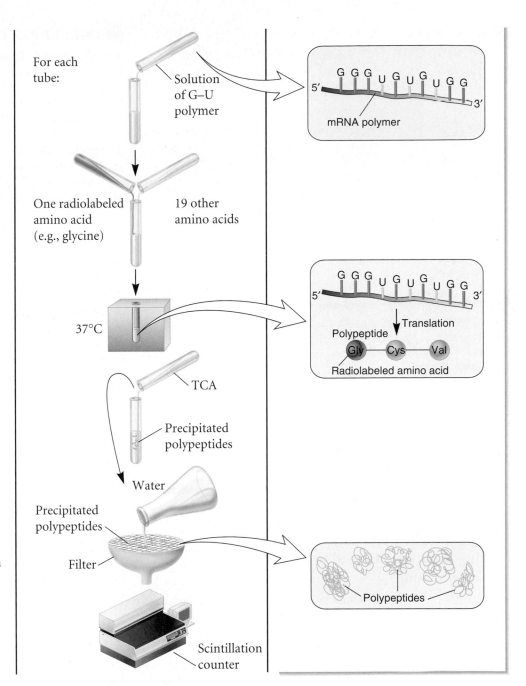

■ THE DATA

Radiolabeled Amino Acid Added	Relative Amount of Radiolabeled Amino Acid Incorporated into Translated Polypeptides (% of total)
Alanine	0
Arginine	0
Asparagine	0
Aspartic acid	0
Cysteine	6
Glutamic acid	0
Glutamine	0
Glycine	49
Histidine	0
Isoleucine	0
Leucine	6
Lysine	0
Methionine	0
Phenylalanine	3
Proline	0
Serine	0
Threonine	0
Tryptophan	15
Tyrosine	0
Valine	21

■ INTERPRETING THE DATA

According to the calculation previously described, codons should occur in the following percentages: 34% GGG, 15% GGU, 6% GUU, 3% UUU, 6% UUG, 15% UGG, 6% UGU, and 15% GUG. In the data shown in Figure 13.3, the value of 49% for glycine is due to two codons: GGG (34%) and GGU (15%). The 6% cysteine is due to UGU, and so on. It is important to realize that the genetic code was not deciphered in a single experiment such as the one described here. Furthermore, this kind of experiment yields information regarding only the nucleotide content of codons, not the specific order of bases within a single codon. For example, this experiment indicates that a cysteine codon contains two U's and one G. However, it does not tell us that a cysteine codon is UGU. Based on these data alone, a cysteine codon could be UUG, GUU, or UGU. However, by comparing many different RNA polymers, the laboratories of Nirenberg and Ochoa established patterns between the specific base sequences of codons and the amino acids they encode. In their first experiments, Nirenberg and Matthaei showed that a random polymer containing only uracil produced a polypeptide containing only phenylalanine. From this result, they inferred that UUU specifies phenylalanine. This idea is consistent with the results shown in the data table. In the random 70% G and 30% U polymer, 3% of the codons will be UUU. Likewise, 3% of the amino acids within the polypeptides were found to be phenylalanine.

A self-help quiz involving this experiment can be found at the Online Learning Center.

The Use of RNA Copolymers and the Triplet Binding Assay Also Helped to Crack the Genetic Code

In the 1960s, H. Gobind Khorana and colleagues developed a novel method to synthesize RNA. They first created short RNA molecules, two to four nucleotides in length, that had a defined sequence. For example, RNA molecules with the sequence 5'–AUC–3' were synthesized chemically. These short RNAs were then linked together enzymatically, in a 5' to 3' manner, to create long copolymers with the sequence

```
5'-AUCAUCAUCAUCAUCAUCAUCAUCAUCAUC-3'
```

This is called a copolymer, because it is made from the linkage of several smaller molecules. Depending on the reading frame, such a copolymer would contain three different codons: AUC, UCA, and CAU. Using a cell-free translation system like the one described in Figure 13.3, such a copolymer produced polypeptides containing isoleucine, serine, and histidine. Table 13.5 summarizes some of the copolymers that were made using this approach and the amino acids that were incorporated into polypeptides.

Finally, another method that helped to decipher the genetic code also involved the chemical synthesis of short RNA molecules. In 1964, Marshall Nirenberg and Philip Leder discovered that RNA molecules containing three nucleotides—a triplet—could

TABLE 13.5

Examples of Copolymers That Were Analyzed by Khorana and Colleagues

Synthetic RNA*	Codon Possibilities	Amino Acids Incorporated into Polypeptides
UC	UCU, CUC	Serine, leucine
AG	AGA, GAG	Arginine, glutamic acid
UG	UGU, GUG	Cysteine, valine
AC	ACA, CAC	Threonine, histidine
UUC	UUC, UCU, CUU	Phenylalanine, serine, leucine
AAG	AAG, AGA, GAA	Lysine, arginine, glutamic acid
UUG	UUG, UGU, GUU	Leucine, cysteine, valine
CAA	CAA, AAC, ACA	Glutamine, asparagine, threonine
UAUC	UAU, AUC, UCU, CUA	Tyrosine, isoleucine, serine, leucine
UUAC	UUA, UAC, ACU, CUU	Leucine, tyrosine, threonine

*The synthetic RNAs were linked together to make copolymers.

stimulate ribosomes to bind a tRNA. In other words, the RNA triplet acted like a codon. Ribosomes were able to bind RNA triplets, and then a tRNA with the appropriate anticodon could subsequently bind to the ribosome. To establish the relationship between triplet sequences and specific amino acids, samples containing ribosomes and a particular triplet were exposed to tRNAs with different radiolabeled amino acids.

As an example, in one experiment the researchers began with a sample of ribosomes that were mixed with 5'–CCC–3' triplets. Portions of this sample were then added to 20 different tubes that had tRNAs with different radiolabeled amino acids. For example, one tube contained radiolabeled histidine, a second tube had radiolabeled proline, a third tube contained radiolabeled glycine, and so on. Only one radiolabeled amino acid was added to each tube. After allowing sufficient time for tRNAs to bind to the ribosomes, the samples were filtered; only the large ribosomes and anything bound to them were trapped on the filter (Figure 13.4). Unbound tRNAs would pass through the filter. Next, the researchers determined the amount of radioactivity trapped on each filter. If the filter contained a large amount of radioactivity, the results indicated that the added triplet encoded the amino acid that was radiolabeled.

Using the triple binding assay, Nirenberg and Leder were able to establish relationships between particular triplet sequences and the binding of tRNAs carrying specific (radiolabeled) amino acids. In the case of the 5'–CCC–3' triplet, they determined that tRNAs carrying radiolabeled proline were bound to the ribosomes. Unfortunately, in some cases, a triplet could not promote sufficient tRNA binding to yield unambiguous results. Nevertheless, the triplet binding assay was an important tool in the identification of the majority of codons.

A Polypeptide Chain Has Directionality from Its Amino Terminal to Its Carboxyl Terminal End

Let's now turn our attention to polypeptide biochemistry. Polypeptide synthesis has a directionality that parallels the order of codons in the mRNA. As a polypeptide is made, a **peptide bond** is formed between the carboxyl group in the last amino acid of the polypeptide chain and the amino group in the amino acid being added. As shown in **Figure 13.5a**, this occurs via a condensation reaction that releases a water molecule. The newest amino acid added to a growing polypeptide always has a free carboxyl group. **Figure 13.5b** compares the sequence of a very short polypeptide with the mRNA that encodes it. The first amino acid is said to be at the **N-terminus,** or **amino terminal end,** of the polypeptide. An amino group (NH_3^+) is found at this site. The term N-terminal refers to the presence of a nitrogen atom (N) at this end. The first amino acid is specified by a codon that is near the 5' end of the mRNA. By comparison, the last amino acid in a completed polypeptide is located at the **C-terminus,** or **carboxyl terminal end.** A carboxyl group (COO–) is always found at this site in the polypeptide chain. This last amino acid is specified by a codon that is closer to the 3' end of the mRNA.

The Amino Acid Sequences of Polypeptides Determine the Structure and Function of Proteins

Now that we understand how mRNAs encode polypeptides, let's consider the structure and function of the gene product, namely,

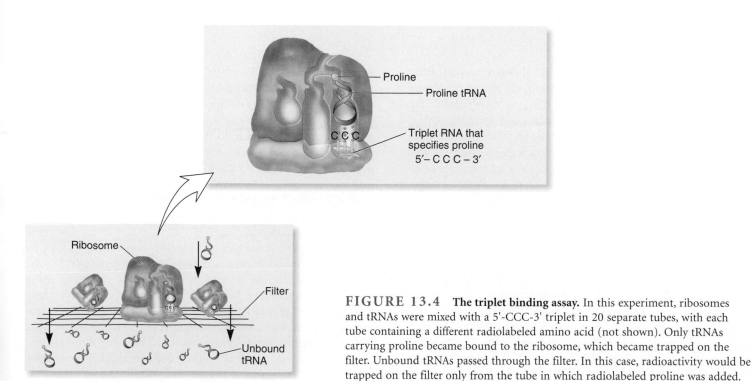

FIGURE 13.4   The triplet binding assay. In this experiment, ribosomes and tRNAs were mixed with a 5'-CCC-3' triplet in 20 separate tubes, with each tube containing a different radiolabeled amino acid (not shown). Only tRNAs carrying proline became bound to the ribosome, which became trapped on the filter. Unbound tRNAs passed through the filter. In this case, radioactivity would be trapped on the filter only from the tube in which radiolabeled proline was added.

(a) Attachment of an amino acid to a peptide chain

FIGURE 13.5   The directionality of polypeptide synthesis. (a) An amino acid is connected to a polypeptide chain via a condensation reaction that releases a water molecule. The letter R is a general designation for an amino acid side chain. (b) The first amino acid in a polypeptide chain (usually methionine) is located at the amino terminal end, and the last amino acid is at the carboxyl terminal end. Thus, the directionality of amino acids in a polypeptide chain is from the amino terminal end to the carboxyl terminal end, which corresponds to the 5' to 3' orientation of codons in mRNA.

(b) Directionality in a polypeptide and mRNA

polypeptides. **Figure 13.6** shows the 20 different amino acids that may be found within polypeptides. Each amino acid contains a unique **side chain,** or **R group,** that has its own particular chemical properties. For example, aliphatic and aromatic amino acids are nonpolar, which means they are less likely to associate with water. These hydrophobic (meaning water-fearing) amino acids are often buried within the interior of a folded protein. In contrast, the polar amino acids are hydrophilic (water-loving) and are more likely to be on the surface of a protein, where they can favorably interact with the surrounding water. The chemical properties of the amino acids and their sequences in a polypeptide are critical factors that determine the unique structure of that polypeptide.

Following gene transcription and mRNA translation, the end result is a polypeptide with a defined amino acid sequence. This sequence is the **primary structure** of a polypeptide. **Figure 13.7** shows the primary structure of an enzyme called lysozyme, a relatively small protein containing 129 amino acids. The primary structure of a typical polypeptide may be a few hundred or even a couple thousand amino acids in length. Within a living cell, a newly made polypeptide is not usually found in a long linear

state for a significant length of time. Rather, to become a functional unit, most polypeptides quickly adopt a compact three-dimensional structure. The folding process begins while the polypeptide is still being translated. The progression from the primary structure of a polypeptide to the three-dimensional structure of a protein is dictated by the amino acid sequence within the polypeptide. In particular, the chemical properties of the amino acid side chains play a central role in determining the folding pattern of a protein. In addition, the folding of some polypeptides is aided by **chaperones**—proteins that bind to polypeptides and facilitate their proper folding.

This folding process of polypeptides is governed by the primary structure and occurs in multiple stages (**Figure 13.8**). The first stage involves the formation of a regular, repeating shape known as a **secondary structure.** The two types of secondary structures are the **α helix** and the **β sheet** (Figure 13.8b). A single polypeptide may have some regions that fold into an α helix and other regions that fold into a β sheet. Because of the geometry of secondary structures, certain amino acids, such as glutamic acid, alanine, and methionine, are good candidates to form an α helix. Other amino acids, such as valine, isoleucine, and tyrosine,

(a) Nonpolar, aliphatic amino acids

(b) Nonpolar, aromatic amino acids

(c) Polar, neutral amino acids

(d) Polar, acidic amino acids

(e) Polar, basic amino acids

FIGURE 13.6 The 20 amino acids found within proteins.

are more likely to be found in a β-sheet conformation. Secondary structures within polypeptides are primarily stabilized by the formation of hydrogen bonds.

The short regions of secondary structure within a polypeptide are folded relative to each other to make the **tertiary structure** of a polypeptide. As shown in Figure 13.8c, α-helical regions and β-sheet regions are connected by irregularly shaped segments to determine the tertiary structure of the polypeptide. The folding of a polypeptide into its secondary and then tertiary conformation can usually occur spontaneously because it is a thermodynamically favorable process. The structure is determined by various interactions, including the tendency of hydrophobic amino acids to avoid water, ionic interactions among charged amino acids, hydrogen bonding among amino acids in the folded polypeptide, and weak bonding known as van der Waals interactions.

A protein is a functional unit that can be composed of one or more polypeptides. Some proteins are composed of a single polypeptide. Many proteins, however, are composed of two or more polypeptides that associate with each other to make a functional protein with a **quaternary structure** (Figure 13.8d). The individual polypeptides are called **subunits** of the protein, each of which has its own tertiary structure. The association of multiple subunits is the quaternary structure of a protein.

Cellular Proteins Are Primarily Responsible for the Characteristics of Living Cells and an Organism's Traits

Why is the genetic material largely devoted to storing the information to make proteins? To a great extent, the characteristics of a cell depend on the types of proteins that it makes. In turn, the traits of multicellular organisms are determined by the properties of their cells. Proteins perform a variety of functions critical to the life of cells and to the morphology and function of organisms (**Table 13.6**). Some proteins are important in determining the shape and structure of a given cell. For example, the protein tubulin assembles into large cytoskeletal structures known as microtubules, which provide eukaryotic cells with internal structure and organization. Some proteins are inserted into the cell membrane and aid in the transport of ions and small molecules across the membrane. An example is a sodium channel that transports sodium ions into nerve cells. Another interesting category of proteins are those that function as biological motors, such as myosin, which is involved in the contractile properties of muscle cells. Within multicellular organisms, certain proteins function in cell signaling and cell surface recognition. For example, proteins,

FIGURE 13.7 **An example of a protein's primary structure.** This is the amino acid sequence of the enzyme lysozyme, which contains 129 amino acids in its primary structure. As you may have noticed, the first amino acid is not methionine; instead, it is lysine. The first methionine residue in this polypeptide sequence is removed after or during translation. The removal of the first methionine occurs in many (but not all) proteins.

TABLE 13.6

Functions of Selected Cellular Proteins

Function	Examples
Cell shape and organization	Tubulin: Forms cytoskeletal structures known as microtubules
	Ankyrin: Anchors cytoskeletal proteins to the plasma membrane
Transport	Sodium channels: Transport sodium ions across the nerve cell membrane
	Lactose permease: Transports lactose across the bacterial cell membrane
	Hemoglobin: Transports oxygen in red blood cells
Movement	Myosin: Involved in muscle cell contraction
	Kinesin: Involved in the movement of chromosomes during cell division
Cell signaling	Insulin: A hormone that influences target cell metabolism and growth
	Epidermal growth factor: A growth factor that promotes cell division
	Insulin receptor: Recognizes insulin and initiates a cell response
Cell surface recognition	Integrins: Bind to large extracellular proteins
Enzymes	Hexokinase: Phosphorylates glucose during the first step in glycolysis
	β-galactosidase: Cleaves lactose into glucose and galactose
	Glycogen synthetase: Uses glucose molecules as building blocks to synthesize a large carbohydrate known as glycogen
	Acyl transferase: Links together fatty acids and glycerol phosphate during the synthesis of phospholipids
	RNA polymerase: Uses ribonucleotides as building blocks to synthesize RNA
	DNA polymerase: Uses deoxyribonucleotides as building blocks to synthesize DNA

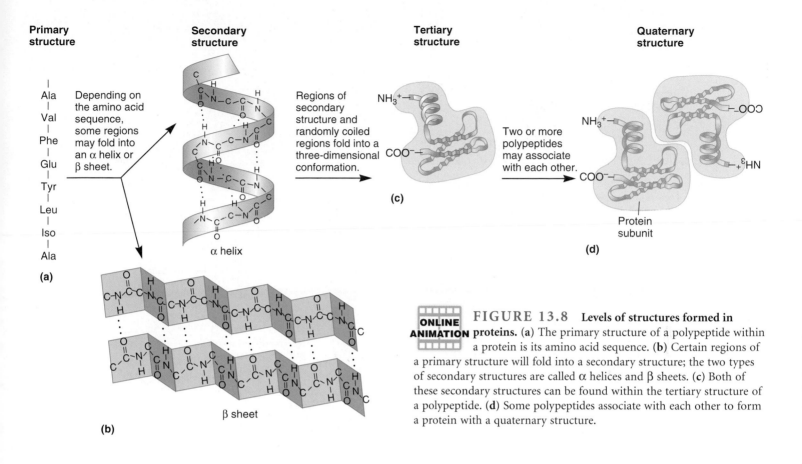

Primary structure

| Ala
| Val
| Phe
| Glu
| Tyr
| Leu
| Iso
| Ala
|

(a)

Depending on the amino acid sequence, some regions may fold into an α helix or β sheet.

Secondary structure

α helix

β sheet

(b)

Regions of secondary structure and randomly coiled regions fold into a three-dimensional conformation.

Tertiary structure

(c)

Two or more polypeptides may associate with each other.

Quaternary structure

Protein subunit

(d)

FIGURE 13.8 **Levels of structures formed in proteins.** (a) The primary structure of a polypeptide within a protein is its amino acid sequence. (b) Certain regions of a primary structure will fold into a secondary structure; the two types of secondary structures are called α helices and β sheets. (c) Both of these secondary structures can be found within the tertiary structure of a polypeptide. (d) Some polypeptides associate with each other to form a protein with a quaternary structure.

ONLINE ANIMATION

such as the hormone insulin, are secreted by endocrine cells and bind to the insulin receptor proteins found within the plasma membrane of target cells.

A key category of proteins are **enzymes,** which function to accelerate chemical reactions within the cell. Some enzymes assist in the breakdown of molecules or macromolecules into smaller units. These are known as catabolic enzymes and are important in generating cellular energy. In contrast, anabolic enzymes function in the synthesis of molecules and macromolecules. Several anabolic enzymes are listed in Table 13.6, including DNA polymerase, which is required for the synthesis of DNA from nucleotide building blocks. Throughout the cell, the synthesis of molecules and macromolecules relies on enzymes and accessory proteins. Ultimately, then, the construction of a cell greatly depends on its anabolic enzymes because these are required to synthesize all cellular macromolecules.

A fundamental question in genetics is, How can we relate protein function to an organism's traits? Let's extend our understanding of protein function by considering an example. In *Drosophila*, the color of the eye provides an interesting case in point (**Figure 13.9**). The eye of the fruit fly is composed of several types of cells. Some of them are pigment cells that are responsible for eye color. In the pigment cells of the red-eyed fly, the X chromosomes contain the red-eye allele (w^+), which codes for a protein that is located in the cell membrane. In the 1990s, the cloning and DNA sequencing of this gene indicated that the encoded protein transports colorless pigment precursor molecules into the cell. In the pigment cells of the red eye, the w^+ genes are transcribed and translated to produce this transport protein. This allows the cells to take up the pigment precursor molecules. Once inside, additional enzymes (that are encoded by different genes) convert the colorless precursors into red pigment. Therefore, the eyes of this fly appear red. In contrast, the white allele (w) of this gene is an example of a defective or loss-of-function allele. In the pigment cells of the white-eyed fly, this gene is not expressed properly, so little or no functional transport protein is made. Without this transport protein, the pigment precursor molecules are not transported into the pigment cells of the eye, so the red pigment cannot be made, and the eyes of the fly remain white.

13.2 STRUCTURE AND FUNCTION OF tRNA

Thus far, we have considered the general features of translation and surveyed the structure and functional significance of cellular proteins. The rest of this chapter is devoted to a molecular understanding of translation as it occurs in living cells. Biochemical studies of protein synthesis and tRNA molecules began in the 1950s. As work progressed toward an understanding of translation, research revealed that different kinds of RNA molecules are involved in the incorporation of amino acids into growing polypeptides. Francis

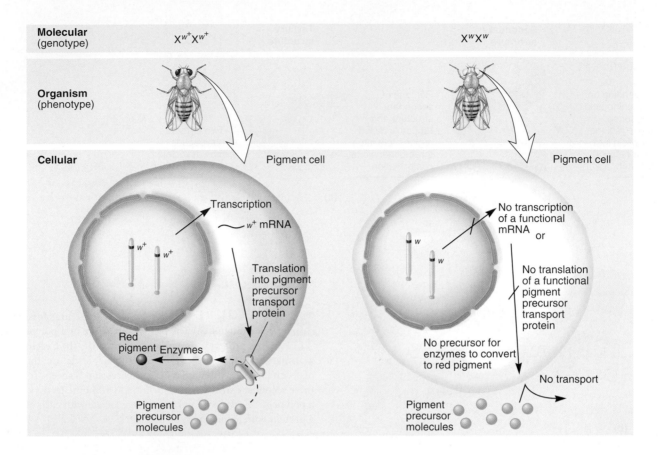

FIGURE 13.9   The relationship between genes and traits at the molecular, organism, and cellular levels.

Genes→Traits The wild-type (w^+) allele encodes a protein that transports a colorless pigment precursor into the pigment cells of the eye. This precursor is converted to a red pigment, so the eyes become red. The mutant w allele cannot produce a functional transporter, so little or none of the precursor is transported into the pigment cells. Therefore, the X^wX^w homozygous fly has colorless (white) eyes.

Crick proposed the **adaptor hypothesis.** According to this idea, the position of an amino acid within a polypeptide chain is determined by the binding between the mRNA and an adaptor molecule carrying a specific amino acid. Later, work by Paul Zamecnik and Mahlon Hoagland suggested that the adaptor molecule is tRNA. During translation, a tRNA has two functions. It recognizes a three-base codon sequence in mRNA, and it carries an amino acid specific for that codon. In this section, we will examine the general function of tRNA molecules. We will begin by considering an experiment that was critical in supporting the adaptor hypothesis and then explore some of the important structural features that underlie tRNA function.

The Function of a tRNA Depends on the Specificity Between the Amino Acid It Carries and Its Anticodon

The adaptor hypothesis proposes that tRNA molecules recognize the codons within mRNA and carry the correct amino acids to the site of polypeptide synthesis. During mRNA-tRNA recognition, the anticodon in a tRNA molecule binds to a codon in mRNA due to their complementary sequences (**Figure 13.10**). Importantly, the anticodon in the tRNA corresponds to the amino acid that it carries. For example, if the anticodon in the tRNA is 3'–AAG–5', it is complementary to a 5'–UUC–3' codon. According to the genetic code, described earlier in this chapter, the UUC codon specifies phenylalanine. Therefore, the tRNA with a 3'–AAG–5' anticodon must carry a phenylalanine. As another example, if the tRNA has a 3'–GGC–5' anticodon, it is complementary to a 5'–CCG–3' codon that specifies proline. This tRNA must carry proline.

Recall that the genetic code has 64 codons. Of these, 61 are sense codons that specify the 20 amino acids. Therefore, to synthesize proteins, a cell must produce many different tRNA molecules having specific anticodon sequences. To do so, the chromosomal DNA contains many distinct tRNA genes that encode tRNA molecules with different sequences. According to the adaptor hypothesis, the anticodon in a tRNA specifies the type of amino acid that it carries. Due to this specificity, tRNA molecules are named according to the type of amino acid that they carry. For example, a tRNA that attaches to phenylalanine is described as tRNA^phe, whereas a tRNA that carries proline is tRNA^pro.

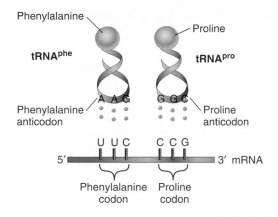

FIGURE 13.10 **Recognition between tRNAs and mRNA.** The anticodon in the tRNA binds to a complementary sequence in the mRNA. At its other end, the tRNA carries the amino acid that corresponds to the codon in the mRNA via the genetic code.

tRNA Functions as the Adaptor Molecule Involved in Codon Recognition

In 1962, François Chapeville and his colleagues conducted experiments aimed at testing the adaptor hypothesis. Their technical strategy was similar to that of the Nirenberg experiments that helped to decipher the genetic code (see Experiment 13A). In this approach, a cell-free translation system was made from cell extracts that contained the components necessary for translation. These components include ribosomes, tRNAs, and other translation factors. A cell-free translation system can synthesize polypeptides in vitro if mRNA and amino acids are added. Such a translation system can be used to investigate the role of specific factors by adding a particular mRNA template and varying individual components required for translation.

According to the adaptor hypothesis, the amino acid attached to a tRNA is not directly involved in codon recognition. Chapeville reasoned that if this were true, the alteration of an amino acid already attached to a tRNA should cause that altered amino acid to be incorporated into the polypeptide instead of the normal amino acid. For example, consider a tRNA^{cys} that carries the amino acid cysteine. If the attached cysteine were changed to an alanine, this tRNA^{cys} should insert an alanine into a polypeptide where it would normally put a cysteine. Fortunately, Chapeville could carry out this strategy because he had a reagent, known as Raney nickel, that can chemically convert cysteine to alanine.

An elegant aspect of the experimental design was the choice of the mRNA template. Chapeville and his colleagues synthesized an mRNA template that contained only U and G. Therefore, this template contained only the following codons (refer back to the genetic code in Table 13.2):

UUU = phenylalanine	GGU = glycine
UUG = leucine	GUU = valine
UGG = tryptophan	GUG = valine
GGG = glycine	UGU = cysteine

Among the eight possible codons, one cysteine codon occurs, but no alanine codons can be formed from a polyUG template.

As shown in the experiment of **Figure 13.11**, Chapeville began with a cell-free translation system that contained tRNA molecules. Amino acids, which would become attached to tRNAs, were added to this mixture. Of the 20 amino acids, only cysteine was radiolabeled. After allowing sufficient time for the amino acids to become attached to the correct tRNAs, the sample was divided into two tubes. One tube was treated with Raney nickel, while the control tube was not. As mentioned, Raney nickel converts cysteine into alanine by removing the –SH (sulfhydryl) group. However, it would not remove the radiolabel, which was a ^{14}C-label within the cysteine amino acid. Next, the polyUG mRNA was added as a template, and the samples were incubated to allow the translation of the mRNA into a polypeptide. In the control tube, we would expect the polypeptide to contain phenylalanine, leucine, tryptophan, glycine, valine, and cysteine, because these are the codons that contain only U and G. However, in the Raney nickel-treated sample, if the tRNA^{cys} was using its anticodon region to recognize the mRNA, we would expect to see alanine instead of cysteine.

Following translation, the polypeptides were isolated and hydrolyzed via a strong acid treatment, and then the individual amino acids were separated by column chromatography. The column separated cysteine from alanine; alanine eluted in a later fraction. The amount of radioactivity in each fraction was determined by liquid scintillation counting.

■ THE HYPOTHESIS

Codon recognition is dictated only by the tRNA anticodon; the chemical structure of the amino acid attached to the tRNA does not play a role.

TESTING THE HYPOTHESIS — FIGURE 13.11   Evidence that tRNA uses its anticodon sequence to recognize mRNA.

Starting material: A cell-free translation system that can synthesize polypeptides if mRNA and amino acids are added.

Experimental level   Conceptual level

1. Place cell-free translation system into a tube. Note: This drawing emphasizes only tRNA^{cys}, even though the cell-free translation system contains all types of tRNAs, and other components, such as ribosomes. In the translation system, a substantial proportion of the tRNAs do not have an attached amino acid. The translation system also contains enzymes that attach amino acids to tRNAs. (These enzymes will be described later in the chapter.)

2. Add amino acids, including radiolabeled cysteine. An enzyme within the translation system will specifically attach the radiolabeled cysteine to tRNA^{cys}. The other tRNAs will have unlabeled amino acids attached to them. Incubate and divide into two tubes.

3. In one tube, treat the tRNAs with Raney nickel. This removes the –SH group from cysteine, converting it to alanine. In the control tube, do not add Raney nickel.

4. Add polyUG mRNA made via polynucleotide phosphorylase as a template. A polyUG mRNA contains cysteine codons but no alanine codons.

5. Allow translation to proceed.

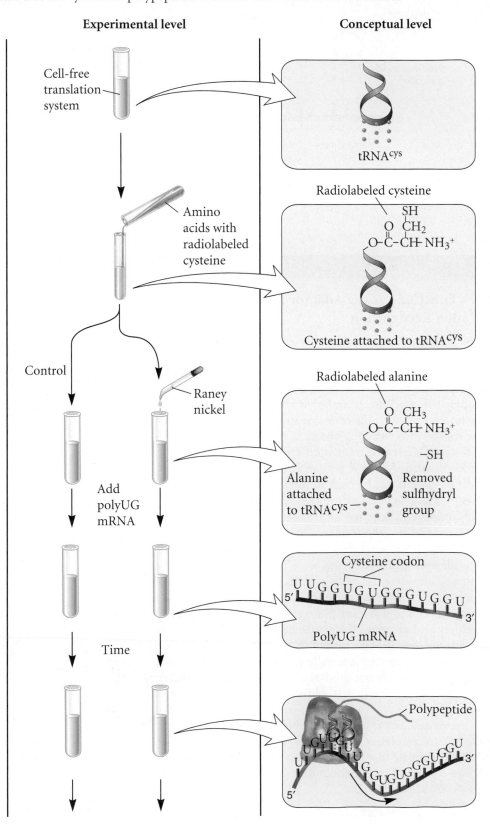

6. Precipitate the newly made polypeptides with trichloroacetic acid and then isolate the polypeptides on a filter.

7. Hydrolyze the polypeptides to their individual amino acids by treatment with a solution containing concentrated hydrochloric acid.

8. Run the sample over a column that separates cysteine and alanine. (See the Appendix for a description of column chromatography.) Separate into fractions. Note: Cysteine runs through the column more quickly and comes out in fraction 3. Alanine comes out later, in fraction 7.

9. Determine the amount of radioactivity in the fractions that contain alanine and cysteine.

Solution with precipitated polypeptides

Filter

Polypeptides

Filter in acid bath

Individual amino acids

Radiolabeled amino acids

1 2 3 4 5 6 7 8

Column fractions

■ THE DATA

	Amount of Radiolabeled Amino Acids Incorporated into Polypeptide (cpm)*		
Conditions	*Cysteine*	*Alanine*	*Total*
Control, untreated tRNA	2,835	83	2,918
Raney nickel-treated tRNA	990	2,020	3,010

*Cpm is the counts per minute of radioactivity in the sample.
Adapted from Chapeville et al. (1962) *PNAS 48*, 1086–1092.

■ INTERPRETING THE DATA

In the control sample, nearly all of the radioactivity was found in the fraction containing cysteine. This result was expected because the only radiolabeled amino acid attached to tRNAs was cysteine. The low radioactivity in the alanine fraction (83 counts per minute) probably represents contamination of this fraction by a small amount of cysteine. By comparison, when the tRNAs were treated with Raney nickel, a substantial amount of radiolabeled alanine became incorporated into polypeptides. This occurred even though the mRNA template did not contain any alanine codons. How do we explain these results? They are consistent with the explanation that a tRNA^{cys}, which carried alanine instead of cysteine, incorporated alanine into the synthesized polypeptide. These observations indicate that the codons in mRNA are identified directly by the tRNA rather than the attached amino acid.

As seen in the data of Figure 13.11, the Raney nickel-treated sample still had 990 cpm of cysteine incorporated into polypeptides. This is about one-third of the total amount of radioactivity (namely, 990/3,010). In other experiments conducted in this study, the researchers showed that the Raney nickel did not react with about one-third of the tRNA^{cys}. Therefore, this proportion of the Raney nickel-treated tRNA^{cys} would still carry cysteine. This observation was consistent with the data shown here. Overall, the results of this experiment supported the adaptor hypothesis, indicating that tRNAs act as adaptors to carry the correct amino acid to the ribosome based on their anticodon sequence.

A self-help quiz involving this experiment can be found at the Online Learning Center.

Common Structural Features Are Shared by All tRNAs

To understand how tRNAs act as carriers of the correct amino acids during translation, researchers have examined the structural characteristics of these molecules in great detail. Though a cell makes many different tRNAs, all tRNAs share common structural features. As originally proposed by Robert Holley in 1965, the secondary structure of tRNAs exhibits a cloverleaf pattern. A tRNA has three stem-loop structures, a few variable sites, and an acceptor stem with a 3' single-stranded region (**Figure 13.12**). The acceptor stem is where an amino acid becomes attached to a tRNA (see inset). A conventional numbering system for the nucleotides within a tRNA molecule begins at the 5' end and proceeds toward the 3' end. Among different types of tRNA molecules, the variable sites (shown in blue) can differ in the number of nucleotides they contain. The anticodon is located in the second loop region.

The actual three-dimensional, or tertiary, structure of tRNA molecules involves additional folding of the secondary structure. In the tertiary structure of tRNA, the stem-loop regions are folded into a much more compact molecule. The ability of RNA molecules to form stem-loop structures and the tertiary folding of tRNA molecules are described in Chapter 9 (see Figure 9.24). Interestingly, in addition to the normal A, U, G, and C nucleotides, tRNA molecules commonly contain modified nucleotides within their primary structures. For example, Figure 13.12 illustrates a tRNA that contains several modified bases. Among many different species, researchers have found that more than 60 different nucleotide modifications can occur in tRNA molecules. We will explore the significance of modified bases in codon recognition later in this chapter.

Aminoacyl-tRNA Synthetases Charge tRNAs by Attaching the Appropriate Amino Acid

To function correctly, each type of tRNA must have the appropriate amino acid attached to its 3' end. How does an amino acid get attached to a tRNA with the correct anticodon? Enzymes in the cell known as **aminoacyl-tRNA synthetases** catalyze the attachment of amino acids to tRNA molecules. Cells produce 20 different aminoacyl-tRNA synthetase enzymes, one for each of the 20 distinct amino acids. Each aminoacyl-tRNA synthetase is named for the specific amino acid it attaches to tRNA. For example, alanyl-tRNA synthetase recognizes a tRNA with an alanine anticodon—tRNA^{ala}—and attaches an alanine to it.

Aminoacyl-tRNA synthetases catalyze a chemical reaction involving three different molecules: an amino acid, a tRNA molecule, and ATP. In the first step of the reaction, a synthetase recognizes a specific amino acid and also ATP (**Figure 13.13**). The ATP is hydrolyzed, and AMP becomes attached to the amino acid; pyrophosphate is released. During the second step, the correct tRNA binds to the synthetase. The amino acid becomes covalently attached to the 3' end of the tRNA molecule at the acceptor stem, and AMP is released. Finally, the tRNA with its attached amino acid is released from the enzyme. At this stage, the tRNA is called a **charged tRNA** or an **aminoacyl tRNA.** In a charged tRNA molecule, the amino acid is attached to the 3' end of the tRNA by a covalent bond (see Figure 13.12 inset).

The ability of the aminoacyl-tRNA synthetases to recognize tRNAs has sometimes been called the "second genetic code." This recognition process is necessary to maintain the fidelity of genetic information. The frequency of error for aminoacyl-tRNA synthetases is less than 10^{-5}. In other words, the wrong amino acid will be attached to a tRNA less than once in 100,000 times!

FIGURE 13.12   Secondary structure of tRNA. The conventional numbering of nucleotides begins at the 5′ end and proceeds toward the 3′ end. In all tRNAs, the nucleotides at the 3′ end contain the sequence CCA. Certain locations can have additional nucleotides not found in all tRNA molecules; these variable sites are shown in blue. The figure also shows the locations of a few modified bases specifically found in a yeast tRNA that carries alanine. The modified bases are as follows: I = inosine, mI = methylinosine, T = ribothymidine, UH_2 = dihydrouridine, m_2G = dimethylguanosine, and P = pseudouridine. The inset shows an amino acid covalently attached to the 3′ end of a tRNA.

As you might expect, the anticodon region of the tRNA is usually important for precise recognition by the correct aminoacyl-tRNA synthetase. In studies of *Escherichia coli* synthetases, 17 of the 20 types of aminoacyl-tRNA synthetases recognize the anticodon region of the tRNA. However, other regions of the tRNA are also important recognition sites. These include the acceptor stem and bases in the stem-loop regions.

As mentioned previously, tRNA molecules frequently contain bases within their structure that have been chemically modified. These modified bases can have important effects on tRNA function. For example, modified bases within tRNA molecules affect the rate of translation and the recognition of tRNAs by aminoacyl-tRNA synthetases. Positions 34 and 37 contain the largest variety of modified nucleotides; position 34 is the first base in the anticodon that matches the third base in the codon

FIGURE 13.13   Catalytic function of aminoacyl-tRNA synthetase. Aminoacyl-tRNA synthetase has binding sites for a specific amino acid, ATP, and a particular tRNA. In the first step, the enzyme catalyzes the covalent attachment of AMP to an amino acid, yielding an activated amino acid. In the second step, the activated amino acid is attached to the appropriate tRNA.

ONLINE ANIMATION

of mRNA. As discussed next, a modified base at position 34 can have important effects on codon-anticodon recognition.

Mismatches That Follow the Wobble Rule Can Occur at the Third Position in Codon-Anticodon Pairing

After considering the structure and function of tRNA molecules, let's reexamine some subtle features of the genetic code. As discussed earlier, the genetic code is degenerate, which means that more than one codon can specify the same amino acid. Degeneracy usually occurs at the third position in the codon. For example, valine is specified by GUU, GUC, GUA, and GUG. In all four cases, the first two bases are G and U. The third base, however, can be U, C, A, or G. To explain this pattern of degeneracy, Francis Crick proposed in 1966 that it is due to "wobble" at the third position in the codon-anticodon recognition process. According to the **wobble rules,** the first two positions pair strictly according to the AU/GC rule. However, the third position can tolerate certain types of mismatches (**Figure 13.14**). This proposal suggested that the base at the third position in the codon does not have to hydrogen bond as precisely with the corresponding base in the anticodon.

Because of the wobble rules, some flexibility is observed in the recognition between a codon and anticodon during the process of translation. When two or more tRNAs that differ at the wobble base are able to recognize the same codon, these are termed **isoacceptor tRNAs.** As an example, tRNAs with an anticodon of 3'–CCA–5' or 3'–CCG–5' would be able to recognize a codon with the sequence of 5'–GGU–3'. In addition, the wobble rules enable a single type of tRNA to recognize more than one codon. For example, a tRNA with an anticodon sequence of 3'–AAG–5' can recognize a 5'–UUC–3' and a 5'–UUU–3' codon. The 5'–UUC–3' codon is a perfect match with this tRNA. The 5'–UUU–3' codon is mismatched according to the standard RNA-RNA hybridization rules (namely, G in the anticodon is mismatched to U in the codon), but the two can fit according to the wobble rules described in Figure 13.14. Likewise, the modification of the wobble base to an inosine allows a tRNA to recognize three different codons. At the cellular level, the ability of a single tRNA to recognize more than one codon makes it unnecessary for a cell to make 61 different tRNA molecules with anticodons that are complementary to the 61 possible sense codons. *E. coli* cells, for example, make a population of tRNA molecules that have just 40 different anticodon sequences.

13.3 RIBOSOME STRUCTURE AND ASSEMBLY

In Section 13.2, we examined how the structure and function of tRNA molecules are important in translation. According to the adaptor hypothesis, tRNAs bind to mRNA due to complementarity between the anticodons and codons. Concurrently, the tRNA molecules have the correct amino acid attached to their 3' ends.

To synthesize a polypeptide, additional events must occur. In particular, the bond between the 3' end of the tRNA and the amino acid must be broken, and a peptide bond must be formed between the adjacent amino acids. To facilitate these events, translation occurs on the surface of a macromolecular complex known as the **ribosome.** The ribosome can be thought of as the macromolecular arena where translation takes place.

In this section, we will begin by outlining the biochemical compositions of ribosomes in bacterial and eukaryotic cells. We will then examine the key functional sites on ribosomes for the translation process.

Bacterial and Eukaryotic Ribosomes Are Assembled from rRNA and Proteins

Bacterial cells have one type of ribosome that is found within the cytoplasm. Eukaryotic cells contain biochemically distinct ribosomes in different cellular locations. The most abundant type of ribosome functions in the cytosol, which is the region of

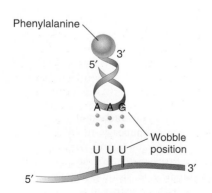

(a) Location of wobble position

Nucleotide of tRNA anticodon	Third nucleotide of mRNA codon
G	C, U
C	G
A	U, C, G, (A)
U	A, U, G, (C)
I	U, C, A
xm^5s^2U	
xm^5Um	
Um	A, (G)
xm^5U	
xo^5U	U, A, G
k^2C	A

(b) Revised wobble rules

FIGURE 13.14   Wobble position and base pairing rules. (a) The wobble position occurs between the first base (meaning the first base in the 5' to 3' direction) in the anticodon and the third base in the mRNA codon. **(b)** The revised wobble rules are slightly different from those originally proposed by Crick. The standard bases found in RNA are G, C, A, and U. In addition, the structures of bases in tRNAs may be modified. Some modified bases that may occur in the wobble position in tRNA are I = inosine; xm^5s^2U = 5-methyl-2-thiouridine; xm5Um = 5-methyl-2'-O-methyluridine; Um = 2'-O-methyluridine; xm^5U = 5-methyluridine; xo^5U = 5-hydroxyuridine; k^2C = lysidine (a cytosine derivative). The mRNA bases in parentheses are recognized very poorly by the tRNA.

the eukaryotic cell that is inside the plasma membrane but outside the organelles. Besides the cytosolic ribosomes, all eukaryotic cells have ribosomes within the mitochondria. In addition, plant cells and algae have ribosomes in their chloroplasts. The compositions of mitochondrial and chloroplast ribosomes are quite different from that of the cytosolic ribosomes. Unless otherwise noted, the term eukaryotic ribosome refers to ribosomes in the cytosol, not to those found within organelles. Likewise, the description of eukaryotic translation refers to translation via cytosolic ribosomes.

Each ribosome is composed of structures called the large and small subunits. This term is perhaps misleading because each ribosomal subunit itself is formed from the assembly of many different proteins and RNA molecules called ribosomal

RNA or rRNA. In bacterial ribosomes, the 30S subunit is formed from the assembly of 21 different ribosomal proteins and a 16S rRNA molecule; the 50S subunit contains 34 different proteins and 23S and 5S rRNA molecules (**Figure 13.15a**). The designations 30S and 50S refer to the rate that these subunits sediment when subjected to a centrifugal force. This rate is described as a sedimentation coefficient in Svedberg units (S), in honor of Theodor Svedberg, who invented the ultracentrifuge. Together, the 30S and 50S subunits form a 70S ribosome. (Note: Svedberg units do not add up linearly.) In bacteria, the ribosomal proteins and rRNA molecules are synthesized in the cytoplasm, and the ribosomal subunits are assembled there.

The synthesis of eukaryotic rRNA occurs within the nucleus, and the ribosomal proteins are made in the cytosol,

(a) Bacterial cell

(b) Eukaryotic cell

FIGURE 13.15 **Composition of bacterial and eukaryotic ribosomes. (a)** Bacterial ribosomes contain two subunits: 30S and 50S, which assemble to form a 70S ribosome. The 30S subunit contains 21 different proteins and 16S rRNA; the 50S subunit contains 34 proteins and 23S and 5S rRNAs. **(b)** Eukaryotic ribosomes found in the cytosol contain two subunits: 40S and 60S. The 40S subunit contains 33 proteins and 18S rRNA; the 60S subunit contains 49 proteins and 5S, 28S, and 5.8S rRNAs. The 40S and 60S subunits are formed in the nucleolus and then are exported into the cytosol, where they assemble to form an 80S ribosome.

where translation takes place. The 40S subunit is composed of 33 proteins and an 18S rRNA; the 60S subunit is made of 49 proteins and 5S, 5.8S, and 28S rRNAs (**Figure 13.15b**). The assembly of the rRNAs and ribosomal proteins to make the 40S and 60S subunits occurs within the **nucleolus,** a region of the nucleus specialized for this purpose. The 40S and 60S subunits are then exported into the cytosol, where they associate to form an 80S ribosome during translation.

Components of Ribosomal Subunits Form Functional Sites for Translation

To understand the structure and function of the ribosome at the molecular level, researchers must determine the locations and functional roles of the individual ribosomal proteins and rRNAs.

In recent years, many advances have been made toward a molecular understanding of ribosomes. Microscopic and biophysical methods have been used to study ribosome structure. An electron micrograph of bacterial ribosomes is shown in **Figure 13.16a.** More recently, a few research groups have succeeded in crystallizing ribosomal subunits, and even intact ribosomes. This is an amazing technical feat, because it is difficult to find the right conditions under which large macromolecules will form highly ordered crystals. **Figure 13.16b** shows the crystal structure of bacterial ribosomal subunits. The overall shape of each subunit is largely determined by the structure of the rRNAs, which constitute most of the mass of the ribosome. The interface between the 30S and 50S subunits is primarily composed of rRNA. Ribosomal proteins cluster on the outer surface of the ribosome and on the periphery of the interface.

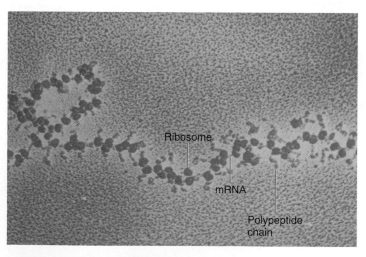

(a) Ribosomes as seen with electron microscope

(b) Bacterial ribosome model based on X–ray diffraction studies

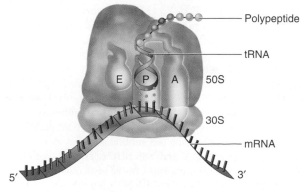

(c) Model for ribosome structure

FIGURE 13.16   Ribosomal structure. (a) Electron micrograph of ribosomes attached to an mRNA molecule. **(b)** Crystal structure of the 50S and 30S subunits in bacteria. This model shows the interface between the two subunits. The rRNA is shown in gray (50S subunit) and turquoise (30S subunit), and proteins are shown in violet (50S subunit) and dark blue (30S subunit). **(c)** A model depicting the sites where tRNA and mRNA bind to an intact ribosome. The mRNA lies on the surface of the 30S subunit. The E, P, and A sites are formed at the interface between the large and small subunits. The growing polypeptide chain exits through a hole in the 50S subunit.

During bacterial translation, the mRNA lies on the surface of the 30S subunit within a space between the 30S and 50S subunits. As the polypeptide is being synthesized, it exits through a channel within the 50S subunit (**Figure 13.16c**). Ribosomes contain discrete sites where tRNAs bind and the polypeptide is synthesized. In 1964, James Watson was the first to propose a two-site model for tRNA binding to the ribosome. These sites are known as the **peptidyl site (P site)** and **aminoacyl site (A site)**. In 1981, Knud Nierhaus, Hans Sternbach, and Hans-Jorg Rheinberger proposed a three-site model. This model incorporated the observation that uncharged tRNA molecules can bind to a site on the ribosome that is distinct from the P and A sites. This third site is now known as the **exit site (E site)**. The locations of the E, P, and A sites are shown in Figure 13.16c. Next, we will examine the roles of these sites during the three stages of translation.

13.4 STAGES OF TRANSLATION

Like transcription, the process of translation can be viewed as occurring in three stages: initiation, elongation, and termination. **Figure 13.17** presents an overview of these stages. During **initiation,** the ribosomal subunits, mRNA, and the first tRNA assemble to form a complex. After the initiation complex is formed, the ribosome slides along the mRNA in the 5' to 3' direction, moving over the codons. This is the **elongation** stage of translation. As the ribosome moves, tRNA molecules sequentially bind to the mRNA at the A site in the ribosome, bringing with them the appropriate amino acids. Therefore, amino acids are linked in the order dictated by the codon sequence in the mRNA. Finally, a stop codon is reached, signaling the **termination** of translation. At this point, disassembly occurs, and the newly made polypeptide is released.

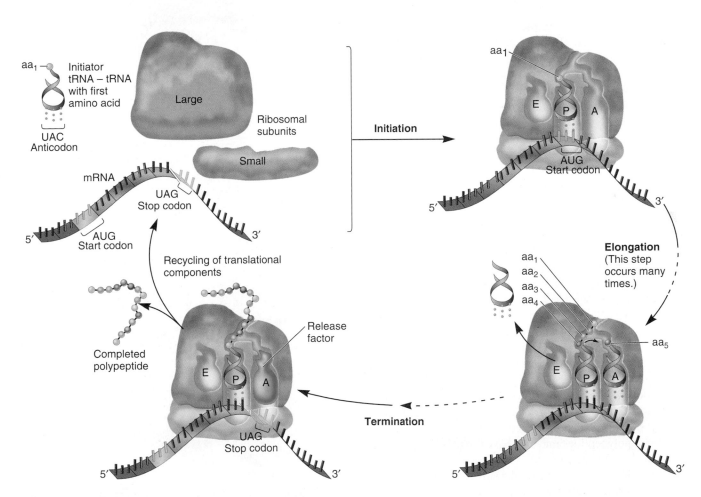

FIGURE 13.17   Overview of the stages of translation. The initiation stage involves the assembly of the ribosomal subunits, mRNA, and the initiator tRNA carrying the first amino acid. During elongation, the ribosome slides along the mRNA and synthesizes a polypeptide chain. Translation ends when a stop codon is reached and the polypeptide is released from the ribosome. (Note: In this and succeeding figures in this chapter, the ribosomes are drawn schematically to emphasize different aspects of the translation process. The structures of ribosomes are described in Figures 13.15 and 13.16.)

Genes→Traits The ability of genes to produce an organism's traits relies on the molecular process of gene expression. During translation, the codon sequence within mRNA (which is derived from a gene sequence during transcription) is translated into a polypeptide sequence. After polypeptides are made within a living cell, they function as proteins to govern an organism's traits. For example, once the β-globin polypeptide is made, it functions within the hemoglobin protein and provides red blood cells with the ability to carry oxygen, a vital trait for survival. Translation allows functional proteins to be made within living cells.

In this section, we will examine the components required for the translation process and consider their functional roles during the three stages of translation.

The Initiation Stage Involves the Binding of mRNA and the Initiator tRNA to the Ribosomal Subunits

During initiation, an mRNA and the first tRNA bind to the ribosomal subunits. A specific tRNA functions as the **initiator tRNA,** which recognizes the start codon in the mRNA. In bacteria, the initiator tRNA, which is also designated tRNA^{fmet}, carries a methionine that has been covalently modified to N-formylmethionine. In this modification, a formyl group (—CHO) is attached to the nitrogen atom in methionine after the methionine has been attached to the tRNA.

Figure 13.18 describes the initiation stage of translation in bacteria during which the mRNA, tRNA^{fmet}, and ribosomal subunits associate with each other to form an initiation complex. The formation of this complex requires the participation of three initiation factors: IF1, IF2, and IF3. First, the mRNA binds to the 30S ribosomal subunit. This binding is facilitated by IF3, which requires GTP for its function. In addition, complementarity between a short region of the mRNA and the 16S rRNA within the 30S subunit plays a key role. A nine-nucleotide sequence within bacterial mRNAs, known as the ribosomal-binding site, or **Shine-Dalgarno sequence,** is involved in the binding of the mRNA to the 30S subunit. The location of this sequence is shown in Figure 13.18 and in more detail in **Figure 13.19.** How does the Shine-Dalgarno sequence facilitate the binding of mRNA to the ribosome? The Shine-Dalgarno sequence is complementary to a short sequence within the 16S rRNA, which promotes the hydrogen bonding of the mRNA to the 30S subunit.

Next, tRNA^{fmet} binds to the mRNA that is already attached to the 30S subunit (Figure 13.18). This step requires the function of IF2, which also uses GTP. The tRNA^{fmet} binds to the start codon, which is typically a few nucleotides downstream from the Shine-Dalgarno sequence. The start codon is usually AUG, but in some cases it can be GUG or UUG. Even when the start codon is GUG (which normally encodes valine) or UUG (which normally encodes leucine), the first amino acid in the polypeptide is still a formylmethionine because only a tRNA^{fmet} can initiate translation. During or after translation of the entire polypeptide, the formyl group or the entire formylmethionine may be removed. Therefore, some polypeptides may not have formylmethionine or methionine as their first amino acid.

The initiation stage of bacterial translation is completed when the 50S ribosomal subunit associates with the 30S subunit. For this to occur, the initiation factors must be released from the 30S subunit. Much later, after translation is completed, IF1 is necessary to dissociate the 50S and 30S ribosomal subunits so that the 30S subunit can reinitiate with another mRNA molecule. For this reason, it is viewed as an initiation factor. As noted in Figure 13.18, the tRNA^{fmet} binds to the P site on the ribosome. During the elongation stage, which is discussed later, all of the other tRNAs initially bind to the A site.

FIGURE 13.18 **The initiation stage of translation in bacteria.** IF3 promotes the binding of mRNA to the 30S subunit. In addition, the Shine-Dalgarno sequence hydrogen bonds with a portion of the 16S rRNA. IF2 then binds to the initiator tRNA, which binds to the start codon in the P site. IF2 and IF3 are released, and the 50S subunit associates with the 30S subunit to form a 70S initiation complex. This marks the end of the initiation stage.

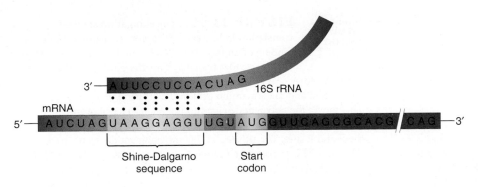

FIGURE 13.19 **The locations of the Shine-Dalgarno sequence and the start codon in bacterial mRNA.** The Shine-Dalgarno sequence is complementary to a sequence in the 16S rRNA. It hydrogen bonds with the 16S rRNA to promote initiation. The start codon is typically a few nucleotides downstream from the Shine-Dalgarno sequence.

In eukaryotes, the assembly of the initiation complex bears similarities to that in bacteria. However, as described in **Table 13.7**, additional factors are required for the initiation process. Note that the initiation factors are designated eIF (eukaryotic Initiation Factor) to distinguish them from bacterial initiation factors. The initiator tRNA in eukaryotes carries methionine rather than formylmethionine, as in bacteria. A eukaryotic initiation factor, eIF2, binds directly to tRNA^{met} to recruit it to the 40S subunit. Eukaryotic mRNAs do not have a Shine-Dalgarno sequence. How then are eukaryotic mRNAs recognized by the ribosome? Several initiation factors (CBPI, eIF4A, eIF4B, eIF4F, and others) bind to the mRNA. Cap-binding protein I (CBPI) recognizes the 7-methylguanosine cap structure and aids in the recruitment of the other initiation factors. Collectively, these initiation factors unwind any secondary structure in the mRNA and promote its binding to the ribosome.

The identification of the correct AUG start codon in eukaryotes differs greatly from that in bacteria. After the initial binding of mRNA to the ribosome, the next step is to locate an AUG start codon that is somewhere downstream from the 5' cap structure. In 1986, Marilyn Kozak proposed that the ribosome begins at the 5' end and then scans along the mRNA in the 3' direction in search of an AUG start codon. In many, but not all, cases, the ribosome uses the first AUG codon that it encounters as a start codon. When a start codon is identified, the 60S subunit assembles with the aid of eIF5.

By analyzing the sequences of many eukaryotic mRNAs, researchers have found that not all AUG codons near the 5' end of mRNA can function as start codons. In some cases, the scanning ribosome passes over the first AUG codon and chooses an AUG farther down the mRNA. The sequence of bases around the AUG codon plays an important role in determining whether or not it will be selected as the start codon by a scanning ribosome. The consensus sequence for optimal start codon recognition is shown here.

				Start Codon					
G	C	C	(A/G)	C	C	A	U	G	G
–6	–5	–4	–3	–2	–1	+1	+2	+3	+4

Aside from an AUG codon itself, a guanine at the +4 position and a purine, preferably an adenine, at the –3 position are the most important sites for start codon selection. These rules for optimal translation initiation are called **Kozak's rules.**

TABLE 13.7

A Comparison of Translational Protein Factors in Bacteria and Eukaryotes

Bacterial Factors	Eukaryotic Factors	Function
Initiation Factors		
IF1	eIF2	Involved in forming the initiation complex
IF2	eIF3	Involved in forming the initiation complex
IF3	eIF4C	Involved in forming the initiation complex
	CBPI	Binds to the 7-methylguanosine cap
	eIF4A, eIF4B, eIF4F	Involved in the search for a start codon
	eIF5	Involved in forming the initiation complex; helps dissociate eIF2, eIF3, and eIF4C
	eIF6	Helps dissociate 60S subunit from inactive ribosomes
Elongation Factors		
EF-Tu	eEF1α	Binding of tRNAs to the A site
EF-Ts	eEF1βγ	Recycling factor for EF-Tu and eEF1α, respectively.
EF-G	eEF2	Required for translocation
Release Factors		
RF1	eRF	Recognizes stop codon and promotes termination
RF2		
RF3		

Polypeptide Synthesis Occurs During the Elongation Stage

During the elongation stage of translation, amino acids are added, one at a time, to the polypeptide chain (**Figure 13.20**). Even though this process is rather complex, it occurs at a remarkable rate. Under normal cellular conditions, a polypeptide chain

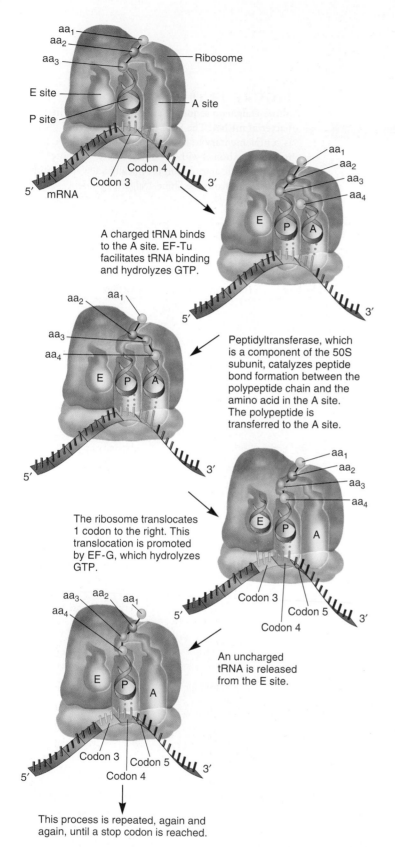

A charged tRNA binds to the A site. EF-Tu facilitates tRNA binding and hydrolyzes GTP.

Peptidyltransferase, which is a component of the 50S subunit, catalyzes peptide bond formation between the polypeptide chain and the amino acid in the A site. The polypeptide is transferred to the A site.

The ribosome translocates 1 codon to the right. This translocation is promoted by EF-G, which hydrolyzes GTP.

An uncharged tRNA is released from the E site.

This process is repeated, again and again, until a stop codon is reached.

FIGURE 13.20   The elongation stage of translation in bacteria. This process begins with the binding of an incoming tRNA. The hydrolysis of GTP by EF-Tu provides the energy for the binding of the tRNA to the A site. A peptide bond is then formed between the incoming amino acid and the last amino acid in the growing polypeptide chain. This moves the polypeptide chain to the A site. The ribosome then translocates in the 3' direction so that the two tRNAs are moved to the E and P sites. The tRNA carrying the polypeptide is now back in the P site. This translocation requires the hydrolysis of GTP via EF-G. The uncharged tRNA in the E site is released from the ribosome. Now the process is ready to begin again. Each cycle of elongation causes the polypeptide chain to grow by one amino acid.

can elongate at a rate of 15 to 18 amino acids per second in bacteria and 6 amino acids per second in eukaryotes!

To begin elongation, a charged tRNA brings a new amino acid to the ribosome so that it can be attached to the end of the growing polypeptide chain. At the top of Figure 13.20, a short polypeptide is attached to the tRNA located at the P site of the ribosome. A charged tRNA carrying a single amino acid binds to the A site. This binding occurs because the anticodon in the tRNA is complementary to the codon in the mRNA. The hydrolysis of GTP by the elongation factor, EF-Tu, provides energy for the binding of a tRNA to the A site. In addition, the 16S rRNA, which is a component of the small 30S ribosomal subunit, plays a key role that ensures the proper recognition between the mRNA and correct tRNA. The 16S rRNA can detect when an incorrect tRNA is bound at the A site and will prevent elongation until the mispaired tRNA is released from the A site. This phenomenon, termed the **decoding function** of the ribosome, is important in maintaining high fidelity of mRNA translation. An incorrect amino acid is incorporated into a growing polypeptide at a rate of approximately one mistake per 10,000 amino acids or 10^{-4}.

The next step of elongation is the **peptidyl transfer** reaction—the polypeptide is removed from the tRNA in the P site and transferred to the amino acid at the A site. This transfer is accompanied by the formation of a peptide bond between the amino acid at the A site and the polypeptide chain, lengthening the chain by one amino acid. The peptidyl transfer reaction is catalyzed by a component of the 50S subunit known as **peptidyltransferase,** which is composed of several proteins and rRNA. Interestingly, based on the crystal structure of the 50S subunit, Thomas Steitz, Peter Moore, and their colleagues concluded that the 23S rRNA—not the ribosomal protein—catalyzes bond formation between adjacent amino acids. In other words, the ribosome is a ribozyme!

After the peptidyl transfer reaction is complete, the ribosome moves, or translocates, to the next codon in the mRNA. This moves the tRNAs at the P and A sites to the E and P sites, respectively. Finally, the uncharged tRNA exits the E site. You should notice that the next codon in the mRNA is now exposed in the unoccupied A site. At this point, a charged tRNA can enter the empty A site, and the same series of steps can add the next amino acid to the polypeptide chain. As you may have realized, the A, P, and E sites are named for the role of the tRNA that is usually found there. The A site binds an aminoacyl-tRNA (also called a charged tRNA), the P site usually contains the peptidyl-tRNA (a tRNA with an attached peptide), and the E site is where the uncharged tRNA exits.

Termination Occurs When a Stop Codon Is Reached in the mRNA

The final stage of translation, known as termination, occurs when a stop codon is reached in the mRNA. In most species, the three stop codons are UAA, UAG, and UGA. These codons are sometimes referred to as ochre (UAA), amber (UAG), and opal (UGA). (Note: The term amber, or brownstone, is the English translation of the name Bernstein, a graduate student who was involved in the discovery of the UAG codon. In keeping with this tradition, the lighthearted names ocher and opal were given to the other two stop codons.) The stop codons are not recognized by a tRNA with a complementary sequence. Instead, they are recognized by proteins known as **release factors** (see Table 13.7). Interestingly, the three-dimensional structure of release factor proteins mimics the structure of tRNAs. In bacteria, RF1 recognizes UAA and UAG, and RF2 recognizes UGA and UAA. A third release factor, RF3, is also required. In eukaryotes, a single release factor, eRF, recognizes all three stop codons.

Figure 13.21 illustrates the termination stage of translation in bacteria. At the top of this figure, the completed polypeptide chain is attached to a tRNA in the P site. A stop codon is located at the A site. In the first step, RF1 or RF2 binds to the stop codon at the A site and RF3 (not shown) binds at a different location on the ribosome. After RF1 (or RF2) and RF3 have bound, the bond between the polypeptide and the tRNA is hydrolyzed. The polypeptide and tRNA are then released from the ribosome. The final step in translational termination is the disassembly of ribosomal subunits, mRNA, and the release factors.

Bacterial Translation Can Begin Before Transcription Is Completed

While most of our knowledge concerning transcription and translation has come from genetic and biochemical studies, electron microscopy has also been an important tool in elucidating the mechanisms of transcription and translation. As described earlier in this chapter, electron microscopy (EM) has been a critical technique in facilitating our understanding of ribosome structure. In addition, EM has been employed to visualize genetic processes such as translation.

The first success in the EM observation of gene expression was achieved by Oscar Miller, Jr., and his colleagues in 1967. **Figure 13.22** shows an electron micrograph of a bacterial gene in the act of gene expression. Prior to this experiment, biochemical and genetic studies had suggested that the translation of a bacterial structural gene begins before the mRNA transcript is completed. In other words, as soon as an mRNA strand is long enough, a ribosome will attach to the 5' end and begin translation, even before RNA polymerase has reached the transcriptional termination site within the gene. This phenomenon is termed the coupling between transcription and translation in bacterial cells. Note that

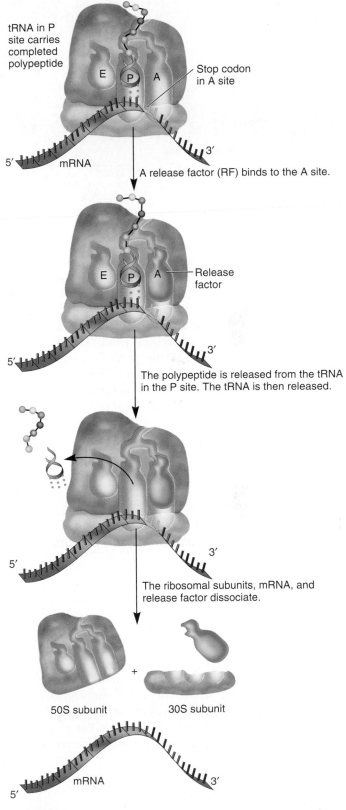

FIGURE 13.21 The termination stage of translation in bacteria. When a stop codon is reached, RF1 or RF2 binds to the A site. (RF3 binds elsewhere and uses GTP to facilitate the termination process.) The polypeptide is cleaved from the tRNA in the P site and released. The tRNA is released, and the rest of the components disassemble.

FIGURE 13.22   Coupling between transcription and translation in bacteria. An electron micrograph showing the simultaneous transcription and translation processes. The DNA is transcribed by many RNA polymerases that move along the DNA from left to right. Note that the RNA transcripts are getting longer as you go from left to right. Ribosomes attach to the mRNA, even before transcription is completed. The complex of many ribosomes bound to the same mRNA is called a polyribosome or a polysome. Several polyribosomes are seen here.

coupling of these processes does not usually occur in eukaryotes, because transcription takes place in the nucleus while translation occurs in the cytosol.

As shown in Figure 13.22, several RNA polymerase enzymes have recognized a gene and begun to transcribe it. Because the transcripts on the right side are longer than those on the left, Miller concluded that transcription was proceeding from left to right in the micrograph. This EM image also shows the process of translation. Relatively small mRNA transcripts, near the left side of the figure, have a few ribosomes attached to them. As the transcripts become longer, additional ribosomes are attached to them. The term **polyribosome, or polysome,** is used to describe an mRNA transcript that has many bound ribosomes in the act of translation. In this electron micrograph, the nascent polypeptide chains were too small for researchers to observe. In later studies, as EM techniques became more refined, the polypeptide chains emerging from the ribosome were also visible (see Figure 13.16a).

The Amino Acid Sequences of Some Proteins Contain Sorting Signals

The amino acid sequence of a protein may contain a sorting signal that will direct the protein to its correct location. In bacteria, for example, certain proteins may contain sorting signals that direct them across the plasma membrane. Also, some bacterial proteins have sorting signals that direct their attachment to the cell wall. In eukaryotes, many more sorting signals are necessary because eukaryotic cells are compartmentalized into many different membrane-bounded organelles, such as mitochondria, lysosomes, the nucleus, and so on. In general, any particular protein is meant to function in only a single location. For example, the ATP-synthase functions in the mitochondrion to make ATP and

is not found in other locations in the cell. Cytoskeletal proteins such as tubulin and actin are found in the cytosol but not within the lumen of cellular organelles. To fulfill its function, each type of protein must be sorted into the correct cellular compartment. In eukaryotes, this sorting can occur either after translation is completed—**posttranslational sorting**—or during translation—**cotranslational sorting.**

Most eukaryotic proteins begin their synthesis on ribosomes in the cytosol (**Figure 13.23**). Translation is completed in the cytosol for those proteins destined for the cytosol, chloroplast, mitochondrion, nucleus, or peroxisome. The uptake of proteins into the chloroplast, mitochondrion, nucleus, and peroxisome then occurs posttranslationally. By comparison, the synthesis of other eukaryotic proteins begins in the cytosol and then is temporarily halted by a protein/RNA complex called the signal recognition particle (SRP). The SRP recognizes an amino acid sequence near the amino-terminal end of a polypeptide as it is being made. Translation resumes when the ribosome has become bound to the membrane of the endoplasmic reticulum (ER) and the polypeptide is synthesized into the ER lumen or ER membrane. This event is termed **cotranslational import** into the ER. Proteins destined for the ER, Golgi complex, lysosome, secretory vesicles, and the plasma membrane are first directed to the ER via this type of cotranslational sorting mechanism.

How is sorting achieved? A vital aspect of gene sequences is that the encoded proteins may contain short amino acid sequences that play a role in protein sorting. These short amino acid sequences are called **sorting signals.** Table 13.8 provides a general description of the sorting signals contained within eukaryotic proteins. Each signal is recognized by specific cellular components that facilitate the sorting of the protein to its correct compartment, either posttranslationally or cotranslationally.

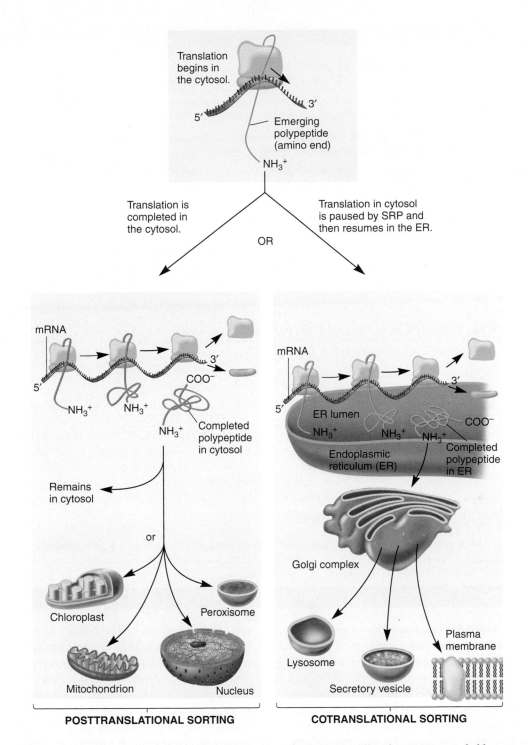

FIGURE 13.23 **Posttranslational and cotranslational protein sorting in eukaryotic cells.** The mRNA encoded by genes in the cell nucleus is exported from the nucleus and begins its translation in the cytosol. Proteins destined for the cytosol, chloroplasts, mitochondria, nucleus, and peroxisomes are completely synthesized in the cytosol and then are sorted posttranslationally. Other polypeptides contain an amino acid sequence at their amino terminal end that is recognized by the signal recognition particle (SRP). SRP directs these proteins to the ER, where they are sorted to the endoplasmic reticulum (ER), Golgi complex, lysosome, secretory vesicles, or plasma membrane.

TABLE **13.8**

Sorting Signals in Eukaryotic Proteins

Type of Signal	Description
Mitochondrial-sorting signal	Usually a short sequence at the amino terminal end of a protein that contains several positively charged amino acids (i.e., lysine and arginine). This signal folds into an α helix in which the positive charges are on one face of the helix.
Nuclear-localization signal	Can be located almost anywhere in the polypeptide sequence. The signal is four to eight amino acids in length and contains several positively charged amino acids and usually one or more prolines.
Peroxisomal-sorting signal	Either a specific sequence of three amino acids (serine, lysine, leucine) that is located near the carboxyl terminal end of the protein or a 26-amino-acid sequence at the amino terminal end.
SRP signal	A sequence of ~20 amino acids near the amino terminal end that is composed of mostly nonpolar amino acids.
ER retention signal	A sequence of four amino acids (lysine, aspartic acid, glutamic acid, leucine) that is located at the carboxyl terminal end of the protein.
Golgi retention signal	A sequence of 20 hydrophobic amino acids that forms a transmembrane domain that is flanked by positively charged amino acids.
Lysosomal-sorting signal	A patch of amino acids within the polypeptide sequence. Positively charged amino acids within this patch are thought to play an important role. This patch causes lysosomal proteins to be covalently modified to contain a mannose-6-phosphate that directs the protein to the lysosome.

Destination of a Cellular Protein	Type of Signal the Protein Contains Within Its Amino Acid Sequence
Cytosol	No signal required.
Chloroplast	Chloroplast-sorting signal
Mitochondrion	Mitochondrial-sorting signal
Nucleus	Nuclear-localization signal
Peroxisome	Peroxisomal-sorting signal
ER	SRP signal and an ER retention signal
Golgi	SRP signal and a Golgi retention signal
Lysosome	SRP signal and a lysosomal-sorting signal
Secretory vesicles	SRP signal. No additional signal required.
Plasma membrane	SRP signal and the protein contains a hydrophobic transmembrane domain that anchors it in the membrane.

CONCEPTUAL SUMMARY

An appreciation for the molecular characteristics of gene structure and function is central to our understanding of genetics. Early ideas of Garrod and of Beadle and Tatum suggested a link between genes and the production of enzymes. Since their time, the molecular nature of that link has become clear. Segments of DNA within the chromosomes are organized into units called genes. Most genes are structural genes, which means that they provide the code for the amino acid sequence of a polypeptide. The function of most genes, therefore, is to store the information for the synthesis of a particular polypeptide—ultimately, for the synthesis of a functional protein. The action of proteins within living cells and organisms directly determines the outcome of an organism's traits.

During translation, the language of mRNA, which is a nucleotide sequence, is translated into the language of proteins, which is an amino acid sequence. The translation process employs a genetic code, which is nearly universal in all spe-cies. The use of this code involves the recognition of three-base codons within the mRNA by complementary anticodons in tRNA molecules. The tRNAs carry the correct amino acid, allowing the synthesis of a polypeptide with a defined amino acid sequence. The sequence of amino acids is known as a polypeptide's primary structure. The chemical properties of the amino acid side chains within a polypeptide cause it to fold into secondary and tertiary structures. Some proteins are composed of two of more polypeptide subunits that associate with each other and create a quaternary structure. The final three-dimensional structure of a protein determines its functional role.

The translation of mRNA into a polypeptide sequence requires many cellular components, including tRNAs, ribosomes, and protein factors. According to the adaptor hypothesis, the anticodons in tRNA molecules act as translators of the genetic code within mRNA. Prior to translation, enzymes known as aminoacyl-tRNA synthetases recognize tRNAs and attach the correct amino acid to their 3' end. At this stage, the tRNA is termed a charged tRNA. During translation, the anticodon in a tRNA binds to a codon in the mRNA due to their complementary sequences. Concurrently, the tRNA carries the correct amino acid, which is then added to the growing polypeptide chain. The third base in the codon of the mRNA is referred to as the wobble position because the complementary base in the tRNA anticodon does not require a precise fit when binding to the mRNA. The wobble rules enable a single type of tRNA to recognize more than one codon. In addition, modified bases in the tRNA can often be recognized by different bases in the mRNA.

The ribosomes are large macromolecular structures that provide a site for translation to occur. Each ribosome is composed of one small and one large subunit; each subunit contains several proteins and rRNAs that assemble together. In bacteria, both ribosomal assembly and translation take place in the cytoplasm. In eukaryotes, this assembly occurs within a region of the nucleus called the nucleolus; the ribosomal subunits are then exported into the cytosol, where translation occurs.

The translation process occurs in three stages, known as initiation, elongation, and termination. During initiation, mRNA and initiator tRNA bind to the ribosomal subunits. In bacteria, the ribosomal-binding site, or Shine-Dalgarno sequence, is located a few nucleotides upstream from the start codon. In eukaryotes, the ribosome binds at the 5' end of the mRNA and then scans along the mRNA until it finds a start codon. After the start codon has been identified, translation can proceed to the elongation phase, in which the polypeptide is made. During elongation, three sites on the ribosome are used for tRNA binding. An appropriate tRNA first binds to the A site, where its anti-codon recognizes the codon in the mRNA. A peptide bond then forms between the amino acid attached to the tRNA at the A site and the growing polypeptide chain attached to a tRNA in the P site. The tRNA at the P site is then released from the ribosome at the E site. Finally, during the termination stage, the stop codon in the mRNA is reached. At this point, release factors bind to the ribosome and cause the disassembly of the completed polypeptide chain, tRNA, ribosomal subunits, and mRNA.

Some interesting differences are found between bacterial and eukaryotic translation due to their differing degrees of cell compartmentalization. Because bacteria contain a single intracellular compartment, translation of mRNA can begin before transcription is completed. This is referred to as coupling between transcription and translation. Coupling of these processes does not occur in eukaryotes because transcription takes place in the nucleus, while translation occurs in the cytosol. Moreover, compartmentalization in eukaryotic cells, due to the presence of different types of organelles, requires a more complicated system of sorting the proteins into their correct locations. This occurs via the presence of sorting signals within the translated polypeptides. The sorting of cytosol, chloroplast, mitochondrion, nucleus, and peroxisome proteins occurs after translation is completed and is called posttranslational sorting. The synthesis of proteins destined for the ER, Golgi, lysosomes, secretory vesicles, or plasma membrane occurs during translation and is called cotranslational sorting. In this case, translation begins in the cytosol and is temporarily halted; it resumes when the ribosome is bound to the membrane of the ER and the polypeptide is synthesized into the ER lumen or membrane.

EXPERIMENTAL SUMMARY

The first biochemical insights into the function of genes came from the experimental observations of Garrod. By making the connection between inherited defects in metabolism and the absence of a specific functional enzyme, he proposed that the function of genes is related to the production of enzymes. Later, Beadle and Tatum analyzed mutant genes in *Neurospora* and concluded that a single gene encodes a single enzyme, which is known as the one gene–one enzyme hypothesis. The experimental strategy used by Nirenberg and Ochoa established the relationship between the mRNA sequence and the polypeptide sequence. By synthesizing RNAs in vitro with a known nucleotide composition, they were able to correlate the predicted percentage of codons with the observed percentage of amino acids within a polypeptide chain. This helped to decipher the genetic code by determining the relationship between a codon's composition and the amino acid it specifies. Similarly, the use of RNA copolymers and the triplet binding assay were instrumental in cracking the genetic code.

Our molecular understanding of translation has come from several lines of experimentation, including biochemistry, genetics, and microscopy. The structure and composition of tRNAs, mRNAs, and ribosomes have been biochemically investigated for many decades. X-ray crystallography has aided our understanding of ribosome structure, and electron microscopy has provided a method to demonstrate that transcription and translation are coupled in bacteria.

The steps of translation have also been investigated at the molecular level. By modifying the amino acid attached to tRNA, Chapeville showed that the tRNA anticodon sequence—not the amino acid attached to the tRNA—is the determining factor during translation. This experimental observation supported the adaptor hypothesis initially proposed by Crick; the function of tRNA is to recognize a codon in mRNA and carry an amino acid specific for that codon. We have also learned a great deal about the translation process by examining genetic sequences. In addition, geneticists have identified the functional roles of sequences, such as the Shine-Dalgarno sequence in bacteria and Kozak's rules in eukaryotes, by analyzing mutations that interfere with translation.

PROBLEM SETS & INSIGHTS

Solved Problems

S1. The first amino acid in a certain bacterial polypeptide chain is methionine. The start codon in the mRNA is GUG, which codes for valine. Why isn't the first amino acid formylmethionine or valine?

Answer: The first amino acid in a polypeptide chain is carried by the initiator tRNA, which always carries formylmethionine. This occurs even when the start codon is GUG (valine) or UUG (leucine). The formyl group can be later removed to yield methionine as the first amino acid.

S2. A tRNA has the anticodon sequence 3'–CAG–5'. What amino acid does it carry?

Answer: Because the anticodon is 3'–CAG–5', it would be complementary to a codon with the sequence 5'–GUC–3'. According to the genetic code, this codon specifies the amino acid valine. Therefore, this tRNA must carry valine at its acceptor stem.

S3. In eukaryotic cells, the assembly of ribosomal subunits occurs in the nucleolus. As discussed in Chapter 12 (see Figure 12.16), a single 45S rRNA transcript is cleaved to produce the three rRNA fragments—18S, 5.8S, and 28S rRNA—that play a key role in creating the structure of the ribosome. The genes that encode the 45S precursor are found in multiple copies (i.e., they are moderately repetitive). The segments of chromosomes that contain the 45S rRNA genes align themselves at the center of the nucleolus. This site is called the nucleolar-organizing center. In this region, active transcription of the 45S gene takes place. Briefly explain how the assembly of the ribosomal subunits occurs.

Answer: In the nucleolar-organizing center, the 45S RNA is cleaved to the 18S, 5.8S, and 28s rRNAs. The other components of the ribosomal subunits, 5S rRNA and ribosomal proteins, must also be imported into the nucleolar region. Because proteins are made in the cytosol, they must enter the nucleus through the nuclear pores. When all the components are present, they assemble into 40S and 60S ribosomal subunits. Following assembly, the ribosomal subunits exit the nucleus through the nuclear pores and enter the cytosol.

S4. Throughout this chapter, we have seen that the general mechanism for bacterial and eukaryotic translation is very similar at the molecular level. In both cases, a polypeptide with a defined amino acid sequence is made on the surface of a ribosome due to the sequential interactions between the anticodons in tRNA molecules and the codons in an mRNA. In addition, we have also seen some interesting distinctions between these processes in bacteria and eukaryotes. Summarize the differences between bacterial and eukaryotic translation.

Answer: See table below.

A Comparison of Bacterial and Eukaryotic Translation

	Bacterial	Eukaryotic

Ribosome composition:	70S ribosomes:
	30S subunit—
	  21 proteins + 1 rRNA
	50S subunit—
	  34 proteins + 2 rRNAs
		80S ribosomes:
		40S subunit—
		  33 proteins + 1 rRNA
		60S subunit—
		  49 proteins + 3 rRNAs

Initiator tRNA:	tRNA^{fmet}	tRNA^{met}
Formation of the initiation complex:	Requires IF1, IF2, and IF3	Requires more initiation factors compared to bacterial initiation
Initial binding of mRNA to the ribosome:	Requires a Shine-Dalgarno sequence	Requires a 7-methylguanosine cap
Selection of a start codon:	AUG, GUG, or UUG located just downstream from the Shine-Dalgarno sequence	According to Kozak's rules
Termination:	Requires RF1, RF2, and RF3	Requires eRF
Coupled to transcription:	Yes	Not usually
Coordinated with cell compartmentalization:	Less complex, due to a lack of internal organelles	Yes, via many different sorting signals, such as the SRP signal

S5. An antibiotic is a drug that kills or inhibits the growth of microorganisms. The use of antibiotics has been of great importance in the battle against many infectious diseases caused by microorganisms. For many antibiotics, their mode of action is to inhibit the translation process within bacterial cells. Certain antibiotics selectively bind to bacterial (70S) ribosomes but do not inhibit eukaryotic (80S) ribosomes. Their ability to inhibit translation can occur at different steps in the translation process. For example, tetracycline prevents the attachment of tRNA to the ribosome, while erythromycin inhibits the translocation of the ribosome along the mRNA. Why would an antibiotic bind to a bacterial ribosome but not to a eukaryotic ribosome? How does inhibition of translation by antibiotics such as tetracycline prevent bacterial growth?

Answer: Because bacterial ribosomes have a different protein and rRNA composition than eukaryotic ribosomes, certain drugs can recognize

these different components, bind specifically to bacterial ribosomes, and thereby interfere with the process of translation. In other words, the surface of a bacterial ribosome must be somewhat different compared to the surface of a eukaryotic ribosome so that the drugs will bind to the surface of only bacterial ribosomes. If a bacterial cell is exposed to tetracycline or other antibiotics, it cannot synthesize new polypeptides. Because polypeptides form functional proteins needed for processes such as cell division, the bacterium will be unable to grow and proliferate.

Conceptual Questions

C1. An mRNA has the following sequence:

5'-GGCGAUGGGCAAUAAACCGGGCCAGUAAGC-3'

Identify the start codon and determine the complete amino acid sequence that would be translated from this mRNA.

C2. What does it mean when we say that the genetic code is degenerate? Discuss the universality of the genetic code.

C3. According to the adaptor hypothesis, are the following statements true or false?

A. The sequence of anticodons in tRNA directly recognizes codon sequences in mRNA, with some room for wobble.

B. The amino acid attached to the tRNA directly recognizes codon sequences in mRNA.

C. The amino acid attached to the tRNA affects the binding of the tRNA to codon sequences in mRNA.

C4. In bacteria, researchers have isolated strains that carry mutations within tRNA genes. These mutations can change the sequence of the anticodon. For example, a normal tRNA^{trp} gene would encode a tRNA with the anticodon 3'-ACC-5'. A mutation could change this sequence to 3'-CCC-5'. When this mutation occurs, the tRNA still carries a tryptophan at its 3' acceptor stem, even though the anticodon sequence has been altered.

A. How would this mutation affect the translation of polypeptides within the bacterium?

B. What does this mutation tell you about the recognition between tryptophanyl-tRNA synthetase and tRNA^{trp}? Does the enzyme primarily recognize the anticodon or not?

C5. The covalent attachment of an amino acid to a tRNA is an endergonic reaction. In other words, it requires an input of energy for the reaction to proceed. Where does the energy come from to attach amino acids to tRNA molecules?

C6. The wobble rules for tRNA-mRNA pairing are shown in Figure 13.14. If we assume that the tRNAs do not contain modified bases, what is the minimum number of tRNAs needed to efficiently recognize the codons for the following types of amino acids?

A. Leucine

B. Methionine

C. Serine

C7. How many different sequences of mRNA could encode a peptide with the sequence proline-glycine-methionine-serine?

C8. If a tRNA molecule carries a glutamic acid, what are the two possible anticodon sequences that it could contain? Be specific about the 5' and 3' ends.

C9. A tRNA has an anticodon sequence 3'-GGU-5'. What amino acid does it carry?

C10. If a tRNA has an anticodon sequence 3'-CCI-5', what codon(s) can it recognize?

C11. Describe the anticodon of a single tRNA that could recognize the codons 5'-AAC-3' and 5'-AAU-3'. How would this tRNA need to be modified for it to also recognize 5'-AAA-3'?

C12. Describe the structural features that all tRNA molecules have in common.

C13. In the tertiary structure of tRNA, where is the anticodon region relative to the attachment site for the amino acid? Are they adjacent to each other?

C14. What is the role of aminoacyl-tRNA synthetase? The ability of the aminoacyl-tRNA synthetases to recognize tRNAs has sometimes been called the "second genetic code." Why has the function of this type of enzyme been described this way?

C15. What is an activated amino acid?

C16. Discuss the significance of modified bases within tRNA molecules.

C17. How and when does formylmethionine become attached to the initiator tRNA in bacteria?

C18. Is it necessary for a cell to make 61 different tRNA molecules, corresponding to the 61 codons for amino acids? Explain your answer.

C19. List the components required for translation. Describe the relative sizes of these different components. In other words, which components are small molecules, macromolecules, or assemblies of macromolecules?

C20. Describe the components of eukaryotic ribosomal subunits and where the assembly of the subunits occurs within living cells.

C21. The term subunit can be used in a variety of ways. Compare the use of the term subunit in proteins versus ribosomal subunit.

C22. Do the following events during bacterial translation occur primarily within the 30S subunit, within the 50S subunit, or at the interface between these two ribosomal subunits?

A. mRNA-tRNA recognition

B. Peptidyl transfer reaction

C. Exit of the polypeptide chain from the ribosome

D. Binding of initiation factors IF1, IF2, and IF3

C23. What are the three stages of translation? Discuss the main events that occur during these three stages.

C24. Describe the sequence in bacterial mRNA that promotes recognition by the 30S subunit.

C25. For each of the following initiation factors, how would eukaryotic initiation of translation be affected if it were missing?

A. eIF2

B. eIF4A

C. eIF5

as the *lac* operon, contains a CAP site; promoter (*lacP*); operator site (*lacO*); three structural genes, *lacZ*, *lacY*, and *lacA*; and a terminator. *LacZ* encodes the enzyme β-galactosidase, an enzyme that cleaves lactose into galactose and glucose. As a side reaction, β-galactosidase also converts a small percentage of lactose into allolactose, a structurally similar sugar (**Figure 14.3b**). As we will see later, allolactose acts as a small effector molecule to regulate the *lac* operon. The *lacY* gene encodes lactose permease, a membrane protein required for the active transport of lactose into the cytoplasm of the bacterium. The *lacA* gene encodes galactoside transacetylase, an enzyme that can covalently modify lactose and lactose analogues. Although the functional necessity of the transacetylase remains unclear, the acetylation of nonmetabolizable lactose analogues may prevent their toxic buildup within the bacterial cytoplasm.

The CAP site and the operator site are short DNA segments that function in gene regulation. The **CAP site** is a DNA sequence recognized by an activator protein called the **catabolite activator protein (CAP)**. The **operator site** (also known simply as the **operator**) is a sequence of bases that provides a binding site for a repressor protein.

A second transcriptional unit involved in genetic regulation is the *lacI* gene (Figure 14.3a). The *lacI* gene is not part of the *lac* operon. The *lacI* gene encodes the **lac repressor,** a protein that is important for the regulation of the *lac* operon. The lac repressor functions as a homotetramer, a protein composed of four identical subunits. The *lacI* gene, which is constitutively expressed at fairly low levels, has its own promoter, the *i* promoter. The amount of lac repressor made is approximately 10 homotetramer proteins per cell. Only a small amount of the lac repressor protein is needed to repress the *lac* operon.

The *lac* Operon Is Regulated by a Repressor Protein

The *lac* operon can be transcriptionally regulated in more than one way. The first mechanism that we will examine is one that is inducible and under negative control. As shown in **Figure 14.4**, this form of regulation involves the lac repressor protein, which binds to the sequence of nucleotides found within the *lac* operator site. Once bound, the lac repressor prevents RNA polymerase from transcribing the *lacZ*, *lacY*, and *lacA* genes (Figure 14.4a). The binding of the repressor to the operator site is a reversible process. In the absence of allolactose, the lac repressor is bound to the operator site most of the time.

The ability of the lac repressor to bind to the operator site depends on whether or not allolactose is bound to it. Each of the repressor protein's four subunits has a single binding site for allolactose, the inducer. How does a small molecule like allolactose exert its effects? When allolactose binds to the repressor, a conformational change occurs in the lac repressor protein that prevents it from binding to the operator site. Under these conditions, RNA polymerase is now free to transcribe the operon (Figure 14.4b). In genetic terms, we would say that the operon has been **induced.** The action of a small effector molecule, such as allolactose, is called **allosteric regulation.** The effector molecule binds to the protein's **allosteric site,** which is a site other than

(a) No lactose in the environment

(b) Lactose present

FIGURE 14.4 **Mechanism of induction of the** *lac* **operon.** (a) In the absence of the inducer allolactose, the repressor protein is tightly bound to the operator site, thereby inhibiting the ability of RNA polymerase to transcribe the operon. (b) When allolactose is available, it binds to the repressor. This alters the conformation of the repressor protein in such a way as to prevent it from binding to the operator site. Therefore, RNA polymerase can transcribe the operon. (The CAP site is not labeled in this drawing.)

the protein's active site. In the case of the lac repressor, the active site is the part of the protein that binds to the DNA.

To better appreciate this form of regulation at the cellular level, let's consider the process as it occurs over time. **Figure 14.5** illustrates the effects of external lactose on the regulation of the *lac* operon. In the absence of lactose, no inducer is available to bind to the lac repressor. Therefore, the lac repressor binds to the

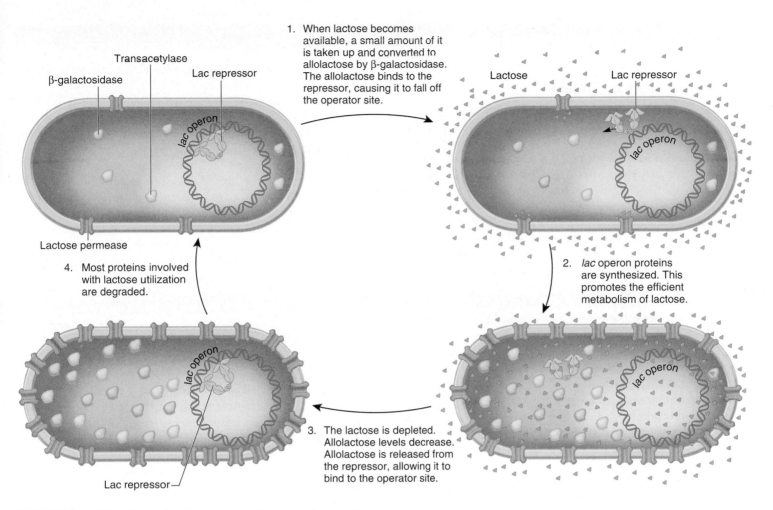

β-galactosidase

Transacetylase

Lac repressor

1. When lactose becomes available, a small amount of it is taken up and converted to allolactose by β-galactosidase. The allolactose binds to the repressor, causing it to fall off the operator site.

Lactose

Lac repressor

lac operon

Lactose permease

4. Most proteins involved with lactose utilization are degraded.

lac operon

2. *lac* operon proteins are synthesized. This promotes the efficient metabolism of lactose.

lac operon

Lac repressor

3. The lactose is depleted. Allolactose levels decrease. Allolactose is released from the repressor, allowing it to bind to the operator site.

FIGURE 14.5   The cycle of *lac* operon induction and repression.

Genes→Traits The genes and the genetic regulation of the *lac* operon provide the bacterium with the trait of being able to metabolize lactose in the environment. When lactose is present, the genes of the *lac* operon are induced, and the bacterial cell can efficiently metabolize this sugar. When lactose is absent, these genes are repressed, so that the bacterium does not waste its energy expressing these genes.

operator site and inhibits transcription. In reality, the repressor does not completely inhibit transcription, so very small amounts of β-galactosidase, lactose permease, and transacetylase are made. However, the levels are far too low for the bacterium to readily use lactose. When the bacterium is exposed to lactose, a small amount can be transported into the cytoplasm via lactose permease, and β-galactosidase converts some of it to allolactose. As this occurs, the cytoplasmic level of allolactose gradually rises; eventually, allolactose binds to the lac repressor. The binding of allolactose promotes a conformational change that prevents the repressor from binding to the *lac* operator site and thereby allows transcription of the *lacZ*, *lacY*, and *lacA* genes to occur. Translation of the encoded polypeptides will produce the proteins needed for lactose uptake and metabolism.

To understand how the induction process is shut off in a lactose-depleted environment, let's consider the interaction between allolactose and the lac repressor. The lac repressor has

a measurable affinity for allolactose. The binding of allolactose to the lac repressor is reversible. The likelihood that allolactose will bind to the repressor depends on the allolactose concentration. During induction of the operon, the concentration of allolactose rises and approaches the affinity for the repressor protein. This makes it likely that allolactose will bind to the lac repressor, thereby causing it to be released from the operator site. Later on, however, the bacterial cell metabolizes the sugars and thereby lowers the concentration of allolactose below its affinity for the repressor. At this point, the lac repressor is unlikely to be bound to allolactose. When allolactose is released, the lac repressor returns to the conformation that binds to the operator site. In this way, the binding of the repressor shuts down the *lac* operon when lactose is depleted from the environment. After repression occurs, the mRNA and proteins encoded by the *lac* operon will be degraded because they have a relatively short half-life (Figure 14.5).

The *lacI* Gene Encodes a Diffusible Repressor Protein

Now that we have an understanding of the *lac* operon, let's consider one of the experimental approaches that was used to elucidate its regulation. In the 1950s, Jacob, Monod, and their colleague Arthur Pardee had identified a few rare mutant strains of bacteria that had abnormal lactose adaptation. One type of mutant, designated *lacI⁻*, resulted in the constitutive expression of the *lac* operon even in the absence of lactose. From this observation, the researchers incorrectly hypothesized that the *lacI⁻* mutation resulted in the synthesis of an internal inducer, making it unnecessary for cells to be exposed to lactose for induction (**Figure 14.6a**). By comparison, **Figure 14.6b** shows the correct explanation. A loss-of-function mutation in the *lacI* gene prevented the lac repressor protein from inhibiting transcription. At the time of their work, however, the function of the lac repressor was not yet known.

To further explore the nature of this mutation, Jacob, Monod, and Pardee applied a genetic approach. In order to understand their approach, let's briefly consider the process of bacterial conjugation (described in Chapter 6). The earliest studies of Jacob, Monod, and Pardee in 1959 involved matings between recipient cells, termed F⁻, and donor cells, which were *Hfr* strains that transferred a portion of the bacterial chromosome. Later experiments in 1961 involved the transfer of circular segments of DNA known as F factors. We will consider the latter type of experiment here. Sometimes an F factor also carries genes that were originally found within the bacterial chromosome. These types of F factors are called F' factors (F prime factors). In their studies, Jacob, Monod, and Pardee identified F' factors that carried the *lacI* gene and portions of the *lac* operon. These F' factors can be transferred from one cell to another by bacterial con-

jugation. A strain of bacteria containing F' factor genes is called a **merozygote,** or partial diploid.

The production of merozygotes was instrumental in allowing Jacob, Monod, and Pardee to elucidate the function of the *lacI* gene. This experimental approach has two key points. First, the two *lacI* genes in a merozygote may be different alleles. For example, the *lacI* gene on the chromosome may be a *lacI⁻* allele that causes constitutive expression, while the *lacI* gene on the F' factor may be normal. Second, the genes on the F' factor and the genes on the bacterial chromosome are not physically adjacent to each other. As we now know, the expression of the *lacI* gene on an F' factor should produce repressor proteins that could diffuse within the cell and eventually bind to the operator site of the *lac* operon located on the chromosome.

Figure 14.7 shows one experiment of Jacob, Monod, and Pardee in which they analyzed a *lacI⁻* mutant strain that was already known to constitutively express the *lac* operon and compared it to the corresponding merozygote. The merozygote had a *lacI⁻* mutant gene on the chromosome and a normal *lacI* gene on an F' factor. These two strains were grown and then divided into two tubes each. In half of the tubes, lactose was omitted. In the other tubes, the strains were incubated with lactose to determine if lactose was needed to induce the expression of the operon. The cells were lysed by sonication, and then a lactose analogue, β-ONPG, was added. This molecule is colorless, but β-galactosidase cleaves it into a product that has a yellow color. Therefore, the amount of yellow color produced in a given amount of time is a measure of the amount of β-galactosidase that is being expressed from the *lac* operon.

■ THE HYPOTHESIS

The *lacI⁻* mutation results in the synthesis of an internal inducer.

(a) Internal inducer hypothesis

(b) Correct explanation

FIGURE 14.6   **Alternative ideas to explain how a *lacI⁻* mutation could cause the constitutive expression of the *lac* operon.** (a) The hypothesis of Jacob, Monod, and Pardee. In this case, the *lacI⁻* mutation would result in the synthesis of an internal inducer that turns on the *lac* operon. (b) The correct explanation in which the *lacI⁻* mutation eliminates the function of the lac repressor protein, which prevents it from repressing the *lac* operon.

TESTING THE HYPOTHESIS — FIGURE 14.7   **Evidence that the *lacI* gene encodes a diffusible repressor protein.**

Starting material: The genotype of the mutant strain was *lacI⁻ lacZ⁺ lacY⁺ lacA⁺*. The merozygote strain had an F' factor that was *lacI⁺ lacZ⁺ lacY⁺ lacA⁺*, which had been introduced into the mutant strain via conjugation.

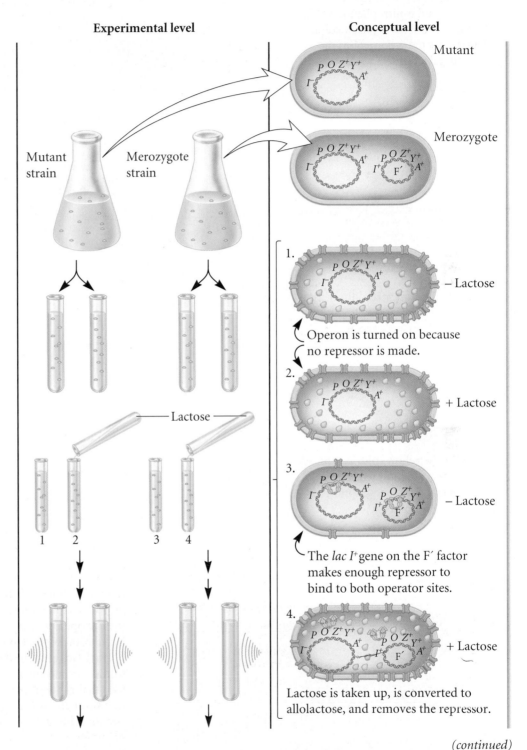

1. Grow mutant strain and merozygote strain separately.

2. Divide each strain into two tubes.

3. In one of the two tubes, add lactose.

4. Incubate the cells long enough to allow *lac* operon induction.

5. Burst the cells with a sonicator. This allows β-galactosidase to escape from the cells.

Experimental level

Mutant strain   Merozygote strain

Lactose

1   2   3   4

Conceptual level

Mutant

Merozygote

1. – Lactose

Operon is turned on because no repressor is made.

2. + Lactose

3. – Lactose

The *lac I⁺* gene on the F' factor makes enough repressor to bind to both operator sites.

4. + Lactose

Lactose is taken up, is converted to allolactose, and removes the repressor.

(continued)

6. Add β-*o*-nitrophenylgalactoside (β-ONPG). This is a colorless compound. β-galactosidase will cleave the compound to produce galactose and *o*-nitrophenol (O-NP). *O*-nitrophenol has a yellow color. The deeper the yellow color, the more β-galactosidase was produced.

7. Incubate the sonicated cells to allow β-galactosidase time to cleave β-*o*-nitrophenylgalactoside.

8. Measure the yellow color produced with a spectrophotometer. (See the Appendix for a description of spectrophotometry.)

■ THE DATA

Strain	Addition of Lactose	Amount of β-Galactosidase (percentage of parent strain)
Mutant	No	100%
Mutant	Yes	100%
Merozygote	No	<1%
Merozygote	Yes	220%

■ INTERPRETING THE DATA

As seen in the data, the yellow production in the original mutant strain was the same in the presence or absence of lactose. This result is expected because the expression of β-galactosidase in the *lacI⁻* mutant strain was already known to be constitutive. In other words, the presence of lactose was not needed to induce the operon due to a defective *lacI* gene. In the merozygote strain, however, a different result was obtained. In the absence of lactose, the *lac* operons were repressed—even the operon on the bacterial chromosome. How do we explain these results? Because the normal *lacI* gene on the F' factor was not physically located next to the chromosomal *lac* operon, this result is consistent with the idea that the *lacI* gene codes for a repressor protein that can diffuse throughout the cell and bind to any *lac* operon. The hypothesis that the *lacI⁻* mutation resulted in the synthesis of an internal inducer was rejected. If that hypothesis had been correct, the merozygote strain would have still made an internal inducer, and the *lac* operons in the merozygote would have been expressed in the absence of lactose. This result was not obtained.

The interactions between regulatory proteins and DNA sequences illustrated in this experiment have led to the definition of two genetic terms. A ***trans*-effect** is a form of genetic regulation that can occur even though two DNA segments are not physically adjacent. The action of the lac repressor on the *lac* operon is a *trans*-effect. A regulatory protein, such as the lac repressor, is called a ***trans*-acting factor.** In contrast, a ***cis*-acting element** is a DNA segment that must be adjacent to the gene(s) that it regulates, and it is said to have a ***cis*-effect** on gene expression. The *lac* operator site is an example of a *cis*-acting element. A *trans*-effect is mediated by genes that encode regulatory proteins, whereas a *cis*-effect is mediated by DNA sequences that are bound by regulatory proteins.

Jacob and Monod also isolated constitutive mutants that affected the operator site, *lacO.* Table 14.1 summarizes the effects of mutations based on their locations in the *lacI* regulatory gene versus *lacO* and their analysis in merozygotes. As seen here, a loss-of-function mutation in a gene encoding a repressor protein has the same effect as a mutation in an operator site that cannot bind a repressor protein. In both cases, the genes of the *lac* operon are constitutively expressed. In a merozygote, however, the results are quite different. When a normal *lacI* gene and a normal *lac* operon are introduced into a cell harboring a defective *lacI* gene, the normal *lacI* gene can regulate both operons. In contrast, when a *lac* operon with a normal operator site is introduced into a cell with a defective operator site, the operon with the defective operator site continues to be expressed without lactose present. Overall, a mutation in a *trans*-acting factor can be complemented by the introduction of a second gene with nor-

mal function. However, a mutation in a *cis*-acting element is not affected by the introduction of another *cis*-acting element with normal function into the cell.

A self-help quiz involving this experiment can be found at the Online Learning Center.

TABLE 14.1

A Comparison of Loss-of-Function Mutations in the *lacI* Gene versus the Operator Site

Chromosome	F' factor	Expression of the *lac* Operon	
		With Lactose	Without Lactose
Wild type	None	100%	<1%
lacI⁻	None	100%	100%
lacO⁻	None	100%	100%
lacI⁻	*lacI⁺* and a normal *lac* operon	200%	<1%
lacO⁻	*lacI⁺* and a normal *lac* operon	200%	100%

The *lac* Operon Is Also Regulated by an Activator Protein

The *lac* operon can be transcriptionally regulated in a second way, known as **catabolite repression** (as we shall see, a somewhat imprecise term). This form of transcriptional regulation is influenced by the presence of glucose, which is a catabolite—a substance that is broken down inside the cell. The presence of glucose ultimately leads to repression of the *lac* operon. When exposed to both glucose and lactose, *E. coli* cells first use glucose, and catabolite repression prevents the use of lactose. Why is this an advantage? The explanation is efficiency. The bacterium does not have to express all the genes necessary for both glucose and lactose metabolism. If the glucose is used up, catabolite repression is alleviated, and the bacterium then expresses the *lac* operon. The sequential use of two sugars by a bacterium, known as **diauxic growth,** is a common phenomenon among many bacterial species. Typically, glucose, a more commonly encountered sugar, is metabolized preferentially, and then a second sugar is metabolized only after glucose has been depleted from the environment.

Glucose, however, is not itself the small effector molecule that binds directly to a genetic regulatory protein. Instead, this form of regulation involves a small effector molecule, **cyclic AMP (cAMP),** which is produced from ATP via an enzyme known as adenylyl cyclase. When a bacterium is exposed to glucose, the transport of glucose into the cell stimulates a signaling pathway that causes the intracellular concentration of cAMP to decrease

because the pathway inhibits adenylyl cyclase, an enzyme needed for cAMP synthesis. The effect of cAMP on the *lac* operon is mediated by an activator protein called the catabolite activator protein (CAP). CAP is composed of two subunits, each of which binds one molecule of cAMP.

Figure 14.8 considers how the interplay between the lac repressor and CAP determines whether the *lac* operon is expressed in the presence or absence of lactose and/or glucose. When only lactose is present, cAMP levels are high (Figure 14.8a). The cAMP binds to CAP, and then CAP binds to the CAP site and stimulates the ability of RNA polymerase to begin transcription. In the presence of lactose, the lac repressor is not bound to the operator site, so transcription can proceed. In the absence of both lactose and glucose, cAMP levels are also high (Figure 14.8b). Under these conditions, however, the binding of the lac repressor inhibits transcription even though CAP is bound to the DNA.

Figure 14.8c considers the situation in which both sugars are present. The presence of lactose causes the lac repressor to be inactive, which prevents it from binding to the operator site. Even so, the presence of glucose decreases cAMP levels so that cAMP is released from CAP, which prevents CAP from binding to the CAP site. Because CAP is not bound to the CAP site, the *lac* operon is not expressed in the presence of both sugars. Finally, Figure 14.8d illustrates what happens when only glucose is present. The *lac* operon is shut off because the lac repressor is bound to the operator site and CAP is not bound to the CAP site due to low cAMP levels.

(a) Lactose, no glucose (high CAMP)

(b) No lactose or glucose (high cAMP)

(c) Lactose and glucose (low cAMP)

(d) Glucose, no lactose (low cAMP)

FIGURE 14.8   **The roles of the lac repressor and catabolite activator protein (CAP) in the regulation of the *lac* operon.** This figure illustrates how the *lac* operon will be regulated depending on its exposure to lactose or glucose.

Genes→Traits   The mechanism of catabolite repression provides the bacterium with the trait of being able to choose between two sugars. When exposed to both glucose and lactose, the bacterium chooses glucose first. After the glucose is used up, it then expresses the genes necessary for lactose metabolism. This trait allows the bacterium to more efficiently use sugars from its environment.

The effect of glucose, called catabolite repression, may seem like a puzzling way to describe this process because this regulation involves the action of an inducer (cAMP) and an activator protein (CAP), not a repressor. The term was coined before the action of the cAMP-CAP complex was understood. At that time, the primary observation was that glucose (a catabolite) inhibited (repressed) lactose metabolism.

Many bacterial promoters that transcribe genes involved in the breakdown of other sugars, such as maltose, arabinose, and melibiose, also have binding sites for CAP. Therefore, when glucose levels are high, these operons are inhibited, thereby promoting diauxic growth.

Further Studies Have Revealed That the *lac* Operon Has Three Operator Sites for the lac Repressor

Our traditional view of the regulation of the *lac* operon has been modified as we have gained a greater molecular understanding of the process. In particular, detailed genetic and crystallography studies have shown that the binding of the lac repressor is more complex than originally realized. The site in the *lac* operon that is commonly called the operator site was first identified by mutations that prevented lac repressor binding. These mutations, called $lacO^-$ or $lacO^C$ mutants, resulted in the constitutive expression of the *lac* operon even in strains that make a normal lac repressor protein. $LacO^C$ mutations were localized in the *lac* operator site, which is now known as O_1. This led to the view that a single operator site was bound by the lac repressor to inhibit transcription, as in Figure 14.4.

In the late 1970s and 1980s, two additional operator sites were identified. As shown at the top of **Figure 14.9**, these sites are called O_2 and O_3. O_1 is the operator site slightly downstream from the promoter. O_2 is located farther downstream in the *lacZ* coding sequence, and O_3 is located slightly upstream from the promoter. The O_2 and O_3 operator sites were initially called pseudo-operators, because substantial repression occurred in the absence of either one of them. However, studies by Benno Müller-Hill and his colleagues revealed a surprising result. As shown in Figure 14.9 (fourth example down), if both O_2 and O_3 are missing, repression is dramatically reduced even when O_1 is present. When O_1 is missing, even in the presence of one of the other operator sites, repression is nearly abolished.

How were these results interpreted? The data of Figure 14.9 supported a hypothesis that the lac repressor must bind to O_1 and either O_2 or O_3 to cause full repression. According to this view, the lac repressor can readily bind to O_1 and O_2, or to O_1 and O_3, but not to O_2 and O_3. If either O_2 or O_3 were missing, maximal repression is not achieved because it is less likely for the repressor to bind when only two operator sites are present. If you take a look at Figure 14.9, you will notice that the operator sites are a fair distance away from each other. For this reason, it was proposed from these studies that the binding of the lac repressor to two operator sites requires the DNA to form a loop. A loop in the DNA would bring the operator sites closer together and thereby facilitate the binding of the repressor protein (**Figure 14.10a**).

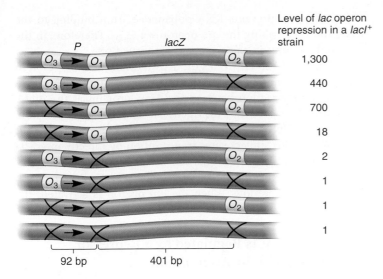

Level of *lac* operon repression in a *lacI*⁺ strain

O_3 → O_1 ... O_2	1,300
O_3 → O_1 ... X	440
X → O_1 ... O_2	700
X → O_1 ... X	18
O_3 → X ... O_2	2
O_3 → X ... X	1
X → X ... O_2	1
X → X ... X	1

92 bp   401 bp

FIGURE 14.9   The identification of three *lac* operator sites. The top of this figure shows the locations of three *lac* operator sites, designated O_1, O_2, and O_3. O_1 is analogous to the *lac* operator site shown in previous figures. The arrows depict the starting site for transcription. Defective operator sites are indicated with an X. When all three operator sites are present, the repression of the *lac* operon is 1,300-fold; this means there is 1/1,300 the level of expression than when lactose is present. This figure also shows the amount of repression when one or more operator sites are removed. A repression value of 1.0 indicates that no repression is occurring. In other words, a value of 1.0 indicates constitutive expression.

In 1996, the proposal that the lac repressor binds to two operator sites was confirmed by studies in which the lac repressor was crystallized by Mitchell Lewis and his colleagues. The crystal structure of the lac repressor has provided exciting insights into its mechanism of action. As mentioned earlier in this chapter, the lac repressor is a tetramer of four identical subunits. The crystal structure revealed that each dimer within the tetramer recognizes one operator site. **Figure 14.10b** is a molecular model illustrating the binding of the lac repressor to the O_1 and O_3 sites. The amino acid side chains in the protein interact directly with bases in the major groove of the DNA helix. This is how genetic regulatory proteins recognize specific DNA sequences. Because each dimer within the tetramer recognizes a single operator site, the association of two dimers to form a tetramer requires that the two operator sites be close to each other. For this to occur, a loop must form in the DNA. The formation of this loop would dramatically inhibit the ability of RNA polymerase to slide past the O_1 site and transcribe the operon.

Figure 14.10b also shows the binding of the cAMP-CAP complex to the CAP site (see the blue protein within the loop). A particularly striking observation is that the binding of the cAMP-CAP complex to the DNA causes a 90° bend in the DNA structure. When the repressor is active—not bound to allolactose—the cAMP-CAP complex facilitates the binding of the lac repressor to the O_1 and O_3 sites. When the repressor is inactive, this bending also appears to be important in the ability of RNA polymerase to initiate transcription slightly downstream from the bend.

(a) DNA loops caused by the binding of the lac repressor

lac repressor tetramer

O_3 O_1

O_1 O_2

(b) Proposed model of the lac repressor binding to O_1 and O_3 based on crystallography studies

CAP

lac repressor

FIGURE 14.10   The binding of the lac repressor to two operator sites. (a) The binding of the lac repressor protein to the O_1 and O_3 or to the O_1 and O_2 operator sites. Because the two sites are far apart, a loop must form in the DNA. **(b)** A molecular model for the binding of the lac repressor to O_1 and O_3. Each repressor dimer binds to one operator site, so the repressor tetramer brings the two operator sites together. This causes the formation of a DNA loop in the intervening region. Note that the DNA loop contains the −35 and −10 regions (shown in green), which are recognized by σ factor of RNA polymerase. This loop also contains the binding site for the cAMP-CAP complex, which is the protein within the loop.

The *ara* Operon Can Be Regulated Positively or Negatively by the Same Regulatory Protein

Now that we have considered the regulation of the *lac* operon, let's compare its regulation to that of other genes in the bacterial chromosome. Another operon in *E. coli* involved in sugar metabolism is the *ara* (arabinose) operon. The sugar arabinose is a constituent of the cell walls of a few types of plants. As shown in **Figure 14.11**, the *ara* operon contains three structural genes, *araB*, *araA*, and *araD*, encoding a polycistronic mRNA for the three enzymes involved in arabinose metabolism. The actions of the three enzymes metabolize arabinose into D-xylulose-5-phosphate.

Like the *lac* operon, the *ara* operon contains a single promoter, designated P_{BAD}. The operon also contains a CAP site for the binding of the catabolite activator protein. The *araC* gene, which has its own promoter (P_C), is adjacent to the *ara* operon. *AraC* encodes a regulatory protein, called the AraC protein, that can bind to operator sites designated *araI*, *araO_1*, and *araO_2*.

As we have seen with the *lac* operon, some regulatory proteins such as the lac repressor inhibit transcription, whereas others such as CAP turn on transcription. AraC is a rather interesting and unusual protein, because it can act as either a negative or positive regulator of transcription, depending on whether or not arabinose is present. How can a protein function in both ways? In the absence of arabinose, the AraC protein binds to the *araI*, *araO_1*, and *araO_2* operator sites (**Figure 14.12a**). An AraC protein dimer is bound to *araO_1*, while monomers are bound at *araO_2* and *araI*. The binding of the AraC proteins to the *araO_1* site inhibits the transcription of the *araC* gene. In other words, the AraC protein is a negative regulator of the *araC* gene. This keeps AraC protein levels fairly low. The AraC proteins bound at *araO_2* and *araI* repress the *ara* operon, but only in the absence of arabinose. As shown in Figure 14.12a, the AraC proteins at *araO_2* and *araI* can bind to each other by causing a loop in the DNA, as originally proposed by Robert Schleif and his colleagues in 1990.

This DNA loop prevents RNA polymerase from binding to the DNA and transcribing the *ara* operon via P_{BAD}. Therefore, in the absence of arabinose, the *ara* operon is turned off.

Figure 14.12b illustrates the activation of the *ara* operon in the presence of arabinose. When arabinose is bound to the AraC protein, the interaction between the AraC proteins at the *araO_2* and *araI* sites is broken. This opens the DNA loop. In addition, a second AraC protein binds at the *araI* site. This AraC dimer at the *araI* operator site activates transcription by directly interacting with RNA polymerase. This activation can occur in conjunction with the activation of the *ara* operon by CAP and cAMP if glucose levels are low. When the *ara* operon is activated, the bacterial cell can efficiently metabolize arabinose.

The *trp* Operon Is Regulated by a Repressor Protein and Also by Attenuation

The *trp* operon (pronounced "trip") encodes enzymes that are needed for the biosynthesis of the amino acid tryptophan. The *trpE*, *trpD*, *trpC*, *trpB*, and *trpA* genes encode enzymes involved in tryptophan biosynthesis. The *trpR* and *trpL* genes are involved in regulating the *trp* operon in two different ways. The *trpR* gene encodes the **trp repressor** protein. When tryptophan levels within the cell are very low, the trp repressor cannot bind to the operator site. Under these conditions, RNA polymerase transcribes the *trp* operon (**Figure 14.13a**). In this way, the cell expresses the genes required for the synthesis of tryptophan. When the tryptophan levels within the cell become high, tryptophan acts as a corepressor that binds to the trp repressor protein. This causes a conformational change in the trp repressor that allows it to bind to the *trp* operator site (**Figure 14.13b**). This inhibits the ability of RNA polymerase to transcribe the operon. Therefore, when a high level of tryptophan is present within the cell—when the cell does not need to make more tryptophan—the *trp* operon is turned off.

FIGURE 14.11   Organization of the *ara* operon and other genes involved in arabinose metabolism. *AraC* encodes a genetic regulatory protein called the AraC protein. This protein can bind to three different regulatory sites, called *araI*, *araO_1*, and *araO_2*. P_C is the promoter for the *araC* gene; P_{BAD} is the promoter for the *ara* operon, which contains three genes (*araB*, *araA*, and *araD*) encoding proteins involved in arabinose metabolism.

(a) Operon inhibited in the absence of arabinose

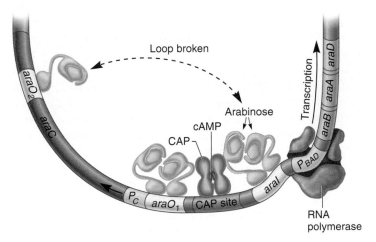

(b) Operon activated in the presence of arabinose

FIGURE 14.12 **DNA looping and unlooping via the AraC protein.** (a) In the absence of arabinose, an AraC protein binds to the $araO_2$ operator site and another to the $araI$ site. These two AraC proteins interact to promote a loop in the DNA. This loop prevents RNA polymerase from transcribing the *ara* operon from the P_{BAD} promoter. (b) In the presence of arabinose, arabinose binds to the AraC proteins and causes a conformational change that breaks the interaction between the AraC proteins bound at $araO_2$ and $araI$. This causes the DNA loop to be broken. Under these conditions, an additional AraC protein binds to $araI$, and RNA polymerase can transcribe the *ara* operon.

(a) Low tryptophan levels, transcription of the entire *trp* operon occurs

(b) High tryptophan levels, repression occurs

(c) High tryptophan levels, attenuation occurs

FIGURE 14.13 **Organization of the *trp* operon and regulation via the trp repressor protein.** (a) When tryptophan levels are low, tryptophan does not bind to the repressor protein, which prevents the repressor protein from binding to the operator site. Under these conditions, RNA polymerase can transcribe the operon, which leads to the expression of the *trpE*, *trpD*, *trpC*, *trpB*, and *trpA* genes. These genes encode enzymes involved in tryptophan biosynthesis. (b) When tryptophan levels are high, tryptophan acts as a corepressor that binds to the trp repressor protein. The tryptophan-trp repressor complex then binds to the operator site to inhibit transcription. (c) Another mechanism of regulation is attenuation. When attenuation occurs, the RNA is transcribed only to the attenuator sequence, and then transcription is terminated.

In the 1970s, after the action of the trp repressor was elucidated, Charles Yanofsky and coworkers made a few unexpected observations. Mutant strains were found that lacked the trp repressor protein. Surprisingly, these mutant strains still could inhibit the expression of the *trp* operon in the presence of tryptophan. In addition, *trp* operon mutations were identified in which a region including the *trpL* gene was missing from the operon. These mutations resulted in higher levels of expression of the other genes in the *trp* operon. As is often the case, unusual observations can lead scientists into productive avenues of study. By pursuing this research further, Yanofsky discovered a second regulatory mechanism in the *trp* operon, called **attenuation,** that is mediated by the region that includes the *trpL* gene (**Figure 14.13c**).

Attenuation can occur in bacteria because the processes of transcription and translation are coupled. As described in Chapter 13, bacterial ribosomes quickly attach to the 5' end of mRNA soon after its synthesis begins via RNA polymerase. During attenuation, transcription actually begins, but it is terminated before the entire mRNA is made. A segment of DNA, termed the **attenuator sequence,** is important in facilitating this termination. When attenuation occurs, the mRNA from the *trp* operon is made as a short piece that terminates shortly past the *trpL* gene (Figure 14.13c). Because this short mRNA has been terminated before RNA polymerase has transcribed the *trpE*, *trpD*, *trpC*, *trpB*, and *trpA* genes, it will not encode the proteins required for tryptophan biosynthesis. In this way, attenuation inhibits the further production of tryptophan in the cell.

The segment of the *trp* operon immediately downstream from the operator site plays a critical role during attenuation. The first gene in the *trp* operon is the *trpL* gene, which encodes a peptide containing 14 amino acids called the leader peptide. As shown in **Figure 14.14**, two features are key in the attenuation mechanism. First, two tryptophan (Trp) codons are found within the mRNA that encodes the *trp* leader peptide. What is the role of these codons? As we will see later, these two codons provide a way to sense whether or not the bacterium has sufficient tryptophan to synthesize its proteins. Second, the mRNA can form stem-loop structures. The type of stem-loop structure that forms underlies attenuation.

Different combinations of stem-loop structures are possible due to interactions among four sequences within the RNA transcript (see the yellow regions in Figure 14.14). Region 2 is complementary to region 1 and also to region 3. Region 3 is complementary to region 2 as well as to region 4. Therefore, three stem-loop structures are possible: 1–2, 2–3, and 3–4. Even so, keep in mind that a particular segment of RNA can participate in the formation of only one stem-loop structure. For example, if region 2 forms a stem-loop with region 1, it cannot (at the same time) form a stem-loop with region 3. Alternatively, if region 2 forms a stem-loop with region 3, then region 3 cannot form a stem-loop with region 4. Though three stem-loop structures are possible, the 3–4 stem-loop structure is functionally unique. The 3–4 stem-loop together with the U-rich attenuator sequence acts as an intrinsic terminator—a ρ-independent terminator, as described in Chapter 12. Therefore, the formation of the 3–4 stem-loop causes RNA polymerase to

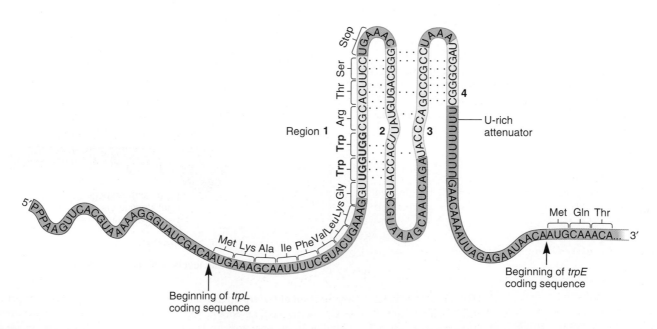

FIGURE 14.14   Sequence of the *trpL* mRNA produced during attenuation. A second method of regulation of the *trp* operon is attenuation. At high tryptophan levels, it may occasionally happen that the trp repressor protein does not bind to the operator site to inhibit transcription. If so, a short mRNA is made that includes the *trpL* region. As shown here, this mRNA has several regions that are complementary to each other. The hydrogen bonding between yellow regions 1 and 2, 2 and 3, and 3 and 4 is also shown. The last U in the purple attenuator sequence is the last nucleotide that would be transcribed during attenuation. At low tryptophan concentrations, however, transcription would occur beyond the end of *trpL* and proceed through the *trpE* gene and the rest of the *trp* operon.

(a) No translation

(b) Low tryptophan levels, 2–3 stem-loop forms

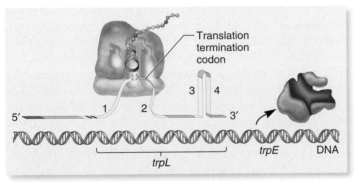

(c) High tryptophan levels, 3–4 stem-loop forms

ONLINE ANIMATION

FIGURE 14.15   Possible stem-loop structures formed from *trpL mRNA* under different conditions of translation. (a) When translation is not coupled with transcription, the most stable form of the mRNA occurs when region 1 hydrogen bonds to region 2 and region 3 hydrogen bonds to region 4. A terminator stem-loop forms, and transcription will be terminated just past the *trpL* gene. **(b)** Coupled transcription and translation have occurred under conditions in which the tryptophan concentration is low. The ribosome pauses at the Trp codons in the *trpL* gene because insufficient amounts of charged tRNA^{trp} are present. This pause blocks region 1 of the mRNA, so region 2 can hydrogen bond only with region 3. When this happens, the 3–4 stem-loop structure cannot form. Transcriptional termination does not occur, and RNA polymerase transcribes the rest of the operon. **(c)** Coupled transcription and translation occurs under conditions in which a sufficient amount of tryptophan is present in the cell. Translation of the *trpL* gene progresses to its stop codon, where the ribosome pauses. This blocks region 2 from hydrogen bonding with any region and thereby enables region 3 to hydrogen bond with region 4. This terminates transcription at the U-rich attenuator.

pause, and the U-rich sequence dissociates from the DNA. This terminates transcription at the U-rich attenuator. In comparison, if region 3 forms a stem-loop with region 2, transcription will not be terminated because a 3–4 stem-loop cannot form.

Conditions that favor the formation of the 3–4 stem-loop ultimately rely on the translation of the *trpL* gene. As shown in **Figure 14.15**, three scenarios are possible. In Figure 14.15a, translation is not coupled with transcription. Region 1 rapidly hydrogen bonds to region 2, and region 3 is left to hydrogen bond to region 4. Therefore, the terminator stem-loop forms, and transcription will be terminated just past the *trpL* gene at the U-rich attenuator.

In Figure 14.15b, coupled transcription and translation occur under conditions in which the tryptophan concentration is low. When tryptophan levels are low, the cell cannot make a sufficient amount of charged tRNA^{trp}. As we see in Figure 14.15b, the ribosome pauses at the Trp codons in the *trpL* mRNA because it is waiting for charged tRNA^{trp}. This pause occurs in such a way that the ribosome shields region 1 of the mRNA. This sterically prevents region 1 from hydrogen bonding to region 2. As an alternative, region 2 hydrogen bonds to region 3. Therefore, because region 3 is already hydrogen bonded to region 2, the 3–4 stem-loop structure cannot form. Under these conditions, transcriptional termination does not occur, and RNA polymerase transcribes the rest of the operon. This ultimately enables the bacterium to make more tryptophan.

Finally, in Figure 14.15c, coupled transcription and translation occur under conditions in which a sufficient amount of tryptophan is present in the cell. In this case, translation of the *trpL* mRNA progresses to its stop codon, where the ribosome pauses. The pausing at the stop codon prevents region 2 from hydrogen bonding with any region and thereby enables region 3 to hydrogen bond with region 4. As in Figure 14.15a, this terminates transcription. Of course, keep in mind that the *trpL* gene contains two tryptophan codons. For the ribosome to smoothly progress to the *trpL* stop codon, enough charged tRNA^{trp} must be available to translate this mRNA. It follows that the bacterium must have a sufficient amount of tryptophan. Under these conditions, the rest of the transcription of the operon is terminated.

Attenuation is a mechanism to regulate transcription that is found in several other operons involved with amino acid biosynthesis. In all cases, the mRNAs that encode the leader peptides are rich in codons for the particular amino acid that is synthesized by the enzymes encoded by the particular operon. For example, the mRNA that encodes the leader peptide of the histidine operon has seven histidine codons in its sequence, and the mRNA for the leader peptide of the leucine operon has four leucine codons. Like the *trp* operon, these other operons have alternative stem-loop structures, one of which is a transcriptional terminator.

Inducible Operons Encode Catabolic Enzymes, and Repressible Operons Usually Encode Anabolic Enzymes

Thus far, we have seen that bacterial genes can be transcriptionally regulated in a positive or negative way—and sometimes both. The *lac* operon and *ara* operon are inducible systems regulated by sugar molecules that activate transcription of these operons. By comparison, the *trp* operon is a repressible operon regulated by tryptophan, a corepressor that binds to the repressor and turns the operon off. In addition, an abundance of charged tRNA^trp in the cytoplasm can turn the *trp* operon off via attenuation.

By studying the genetic regulation of many operons, geneticists have discovered a general trend concerning inducible versus repressible regulation. When the genes in an operon encode proteins that function in the catabolism or breakdown of a substance, they are usually regulated in an inducible manner. The substance to be broken down or a related compound often acts as the inducer. For example, allolactose and arabinose act as inducers of the *lac* and *ara* operons, respectively. An inducible form of regulation allows the bacterium to phenotypically express the appropriate genes only when they are needed to catabolize these sugars.

In contrast, other enzymes are important for the anabolism or synthesis of small molecules. The genes that encode these anabolic enzymes tend to be regulated by a repressible mechanism. The corepressor or inhibitor is commonly the small molecule that is the product of the enzymes' biosynthetic activities. For example, tryptophan is produced by the sequential action of several enzymes that are encoded by the *trp* operon. Tryptophan itself acts as a corepressor that can bind to the trp repressor protein when the intracellular levels of tryptophan become relatively high. This mechanism turns off the genes required for tryptophan biosynthesis when enough of this amino acid has been made. Therefore, genetic regulation via repression provides the bacterium with a way to prevent the overproduction of the product of a biosynthetic pathway.

14.2 TRANSLATIONAL AND POSTTRANSLATIONAL REGULATION

Though genetic regulation in bacteria is exercised predominantly at transcription, many examples are known in which regulation is exerted at a later stage in gene expression. In some cases, specialized mechanisms have evolved to regulate the translation of certain mRNAs. Recall that the translation of mRNA occurs in three stages: initiation, elongation, and termination. Genetic regulation of translation is usually aimed at preventing the initiation step.

The net result of translation is the synthesis of a protein. The activities of proteins within living cells and organisms ultimately determine an individual's traits. Therefore, to fully understand how proteins influence an organism's traits, researchers have investigated how the functions of proteins are regulated. The term **posttranslational** regulation refers to the functional control of proteins that are already present in the cell rather than regulation of transcription or translation. Posttranslational control can either activate or inhibit the function of a protein. Compared to transcriptional or translational regulation, posttranslational control can be relatively fast, occurring in a matter of seconds, which is an important advantage. In contrast, transcriptional and translational regulation typically require several minutes or even

hours to take effect because these two mechanisms involve the synthesis and turnover of mRNA and polypeptides. In this section, we will examine some of the ways that bacteria can regulate the initiation of translation, as well as ways that protein function can be regulated posttranslationally.

Repressor Proteins and Antisense RNA Can Inhibit Translation

For some bacterial genes, the translation of mRNA is regulated by the binding of proteins or other RNA molecules that influence the ability of ribosomes to translate the mRNA into a polypeptide. A **translational regulatory protein** recognizes sequences within the mRNA, much as transcription factors recognize DNA sequences. In most cases, translational regulatory proteins act to inhibit translation. These are known as **translational repressors.** When a translational repressor protein binds to the mRNA, it can inhibit translational initiation in one of two ways. One possibility is that it can bind in the vicinity of the Shine-Dalgarno sequence and/or the start codon and thereby sterically block the ribosome's ability to initiate translation in this region. Alternatively, the repressor protein may bind outside the Shine-Dalgarno/start codon region but stabilize an mRNA secondary structure that prevents initiation. Translational repression is also a form of genetic regulation found in eukaryotic species, and we will consider specific examples in Chapter 15.

A second way to regulate translation is via the synthesis of **antisense RNA,** an RNA strand that is complementary to a strand of mRNA. (The mRNA strand has the same sequence as the DNA sense strand.) To understand this form of genetic regulation, let's consider a trait known as osmoregulation, which is essential for the survival of most bacteria. Osmoregulation refers to the ability to control the amount of water inside the cell. Because the solute concentrations in the external environment may rapidly change between hypotonic and hypertonic conditions, bacteria must have an osmoregulation mechanism to maintain their internal cell volume. Otherwise, bacteria would be susceptible to the harmful effects of lysis or shrinking.

In *E. coli*, an outer membrane protein encoded by the *ompF* gene is important in osmoregulation. At low osmolarity, the ompF protein is preferentially produced, whereas at high osmolarity, its synthesis is decreased. The expression of another gene, known as *micF*, is responsible for inhibiting the expression of the *ompF* gene at high osmolarity. As shown in **Figure 14.16**, the inhibition occurs because the *micF* RNA is complementary to the *ompF* mRNA; it is an antisense strand of RNA. When the *micF* gene is transcribed, its RNA product binds to the *ompF* mRNA via hydrogen bonding between their complementary regions. The binding of the *micF* RNA to the *ompF* mRNA prevents the *ompF* mRNA from being translated. The RNA transcribed from the *micF* gene is called antisense RNA because it is complementary to the *ompF* mRNA, which is a sense strand of mRNA that encodes a protein. The *micF* RNA does not encode a protein.

FIGURE 14.16 **The double-stranded RNA structure formed between the *micF* antisense RNA and the *ompF* mRNA.** Because they have regions that are complementary to each other, the *micF* antisense RNA binds to the *ompF* mRNA to form a double-stranded structure in which the *ompF* mRNA cannot be translated.

Posttranslational Regulation Can Occur Via Feedback Inhibition and Covalent Modifications

Let's now turn our attention to ways that protein function is regulated posttranslationally. A common mechanism to regulate the activity of metabolic enzymes is **feedback inhibition.** The synthesis of many cellular molecules such as amino acids, vitamins, and nucleotides occurs via the action of a series of enzymes that convert precursor molecules to a particular product. The final product in a metabolic pathway then inhibits an enzyme that acts early in the pathway.

Figure 14.17 depicts feedback inhibition in a metabolic pathway. Enzyme 1 is an example of an **allosteric enzyme,** an enzyme that contains two different binding sites. (The lac repressor is an allosteric protein, but not an enzyme.) The catalytic site is responsible for the binding of the substrate and its conversion to intermediate 1. The second site is a regulatory or allosteric site. This site binds the final product of the metabolic pathway. When bound to the regulatory site, the final product inhibits the catalytic ability of enzyme 1.

To appreciate feedback inhibition at the cellular level, we can consider the relationship between the product concentration and the regulatory site on enzyme 1. As the final product is made within the cell, its concentration will gradually increase. Once the final product concentration has reached a level that is similar to its affinity for enzyme 1, the final product is likely to bind to the regulatory site on enzyme 1 and inhibit its function. In this way, the net result is that the final product of a metabolic pathway inhibits the further synthesis of more product. Under these conditions, the concentration of the final product has reached a level sufficient for the purpose of the bacterium.

A second strategy to control the function of proteins is by the covalent modification of their structure, a process called **posttranslational covalent modification.** Certain types of modifications are involved primarily in the assembly and construction of a functional protein. These alterations include proteolytic processing; disulfide bond formation; and the attachment of prosthetic groups, sugars, or lipids. These are typically irreversible

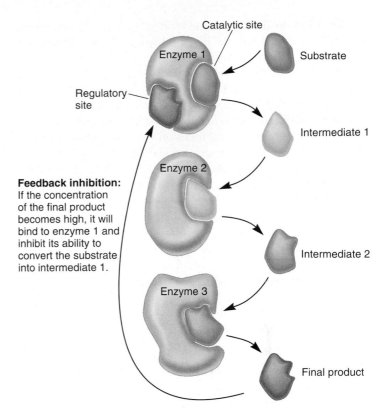

Feedback inhibition: If the concentration of the final product becomes high, it will bind to enzyme 1 and inhibit its ability to convert the substrate into intermediate 1.

FIGURE 14.17 **Feedback inhibition in a metabolic pathway.** The substrate is converted to a product by the sequential action of three different enzymes. Enzyme 1 has a catalytic site that recognizes the substrate, and it also has a regulatory site that recognizes the final product. When the final product binds to the regulatory site, it inhibits enzyme 1.

changes required to produce a functional protein. In contrast, other types of modifications, such as phosphorylation ($-PO_4$), acetylation ($-COCH_3$), and methylation ($-CH_3$), are often reversible modifications that transiently affect the function of a protein.

14.3 GENE REGULATION IN THE BACTERIOPHAGE REPRODUCTIVE CYCLE

Viruses are small particles that contain genetic material surrounded by a protein coat. They can infect a living cell and then propagate themselves by using the energy and metabolic machinery of the host cell. For this to occur, the genetic material of viruses orchestrates an intricate series of steps, involving the expression of many viral genes. During the past several decades, the reproduction of viruses has presented an interesting and challenging problem for geneticists to investigate. The study of bacteriophages, which are viruses that infect bacteria, has greatly advanced our basic knowledge of how genetic regulatory proteins work. In addition, the study of viruses has been instrumental in our ability to devise medical strategies aimed at combating viral diseases. For example, our knowledge of the reproductive cycles of human viruses has led to the development of drugs that inhibit viral growth. Azidothymidine (AZT), which is used to combat human immunodeficiency virus (HIV), suppresses the production of viral DNA by inhibiting a viral gene product called reverse transcriptase that is involved in viral DNA synthesis.

In this section, we will focus on the function of bacteriophage genes that encode genetic regulatory proteins. The structural genes of bacteriophages are often arranged in operons. This enables all of the genes within an operon to be controlled by regulatory proteins that bind to operator sites and influence the function of nearby promoters. Like bacterial operons, phage operons can be controlled by repressor proteins or activator proteins. To understand how this works, we will carefully examine the two reproductive cycles—lytic and lysogenic—of a virus called phage λ (lambda), which was discovered by André Lwoff and his colleagues in the 1940s. Since its discovery, phage λ has been investigated extensively and has provided geneticists with a model on which to base our understanding of viral proliferation.

Phage λ Can Follow a Lytic or Lysogenic Cycle

Phage λ can bind to the surface of a bacterium and inject its genetic material into the bacterial cytoplasm. After this occurs, the phage proceeds along only one of two alternative cycles,

known as the **lytic cycle** and the **lysogenic cycle.** This topic is also discussed in Chapter 6 (see Figure 6.9). During the lytic cycle, the genetic instructions of the bacteriophage direct the synthesis of many copies of the phage genetic material and coat proteins that are then assembled to make new phages. When synthesis and assembly are completed, the bacterial host cell is lysed, and the newly made phages are released into the environment.

Alternatively, in the lysogenic cycle, the phage can act as a **temperate phage,** which will usually not produce new phages and will not kill the bacterial cell that acts as its host. During the lysogenic cycle, phage λ integrates its genetic material into the chromosome of the bacterium. This integrated phage DNA is known as a **prophage.** A prophage can exist in a dormant state for a long time, during which no new bacteriophages are made. When a bacterium containing a lysogenic prophage divides to produce two daughter cells, it will copy the prophage's genetic material along with its own chromosome. Therefore, both daughter cells inherit the prophage. At some later time, in a process called induction, a prophage may become activated to excise itself from the bacterial chromosome and enter the lytic cycle. During induction, the phage promotes the synthesis of new phages and eventually lyses the host cell.

Inside the virus's head, phage λ DNA is linear. After injection into the bacterium, the two ends of the DNA become covalently attached to each other to form a circular piece of DNA. **Figure 14.18** shows the genome of phage λ. The organization of

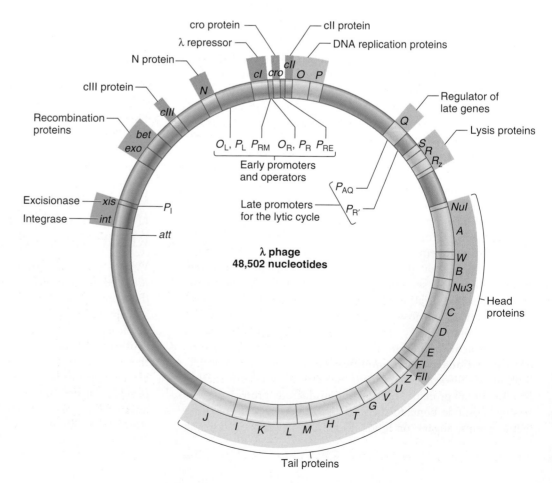

FIGURE 14.18   The genome of phage λ. The genes shaded in orange encode genetic regulatory proteins that determine whether the lysogenic or lytic cycle prevails. Genes shaded in green encode proteins necessary for the lysogenic cycle. The genes shaded in dark and light tan encode proteins required for the lytic cycle.

the genes within this circular structure reflects the two alternative cycles of this virus. The genes in the top center of the figure are transcribed very soon after infection. This occurs at the beginning of either reproductive cycle. As we will see, the expression pattern of these early genes determines whether the lytic or lysogenic cycle prevails.

The genes on the upper left side of Figure 14.18 encode proteins needed for the lysogenic cycle, while the genes on the right side and bottom promote the lytic cycle. If the lysogenic cycle prevails, the integrase (*int*) gene is subsequently turned on. The integrase gene encodes an enzyme that integrates the λ DNA into the bacterial chromosome. (The mechanism of phage λ integration is described in Chapter 17.) If the lysogenic cycle is not chosen, the genes on the right side and bottom of the figure will be transcribed. These genes are necessary for λ assembly and release. They encode replication proteins, coat proteins, proteins involved in coat assembly, proteins involved in packaging the DNA into the phage head, and enzymes that cause the bacterium to lyse.

During the Lysogenic Cycle, the cII Protein Activates Expression of the λ Repressor

Now that we have an understanding of the phage λ reproductive cycles and genome organization, let's examine how the decision is made between the lytic and lysogenic cycles. This choice depends on the actions of several genetic regulatory proteins. Our molecular understanding of the phage λ cycles represents an extraordinary accomplishment in the field of genetic regulation. This process is quite detailed because it involves a series of intricate steps in which regulatory proteins bind to several different sites in the λ genome. To simplify these events, we will begin by considering the steps that occur when the lysogenic cycle prevails.

Soon after the λ DNA enters the bacterial cell, two promoters, designated P_L and P_R, are used for transcription. This initiates a competition between the lytic and lysogenic cycles (**Figure 14.19**). Initially, transcription from P_L and P_R results in the synthesis of two short RNA transcripts that encode two proteins called the N protein and the cro protein, both of which are genetic regulatory proteins. The N protein is a genetic regulatory protein with an interesting function that we have not yet considered. Its function, known as **antitermination,** is to prevent transcriptional termination. The N protein inhibits termination at three sites, designated t_L, t_{R1}, and t_{R2}. The N protein actually binds to RNA polymerase and prevents transcriptional termination when these sites are being transcribed. When the N protein prevents termination at t_{R1} and t_{R2}, the transcript from P_R is extended to include the *cII*, *O*, *P*, and *Q* genes. The *cII* gene encodes an activator protein, the *O* and *P* genes encode enzymes needed for the initiation of λ DNA synthesis, and the *Q* gene encodes another antiterminator that is required for the lytic cycle. When the N protein prevents termination at t_L, the transcript from P_L is extended to include the *int*, *xis*, and *cIII* genes. The *int* gene encodes integrase, which is involved with integrating λ DNA into the *E. coli* chromosome, and the *xis* gene encodes excisionase, which can excise the λ DNA if a switch is made from the lysogenic to the lytic cycle. The *cIII* gene encodes the cIII protein, which forms a complex with

cII that helps to stabilize the cII activator protein and make it less vulnerable to protease digestion.

As shown on the left side of Figure 14.19, if the cII-cIII complex accumulates to sufficient levels, the lysogenic cycle is favored. Once it is made, the cII protein activates two different promoters in the λ genome. When the cII protein binds to the promoter P_{RE}, it turns on the transcription of *cI*, a gene that encodes the λ repressor. The cII protein also activates the *int* gene by binding to the promoter P_I. The λ repressor and integrase proteins play central roles in promoting the lysogenic cycle. When the λ repressor is made in sufficient quantities, it binds to operator sites (O_L and O_R) that are adjacent to P_L and P_R. When the λ repressor is bound to O_R, it inhibits the expression of the genes required for the lytic cycle.

As you are looking at the left side of Figure 14.19, you may have noticed that the binding of the λ repressor to O_R will inhibit the expression of *cII*. This may seem counterintuitive, because the cII protein was initially required to activate the *cI* gene that encodes the λ repressor. You may be thinking that the inhibition of the *cII* gene will eventually prevent the expression of the *cI* gene and ultimately stop the synthesis of the λ repressor protein. What prevents the inhibition of *cI* gene expression? The explanation is that the *cI* gene actually has two promoters: P_{RE}, which is activated by the cII protein, and a second promoter called P_{RM}. Transcription from P_{RE} occurs early in the lysogenic cycle. P_{RE} gets its name because the use of this promoter results in the expression of the λ Repressor during the Establishment of the lysogenic cycle. The transcript made from the use of P_{RE} is very stable and quickly leads to a buildup of the λ repressor protein. This causes an abrupt inhibition of the lytic cycle because the binding of the λ repressor protein to O_R blocks the P_R promoter. Later in the lysogenic cycle, it is no longer necessary to make a large amount of the λ repressor. At this point, the use of the P_{RM} promoter is sufficient to make enough Repressor protein to Maintain the lysogenic cycle. Interestingly, the P_{RM} promoter is activated by the λ repressor protein. The λ repressor was named when it was understood that it repressed the lytic cycle. Later studies revealed that it also activates its own transcription from P_{RM}.

The Lytic Cycle Depends on the Action of the Cro Protein

As we have just seen, the λ repressor protein binds to O_R and prevents the expression of the operons needed for the lytic cycle. For the lytic cycle to occur, the λ repressor must be prevented from inhibiting P_R. This is the role of the cro protein. If the activity of the cro protein exceeds the activity of the cII protein, the lytic cycle prevails (see the right side of Figure 14.19). As was mentioned, an early step in the expression of λ genes is the transcription from P_R to produce the cro protein. If the concentration of the cro protein builds to sufficient levels, it will bind to two operator regions, O_L and O_R. The binding of cro to O_L inhibits transcription from P_L; the binding of cro to O_R has several effects. When the cro protein binds to O_R, it inhibits transcription from P_{RM} in the leftward direction. This inhibition prevents the expression of the *cI* gene, which encodes the λ repressor; the λ repressor is

FIGURE 14.19 **The sequence of events that occur during the lysogenic and lytic cycles of phage λ.** The top part of this figure shows the region of the phage λ genome that regulates the choice between the lytic and lysogenic cycles. DNA is shown in blue. The names of genes that encode proteins are shown above the DNA. Promoters and operator sites are shown below the DNA. RNA transcripts are shown in red or green. The key regulatory proteins are indicated as spheres.

Immediately after infection, P_L and P_R are used to make two short mRNAs. These mRNAs encode two early proteins, designated N and cro. The N protein prevents transcriptional termination at three sites in the RNA (t_L, t_{R1}, and t_{R2}). This allows the transcription of several genes, which include *cIII*, *cII*, *O*, *P*, and *Q*. The expression of the *O*, *P*, and *Q* genes is necessary only for the lytic cycle. If the lysogenic cycle is chosen, transcription of these genes is abruptly inhibited.

The left side shows the steps that lead to the lysogenic cycle. The cII-cIII protein complex binds to P_{RE}. This activates the transcription of the *cI* gene, which encodes the λ repressor. The cII-cIII complex also binds to P_I to activate the transcription of the *int* gene. The λ repressor binds to O_L and O_R to inhibit transcription from P_L and P_R. This prevents the lytic cycle. The integrase protein catalyzes the integration of the λ DNA into the *E. coli* chromosome. At a later stage in the lysogenic cycle, the *cI* gene is transcribed from the P_{RM} promoter. This results in a low but steady synthesis of the λ repressor, which is necessary to further maintain the lysogenic state.

The right side shows the steps that lead to the lytic cycle. The cro protein binds to O_L and blocks transcription from P_L. Cro also binds to O_R and blocks transcription from P_{RM}. This prevents the synthesis of the λ repressor. Transcription from P_R is still allowed to occur at a low level, which leads to a buildup of the O, P, and Q proteins. The O and P proteins catalyze the replication of additional λ DNA. The Q protein allows transcription through $P_{R'}$ (not shown in this figure, but see Figure 14.18). This leads to the synthesis of many proteins that are necessary to make new λ phages and to lyse the cell.

Genes→Traits The ability to choose between two alternative reproductive cycles can be viewed as a trait of this bacteriophage. As described here, the choice between the two cycles depends on the pattern of gene regulation.

needed to maintain the lysogenic state. Therefore, the λ repressor cannot successfully shut down transcription from P_R.

The binding of the cro protein to O_R also allows a low level of transcription from P_R in the rightward direction. This enables the transcription of the O, P, and Q genes. The O and P proteins are necessary for the replication of the λ DNA. The Q protein is an antiterminator protein that permits transcription through another promoter, designated $P_{R'}$ (not to be confused with P_R). The $P_{R'}$ promoter controls a very large operon that encodes the proteins necessary for the phage coat, the assembly of the coat proteins, the packaging of the λ DNA, and the lysis of the bacterial cell (refer back to Figure 14.18). These proteins are made toward the end of the lytic cycle. The expression of these late genes leads to the synthesis and assembly of many new λ phages that are released from the bacterial cell when it lyses.

Cellular Proteases Influence the Choice Between the Lytic and Lysogenic Cycle

What factors determine whether the lytic or lysogenic cycle prevails? As you may have noticed from Figure 14.19, the first two steps of both cycles are identical. Whether the lytic or lysogenic cycle prevails depends on the steps that occur after the early genes are transcribed. In particular, the activity of the cII protein plays a key role in directing λ to one of the two cycles. A critical physiological issue is that the cII protein is easily degraded by cellular proteases that are produced by *E. coli*. Whether or not these proteases are made depends on the environmental conditions. If the growth conditions are very favorable, such as a rich growth medium, the intracellular protease levels are relatively high, and the cII protein tends to be degraded. When cII protein is degraded, P_{RE} cannot be activated, and therefore, the λ repressor is not made. Instead, the cro protein slowly accumulates to sufficient levels, as in the right side of Figure 14.19. The binding of the cro protein to O_R prevents transcription of the λ repressor gene from P_{RM} and, at the same time, allows the lytic cycle to proceed. In this way, environmental conditions that are favorable

for growth promote the lytic cycle. This makes sense, because a sufficient supply of nutrients is necessary to synthesize new bacteriophages.

Alternatively, starvation conditions favor the lysogenic cycle. When nutrients are limiting, cellular proteases are relatively inactive. Under these conditions, the cII protein will build up much more quickly than the cro protein. Therefore, the cII protein will turn on P_{RE} and lead to the transcription of the λ repressor (as in the left side of Figure 14.19). This event favors the lysogenic cycle. From the perspective of the bacteriophage, lysogeny may be a better choice under starvation conditions because nutrients may be insufficient for the production of new λ phages.

After lysogeny is established, certain environmental conditions can also favor induction to the lytic cycle at some later time. For example, exposure to UV light promotes induction. This also is caused by the activation of cellular proteases. In this case, a cellular protein known as recA (a protein ordinarily involved in facilitating recombination between DNA molecules) detects the DNA damage from UV light and is activated to become a mediator of protein cleavage. RecA protein mediates cleavage of the λ repressor and thereby inactivates it. This allows transcription from P_R and eventually leads to the accumulation of the cro protein and thereby favors the lytic cycle. Under these conditions, it may be advantageous for λ to make new phages and lyse the cell, because the exposure to UV light may have already damaged the bacterium to the point where further bacterial growth and division are prevented.

The O_R Region Provides a Genetic Switch Between the Lytic and Lysogenic Cycles

Before we end this section on the λ reproductive cycles, let's consider how the O_R region acts as a genetic switch between the two cycles. Depending on the binding of genetic regulatory proteins to this region, the switch can be turned to favor the lytic or lysogenic cycle. How does this switch work? To understand how, we need to take a closer look at the O_R region (**Figure 14.20**).

FIGURE 14.20 **The O_R region, the genetic switch between the lysogenic and lytic cycles.** The O_R region contains three operator sites and two promoters, P_{RM} and P_R, which transcribe in opposite directions. The left side of the figure depicts the events that promote the lysogenic cycle. The λ repressor protein is a dimer; each subunit contains two globular domains connected by a short link. The λ repressor protein dimer first binds to O_{R1} because it has the highest affinity for this site. A second λ repressor dimer binds to O_{R2}. This occurs very rapidly, because the binding of the first dimer to O_{R1} favors the binding of a second dimer to O_{R2}. The binding of the λ repressor to O_{R1} and O_{R2} inhibits transcription from P_R and thereby switches off the lytic cycle. Early in the lysogenic cycle, the λ repressor protein concentration may become so high that it occupies O_{R3}. Later, the λ repressor concentration begins to drop, because the inhibition of P_R decreases the synthesis of cII protein, which activates the λ repressor gene (from P_{RE}). As the λ repressor concentration gradually falls, it is first removed from O_{R3}. This allows transcription from P_{RM} and maintains the lysogenic cycle.

The right side depicts events at the O_R region that promote the lytic cycle. The cro repressor protein is a small globular protein that also binds to each operator site as a dimer. The cro protein has its highest affinity for O_{R3}, and so it binds there first. This blocks transcription from P_{RM} and thereby switches off the lysogenic cycle. The cro protein has a similar affinity for O_{R2} and O_{R1}, and so it may occupy either of these sites next. Later in the lytic cycle, the cro protein concentration will continue to rise, so eventually it binds to both O_{R2} and O_{R1}. This turns down the expression from P_R, which is not needed in the later stages of the lytic cycle.

The O_R region contains three operator sites, designated O_{R1}, O_{R2}, and O_{R3}. These operator sites control two promoters called P_{RM} and P_R that transcribe in opposite directions. The λ repressor protein or the cro protein can bind to any or all of the three operator sites. The binding of these two proteins at these sites governs the switch between the lysogenic and lytic cycles. Two critical issues influence this binding event. The first is the relative affinities that the regulatory proteins have for these operator sites. The second is the concentrations of the λ repressor protein and the cro protein within the cell.

Let's consider how an increasing concentration of the λ repressor protein can switch on the lysogenic cycle and switch off the lytic cycle (Figure 14.20, left side). This protein was first isolated by Mark Ptashne and his colleagues in 1967. Their studies showed that λ repressor binds with highest affinity to O_{R1}, followed by O_{R2} and O_{R3}. As the concentration of the λ repressor builds within the cell, a dimer of the λ repressor protein first binds to O_{R1} because it has the highest affinity for this site. Next, a second λ repressor dimer binds to O_{R2}. This occurs very rapidly, because the binding of the first dimer to O_{R1} favors the binding of a second dimer to O_{R2}. This is called a cooperative interaction. The binding of the λ repressor to O_{R1} and O_{R2} inhibits transcription from P_R and thereby switches off the lytic cycle.

Early in the lysogenic cycle, the λ repressor protein concentration may become so high that it occupies O_{R3}. Eventually, however, the λ repressor concentration begins to drop, because the inhibition of P_R decreases the synthesis of cII, which activates the λ repressor gene from P_{RE}. As the λ repressor concentration gradually falls, it is first removed from O_{R3}. This allows transcription from P_{RM}. As mentioned earlier, the term λ repressor is somewhat misleading because the binding of the λ repressor at only O_{R1} and O_{R2} acts as an activator of P_{RM}. The ability of the λ repressor to activate its own transcription allows the switch to the lysogenic cycle to be maintained.

In the lytic cycle (Figure 14.20, right side), the binding of the cro protein controls the switch. The cro protein has its highest affinity for O_{R3} and has a similar affinity for O_{R2} and O_{R1}. Under conditions that favor the lytic cycle, the cro protein accumulates, and a cro dimer first binds to O_{R3}. This blocks transcription from P_{RM} and thereby switches off the lysogenic cycle. Later in the lytic cycle, the cro protein concentration continues to rise, so eventually it binds to O_{R2} and O_{R1}. This turns down expression from P_R, which is not needed in the later stages of the lytic cycle.

Genetic switches, like the one just described for phage λ, represent an important form of genetic regulation. As we have just seen, a genetic switch can be used to control two alternative reproductive cycles of a bacteriophage. In addition, genetic switches are also important in the developmental pathways for bacteria and eukaryotes. For example, certain species of bacteria can grow in a vegetative state when nutrients are abundant but will produce spores when conditions are unfavorable for growth. The choice between vegetative growth and sporulation involves genetic switches. Likewise, genetic switches operate in the developmental pathways in eukaryotes. As we will discover in Chapter 23, they are key events in the initiation of cell differentiation during development. Studies of the phage λ life cycle have provided fundamental information with which to understand how these other switches can operate at the molecular level.

CONCEPTUAL SUMMARY

Genes can be regulated at any step in the pathway of gene expression. In bacterial cells, it is common for structural genes to be organized into a multigene unit called an operon. An operon typically consists of a promoter, an operator site, two or more structural genes, and a terminator. The organization of an operon allows two or more genes to be regulated as a single unit. In most cases, transcriptional regulation involves the actions of regulatory proteins that bind to the operon and can greatly influence the rate of transcription. A repressor is a protein that inhibits transcription, while an activator is a protein that increases the rate of transcription. Furthermore, the binding of small effector molecules to regulatory proteins can influence whether or not the regulatory proteins can bind to the DNA. Some regulatory proteins that exert negative control, such as the lac repressor and trp repressor, inhibit transcription. Others, such as the catabolite activator protein (CAP), exert positive control by enhancing transcription. The AraC protein is interesting because it can act as both a repressor and an activator, depending on the presence or absence of arabinose. Experiments involving the use of F' factors showed that regulatory proteins are synthesized and can diffuse within the cell to ultimately bind to a distant operator site and cause repression; this form of genetic regulation is called a *trans*-effect. In contrast, a *cis*-effect on gene expression is due to DNA sequences adjacent to the gene.

Our understanding of transcriptional regulation has been greatly aided by studying the reproductive cycles of viruses. In this chapter, we considered the ability of phage λ to proceed along one of two alternative cycles: the lytic and lysogenic cycles. This choice is determined by the actions of several genetic regulatory proteins that can bind to the λ DNA and influence the transcription of nearby genes. The O_R region provides a genetic switch for the lytic or lysogenic cycle. If the λ repressor controls the switch, the lysogenic cycle is favored, whereas binding of the cro protein to this switch favors the lytic cycle. Cellular proteases, which are influenced by environmental conditions, play a key role in the choice between the lytic and lysogenic cycles.

In addition to direct transcriptional regulation, gene expression in bacteria can be affected at later stages in the expression process. During attenuation of the *trp* operon, for example, transcription actually begins, but it is terminated before the entire mRNA is made; the length of the transcript is determined by the translation of the *trpL* gene. This form of regulation is unique to bacteria because they couple the processes of transcription and translation. The translation of a complete mRNA transcript can also be regulated. One form of translational regulation involves the binding of translational repressor proteins that prevent translational initiation. A second form involves the binding of antisense RNA to an mRNA to block translation. Finally, the function of proteins that have already been made can be influenced posttranslationally. Feedback inhibition involves the noncovalent binding of molecules to an allosteric site on a protein. This inhibition blocks the function of an enzyme that is required during an early step in a metabolic pathway. In addition, the function of proteins can be controlled by irreversible and reversible covalent modifications.

EXPERIMENTAL SUMMARY

Early insights into the molecular mechanisms of gene expression came from experiments involving lactose metabolism in *E. coli*. Jacob and Monod examined the phenomenon of enzyme adaptation and concluded that the proteins involved in lactose metabolism were made after the bacterium was exposed to lactose. This observation led them to investigate the genes involved in this regulation process. By constructing merozygotes, strains of bacteria containing F′ factor genes, they discovered that the *lacI* gene encodes a repressor protein that in the absence of lactose is bound to the operator site. More recently, studies by Müller-Hill revealed the existence of three operator sites in the *lac* operon. The crystal structure of the lac repressor tetramer bound to two operator sites has been determined.

A different form of genetic regulation, known as attenuation, was discovered by Yanofsky and colleagues. They identified mutant strains of bacteria that were missing the trp repressor yet still could repress the *trp* operon when tryptophan levels were high. Additional mutations in the *trpL* region destroyed the ability to attenuate transcription. An analysis of the DNA sequence in this region revealed the ability of the mRNA to form alternative stem-loop structures. Of the three possible stem-loop structures, only the 3–4 stem-loop structure acts as a transcription terminator and thereby prevents the further transcription of the *trp* operon.

An analysis of genetic sequences also revealed a translational method of gene regulation involving the synthesis of antisense RNA. In *E. coli*, a gene known as *micF* inhibits expression of the *ompF* gene. Researchers discovered that the *micF* RNA is complementary to the *ompF* mRNA. The binding of the *micF* RNA (the antisense RNA) to *ompF* mRNA (the sense strand of mRNA that encodes a protein) inhibits the translation of the *ompF* mRNA.

We concluded the chapter with a description of gene regulation in bacteriophage λ. Many of the same kinds of experimental approaches were used to elucidate the molecular mechanisms that underlie the choice between the lysogenic and lytic cycles. In particular, many researchers have studied how mutations in particular genes in the λ genome favor the lysogenic or lytic cycles. In addition, biochemical and biophysical studies have analyzed the detailed interactions between λ regulatory proteins and the operator sites that they recognize.

PROBLEM SETS & INSIGHTS

Solved Problems

S1. Researchers have identified mutations in the promoter region of the *lacI* gene that make it more difficult for the *lac* operon to be induced. These are called *lacI*^{Q} mutants, because a greater Quantity of lac repressor protein is made. Explain why an increased transcription of the *lacI* gene makes it more difficult to induce the *lac* operon.

Answer: An increase in the amount of lac repressor protein makes it easier to repress the *lac* operon. When the cells become exposed to lactose, allolactose levels slowly rise. Some of the allolactose binds to the lac repressor protein and causes it to be released from the operator site. If many more lac repressor proteins are found within the cell, more allolactose is needed to ensure that no unoccupied repressor proteins can repress the operon.

S2. Explain how the pausing of the ribosome in the presence or absence of tryptophan affects the formation of a terminator stem-loop.

Answer: The key issue is the location where the ribosome stalls. In the absence of tryptophan, it stalls over the Trp codons in the *trpL* mRNA. Stalling at this site shields region 1 in the attenuator region. Because region 1 is unavailable to hydrogen bond with region 2, region 2 hydrogen bonds with region 3. Therefore, region 3 cannot form a terminator stem-loop with region 4. Alternatively, if tryptophan levels in the cell are sufficient, the ribosome pauses over the stop codon in the *trpL* RNA. In this case, the ribosome shields region 2. Therefore, regions 3 and 4 hydrogen bond with each other to form a terminator stem-loop, which abruptly halts the continued transcription of the *trp* operon.

S3. With regard to the key proteins that affect the choice between the lytic and lysogenic cycles, which one(s) may be degraded by cellular proteases?

Answer: After infection, the key protein affected by cellular proteases is cII. If protease levels are high, as under good growth conditions, cII is degraded. This promotes the lytic cycle. Under starvation conditions, the protease levels are low. This prevents the degradation of cII and thereby promotes the lysogenic cycle. After lysogeny has been established, the key protein affected by cellular proteases is the λ repressor. Agents such as UV light activate cellular proteases that digest the λ repressor. This permits induction of the lytic cycle.

S4. In bacteria, it is common for two or more structural genes to be arranged together in an operon. Discuss the arrangement of genetic sequences within an operon. What is the biological advantage of an operon organization?

Answer: An operon contains several different DNA sequences that play specific roles. For transcription to take place, an operon is flanked by a promoter to signal the beginning of transcription and a terminator to signal the end of transcription. Two or more structural genes that encode different proteins are found between these two sequences. A key feature of an operon is that the expression of the structural genes occurs as a unit. When transcription takes place, a polycistronic mRNA is made that encodes all of the structural genes.

In order to control the ability of RNA polymerase to transcribe an operon, an additional DNA sequence, known as the operator site, is usually present. The base sequence within the operator site can serve as a binding site for genetic regulatory proteins called activators or repressors. The advantage of an operon organization is that it allows a bacterium to coordinately regulate a group of genes whose encoded proteins have a common function. For example, an operon may contain a group of genes involved in lactose breakdown or a group of genes involved in tryptophan synthesis. The genes within an operon usually encode proteins within a common metabolic pathway or cellular function.

S5. The sequential use of two sugars by a bacterium is known as diauxic growth. It is a common phenomenon among many bacterial species. When glucose is one of the two sugars available, it is typical that the bacterium metabolizes glucose first, and then a second sugar after the glucose has been used up. Among *E. coli* and related species, diauxic growth is regulated by intracellular cAMP levels and the catabolite activator protein (CAP). Summarize the effects of glucose and lactose on the ability of the lac repressor and the cAMP-CAP complex to regulate the *lac* operon.

Answer: In the absence of lactose, the lac repressor has the dominating effect of shutting off the *lac* operon. Even when glucose is also absent and the cAMP-CAP complex is formed, the presence of the bound lac repressor prevents the expression of the *lac* operon. The effects of the cAMP-CAP complex are exerted only in the presence of lactose. When lactose is present and glucose is absent, the cAMP-CAP complex acts to enhance the rate of transcription. However, when both lactose and glucose are present, the inability of CAP to bind to the *lac* operon decreases the rate of transcription. The table shown here summarizes these effects.

lac Operon Regulation

Sugar Present	Transcription of the *lac* Operon
None	The operon is turned off due to the dominating effect of the lac repressor protein.
Lactose	The operon is maximally turned on. The repressor protein is removed from the operator site, and the cAMP-CAP complex is bound to the CAP site.
Glucose	The operon is turned off due to the dominating effect of the lac repressor protein.
Lactose and glucose	The expression of the *lac* operon is greatly decreased. The lac repressor is removed from the operator site, and the catabolite activator protein is not bound to the CAP site. The absence of CAP at the CAP site makes it difficult for RNA polymerase to begin transcription. However, a little more transcription occurs under these conditions than in the absence of lactose, when the repressor is bound.

S6. The ability of DNA-binding proteins to promote a loop in DNA structure is an interesting phenomenon that is important in the structure and function of DNA. Besides the regulation of genes and operons, DNA looping is required in the compaction of DNA within the nucleoid of a bacterium and the nucleus of a eukaryotic cell (see Chapter 10). In addition, DNA looping is frequently involved in the expression of eukaryotic genes (see Chapter 15). In this solved problem, we will examine an experimental approach that made it possible for Robert Schleif and his colleagues to determine that the AraC protein causes a loop to form in the DNA. This work relied on the mobility of DNA in an acrylamide gel. A segment of DNA that contains a loop is more compact than the same DNA segment without a loop. Therefore, when these two alternative structures (looped versus unlooped) are run through a gel, the looped structure migrates more quickly to the bottom of the gel because it can more easily penetrate the gel matrix. As a starting material, Schleif had a sample of DNA that contained a portion of the *ara* operon including both the *araI* and *araO₂* sites. In the gel show here, this segment of DNA was exposed to the following conditions before it was run on the gel:

Lane 1. No further additions
Lane 2. Add AraC protein
Lane 3. Add arabinose
Lane 4. Add AraC protein and arabinose

Explain these results.

Answer: In lane 2, AraC protein was added and arabinose was not. As expected from the DNA looping hypothesis, a DNA loop was formed as evidenced from the faster mobility on the acrylamide gel. In lane 1, no AraC protein was added, so no DNA loop was able to form. These results confirm the idea that the AraC protein causes a loop to form in the sample loaded into lane 2. In lane 3, only the sugar arabinose is present but not the AraC protein. Because the DNA remains unlooped, the sugar by itself has no effect. In the sample loaded into lane 4, AraC protein was added, and arabinose was added as well. In this case, no DNA loop is observed. These results are consistent with the idea that arabinose binds to the AraC protein and breaks the DNA loop that is promoted by the AraC protein.

Conceptual Questions

C1. What is the difference between a constitutive gene and a regulated gene?

C2. In general, why is it important to regulate genes? Discuss examples of situations in which it would be advantageous for a bacterial cell to regulate genes.

C3. If a gene is repressible and under positive control, describe what kind of effector molecule and regulatory protein are involved. Explain how the binding of the effector molecule affects the regulatory protein.

C4. Transcriptional regulation often involves a regulatory protein that binds to a segment of DNA and a small effector molecule that binds to the regulatory protein. Do the following terms apply to a regulatory protein, a segment of DNA, or a small effector molecule?

A. Repressor   E. Activator

B. Inducer   F. Attenuator

C. Operator site   G. Inhibitor

D. Corepressor

C5. An operon is repressible—a small effector molecule turns off transcription. Which combinations of small effector molecules and regulatory proteins could be involved?

A. An inducer plus a repressor

B. A corepressor plus a repressor

C. An inhibitor plus an activator

D. An inducer plus an activator

C6. Some mutations have a *cis*-effect, whereas others have a *trans*-effect. Explain the molecular differences between *cis*- and *trans*-mutations. Which type of mutation (*cis* or *trans*) can be complemented in a merozygote experiment?

C7. What is enzyme adaptation? From a genetic point of view, how does it occur?

C8. In the *lac* operon, how would gene expression be affected if one of the following segments were missing?

A. *lac* operon promoter

B. Operator site

C. *lacA* gene

C9. If an abnormal repressor protein could still bind allolactose but the binding of allolactose did not alter the conformation of the repressor protein, how would this affect the expression of the *lac* operon?

C10. What is diauxic growth? Explain the roles of cAMP and the catabolite activator protein in this process.

C11. Mutations may have an effect on the expression of the *lac* operon, the *ara* operon, and the *trp* operon. Would the following mutations have a *cis*- or *trans*-effect on the expression of the structural genes in the operon?

A. A mutation in the operator site that prevents the lac repressor from binding to it

B. A mutation in the *lacI* gene that prevents the lac repressor from binding to DNA

C. A mutation in the *araC* gene that prevents two AraC proteins from binding to each other and forming a loop

D. A mutation in *trpL* that prevents attenuation

C12. Would a mutation that inactivated the lac repressor and prevented it from binding to the *lac* operator site result in the constitutive expression of the *lac* operon under all conditions? Explain. What is the disadvantage to the bacterium of having a constitutive *lac* operon?

C13. Describe the function of the AraC protein. How does it positively and negatively regulate the *ara* operon?

C14. Explain how a mutation would affect the regulation of the *ara* operon if the mutation prevented AraC protein from binding to the following sites:

A. *araO$_2$*

B. *araO$_1$*

C. *araI*

D. *araO$_2$* and *araI*

C15. What is meant by the term attenuation? Is it an example of gene regulation at the level of transcription or translation? Explain your answer.

C16. As described in Figure 14.14, four regions within the *trpL* gene can form stem-loop structures. Let's suppose that mutations have been previously identified that prevent the ability of a particular region to form a stem-loop structure with a complementary region. For example, a region 1 mutant cannot form a 1–2 stem-loop structure, but it can still form a 2–3 or 3–4 structure. Likewise, a region 4 mutant can form a 1–2 or 2–3 stem-loop but not a 3–4 stem-loop. Under the following conditions, would attenuation occur?

A. Region 1 is mutant, tryptophan is high, and translation is not occurring.

B. Region 2 is mutant, tryptophan is low, and translation is occurring.

C. Region 3 is mutant, tryptophan is high, and translation is not occurring.

D. Region 4 is mutant, tryptophan is low, and translation is not occurring.

C17. As described in Chapter 13, enzymes known as aminoacyl-tRNA synthetases are responsible for attaching amino acids to tRNAs. Let's suppose that tryptophanyl-tRNA synthetase was partially defective at attaching tryptophan to tRNA; its activity was only 10% of that found in a normal bacterium. How would that affect attenuation of the *trp* operon? Would it be more or less likely to be attenuated? Explain your answer.

C18. The 3–4 stem-loop and U-rich attenuator found in the *trp* operon (see Figure 14.14) is an example of ρ-independent termination. The function of ρ-independent terminators is described in Chapter 12. Would you expect attenuation to occur if the tryptophan levels were high and mutations at the end of the *trpL* gene changed the UUUUUUUU sequence to UGGUUGUC? Explain why or why not.

C19. Mutations in tRNA genes can create tRNAs that recognize stop codons. Because stop codons are sometimes called nonsense codons, these types of mutations that affect tRNAs are called nonsense suppressors. For example, a normal tRNA^{gly} has an anticodon sequence CCU that recognizes a glycine codon in mRNA (GGA) and puts in a glycine during translation. However, a mutation in the gene that encodes tRNA^{gly} could change the anticodon to ACU. This mutant tRNA^{gly} would still carry glycine, but it would recognize the stop codon UGA. Would this mutation affect attenuation of the *trp* operon? Explain why or why not. Note: To answer this question, you need to look carefully at Figure 14.14 and see if you can identify any stop codons that may exist beyond the UGA stop codon that is found after region 1.

C20. Translational control is usually aimed at preventing the initiation of translation. With regard to cellular efficiency, why do you think this is the case?

C21. What is antisense RNA? How does it affect the translation of a complementary mRNA?

C22. A species of bacteria can synthesize the amino acid histidine so that it does not require histidine in its growth medium. A key enzyme, which we will call histidine synthetase, is necessary for histidine biosynthesis. When these bacteria are given histidine in their growth media, they stop synthesizing histidine intracellularly. Based on this observation alone, propose three different regulatory mechanisms to explain why histidine biosynthesis ceases when histidine is in the growth medium. To explore this phenomenon further, you measure the amount of intracellular histidine synthetase protein when cells are grown in the presence and absence of histidine. In both conditions, the amount of this protein is identical. Which mechanism of regulation would be consistent with this observation?

C23. Using three examples, describe how allosteric sites are important in the function of genetic regulatory proteins.

C24. In what ways are the actions of the lac repressor and trp repressor similar and different? In other words, discuss the similarities and differences with regard to their binding to operator sites, their effects on transcription, and the influences of small effector molecules.

C25. Transcriptional repressor proteins (e.g., lac repressor), antisense RNA, and feedback inhibition are three different mechanisms that can turn off the expression of genes and gene products. Which of these three mechanisms would be most effective in each of the following situations?

A. Shutting down the synthesis of a polypeptide

B. Shutting down the synthesis of mRNA

C. Shutting off the function of a protein

For your answers in parts A–C that have more than one mechanism, which mechanism would be the fastest or the most efficient?

C26. What are key features that distinguish the lytic and lysogenic cycles?

C27. With regard to promoting the lytic or lysogenic cycle, what would happen if the following genes were missing from the λ genome?

A. *cro*

B. *cI*

C. *cII*

D. *int*

E. *cII* and *cro*

C28. How do the λ repressor and the cro protein affect the transcription from P_R and P_{RM}? Explain where these proteins are binding to cause their effects.

C29. In your own words, explain why it is necessary for the *cI* gene to have two promoters. What would happen if it had only P_{RE}?

C30. A mutation in P_R causes its transcription rate to be increased 10-fold. Do you think that this mutation would favor the lytic or lysogenic cycle? Explain your answer.

C31. When an *E. coli* bacterium already has a λ prophage integrated into its chromosome, another λ phage cannot usually infect the cell and establish the lysogenic or lytic cycle. Based on your understanding of the genetic regulation of the λ life cycles, why do you think the other phage would be unsuccessful?

C32. If a bacterium were exposed to a drug that inhibited the N protein, what would you expect to happen if the bacterium was later infected by phage λ? Would phage λ follow the lytic cycle, the lysogenic cycle, or neither? Explain your answer.

C33. Figure 14.20 shows a genetic switch that controls the choice between the lytic and lysogenic cycles of phage λ. What is a genetic switch? Compare the roles of a genetic switch and a simple operator site (like the one found in the *lac* operon) in gene regulation.

C34. This question combines your knowledge of conjugation (described in Chapter 6) and the genetic regulation that directs the phage λ reproductive cycles. When donor *Hfr* strains conjugate with recipient F⁻ bacteria that are lysogenic for phage λ, the conjugated cells survive normally. However, if donor *Hfr* strains that are lysogenic for phage λ conjugate with recipient F⁻ bacteria that do not contain any phage λ, the conjugated cells often lyse, due to the induction of λ into the lytic cycle. Based on your knowledge of the regulation of the two λ cycles, explain this observation.

Experimental Questions

E1. Answer the following questions that pertain to the experiment of Figure 14.7.

A. Why was β-ONPG used? Why was no yellow color observed in one of the four tubes? Can you propose alternative methods to measure the level of expression of the *lac* operon?

B. The optical density values were twice as high for the mated strain as for the parent strain. Why was this result obtained?

E2. Chapter 18 describes a blotting method known as Northern blotting, which can be used to detect RNA transcribed from a particular gene or a particular operon. In this method, a specific RNA is detected by using a short segment of cloned DNA as a probe. The DNA probe, which is radioactive, is complementary to the RNA that the researcher wishes to detect. After the radioactive probe DNA binds to the RNA within a blot of a gel, the RNA is visualized as a dark (radioactive) band on an X-ray film. For example, a DNA probe complementary to the mRNA of the *lac* operon could be used to specifically detect the *lac* operon mRNA on a gel blot. As shown here, the method of Northern blotting can be used to determine the amount of a particular RNA transcribed under different types of growth conditions. In this Northern blot, bacteria containing a normal *lac* operon were grown under different types of conditions, and then the mRNA was isolated from the cells and subjected to a Northern blot, using a probe that is complementary to the mRNA of the *lac* operon.

Lane 1. Growth in media containing glucose
Lane 2. Growth in media containing lactose
Lane 3. Growth in media containing glucose and lactose
Lane 4. Growth in media that doesn't contain glucose or lactose

Based on your understanding of the regulation of the *lac* operon, explain these results. Which is more effective at shutting down the *lac* operon, the binding of the *lac* repressor or the removal of CAP? Explain your answer based on the results shown in the Northern blot.

E3. As described in experimental question E2 and also in Chapter 18, the technique of Northern blotting can be used to detect the transcription of RNA. Draw the results you would expect from a Northern blot if bacteria were grown in media containing lactose (and no glucose) but had the following mutations:

Lane 1. Normal strain

Lane 2. Strain with a mutation that inactivates the lac repressor

Lane 3. Strain with a mutation that prevents allolactose from binding to the lac repressor

Lane 4. Strain with a mutation that inactivates CAP

How would your results differ if these bacterial strains were grown in media that did not contain lactose or glucose?

E4. An absentminded researcher follows the protocol described in Figure 14.7 and (at the end of the experiment) does not observe any yellow color in any of the tubes. Yikes! Which of the following mistakes could account for this observation?

A. Forgot to sonicate the cells

B. Forgot to add lactose to two of the tubes

C. Forgot to add β-ONPG to the four tubes

E5. Explain how the data shown in Figure 14.9 indicate that two operator sites are necessary for repression of the *lac* operon. What would the results have been if all three operator sites were required for the binding of the lac repressor?

E6. A mutant strain has a defective *lac* operator site that results in the constitutive expression of the *lac* operon. Outline an experiment you would carry out to demonstrate that the operator site must be physically adjacent to the genes that it influences. Based on your knowledge of the *lac* operon, describe the results you would expect.

E7. Let's suppose you have isolated a mutant strain of *E. coli* in which the *lac* operon is constitutively expressed. To understand the nature of this defect, you create a merozygote in which the mutant strain contains an F' factor with a normal lac operon and a normal *lacI* gene. You then compare the mutant strain and the merozygote with regard to their β-galactosidase activities in the presence and absence of lactose. You obtain the following results:

	Addition of Lactose	Amount of β-Galactosidase (percentage of mutant strain in the presence of lactose)
Mutant	No	100
Mutant	Yes	100
Merozygote	No	100
Merozygote	Yes	200

Explain the nature of the defect in the mutant strain.

E8. In the experiment of Figure 14.7, a *lacI⁻* mutant was conjugated to a strain that had a functional *lacI* gene on an F' factor. The results of this experiment were important in determining the action of the lac repressor protein. What results would you expect if you used the same approach to investigate the regulation of the *ara* operon via AraC? In other words, what would be the level of expression of the *ara* operon in an araC⁻ strain in the presence and absence of arabinose, and how would the level of expression change when a functional *araC* gene was introduced into the strain via conjugation?

E9. A segment of DNA that contains a loop is more compact than the same DNA segment without a loop. Therefore, when these two alternative structures (looped versus unlooped) are electrophoresed through a gel, the looped structure migrates more quickly to the bottom of the gel because it can more easily penetrate the gel matrix (see solved problem S6). Let's suppose that a mutant *E. coli* strain has been identified in which the AraC protein represses the *ara* operon, even in the presence of arabinose. In the experiment shown here, mutant or normal AraC protein was mixed with a segment of DNA containing the *ara* operon in the absence or presence of arabinose and then run on a gel.

Describe the defect in this mutant AraC protein.

Questions for Student Discussion/Collaboration

1. Discuss the advantages and disadvantages of genetic regulation at the different levels described in Figure 14.1.

2. As you look at Figure 14.10, discuss possible "molecular ways" that the cAMP-CAP complex and lac repressor may influence RNA polymerase function. In other words, try to explain how the bending and looping in DNA may affect the ability of RNA polymerase to initiate transcription.

3. Certain environmental conditions such as UV light are known to activate lysogenic λ prophages and cause them to progress into the lytic cycle. UV light initially causes the repressor protein to be proteolytically degraded. Make a flow diagram that describes the subsequent events that would lead to the lytic cycle. Note: The *xis* gene codes for an enzyme that is necessary to excise the λ prophage from the *E. coli* chromosome. The integrase enzyme is also necessary to excise the λ prophage.

Note: All answers appear at the website for this textbook; the answers to even-numbered questions are in the back of the textbook.

www.mhhe.com/brookergenetics3e

Visit the Online Learning Center for practice tests, answer keys, and other learning aids for this chapter. Enhance your understanding of genetics with our interactive exercises, quizzes, animations, and much more.

GENE REGULATION IN EUKARYOTES

Gene regulation provides many benefits to eukaryotic organisms, a category that includes protists, fungi, plants, and animals. Like their prokaryotic counterparts, eukaryotic cells need to adapt to changes in their environment. For example, eukaryotic cells can respond to changes in nutrient availability by enzyme adaptation, much as prokaryotic cells do. Eukaryotic cells also respond to environmental stresses such as ultraviolet (UV) radiation by inducing genes that provide protection against harmful environmental agents. An example is the ability of humans to develop a tan. The tanning response helps to protect a person's cells against the damaging effects of UV rays.

Among plants and animals, multicellularity and a more complex cell structure also demand a much greater level of gene regulation. The life cycle of complex eukaryotic organisms involves the progression through several developmental stages to achieve a mature organism. Some genes are expressed only during early stages of development, such as the embryonic stage, whereas others are expressed in the adult. In addition, complex eukaryotic species are composed of many different tissues that contain a variety of cell types. Gene regulation is necessary to ensure the differences in structure and function among distinct cell types. It is amazing that the various cells within a multicellular organism usually contain the same genetic material, yet phenotypically may look quite different. For example, the appearance of a human nerve cell seems about as similar to a muscle cell as an amoeba is to a paramecium. In spite of these phenotypic differences, a human nerve cell and muscle cell actually contain the same complement of human chromosomes. Nerve and muscle cells are strikingly different because of gene regulation rather than differences in DNA content. Many genes are expressed in the nerve cell and not the muscle cell, and vice versa.

The molecular mechanisms that underlie gene regulation in eukaryotes bear many similarities to the ways that bacteria regulate their genes. As in prokaryotes, regulation in eukaryotes can occur at any step in the pathway of gene expression (**Figure 15.1**). We will begin this chapter with an exploration of gene regulation at the level of transcription, an important form of control. In addition, research in the past few decades has revealed that eukaryotic organisms frequently regulate gene expression at points other than transcription. We will discuss some well-studied examples in which genes are regulated at many of these other control points.

Certain proteins, known as regulatory transcription factors, have the ability to bind into the major groove of DNA and regulate gene transcription.

**REGULATION OF
GENE EXPRESSION**

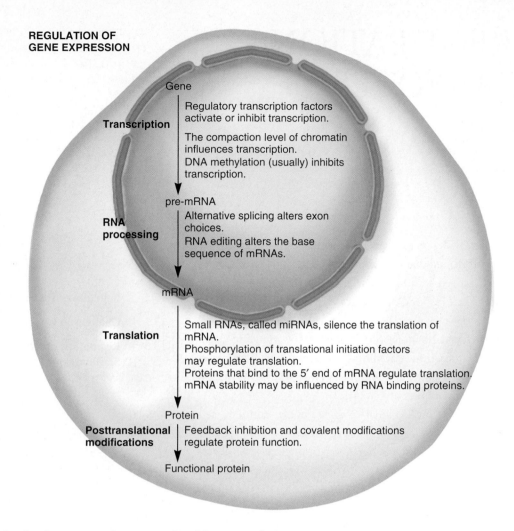

Gene

Transcription
Regulatory transcription factors
activate or inhibit transcription.

The compaction level of chromatin
influences transcription.
DNA methylation (usually) inhibits
transcription.

pre-mRNA

**RNA
processing**
Alternative splicing alters exon
choices.
RNA editing alters the base
sequence of mRNAs.

mRNA

Translation
Small RNAs, called miRNAs, silence the translation of
mRNA.
Phosphorylation of translational initiation factors
may regulate translation.
Proteins that bind to the 5′ end of mRNA regulate translation.
mRNA stability may be influenced by RNA binding proteins.

Protein

**Posttranslational
modifications**
Feedback inhibition and covalent modifications
regulate protein function.

Functional protein

FIGURE 15.1   **Levels of gene expression commonly subject to regulation.**

15.1 REGULATORY TRANSCRIPTION FACTORS

The term **transcription factor** is broadly used to describe proteins that influence the ability of RNA polymerase to transcribe a given gene. We will focus our attention on transcription factors that affect the ability of RNA polymerase to begin the transcription process. Such transcription factors can regulate the binding of the transcriptional apparatus to the core promoter and/or control the switch from the initiation to the elongation stage of transcription. Two categories of transcription factors play a key role in these processes. In Chapter 12, we considered **general transcription factors,** which are required for the binding of RNA polymerase to the core promoter and its progression to the elongation stage. General transcription factors are necessary for a basal level of transcription. In addition, eukaryotic cells possess a diverse array of **regulatory transcription factors** that serve to regulate the rate of transcription of target genes. The importance of transcription factors is underscored by the number of genes that encode this category of proteins. A sizeable portion of the genomes of complex organisms such as plants and animals is devoted to the process of gene regulation. For example, in *Arabidopsis thaliana,* a plant that is used as a model organism by plant geneticists, approximately 5% of its genome encodes transcription factors. This species has more than 1,500 different genes that encode proteins that regulate the transcription of other genes. Similarly, 2 to 3% of the genes in the human genome encode transcription factors.

As discussed in this section, regulatory transcription factors exert their effects by influencing the ability of RNA polymerase to begin transcription of a particular gene. They typically recognize *cis*-acting elements that are located in the vicinity of the core promoter. These DNA sequences are analogous to the operator sites found near bacterial promoters. In eukaryotes, these DNA sequences are generally known as **control elements** or **regulatory elements.** When a regulatory transcription factor binds to a regulatory element, it affects the transcription of an associated gene. For example, the binding of regulatory transcription factors may enhance the rate of transcription (**Figure 15.2a**). Such a transcription factor is termed an **activator,** and the sequence it binds to is called an **enhancer.** Alternatively, regulatory transcription factors may act as **repressors** by binding to elements called **silencers** and preventing transcription from occurring (**Figure 15.2b**).

(a) Gene activation

(b) Gene repression

FIGURE 15.2 **Overview of transcriptional regulation by regulatory transcription factors.** These proteins can act as either **(a)** an activator to increase the rate of transcription or **(b)** a repressor to decrease the rate of transcription.

By studying transcriptional regulation, researchers have discovered that most eukaryotic genes, particularly those found in multicellular species, are regulated by many factors. This phenomenon is called **combinatorial control** because the combination of many factors determines the expression of any given gene. At the level of transcription, the following are common factors that contribute to combinatorial control:

1. One or more activator proteins may stimulate the ability of RNA polymerase to initiate transcription.
2. One or more repressor proteins may inhibit the ability of RNA polymerase to initiate transcription.
3. The function of activators and repressors may be modulated in a variety of ways. These include the binding of small effector molecules, protein–protein interactions, and covalent modifications.
4. Activator proteins may promote the loosening of chromatin compaction in the chromosome where a gene is located, thereby making it easier for the gene to be recognized and transcribed by RNA polymerase.
5. DNA methylation may inhibit transcription, either by preventing the binding of an activator protein or by recruiting proteins that cause the chromatin to become more compact.

All five of these factors can contribute to the regulation of a single gene, or possibly only three or four will play a role. In most cases, transcriptional regulation is aimed at controlling the initiation of transcription at the promoter. In this section, we will survey the first three factors that affect transcription. In the following section, we will consider the fourth and fifth factors.

Structural Features of Regulatory Transcription Factors Allow Them to Bind to DNA

Genes that encode general and regulatory transcription factor proteins have been identified and sequenced from a wide variety of eukaryotic species, including yeast, plants, and animals. Several different families of evolutionarily related transcription factors have been discovered. In recent years, the molecular structures of transcription factor proteins have become an area of intense research. Transcription factor proteins contain regions, called **domains,** that have specific functions. For example, one domain of a transcription factor may have a DNA-binding function, while another may provide a binding site for a small effector molecule. When a domain or portion of a domain has a very similar structure in many different proteins, such a structure is called a **motif.**

Figure 15.3 depicts several different domain structures found in transcription factor proteins. The protein secondary structure known as an α helix is frequently found in transcription factors. Why is the α helix common in such proteins? The explanation is that the α helix is the proper width to bind into the major groove of the DNA double helix. In helix-turn-helix and helix-loop-helix motifs, an α helix called the recognition helix makes contact with and recognizes a base sequence along the major groove of the DNA (Figure 15.3a and b). Recall that the major groove is a region of the DNA double helix where the bases contact the surrounding water in the cell. Hydrogen bonding between an α helix and nucleotide bases is one way that a transcription factor can bind to a specific DNA sequence. In addition, the recognition helix often contains many positively charged amino acids (e.g., arginine and lysine) that favorably interact with the DNA backbone, which is negatively charged. Such basic domains are a common feature of many DNA-binding proteins.

A zinc finger motif is composed of one α helix and two β-sheet structures that are held together by a zinc (Zn^{2+}) metal ion (Figure 15.3c). The zinc finger also can recognize DNA sequences within the major groove.

A second interesting feature of certain motifs is that they promote protein dimerization. The leucine zipper (Figure 15.3d) and helix-loop-helix motif (Figure 15.3b) mediate protein dimerization. For example, Figure 15.3d depicts the dimerization and DNA binding of two proteins that have several leucine amino acids (a zipper). Alternating leucines in both proteins interact ("zip up"), resulting in protein dimerization. Two identical transcription factor proteins will come together to form a **homodimer,** or two different transcription factors can form a **heterodimer.** As discussed later, the dimerization of transcription factors can be an important way to modulate their function.

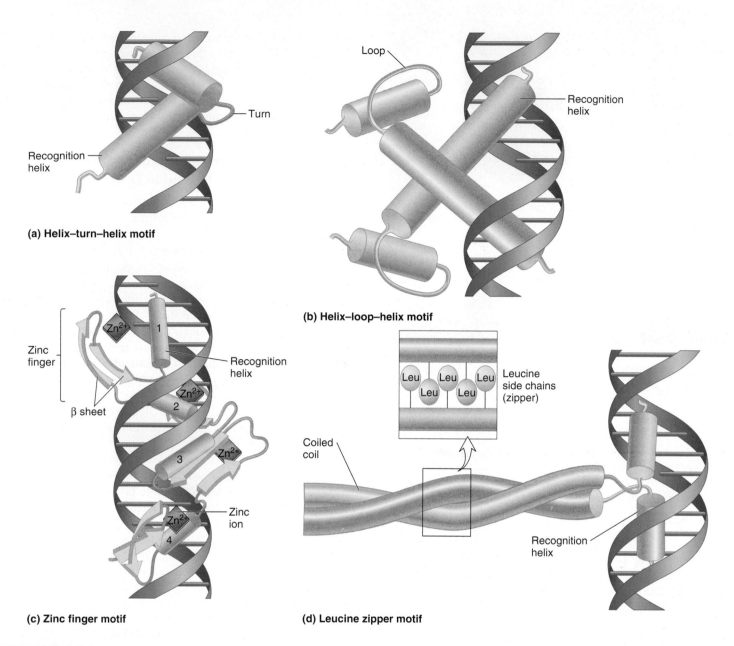

(a) Helix–turn–helix motif

(b) Helix–loop–helix motif

(c) Zinc finger motif

(d) Leucine zipper motif

FIGURE 15.3 **Structural motifs found in transcription factor proteins.** Certain types of protein secondary structure are found in many different transcription factors. In this figure, α helices are shown as cylinders and β sheets as flattened arrows. (**a**) Helix-turn-helix motif: Two α helices are connected by a turn. The α helices lie in the major groove of the DNA. (**b**) Helix-loop-helix motif: A short α helix is connected to a longer α helix by a loop. In this illustration, a dimer is formed from the interactions of two helix-loop-helix motifs, and the longer helices are binding to the DNA. (**c**) Zinc finger motif: Each zinc finger is composed of one α helix and two antiparallel β sheets. A zinc ion (Zn^{2+}), shown in red, holds the zinc finger together. This illustration shows four zinc fingers in a row. (**d**) Leucine zipper motif: The leucine zipper promotes the dimerization of two transcription factor proteins. Two α helices (termed a coiled coil) are intertwined via the leucines (see inset).

Regulatory Transcription Factors Recognize Regulatory Elements That Function as Enhancers or Silencers

As mentioned previously, when the binding of a regulatory transcription factor to a regulatory element increases transcription, the regulatory element is known as an enhancer. Such elements can stimulate transcription 10- to 1,000-fold; this is termed **up regulation.** Alternatively, regulatory elements that serve to inhibit transcription are called silencers, and their action is called **down regulation.**

Many regulatory elements are **orientation independent** or **bidirectional.** This means that the regulatory element can function in the forward or reverse direction. For example, if the forward orientation of an enhancer is

$$5'-GATA-3'$$
$$3'-CTAT-5'$$

this enhancer will also be bound by a regulatory transcription factor and enhance transcription even when it is oriented in the reverse direction:

$$5'-TATC-3'$$
$$3'-ATAG-5'$$

Striking variation is also found in the location of regulatory elements relative to a gene's promoter. Regulatory elements are often located in a region within a few hundred base pairs upstream from the promoter site. However, they can be quite distant from the promoter, even 100,000 base pairs away, yet exert strong effects on the ability of RNA polymerase to initiate transcription at the core promoter! Regulatory elements were first discovered by Susumu Tonegawa and coworkers in the 1980s. While studying genes that play a role in immunity, they identified a region far away from the core promoter, but needed for high levels of transcription to take place. In some cases, regulatory elements are located downstream from the promoter site and may even be found within introns, the noncoding parts of genes. As you may imagine, the variation in regulatory element orientation and location profoundly complicates the efforts of geneticists to identify the regulatory elements that affect the expression of any given gene.

Regulatory Transcription Factors May Exert Their Effects Through TFIID and Mediator

Different mechanisms have been discovered that explain how a regulatory transcription factor can bind to a regulatory element and thereby affect gene transcription. Indeed, more than one mechanism is typically involved. The net effect of a regulatory transcription factor is to influence the ability of RNA polymerase to transcribe a given gene. However, most regulatory transcription factors do not bind directly to RNA polymerase. How then do most regulatory transcription factors exert their effects? In many cases, their mechanism of action is to influence the function of RNA polymerase by interacting with other proteins that directly bind to RNA polymerase. Two protein complexes that communicate the effects of regulatory transcription factors are TFIID and mediator.

Figure 15.4 depicts how regulatory transcription factors may control transcription. In some cases, regulatory transcription factors bind to a regulatory element and then influence the function of TFIID (Figure 15.4a). As discussed in Chapter 12, **TFIID** is a general transcription factor that binds to the TATA box and is needed to recruit RNA polymerase to the core promoter. Activator proteins are expected to enhance the ability of TFIID to initiate transcription. One possibility is that activator proteins could help recruit TFIID to the TATA box or they could enhance the function of TFIID in a way that facilitates its ability to bind RNA polymerase. In contrast, repressors inhibit the function of TFIID. They could exert their effects by preventing the binding of TFIID to the TATA box or by inhibiting the ability of TFIID to recruit RNA polymerase to the core promoter.

A second way that regulatory transcription factors control RNA polymerase is via mediator, a protein complex discovered by Roger Kornberg and colleagues in 1990 (Figure 15.4b). The term **mediator** refers to the observation that it mediates the interaction between RNA polymerase and regulatory transcription factors. As discussed in Chapter 12, mediator controls the ability of RNA polymerase to progress to the elongation stage of transcription. Transcriptional activators stimulate the ability of mediator to facilitate the switch between the initiation and elongation stages, whereas repressors have the opposite effect. When a repressor protein interacts with mediator, RNA polymerase cannot progress to the elongation stage of transcription.

A third way that regulatory transcription factors can influence transcription is by recruiting proteins to the promoter region that affect chromatin compaction. For example, certain transcriptional activators can recruit proteins to the promoter region and thereby promote the conversion of chromatin from a closed to an open conformation. We will return to this topic later in this chapter.

The Function of Regulatory Transcription Factor Proteins Can Be Modulated in Three Ways

Thus far, we have considered the structures of regulatory transcription factors and the molecular mechanisms that account for their abilities to control transcription. The functions of the regulatory transcription factors themselves must also be modulated. Why is this necessary? The answer is that the genes they control must be turned on at the proper time, in the correct cell type, and under the appropriate environmental conditions. Therefore, eukaryotes have evolved different ways to modulate the functions of these proteins.

The functions of regulatory transcription factor proteins are controlled in three common ways, through (1) the binding of a small effector molecule, (2) protein–protein interactions, and (3) covalent modifications. **Figure 15.5** depicts these three mechanisms of modulating regulatory transcription factor function. Usually, one or more of these modulating effects are important in determining whether a transcription factor can bind to the DNA and/or influence transcription by RNA polymerase. For example, a small effector molecule may bind to a regulatory transcription factor and promote its binding to DNA (Figure 15.5a). We will see that steroid hormones function in this manner. Another important way is via protein–protein interactions (Figure 15.5b). The formation of homodimers and heterodimers is a fairly common means of controlling transcription. Finally, the function of a regulatory transcription factor can be affected by covalent modifications such as the attachment of a phosphate group (Figure 15.5c). As discussed later, the phosphorylation of activators can control their ability to stimulate transcription.

Steroid Hormones Exert Their Effects by Binding to a Regulatory Transcription Factor

Now that we have a general understanding regarding the structure and function of transcription factors, let's turn our attention to specific examples that illustrate how regulatory transcription factors carry out their roles within living cells. Our first example

The activator protein recruits TFIID to the core promoter and/or activates its function. Transcription will be activated.

The repressor protein inhibits the binding of TFIID or inhibits its function. Transcription is repressed.

(a) Regulatory transcription factors and TFIID

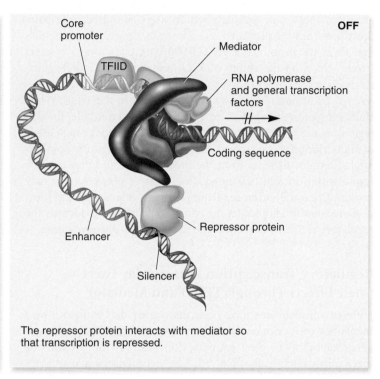

The activator protein interacts with mediator. This enables RNA polymerase to form a preinitiation complex that can proceed to the elongation phase of transcription.

The repressor protein interacts with mediator so that transcription is repressed.

(b) Regulatory transcription factors and mediator

FIGURE 15.4 **Ability of regulatory transcription factors to affect transcription.** (a) Some regulatory transcription factors exert their effects through TFIID. Transcriptional activators may recruit TFIID to the core promoter and/or activate its function (left), while repressors inhibit TFIID binding or its activity (right). (b) Regulatory transcription factors may also exert their effects via mediator. By interacting with mediator in different ways, activators stimulate transcription (left), while repressors inhibit transcription (right). In this example, the enhancer and silencer are relatively far away from the core promoter. Therefore, a loop must form in the DNA so that the regulatory transcription factor can interact with mediator.

is a category that responds to steroid hormones. This type of regulatory transcription factor is known as a **steroid receptor,** because the steroid hormone binds directly to the protein.

The ultimate action of a steroid hormone is to affect gene transcription. Steroid hormones act as signaling molecules that are synthesized by endocrine glands of animals and secreted into the bloodstream. The hormones are then taken up by cells that can respond to the hormones in different ways. For example, glucocorticoid hormones influence nutrient metabolism in most body cells. Other steroid hormones, such as estrogen and testosterone, are called gonadocorticoids because they influence the growth and function of the gonads.

Figure 15.6 shows the stepwise action of glucocorticoid hormones, which are produced in mammals. In this example, the hormone enters the cytosol of a cell by diffusing through the plasma membrane. Once inside, the hormone specifically binds to **glucocorticoid receptors.** Prior to hormone binding, the glucocorticoid receptor is complexed with proteins known as heat shock proteins (HSP), one example being HSP90. After the hormone binds to the glucocorticoid receptor, HSP90 is released. This exposes a nuclear localization signal (NLS)—a signal that directs a protein into the nucleus. Two glucocorticoid receptors form a homodimer and then travel through a nuclear pore into the nucleus.

(a) Binding of a small effector molecule such as a hormone

(b) Protein–protein interaction

(c) Covalent modification such as phosphorylation

FIGURE 15.5 Common ways to modulate the function of regulatory transcription factors. (**a**) The binding of an effector molecule such as a hormone may influence the ability of a transcription factor to bind to the DNA. (**b**) Protein–protein interactions among transcription factor proteins may influence their functions. (**c**) Covalent modifications such as phosphorylation may alter transcription factor function.

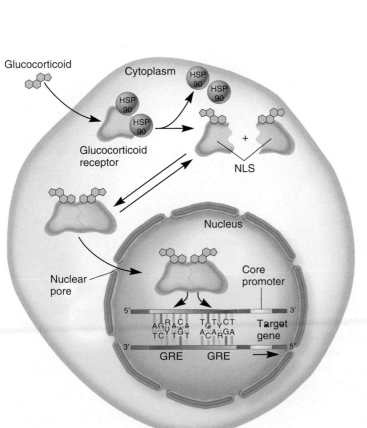

FIGURE 15.6 The action of glucocorticoid hormones. Once inside the cell, the glucocorticoid hormone binds to the glucocorticoid receptor, releasing it from a protein known as HSP90. This exposes a nuclear localization signal (NLS). Two glucocorticoid receptors then form a dimer and travel into the nucleus, where the dimer binds to glucocorticoid response elements (GREs) that are next to particular genes. The binding of the glucocorticoid receptors to the GREs activates the transcription of the adjacent target gene.

Genes→Traits Glucocorticoid hormones are produced by the endocrine glands in response to fasting and activity. They enable the body to regulate its metabolism properly. When glucocorticoids are produced, they are taken into cells and bind to the glucocorticoid receptors. This eventually leads to the activation of genes that encode proteins involved in the synthesis of glucose, the breakdown of proteins, and the mobilization of fats.

How does the glucocorticoid receptor regulate the expression of particular genes? In the nucleus, the glucocorticoid receptor homodimer binds to glucocorticoid response elements (GREs) with the following consensus sequence:

$$5'-AGRACA-3'$$
$$3'-TCYTGT-5'$$

where R is a purine and Y is pyrimidine. This sequence is found next to many genes and functions as an enhancer. The binding of the glucocorticoid receptor homodimer to adjacent GREs activates the transcription of the nearby gene, eventually leading to the synthesis of the encoded protein.

Mammalian cells usually have a large number of glucocorticoid receptors within the cytoplasm. Because GREs are located near dozens of different genes, the uptake of many hormone molecules can activate many glucocorticoid receptors and thereby stimulate the transcription of many different genes. For this reason, a cell can respond to the presence of the hormone in a very complex way. Glucocorticoid hormones stimulate many genes that encode proteins involved in several different cellular processes, including the synthesis of glucose, the breakdown of proteins, and the mobilization of fats. Although the genes are not physically connected to each other, the regulation of multiple genes via glucocorticoid hormones is much like the ability of bacterial operons to simultaneously control the expression of several genes.

The CREB Protein Is an Example of a Regulatory Transcription Factor Modulated by Covalent Modification

As we have just seen, steroid hormones function as signaling molecules that bind directly to regulatory transcription factors to alter their function. This enables a cell to respond to a hormone by up regulating a particular set of genes. Most extracellular signaling molecules, however, do not enter the cell or bind directly to transcription factors. Instead, most signaling molecules must bind to receptors in the plasma membrane. This binding activates the receptor and may lead to the synthesis of an intracellular signal that causes a cellular response. One type of cellular response is to affect the transcription of particular genes within the cell.

As our second example of regulatory transcription factor function within living cells, we will examine the **cAMP response element–binding (CREB) protein.** The CREB protein is a regulatory transcription factor that becomes activated in response to cell-signaling molecules that cause an increase in the cytoplasmic concentration of the molecule cyclic adenosine monophosphate (cAMP). This transcription factor recognizes a response element with the following consensus sequence.

$$5'-TGACGTCA-3'$$
$$3'-ACTGCAGT-5'$$

This response element, which is found near many different genes, has been termed a **cAMP response element (CRE).**

Figure 15.7 shows the steps leading to the activation of the CREB protein. A wide variety of hormones, growth factors, neurotransmitters, and other signaling molecules can bind to plasma membrane receptors to initiate an intracellular response. In this case, the response involves the production of a second messenger, cAMP. The extracellular signaling molecule itself is considered the primary messenger. When the signaling molecule binds to the receptor, it activates a G protein that subsequently activates the enzyme adenylyl cyclase. The activated adenylyl cyclase catalyzes the synthesis of cAMP. The cAMP molecule then binds to a second enzyme, protein kinase A, and activates it. This enzyme travels into the nucleus and phosphorylates several different cellular proteins, including the CREB protein. When phosphorylated,

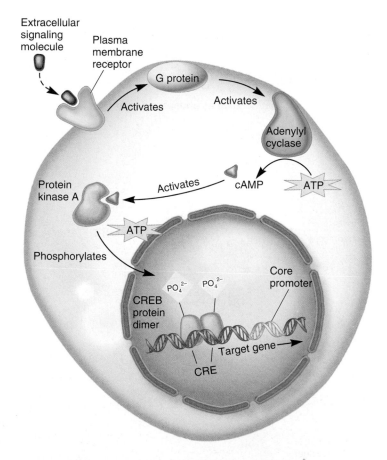

FIGURE 15.7   The activity of the CREB protein. When an extracellular signaling molecule binds to a receptor in the plasma membrane, this activates a G protein, which activates adenylyl cyclase, leading to the synthesis of cAMP. Next, cAMP binds to protein kinase A, which activates it. Protein kinase A then travels into the nucleus and phosphorylates the CREB protein. Once phosphorylated, the CREB protein acts as a transcriptional activator.

CREB proteins stimulate transcription. In contrast, unphosphorylated CREB proteins can still bind to CREs but do not activate RNA polymerase.

15.2 CHANGES IN CHROMATIN STRUCTURE

Changes in chromatin structure can involve changes in the level of chromosomal compaction and/or direct alterations in the structure of DNA. Because genes are segments of DNA, it is not surprising that such alterations can affect gene expression in a variety of ways (**Table 15.1**).

In eukaryotes, changes in chromatin compaction are required for gene transcription to take place. For this reason, the regulation of gene transcription is not simply dependent on the activity of regulatory transcription factors that influence RNA polymerase via TFIID and mediator to turn genes on or off. Transcription must also involve changes in chromatin structure that affect the ability of transcription factors to gain access to and bind their target sequences in the promoter region. If the chromatin is very tightly packed or in a **closed conformation,** transcription may be difficult or impossible. By comparison, chromatin that is in an **open conformation** is more easily accessible to transcription factors and RNA polymerase so that transcription can take place.

Gene regulation may also involve changes in DNA structure. **Gene amplification,** an increase in the number of copies of a gene, is a possible but not common method of gene regulation. Another change in DNA that affects gene expression is the rearrangement of the DNA structure. In Chapter 17, we will see that **gene rearrangement** occurs in the DNA of immune system cells to generate a diverse array of antibody proteins. A more common way that alterations of DNA structure affect gene regulation is **DNA methylation.**

In this section, we will begin by considering how the conversion between the closed and open conformations of chromatin is associated with the expression of genes. We will then examine molecular mechanisms that explain how chromosome compaction in the vicinity of specific genes is altered. Finally, we will examine how DNA methylation occurs and how it can regulate genes in a tissue-specific manner.

TABLE 15.1

Gene Regulation That Occurs via Changes in Chromatin or DNA Structure

Modification	Description
Chromatin compaction	This is a common mechanism of eukaryotic gene regulation. For a gene to be transcribed, it must be converted from a closed (highly compacted) conformation to an open conformation through the actions of enzymes such as histone acetyltransferases and ATP-dependent remodeling enzymes. Transcriptional activators recruit these enzymes to promoter regions to activate gene transcription.
Gene amplification	Though not a common mechanism of gene regulation, gene amplification involves an increase in gene number. For example, in the oocytes of *Xenopus laevis* (a South African frog), the number of rRNA genes is increased through gene amplification. The mechanism involves an unusual DNA replication event in which the rRNA genes are replicated to form many copies of circular DNA molecules called minichromosomes.
Gene rearrangement	This is not a common mechanism of gene regulation. As discussed in Chapter 17, gene rearrangement occurs within immunoglobulin genes to generate a diverse array of genes that encode antibody proteins.
DNA methylation	This is a common DNA modification in vertebrates and plants that involves the attachment of methyl groups to cytosine bases. When DNA methylation occurs in the promoter region, it usually inhibits transcription. This inhibition may be due to the inability of transcription factors to recognize the promoter region and/or a conversion of the chromatin to the closed conformation.

Gene Accessibility Can Be Controlled by Changes in Chromatin Compaction

Chromatin is composed of DNA and proteins that are organized into a compact structure that fits inside the nucleus of the cell. As discussed in Chapter 10, the DNA is wound around histone proteins to form nucleosomes that are 11 nm in diameter. This 11 nm fiber is condensed further to a 30 nm fiber, which is the predominant form of euchromatin found within the nucleus during interphase, when gene expression primarily occurs. Nevertheless, chromatin is a very dynamic structure that can alternate between highly condensed and highly extended conformations. The dynamic nature of chromatin is important in regulating gene transcription.

Variations in the degree of chromatin packing occur along the length of eukaryotic chromosomes during interphase. Tightly packed chromatin in a closed conformation cannot be transcribed. During gene activation, such chromatin must be converted to an open conformation that is less tightly packed than a 30 nm fiber. In certain cells, researchers can microscopically observe the decondensation of the 30 nm fiber when transcription is occurring. **Figure 15.8** shows photomicrographs of a chromosome from an amphibian oocyte, the genes of which are being actively transcribed. This chromosome does not form a uniform, compact 30 nm fiber. Instead, many decondensed loops radiate from the central axis of the chromosome. These loops are regions of DNA in which the genes are being actively transcribed. These chromosomes have been named lampbrush chromosomes because their feathery appearance resembles the brushes that were once used to clean kerosene lamps.

FIGURE 15.8
A lampbrush chromosome from an amphibian oocyte. The loops that radiate from the chromosome are regions of DNA containing genes that are being actively transcribed.

20 μm

2 μm

EXPERIMENT 15A

DNase I Sensitivity Can Be Used to Study Changes in Chromatin Compaction

To understand the interconversions between closed and open chromatin conformations, researchers need methods to evaluate the degree of chromatin packing as it occurs in living cells. An important technique is the use of DNase I to monitor DNA conformation. Recall from Chapter 10 (see Figure 10.15) that DNase I is an endonuclease that cleaves DNA. DNase I is much more likely to cleave DNA in an open conformation than a closed conformation because a looser conformation allows greater accessibility of DNase I to the DNA. When DNA is in an open conformation and able to be digested by DNase I, such DNA is said to exhibit **DNase I sensitivity.**

In 1976, Harold Weintraub and Mark Groudine used DNase I sensitivity as a tool to evaluate differences in chromatin structure that occur when a gene is being actively transcribed. In particular, they focused their attention on the β-globin gene. Humans possess several different genes that encode globin polypeptides. These polypeptides are the subunits of the oxygen-carrying protein hemoglobin. Prior to this work, it had been well established that globin genes are specifically expressed in reticulocytes (immature red blood cells) but not in other cell types such as brain cells and fibroblasts. Weintraub and Groudine asked the question, Is there a difference in the chromatin packing of globin genes in cells that can actively transcribe the globin genes compared with cells in which the globin genes are turned off? To answer this question, they used DNase I sensitivity to compare the degree of globin gene packing in reticulocytes versus brain cells and fibroblasts.

Before discussing the steps in their experimental protocol, let's consider the rationale behind their experimental approach (**Figure 15.9**). Because the globin genes are only a small part of the total chromosomal DNA, having a way to specifically monitor the DNase I digestion of the β-globin gene was a vital aspect of their protocol. They accomplished this by using a probe—a segment of DNA complementary to the β-globin gene. In Weintraub and Groudine's experiments, the DNA probe was radiolabeled to allow its detection. Following the exposure of the chromatin to DNase I, the chromosomal DNA was denatured and the probe was added. If DNase I had digested the chromosomal β-globin gene, the probe DNA would not be able to hybridize with the β-globin gene (Figure 15.9a). Alternatively, if the chromosomal β-globin gene had not been digested by DNase I, the probe would hybridize to it (Figure 15.9b).

After hybridization, the samples were then exposed to another enzyme, known as S1 nuclease. This enzyme digests DNA, but only when it is single stranded, not when it is double stranded. As shown in Figure 15.9a, S1 nuclease digests the radiolabeled DNA probe because it had not hybridized with the complementary chromosomal strand. In contrast, S1 nuclease does not digest the probe if the chromosomal DNA remains intact and hybridizes with the probe (Figure 15.9b). In this way, an intact strand of chromosomal DNA protects the probe from S1 nuclease digestion.

After S1 incubation, Weintraub and Groudine reasoned that their sample would contain an intact radiolabeled DNA probe if

(a)

(b)

FIGURE 15.9 The use of DNase I and S1 nuclease to probe the chromatin structure of the β-globin gene. (a) Accessible DNA has been digested into small pieces via DNase I. These small pieces cannot hybridize with the probe DNA. When S1 nuclease is added, the single-stranded probe DNA is digested. (b) Due to compaction, the DNA is not cut with DNase I. The probe DNA can hybridize with a strand of chromosomal DNA after it has been denatured. When S1 nuclease is added, it will digest only the single-stranded regions of DNA, not the double-stranded region where the probe DNA and chromosomal DNA are bound to each other.

the chromosomal DNA had been in a closed conformation. This is because DNase I would not digest the chromosomal gene, and therefore, the β-globin gene would be available to hybridize to the radiolabeled DNA probe. However, if the chromosomal globin gene was in an open conformation, DNase I would digest the chromosomal DNA, preventing it from hybridizing with the radiolabeled DNA probe. In this case, the radiolabeled DNA strand would be digested by S1 nuclease. Therefore, the susceptibility of the radiolabeled DNA to S1 nuclease digestion allowed them to evaluate whether the chromosomal globin gene was in an open or closed conformation.

With these concepts in mind, let's examine the steps in their experimental procedure (**Figure 15.10**). They began with three cell types: reticulocytes, brain cells, and fibroblasts. Only the reticulocytes express the β-globin gene. They hypothesized that if the globin gene was expressed, the DNA would be in an open conformation and accessible to digestion. The nuclei were extracted from these cells and incubated with DNase I. The DNA was then isolated from the three types of nuclei. Following the isolation procedure, the chromosomal DNA was broken into fragments by sonication, with the average fragment size larger than the probe DNA. The DNA fragments and probe were mixed together, denatured, and then allowed to hybridize. The cooled samples were then divided into two tubes each. Into one of the two tubes, S1 nuclease was added. A key point is that the probe would not be digested by S1 nuclease if it had hybridized to a complementary strand from the globin gene and thus was double stranded. The DNA fragments were then precipitated using trichloroacetic acid. The samples were subjected to centrifugation, and the amounts of radioactivity in the pellets were determined by scintillation counting.

■ THE HYPOTHESIS

A loosening of chromatin structure occurs when the β-globin gene is transcriptionally active.

■ TESTING THE HYPOTHESIS — FIGURE 15.10   **The use of DNase I sensitivity to study the compaction of the β-globin gene.**

Starting material: Nuclei were isolated from three different cell types in chicken: reticulocytes (immature red blood cells), brain cells, and fibroblasts. The globin genes are expressed in red blood cells but not in brain cells and fibroblasts.

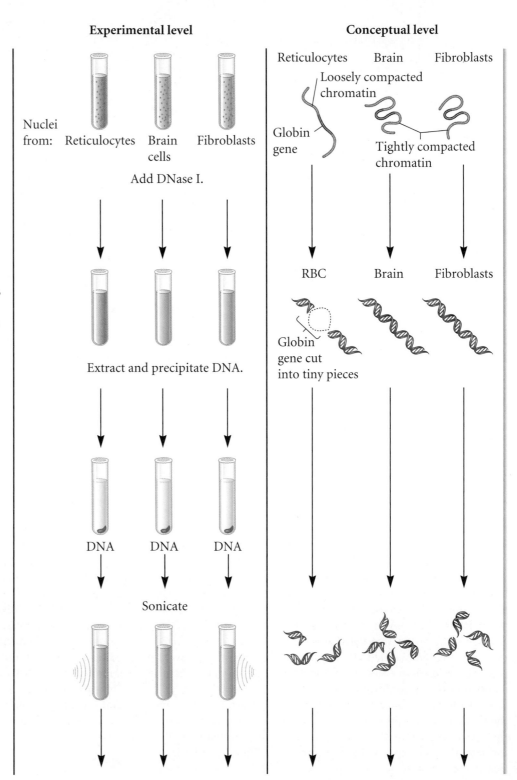

Experimental level

1. Treat nuclei from three types of cells with the same amount of DNase I.

Nuclei from: Reticulocytes   Brain cells   Fibroblasts

Add DNase I.

2. Extract the DNA from the nuclei. This involves lysing the nuclei with detergent, removing the protein by treatment with a phenol–chloroform mixture, and then precipitating the DNA by adding ethanol. The precipitated DNA forms a pellet at the bottom of a test tube following centrifugation. The DNA at the bottom of the tube can then be resuspended in an appropriate solution for the next step.

Extract and precipitate DNA.

DNA   DNA   DNA

Sonicate

3. Subject the DNA to sound waves (i.e., sonication) to break the DNA into fragments of an average length of 500 bp.

Conceptual level

Reticulocytes   Brain   Fibroblasts

Loosely compacted chromatin

Globin gene

Tightly compacted chromatin

RBC   Brain   Fibroblasts

Globin gene cut into tiny pieces

4. Add a radiolabeled DNA probe that is complementary to the β-globin gene.

5. Denature the DNA into single strands by treatment with high temperature. Then cool it to 65°C to allow the complementary DNA strands to hybridize with each other.

6. Reduce the temperature further and divide each sample into two tubes. Add S1 nuclease to one tube of each sample. Omit S1 nuclease from the second tube.

7. Precipitate the DNA with trichloroacetic acid. Subject to centrifugation.

8. Count the amount of radioactivity in the pellet. (The technique of scintillation counting is described in the Appendix.) The amount of radioactivity in the presence of S1 nuclease divided by the amount of radioactivity in the absence of S1 nuclease provides a measure of the percentage of radiolabeled DNA that has hybridized to the chromosomal DNA.

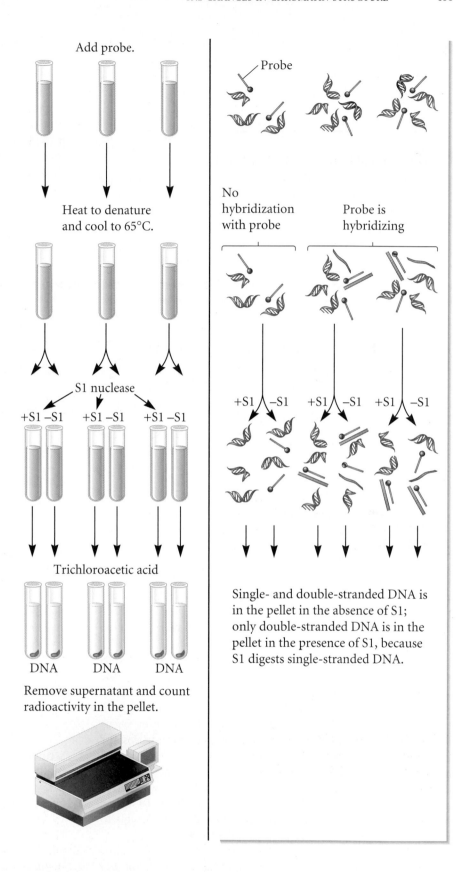

Add probe.

Heat to denature and cool to 65°C.

S1 nuclease

+S1 −S1   +S1 −S1   +S1 −S1

Trichloroacetic acid

DNA   DNA   DNA

Remove supernatant and count radioactivity in the pellet.

Probe

No hybridization with probe

Probe is hybridizing

+S1 / −S1   +S1 / −S1   +S1 / −S1

Single- and double-stranded DNA is in the pellet in the absence of S1; only double-stranded DNA is in the pellet in the presence of S1, because S1 digests single-stranded DNA.

THE DATA

Source of Nuclei	% Hybridization of DNA Probe
Reticulocytes	25%
Brain cells	>94%
Fibroblasts	>94%

INTERPRETING THE DATA

As shown in the data of Figure 15.10, a much smaller percentage of the radiolabeled DNA probe hybridized to the chromosomal DNA in reticulocytes compared to brain cells and fibroblasts. What do these results mean? The data are consistent with the idea that DNase I had digested the globin gene in the chromatin of reticulocytes into small fragments that were too small to hybridize to the radiolabeled DNA probe. In other words, the globin genes in reticulocytes are more sensitive to DNase I digestion. By comparison, the globin genes in brain cells and fibroblasts are relatively resistant. Because the globin genes are expressed in reticulocytes but not in brain cells and fibroblasts, these results are consistent with the hypothesis that the globin gene is less tightly packed when it is being expressed. In cells where the globin genes should not be expressed, such as brain cells and fibroblasts, the chromatin containing the β-globin gene is tightly compacted. Changes in chromatin compaction provide one way to regulate globin gene expression among different cell types.

A self-help quiz involving this experiment can be found at the Online Learning Center.

Transcriptional Activators Recruit Chromatin-Remodeling Enzymes to the Promoter Region

In recent years, geneticists have been trying to identify the steps that promote the interconversion between the closed and open conformations of chromatin. As we saw for globin genes, changes in chromatin structure are associated with the level of gene transcription. Based on the analysis of many genes, researchers have discovered that a key role of some transcriptional activators is to orchestrate changes in chromatin compaction from the closed to the open conformation.

As discussed in Chapter 12, chromatin structure is altered in eukaryotes in two common ways: ATP-dependent chromatin remodeling and covalent modification of histones. The SWI/SNF family is a major category of proteins found in all eukaryotic species that catalyze ATP-dependent chromatin remodeling. The net result of ATP-dependent chromatin remodeling is a change in the locations of nucleosomes. This may involve a shift in nucleosomes to a new location or a change in the spacing of nucleosomes over a long stretch of DNA, both of which may loosen the level of chromatin compaction.

Nucleosomes have been shown to change position in cells that normally express a particular gene but not in cells where the gene is inactive. For example, in fibroblasts that do not express the β-globin gene, nucleosomes are positioned at regular intervals from nucleotides −3,000 to +1,500 (**Figure 15.11a**). However, in reticulocytes that express the β-globin gene, a disruption in nucleosome positioning occurs in the region from nucleotide −500 to +200 (**Figure 15.11b**). This disruption is a key step in gene activation. The position of a nucleosome may greatly influence whether or not a gene can be transcribed. For example, if the TATA box is tightly bound to a core histone protein, it may be inaccessible to general transcription factors and RNA polymerase.

A second mechanism that leads to an alteration in chromatin structure involves the covalent modification of histones. A histone code appears to play a key role in controlling the level of chromatin compaction. The amino terminal ends of histone proteins are covalently modified in several ways, including the acetylation of lysines, methylation of lysines, and phosphorylation of

(a) Nucleosomes in fibroblasts, which do not express the β-globin gene

(b) Nucleosomes in reticulocytes, which express the β-globin gene

FIGURE 15.11   Changes in nucleosome position during the activation of the β-globin gene. (**a**) In fibroblasts, which do not express β globin, nucleosome positioning is uninterrupted. (**b**) In reticulocytes, which express this gene, disruption occurs in the positions of nucleosomes from the −500 to +200 region. (Position +1 is the beginning of transcription.) It is not yet clear whether the histones are removed, the histones are partially displaced, and/or other proteins are bound to this region.

serines (see Chapter 10, Figure 10.20). Because these covalent modifications influence the compaction of chromatin, they play a key role in transcriptional regulation. For example, positively charged lysines within the core histone proteins can be acetylated by enzymes called **histone acetyltransferases.** The attachment of the acetyl group ($-COCH_3$) may have two effects. First, it eliminates the positive charge on the lysine side chain and thereby disrupts the electrostatic attraction between the histone protein and the negatively charged DNA backbone. Secondly, covalently modified histones are recognized by DNA-binding proteins that promote changes in the compaction level of the chromatin. In the case of acetylation, the net effect is to diminish the interactions between DNA and nucleosomes and thereby loosen the compaction of the 30 nm fiber. This would facilitate the ability of RNA polymerase to transcribe a gene. Some studies suggest that histones are completely displaced, whereas others suggest they are loosened but remain attached to the DNA.

An important role of certain transcriptional activators is to recruit ATP-dependent remodeling enzymes and histone-modifying enzymes to the promoter region. Though the order of recruitment may differ among specific transcriptional activators, this appears to be a critical step that is necessary to initiate transcription. A well-studied example involves a gene in yeast that is involved in mating. Yeast can exist in two mating types, termed *a* and α. The yeast *HO* gene encodes an enzyme that is required for the switch from one mating type to the other.

Prior to transcription, much of the region surrounding the core promoter of the *HO* gene is in a closed conformation. How is this region loosened up so that transcription can occur? A simplified mechanism for the transcription of this gene is shown in **Figure 15.12**. A regulatory transcription factor (SWI5P) binds to an accessible enhancer in the vicinity of the core promoter. This transcription factor then recruits an ATP-dependent remodeling enzyme (SWI/SNF) to the region, which promotes the conversion of the chromatin from the closed to the open conformation. Next, a histone acetyltransferase (SAGA) is recruited to this region, which serves to further open the chromatin conformation. Overall, the actions of the ATP-dependent remodeling enzyme and the histone acetyltransferase remodel a region around the promoter that is approximately 1,000 bp in length. A second regulatory transcription factor (SBP) is then able to bind to an enhancer near the core promoter and thereby initiate the binding of general transcription factors and RNA polymerase.

DNA Methylation Usually Inhibits Gene Transcription

We now turn our attention to a change in chromatin structure that silences gene expression. As discussed in previous chapters, DNA structure can be modified by the covalent attachment of methyl groups. DNA methylation is common in some but not all eukaryotic species. For example, yeast and *Drosophila* have little or no detectable methylation of their DNA, whereas DNA methylation in vertebrates and plants is relatively abundant. In mammals, approximately 2 to 7% of the DNA is methylated. As shown in **Figure 15.13**, eukaryotic DNA methylation occurs via an enzyme called **DNA methyltransferase,** which attaches

ONLINE ANIMATION

FIGURE 15.12   An example of chromatin remodeling of a promoter region that leads to transcriptional activation. The *HO* gene is involved in yeast mating type switching. For this gene to be transcribed, the promoter region must be converted from a closed to an open conformation. A transcription factor termed SWI5P (SWI refers to mating type <u>swi</u>tching) binds to an enhancer in the promoter region and recruits an ATP-dependent chromatin remodeling enzyme (of the SWI/SNF family) to the region. This begins the opening process. A histone acetyltransferase termed SAGA is then recruited to the region and further decompacts the chromatin. SAGA is an acronym for Spt/<u>A</u>da/<u>GCN5</u>/<u>A</u>cetyltransferase. *Spt, Ada,* and *GCN5* are the names of three genes that had been shown to be transcriptionally regulated by this histone acetyltransferase. A second transcription factor termed SBP (an acronym for a mating type <u>s</u>witching cell cycle <u>b</u>ox <u>p</u>rotein) binds to an enhancer in the promoter region and initiates transcription by recruiting general transcription factors and RNA polymerase.

Within the figure:

Enhancer for SWI5P · Enhancer for SBP · Core promoter · HO coding sequence

A transcriptional activator (SWI5P) binds to its enhancer.

SWI5P

An ATP-dependent remodeling enzyme (SWI/SNF) is recruited to this region.

A histone acetyltransferase (SAGA) is recruited to this region. The recruitment of both enzymes leads to an open conformation in the promotor region.

The open conformation allows a second transcriptional activator (SBP) to bind to its enhancer element.

SBP

This second transcriptional activator recruits RNA polymerase and general transcription factors to the core promoter. This initiates transcription.

TFIID

RNA polymerase

(a) The methylation of cytosine

Cytosine

DNA methyltransferase

5-methylcytosine

(a) The methylation of cytosine

(b) Unmethylated

(c) Hemimethylated

(d) Fully methylated

FIGURE 15.13   **DNA methylation on cytosine bases.**
(**a**) Methylation occurs via an enzyme known as DNA methyltransferase, which attaches a methyl group to the number 5 carbon on cytosine. The CG sequence can be (**b**) unmethylated, (**c**) hemimethylated, or (**d**) fully methylated.

a methyl group to the number 5 position of the cytosine base, forming 5-methylcytosine. The sequence that is methylated is shown here.

Note that this sequence contains cytosines in both strands. Methylation of the cytosine in both strands is termed full methylation, whereas methylation of only one strand is called hemimethylation.

DNA methylation usually inhibits the transcription of eukaryotic genes, particularly when it occurs in the vicinity of the promoter. In vertebrates and plants, **CpG islands** occur near many promoters of genes. (Note: CpG refers to a dinucleotide of C and G in DNA that is connected by a phosphodiester linkage.) These CpG islands are commonly 1,000 to 2,000 base pairs in length and contain a high number of CpG sites. In the case of **housekeeping genes**—genes that encode proteins required in most cells of a multicellular organism—the cytosine bases in the CpG islands are unmethylated. Therefore, housekeeping genes tend to be expressed in most cell types. By comparison, other genes are highly regulated and may be expressed only in a particular cell type. These are **tissue-specific genes.** In some cases, it has been found that the expression of such genes may be silenced by the methylation of CpG islands. Overall, evidence is accumulating that unmethylated CpG islands are correlated with active genes, whereas suppressed genes contain methylated CpG islands. In this way, DNA methylation is thought to play an important role in the silencing of tissue-specific genes so they are not expressed in the wrong tissue.

Methylation can inhibit transcription in two general ways. First, methylation of CpG islands may prevent the binding of transcription factors to the promoter region. For example, methylated CG sequences could prevent the binding of an activator protein to an enhancer element, presumably by the methyl group protruding into the major groove of the DNA (**Figure 15.14a**). The inability of an activator protein to bind to the DNA would inhibit the initiation of transcription. However, CG methylation does not slow down the movement of RNA polymerase along a gene. In vertebrates and plants, coding regions downstream from the core promoter usually contain methylated CG sequences, but these do not hinder the elongation phase of transcription. This suggests that methylation must occur in the vicinity of the promoter to have an inhibitory effect on transcription.

A second way that methylation inhibits transcription is by converting chromatin from an open to a closed conformation (**Figure 15.14b**). Proteins known as **methyl-CpG-binding proteins** bind methylated sequences. These proteins contain a domain called the methyl-binding domain that specifically recognizes a methylated CG sequence. Once bound to the DNA, the methyl-CpG-binding protein recruits other proteins to the region that cause the chromatin to become very compact. For example, methyl-CpG-binding proteins may recruit histone deacetylase to a methylated CpG island near a promoter. Histone deacetylation removes acetyl groups from the histone proteins and thereby favors a chromatin conformation that is very compact.

DNA Methylation Is Heritable

Methylated DNA sequences are inherited during cell division. Experimentally, if fully methylated DNA is introduced into a plant or vertebrate cell, the DNA will remain fully methylated even in subsequently produced daughter cells. However, if the same

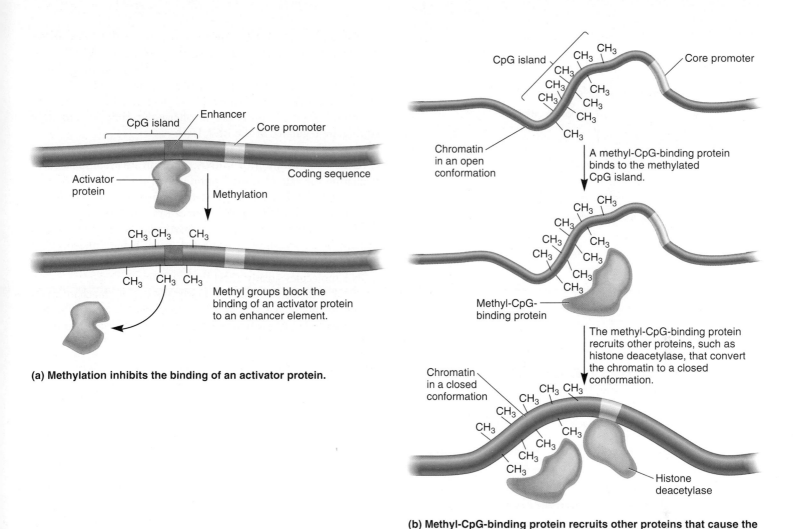

(a) **Methylation inhibits the binding of an activator protein.**

(b) **Methyl-CpG-binding protein recruits other proteins that cause the region to become more compact.**

FIGURE 15.14 **Transcriptional silencing via methylation.** (a) The methylation of a CpG island may inhibit the binding of transcriptional activators to the promoter region. (b) The binding of a methyl-CpG-binding protein to a CpG island may lead to the recruitment of proteins, such as histone deacetylase, that convert chromatin to a closed conformation and thus suppress transcription.

sequence of nonmethylated DNA is introduced into a cell, it will remain nonmethylated in the daughter cells. These observations indicate that the pattern of methylation is retained following DNA replication and, therefore, is inherited in future daughter cells.

How can methylation be inherited from cell to cell? **Figure 15.15** illustrates a molecular model that explains this process, which was originally proposed by Arthur Riggs, Robin Holliday, and J. E. Pugh. The DNA in a particular cell may become methylated by *de novo* **methylation**—the methylation of DNA that was previously unmethylated. When a fully methylated segment of DNA replicates in preparation for cell division, the newly made daughter strands contain unmethylated cytosines. Because only one strand is methylated, such DNA is called hemimethylated. This hemimethylated DNA is efficiently recognized by DNA methyltransferase, which makes it fully methylated. This process is called **maintenance methylation,** because it preserves the methylated condition in future cells. However, maintenance

methylation does not act on unmethylated DNA. Overall, maintenance methylation appears to be an efficient process that routinely occurs within vertebrate and plant cells. By comparison, *de novo* methylation and demethylation are infrequent and highly regulated events. According to this view, the initial methylation or demethylation of a given gene can be regulated so that it occurs in a specific cell type or stage of development. Once methylation has occurred, it can then be transmitted from mother to daughter cells via maintenance methylation.

The methylation mechanism shown in Figure 15.15 can explain the phenomenon of genomic imprinting, which is described in Chapter 7. In this case, specific genes are methylated during oogenesis or spermatogenesis, but not both. Following fertilization, the pattern of methylation is maintained in the offspring. For example, if a gene is methylated only during spermatogenesis, the allele that is inherited from the father will be methylated in the somatic cells of the offspring, while the maternal allele

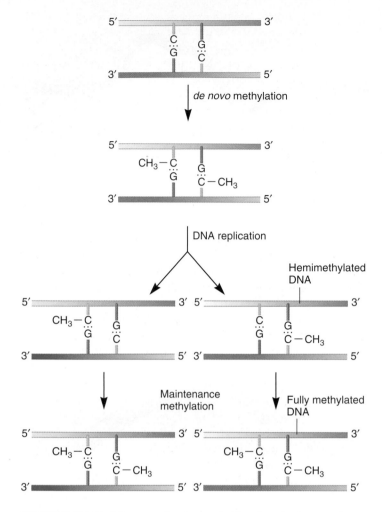

FIGURE 15.15   A molecular model for the inheritance of DNA methylation. The DNA initially undergoes *de novo* methylation, which is a rare, highly regulated event. Once this occurs, DNA replication produces hemimethylated DNA molecules, which are then fully methylated by DNA methyltransferase. This process, called maintenance methylation, is a routine event that is expected to occur for all hemimethylated DNA.

will remain unmethylated. Along these lines, geneticists are also eager to determine how variations in DNA methylation patterns may be important for cell differentiation. It may be a key way to silence genes in different cell types. However, additional research will be necessary to understand how specific genes may be targeted for *de novo* methylation or demethylation during different developmental stages or in specific cell types.

15.3 REGULATION OF RNA PROCESSING, RNA STABILITY, AND TRANSLATION

Thus far, we have considered a variety of mechanisms that regulate the level of gene transcription. These mechanisms control the amount of RNA transcribed from a given gene. In addition, eukaryotic gene expression is commonly regulated after the RNA

is made (**Table 15.2**). One mechanism is pre-mRNA processing. Following transcription, a pre-mRNA transcript is processed before it becomes a functional mRNA. These processing events, which include splicing, capping, polyA tailing, and RNA editing, were described in Chapter 12. We will begin this section by examining how the processes of alternative splicing and RNA editing are regulated at the RNA level.

Another strategy for regulating gene expression is to influence the concentration of mRNA. This can be accomplished by regulating the rate of transcription. When the transcription of a gene is increased, a higher concentration of the corresponding RNA results. In addition, RNA concentration is greatly affected by the stability or half-life of a particular RNA. Factors that increase RNA stability are expected to raise the concentration of that RNA molecule. We will explore how sequences within mRNA molecules greatly affect their stability.

We will also consider a newly discovered mechanism of mRNA silencing, known as RNA interference, which involves double-stranded RNAs that may direct the breakdown of specific

TABLE **15.2**	
Gene Regulation via RNA Processing and Translation	
Effect	**Description**
Alternative splicing	Certain pre-mRNAs can be spliced in more than one way, leading to polypeptides that have different amino acid sequences. Alternative splicing is often cell specific so that a protein can be fine-tuned to function in a particular cell type. It is an important form of gene regulation in multicellular eukaryotic species.
RNA editing	The sequence of RNAs can be altered after the RNA is made. This usually involves the addition or deletion of one or a few bases, or a change of C to U, or A to I. This does not appear to be a widespread phenomenon, though the study of RNA editing is still relatively recent.
RNA stability	The amount of RNA is greatly influenced by the half-life of RNA transcripts. A long polyA tail on mRNAs promotes their stability due to the binding of polyA-binding protein. Some RNAs with a relatively short half-life contain sequences that target them for rapid destruction. Some RNAs are stabilized by specific RNA-binding proteins that usually bind near the 3' end.
RNA interference	Double-stranded RNA can mediate the degradation of homologous mRNAs in the cell. This is a mechanism of gene regulation. Also, it probably provides eukaryotic cells with protection from invasion by certain types of viruses and may prevent the movement of transposable elements.
General regulation of translation	The function of translational initiation factors may be regulated to permit or inhibit translation. This regulation affects the translation of all cellular mRNAs. Inhibition of translation is desirable if a cell has been exposed to a virus or to toxic materials.
Translational regulation of specific mRNAs	Some mRNAs are regulated via binding proteins that inhibit the ability of the ribosomes to initiate translation. These proteins usually bind at the 5' end of the mRNA and thereby prevent the ribosome from binding.

RNAs or inhibit their translation. In addition to RNA interference, the process of mRNA translation may be regulated at the level of the translational machinery such as ribosomes. In eukaryotes, this can occur by directly controlling the function of translational initiation factors. Also, RNA-binding proteins can prevent ribosomes from initiating the translation process for specific mRNAs. In this section, we will examine these interesting mechanisms for regulating mRNA translation.

Alternative Splicing Regulates Which Exons Occur in an RNA Transcript, Allowing Different Polypeptides to Be Made from the Same Structural Gene

When it was first discovered, the phenomenon of splicing seemed like a rather wasteful process. During transcription, energy is used to synthesize intron sequences. Likewise, energy is also used to remove introns via a large spliceosome complex. This observation intrigued many geneticists, because natural selection tends to eliminate wasteful processes. Therefore, instead of simply viewing splicing as a wasteful process, many geneticists expected to find that pre-mRNA splicing has one or more important biological roles. In recent years, one very important biological advantage has become apparent. This is **alternative splicing,** which refers to the phenomenon that a pre-mRNA can be spliced in more than one way.

What is the advantage of alternative splicing? To understand the biological effects of alternative splicing, remember that the sequence of amino acids within a polypeptide determines the structure and function of a protein. Alternative splicing produces two or more polypeptides with differences in their amino acid sequences, leading to possible changes in their functions. In most cases, the alternative versions of the protein will have similar functions, because much of their amino acid sequences are iden-

tical to each other. Nevertheless, alternative splicing produces differences in amino acid sequences that will provide each polypeptide with its own unique characteristics. Because alternative splicing allows two or more different polypeptide sequences to be derived from a single gene, some geneticists have speculated that an important advantage of this process is that it allows an organism to carry fewer genes in its genome.

The degree of splicing and alternative splicing varies greatly among different species. Baker's yeast, for example, contains about 6,300 genes and approximately 300 (i.e., approximately 5%) encode pre-mRNAs that are spliced. Of these, only a few have been shown to be alternatively spliced. Therefore, in this unicellular eukaryote, alternative splicing is not a major mechanism to generate protein diversity. In comparison, complex multicellular organisms seem to rely on alternative splicing to a great extent. Humans contain approximately 20,000 to 25,000 different genes, and most of these contain one or more introns. Recent estimates suggest that about 70% of all human pre-mRNAs are alternatively spliced. Furthermore, certain pre-mRNAs are alternatively spliced to an extraordinary extent. In fact, some pre-mRNAs can be alternatively spliced to produce dozens of different mRNAs. This provides a much greater potential for human cells to create protein diversity.

Figure 15.16 considers an example of alternative splicing for a gene that encodes a protein known as α-tropomyosin. This protein functions in the regulation of cell contraction. It is located along the thin filaments found in smooth muscle cells, such as those in the uterus and small intestine, and in striated muscle cells that are found in cardiac and skeletal muscle. Alpha-tropomyosin is also synthesized in many types of nonmuscle cells but in lower amounts. Within a multicellular organism, different types of cells must regulate their contractibility in subtly different ways. One way that this may be accomplished is by the production of different forms of α-tropomyosin.

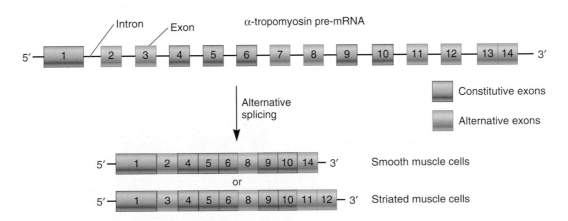

FIGURE 15.16 **Alternative ways that the rat α-tropomyosin pre-mRNA can be spliced.** The top part of this figure depicts the structure of the rat α-tropomyosin pre-mRNA. Exons are shown as colored boxes, and introns are illustrated as connecting black lines. The lower part of the figure describes the final mRNA products in smooth and striated muscle cells. Note: Exon 8 is found in the final mRNA of smooth and striated muscle cells, but not in the mRNA of certain other cell types. The junction between exons 13 and 14 contains a 3' splice so that exon 13 can be separated from exon 14.

Genes→Traits α-tropomyosin functions in the regulation of cell contraction in muscle and nonmuscle cells. Alternative splicing of the pre-mRNA provides a way to vary contractibility in different types of cells by modifying the function of α-tropomyosin. As shown here, the alternatively spliced versions of the pre-mRNA produce α-tropomyosin proteins that differ slightly from each other in their structure (i.e., amino acid sequence). These alternatively spliced versions vary in function to meet the needs of the cell type in which they are found. For example, the sequence of exons 1–2–4–5–6–8–9–10–14 produces an α-tropomyosin protein that functions suitably in smooth muscle cells. Overall, alternative splicing affects the traits of an organism by allowing a single gene to encode several versions of a protein, each optimally suited to the cell type in which it is made.

The intron-exon structure of the rat α-tropomyosin pre-mRNA and two alternative ways that the pre-mRNA can be spliced are described in Figure 15.16. The pre-mRNA contains 14 exons, six of which are **constitutive exons** (shown in red), which are always found in the mature mRNA from all cell types. Presumably, constitutive exons encode polypeptide segments of the α-tropomyosin protein that are necessary for its general structure and function. By comparison, **alternative exons** (shown in green) are not always found in the mRNA after splicing has occurred. The polypeptide sequences encoded by alternative exons may subtly change the function of α-tropomyosin to meet the needs of the cell type in which it is found. For example, Figure 15.16 shows the predominant splicing products found in smooth muscle cells and striated muscle cells. Exon 2 encodes a segment of the α-tropomyosin protein that alters its function to make it suitable for smooth muscle cells. By comparison, the α-tropomyosin mRNA found in striated muscle cells does not include exon 2. Instead, this mRNA contains exon 3, which is more suitable for that cell type.

Alternative splicing is not a random event. Rather, the specific pattern of splicing is regulated in any given cell. The molecular mechanism for the regulation of alternative splicing involves proteins known as **splicing factors.** Such splicing factors play a key role in the choice of particular splice sites. **SR proteins** are an example of a type of splicing factor. SR proteins contain a domain at their carboxyl terminal end that is rich in serines (S) and arginines (R) and is involved in protein-protein recognition. They also contain an RNA-binding domain at their amino terminal end.

As discussed in Chapter 12, components of the spliceosome recognize the 5' and 3' splice sites and then remove the intervening intron. The key effect of splicing factors is to modulate the ability of the spliceosome to choose 5' and 3' splice sites. This can occur in two ways. Some splicing factors act as repressors that inhibit the ability of the spliceosome to recognize a splice site. In **Figure 15.17a**, a splicing repressor binds to a 3' splice site and prevents the spliceosome from recognizing the site. Instead, the spliceosome binds to the next available 3' splice site. The splicing

(a) Splicing repressors

(b) Splicing enhancers

FIGURE 15.17   The roles of splicing factors during alternative splicing. (a) Splicing factors can act as repressors to prevent the recognition of splice sites. In this example, the presence of the splicing repressor causes exon 2 to be skipped and thus not included in the mRNA. (b) Other splicing factors can enhance the recognition of splice sites. In this example, the splicing enhancers promote the recognition of sites that flank exon 3, thereby causing its inclusion in the mRNA.

repressor causes exon 2 to be spliced out of the mature mRNA, an event called **exon skipping.** Alternatively, other splicing factors enhance the ability of the spliceosome to recognize particular splice sites. In **Figure 15.17b**, splicing enhancers bind to the 3' and 5' splice sites that flank exon 3, which results in the inclusion of exon 3 in the mature mRNA.

Alternative splicing in different tissues is thought to occur because each cell type has its own characteristic concentration of many kinds of splicing factors. Furthermore, much like transcription factors, splicing factors may be regulated by the binding of small effector molecules, protein–protein interactions, and covalent modifications. Overall, the differences in the composition of splicing factors, and the regulation of their activities, form the basis for alternative splicing decisions.

The Nucleotide Sequence of RNA Can Be Modified by RNA Editing

The term **RNA editing** refers to a change in the nucleotide sequence of an RNA molecule that involves additions or deletions of particular bases, or a conversion of one type of base to another, such as a cytosine to a uracil. In the case of mRNAs, editing can have various effects, such as generating start codons, generating stop codons, and changing the coding sequence for a polypeptide.

The phenomenon of RNA editing was first identified in trypanosomes, the protists that cause sleeping sickness. As with the discovery of RNA splicing, the initial finding of RNA editing was met with great skepticism. Since that time, however, RNA editing has been shown to occur in various organisms and in a variety of ways, although its functional significance is slowly emerging (**Table 15.3**).

The molecular mechanisms of RNA editing have been the subject of many investigations. In the specific case of trypanosomes, which are single-celled protists, the editing process involves the participation of a **guide RNA.** This guide RNA directs the addition of one or more uracil nucleotides into an RNA or the deletion of one or more uracils. As shown in **Figure 15.18**, the guide RNA has two important characteristics. First, its 5' anchor is complementary to the RNA that is to be edited. Second, it has uracil nucleotides at its 3' end. During the editing process, the 5' anchor binds to the target RNA. The RNA is cleaved at a defined location by an endonuclease, and the 3' end of the guide RNA becomes displaced from the target RNA. For additions, an enzyme called terminal U-transferase then attaches uracil-containing nucleotides. For deletions, an enzyme called 3'-U-exonuclease removes one or more uracils. Finally, RNA ligase rejoins the two pieces of RNA.

A more widespread mechanism for RNA editing involves changes of one type of base to another. In this form of editing, a base in the RNA is deaminated—an amino group is removed from the base. When cytosine is deaminated, uracil is formed, and when adenine is deaminated, inosine is formed (**Figure 15.19**). Inosine is recognized as guanine during translation.

An example of RNA editing occurs in mammals involving an mRNA that encodes a protein called apolipoprotein B. In the liver, the RNA editing process produces apolipoprotein B-100, a protein essential for the transport of cholesterol in the blood. In intestinal cells, the mRNA may be edited so that a single C is changed to a U. What is the significance of this base substitution? This change converts a glutamine codon (CAA) to a stop codon (UAA) and thereby results in a shorter apolipoprotein. In this case, RNA editing produces an apolipoprotein B with an altered structure. Therefore, RNA editing can produce two proteins from the same gene, much like the phenomenon of alternative splicing described earlier in this chapter.

The Stability of mRNA Influences mRNA Concentration

In eukaryotes, the stability of mRNAs can vary considerably. Certain mRNAs have very short half-lives, such as several minutes, whereas others can persist for many days or even months. In some cases, the stability of an mRNA can be regulated so that its half-life is shortened or lengthened. A change in the stability of mRNA can greatly influence the cellular concentration of that mRNA molecule. In this way, mechanisms that control mRNA stability can dramatically affect gene expression.

Various factors can play a role in mRNA stability. One important structural feature is the length of the polyA tail. As you may recall from Chapter 12, most newly made mRNAs contain a polyA tail that averages 200 nucleotides in length. The polyA tail is recognized by the **polyA-binding protein.** The binding of this protein enhances RNA stability. However, as an mRNA ages, its polyA tail tends to be shortened by the action of cellular exonucleases. Once it becomes less than 10 to 30 adenosines in length, the polyA-binding protein can no longer bind, and the mRNA is rapidly degraded by exo- and endonucleases.

Certain mRNAs, particularly those with short half-lives, contain sequences that act as destabilizing elements. While these destabilizing elements can be located anywhere within the mRNA,

TABLE 15.3

Examples of RNA Editing

Organism	Type of Editing	Found in
Trypanosomes (protozoa)	Primarily additions but occasionally deletions of uracil nucleotides	Many mitochondrial mRNAs
Land plants	C-to-U conversion	Many mitochondrial and chloroplast mRNAs, tRNAs, and rRNAs
Slime mold	C additions	Many mitochondrial mRNAs
Mammals	C-to-U conversion	Apolipoprotein B mRNA, and NF1 mRNA, which encodes a tumor-suppressor protein
	A-to-I conversion	Glutamate receptor mRNA, many tRNAs
Drosophila	A-to-I conversion	mRNA for calcium and sodium channels

FIGURE 15.18   RNA editing in trypanosomes. The 5' end of the guide RNA (the anchor) binds to the target mRNA. The mRNA is then cleaved at the editing site by an endonuclease. On the left side, uracil residues are added by terminal U-transferase. On the right side, uracils are removed by 3'-U-exonuclease. In both cases, the ends of the target mRNA are rejoined by RNA ligase.

FIGURE 15.19   RNA editing by deamination. A cytidine deaminase can remove an amino group from cytosine, thereby creating uracil. An adenine deaminase can remove an amino group from adenine to make inosine.

they are most commonly located near the 3' end between the stop codon and the polyA tail. This region of the mRNA is known as the **3'-<u>un</u>translated <u>r</u>egion (3'-UTR)**. An example of a destabilizing element is the <u>AU</u>-<u>r</u>ich <u>e</u>lement (ARE) that is found in many short-lived mRNAs (**Figure 15.20**). This element, which contains the consensus sequence AUUUA, is recognized by cellular proteins that bind to the ARE and thereby influence whether or not the mRNA is rapidly degraded.

Double-Stranded RNA Can Silence the Expression of mRNA

Another way that specific RNAs can be targeted for degradation or translational inhibition is via a recently discovered mechanism involving double-stranded RNA. Research in plants and the nematode *Caenorhabditis elegans* led to the discovery that double-stranded RNA can silence the expression of particular genes. Studies of plant viruses were one avenue of research that identified this mechanism of gene regulation. Certain plant viruses produce double-stranded RNA as part of their life cycle. In addition, these plant viruses may carry genes very similar to genes that already exist in the genome of the plant cell. When such viruses infect plant cells, they silence the expression of the plant gene that is similar to the viral gene.

Plant research that involved the production of transgenic plants also was consistent with the idea that double-stranded RNA can cause mRNA to be degraded or inhibited. Using molecular techniques described in Chapter 19, cloned genes can be introduced into the genome of plants. Surprisingly, researchers observed that when cloned genes were introduced in multiple copies, the expression of the gene was often silenced. How were these results explained? We now know that this may be due to the formation of double-stranded RNA. When a cloned gene randomly inserts into a genome, it may, as a matter of random chance, happen to insert itself next to a promoter for a plant gene that is already present in the genome (**Figure 15.21**). The cloned gene itself has a promoter to make a copy of the sense strand. If it randomly inserts next to a plant gene promoter that is oriented in the opposite direction, both strands of the cloned gene would be transcribed, thereby generating double-stranded RNA. As more copies of a cloned gene are introduced into a genome, the likelihood is greater that the scenario described in Figure 15.21 may occur. This event silences the expression of the cloned gene even though it is present in multiple copies. Furthermore, if the cloned gene is homologous to a plant gene that is already present in the plant cell, this phenomenon will also silence the endogenous plant gene. Therefore, the curious observation made by researchers was that increasing the number of copies of a cloned gene frequently led to gene silencing.

FIGURE 15.20   **The location of AU-rich elements (AREs) within mRNAs.** One or more AREs are commonly found within the 3'-UTRs (untranslated regions) of mRNAs with short half-lives. The 5'-UTR is the untranslated region of the mRNA that precedes the start codon. AREs are recognized by cellular proteins that bind and thereby influence whether or not the mRNA is rapidly degraded.

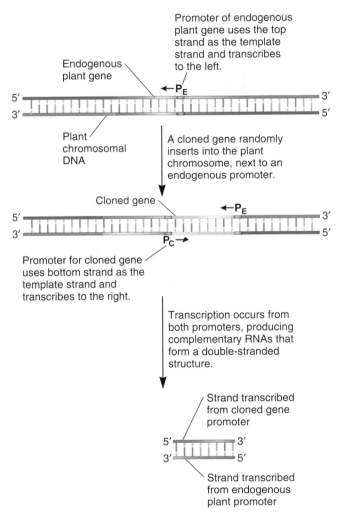

FIGURE 15.21   **Gene insertion leading to the production of double-stranded RNA.** When a cloned gene with its own promoter is introduced into a plant cell, it will randomly insert into the chromosomal DNA. In some cases, the cloned gene may insert next to a promoter for an endogenous plant gene, so the promoter for the endogenous gene will transcribe the antisense strand of the cloned gene. Therefore, the two promoters will transcribe the sense strand and the antisense strand to produce complementary RNAs that will form double-stranded RNA.

Fire and Mello Show That Double-Stranded RNA Is More Potent Than Antisense RNA at Silencing mRNA

The discovery that double-stranded RNA inhibits mRNA function came about as a result of studies involving gene expression. As a research tool to investigate the function of particular genes, researchers had often introduced antisense RNA (RNA that is complementary to mRNA) into cells as a way to inhibit mRNA translation. Because antisense RNA is complementary to mRNA, the antisense RNA would bind to the mRNA and thereby prevent translation. Oddly, in some experiments, researchers introduced sense RNA (RNA with the same sequence as mRNA) into cells, and this also inhibited mRNA translation. Another curious observation was that the effects of antisense RNA often persisted for a very long time, much longer than would have been predicted by the relatively short half-lives of most RNA molecules in the cell. These two unusual observations caused Andrew Fire, Craig Mello, and colleagues to investigate how the injection of RNA into cells inhibits mRNA.

They used *C. elegans* as their experimental organism because it was relatively easy to inject with RNA and the expression of many genes had already been established. In 1998, Fire and Mello investigated the effects of several injected RNAs known to be complementary to cellular mRNAs. In the investigation of **Figure 15.22**, we will focus on one of their experiments involving a gene called *mex-3*, which had already been shown to be highly expressed in early embryos. They started with the cloned gene for *mex-3*. This gene was genetically engineered using techniques described in Chapter 18 so that one version had a promoter that would result in the synthesis of the normal mRNA, which is the sense RNA. They made another version of *mex-3* in which a promoter directed the synthesis of the opposite strand, which produced antisense RNA. To make RNA in vitro, they added RNA polymerase and nucleotides to these cloned genes to make sense or antisense RNA.

Next, they injected RNA into *C. elegans* eggs and observed the effects in developing embryos. They either injected antisense RNA or they mixed sense and antisense RNA and injected double-stranded RNA. They also used uninjected eggs as controls. To determine the expression of *mex-3*, they incubated the resulting embryos with a probe complementary to the *mex-3* mRNA. The probe was labeled so it could be observed under the microscope. After this incubation step, any probe that was not bound to this mRNA was washed away.

■ THE GOAL

The goal was to further understand how the experimental injection of RNA was responsible for the silencing of particular mRNAs.

■ ACHIEVING THE GOAL — FIGURE 15.22   **Injection of antisense and double-stranded RNA into *C. elegans* to compare their effects on mRNA silencing.**

Starting material: The researchers used eggs from *C. elegans*. They also had the cloned *mex-3* gene, which had been previously shown to be highly expressed in the embryo.

	Experimental level	Conceptual level

1. Make sense and antisense *mex-3* RNA in vitro using cloned genes for *mex-3* with promoters on either side of the gene. RNA polymerase and nucleotides are added to synthesize the RNAs.

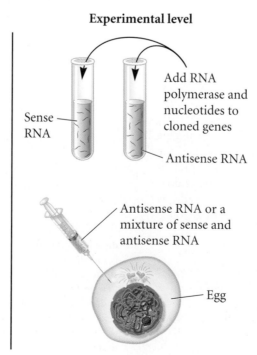

Sense RNA

Add RNA polymerase and nucleotides to cloned genes

Antisense RNA

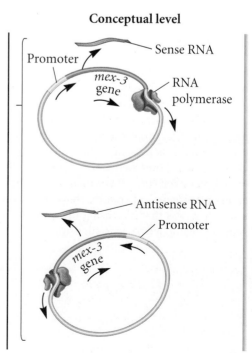

Promoter

Sense RNA

mex-3 gene

RNA polymerase

2. Inject *C. elegans* eggs (or early embryos) with either *mex-3* antisense RNA or a mixture of *mex-3* sense and antisense RNA. As a control, do not inject any RNA.

Antisense RNA or a mixture of sense and antisense RNA

Egg

Antisense RNA

Promoter

mex-3 gene

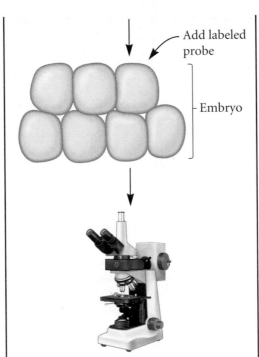

3. Incubate and then subject early embryos to *in situ* hybridization. In this method, a labeled probe is added that is complementary to *mex-3* mRNA. If cells express *mex-3*, the mRNA in the cells will bind to the probe and become labeled. After incubation with a labeled probe, the cells are washed to remove unbound probe.

4. Observe embryos under the microscope.

■ THE DATA

Control

Injected with *mex-3* antisense RNA

Injected with both *mex-3* sense and antisense RNA

Photos reprinted by permission from Macmillan Publishers Ltd. *Nature*. Potent and specific genetic interference by double-stranded RNA in *Caenohabditis elegans*. Andrew Fire et al. 391:6669, 806–811, 1998.

■ INTERPRETING THE DATA

As seen in the data of Figure 15.22, the control embryos were very darkly staining. These results indicated that the *mex-3* gene is highly expressed, which was known from previous research. In the embryos injected with antisense RNA, *mex-3* mRNA levels were decreased, but detectable. Remarkably, in samples injected with double-stranded RNA, no *mex-3* mRNA was detected! These results indicated that double-stranded RNA is more potent at silencing mRNA than is antisense RNA. In this case, the double-stranded RNA caused the *mex-3* mRNA to be degraded. They used the term **RNA interference** (**RNAi**) to describe the phenomenon in which double-stranded RNA causes the silencing of mRNA. This surprising observation led researchers to investigate the underlying molecular mechanism that accounts for this phenomenon, as described next.

A self-help quiz involving this experiment can be found at the Online Learning Center.

RNA Interference Is Mediated by MicroRNAs and the RNA-Induced Silencing Complex

RNA interference is widely found in eukaryotic species. Double-stranded RNA can come from a variety of sources. As mentioned, viruses sometimes produce double-stranded RNA as part of their reproductive cycle, and the insertion of multiple copies of a gene into the genome can also result in double-stranded RNA. In addition, the genomes of eukaryotic organisms have been found to contain genes that encode RNA molecules that cause RNA interference. These RNA molecules, called **microRNAs** (abbreviated **miRNAs**), are small RNA molecules, typically 21 to 23 nucleotides in length, that silence the expression of specific mRNAs. In 1993, Victor Ambros and his colleagues, who were interested in the developmental stages that occur in the worm *C. elegans*, determined that the transcription of a particular gene produced a small RNA, now called a microRNA, that does not encode a protein. Instead, this miRNA was found to be complementary to an mRNA, and it inhibited the translation of the mRNA. Since this study, researchers have discovered that genes encoding miRNAs are widely found in animals and plants. In humans, for example, approximately 200 different genes encode miRNAs.

MiRNAs represent an important mechanism of mRNA silencing. How do miRNAs cause the silencing of specific mRNAs? **Figure 15.23** shows how an miRNA, which is encoded by a gene,

Pre-miRNA transcribed from a gene

Hairpin structure

5′

3′

Dicer

The double-stranded RNA is cut by dicer to yield a double-stranded RNA about 21 to 23 bp long.

The double-stranded RNA is recognized by a protein that associates with other proteins to form the RNA-induced silencing complex (RISC). One of the RNA strands is degraded.

RISC

Complementary region between cellular mRNA and miRNA

The RISC recognizes specific cellular mRNAs, due to complementarity.

RISC

OR

The cellular mRNA is degraded.
(High complementarity)

The mRNA is unable to be translated.
(Low complementarity)

ONLINE ANIMATION

FIGURE 15.23 **Mechanism of RNA interference.** A double-stranded region of RNA within a pre-miRNA is recognized by an endonuclease called dicer that digests the RNA to produce a small piece, 21 to 23 bp in length. This small RNA binds with proteins to form an RNA-induced silencing complex (RISC). One strand of the RNA will be degraded. The remaining RNA strand in RISC will bind to complementary mRNAs. After RISC binding, endonucleases within the complex may cut the mRNA and thereby inactivate it. It should be emphasized that the miRNAs are mRNA specific. For example, if the miRNA within RISC was complementary to the β-globin mRNA, only the β-globin mRNA would be silenced.

leads to RNA interference. In this example, the miRNA is first synthesized as a pre-miRNA, which forms a hairpin structure that is cut by an endonuclease called **dicer.** This releases a double-stranded RNA molecule that associates with cellular proteins to form a complex called the **RNA-induced silencing complex (RISC).** One of the RNA strands is degraded. The remaining single-stranded miRNA, which is complementary to specific cellular mRNAs, will allow the RISC to specifically recognize and bind to those mRNA molecules.

Upon binding with the mRNA, two different things may happen. In some cases, the RISC may direct the degradation of the mRNA. This tends to occur when the miRNA and mRNA are a perfect match or are highly complementary. Alternatively, the RISC may inhibit translation. This is more common when the miRNA and mRNA are not a perfect match or are only partially complementary. In either case, the expression of the mRNA is silenced. The effect is termed RNA interference because the miRNA interferes with the proper expression of an mRNA. Likewise, dicer and RISC may act upon double-stranded RNA that comes from viruses or from the insertion of multiple copies of a gene into the genome (as shown at the bottom of Figure 15.21). In 2006, Fire and Mello received the Nobel Prize for their discovery of this phenomenon.

RNA interference is believed to have at least three benefits. First, this phenomenon represents a newly identified form of gene regulation. When genes encoding pre-miRNAs are turned on, the production of miRNAs will silence the expression of other genes. Second, RNAi may offer a defense mechanism against certain viruses. In particular, this mechanism may inhibit the proliferation of viruses that have a double-stranded RNA genome and viruses that produce double-stranded RNAs as part of their reproductive cycles. After entering the host cell, the viral RNA would be degraded, as in Figure 15.23, and the cell would survive the infection. Third, researchers have speculated that RNAi may also play a role in the silencing of certain transposable elements. As discussed in Chapter 17, transposable elements are DNA segments that have the capacity to move throughout the genome, an event termed transposition. Transposable elements carry genes, such as transposase, that are needed in the transposition process. The random insertion of many transposable elements in a cell may ultimately lead to gene silencing due to the scenario described in Figure 15.21. The silencing of genes within transposable elements via RNAi would protect the organism against the potentially harmful effects of transposition.

Phosphorylation of Ribosomal Initiation Factors Can Alter the Rate of Translation

Let's now turn our attention to another regulatory mechanism that affects translation. Modulation of translational initiation factors is widely used to control fundamental cellular processes. Under certain conditions, it is advantageous for a cell to stop synthesizing proteins. For example, if a virus infects a cell, the inhibition of protein synthesis can prevent viral proliferation by inhibiting the production of new viral proteins. Likewise, if critical nutrients are in short supply, it is beneficial for a cell to conserve its resources by inhibiting unnecessary protein synthesis.

Translational initiation factors are required to begin protein synthesis. The phosphorylation of many different initiation factors has been found to affect translation in eukaryotic cells. Two factors, eIF2 and eIF4F, appear to play a central role in controlling the initiation of translation. The functions of these two translational initiation factors are modulated by phosphorylation in opposite ways. When the α subunit of eIF2 (known as eIF2α) is phosphorylated, translation is inhibited, whereas the phosphorylation of eIF4F increases the rate of translation.

Figure 15.24 shows the events leading to translational inhibition by eIF2α. A variety of conditions can lead to a shutdown of protein synthesis, including viral infection, nutrient deprivation, heat shock, and the presence of toxic heavy metals. These conditions promote the activation of protein kinases known as eIF2α protein kinases. Several eIF2α protein kinases have been identified. Once activated, an eIF2α protein kinase can phosphorylate eIF2α. The phosphorylation of eIF2α causes it to bind tightly to another initiation factor subunit called eIF2B. Functional eIF2B is necessary so that eIF2 can promote the binding of the initiator tRNA^{met} to the 40S ribosomal subunit. However, when the phosphorylated eIF2α binds to eIF2B, it prevents eIF2B from functioning. Therefore, the initiator tRNA^{met} does not bind to the 40S subunit, and translation is inhibited.

A second important way to control translation is via the eIF4F translation factor that modulates the binding of mRNA to the ribosomal initiation complex. The function of eIF4F is stimulated by phosphorylation. A variety of conditions have been shown to cause eIF4F to become phosphorylated. These include the presence of growth factors, insulin, and other signaling molecules that promote cell proliferation. Conversely, conditions such as heat shock and viral infection decrease the level of eIF4F phosphorylation and thereby inhibit translation.

The Regulation of Iron Assimilation Is an Example of the Regulatory Effect of RNA-Binding Proteins on Translation

As we have just seen, the phosphorylation of translational initiation factors can modulate the translation of mRNA. Because these initiation factors are necessary to translate all of a cell's mRNA, this form of regulation affects the expression of many mRNAs. By comparison, specific mRNAs are sometimes regulated by RNA-binding proteins that directly affect translational initiation or RNA stability. The regulation of iron assimilation provides a well-studied example in which both of these phenomena occur. Before discussing this form of translational control, let's consider the biology of iron metabolism.

Iron is an essential element for the survival of living organisms because it is required for the function of many different enzymes. The pathway by which mammalian cells take up iron is depicted in **Figure 15.25**. Iron ingested by an animal is absorbed

FIGURE 15.24 **The pathway that leads to the phosphorylation of eIF2α (eukaryotic initiation factor α) and the inhibition of translation.**

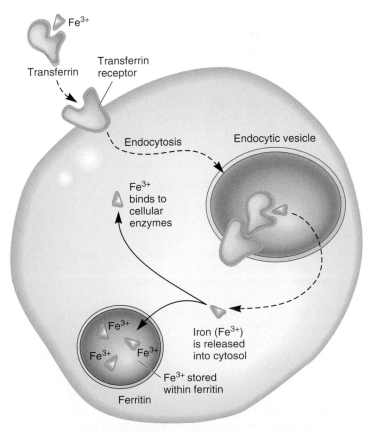

FIGURE 15.25 **The uptake of iron (Fe³⁺) into mammalian cells.**

into the bloodstream and becomes bound to transferrin, a protein that carries iron through the bloodstream. The transferrin-Fe^{3+} complex is recognized by a transferrin receptor on the surface of cells; the complex binds to the receptor and then is transported into the cytosol by endocytosis. Once inside, the iron is then released from transferrin. At this stage, Fe^{3+} may bind to cellular enzymes that require iron for their activity. Alternatively, if too much iron is present, the excess iron is stored within a hollow, spherical protein known as ferritin. The storage of excess iron within ferritin helps to prevent the toxic buildup of too much iron within the cell.

Because iron is a vital yet potentially toxic substance, mammalian cells have evolved an interesting way to regulate iron assimilation. The two mRNAs that encode ferritin and the transferrin receptor are both influenced by an RNA-binding protein known as the **iron regulatory protein (IRP)**. How does IRP exert its effects? This protein binds to a regulatory element within these two mRNAs known as the **iron response element (IRE)**. The ferritin mRNA has an IRE in its 5'-untranslated region (5'-UTR). When IRP binds to this IRE, it inhibits the translation of the ferritin mRNA (**Figure 15.26a**, left). However, when iron is abundant in the cytosol, the iron binds directly to IRP and prevents it from binding to the IRE. Under these conditions, the ferritin mRNA is translated to make more ferritin protein (Figure 15.26a, right), which prevents the toxic buildup of iron within the cytosol.

The transferrin receptor mRNA also contains an iron response element, but it is located in the 3'-UTR. When IRP binds to this IRE, it does not inhibit translation. Instead, the binding of IRP increases the stability of the mRNA by blocking the action

(a) Regulation of ferritin mRNA

(b) Regulation of transferrin receptor mRNA

FIGURE 15.26   The regulation of iron assimilation genes by IRP and IRE. (a) When Fe^{3+} concentrations are low, the binding of the iron regulatory protein (IRP) to the iron response element (IRE) in the 5'-UTR of ferritin mRNA inhibits translation (left). When Fe^{3+} concentrations are high and Fe^{3+} binds to IRP, the IRP is removed from the ferritin mRNA, so translation can proceed (right). **(b)** The binding of IRP to IRE in the 3'-UTR of the transferrin receptor mRNA enhances the stability of the mRNA and leads to a higher concentration of this mRNA. Therefore, more transferrin receptor protein is made when Fe^{3+} concentration is low (left). When Fe^{3+} levels are high and this metal binds to IRP, the IRP dissociates from the IRE, and the transferrin receptor mRNA is rapidly degraded (right).

Genes→Traits This form of translational control allows cells to use iron appropriately. When the cellular concentration of iron is low, the translation of the transferrin receptor is increased, thereby enhancing the ability of cells to take up more iron. Also, the translation of ferritin mRNA is inhibited, because ferritin is not needed to store excess iron. By comparison, when the cellular concentration of iron is high, the translation of ferritin mRNA is enhanced. This leads to the synthesis of ferritin and prevents the toxic buildup of iron. When iron is high, the transferrin receptor mRNA is degraded, which decreases further uptake of iron.

of RNA-degrading enzymes. This leads to increased amounts of transferrin receptor mRNA within the cell when the cytosolic levels of iron are very low (**Figure 15.26b**, left). Under these conditions, more transferrin receptor is made. This promotes the uptake of iron when it is in short supply. In contrast, when iron is abundant within the cytosol, IRP is removed from the transferrin receptor mRNA, and the mRNA becomes rapidly degraded (Figure 15.26b, right). This leads to a decrease in the amount of transferrin receptor and thereby helps to prevent the uptake of too much iron into the cell.

CONCEPTUAL SUMMARY

In this chapter, we have surveyed a wide variety of molecular mechanisms through which eukaryotic organisms regulate gene expression. Eukaryotic cells have an array of regulatory transcription factors that influence the ability of RNA polymerase to begin transcription of a particular gene. Such transcription factors recognize and bind to short DNA sequences called regulatory elements that control a particular gene. The binding of regulatory transcription factors may enhance (up regulate) the rate of transcription, in which case the transcription factor is termed an activator and the sequence it binds to is called an enhancer. Conversely, the binding may prevent transcription from occurring (down regulate), in which case the transcription factor is termed a repressor and the sequence, a silencer. Regulatory elements can be bidirectional and function at a fairly large distance away from the core promoter site.

Regulatory transcription factors may influence the function of RNA polymerase by interacting with protein complexes called TFIID and mediator. We considered common ways that the functions of regulatory transcription factors themselves are modulated. The glucocorticoid hormone is an example of an effector molecule that binds to a regulatory transcription factor (the glucocorticoid receptor) and alters its function. Once the hormone is bound, it activates several different genes by binding to glucocorticoid response elements (GREs) that are next to the genes. By comparison, the CREB protein is a regulatory transcription factor that is activated in response to cell-signaling molecules that cause increased intracellular levels of cAMP. The CREB protein binds to a response element known as CRE. When phosphorylated, CREB proteins stimulate transcription.

Gene regulation may also involve changes in the level of chromatin compaction or changes in DNA structure. For transcription to occur, the chromatin cannot be in a closed conformation. Instead, it must be in a loosely packed or open conformation. Transcriptional activators play a role in the recruitment of enzymes to the promoter region that convert the chromatin from a closed to an open conformation. These include ATP-dependent chromatin remodeling enzymes and histone acetyltransferase. The DNA itself can be altered by gene amplification, gene rearrangement, and DNA methylation. The first two mechanisms are uncommon and not examined in this chapter. In contrast, DNA methylation in vertebrates and higher plants appears to be an important mechanism to turn genes off in a tissue-specific or developmentally specific manner. CpG islands, regions that are commonly 1,000 to 2,000 base pairs in length and contain a high concentration of CpG sites, may be found in the vicinity of promoters. The methylation of CpG islands is one way to inhibit transcription. Methylation may prevent the binding of transcription factors and/or promote compaction. Methylated DNA sequences are passed from parent to daughter cell during cell division, a process called maintenance methylation. This preserves the methylated condition in future cells and appears to be a process that routinely occurs within vertebrate and plant cells.

Gene regulation can also occur at the level of RNA processing and translation. For example, a pre-mRNA transcript can be regulated by alternative splicing, an RNA processing event that allows different proteins to be made from a single type of pre-mRNA. RNA editing involves an addition or deletion of bases or a conversion of one type of base to another. The stability of RNA can also be affected by particular sequences within the mRNA. One important feature influencing stability is the length of the polyA tail; one example of a destabilizing element is the AU-rich element found in mRNAs. RNA can also be a target for degradation by double-stranded RNA, a mechanism termed RNA interference (RNAi).

We explored two ways in which the rate of mRNA translation can be controlled. First, the phosphorylation of two initiation factors, eIF2α and eIF4F, can affect translation in eukaryotic cells. The functions of these factors are modulated in opposite ways. When eIF2α is phosphorylated, translation is inhibited, whereas the phosphorylation of eIF4F increases the rate of translation. Second, specific RNA-binding proteins can directly affect the ability of certain mRNAs to be translated. The regulation of iron assimilation genes by iron regulatory proteins (IRPs) and iron response elements (IREs) is a form of translational control that allows cells to respond to changing levels of iron in the cell.

EXPERIMENTAL SUMMARY

The regulation of gene expression in eukaryotes has been studied via many different techniques. In Chapter 15, we have considered a few approaches to study this phenomenon. In one example, we saw how Weintraub and Groudine used DNase I sensitivity to evaluate the degree of globin gene packing in reticulocytes versus brain cells and fibroblasts. This approach exploits the fact that tightly packed chromatin is less susceptible to DNase I digestion than is DNA in an open conformation.

Other studies have focused on the identification of regulatory transcription factors that influence the expression of particular genes. By comparing the structures of many different transcription factors, researchers have found that they tend to have common domains that either interact with small effector molecules or bind to DNA. The sequences of many regulatory elements have also been determined. In a few cases, the combined efforts of many research groups have elucidated the detailed molecular mechanisms for the regulation of particular genes. For example, we examined how the glucocorticoid receptor and CREB protein regulate genes at the level of transcription.

Insights into how specific RNAs can be targeted for degradation or translational inhibition came from a study by Fire and Mello that showed how double-stranded RNA is more potent than antisense RNA at silencing mRNA. Their research led to the discovery that this phenomenon, termed RNA interference, is mediated by microRNAs and an RNA-induced silencing complex (RISC). The solved problems and experimental questions at the end of Chapter 15 also serve to illustrate several techniques that have been used to understand the molecular mechanisms of eukaryotic gene regulation.

PROBLEM SETS & INSIGHTS

Solved Problems

S1. Describe how the tight packing of chromatin in a closed conformation may prevent the initiation of gene transcription.

Answer: The tight packing of chromatin may physically inhibit transcription in a few ways. First, it may prevent transcription factors and/or RNA polymerase from binding to the major groove of the DNA. Second, tight packing may prevent RNA polymerase from forming an open complex, which is necessary to begin transcription.

S2. What are the two alternative ways that IRP can affect gene expression at the RNA level?

Answer: The ferritin mRNA has an IRE in its 5'-UTR. When IRP binds to this IRE, it inhibits the translation of the ferritin mRNA. This decreases the amount of ferritin protein, which is not needed when iron levels are low. However, when iron is abundant in the cytosol, the iron binds directly to IRP and prevents it from binding to the IRE. This allows the ferritin mRNA to be translated, producing more ferritin protein. An IRE is also located in the transferrin receptor mRNA in the 3'-UTR. When IRP binds to this IRE, it increases the stability of the mRNA, which leads to an increase in the amount of transferrin receptor mRNA within the cell when the cytosolic levels of iron are very low. Under these conditions, more transferrin receptor is made to promote the uptake of iron, which is in short supply. When the iron is found in abundance within the cytosol, IRP is removed from the mRNA, and the mRNA becomes rapidly degraded. This leads to a decrease in the amount of the transferrin receptor.

S3. Eukaryotic regulatory elements are often orientation independent and can function in a variety of locations. Explain the meaning of this statement.

Answer: Orientation independence means that the regulatory element can function in the forward or reverse direction. In addition, regulatory elements can function at a variety of locations that may be upstream or downstream from the core promoter. A loop in the DNA must form to bring the regulatory transcription factors bound at the regulatory elements and the core promoter in close proximity with one another.

S4. To gain a molecular understanding of how the glucocorticoid receptor works, geneticists have attempted to "dissect" the protein to identify smaller domains that play specific functional roles. Using recombinant DNA techniques described in Chapter 18, particular segments in the coding region of the glucocorticoid receptor gene can be removed. The altered gene can then be expressed in a living cell to see if functional aspects of the receptor have been changed or lost. For example, the removal of

the portion of the gene encoding the carboxyl terminal half of the protein causes a loss of glucocorticoid binding. These results indicate that the carboxyl terminal portion contains a domain that functions as a glucocorticoid-binding site. The figure shown here illustrates the locations of several functional domains within the glucocorticoid receptor relative to the entire amino acid sequence.

Based on your understanding of the mechanism of glucocorticoid receptor function described in Figure 15.6, explain the functional roles of the different domains in the glucocorticoid receptor.

Answer: The hormone-binding domain is located in the carboxyl terminal half of the protein. This part of the protein also contains the region necessary for HSP90 binding and receptor dimerization. A nuclear localization sequence (NLS) is located near the center of the protein. After hormone binding, the NLS is exposed on the surface of the protein and allows it to be targeted to the nucleus. The DNA-binding domain, which contains zinc fingers, is also centrally located in the primary amino acid sequence. Zinc fingers promote DNA binding to the major groove. Finally, two separate regions of the protein, one in the amino terminal half and one in the carboxyl terminal half, are necessary for the transactivation of RNA polymerase. If these domains are removed from the receptor, it can still bind to the DNA, but it cannot activate transcription.

S5. A common approach to identify genetic sequences that play a role in the transcriptional regulation of a gene is the strategy that is sometimes called **promoter bashing.** This approach requires gene cloning methods, which are described in Chapter 18. A clone is obtained that has the coding region for a structural gene as well as the region that is upstream from the core promoter. This upstream region is likely to contain genetic regulatory elements such as enhancers and silencers. The diagram shown here depicts a cloned DNA region that contains the upstream region, the core promoter, and the coding sequence for a protein that is expressed in human liver cells. The upstream region may be several thousand base pairs in length.

Upstream region Core promoter Coding sequence of liver-specific gene

To determine if promoter bashing has an effect on transcription, it is helpful to have an easy way to measure the level of gene expression. One way to accomplish this is to swap the coding sequence of the gene of interest with the coding sequence of another gene. For example, the coding sequence of the *lacZ* gene, which encodes β-galactosidase, is frequently swapped because it is easy to measure the activity of β-galactosidase using an assay for its enzymatic activity. The *lacZ* gene is called a "reporter gene" because it is easy to measure its activity. As shown here, the coding sequence of the *lacZ* gene has been swapped with the coding sequence of the liver-specific gene. In this new genetic construct, the expression and transcriptional regulation of the *lacZ* gene is under the control of the core promoter and upstream region of the liver-specific gene.

Upstream region Core promoter   *lacZ* gene coding sequence

Now comes the "bashing" part of the experiment. Different segments of the upstream region are deleted, and then the DNA is transformed into living cells. In this case, the researcher would probably transform the DNA into liver cells, because those are the cells where the gene is normally expressed. The last step is to measure the β-galactosidase activity in the transformed liver cells

In the diagram shown here, the upstream region and the core promoter have been divided into five regions, labeled A–E.

Upstream region Core promoter   *lacZ* gene coding sequence

One of these regions was deleted (i.e., bashed out), and the rest of the DNA segment was transformed into liver cells. The data shown here are from this experiment.

Region Deleted	Percentage of β-Galactosidase Activity*
None	100
A	100
B	330
C	100
D	5
E	<1

*The amount of β-galactosidase activity in the cells carrying an undeleted upstream and promoter region was assigned a value of 100%. The amounts of activity in the cells carrying a deletion were expressed relative to this 100% value.

Explain what these results mean.

Answer: The amount of β-galactosidase activity found in liver cells that do not carry a deletion reflects the amount of expression under normal circumstances. If the core promoter (region E) is deleted, very little expression is observed. This is expected because a core promoter is needed for transcription. If enhancers are deleted, the activity should be less than 100%. It appears that one or more enhancers are found in region D. If a silencer is deleted, the activity should be above 100%. From the data shown it appears that one or more silencers are found in region B. Finally, if a deletion has no effect, there may not be any regulatory elements there. This was observed for regions A and C.

Note: The deletion of an enhancer will have an effect on β-galactosidase activity only if the cell is expressing the regulatory transcription factor that binds to the enhancer and activates transcription. Likewise, the deletion of a silencer will have an effect only if the cell is expressing the repressor protein that binds to the silencer and inhibits transcription. In the problem described, the liver cells must be expressing the activators and repressors that recognize the regulatory elements found in regions B and D, respectively.

Conceptual Questions

C1. Discuss the common points of control in eukaryotic gene regulation.

C2. Discuss the structure and function of regulatory elements. Where are they located relative to the core promoter?

C3. What is meant by the term transcription factor modulation? List three general ways that this can occur.

C4. What are the functions of transcriptional activator proteins and repressor proteins? Explain how they work at the molecular level.

C5. Are the following statements true or false?

A. An enhancer is a type of regulatory element.

B. A core promoter is a type of regulatory element.

C. Regulatory transcription factors bind to regulatory elements.

D. An enhancer may cause the down regulation of transcription.

C6. Transcription factors usually contain one or more motifs that play key roles in their function. What is the function of the following motifs?

A. Helix-turn-helix

B. Zinc finger

C. Leucine zipper

C7. The binding of an effector molecule, protein-protein interactions, and covalent modifications are three common ways to modulate the activities of transcription factors. Which of these three mechanisms are used by steroid receptors and the CREB protein?

C8. Describe the steps that occur for the glucocorticoid receptor to bind to a GRE.

C9. Let's suppose a mutation in the glucocorticoid receptor does not prevent the binding of the glucocorticoid hormone to the protein but prevents the ability of the receptor to activate transcription. Make a list of all the possible defects that may explain why transcription cannot be activated.

C10. Explain how phosphorylation affects the function of the CREB protein.

C11. A particular drug inhibits the protein kinase that is responsible for phosphorylating the CREB protein. How would this drug affect the following events?

A. The ability of the CREB protein to bind to CREs

B. The ability of extracellular hormone to enhance cAMP levels

C. The ability of the CREB protein to stimulate transcription

D. The ability of the CREB protein to dimerize

C12. The glucocorticoid receptor and the CREB protein are two examples of transcriptional activators. These proteins bind to response elements and activate transcription. (Note: The answer to this question is not directly described in Chapter 15. You have to rely on your understanding of the functioning of other proteins that are modulated by the binding of effector molecules, such as the lac repressor.)

A. How would the function of the glucocorticoid receptor be shut off?

B. What type of enzyme would be needed to shut off the activation of transcription by the CREB protein?

C13. Transcription factors such as the glucocorticoid receptor and the CREB protein form homodimers and activate transcription. Other transcription factors form heterodimers. For example, a transcription factor known as myogenic bHLH forms a heterodimer with a protein called the E protein. This heterodimer activates the transcription of genes that promote muscle cell differentiation. However, when myogenic bHLH forms a heterodimer with a protein called the Id protein, transcriptional activation does not occur. (Note: Id stands for Inhibitor of differentiation.) Which of the following possibilities would best explain this observation? Only one possibility is correct.

| | **Myogenic** | | |
	bHLH	**E Protein**	**Id Protein**
Possibility 1			
DNA-binding domain:	Yes	No	No
Leucine zipper:	Yes	No	Yes
Possibility 2			
DNA-binding domain:	Yes	Yes	No
Leucine zipper:	Yes	Yes	Yes
Possibility 3			
DNA-binding domain:	Yes	No	Yes
Leucine zipper:	Yes	No	No

C14. An enhancer, located upstream from a gene, has the following sequence:

5'–GTAG–3'
3'–CATC–5'

This enhancer is orientation independent. Which of the following sequences would also work as an enhancer?

A. 5'–CTAC–3'
   3'–GATG–5'

B. 5'–GATG–3'
   3'–CTAC–5'

C. 5'–CATC–3'
   3'–GTAG–5'

C15. The DNA-binding domain of each CREB protein recognizes the sequence 5'–TGACGTCA–3'. As a matter of random chance, how often would you expect this sequence to occur in the human genome, which contains approximately 3 billion base pairs? Actually, only a few dozen genes are activated by the CREB protein. Does the value of a few dozen agree with the number of random sites found in the human genome? If the number of random sites in the human genome is much higher than a few dozen, provide at least one explanation why the CREB protein is not activating more than a few dozen genes.

C16. Solved problem S4 shows the locations of domains in the glucocorticoid receptor relative to the amino and carboxyl terminal ends of the protein. Make a drawing that illustrates the binding of a glucocorticoid receptor dimer to the DNA. In your drawing, label the amino and carboxyl ends, the hormone-binding domains, the DNA-binding domains, the dimerization domains, and the transactivation domains.

C17. The gene that encodes the enzyme called tyrosine hydroxylase is known to be up regulated by the CREB protein. Tyrosine hydroxylase is expressed in nerve cells and is involved in the synthesis of catecholamine, a neurotransmitter. The exposure of cells to adrenaline normally up regulates the transcription of the tyrosine hydroxylase gene. A mutant cell line was identified in which the tyrosine hydroxylase gene was not up regulated when exposed to adrenaline. List all the possible mutations that could explain this defect. How would you explain the defect if only the tyrosine hydroxylase gene was not up regulated by the CREB protein and other genes having CREs were properly up regulated in response to adrenaline in this cell line?

C18. What is the predominant form of chromatin in the eukaryotic nucleus? What must happen to its structure for transcription to take place? Discuss what is believed to cause this change in chromatin structure.

C19. Explain how the acetylation of core histones may loosen chromatin packing.

C20. Figure 15.11 shows the nucleosome structure of the β-globin gene in an active and inactive state. Suggest two or more possible reasons why the arrangement shown in Figure 15.11a would be transcriptionally inactive.

C21. What is DNA methylation? When we say that DNA methylation is heritable, what do we mean? How is it passed from a mother to a daughter cell?

C22. Let's suppose that a vertebrate organism carries a mutation that causes some cells that would normally differentiate into nerve cells to differentiate into muscle cells. A molecular analysis of this mutation revealed that it was in a gene that encodes a methyltransferase. Explain how an alteration in a methyltransferase could produce this phenotype.

C23. What is a CpG island? Where would you expect one to be located? How does the methylation of CpG islands affect gene expression?

C24. What is the function of a splicing factor? Explain how splicing factors can regulate the tissue-specific splicing of mRNAs.

C25. Figure 15.16 shows the products of alternative splicing for the α-tropomyosin pre-mRNA. Let's suppose that smooth muscle cells produce splicing factors that are not produced in other cell types. Explain where you think such splicing factors bind and how they influence the splicing of the α-tropomyosin pre-mRNA.

C26. Let's suppose a person is homozygous for a mutation in the IRP gene that changed the structure of the iron regulatory protein in such a way that it could not bind iron, but it could still bind to IREs. How would this mutation affect the regulation of ferritin and transferrin receptor mRNAs? Do you think such a person would need more iron in his/her diet compared to normal individuals? Do you think that excess iron in his/her diet would be more toxic compared to normal individuals? Explain your answer.

C27. In response to potentially toxic substances (e.g., high levels of iron), eukaryotic cells often use translational or posttranslational regulatory mechanisms to prevent cell death, rather than using transcriptional regulatory mechanisms. Explain why.

C28. What are the advantages and disadvantages of mRNAs with a short half-life compared to mRNAs with a long half-life?

C29. What is the phenomenon of RNA interference (RNAi)? During RNAi, explain how the double-stranded RNA is processed and how it leads to the silencing of a complementary mRNA.

C30. With regard to RNAi, what are three possible sources for double-stranded RNA?

C31. What conditions lead to the phosphorylation of eIF2α? Discuss why a cell would want to shut down translation when these conditions occur.

C32. What is the relationship between mRNA stability and mRNA concentration? What factors affect mRNA stability?

C33. Describe how the binding of the iron regulatory protein affects the mRNAs for ferritin and the transferrin receptor. How does iron influence this process?

Experimental Questions

E1. Why is a DNA site able to be cleaved by DNase I? With regard to DNase I cleavage, what do you expect will happen when a gene is converted from an inactive to a transcriptionally active state?

E2. In the experiment of Figure 15.10, explain how the use of S1 nuclease makes it possible to determine whether or not the β-globin gene is DNase I sensitive. In this experiment, why is it necessary to precipitate the DNA in step 7?

E3. Researchers can isolate a sample of cells, such as skin fibroblasts, and grow them in the laboratory. This procedure is called a cell culture. A cell culture can be exposed to a sample of DNA. If the cells are treated with agents that make their membranes permeable to DNA, the cells may take up the DNA and incorporate the DNA into their chromosomes. This process is called transformation or transfection. Experimenters have transformed human skin fibroblasts with methylated DNA and then allowed the fibroblasts to divide for several cellular generations. The DNA in the daughter cells was then isolated, and the segment that corresponded to the transformed DNA was examined. This DNA segment in the daughter cells was also found to be methylated. However, if the original skin fibroblasts were transformed with unmethylated DNA, the DNA found in the daughter cells was also unmethylated. Do fibroblasts undergo *de novo* methylation, maintenance methylation, or both? Explain your answer.

E4. Restriction enzymes, described in Chapter 18, are enzymes that recognize a particular DNA sequence and cleave the DNA (along the DNA backbone) at that site. The restriction enzyme known as *Not*I recognizes the sequence

5'–GCGGCCGC–3'
3'–CGCCGGCG–5'

However, if the cytosines in this sequence have been methylated, *Not*I will not cleave the DNA at this site. For this reason, *Not*I is commonly used to investigate the methylation state of CpG islands.

A researcher has studied a gene, which we will call gene *T*, that is found in corn. This gene encodes a transporter involved in the uptake of phosphate from the soil. A CpG island is located near the core promoter of gene *T*. The CpG island has a single *Not*I site. The arrangement of gene *T* is shown here.

	CpG island	Core promoter	Coding sequence for gene *T*
Sal I	1,500 bp	*Not* I	3,800 bp	*Eco*RI

A *Sal*I restriction site is located upstream from the CpG island, and an *Eco*RI restriction site is located near the end of the coding sequence for gene *T*. The distance between the *Sal*I and *Not*I sites is 1,500 bp, and the distance between the *Not*I and *Eco*RI sites is 3,800 bp. No other sites for *Sal*I, *Not*I, or *Eco*RI are found in this region.

Here is the question. Let's suppose a researcher has isolated DNA samples from four different tissues in a corn plant. These include the leaf, the tassel, a section of stem, and a section of root. The DNA was then digested with all three restriction enzymes, separated by gel electrophoresis, and then probed with a DNA fragment complementary to the gene *T* coding sequence. The results are shown here.

In which type of tissue is the CpG island methylated? Does this make sense based on the function of the protein encoded by gene *T*?

E5. You will need to understand solved problem S5 before answering this question. A muscle-specific gene was cloned and then subjected to promoter bashing as described in solved problem S5. As shown here, six regions, labeled A–F, were deleted, and then the DNA was transformed into muscle cells.

The data shown here are from this experiment.

Region Deleted	Percentage of β-Galactosidase Activity
None	100
A	20
B	330
C	100
D	5
E	15
F	<1

Explain these results.

E6. You will need to understand solved problem S5 before answering this question. A gene that is normally expressed in pancreatic cells was cloned and then subjected to promoter bashing as described in solved problem S5. As shown here, four regions, labeled A–D, were individually deleted, and then the DNA was transformed into pancreatic cells or into kidney cells.

The data shown here are from this experiment.

Region Deleted	Cell Type Transformed	Percentage of β-Galactosidase Activity
None	Pancreatic	100
A	Pancreatic	5
B	Pancreatic	100
C	Pancreatic	100
D	Pancreatic	<1
None	Kidney	<1
A	Kidney	<1
B	Kidney	100
C	Kidney	<1
D	Kidney	<1

If we assume that the upstream region has one silencer and one enhancer, answer the following questions:

A. Where are the silencer and enhancer located?

B. Why don't we detect the presence of the silencer in the pancreatic cells?

C. Why isn't this gene normally expressed in kidney cells?

E7. A gel retardation assay can be used to determine if a protein binds to a segment of DNA. When a segment of DNA is bound by a protein, its mobility will be retarded, and the DNA band will appear higher in the gel. In the gel retardation assay shown here, a cloned gene fragment that is 750 bp in length contains a regulatory element that is recognized by a transcription factor called protein X. Previous experiments have shown that the presence of hormone X results in transcriptional activation by protein X. The results of a gel retardation assay are shown here.

Explain the action of hormone X.

E8. Explain how the data of Fire and Mellow suggested that double-stranded RNA is responsible for the silencing of the *mex-3* gene.

E9. Chapter 18 describes a blotting method known as Northern blotting, in which a short segment of cloned DNA is used as a probe to detect RNA that is transcribed from a particular gene. The DNA probe, which is radioactive, is complementary to the RNA that the researcher wishes to detect. After the radioactive probe DNA binds to the RNA within a blot of a gel, the RNA is visualized as a dark (radioactive) band on an X-ray film. The method of Northern blotting can be used to determine the amount of a particular RNA transcribed in a given cell type. If one type of cell produces twice as much of a particular mRNA compared to another cell, the band appears twice as intense.

For this question, a researcher has a DNA probe complementary to the ferritin mRNA. This probe can be used to specifically detect the amount of ferritin mRNA on a gel. A researcher began with two flasks of human skin cells. One flask contained a very low concentration of iron, and the other flask had a high concentration of iron. The mRNA was isolated from these cells and then subjected to Northern blotting, using a probe complementary to the ferritin mRNA. The sample loaded in lane 1 was from the cells grown in a low concentration of iron, and the sample in lane 2 was from the cells grown in a high concentration of iron. Three Northern blots are shown here, but only one of them is correct. Based on your understanding of ferritin mRNA regulation, which blot (a, b, or c) would be your expected result? Explain. Which blot (a, b, or c)

would be your expected result if the gel had been probed with a DNA segment complementary to the transferrin receptor mRNA?

(a)   (b)   (c)

Questions for Student Discussion/Collaboration

1. Explain how DNA methylation could be used to regulate gene expression in a tissue-specific way. When and where would *de novo* methylation occur, and when would demethylation occur? What would occur in the germ line?

2. Enhancers can be almost anywhere and affect the transcription of a gene. Let's suppose you have a gene cloned on a piece of DNA, and the DNA fragment is 50,000 bp in length. Using cloning methods described in Chapter 18, you can cut out short segments from this 50,000 bp fragment and then reintroduce the smaller fragments into a cell that can express the gene. You would like to know if any enhancers are within the 50,000 bp region that may affect the expression of the gene. Discuss the most efficient strategy you can think of to trim your 50,000 bp fragment and thereby locate

enhancers. You can assume that the coding sequence of the gene is in the center of the 50,000 bp fragment and that you can trim the 50,000 bp fragment into any size piece you want using molecular techniques described in Chapter 18.

3. How are regulatory transcription factors and regulatory splicing factors similar in their mechanism of action? In your discussion, consider the domain structures of both types of proteins. How are they different?

Note: All answers appear at the website for this textbook; the answers to even-numbered questions are in the back of the textbook.

---

www.mhhe.com/brookergenetics3e

Visit the Online Learning Center for practice tests, answer keys, and other learning aids for this chapter. Enhance your understanding of genetics with our interactive exercises, quizzes, animations, and much more.

16

GENE MUTATION AND DNA REPAIR

The effects of a mutation. A mutation during embryonic development has caused this sheep to have a black spot on its side.

The primary function of DNA is to store information for the synthesis of cellular proteins. A key aspect of the gene expression process is that the DNA itself does not normally change. This allows DNA to function as a permanent storage unit. However, on relatively rare occasions, a mutation can occur. The term **mutation** refers to a heritable change in the genetic material. This means that the structure of DNA has been changed permanently, and this alteration can be passed from mother to daughter cells during cell division. If a mutation occurs in reproductive cells, it may also be passed from parent to offspring.

The topic of mutation is centrally important in all fields of genetics, including molecular genetics, Mendelian inheritance, and population genetics. Mutations provide the allelic variation that we have discussed throughout this textbook. For example, phenotypic differences, such as tall versus dwarf pea plants, are due to mutations that alter the expression of particular genes. With regard to their phenotypic effects, mutations can be beneficial, neutral, or detrimental. On the positive side, mutations are essential to the continuity of life. They provide the variation that enables species to change and adapt to their environment. Mutations are the foundation for evolutionary change. On the negative side, however, new mutations are much more likely to be harmful rather than beneficial to the individual. The genes within each species have evolved to work properly. They have functional promoters, coding sequences, terminators, and so on, that allow the genes to be expressed. Random mutations are more likely to disrupt these sequences rather than improve their function. For example, many inherited human diseases result from mutated genes. In addition, diseases such as skin and lung cancer can be caused by environmental agents that are known to cause DNA mutations. For these and many other reasons, understanding the molecular nature of mutations is a deeply compelling area of research. In this chapter, we will consider the nature of mutations and their consequences on gene expression at the molecular level.

Because mutations can be quite harmful, organisms have developed several ways to repair damaged DNA. DNA repair systems reverse DNA damage before it results in a mutation that could potentially have negative consequences. DNA repair systems have been studied extensively in many organisms, particularly *Escherichia coli*, yeast, mammals, and plants. A variety of systems repair different types of DNA lesions. In this chapter, we will examine the ways that several of these DNA repair systems operate.

16.1 CONSEQUENCES OF MUTATION

How do mutations affect phenotype? To answer this question, we must appreciate how changes in DNA structure can ultimately affect DNA function. Much of our understanding of mutation has come from the study of experimental organisms, such as bacteria, yeast, and *Drosophila*. Researchers can expose these organisms to environmental agents that cause mutation and then study the consequences of the induced mutations. In addition, because these organisms have a short generation time, researchers can investigate the effects of mutation when they are passed from cell to cell and from parent to offspring.

As discussed in Chapter 8, changes in chromosome structure and number are important occurrences within natural populations of eukaryotic organisms. These types of changes are considered to be mutations because the genetic material has been altered in a way that can be inherited. In comparison, a gene mutation is a relatively small change in DNA structure that affects a single gene. In this section, we will be primarily concerned with the ways that mutations may affect the molecular and phenotypic expression of single genes. We will also consider how the timing of mutations during an organism's development has important consequences.

Gene Mutations Are Molecular Changes in the DNA Sequence of a Gene

A gene mutation occurs when the sequence of the DNA within a gene is altered in a permanent way. A gene mutation can involve a base substitution in the sequence within a gene or a removal or addition of one or more base pairs.

A **point mutation** is a change in a single base pair within the DNA. For example, the DNA sequence shown here has been altered by a **base substitution,** in which one base is substituted for another base.

$$\downarrow$$

5'-AACGCTAGATC-3' → 5'-AACGCGAGATC-3'
3'-TTGCGATCTAG-5'   3'-TTGCGCTCTAG-5'

A change of a pyrimidine to another pyrimidine, such as C to T, or a purine to another purine, such as A to G, is called a **transition.** This type of mutation is more common than a **transversion,** in which purines and pyrimidine are interchanged. The example just shown is a transversion (a T to G change), not a transition.

Besides base substitutions, a short sequence of DNA may be deleted from or added to the chromosomal DNA:

$$\downarrow$$

5'-AACGCTAGATC-3' → 5'-AACGCTC-3'
3'-TTGCGATCTAG-5'   3'-TTGCGAG-5'

(Deletion of 4 bp)

$$\underline{\ \ }\downarrow\underline{\ \ }$$

5'-AACGCTAGATC-3' → 5'-AACAGTCGCTAGATC-3'
3'-TTGCGATCTAG-5'   3'-TTGTCAGCGATCTAG-5'

(Addition of 4 bp)

As we will see next, small deletions or additions to the sequence of a gene can significantly affect its function.

Gene Mutations Can Alter the Coding Sequence Within a Gene

How might a mutation within the coding sequence of a structural gene affect the amino acid sequence of the polypeptide that is encoded by the gene? **Table 16.1** describes the possible effects of point mutations. **Silent mutations** are those that do not alter the amino acid sequence of the polypeptide even though the nucleotide sequence has changed. Because the genetic code is degenerate, silent mutations can occur in certain bases within a codon, such as the third base, so that the specific amino acid is not changed. In contrast, **missense mutations** are base substitutions in which an amino acid change does occur. An example of a missense mutation occurs in the human disease known as sickle-cell anemia. This disease involves a mutation in the β-globin gene, which alters the polypeptide sequence so that the sixth amino acid is changed from a glutamic acid to valine. This single amino acid substitution alters the structure and function of the hemoglobin protein. One consequence of this alteration is that the red blood cells sickle under conditions of low oxygen (**Figure 16.1**). In this case, a single amino acid substitution has a profound effect on the phenotype of cells and even causes a serious disease.

Nonsense mutations involve a change from a normal codon to a stop codon. This terminates the translation of the polypeptide earlier than expected, producing a truncated polypeptide (see Table 16.1). When a nonsense mutation occurs in a bacterial operon, it may also inhibit the expression of downstream genes. This phenomenon, termed **polarity,** is described in solved problem S4 at the end of this chapter. Finally, **frameshift mutations** involve the addition or deletion of a number of nucleotides that is not divisible by three. Because the codons are read in multiples of three, this shifts the reading frame, Therefore, translation of the mRNA will result in a completely different amino acid sequence downstream from the mutation.

Except for silent mutations, new mutations are more likely to produce polypeptides that have reduced rather than enhanced function. For example, nonsense mutations will produce polypeptides that are substantially shorter and, therefore, unlikely to function properly. Likewise, frameshift mutations dramatically alter the amino acid sequence of polypeptides and are thereby likely to disrupt function. Missense mutations are less likely to alter function because they involve a change of a single amino acid within polypeptides that typically contain hundreds of amino acids. When a missense mutation has no detectable effect on protein function, it is referred to as a **neutral mutation.** A missense mutation that substitutes an amino acid with a similar chemistry as the original amino acid is likely to be neutral or nearly neutral. For example, a missense mutation that substitutes a glutamic acid for an aspartic acid is likely to be neutral because both amino acids are negatively charged and have similar side chain structures. Silent mutations are also considered a type of neutral mutation.

TABLE 16.1

Consequences of Point Mutations Within the Coding Sequence

Type of Change	Mutation in the DNA	Example*	Amino Acids Altered	Likely Effect on Protein Function
None	None	5′–A–T–G–A–C–C–G–A–C–C–C–G–A–A–A–G–G–G–A–C–C–3′ Met – Thr – Asp – Pro – Lys – Gly – Thr –	None	None
Silent	Base substitution	↓ 5′–A–T–G–A–C–C–G–A–C–C–C–C–A–A–A–G–G–G–A–C–C–3′ Met – Thr – Asp – Pro – Lys – Gly – Thr –	None	None
Missense	Base substitution	↓ 5′–A–T–G–C–C–C–G–A–C–C–C–G–A–A–A–G–G–G–A–C–C–3′ Met – Pro – Asp – Pro – Lys – Gly – Thr –	One	Neutral or inhibitory
Nonsense	Base substitution	↓ 5′–A–T–G–A–C–C–G–A–C–C–C–G–T–A–A–G–G–G–A–C–C–3′ Met – Thr – Asp – Pro – STOP!	Many	Inhibitory
Frameshift	Addition/deletion	↓ 5′–A–T–G–A–C–C–G–A–C–G–C–C–G–A–A–A–G–G–G–A–C–C–3′ Met – Thr – Asp – Ala – Glu – Arg – Asp –	Many	Inhibitory

*DNA sequence in the coding strand. Note that this sequence is the same as the mRNA sequence except that the RNA contains uracil (U) instead of thymine (T). The 3-base codons are shown in alternating black and red colors. Mutations are shown in green.

Normal red blood cells   ⊢——— 10 µm ———⊣   Sickled red blood cells   ⊢——— 10 µm ———⊣

(a) Micrographs of red blood cells

NORMAL: NH₂ – VALINE – HISTIDINE – LEUCINE – THREONINE – PROLINE – GLUTAMIC ACID – GLUTAMIC ACID...

SICKLE CELL: NH₂ – VALINE – HISTIDINE – LEUCINE – THREONINE – PROLINE – VALINE – GLUTAMIC ACID...

(b) A comparison of the amino acid sequence between normal β-globin and sickle-cell β-globin

FIGURE 16.1 **Missense mutation in sickle-cell anemia.** (a) Normal red blood cells (left) and sickled red blood cells (right). (b) A comparison of the amino acid sequence of the normal β-globin polypeptide and the polypeptide encoded by the sickle-cell allele. This figure shows only a portion of the polypeptide sequence, which is 146 amino acids long. As seen here, a missense mutation changes the sixth amino acid from a glutamic acid to a valine.

Genes→Traits A missense mutation alters the structure of β globin, which is a subunit of hemoglobin, the oxygen-carrying protein in the red blood cells. When an individual is homozygous for this allele, this missense mutation causes the red blood cells to sickle under conditions of low oxygen concentration. The sickling phenomenon is a description of the trait at the cellular level. At the organism level, the sickled cells can clog the capillaries, thereby causing painful episodes, called crises, that can result in organ damage. The shortened life span of the red blood cells leads to symptoms of anemia.

Mutations can occasionally produce a polypeptide that has an enhanced ability to function. While these favorable mutations are relatively rare, they may result in an organism with a greater likelihood to survive and reproduce. If this is the case, natural selection may cause such a favorable mutation to increase in frequency within a population. This topic will be discussed later in this chapter and also in Chapter 24.

Gene Mutations Are Also Given Names That Describe How They Affect the Wild-Type Genotype and Phenotype

Thus far, several genetic terms have been introduced that describe the molecular effects of mutations. Genetic terms are also used to describe the effects of mutations relative to a wild-type genotype or phenotype. In a natural population, the **wild type** is a relatively prevalent genotype. For some genes with multiple alleles, a population may have two or more wild-type alleles. A mutation may change the wild-type genotype by altering the DNA sequence of a gene. When such a mutation is rare in a population, it is generally referred to as a **mutant allele.** A reverse mutation, more commonly called a **reversion,** changes a mutant allele back to a wild-type allele.

Another way to describe a mutation is based on its influence on the wild-type phenotype. Mutants are often characterized by their differential ability to survive. As mentioned, a neutral mutation does not alter protein function, so it does not affect survival or reproductive success. A **deleterious mutation** will decrease the chances of survival and reproduction. The extreme example of a deleterious mutation is a **lethal mutation,** which results in death to the cell or organism. On the other hand, a **beneficial mutation** will enhance the survival or reproductive success of an organism. In some cases, an allele may be either deleterious or beneficial depending on the genotype and/or the environmental conditions. An example is the sickle-cell allele. In the homozygous state, the sickle-cell allele lessens the chances of survival. However, when an individual is heterozygous for the sickle-cell allele and wild-type allele, this increases the chances of survival due to malarial resistance. Finally, some mutations are called **conditional mutants** because they affect the phenotype only under a defined set of conditions. Geneticists often study conditional mutants in microorganisms; a common example is a temperature-sensitive (*ts*) mutant. A bacterium that has a *ts* mutation grows normally in one temperature range—the permissive temperature range—but exhibits defective growth at a different temperature range—the nonpermissive temperature range. For example, an *E. coli* strain carrying a *ts* mutation may be able to grow from 33 to 38°C but not between 40 to 42°C, whereas the wild-type strain can grow at either temperature range.

A second mutation will sometimes affect the phenotypic expression of a first mutation. As an example, let's consider a mutation that causes an organism to grow very slowly. A second mutation at another site in the organism's DNA may restore the normal growth rate, converting the mutant back to the wild-type phenotype. Geneticists call these second-site mutations **suppressors** or **suppressor mutations.** This name is meant to indicate that a suppressor mutation acts to suppress the phenotypic effects of another mutation. A suppressor mutation differs from a reversion, because it occurs at a DNA site that is distinct from the first mutation.

Suppressor mutations are classified according to their relative locations with regard to the mutation they suppress (**Table 16.2**). When the second mutant site is within the same gene as the first mutation, the mutation is termed an **intragenic suppressor.** This type of suppressor often involves a change in protein structure that compensates for an abnormality in protein structure caused by the first mutation. Researchers often isolate suppressor mutations to obtain information about protein structure and function. For example, Robert Brooker and colleagues have isolated many intragenic suppressors in the *lacY* gene of *E. coli*, which encodes the lactose permease described in Chapter 15. This protein must undergo conformational changes to transport lactose across the cell membrane. They began with single mutations that altered amino acids on transmembrane regions, which inhibited this conformational change and thereby prevented growth on lactose. Suppressor mutations were then isolated that allowed growth on lactose by restoring transport function. By analyzing the locations of these suppressor mutations, the researchers were able to determine that certain transmembrane regions in the protein are critical for conformational changes required for lactose transport.

Alternatively, a suppressor mutation can be in a different gene from the first mutation—an **intergenic suppressor.** Researchers often study intergenic suppressors as a way to gain information about proteins that have similar or redundant functional roles, proteins that participate in a common pathway, multimeric proteins with two or more subunits, and the regulation of protein expression by transcription factors. How do intergenic suppressors work? These suppressor mutations usually involve a change in the expression of one gene that compensates for a loss-of-function mutation affecting another gene (see Table 16.2). For example, a first mutation may cause one protein to be partially or completely defective. An intergenic suppressor mutation in a different structural gene might overcome this defect by altering the structure of a second protein so that it could take over the functional role the first protein cannot perform. Alternatively, intergenic suppressors may involve proteins that participate in a common cellular pathway. When a first mutation decreases the activity of a protein, a suppressor mutation could enhance the function of a second protein involved in this pathway and thereby overcome the defect in the first protein.

In some cases, intergenic suppressors involve multimeric proteins, with each subunit encoded by a different gene. A mutation in one subunit that inhibits function may be compensated for by a mutation in another subunit. Another type of intergenic suppressor is one that involves mutations in genetic regulatory proteins such as transcription factors. When a first mutation causes a protein to be defective, a suppressor mutation may occur in a gene that encodes a transcription factor. The mutant transcription factor transcriptionally activates another gene that can compensate for the loss-of-function mutation in the first gene.

TABLE 16.2

Examples of Suppressor Mutations

Type	No Mutation	First Mutation	Second Mutation	Description
Intragenic	Transport can occur	Transport inhibited	Transport can occur	A first mutation disrupts normal protein function and a suppressor mutation affecting the same protein restores function. In this example, the first mutation inhibits lactose transport function, and the second mutation restores lactose transport.
Intergenic Redundant function	Enzymatic function	Loss of enzymatic function	Gain of a new enzymatic function	A first mutation inhibits the function of a protein, and a second mutation alters a different protein to carry out that function. In this example, the proteins function as enzymes.
Common pathway	Precursor → Fast → Intermediate → Slow → Product	Precursor → Slow → Intermediate → Slow → Little product	Precursor → Slow → Intermediate → Fast → Product	Two or more different proteins may be involved in a common pathway. A mutation that causes a defect in one protein may be compensated for by a mutation that alters the function of a different protein in the same pathway.
Multimeric protein	Active	Inactive	Active	A mutation in a gene encoding one protein subunit that inhibits function may be suppressed by a mutation in a gene that encodes a different subunit. The double mutant has restored function.
Transcription factor	Normal function	Loss of function	Loss of function	A first mutation causes loss of function of a particular protein. A second mutation may alter a transcription factor and cause it to activate the expression of another gene. This other gene encodes a protein that can compensate for the loss of function caused by the first mutation.

Mutant transcription factor turns on a gene that compensates for the loss of function

Transcription factor

Compensates for inactive protein

Causes expression of this protein.

Less commonly, intergenic suppression involves mutations in nonstructural genes that enable the cell to defy the genetic code. Examples are suppressor tRNA mutants, which have been identified in microorganisms. Suppressor tRNA mutants have a change in the anticodon region that causes the tRNA to behave contrary to the genetic code. Nonsense suppressors are mutant tRNAs that recognize a stop codon, and instead of stopping translation, put an amino acid into the growing polypeptide chain. This type of mutant tRNA can suppress a nonsense mutation in a structural gene. However, such bacterial strains grow poorly because they may also suppress stop codons in normal genes.

Gene Mutations Can Occur Outside of the Coding Sequence and Still Influence Gene Expression

Thus far, we have focused our attention primarily on mutations in the coding regions of genes and their effects on polypeptide structure and protein function. In previous chapters, we learned how various sequences outside the coding sequence play important roles during the process of gene expression. A mutation can occur within noncoding sequences and thereby affect gene expression (Table 16.3). For example, a mutation may alter the sequence within the core promoter of a gene. Promoter mutations that increase transcription are termed **up promoter mutations.** Mutations that make a sequence more like the consensus sequence are likely to be up promoter mutations. In contrast, in a **down promoter mutation,** a mutation causes the promoter to become less like the consensus sequence, decreasing its affinity for regulatory factors and decreasing the transcription rate.

In Chapter 14, we considered how mutations could affect regulatory elements. For example, mutations in the *lac* operator site, called *lacO^C* mutations, prevent the binding of the lac repressor protein. This causes the *lac* operon to be constitutively expressed even in the absence of lactose. Bacteria strains with *lacO^C* mutations are at a selective disadvantage compared to wild-type *E. coli* strains because they waste their energy expressing the *lac* operon even when these proteins are not needed. As noted in Table 16.3, mutations can also occur in other noncoding regions of a gene and alter gene expression in a way that may

affect phenotype. For example, mutations that affect the untranslated regions of mRNA—the 5'-UTR and 3'-UTR—may affect gene expression if they alter its ability to be translated or its stability. In addition, mutations in eukaryotic genes can alter splice junctions and affect the order and/or number of exons that are contained within mRNA.

DNA Sequences Known as Trinucleotide Repeats Are Hotspots for Mutation

Researchers have discovered several human genetic diseases caused by an unusual form of mutation known as **trinucleotide repeat expansion (TNRE).** The term refers to the phenomenon in which a repeated sequence of three nucleotides can readily increase in number from one generation to the next. In normal individuals, certain genes and chromosomal locations contain regions where trinucleotide sequences are repeated in tandem. These sequences are transmitted normally from parent to offspring without mutation. However, in persons with TNRE disorders, the length of a trinucleotide repeat has increased above a certain critical size and becomes prone to frequent expansion. This phenomenon is depicted here, where the trinucleotide repeat of CAG has expanded from 11 tandem copies to 18 copies.

CAGCAGCAGCAGCAGCAGCAGCAGCAGCAGCAG   n = 11

to

CAGCAGCAGCAGCAGCAGCAGCAGCAGCAGCAGCAGCAG-
CAGCAGCAGCAG   n = 18

Several human diseases have been discovered that involve these types of expansions, including spinal and bulbar muscular atrophy (SBMA), Huntington disease (HD), spinocerebellar ataxia (SCA1), fragile X syndromes (FRAXA and FRAXE), and myotonic muscular dystrophy (DM) (Table 16.4). In some cases, the expansion is within the coding sequence of the gene. Typically, such an expansion is a CAG repeat. Because CAG encodes a glutamine codon, these repeats cause the encoded proteins to contain long tracks of glutamine. Although the reasons for the disease symptoms in TNRE disorders are not well understood, the presence of glutamine tracts causes the proteins to aggregate. This aggregation of proteins or protein fragments carrying glutamine repeats is correlated with the progression of the disease. In other TNRE disorders, the expansions are located in the noncoding regions of genes. In the case of the two fragile X syndromes, the repeat produces CpG islands that become methylated. As discussed in Chapter 15, methylation can lead to chromatin compaction and thereby silence gene transcription. For myotonic muscular dystrophy, it has been hypothesized that these expansions cause abnormal changes in RNA structure and thereby produce disease symptoms.

TNRE disorders have the unusual feature that their severity tends to worsen in future generations—a phenomenon called **anticipation.** However, anticipation does not occur with all TNRE disorders and usually depends on whether the disease is inherited from the mother or father. In the case of Huntington disease, anticipation is likely to occur if the mutant gene is inherited from the father. In contrast, myotonic muscular dystrophy is

---

TABLE 16.3

Possible Consequences of Gene Mutations Outside of the Coding Sequence

Sequence	Effect of Mutation
Promoter	May increase or decrease the rate of transcription
Regulatory element/operator site	May disrupt the ability of the gene to be properly regulated
5'-UTR/3'-UTR	May alter the ability of mRNA to be translated; may alter mRNA stability
Splice recognition sequence	May alter the ability of pre-mRNA to be properly spliced

TABLE **16.4**

TNRE Disorders

Disease	SBMA	HD	SCA1	FRAXA	FRAXE	DM
Repeat Sequence	CAG	CAG	CAG	CGG	GCC	CTG
Location of Repeat	Coding sequence	Coding sequence	Coding sequence	5'-UTR	5'-UTR	3'-UTR
Number of Repeats in Unaffected Individuals	11–33	6–37	6–44	6–53	6–35	5–37
Number of Repeats in Affected Individuals	36–62	27–121	43–81	>200	>200	>200
Pattern of Inheritance	X linked	Autosomal dominant	Autosomal dominant	X linked	X linked	Autosomal dominant
Disease Symptoms	Neuro-degenerative	Neuro-degenerative	Neuro-degenerative	Mental impairment	Mental impairment	Muscle disease
Anticipation *	None	Male	Male	Female	None	Female

*Indicates the parent in which anticipation occurs most prevalently.

more likely to get worse if the gene is inherited from the mother. These results suggest that TNRE can happen more frequently during oogenesis or spermatogenesis, depending on the particular gene involved. Overall, TNRE is a newly discovered form of mutation that is receiving a lot of attention by the research community. It poses many challenging questions in molecular genetics. The phenomenon of TNRE also makes it particularly difficult for genetic counselors to advise couples as to the severity of these diseases if they are passed to their children.

At the DNA level, the cause of TNRE is not well understood. Researchers have speculated that tandem repeats, such as trinucleotide repeats, produce alterations in DNA structure, including stem-loop formation and DNA strand slippage, and this may lead to errors in DNA replication. However, future research will be necessary to understand the underlying mechanism that causes TNRE. Nevertheless, it is well established that TNRE within certain genes alters the expression of the gene and thereby produces the disease symptoms.

Changes in Chromosome Structure Can Affect the Expression of a Gene

Thus far, we have considered small changes in the DNA sequence of particular genes. A change in chromosome structure can also be associated with an alteration in the expression of specific genes. Quite commonly, an inversion or translocation has no obvious phenotypic consequence. However, in 1925, Alfred Sturtevant was the first to recognize that chromosomal rearrangements in *Drosophila melanogaster* can influence phenotypic expression (namely, eye morphology). In some cases, a chromosomal rearrangement may affect a gene because a chromosomal **breakpoint**—the region where two chromosome pieces break and rejoin with other chromosome pieces—occurs within a gene. A breakpoint within the middle of a gene is very likely to inhibit gene function because it will separate the gene into two pieces.

In other cases, a gene may be left intact, but its expression may be altered when it is moved to a new location. When this occurs, the change in gene location is said to have a **position effect.** How do position effects alter gene expression? Researchers have discovered two common explanations. **Figure 16.2** depicts a schematic example in which a piece of one chromosome has been inverted or translocated to a different chromosome. One possibility is that a gene may be moved next to regulatory sequences, such as silencers or enhancers, that influence the expression of the relocated gene (Figure 16.2a). Alternatively, a chromosomal rearrangement may reposition a gene from a less condensed or euchromatic chromosome to a very highly condensed or heterochromatic chromosome. When the gene is moved to a heterochromatic region, its expression may be turned off (Figure 16.2b). This second type of position effect may produce a variegated phenotype in which the expression of the gene is variable. For genes that affect pigmentation, this produces a mottled appearance rather than an even color. **Figure 16.3** shows a position effect that alters eye color in *Drosophila.* Figure 16.3a depicts a normal red-eyed fruit fly, and Figure 16.3b shows a mutant fly that has inherited a chromosomal rearrangement in which a gene affecting eye color has been relocated to a heterochromatic chromosome. The variegated appearance of the eye occurs because the degree of heterochromatin formation varies across different regions of the eye. In cells where heterochromatin formation has turned off the eye color gene, a white phenotype occurs, while other cells allow this same region to remain euchromatic and produce a red phenotype.

Mutations Can Occur in Germ-Line or Somatic Cells

In this section, we have considered many different ways that mutations affect gene expression. For multicellular organisms, the timing of mutations also plays an important role. A mutation can occur very early in life, such as in a gamete or a fertilized egg, or it may occur later in life, such as in the embryonic or adult stages. The exact time when mutations occur can be important with regard to the severity of the genetic effect and whether they are passed from parent to offspring.

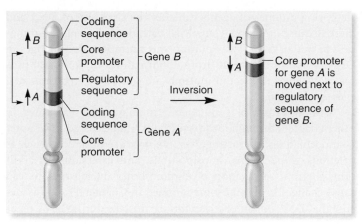

(a) Position effect due to regulatory sequences

(b) Position effect due to translocation to a heterochromatic chromosome

FIGURE 16.2   Causes of position effects. (a) A chromosomal inversion has repositioned the core promoter of gene *A* next to the regulatory sequences for gene *B*. Because regulatory sequences are often bidirectional, the regulatory sequences for gene *B* may regulate the transcription of gene *A*. **(b)** A translocation has moved a gene from a euchromatic to a heterochromatic chromosome. This type of position effect prevents the expression of the relocated gene.

Geneticists classify the cells of animals into two types—the germ line and the somatic cells. The term **germ line** refers to cells that give rise to the gametes such as eggs and sperm. A **germ-line mutation** can occur directly in a sperm or egg cell, or it can occur in a precursor cell that produces the gametes. If a mutant gamete participates in fertilization, all cells of the resulting offspring will contain the mutation (**Figure 16.4a**). Likewise, when an individual with a germ-line mutation produces gametes, the mutation may be passed along to future generations of offspring.

The **somatic cells** comprise all cells of the body excluding the germ-line cells. Examples include muscle cells, nerve cells, and skin cells. Mutations can also happen within somatic cells at early or late stages of development. **Figure 16.4b** illustrates the consequences of a mutation that took place during the embryonic stage. In this example, a **somatic mutation** has occurred

FIGURE 16.3   A position effect that alters eye color in *Drosophila*. (a) A normal red eye. **(b)** An eye in which an eye color gene has been relocated to a heterochromatic chromosome. This can inactivate the gene in some cells and produces a variegated phenotype.

Genes→Traits Variegated eye color occurs because the degree of heterochromatin formation varies throughout different regions of the eye. In some cells, heterochromatin formation occurs and turns off the eye color gene, thereby leading to the white phenotype. In other patches of cells, the region containing the eye color allele remains euchromatic, yielding a red phenotype.

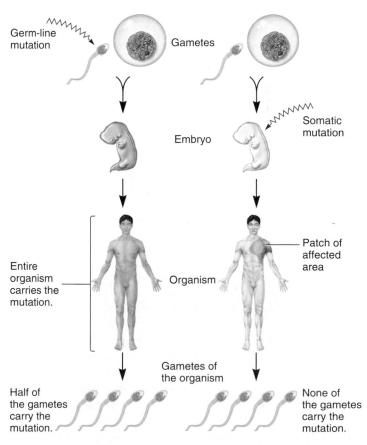

FIGURE 16.4   The effects of germ-line versus somatic mutations.

within a single embryonic cell. As the embryo grows, this single cell will be the precursor for many cells of the adult organism. Therefore, in the adult, a portion of the body will contain the mutation. The size of the affected region will depend on the timing of the mutation. In general, the earlier the mutation occurs during development, the larger the region. An individual that has somatic regions that are genotypically different from each other is called a **genetic mosaic.**

Figure 16.5 illustrates an individual who had a somatic mutation occur during an early stage of development. In this case, the person has a patch of white hair, but the rest of the hair is pigmented. Presumably, this individual initially had a single mutation occur in an embryonic cell that ultimately gave rise to a patch of scalp that produced the white hair. Although a patch of white hair is not a harmful phenotypic effect, mutations during early stages of life can be quite detrimental, especially if they disrupt essential developmental processes. Therefore, even though it is smart to avoid environmental agents that cause mutations during all stages of life, the possibility of somatic mutations is a rather compelling reason to avoid them during the very early stages of life such as fetal development, infancy, and early childhood. For example, the possibility of somatic mutations in an embryo is a reason why women are advised to avoid getting X-rays during pregnancy.

16.2 OCCURRENCE AND CAUSES OF MUTATION

As we have seen, mutations can have a wide variety of effects on the phenotypic expression of genes. For this reason, geneticists have expended a great deal of effort identifying the causes of mutations. This task has been truly challenging, because a

© Scott Aitken, www.scottpix.com

FIGURE 16.5   Example of a somatic mutation.
Genes→Traits This person has a patch of white hair because a somatic mutation occurred in a single cell during embryonic development that prevented pigmentation of the hair. This cell continued to divide to produce a patch of white hair.

staggering number of agents can alter the structure of DNA and thereby cause mutation. Geneticists categorize the cause of mutation in one of two ways. **Spontaneous mutations** are changes in DNA structure that result from abnormalities in biological processes, whereas **induced mutations** are caused by environmental agents (**Table 16.5**).

Many causes of spontaneous mutations are examined in other chapters throughout this textbook. As discussed in Chapter 8, abnormalities in crossing over can produce mutations such as deletions, duplications, translocations, and inversions. In addition, aberrant segregation of chromosomes during meiosis can cause changes in chromosome number. In Chapter 11, we discovered that DNA polymerase can make a mistake during DNA replication by putting the wrong base in a newly synthesized daughter strand. Errors in DNA replication are usually infrequent except in certain viruses, such as HIV, that have relatively high rates of spontaneous mutations. Also, normal metabolic processes may produce chemicals within the cell that can react directly with the DNA and alter its structure. As discussed in Chapter 17, transposable genetic elements can alter gene sequences by inserting themselves into

TABLE **16.5**	
Causes of Mutations	
Common Causes of Mutations	**Description**
Spontaneous	
Aberrant recombination	Abnormal crossing over may cause deletions, duplications, translocations, and inversions (see Chapter 8).
Aberrant segregation	Abnormal chromosomal segregation may cause aneuploidy or polyploidy (see Chapter 8).
Errors in DNA replication	A mistake by DNA polymerase may cause a point mutation (see Chapter 11).
Toxic metabolic products	The products of normal metabolic processes may be chemically reactive agents that can alter the structure of DNA.
Transposable elements	Transposable elements can insert themselves into the sequence of a gene (see Chapter 17).
Depurination	On rare occasions, the linkage between purines (i.e., adenine and guanine) and deoxyribose can spontaneously break. If not repaired, it can lead to mutation.
Deamination	Cytosine and 5-methylcytosine can spontaneously deaminate to create uracil or thymine.
Tautomeric shifts	Spontaneous changes in base structure can cause mutations if they occur immediately prior to DNA replication.
Induced	
Chemical agents	Chemical substances may cause changes in the structure of DNA.
Physical agents	Physical phenomena such as UV light and X-rays can damage the DNA.

genes. In this section, we will examine how spontaneous changes in nucleotide structure, such as depurination, deamination, and tautomeric shifts, can cause mutation. Overall, a distinguishing feature of spontaneous mutations is that their underlying cause originates within the cell. By comparison, the cause of induced mutations originates outside the cell. Induced mutations are produced by environmental agents, either chemical or physical, that enter the cell and lead to changes in DNA structure. Agents known to alter the structure of DNA are called **mutagens.**

We will begin by examining the random nature of spontaneous mutations and general features of the mutation rate. Then we will explore several mechanisms by which mutagens can alter the structure of DNA. Finally, laboratory tests that can identify potential mutagens will be described.

Spontaneous Mutations Are Random Events

For a couple of centuries, biologists had questioned whether heritable changes occur purposefully as a result of behavior or exposure to particular environmental conditions or whether they are spontaneous events that may happen randomly. In the nineteenth century, the naturalist Jean Baptiste Lamarck proposed that physiological events—such as the use or disuse of muscles—determine whether traits are passed along to offspring. For example, his hypothesis suggested that an individual who practiced and became adept at a physical activity and developed muscular legs would pass that characteristic on to the next generation. The alternative point of view is that genetic variation exists in a population as a matter of random chance and natural selection results in the differential reproductive success of organisms that are better adapted to their environments. Those individuals who happen to have beneficial mutations will be more likely to survive and pass these genes to their offspring. These opposing ideas of the nineteenth century—one termed physiological adaptation and the other termed random mutation—were tested in bacterial studies in the 1940s and 1950s, two of which are described here.

Salvadore Luria and Max Delbrück were interested in the ability of bacteria to become resistant to infection by a bacteriophage called T1. When a population of *E. coli* cells is exposed to T1, a small percentage of bacteria are found to be resistant to T1 infection and pass this trait to their progeny. Luria and Delbrück were interested in whether such resistance, called *ton^r* (T one resistance), is due to the occurrence of random mutations or whether it is a physiological adaptation that occurs at a low rate within the bacterial population.

According to the physiological adaptation hypothesis, the rate of mutation should be a relatively constant value and depend on the exposure to the bacteriophage. Therefore, when comparing different populations of bacteria, the number of *ton^r* bacteria should be an essentially constant proportion of the total population. In contrast, the random mutation hypothesis depends on the timing of mutation. If a *ton^r* mutation occurs early within the proliferation of a bacterial population, many *ton^r* bacteria will be found within that population because mutant bacteria will have time to grow and multiply. However, if it occurs much later in

population growth, fewer *ton^r* bacteria will be observed. In general, a random mutation hypothesis predicts a much greater fluctuation in the number of *ton^r* bacteria among different populations.

To distinguish between the physiological adaptation and random mutation hypotheses, Luria and Delbrück inoculated one large flask and 20 individual tubes with *E. coli* cells and grew them in the absence of T1 phage. The flask was grown to produce a very large population of cells, while each individual culture was grown to a smaller population of approximately 20 million cells. They then plated the individual cultures onto media containing T1 phage. Likewise, 10 subsamples, each consisting of 20 million bacteria, were removed from the large flask and plated onto media with T1 phage.

The results of the Luria and Delbrück experiment are shown in **Figure 16.6.** Within the smaller individual cultures, a great fluctuation was observed in the number of *ton^r* mutants. This test, therefore, has become known as the **fluctuation test.** Which hypothesis do these results support? They are consistent with a random mutation hypothesis in which the timing of a mutation during the growth of a culture greatly affects the number of mutant cells. For example, in tube 14, many *ton^r* bacteria were observed. Luria and Delbrück reasoned that a mutation occurred randomly in one bacterium at an early stage of the population growth, before the bacteria were exposed to T1 on plates. This mutant bacterium then divided to produce many daughter cells that inherited the *ton^r* trait. In other tubes, such as 1 and 3, this spontaneous mutation did not occur, so none of the bacteria had a *ton^r* phenotype. By comparison, the cells plated from the large flask tended toward a relatively constant and intermediate number of *ton^r* bacteria. Because the large flask had so many cells, several independent *ton^r* mutations were likely to have occurred during different stages of its growth. In a single flask, however, these independent events would be mixed together to give an average value of *ton^r* cells.

Several years later, Joshua and Esther Lederberg were also interested in the relationship between mutation and the environmental conditions that select for mutation. To distinguish between the physiological adaptation and random mutation hypotheses, they developed a technique known as **replica plating** in the 1950s. As shown in **Figure 16.7,** they plated a large number of bacteria onto a master plate that did not contain any selective agent (namely, the T1 phage). A sterile piece of velvet cloth was lightly touched to this plate in order to pick up a few bacterial cells from each colony. This replica was then transferred to two secondary plates that contained an agent that selected for the growth of bacterial cells with a particular genotype.

In the example shown in Figure 16.7, the secondary plates contained T1 bacteriophages. On these plates, only those mutant cells that are *ton^r* could grow. On the secondary plates, a few colonies were observed. Strikingly, they occupied the same location on each plate. These results indicated that the mutations conferring *ton^r* occurred randomly while the cells were growing on the nonselective master plate. The presence of the T1 phage in the secondary plates simply selected for the growth of previously occurring *ton^r* mutants. These results supported the random mutation hypothesis. In contrast, the physiological adaptation hypothesis

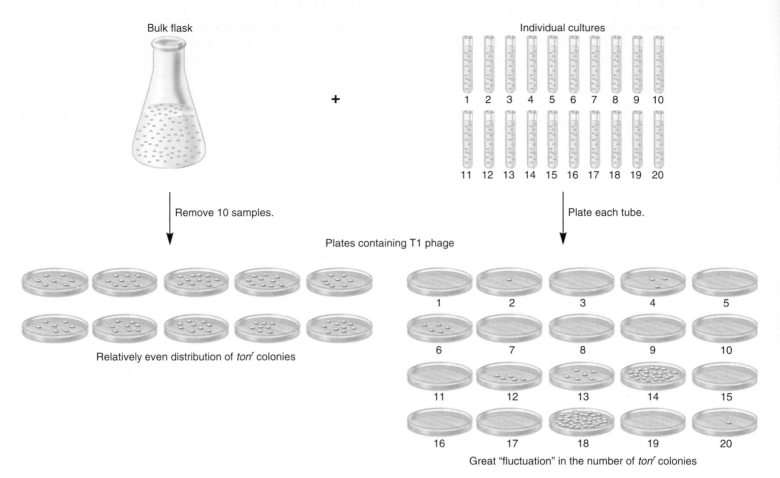

FIGURE 16.6   **The Luria-Delbrück fluctuation test.**

would have predicted that *ton^r* bacterial mutants would occur after exposure to the selective agent. If that had been the case, the colonies would not be expected to arise in identical locations on different secondary plates but rather in random patterns.

Taken together, the results of Luria and Delbrück and those of the Lederbergs supported the random mutation hypothesis, now known as the **random mutation theory.** According to this theory, mutations are a random process—they can occur in any gene and do not involve exposure of an organism to a particular condition that selects for specific types of mutations. In some cases, a random mutation may provide a mutant organism with an advantage, such as resistance to T1 phage. Although such mutations occur as a matter of random chance, growth conditions may select for organisms that happen to carry them.

As researchers have learned more about mutation at the molecular level, the view that mutations are a totally random process has required some modification. Within the same individual, some genes mutate at a much higher rate than other genes. Why does this happen? Some genes are larger than others, which provides a greater chance for mutation. Also, the relative locations of genes within a chromosome may cause some genes to be more susceptible to mutation compared to others. Even within a single gene, **hot spots** are usually found—certain regions of a gene that

are more likely to mutate compared to other regions (refer back to Chapter 6, Figure 6.20).

Mutation Rates and Frequencies Are Ways to Quantitatively Assess Mutation in a Population

Because mutations occur spontaneously among populations of living organisms, geneticists are greatly interested in learning how prevalent they are. The term **mutation rate** is the likelihood that a gene will be altered by a new mutation. This rate is commonly expressed as the number of new mutations in a given gene per cell generation. In general, the spontaneous mutation rate for a particular gene is in the range of 1 in 100,000 to 1 in 1 billion, or 10^{-5} to 10^{-9} per cell generation. These numbers tell us that it is very unlikely that a particular gene will mutate due to natural causes. However, the mutation rate is not a constant number. The presence of certain environment agents, such as X-rays, can increase the rate of induced mutations to a much higher value than the spontaneous mutation rate. In addition, mutation rates vary substantially from species to species and even within different strains of the same species. One explanation for this variation is that there are many different causes of mutations (refer back to Table 16.5).

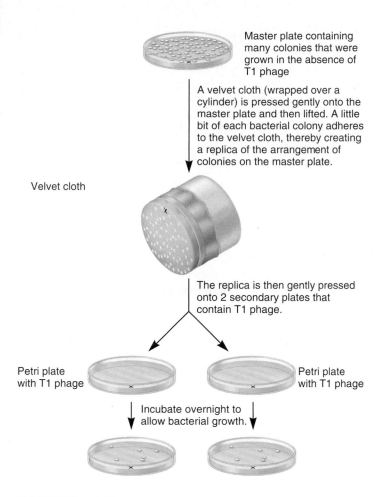

FIGURE 16.7   Replica plating. Bacteria were first plated on a master plate under nonselective conditions. A sterile velvet cloth was used to make a replica of the master plate. This replica was gently pressed onto two secondary plates that contained a selective agent. In this case, the two secondary plates contained T1 bacteriophage. Only those mutant cells that are *ton^r* (resistant to T1) could grow to form visible colonies. Note: The black x indicates the alignment of the velvet and the plates.

Before we end our discussion of mutation rate, let's distinguish the rate of new mutation from the concept of **mutation frequency.** The mutation frequency for a gene is the number of mutant genes divided by the total number of genes within a population. If 1 million bacteria were plated and 10 were found to be mutant, the mutation frequency would be 1 in 100,000, or 10^{-5}. As we have seen, Luria and Delbrück showed that among the bacteria in the 20 tubes, the timing of mutations influenced the mutation frequency within any particular tube. Some tubes had a high frequency of mutation, while others did not. Therefore, the mutation frequency depends not only on the mutation rate but also on the timing of mutation, and on the likelihood that a mutation will be passed to future generations. The mutation frequency is an important genetic concept, particularly in the field of population genetics. As we will see in Chapter 24, mutation frequencies may rise above the mutation rate due to evolutionary forces such as natural selection and genetic drift.

Spontaneous Mutations Can Arise by Depurination, Deamination, and Tautomeric Shifts

Thus far, we have considered the random nature of mutation and the quantitative difference between mutation rate and mutation frequency. We now turn our attention to the molecular changes in DNA structure that can cause mutation. Our first examples concern changes that can occur spontaneously, albeit at a low rate. The most common type of chemical change that occurs naturally is **depurination,** which involves the removal of a purine (adenine or guanine) from the DNA. The covalent bond between deoxyribose and a purine base is somewhat unstable and occasionally undergoes a spontaneous reaction with water that releases the base from the sugar, thereby creating an **apurinic site (Figure 16.8a)**. In a typical mammalian cell, approximately 10,000 purines are lost from the DNA in a 20-hour period at 37°C. The rate of loss is higher if the DNA is exposed to agents that cause certain types of base modifications such as the attachment of alkyl groups (methyl or ethyl groups). Fortunately, as discussed later in this chapter, apurinic sites are recognized by DNA base excision repair enzymes that repair the site. If the repair system fails, however, a mutation may result during subsequent rounds of DNA replication. What happens at an apurinic site during DNA replication? Because a complementary base is not present to specify the incoming base for the new strand, any of the four bases will be added to the

(a) Depurination

(b) Replication over an apurinic site

FIGURE 16.8   Spontaneous depurination. (a) The bond between guanine and deoxyribose is broken, thereby releasing the base. This leaves an apurinic site in the DNA. **(b)** If an apurinic site remains in the DNA as it is being replicated, any of the four nucleotides can be added to the newly made strand. Because three out of four (A, T, and G) are the incorrect base, the chance of causing a mutation is 75%.

new strand in the region that is opposite the apurinic site (**Figure 16.8b**). This may lead to a new mutation.

A second spontaneous lesion that may occur in DNA is the **deamination** of cytosines. The other bases are not readily deaminated. As shown in **Figure 16.9a**, deamination involves the removal of an amino group from the cytosine base. This produces uracil. As discussed later, DNA repair enzymes can recognize uracil as an inappropriate base within DNA and subsequently remove it. However, if such repair does not take place, a mutation may result because uracil hydrogen bonds with adenine during DNA replication. Therefore, if a DNA template strand has uracil instead of cytosine, a newly made strand will incorporate adenine into the daughter strand instead of guanine.

Figure 16.9b shows the deamination of 5-methylcytosine. As discussed in Chapter 15, the methylation of cytosine occurs in many eukaryotic species. It also occurs in prokaryotes. If 5-methylcytosine is deaminated, the resulting base is thymine,

which is a normal constituent of DNA. Therefore, this poses a problem for DNA repair enzymes because they cannot distinguish which is the incorrect base—the thymine that was produced by deamination or the guanine in the opposite strand that originally base-paired with the methylated cytosine. For this reason, methylated cytosine bases tend to create hot spots for mutation. As an example, researchers analyzed 55 spontaneous mutations that occurred within the *lacI* gene of *E. coli* and determined that 44 of them involved changes at sites that were originally occupied by a methylated cytosine base.

A third way that mutations may arise spontaneously involves a temporary change in base structure called a **tautomeric shift.** In this case, the **tautomers** are bases, which exist in keto and enol or amino and imino forms. These forms can interconvert by a chemical reaction that involves the migration of a hydrogen atom and a switch of a single bond and an adjacent double bond. The common, stable form of guanine and thymine is the keto form; the common form of adenine and cytosine is the amino form (**Figure 16.10a**). At a low rate, G and T can interconvert to an enol form, and A and C can change to an imino form. Though the relative amounts of the enol and imino forms of these bases are relatively small, they can cause a mutation because these rare forms of the bases do not conform to the AT/GC rule of base pairing. Instead, if one of the bases is in the enol or imino form, hydrogen bonding will promote TG and CA base pairs, as shown in **Figure 16.10b**.

How does a tautomeric shift cause a mutation? The answer is that it must occur immediately prior to DNA replication. When DNA is in a double-stranded condition, the base pairing usually holds the bases in their more stable forms. After the strands unwind, however, a tautomeric shift may occur. In the example shown in **Figure 16.10c**, a thymine base in the template strand has undergone a tautomeric shift just prior to the replication of the complementary daughter strand. During replication, the daughter strand will incorporate a guanine opposite this thymine, creating a base mismatch. This mismatch could be repaired via the proofreading function of DNA polymerase or via a mismatch repair system (discussed later in this chapter). However, if these repair mechanisms fail, the next round of DNA replication will create a double helix with a CG base pair, while the correct base pair should be TA. As shown in the right side of Figure 16.10c, one of four daughter cells will inherit this CG mutation.

(a) Deamination of cytosine

(b) Deamination of 5-methylcytosine

FIGURE 16.9   Spontaneous deamination of cytosine and 5-methylcytosine. (a) The deamination of cytosine produces uracil. **(b)** The deamination of 5-methylcytosine produces thymine.

(a) Tautomeric shifts that occur in the 4 bases found in DNA

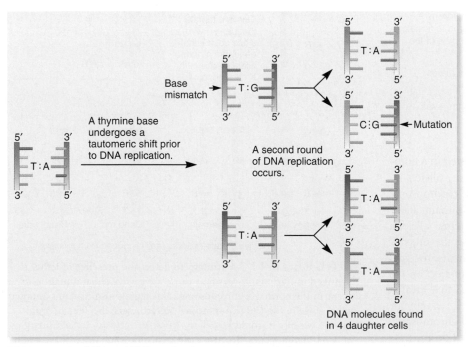

(b) Mis–base pairing due to tautomeric shifts

(c) Tautomeric shifts and DNA replication can cause mutation

FIGURE 16.10   Tautomeric shifts and their ability to cause mutation. (a) The common forms of the bases are shown on the left, and the rare forms produced by a tautomeric shift are shown on the right. (b) On the left, the rare enol form of thymine pairs with the common keto form of guanine (instead of adenine); on the right, the rare imino form of cytosine pairs with the common amino form of adenine (instead of guanine). (c) A tautomeric shift occurred in a thymine base just prior to replication, causing the formation of a TG base pair. If not repaired, a second round of replication will lead to the formation of a permanent CG mutation. Note: A tautomeric shift is a very temporary situation. During the second round of replication, the thymine base that shifted prior to the first round of DNA replication is likely to have shifted back to its normal form. Therefore, during the second round of replication, an adenine base will be found opposite this thymine.

X-Rays Were the First Environmental Agent Shown to Cause Induced Mutations

As shown in Table 16.5, changes in DNA structure can also be caused by environmental agents, either chemical or physical agents. These agents are called mutagens, and the mutations they cause are referred to as induced mutations. In 1927, Hermann Müller devised an approach to show that X-rays can cause induced mutations in *Drosophila melanogaster*. Müller reasoned that a mutagenic agent might cause some genes to become defective. His experimental approach focused on the ability of a mutagen to cause defects in X-linked genes that result in a recessive lethal phenotype.

To determine if X-rays increase the rate of recessive, X-linked lethal mutations, Müller sought an easy way to detect the occurrence of such mutations. He cleverly realized that he had a laboratory strain of fruit flies that could make this possible. In particular, he conducted his crosses in such a way that a female fly that inherited a new mutation causing a recessive X-linked lethal allele would not be able to produce any male offspring. This made it very easy for him to detect lethal mutations; he had to count only the number of female flies that could not produce sons. To understand Müller's crosses, we need to take a closer look at a peculiar version of one of the X chromosomes in a strain of flies that he used in his crosses. This X chromosome, designated *ClB*, had three important genetic alterations.

C: Contained a large inversion that prevents it from **C**rossing over with the other X chromosome in female flies. The letter *C* is a reminder that this region of the chromosome cannot cross over.

l: Carried a **l**ethal recessive X-linked gene. If males (XY) inherit this chromosome, they will die.

B: Carried a dominant mutation that causes the eyes of the fly to have a **B**ar shape.

(Note: *C* and *B* are uppercase because they are inherited in a dominant manner, and *l* is lowercase because it is a recessive allele.)

A female fly that has one copy of this X chromosome would have bar-shaped eyes, because bar is a dominant allele. Even though this X chromosome has a recessive lethal allele, a female fly can survive if the corresponding gene on the other X chromosome is a normal allele. In Müller's experiments, the goal was to determine if exposure to X-rays caused a mutation on the normal X chromosome (not the *ClB* chromosome) that created a recessive lethal allele in any essential X-linked gene except for the gene that already had a lethal allele on the *ClB* chromosome (**Figure 16.11**). If a recessive lethal mutation occurred on the normal X chromosome, this female could survive because it would be

heterozygous for recessive lethal mutations in two different genes. However, because each X chromosome would have a lethal mutation, this female would not be able to produce any living sons.

The steps in Müller's protocol are shown in **Figure 16.12**. He began with wild-type males and exposed them to X-rays. These X-rays may mutate the X chromosome in sperm cells, resulting in a recessive lethal allele. These males, and a control group of males that were not exposed to X-rays, were then mated to females carrying the *ClB* chromosome. Daughters with bar eyes were saved from this cross and mated to nonirradiated males. You should look carefully at this cross and realize that if these daughters also contained a lethal allele on the X chromosome they inherited from their father (e.g., an irradiated male in step 1), they would not be able to produce living sons.

■ THE HYPOTHESIS

The exposure of flies to X-rays will increase the rate of mutation.

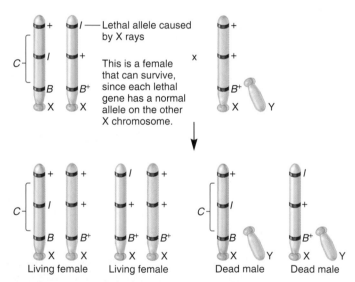

FIGURE 16.11   A strategy to detect the presence of lethal X-linked mutations. X-rays may cause a recessive lethal mutation to occur in the normal X chromosome. This female also contains another lethal allele in the *ClB* chromosome. Nevertheless, this female could survive because it would be heterozygous for recessive lethal mutations in two different genes. Because each X chromosome would have a lethal mutation, this female would not be able to produce any living sons.

■ TESTING THE HYPOTHESIS — FIGURE 16.12   **Evidence that X-rays cause mutation.**

Starting material: The female flies used in this study had one normal X chromosome and a *ClB* X chromosome. The male flies had a normal X chromosome.

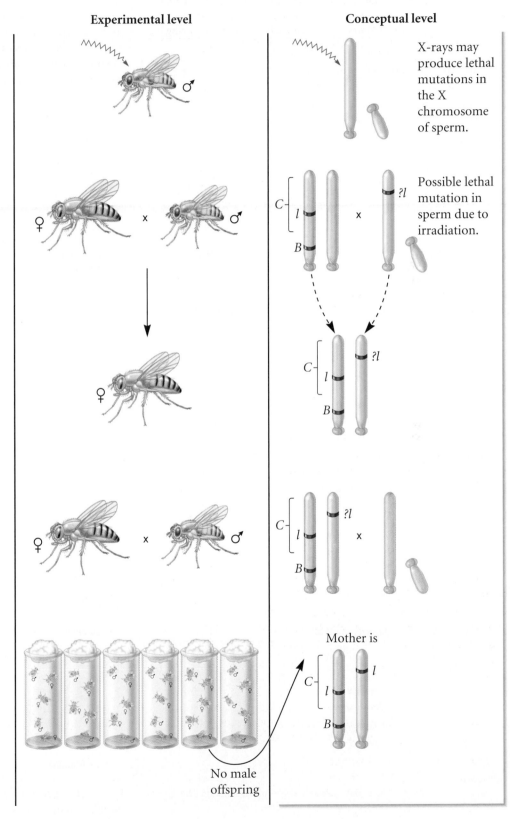

	Experimental level	**Conceptual level**

1. Expose male flies to X-rays. Also, have a control group that is not exposed to X-rays.

   X-rays may produce lethal mutations in the X chromosome of sperm.

2. Mate the male flies to female flies carrying one normal X chromosome and one *ClB* X chromosome.

   Possible lethal mutation in sperm due to irradiation.

3. Save about 1,000 daughters with bar eyes. Note: These females contain a *ClB* X chromosome from their mothers and an X chromosome from their fathers that may (or may not) have a recessive lethal mutation.

4. Mate each bar-eyed daughter with normal (nonirradiated) males. Note: This is done in (1,000) individual tubes. (Only six are shown.)

5. Count the number of crosses that do not contain any male offspring. These crosses indicate that the bar-eyed female parent contained an X-linked lethal recessive mutation on the non-*ClB* X chromosome.

   Mother is

   No male offspring

THE DATA

Treatment of Fathers of the *ClB* Daughters	Number of *ClB* Daughters Crossed to Normal Males*	Number of Tubes Containing Any Offspring†	Number of Tubes with Female Offspring but Lacking Male Offspring
Control	1,011	947	1
X-ray treated	1,015	783	91

*See step 4.
†The reason why these values are less than the numbers of mated *ClB* daughters is because some crosses did not produce living offspring.

INTERPRETING THE DATA

As shown in the data of Figure 16.12, in the absence of X-ray treatment, only 1 cross in approximately 1,000 was unable to produce male offspring. This means that the spontaneous rate for any X-linked lethal mutation was relatively low. By comparison, X-ray treatment of the fathers that gave rise to these *ClB* females resulted in 91 crosses without male offspring. Because these females inherited their non-*ClB* chromosome from irradiated fathers, these results indicate that X-rays greatly increase the rate of X-linked, recessive lethal mutations. This conclusion has been confirmed in many subsequent studies, which have shown that the increase in mutation rate is correlated with the amount of exposure to X-rays.

A self-help quiz involving this experiment can be found at the Online Learning Center.

Mutagens Alter DNA Structure in Different Ways

Since this pioneering study of Müller, researchers have found that an enormous array of agents can act as mutagens to permanently alter the structure of DNA. We often hear in the news media that we should avoid these agents in our foods and living environment. For example, we use products such as sunscreens to help us avoid the mutagenic effects of ultraviolet (UV) rays. The public is concerned about mutagens for two important reasons. First, mutagenic agents are often involved in the development of human cancers. In addition, because new mutations may be deleterious, people want to avoid mutagens to prevent gene mutations that may have harmful effects in their future offspring.

Mutagenic agents are usually classified as chemical or physical mutagens. Examples of both types of agents are listed in **Table 16.6**. In some cases, chemicals that are not mutagenic can be altered to a mutagenically active form after they have been ingested into the body. Cellular enzymes such as oxidases have been shown to activate some mutagens. Certain foods contain chemicals that act as antioxidants. Many scientists are investigating whether particular antioxidants found in food can counteract the effects of oxidases and thereby lower the cancer rate. Colorful fruits and vegetables, including prunes, grapes, blueberries, cranberries, citrus fruits, spinach, broccoli, beets, beans, red peppers, carrots, and strawberries are usually high in antioxidants.

Mutagens can alter the structure of DNA in various ways. Some mutagens act by covalently modifying the structure of bases. For example, **nitrous acid** (HNO_2) replaces amino groups with keto groups ($=NH_2$ to $=O$), a process called deamination. This can change cytosine to uracil and adenine to hypoxanthine. When this altered DNA replicates, the modified bases do not pair with the appropriate bases in the newly made strand. Instead, uracil pairs with adenine, and hypoxanthine pairs with cytosine (**Figure 16.13**).

TABLE 16.6

Examples of Mutagens

Mutagen	Effect(s) on DNA Structure
Chemical	
Nitrous acid	Deaminates bases
Nitrogen mustard	Alkylating agent
Ethyl methanesulfonate	Alkylating agent
Proflavin	Intercalates within DNA helix
5-bromouracil	Base analogue
2-aminopurine	Base analogue
Physical	
X-rays	Cause base deletions, single-stranded nicks in DNA, cross-linking, and chromosomal breaks
UV light	Promotes pyrimidine dimer formation, such as thymine dimers

FIGURE 16.13   Mispairing of modified bases that have been deaminated by nitrous acid. Nitrous acid converts cytosine to uracil, and adenine to hypoxanthine. During DNA replication, uracil pairs with adenine, and hypoxanthine pairs with cytosine. This will create mutations in the newly replicated strand during DNA replication.

Other chemical mutagens can also disrupt the appropriate pairing between nucleotides by alkylating bases within the DNA. During alkylation, methyl or ethyl groups are covalently attached to the bases. Examples of alkylating agents include **nitrogen mustard** (a type of mustard gas) and **ethyl methanesulfonate (EMS)**. Mustard gas was used as a chemical weapon during World War I. Such agents damage the skin, eyes, mucous membranes, lungs, and blood-forming organs.

Some mutagens exert their effects by directly interfering with the DNA replication process. For example, **acridine dyes** such as **proflavin** contain flat structures that intercalate, or insert themselves, between adjacent base pairs, thereby distorting the helical structure. When DNA containing these mutagens is replicated, single-nucleotide additions and/or deletions can be incorporated into the newly made daughter strands, creating a frameshift mutation.

Compounds such as **5-bromouracil (5BU)** and **2-aminopurine** are base analogues that become incorporated into daughter strands during DNA replication. 5BU is a thymine analogue that can be incorporated into DNA instead of thymine. Like thymine, 5BU base-pairs with adenine. However, at a relatively high rate, 5BU will undergo a tautomeric shift and base-pair with guanine (**Figure 16.14a**). When this occurs during DNA replication, 5BU causes a mutation in which a TA base pair is changed to a 5BU-G base pair (**Figure 16.14b**). This is a transition, because the adenine has been changed to a guanine, both of which are purines. During the next round of DNA replication, the template strand containing the guanine base will create a GC base pair. In this way, 5-bromouracil can promote a change of an AT base pair into a GC base pair.

Compounds like 5BU are sometimes used in chemotherapy for cancer. The rationale is that these compounds will be incorporated only into the DNA of actively replicating cells such as cancer cells. When incorporated, these compounds tend to cause many mutations in the cells, hopefully leading to the death of cancer cells. Unfortunately, other actively dividing cells, such as those in the skin and the lining of the digestive tract, will also incorporate 5BU. This leads to unwanted side effects of chemotherapy such as hair loss and a diminished appetite.

DNA molecules are also sensitive to physical agents such as radiation. In particular, radiation of short wavelength and high energy, known as ionizing radiation, can alter DNA structure. This type of radiation includes X-rays and gamma rays. Ionizing radiation can penetrate deeply into biological materials, where it creates chemically reactive molecules known as free radicals. These molecules can alter the structure of DNA in a variety of ways. Exposure to high doses of ionizing radiation can cause base deletions, single nicks in DNA strands, cross-linking, and even chromosomal breaks.

Nonionizing radiation, such as UV light, contains less energy, and so it penetrates only the surface of an organism, such as the skin. Nevertheless, UV light is known to cause DNA mutations. As shown in **Figure 16.15**, UV light causes the formation of cross-linked **thymine dimers.** Thymine dimers interfere with transcription and DNA replication. This can result in a mutation when that DNA strand is replicated. Plants, in particular, must have effective ways to prevent UV damage because they are exposed to sunlight throughout the day. Also, sun tanning greatly increases a person's exposure to UV light, raising the potential

for thymine dimers and mutation. This explains the higher incidence of skin cancer among people who have been exposed to large amounts of sun during their lifetime. Because of the link between skin cancer and sun exposure, people now apply sunscreen to their skin to prevent the harmful effects of UV light. Most sunscreens contain organic compounds that absorb UV light, such as oxybenzone, and/or opaque ingredients that reflect UV light, such as zinc oxide.

Testing Methods Can Determine If an Agent Is a Mutagen

To determine if an agent is mutagenic, researchers use testing methods that can monitor whether or not an agent increases the rate of mutation. Many different kinds of tests have been used to evaluate mutagenicity. One commonly used test is the **Ames**

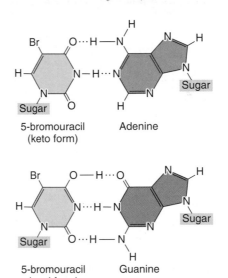

(a) Base pairing of 5BU with adenine or guanine

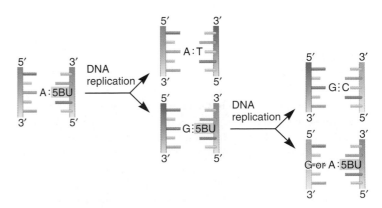

(b) How 5BU causes a mutation in a base pair during DNA replication

FIGURE 16.14   Structure and effects of the mispairing of 5-bromouracil with guanine. (a) In its keto form, 5BU bonds with adenine; in its enol form, it bonds with guanine. **(b)** During DNA replication, guanine may be incorporated into a newly made strand by pairing with 5BU. After a second round of replication, the DNA will contain a GC base pair instead of the original AT base pair.

Thymine

Thymine

Ultraviolet light

Thymine dimer

FIGURE 16.15   Formation and structure of a thymine dimer.

Mix together the *Salmonella* strain, rat liver extract, and suspected mutagen. The suspected mutagen is omitted from the control sample.

Control

Suspected mutagen — Rat liver extract

Salmonella strain (requires histidine)

Rat liver extract

Salmonella strain (requires histidine)

Plate the mixtures onto petri plates that lack histidine.

Incubate overnight to allow bacterial growth.

A large number of colonies suggests that the suspected mutagen causes mutation.

FIGURE 16.16   The Ames test for mutagenicity.

test, which was developed by Bruce Ames. This test uses strains of a bacterium, *Salmonella typhimurium*, that cannot synthesize the amino acid histidine. These strains contain a point mutation within a gene that encodes an enzyme required for histidine biosynthesis. The mutation renders the enzyme inactive. Therefore, the bacteria cannot grow on petri plates unless histidine has been added to the growth medium. However, a second mutation—a reversion—may occur that restores the ability to synthesize histidine. In other words, a second mutation can cause a reversion back to the wild-type condition. The Ames test monitors the rate at which this second mutation occurs and thereby indicates whether an agent increases the mutation rate above the spontaneous rate.

Figure 16.16 outlines the steps in the Ames test. The suspected mutagen is mixed with a rat liver extract and a strain of *Salmonella* that cannot synthesize histidine. A mutagen may require activation by cellular enzymes, which are provided by the rat liver extract. This step improves the ability of the test to identify agents that may cause mutation in mammals. After the incubation period, a large number of bacteria are then plated on a growth medium that does not contain histidine. The *Salmonella* strain is not expected to grow on these plates. However,

if a mutation has occurred that allows the strain to synthesize histidine, it can grow on these plates to form a visible bacterial colony. To estimate the mutation rate, the colonies that grow on the media are counted and compared to the total number of bacterial cells that were originally streaked on the plate. For example, if 10,000,000 bacteria were plated and 10 growing colonies were observed, the rate of mutation is 10 out of 10,000,000; this equals 1 in 10^6, or simply 10^{-6}. As a control, bacteria that have not been exposed to the mutagen are also tested, because a low level of spontaneous mutations is expected to occur.

How do we judge if an agent is a mutagen? Researchers compare the mutation rate in the presence and absence of the suspected mutagen. The experimental approach shown in Figure 16.16 is conducted several times. If statistics reveal that the mutation rate in the experimental and control samples are significantly different, researchers may tentatively conclude that the agent may be a mutagen. Interestingly, many studies have been conducted in which researchers used the Ames test to compare the urine from cigarette smokers to that from nonsmokers. This research has shown that the urine from smokers contains much higher levels of mutagens. Furthermore, heavy smokers have more mutagens in their urine compared to light smokers.

As discussed in solved problem S3 at the end of this chapter, several *his*⁻ strains of Salmonella are available that carry different kinds of mutations within the coding sequence. This makes it possible to determine if a mutagen causes transitions, transversions, or frameshift mutations.

16.3 DNA REPAIR

Because most mutations are deleterious, DNA repair systems are vital to the survival of all organisms. If DNA repair systems did not exist, spontaneous and environmentally induced mutations would be so prevalent that few species, if any, would survive. The necessity of DNA repair systems becomes evident when they are missing. Bacteria contain several different DNA repair systems. Yet, when even a single system is absent, the bacteria have a much higher rate of mutation. In fact, the rate of mutation is so high that these bacterial strains are sometimes called mutator strains. Likewise, in humans, an individual who is defective in only a single DNA repair system may manifest various disease symptoms, including a higher risk of skin cancer. This increased risk is due to the inability to repair UV-induced mutations.

Living cells contain several DNA repair systems that can fix different types of DNA alterations (Table 16.7). Each repair system is composed of one or more proteins that play specific roles in the repair mechanism. In most cases, DNA repair is a multistep process. First, one or more proteins in the DNA repair system detect an irregularity in DNA structure. Next, the abnormality is removed by the action of DNA repair enzymes. Finally, normal DNA is synthesized via DNA replication enzymes. In this section, we will examine several different repair systems that have

been characterized in bacteria, yeast, mammals, and plants. Their diverse ways of repairing DNA underscore the extreme necessity for the structure of DNA to be maintained properly.

Damaged Bases Can Be Directly Repaired

In a few cases, the covalent modification of nucleotides by mutagens can be reversed by specific cellular enzymes. As discussed earlier in this chapter, UV light causes the formation of thymine dimers. Yeast, such as *Saccharomyces cerevisiae*, and most plants produce an enzyme called **photolyase** that can repair thymine dimers by splitting the dimers, which returns the DNA to its original condition (**Figure 16.17a**). This process directly restores the structure of DNA. Because plants are exposed to sunlight throughout the day, photolyase is a critical DNA repair enzyme for many plant species.

A protein known as **alkyltransferase** can remove methyl or ethyl groups from guanine bases that have been mutagenized by alkylating agents such as nitrogen mustard and ethyl methanesulfonate. This protein is called alkyltransferase because it transfers the methyl or ethyl group from the base to a cysteine side chain within the alkyltransferase protein (**Figure 16.17b**). Surprisingly, this permanently inactivates alkyltransferase, which means it can be used only once!

Base Excision Repair Removes a Damaged Base

A second type of repair system, called **base excision repair (BER),** involves the function of a category of enzymes known as **DNA N-glycosylases.** This type of enzyme can recognize an abnormal base and cleave the bond between it and the sugar in the DNA backbone, creating an apurinic or apyrimidinic site (**Figure 16.18**). Depending on the species, this repair system can eliminate abnormal bases such as uracil, 3-methyladenine, 7-methylguanine, and pyrimidine dimers.

Figure 16.18 illustrates the general steps involved in DNA repair via *N*-glycosylase. In this example, the DNA contains a uracil in its sequence. This could have happened spontaneously or by the action of a chemical mutagen. *N*-glycosylase recognizes a uracil within the DNA and cleaves (nicks) the bond between the sugar and base. This releases the uracil base and leaves behind an apyrimidinic nucleotide. This abnormal nucleotide is recognized by a second enzyme, **AP endonuclease,** which makes a cut on the 5' side. DNA polymerase, which has a 5' to 3' exonuclease activity, removes the abnormal region and, at the same time, replaces it with normal nucleotides. This process is called **nick translation** (although DNA replication, not mRNA translation, actually occurs). Finally, DNA ligase closes the nick.

Nucleotide Excision Repair Systems Remove Segments of Damaged DNA

An important general process for DNA repair is the **nucleotide excision repair (NER)** system. This type of system can repair many different types of DNA damage, including thymine dimers, chemically modified bases, missing bases, and certain types of cross-links. In NER, several nucleotides in the damaged strand are removed from the DNA, and the intact strand is used as a

TABLE 16.7

Common Types of DNA Repair Systems

System	Description
Direct repair	An enzyme recognizes an incorrect alteration in DNA structure and directly converts it back to a correct structure.
Base excision repair and nucleotide excision repair	An abnormal base or nucleotide is first recognized and removed from the DNA. A segment of DNA in this region is excised, and then the complementary DNA strand is used as a template to synthesize a normal DNA strand.
Mismatch repair	Similar to excision repair except that the DNA defect is a base pair mismatch in the DNA, not an abnormal nucleotide. The mismatch is recognized, and a segment of DNA in this region is removed. The parental strand is used to synthesize a normal daughter strand of DNA.
Homologous recombination repair	Occurs at double-strand breaks or when DNA damage causes a gap in synthesis during DNA replication. The strands of a normal sister chromatid are used to repair a damaged sister chromatid.
Nonhomologous end joining	Occurs at double-strand breaks. The broken ends are recognized by proteins that keep the ends together; the broken ends are eventually rejoined.

Thymine dimer

DNA backbone

DNA photolyase cleaves the 2 bonds between the thymine dimer.

The normal structure of the 2 thymines is restored.

(a) Direct repair of a thymine dimer

O^6-methylguanine

Alkyltransferase

DNA backbone

Alkyltransferase catalyzes the removal of the methyl group onto itself.

Guanine

The normal structure of guanine is restored.

(b) Direct repair of a methylated base

 FIGURE 16.17 **Direct repair of damaged bases in DNA.** (a) The repair of thymine dimers by photolyase. (b) The repair of methylguanine by the transfer of the methyl group to alkyltransferase.

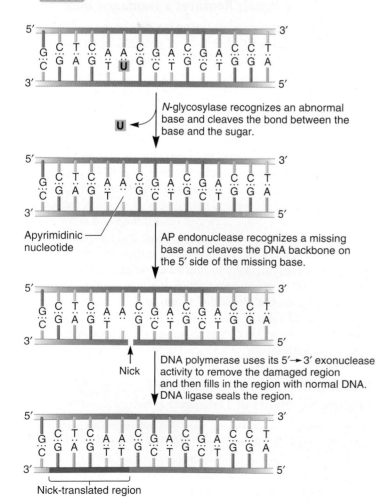

N-glycosylase recognizes an abnormal base and cleaves the bond between the base and the sugar.

Apyrimidinic nucleotide

AP endonuclease recognizes a missing base and cleaves the DNA backbone on the 5′ side of the missing base.

Nick

DNA polymerase uses its 5′→3′ exonuclease activity to remove the damaged region and then fills in the region with normal DNA. DNA ligase seals the region.

Nick-translated region

template for resynthesis of a normal complementary strand. NER is found in all eukaryotes and prokaryotes, although its molecular mechanism is better understood in prokaryotic species.

In *E. coli*, the NER system requires four key proteins, designated UvrA, UvrB, UvrC, and UvrD, plus the help of DNA polymerase and DNA ligase. UvrA, B, C, and D recognize and remove a short segment of a damaged DNA strand. They are named Uvr because they are involved in Ultraviolet light repair of pyrimidine dimers, although the UvrA–D proteins are also important in repairing chemically damaged DNA.

Figure 16.19 outlines the steps involved in the *E. coli* NER system. A protein complex consisting of two UvrA molecules and one UvrB molecule tracks along the DNA in search of damaged DNA. Such DNA will have a distorted double helix, which is sensed by the UvrA/UvrB complex. When a damaged segment is identified, the two UvrA proteins are released, and UvrC binds to the site. The UvrC protein makes cuts in the damaged strand on both sides of the damaged site. Typically, the damaged strand is cut four to five nucleotides away from the 3′ side of the damage and eight nucleotides from the 5′ end. After this process, UvrD, which is a helicase, recognizes the region and separates the two

FIGURE 16.18 **Base excision repair.** An abnormal base is initially recognized by an enzyme known as N-glycosylase, which cleaves the bond between the base and the sugar. Depending on whether a purine or pyrimidine is removed, this creates an apurinic or apyrimidinic site, respectively. In either case, the region is recognized by AP endonuclease, which makes a cut at the 5′ side of the abnormal nucleotide. DNA polymerase then removes the abnormal region and fills it in with normal DNA. Finally, DNA ligase seals the repaired strand.

strands of DNA. This releases a short DNA segment that contains the damaged region, and UvrB and UvrC are also released. Following the excision of the damaged DNA, DNA polymerase fills in the gap, using the undamaged strand as a template. Finally,

Thymine dimer

The UvrA/UvrB complex tracks along the DNA in search of damaged DNA.

After damage is detected, UvrA is released, and UvrC binds.

UvrC makes cuts on both sides of the thymine dimer.

Cut   Cut

UvrD, which is a helicase, removes the damaged region. UvrB and UvrC are also released.

DNA polymerase fills in the gap, and DNA ligase seals the gap.

No thymine dimer

ONLINE ANIMATION

FIGURE 16.19   Nucleotide excision repair in *E. coli.* UvrA/UvrB complex scans along the DNA. When the complex encounters a thymine dimer in the DNA, the UvrA dimer is released. UvrC binds to UvrB. UvrC make two cuts in the damaged DNA strand. One cut is made eight nucleotides 5' to the site, the other approximately five nucleotides 3' from the site. UvrD (not shown in this figure) binds to the site. UvrD is a helicase that causes the damaged strand to be removed. During this process, UvrB and UvrC are released. DNA polymerase resynthesizes a complementary strand. DNA ligase makes the final connection.

DNA ligase makes the final covalent connection between the newly made DNA and the original DNA strand.

Several human diseases are due to inherited defects in genes involved in nucleotide excision repair. These include xeroderma pigmentosum (XP), Cockayne syndrome (CS), and PIBIDS. (PIBIDS is an acronym for a syndrome with symptoms that include photosensitivity; ichthyosis, a skin abnormality; brittle hair; impaired intelligence; decreased fertility; and short stature.) A common characteristic in all three syndromes is an increased sensitivity to sunlight because of an inability to repair UV-induced lesions. **Figure 16.20** shows a photograph of an individual with xeroderma pigmentosum. Such individuals have pigmentation abnormalities, many premalignant lesions, and a high predisposition to skin cancer. They may also develop early degeneration of the nervous system.

Genetic analyses of patients with XP, CS, and PIBIDS have revealed that these syndromes result from defects in a variety of different genes that encode NER proteins. For example, xeroderma pigmentosum can be caused by defects in any of seven different NER genes. In all cases, individuals have a defective

FIGURE 16.20   An individual affected with xeroderma pigmentosum.

Genes→Traits This disease involves a defect in genes that are involved with nucleotide excision repair. Xeroderma pigmentosum can be caused by defects in seven different NER genes. Affected individuals have an increased sensitivity to sunlight because of an inability to repair UV-induced DNA lesions. In addition, they may also have pigmentation abnormalities, many premalignant lesions, and a high predisposition to skin cancer.

NER pathway. In recent years, several human NER genes have been successfully cloned and sequenced. Although more research is needed to completely understand the mechanisms of DNA repair, the identification of NER genes has helped unravel the complexities of NER pathways in human cells.

Mismatch Repair Systems Recognize and Correct a Base Pair Mismatch

Thus far, we have considered several DNA repair systems that recognize abnormal nucleotide structures within DNA, including thymine dimers, alkylated bases, and the presence of uracil in the DNA. Another type of abnormality that should not occur in DNA is a **base mismatch.** The structure of the DNA double helix obeys the AT/GC rule of base pairing. During the normal course of DNA replication, however, an incorrect nucleotide may be added to the growing strand by mistake. This creates a mismatch between a nucleotide in the parental and newly made strand. Various DNA repair mechanisms can recognize and remove this mismatch. For example, as described in Chapter 11, DNA polymerase has a 3' to 5' proofreading ability that can detect mismatches and remove them. However, if this proofreading ability fails, cells contain additional DNA repair systems that can detect base mismatches and fix them. An interesting DNA repair system that exists in all species is the **mismatch repair system.**

In the case of a base mismatch, how does a DNA repair system determine which base to remove? If the mismatch is due to an error in DNA replication, the newly made daughter strand contains the incorrect base, while the parental strand is normal. Therefore, an important aspect of mismatch repair is that it specifically repairs the newly made strand rather than the parental template strand. Prior to DNA replication, the parental DNA has already been methylated. Immediately after DNA replication, some time must pass before a newly made strand is methylated. Therefore, newly replicated DNA is hemimethylated—only the parental DNA strand is methylated. Hemimethylation provides a way for a DNA repair system to distinguish between the parental DNA strand and the daughter strand.

The molecular mechanism of mismatch repair has been studied extensively in *E. coli.* As shown in **Figure 16.21,** three proteins, designated MutS, MutL, and MutH, detect the mismatch and direct the removal of the mismatched base from the newly made strand. These proteins are named Mut because their absence leads to a much higher mutation rate compared to normal strains of *E. coli.* The role of MutS is to locate mismatches. Once a mismatch is detected, MutS forms a complex with MutL. MutL acts as a linker that binds to MutH by a looping mechanism. This stimulates MutH, which is bound to a hemimethylated site, to make a cut in the newly made, nonmethylated DNA strand. After the strand is cut, MutU, which functions as helicase, separates the strands, and an exonuclease then digests the nonmethylated DNA strand in the direction of the mismatch and proceeds beyond the mismatch site. This leaves a gap in the daughter strand that is repaired by DNA polymerase and DNA ligase. The net result is that the mismatch has been corrected by removing the incorrect region in the daughter strand and then

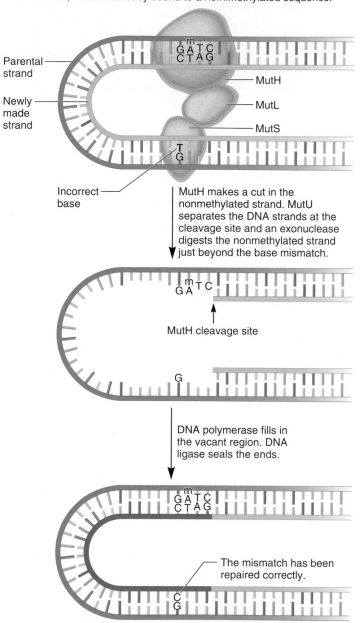

The MutS protein finds a mismatch. The MutS/MutL complex binds to MutH, which is already bound to a hemimethylated sequence.

Parental strand

Newly made strand

MutH

MutL

MutS

Incorrect base

MutH makes a cut in the nonmethylated strand. MutU separates the DNA strands at the cleavage site and an exonuclease digests the nonmethylated strand just beyond the base mismatch.

MutH cleavage site

DNA polymerase fills in the vacant region. DNA ligase seals the ends.

The mismatch has been repaired correctly.

ONLINE ANIMATION

FIGURE 16.21   Mismatch repair in *E. coli.* MutS slides along the DNA and recognizes base mismatches in the double helix. MutL binds to MutS and acts as a linker between MutS and MutH. The DNA must loop for this interaction to occur. The role of MutH is to identify the methylated strand of DNA, which is the nonmutated parental strand. MutH can be bound to a methylated site on either side of the base mismatch. In this example, the methylated adenine is designated with an m. MutH makes a cut in the nonmethylated strand. MutU (not shown in the figure) begins to separate the strands at the cleavage site, and then an exonuclease digests the nonmethylated strand until it passes the MutS/MutL region. This leaves a gap, which is filled in by DNA polymerase and sealed by DNA ligase.

resynthesizing the correct sequence using the parental DNA as a template.

Eukaryotic species also have homologs to MutS and MutL, along with many other proteins that are needed for mismatch repair. However, a MutH homolog has not been identified in eukaryotes, and the mechanism by which the eukaryotic mismatch repair system distinguishes between the parental and daughter strands is not well understood. As with defects in nucleotide excision repair systems, mutations in the human mismatch repair system are associated with particular types of cancer. For example, mutations in two human mismatch repair genes, *hMSH2* and *hMLH1*, play a role in the development of a type of colon cancer known as hereditary nonpolyposis colorectal cancer.

Double-Strand Breaks Can Be Repaired by Homologous Recombination Repair and by Nonhomologous End Joining

Of the many types of DNA damage that can occur within living cells, the breakage of chromosomes—called a DNA double-strand break (DSB)—is perhaps the most dangerous. DSBs can be caused by ionizing radiation (X-rays or gamma rays), chemical mutagens, and certain drugs used for chemotherapy. In addition, free radicals that are the byproducts of cellular metabolism can cause double-strand breaks. Surprisingly, researchers estimate that naturally occurring double-strand breaks in a typical human cell occur at a rate of between 10 and 100 breaks per cell per day! Such breaks can be harmful in a variety of ways. First, DSBs can result in chromosomal rearrangements such as inversions and translocations (see Figure 8.2). In addition, DSBs can lead to terminal or interstitial deficiencies (see Figure 8.3). Such genetic changes have the potential to result in detrimental phenotypic effects.

How are DSBs repaired? The two main mechanisms are **homologous recombination repair (HRR)** and **nonhomologous end joining (NHEJ)**. Homologous recombination repair, also called homology-directed repair, occurs when homologous DNA strands, usually from a sister chromatid, are used to repair a DSB in the other sister chromatid (**Figure 16.22**). First, the DSB is processed by the short digestion of DNA strands at the break site. This processing event is followed by the exchange of DNA strands between the broken and unbroken sister chromatids. The unbroken strands are then used as templates to synthesize DNA in the region where the break occurred. Finally, the criss-crossed strands are resolved, which means that they are broken and then rejoined in a way that produces separate chromatids. Because sister chromatids are genetically identical, an advantage is that homologous recombination repair can be an error-free mechanism to repair a DSB. A disadvantage is that sister chromatids are available only during the S and G_2 phases of the cell cycle in eukaryotes or following DNA replication in bacteria. Although sister chromatids are strongly preferred, HRR may also occur between homologous regions that are not identical. Therefore, HRR may occasionally happen when sister chromatids are unavailable. The proteins involved in homologous recombination repair are described in Chapter 17.

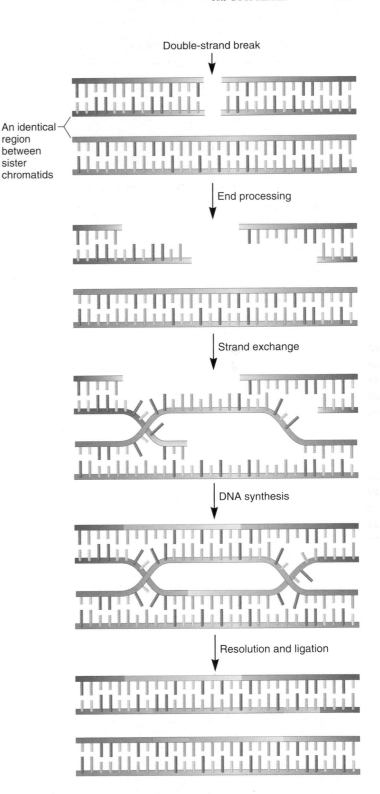

FIGURE 16.22 **DNA repair of double-strand breaks (DSBs) via homologous recombination repair.**

During nonhomologous end joining, the two broken ends of DNA are simply pieced back together (**Figure 16.23**). This mechanism requires the participation of several proteins that play key roles in the process. First, the DSB is recognized by end-binding proteins. These proteins then recognize additional

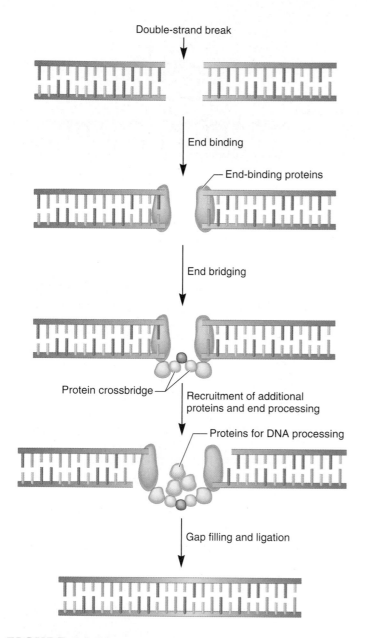

FIGURE 16.23 **DNA repair of double-strand breaks via nonhomologous end joining.**

*(Figure labels, top to bottom:)*

Double-strand break

End binding

End-binding proteins

End bridging

Protein crossbridge

Recruitment of additional proteins and end processing

Proteins for DNA processing

Gap filling and ligation

Actively Transcribed DNA Is Repaired More Efficiently Than Nontranscribed DNA

Not all DNA is repaired at the same rate. In the 1980s, Philip Hanawalt and colleagues conducted experiments showing that actively transcribed genes in eukaryotes and prokaryotes are more efficiently repaired following radiation damage than is nontranscribed DNA. The targeting of DNA repair enzymes to actively transcribing genes may have several biological advantages. First, active genes are loosely packed and may be more vulnerable to DNA damage. Likewise, the process of transcription itself may make the DNA more susceptible to agents that cause damage. In addition, DNA regions that contain actively transcribed genes are more likely to be important for survival than nontranscribed regions. This may be particularly true in differentiated cells, such as nerve cells, which no longer divide but actively transcribe genes for many years. In nondividing cells, gene transcription rather than DNA replication is of utmost importance.

How are actively transcribed regions targeted by DNA repair systems? In *E. coli*, a protein known as **transcription-repair coupling factor** (**TRCF**) is responsible for targeting the nucleotide excision repair system to actively transcribed genes having damaged DNA. **Figure 16.24** illustrates a case where RNA polymerase had been transcribing a gene until it encountered a thymine dimer and became stalled. TRCF is a helicase that displaces the stalled RNA polymerase from the damaged template strand. TRCF also has a binding site for UvrA. Therefore, after RNA polymerase has been removed, TRCF recruits the nucleotide excision repair system to the damaged site by recognizing the UvrA/UvrB complex. After the UvrA/UvrB complex has bound to this site, TRCF is released from the damaged site. At this stage, the UvrA/UvrB complex can begin nucleotide excision repair, as described earlier in Figure 16.19.

In eukaryotes, the mechanism that couples DNA repair and transcription is not completely understood. Several different proteins have been shown to act as transcription-repair coupling factors. Some of these have been identified in people with high rates of mutation. As mentioned previously, Cockayne syndrome (CS) involves defects in genes that play a role in nucleotide excision repair. Two genes that are defective in CS, termed *CS-A* and *CS-B*, encode proteins that function as transcription-repair coupling factors. In addition, certain general transcription factors, such as TFIIH, play a role in both transcription and nucleotide excision repair.

Damaged DNA May Be Replicated by Translesion DNA Polymerases

Despite the efficient action of numerous repair systems that remove lesions in DNA in an error-free manner, it is inevitable that some lesions may escape these repair mechanisms. Such lesions may be present when DNA is being replicated. If so, replicative DNA polymerases, such as polIII in *E. coli*, are highly sensitive to geometric distortions in DNA and are unable to replicate through DNA lesions. During the past decade, researchers have discovered that cells are equipped with specialized DNA polymerases that

proteins that form a crossbridge that prevents the two ends from drifting apart. Next, additional proteins are recruited to the region that may process the ends of the broken chromosome by digesting particular DNA strands. This processing may result in the deletion of a small amount of genetic material from the region. Finally, any gaps are filled in via DNA polymerase, and the DNA ends are ligated together. One advantage of NHEJ is that it doesn't involve the participation of a sister chromatid, so it can occur at any stage of the cell cycle. However, a disadvantage is that NHEJ can result in small deletions in the region that has been repaired.

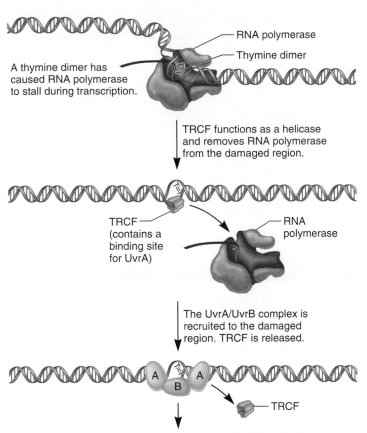

A thymine dimer has caused RNA polymerase to stall during transcription.

— RNA polymerase
— Thymine dimer

TRCF functions as a helicase and removes RNA polymerase from the damaged region.

TRCF (contains a binding site for UvrA)

RNA polymerase

The UvrA/UvrB complex is recruited to the damaged region. TRCF is released.

A   B   A

— TRCF

The region is repaired as described in Figure 16.19.

FIGURE 16.24 **Targeting of the nucleotide excision repair system to an actively transcribing gene containing a DNA alteration.** When RNA polymerase encounters damaged DNA, it will stall over the damaged site. TRCF, which is a helicase, recognizes a stalled RNA polymerase and removes it from the damaged site. TRCF has a binding site for UvrA and thereby recruits the UvrA/UvrB complex to the region. This initiates the process of nucleotide excision repair.

will assist the replicative DNA polymerase during the process of **translesion synthesis (TLS)**—the synthesis of DNA over a template strand that harbors some type of DNA damage. These lesion-replicating polymerases, which are also described in Chapter 11 (see Table 11.4), contain an active site with a loose, flexible pocket that can accommodate aberrant structures in the template strand. When a replicative DNA polymerase encounters a damaged region, it is swapped with a lesion-replicating polymerase.

A negative consequence of translesion synthesis is low fidelity. Due to their flexible active site, lesion-replicating polymerases are much more likely to incorporate the wrong nucleotide in a newly made daughter strand. The mutation rate is typically in the range of 10^{-2} to 10^{-3}. By comparison, replicative DNA polymerases are highly intolerant of the geometric distortions imposed on DNA by the incorporation of incorrect nucleotides, and consequently, they incorporate wrong nucleotides with a very low frequency. In other words, they copy DNA with a high degree of fidelity.

In *E. coli*, translesion synthesis has been shown to occur under extreme conditions that promote DNA damage, an event termed the **SOS response.** Causes of the SOS response include high doses of UV light and other types of mutagens. Such environmental factors result in the up regulation of several genes that function to repair the DNA lesions, restore replication, and prevent premature cell division. When the SOS response occurs, the damaged DNA that has not been repaired is replicated by DNA polymerases II, IV, and V, which can replicate over damaged regions. As previously mentioned, this translesion DNA synthesis results in a high rate of mutation. Even though this result may be harmful, it nevertheless allows the bacteria to survive under conditions of extreme environmental stress. Furthermore, the high rate of mutation may provide genetic variability within the bacterial population, so certain cells may become resistant to the harsh conditions.

CONCEPTUAL SUMMARY

Mutations are heritable changes in the genetic material. They can result from small changes in the genetic material (single-gene mutations), alterations of chromosome structure, or changes in chromosome number. Mutations can be beneficial, neutral, or detrimental. While they provide the variation that enables species to change and adapt to their environment, mutations are more likely to disrupt sequences within genes rather than improve their function.

Mutations can have various consequences on gene expression. If a point mutation—a change in a single base pair—occurs within the coding sequence, it may alter the amino acid sequence of the encoded polypeptide. Missense mutations are base substitutions that cause a change in a single amino acid; nonsense mutations are base substitutions that change a normal codon to a stop codon, resulting in a truncated polypeptide; and frameshift mutations involve the addition or deletion of nucleotides and thus produce a shift in the mRNA codon reading frame.

Mutations within promoters, regulatory elements, splicing sequences, and noncoding regions of genes can have major consequences on the level of gene expression. If a mutation occurs within the germ line—those cells that give rise to the eggs and sperm—it will occur in all the cells of an organism and can be passed from parent to offspring. By comparison, mutations in somatic cells—all cells of the body excluding the germ-line cells—affect only a portion of the organism.

Geneticists categorize mutations as spontaneous mutations—changes in DNA structure that result from abnormalities in biological processes—and induced mutations—changes in DNA structure that are caused by environmental agents. Spontaneous mutations occur during normal cellular conditions. These can arise from depurination, the removal of a purine (guanine or adenine) from the DNA; deamination, the removal of an amino group from the cytosine base; and tautomeric shifts, a temporary change of a nitrogen base structure to an alternative form. In addition, a variety of environmental agents, known as chemical or physical mutagens, are known to induce mutations. Mutagens can alter DNA structure by modifying or removing bases, causing errors in DNA replication, forming thymine dimers, and producing breaks in DNA strands. Testing methods such as the Ames test can determine whether an agent is mutagenic.

Due to the prevalence of mutagens in the environment, organisms have evolved DNA repair systems that can detect and repair different types of DNA lesions. Some DNA repair systems can recognize altered bases and repair them directly. For example, photolyase is a critical DNA repair enzyme for many plant species that removes thymine dimers, and alkyltransferase is a protein that can remove methyl or ethyl groups from guanine bases that have been mutagenized by alkylating agents. Base

excision repair and nucleotide excision repair systems recognize alterations in DNA structure and excise the abnormal base or region, respectively. Certain proteins, such as the transcription-repair coupling factor (TRCF) of *E. coli*, can target the nucleotide excision repair system to actively transcribing genes that have damaged DNA. Mismatch repair systems detect and repair base mismatches between nucleotides in parental and newly made strands. The breakage of chromosomes—called a DNA double-strand break (DSB)—is a dangerous type of DNA damage. The two main mechanisms for repair of DSBs are homologous recombination repair (HRR) and nonhomologous end joining (NHEJ). Finally, in translesion synthesis (TLS), specialized DNA polymerases are able to replicate over damaged DNA.

EXPERIMENTAL SUMMARY

Experimentally, the occurrence, causes, and consequences of mutation have been studied in many ways. Luria and Delbrück conducted a fluctuation test to explore whether mutations occur randomly in a population or whether they are a physiological adaptation. Similarly, the Lederbergs used replica plating to distinguish between the random mutation hypothesis and the physiological adaptation hypothesis. Results from both studies supported what is now known as the random mutation theory. According to this theory, mutations are a random process that can occur in any gene and do not involve exposure of an organism to particular conditions that select for specific types of muta-

tions. However, certain regions of genes, called hot spots, occur where mutations are more frequent.

Müller was the first scientist to establish that environmental agents such as X-rays can cause induced mutations. His work showed that X-rays dramatically increased the likelihood of X-linked, recessive lethal mutations. Since these studies, biochemists, microbiologists, and geneticists have discovered an enormous array of agents that can act as mutagens that permanently alter the structure of DNA. Many kinds of testing methods, such as the commonly used Ames test, can ascertain whether or not an agent is a mutagen.

PROBLEM SETS & INSIGHTS

Solved Problems

S1. Mutant tRNAs may act as nonsense and missense suppressors. At the molecular level, explain how you think these suppressors work.

Answer: A suppressor is a second-site mutation that suppresses the phenotypic effects of a first mutation. Intergenic suppressor mutations in tRNA genes can act as nonsense or missense suppressors. For example, let's suppose a first mutation puts a stop codon into a structural gene. A second mutation in a tRNA gene can alter the anticodon region of a tRNA so that the anticodon recognizes a stop codon but inserts an amino acid at this site. A missense suppressor is a mutation in a tRNA gene that changes the anticodon so that it puts in the wrong amino acid at a normal codon that is not a stop codon. These mutant tRNAs are termed missense tRNAs. For example, a tRNA that normally recognizes glutamic acid may incur a mutation that changes its anticodon sequence so that it recognizes a glycine codon instead. Like nonsense suppressors, missense suppressors can be produced by mutations in the anticodon region of tRNAs so that the tRNA recognizes an incorrect codon. Alternatively, missense suppressors can also be produced by mutations in aminoacyl-tRNA synthetases that cause them to attach the incorrect amino acid to a tRNA.

S2. If the rate of mutation is 10^{-5} per gene, how many new mutations per gene would you expect in a population of 1 million bacteria?

Answer: If we multiply the mutation rate times the number of bacteria ($10^{-5} \times 10^6$), we obtain a value of 10 new mutations per gene in this

population. This answer is correct, but it is an oversimplification of mutation rate. For any given gene, the mutation rate is based on a probability that an event will occur. Therefore, when we consider a particular population of bacteria, we should be aware that the actual rate of new mutation would vary. Even though the rate may be 10^{-5}, we would not be surprised if a population of 1 million bacteria had 9 or 11 new mutations per gene instead of the expected number of 10. We would be surprised if it had 5,000 new mutations per gene, because this value would deviate much too far from our expected number.

S3. In the Ames test, several *Salmonella* strains are used that contain different types of mutations within the gene that encodes an enzyme necessary for histidine biosynthesis. These mutations include transversions, transitions, and frameshift mutations. Why do you think it would be informative to test a mutagen with these different strains?

Answer: Different types of mutagens have different effects on DNA structure. For example, if a mutagen caused transversions, an experimenter would want to use a *Salmonella* strain in which a transversion would convert a *his⁻* strain into a *his⁺* strain. This type of strain would make it possible to detect the effects of the mutagen.

S4. In Chapter 14, we discussed how bacterial genes can be arranged in an operon structure in which a polycistronic mRNA contains the coding sequences for two or more genes. For genes in an operon, a relatively short distance occurs between the stop codon in the first gene and the start codon in the next gene. In addition to the ribosomal-binding sequence at the beginning of the first gene, a ribosomal-binding sequence is also found at the beginning of the second gene and at the beginning of all subsequent genes. After the ribosome has moved past the stop codon in the first gene, it quickly encounters a ribosomal-binding site in the next gene. This prevents the complete disassembly of the ribosome as would normally occur after a stop codon and leads to the efficient translation of the second gene.

Jacob and Monod, who studied mutations in the *lac* operon, discovered a mutation in the *lacZ* gene that had an unusual effect. This mutation resulted in a shorter version of the β-galactosidase protein, and it also prevented the expression of the *lacY* gene, even though the *lacY* gene was perfectly normal. Explain this mutation.

Answer: This mutation introduced a stop codon much earlier in the *lacZ* gene sequence. This type of mutation, which affects the translation of a gene downstream in an operon, is referred to as a **polar mutation,** and the phenomenon is called polarity. The explanation is that the early nonsense codon causes the ribosome to disassemble before it has a chance to reach the next gene sequence in the operon. The ribosomal-binding sequence that precedes the second gene is efficiently recognized only by a ribosome that has traveled to the normal stop codon in the first gene.

S5. A reverse mutation or a reversion is a mutation that returns a mutant codon back to a codon that gives a wild-type phenotype. At the DNA level, this can be an exact reversion or an equivalent reversion.

GAG (glutamic acid)	First → GTG mutation (valine)	Exact → GAG reversion (glutamic acid)
GAG (glutamic acid)	First → GTG mutation (valine)	Equivalent → GAA reversion (glutamic acid)
GAG (glutamic acid)	First → GTG mutation (valine)	Equivalent → GAT reversion (aspartic acid)

An equivalent reversion produces a protein equivalent to the wild type with regard to structure and function. This can occur in two ways. In some cases, the reversion produces the wild-type amino

acid (in this case, glutamic acid), but it uses a different codon than the wild-type gene. Alternatively, an equivalent reversion may substitute an amino acid structurally similar to the wild-type amino acid. In our example, an equivalent reversion has changed valine to an aspartic acid. Because aspartic and glutamic acids are structurally similar—they are acidic amino acids—this type of reversion can restore the wild-type structure and function. Now here is the question. The template strand within the coding sequence of a gene has the following sequence:

3'-TACCCCTTCGACCCCGGA-5'

This template produces an mRNA, 5'–AUGGGGAAGCUGGGG CCA–3', that encodes a polypeptide with the sequence methionine–glycine–lysine–leucine–glycine–proline.

A mutation changes the template strand to 3'–TACCCCT<u>A</u> CGACCCCGGA-5'.

After the first mutation, another mutation occurs to change this sequence again. Would the following second mutations be an exact reversion, an equivalent reversion, or neither?

A. 3'-TACCCCT<u>CC</u>GACCCCGGA-5'

B. 3'-TACCCCT<u>T</u>CGACCCCGGA-5'

C. 3'-TACCCCG<u>A</u>CGACCCCGGA-5'

Answer:

A. This is probably an equivalent reversion. The third codon, which encodes a lysine in the normal gene, is now an arginine codon. Arginine and lysine are both basic amino acids, so the polypeptide would probably function normally.

B. This is an exact reversion.

C. The third codon, which is a lysine in the normal gene, has been changed to a leucine codon. It is difficult to say if this would be an equivalent reversion or not. Lysine is a basic amino acid and leucine is a nonpolar, aliphatic amino acid. The protein may still function normally with a leucine at the third codon, or it may function abnormally. You would need to test the function of the protein to determine if this was an equivalent reversion or not.

Conceptual Questions

C1. For each of the following mutations, is it a transition, transversion, addition, or deletion? The original DNA strand is 5'–GGACTAGATAC–3'. (Note: Only the coding DNA strand is shown.)

A. 5'-GAACTAGATAC-3'

B. 5'-GGACTAGAGAC-3'

C. 5'-GGACTAGTAC-3'

D. 5'-GGAGTAGATAC-3'

C2. A gene mutation changes an AT base pair to a GC pair. This causes a gene to encode a truncated protein that is nonfunctional. An organism that carries this mutation cannot survive at high temperatures. Make a list of all the genetic terms that could be used to describe this type of mutation.

C3. What does a suppressor mutation suppress? What is the difference between an intragenic and intergenic suppressor?

C4. How would each of the following types of mutations affect the amount of functional protein that is expressed from a gene?

A. Nonsense

B. Missense

C. Up promoter mutation

D. Mutation that affects splicing

C5. X-rays strike a chromosome in a living cell and ultimately cause the cell to die. Did the X-rays produce a mutation? Explain why or why not.

C6. The lactose permease is encoded by the *lacY* gene of the lac operon. A mutation occurred at codon 64 that changed the normal glycine codon into a valine codon. The mutant lactose permease is unable to function. However, a second mutation, which changes codon 50 from an alanine codon to a threonine codon, is able to restore function. Are the following terms appropriate or inappropriate to describe this second mutation?

A. Reversion

B. Intragenic suppressor

C. Intergenic suppressor

D. Missense mutation

C7. Nonsense suppressors tend to be very inefficient at their job of allowing readthrough of a stop codon. How would it affect the cell if they were efficient at their job?

C8. Are each of the following mutations silent, missense, nonsense, or frameshift mutations? The original DNA strand is 5'–ATGGGACTAGATACC–3'. (Note: Only the coding strand is shown; the first codon is methionine.)

A. 5'-ATGGGTCTAGATACC-3'

B. 5'-ATGCGACTAGATACC-3'

C. 5'-ATGGGACTAGTTACC-3'

D. 5'-ATGGGACTAAGATACC-3'

C9. In Chapters 12 through 15, we discussed many sequences that are outside the coding sequence and are important for gene expression. Look up two of these sequences and write them out. Explain how a mutation could change these sequences and thereby alter gene expression.

C10. Explain two ways that a chromosomal rearrangement can cause a position effect.

C11. Is a random mutation more likely to be beneficial or harmful? Explain your answer.

C12. Which of the following mutations could be appropriately described as a position effect?

A. A point mutation at the −10 position in the promoter region prevents transcription.

B. A translocation places the coding sequence for a muscle-specific gene next to an enhancer that is turned on in nerve cells.

C. An inversion flips a gene from the long arm of chromosome 17 (which is euchromatic) to the short arm (which is heterochromatic).

C13. As discussed in Chapter 22, most forms of cancer are caused by environmental agents that produce mutations in somatic cells. Is an individual with cancer considered a genetic mosaic? Explain why or why not.

C14. Discuss the consequences of a germ-line versus a somatic mutation.

C15. Draw and explain how alkylating agents alter the structure of DNA.

C16. Explain how a mutagen can interfere with DNA replication to cause a mutation. Give two examples.

C17. What type of mutation (transition, transversion, and/or frameshift) would you expect each of the following mutagens to cause?

A. Nitrous acid

B. 5-bromouracil

C. Proflavin

C18. Explain what happens to the sequence of DNA during trinucleotide repeat expansion. If someone was mildly affected with a trinucleotide repeat expansion (TNRE) disorder, what issues would be important when considering whether to have offspring?

C19. Distinguish between spontaneous and induced mutations. Which are more harmful? Which are avoidable?

C20. Are mutations random events? Explain your answer.

C21. Give an example of a mutagen that can change cytosine to uracil. Which DNA repair system(s) would be able to repair this defect?

C22. If a mutagen causes bases to be removed from nucleotides within DNA, what repair system would fix this damage?

C23. Trinucleotide repeat expansions (TNREs) are associated with several different human inherited diseases. Certain types of TNREs produce a long stretch of glutamines (an amino acid) within the encoded protein. This long stretch of glutamines somehow inhibits the function of the protein and thereby causes a disorder. In cases where a TNRE exerts its detrimental effect by producing a glutamine stretch, are the following statements true or false?

A. The TNRE is within the coding sequence of the gene.

B. The TNRE prevents RNA polymerase from transcribing the gene properly.

C. The trinucleotide sequence is CAG.

D. The trinucleotide sequence is CCG.

C24. With regard to TNRE, what is meant by the term anticipation?

C25. What is the difference between the mutation rate and the mutation frequency?

C26. Achondroplasia is a rare form of dwarfism. It is caused by an autosomal dominant mutation within a single gene. Among 1,422,000 live births, the number of babies born with achondroplasia was 31. Among those 31 babies, 18 of them had one parent with achondroplasia. The remaining babies had two unaffected parents. What is the mutation frequency for this disorder among these 1,422,000 babies? What is the mutation rate for achondroplasia?

C27. A segment of DNA has the following sequence:

TTGGATGCTGG
AACCTACGACC

A. What would be the sequence immediately after reaction with nitrous acid? Let the letters H represent hypoxanthine and U represent uracil.

B. Let's suppose this DNA was reacted with nitrous acid. The nitrous acid was then removed, and the DNA was replicated for two generations. What would be the sequences of the DNA products after the DNA had replicated two times? Your answer should contain the sequences of four double helices.

C28. In the treatment of cancer, the basis for many types of chemotherapy and radiation therapy is that mutagens are more effective at killing dividing cells compared to nondividing cells. Explain why. What are possible harmful side effects of chemotherapy and radiation therapy?

C29. An individual contains a somatic mutation that changes a lysine codon into a glutamic acid codon. Prior to acquiring this mutation, the individual had been exposed to UV light, proflavin, and 5-bromouracil. Which of these three agents would be the most likely to have caused this somatic mutation? Explain your answer.

C30. Which of the following examples is likely to be caused by a somatic mutation?

A. A purple flower has a small patch of white tissue.

B. One child, in a family of seven, is an albino.

C. One apple tree, in a very large orchard, produces its apples two weeks earlier than any of the other trees.

D. A 60-year-old smoker develops lung cancer.

C31. How would nucleotide excision repair be affected if one of the following proteins were missing? Describe the condition of the DNA that had been repaired in the absence of the protein.

A. UvrA

B. UvrC

C. UvrD

D. DNA polymerase

C32. During mismatch repair, why is it necessary to distinguish between the template strand and the newly made daughter strand? How is this accomplished?

C33. What are the two main mechanisms by which cells repair double-strand breaks? Briefly describe each one.

C34. With regard to the repair of double-strand breaks, what are the advantages and disadvantages of homologous recombination repair versus nonhomologous end joining?

C35. When DNA N-glycosylase recognizes thymine dimers, it detects only the thymine located on the 5' side of the thymine dimer as being abnormal. Draw and explain the steps whereby a thymine dimer would be removed by the consecutive actions of DNA N-glycosylase, AP endonuclease, and DNA polymerase.

C36. Discuss the relationship between transcription and DNA repair. Why is it beneficial to repair actively transcribed DNA more efficiently than nontranscribed DNA?

C37. Three common ways to repair changes in DNA structure are nucleotide excision repair, mismatch repair, and homologous recombination repair. Which of these three mechanisms would be used to fix the following types of DNA changes?

A. A change in the structure of a base caused by a mutagen in a nondividing eukaryotic cell

B. A change in DNA sequence caused by a mistake made by DNA polymerase

C. A thymine dimer in the DNA of an actively dividing bacterial cell

C38. What is the underlying genetic defect that causes xeroderma pigmentosum? How can the symptoms of this disease be explained by the genetic defect?

C39. In E. coli, a methyltransferase enzyme encoded by the dam gene recognizes the sequence 5'–GATC–3' and attaches a methyl group to the N^6 position of adenine. E. coli strains that have the dam gene deleted are known to have a higher spontaneous mutation rate compared to normal strains. Explain why.

C40. Discuss the similarities and differences between the nucleotide excision repair and mismatch repair systems.

Experimental Questions

E1. The Luria-Delbrück fluctuation test is consistent with the random mutation theory. How would the results have been different if the physiological adaptation hypothesis had been correct?

E2. Explain how the technique of replica plating supports a random mutation theory but conflicts with the physiological adaptation hypothesis. Outline how you would use this technique to show that antibiotic resistance is due to random mutations.

E3. In the experiment of Figure 16.12, the ClB chromosome carries a large paracentric inversion, in which the centromere is found outside of the inverted region. This inversion prevents crossing over with the other X chromosome (which may carry an X-ray-induced lethal mutation). Does a paracentric inversion really prevent crossing over? Explain. You may wish to review Chapter 8 to answer this question.

E4. As you may have noticed in the data table of Figure 16.12, the daughters whose fathers were exposed to X-rays were less likely to produce any offspring compared to the control daughters (whose fathers were not exposed to X-rays). Among 1,015 daughters whose fathers had been exposed to X-rays, only 783 produced offspring. By comparison, among 1,011 control daughters, 947 were able to produce offspring. Suggest reasons why the daughters of irradiated fathers appear less likely to produce any offspring compared to the control daughters.

E5. In Müller's experiment with ClB chromosomes, is the experiment measuring the mutation rate within a single gene? Explain. If we divide 91 by 783, we obtain a mutation rate of 11.6%. In your own words, explain what this value means.

E6. Researchers were interested in whether two different chemicals were mutagens. Let's call them chemical A and chemical B. They followed the protocol described in the experiment of Figure 16.12 and obtained the following results:

Treatment	Number of ClB Daughters Crossed to Normal Males	Number of Tubes Containing Offspring	Number of Tubes Lacking Male Offspring
Control	2,108	2,077	3
Chemical A	1,402	1,378	2
Chemical B	4,203	3,100	77

Would you conclude that chemical A and/or chemical B is a mutagen? Explain.

E7. From an experimental point of view, is it better to use haploid or diploid organisms for mutagen testing? Consider both the Ames test and the experiment of Figure 16.12 when preparing your answer.

E8. How would you modify the Ames test to discover physical mutagens? Would it be necessary to add the rat liver extract? Explain why or why not.

E9. During an Ames test, bacteria were exposed to a potential mutagen. Also, as a control, another sample of bacteria was not exposed to the mutagen. In both cases, 10 million bacteria were plated and the following results were obtained:

No mutagen: 17 colonies

With mutagen: 2,017 colonies

Calculate the mutation rate in the presence and absence of the mutagen. How much does the mutagen increase the rate of mutation?

E10. Richard Boyce and Paul Howard-Flanders conducted an experiment that provided biochemical evidence that thymine dimers are removed from the DNA by a DNA repair system. In their studies, bacterial DNA was radiolabeled so that the amount of radioactivity reflected the amount of thymine dimers. The DNA was then subjected to UV light, causing the formation of thymine dimers. When radioactivity was found in the soluble fraction, thymine dimers had been excised from the DNA by a DNA repair system. But when the radioactivity was in the insoluble fraction, the thymine dimers had been retained within the DNA. The following table illustrates some of their results involving a normal strain of *E. coli* and a second strain that was very sensitive to killing by UV light:

Strain	Treatment	Radioactivity in the Insoluble Fraction (cpm)	Radioactivity in the Soluble Fraction (cpm)
Normal	No UV	<100	<40
Normal	UV-treated, incubated 2 hours at 37°C	357	940
Mutant	No UV	<100	<40
Mutant	UV-treated, incubated 2 hours at 37°C	890	<40

(Adapted from: Boyce, R. P., and Howard-Flanders, P. Release of ultraviolet light-induced thymine dimers from DNA in *E. coli* K-12. *Proc. Natl. Acad. Sci. USA* 51, 293–300.)

Explain the results found in this table. Why is the mutant strain sensitive to UV light?

Questions for Student Discussion/Collaboration

1. In *E. coli*, a variety of mutator strains have been identified in which the spontaneous rate of mutation is much higher than normal strains. Make a list of the types of abnormalities that could cause a strain of bacteria to become a mutator strain. Which abnormalities do you think would give the highest rate of spontaneous mutation?

2. Discuss the times in a person's life when it would be most important to avoid mutagens. Which parts of a person's body should be most protected from mutagens?

3. A large amount of research is aimed at studying mutation. However, there is not an infinite amount of research dollars. Where would you put your money for mutation research?

   A. Testing of potential mutagens

   B. Investigating molecular effects of mutagens

   C. Investigating DNA repair mechanisms

   D. Or some other place?

Note: All answers appear at the website for this textbook; the answers to even-numbered questions are in the back of the textbook.

www.mhhe.com/brookergenetics3e

Visit the Online Learning Center for practice tests, answer keys, and other learning aids for this chapter. Enhance your understanding of genetics with our interactive exercises, quizzes, animations, and much more.

17

RECOMBINATION AND TRANSPOSITION AT THE MOLECULAR LEVEL

In this chapter, we will examine a variety of molecular processes in which segments of DNA, such as those found in chromosomes, become rearranged. **Homologous recombination** is the process whereby DNA segments, which are similar or identical to each other, break and rejoin to form a new combination. As described in Chapter 3, homologous recombination occurs when chromosomes cross over during meiosis. Not only does homologous recombination enhance genetic diversity, it also helps to repair DNA and ensures the proper segregation of chromosomes.

We will also examine ways that nonhomologous segments of DNA may recombine with each other. During **site-specific recombination,** nonhomologous DNA segments are recombined at specific sites. This type of recombination occurs when certain viruses integrate their genomes into host cell DNA and also happens within genes that encode antibody polypeptides. In addition, we will consider a widespread form of recombination known as **transposition.** As you will learn, small segments of DNA called transposable elements can move themselves to multiple locations within the chromosomal DNA.

From a molecular viewpoint, homologous recombination, site-specific recombination, and transposition are important mechanisms for DNA rearrangement. These processes involve a series of steps that direct the breakage and rejoining of DNA fragments. Various cellular proteins are necessary for these steps to occur properly. The past few decades have seen many exciting advances in our understanding of recombination at the molecular level. In this chapter, we will consider the general concepts of recombination and examine molecular models that explain how recombination occurs.

Speckled corn kernels. This speckling phenotype is due to the movement of DNA segments called transposable elements.

17.1 HOMOLOGOUS RECOMBINATION

Homologous recombination involves an exchange between DNA segments that are similar or identical. Eukaryotic chromosomes that have similar or identical sequences frequently participate in crossing over during meiosis I and occasionally during mitosis. Crossing over involves the alignment of a pair of homologous chromosomes, followed by the breakage of two chromatids at analogous locations, and the subsequent exchange of the corresponding segments (refer to Figure 3.10).

Figure 17.1 shows two types of crossing over that may occur between replicated chromosomes in a diploid species. When crossing over takes place between sister chromatids, the process is called **sister chromatid exchange (SCE).** Because sister chromatids are genetically identical to each other, SCE

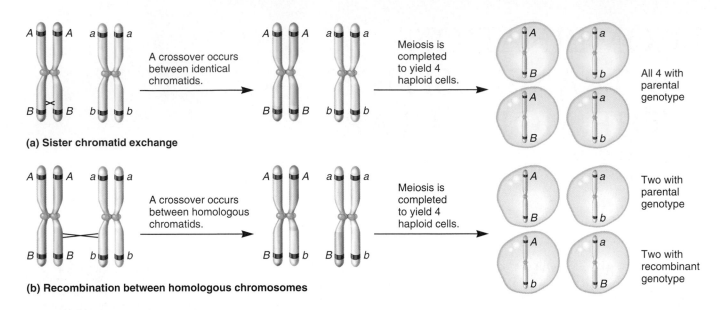

(a) Sister chromatid exchange

A crossover occurs between identical chromatids.

Meiosis is completed to yield 4 haploid cells.

All 4 with parental genotype

(b) Recombination between homologous chromosomes

A crossover occurs between homologous chromatids.

Meiosis is completed to yield 4 haploid cells.

Two with parental genotype

Two with recombinant genotype

FIGURE 17.1 **Crossing over between eukaryotic chromosomes.** (a) Sister chromatid exchange occurs when genetically identical chromatids cross over. (b) Homologous recombination can also occur when homologous chromosomes cross over. This form of homologous recombination may lead to a new combination of alleles, which is called a recombinant (or nonparental) genotype.

Genes→Traits Homologous recombination is particularly important when we consider the relationships between multiple genes and multiple traits. For example, if the X chromosome in a female fruit fly carried alleles for red eyes and gray body and its homologue carried alleles for white eyes and yellow body, homologous recombination could produce recombinant chromosomes that carry alleles for red eyes and yellow body, or alleles for white eyes and gray body. Therefore, new combinations of two or more alleles can arise when homologous recombination takes place.

does not produce a new combination of alleles (Figure 17.1a). By comparison, it is common for crossing over to occur between homologous chromosomes. As shown in Figure 17.1b, this form of homologous recombination may produce a new combination of alleles in the resulting chromosomes. In this second case, homologous recombination has resulted in **genetic recombination,** which refers to the shuffling of genetic material to create a new genetic combination that differs from the original. Homologous recombination is an important mechanism to foster genetic recombination.

Although bacteria are usually haploid—having only one type of chromosome—they also can undergo homologous recombination. How can the exchange of DNA segments occur in a

haploid organism? First, bacteria may have more than one copy of a chromosome per cell, though the copies are genetically identical. These copies can exchange genetic material via homologous recombination. Second, during DNA replication, the replicated regions may also undergo homologous recombination. In bacteria, homologous recombination is particularly important in the repair of DNA segments that have been damaged.

In this section, we will begin with an experimental approach to detect SCE, overcoming the obstacle that the exchanged chromosomes are genetically identical to each other. We will then focus our attention on the molecular mechanisms that underlie homologous recombination.

EXPERIMENT 17A

The Staining of Harlequin Chromosomes Can Reveal Recombination Between Sister Chromatids

Our understanding of crossing over and homologous recombination has come from a variety of experimental approaches, including genetic, biochemical, and cytological analyses. Chromosomal staining methods have allowed researchers to visualize the genetic exchange between eukaryotic chromosomes. In the 1970s, the Russian cytogeneticist A. F. Zakharov and colleagues developed methods that improved our ability to identify chromosomes. They made the interesting observation that chromosomes labeled with the nucleotide analog **5-bromodeoxyuridine (BrdU)** bind certain types of stain to a different degree compared to normal chromosomes. In 1974, Paul Perry and Sheldon Wolff

extended this approach to differentially stain sister chromatids and microscopically identify SCEs.

Before we consider the experiment of Perry and Wolff, let's examine how their staining procedure allowed them to accurately distinguish the two sister chromatids. In their approach, eukaryotic cells were grown in a laboratory and exposed to BrdU for two rounds of DNA replication. After the second round of DNA replication, one of the sister chromatids in each pair contained one unlabeled strand and one BrdU-labeled strand. The other sister chromatid had two BrdU-labeled strands (**Figure 17.2**). When treated with two dyes, Hoechst 33258 and Giemsa, the sister chromatid containing two strands with BrdU stains very weakly and appears light, whereas the sister chromatid with only one strand containing BrdU stains much more strongly and

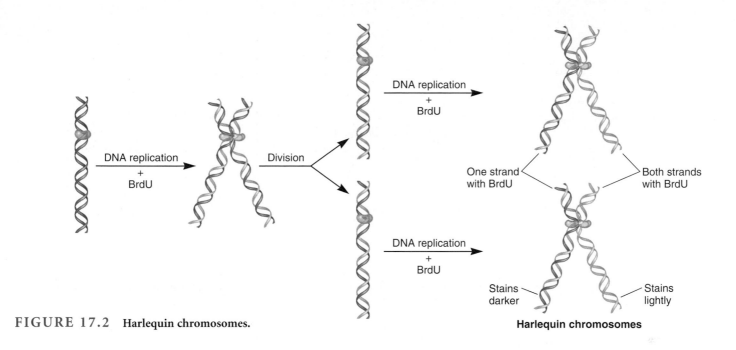

FIGURE 17.2 **Harlequin chromosomes.**

One strand with BrdU / Both strands with BrdU

Stains darker / Stains lightly

Harlequin chromosomes

appears very dark. In this way, the two sister chromatids can be distinguished microscopically. Chromosomes stained in this way are referred to as harlequin chromosomes, because they are reminiscent of a harlequin character's costume with its variegated pattern of light and dark patches. In these chromosomes, SCEs can be clearly identified as exchanges between light and dark chromatids.

The steps in Perry and Wolff's protocol are shown in **Figure 17.3**. They began with Chinese hamster ovary cells, a com-

monly used mammalian cell line, and exposed the cells to BrdU for two rounds of DNA replication. Near the end of the second round, colcemid was added to prevent the completion of mitosis. The cells were treated with KCl to spread out the chromosomes, which were subsequently fixed and then stained with Hoechst 33258 and Giemsa.

■ **THE HYPOTHESIS**

Crossing over may occur between sister chromatids.

■ **TESTING THE HYPOTHESIS — FIGURE 17.3** **The staining of harlequin chromosomes reveals sister chromatid exchange.**

Starting material: A laboratory cell line of Chinese hamster ovary (CHO) cells.

Experimental level   **Conceptual level**

1. Expose CHO cells to BrdU for two cell generations (approximately 24 hours). Note: CHO cells are Chinese hamster ovary cells, which is a commonly used mammalian cell line.

CHO cells

BrdU

24 hours

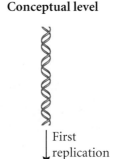

First replication

2. Near the end of the growth, expose the cells to colcemid. This prevents the cells from completing mitosis following the second round of DNA replication.

Colcemid

Second replication

3. Add 0.075 M KCl to spread the chromosomes and then methanol/acetic acid to fix the cells.

Add KCl.
Add methanol/acetic acid.

Occasional sister chromatid crossover

Crossing over

4. Stain with Hoechst 33258, rinse, and later stain with Giemsa. Note: This refinement in the staining procedure greatly improved the ability to discern the sister chromatids.

Treat with stains.

Stain

5. View under a microscope.

■ THE DATA

Reprinted by permission from Macmillan Publishers Ltd.
Nature, New Giemsa method for the differential staining of sister chromatids. Perry P. & Wolff S. 251:5471, 156–158, 1974.

■ INTERPRETING THE DATA

A micrograph of their results is shown in the data of Figure 17.3. As seen here, the chromosomes show the classic harlequin appearance due to the differential staining of the sister chromatids. Furthermore, examples of SCE are clearly visible. The arrows depict regions where crossing over has taken place. In this study, Perry and Wolff found that SCEs occurred at a frequency of approximately 0.67 per chromosome. This method has provided an accurate (and dramatic) way to visualize genetic exchange between sister chromatids.

Many subsequent studies have used the harlequin staining method to study the effects of agents that may influence the frequency of genetic exchanges. Researchers have found that DNA damage caused by radiation and chemical mutagens tends to increase the level of genetic exchange. When cells are exposed to these types of mutagens, the technique of harlequin staining has revealed a substantial increase in the frequency of SCEs. In addition, certain genetic disorders that result in higher levels of chromosome breakage also show elevated SCEs. For example, Bloom syndrome is a rare autosomal recessive disorder characterized by short stature, skin abnormalities, and a predisposition for developing certain forms of cancer. The defect is associated

with a gene that is involved with DNA replication. In the cells of Bloom syndrome patients, chromosome breaks are more frequent during DNA replication. Likewise, SCE is typically 10- to 15-fold more frequent in Bloom syndrome patients compared to unaffected individuals.

A self-help quiz involving this experiment can be found at the Online Learning Center.

The Holliday Model Describes a Molecular Mechanism for the Recombination Process

We now turn our attention to genetic exchange that occurs between homologous chromosomes. Perhaps it is surprising that the first molecular model of homologous recombination did not come from a biochemical analysis of DNA or from electron microscopy studies. Instead, it was deduced from the outcome of genetic crosses in fungi.

As discussed in Chapter 5, geneticists have learned a great deal from the analysis of fungal asci, because an ascus is a sac that contains the products of a single meiosis. When two haploid fungi that differ at a single gene are crossed to each other, the ascus is expected to contain an equal proportion of each genotype. For example, if a pigmented strain of *Neurospora* producing orange spores is crossed to an albino strain producing white spores, the resulting group of eight cells, or octad, should contain four orange spores and four white spores (refer to Figure 5.14). As early as 1934, H. Zickler noticed that unequal proportions of the spores sometimes occurred within asci. He occasionally observed octads with six orange spores and two white spores, or six white spores and two orange spores.

Zickler used the term **gene conversion** to describe the phenomenon in which one allele is converted to the allele on the homologous chromosome. Subsequent studies by several researchers confirmed this phenomenon in yeast and *Neurospora*. Gene conversion occurred at too high a rate to be explained by new mutations. In addition, research showed that gene conversion often occurs in a chromosomal region where a crossover has taken place.

Based on studies involving gene conversion, Robin Holliday proposed a model in 1964 to explain the molecular steps that occur during homologous recombination. We will first consider the steps in the Holliday model and then consider more recent models. Later, we will examine how the Holliday model can explain the phenomenon of gene conversion.

The **Holliday model** is shown in **Figure 17.4a**. At the beginning of the process, two homologous chromatids are aligned with each other. According to the model, a break or nick occurs at identical sites in one strand of each of the two homologous chromatids. The strands then invade the opposite helices and base pair with the complementary strands. This event is followed by a covalent linkage to create a **Holliday junction.** The cross in the Holliday junction can migrate in a lateral direction. As it does so, a DNA strand in one helix is swapped for a DNA strand in the other helix. This process is called **branch migration**

because the branch connecting the two double helices migrates laterally. Because the DNA sequences in the homologous chromosomes are similar but not identical, the swapping of the DNA strands during branch migration may produce a **heteroduplex,** a region in the double-stranded DNA that contains base mismatches. In other words, because the DNA strands in this region are from homologous chromosomes, their sequences are not perfectly complementary, yielding mismatches.

The final two steps in the recombination process are collectively called **resolution** because they involve the breakage and rejoining of two DNA strands to create two separate chromosomes. In other words, the entangled DNA strands become resolved into two separate structures. The bottom left side of Figure 17.4a shows the Holliday junction viewed from two different planes. If breakage occurs in the same two DNA strands that were originally nicked at the beginning of this process, the subsequent joining of strands produces nonrecombinant chromosomes with a heteroduplex region. Alternatively, if breakage occurs in the strands that were not originally nicked, the rejoining process results in recombinant chromosomes, also with a heteroduplex region.

The Holliday model can account for the general properties of recombinant chromosomes formed during eukaryotic meiosis. As mentioned, the original model was based on the results of crosses in fungi where the products of meiosis are contained within a single ascus. Nevertheless, molecular research in many other organisms has supported the central tenets of the model. Particularly convincing evidence came from electron microscopy studies in which recombination structures could be visualized. **Figure 17.4b** shows an electron micrograph of two DNA fragments in the process of recombination. This structure has been called a chi form because its shape is similar to the Greek letter χ (chi).

More Recent Models Have Refined the Molecular Steps of Homologous Recombination

As more detailed studies of homologous recombination have become available, certain steps in the Holliday model have been reconsidered. In particular, more recent models have modified the initiation phase of recombination. Researchers now suggest that homologous recombination is not likely to involve nicks at identical sites in one strand of each homologous chromatid. Instead, it is more likely for a DNA helix to incur a break in both strands of one chromatid or a single nick. Both of these types of changes have been shown to initiate homologous recombination. Therefore, newer models have tried to incorporate these experimental

ONLINE ANIMATION

FIGURE 17.4   The Holliday model for homologous recombination. The Holliday model is adapted from Holliday, R. (1964). A mechanism for gene conversion in fungi. *Genet. Res.* 5, 282–304.

Sister chromatids

Homologous chromatids

Both chromatids are nicked at identical locations.

Nicks

The DNA strands to the left of the nicks invade the homologous chromosomes and attach to the strands to the right of the nicks.

Holliday junction

The Holliday junction migrates from left to right. This is called branch migration. It creates 2 heteroduplex regions.

Two heteroduplex regions that have a few base mismatches

In this next step, the figure is simply redrawn by bending the ends labeled *A* and *B* upwards, and bending the ends labeled *a* and *b* downwards. This makes it look more like a true Holliday junction. The Holliday junction is viewed in two different planes.

The strands that were originally nicked are broken.

The strands are connected to create nonrecombinant chromosomes with a short heteroduplex region.

Viewed in a different plane (when rotated 180°)

The strands that were not originally nicked are broken.

The strands are connected to create recombinant chromosomes with a short heteroduplex region.

(b) Micrograph of a Holliday junction

End result

Short heteroduplex region

Nonrecombinant chromosomes with a heteroduplex region

Short heteroduplex region

Recombinant chromosomes with a heteroduplex region

(a) The Holliday model for homologous recombination

observations. In 1975, a model proposed by Matthew Meselson and Charles Radding hypothesized that a single nick in one DNA strand initiates recombination. A second model, proposed by Jack Szostak, Terry Orr-Weaver, Rodney Rothstein, and Franklin Stahl, suggests that a double-strand break initiates the recombination process. This is called the **double-strand break model.** Though recombination may occur via more than one mechanism, recent evidence suggests that double-strand breaks commonly promote homologous recombination during meiosis and during DNA repair.

Figure 17.5 shows the general steps in the double-strand break model. As seen here, the top chromosome has experienced

a double-strand break. A small region near the break is degraded, which generates a single-stranded DNA segment that can invade the intact double helix. The strand displaced by the invading segment forms a structure called a displacement loop (D-loop). After the D-loop is formed, two regions have a gap in the DNA. How is the problem fixed? DNA synthesis occurs in the relatively short gaps where a DNA strand is missing. This DNA synthesis is called **DNA gap repair synthesis.** Once this is completed, two Holliday junctions are produced. Depending on the way these are resolved, the end result is nonrecombinant or recombinant chromosomes containing a short heteroduplex. In eukaryotes such as yeast, recent evidence suggests that certain proteins bound to Holliday junctions may regulate the resolution step in a way that favors the formation of recombinant chromosomes rather than nonrecombinant chromosomes.

Various Proteins Are Necessary to Facilitate Homologous Recombination

The homologous recombination process requires the participation of many proteins that catalyze different steps in the recombination pathway. Homologous recombination is found in all species, and the types of proteins that participate in the steps outlined in Figure 17.5 are very similar. The cells of any given species may have more than one molecular mechanism to carry out homologous recombination. This process is best understood in *Escherichia coli.* **Table 17.1** summarizes some of the *E. coli* proteins that play critical roles in one recombination pathway found in this species. Though it is beyond the scope of this textbook, *E. coli* has other pathways to carry out homologous recombination.

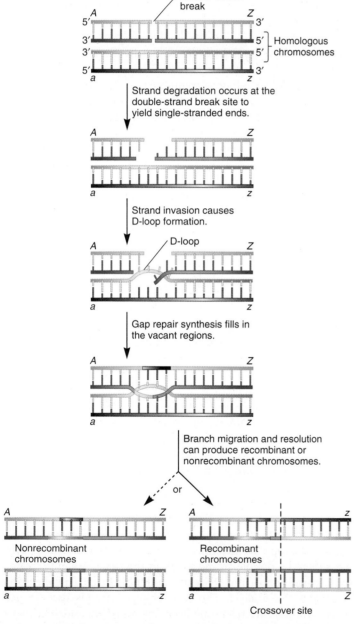

FIGURE 17.5 A simplified version of the double-strand break model. For simplicity, this illustration does not include the formation of heteroduplexes. The dashed arrow indicates that the pathway to the left may be less favored.

TABLE 17.1

E. coli Proteins That Play a Role in Homologous Recombination

Protein	Description
RecBCD	A complex of three proteins that tracks along the DNA and recognizes double-strand breaks. The complex partially degrades the double-stranded regions to generate single-stranded regions that can participate in strand invasion. RecBCD is also involved in loading RecA onto single-stranded DNA. In addition, RecBCD can create single-strand breaks at chi sequences.
Single-stranded binding protein	Coats broken ends of chromosomes and prevents excessive strand degradation.
RecA	Binds to single-stranded DNA and promotes strand invasion, which enables homologous strands to find each other. It also promotes the displacement of the complementary strand to generate a D-loop.
RuvABC	This protein complex binds to Holliday junctions. RuvAB promotes branch migration. RuvC is an endonuclease that cuts the crossed or uncrossed strands to resolve Holliday junctions into separate chromosomes.
RecG	RecG protein can also promote branch migration of Holliday junctions.

RecBCD is a protein complex composed of the RecB, RecC, and RecD proteins. (The term Rec indicates that these proteins are involved with recombination.) The RecBCD complex plays an important role in the initiation of recombination involving double-strand breaks, as outlined in Figure 17.5. In this process, RecBCD recognizes a double-strand break within DNA and catalyzes DNA unwinding and strand degradation. The action of RecBCD produces single-stranded DNA ends that can participate in strand invasion and exchange. The single-stranded DNA ends are coated with single-stranded binding protein to prevent their further degradation. The RecBCD complex can also create breaks in the DNA at sites known as chi sequences. In *E. coli*, the chi sequence is 5'–GCTGGTGG–3'. As RecBCD tracks along the DNA, when it encounters a chi sequence, it cuts one of the DNA strands to the 3' side of this sequence. This generates a single nick that can initiate homologous recombination.

The function of the RecA protein is to promote strand invasion. To accomplish this task, it binds to the single-stranded ends of DNA molecules generated from the activity of RecBCD. A large number of RecA proteins bind to single-stranded DNA, forming a structure called a filament. During strand invasion, this filament makes contact with the unbroken chromosome. Initially, this contact is most likely to occur at nonhomologous regions. The contact point slides along the DNA until it reaches a homologous region. Once a homologous site is located, RecA catalyzes the displacement of one DNA strand, and the invading single-stranded DNA quickly forms a double helix with the other strand. This results in a D-loop, as shown in Figure 17.5. RecA proteins mediate the movement of the invading strand and the displacement of the complementary strand. This occurs in such a way that the displaced strand invades the vacant region of the broken chromosome.

Proteins that bind specifically to Holliday junctions have also been identified. These include a complex of proteins termed RuvABC and a protein called RecG. RuvA is a tetramer that forms a platform on which the Holliday junction is held in a square planar configuration. This platform also contains two hexameric rings of RuvB. Together, RuvA and RuvB act as a helicase that catalyzes an ATP-dependent migration of a four-way branched DNA junction in either the 5' → 3' or 3' → 5' direction. RuvC is an endonuclease that binds as a dimer to Holliday junctions and resolves these structures by making cuts in the DNA (see bottom of Figure 17.4). RecG is a helicase that also catalyzes branch migration of Holliday junctions.

Before ending this discussion regarding the molecular mechanism of homologous recombination, let's consider recombinational events during meiosis in eukaryotic cells. As described in Chapter 3, crossing over between homologous chromosomes is an important event during prophase of meiosis I. An intriguing question is, How are crossover sites chosen between two homologous chromosomes? While the answer is not entirely understood, molecular studies in two different yeast species, *Saccharomyces cerevisiae* and *Schizosaccharomyces pombe*, suggest that double-strand breaks initiate the homologous recombination that occurs during meiosis. In other words, double-strand breaks create sites where a crossover will occur. In *S. cerevisiae*, the formation of DNA double-strand breaks that initiate meiotic recombination requires at least 10 different proteins. One particular protein, termed Spo11 protein, is thought to be instrumental in cleaving the DNA and thereby creating a double-strand break. However, the roles of the other proteins, and the interactions among them, are not well understood. Once a double-strand break is made, homologous recombination can then occur according to the model described in Figure 17.5.

Gene Conversion May Result from DNA Mismatch Repair or DNA Gap Repair

As mentioned earlier, homologous recombination can lead to an event where one allele is converted to the allele on the homologous chromosome, a process known as gene conversion. The original Holliday model was based on the phenomenon of gene conversion.

How can homologous recombination account for gene conversion? Researchers have identified two possible ways that gene conversion can occur. One mechanism involves DNA mismatch repair, a topic that was described in Chapter 16. To understand how this works, let's take a closer look at the heteroduplexes formed during branch migration of a Holliday junction (see Figure 17.4a). A heteroduplex contains a DNA strand from each of the two original parental chromosomes. The two parental chromosomes may contain an allelic difference within this region. In other words, this short region may contain DNA sequence differences. If this is the case, the heteroduplex formed after branch migration will contain an area of base mismatch. Gene conversion occurs when recombinant chromosomes are repaired and result in two copies of the same allele.

As shown in **Figure 17.6**, mismatch repair of a heteroduplex may result in gene conversion. In this example, the two parental chromosomes had different alleles due to a single base-pair difference in their DNA sequences, as shown at the top of the figure. During recombination, branch migration has occurred across this region, thereby creating two heteroduplexes with base mismatches. As described in Chapter 16, DNA mismatches will be recognized by DNA repair systems and repaired to a double helix that obeys the AT/GC rule. These two mismatches can be repaired in four possible ways. As shown here, two possibilities produce no gene conversion, whereas the other two lead to gene conversion.

A second mechanism for gene conversion is via DNA gap repair synthesis. **Figure 17.7** illustrates how gap repair synthesis can lead to gene conversion according to the double-strand break model. The top chromosome, which carries the recessive *b* allele, has suffered a double-strand break in this gene. A gap is created by the digestion of the DNA in the double helix. This digestion eliminates the *b* allele. The two template strands used in gap repair synthesis are from one homologous chromatid. This helix carries the dominant *B* allele. Therefore, after gap repair synthesis takes place, the top chromosome will contain the *B* allele, as will the bottom chromosome. Gene conversion has changed the recessive *b* allele to a dominant *B* allele.

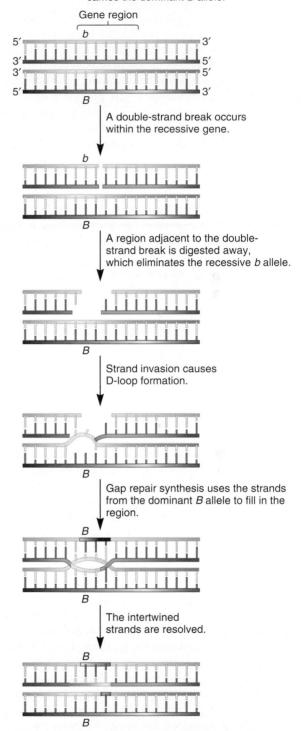

FIGURE 17.6 **Gene conversion by DNA mismatch repair.** A branch migrates past a homologous region that contains slightly different DNA sequences. This produces two heteroduplexes—DNA double helices with mismatches. The mismatches can be repaired in four possible ways by the mismatch repair system described in Chapter 16. Two of these ways result in gene conversion. The repaired base is shown in red.

FIGURE 17.7 **Gene conversion by gap repair synthesis in the double-strand break model.** A gene is found in two alleles, designated *B* and *b*. A double-strand break occurs in the DNA encoding the *b* allele. Both of these DNA strands are digested away, thereby eliminating the *b* allele. A complementary DNA strand encoding the *B* allele migrates to this region and provides the template to synthesize a double-stranded region. Following resolution, both DNA double helices carry the *B* allele.

17.2 SITE-SPECIFIC RECOMBINATION

Thus far, we have examined recombination between segments of DNA similar or identical to each other. **Site-specific recombination** is another mechanism by which DNA fragments can recombine to make new genetic combinations. During this process, two DNA segments with little or no homology align themselves at specific sites. The sites are relatively short DNA sequences (a dozen or so nucleotides in length) that provide a specific location where recombination will occur. Chromosome breakage and reunion occur at these defined locations to create a recombinant chromosome. For this to occur, these sites are recognized by specialized enzymes that catalyze the breakage and rejoining of DNA fragments within the sites.

Certain viruses use site-specific recombination to insert their viral DNA into their host cell's chromosome. This process has been examined extensively in bacteriophage λ. In addition, mammalian genes that encode antibody polypeptides are rearranged by site-specific recombination that enables the generation of a diverse array of antibodies. In this section, we will consider both mechanisms.

The Integration of Viral Genomes Can Occur by Site-Specific Recombination

The reproductive cycle of some viruses involves the integration of viral DNA into host cell DNA. Certain bacteriophages, for example, can integrate their viral DNA into the bacterial chromosome, creating a prophage. The prophage can exist in a latent, or **lysogenic,** state for many generations. The two cycles of phage λ—lytic and lysogenic—are described in Chapter 14. Phage λ infects *E. coli*. The integration of this phage into the host chromosome is well understood and occurs by a mechanism involving site-specific recombination (**Figure 17.8**).

Integration of the λ DNA into the *E. coli* chromosome requires sequences known as **attachment sites.** As shown at the top of Figure 17.8, a common core sequence within the attachment site sequence is identical in the λ DNA and the *E. coli* chromosome. An enzyme known as integrase is encoded by a gene in the λ DNA. Several molecules of integrase recognize the core sequences and bring them close together. Integrase then makes staggered cuts in both the λ and *E. coli* attachment sites. The strands are then exchanged, and the ends are ligated together. In this way, the phage DNA is integrated into the host cell chromosome. As a prophage, the λ DNA may remain latent for many generations. Certain environmental conditions, such as exposure to UV light, may act to stimulate the excision of the prophage from the host DNA, thereby reactivating the virus. Excision also requires integrase, which catalyzes the reverse reaction, as well as a second protein known as excisionase.

Antibody Diversity in the Immune System Is Produced by Site-Specific Recombination

As we have just seen, viruses can integrate their DNA into the host cell chromosome by site-specific recombination. This process

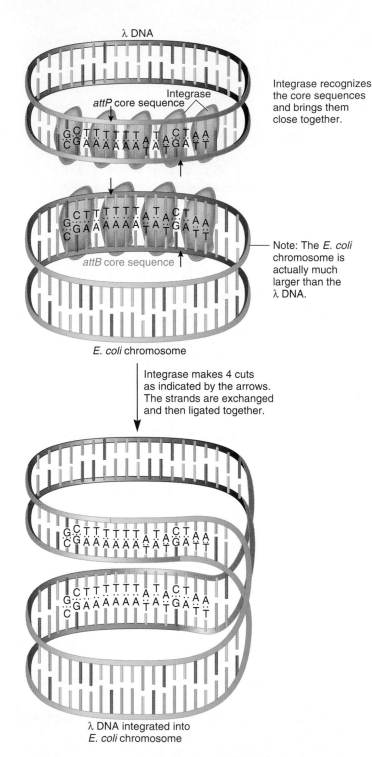

FIGURE 17.8   The integration of λ DNA into the *E. coli* chromosome. The core sequence within *attP* in the λ DNA attaches to the core sequence within *attB* in the *E. coli* chromosome. As noted here, the core sequences of *attP* and *attB* are identical to each other and thereby provide recognition sites for site-specific recombination.

requires a viral enzyme, integrase, that recognizes the sites and catalyzes the recombination reaction. A similar process occurs in certain cells of the immune system. The DNA sequences within antibody genes are rearranged by enzymes that recognize specific sites within those genes and catalyze the breakage and reunion of DNA segments. Before we discuss the details of this mechanism, let's first consider the biology of antibodies.

Antibodies, or **immunoglobulins (Igs),** are proteins produced by the B cells of the immune system. Their function is to recognize foreign material, such as viruses and bacteria, and target them for destruction. A foreign material that elicits an immune response is called an **antigen.** Antibodies recognize sites within antigens. The recognition between an antibody and antigen is very specific. Within the immune system, each B cell produces a single type of antibody. However, our bodies have millions of B cells. Site-specific recombination allows each B cell to produce an antibody with a different amino acid sequence. These differences in the amino acid sequences of antibody proteins enable them to recognize different antigens. In this way, site-specific recombination plays an important role in the ability of the immune system to identify an impressive variety of foreign materials as being antigens and thereby target them for destruction.

From a genetic viewpoint, the production of millions of different antibodies poses an interesting problem. If a distinct gene was needed to produce each different antibody polypeptide, the genome would need to contain millions of different antibody genes. By comparison, consider that the entire human genome contains only about 20,000 to 25,000 different genes. How then is it possible for the genome to produce millions of different antibody proteins, which are encoded by genes? The answer to this question baffled geneticists for several decades until research revealed that antibodies with different polypeptide sequences are generated by an unusual mechanism in which the DNA is cut and reconnected by site-specific recombination. We might call this "DNA splicing," although the term splicing is usually reserved for the cutting and rejoining of RNA molecules. With this mechanism, only a few large antibody precursor genes are needed to produce millions of different antibodies. These precursor genes are spliced in many different ways to produce a vast array of polypeptides with differing amino acid sequences.

Figure 17.9 shows the site-specific recombination of an antibody precursor gene. Antibodies are tetrameric proteins composed of two heavy polypeptide chains and two light chains. One type of light chain is the κ (kappa) light chain, which is a component of a class of antibodies known as immunoglobulin G (IgG). The organization of the precursor gene for the κ light chain is shown at the top of Figure 17.9. At the left side, the gene has approximately 300 regions known as variable (V) sequences or domains. In addition, the gene contains four different joining (J) sequences and a single constant (C) sequence. Each variable domain or joining domain encodes a different amino acid sequence.

During the maturation of B cells, the κ light-chain precursor gene is cut and rejoined so that one variable domain becomes adjacent to a joining domain. At the end of every V domain and the beginning of every J domain is located a **recombination signal sequence** that functions as a recognition site for site-specific recombination between the V and J regions. The recombination event is initiated by two proteins called **RAG1** and **RAG2.** RAG is an acronym for recombination-activating gene. These proteins recognize recombination signal sequences and generate two

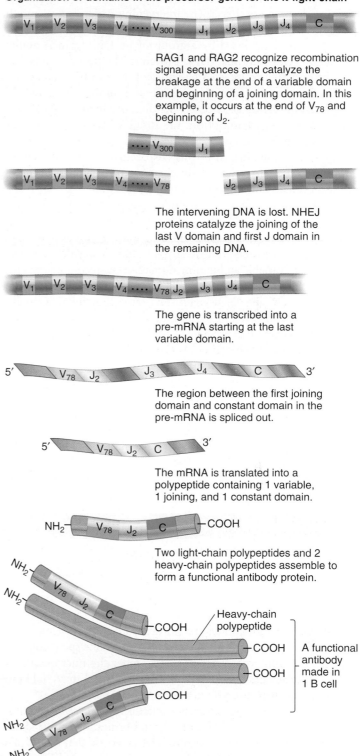

Organization of domains in the precursor gene for the κ light-chain

RAG1 and RAG2 recognize recombination signal sequences and catalyze the breakage at the end of a variable domain and beginning of a joining domain. In this example, it occurs at the end of V_{78} and beginning of J_2.

The intervening DNA is lost. NHEJ proteins catalyze the joining of the last V domain and first J domain in the remaining DNA.

The gene is transcribed into a pre-mRNA starting at the last variable domain.

The region between the first joining domain and constant domain in the pre-mRNA is spliced out.

The mRNA is translated into a polypeptide containing 1 variable, 1 joining, and 1 constant domain.

Two light-chain polypeptides and 2 heavy-chain polypeptides assemble to form a functional antibody protein.

Heavy-chain polypeptide

A functional antibody made in 1 B cell

FIGURE 17.9   Site-specific recombination within the precursor gene that encodes the κ light chain for immunoglobulin G (IgG) proteins.

double-strand breaks, one at the end of a V domain and one at the beginning of a J domain. For example, in the recombination event shown in Figure 17.9, RAG1 and RAG2 have made cuts at the end of variable domain number 78 and the beginning of joining domain number 2. The intervening region is lost, and the two ends are then joined to each other. The connection phase of this process is catalyzed by a group of proteins termed **nonhomologous end-joining (NHEJ) proteins.** The fusion process may not be entirely precise, so a few nucleotides can be added or lost at the junction between the variable and joining domains. This imprecision further accentuates the diversity in antibody genes.

Following transcription, the fused VJ region is contained within a pre-mRNA transcript that is then spliced to connect the J and C domains. After this has occurred, a B cell will produce only the particular κ light chain encoded by the specific VJ fusion domain and the constant domain.

The heavy-chain polypeptides are produced by a similar recombination mechanism. In this case, the heavy-chain gene has about 500 variable domains and four joining domains. In addition, the gene encodes 12 diversity (D) domains, which are found between the variable and joining domains. The recombination first involves the connection of a D and J domain, followed by the connection of a V and DJ domain. The same proteins that catalyze VJ fusion are involved in the recombination with the heavy-chain gene. Collectively, this process is called **V(D)J recombination.** The D is in parentheses because this type of domain is found only in the heavy-chain genes, not in the light-chain genes.

The recombination process within immunoglobulin genes produces an enormous diversity in polypeptides. Even though it occurs at specific junctions within the antibody gene, the recombination is fairly random with regard to the particular V and J domains that can be joined. Overall, the possible number of functional antibodies that can be produced by V(D)J recombination is rather staggering. If we assume that any of the 300

different variable sequences can be spliced next to any of the four joining sequences, this results in 1,200 possible light-chain combinations. The number of heavy-chain possibilities is $500 \times 12 \times 4 = 24,000$. Because any light-chain–heavy-chain combination is possible, this yields $1,200 \times 24,000 = 28,800,000$ possible antibody molecules from the random recombination within two precursor genes!

17.3 TRANSPOSITION

The last form of recombination we will consider is transposition. In some ways, transposition resembles the site-specific recombination we examined for phage λ. In that case, a segment of λ DNA was able to integrate itself into the *E. coli* chromosome. Transposition also involves the integration of small segments of DNA into the chromosome. Transposition, though, can occur at many different locations within the genome. The DNA segments that transpose themselves are known as **transposable elements (TEs).** TEs have sometimes been referred to as "jumping genes" because they are inherently mobile.

Transposable elements were first identified by Barbara McClintock in the early 1950s from her classic studies with corn plants. Since that time, geneticists have discovered many different types of TEs in organisms as diverse as bacteria, fungi, plants, and animals. The advent of molecular technology has allowed scientists to understand more about the characteristics of TEs that enable them to be mobile. In this section, we will examine the characteristics of TEs and explore the mechanisms that explain how they move. We will also discuss the biological significance of TEs and their uses as experimental tools.

EXPERIMENT 17B

McClintock Found That Chromosomes of Corn Plants Contain Loci That Can Move

Barbara McClintock began her scientific career as a student at Cornell University. Her interests quickly became focused on the structure and function of the chromosomes of corn plants, an interest that continued for the rest of her life. She spent countless hours examining corn chromosomes under the microscope. She was technically gifted and had a theoretical mind that could propose ideas that conflicted with conventional wisdom.

During her long career as a scientist, McClintock identified many unusual features of corn chromosomes. She noticed that one strain of corn had the strange characteristic that a particular chromosome, number 9, tended to break at a fairly high rate at the same site. McClintock termed this a **mutable site** or locus. This observation initiated a six-year study concerned with highly unstable chromosomal locations. In 1951, at the end of her study, McClintock proposed that these sites are actually locations where transposable elements have been inserted into the chromosomes. At the time of McClintock's studies, such an idea was entirely unorthodox.

McClintock focused her efforts on the relationship between a mutable locus and its phenotypic effects on corn kernels. Chromosome 9 with a mutable locus also carried several genes that affected the phenotype of corn kernels. Each gene existed in (at least) one dominant and one recessive allele. The mutable locus was termed *Ds* (for dissociation), because the locus was known to frequently cause chromosomal breaks. In the chromosome shown here, the *Ds* locus is located next to several genes affecting kernel traits.

In this case, there are three genes that exist as two or more alleles:

1. *C* is an allele for normal kernel color (dark red), *c* is a recessive allele of the same gene that causes a colorless kernel, while C^I is a third allele of this gene that is dominant to both *C* and *c* and causes a colorless kernel.
2. *Sh* is an allele that produces normal endosperm, while *sh* is a recessive allele that causes shrunken endosperm. (Note:

The endosperm is the storage material in the kernel that is used by the plant embryo to provide energy for growth.)

3. *Wx* is the allele that produces normal starch in the endosperm, while *wx* is a recessive allele that produces a waxy-appearing phenotype.

During her intensive work, in which she studied corn chromosomes under the microscope, McClintock identified strains of corn in which the *Ds* locus was found in different locations within the corn genome. She could determine the location of *Ds* because the movement of *Ds* occasionally causes chromosome breakage (**Figure 17.10**). Keep in mind that the endosperm of a kernel is triploid because it is derived from the fusion of two maternal haploid nuclei and one paternal haploid nucleus (refer back to Chapter 2, Figure 2.2c). The kernel shown in Figure 17.10 was produced by a cross in which the pollen carried the top chromosome—*C^I Sh Wx Ds*—while the two maternal chromosomes are *C sh wx*. This kernel is expected to be colorless, because the *C^I* allele is dominant and causes a colorless phenotype. However, as the kernel grows by cell division, the movement of the *Ds* locus out of its original location may occasionally cause a chromosome to break, and the distal part of this chromosome is lost. This chromosome breakage may happen in several cells, which continue to divide and grow as the kernel becomes larger. This process produces a sectoring phenotype—patches of cells occur in the kernel that are red, shrunken, and waxy.

By analyzing many kernels, McClintock was also able to identify cases in which *Ds* had moved to a new location. For example, if *Ds* had moved out of its original location and inserted

between *Sh* and *Wx*, a break at *Ds* would produce the following combination:

This genotype would produce patches on the kernel that are red and shrunken but not waxy. In this way, McClintock identified 20 independent cases in which the *Ds* locus had moved to a new location within this chromosome. Overall, the results from many crosses were consistent with the idea that *Ds* can transpose itself throughout the corn genome. McClintock also found that a second locus, termed *Ac* (for activator), was necessary for the *Ds* locus to move. Researchers later discovered that the *Ac* locus contains a gene that encodes an enzyme called transposase, which is necessary for *Ds* to move. We will discuss the function of transposase later in this chapter. Some strains of McClintock's corn contained the *Ac* locus, while others did not.

During her studies, McClintock noticed a particularly exciting and unusual event. By making the appropriate cross, she sought to produce kernels with the following genotype:

The kernels were expected to be red. Because the strain also contained the *Ac* locus, breakage would occur occasionally at the *Ds* locus to produce colorless patches. Among 4,000 kernels, she noticed one kernel with the opposite phenotype—a colorless kernel with red patches. This observation suggested that the inherited genotype in this case, which produced a colorless phenotype, was mutable to become *C*.

How did McClintock explain these results? She postulated that the colorless phenotype was due to a transposition of *Ds* into the *C* gene:

When *Ds* was located within the *C* gene, it inactivated the *C* gene, thereby resulting in the colorless phenotype. However, McClintock proposed that when *Ds* occasionally transposed out of the gene during kernel growth, the *C* allele would be restored and a red patch would result. In this case, the formation of patches, or sectoring, was due to the movement of *Ds* out of its

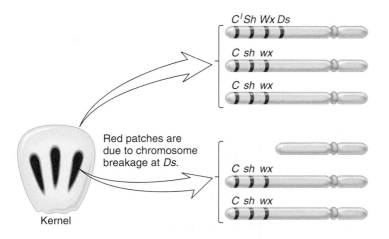

FIGURE 17.10   The sectoring trait in corn kernels.

Genes→Traits This kernel is expected to be colorless, because the *C^I* allele is dominant. On occasion, though, the movement of *Ds* may cause a chromosome break, thereby losing the *C^I*, *Sh*, and *Wx* alleles. As such a cell continues to divide, it will produce a patch of daughter cells that are red, shrunken, and waxy. Therefore, this sectoring trait arises from the loss of genes that occurs when the movement of *Ds* causes chromosome breakage. Note: The movement of *Ds* does not usually cause chromosome breakage. In most cases, the movement of *Ds* out of a site is followed by nonhomologous end joining (described in Chapter 16) in which the chromosome pieces are connected together.

location within the *C* gene and the rejoining of the two ends, not simply due to chromosome breakage. According to this hypothesis, the red phenotype should be associated with two observations. First, *Ds* should have moved to a new location. Second, the restored *C* allele should no longer be mutable because the mutable locus had been removed.

To further study this phenomenon, McClintock carried out the experiment shown in **Figure 17.11**. As seen here, a cross produced kernels that were not entirely colorless. In most cases, red sectoring occurred because the *Ds* element left the *C* gene in a few cells during the growth of the kernel. However, in the ear of corn shown in Figure 17.11, one kernel was completely red. This suggests that the *Ds* element transposed out of the *C* gene prior to kernel growth. In this case, the *Ds* element transposed out of

the *C* gene during the formation of the haploid male gametophyte that produced the pollen nuclei that fertilized this particular red kernel. Therefore, all of the cells in this kernel would have a red phenotype. (Note: Every kernel in an ear of corn is equivalent to a distinct offspring, because each is produced from the union of haploid cells from the male and female gametophytes.)

▆ THE HYPOTHESIS

The transposition of the *Ds* element into the normal *C* gene prevents kernel pigmentation. When the *Ds* element transposes back out of the *C* gene, the normal *C* allele is restored, which gives a red phenotype.

▆ TESTING THE HYPOTHESIS — FIGURE 17.11   Evidence for transposable elements in corn.

Starting material: The male pollen of the corn plant was homozygous for a chromosome in which the *Ds* element had moved into the *C* gene: *CDsC Sh wx*. The male plant also contained the *Ac* locus. The plant contributing the female gamete was homozygous for a chromosome carrying *c sh Wx*.

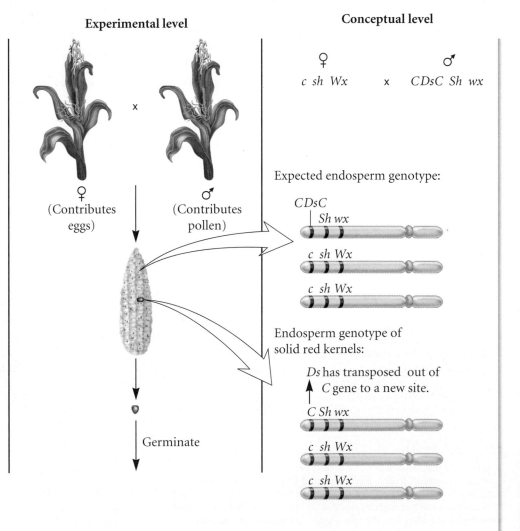

1. Cross a plant that is homozygous for *c*, *sh*, and *Wx* to a plant that is homozygous for *CDsC*, *Sh*, *wx*.

2. In this cross, most kernels will have a colorless background with red sectoring. On rare occasions, *Ds* may have transposed during male gametophyte formation, producing pollen that gives rise to a completely red kernel.

3. Identify those occasional kernels that are completely red.

4. Germinate the solid red kernels.

Experimental level

Conceptual level

♀   ♂
c sh Wx   x   *CDsC Sh wx*

(Contributes eggs)   (Contributes pollen)

Expected endosperm genotype:

CDsC
  Sh wx

c sh Wx

c sh Wx

Endosperm genotype of solid red kernels:

Ds has transposed out of *C* gene to a new site.

C Sh wx

c sh Wx

c sh Wx

Germinate

5. Conduct crosses to determine the location of *Ds* in the plants derived from the solid red kernels. Note: This can be done using the chromosomal markers and the strategy that was described at the beginning of this experiment. When observing the results of these crosses, determine whether the *C* gene is still mutable. (Is red sectoring occurring?)

Make crosses and observe kernels.

◼ THE DATA

Strain	Kernel Phenotype	Location of *Ds*	Mutability?
From parental cross (see step 2)	Colorless background with red sectoring	Within the *C* gene	Yes, red sectoring occurred in strains containing *Ac*.
From red kernels (see step 3)	Red kernels	*Ds* had moved out of the *C* gene to another location.	No, the *C* gene was stable; no sectoring was observed.

◼ INTERPRETING THE DATA

By conducting the appropriate crosses, McClintock found that in the progeny of a solid red kernel, the *Ds* locus had moved out of the *C* gene to another location (see the data of Figure 17.11).

In addition, the "restored" *C* gene behaved normally. In other words, the *C* gene was no longer highly mutable, and the kernels did not show a sectoring phenotype. Taken together, the results are consistent with the hypothesis that the *Ds* locus can move around the corn genome by transposition.

When McClintock published these results in 1951, they were met with great skepticism. Some geneticists of that time were unable to accept the idea that the genetic material was susceptible to frequent rearrangement. Instead, they believed that the genetic material was always very stable and permanent in its structure. Over the next several decades, the scientific community came to realize that transposable elements are a widespread phenomenon. Much like Gregor Mendel and Charles Darwin, Barbara McClintock was clearly ahead of her time. She was awarded the Nobel Prize in 1983, more than 30 years after her original discovery.

A self-help quiz involving this experiment can be found at the Online Learning Center.

Transposable Elements and Retroelements Move Via One of Three Transposition Pathways

Since the pioneering studies of Barbara McClintock, many different transposable elements (TEs) have been found in bacteria, fungi, plants, and animals. Three general types of transposition pathways have been identified (**Figure 17.12**). In **simple,** or **conservative transposition,** the TE is removed from its original site and transferred to a new target site. This mechanism is also called a cut-and-paste mechanism because the element is cut out of its original site and pasted into a new site. Transposable elements that move via simple transposition are widely found in bacterial and eukaryotic species. By comparison, **replicative transposition** involves the replication of the TE and insertion of the newly made copy into a second site. In this case, one of the TEs remains in its original location, and the other is inserted at another location. Replicative transposition is relatively uncommon and found only in bacterial species.

The first two categories of TEs we have considered thus far move as a DNA molecule from one site to another. Such TEs are called **transposons.** A third category of elements move via an RNA intermediate. This form of transposition, termed **retrotransposition,** is found in eukaryotic species, where it is very common. Transposable elements that move via retrotransposition are known as **retroelements, retrotransposons,** or **retroposons.** In retrotransposition, the element is transcribed into RNA. An enzyme called reverse transcriptase uses the RNA as a template to synthesize a DNA molecule that is integrated into a new region of the genome. Like replicative transposons, retroelements increase in number during retrotransposition.

(a) Simple transposition

(b) Replicative transposition

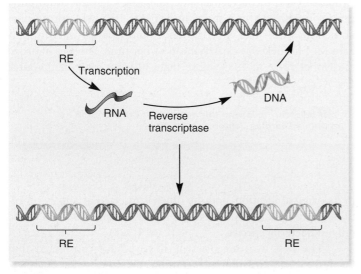

(c) Retrotransposition

FIGURE 17.12   Three mechanisms of transposition.

Each Type of Transposable Element Has a Characteristic Pattern of DNA Sequences

As researchers have studied TEs from many species, they have found that DNA sequences within transposable elements are organized in several different ways. **Figure 17.13** describes a

(a) Elements that move by simple transposition

(b) An element that moves by replicative transposition

(c) Elements that move by retrotransposition (via an RNA intermediate)

FIGURE 17.13   **Common organizations of transposable elements.** Direct repeats (DRs) are found within the host DNA. Inverted repeats (IRs) are at the ends of most transposable elements. Long terminal repeats (LTRs) are regions containing a large number of tandem repeats.

few ways that TEs are organized, although many variations are possible. All TEs are flanked by **direct repeats (DR),** which are identical nucleotide sequences that are in the same <u>direction</u> and <u>repeat</u>ed. Direct repeats are adjacent to both ends of the element. The simplest TE, which is commonly found in bacteria, is known as an **insertion sequence.** As shown in Figure 17.13a, an insertion sequence has two important characteristics. First, both ends of the insertion sequence contain **inverted repeats (IRs).** Inverted repeats are DNA sequences that are identical (or very similar) but run in opposite directions, such as the following:

```
5'-CTGACTCTT-3'    and    5'-AAGAGTCAG-3'
3'-GACTGAGAA-5'           3'-TTCTCAGTC-5'
```

Depending on the particular element, the lengths of inverted repeats range from 9 to 40 bp in length. In addition, insertion sequences may contain a central region that encodes the enzyme **transposase,** which catalyzes the transposition event.

Composite transposons contain additional genes that are not necessary for transposition per se. They commonly contain genes that confer a selective advantage to the organism. Composite transposons are prevalent in bacteria, where they often contain genes that provide resistance to antibiotics or toxic heavy metals. For example, the composite transposon shown in Figure 17.13a contains two insertion sequences flanking a gene that confers antibiotic resistance. During transposition of a composite transposon, only the inverted repeats at the ends of the transposon are involved in the transpositional event. Whenever insertion sequences are found at both ends of a gene, they create a composite transposon. Both insertion sequences and composite transposons are elements that move via simple transposition, also called cut-and-paste transposition.

Replicative transposons, elements that move by replicative transposition, have a sequence organization that is similar to insertion sequences except that replicative transposons have a resolvase gene that is found between the inverted repeats (Figure 17.13b). As discussed later in this chapter, both transposase and resolvase are needed to catalyze the transposition of replicative transposons.

The organization of retroelements can be quite variable, and they are categorized based on their evolutionary relationship to retroviral sequences. Retroviruses are RNA viruses that make a DNA copy that integrates into the host's genome. The **viral-like retroelements** are evolutionarily related to known retroviruses. These transposable elements have retained the ability to move around the genome, though, in most cases, they do not produce mature viral particles. Viral-like retroelements contain **long terminal repeats (LTRs)** at both ends of the element (Figure 17.13c). The LTRs are typically a few hundred nucleotides in length. Like their viral counterparts, viral-like retroelements encode virally related proteins such as reverse transcriptase and integrase that are needed for the transposition process.

By comparison, **nonviral-like retroelements** appear less like retroviruses in their sequence, although some similarity, such as the occurrence of a reverse transcriptase gene, may be present (Figure 17.13c). Many nonviral-like retroelements, however, do not share any sequence similarity with known viruses. Instead, some nonviral-like retroelements are evolutionarily derived from normal eukaryotic genes. For example, the *Alu* family of repetitive sequences found in humans is derived from a single ancestral gene known as the *7SL RNA* gene (a component of the complex called signal recognition particle, which is described in Chapter 13). This gene sequence has been copied by retrotransposition to achieve the current number of approximately 1,000,000 copies.

Transposable elements are considered to be complete or **autonomous elements** when they contain all the information necessary for transposition or retrotransposition to take place.

However, TEs are often incomplete or nonautonomous. A **nonautonomous element** typically lacks a gene such as transposase or reverse transcriptase that is necessary for transposition. The *Ds* locus described in the experiment of Figure 17.11 is a nonautonomous element, because it lacks a transposase gene. An element that is similar to *Ds* but contains a functional transposase gene is called the *Ac* locus or *Ac* element, which stands for <u>Ac</u>tivator element. As mentioned earlier, an *Ac* locus provides a transposase gene that enables *Ds* to transpose. Therefore, nonautonomous TEs such as *Ds* can transpose only when the *Ac* locus is present at another region in the genome.

Transposase Catalyzes the Excision and Insertion of Transposable Elements

Now that we have an understanding of the typical organization of transposable elements, let's examine the steps of the transposition process. The enzyme transposase catalyzes the removal of a TE from its original site in the chromosome and its subsequent insertion at another location. A general scheme for simple transposition is shown in **Figure 17.14**. Transposase binds to the inverted repeat sequences at the ends of the TE and brings them close together. The DNA is cleaved at the ends of the TE, excising it from its original site within the chromosome. The transposase carries the TE to a new site and cleaves the target DNA sequence at staggered recognition sites. The TE is then inserted and ligated to the target DNA.

As noted in Figure 17.14, the ligation of the transposable element into its new site initially leaves short gaps in the target DNA. Notice that the DNA sequences in these gaps are complementary to each other (in this case, ATGCT and TACGA). Therefore, when they are filled in by DNA gap repair synthesis, the DNA base pair sequences that flank both ends of the TE are identical. These direct repeats are common features found adjacent to all TEs (see Figure 17.13).

Although the transposition process depicted in Figure 17.14 does not directly alter the number of transposable elements, simple transposition is known to increase the number of TEs in genomes, in some cases to fairly high levels. How can this happen? The answer is that transposition often occurs around the time of DNA replication (**Figure 17.15**). After a replication fork has passed a region containing a TE, two TEs will be found behind the fork—one in each of the replicated regions. One of these TEs could then transpose from its original location into a region ahead of the replication fork. After the replication fork has passed this second region and DNA replication is completed, two TEs will be found in one of the chromosomes and one TE in the other chromosome. In this way, simple transposition can lead to an increase in TEs. We will discuss the biological significance of transposon proliferation later in this chapter.

Replicative Transposition Requires Both Transposase and Resolvase

Replicative transposition has been studied in several bacterial transposons and in bacteriophage μ (mu), which behaves like a

FIGURE 17.14   **Simple transposition.**

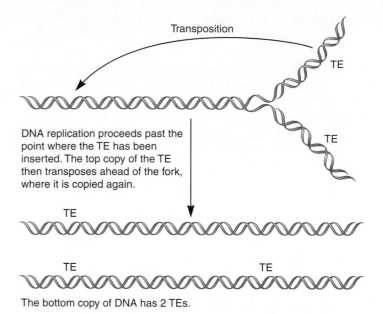

The bottom copy of DNA has 2 TEs.

FIGURE 17.15   **Increase in TE copy number via simple transposition.**   In this example, a TE that has already been replicated transposes to a new site that has not yet replicated. Following the completion of DNA replication, the TE has increased in number.

transposon. The net result of replicative transposition is that a transposable element occurs at a new site, and a TE also remains in its original location. **Figure 17.16** describes a model for replicative transposition between two circular DNA molecules. One DNA molecule already has a TE, whereas the other does not. In this mechanism, transposase initially makes one cut at each end of the TE and two cuts in the target DNA. Note that this differs from simple transposition, in which the transposase makes four cuts and completely removes the TE from its original site (refer back to Figure 17.14). In replicative transposition, the TE is left at its original location.

Following ligation, both the target DNA and the transposable element have a long gap. DNA gap repair synthesis copies the target DNA gap as well as the TE. This creates two copies of the TE within a large circular molecule known as a cointegrant. The enzyme resolvase catalyzes homologous recombination within the TEs so that the cointegrant can be resolved into two separate DNA molecules. One of these molecules contains the TE in its original location, and the other has a TE at a new location.

Retroelements Use Reverse Transcriptase and Integrase for Retrotransposition

Thus far, we have considered how DNA elements—transposons—can move throughout the genome. By comparison, retroelements use an RNA intermediate in their transposition mechanism. As shown in **Figure 17.17**, the movement of retroelements also requires two key enzymes: reverse transcriptase and integrase. In this example, the cell already contains a retroelement known as the *Alu* sequence within its genome. This retroelement, which

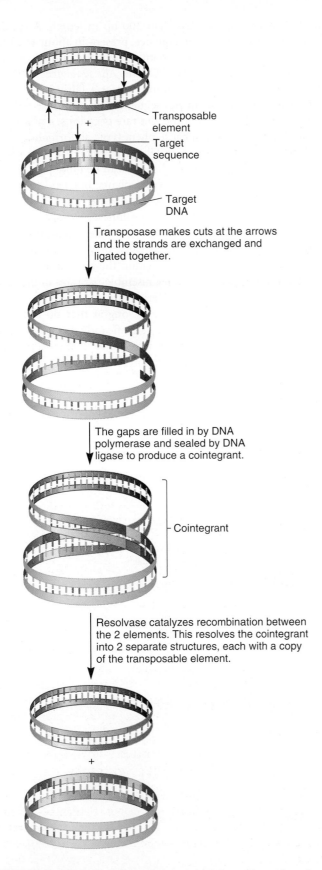

Transposable element

Target sequence

Target DNA

Transposase makes cuts at the arrows and the strands are exchanged and ligated together.

The gaps are filled in by DNA polymerase and sealed by DNA ligase to produce a cointegrant.

Cointegrant

Resolvase catalyzes recombination between the 2 elements. This resolves the cointegrant into 2 separate structures, each with a copy of the transposable element.

FIGURE 17.16   Replicative transposition. The end result of this process is that one TE remains in its original site and another TE is inserted at a new site.

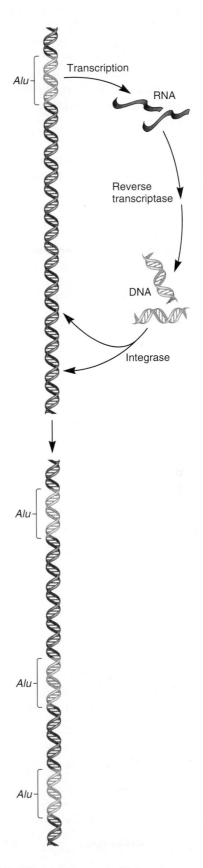

Alu

Transcription

RNA

Reverse transcriptase

DNA

Integrase

Alu

Alu

Alu

FIGURE 17.17   Retrotransposition.

behaves like a gene, is transcribed into RNA. In a series of steps, **reverse transcriptase** uses this RNA as a template to synthesize a double-stranded DNA molecule. The ends of the double-stranded DNA are then recognized by **integrase**, which catalyzes the insertion of the DNA into the target chromosomal DNA. The integration of retroelements can occur at many locations within the genome. Furthermore, because a single retroelement can be copied into many RNA transcripts, retroelements may accumulate rapidly within a genome.

Transposable Elements May Have Important Influences on Mutation and Evolution

Over the past few decades, researchers have found that transposable elements probably occur in the genomes of all species. Table 17.2 describes a few TEs that have been studied in great detail. As discussed in Chapter 10, the genomes of eukaryotic species typically contain moderately and highly repetitive sequences. In some cases, these repetitive sequences are due to the proliferation of TEs. In mammals, for example, **LINEs** are long interspersed elements that are usually 1,000 to 5,000 bp in length and found in 20,000 to 100,000 copies per genome. **SINEs** are short inter-spersed elements that are less than 500 bp in length. A specific example of a SINE is the *Alu* sequence, present in about 1 million copies in the human genome. About 10% of the human genome is composed of this particular TE! The *Alu* sequence continues to proliferate in the human genome but at a fairly low rate. In about 1 live birth in 200, an *Alu* sequence has been inserted into a new site in the human genome. On rare occasions, a new insertion can disrupt a gene and cause phenotypic abnormalities.

The relative abundance of TEs varies widely among different species. As shown in **Table 17.3**, TEs can be quite prevalent in amphibians, mammals, and flowering plants, but tend to be less abundant is simpler organisms such as bacteria and yeast. The biological significance of TEs in the evolution of prokaryotic and eukaryotic species remains a matter of debate. According to the **selfish DNA hypothesis**, TEs exist because they contain characteristics that allow them to multiply within the chromosomal DNA of living cells. In other words, they resemble parasites in the sense that they inhabit a cell without offering any selective advantage to the organism. They can proliferate as long as they do not harm the organism to the extent that they significantly disrupt survival.

Alternatively, other geneticists have argued that most transpositional events are deleterious. Therefore, TEs would be eliminated

TABLE **17.2**

Examples of Transposable Elements

Element	Type	Approximate Length (bp)	Description
Bacterial			
IS1	Insertion sequence	768	An insertion sequence that is commonly found in 5–8 copies in *E. coli*.
Mu	Replicative transposon	36,000	A true virus that can insert itself anywhere in the *E. coli* chromosome. Its name, *Mu*, is derived from its ability to insert into genes and mutate them.
Tn10	Composite transposon	9,300	One of many different bacterial transposons that carries antibiotic resistance.
Tn951	Composite transposon	16,600	A transposon that provides bacteria with genes that allow them to metabolize lactose.
Yeast			
Ty elements	Viral-like retroelement	6,200	A retroelement found in *S. cerevisiae* at about 35 copies per genome.
Drosophila			
P elements	Simple transposon	500–3,000	A transposon that may be found in 30–50 copies in P strains of *Drosophila*. It is absent from M strains.
Copia-like elements	Viral-like retroelement	5,000–8,000	A family of *copia*-like elements found in *Drosophila*, which vary slightly in their lengths and sequences. Typically, each family member is found at about 5–100 copies per genome.
Humans			
Alu sequence	Nonviral-like retroelement	300	A SINE that is abundantly interspersed throughout the human genome.
L1	Viral-like retroelement	6,500	A LINE found in 50,000–100,000 copies in the human genome.
Plants			
Ac/Ds	Simple transposon	4,500	*Ac* is an autonomous transposon found in corn and other plant species. It carries a transposase gene. *Ds* is a nonautonomous version that lacks a functional transposase gene.

TABLE 17.3

Abundance of TEs in the Genomes of Selected Species

Species	Percentage of the Total Genome Composed of Transposable Elements*
Frog (*Xenopus laevis*)	77
Corn (*Zea mays*)	60
Human (*Homo sapiens*)	45
Mouse (*Mus musculus*)	40
Fruit fly (*Drosophila melanogaster*)	20
Nematode (*Caenorhabditis elegans*)	12
Yeast (*Saccharomyces cerevisiae*)	4
Bacterium (*Escherichia coli*)	0.3

*In some cases, the abundance of TEs may vary somewhat among different strains of the same species. The values reported here are typical values.

TABLE 17.4

Possible Consequences of Transposition

Consequence	Cause
Chromosome Structure	
Chromosome breakage	Excision of a TE.
Chromosomal rearrangements	Homologous recombination between TEs located at different positions in the genome.
Gene Expression	
Mutation	Incorrect excision of TEs.
Gene inactivation	Insertion of a TE into a gene.
Alteration in gene regulation	Transposition of a gene next to regulatory sequences or the transposition of regulatory sequences next to a gene.
Alteration in the exon content of a gene	Insertion of exons into the coding sequence of a gene via TEs. This phenomenon is called exon shuffling.
Gene duplications	Creation of a composite transposon that transposes to another site in the genome.

from the genome by natural selection if they did not also offer a compensating advantage. Several potential advantages have been suggested. For example, TEs may cause greater genetic variability by promoting recombination. In addition, bacterial TEs often carry an antibiotic-resistance gene that provides the organism with a survival advantage (refer back to Figure 17.13a). Researchers have also suggested that transposition may cause the insertion of exons into the coding sequences of structural genes. This phenomenon, called **exon shuffling,** may lead to the evolution of genes with more diverse functions, as described in Chapter 24.

While this controversy remains unresolved, it is clear that transposable elements can rapidly enter the genome of an organism and proliferate quickly. In *Drosophila melanogaster*, for example, a TE known as the P element was probably introduced into this species in the 1950s. Laboratory stocks of *D. melanogaster* collected prior to this time do not contain P elements. Remarkably, in the last 50 years, the P element has expanded throughout *D. melanogaster* populations worldwide. The only strains without the P element are laboratory strains collected prior to the 1950s. This observation underscores the surprising ability of TEs to infiltrate a population of organisms.

Transposable elements have a variety of effects on chromosome structure and gene expression (**Table 17.4**). Because many of these outcomes are likely to be harmful, transposition is usually a highly regulated phenomenon that occurs only in a few individuals under certain conditions. Agents such as radiation, chemical mutagens, and hormones stimulate the movement of TEs. When it is not carefully regulated, transposition is likely to be potently detrimental. For example, in *D. melanogaster*, if females that lack P elements (M strain females) are crossed with males that contain numerous P elements (P strain males), the egg cells allow the P elements inherited via the sperm to transpose at a high rate. The resulting hybrid offspring exhibit a variety of abnormalities, which include a high rate of sterility, mutation, and chromosome breakage. This deleterious outcome, which is called **hybrid dysgenesis,** occurs because the P elements were able to insert into a variety of locations in the genome.

Transposons Have Become Important Tools in Molecular Biology

The unique and unusual features of transposons have made them an important experimental tool in molecular biology. For researchers, the introduction of a transposon into a cell is a convenient way to alter the expression of a particular gene. If a transposon "hops" into a gene, it is likely to inactivate the gene's function. This phenomenon can be used to clone a particular gene in an approach known as **transposon tagging.** In this strategy, researchers use transposons in an attempt to clone genes.

An early example of transposon tagging involved an X-linked gene in *Drosophila* that affects eye color. This X-linked gene can exist in the wild-type (red) allele and a loss-of-function allele that causes a white-eye phenotype. In 1981, Paul Bingham, in collaboration with Robert Levis and Gerald Rubin, used transposon tagging to clone this gene (**Figure 17.18**). Prior to their cloning work, a wild-type strain of *Drosophila* had been characterized that carried a transposable element called *copia*. From this red-eyed strain, a white-eyed strain was obtained in which the *copia* element had transposed into a region on the X chromosome that corresponded to where the eye color gene mapped.

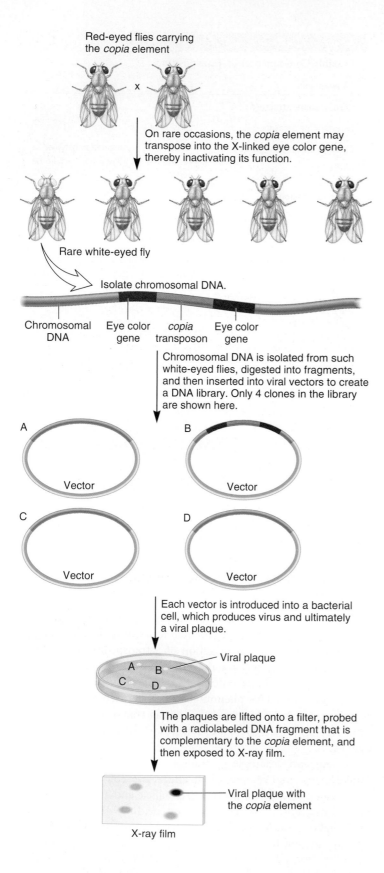

Red-eyed flies carrying
the *copia* element

X

On rare occasions, the *copia* element may
transpose into the X-linked eye color gene,
thereby inactivating its function.

Rare white-eyed fly

Isolate chromosomal DNA.

Chromosomal | Eye color | *copia* | Eye color
DNA | gene | transposon | gene

Chromosomal DNA is isolated from such
white-eyed flies, digested into fragments,
and then inserted into viral vectors to create
a DNA library. Only 4 clones in the library
are shown here.

A   B

Vector   Vector

C   D

Vector   Vector

Each vector is introduced into a bacterial
cell, which produces virus and ultimately
a viral plaque.

A   B
C   D

Viral plaque

The plaques are lifted onto a filter, probed
with a radiolabeled DNA fragment that is
complementary to the *copia* element, and
then exposed to X-ray film.

Viral plaque with
the *copia* element

X-ray film

The researchers reasoned that the white-eye phenotype could be due to the insertion of the *copia* element into the wild-type gene, thereby inactivating it.

To clone the eye color gene, chromosomal DNA from this white-eyed strain was isolated, digested with restriction enzymes, and cloned into vectors that were derived from a naturally occurring virus (viral vectors). This procedure creates a DNA library, a collection of vectors that contain different pieces of chromosomal DNA (described in Chapter 18). If a transposon had "jumped" into the eye color gene, vectors that contain this gene will also contain the transposon sequence. In other words, the presence of the transposon tags the eye color gene. Each vector is introduced into a bacterial cell, which produces a virus and ultimately a viral plaque, a clear area where the bacteria have been lysed. A radiolabeled fragment of DNA that is complementary to the transposon sequence can be used as a probe to identify plaques that also contain the eye color gene. The method of using a probe to screen a DNA library is also described in Chapter 18. In the example of Figure 17.18, the method of transposon tagging was successful at cloning an eye color gene in *Drosophila*.

FIGURE 17.18   The procedure of transposon tagging.

Genes→Traits A white-eyed fly may occur due to the insertion of a transposable element into a gene that confers red eye color. As discussed in Chapter 13, the wild-type eye color gene encodes a protein that is necessary for red pigment production. When a TE inserts into this gene, it disrupts the coding sequence and thereby causes the gene to produce a nonfunctional protein. Therefore, no red pigment can be made, and a white-eye phenotype results. In many cases, transposons affect the phenotypes of organisms by inactivating individual genes.

CONCEPTUAL SUMMARY

Homologous recombination, an essential feature of all species, is a major category of genetic recombination. Sister chromatid exchange (SCE) is a type of homologous recombination that involves crossing over between sister chromatids and the exchange of identical genetic material. By comparison, homologous recombination can also occur when homologous chromosomes cross over. During this form of homologous recombination, homologous DNA regions align and exchange DNA segments. This enhances genetic variability by producing chromosomes with new combinations of alleles.

At the molecular level, several models have been proposed to explain the steps in homologous recombination. The Holliday model was the first example of a molecular explanation for the recombination process. This model was able to account for the phenomenon of gene conversion, in which one allele is converted to the allele on the homologous chromosome. More recently, other models, such as that proposed by Meselson and Radding and the double-strand break model, have more accurately described steps in the recombination pathway. In addition, much progress has been made toward identifying the many proteins that play important roles in recombination.

A second form of recombination is known as site-specific recombination, because the breakage and rejoining of nonho-mologous DNA segments occur at particular DNA sequences. Site-specific recombination is responsible for the integration of certain bacteriophages such as λ into the host cell DNA. In addition, site-specific recombination within immunoglobulin genes is important in generating an astounding diversity in antibody polypeptides. During this process, regions in precursor immunoglobulin genes known as V, D, and J domains are randomly connected to each other via recombination. This mechanism enables the production of a very diverse array of antibody proteins.

In transposition, a third form of recombination, short segments of DNA known as transposable elements (TEs) integrate themselves into different locations within the genome. During simple transposition, a cut-and-paste mechanism occurs in which the TE is cut out of its original site and transferred to a new site. This process is catalyzed by transposase. In replicative transposition, a TE is duplicated, with one TE remaining in its original location and a new TE located at a new site. The process is catalyzed by the enzymes transposase and resolvase. In retrotransposition, a retroelement is transcribed into RNA. Reverse transcriptase uses this RNA as a template to synthesize DNA, which is then integrated into the chromosome by integrase. From an evolutionary viewpoint, all mechanisms of transposition can cause many different types of mutations.

EXPERIMENTAL SUMMARY

Homologous recombination, which may occur between sister chromatids or homologous chromosomes, can be detected by several techniques. Sister chromatid exchange (SCE) can be observed using staining methods that produce harlequin chromosomes, which distinguish the sister chromatids in which crossing over has occurred. At the molecular level, researchers have studied recombination that occurs between homologous chromosomes by identifying and characterizing the proteins and intermediates that facilitate the process. The Holliday model describes a molecular mechanism for the recombination process.

Likewise, researchers have elucidated the mechanisms of site-specific recombination and transposition using genetic and molecular techniques. McClintock conducted genetic crosses in corn that provided compelling evidence for the existence of transposable elements (TEs). The sequencing of TEs has revealed certain patterns of DNA sequences that are required for simple transposition, replicative transposition, and retrotransposition. Researchers have also identified the proteins necessary to promote the movement of TEs. In addition, molecular biologists have taken advantage of our knowledge of transposition and routinely use transposons to inactivate genes and to clone genes via transposon tagging.

PROBLEM SETS & INSIGHTS

Solved Problems

S1. Zickler was the first person to demonstrate gene conversion by observing unusual ratios in *Neurospora* octads. At first, it was difficult for geneticists to believe these results because they seemed to contradict the Mendelian concept that alleles do not physically interact with each other. However, work by Mary Mitchell provided convincing evidence that gene conversion actually takes place. She investigated three different genes in *Neurospora*. One *Neurospora* strain had three mutant alleles: *pdx-1* (pyridoxine-requiring), *pyr-1* (pyrimidine-requiring), and *col-4* (a mutation that affected growth morphology). The *pdx-l* gene had been previously shown to map between the *pyr-1* and *col-4* genes. As shown here, Mitchell crossed this strain to a wild-type *Neurospora* strain:

pyr-1 pdx-1 col-4 × pyr-1⁺ pdx-1⁺ col-4⁺

She first analyzed many octads with regard to their requirement for pyridoxine. Out of 246 octads, two of them had an aberrant ratio in which two spores were *pdx-1*, and six were *pdx-1⁺*. These same spores were then analyzed with regard to the other two genes. In both cases, the aberrant asci gave a normal 4:4 ratio of *pyr-1:pyr-1⁺* and *col-4:col-4⁺*. Explain these results.

Answer: These results can be explained by gene conversion. The gene conversion took place in a limited region of the chromosome (within the *pdx-1* gene), but it did not affect the flanking genes (*pyr-1* and *col-4*) located on either side of the *pdx-1* gene. In the asci containing two *pdx-1* alleles and six *pdx-1⁺* alleles, a crossover occurred during meiosis I in the region of the *pdx-1* gene. Gene conversion changed the *pdx-1* allele into the *pdx-1⁺* allele. This gene conversion could have occurred by two mechanisms. If branch migration occurred across the *pdx-1* gene, a heteroduplex may have formed, and this could be repaired by mismatch DNA repair, as described in Figure 17.6. In the aberrant asci with two *pdx-1* and six *pdx-1⁺* alleles, the *pdx-1* allele was converted to *pdx-1⁺*. Alternatively, gene conversion of *pdx-1* into *pdx-1⁺* could have taken place via gap repair synthesis, as described in Figure 17.7. In this case, the *pdx-1* allele would have been digested away, and the DNA encoding the *pdx-1⁺* allele would have migrated into the digested region and provided a template to make a copy of the *pdx-1⁺* allele. (Note: Since this pioneering work, additional studies have shed considerable light concerning the phenomenon of gene conversion. It occurs at a fairly high rate in fungi, approximately 0.1 to 1% of the time. It is not due to new mutations occurring during meiosis.)

S2. Recombination involves the pairing of identical or similar sequences, followed by crossing over and the resolution of the intertwined helices. On rare occasions, the direct repeats or the inverted repeats within a single transposable element can align and undergo homologous recombination. What are the consequences when the direct repeats recombine? What are the consequences when the inverted repeats recombine?

Answer:

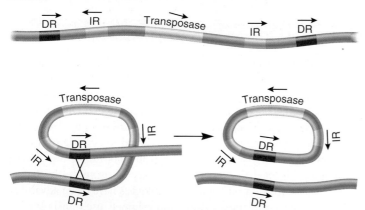

Most of the transposable element has been excised.

(a) Recombination between direct repeats

The sequence within the transposable element has been inverted. Note that the transposase gene has changed to the opposite direction.

(b) Recombination between inverted repeats

S3. A schematic drawing of an uncrossed Holliday junction is shown here. One chromatid is shown in red, and the homologous chromatid is shown in blue. The red chromatid carries a dominant allele labeled *A* and a recessive allele labeled *b*, whereas the blue chromatid carries a recessive allele labeled *a* and a dominant allele labeled *B*.

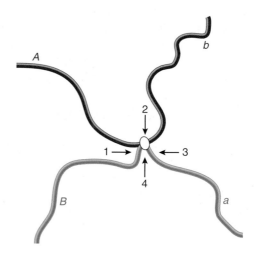

Where would the breakage of the crossed strands have to occur to get recombinant chromosomes? Would it have to occur at sites 1 and 3, or at sites 2 and 4? What would be the genotypes of the two recombinant chromosomes?

Answer:

Breakage would have to occur at the arrows labeled 2 and 4. This would connect the *A* allele with the *B* allele. The *a* allele in the other homologue would become connected with the *b* allele. In other words, one chromosome would be *AB*, and the homologue would be *ab*.

Conceptual Questions

C1. Describe the similarities and differences between homologous recombination involving sister chromatid exchange (SCE) versus that involving homologues. Would you expect the same types of proteins to be involved in both processes? Explain.

C2. The molecular mechanism of SCE is similar to homologous recombination between homologues except that the two segments of DNA are sister chromatids instead of homologous chromatids. If branch migration occurs during SCE, will a heteroduplex be formed? Explain why or why not. Can gene conversion occur during sister chromatid exchange?

C3. Which steps in the double-strand break model for recombination would be inhibited if the following proteins were missing? Explain the function of each protein required for the step that is inhibited.

A. RecBCD

B. RecA

C. RecG

D. RuvABC

C4. What are the two molecular mechanisms that can explain the phenomenon of gene conversion? Would both of these mechanisms occur in the double-strand break model?

C5. Is homologous recombination an example of mutation? Explain.

C6. What are recombinant chromosomes? How do they differ from the original parental chromosomes from which they are derived?

C7. In the Holliday model for homologous recombination (see Figure 17.4), the resolution steps can produce recombinant or nonrecombinant chromosomes. Explain how this can occur.

C8. What is gene conversion?

C9. Make a list of the differences between the Holliday model and the double-strand break model.

C10. In recombinant chromosomes, where is gene conversion likely to have taken place: near the breakpoint or far away from the breakpoint? Explain.

C11. What are the events that RecA protein facilitates?

C12. According to the double-strand break model, does gene conversion necessarily involve DNA mismatch repair? Explain.

C13. What type of DNA structure is recognized by RecG and RuvABC? Do you think these proteins recognize DNA sequences? Be specific about what type(s) of molecular recognition these proteins can perform.

C14. Briefly describe three ways that antibody diversity is produced.

C15. Describe the function of RAG1 and RAG2 proteins and NHEJ proteins.

C16. According to the scenario shown in Figure 17.9, how many segments of DNA (one, two, or three) are removed during site-specific recombination within the gene that encodes the κ (kappa) light chain for IgG proteins? How many segments are spliced out of the pre-mRNA?

C17. Describe the role that integrase plays during the insertion of λ DNA into the host chromosome.

C18. If you were examining a sequence of chromosomal DNA, what characteristics would cause you to believe that the DNA contained a transposable element?

C19. According to the model for replicative transposition shown in Figure 17.16, does the transposable element replicate before or after it transposes? Explain your answer.

C20. Why does transposition always produce direct repeats in the chromosomal DNA?

C21. Which types of transposable elements have the greatest potential for proliferation: insertion sequences, replicative TEs, or retroelements? Explain your choice.

C22. Do you consider transposable elements to be mutagens? Explain.

C23. Let's suppose that a species of mosquito has two different types of simple transposons that we will call X elements and Z elements. The X elements appear quite stable. When analyzing a population of 100 mosquitoes, every mosquito has six X elements, and they are always located in the same chromosomal locations among different individuals. In contrast, the Z elements seem to "move around" quite a bit. Within the same 100 mosquitoes, the number of Z elements ranges from 2 to 14, and the locations of the Z elements tend to vary considerably among different individuals. Explain how one simple transposon can be stable and another simple transposon can be mobile, within the same group of individuals.

C24. This chapter describes five different types of transposable elements including insertion sequences, composite transposons, replicative transposons, viral-like retroelements, and nonviral-like retroelements. Which of these five types of TEs would have the following features?

A. Require reverse transcriptase to transpose

B. Require transposase to transpose

C. Are flanked by direct repeats

D. Have inverted repeats

C25. What features distinguish a transposon from a retroelement? How are their sequences different and how are their mechanisms of transposition different?

C26. Solved problem S2 illustrates the consequences of crossing over between the direct and inverted repeats within a single transposable element. The drawing here shows the locations of two copies of the same transposable element within a single chromosome. The chromosome is depicted according to its G banding pattern. (Note: G bands are illustrated in Figure 8.1.) The direct and inverted repeats are labeled 1, 2, 3, and 4, from left to right.

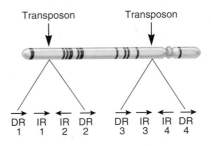

Draw the end result of recombination between the following sequences. Your drawing should include the banding pattern of the resulting chromosome.

A. DR-1 and DR-4

B. IR-1 and IR-4

C27. What is the difference between an autonomous and a nonautonomous transposable element? Is it possible for nonautonomous TEs to move? If yes, explain how.

C28. An operon in the bacterium *Salmonella typhimurium* has the following arrangement:

DR IR   IR DR

↓ ↓ promoter→ ↓ ↓ *H2 rH1*

The promoter for this operon is contained within a transposable element. The *H2* gene encodes a protein that is part of the bacterial flagellum. The *rH1* gene encodes a repressor protein that represses the *H1* gene, which is found at another location in the bacterial chromosome. The *H1* gene also encodes a flagellar protein. When the promoter is found in the arrangement shown here, the *H2/rH1* operon is turned on. This results in flagella that contain the *H2* protein. The *H1* protein is not made because the *rH1* repressor prevents the transcription of the *H1*

gene. At a frequency of approximately 1 in 10,000 (which is much higher than the spontaneous mutation rate), this strain of bacterium can "switch" its expression so that *H2* is turned off and *H1* is turned on. Bacteria that have *H1* turned on and *H2* turned off can also switch back to having *H2* turned on and *H1* turned off. This switch also occurs at a frequency of about 1 in 10,000. Based on your understanding of transposons and recombination, explain how switching occurs in *Salmonella typhimurium*. Hint: Take a look at solved problem S2.

C29. The occurrence of multiple transposons within the genome of organisms has been suggested as a possible cause of chromosomal rearrangements such as deletions, translocations, and inversions. How could the occurrence of transposons promote these kinds of structural rearrangements?

Experimental Questions

E1. With the harlequin staining technique, one sister chromatid appears to fluoresce more brightly than the other. Why?

E2. In the data shown here, harlequin staining was used to determine the frequency of sister chromatid exchanges (SCEs) in the presence of a suspected mutagen.

Frequency of SCEs/Chromosome

No mutagen   0.67

With suspected mutagen   14.7

Would you conclude that this substance is a mutagen?

E3. In the experiment described in experimental question E2, at what point would you need to add the mutagen: before the first round of DNA replication, after the first round but before the second round, or after the second round?

E4. Let's suppose that a researcher followed the protocol described in the experiment of Figure 17.3 but exposed the cells to BrdU for three cell generations instead of two. Near the end of growth, the cells were exposed to colcemid to prevent them from completing mitosis following the third round of DNA replication. What would be the expected results if a parental cell contained a total of four chromosomes (two homologous pairs) and a single sister chromatid exchange occurred after the second replication? Your drawing should show four cells that contain four condensed chromosomes in each cell.

E5. Based on your understanding of the experiment of Figure 17.3, does BrdU enhance the binding of Giemsa to the chromatids or inhibit the binding of Giemsa? Explain.

E6. Briefly explain how McClintock determined that *Ds* was occasionally moving from one chromosomal location to another. Discuss the type of data she examined to arrive at this conclusion.

E7. In the data of Figure 17.3, is the solid red phenotype due to chromosome breakage or the excision of a transposable element? Explain how you have arrived at your conclusion.

E8. In your own words, explain the term transposon tagging.

E9. Tumor-suppressor genes are normal human genes that prevent uncontrollable cell growth. Starting with a normal laboratory human cell line, describe how you could use transposon tagging to identify tumor-suppressor genes. Note: When a transposable element hops into a tumor-suppressor gene, it may cause uncontrolled cell growth. This is detected as a large clump of cells among a normal monolayer of cells.

E10. As discussed in the experiment of Figure 17.11, the presence of a transposon can create a mutable site or locus that is subject to frequent chromosome breakage. Why do you think a transposon creates a mutable site? If chromosome breakage occurs, do you think the transposon has moved somewhere else? How would you experimentally determine if it has?

E11. Gerald Rubin and Allan Spradling devised a method of introducing a transposon into *Drosophila*. This approach has been important for the transposon tagging of many *Drosophila* genes. They began with a P element that had been cloned on a plasmid. (Note: Methods of cloning are described in Chapter 18.) Using cloning methods, they inserted the wild-type allele for the *rosy* gene into the P element in this plasmid. The recessive allele, *rosy*, results in a rosy eye color, while the wild-type allele, *rosy*+, produces red eyes. This plasmid also has an intact transposase gene. The cloned DNA is shown here:

They used a micropipette to inject this DNA into regions of embryos that would later become reproductive cells. These embryos were originally homozygous for the recessive *rosy* allele. However, the P element carrying the *rosy*+ allele could "hop" out of the plasmid and into a chromosome of the cells that were destined to become germ cells (i.e., sperm or egg cells). After they had matured to adults, these flies were then mated to flies that were homozygous for the recessive rosy allele. If offspring inherited a chromosome carrying the P element with the *rosy*+ gene, such offspring would have red eyes. Therefore, the phenotype of red eyes provided a way to identify offspring that had a P element insertion.

Now here is the question. Let's suppose you were interested in identifying genes that play a role in wing development. Outline the experimental steps you would follow, using the plasmid with the P element containing the *rosy*+ gene, as a way to transposon tag genes that play a role in wing development. Note: You should assume that the inactivation of a gene involved in wing development would cause an abnormality in wing shape. Also keep in mind that most P element insertions inactivate genes and may be inherited in a recessive manner.

Questions for Student Discussion/Collaboration

1. Make a list of the similarities and differences among homologous recombination, site-specific recombination, and transposition.

2. If no homologous recombination of any kind could occur, what would be the harmful and beneficial consequences?

3. Based on your current knowledge of genetics, discuss whether or not you think the selfish DNA hypothesis is correct.

Note: All answers appear at the website for this textbook; the answers to even-numbered questions are in the back of the textbook.

www.mhhe.com/brookergenetics3e

Visit the Online Learning Center for practice tests, answer keys, and other learning aids for this chapter. Enhance your understanding of genetics with our interactive exercises, quizzes, animations, and much more.

18

RECOMBINANT DNA TECHNOLOGY

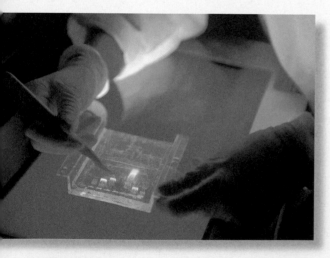

Visualization of DNA bands on a gel. DNA can be cut into fragments that can be separated via gel electrophoresis and visualized by staining. The cutting and pasting of DNA fragments allows researchers to clone genes.

PART V
Genetic Technologies

Recombinant DNA technology is the use of in vitro molecular techniques to isolate and manipulate fragments of DNA. In the early 1970s, the first successes in making recombinant DNA molecules were accomplished independently by two groups at Stanford University: David Jackson, Robert Symons, and Paul Berg; and Peter Lobban and A. Dale Kaiser. Both groups were able to isolate and purify pieces of DNA in a test tube and then covalently link DNA fragments from two different sources. In other words, they constructed molecules called **recombinant DNA molecules.** Shortly thereafter, researchers were able to introduce such recombinant DNA molecules into living cells. Once inside a host cell, the recombinant molecules can be replicated to produce many identical copies of a gene—a process called gene cloning.

Recombinant DNA technology and gene cloning have enabled geneticists to probe relationships between gene sequences and phenotypic consequences, and thereby have been fundamental to our understanding of gene structure and function. Most researchers in molecular genetics are familiar with recombinant DNA technology and apply it frequently in their work. Significant practical applications of recombinant DNA technology also have been developed. These include exciting advances such as gene therapy, screening for human diseases, recombinant vaccines, and the production of transgenic plants and animals in agriculture, in which a cloned gene from one species is transferred to some other species. Transgenic organisms have also been important in basic research.

In this chapter, we will focus primarily on the use of recombinant DNA technology as a way to further our understanding of gene structure and function. We will look at the materials and molecular techniques used in gene cloning, examine the first gene cloning experiment, and explore polymerase chain reaction (PCR), which can make many copies of DNA in a defined region. We then explore techniques to identify a specific gene or gene product, and methods to detect the binding of proteins to DNA sequences. Finally, we will examine how scientists analyze and alter DNA sequences through the techniques of DNA sequencing, a method that enables researchers to determine the base sequence of a DNA strand, and site-directed mutagenesis, a procedure that allows researchers to make mutations within a cloned segment of DNA.

In Chapter 19, we will consider many of the practical applications that have arisen as a result of these technologies. Chapters 20 and 21 are devoted to genomics, the molecular analysis of many genes and even the entire genome of a species.

18.1 GENE CLONING

Molecular biologists want to understand how the molecules within living cells contribute to cell structure and function. Because proteins are the workhorses of cells and because they are the products of genes, most molecular biologists focus their attention on the structure and function of proteins or the genes that encode them. Researchers may focus their efforts on the study of just one or perhaps a few different genes or proteins. At the molecular level, this poses a daunting task. In eukaryotic species, any given cell can express thousands of different proteins, making the study of any single gene or protein akin to a "needle in a haystack" exploration. To overcome this truly formidable obstacle, researchers frequently take the approach of cloning the genes that encode their proteins of interest. The term **gene cloning** refers to the phenomenon of isolating and making many copies of a gene in order to study it in detail. The laboratory methods necessary to clone a gene were devised during the early 1970s. Since then, many technical advances have enabled gene cloning to become a widely used procedure among scientists, including geneticists, cell biologists, biochemists, plant biologists, microbiologists, evolutionary biologists, clinicians, and biotechnologists.

Table 18.1 summarizes some of the common uses of gene cloning. In modern molecular biology, the myriad of uses for gene cloning is remarkable. For this reason, gene cloning has provided the foundation for critical technical advances in a variety of disciplines including molecular biology, genetics, cell biology, biochemistry, and medicine. In this section, we will examine the two general strategies used to make copies of a gene—the insertion of a gene into a vector that is then propagated in living cells, and cloning via polymerase chain reaction. Later sections in this chapter, as well as Chapters 19 through 21, will consider many of the uses of gene cloning that are described in Table 18.1.

Cloning Experiments May Involve Two Kinds of DNA Molecules: Chromosomal DNA and Vector DNA

If a scientist wants to clone a particular gene, the source of the gene is the chromosomal DNA of the species that carries the gene. For example, if the goal is to clone the rat β-globin gene, this gene is found within the chromosomal DNA of rat cells. In this case, therefore, the rat's chromosomal DNA is one type of DNA needed in a cloning experiment. To prepare chromosomal DNA, an experimenter first obtains cellular tissue from the organism of interest. The preparation of chromosomal DNA then involves the breaking open of cells and the extraction and purification of the DNA using biochemical techniques such as chromatography and centrifugation (see the Appendix for a description of these techniques).

Let's begin our discussion of gene cloning by considering a recombinant DNA technology in which a gene is removed from its native site within a chromosome and inserted into a smaller segment of DNA known as a **vector**—a small DNA molecule that can replicate independently of host cell chromosomal DNA and

TABLE 18.1

Some Uses of Gene Cloning

Technique	Description
Gene sequencing	Cloned genes provide enough DNA to subject the gene to DNA sequencing (described later in this chapter). The sequence of the gene can reveal the gene's promoter, regulatory sequences, and coding sequence. Gene sequencing is also important in the identification of alleles that cause cancer and inherited human diseases.
Site-directed mutagenesis	A cloned gene can be manipulated to change its DNA sequence. Mutations within genes can help to identify gene sequences such as promoters and regulatory elements. The study of a mutant gene can also help to elucidate its normal function and how its expression may affect the roles of other genes. Mutations in the coding sequence can reveal which amino acids are important for a protein's structure and function.
Gene probes	Labeled DNA strands from a cloned gene can be used as probes to identify similar or identical genes or RNA. These methods, known as Southern and Northern blot analysis, are described later in this chapter. Probes can also be used to localize genes within intact chromosomes (see Chapter 20). DNA probes are used in techniques such as DNA fingerprinting to identify criminals (see Chapter 24).
Expression of cloned genes	Cloned genes can be introduced into a different cell type or different species. The expression of cloned genes has many uses: *Research* 1. The expression of a cloned gene can help to elucidate its cellular function. 2. The coding sequence of a gene can be placed next to an active promoter and then introduced into a culture of cells that will express a large amount of the protein. This greatly aids in the purification of large amounts of protein that may be needed for biochemical or biophysical studies. *Biotechnology* 1. Cloned genes can be introduced into bacteria to make pharmaceutical products such as insulin (see Chapter 19). 2. Cloned genes can be introduced into plants and animals to make transgenic species with desirable traits (see Chapter 19). *Clinical trials* 1. Cloned genes have been used in clinical trials involving gene therapy (see Chapter 19).

produce many identical copies of an inserted gene. The purpose of vector DNA is to act as a carrier of the DNA segment that is to be cloned. (The term vector comes from a Latin term meaning carrier.) In cloning experiments, a vector may carry a small segment of chromosomal DNA, perhaps only a single gene. By comparison, a chromosome carries many more genes, perhaps a few thousand. Like a chromosome, a vector is replicated when it resides within a living cell; a cell that harbors a vector is called a **host cell.** When a vector is replicated within a host cell, the DNA that it carries is also replicated.

The vectors commonly used in gene cloning experiments were derived originally from two natural sources: plasmids or viruses. Some vectors are **plasmids,** which are small circular pieces of DNA. As discussed in Chapter 6, plasmids are found naturally in many strains of bacteria and occasionally in eukaryotic cells. Many naturally occurring plasmids carry genes that confer resistance to antibiotics or other toxic substances. These plasmids are called **R factors.** Some of the plasmids used in modern cloning experiments were derived from R factors.

Plasmids also contain a DNA sequence, known as an **origin of replication,** that is recognized by the replication enzymes of the host cell and allow it to be replicated. The sequence of the origin of replication determines whether or not the vector can replicate in a particular type of host cell. Some plasmids have origins of replications with a broad host range. Such a plasmid can replicate in the cells of many different species. Alternatively, many vectors used in cloning experiments have a limited host cell range. In cloning experiments, researchers must choose a vector that replicates in the appropriate cell types for their experiments. For example, if researchers want a cloned gene to be propagated in *Escherichia coli,* the vector they employ must have an origin of replication that is recognized by this species of bacterium. The origin of replication also determines the copy number of a plasmid. Some plasmids are said to have strong origins because they can achieve high copy number—perhaps 100 to 200 copies of the plasmid per cell. Others have weaker origins whereby only one or two plasmids are found per cell.

Commercially available plasmids have been genetically engineered for effective use in cloning experiments. They contain unique sites where geneticists can easily insert pieces of DNA. Another useful feature of cloning vectors is that they often contain resistance genes that provide host bacteria with the ability to grow in the presence of a toxic substance. Such a gene is called a **selectable marker,** because the expression of the gene selects for the growth of the bacterial cells. Many selectable markers are genes that confer antibiotic resistance to the host cell. For example, the gene amp^R encodes an enzyme known as β-lactamase. This enzyme degrades ampicillin, an antibiotic that normally kills bacteria. Bacteria containing the amp^R gene can grow on media containing ampicillin because they can degrade it. In a cloning experiment where the amp^R gene is found within the plasmid, the growth of cells in the presence of ampicillin identifies bacteria that contain the plasmid. These bacteria can grow and form visible colonies on solid growth media. In contrast, those cells that do not contain the plasmid are ampicillin sensitive and will not grow.

An alternative type of vector used in cloning experiments is a viral vector. As discussed in Chapter 6, viruses can infect living cells and propagate themselves by taking control of the host cell's metabolic machinery. When a chromosomal gene is inserted into a viral genome, the gene will be replicated whenever the viral DNA is replicated. Therefore, viruses can be used as vectors to carry other pieces of DNA. When a virus is used as a vector, the researcher may analyze viral plaques rather than bacterial colonies. The characteristics of viral plaques are described in Chapter 6 (see Figure 6.14).

Molecular biologists use hundreds of different vectors in cloning experiments. Table 18.2 provides a general description of several different types of vectors that are commonly used to clone small segments of DNA. In addition, other types of vectors, such as cosmids, BACs, and YACs, are used to clone large pieces of DNA. These vectors are described in detail in Chapter 20. Vectors designed to introduce genes into plants and animals are discussed in Chapter 19.

Enzymes Are Used to Cut DNA into Pieces and Join the Pieces Together

A key step in a cloning experiment is the insertion of chromosomal DNA into a plasmid or viral vector. This requires the cutting and pasting of DNA fragments. To cut DNA, researchers use enzymes known as **restriction endonucleases,** or **restriction enzymes.** The restriction enzymes used in cloning experiments

TABLE 18.2

Some Vectors Used in Cloning Experiments

Example	Type	Description
pBluescript	Plasmid	A type of vector like the one shown in Figure 18.2. It is used to clone small segments of DNA and propagate them in *E. coli*.
YEp24	Plasmid	This plasmid is an example of a **shuttle vector,** which can replicate in two different host species, *E. coli* and *Saccharomyces cerevisiae*. It carries an origin of replication for both species.
λgt11	Viral	This vector is derived from the bacteriophage λ, which is described in Chapter 14. λgt11 also contains a promoter from the *lac* operon. When fragments of DNA are cloned next to this promoter, the DNA is expressed in *E. coli*. This is an example of an **expression vector.** An expression vector is designed to clone the coding sequence of genes so they will be transcribed and translated correctly.
SV40	Viral	This virus naturally infects mammalian cells. Genetically altered derivatives of the SV40 viral DNA are used as vectors for the cloning and expression of genes in mammalian cells that are grown in the laboratory.
Baculovirus	Viral	This virus naturally infects insect cells. In a laboratory, insect cells can be grown in liquid media. Unlike many other types of cells, insect cells often express large amounts of proteins that are encoded by cloned genes. When researchers want to make a large amount of a protein, they can clone the gene that encodes the protein into baculovirus and then purify the protein from insect cells.

bind to a specific base sequence and then cleave the DNA backbone at two defined locations, one in each strand. Proposed by Werner Arber in the 1960s and discovered by Hamilton Smith and Daniel Nathans in the 1970s, restriction enzymes are made naturally by many different species of bacteria and protect bacterial cells from invasion by foreign DNA, particularly that of bacteriophages. Researchers have isolated and purified restriction enzymes from many bacterial species and now use them in their cloning experiments.

Figure 18.1 shows the role of a restriction enzyme, called *Eco*RI, in producing a recombinant DNA molecule. Certain types of restriction enzymes are useful in cloning because they digest DNA into fragments with "sticky ends." As shown in Figure 18.1, the sticky ends are single-stranded regions of DNA that can hydrogen bond to a complementary sequence of DNA from a different source. The ends of two different DNA pieces will hydrogen bond to each other because of their complementary sticky ends.

The hydrogen bonding between the sticky ends of DNA fragments promotes a temporary interaction between the two fragments. However, this interaction is not stable because it involves only a few hydrogen bonds between complementary bases. How can this interaction be made more permanent? The answer is that the sugar–phosphate backbones within the DNA

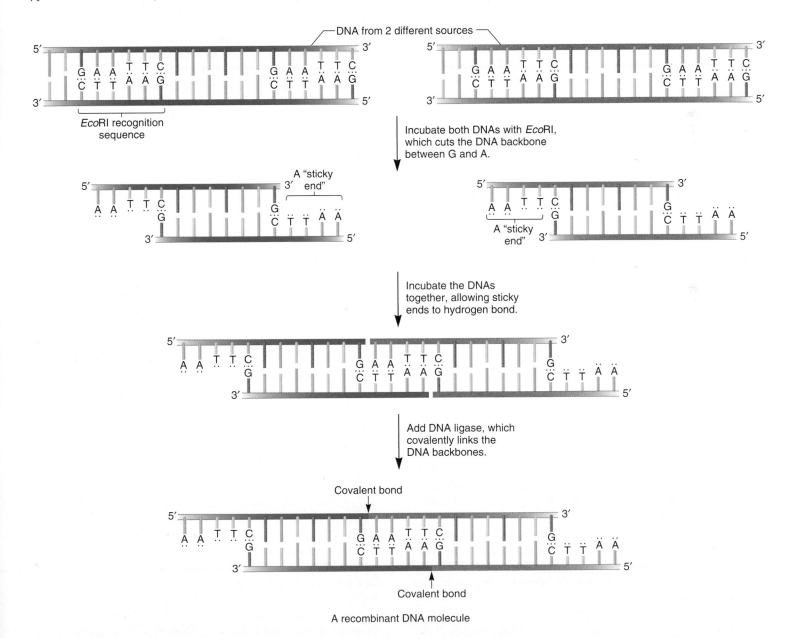

FIGURE 18.1   The action of a restriction enzyme and the production of recombinant DNA. The restriction enzyme *Eco*RI binds to a specific sequence, in this case 5'–GAATTC–3'. It then cleaves the DNA backbone between G and A, producing DNA fragments. The single-stranded ends of the DNA fragments can hydrogen bond with each other, because they have complementary sequences. The enzyme DNA ligase can then catalyze the formation of covalent bonds in the DNA backbones of the fragments.

TABLE 18.3

Some Restriction Enzymes Used in Gene Cloning

Restriction Enzyme*	Bacterial Source	Sequence Recognized†
BamHI	Bacillus amyloliquefaciens H	↓ 5'–GGATCC–3' 3'–CCTAGG–5' ↑
ClaI	Caryophanon latum	↓ 5'–ATCGAT–3' 3'–TAGCTA–5' ↑
EcoRI	E. coli RY13	↓ 5'–GAATTC–3' 3'–CTTAAG–5' ↑
NaeI	Nocardia aerocolonigenes	↓ 5'–GCCGGC–3' 3'–CGGCCG–5' ↑
PstI	Providencia stuartii	↓ 5'–CTGCAG–3' 3'–GACGTC–5' ↑
SacI	Streptomyces achromogenes	↓ 5'–GAGCTC–3' 3'–CTCGAG–5' ↑

*Restriction enzymes are named according to the species in which they are found. The first three letters are italicized because they indicate the genus and species names. Because a species may produce more than one restriction enzyme, the enzymes are designated I, II, III, and so on, to indicate the order in which they were discovered in a given species. Some restriction enzymes, like EcoRI, produce a sticky end with a 5' overhang (see Figure 18.1), whereas others, such as PstI, produce a 3' overhang. However, not all restriction enzymes cut DNA to produce sticky ends. For example, the enzyme NaeI cuts DNA to produce blunt ends.

†The arrows show the locations in the upper and lower DNA strands where the restriction enzymes cleave the DNA backbone. A complete list of restriction enzymes can be found at http://rebase.neb.com/rebase/rebase.html.

strands must be covalently linked together. This linkage is catalyzed by **DNA ligase.** Figure 18.1 illustrates the action of DNA ligase, which catalyzes covalent bond formation in the sugar–phosphate backbones of both DNA strands after the sticky ends have hydrogen bonded with each other.

Currently, several hundred different restriction enzymes from various bacterial species have been identified and are available commercially to molecular biologists. Table 18.3 gives a few examples. Restriction enzymes usually recognize sequences that are **palindromic;** that is, the sequence in one strand is identical when read in the opposite direction in the complementary strand. For example, the sequence recognized by EcoRI is 5'–GAATTC–3' in the top strand. Read in the opposite direction in the bottom strand, this sequence is also 5'–GAATTC–3'.

Gene Cloning Involves the Insertion of DNA Fragments into Vectors, Which Are Then Propagated Within Host Cells

Now that we are familiar with the materials, let's outline the general strategy that is followed in a typical cloning experiment. In the procedure shown in **Figure 18.2**, the goal is to clone a chromosomal gene of interest into a plasmid vector that already carries the amp^R gene. To begin this experiment, the chromosomal DNA is isolated and digested with a restriction enzyme. This enzyme will cut the chromosomes into many small fragments. The plasmid DNA is also cut at one site with the same restriction enzyme, so the vector will have the same sticky ends as the chromosomal DNA fragments. The digested chromosomal DNA and plasmid DNA are mixed together and incubated under conditions that promote the binding of complementary sticky ends.

DNA ligase is then added to catalyze the covalent linkage between DNA fragments. In some cases, the two ends of the vector will simply ligate back together, restoring the vector to its original structure. This is called a recircularized vector. In other cases, a fragment of chromosomal DNA may become ligated to both ends of the vector. In this way, a segment of chromosomal DNA has been inserted into the vector. The vector containing a piece of chromosomal DNA is referred to as a **recombinant vector.**

Following ligation, the DNA is introduced into living cells treated with agents that render them permeable to DNA molecules. Cells that can take up DNA from the extracellular medium are called **competent cells.** This step in the procedure is commonly called **transformation** (see Chapter 6). In the experiment shown in Figure 18.2, a plasmid is introduced into bacterial cells that were originally sensitive to ampicillin. The bacteria are then streaked onto plates containing bacterial growth media and ampicillin. A bacterium that has taken up a plasmid carrying the amp^R gene will continue to divide and form a bacterial colony containing tens of millions of cells. Because each cell within a single colony is derived from the same original cell, all cells within a colony contain the same type of plasmid DNA.

In the experiment shown in Figure 18.2, how can the experimenter distinguish between bacterial colonies that contain a recircularized vector versus those with a recombinant vector carrying a piece of chromosomal DNA? As shown here, the chromosomal DNA has been inserted into a region of the vector that contains the lacZ gene, which encodes the enzyme β-galactosidase (see Chapter 14). The insertion of chromosomal DNA into the vector disrupts the lacZ gene so that it is no longer able to produce a functional enzyme. By comparison, a recircularized vector has a functional lacZ gene. The functionality of lacZ can be determined by providing the growth media with a colorless compound, X-Gal, which is cleaved by β-galactosidase into a blue dye. Bacteria grown in the presence of X-Gal and IPTG (an inducer of the lacZ gene) will form blue colonies if they have a functional lacZ gene and white colonies if they do not. In this experiment, therefore, bacterial colonies

FIGURE 18.2   **The steps in gene cloning.** Note: X-Gal refers to the colorless compound 5-bromo-4-chloro-3-indolyl-β-D-galactoside. IPTG is an acronym for isopropyl-β-D-thiogalactopyranoside, which is a nonmetabolizable lactose analogue that can induce the *lac* promoter.

containing recircularized vectors will form blue colonies, while colonies containing recombinant vectors will be white.

In the example of Figure 18.2, one of the white colonies contains cells with a recombinant vector that carries a human gene of interest; the segment containing the human gene is shown in red. The net result of gene cloning is to produce an enormous number of copies of a recombinant vector. During transformation, a single bacterial cell usually takes up a single copy of a recombinant vector. However, two subsequent events lead to the amplification of the cloned gene. First, because the vector has an origin of replication, the bacterial host cell replicates the recombinant vector to produce many copies per cell. Second, the bacterial cells divide approximately every 20 minutes. Following overnight growth, a population of many millions of bacteria will be obtained. Each of these bacterial cells will contain many copies of the cloned gene. For example, a bacterial colony may be composed of 10 million cells, with each cell containing 50 copies of the recombinant vector. Therefore, this bacterial colony would contain 500 million copies of the cloned gene!

The preceding description has acquainted you with the steps required to clone a gene. A misleading aspect of Figure 18.2 is that the digestion of the chromosomal DNA with restriction enzymes appears to yield only a few DNA fragments, one of which contains the gene of interest. In an actual cloning experiment, however, the digestion of the chromosomal DNA with a restriction enzyme produces tens of thousands of different pieces of chromosomal DNA, not just a single piece of chromosomal DNA that happens to be the gene of interest. Later, we will consider methods to identify bacterial colonies containing the specific gene that a researcher wants to clone.

Recombinant DNA technology can also be used to clone fragments of DNA that do not code for genes. For example, sequences such as telomeres, centromeres, and highly repetitive sequences have been cloned by this procedure.

EXPERIMENT 18A

In the First Gene Cloning Experiment, Cohen, Chang, Boyer, and Helling Inserted a *kanamycin^R* Gene into a Plasmid Vector

Now that we are familiar with the basic procedures followed in a cloning experiment, let's consider the first successful attempt at creating a recombinant DNA molecule and propagating that molecule in bacterial cells. This was accomplished in a collaboration between Stanley Cohen, Annie Chang, Herbert Boyer, and Robert Helling in 1973. Prior discoveries had led to their ability to clone a gene. In 1970, H. Gobind Khorana found that DNA ligase could covalently link DNA fragments together. In 1972, Janet Mertz, Ronald Davis, and Vittorio Sgaramella discovered that the digestion of DNA with the restriction enzyme *Eco*RI produced sticky ends that enabled the DNA fragments to hydrogen bond. With these two observations in mind, Cohen, Chang, Boyer, and Helling realized it might be possible to create recombinant DNA molecules by digesting DNA with *Eco*RI, allowing the fragments to hydrogen bond with each other, and then covalently linking these fragments together with DNA ligase.

One last condition was necessary for the researchers to succeed in cloning a gene. They needed to identify a vector that could independently replicate itself once inside a host cell. They chose a plasmid vector as their vehicle for gene cloning. In their collection of plasmids, they found one small plasmid, designated pSC101, that had a single *Eco*RI site and carried a gene for tetracycline resistance (*tet^R*).

They reasoned that it should be possible to open up this plasmid with *Eco*RI and then insert another piece of DNA into this *Eco*RI site via the hydrogen bonding of the *Eco*RI sticky ends.

As a source of a gene to insert into pSC101, they obtained a second plasmid, which they called pSC102.

This plasmid carried a gene, *kan^R*, that provided resistance to the antibiotic kanamycin. However, this second plasmid was cut at three locations by *Eco*RI, yielding three DNA fragments. One of these DNA fragments would be expected to carry the kanamycin-resistance gene (unless, unluckily, *Eco*RI happened to cut in the middle of the *kan^R* gene). Their goal in this experiment was to clone the *kan^R* gene by inserting the DNA fragment carrying this gene into the *Eco*RI site of pSC101 and then introduce the recombinant plasmid into bacterial cells.

As shown in **Figure 18.3**, the researchers isolated plasmid DNA from bacterial strains carrying pSC101 or pSC102 (step 1). The purified plasmids were digested with *Eco*RI and then mixed together to allow the sticky ends to hydrogen bond with each other. Ligase was added to covalently seal the plasmids (step 4). Their goal was to obtain a recombinant vector in which pSC101 carried the kanamycin-resistance gene from pSC102. Such a plasmid would confer resistance to both tetracycline and kanamycin. To identify such a plasmid, the DNA sample was exposed to bacteria that were treated in such a way as to promote the uptake of DNA. Some colonies were obtained that were resistant to both antibiotics (step 7). Four of these colonies were picked, and DNA was isolated from them. How did the researchers confirm that they had constructed recombinant plasmids? The plasmid DNA from these colonies, termed pSC105, was then subjected to two different procedures. In step 9, some of the DNA was analyzed by cesium chloride density gradient centrifugation and scintillation counting. In step 10, some of the DNA was digested with *Eco*RI and subjected to gel electrophoresis. (See the Appendix for a description of the techniques of density gradient centrifugation, scintillation counting, and gel electrophoresis.) As a control, samples of pSC101 and pSC102 also underwent these last two steps.

■ THE HYPOTHESIS

A piece of DNA carrying a gene can be inserted into a plasmid vector using recombinant DNA techniques. If this recombinant plasmid is introduced into a bacterial host cell, it will be replicated and transmitted to daughter cells, producing many copies of the recombinant plasmid.

■ TESTING THE HYPOTHESIS — FIGURE 18.3   The first success at gene cloning.

Starting material: Three different strains of *E. coli*: one strain that did not carry any plasmid and two strains that carried pSC101 or pSC102.

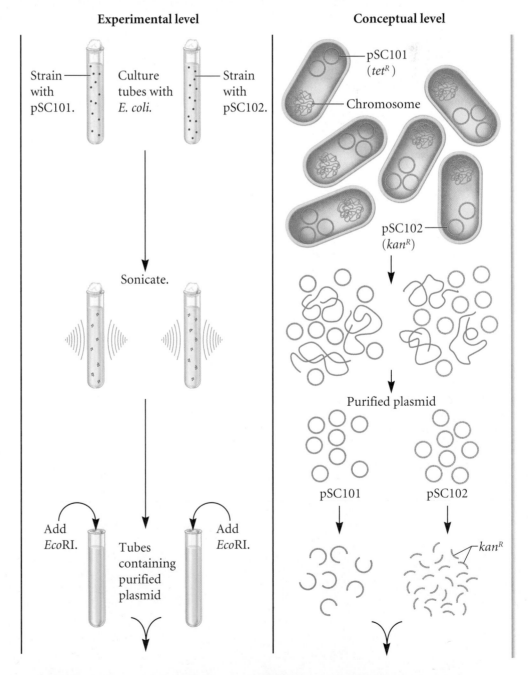

Experimental level   Conceptual level

1. Isolate and purify the two types of plasmid DNA:

Grow the bacterial cells containing the plasmids.

Strain with pSC101.   Culture tubes with *E. coli*.   Strain with pSC102.

pSC101 (*tet*^{R})
Chromosome
pSC102 (*kan*^{R})

Break open the cells. One way to break open cells is to subject them to harsh sound waves, or sonication.

Sonicate.

Isolate each type of plasmid DNA by density gradient centrifugation (see the Appendix for a description of this procedure).

Purified plasmid

pSC101   pSC102

2. Digest the plasmid DNAs with *Eco*RI.

Add *Eco*RI.   Tubes containing purified plasmid   Add *Eco*RI.

kan^R

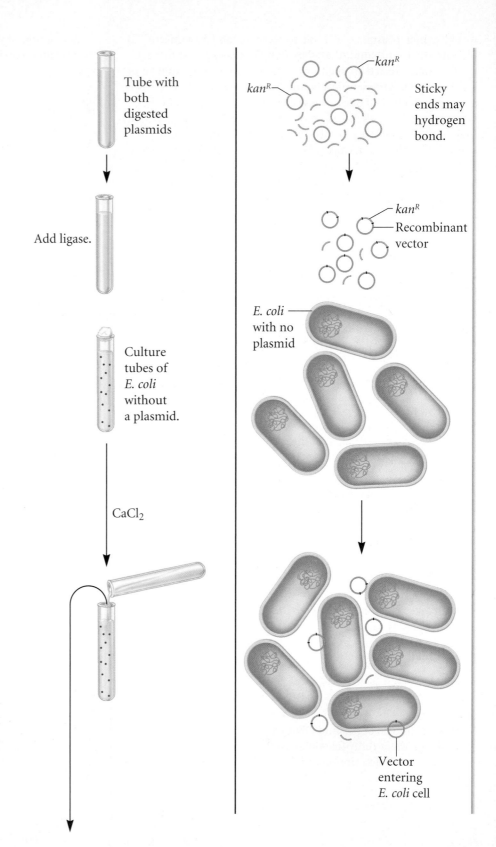

3. Mix together the two samples.

4. Add DNA ligase to covalently link DNA pieces.

5. Grow an *E. coli* strain that does not carry a plasmid. Treat the cells with $CaCl_2$ to make them permeable to DNA.

6. Add the ligated DNA samples from step 4 to the bacterial cells. Most cells do not take up any plasmid but an occasional cell will take up one plasmid.

Tube with both digested plasmids

Add ligase.

Culture tubes of *E. coli* without a plasmid.

$CaCl_2$

kan[R]

kan[R]

Sticky ends may hydrogen bond.

kan[R]
Recombinant vector

E. coli with no plasmid

Vector entering *E. coli* cell

7. Plate the cells on growth media containing both tetracycline and kanamycin. Grow overnight to allow growth of visible bacterial colonies.

8. Pick four colonies from the plates. The plasmid found in these colonies is designated pSC105. Grow the colonies in liquid culture containing radiolabeled deoxyribonucleotides.

9. To isolate the radiolabeled plasmid DNA, break open the cells and subject the DNA to cesium chloride density gradient centrifugation. Collect fractions and count the amount of radioactivity in each fraction using a scintillation counter. (See data on the next page that is shown on the top left.) As a control, subject pSC101 and pSC102 to the same procedure. (See data on the bottom left.)

10. Digest the plasmid DNAs with *Eco*RI, and subject them to gel electrophoresis. (See Appendix for a description of gel electrophoresis.) (See data shown on the next page that is on the right.)

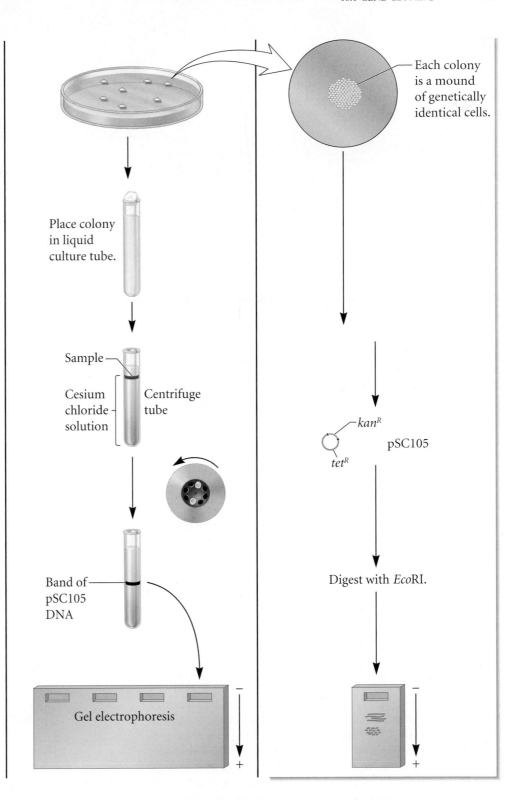

Each colony is a mound of genetically identical cells.

Place colony in liquid culture tube.

Sample

Cesium chloride solution

Centrifuge tube

Band of pSC105 DNA

Gel electrophoresis

kan^R

pSC105

tet^R

Digest with *Eco*RI.

■ THE DATA

Results from step 9:

Results from step 10:

Lane 1: pSC101

2: pSC102

3: pSC101 + pSC102

4: pSC105

■ INTERPRETING THE DATA

With regard to their data, let's first consider the results of density gradient centrifugation. In the control experiment, the pSC101 and pSC102 plasmids were mixed together and then subjected to density gradient centrifugation. As shown in the data at the bottom left, this yielded two peaks at 27 S and 39.5 S. The pSC102 plasmid is larger than pSC101 and sediments at a density of 39.5 S, whereas the latter plasmid has a density of 27 S. (Note: The letter S refers to Svedberg units, a unit of centrifugation velocity; see the Appendix.) The top graph shows the results for pSC105 obtained from a bacterial colony that was resistant to both tetracycline and kanamycin. In this case, the plasmid had an intermediate density of 32 S. These results indicate that this bacterial colony contained a recombinant plasmid, not a mixture of pSC101 and pSC102 plasmids.

This conclusion was confirmed in the gel electrophoresis analysis. When digested with *Eco*RI, pSC101 yielded a single band, whereas pSC102 yielded three bands (see lanes 1 and 2). In lane 3, the experimenters had mixed together samples of pSC101 and pSC102 (isolated in step 1) and digested them with *Eco*RI. As expected, this produced four bands. Lane 4 shows the plasmid DNA, pSC105, from a bacterial colony that was tetracycline and kanamycin resistant. This plasmid showed the pSC101 band plus one other band that was found in pSC102. The results of lane 4 are consistent with the idea that the pSC105 plasmid is formed

by the insertion of one fragment from pSC102 into the single *Eco*RI site of pSC101. This recombinant plasmid is shown here.

kan^R

pSC102 fragment

*Eco*RI site — — *Eco*RI site

*Eco*RI site

pSC101 — — *tet^R*

pSC105

This study showed the scientific community that it is possible to create recombinant DNA molecules and then propagate them in bacterial cells. In other words, researchers were able to clone genes. This hallmark achievement ushered in the era of gene cloning.

A self-help quiz involving this experiment can be found at the Online Learning Center.

cDNA Can Be Made from mRNA Via Reverse Transcriptase

In the discussion of gene cloning described earlier in Figure 18.2, chromosomal DNA and plasmid DNA were used as the material to clone genes. Alternatively, a sample of RNA can provide a starting point to clone DNA. As mentioned in Chapter 17, the enzyme **reverse transcriptase** can use RNA as a template to make a complementary strand of DNA. This enzyme is encoded in the genome of retroviruses and provides a way for retroviruses to copy their RNA genome into DNA molecules that integrate into the host cell's chromosomes. Likewise, reverse transcriptase is encoded in viral-like retroelements and is needed in the retrotransposition of such elements.

Researchers can use purified reverse transcriptase in a strategy to clone genes, using mRNA as the starting material (**Figure 18.4**). To begin this experiment, RNA is purified from a sample of cells. The RNA is mixed with primers composed of a string of thymine-containing nucleotides. This short strand of DNA, or **oligonucleotide,** is called a poly-dT primer. Because eukaryotic mRNAs contain a polyA tail, poly-dT primers will be complementary to the 3' end of mRNAs. Reverse transcriptase and deoxyribonucleotides (dNTPs) are then added to make a DNA strand that is complementary to the mRNA. Next, RNaseH, DNA polymerase, and DNA ligase are added. One way to make the other DNA strand is to use RNaseH, which partially digests the RNA, generating short RNAs that are used as primers by DNA polymerase to make a second DNA strand that is complementary to the strand made by reverse transcriptase. Finally, DNA ligase will seal any nicks in this second DNA strand. When DNA is made from RNA as the starting material, the DNA is called **complementary DNA (cDNA).** The term originally referred to the single strand of DNA that is complementary to the RNA template. However, cDNA now refers to any DNA, whether it is single or double stranded, that is made using RNA as the starting material.

Why is cDNA cloning useful? From a research perspective, an important advantage of cDNA is that it lacks introns. Because introns can be quite large, it is much simpler to insert cDNAs into vectors if researchers want to focus their attention on the coding sequence of a gene. For example, if the primary goal was to determine the coding sequence of a structural gene, a researcher would insert cDNA into a vector and then determine the DNA sequence of the insert, as described later in this chapter. Similarly, if a scientist wanted to express an encoded protein of interest in a cell that would not splice out the introns properly (e.g., in a bacterial cell), it would be necessary to make cDNA clones of the respective gene.

Restriction Mapping Is Used to Locate the Restriction Sites Within a Vector

As we have seen, DNA or gene cloning involves the digestion of vector and chromosomal DNA with restriction enzymes and the subsequent ligation of DNA fragments into vectors. In this type of procedure, the locations of restriction enzyme sites, or simply restriction sites, are important for the design of experiments. In the vector, for example, it is desirable to have unique restriction sites for the insertion of chromosomal DNA.

FIGURE 18.4   Synthesis of cDNA. A poly-dT primer anneals to the 3' end of mRNAs. Reverse transcriptase then catalyzes the synthesis of a complementary DNA strand (cDNA). RNaseH digests the mRNA into short pieces that are used as primers by DNA polymerase I to synthesize the second DNA strand. The 5' to 3' exonuclease function of DNA polymerase I removes all of the RNA primers except the one at the 5' end (because there is no primer upstream from this site). This RNA primer can be removed by the subsequent addition of an RNase. After the double-stranded cDNA is made, it can then be inserted into vectors as described in Figure 18.2. Prior to vector insertion, short pieces of DNA, called linkers, are covalently attached to both ends of the cDNA. The DNA sequence of the linkers contains unique restriction sites, making it easy to cut the cDNA at both ends and insert it into a vector.

A common approach to determine the locations of restriction sites is known as **restriction mapping. Figure 18.5** outlines the restriction mapping of a bacterial plasmid. To begin this experiment, the small circular plasmid DNA is isolated and purified

FIGURE 18.5 **Restriction mapping of the pBR322 plasmid.**

from host cells. Samples of the purified DNA are then placed in separate test tubes that contain a particular restriction enzyme or combination of enzymes. The plasmid DNA is incubated with the restriction enzymes long enough for digestion to occur. The DNA fragments are then separated by gel electrophoresis. In lane 1 of this experiment, a different DNA sample, known as a set of molecular markers, is also subjected to gel electrophoresis. This sample contains a mixture of DNA fragments with molecular masses that are known from previous experiments. (These markers are obtained from commercial sources or can be prepared in the laboratory.) To determine the sizes of the fragments obtained by digesting the plasmid, the fragments in lanes 2 to 8 are compared to the known markers in lane 1.

The restriction map shown in Figure 18.5 was deduced by comparing the sizes of fragments obtained from digestions with one, two, or all three of the restriction enzymes. The starting plasmid is a circular molecule 4,363 bp in length. A single digestion with any of the three enzymes yields a single linear fragment of size 4,363 bp. This means that *Eco*RI, *Bam*HI, and *Pst*I cut the plasmid at a single site. The double digestion with *Eco*RI and *Bam*HI yields two fragments of about 380 bp and 3,980 bp. This result indicates that the *Eco*RI and *Bam*HI sites are approximately 380 bp apart in one direction along the circle and 3,980 bp apart along the circle in the opposite direction. Likewise, the pairwise combinations of *Eco*RI/*Pst*I and *Bam*HI/*Pst*I indicate how far apart these sites are along the circular plasmid. Finally, the triple digestion confirms the locations of the single sites for *Eco*RI, *Bam*HI, and *Pst*I. Taken together, these results provide a map of the restriction sites within this plasmid. A similar approach can be used on a recombinant vector to determine the locations of restriction sites within a fragment of DNA that has been inserted into a plasmid.

Alternatively, another way to obtain a restriction map is via DNA sequencing, a technique described later in the chapter. If the DNA sequence of a recombinant vector has been determined, computer programs can scan the sequence and identify sites that are recognized by particular restriction enzymes. For example, *Eco*RI recognizes the sequence 5'–GAATTC–3'. If a recombinant vector contained this sequence, a computer program would identify this as an *Eco*RI site and place it on a map. Such a program would scan the DNA sequence for the recognition sequences of many different restriction enzymes and generate a detailed restriction map of a recombinant vector.

A second use of restriction enzymes is gene mapping. The technique known as restriction fragment length polymorphism (RFLP) enables researchers to map particular genes within a species' genome. This approach is discussed in Chapter 20.

Polymerase Chain Reaction (PCR) Can Also Be Used to Make Many Copies of DNA

In our previous discussions of gene cloning, the DNA of interest was inserted into a vector, which then was introduced into a host cell. The replication of the vector within the host cell, and the proliferation of the host cells, led to the production of many copies of the DNA. Another way to copy DNA, without the aid of vectors and host cells, is a technique called **polymerase chain reaction (PCR),** which was developed by Kary Mullis in 1985.

The PCR method is outlined in **Figure 18.6.** In this example, the starting material contains a sample of DNA that someone wants to clone. This DNA is called **template DNA.** The goal of PCR is to make many copies of the template DNA in a defined region. Several reagents are added to facilitate the synthesis of DNA. These include a high concentration of two oligonucleotide primers, which are complementary to sequences at the ends of the DNA fragment to be amplified; deoxyribonucleoside triphosphates (dNTPs); and a thermostable form of DNA polymerase such as *Taq* **polymerase,** isolated from the bacterium *Thermus aquaticus*. A thermostable form of DNA polymerase is necessary because PCR involves heating steps that would inactivate most other natural forms of DNA polymerase (which are thermolabile, or readily denatured by heat).

To make copies of the DNA, the template DNA is first denatured by heat treatment, causing the strands to separate (see Online Animation). As the temperature is lowered, oligonucleotide primers bind to the DNA in a process called **annealing.** Once the primers have annealed, *Taq* polymerase catalyzes the synthesis of complementary DNA strands, starting at the primers. This doubles the amount of the template DNA. The sequential process of denaturation—annealing—synthesis is then repeated to double the amount of template DNA many times in a row. This method is called a chain reaction because the products of each previous reaction (the newly made DNA strands) are used as reactants (the template strands) in subsequent reactions.

PCR is carried out in a thermal reactor, known as a **thermocycler,** that automates the timing of each cycle. The experimenter mixes the DNA sample, dNTPs, *Taq* polymerase, and an excess amount of primers together in a single tube. The tube is placed in a thermocycler, and the experimenter sets the machine to operate within a defined temperature range and number of cycles. During each cycle, the thermocycler increases the temperature to denature the DNA strands and then lowers the temperature to allow annealing and DNA synthesis to take place. Typically, each cycle lasts 2 to 3 minutes and is then repeated. A typical PCR run is likely to involve 20 to 30 cycles of replication and takes a couple of hours to complete. The PCR technique can amplify the amount of DNA by a staggering amount. After 30 cycles of amplification, the intervening region between the two primers will increase 2^{30}-fold, which is approximately a billion-fold!

In the PCR reaction shown in Figure 18.6, the outcome of PCR is the synthesis of many copies of a DNA sequence that is flanked by two primers. The sequences of the PCR primers are complementary to two specific sequences within the template DNA. How are the primers designed, and how are they made? To conduct this type of PCR experiment, a researcher must have prior knowledge about the sequence of the template DNA in order to design oligonucleotide primers complementary to two sites in the template sequence. This information might come from the sequencing of a species' genome. As discussed in Chapter 20, the DNA sequence of the entire genome of many organisms is

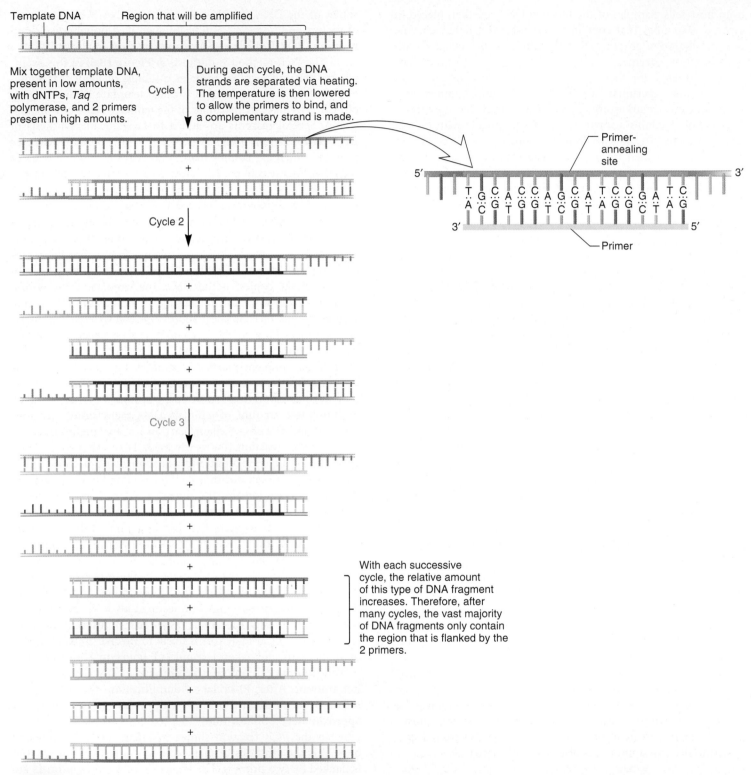

FIGURE 18.6 **The technique of polymerase chain reaction (PCR).** During each cycle, oligonucleotides that are complementary to the ends of the targeted DNA sequence bind to the DNA and act as primers for the synthesis of this DNA region. The primers used in actual PCR experiments are usually about 20 nucleotides in length. The region between the two primers is typically hundreds of nucleotides in length, not just several nucleotides as shown here. The net result of PCR is the synthesis of many copies of DNA in the region flanked by the two primers. Note: The DNA template strands are purple, and those strands made during the first, second, and third cycles are blue, red, and gray, respectively.

ONLINE ANIMATION

known. Complete genome sequences have been obtained for many species of bacteria, yeast, and several plants and animals. Once a researcher decides which gene or DNA region to amplify, she/he must choose DNA sites that flank both sides of this region and specify short DNA primer sequences, typically about 20 nucleotides in length, that are complementary to those sites. After the sequences are chosen, researchers usually obtain such primers via biotechnology companies or academic facilities that synthesize them chemically.

When specific primers can be constructed, PCR can amplify a particular region of DNA from a very complex mixture of template DNA. For example, if a researcher uses two primers that anneal to the human β-globin gene, PCR can amplify just the β-globin gene from a DNA sample that contains all of the human chromosomes!

Alternatively, PCR can be used to amplify a sample of chromosomal DNA semispecifically or nonspecifically. As discussed in Chapter 24, this approach is used in DNA fingerprinting analysis. In a semispecific PCR experiment, the primers recognize a known repetitive DNA sequence found at several sites within the genome. When chromosomal DNA is used as a template, this will amplify many different DNA fragments. In a nonspecific approach, a mixture of short PCR primers with many different random sequences is used. These primers will anneal randomly throughout the genome and amplify most of the chromosomal DNA. Nonspecific DNA amplification is used to increase the total amount of DNA in very small samples, such as blood stains found at crime scenes.

PCR is also used to detect and quantitate the amount of specific RNAs in living cells. To accomplish this goal, RNA is isolated from a sample and mixed with reverse transcriptase and a primer that will bind near the 3' end of the RNA of interest. This generates a single-stranded cDNA, which then can be used as template DNA in a conventional PCR reaction. This method, called **reverse transcriptase PCR (RT-PCR),** is extraordinarily sensitive. RT-PCR can detect the expression of small amounts of RNA from a single cell!

18.2 DETECTION OF GENES, GENE PRODUCTS, AND PROTEIN-DNA INTERACTIONS

The advent of gene cloning has enabled scientists to investigate gene structure and function at the molecular level. However, unlike the first gene cloning experiment, in which a single gene was removed from one plasmid (pSC102) and cloned into another plasmid (pSC101), molecular geneticists usually want to study genes that are originally within the chromosomes of living species. This presents a problem, because chromosomal DNA contains thousands of different genes. For this reason, researchers need methods to specifically identify a gene within a mixture of many other genes or DNA fragments. Similarly, when studying gene expression at the molecular level, scientists also need techniques for the identification of gene products, such as the RNA that is transcribed from a particular gene or the protein that is encoded by an mRNA. In this section, we will consider the methodology and uses of several common techniques used to detect DNA, RNA, proteins, and protein-DNA interactions.

A DNA Library Is Constructed and Then Screened by Colony Hybridization to Identify a Cloned Gene

In a typical cloning experiment (refer back to Figure 18.2), the treatment of the chromosomal DNA with restriction enzymes yields tens of thousands of different DNA fragments. Therefore, after the DNA fragments are ligated individually to vectors, the researcher has a collection of recombinant vectors, with each vector containing a particular fragment of chromosomal DNA. A collection of recombinant vectors is known as a **DNA library.** When the starting material is chromosomal DNA, the library is called a **genomic library** (**Figure 18.7**). The library shown here uses a plasmid vector. Alternatively, a viral vector could be used, which would result in viral plaques rather than bacterial colonies.

It is also common for researchers to make a **cDNA library** that contains recombinant vectors with cDNA inserts. Recall that because cDNA is produced from mature mRNA via reverse transcriptase, it lacks any introns. A cDNA library could be made because a researcher wanted to express the encoded protein of interest in a cell that would not splice out the introns properly.

In many cloning experiments, the ultimate goal is to clone a specific gene. For example, let's suppose that a geneticist wishes to clone the rat β-globin gene. To begin a cloning experiment, chromosomal DNA would be isolated from rat cells. This chromosomal DNA would be digested with a restriction enzyme, yielding thousands of DNA fragments. The chromosomal fragments would then be ligated to vector DNA and transformed into bacterial cells. Unfortunately, only a small percentage of the recombinant vectors, perhaps one in a few thousand, would actually contain the rat β-globin gene. For this reason, researchers must have some way to identify those rare bacterial colonies that happen to contain the cloned gene of interest, in this case, a colony that contains the rat β-globin gene.

Figure 18.8 describes a method for identifying a bacterial colony that contains the gene of interest. This procedure is referred to as **colony hybridization.** The master plate shown at the top of Figure 18.8 has many bacterial colonies. Each bacterial colony is composed of bacterial cells containing a recombinant vector with a different piece of rat chromosomal DNA. The goal is to identify a colony that contains the gene of interest, in this case, the rat β-globin gene. To do so, a nylon membrane is laid gently onto the master plate containing many bacterial colonies. After the membrane is lifted, some cells from each colony are attached to it. In this way, the membrane contains a replica of the colonies on the master plate.

In this procedure, cells are treated with detergent, which dissolves the cell membrane and makes the DNA accessible. The DNA within the cells is then fixed (i.e., adhered) to the nylon membrane and denatured with NaOH. The membrane is then submerged in a solution containing a **probe,** which can be a oligonucleotide, a fragment of DNA from a cloned gene, or a specific RNA. In this case, the probe is single-stranded DNA with a base sequence that is complementary to one of the DNA strands of the rat β-globin gene. It is radiolabeled for ease of detection. The probe is given time to hybridize to the DNA on the membrane, which has already

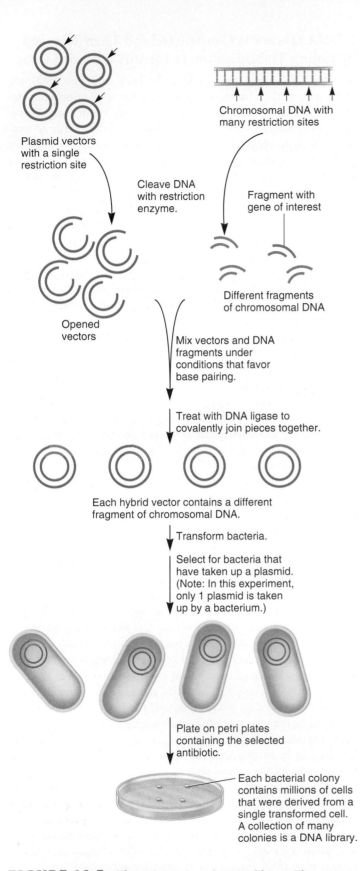

FIGURE 18.7   The construction of a DNA library. The digestion of chromosomal DNA produces many fragments. The fragment containing the gene of interest is highlighted in red. Following ligation, each vector contains a different piece of chromosomal DNA.

FIGURE 18.8   A colony hybridization experiment. Note: If the vector used was a viral vector, the master plate would have viral plaques. The steps shown here could also be followed, in which case the experiment would be called plaque hybridization.

been denatured into single strands. If a bacterial colony contains the rat β-globin gene, the probe will hybridize to the DNA in this colony. Most bacterial colonies are not expected to contain this gene. The unbound probe is then washed away, and the membrane is placed next to X-ray film. If the DNA within a bacterial colony did hybridize to the probe, a dark spot will appear on the film in the corresponding location. The developed film is compared to the master plate to identify bacterial colonies carrying the rat β-globin gene. Following the identification of "hot" colonies, the experimenter can pick them from the plate, and grow bacteria containing the cloned gene.

A scientist might follow several different strategies to obtain a probe for a colony hybridization experiment. In some cases, the gene of interest may have already been cloned. A piece of the cloned gene then can be used as a radiolabeled probe. However, in many circumstances, a scientist may attempt to clone a gene that has never been cloned before. In such a case, one strategy is to use a probe that likely has a sequence similar to that of the gene of interest. For example, if the goal is to clone the β-globin gene from a South American rodent, it is expected that this gene is similar to the rat β-globin gene, which has already been cloned. Therefore, the cloned rat β-globin gene can be used as a DNA probe to "fish out" the β-globin gene from a similar species.

Can a researcher design a DNA probe for a novel type of gene that no one else has ever cloned before from any species? If the protein of interest has been previously isolated from living cells, fragments of the protein may be subjected to amino acid sequence analysis to obtain short amino acid sequences. Because the genetic code is known, a researcher can use this amino acid sequence information to design oligonucleotide probes that would correspond to the coding sequence of the gene that encodes the protein. Alternatively, the purified protein could be used to generate antibodies. In that case, the antibody would recognize clones that express the protein. Like DNA probes, labeled antibodies can be used as probes in a colony hybridization experiment. The use of antibodies as labeling tools in the technique known as Western blotting will be discussed later in this chapter.

Southern Blotting Is Used to Detect DNA Sequences

The technique of **Southern blotting** is used to detect the presence of a particular gene sequence within a mixture of many chromosomal DNA fragments. This method, developed by Edwin Southern in 1975, has several uses. It can determine the copy number of a gene within the genome of an organism. For example, Southern blotting has revealed that rRNA genes are found in multiple copies within a genome, whereas many structural genes are unique. In the study of human genetic diseases, Southern blotting can also detect small gene deletions that cannot be distinguished under the light microscope (see solved problem S4 at the end of this chapter). A common use of Southern blotting is to identify gene families. As discussed in Chapter 8, a gene family is a group of two or more genes derived from the same ancestral gene. The members of a gene family have similar but not identical DNA sequences; they are homologous genes. As we will see, Southern blotting can distinguish the homologous members of a gene family. Similarly, Southern blotting can identify homologous genes in different species. This is called a zoo blot.

Prior to a Southern blotting experiment, a gene or fragment of a gene of interest has been cloned. This cloned DNA is labeled (e.g., radiolabeled) in vitro, and then the labeled DNA is used as a probe to detect the presence of the gene or a homologous gene within a mixture of many DNA fragments. The basis for a Southern blotting experiment is that two DNA fragments will bind to each other only if they have complementary sequences. In such an experiment, the labeled strands from the cloned gene will pair specifically with complementary DNA strands, even if the complementary strands are found within a mixture of many other DNA pieces.

Figure 18.9a shows the Southern blotting procedure. The goal of this experiment is to determine if the chromosomal DNA contains a base sequence complementary to a specific probe. To begin the experiment, the chromosomal DNA is isolated and digested with a restriction enzyme. Because the restriction enzyme cuts the chromosomal DNA at many different sites within the chromosomes, this step produces thousands of DNA pieces of different sizes. Depending on the location of the restriction sites, each gene will be contained on one or more DNA pieces with specific masses. The chromosomal pieces are separated according to their masses by gel electrophoresis. The DNA pieces within the gel are denatured by soaking the gel in a NaOH solution, and then the DNA is transferred (i.e., blotted) to a nylon membrane.

As shown in **Figure 18.9b**, the traditional way to transfer the DNA to the membrane is to lay the gel on blotting paper in a transfer solution. A nylon membrane is placed on top of the gel, and then additional blotting paper and dry paper towels are layered above the nylon membrane. A glass plate and heavy weight are placed on the very top. The liquid from the transfer solution is drawn upward, carrying the DNA from the gel to the nylon membrane. When the DNA enters the nylon membrane, it remains bound to it. Alternatively, another way to transfer DNA is via a transfer apparatus in which the gel is placed on top of a nylon membrane and then an electric field causes the DNA fragments to electrophoretically move from the gel into the membrane (**Figure 18.9c**).

The procedures shown in Figure 18.9b and c result in a nylon membrane that contains many unlabeled DNA fragments that have been denatured and separated according to size. The next step is to determine if any of these unlabeled fragments from the chromosomal DNA contain sequences complementary to a labeled probe. One labeling method is to incorporate the radioisotope ^{32}P into the probe DNA, which labels the phosphate group in the DNA backbone. The nylon membrane, which has the unlabeled chromosomal DNA attached to it, is submerged in a solution containing the radiolabeled probe. If the radiolabeled probe and a fragment of chromosomal DNA are complementary, they will hydrogen bond to each other. Any unbound radiolabeled DNA is then washed away, and the membrane is exposed to X-ray film. Locations where radiolabeled probe has bound will appear as dark bands on the X-ray film.

Important variables in the Southern blotting procedure are the temperature and ionic strength of the hybridization and wash steps. If the procedure is done at very high temperatures and/or at low salt concentrations, then the probe DNA and chromosomal fragment must be very complementary—nearly a perfect match—in order to hybridize. This condition is called **high stringency.** Conditions of high stringency are used to detect close or exact complementarity between the probe and a chromosomal DNA fragment. However, if the temperature is lower and/or the salt concentration is higher, DNA sequences that are not perfectly complementary may hybridize to the probe. This is called **low stringency.** Conditions of low stringency are used to detect homologous genes with DNA sequences that are similar

(a) The steps in Southern blotting

A sample of chromosomal DNA is digested into small fragments with a restriction enzyme.

The fragments are separated by gel electrophoresis, and then denatured.

Gel

As shown in parts b and c, the DNA bands are transferred (blotted) to a nylon membrane.

Nylon membrane

The membrane is placed in a solution containing a radiolabeled probe. The binding can be done under conditions of low or high stringency. Excess probe is washed away, and the membrane is exposed to X-ray film.

High stringency   Low stringency

X-ray film

(b) Transfer step (Traditional method)

Weight
Glass plate
Dry paper towels
Blotting paper
Nylon membrane
Gel
Support for blotting paper and gel
Transfer solution

(c) The transfer step via electrophoresis

Lid
Cathode plate
Blotting paper
Gel
Nylon membrane
Blotting paper
Anode plate
Base

FIGURE 18.9   The technique of Southern blotting. (a) Chromosomal DNA is isolated from a sample of cells and then digested with a restriction enzyme to yield many DNA fragments. The fragments are then separated by gel electrophoresis according to molecular mass and denatured with NaOH. The bands within the gel are transferred to a nylon membrane, a step known as blotting. The blot is then placed in a solution containing a radiolabeled DNA probe. When the DNA probe is hybridized to the mixture of chromosomal DNA fragments at high temperature and/or low ionic strength (i.e., high stringency), the probe recognizes only a single band. When hybridization is performed at lower temperature and/or higher ionic strength (i.e., low stringency), this band and two additional bands are produced. **(b)** The transfer step of DNA blotting (traditional method). **(c)** The transfer step of DNA blotting via electrophoresis.

but not identical to the cloned gene being used as a probe. In the results shown in Figure 18.9a, conditions of high stringency reveal that the gene of interest is unique in the genome. At low stringency, however, two other bands are detected. These results suggest that this gene is a member of a gene family composed of three distinct members.

Northern Blotting Is Used to Detect RNA

Let's now turn our attention to the technique known as Northern blotting, which is used to identify a specific RNA within a mixture of many RNA molecules. (Note: Even though Southern blotting was named after Edwin Southern, Northern blotting was not named after anyone called Northern! It was originally termed reverse-Southern blotting and later became known as Northern blotting.) Northern blotting is used to investigate the transcription of genes at the molecular level. This method can determine if a specific gene is transcribed in a particular cell type, such as nerve or muscle cells, or at a particular stage of development, such as fetal or adult cells. Also, Northern blotting can reveal if a pre-mRNA transcript is alternatively spliced into two or more mRNAs of different sizes.

From a technical viewpoint, Northern blotting is rather similar to Southern blotting with a few important differences. RNA is the starting material and is extracted and purified from living cells. This RNA can be isolated from a particular cell type under a given set of conditions or during a particular stage of development. Any given cell will produce thousands of different types of RNA molecules, because cells express many genes at any given time. After the RNA is extracted from cells and purified, it is loaded onto a gel that separates the RNA transcripts according to their size. The RNAs within the gel are then blotted onto a nylon membrane and probed with a radiolabeled fragment of DNA from a cloned gene. Using this method, RNAs that are complementary to the radiolabeled DNA fragment are detected as dark bands on an X-ray film. In **Figure 18.10**, the probe was complementary to an mRNA that encodes a protein called tropomyosin. In lane 1, the RNA was isolated from smooth muscle cells, in lane 2 from striated muscle cells, and in lane 3 from brain cells. As seen here, smooth and striated muscle cells contain a large amount of this mRNA. This result is expected because tropomyosin plays a role in the regulation of cell contraction. By comparison, brain cells have much less of this mRNA. In addition, we see that the molecular masses of the three mRNAs are slightly different among the three cell types. This observation indicates that the pre-mRNA is alternatively spliced to contain different combinations of exons.

As an alternative to Northern blotting, another way to detect a specific RNA is via RT-PCR, a method discussed previously in the chapter.

Western Blotting Is Used to Detect Proteins

For structural genes, the end result of gene expression is the synthesis of proteins. A particular protein within a mixture of many different protein molecules can be identified by **Western blotting,** a third detection procedure. This method can determine if a specific protein is made in a particular cell type or at a particular stage of development. Technically, this procedure is also similar to Southern and Northern blotting.

In a Western blotting experiment, proteins are extracted from living cells. As with RNA, any given cell will produce many different proteins at any time, because it is expressing many structural genes. After the proteins have been extracted from the cells, they are loaded onto a gel that separates them by molecular mass. To perform the separation step, the proteins are first dissolved in s̲odium d̲odecyl s̲ulfate (SDS), a detergent that denatures proteins and coats them with negative charges. The negatively charged proteins are then separated in a gel made of polyacrylamide. This method of separating proteins is called SDS-PAGE (p̲oly̲a̲crylamide g̲el e̲lectrophoresis).

Following SDS-PAGE, the proteins within the gel are blotted to a nylon membrane. The next step is to use a probe that will recognize a specific protein of interest. An important difference between Western blotting and either Southern or Northern blotting is the use of an **antibody** as a probe, rather than a labeled DNA strand. Antibodies bind to sites known as **epitopes.** An epitope has a three-dimensional structure that is recognized by an antibody. The term **antigen** refers to any molecule that is recognized by an antibody. An antigen contains one or more epitopes. In the case of proteins, an epitope is a short sequence of amino acids. Because the amino acid sequence is a unique feature of each protein, any given antibody will specifically recognize a particular protein. In a Western blot, this is called the primary antibody for that protein.

After the primary antibody has been given sufficient time to recognize the protein of interest, any unbound primary antibody is washed away, and a secondary antibody is added. A secondary antibody is an antibody that binds to the primary antibody. Secondary antibodies, which may be radiolabeled or conjugated to an enzyme, are used for convenience, as secondary antibodies are available commercially. In general, it is easier for researchers to obtain these antibodies from commercial sources rather than label their own primary antibodies. In a Western blotting experiment, the secondary antibody provides a way to detect

Lane 1: Smooth muscle cells

Lane 2: Striated muscle cells

Lane 3: Brain cells

FIGURE 18.10 **The results of Northern blotting.** RNA is isolated from a sample of cells and then separated by gel electrophoresis. The separated RNA bands are blotted to a nylon membrane and then placed in a solution containing a radioactive probe. Following autoradiography, RNA molecules that are complementary to the probe appear as dark bands on the X-ray film.

Lane 1: Red blood cells

Lane 2: Brain cells

Lane 3: Intestinal cells

FIGURE 18.11   **The results of Western blotting.** The black band indicates where the primary antibody has recognized the protein of interest. In lane 1, proteins were isolated from mouse red blood cells. As seen here, the β-globin polypeptide is made in these cells. By comparison, lanes 2 and 3 were samples of protein from brain cells and intestinal cells, which do not synthesize β globin.

FIGURE 18.12   **The results of a gel retardation assay.** The binding of protein to a labeled fragment of DNA retards its rate of movement through a gel. For the results shown in the lane on the right, if the concentration of the DNA fragment was higher than the concentration of the protein, there would be two bands: one band with protein bound (at a higher molecular mass) and one band without protein bound (corresponding to the band found in the left lane).

the protein of interest in a gel blot. For example, it is common for the secondary antibody to be linked to the enzyme alkaline phosphatase. When the colorless dye, XP (5-bromo-4-chloro-3-indolyl-phosphate), is added to the blotting solution, alkaline phosphatase converts the dye to a black compound. Because the secondary antibody binds to the primary antibody, a protein band that is recognized by the primary antibody will become black. In the example shown in **Figure 18.11**, the primary antibody recognizes β globin. In lane 1, proteins were isolated from mouse red blood cells. The black band indicates that the β-globin polypeptide is made in these cells. By comparison, the proteins loaded into lanes 2 and 3 were from brain cells and intestinal cells, respectively. The absence of any bands indicates that these cell types do not synthesize β globin.

Techniques Can Be Used to Detect the Binding of Proteins to DNA Sequences

In addition to detecting the presence of genes and gene products using blotting techniques, researchers often want to study the binding of proteins to specific sites on a DNA molecule. For example, the molecular investigation of transcription factors requires methods that can identify interactions between transcription factor proteins and specific DNA sequences. A technically simple, widely used method for identifying this type of interaction is the **gel retardation assay,** also known as the **band shift assay.** This technique was used originally to study rRNA–protein interactions and quickly became popular after its success in studying protein–DNA interactions in the *lac* operon. Now it is commonly used as a technique to detect interactions between eukaryotic transcription factors and DNA regulatory elements.

The technical basis for a gel retardation assay is that the binding of a protein to a DNA fragment will retard the fragment's ability to move within a polyacrylamide or agarose gel. During electrophoresis, DNA fragments are pulled through the gel matrix toward the bottom of the gel by a voltage gradient. Smaller fragments of DNA migrate more quickly through a gel

matrix than do larger fragments. As you might expect, therefore, the binding of a protein to a DNA fragment will retard the DNA's rate of movement through the gel matrix, because a protein–DNA complex has a higher mass. When comparing a DNA fragment and a protein–DNA complex after electrophoresis, the complex is shifted to a higher band than the DNA alone, because the complex migrates more slowly to the bottom of the gel (**Figure 18.12**).

A gel retardation assay must be carried out under nondenaturing conditions. This means that the buffers and gel cannot cause the unfolding of proteins or the separation of the DNA double helix. This is necessary so that the proteins and DNA retain their proper structure and thereby can bind to each other. The nondenaturing conditions of a gel retardation assay differ from the more common SDS–gel electrophoresis, in which the proteins are denatured by the detergent SDS.

A second method for studying protein–DNA interactions is **DNase I footprinting,** a technique described originally by David Galas and Albert Schmitz in 1978. A DNase I footprinting experiment attempts to identify one or more regions of DNA that interact with a DNA-binding protein. In their original study, Galas and Schmitz identified a site in the *lac* operon, known as the *lac* operator site, that is bound by a DNA-binding protein called the lac repressor.

To understand the basis of a DNase I footprinting experiment, we need to consider the molecular interactions among three things: a fragment of DNA, DNA-binding proteins, and agents that can alter DNA structure. As an example, let's examine the binding of RNA polymerase to a bacterial promoter, a topic discussed in Chapter 12. When RNA polymerase holoenzyme binds to the promoter to form a closed complex, it binds tightly to the −35 and −10 promoter regions, but the protein covers up an even larger region of the DNA. Therefore, holoenzyme bound at the promoter prevents other molecules from gaining access to this region of the DNA. The enzyme DNase I, which can cleave covalent bonds in the DNA backbone, is used as a reagent to determine if a DNA region has a protein bound to it. Galas and Schmitz reasoned that DNase I cannot cleave the DNA at locations where a protein is bound. In this example, it is expected that RNA polymerase holoenzyme will bind to a promoter and protect this DNA region from DNase I cleavage.

Figure 18.13 shows the results of a DNase I footprinting experiment. In this experiment, a sample of many identical DNA fragments, which are 150 bp in length, were radiolabeled at only one end. The sample of fragments was then divided into two tubes: tube A, which did not contain any holoenzyme, and tube B, which contained RNA polymerase holoenzyme. DNase I was then added to both tubes. The tubes were incubated long enough for DNase I to cleave the DNA at a single site in each DNA fragment. Each tube contained many 150 bp DNA fragments, and the cutting in any DNA strand by DNase I occurred randomly. Therefore, the DNase I treatment should produce a mixture of many smaller DNA fragments. A key point, however, is that DNase I cannot cleave the DNA in a region where RNA polymerase holoenzyme is bound. After DNase I treatment, the DNA fragments within the two tubes were separated by gel electrophoresis, and DNA fragments containing the labeled end were detected by autoradiography.

In the absence of RNA polymerase holoenzyme (tube A), DNase I should cleave the 150 bp fragments randomly at any single location. Therefore, a continuous range of sizes occurs (Figure 18.13). However, if we look at the gel lane from tube B, no bands are observed in the size range from 25 to 105 nucleotides. Why are these bands missing? The answer is that DNase I cannot cleave the DNA within the region where the holoenzyme is bound. The middle portion of the 150 bp fragment contains a promoter sequence that binds the RNA polymerase holoenzyme. Along the right side of the gel, the bases are numbered according to their position within the gene. (The site labeled +1 is where transcription begins.) As seen here, RNA polymerase covers up a fairly large region (its "footprint") of about 80 nucleotides, from the −50 region to the +30 region.

As illustrated in this experiment, DNase I footprinting can identify the DNA region that interacts with a DNA-binding protein. In addition to RNA polymerase–promoter binding, DNase I footprinting has been used to identify the binding sites for many other types of DNA-binding proteins, such as eukaryotic transcription factors and histones. This technique has greatly facilitated our understanding of protein-DNA interactions.

A newer method, called chromatin immunoprecipitation (ChIP), can determine if proteins bind to particular DNA sequences in a living cell. Because this method is sometimes done in conjunction with DNA microarrays, we will consider this approach in Chapter 21.

18.3 ANALYSIS AND ALTERATION OF DNA SEQUENCES

As we have seen throughout this textbook, our knowledge of genetics can be largely attributed to an understanding of DNA structure and function. The feature that underlies all aspects of

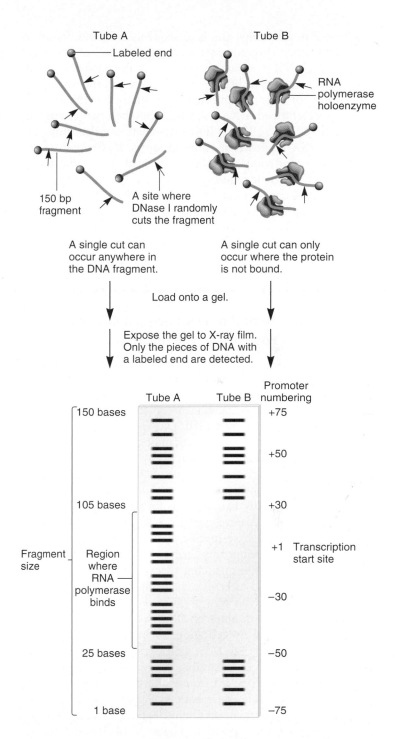

FIGURE 18.13   A DNase I footprinting experiment. Both tubes contained 150 bp fragments of DNA that were incubated with DNase I. Tube B also contained RNA polymerase holoenzyme. The binding of RNA polymerase holoenzyme protected a region of about 80 nucleotides (namely, the −50 region to the +30 region) from DNase I digestion. Note: The promoter numbering convention shown here is the same as that described previously in Figure 12.3.

inherited traits is the DNA sequence. For this reason, analyzing and altering DNA sequences is a powerful approach in understanding genetics. In this last section, we will begin by examining a technique called **DNA sequencing.** This method enables researchers to determine the base sequence of DNA found in genes and other chromosomal regions. It is one of the most important tools for exploring genetics at the molecular level. DNA sequencing is practiced by scientists around the world, and the amount of scientific information contained within experimentally determined DNA sequences has become enormous. In Chapter 20, we will learn how researchers can determine the complete DNA sequence of entire genomes. In Chapter 21, we will consider how computers play an essential role in the storage and analysis of genetic sequences.

Not only can researchers determine DNA sequences, they can also use another technique, known as site-directed mutagenesis, to change the sequence of cloned DNA segments. At the end of this section, we will examine how site-directed mutagenesis is conducted and how it provides information regarding the function of genes.

The Dideoxy Method of DNA Sequencing Is Based on Our Knowledge of DNA Replication

Molecular geneticists often seek to determine base sequences as a first step toward understanding the function and expression of genes. For example, the investigation of genetic sequences has been vital to our understanding of promoters, regulatory elements, and the genetic code itself. Likewise, an examination of sequences has facilitated our understanding of origins of replication, centromeres, telomeres, and transposable elements. In this section, we will examine the technique of DNA sequencing, which is used to determine the base sequence within a DNA strand.

During the 1970s, two methods for DNA sequencing were devised. One method, developed by Allan Maxam and Walter Gilbert, involves the base-specific chemical cleavage of DNA. Another method, developed by Frederick Sanger and colleagues, is known as **dideoxy sequencing.** Because it has become the more popular method of DNA sequencing, we will consider the dideoxy method here.

The dideoxy procedure of DNA sequencing is based on our knowledge of DNA replication but uses a clever twist. As described in Chapter 11, DNA polymerase connects adjacent deoxyribonucleotides by catalyzing a covalent linkage between the 5' phosphate on one nucleotide and the 3' –OH group on the previous nucleotide (refer back to Figure 11.11). Chemists, though, can synthesize deoxyribonucleotides that are missing the –OH group at the 3' position (**Figure 18.14**). These synthetic nucleotides are called **dideoxyribonucleotides.** (Note: The prefix dideoxy- indicates that this sugar has two (di) removed (de) oxygens (oxy) compared to ribose; ribose has –OH groups at both the 2' and 3' positions.) Sanger reasoned that if a dideoxyribonucleotide is added to a growing DNA strand, the strand can no longer grow because the dideoxyribonucleotide is missing the 3' –OH group. The incorporation of a dideoxyribonucleotide into a growing strand is therefore referred to as **chain termination.**

Before describing the steps of this DNA-sequencing protocol, we need to become acquainted with the DNA segments that are used in a sequencing experiment. Prior to DNA sequencing, the segment of DNA to be sequenced must be obtained in large

amounts. This is accomplished using gene cloning, which was described earlier in this chapter. In **Figure 18.15**, the segment of DNA to be sequenced, which we will call the target DNA, was cloned into a vector at a defined location. The target DNA was inserted next to a site in the vector where a primer will bind, which is called the primer-annealing site. The aim of the experiment is to determine the base sequence of the target DNA that has been inserted next to the primer-annealing site. In the experiment shown in Figure 18.15, the recombinant vector DNA has been previously denatured into single strands, usually via heat treatment. Only the strand needed for DNA sequencing is shown here.

With these ideas in mind, we can now consider the steps involved in DNA sequencing (Figure 18.15). First, a sample containing many copies of the single-stranded DNA of the recombinant vector is mixed with primers that will bind to the primer-annealing site. This annealing process is identical to hybridization, because the primer and primer-annealing site are complementary to each other. All four types of deoxyribonucleotides and DNA polymerase are then added to the annealed DNA fragments, and the mixture is divided into four separate tubes. Each of the four tubes has a low concentration of one dideoxyribonucleotide (ddGTP, ddATP, ddTTP, or ddCTP). The tubes are then incubated to allow DNA polymerase to make strands complementary to the target DNA sequence. In the third tube in this example, which contains ddTTP, chain termination can occasionally occur at the sixth or thirteenth position of the newly synthesized DNA strand if a ddT becomes incorporated at either of these sites. Note that the complementary A base is found at the sixth and thirteenth position in the target DNA. Therefore, in this tube, we would expect to make DNA strands that will terminate at the sixth or thirteenth positions. Likewise, in the second tube, which contains ddATP, chain termination will occur only at the second, seventh, eighth, or eleventh positions because a complementary T base is found at the corresponding positions in the target strand.

Within the four tubes, mixtures of DNA strands of different lengths will be made, depending on the number of nucleotides attached to the primer. These DNA strands can be separated according to their lengths by running them on an acrylamide gel. The shorter strands move to the bottom of the gel more quickly than the longer strands. To detect the newly made DNA strands, small amounts of radiolabeled deoxyribonucleotides are added to each reaction. This enables the strands to be visualized as bands when the gel is exposed to X-ray film. In Figure 18.15,

2′, 3′-Dideoxyadenosine triphosphate

FIGURE 18.14   The structure of a dideoxyribonucleotide. Note that the 3' group is a hydrogen rather than an –OH group. For this reason, another nucleotide cannot be attached at the 3' position.

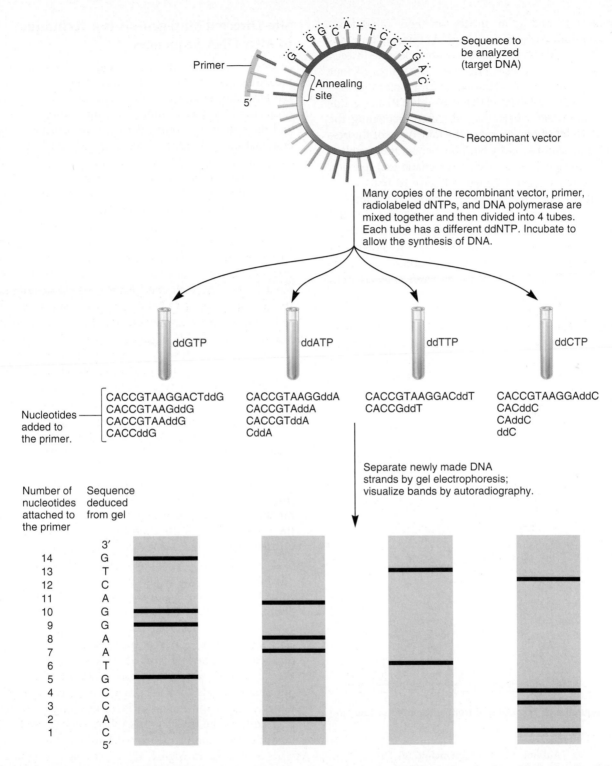

FIGURE 18.15 **The protocol for DNA sequencing by the dideoxy method.** The method shown here begins with single-stranded DNA in which the target DNA has been inserted into a viral vector. Alternatively, double-stranded DNA can be used as the template DNA. However, for the primer to bind, the template must be denatured into single-stranded DNA at the beginning of the experiment.

the DNA strands in the four tubes were run in separate lanes on an acrylamide gel. Because we know which dideoxyribonucleotide was added to each tube, we also know which base is at the very end of each DNA strand separated on this gel. Therefore, we can deduce the DNA sequence that is complementary to the target

DNA by "reading" which base is at the end of every DNA strand and matching this sequence with the length of the strand. Reading the base sequence, from bottom to top, is much like climbing a ladder of bands. For this reason, the sequence obtained by this method is referred to as a **sequencing ladder.**

An important innovation in the method of dideoxy sequencing is **automated sequencing.** Instead of having four separate tubes, with a single type of dideoxyribonucleotide in each tube, automated sequencing uses one tube containing all four types of dideoxyribonucleotides. However, each type of dideoxyribonucleotide (ddGTP, ddATP, ddTTP, and ddCTP) has a different-colored fluorescent label attached. After incubating the target DNA with deoxyribonucleotides, the four types of fluorescent dideoxyribonucleotides, and DNA polymerase, the sample is then loaded into a single lane of a gel. A schematic example of a lane of automated sequencing is shown in **Figure 18.16a.**

In automated sequencing, a sample is loaded at the top of a gel, and then the fragments are separated by electrophoresis. Theoretically, it would be possible to read this sequence directly from the gel. From a practical perspective, however, it is more efficient to automate the procedure using a laser and fluorescence detector. Electrophoresis is continued until each band emerges from the bottom of the gel. As each band comes off, a laser excites the fluorescent dye, and a fluorescence detector records the amount of fluorescence emission. The detector reads the level of fluorescence at four wavelengths, corresponding to the four dyes. An example of the printout from the fluorescence detector is shown in **Figure 18.16b.** As seen here, the peaks of fluorescence correspond to the DNA sequence that is complementary to the target DNA.

(a) Automated sequencing gel

(b) Output from automated sequencing

FIGURE 18.16   Automated DNA sequencing. (a) This diagram schematically depicts a series of bands on a gel; the four colors of the bands occur because each type of dideoxynucleotide is labeled with a different colored fluorescent molecule. **(b)** The mixture of DNA fragments is electrophoresed off the end of the gel. As each band is released from the bottom of the gel, the fluorescent dye is excited by a laser, and the fluorescence emission is recorded by a fluorescence detector. The detector reads the level of fluorescence at four wavelengths, corresponding to the four dyes. As shown in the printout, the peaks of fluorescence correspond to the DNA sequence that is complementary to the target DNA described in Figure 18.15.

Site-Directed Mutagenesis Is a Technique to Alter DNA Sequences

As we have seen, dideoxy sequencing provides a way to determine the base sequence of DNA. To understand how the genetic material functions, researchers often analyze mutations that alter the normal DNA sequence and thereby affect the expression of genes and the outcome of traits. For example, geneticists have discovered that many inherited human diseases, such as sickle-cell disease and hemophilia, involve mutations within specific genes. These mutations provide insight into the function of the genes in unaffected individuals. Hemophilia, for example, involves deleterious mutations in genes that encode blood clotting factors.

Because the analysis of mutations can provide important information about normal genetic processes, researchers often wish to obtain mutant organisms. As we discussed in Chapter 16, mutations can arise spontaneously or can be induced by environmental agents. Mendel's pea plants are a classic example of allelic strains with different phenotypes that arose from spontaneous mutations. X-rays and UV light are physical agents that can cause induced mutations. In addition, experimental organisms can be treated with mutagens that increase the rate of mutations.

More recently, researchers have developed molecular techniques to make mutations within cloned genes or other DNA segments. One widely used method, known as **site-directed mutagenesis,** allows a researcher to produce a mutation at a specific site within a cloned DNA segment. For example, if a DNA sequence is 5'–AAATTTCTTTAAA–3', a researcher can use site-directed mutagenesis to change it to 5'–AAATTTGTTTAAA–3'. In this case, the researcher deliberately changed the seventh base from a C to a G. Why is this method useful? The site-directed mutant can then be introduced into a living organism to see how the mutation affects the expression of a gene, the function of a protein, and the phenotype of an organism.

The first successful attempts at site-directed mutagenesis involved changes in the sequences of viral genomes. These studies were conducted in the 1970s. Mark Zoller and Michael Smith also developed a protocol for the site-directed mutagenesis of DNA that has been cloned into a viral vector. Since these early studies, many approaches have been used to achieve site-directed mutagenesis. **Figure 18.17** describes the general steps in the procedure. Prior to this experiment, the DNA was denatured into single strands; only the single strand needed for site-directed mutagenesis is shown. As in PCR, this single-stranded DNA is referred to as the template DNA, because it is used as a template to synthesize a complementary strand.

As shown in Figure 18.17, an oligonucleotide primer is allowed to hybridize or anneal to the template DNA. The primer, typically 20 or so nucleotides in length, is synthesized chemically. (A shorter version of the primer is shown in Figure 18.17 for simplicity.) The scientist designs the base sequence of the primer. The primer has two important characteristics. First, most of the sequence of the primer is complementary to the site in the DNA where the mutation is to be made. However, a second feature is that the primer contains a region of mismatch where the primer and template DNA are not complementary. The mutation will

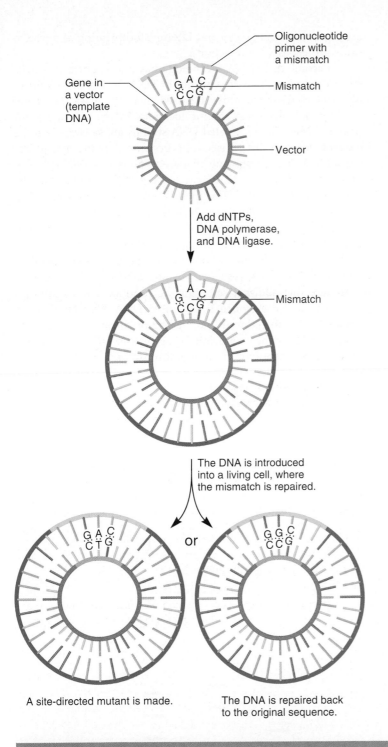

Oligonucleotide primer with a mismatch

Gene in a vector (template DNA)

Mismatch

Vector

Add dNTPs, DNA polymerase, and DNA ligase.

Mismatch

The DNA is introduced into a living cell, where the mismatch is repaired.

or

A site-directed mutant is made.

The DNA is repaired back to the original sequence.

occur in this mismatched region. For this reason, site-directed mutagenesis is sometimes referred to as oligonucleotide-directed mutagenesis.

After the primer and template have annealed, the complementary strand is synthesized by adding deoxyribonucleoside triphosphates (dNTPs), DNA polymerase, and DNA ligase. This yields a double-stranded molecule that contains a mismatch only at the desired location. This double-stranded DNA is then introduced into a bacterial cell. Within the cell, the DNA mismatch will likely be repaired (see Chapter 16). Depending on which base is replaced, this may produce the mutant sequence or the original sequence. Clones containing the desired mutation can be identified by DNA sequencing and used for further studies.

After a site-directed mutation has been made within a cloned gene, its consequences are analyzed by introducing the mutant gene into a living cell or organism. As described earlier, recombinant vectors containing cloned genes can be introduced into bacterial cells. Following transformation into a bacterium, a researcher can study the differences in function between the mutant and wild-type genes and the proteins they encode. Similarly, mutant genes made via site-directed mutagenesis can be introduced into plants and animals, as discussed in Chapter 19.

FIGURE 18.17   The method of site-directed mutagenesis.

Genes→Traits To examine the relationship between genes and traits, researchers can alter gene sequences via site-directed mutagenesis. The altered gene can then be introduced into a living organism to examine how the mutation affects the organism's traits. For example, a researcher could introduce a nonsense mutation into the middle of the *lacY* gene in the *lac* operon. If this site-directed mutant was introduced into an *E. coli* bacterium that did not a have a normal copy of the *lacY* gene, the bacterium would be unable to use lactose. These results indicate that a functional *lacY* gene is necessary for bacteria to have the trait of lactose utilization.

CONCEPTUAL AND EXPERIMENTAL SUMMARY

Recombinant DNA technology, the development of which began in the early 1970s, has revolutionized our understanding of molecular genetics. Researchers can now use restriction enzymes to cut DNA fragments out of their native sites within a chromosome. The DNA fragments can then be ligated to a vector, either a plasmid or viral vector, which propagates within a living host cell. This technology, known as gene cloning, refers to the phenomenon of isolating and making many copies of a gene in order to study it in detail. Similarly, the goal of polymerase chain

reaction (PCR) is to make many copies of the template DNA in a defined region. PCR can achieve gene cloning without the need for vectors and host cells and can amplify the amount of DNA by a staggering amount. Gene cloning has many practical applications, which will be described in Chapter 19.

In some cases, a researcher needs to detect a specific gene or gene product. For example, gene detection procedures, such as colony hybridization, are usually needed to identify particular cloned genes in a DNA library. In addition, detection of gene products can provide information regarding the expression pattern of a gene in particular cell types or during specific stages of development. Three common methods of detection are Southern blotting, Northern blotting, and Western blotting. In Southern blotting, a DNA probe is used to detect the presence of a gene or other DNA sequence within a mixture of many DNA fragments. Under conditions of high stringency, this technique can detect close or exact complementarity between the probe and a chromosomal DNA fragment. Under conditions of low stringency, it can detect homologous genes with DNA sequences that are similar but not identical to the DNA probe. The Northern blotting procedure is used to detect the amount of RNA from a specific gene. In this technique, a DNA probe is used to detect RNA. Both Southern and Northern blotting rely on sequence homology between the DNA probe and the DNA or RNA in the sample. They are called hybridization techniques because the labeled DNA probe forms a hybrid with a specific molecule in the sample. Finally, Western blotting is used to detect the protein product from a particular gene. In this technique, an antibody is used as a probe because antibodies bind to proteins very specifically. In addition to these detection methods, researchers can use a gel retardation assay and DNase I footprinting as a way to study protein-DNA interactions.

Recombinant DNA technology has also enabled researchers to determine the base sequence of DNA. In dideoxy sequencing, dideoxyribonucleotides are used that terminate the action of DNA polymerase at specific locations in the growing DNA strands. When the terminated DNA strands are separated on the basis of their size by gel electrophoresis, this creates a sequencing ladder of DNA fragments that are terminated at defined locations. The DNA sequence can then be determined by reading the bases along the ladder. The method of automated sequencing uses four different fluorescently labeled dideoxyribonucleotides and a fluorescence detector that reads the level of fluorescence at four wavelengths, corresponding to the four dyes. In addition, researchers can answer questions concerning the functional importance of DNA sequences by site-directed mutagenesis, a procedure that allows a researcher to produce a mutation at a particular site in a cloned fragment of DNA. The site-directed mutant can be introduced into a living organism to see how the mutation affects the expression of a gene, the function of a protein, and the phenotype of an organism.

PROBLEM SETS & INSIGHTS

Solved Problems

S1. RNA was isolated from four different cell types and probed with radiolabeled DNA strands from a cloned gene that is called gene *X*. The results are shown here.

Lane 1: Muscle cells
Lane 2: Liver cells
Lane 3: Spleen cells
Lane 4: Nerve cells

Explain the results of this experiment.

Answer: In this Northern blot, a dark band appears in those lanes where RNA was isolated from muscle and spleen cells but not from liver and nerve cells. These results indicate that the muscle and spleen cells contain a significant amount of RNA from gene *X*, but the liver and nerve cells do not. The muscle cells show a single band, whereas the spleen cells show this band plus a second band of lower molecular mass. An interpretation of these results is that the spleen cells can alternatively splice the RNA to produce a second RNA containing fewer exons.

S2. In the Western blotting experiment shown here, proteins were extracted from red blood cells obtained from tissue samples at different stages of human development. An equal amount of total cellular proteins was added to each lane. The primary antibody recognizes the β-globin polypeptide that is found in the hemoglobin protein.

Lane 1: Embryonic red blood cells
Lane 2: Fetal red blood cells
Lane 3: Newborn red blood cells
Lane 4: Adult red blood cells

Explain these results.

Answer: As shown here, the amount of β globin increases during development. Little detectable β globin is produced during embryonic development. The amount increases significantly during fetal development and becomes maximal in the adult. These results indicate that the β-globin gene is "turned on" in later stages of development, leading to the synthesis of the β-globin polypeptide. This experiment illustrates how a Western blot can provide information concerning the relative amount of a specific protein within living cells.

S3. A DNA strand is 3'–ATACGACTAGTCGGGACCATATC–5'. If the primer in a dideoxy sequencing reaction anneals just to the left of this sequence, draw what the sequencing ladder would look like.

Answer:

S4. The human genetic disease phenylketonuria (PKU) involves a defect in a gene that encodes the enzyme phenylalanine hydroxylase. It is inherited as a recessive autosomal disorder. Using a strand of DNA from the nonmutant phenylalanine hydroxylase gene as a probe, a Southern blot was carried out on a PKU patient, one of her parents, and an unaffected, unrelated person. In this example, the DNA fragments were subjected to acrylamide gel electrophoresis, rather than agarose gel electrophoresis, because acrylamide gel electrophoresis is better able to detect small deletions within genes. The following results were obtained:

Suggest an explanation for these results.

Answer: In the affected person, the PKU defect is caused by a small deletion within the PKU gene. The parents are heterozygous for the nonmutant gene and the deletion. The PKU-affected person carries only the deletion, which runs at a lower molecular mass than the nonmutant gene. The unrelated person carries two copies of the nonmutant (nondeleted) gene.

Conceptual Questions

C1. Discuss three important advances that have resulted from gene cloning.

C2. What is a restriction enzyme? What structure does it recognize? What type of chemical bond does it cleave? Be as specific as possible.

C3. Write a double-stranded sequence that is 20 bases long and is palindromic.

C4. What is cDNA? In eukaryotes, how would cDNA differ from genomic DNA?

C5. Explain and draw the structural feature of a dideoxyribonucleotide that causes chain termination.

Experimental Questions

E1. What is the functional significance of sticky ends in a cloning experiment? What type of bonding makes the ends sticky?

E2. Table 18.3 describes the cleavage sites of six different restriction enzymes. After these restriction enzymes have cleaved the DNA, five of them produce sticky ends that can hydrogen bond with complementary sticky ends, as shown in Figure 18.1. The efficiency of sticky ends binding together depends on the number of hydrogen bonds; more hydrogen bonds makes the ends "stickier" and more likely to stay attached. Rank these five restriction enzymes in Table 18.3 (from best to worst) with regard to the efficiency of their sticky ends binding to each other.

E3. Describe the important features of cloning vectors. Explain the purpose of selectable marker genes in cloning experiments.

E4. How does gene cloning produce many copies of a gene?

E5. In your own words, describe the series of steps necessary to clone a gene. Your answer should include the use of a probe to identify a bacterial colony that contains the cloned gene of interest.

E6. What is a recombinant vector? How is a recombinant vector constructed? Explain how X-Gal can be used in a method to identify recombinant vectors that contain segments of chromosomal DNA.

E7. In the experiment of Figure 18.3, would the researchers have been successful if pSC102 had an *Eco*RI site in the middle of the *kan^R* gene? Explain why or why not.

E8. In the experiment of Figure 18.3, would the results have been entirely convincing if they had done only gel electrophoresis (step 10) but not the density gradient centrifugation experiment (step 9)? Can you think of an alternative explanation for the results shown in step 10 that would not require the insertion of the *kan^R* gene into pSC101? In other words, can you think of an explanation for the results shown in step 10 that would not require the construction of a recombinant plasmid? Do the results shown in step 9 rule out your alternative explanation?

E9. A circular plasmid was digested with one or more restriction enzymes, run on a gel, and the following results were obtained:

Construct a restriction map for this plasmid.

E10. If a researcher began with a sample that contained three copies of double-stranded DNA, how many copies would be present after 27 cycles of PCR?

E11. Why is a thermostable form of DNA polymerase (e.g., *Taq* polymerase) used in PCR? Is it necessary to use a thermostable form of DNA polymerase in the techniques of dideoxy DNA sequencing or site-directed mutagenesis?

E12. Reverse transcriptase is an enzyme that uses RNA as template to make a complementary strand of DNA. Experimentally, it is used to make cDNA. Reverse transcriptase can also be used in conjunction with PCR to amplify RNAs. This method is called reverse transcriptase PCR, or RT-PCR. In other words, it is possible to make many copies of double-stranded DNA using RNA as a template. Starting with a sample of RNA that contained the mRNA for the β-globin gene, explain how you could create many copies of the β-globin cDNA using RT-PCR.

E13. Let's suppose you have recently cloned a gene, which we will call gene *X*, from corn. You use a labeled DNA strand from this cloned gene to probe genomic DNA from corn in a Southern blot experiment under conditions of low and high stringency. The following results were obtained:

High stringency   Low stringency

What do these results mean?

E14. What is a DNA library? Do you think this is an appropriate name?

E15. Some vectors used in cloning experiments contain bacterial promoters that are adjacent to unique cloning sites. This makes it possible to insert a gene sequence next to the bacterial promoter and express the gene in bacterial cells. These are called expression vectors. If you wanted to express a eukaryotic protein in bacterial cells, would you clone genomic DNA or cDNA into the expression vector? Explain your choice.

E16. Southern and Northern blotting depend on the phenomenon of hybridization. In these two techniques, explain why hybridization occurs. Which member of the hybrid is labeled?

E17. In Southern, Northern, and Western blotting, what is the purpose of gel electrophoresis?

E18. What is the purpose of a Northern blotting experiment? What types of information can it tell you about the transcription of a gene?

E19. Let's suppose an X-linked gene in mice exists as two alleles, which we will call *B* and *b*. X inactivation, a process in which one X chromosome is turned off, occurs in the somatic cells of female mammals (see Chapter 7). Allele *B* encodes an mRNA that is 900 nucleotides long, while allele *b* contains a small deletion that shortens the mRNA to a length of 825 nucleotides. Draw the expected results of a Northern blot using mRNA isolated from somatic tissue of the following mice:

Lane 1. mRNA from an X^bY male mouse

Lane 2. mRNA from an X^bX^b female mouse

Lane 3. mRNA from an X^BX^b female mouse. Note: The sample taken from the female mouse is not from a clone of cells. It is from a tissue sample, like the one shown at the beginning of the experiment of Figure 7.6.

E20. The method of Northern blotting can be used to determine the amount and size of a particular RNA transcribed in a given cell type. Alternative splicing (discussed in Chapter 15) can produce mRNAs from the same gene that have different lengths. A Northern blot is shown here using a DNA probe that is complementary to the mRNA encoded by a particular gene. The mRNA in lanes 1–4 was isolated from different cell types, and equal amounts of total cellular mRNA were added to each lane.

Lane 1: mRNA isolated from nerve cells
Lane 2: mRNA isolated from kidney cells
Lane 3: mRNA isolated from spleen cells
Lane 4: mRNA isolated from muscle cells

Explain these results.

E21. Southern blotting can be used to detect the presence of repetitive sequences, such as transposable elements, that are present in multiple copies within the chromosomal DNA of an organism. (Note: Transposable elements are described in Chapter 17.) In the Southern blot shown here, chromosomal DNA was isolated from three different strains of baker's yeast, digested with a restriction enzyme, run on a gel, blotted, and then probed with a radioactive DNA probe that is complementary to a transposable element called the *Ty* element.

Southern blot

Explain, in a general way, why the banding patterns are not the same in lanes 1, 2, and 3.

E22. In Chapter 8, Figure 8.10 describes the evolution of the globin gene family. All of the genes in this family are homologous to each other, though the degree of sequence similarity varies depending on the time of divergence. Genes that have diverged more recently have sequences that are more similar. For example, the α_1 and α_2 genes have DNA sequences that are more similar to each other compared to the α_1 and ξ genes. In a Southern blotting experiment, the degree of sequence similarity can be discerned by varying the stringency of hybridization. At high temperature (i.e., high stringency), the probe will recognize genes that are only a perfect or very close match. At a lower temperature, however, homologous genes with lower degrees of similarities can be detected because slight mismatches are tolerated. If a Southern blot was conducted on a sample of human chromosomal DNA and a probe was used that was a perfect match to the β-globin gene, rank the following genes (from those that would be detected at high stringency down to those that would only be detected at low stringency) as they would appear in a Southern blot experiment: Mb, α_1, β, γ_A, δ, and ϵ.

E23. In the Western blot shown here, polypeptides were isolated from red blood cells and muscle cells from two different individuals. One individual was unaffected, and the other individual suffered from a disease known as thalassemia, which involves a defect in hemoglobin. In the Western blot, the gel blot was exposed to an antibody that recognizes β globin, which is one of the polypeptides that constitute hemoglobin. Equal amounts of total cellular proteins were added to each lane.

Lane 1: Proteins isolated from normal red blood cells

Lane 2: Proteins isolated from the red blood cells of a thalassemia patient

Lane 3: Proteins isolated from normal muscle cells

Lane 4: Proteins isolated from the muscle cells of a thalassemia patient

Explain these results.

E24. Let's suppose a researcher was interested in the effects of mutations on the expression of a structural gene that encodes a polypeptide 472 amino acids in length. This polypeptide is expressed in leaf cells of *Arabidopsis thaliana*. Because the average molecular mass of an amino acid is 120 Daltons, this protein has a molecular mass of approximately 56,640 Daltons. Make a drawing that shows the expected results of a Western blot using polypeptides isolated from the leaf cells that were obtained from the following individuals:

Lane 1. A plant homozygous for a nonmutant gene

Lane 2. A plant homozygous for a deletion that removes the promoter for this gene

Lane 3. A heterozygous plant in which one gene is nonmutant and the other gene has a mutation that introduces an early stop codon at codon 112

Lane 4. A plant homozygous for a mutation that introduces an early stop codon at codon 112

Lane 5. A plant homozygous for a mutation that changes codon 108 from a phenylalanine codon into a leucine codon

E25. If you wanted to know if a protein was made during a particular stage of development, what technique would you choose?

E26. Explain the basis for using an antibody as a probe in a Western blotting experiment.

E27. Starting with pig cells and a probe that is a labeled DNA strand from the human β-globin gene, describe how you would clone the β-globin gene from pigs. You may assume that you have available all the materials needed in a cloning experiment. How would you confirm that a putative clone really contained a β-globin gene?

E28. A cloned gene fragment contains a response element that is recognized by a regulatory transcription factor. Previous experiments have shown that the presence of a hormone results in transcriptional activation by this transcription factor. To study this effect, you conduct a gel retardation assay and obtain the following results:

Explain the action of the hormone.

E29. Describe the rationale behind a gel retardation assay.

E30. Certain hormones, such as epinephrine, can increase the levels of cAMP within cells. Let's suppose you can pretreat cells with or without epinephrine and then prepare a cell extract that contains the CREB protein (see Chapter 15 for a description of the CREB protein). You then use a gel retardation assay to analyze the ability of the CREB protein to bind to a DNA fragment containing a cAMP response element (CRE). Describe what the expected results would be.

E31. A gel retardation assay can be used to study the binding of proteins to a segment of DNA. In the experiment shown here, a gel retardation assay was used to examine the requirements for the binding of RNA polymerase II (from eukaryotic cells) to the promoter of a structural gene. The assembly of RNA polymerase II at the core promoter is described in Chapter 12 (Figure 12.13). In this experiment, the segment of DNA containing a promoter sequence was 1,100 bp in length. The fragment was mixed with various combinations of proteins and then subjected to a gel retardation assay.

Lane 1: No proteins added
Lane 2: TFIID
Lane 3: TFIIB
Lane 4: RNA polymerase II
Lane 5: TFIID + TFIIB
Lane 6: TFIID + RNA polymerase II
Lane 7: TFIID + TFIIB + RNA polymerase II

Explain which proteins (TFIID, TFIIB, and/or RNA polymerase II) are able to bind to this DNA fragment by themselves. Which transcription factors (i.e., TFIID and/or TFIIB) are needed for the binding of RNA polymerase II?

E32. As described in Chapter 15 (Figure 15.6), certain regulatory transcription factors bind to the DNA and activate RNA polymerase II. When glucocorticoid binds to the glucocorticoid receptor (a regulatory transcription factor), this changes the conformation of the receptor and allows it ultimately to bind to the DNA. The glucocorticoid receptor binds to a DNA sequence called a glucocorticoid response element (GRE). In contrast, other regulatory transcription factors, such as the CREB protein, do not require hormone binding in order to bind to DNA. The CREB protein can bind to the DNA in the absence of any hormone, but it will not activate RNA polymerase II unless the CREB protein is phosphorylated. (Phosphorylation is stimulated by certain hormones.) The CREB protein binds to a DNA sequence called a cAMP response element (CRE). With these ideas in mind, draw the expected results of a gel retardation assay conducted on the following samples:

Lane 1. A 600 bp fragment containing a GRE, plus the glucocorticoid receptor

Lane 2. A 600 bp fragment containing a GRE, plus the glucocorticoid receptor, plus glucocorticoid hormone

Lane 3. A 600 bp fragment containing a GRE, plus the CREB protein

Lane 4. A 700 bp fragment containing a CRE, plus the CREB protein

Lane 5. A 700 bp fragment containing a CRE, plus the CREB protein, plus a hormone (such as epinephrine) that causes the phosphorylation of the CREB protein

Lane 6. A 700 bp fragment containing a CRE, plus the glucocorticoid receptor, plus glucocorticoid hormone.

E33. In the technique of DNase I footprinting, the binding of a protein to a region of DNA will protect that region from digestion by DNase I by blocking the ability of DNase I to gain access to the phosphodiester linkages in the DNA. In the DNase I footprinting experiment shown here, a researcher began with a sample of cloned DNA 400 bp in length. This DNA contained a eukaryotic promoter for RNA polymerase II. The assembly of RNA polymerase II at the core promoter is described in Chapter 12. For the sample loaded in lane 1, no proteins were added. For the sample loaded in lane 2, the 400 bp fragment was mixed with RNA polymerase II plus TFIID and TFIIB.

Which region of this 400 bp fragment of DNA is bound by RNA polymerase II and TFIID and TFIIB?

E34. Explain the rationale behind a DNase I footprinting experiment.

E35. DNA sequencing can help us to identify mutations within genes. The following data are derived from an experiment in which a normal gene and a mutant gene have been sequenced:

Locate and describe the mutation.

E36. A sample of DNA was subjected to automated DNA sequencing as shown here.

G = Black   T = Red
A = Green   C = Blue

A. What is the sequence of this DNA segment?

B. Discuss the advantages of automated sequencing over conventional sequencing that uses radiolabeled nucleotides.

E37. A portion of the coding sequence of a cloned gene is shown here:

5'-GCCCCCGATCTACATCATTACGGCGAT-3'
3'-CGGGGGCTAGATGTAGTAATGCCGCTA-5'

This portion of the gene encodes a polypeptide with the amino acid sequence alanine–proline–aspartic acid–leucine–histidine–histidine–tyrosine–glycine–aspartic acid. Using the method of site-directed mutagenesis, a researcher wants to change the leucine codon into an arginine codon, with an oligonucleotide that is 19 nucleotides long. What is the sequence of the oligonucleotide that should be made? You should designate the 5' and 3' ends of the oligonucleotide in your answer. Note: The mismatch should be in the middle of the oligonucleotide, and a one-base mismatch is preferable over a two- or three-base mismatch. Use the bottom strand as the template strand for this site-directed mutagenesis experiment.

E38. Let's suppose you want to use site-directed mutagenesis to investigate a DNA sequence that functions as a response element for hormone binding. From previous work, you have narrowed down the response element to a sequence of DNA that is 20 bp in length with the following sequence:

```
5'-GGACTGACTTATCCATCGGT-3'
3'-CCTGACTGAATAGGTAGCCA-5'
```

As a strategy to pinpoint the actual response element sequence, you decide to make 10 different site-directed mutants and then analyze their effects by a gel retardation assay. What mutations would you make? What results would you expect to obtain?

E39. Site-directed mutagenesis can also be used to explore the structure and function of proteins. For example, changes can be made to the coding sequence of a gene to determine how alterations in the amino acid sequence affect the function of a protein. Let's suppose that you are interested in the functional importance of one glutamic acid residue within a protein you are studying. By site-directed mutagenesis, you make mutant proteins in which this glutamic acid codon has been changed to other codons. You then test the encoded mutant proteins for functionality. The results are as follows:

	Functionality
Normal protein	100%
Mutant proteins containing	
Tyrosine	5%
Phenylalanine	3%
Aspartic acid	94%
Glycine	4%

From these results, what would you conclude about the functional significance of the glutamic acid residue within the protein?

Questions for Student Discussion/Collaboration

1. Discuss and make a list of some of the reasons why it would be informative for a geneticist to determine the amount of a gene product. Use specific examples of known genes (e.g., β-globin and other genes) when making your list.

2. Make a list of all the possible genetic questions that could be answered using site-directed mutagenesis.

Note: All answers appear at the website for this textbook; the answers to even-numbered questions are in the back of the textbook.

www.mhhe.com/brookergenetics3e

Visit the Online Learning Center for practice tests, answer keys, and other learning aids for this chapter. Enhance your understanding of genetics with our interactive exercises, quizzes, animations, and much more.

BIOTECHNOLOGY

Biotechnology is broadly defined as the application of technologies that involve the use of living organisms, or products from living organisms, for the development of products that benefit humans. Biotechnology is not a new topic. It began about 12,000 years ago when humans began to domesticate animals and plants for the production of food. Since that time, many species of microorganisms, animals, and plants have become routinely used by people. More recently, the term biotechnology has become associated with molecular genetics. Since the 1970s, molecular genetic tools have provided novel ways to make use of living organisms for products and services. As discussed in Chapter 18, recombinant DNA techniques can be used to genetically engineer microorganisms. In addition, recombinant methods enable the introduction of genetic material into animals and plants. **Genetically modified organisms (GMOs)** have received genetic material via recombinant DNA technology. If an organism has received genetic material from a different species, it is called a **transgenic organism.** A gene from one species that is introduced into another species is called a **transgene.**

In the 1980s, court rulings made it possible to patent recombinant organisms such as transgenic animals and plants. This was one factor that contributed to the growth of many biotechnology industries. In this chapter, we will examine how molecular techniques have expanded our knowledge of the genetic characteristics of commercially important species. We will also discuss examples in which recombinant microorganisms and transgenic animals and plants have been given characteristics that are useful in the treatment of disease or in agricultural production. These include recombinant bacteria that make human insulin, transgenic livestock that produce human proteins in their milk, and transgenic tomatoes with a longer shelf life. In addition, the topics of mammalian cloning and stem cell research will be examined from a technical point of view. Likewise, the current and potential use of human gene therapy—the introduction of cloned genes into living cells in the treatment of a disease—will be addressed. In the process, we will also touch upon some of the ethical issues associated with these technologies.

The sheep named Dolly, which was cloned using genetic material from a somatic cell.

19.1 THE USES OF MICROORGANISMS IN BIOTECHNOLOGY

Microorganisms are used to benefit humans in various ways (Table 19.1). In this section, we will examine how molecular genetic tools have become increasingly important for improving our use of microorganisms. Such tools can produce recombinant microorganisms with genes that have been manipulated in vitro. Why are recombinant organisms useful? Recombinant techniques can improve strains of microorganisms and have even yielded strains that make products not normally produced by microorganisms. For example, human genes have been introduced into bacteria to produce medically important products such as insulin and human growth hormone. As discussed in this section, several recombinant strains are in widespread use. However, in some areas of biotechnology and in some parts of the world, the commercialization of recombinant strains has proceeded very slowly. This is particularly true for applications in which recombinant microorganisms may be used to produce food products or where they are released into the environment. In such cases, safety and environmental concerns and negative public perceptions have slowed or even halted the commercial use of recombinant microorganisms. Nevertheless, molecular genetic research continues, and many biotechnologists expect an expanding use of recombinant microbes in the future.

TABLE 19.1

Common Uses of Microorganisms

Application	Examples
Production of medicines	Antibiotics
	Synthesis of human insulin in recombinant *E. coli*
Food fermentation	Cheese, yogurt, vinegar, wine, and beer
Biological control	Control of plant diseases, insect pests, and weeds
	Symbiotic nitrogen fixation
	Prevention of frost formation
Bioremediation	Cleanup of environmental pollutants such as petroleum hydrocarbons and synthetics that are difficult to degrade

EXPERIMENT 19A

Somatostatin Was the First Human Peptide Hormone Produced by Recombinant Bacteria

During the 1970s, geneticists became aware of the great potential of recombinant DNA technology to produce therapeutic agents to treat certain human diseases. Healthy individuals possess many different genes that encode short peptide and longer polypeptide hormones. Diseases can result when an individual is unable to produce these hormones.

In 1976, Robert Swanson and Herbert Boyer formed Genentech Inc. The aspiration of this company was to engineer bacteria to synthesize useful products, particularly peptide and polypeptide hormones. Their first contract was with researchers Keiichi Itakura and Arthur Riggs. Their intent was to engineer a bacterial strain that would produce somatostatin, a human hormone that inhibits the secretion of a number of other hormones, including growth hormone, insulin, and glucagon. Somatostatin was not chosen for its commercial potential. Instead, it was chosen because the researchers thought it would be technically less difficult than other hormones. Somatostatin is very small (only 14 amino acids long), which requires a short coding sequence, and it can be detected easily.

Before discussing the details of this experiment, let's consider the researchers' approach to constructing the somatostatin gene. To express somatostatin in bacteria, the coding sequence for somatostatin must be inserted next to a bacterial promoter that is contained within a plasmid. Rather than obtaining the gene from the human genome that encodes this 14-amino acid hormone, the researchers took a different approach. As shown below, they chemically synthesized short (single-stranded) oligonucleotides that would hydrogen bond with each other to form the coding sequence for this gene:

Eight separate oligonucleotides (labeled A through H) were synthesized chemically. Due to base complementarity within their sequences, the oligonucleotides hydrogen bonded to each other forming a longer double-stranded DNA fragment with two important characteristics. First, its single-stranded ends (i.e., overhangs) allowed it to be inserted into *Eco*RI and *Bam*HI restriction sites within plasmid DNA. Second, the middle of this DNA fragment encodes the amino acid sequence of the somatostatin peptide hormone. (Today, oligonucleotide synthesis methods have greatly improved, making it unnecessary to synthesize several, short oligonucleotides. Instead, researchers now could make an oligonucleotide that would span the entire length of the somatostatin coding sequence.)

The coding sequence was constructed so that an extra methionine would be located at the amino terminal end of somatostatin. This methionine provided a link between somatostatin and a bacterial protein, β-galactosidase. As discussed in Chapter 14, this enzyme is encoded by the *lacZ* gene. Why was this link necessary? During the course of their experiments, the researchers learned that somatostatin made in bacteria is rapidly degraded by cellular proteases. To prevent this from happening, they linked the somatostatin sequence to the *lacZ* gene encoding β-galactosidase. When this linked gene is expressed in bacteria, a fusion protein is made between somatostatin and β-galactosidase. The fusion protein is not rapidly degraded. The researchers could then separate somatostatin from β-galactosidase by treatment with cyanogen bromide (CNBr), which cleaves polypeptides at the carboxyl terminal side of methionine. Because no methionines are found within the somatostatin sequence itself, this treatment does not degrade somatostatin.

β-galactosidase

Somatostatin

Somatostatin

The steps in their protocol are shown in **Figure 19.1**. As described here, the researchers made a synthetic somatostatin gene that was flanked by unique restriction sites and had a methionine codon at the beginning of the somatostatin-coding sequence. This gene was then inserted into a plasmid at the end of the *lacZ* gene. As a control, they also inserted the somatostatin gene in the wrong orientation (shown on the right of step 3). The plasmid with the wrong orientation should not make any somatostatin. The plasmids were then introduced into *E. coli* cells. The plasmid contained the *lac* promoter (from the *lac* operon), which was induced with IPTG (a nonmetabolizable lactose analogue). In the cells harboring the somatostatin gene in the correct orientation, this would induce the synthesis of a fusion protein containing β-galactosidase and somatostatin. The bacterial cells were then collected by centrifugation and exposed to formic acid and cyanogen bromide (CNBr). As mentioned, the CNBr cleaves polypeptides next to methionine residues. Therefore, this treatment would break the link between β-galactosidase and somatostatin. The amount of somatostatin was then determined by a radioimmunoassay (RIA). (See the Appendix for a description of radioimmunoassay.)

■ THE GOAL

The researchers wanted to produce human somatostatin in a recombinant bacterium.

■ ACHIEVING THE GOAL — FIGURE 19.1   The production of human somatostatin in *E. coli.*

Starting material: A normal *E. coli* strain that was unable to synthesize somatostatin, and bacterial plasmids that carry the *amp^R* gene along with bacterial promoters.

Experimental level	**Conceptual level**

1. Chemically synthesize eight oligonucleotides. When they are mixed together, the complementary oligonucleotides will hybridize to each other.

2. Treat with DNA ligase to covalently link the oligonucleotides, which will form the molecule shown at the right.

3. Using recombinant techniques described in Chapter 18 (Figure 18.2), insert this fragment into a plasmid by digesting the plasmid at unique sites for *Eco*RI and *Bam*HI.

4. Transform the plasmids into *E. coli* by treatment with CaCl₂. The transformed cells are spread on plates containing ampicillin. Grow overnight to obtain bacterial colonies that are ampicillin resistant because they contain the plasmid.

5. Pick cells from a bacterial colony and grow these recombinant bacteria in liquid media; activate the *lac* promoter with isopropyl thiodigalactoside (IPTG).

Note: The *lac* promoter in this plasmid is controlled by the lac repressor (see Chapter 15). IPTG is an inducer that activates transcription of the *lacZ* gene by removing the lac repressor.

6. Place in a tube and centrifuge to obtain a bacterial cell pellet.

7. Resuspend the pellet in 70% formic acid and cyanogen bromide (5 mg/ml).

Note: This breaks open the cells and cleaves polypeptides at methionine residues.

8. Determine the amount of somatostatin using a radioimmunoassay. (See the Appendix for a description of this procedure.)

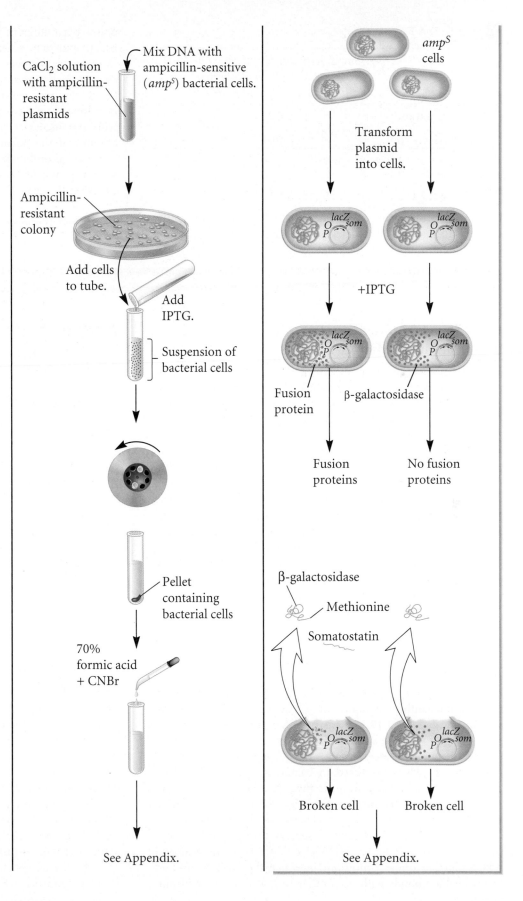

CaCl₂ solution with ampicillin-resistant plasmids

Mix DNA with ampicillin-sensitive (*amp*ˢ) bacterial cells.

Ampicillin-resistant colony

Add cells to tube.

Add IPTG.

Suspension of bacterial cells

Pellet containing bacterial cells

70% formic acid + CNBr

See Appendix.

*amp*ˢ cells

Transform plasmid into cells.

+IPTG

Fusion protein

β-galactosidase

Fusion proteins

No fusion proteins

β-galactosidase

Methionine

Somatostatin

Broken cell

Broken cell

See Appendix.

■ THE DATA

Plasmid Strain	Amount of Somatostatin (detected by RIA) (picograms of somatostatin/ milligram of bacterial proteins)
Correct orientation	8–320*
Incorrect orientation	<0.4

*The amount of somatostatin was determined in several independent experiments.

■ INTERPRETING THE DATA

As shown in the data table, recombinant bacteria carrying the somatostatin gene in the correct orientation produced this hormone. The amount of somatostatin varied from 8 to 320 pico-grams per milligram of bacterial proteins. This variability could be attributed to several factors, including protein degradation, incomplete cyanogen bromide cleavage, and unknown genetic changes in the plasmids during bacterial cell growth. In spite of this variability, the exciting result was the production of a human hormone in recombinant bacteria. By comparison, the plasmid with the incorrect orientation did not produce a significant amount of the hormone. This study was the first demonstration that recombinant bacteria could make products encoded by human genes. At the time, this was a major breakthrough that catalyzed the growth of the biotechnology industry!

A self-help quiz involving this experiment can be found at the Online Learning Center.

Many Important Medicines Are Produced by Recombinant Microorganisms

Since the pioneering study described in the experiment of Figure 19.1, recombinant DNA technology has been used to develop bacterial strains that synthesize several other medical agents, a few of which are described in **Table 19.2**.

In 1982, the U.S. Food and Drug Administration approved the sale of the first genetically engineered drug, human insulin, which was produced by Genentech and marketed by Eli Lilly. In nondiabetic individuals, insulin is produced by the β cells of the pancreas. Insulin functions to regulate several physiological processes, particularly the uptake of glucose into fat and muscle cells. Persons with insulin-dependent diabetes cannot synthesize an adequate amount of insulin due to a loss of their β cells. Prior to 1982, insulin was isolated from the pancreases removed from cattle and pigs. Unfortunately, in some cases, diabetic individuals became allergic to such insulin and had to use expensive combinations of insulin from other animals and human cadavers. Today, people with diabetes can use genetically engineered human insulin to treat their disease.

Insulin is a hormone composed of two polypeptide chains, called the A and B chains. To make this hormone using bacteria, the coding sequences of the A and B chains are placed next to the coding sequence of a native *E. coli* protein, β-galactosidase (**Figure 19.2**). This creates a fusion protein comprising β-galactosidase and the A or B chain. As with somatostatin, this step is necessary because the A and B chains are rapidly degraded when expressed in bacterial cells by themselves. The fusion proteins, however, are not. How are the two fusion proteins used to make human insulin? After the fusion proteins are expressed in bacteria, they can be purified and then treated with cyanogen bromide (CNBr) to separate β-galactosidase from the A or B chain. The A and B chains are then purified and mixed together under conditions in which they will refold and form disulfide bonds with each other to make an active insulin hormone.

TABLE 19.2

Examples of Medical Agents Produced by Recombinant Microorganisms

Drug	Action	Treatment
Insulin	A hormone that promotes glucose uptake	For diabetic patients
Tissue plasminogen activator (TPA)	Dissolves blood clots	For heart attack victims and other arterial occlusions
Superoxide dismutase	Antioxidant	For heart attack victims to minimize tissue damage
Factor VIII	Blood clotting factor	For certain types of hemophilia patients
Renin inhibitor	Lowers blood pressure	For hypertension
Erythropoietin	Stimulates the production of red blood cells	For anemia

Bacterial Species Can Be Used as Biological Control Agents

The term **biological control** refers to the use of living organisms or their products to alleviate plant diseases or damage from environmental conditions. During the past 20 years, interest in the biological control of plant diseases and insect pests as an alternative to chemical pesticides has increased. Biological control agents can prevent disease in several ways. In some cases, nonpathogenic microorganisms are used to compete effectively against pathogenic strains for nutrients or space. In other cases, microorganisms may produce a toxin that inhibits other pathogenic microorganisms or insects without harming the plant.

Biological control can also involve the use of microorganisms living in the field. A successful example is the use of *Agrobacterium radiobacter* to prevent crown gall disease caused by

FIGURE 19.2 **The use of bacteria to make human insulin.**

Genes → Traits The synthesis of human insulin is not a trait that bacteria normally possess. However, genetic engineers can introduce the genetic sequences that encode the A and B chains of human insulin via recombinant DNA technology, yielding bacteria that make these polypeptides as fusion proteins with β-galactosidase. CNBr treatment releases the A and B polypeptides, which are then purified and mixed together under conditions in which they refold and form functional human insulin.

Agrobacterium tumefaciens. The disease gets its name from the large swellings (galls) produced by the plant in response to the bacteria. *A. radiobacter* produces agrocin 84, an antibiotic that kills *A. tumefaciens.* Molecular geneticists have determined that *A. radiobacter* contains a plasmid with genes responsible for agrocin 84 synthesis and resistance. Unfortunately, this plasmid is occasionally transferred from *A. radiobacter* to *A. tumefaciens* during interspecies conjugation. When this occurs, *A. tumefaciens* can gain resistance to agrocin 84. Researchers have identified *A. radiobacter* strains in which this plasmid has been altered genetically to prevent its transfer during conjugation. This conjugation-deficient strain is now used commercially worldwide to prevent crown gall disease.

Another biological control agent is *Bacillus thuringiensis,* usually referred to as Bt (pronounced "bee-tee"). This naturally occurring bacterium produces toxins that are lethal to many caterpillars and beetles that feed on trees, shrubs, flowers, and fruits. Bt is generally harmless to plants and other animals, such as humans, and does not usually harm beneficial insects that act as pollinators. Therefore, it is viewed as an environmentally friendly pesticide. Commercially, Bt is sold in a powder form that is used as a dust or mixed with water as a foliar spray. Bt is then dusted or sprayed on plants that are under attack by caterpillars or beetles so that the pests will ingest the bacteria as they eat the leaves, flowers, or fruits. The toxins produced by Bt bring about paralysis of the insect's digestive tract, causing it to stop feeding within hours and die within a few days. Geneticists have cloned the genes that encode Bt toxins, which are proteins. As discussed later, such genes have been introduced into crops, such as corn, to produce transgenic plants resistant to insect attack.

The Release of Recombinant Microorganisms into the Environment Is Sometimes Controversial

As we have seen, genetically altered strains can have commercial applications in the field. Whether or not a microorganism is recombinant has become an important issue in the use of biological control agents that are released into the environment. Alterations in the genetic characteristics of a microorganism, such as the acquisition of naturally occurring plasmids or mutagenesis by chemical agents and radiation, do not produce bacteria that are classified as recombinant strains. The *A. radiobacter* strain and Bt bacterium fall into this category. This distinction is important from the perspective of both governmental regulation and public perception, even though the organisms produced by nonrecombinant and recombinant approaches may be genetically identical or nearly identical.

Knowledge from molecular genetic research is used to develop both nonrecombinant and recombinant strains with desirable characteristics. Each year, many new strains of nonrecombinant microorganisms are analyzed in field tests for the biological control of plant diseases and insect pests. By comparison, the use of recombinant microorganisms in field tests has proceeded much more slowly. This slow progress is related to increased levels of governmental regulation and, in some cases, to negative public perception of recombinant microorganisms.

As an example of the controversial nature of this topic, let's consider the first field test of a recombinant bacterium, which involved the use of a genetically engineered strain of *Pseudomonas syringae* to control frost damage. Experiments by Steven Lindow and colleagues showed that the formation of ice on the surface of plants is enhanced by the presence of certain bacterial species. These *Ice*⁺ species synthesize cellular proteins that promote ice nucleation—the initiation of ice crystals. Using recombinant DNA technology, Lindow constructed an *Ice*⁻ strain of *P. syringae* in the early 1980s, which lacks the gene responsible for the production of ice-nucleation proteins. When applied to the surface of plants, an *Ice*⁻ strain can compete with and thereby reduce the proliferation of *Ice*⁺ bacteria, thus inhibiting the formation of frost.

Lindow sought approval for field tests of an *Ice*⁻ recombinant strain in Tulelake, California. For several years, these tests were delayed because of a lawsuit from the Foundation on Economic Trends. During that time, Lindow made great efforts to ensure the safety of this project by studying the local environment. He also consulted with local townspeople where the field test was to take place. Initially, the idea was well received by the local residents. However, another company tested similar bacteria on the roof of an Oakland facility without Environmental Protection Agency (EPA) approval. The media reported this incident, and it caused many Tulelake townspeople to become apprehensive about the release of recombinant bacteria. Nevertheless, in 1987, approval was finally granted for the field testing of the recombinant *P. syringae* (**Figure 19.3**).

During the first test on several thousand strawberry plants, the plants were ripped out by vandals. In a second field test, the ability of *Ice*⁻ bacteria to protect potato plants was tested. While some of the plants were destroyed by vandals, the results of this field experiment showed that the *Ice*⁻ bacteria did protect potato plants from frost damage. In addition, soil sampling showed that the recombinant bacteria were contained at the field site and did not proliferate into surrounding areas. Even so, the release of recombinant microorganisms into the environment remains controversial. Since this first test, relatively few recombinant strains have been released. In the case of *Ice*⁻ bacteria, further research was discouraged by a variety of factors, including governmental regulation and the expense of doing additional experiments. The recombinant *Ice*⁻ bacteria were never commercialized.

FIGURE 19.3   **The release of recombinant *P. syringae* in a field test.**

Microorganisms Can Reduce Environmental Pollutants

The term **bioremediation** refers to the use of living organisms or their products to decrease pollutants in the environment. As its name suggests, this is a biological remedy for pollution. During bioremediation via microorganisms, enzymes produced by a microorganism modify a toxic pollutant by altering or transforming its structure. This event is called **biotransformation.** In many cases, biotransformation results in **biodegradation,** in which the toxic pollutant is degraded, yielding less complex, nontoxic metabolites. Alternatively, biotransformation without biodegradation can also occur. For example, toxic heavy metals can often be rendered less toxic by oxidation or reduction reactions carried out by microorganisms. Another way to alter the toxicity of organic pollutants is by promoting polymerization. In many cases, polymerized toxic compounds are less likely to leach from the soil and, therefore, are less environmentally toxic than their parent compounds.

Since the early 1900s, microorganisms have been used in the treatment and degradation of sewage. More recently, the field of bioremediation has expanded into the treatment of hazardous and refractory chemical wastes—chemicals that are difficult to degrade and usually associated with industrial activity. These pollutants include petroleum hydrocarbons, halogenated organic compounds, pesticides, herbicides, and organic solvents. Many new applications that use microorganisms to degrade these pollutants are being tested. The field of bioremediation has been fostered, to a large extent, by better knowledge of how pollutants are degraded by microorganisms, the identification of new and useful strains of microbes, and the ability to enhance bioremediation through genetic engineering.

Molecular genetic technology is key in identifying genes that encode enzymes involved in bioremediation. The characterization of the relevant genes greatly enhances our understanding of how microbes can modify toxic pollutants. In addition, recombinant strains created in the laboratory can be more efficient at degrading certain types of pollutants.

In 1980, in a landmark case (*Diamond v. Chakrabarty*), the U.S. Supreme Court ruled that a live, recombinant microorganism is patentable as a "manufacture or composition of matter." The first recombinant microorganism to be patented was an "oil-eating" bacterium that contained a laboratory-constructed plasmid. This strain can oxidize the hydrocarbons commonly found in petroleum. It grew faster on crude oil than did any of the natural isolates tested. However, it has not been a commercial success because this recombinant strain metabolizes only a limited number of toxic compounds; the number of compounds actually present in crude oil is over 3,000. Unfortunately, the recombinant strain did not degrade many higher-molecular-weight compounds, which tend to persist in the environment.

Currently, bioremediation should be considered a developing industry. This field will need well-trained molecular geneticists to conduct research aimed at elucidating the mechanisms whereby microorganisms degrade toxic pollutants. In the future, recombinant microorganisms may provide an effective way to decrease the levels of toxic chemicals within our environment. However, this approach will require careful studies to demonstrate that recombinant organisms are effective at reducing pollutants and are safe when released into the environment.

19.2 GENETICALLY MODIFIED ANIMALS

As mentioned at the beginning of this chapter, transgenic organisms contain recombinant DNA from another species that has been integrated into their genome. A dramatic example of this is shown in **Figure 19.4**. In this case, the gene that encodes the human growth hormone was introduced into the genome of a mouse. The larger mouse on the right is a transgenic mouse that expresses the human growth hormone gene to a very high level.

The production of transgenic animals is a relatively new area of biotechnology. In recent years, a few transgenic species have reached the stage of commercialization. Many researchers believe that this technology holds great promise for innovations in biotechnology. However, the degree to which this potential may be realized will depend, in part, on the public's concern about the creation and consumption of transgenic species.

In this section, we will begin by examining the mechanisms through which cloned DNA becomes integrated into the chromosomal DNA of animal cells. We will then explore the current techniques used to create transgenic animals and their potential uses. In addition, we will consider mammalian cloning and stem cell research. These topics have received enormous public attention due to the complex ethical issues that they raise.

The Integration of a Cloned Gene into a Chromosome Can Result in Gene Replacement or Gene Addition

In Chapter 18, we considered methods to clone genes. A common approach is to insert a chromosomal gene into a vector and then propagate the vector in living microorganisms such as bacteria or yeast cells. Cloned genes can also be introduced into animal and plant cells. However, to be inherited stably from generation to generation, the cloned gene must become integrated into one (or more) of the chromosomes that reside in the cell nucleus. The integration of cloned DNA into a chromosome occurs by recombination, which is described in Chapter 17.

Figure 19.5 illustrates how a cloned gene can integrate into a chromosome by recombination. If the genome of the host cell carries the same type of gene and if the cloned gene is swapped with the normal chromosomal gene by homologous recombination, then the cloned gene will replace the normal gene within the chromosome (Figure 19.5a). This process is termed **gene replacement.** If the cloned gene has been rendered inactive by mutation and replaces the normal gene, researchers can study how the loss of normal gene function affects the organism. This is called a **gene knockout.** More commonly, a cloned gene may be introduced into a cell and become integrated into the genome by nonhomologous recombination (Figure 19.5b) in a process known as **gene addition.** As shown here, both the cloned and normal genes will be present following gene addition.

Researchers may also introduce a gene that is not already present in a particular species' genome. An interesting example of this type of gene addition involves the production of aquarium fish that "glow," which are aptly named **GloFish®**. The GloFish is a brand of transgenic zebrafish (*Danio rerio*) that glows with bright green, red, or yellow fluorescent color (**Figure 19.6**). How were these fish produced? In 1999, Dr. Zhiyuan Gong and his colleagues started with a gene from jellyfish that encodes a green fluorescent protein and inserted it into the zebrafish genome by gene addition, causing the zebrafish to glow green. By placing the gene next to a gene promoter that would turn the gene on in the presence of certain environmental toxins, their goal was to eventually develop a fish that could be used to detect water pollution. The researchers subsequently collaborated with a company to market the fish for aquarium use. They developed a red fluorescent zebrafish by adding a gene from a sea coral, and a yellow fluorescent zebrafish by adding a variant of the jellyfish gene. In 2003, the GloFish became the first genetically modified organism to be sold as a pet. GloFish have been successfully marketed in several countries including the United States, although the sale of GloFish is banned in California.

FIGURE 19.4 **A comparison between a normal mouse and a transgenic mouse that carries a growth hormone gene.**

Genes → Traits The transgenic mouse (on the right) carries a gene encoding human growth hormone. The introduction of the human gene into the mouse's genome has caused it to grow larger.

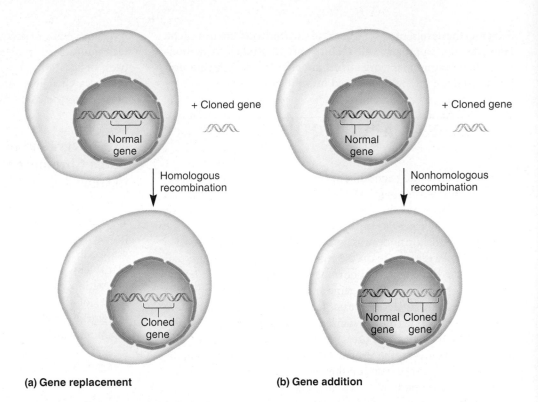

FIGURE 19.5 **The introduction of a cloned gene into a cell can lead to gene replacement or gene addition.** In these two examples, the cloned gene is similar to a normal gene already present in the genome of a cell. (**a**) If the cloned gene undergoes homologous recombination and replaces the normal gene, this event is called gene replacement. (**b**) Alternatively, the cloned gene may recombine nonhomologously at some other chromosomal location, leading to gene addition.

FIGURE 19.6 **The use of gene addition to produce fish that glow.** The aquarium fish shown here, which are named GloFish®, are transgenic organisms that have received a gene from jellyfish or sea corals that encodes a fluorescent protein, causing them to glow green, red, or yellow.

Molecular Biologists Can Produce Mice That Contain Gene Replacements

In bacteria and yeast, which have relatively small genomes, homologous recombination between cloned genes and the host cell chromosome occurs at a relatively high rate, so gene replacement is commonly achieved. Gene replacement is useful when a researcher or biotechnologist wants to make a gene mutation in the laboratory and then introduce the mutant gene into a living organism to compare the effects of the normal and mutant genes on the phenotype. However, in more complex eukaryotes with very large genomes, the introduction of cloned genes into cells is much more likely to result in gene addition rather than gene replacement. For example, when a cloned gene is introduced into a mouse cell, it will undergo homologous recombination less than 0.1% of the time. Most of the time (over 99.9%), gene addition occurs.

To produce mice with gene replacements, molecular biologists have devised laboratory procedures to preferentially select cells in which homologous recombination has occurred. One approach is shown in **Figure 19.7**. The gene of interest is found in a mouse chromosome and is cloned so that it can be manipulated in vitro. The cloned gene is shown at the top of Figure 19.7. The cloning procedure involves two selectable marker genes that influence whether or not mouse embryonic cells can grow in the presence of certain drugs. First, the gene of interest is inactivated by inserting a neomycin-resistance gene (called Neo^R) into the center of its coding sequence. Neo^R provides cells with resistance to neomycin. Next, a thymidine kinase gene, designated TK, is inserted adjacent to the gene of interest but not within the gene itself. The TK gene renders cells sensitive to killing by a drug called gancyclovir.

After the cloned gene has been modified in vitro, it is introduced into mouse embryonic stem cells. How do researchers identify the cells in which homologous recombination has occurred? When the cells are grown in the presence of neomycin and gancyclovir, most nonhomologous recombinants will be killed because they will also carry the TK gene. In contrast, homologous recombinants in which the normal gene has been partially replaced with the cloned gene contain only the Neo^R gene, and so they will be resistant to both drugs. The surviving embryonic cells can then be injected into blastocysts, early embryos that are obtained from a pregnant mouse. In the example shown in Figure 19.7, the embryonic cells are from a mouse with dark fur, and the blastocysts are from a mouse with white fur. The embryonic cells can mix with the blastocyst cells to create a **chimera,** an organism that contains cells from two different individuals. To identify chimeras, the injected blastocysts are reimplanted into the uterus of a female mouse and allowed to develop. When this mouse gives birth, chimeras are identified easily because they contain patches of white and dark fur (Figure 19.7).

Chimeric animals that contain a single-gene replacement can then be mated to other mice to produce offspring that carry the mutant gene. Because mice are diploid, researchers must make two or more subsequent crosses to create a strain of mice that contains both copies of the mutant target gene. However, homozygous strains for a gene knockout cannot be produced if the mutant gene is lethal in the homozygous state.

An alternative method for producing genetically modified mice is to inject the desired gene into a fertilized egg. To conduct this type of experiment, researchers obtain mouse eggs and fertilize them in vitro. Immediately following fertilization, the cloned gene is injected into the sperm pronucleus—the haploid nucleus that has not yet fused with the egg nucleus. The cloned DNA then integrates into the genome, and the two pronuclei fuse to form the diploid nucleus of a zygote. The zygote begins to divide and is introduced into the uterus of a female mouse, where it becomes implanted and grows. As discussed next, genetically modified mice are used in basic research and to study human diseases.

Gene Knockouts and Knockins Are Produced in Mice to Understand Gene Function and Human Disease

As we have seen, researchers can replace a normal mouse gene with one that has been inactivated by the insertion of an antibiotic-resistance gene. As mentioned, when a mouse is homozygous for an inactivated gene, this is called a gene knockout. The inactive mutant gene has replaced both copies of the normal gene. In other words, the function of the normal gene has been "knocked out." By creating gene knockouts, researchers can study how the loss of normal gene function affects the organism. Gene knockouts frequently have specific effects on the phenotype of a mouse, which helps researchers to determine that the function of a gene is critical within a particular tissue or during a specific stage of development. In many cases, however, a gene knockout produces no obvious phenotypic effect at all. One explanation is that a single gene may make such a small contribution to an organism's phenotype that its loss may be difficult to detect. Alternatively, another possible explanation for a lack of observable phenotypic change in a knockout mouse may involve **gene redundancy.** This means that when one type of gene is inactivated, another gene with a similar function may be able to compensate for the inactive gene.

A particularly exciting avenue of gene knockout research is its application in the study and treatment of human disease. How is this useful? Knocking out the function of a gene may provide clues about what that gene normally does. Because humans share many genes with mice, observing the characteristics of knockout mice gives researchers information that can be used to better understand how a similar gene may cause or contribute to a disease in humans. Examples of research areas in which knockout mice have been useful include cancer, obesity, heart disease, diabetes, and many inherited disorders. In 2006, the National Institutes of Health (NIH) launched the Knockout Mouse Project. The goal of this program is to build a comprehensive and publicly available resource of knockout mutations in mice. The NIH Knockout Mouse Project will collaborate with other large-scale efforts to produce mouse knockouts that are under way in Canada, called the North American Conditional Mouse Mutagenesis Project (NorCOMM), and in Europe, called the European Conditional Mouse Mutagenesis Program (EUCOMM). The collective goal of these programs is to create at least one loss-of-function mutation in each of the approximately 20,000 genes in the mouse genome.

In contrast to knockouts, researchers may introduce genes into the mouse genome to study the effects of gene overexpression or to examine the effects of particular alleles, such as those that may cause disease in humans. To accomplish this, researchers can produce a **gene knockin.** A gene knockin is a gene addition in which a gene of interest has been added to a particular site in the mouse genome (**Figure 19.8**). In this example, the cloned gene is inserted into the middle of a segment of DNA from a noncritical site in the mouse genome. The noncritical segment is very long, which allows the transgene to be targeted to that specific, noncritical integration site by homologous recombination after it is

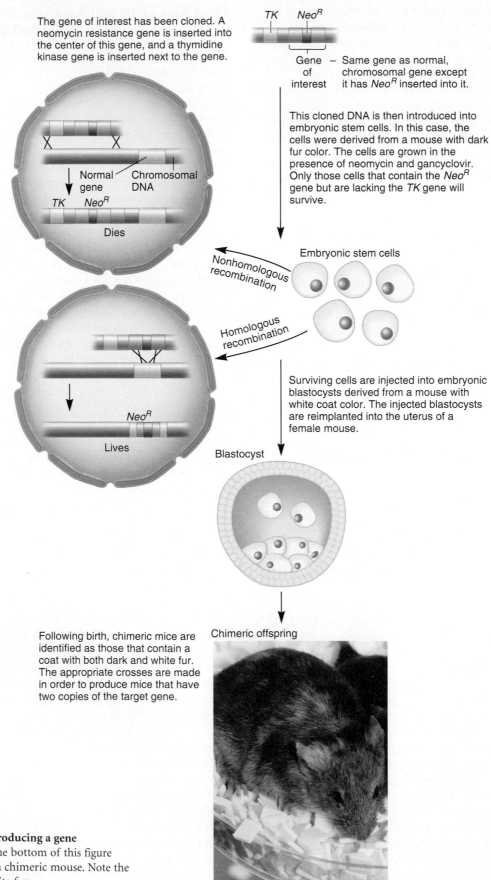

The gene of interest has been cloned. A neomycin resistance gene is inserted into the center of this gene, and a thymidine kinase gene is inserted next to the gene.

TK   *Neo^R*

Gene of interest — Same gene as normal, chromosomal gene except it has *Neo^R* inserted into it.

This cloned DNA is then introduced into embryonic stem cells. In this case, the cells were derived from a mouse with dark fur color. The cells are grown in the presence of neomycin and gancyclovir. Only those cells that contain the *Neo^R* gene but are lacking the *TK* gene will survive.

Normal gene   Chromosomal DNA

TK   *Neo^R*

Dies

Nonhomologous recombination

Embryonic stem cells

Homologous recombination

Neo^R

Lives

Surviving cells are injected into embryonic blastocysts derived from a mouse with white coat color. The injected blastocysts are reimplanted into the uterus of a female mouse.

Blastocyst

Following birth, chimeric mice are identified as those that contain a coat with both dark and white fur. The appropriate crosses are made in order to produce mice that have two copies of the target gene.

Chimeric offspring

FIGURE 19.7   Producing a gene replacement in mice. The bottom of this figure shows a photograph of a chimeric mouse. Note the patches of black and white fur.

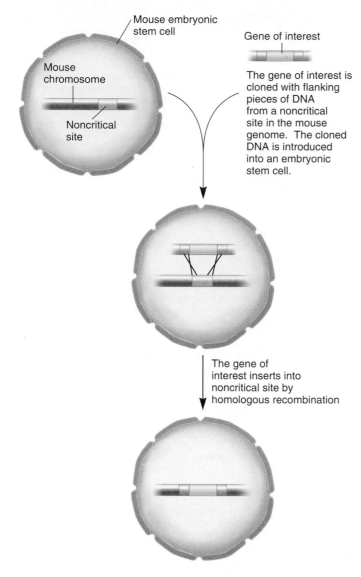

Mouse embryonic stem cell

Mouse chromosome

Noncritical site

Gene of interest

The gene of interest is cloned with flanking pieces of DNA from a noncritical site in the mouse genome. The cloned DNA is introduced into an embryonic stem cell.

The gene of interest inserts into noncritical site by homologous recombination

FIGURE 19.8   Producing a gene knockin in mice. This simplified diagram illustrates the strategy of producing a gene knockin. The gene of interest is cloned, and the cloned gene is flanked by DNA sequences that are homologous to a noncritical site in the mouse genome. When introduced into a mouse embryonic stem cell, the gene will insert into the genome by homologous recombination, which is an example of a targeted gene addition.

introduced into mouse cells. Such gene knockins tend to result in a more consistent level of expression of the transgene compared to gene additions that may occur randomly in another place in the genome. Also, because a targeted transgene is not interfering with a critical locus, the researcher can be more certain that any resulting phenotypic effect is due to the expression of the transgene.

To study human diseases, researchers have produced strains of transgenic mice that harbor both gene knockouts and gene knockins. A strain of mice engineered to carry a mutation that is analogous to a disease-causing mutation in a human gene is termed a mouse model. As an example, let's consider sickle-cell disease, which is due to a mutation in the human

β-globin gene (refer back to Figure 4.7). This gene encodes a polypeptide called β globin; adult hemoglobin is composed of both α-globin and β-globin polypeptides. When researchers produced a gene knockin by introducing the mutant human β-globin gene into mice, the resulting mice showed only mild symptoms of the disease. However, Chris Paszty and Edward Rubin produced a mouse model with multiple gene knockins and gene knockouts. In particular, the mice had gene knockins for the normal human α-globin gene and the mutant β-globin gene from patients with sickle-cell anemia. The strain also had gene knockouts of the mouse α-globin gene and β-globin gene. Therefore, these mice made adult hemoglobin just like people with sickle-cell disease, but they did not produce any normal mouse hemoglobin. These transgenic mice exhibit the major features of sickle-cell disease—sickled red blood cells, anemia, and multiorgan pathology. They have been useful as a model to study the disease and to test potential therapies for treatment.

Biotechnology Holds Promise in Producing Transgenic Livestock

The technology for creating transgenic mice has been extended to other animals, and much research is under way to develop transgenic species of livestock, including fish, sheep, pigs, goats, and cattle. For some farmers, the ability to modify the characteristics of livestock via the introduction of cloned genes is an exciting prospect. In addition, work is currently under way to produce genetically modified pigs that are expected to be resistant to rejection mechanisms that occur following organ transplantation to humans. These strains may in time become a source of organs or cells for patients.

A novel avenue of research involves the production of medically important proteins in the mammary glands of livestock. This approach is sometimes called **molecular pharming.** (The term is also used to describe the manufacture of medical products by agricultural plants.) As shown in **Table 19.3,** several human proteins have been successfully produced in the milk of domestic livestock. Compared to the production of proteins in bacteria, one advantage is that certain proteins are more likely to function properly when expressed in mammals. This may be due to covalent modifications, such as the attachment of carbohydrate groups, which occur in eukaryotes but not in bacteria. In addition, certain proteins may be degraded rapidly or folded improperly when expressed in bacteria. Furthermore, the yield of recombinant proteins in milk can be quite large. Dairy cows, for example, produce about 10,000 liters of milk per year per cow. In most cases, a transgenic cow can produce approximately 1 g of the transgenic protein per liter of milk.

To introduce a human gene into an animal so that the encoded protein will be secreted into its milk, the strategy is to insert the gene next to a milk-specific promoter. Eukaryotic genes often are expressed in a tissue-specific fashion. In mammals, certain genes are expressed specifically within the mammary gland so that their protein product will be secreted into the milk. Examples of milk-specific genes include genes that encode milk proteins such as β-lactoglobulin, casein, and whey acidic protein.

TABLE 19.3

Proteins That Can Be Produced in the Milk of Domestic Animals

Protein	Host	Use
Lactoferrin	Cattle	Used as an iron supplement in infant formula
Tissue plasminogen activator (TPA)	Goat	Dissolves blood clots
Antibodies	Cattle	Used to combat specific infectious diseases
α-1-antitrypsin	Sheep	Treatment of emphysema
Factor IX	Sheep	Treatment of certain inherited forms of hemophilia
Insulin-like growth factor	Cattle	Treatment of diabetes

To express a human gene that encodes a protein hormone into a domestic animal's milk, the promoter for a milk-specific gene is linked to the coding sequence for the human gene (**Figure 19.9**). The DNA is then injected into an oocyte, where it will be integrated into the genome. The fertilized oocyte is then implanted into the uterus of a female animal, which later gives birth to a transgenic offspring. If the offspring is a female, the protein hormone encoded by the human gene will be expressed within the mammary gland and secreted into the milk. The milk can then be obtained from the animal, and the human hormone isolated.

Researchers Have Succeeded in Cloning Mammals from Somatic Cells

We now turn our attention to cloning as a way to genetically manipulate mammals. The term cloning has many different meanings. In Chapter 18, we discussed gene cloning, which involves methods that produce many copies of a gene. The cloning of an entire organism is a different matter. **Reproductive cloning** refers to methods that produce two or more genetically identical individuals. This happens occasionally in nature; identical twins are genetic clones that began from the same fertilized egg. Similarly, researchers can take mammalian embryos at an early stage of development (e.g., the two-cell to eight-cell stage), separate the cells, implant them into the uterus, and obtain multiple births of genetically identical individuals.

In the case of plants, cloning is an easier undertaking, as we will explore later in the chapter. Plants can be cloned from somatic cells. In most cases, it is easy to take a cutting from a plant, expose it to growth hormones, and obtain a separate plant that is genetically identical to the original. However, this approach has not been possible with mammals. For several decades, scientists believed that chromosomes within the somatic cells of mammals had incurred irreversible genetic changes that render them unsuitable for cloning. However, this hypothesis has proven to be incorrect. In 1997, Ian Wilmut and his colleagues at the Roslin Institute in Scotland announced that a sheep, named Dolly, had been cloned using the genetic material from somatic cells.

Human hormone gene

Using recombinant DNA technology (described in Chapter 18), clone a human hormone gene next to a sheep β-lactoglobulin promoter. This promoter is functional only in mammary cells so that the protein product is secreted into the milk.

β-lactoglobulin promoter

Plasmid vector

Inject this DNA into a sheep oocyte. The plasmid DNA will integrate into the chromosomal DNA, resulting in the addition of the human hormone gene into the sheep's genome.

Implant the fertilized oocyte into a female sheep, which then gives birth to a transgenic sheep offspring.

Transgenic sheep

Obtain milk from female transgenic sheep. The milk contains a human hormone.

Purify the hormone from the milk.

FIGURE 19.9   Strategy for expressing human genes in a domestic animal's milk. The β-lactoglobulin gene is normally expressed in mammary cells, whereas the human hormone gene is not. To express the human hormone gene in milk, the promoter from the milk-specific gene in sheep is linked to the coding sequence of the human hormone gene. In addition to the promoter, a short signal sequence may also be necessary so that the protein will be secreted from the mammary cells and into the milk.

Genes → Traits By using genetic engineering, researchers can give sheep the trait of producing a human hormone in their milk. This hormone can be purified from the milk and used to treat humans.

How was Dolly created? As shown in **Figure 19.10**, the researchers removed mammary cells from an adult female sheep and grew them in the laboratory. The researchers then extracted the nucleus from an egg cell of a different sheep and used electrical pulses to fuse the diploid mammary cell with the enucleated egg cell. After fusion, the zygote began embryonic development and the resulting embryo was implanted into the uterus of a surrogate mother sheep. One hundred and forty-eight days later, Dolly was born.

While Dolly was clearly a clone of the initial adult female sheep, tests conducted when she was three years old suggested that she was "genetically older" than her actual age indicated. As mammals age, chromosomes in somatic cells tend to shorten from the telomeres—the ends of eukaryotic chromosomes. Therefore, older individuals have shorter chromosomes in their somatic cells compared to younger ones. This shortening does not seem to occur in the cells of the germ line, however. When researchers analyzed the chromosomes in Dolly's somatic cells when she was about three years old, the lengths of her chromosomes were consistent with a sheep that was significantly older, say, nine or 10 years old. The sheep that donated the somatic cell that produced Dolly was six years old, and her mammary cells had been grown in culture for several cell doublings before a mammary cell was fused with an oocyte. This led researchers to suspect that Dolly's shorter telomeres were a result of chromosome shortening in the somatic cells of the sheep that donated the nucleus. In 2003, the Roslin Institute announced the decision to euthanize six-year-old Dolly after an examination showed progressive lung disease. Her death has raised concerns among experts that the techniques used to produce Dolly could have caused premature aging.

With regard to telomere length, research in mice and cattle has shown different results; the telomeres of these cloned animals appear to be the correct length. For example, cloning was conducted on mice via the method described in Figure 19.10 for six consecutive generations. The cloned mice of the sixth generation had normal telomeres. Further research will be necessary to determine if cloning via somatic cells has an effect on the length of telomeres in subsequent generations. However, other studies in mice point to various types of genetic flaws in cloned animals. For example, Rudolf Jaenisch and his colleagues used DNA microarray technology (described in Chapter 21) and analyzed the transcription patterns of over 10,000 genes in cloned mice. Up to 4% of those genes were not expressed normally. Furthermore, research has shown that cloned mice die at a younger age than their naturally bred counterparts.

Mammalian cloning is still at an early stage of development. Nevertheless, the breakthrough of creating Dolly has shown that it is technically possible. In recent years, cloning from somatic cells has been achieved in several mammalian species, including sheep, cattle, mice, goats, and pigs. In 2002, the first pet was cloned, which was named Carbon Copy, also called Copy Cat (**Figure 19.11**). Mammalian cloning has the potential to provide many practical applications. With regard to livestock, cloning would enable farmers to use the somatic cells from their best individuals to create genetically homogeneous herds. This could

FIGURE 19.10 **Protocol for the successful cloning of sheep.**
Genes → Traits Dolly was (almost) genetically identical to the sheep that donated a mammary cell to create her. Dolly and the donor sheep were (almost) genetically identical in the same way that identical twins are; they carried the same set of genes and looked remarkably similar. However, they may have had minor genetic differences due to possible variation in their mitochondrial DNA and may have exhibited some phenotypic differences due to maternal effect or imprinted genes.

FIGURE 19.11   Carbon Copy, the first cloned pet. The animal shown here was produced using a procedure similar to the one shown in Figure 19.10.

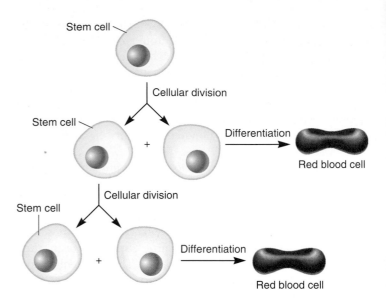

FIGURE 19.12   Growth pattern of stem cells. The two main traits that stem cells exhibit are an ability to divide and an ability to differentiate. When a stem cell divides, one of the two cells remains a stem cell, and the other daughter cell differentiates into a specialized cell type.

be advantageous with regard to agricultural yield, although such a genetically homogeneous herd may be more susceptible to rare diseases.

Though some people are concerned about the practical uses of cloning agricultural species, a majority have become very concerned with the possibility of human cloning. This prospect has raised a host of serious ethical questions. For example, some people feel that it is morally wrong and threatens the basic fabric of parenthood and family. Others feel that it is a technology that could offer a new avenue for reproduction, one that could be offered to infertile couples, for example. In the public sector, the sentiment toward human cloning has been generally negative. Many countries have issued an all-out ban on human cloning, while others permit limited research in this area. Because the technology for cloning exists, our society will continue to wrestle with the legal and ethical aspects of cloning as it applies not only to animals but also to people.

Stem Cells Have the Ability to Divide and Differentiate into Different Cell Types

Stem cells supply the cells that construct our bodies from a fertilized egg. In adults, stem cells also replenish worn-out or damaged cells. To accomplish this task, stem cells have two common characteristics. First, they have the capacity to divide, and second, they can differentiate into one or more specialized cell types. As

shown in **Figure 19.12**, the two daughter cells produced from the division of a stem cell can have different fates. One of the cells may remain an undifferentiated stem cell, while the other daughter cell can differentiate into a specialized cell type. With this type of asymmetric division/differentiation pattern, the population of stem cells remains constant, yet the stem cells provide a population of specialized cells. In the adult, this type of mechanism is needed to replenish cells that have a finite life span, such as skin epithelial cells and red blood cells.

In mammals, stem cells are commonly categorized according to their developmental stage and their ability to differentiate (**Figure 19.13**). The ultimate stem cell is the fertilized egg, which, via multiple cellular divisions, can give rise to an entire organism. A fertilized egg is considered **totipotent,** because it can give rise to all the cell types in the adult organism. The early mammalian embryo contains **embryonic stem cells (ES cells),** which are found in the inner cell mass of the blastocyst. The blastocyst is the stage of embryonic development prior to uterine implantation—the preimplantation embryo. Embryonic stem cells are **pluripotent,** which means they can differentiate into almost every cell type of the body. However, a single embryonic stem cell has lost the ability to produce an entire, intact individual.

During the early fetal stage of development, the germ-line cells found in the gonads also are pluripotent. These cells are called **embryonic germ cells (EG cells).** Interestingly, certain types of human cancers called teratocarcinomas arise from cells that are pluripotent. These bizarre tumors contain a variety of tissues including cartilage, neuroectoderm, muscle, bone, skin, ganglionic structures, and primitive glands. Due to the seemingly embryonic origin of teratocarcinoma cells, these cells are termed **embryonic carcinoma cells (EC cells).**

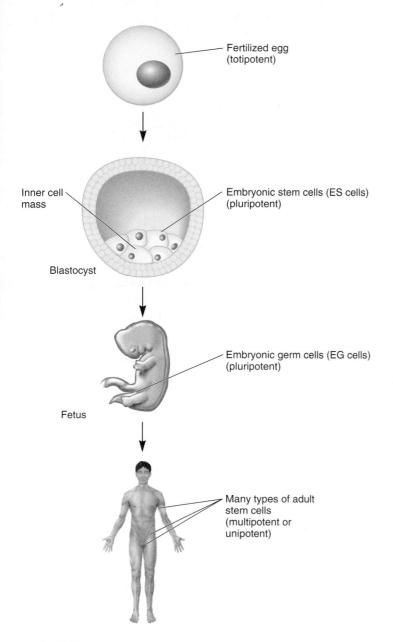

FIGURE 19.13 **Occurrence of stem cells at different stages of development.**

TABLE 19.4

Potential Uses of Stem Cells to Treat Diseases

Cell/Tissue Type	Disease Treatment
Neural	Implantation of cells into the brain to treat Parkinson disease
	Treatment of injuries such as spinal cord injuries
Skin	Treatment of burn victims and other types of skin disorders
Cardiac	Repair of heart damage associated with heart attacks
Cartilage	Repair of joints damaged by injury or arthritis
Bone	Repair of damaged bone or replacement with new bone
Liver	Repair or replacement of liver tissue that has been damaged by injury or disease
Skeletal muscle	Repair or replacement of damaged muscle

into a T cell, B cell, natural killer cell, or dendritic cell. Other stem cells found in the adult seem to be **unipotent.** For example, primordial germ cells in the testis differentiate only into a single cell type, the sperm.

What are the potential uses of stem cells? Interest in stem cells centers around two main areas. Because stem cells have the capacity to differentiate into multiple cell types, the study of stem cells may help us to understand basic genetic mechanisms that underlie the process of development, the details of which are described in Chapter 23. A second compelling reason why people have become interested in stem cells is the potential to treat human diseases or injuries that cause cell and tissue damage. This application has already become a reality in certain cases. For example, bone marrow transplants are used to treat patients with certain forms of cancers. Such patients may be given radiation treatments that destroy their immune systems. When these patients are injected with bone marrow from a healthy person, the stem cells within the transplanted marrow have the ability to proliferate and differentiate within their bodies and provide them with a functioning immune system.

Renewed interest in the use of stem cells in the potential treatment of many other diseases was fostered in 1998 by studies of two separate teams, headed by James Thomson and John Gearhart, showing that embryonic cells, either ES or EG cells, can be successfully propagated in the laboratory. As mentioned, ES and EG cells are pluripotent and therefore have the capacity to produce many different kinds of tissue. As shown in **Table 19.4,** embryonic cells could potentially be used to treat a wide variety of diseases associated with cell and tissue damage. By comparison, it would be difficult, based on our modern knowledge, to treat these diseases with adult stem cells because of the inability to locate most types of adult stem cells within the body and successfully grow them in the laboratory. Even hematopoietic stem cells are elusive. In the bone marrow, about one cell in 10,000 is a stem cell, yet that is enough to populate all of the blood and

As mentioned, adults also contain stem cells, but these are thought to be multipotent or unipotent. A **multipotent** stem cell can differentiate into several cell types but far fewer than an embryonic stem cell. For example, hematopoietic stem cells (HSCs) found in the bone marrow can supply cells that populate two different tissues, namely, the blood and lymphoid tissues (**Figure 19.14**). Furthermore, each of these tissues contains several cell types. Multipotent hematopoietic stems cells can follow a pathway in which cell division produces a myeloid progenitor cell, which can then differentiate into a red blood cell, megakaryocyte, basophil, monocyte, eosinophil, neutrophil, or dendritic cell. Alternatively, an HSC can follow a path in which it becomes a lymphoid progenitor cell, which then differentiates

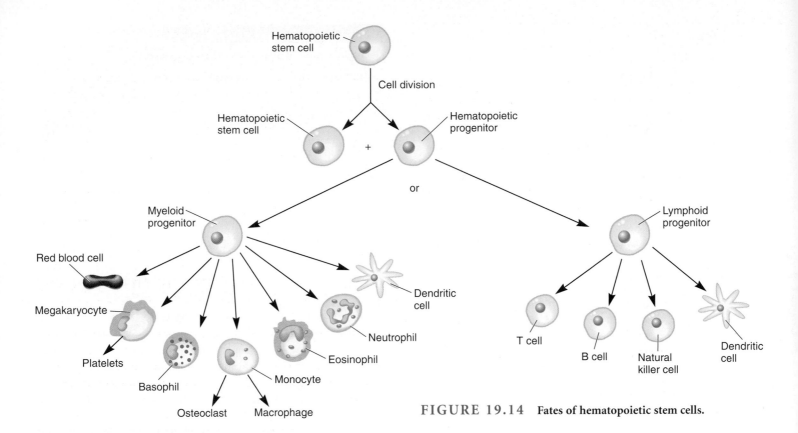

FIGURE 19.14 Fates of hematopoietic stem cells.

lymphoid cells of the body. The stem cells of most other adult tissues are equally difficult to locate, if not more so. In addition, with the exception of stem cells in the blood, other types of stem cells in the adult body are difficult to remove in sufficient numbers for transplantation. By comparison, ES and EG cells are easy to identify and have the great advantage of rapid growth in the laboratory. For these reasons, ES and EG cells offer a greater potential for transplantation, based on our current knowledge of stem cell biology.

For ES or EG cells to be used in transplantation, researchers will need to derive methods that cause them to differentiate into the appropriate type of tissue. For example, if the goal was to repair a spinal cord injury, ES or EG cells would need the appropriate cues that would cause them to differentiate into neural tissue. At present, much research is needed to understand and potentially control the fate of ES or EG cells. Currently, researchers speculate that a complex variety of factors determines the developmental fates of stem cells. These include internal factors within the stem cells themselves, as well as external factors such as the properties of neighboring cells and the presence of hormones and growth factors in the environment.

From an ethical perspective, the primary issue that raises debate is the source of the stem cells for research and potential treatments. Most ES cells have been derived from human embryos that were produced from in vitro fertilization and were subsequently not used. Most EG cells are obtained from aborted fetuses. Some feel that it is morally wrong to use such tissue in research and/or the treatment of disease or fear that this technology could lead to intentional abortions for the sole purpose

of obtaining fetal tissues for transplantation. Alternatively, others feel that the embryos and fetuses that provide the ES and EG cells are not going to become living individuals, and therefore, it is beneficial to study these cells and use them in a positive way to treat human diseases and injury. It is not clear whether these two opposing viewpoints can reach a common ground.

As a compromise, many governments have enacted laws that limit or prohibit the use of embryos or fetuses to obtain stem cells, yet permit the use of stem cell lines that are already available in research laboratories. In the United States, for example, recent interpretations of federal laws prohibit the use of government funding for research projects that involve the destruction of embryos to obtain stem cells. As of 2007, government-sponsored research can be done on the 60 or so stem cell lines that were created prior to this legislation, and nongovernment-sponsored research is not subject to this limitation.

If stem cells could be obtained from adult cells and propagated in the laboratory, an ethical dilemma may be avoided because most people do not have serious moral objections to current procedures such as bone marrow transplantation. In 2006, work by Shinya Yamanaka and colleagues showed that adult mouse fibroblasts (a type of connective tissue cell) could become pluripotent via the injection of four different genes that encode transcription factors. In 2007, Yamanaka's laboratory and two other research groups were able to show that such induced pluripotent stem cells (iPS) can differentiate into all cell types when injected into mouse blastocysts and grown into baby mice. Though further research is still needed, these recent results indicate that adult cells can be reprogrammed to become embryonic stem cells.

19.3 GENETICALLY MODIFIED PLANTS

As we have seen, researchers have succeeded in making genetically modified animals for a variety of reasons. In this section, we will examine the methods that scientists follow to make transgenic plants.

A relatively new area of research is the use of transgenic species in agriculture. For centuries, agriculture has relied on selective breeding programs to produce plants and animals with desirable characteristics. For agriculturally important species, this often means the production of strains that are larger, have disease resistance, and yield high-quality food. Agricultural scientists can now complement traditional breeding strategies with modern molecular genetic approaches. In the mid-1990s, genetically modified crops first became commercialized. Since that time, their use has progressively increased. In 2006, roughly 20% of all agricultural crops were transgenic. Worldwide, more than 100 million hectares (247 million acres) of transgenic crops were planted. In this section, we will discuss some current and potential uses of transgenic plants in agriculture.

Agrobacterium tumefaciens and Other Methods Can Be Used to Make Transgenic Plants

As we have seen, the introduction of cloned genes into embryonic cells can produce transgenic animals. The production of transgenic plants is somewhat easier, because certain plant cells are totipotent, which means they are capable of developing into an entire organism. Therefore, a transgenic plant can be made by the introduction of cloned genes into somatic tissue, such as the tissue of a leaf. After the cells of a leaf have become transgenic, an entire plant can be regenerated by the treatment of the leaf with plant growth hormones, which cause it to form roots and shoots.

Molecular biologists can use the bacterium *Agrobacterium tumefaciens*, which naturally infects plant cells, to produce transgenic plants. A plasmid from the bacterium, known as the **Ti plasmid** (Tumor-inducing plasmid), naturally induces tumor formation after a plant has been infected (**Figure 19.15a**). A segment of the plasmid DNA, known as **T DNA** (for transferred DNA), is transferred from the bacterium to the infected plant cells. The T DNA from the Ti plasmid becomes integrated into the chromosomal DNA of the plant cell by recombination. After this occurs, genes within the T DNA that encode plant growth hormones cause uncontrolled plant cell growth. This produces a cancerous plant growth known as a crown gall tumor (**Figure 19.15b**).

Because *A. tumefaciens* inserts its T DNA into the chromosomal DNA of plant cells, it can be used as a vector to introduce cloned genes into plants. Molecular geneticists have been able to modify the Ti plasmid to make this an efficient process. The T DNA genes that cause the development of a gall have been identified. Fortunately for genetic engineers, when these genes are deleted, the T DNA is still taken up into plant cells and integrated within the plant chromosomal DNA. However, a gall does not form. In addition, geneticists have inserted selectable marker genes into the T DNA to allow selection of plant cells that have taken up the T DNA. A gene that provides resistance to the antibiotic kanamycin is a commonly used selectable marker. The Ti plasmids used in cloning experiments are also modified to contain unique restriction sites for the convenient insertion of any gene.

Figure 19.16 shows the general strategy for producing transgenic plants via T DNA-mediated gene transfer. A gene of interest is inserted into a genetically engineered Ti plasmid and then transformed into *A. tumefaciens*. Plant cells are exposed to the transformed *A. tumefaciens*. After allowing time for infection, the plant cells are exposed to the antibiotics kanamycin and carbenicillin. Carbenicillin kills *A. tumefaciens*, and kanamycin kills any plant cells that have not taken up the T DNA with the antibiotic-resistance gene. Therefore, the only surviving cells are those plant cells that have integrated the T DNA into their genome. Because the T DNA also contains the cloned gene of interest, the selected plant cells are expected to have received this cloned gene as well. The cells are then transferred to a medium that contains the plant growth hormones necessary for the regeneration of entire plants. These plants can then be analyzed to verify that they are transgenic plants containing the cloned gene.

A. tumefaciens infects a wide range of plant species, including most dicotyledonous plants, most gymnosperms, and some monocotyledonous plants. However, not all plant species are infected by this bacterium. Fortunately, other methods are available for introducing genes into plant cells. Another common way to produce transgenic plants is an approach known as **biolistic gene transfer.** In this method, plant cells are bombarded with high-velocity microprojectiles coated with DNA. When fired upon by this "gene gun," the microprojectiles penetrate the cell wall and membrane and thereby enter the plant cell. The cells that take up the DNA are identified with a selectable marker and regenerated into new plants.

Other methods are also available to introduce DNA into plant cells (and also animal cells). For example, DNA can enter plant cells by **microinjection**—the use of microscopic-sized needles—or by **electroporation**—the use of electrical current to create temporary pores in the plasma membrane. Because the rigid plant cell wall is a difficult barrier for DNA entry, other approaches involve the use of protoplasts, which are plant cells that have had their cell walls removed. DNA can be introduced into protoplasts using a variety of methods, including treatment with polyethylene glycol and calcium phosphate.

The production of transgenic plants has been achieved for many agriculturally important plant species. These include alfalfa, corn, cotton, soybean, tobacco, and tomato. Some of the applications of transgenic plants are described next.

Transgenic Plants Can Be Given Characteristics That Are Agriculturally Useful

Various traits can be modified in transgenic plants (**Table 19.5**). Frequently, transgenic research has sought to produce plant strains resistant to insects, disease, and herbicides. For example, transgenic plants highly tolerant of particular herbicides have

Agrobacterium tumefaciens is found within the soil. A wound on the plant enables the bacterium to infect the plant cells.

During infection, the T DNA within the Ti plasmid is transferred to the plant cell. The T DNA becomes integrated into the plant cell's DNA. Genes within the T DNA promote uncontrolled plant cell growth.

The growth of the recombinant plant cells produces a crown gall tumor.

(a) The production of a crown gall tumor by *A. tumefaciens* infection

(b) A crown gall tumor on a pecan tree

FIGURE 19.15   *Agrobacterium tumefaciens* infecting a plant and causing a crown gall tumor.

been made. The Monsanto Company has produced transgenic plant strains tolerant of glyphosate, the active agent in the herbicide Roundup™. The herbicide remains effective against weeds, but the herbicide-resistant crop is spared (**Figure 19.17**).

Another important approach is to make plant strains that are disease resistant. In many cases, virus-resistant plants have been developed by introducing a gene that encodes a viral coat protein. When the plant cells express the viral coat protein, they become resistant to infection by that pathogenic virus.

Many transgenic plants have been approved for human consumption. The first example was the Flavr Savr tomato (**Figure 19.18**), which was developed by Calgene Inc., which is now part of Monsanto. In this technique, a tomato plant was given a gene that encodes an antisense RNA complementary to the mRNA that encodes the enzyme polygalacturonase. This enzyme, which is expressed during ripening, digests sugar linkages within the pectin found in plant cell walls and thus softens the tomato. The antisense RNA binds to the polygalacturonase mRNA, preventing it from being translated. In

T DNA

Site where restriction enzyme cuts

Gene of interest

Gene of interest is inserted into the T DNA of the Ti plasmid.

Recombinant Ti plasmid

The recombinant Ti plasmid is transformed into *A. tumefaciens*.

A. tumefaciens

Recombinant Ti plasmid

Plant cells are exposed to *A. tumefaciens*. The T DNA is transferred and incorporated into the plant cell chromosome. Carbenicillin and kanamycin are then added to kill *A. tumefaciens* cells and plant cells that have not taken up T DNA, respectively.

Plant cell

Inserted T DNA carrying new gene

The plant cells are transferred to a medium containing growth hormones to regenerate an entire plant.

Plant with cloned gene

FIGURE 19.16   **The transfer of genes into plants using the Ti plasmid from *A. tumefaciens* as a vector.**

TABLE **19.5**	
Traits That Have Been Modified in Transgenic Plants	
Trait	**Examples**
Plant Protection	
Resistance to viral, bacterial, and fungal pathogens	Transgenic plants that express the pokeweed antiviral protein are resistant to a variety of viral pathogens.
Resistance to insects	Transgenic plants that express the CryIA protein from *Bacillus thuringiensis* are resistant to a variety of insects (see Figure 19.19).
Resistance to herbicides	Transgenic plants can express proteins that render them resistant to particular herbicides (see Figure 19.17).
Plant Quality	
Improvement in storage	Transgenic plants can express antisense RNA that silences a gene involved in fruit softening (see Figure 19.18).
Change in plant composition	Transgenic strains of canola have been altered with regard to oil composition; the seeds of the Brazil nut have been rendered methionine-rich via transgenic technology.
New Products	
Biodegradable plastics	Transgenic plants have been made that can synthesize polyhydroxyalkanoates, which are used as biodegradable plastics.
Vaccines	Transgenic plants have been modified to produce vaccines in their leaves against many human and animal diseases including hepatitis B, cholera, and malaria.
Pharmaceuticals	Transgenic plants have been made that produce a variety of medicines including human interferon-α (to fight viral diseases and cancer), human epidermal growth factor (for wound repair), and human aprotinin (for transplantation surgery).
Antibodies	Human antibodies have been made in transgenic plants to battle various diseases such as non-Hodgkin's lymphoma.

addition, the double-stranded RNA is targeted for degradation (RNA interference), as discussed in Chapter 15 (Figure 15.23). Because the expression of polygalacturonase is silenced, the practical advantage is that the tomatoes do not soften as quickly as unmodified tomatoes. Therefore, these tomatoes can be allowed to ripen on the vine for a longer period of time, enhancing their flavor and extending their shelf life. Improved taste is an important consideration in the $5-billion annual U.S. tomato market. By comparison, other commercial tomatoes commonly are picked while they are green and ripen later in order to maintain their firmness longer.

The Flavr Savr tomato was not a commercial success, and its sales were eventually discontinued. The failure of the Flavr Savr has been attributed to a variety of issues. In particular, the variety of tomato that was genetically engineered may have not been the best choice, and the antirotting trait was not as helpful

FIGURE 19.17   **Transgenic plants that are resistant to glyphosate.**

Genes → Traits This field of soybean plants has been treated with glyphosate. The plants on the left have been genetically engineered to contain a herbicide-resistance gene. They are resistant to killing by glyphosate. By comparison, the dead or stunted plants in the row with the orange stick do not contain this gene.

FIGURE 19.18   **The genetically engineered Flavr Savr tomato.**

Genes → Traits This transgenic tomato plant has been genetically engineered to contain an artificial gene that encodes an antisense RNA that will bind to the mRNA that encodes polygalacturonase. This prevents the mRNA from being translated, which inhibits the synthesis of polygalacturonase. Without this enzyme, the Flavr Savr tomato can be left on the vine to ripen longer than traditional tomatoes, thereby improving its flavor.

for the tomato business as Calgene had anticipated. Compared to unmodified, green tomatoes, another factor was that the ripe Flavr Savr tomatoes were more delicate and required the purchase of expensive handling equipment.

A much more successful example of the use of transgenic plants has involved the introduction of genes from *Bacillus thuringiensis* (Bt). As discussed earlier in this chapter, this bacterium produces toxins that kill certain types of caterpillars and beetles and has been widely used as a biological control agent for several decades. These toxins are proteins encoded in the genome of *B. thuringiensis.* Researchers have succeeded in cloning toxin genes from *B. thuringiensis* and transferring those genes into plants. Such Bt varieties of plants produce the toxins themselves and therefore are resistant to many types of caterpillars and beetles. Examples of commercialized crops include Bt corn (**Figure 19.19a**) and Bt cotton. Since their introduction in 1996, the commercial use of these two Bt crops has steadily increased (**Figure 19.19b**).

The introduction of transgenic plants into agriculture has been strongly opposed by some people. What are the perceived risks? One potential risk is that transgenes in commercial crops could endanger native species. For example, Bt crops may kill pollinators of native species. Another worry is that the planting of transgenic crops could potentially lead to the proliferation of resistant insects. To prevent this from happening, researchers

(a) A field of Bt corn

FIGURE 19.19   **The production of Bt crops.** (**a**) A field of Bt corn. These corn plants carry an endotoxin gene from *Bacillus thuringiensis* that provides them with resistance to insects such as corn borers, which are a major pest of corn plants. (**b**) A graph showing the increase in usage of Bt corn and Bt cotton since their commercial introduction in 1996.

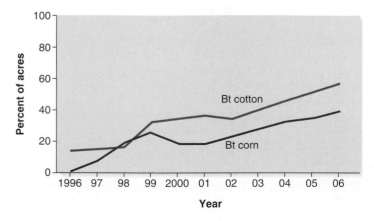

(b) Bt corn and Bt cotton usage since 1996

are producing transgenic strains that carry more than one toxin gene, which makes it more difficult for insect resistance to arise. Despite these and other concerns, many farmers are embracing transgenic crops, and their use continues to rise.

19.4 HUMAN GENE THERAPY

Throughout this textbook, we have considered examples in which mutant genes cause human diseases. As discussed in Chapter 22, these include rare inherited disorders and more common diseases such as cancer. Because mutant genes cause disease, geneticists are actively pursuing the goal of using normal, cloned genes to compensate for defects in mutant genes. Gene therapy is the introduction of cloned genes into living cells in the treatment of disease. It is a potential method of treating a wide variety of illnesses.

Many current research efforts in gene therapy are aimed at alleviating inherited human diseases. Over 4,000 human genetic diseases are known to involve a single gene abnormality. Familiar examples include cystic fibrosis, sickle-cell disease, and hemophilia. In addition, gene therapies have also been aimed at treating diseases such as cancer and cardiovascular disease, which may occur later in life. Some scientists are even pursuing research that will use gene therapy to combat infectious diseases such as AIDS. Even though gene therapy is still at an early stage of development, a large amount of research has already been conducted. Unfortunately, success has been limited, and relatively few patients have been treated with gene therapy. Nevertheless, a few results have been somewhat promising. Table 19.6 describes several types of diseases that are being investigated as potential targets for gene therapy. In this section, we will examine the approaches to gene therapy and how it may be used to treat human disease.

TABLE 19.6

Future Prospects in Gene Therapy

Type of Disease	Treatment of
Blood	Sickle-cell disease, hemophilia, severe combined immunodeficiency disease (SCID)
Metabolic	Glycogen storage diseases, lysosomal storage diseases, and phenylketonuria
Muscular	Duchenne muscular dystrophy, myotonic muscular dystrophy
Lung	Cystic fibrosis
Cancer	Brain tumors, breast cancer, colorectal cancer, malignant melanoma, ovarian cancer, and several other types of malignancies
Cardiovascular	Atherosclerosis, essential hypertension
Infectious	AIDS, possibly other viral diseases that involve latent infections

Gene Therapy Involves the Introduction of Cloned Genes into Human Cells

A key step in gene therapy is the introduction of a cloned gene into the cells of people. The techniques to transfer a cloned gene into human cells can be categorized as nonviral and viral gene transfer methods. The most common nonviral technique involves the use of **liposomes,** which are lipid vesicles (**Figure 19.20a**). The DNA containing the gene of interest is complexed with liposomes that carry a positive charge (i.e., cationic liposomes). The DNA-liposome complexes are taken into cells via endocytosis, in which a portion of the plasma membrane invaginates and creates an intracellular vesicle known as an endosome; the liposome is degraded within the endosome. The DNA is then released into the cytosol, imported into the nucleus, and then integrated into a chromosome of the target cell. An advantage of gene transfer via liposomes is that the liposomes do not elicit an immune response. A disadvantage is that the efficiency of gene transfer may be very low.

A second way to transfer genes into human cells is via viruses. Commonly used viruses for gene therapy include retroviruses, adenoviruses, and parvoviruses. The genetic modification of these viral genomes has led to the development of gene therapy vectors with a capacity to infect cells or tissues, much like the ability of wild-type viruses to infect cells. However, in contrast to wild-type viruses, gene therapy viral vectors have been genetically engineered so they have lost their capacity for replication in target cells. Nevertheless, the genetically engineered viruses are naturally taken up by cells via endocytosis (**Figure 19.20b**). The viral coat disassembles, and the viral genome is released into the cytosol. In the case of retroviruses, the genome is RNA, which is reverse transcribed into DNA. The viral DNA, which carries a gene of interest, travels into the nucleus and is then integrated into a chromosome of the target cell.

A key advantage of viral vectors is their ability to efficiently transfer cloned genes to a variety of human cell types. However, a major disadvantage of viral-mediated gene therapy is the potential to evoke an undesirable immune response when injected into a patient. The inflammatory responses induced by adenovirus particles, for example, can be very strong and even fatal. Therefore, much effort has been aimed at preventing the inflammatory responses mediated by virus particles. These include the application of immunosuppressive drugs and the generation of less immunogenic viral vectors by further genetic modification within the viral genome.

Adenosine Deaminase Deficiency Was the First Inherited Disease Treated with Gene Therapy

Adenosine deaminase (ADA) is an enzyme involved in purine metabolism. If both copies of the *ADA* gene are defective, deoxyadenosine will accumulate within the cells of the individual. At high concentrations, deoxyadenosine is particularly toxic to lymphocytes in the immune system, namely, T cells and B cells. In affected individuals, the destruction of T and B cells leads to a form of severe combined immunodeficiency disease (SCID). If

FIGURE 19.20 **Methods of gene transfer used in gene therapy.** (**a**) In this example, the DNA containing the gene of interest is complexed with cationic liposomes. These complexes are taken into cells by endocytosis, in which a portion of the plasma membrane invaginates and creates an intracellular vesicle known as an endosome. After it is released from the endosome, the DNA may then integrate into the chromosomal DNA via recombination. (**b**) In this example, the gene of interest is cloned into a retrovirus. When the retrovirus infects a cell, the RNA genome is reverse transcribed into double-stranded DNA, which then integrates into the chromosome. Viruses used in gene therapy have been genetically altered so they cannot proliferate after entry into the target cell.

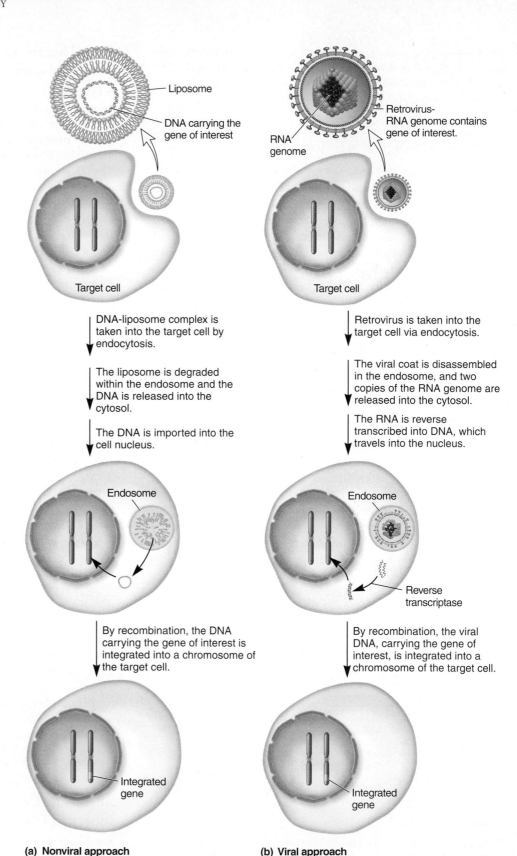

left untreated, SCID is typically fatal at an early age (generally, one to two years old), because the immune system of these individuals is severely compromised and cannot fight infections.

Three approaches can be used to treat ADA deficiency. In some cases, a patient may receive a bone marrow transplant from a compatible donor. A second method is to treat SCID patients with purified ADA that is coupled to polyethylene glycol (PEG). This PEG-ADA is taken up by lymphocytes and can correct the ADA deficiency. Unfortunately, these two approaches are not always available and/or successful. A third, more recent approach is to treat ADA patients with gene therapy.

On September 14, 1990, the first human gene therapy was approved for a young girl suffering from ADA deficiency. This work was carried out by a large team of researchers composed of R. Michael Blaese, Kenneth Culver, W. French Anderson, and colleagues. Prior to this clinical trial, the normal gene for ADA had been cloned into a retroviral vector that can infect lymphocytes. The general aim of this therapy was to remove lymphocytes from the blood of the young girl with SCID, introduce the normal *ADA* gene into her cells, and then return them to her bloodstream.

Figure 19.21 outlines the protocol for the experimental treatment. Lymphocytes (i.e., T cells) were removed and cultured in a laboratory. The lymphocytes were then transfected with a nonpathogenic retrovirus that had been genetically engineered to contain the normal *ADA* gene. During the life cycle of a retrovirus, the retroviral genetic material is inserted into the host cell's DNA. Therefore, because this retrovirus contained the normal ADA gene, this gene also was inserted into the chromosomal DNA of the girl's lymphocytes. After this had occurred in the laboratory, the cells were reintroduced back into the patient. This approach is called an **ex vivo approach** because the genetic manipulations occur outside the body, yet the products are reintroduced into the body.

In this clinical trial, two patients were enrolled, and a third patient was later treated in Japan. Was the treatment a success? The results of this trial showed that the transfer of DNA into a large number of human cells is feasible. In at least one patient, T cells carrying the cloned gene were still detectable 8 to 10 years after they had been transferred. However, most of the circulating T cells were not found to contain the cloned gene. Because the individuals also received a low dose of PEG-ADA treatment, researchers could not determine whether or not gene transfer into T cells in itself was of significant clinical benefit.

Another form of SCID, termed SCID-X1, is inherited as an X-linked trait. SCID-X1 is characterized by a block in T cell growth and differentiation. This block is caused by mutations in the gene encoding the γ_c cytokine receptor, which plays a key role in the recognition of signals that are needed to promote the growth, survival, and differentiation of T cells. A gene therapy

Lymphocytes

Remove ADA-deficient lymphocytes from the SCID patient.

Culture the cells in a laboratory.

Infect the cells with a retrovirus that contains the normal *ADA* gene.

Reinfuse the *ADA*-gene-corrected lymphocytes back into the SCID patient.

FIGURE 19.21   **The first human gene therapy for adenosine deaminase (ADA) deficiency.**

trial for SCID-X1 similar to the trial shown in Figure 19.21 was initiated in 2000 in which a normal γ_c cytokine receptor gene was cloned into a retroviral vector and then introduced into SCID-X1 patients' lymphocytes. The lymphocytes were then reintroduced back into their bodies. At a 10-month follow-up, T cells expressing the normal γ_c cytokine receptor were detected in two patients. Most importantly, the T cell counts in these two patients had risen to levels that were comparable to normal individuals. This clinical trial was the first clear demonstration that gene therapy can offer clinical benefit, providing in these cases what seemed to be a complete correction of the disease phenotype. However, in a French study involving 10 SCID-X1 patients, an unexpected and serious side effect occurred. Within three years of gene therapy treatment, three out of the 10 treated children developed leukemia—a form of cancer involving the proliferation of white blood cells. In these cases, the disease was caused by the integration of the retroviral vector next to a particular gene in the patients' genomes. The development of leukemia in these patients has halted many clinical trials involving gene therapy.

Aerosol Sprays May Be Used to Treat Cystic Fibrosis

Cystic fibrosis (CF) is a rare recessive disorder with debilitating consequences. About 1 in 3,000 babies whose parents are of northern European descent are affected with this disorder. CF is caused by a defect in a gene termed the cystic fibrosis transmembrane regulator (*CFTR*), which encodes a protein that functions in the transport of chloride ions across the plasma membrane of epithelial cells, such as cells lining the respiratory tract and the intestinal tract. A defect in membrane transport leads to an abnormality in salt and water balance, which causes a variety of symptoms, particularly an overaccumulation of mucus in the lungs. Even though great strides have been made in the treatment of the symptoms of CF, this disease remains associated with repeated lung infections and a shortened life span. In most cases, mortality results from chronic lung infections.

CF has been the subject of much gene therapy research. Clinical trials have tested the ability of gene therapy to improve the condition of patients suffering from CF. To implement CF gene therapy, it is necessary to deliver the normal *CFTR* gene to the lung cells. Unlike *ADA* gene therapy, in which the lymphocytes can be treated ex vivo, lung epithelial cells cannot be removed and then put back into the individual. Instead, researchers must design innovative approaches that can target the *CFTR* gene directly to the lung cells.

To achieve this goal, CF gene therapy methods have involved the use of an inhaled aerosol spray. In one protocol, the normal *CFTR* gene is cloned into an adenovirus, a virus that normally infects lung epithelial cells and causes a lung infection. This adenovirus, however, has been engineered so that it will gain entry into the epithelial cells but not cause a lung infection. In addition, the adenovirus has been engineered to contain the normal *CFTR* gene. In a second approach, the normal *CFTR* gene is complexed with liposomes. When inhaled by the patient via an aerosol spray, the lung epithelial cells take up this liposome complex.

Like *ADA* gene therapy, CF gene therapy is at an early stage of development. Researchers hope that gene therapy eventually will become an effective method to alleviate the symptoms associated with this disease.

CONCEPTUAL AND EXPERIMENTAL SUMMARY

Biotechnology, the application of technologies involving the use of living organisms in the development of products and services that benefit humans, has increasingly become associated with molecular genetics. For example, molecular genetic tools can engineer recombinant microorganisms used in the production of medically important products. The production of somatostatin by recombinant bacteria paved the way for the creation of genetically engineered medicines, including insulin and human growth hormone. Recombinant microorganisms have also been used as agents of biological control of plant diseases and for bioremediation to modify toxic pollutants.

Molecular tools are also being used to create transgenic organisms, which contain recombinant DNA from another species that has been integrated into their genome. We explored the current techniques employed to create transgenic animals and their potential uses. The introduction of cloned DNA into a cell can lead to gene replacement or gene addition. In this way, researchers can produce gene knockouts and knockins, which are powerful tools to explore gene function and human disease. Research is developing transgenic species of livestock that produce medically important proteins. In recent years, cloning from somatic cells has been achieved in several mammalian species,

beginning with the first cloned sheep, Dolly. Research studies of stem cells can help us to understand basic genetic mechanisms underlying the process of development. Furthermore, stem cells have the potential to treat human diseases or injuries that cause cell and tissue damage.

A relatively new area of research is the use of transgenic species in agriculture. Transgenic plants can be produced through the use of the bacterium *Agrobacterium tumefaciens* as a vector or through the technique of biolistic gene transfer. An important goal is to make plant strains that are disease or herbicide resistant. While many people have concerns over the use of genetically modified organisms in agriculture, their use continues to rise. In the future, it is expected that the production and commercial success of transgenic organisms will become more prevalent.

Much scientific effort is going into gene therapy, the introduction of cloned genes into living cells in order to treat a wide variety of human diseases. In this approach, a cloned gene is introduced via liposomes or viruses into somatic cells as a way to compensate for a defective gene. While success has been limited and sobering risks have been identified, a few results have been somewhat promising. In the future, this may become an important way to alleviate many human disorders.

PROBLEM SETS & INSIGHTS

Solved Problems

S1. The protocol that was followed to produce somatostatin in recombinant *E. coli* cells would have been difficult to carry out if somatostatin was a long polypeptide. Explain why.

Answer: The researchers constructed the coding sequence for somatostatin by making several short oligonucleotides that were complementary to each other and formed the somatostatin coding sequence. This would have been difficult to achieve if the polypeptide sequence was much longer, say, 100 amino acids in length, because it would have required the synthesis of many short oligonucleotides.

S2. Which of the following would appropriately be described as a transgenic organism?

   A. The sheep "Dolly," which was produced by cloning

   B. A sheep that produces human α-1-antitrypsin in its milk

   C. The Flavr Savr strain of tomato

   D. A hybrid strain of corn produced from crossing two inbred strains of corn (The inbred strains were not transgenic.)

Answer:

   A. No, Dolly was not produced using recombinant techniques. Pieces of DNA were not cut and combined in a new way.

   B. Yes

   C. Yes

   D. No, the hybrids simply contain chromosomal genes from two different parental strains.

S3. Describe the strategy for producing human proteins in the milk of livestock.

Answer: Milk proteins are encoded by genes with promoters and regulatory sequences that direct the expression of these genes within the cells of the mammary gland. To get other proteins expressed in the mammary gland, the strategy is to link the promoter and regulatory sequences from a milk-specific gene to the coding sequence of the gene that encodes the human protein of interest. In some cases, it is also necessary to add a signal sequence to the amino terminal end of the target protein. A signal sequence is a short polypeptide that directs the secretion of a protein from a cell. If the target protein does not already have a signal sequence, it is possible to use a signal sequence from a milk-specific gene to promote the secretion of the target protein from the mammary cells and into the milk. During this process, the signal sequence is cleaved from the secreted protein.

S4. With regard to genetically modified organisms, describe two that have not been successful and two that have.

Answer: Two unsuccessful examples are oil-eating bacteria and the Flavr Savr tomato. Two very successful examples include bacteria that make human insulin and Bt varieties of agricultural crops.

Conceptual Questions

C1. What is a recombinant microorganism? Discuss examples of recombinant microbes.

C2. A conjugation-deficient strain of *A. radiobacter* is used to combat crown gall disease. Explain how this bacterium prevents the disease and the advantage of a conjugation-deficient strain.

C3. What is bioremediation? What is the difference between biotransformation and biodegradation?

C4. What is a biological control agent? Briefly describe three examples.

C5. As described in Table 19.2, several medical agents are now commercially produced by genetically engineered microorganisms. Discuss the advantages and disadvantages of making these agents this way.

C6. What is a mouse model for human disease?

C7. What is a transgenic organism? Describe three examples.

C8. What part of the *A. tumefaciens* DNA gets transferred to the genome of a plant cell during infection?

C9. Explain the difference between gene addition and gene replacement. Would the following descriptions be examples of gene addition or gene replacement?

   A. A mouse model to study cystic fibrosis

   B. Introduction of a pesticide-resistance gene into corn using the Ti plasmid of *A. tumefaciens*

C10. As described in Chapter 7, not all inherited traits are determined by nuclear genes (i.e., genes located in the cell nucleus) that are expressed during the life of an individual. In particular, maternal effect genes and mitochondrial genes are notable exceptions. With these ideas in mind, let's consider the cloning of sheep (e.g., Dolly).

   A. With regard to maternal effect genes, would the phenotype of such a cloned animal be determined by the animal that donated the enucleated egg or by the animal that donated the somatic cell nucleus? Explain.

   B. Would the cloned animal inherit extranuclear traits from the animal that donated the egg or by the animal that donated the somatic cell? Explain.

   C. In what ways would you expect this cloned animal to be similar to or different from the animal that donated the somatic cell? Is it fair to call such an animal a "clone" of the animal that donated the nucleus?

C11. Discuss some of the worthwhile traits that can be modified in transgenic plants.

C12. Discuss the concerns that some people have with regard to the uses of genetically engineered organisms.

Experimental Questions

E1. Recombinant bacteria can produce hormones that are normally produced in humans. Briefly describe how this is accomplished.

E2. In the experiment of Figure 19.1, why did the plasmid with somatostatin in the incorrect orientation fail to produce somatostatin?

E3. *Bacillus thuringiensis* can make toxins that kill insects. This toxin must be applied several times during the growth season to prevent insect damage. As an alternative to repeated applications, one strategy is to apply bacteria directly to leaves. However, *B. thuringiensis* does not survive very long in the field. Other bacteria, such as *Pseudomonas syringae*, do. Propose a way to alter *P. syringae* so that it could be used as an insecticide. Discuss advantages and disadvantages of this approach compared to the repeated applications of the insecticide from *B. thuringiensis*.

E4. In the experiment of Figure 19.1, explain how the coding sequence for somatostatin was constructed. Why was it necessary to link this coding sequence to the sequence for β-galactosidase? How were these two polypeptides separated after the fusion protein was synthesized in *E. coli*?

E5. Explain how it is possible to select for homologous recombination in mice. What phenotypic marker is used to readily identify chimeric mice?

E6. To produce transgenic plants, plant tissue is exposed to *A. tumefaciens* and then grown in media containing kanamycin, carbenicillin, and plant growth hormones. Explain the purpose behind each of these three agents. What would happen if you left out the kanamycin?

E7. List and briefly describe five methods for the introduction of cloned genes into plants.

E8. What is a gene knockout? Is an animal or plant with a gene knockout a heterozygote or homozygote? What might you conclude if a gene knockout does not have a phenotypic effect?

E9. Nowadays, it is common for researchers to identify genes using cloning methods described in Chapters 18 and 19. A gene can be identified according to its molecular features. For example, a segment of DNA can be identified as a gene because it contains the right combination of sequences: a promoter, exons, introns, and a terminator. Or a gene can be identified because it is transcribed into mRNA. In the study of plants and animals, it is relatively common for researchers to identify genes using molecular techniques without knowing the function of the gene. In the case of mice, the function of the gene can be investigated by making a gene knockout. If the knockout causes a phenotypic change in the mouse, this may provide an important clue regarding the function of a gene. For example, if a gene knockout produced an albino mouse, this would indicate the gene knocked out probably plays a role in pigment formation. The experimental strategy of first identifying a gene based on its molecular properties and then investigating its function by making a knockout is called reverse genetics. Explain how this approach is opposite (or "in reverse") to the conventional way that geneticists study the function of genes.

E10. According to the methods described in Figure 19.7, can homologous recombination that results in gene replacement cause the integration of both the *TK* and *Neo*^{R} genes? Explain why or why not. Describe how the *TK* gene and *Neo*^{R} gene are used in a selection scheme that favors gene replacement.

E11. What is a chimera? How are chimeras made?

E12. Evidence (see Shiels et al., "Analysis of telomere lengths in cloned sheep," *Nature* 399, 316–17) suggested that Dolly may have been "genetically older" than her actual age would have suggested. As mammals age, the chromosomes in somatic cells tend to shorten from the telomeres. Therefore, older individuals have shorter chromosomes in their somatic cells compared to younger ones. When researchers analyzed the chromosomes in the somatic cells of Dolly when she was about three years old, the lengths of her chromosomes were consistent with a sheep that was significantly older, say, nine or ten years old. (Note: As described in the chapter, the sheep that donated the somatic cell that produced Dolly was six years old, and her mammary cells had been grown in culture for several cell doublings before a mammary cell was fused with an oocyte.)

A. Suggest an explanation why Dolly's chromosomes seemed so old.

B. Let's suppose that Dolly at age 11 gave birth to a lamb named Molly; Molly was produced naturally (by mating Dolly with a normal male). When Molly was eight years old, a sample of somatic cells was analyzed. How old would you expect Molly's chromosomes to appear, based on the phenomenon of telomere shortening? Explain your answer.

C. Discuss how the observation of chromosome shortening, which was observed in Dolly, might affect the popularity of reproductive cloning.

E13. When transgenic organisms are made, the transgene may integrate into multiple sites within the genome. Furthermore, the integration site may influence the expression of the gene. For example, if a transgene integrates into a heterochromatic (i.e., highly condensed) region of a chromosome, the transgene may not be expressed. For these reasons, it is important for geneticists to analyze transgenic organisms with regard to the number of transgene insertions and the expression levels of the transgenes. Chapter 18 describes three methods (Southern blotting, Northern blotting, and Western blotting) that can be used to detect genes and gene products. Which of these techniques would you use to determine the number of transgenes in a transgenic plant or animal? Why is it important to know the number of transgenes? (Hint: You may want to use a transgenic animal or plant as breeding stock to produce many more transgenic animals or plants.) Which technique would you use to determine the expression levels of transgenes?

E14. What is molecular pharming? Compared to the production of proteins by bacteria, why might it be advantageous?

E15. What is reproductive cloning? Are identical twins in humans considered to be clones? With regard to agricultural species, what are some potential advantages to reproductive cloning?

E16. Researchers have identified a gene in humans that (when mutant) causes severe dwarfism and mental retardation. This disorder is inherited in an autosomal recessive manner, and the mutant allele is known to be a loss-of-function mutation. The same gene has been found in mice, although a mutant version of the gene has not been discovered in mice. To develop drugs and an effective therapy to treat this disorder in humans, it would be experimentally useful to have a mouse model. In other words, it would be desirable to develop a strain of mice that carry the mutant allele in the homozygous condition. Experimentally, how would you develop such a strain?

E17. Treatment of adenosine deaminase (ADA) deficiency is an example of ex vivo gene therapy. Why is this therapy called ex vivo? Can ex vivo gene therapy be used to treat all inherited diseases? Explain.

E18. Describe the targeting methods used in cystic fibrosis gene therapy. Provided the *CFTR* gene gets to the patient's lung cells, would you expect this to be a permanent cure for the patient, or would it be necessary to perform this gene therapy on a regular basis (say, monthly)?

E19. Several research studies are under way that involve the use of gene therapies to inhibit the growth of cancer cells. As discussed in Chapter 22, oncogenes are mutant genes that are overexpressed and cause cancer. New gene therapies are aimed at silencing an oncogene by producing antisense RNA that recognizes the mRNA transcribed from an oncogene. Based on your understanding of antisense RNA (described in Chapters 14 and 15), explain how this strategy would prevent the growth of cancer cells.

Questions for Student Discussion/Collaboration

1. Discuss the advantages and disadvantages of gene therapy. Because a limited amount of funding is available for gene therapy research, make a priority list of the three top diseases that you would fund. Discuss your choices.

2. A commercially available strain of *P. syringae* marketed as Frostban B is used to combat frost damage. This is a naturally occurring *Ice⁻* strain. Discuss the advantages and disadvantages of using this strain compared with a recombinant version.

3. Make a list of the types of traits you would like to see altered in transgenic plants and animals. Suggest ways (i.e., what genes would you use?) to accomplish these alterations.

Note: All answers appear at the website for this textbook; the answers to even-numbered questions are in the back of the textbook.

www.mhhe.com/brookergenetics3e

Visit the Online Learning Center for practice tests, answer keys, and other learning aids for this chapter. Enhance your understanding of genetics with our interactive exercises, quizzes, animations, and much more.

20

GENOMICS I: ANALYSIS OF DNA

Labeling the ends of chromosomes. In this micrograph, the telomeric sequences at ends of chromosomes are labeled with an orange fluorescent probe, while the rest of the chromosomes are labeled in blue. This method, called fluorescent *in situ* hybridization, allows geneticists to identify particular sequences within intact chromosomes.

The term **genome** refers to the total genetic composition of an organism or species. For example, the nuclear genome of humans is composed of 22 different autosomes and an X and Y chromosome. In addition, humans have a mitochondrial genome composed of a single circular chromosome.

As genetic technology has progressed over the past few decades, researchers have gained an increasing ability to analyze the composition of genomes as a whole unit. The term **genomics** refers to the molecular analysis of the entire genome of a species. Genome analysis is a molecular dissection process applied to a complete set of chromosomes. Segments of chromosomes are analyzed in progressively smaller pieces, the locations of which are known on the intact chromosomes. This is the mapping phase of genome analysis. The mapping of the genome ultimately progresses to the determination of the complete DNA sequence, which provides the most detailed description of an organism's genome at the molecular level.

In 1995, a team of researchers headed by Craig Venter and Hamilton Smith obtained the first complete DNA sequence of the bacterial genome from *Haemophilus influenzae*. Its genome is composed of a single circular chromosome 1.83 million base pairs (bp) in length and contains approximately 1,743 genes (**Figure 20.1**). In 1996, the first entire DNA sequence of a eukaryote, *Saccharomyces cerevisiae* (baker's yeast), was completed. This work was carried out by a European-led consortium of more than 100 laboratories, including some in the United States, Japan, and Canada. The overall coordinator was Andre Goffeau in Belgium. The yeast genome contains 16 linear chromosomes; these 16 chromosomes have a combined length of about 12.1 million bp and contain approximately 6,300 genes. Since that time, genome sequences from many prokaryotes and eukaryotes have been completed.

In this chapter, we will focus on methods aimed at elucidating the organization of the sequences within a species' genome. This process begins with the mapping of regions along an organism's chromosomes and ends with the determination of the complete DNA sequence. We will consider three mapping strategies—cytogenetic, linkage, and physical mapping—and the approaches used to carry them out. We will then turn to exploring genome sequencing projects, research endeavors that have the ultimate goal of determining the sequence of DNA bases of the entire genome of a given species. We will consider the methods, goals, and results of these large undertakings, which include the Human Genome Project.

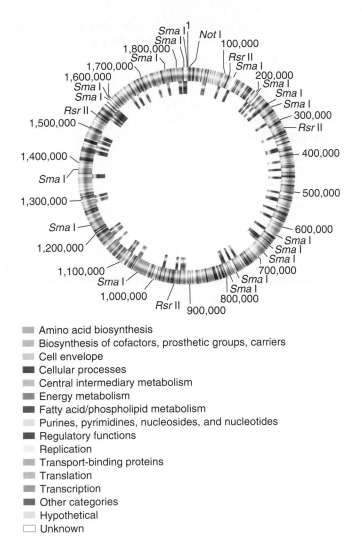

Amino acid biosynthesis
Biosynthesis of cofactors, prosthetic groups, carriers
Cell envelope
Cellular processes
Central intermediary metabolism
Energy metabolism
Fatty acid/phospholipid metabolism
Purines, pyrimidines, nucleosides, and nucleotides
Regulatory functions
Replication
Transport-binding proteins
Translation
Transcription
Other categories
Hypothetical
Unknown

FIGURE 20.1 **A complete map of the genome of the bacterium** *Haemophilus influenzae.* On the outside of the circular chromosome, the numbering of nucleotides begins at the top of the circle and proceeds clockwise in 100,000-nucleotide increments. The map also shows the locations of restriction enzyme sites for *Not*I, *Sma*I, and *Rsr*II. As shown in the key, the general categories of genes are color coded.

Genes→Traits By mapping and DNA sequencing the entire genome of a species, researchers can identify most of the genes it possesses. It is the expression of these many genes that ultimately determines an organism's inherited traits.

Once a genome sequence is known, researchers can examine, at the level of many genes, how the components of a genome interact to produce the traits of an organism. This approach is called **functional genomics.** Ultimately, a long-term goal of researchers is to determine the roles of all cellular proteins, as well as the interactions that these proteins experience, to produce the characteristics of particular cell types and the traits of complete organisms. This is the research area known as **proteomics.** In this chapter, we will explore genomics at the level of DNA seg-

ments and sequences. In Chapter 21, we will consider the exciting prospects that functional genomics and proteomics have to offer regarding our future understanding of genetics.

20.1 CYTOGENETIC AND LINKAGE MAPPING

In genetics, the term **mapping** refers to the experimental process of determining the relative locations of genes or other segments of DNA along individual chromosomes. Researchers may follow three general approaches to map a chromosome: cytogenetic, linkage, and physical mapping strategies. Before we discuss these approaches in detail, let's compare them to each other. **Cytogenetic mapping** (also called cytological mapping) relies on the localization of gene sequences within chromosomes that are viewed microscopically. When stained, each chromosome of a given species has a characteristic banding pattern, and genes are mapped cytogenetically relative to a band location. By comparison, in Chapter 5, we considered how genetic crosses are conducted to map the relative locations of genes within a chromosome. Such genetic studies, which are called **linkage mapping** or genetic mapping, use the frequency of genetic recombination between different genes to determine their relative spacing and order along a chromosome. In eukaryotes, linkage mapping involves crosses among organisms that are heterozygous for two or more genes. The number of recombinant offspring provides a relative measure of the distance between genes, which is computed in map units (or centiMorgans). Finally, a third approach is **physical mapping.** DNA cloning techniques are used to determine the location of and distance between genes and other DNA regions. In a physical map, the distances are computed as the number of nucleotide base pairs between genes.

A **genetic map** is a chart that describes the relative locations of genes or other DNA segments along a chromosome. The term **locus** (plural, **loci**) refers to the site within a genetic map where a specific gene or other DNA segment is found. **Figure 20.2** compares genetic maps that show the loci for two X-linked genes, *sc* (scute, a gene affecting bristle morphology) and *w* (a gene affecting eye color), in *Drosophila melanogaster*. In the cytogenetic map (top), the *sc* gene is located at band 1A8, and the *w* gene is located at band 3B6. In the linkage map, genetic crosses indicate that the two genes are approximately 1.5 map units (mu) apart. The physical map shows that the two genes are approximately 2.4×10^6 bp apart from each other along the X chromosome. Correlations between cytogenetic, linkage, and physical maps often vary from species to species and from one region of the chromosome to another. For example, a distance of 1 mu may correspond to 1 to 2 million base pairs in one region of the chromosome, but other regions may recombine at a much lower rate, so a distance of 1 mu may be a much longer physical segment of DNA.

In this section, we will explore several techniques aimed at producing cytogenetic and linkage maps and consider physical mapping in the section that follows.

Cytogenetic map:

Linkage map:

Physical map:

FIGURE 20.2   **A comparison of cytogenetic, linkage, and physical maps.** Each of these maps shows the distance between the *sc* and *w* genes along the X chromosome in *Drosophila melanogaster*. The cytogenetic map is that of the polytene chromosome.

A Goal of Cytogenetic Mapping Is to Determine the Location of a Gene Along an Intact Chromosome

Cytogenetic mapping is commonly used in eukaryotes, which have very large chromosomes compared to those of bacteria. Microscopically, eukaryotic chromosomes can be distinguished from one another by their size, centromeric location, and banding patterns (refer to Chapter 8, Figure 8.1). By treating chromosomal preparations with particular dyes, a discrete banding pattern is obtained for each chromosome. Cytogeneticists use this banding pattern as a way to describe specific regions along a chromosome.

Cytogenetic mapping attempts to determine the locations of particular genes relative to a banding pattern of a chromosome. For example, the human gene that encodes the cystic fibrosis transmembrane regulator, the protein that is defective in people with cystic fibrosis, is located on chromosome 7, at a specific site in the q3 region.

Cytogenetic mapping may be used as a first step in the localization of genes in plants and animals. However, because it relies on light microscopy, cytogenetic analysis has a fairly crude limit of resolution. In most species, cytogenetic mapping is accurate only within limits of approximately 5 million bp along a chromosome. In species that have large polytene chromosomes, such as *Drosophila*, the resolution is much better. A common strategy used by geneticists is to roughly locate a gene by cytogenetic analysis and then determine its location more precisely by the physical mapping methods described later.

In situ Hybridization Can Localize Genes Along Particular Chromosomes

The technique of *in situ* **hybridization** is widely used to cytogenetically map the locations of genes or other DNA sequences within large eukaryotic chromosomes. The term *in situ* (from the Latin for "in place") indicates that the procedure is conducted on chromosomes that are being held in place—adhered to a surface.

To map a gene via *in situ* hybridization, researchers use a probe to detect the location of the gene within a set of chromosomes. If the gene of interest has been cloned previously, as described in Chapter 18, the DNA of the cloned gene can be used as a probe. Because a DNA strand from a cloned gene, which is a very small piece of DNA relative to a chromosome, will hybridize only to its complementary sequence on a particular chromosome, this technique provides the ability to localize the gene of interest. For example, let's consider the gene that causes the white-eye phenotype in *Drosophila* when it carries a loss-of-function mutation. This gene has already been cloned. If a single-stranded piece of this cloned DNA is mixed with *Drosophila* chromosomes in which the DNA has been denatured, it will bind only to the X chromosome at the location corresponding to the site of the eye color gene.

The most common method of *in situ* hybridization uses fluorescently labeled DNA probes and is referred to as **fluorescence *in situ* hybridization** (**FISH**). **Figure 20.3** describes the steps of the FISH procedure. The cells are prepared using a technique that keeps the chromosomes intact. The cells are treated with agents that cause them to swell, and their contents are fixed to the slide. The chromosomal DNA is then denatured, and a DNA probe is added. For example, the added DNA probe might be single-stranded DNA that is complementary to a specific gene. In this case, the goal of a FISH experiment is to determine the location of the gene within a set of chromosomes. The probe will bind to a site in the chromosomes where the gene is located because the probe and chromosomal gene will line up and hydrogen bond with each other. To detect where the probe has bound to a chromosome, the probe is subsequently tagged with a fluorescent molecule. This is usually accomplished by first incorporating biotin-labeled nucleotides into the probe. Biotin, a small, nonfluorescent molecule, has a very high affinity for a protein called avidin. Fluorescently labeled avidin is added, which binds tightly to the biotin and thereby labels the probe as well.

How is the fluorescently labeled probe detected? A fluorescent molecule is one that absorbs light at a particular wavelength and then emits light at a longer wavelength. To detect the light emitted by a fluorescently labeled probe, a fluorescence microscope is

FIGURE 20.3 **The technique of fluorescence *in situ* hybridization (FISH).** The probe hybridizes to the denatured chromosomal DNA only at specific, complementary sites in the genome. Note that the chromosomes are highly condensed metaphase chromosomes that have already replicated. Therefore, each X-shaped chromosome actually contains two copies of a particular gene. These are sister chromatids. Because the sister chromatids are identical, a probe that recognizes one sister chromatid will also bind to the other.

used. Such a microscope contains filters that allow the passage of light only within a defined wavelength range. The sample is illuminated at the wavelength of light that is absorbed by the fluorescent molecule. The fluorescent molecule then emits light at a longer wavelength. The fluorescence microscope has a filter that allows the transmission of the emitted light. Because only the emitted light is viewed, the background of the sample is dark, and the fluorescence is seen as a brightly glowing color on a dark background. For most FISH experiments, chromosomes are generally counterstained by a fluorescent dye that is specific for DNA. The most common dye is DAPI (4', 6-diamidino-2-phenyl-indol) that is excited by UV light. This provides all of the DNA with a blue background. The results of a FISH experiment are then compared to a sample of chromosomes that have been stained with Giemsa to produce banding, so the location of a probe can be mapped relative to the banding pattern.

Figure 20.4 illustrates the results of an experiment involving six different DNA probes. The six probes were strands of DNA corresponding to six different DNA segments of human chromosome 5. In this experiment, each probe was labeled with a different fluorescent molecule. This enabled researchers to distinguish the probes when they became bound to their corresponding locations on chromosome 5. In this experiment, computer-imaging methods were used to assign each fluorescently labeled

FIGURE 20.4 **The results of a fluorescence *in situ* hybridization experiment.** In this experiment, six different probes were used to locate six different sites along chromosome 5. The colors are due to computer imaging of fluorescence emission; they are not the actual colors of the fluorescent labels. Two spots are usually seen at each site because the probe binds to both sister chromatids.

probe a different color. In this way, fluorescence *in situ* hybridization discerns the sites along chromosome 5 corresponding to the six different probes. In a visual, colorful way, this technique was used here to determine the order and relative distances between several specific sites along a single chromosome. FISH is commonly used in genetics and cell biology research, and its use has become more widespread in clinical applications. For example, clinicians may use FISH to detect small changes in chromosome structure such as deletions and duplications, which may occur in patients with genetic disorders.

Linkage Mapping Can Use Molecular Markers

Let's now turn to linkage mapping, which relies on the frequency of recombinant offspring to determine the distance between sites located along the same chromosome. The linkage mapping methods described in Chapters 5 and 6 used allelic differences between genes to map the relative locations of those genes along a chromosome. As an alternative to gene mapping, geneticists have realized that regions of DNA that do not encode genes can be used as markers along a chromosome. A **molecular marker** is a segment of DNA found at a specific site along a chromosome and has properties that enable it to be uniquely recognized using molecular tools such as polymerase chain reaction (PCR) and gel electrophoresis. As with alleles, the molecular markers may be **polymorphic;** that is, within a population, they may vary from individual to individual. Therefore, the distances between linked molecular markers can be determined from the outcomes of crosses. Using molecular techniques, researchers have found it easier to identify many molecular markers within a given species' genome rather than identify many allelic differences among individuals. For this reason, geneticists have increasingly turned to molecular markers as points of reference along genetic maps. As described in **Table 20.1**, many different kinds of molecular markers are used by geneticists.

Researchers have constructed detailed genetic maps in which a series of many molecular markers have been identified along each chromosome of certain species. These species include humans, model organisms, agricultural species, and many others. Why are molecular markers useful? One key reason is that molecular markers can be used to determine the approximate location of an unknown gene that causes a human disease. As we will see in Experiment 20A, a particular molecular marker (an RFLP) was linked to the *Hb^S* allele that causes sickle-cell disease. Similarly, for inherited diseases in which the mutant genes have yet to be identified, human geneticists can sometimes follow the transmission patterns of polymorphic molecular markers in family pedigrees. The discovery of a particular marker in those who have the disease can indicate that the marker is close to the disease-causing allele. This may help researchers identify the gene by cloning methods, such as chromosome walking, which we will examine later in this chapter.

In addition, molecular markers may help researchers identify the locations of genes involved in quantitative traits, such as fruit yield and meat weight, that are valuable in agriculture. The use of molecular markers to identify such genes is described in Chapter 25 (see Figure 25.7). Genetic maps with a large number

---

TABLE 20.1

Common Types of Molecular Markers

Marker	Description
Restriction fragment length polymorphism (RFLP)	A site in a genome where the distance between two restriction sites varies among different individuals. These sites are identified by restriction enzyme digestion of chromosomal DNA and the use of Southern blotting.
Amplified restriction fragment length polymorphism (AFLP)	The same as an RFLP except that the fragment is amplified via PCR instead of isolating the chromosomal DNA
Minisatellite, also called a site with a variable number of tandem repeats (VNTR)	A site in the genome that contains many repeat sequences. The total length is usually in the size range of several hundred to a few thousand base pairs. Because their total lengths are usually polymorphic, they were once used in DNA fingerprinting but have been largely superseded by microsatellites (see Chapter 24).
Microsatellite, also called a short tandem repeat (STR)	A site in the genome that contains many short tandem repeat sequences. The total length is usually in the size range of 100 to 500 bp, and their lengths may be polymorphic within a population. They are isolated via PCR.
Single-nucleotide polymorphism (SNP)	A site in the genome where a single nucleotide is polymorphic among different individuals. These sites occur commonly in all genomes, and they are gaining greater use in the mapping of disease-causing alleles and in the mapping of genes that contribute to quantitative traits that are valuable in agriculture (see Chapter 24).
Sequence-tagged site (STS)	This is a general term to describe any molecular marker that is found at a unique site in the genome and is amplified by PCR. AFLPs, microsatellites, and SNPs can provide sequence-tagged sites within a genome.

---

of markers can also be used by evolutionary biologists to determine patterns of genetic variation within a species and the evolutionary relatedness of different species.

All of the types of markers described in Table 20.1 have been used in linkage mapping studies. To exemplify their use, we will first consider **restriction fragment length polymorphisms (RFLPs)**, but similar mapping strategies can be followed for other types of markers. As discussed in Chapter 18, restriction enzymes recognize specific DNA sequences and cleave the DNA at those sequences. Along a very long chromosome, a particular restriction enzyme will recognize many sites. For example, a commonly used restriction enzyme, *Eco*RI, recognizes 5'-GAATTC-3'. Simply by chance, this six-nucleotide sequence is expected to occur (on average) every 4^{-6} bases, or once in every 4,096 bases. A chromosome composed of millions of nucleotides contains many *Eco*RI sites, which will be randomly distributed along the chromosome. Therefore, *Eco*RI will digest a chromosome into many smaller pieces of different lengths.

When comparing two different individuals, the digestion of chromosomal DNA by a given restriction enzyme may produce certain fragments that differ in their length, even though such fragments are found at the same chromosomal locations in the two individuals (**Figure 20.5**). In this case, geneticists would say there is polymorphism in the population with regard to the length of a particular DNA fragment. This variation can arise for several reasons. Among different individuals, genetic changes such as small deletions and duplications may subtract or add segments of DNA to a particular region of the chromosome. This tends to occur more frequently if a region contains a repetitive sequence. Alternatively, a mutation may change the DNA sequence in a way that alters a recognition site for a restriction enzyme. For example, in Figure 20.5, the elimination of a restriction site has occurred in the chromosome of individual 2. This has altered the sizes of certain DNA fragments that result from the digestion with *Eco*RI.

Figure 20.6 considers the application of RFLP analysis to a short chromosomal region among three diploid individuals. Due to slight differences in their DNA sequences, these individuals vary with regard to the existence of an *Eco*RI site shown in red. In certain individuals, a mutation may change the sequence at this site so that it is no longer recognized by *Eco*RI. In a diploid

species, each individual has two copies of this region. In individual 1, both of these *Eco*RI sites are present, while in individual 2, both are absent. By comparison, individual 3 is a heterozygote. In one of the chromosomes, the site is present, but in the other, it is absent. In this experiment, the DNA was isolated and then cut by *Eco*RI. The DNA from individual 1 is cut at six sites by *Eco*RI, whereas individual 2 has only five locations where the DNA is cut. In individual 2, an *Eco*RI site is missing because the sequence of DNA in this region has been changed so that it is no longer GAATTC and is not recognized by *Eco*RI.

When subjected to gel electrophoresis, the change in the DNA sequence at a single *Eco*RI restriction site produces a variation in the sizes of homologous DNA fragments. The term polymorphism, meaning many forms, refers to the idea that the individuals within a population differ with regard to these particular DNA fragments. As noted in Figure 20.6, not all restriction sites are polymorphic. The three individuals share many DNA fragments that are identical in size. When a DNA segment is identical among all members of a population, it is said to be **monomorphic** (meaning one form). As a practical rule of thumb, a DNA segment is considered monomorphic when over 99% of the individuals in the population have identical sequences at that segment.

The preceding discussion was meant as an overview of the phenomenon of RFLPs. However, the experiment of Figure 20.6 is a technical oversimplification, because it considers only a very short region of chromosomal DNA. In an actual RFLP analysis,

FIGURE 20.5 **Restriction fragment length polymorphisms (RFLPs).** For simplicity, only one chromosome is shown for each individual. In a diploid species, a cell would contain two copies of each chromosome, which may or may not have the same RFLP.

Cut the DNA from all 3 individuals with *Eco*RI.

Separate the DNA fragments by gel electrophoresis.

Polymorphic bands are indicated at the arrows.

FIGURE 20.6   An RFLP analysis of chromosomal DNA from three different individuals. The *Eco*RI site shown in red is sometimes missing because a mutation has changed the DNA sequence in this region.

DNA samples containing all of the chromosomal DNA would be isolated from these three individuals. The digestion of the chromosomal DNA with *Eco*RI would then yield so many fragments that the results would be very difficult to analyze. To circumvent this problem, Southern blotting is used to identify specific RFLPs.

Figure 20.7 illustrates such an experiment that reconsiders the chromosomes from the three individuals described in Figure 20.6. A DNA strand from a cloned piece of DNA that corresponds to the region near the fourth *Eco*RI site is used as a radiolabeled probe in a Southern blot of the chromosomal DNA. When the blot is exposed to X-ray film, we will see only the DNA bands that can hybridize to the radiolabeled probe. As shown in Figure 20.7, individual 1 shows one band of 3,000 bp, and individual 2 has a band 4,500 bp long. As discussed in regard to Figure 20.6, this difference is due to the absence of an *Eco*RI site in individual 2. Individual 3 has both bands because of heterozygosity for this *Eco*RI site. In other words, the heterozygote will have two thinner bands of different lengths, whereas homozygotes will display only one band, because the RFLP procedure detects the molecular products of each chromosome. In an actual RFLP analysis, researchers would analyze hundreds or thousands of RFLPs by using a collection of probes that recognize RFLPs throughout the genome.

The Distance Between Two Linked RFLPs Can Be Determined

Now that we understand what an RFLP is, let's consider how RFLPs are mapped and used as chromosomal markers. In the linkage mapping experiments described in Chapter 5, we learned how the results of genetic crosses are used to map the distance

Autoradiograph following Southern blotting:

X-ray film

FIGURE 20.7   Southern blotting of a specific RFLP. In this experiment, a Southern blot was conducted on the chromosomal DNA from the same three individuals described in Figure 20.6. A labeled strand of DNA, just to the right of the fourth *Eco*RI site, was used as a probe.

between two genes. In that type of analysis, the proportions of offspring with recombinant phenotypes were used to calculate the map distance between genes. Likewise, we can map the distance between two RFLPs by making crosses and analyzing the offspring. In an RFLP analysis, however, we do not look at the phenotypic characteristics of the offspring (e.g., white eyes or miniature wings). Instead, we isolate DNA from a tissue sample and look at the DNA bands on a gel.

Figure 20.8 shows an example of how RFLP mapping is done. Let's suppose two RFLPs are located at different sites in the genome. One RFLP is detected with probe 1 and yields either a 4,500 bp band or a 6,500 bp band. A second RFLP is detected with probe 2 and yields either a 2,000 bp band or a 1,500 bp band. In the experiment of Figure 20.8, the goal is to determine whether or not the 4,500/6,500 and 2,000/1,500 sites are linked along the same chromosome and, if so, to find the map distance between them. The researcher begins with two strains that are homozygous: 4,500 and 2,000 versus 6,500 and 1,500. These strains are crossed to each other to produce a heterozygote. The heterozygote can then be crossed to a homozygote that carries only the 4,500 and 2,000 bands. This is analogous to a dihybrid test cross described in Chapter 5, in which an F_1 heterozygote is crossed to a homozygote (see Figure 5.9). The results of this cross depend on whether the RFLPs are closely linked or not. If an offspring inherits pairs of fragments like the parents (4,500 bp and 2,000 bp; 6,500 bp and 1,500 bp), such offspring will show a parental phenotype consisting of four bands of 4,500 bp, 2,000 bp, 6,500 bp, and 1,500 bp or two bands of 4,500 and 2,000 bp. The recombinant phenotypes are 4,500, 2,000 bp, and 6,500 bp or 4,500, 2,000, and 1,500 bp. If the RFLPs are not linked, we would expect a 1:1:1:1 ratio of the four types among the offspring. If the RFLPs are linked, we expect a higher percentage of offspring with a parental combination of fragments. Figure 20.8 describes the results of 100 offspring. Many more parental offspring were observed, which indicates that the RFLPs are closely linked.

In an actual analysis of RFLP data, researchers simultaneously analyze many RFLPs and decide on the likelihood of linkage between two RFLPs by using a statistical test called the **lod** (**l**ogarithm of the **od**ds) **score method.** This method was devised by Newton Morton in 1955. Although the theoretical basis of this approach is beyond the scope of this textbook, computer programs analyze pooled data from a large number of pedigrees or crosses involving many RFLPs. The programs determine the probability, or odds, that any two markers exhibit a certain degree of linkage (that is, are within a particular number of map units apart) and also the probability that the data would have been obtained if the two markers were unlinked. The lod score is then calculated as the ratio of these values:

$$\text{Lod score} = \log_{10} \frac{\text{Probability of a certain degree of linkage}}{\text{Probability of independent assortment}}$$

For example, a lod score of 3 reflects a $\log_{10}$ of 1,000. This means the probability is 1,000-fold greater that the two markers are linked rather than assort independently. Traditionally, geneticists accept that two markers are linked if the lod score is 3 or greater.

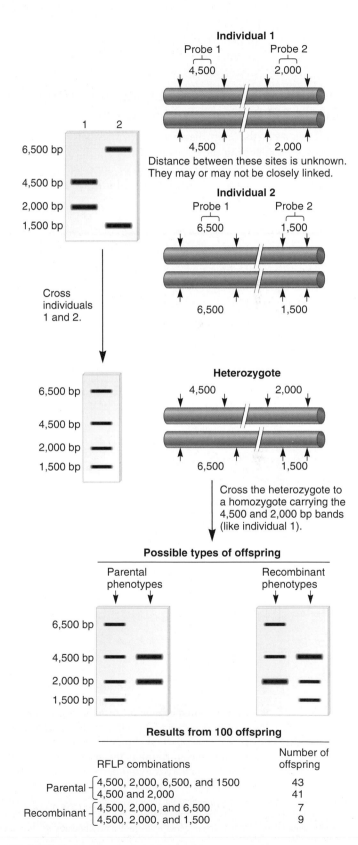

Results from 100 offspring

RFLP combinations		Number of offspring
Parental	4,500, 2,000, 6,500, and 1500	43
	4,500 and 2,000	41
Recombinant	4,500, 2,000, and 6,500	7
	4,500, 2,000, and 1,500	9

FIGURE 20.8   Linkage analysis of RFLP markers.

Figure 20.8 illustrates the general type of approach that researchers can follow to map RFLP markers. If a lod score suggested that these two markers were linked, they would calculate the map distance by dividing the recombinant offspring (7 + 9 = 16) by the total number (100) and multiplying by 100, to obtain a map distance of 16 mu.

An RFLP Map Describes the Locations of Many Different RFLPs Throughout a Genome

As we have seen, the map distance between two RFLPs can be determined by analyzing the transmission of RFLP markers from parents to offspring. Such an analysis can be conducted on many different RFLPs to determine their relative locations throughout

a genome. RFLPs are quite common in virtually all species, so geneticists can easily map the locations of many RFLPs within a genome. A linkage map composed of many RFLP markers is called an **RFLP map**. **Figure 20.9** shows a simplified RFLP map of the plant *Arabidopsis thaliana* (a small plant in the mustard family), which is one of the favorite model organisms of plant molecular geneticists. Many RFLPs have been mapped to different locations along the five *Arabidopsis* chromosomes.

An RFLP map, like the one shown in Figure 20.9, can be used to locate functional genes within the genome. As an example, solved problem S2 illustrates how RFLP analysis can be used to map a gene that confers herbicide resistance. As described next, RFLP analysis can also be used to determine the likelihood that a person may carry a disease-causing allele.

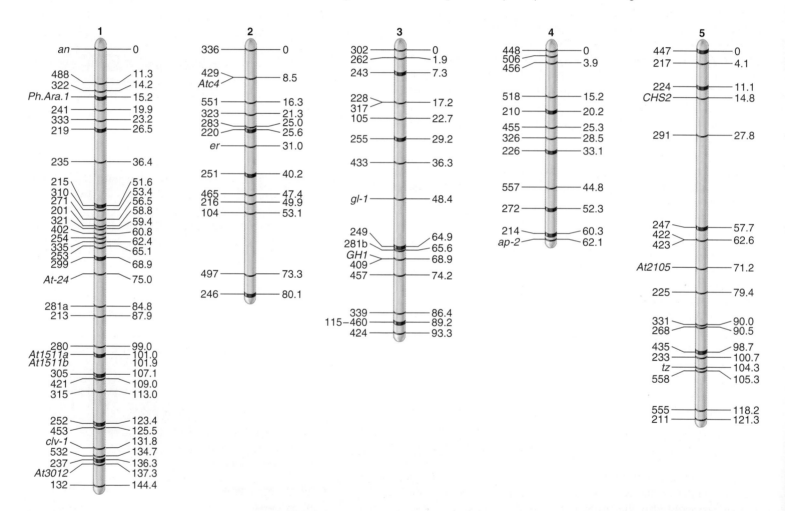

FIGURE 20.9   An RFLP linkage map of *Arabidopsis thaliana*. This plant has five different chromosomes. The left side of each chromosome describes the locations of RFLP markers. The numbers along the right side of each chromosome are the map distances in map units. For example, 219 and 235 are RFLPs located 9.9 mu apart on chromosome 1. The top marker at the end of each chromosome was arbitrarily assigned as the starting point (zero) for each chromosome. In addition, the map shows the locations of a few known genes (shown in red): *Ph.Ara.1* = phytochrome, *At-24* = nitrate reductase, *At1511a/b* = small RNA found in seeds, *clv-1* = clavata-1, *At3012* = alcohol dehydrogenase, *Atc4* = actin, *er* = erecta, *gl-1* = glabra-1, *GH1* = acetolactate synthase, *ap-2* = apelata-2, *CHS2* = chalcone synthase, *At2105* = 12S seed storage protein, and *tz* = thiazole-requiring allele.

Genes→Traits As a first step in mapping the locations of an organism's genes, researchers may initially determine the sites of RFLPs along the chromosomes. In this case, researchers determined the locations of many of these sites along the *Arabidopsis* chromosomes. By mapping these RFLP sites, it becomes easier to locate genes within the *Arabidopsis* genome. The identification of genes helps researchers to elucidate the relationship between genes and traits.

RFLP Analysis Can Be Used to Follow the Inheritance of Disease-Causing Alleles

As mentioned earlier, RFLP analysis can locate particular genes relative to an RFLP marker. This method may also be used to inform prospective parents of the likelihood that they are heterozygous for a recessive allele that may cause a genetic disease. This information would allow them to predict the likelihood of having children affected with the disorder. This genetic testing method is particularly useful for diseases in which the phenotype of the heterozygote may not differ from that of the unaffected homozygote.

The assumption behind this approach is that a disease-causing allele had its origin in a single individual known as a **founder,** who lived many generations ago. Since that time, the allele has spread throughout portions of the human population. In the case of recessive disorders, descendants of the founder who are affected with the disease have inherited two copies of the mutant allele. A second assumption is that the founder is likely to have had a polymorphic molecular marker that lies somewhere near the mutant allele. This is a reasonable assumption, because all people carry many polymorphic markers throughout their genomes. If a polymorphic marker lies very close to the disease-causing allele, it is unlikely that a crossover will occur in the intervening region. Therefore, such a polymorphic marker may be linked to the disease-causing allele for many generations. For this reason, an association between a particular polymorphic marker and a disease-causing allele may help to predict whether a person is a heterozygote for a recessive disease-causing allele.

As an example, let's consider the allele that causes sickle-cell disease. A couple may want to know if they are heterozygous carriers of the recessive allele that causes sickle-cell disease (see Chapter 4 for a description of this disease). While other techniques are now used to distinguish sickle-cell homozygotes and heterozygotes, the experiment described in **Figure 20.10** is a classic study conducted in 1978 that confirmed that RFLP markers can be used to predict the likelihood an individual carries a disease-causing allele. In this study, researchers Yuet Kan and Andree Dozy conducted an RFLP analysis in the region of the β-globin gene. This gene encodes the β-globin polypeptide, which is a subunit of hemoglobin. The common allele (Hb^A) results in the production of hemoglobin A, while a mutant form of the β-globin gene (Hb^S) encodes an abnormal polypeptide that results in the production of an altered hemoglobin form, hemoglobin S. Individuals who are homozygous for this mutant allele develop sickle-cell disease. Kan and Dozy set out to determine if RFLP differences could be detected between chromosomes containing these forms of the β-globin gene. In other words, they wanted to discover if the restriction digestion of chromosomal DNA in the vicinity of the β-globin gene was different in the Hb^A and Hb^S forms of the gene.

In the study described in Figure 20.10, the genotypes of 73 individuals were already known from a biochemical analysis of their hemoglobin proteins. The goal of this research was to see if an RFLP analysis could also distinguish individuals carrying the Hb^S allele from those carrying the Hb^A allele. As shown in Figure 20.10, they began with samples of white blood cells from these individuals and isolated the chromosomal DNA. Next, the chromosomal DNA was digested with the restriction enzyme *Hpa*I. The DNA fragments were separated by gel electrophoresis during the procedure of Southern blotting; strands that are complementary to the β-globin gene were used as probes. An RFLP was identified that produced three different sized fragments among a population of 73 individuals. The sizes of the fragments were 7,000 bp, 7,600 bp, and 13,000 bp, or 7.0, 7.6, and 13 kbp (kilobase pairs).

■ THE HYPOTHESIS

A mutant allele may be closely linked to a particular RFLP. People who carry the mutant allele can be identified by which sized fragment(s) they carry.

■ TESTING THE HYPOTHESIS — FIGURE 20.10   RFLP linkage to a disease-causing allele.

Starting materials: The β-globin gene had been cloned previously, and radiolabeled strands from this gene were used as probes. The researchers obtained blood samples from 73 individuals, some of whom were known to be affected with sickle-cell disease.

Experimental level	Conceptual level

1. Take blood samples from 73 people. (Note: The *Hb* genotype was already known from a biochemical analysis of the subjects' hemoglobin.)

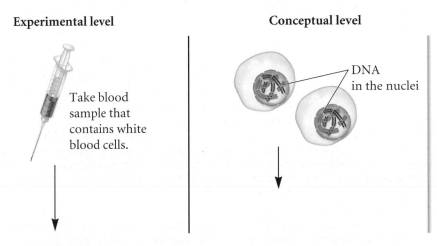

Take blood sample that contains white blood cells.

DNA in the nuclei

2. Isolate chromosomal DNA from the cells. This is similar to plasmid isolation. It involves breaking the cells and then using chromatography or centrifugation to purify the DNA. See Figure 19.1.

3. Digest the chromosomal DNA with the restriction enzyme *Hpa*I.

4. Use radiolabeled DNA strands from the β-globin gene as probes to conduct a Southern blot.

Tube with purified chromosomal DNA

Add *Hpa*I.

See Figure 20.7 for a description of this approach.

Long chromosomal DNA

DNA fragments following digestion

Radiolabeled probe recognizing a DNA fragment

X-ray film

■ THE DATA

Results from 5 different individuals

Summary of data from 73 people

Hemoglobin* Composition	*Hpa*I β-globin RFLP						Frequency of the 13 kbp Fragment
	7.6/7.6	7.6/7.0	7.0/13	7.6/13	13/13	Total	
African**							
AA	8	6	0	1	0	15	3%
AS	5	1	1	9	0	16	31%
SS	0	0	0	4	11	15	87%
Caucasian**							
AA	12	0	0	0	0	12	0%
Asian**							
AA	15	0	0	0	0	15⁄73	0%

*A is β-globin from *Hb*^{A}, and S is β-globin from *Hb*^{S}.

**Racial origin of the individuals used in this study.

Data from: Kan, Y.W., and Dozy, A.M. (1978) Polymorphism of DNA sequence adjacent to human β-globin structural gene: Relationship to sickle mutation. *Proc. Natl. Acad. Sci. USA* 75, 5631–5635.

INTERPRETING THE DATA

With regard to the Hb^A and Hb^S alleles, the occurrence of an RFLP existing in three different fragments was not random. Instead, the 13 kbp fragment was usually found in persons who were known to have at least one copy of the Hb^S allele. For example, in approximately 87% of the homozygotes ($Hb^S\ Hb^S$), the Hb^S allele was associated with the 13 kbp fragment. This observation is consistent with the diagram shown here.

In most people with the Hb^A allele, the site where the normal β-globin gene is located has a distance of 7.0 or 7.6 kbp between the HpaI sites. For people carrying the Hb^S allele, the mutant β-globin gene is found in a chromosomal region where the HpaI sites are usually 13 kbp apart.

In most, but not all, cases, the Hb^S allele was found to be associated with a 13 kbp fragment. This type of information can be used as a predictive tool in human genetic analysis. For example, if an individual was found to have both 7.6 kbp and 13 kbp fragments, it would be fairly likely that the individual is a heterozygous (Hb^AHb^S) carrier. By comparison, if an RFLP analysis revealed that an individual had only a 7.6 kbp fragment, that person would more likely be homozygous for the Hb^A allele.

> A self-help quiz involving this experiment can be found at the Online Learning Center.

Linkage Mapping Also Uses Molecular Markers Called Microsatellites

To make a highly refined map of a genome, many different polymorphic sites must be identified and their transmission followed from parent to offspring over many generations. RFLPs were among the first molecular markers studied by geneticists. More recently, the use of other molecular markers has become more prevalent because they are easier to generate via PCR. As an example, let's consider **microsatellites,** short repetitive sequences that are abundantly interspersed throughout a species' genome and are quite variable in length among different individuals. For example, the most common microsatellite encountered in humans is a sequence $(CA)_n$, where n may range from 5 to more than 50. In other words, this dinucleotide sequence can be tandemly repeated 5 to 50 or more times. The $(CA)_n$ microsatellite is found, on average, about every 10,000 bases in the human genome. Researchers have identified thousands of different DNA segments that contain $(CA)_n$ microsatellites, located at many distinct sites within the human genome. Using primers complementary to the unique DNA sequences that flank a specific $(CA)_n$ region, a particular microsatellite can be amplified by PCR. In other words, the PCR primers will amplify only a particular microsatellite, but not the thousands of others that are interspersed throughout the genome (**Figure 20.11**).

When a pair of PCR primers amplifies a single site within a set of chromosomes, the amplified region is called a **sequence-tagged site (STS)**. In a diploid species, an individual will have two copies of a given STS. When an STS contains a microsatellite, the two PCR products will be identical only if the region is the same length in both copies (i.e., if the individual is homozygous for the microsatellite). However, if an individual has two copies that differ in the number of repeats in the microsatellite sequence (i.e., if the individual is heterozygous for the microsatellite), the two PCR products obtained will be slightly different in length (as in Figure 20.11).

Like RFLPs, microsatellites that have length polymorphisms allow researchers to follow their transmission from parent to offspring. PCR amplification of particular microsatellites provides an important strategy in the genetic analysis of human pedigrees, as shown in **Figure 20.12**. Prior to this analysis, a unique segment of DNA containing a microsatellite had been identified. Using PCR primers complementary to this microsatellite's unique flanking segments, two parents and their three offspring were tested for the inheritance of this microsatellite. A small sample of cells was obtained from each individual and subjected to PCR amplification, as described earlier in Figure 20.11. The amplified PCR products were then analyzed by high-resolution gel electrophoresis, which can detect small differences in the lengths of DNA fragments. The mother's PCR products were 154 and 150 bp in length; the father's were 146 and 140 bp. Their first offspring inherited the 154 bp product from the mother and the 146 bp from the father, the second inherited the 150 bp from the mother and the 146 bp from the father, and the third inherited the 150 bp from the mother and the 140 bp from the father. As shown in the figure, the transmission of polymorphic microsatellites is relatively easy to follow from generation to generation.

The simple pedigree analysis shown in Figure 20.12 illustrates the general method used to follow the transmission of a single microsatellite that is polymorphic in length. In linkage studies, the goal is to follow the transmission of many different microsatellites to determine those that are linked along the same chromosome versus those that are not. Those that are not

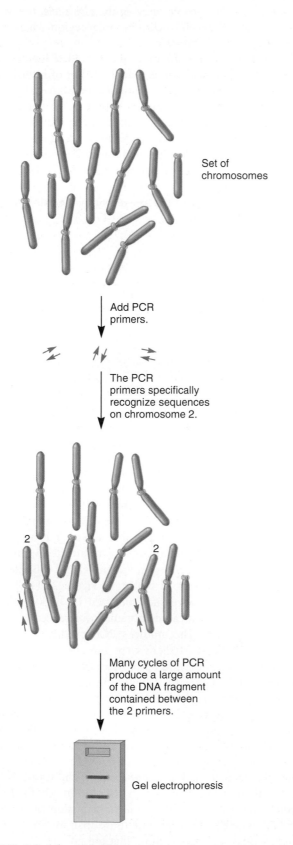

FIGURE 20.11 Identifying a microsatellite using PCR primers.

(a) Pedigree

(b) Electrophoretic gel of PCR products for a polymorphic microsatellite found in the family in (a).

FIGURE 20.12 Inheritance pattern of a polymorphic microsatellite in a human pedigree.

linked will independently assort from generation to generation. Those that are linked tend to be transmitted together to the same offspring. In a large pedigree, it is possible to identify cases where linked microsatellites have segregated due to crossing over. The frequency of crossing over provides a measure of the map distance, in this case between different microsatellites. This approach can help researchers to obtain a finely detailed genetic linkage map of the human chromosomes.

How can pedigree analysis involving STSs, such as polymorphic microsatellites, help researchers to identify the location of disease-causing alleles? By following the transmission of many polymorphic markers within large family pedigrees, it may be possible to determine that particular markers are found in people who carry specific disease-causing alleles. (An example is shown in Chapter 22, Figure 22.5.) As mentioned earlier, this is based on the assumption that the disease-causing allele had its origin in a founder and is likely to have a polymorphic marker nearby. After the identification of a closely linked marker, a disease-causing allele can be identified using a technique called chromosome walking, which is described later in this chapter.

20.2 PHYSICAL MAPPING

We now turn our attention to methods aimed at establishing a physical map of a species' genome. Physical mapping requires the cloning of many pieces of chromosomal DNA. The cloned DNA fragments are then characterized by size (that is, their length in base pairs), as well as the genes they contain and their relative locations along a chromosome.

As mentioned in Chapter 10, eukaryotic genomes are very large; the *Drosophila* genome is roughly 175 million bp long, and the human genome is approximately 3 billion bp in length. When making a physical map of a genome, researchers must characterize many DNA clones that contain much smaller pieces of the genome. For species with very large genomes, the physical mapping of the entire genome of a species is a massive undertaking carried out as a collaborative project by many genome research laboratories. In this section, we will examine the general strategies used in creating a physical map of a species' genome. We will also consider how physical mapping information can be used to clone genes.

A Physical Map of a Chromosome Is Constructed by Creating a Contiguous Series of Clones That Span a Chromosome

As discussed in Chapter 18, a DNA library contains a collection of hybrid vectors, with each vector containing a particular fragment of chromosomal DNA. In physical mapping studies, the goal is to determine the relative locations of the cloned chromosomal fragments from a DNA library, as they would occur in an intact chromosome. In other words, the members of the library must be organized according to their actual locations along a chromosome. To obtain a complete physical map of a chromosome, researchers need a series of clones that contain contiguous, overlapping pieces of chromosomal DNA. Such a collection of clones is known as a **contig** (**Figure 20.13**). A contig represents a physical map of a chromosome. As discussed later, cloning vectors known as BACs and cosmids are commonly used in the construction of a contig.

Different experimental strategies can be used to align the members of a contig. The general approach is to identify clones that contain overlapping regions. Historically, Southern blotting was used to determine if two different clones contain an overlapping region. In the example shown in Figure 20.13, the DNA

from clone 1 could be radiolabeled in vitro and then used as a probe in a Southern blot of the other clones shown in this figure. Clone 1 would hybridize to clone 2, because they share identical DNA sequences in the overlapping region. Similarly, clone 2 could be used as a probe to show that it will hybridize to clone 1 and clone 3. By conducting Southern blots between many combinations of clones, researchers can determine which clones have common overlapping regions and thereby order them as they would occur along the chromosome.

Alternatively, other methods to order the members of a contig involve the use of molecular markers. For example, if many STSs have already been identified along a chromosome via linkage and/or cytogenetic analysis, the STSs can be used as probes to order the members of a contig. Another approach is to subject the members of a library to a restriction enzyme to produce a restriction digest pattern for each member. The patterns of fragments are then analyzed by computer programs, which can identify regions that are potentially overlapping.

An ultimate goal of physical mapping procedures is to obtain a complete contig for each type of chromosome within a full set. For example, in the case of humans, a complete physical map requires a contig for each of the 22 autosomes and for the X and Y chromosomes. Geneticists can also correlate cloned DNA fragments in a contig with locations along a chromosome obtained from linkage or cytogenetic mapping. For example, a member of a contig may contain a gene, RFLP, or STS that previously has been mapped by linkage analysis. **Figure 20.14** considers a situation in which two members of a contig carry genes previously mapped by linkage analysis to be approximately 1.5 mu apart on chromosome 11. In this example, clone 2 has an insert that carries gene *A*, while clone 7 has an insert that carries

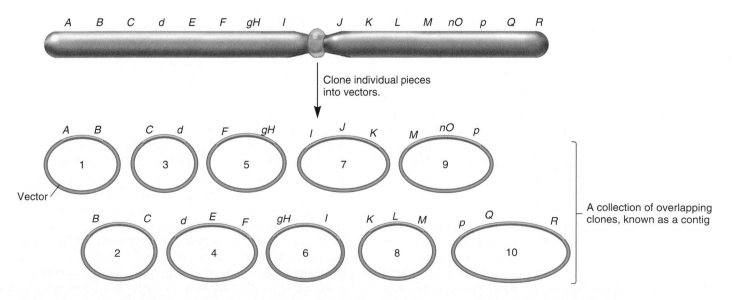

FIGURE 20.13   The construction of a contig. Large pieces of chromosomal DNA are cloned into vectors, and their order is determined by the identification of overlapping regions. For example, clone 3 ends with gene *d*, and clone 4 begins with gene *d*; clone 4 ends with gene *F*, and clone 5 begins with gene *F*.

FIGURE 20.14   The use of genetic markers to align a contig. In this example, gene *A* and gene *B* had been mapped previously to specific regions of chromosome 11. Gene *A* was found within the insert of clone 2, gene *B* within the insert of clone 7. This made it possible to align the contig using gene *A* and gene *B* as genetic markers (i.e., reference points) along chromosome 11.

gene *B*. Because a contig is composed of overlapping members, a researcher can align the contig along chromosome 11, starting with gene *A* and gene *B* as reference points. In this example, genes *A* and *B* serve as markers that identify the location of specific clones within the contig.

YAC, BAC, and PAC Cloning Vectors Are Used to Make Contigs of Eukaryotic Chromosomes

For large eukaryotic genomes, researchers often begin with cloning vectors that can accept chromosomal DNA inserts of very large size. By having large insert sizes, a contig is more easily constructed and aligned because fewer recombinant vectors are needed. In general, most plasmid and viral vectors can accommodate inserts only a few thousand to perhaps tens of thousands of nucleotides in length. If a plasmid or viral vector has a DNA insert that is too large, it will have difficulty with DNA replication and is likely to suffer deletions in the insert.

By comparison, other cloning vectors, known as **artificial chromosomes,** can accommodate much larger sizes of DNA inserts. As their name suggests, they behave like chromosomes when inside of living cells. The first type to be made was the **yeast artificial chromosome (YAC),** which was developed by David Burke, Georges Carle, and Maynard Olson in 1987. An insert within a YAC can be several hundred thousand to perhaps 2 million base pairs in length. For an average human chromosome, a few hundred YACs are sufficient to create a contig with fragments that span the entire length of the chromosome. By comparison, it would take thousands or even tens of thousands of recombinant plasmid vectors to create such a contig.

At the molecular level, YACs have structural similarities to normal eukaryotic chromosomes, yet have characteristics that make them suitable for cloning. The general structure of a YAC vector is shown at the top left of **Figure 20.15.** The YAC vector contains two telomeres (*TEL*), a centromere (*CEN*), a bacterial

origin of replication (*ORI*), a yeast origin of replication (known as an *ARS*, for autonomous replication sequence), selectable markers, and unique cloning sites that are each recognized by a single restriction enzyme. Without an insert, the circular form of this vector can replicate in *E. coli*. After a large fragment has been inserted, the linear form of the vector can replicate in yeast.

In the experiment shown in Figure 20.15, chromosomal DNA is digested with the restriction enzyme *Eco*RI at a low concentration so that only some of the restriction sites are cut. This partial digestion will result in only occasional cleavage of the chromosomal DNA to yield very large DNA fragments.

The circular YAC vector is also digested with *Eco*RI and a second restriction enzyme, *Bam*HI, to yield two arms of the YAC. The YAC arms are then ligated to the large fragments of chromosomal DNA and transformed into yeast cells. When ligation occurs in the desired way, a large piece of chromosomal DNA becomes ligated to both arms of the YAC. Because each arm contains a different selectable marker, it is possible to select for the growth of yeast cells that carry a YAC construct having both arms.

Newer types of cloning vectors called **bacterial artificial chromosomes (BACs)** and **P1 artificial chromosomes (PACs)** have been constructed. BACs were developed from bacterial F factors, which are described in Chapter 6, and PACs were developed from P1 bacteriophage chromosomes. BACs and PACs typically can contain inserts up to 300,000 bp and sometimes larger. These vectors are somewhat easier to use than YACs because the DNA is inserted into a circular molecule and transformed into *E. coli*. BACs and PACs have largely replaced YACs for the cloning of large DNA fragments.

Figure 20.16 shows a simplified drawing of a BAC cloning vector. The vector contains several genes that function in vector replication and segregation. The origin of replication is designated *oriS*, and the *repE* gene encodes a protein essential for replication at *oriS*. The *parA*, *parB*, and *parC* genes encode proteins required for the proper segregation of the vector into daughter cells. A chloramphenicol resistance gene, *cm^R*, provides a way to select for cells that have taken up the vector based on their ability to grow in the presence of the antibiotic chloramphenicol. The vector also contains unique restriction enzyme sites, such as *Hin*dIII, *Bam*HI, and *Sph*I, for the insertion of large fragments of DNA. These sites are located within the *lacZ* gene, which encodes the enzyme β–galactosidase. Vectors with DNA inserts can be determined by plating cells on media containing the compound X-Gal (as described for plasmid vectors in Chapter 18, Figure 18.2).

YAC, BAC, and PAC cloning vectors have been very useful in the construction of contigs that span long segments of chromosomes. They have been used as the first step in creating a rough physical map of a genome. While this is an important step in physical mapping, the large insert sizes make them difficult to use in gene cloning and sequencing experiments. Therefore, libraries containing hybrid vectors with smaller insert sizes are needed. Most commonly, a type of cloning vector called a **cosmid** is used. A cosmid is a hybrid between a plasmid vector and phage λ; its DNA can replicate in a cell like a plasmid or be packaged into a protein coat like a phage. Cosmid vectors typically can accept DNA fragments that are tens of thousands of bp in length.

FIGURE 20.15   **The use of a YAC vector in DNA cloning.** Note: *Eco*RI and *Bam*HI both cut DNA to yield sticky ends. In this diagram, only the *Eco*RI ends are shown with a notch to emphasize that these are the sticky ends that bind to the chromosomal DNA.

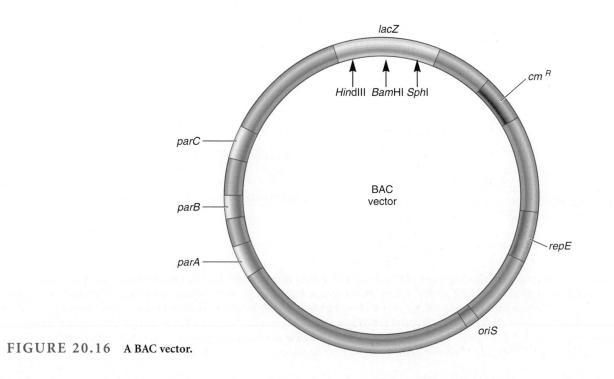

FIGURE 20.16   **A BAC vector.**

Figure 20.17 illustrates a comparison between cytogenetic, linkage, and physical maps of human chromosome 16. Actually, this is a very simplified map of chromosome 16. A much more detailed map is available, although it would take well over 10 pages of this textbook to print it! The top of this figure shows the banding pattern of this chromosome. Underneath the banded chromosome are molecular markers that have been mapped by linkage analysis. These same markers have been mapped cytogenetically. A complete contig of this chromosome has also been produced by generating a series of overlapping YACs. However, Figure 20.17 shows the location of only one YAC within the contig. Cosmids within this region are shown below the YAC. In addition, an STS is found within the region and provides a molecular marker for cosmids N16Y1-19 and N16Y1-10, as well as YAC N16Y1.

FIGURE 20.17   A correlation of cytogenetic, linkage, and physical maps of human chromosome 16. The top part shows the cytogenetic map of human chromosome 16 according to its G banding pattern. A very simple linkage map of molecular markers (D16S85, D16S60, etc.) is aligned below the cytogenetic map. A correlation between the linkage map and a segment of the physical map is shown below the linkage map. A YAC clone designated YAC N16Y1 is located between markers D16AC6.5 and D16S150 on the linkage map. Pieces of DNA from this YAC were subcloned into cosmid vectors, and the cosmids (310C4, N16Y1-29, N16Y1-18, etc.) were aligned relative to each other. One of the cosmids (N16Y1-10) was sequenced, and this sequence was used to generate an STS shown at the bottom of the figure.

Positional Cloning Can Be Achieved Using Chromosome Walking

The creation of a contig bears many similarities to a gene cloning strategy known as **positional cloning,** a strategy in which a gene is cloned based on its mapped position along a chromosome. This approach has been successful in the cloning of many human genes, particularly those that cause genetic diseases when mutated. These include genes involved in cystic fibrosis, Huntington disease, and Duchenne muscular dystrophy.

A common method used in positional cloning is known as **chromosome walking.** To initiate this type of experiment, a gene's position relative to a marker must be known from mapping studies. For example, a gene may be known to be fairly close to a previously mapped gene or RFLP marker. This provides a starting point to molecularly "walk" toward the gene of interest.

Figure 20.18 considers a chromosome walk in which the goal is to locate a gene that we will call gene A. In this example, linkage mapping studies have revealed that gene A is relatively

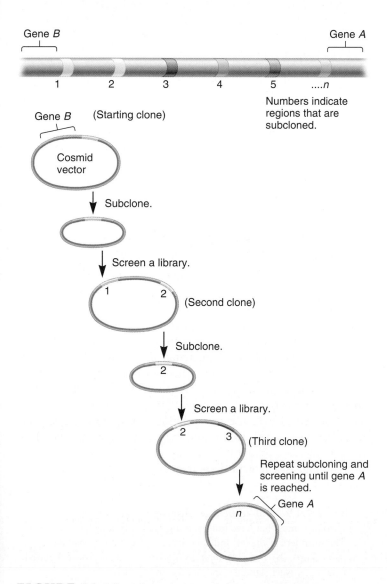

Gene B   (Starting clone)

Numbers indicate regions that are subcloned.

FIGURE 20.18   **The technique of chromosome walking.**

close to another gene, called gene B, that has been previously cloned. Gene A and gene B have been deduced from genetic crosses to be approximately 1 mu apart. To begin this chromosome walk, a cloned DNA fragment that contains gene B and flanking sequences can be used as a starting point.

To walk from gene B to gene A, a series of library screening methods are followed. In this example, the starting materials are a cosmid library and a clone containing gene B. A small piece of DNA from the first cosmid vector containing gene B is inserted into another vector. This is called **subcloning.** The subclone is radiolabeled and used as a probe to screen a cosmid library. This will enable the researchers to identify a second clone that extends into the region that is closer to gene A. A subclone from this second clone is then used to screen the library a second time. This allows the researchers to identify a clone that is even closer to gene A. This repeated pattern of subcloning and library screening is used to reach gene A. The term chromosome walking is an appropriate description of this technique, because each clone takes you a step closer to the gene of interest. When starting at gene B in Figure 20.18, researchers would also want to have markers to the left of gene B to ensure they were not walking in the wrong direction.

The number of steps required to reach the gene of interest depends on the distance between the starting and ending points and on the sizes of the DNA inserts in the library. If the two points are 1 mu apart, they are expected to be approximately 1 million bp apart, although the correlation between map units and physical distances can be quite variable. In a typical walking experiment, each clone might have an average insert size of 50,000 bp. Therefore, it will take about 20 walking steps to reach the gene of interest. Researchers want to locate starting points in a chromosome walking experiment that are as close as possible to the gene they wish to identify.

How do researchers know when they have reached a gene of interest? In the case of a gene that causes a disease when mutant, researchers would conduct their walking steps on DNA from both an unaffected individual and an affected individual. Each set of clones would be subjected to DNA sequencing, and those DNA sequences would be compared to each other. When the researchers reach a spot where the DNA sequences differ between an unaffected individual and an affected individual, such a site may be within the gene of interest. However, this has to be confirmed by sequencing the region from several unaffected and affected individuals to be certain the change in DNA sequence is correlated with the disease.

Genomes Can Be Compared by Pulsed-Field Gel Electrophoresis

Researchers may also compare genomes using techniques that separate large DNA fragments and/or small chromosomes. One technique that can accomplish this goal is **pulsed-field gel electrophoresis (PFGE),** developed in 1984 by David Schwartz and Charles Cantor. Using this method, large pieces of chromosomes or entire small chromosomes can be separated and identified. Pulsed-field gel electrophoresis differs from conventional electrophoresis in that the DNA is driven through the gel using alternating pulses of electrical current at different angles. This can be achieved using electrophoresis devices that have two sets of electrodes. The alternating pulses can separate larger DNA fragments that vary in length from 50,000 bp up to 10 million bp.

For PFGE, large fragments of chromosomes or individual chromosomes must be isolated very carefully. To avoid mechanical breakage of the chromosomes, cells are embedded in agarose blocks (also called plugs). The agarose protects the chromosomal DNA from mechanical breakage. Once inside the plug, the cells are lysed, and if desired, the chromosomal DNA can be subjected to restriction digestion to obtain large DNA fragments. This can be achieved using a restriction enzyme that cuts very infrequently. To separate the large pieces of DNA, the plug is inserted into a well in an agarose gel and then subjected to PFGE.

Figure 20.19a illustrates a separation of yeast chromosomes that have not been subjected to restriction digestion. **Figure 20.19b** shows fragments of DNA that have been obtained by digesting a bacterial chromosome with three different restriction enzymes that cut the chromosome infrequently.

What are the applications of PFGE? In the past, PFGE was often used to isolate large pieces of chromosomes to construct contigs. However, in recent years, PFGE has been most commonly used to identify species of microorganisms or strains

of the same species. For example, some strains of *E. coli* are pathogenic, while others are not. When subjected to PFGE, such strains may differ in their banding patterns. Therefore, this method has gained widespread use to distinguish different strains of microorganisms.

20.3 GENOME SEQUENCING PROJECTS

Genome sequencing projects are research endeavors that have the ultimate goal of determining the sequence of DNA bases of the entire genome of a given species. Such projects involve many participants, including scientists who isolate DNA and perform DNA sequencing reactions, as well as theoreticians who gather the DNA sequence information and assemble it into a long DNA sequence for each chromosome. For bacteria and archaea, which usually have just one chromosome, the genome sequence is that of a single chromosome. For eukaryotes, each chromosome must be sequenced. Thus, for humans, the genome sequence includes sequences of 22 autosomes, 2 sex chromosomes, and the mitochondrial genome.

In just a couple of decades, our ability to map and sequence genomes has improved dramatically. By 2007, the complete genome sequences have been obtained from hundreds of different species, including over 400 bacteria, 30 archaea, 20 protists, 20 fungi, 6 plants, and 20 animals. Considering that the first genome sequence was generated in 1995, the progress of genome sequencing projects since then has been truly remarkable! In this section, we will examine the approaches that researchers follow when tackling such a large project. We will also survey some of the general goals of the Human Genome Project, the largest of its kind, and compare the results from the genome sequencing of various species.

Shotgun Sequencing Is Used to Determine the DNA Sequence of Entire Genomes

When sequencing an entire genome, researchers must consider factors such as genome size, the efficiency of the methods used to sequence DNA, and the costs of the project. Since genome sequencing projects began in the 1990s, researchers have learned that the most efficient and inexpensive way to sequence genomes is via an approach called **shotgun sequencing,** in which DNA fragments to be sequenced are randomly generated from larger DNA fragments. In this method, genomic DNA is isolated and sheared into smaller DNA fragments, typically 1,500 bp or longer in length. The researchers then use the technique of dideoxy sequencing, described in Chapter 18, to randomly sequence fragments from the genome. As a matter of chance, some of the fragments will be overlapping, as schematically shown at the top of the next page.

(a) Each band is a different yeast chromosome

(b) Each band is a fragment of a bacterial chromosome

FIGURE 20.19 Pulsed-field gel electrophoresis (PFGE).
(a) Yeast chromosomes that have not been digested with a restriction enzyme. The numbers at the left indicate the size of the chromosomes in kilobase pairs (kbp). **(b)** The *Haemophilus influenzae* chromosome digested with *Eag*I (lane A), *Nae*I (lane B), and *Sma*I (lane C). Lane D shows molecular weight markers.

Overlapping Region

```
TTACGGTACCAGTTACAAATTCCAGACCTAGTACC
AATGCCATGGTCAATGTTTAAGGTCTGGATCATGG
              GACCTAGTACCGGACTTATTCGATCCCCAATTTTGCAT
              CTGGATCATGGCCTGAATAAGCTAGGGGTTAAAACGTA
```

The DNA sequence in two different fragments will be identical in the overlapping region. This allows researchers to order them as they are found in the intact chromosome.

An advantage of shotgun DNA sequencing is that it does not require extensive mapping, which can be very time consuming and expensive. A disadvantage is that researchers will waste some time sequencing the same region of DNA more times than needed.

To obtain a complete sequence of a genome with the shotgun approach, how do researchers decide how many fragments to sequence? We can calculate the probability that a base will not be sequenced using this approach with the following equation:

$$P = e^{-m}$$

where

P is the probability that a base will be left unsequenced

e is the base of the natural logarithm; e = 2.72

m is the number of bases sequenced divided by the total genome size

For example, in the case of a bacterial species with a genome size of 1.8 Mb (i.e., 1.8 million bp), if researchers sequenced 9 Mb, m = 5 (i.e., 9.0 Mb divided by 1.8 Mb):

$$P = e^{-m} = e^{-5} = 0.0067, \text{ or } 0.67\%$$

This means that if we randomly sequence 9.0 Mb, which is five times the length of a single genome, we are likely to miss only 0.67% of the genome. With a genome size of 1.8 Mb, we would miss about 12,000 nucleotides out of approximately 1,800,000. Such missed sequences are typically on small DNA fragments that—as a matter of random chance—did not happen to be sequenced. The missing links in the genome can be sequenced later using mapping methods such as chromosome walking.

For a genome sequencing project, researchers have typically followed two types of shotgun strategies—hierarchical shotgun sequencing or whole genome shotgun sequencing. In the late 1990s, when researchers first began to sequence the larger genomes from eukaryotic species, they first took the approach of **hierarchical shotgun sequencing** (**Figure 20.20a**). To begin this process, researchers clone large (e.g., 150,000 bp long) DNA fragments into BACs (or YACs) to generate a contig for each chromosome. For DNA sequencing, researchers choose a set of clones from each chromosome contig that have relatively short overlapping regions. Each BAC clone in a chosen set is then subjected to shotgun sequencing. A clone is sheared into smaller pieces (e.g., 2,000 bp), and those small pieces are cloned into vectors. The recombinant vectors are then randomly sequenced using a primer that anneals at a site next to the inserted DNA (see Figure 18.15). The method of dideoxy sequencing yields a DNA sequence, usually 500 to 1,000 bp in length, at one end of each insert. The resulting DNA sequences are analyzed by a computer program that identifies overlapping sequences between two DNA fragments and assembles the DNA sequence into one long sequence along a single chromosome.

In the hierarchical method, the nucleotide sequence is determined clone by clone. For this reason, the method is also called the clone-by-clone or BAC-to-BAC approach. This strategy has the advantage of making it easier to align the overlapping sequences because the DNA sequences are grouped according to the BAC clone from which the sequence was obtained. Another issue is that it minimizes the amount of DNA sequencing by reducing the amount of sequencing in overlapping regions. This was an advantage in the 1990s when DNA sequencing was an expensive and labor-intensive procedure. However, a disadvantage of this approach is that BAC mapping can be time-consuming and expensive.

In more recent years, advances in DNA sequencing technology have greatly diminished the time and expense of dideoxy sequencing. Modern genome sequencing projects more commonly follow a strategy called **whole genome shotgun sequencing.** In this method, the physical mapping step used in the hierarchical approach is bypassed. The DNA from the entire genome is isolated and sheared into small (e.g., 2,000 or 10,000 bp) and large (50,000 or 150,000 bp) fragments and cloned into vectors (**Figure 20.20b**). Each insert is usually sequenced at both ends, an approach called **double-barrel shotgun sequencing.** As in the hierarchical approach, the DNA sequences, which are 500 bp to 1,000 bp in length, are then analyzed by a computer program that identifies overlapping regions between two DNA fragments and assembles the data into one long sequence. The double-barrel strategy improves the ability of researchers to align their DNA sequences. Most of the DNA sequencing in a whole genome shotgun approach is carried out on small DNA fragments. However, due to the prevalence of repetitive sequences, particularly within the genomes of eukaryotic species, the alignment of small pieces may be difficult. The sequencing of both ends of longer fragments makes alignment easier, because the relative spacing of particular DNA sequences is known.

The Human Genome Project Was the Largest Genome Sequence Project in History

Due to its large size, the sequencing of the human genome was an enormous undertaking. Scientists had been discussing how to undertake this project since the mid-1980s. In 1988, the National Institutes of Health established an Office of Human Genome Research, with James Watson as its first director. The **Human Genome Project,** which officially began on October 1, 1990, was a 13-year effort coordinated by the U.S. Department of Energy

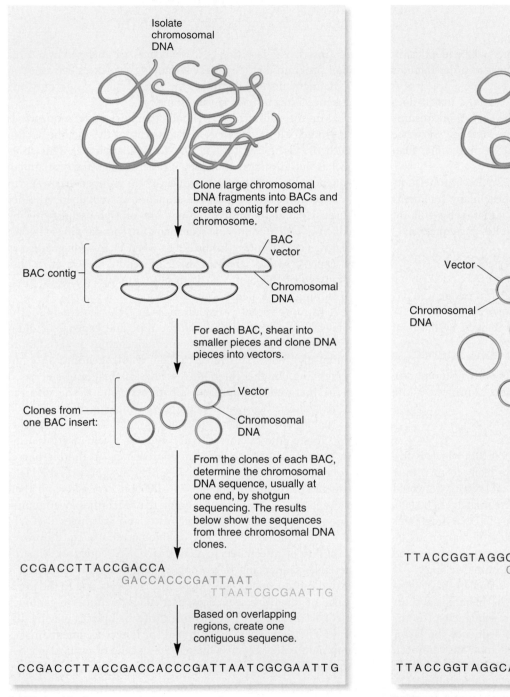

(a) Hierarchical genome shotgun sequencing

(b) Whole genome shotgun sequencing

FIGURE 20.20 **Two approaches of shotgun sequencing of genomes.** Note: the BAC vector and chromosomal DNA are not drawn to scale. The chromosomal DNA would be much longer than the BAC vector DNA.

and the National Institutes of Health. From its outset, the Human Genome Project had the following goals:

1. *To obtain a genetic linkage map of the human genome.* This has involved the identification of thousands of genetic markers and their localization along the autosomes and sex chromosomes.

2. *To obtain a physical map of the human genome.* This has required the cloning of many segments of chromosomal DNA into BACs, YACs, and cosmids.

3. *To obtain the DNA sequence of the entire human genome.* The first (nearly complete) sequence was published in February 2001. This was considered a first draft. A second draft was published in 2003, and the completed maps and sequences for all of the human chromosomes were published by 2006. The entire genome is approximately 3 billion nucleotide base pairs in length. If the entire human genome were typed in a textbook like this, it would be nearly 1 million pages long!

4. *To develop technology for the management of human genome information.* The amount of information obtained from this project is staggering, to say the least. The Human Genome Project developed user-friendly tools to provide scientists easy access to up-to-date information obtained from the project. The Human Genome Project also developed analytical tools to interpret genome information.

5. *To analyze the genomes of other model organisms.* These include bacterial species (e.g., *Escherichia coli* and *Bacillus subtilis*), *Drosophila melanogaster* (fruit fly), *Caenorhabditis elegans* (a nematode), *Arabidopsis thaliana* (a simple plant), and *Mus musculus* (mouse).

6. *To develop programs focused on understanding and addressing the ethical, legal, and social implications of the results obtained from the Human Genome Project.* The Human Genome Project has sought to identify the major genetic issues that will affect members of our society and to develop policies to address these issues. For example, what is an individual's right to privacy regarding genetic information? Some people are worried that their medical insurance company may discriminate against them if it is found that they carry a disease-causing or otherwise deleterious gene.

7. *To develop technological advances in genetic methodologies.* Some of the efforts of the Human Genome Project have involved improvements in molecular genetic technology such as gene cloning, contig construction, DNA sequencing, and so forth. The project has also been aimed at developing computer technology for data processing, storage, and the analysis of sequence information (see Chapter 21).

A great benefit expected from the characterization of the human genome is the ability to identify and study the sequences of our genes. Mutations in many different genes are known to be correlated with human diseases, which include cancer, heart disease, and many other abnormalities. The identification of mutant genes that cause inherited diseases was a strong motivation for the Human Genome Project. A detailed genetic and physical map has made it profoundly easier for researchers to locate such genes. Furthermore, a complete DNA sequence of the human genome provides researchers with insight into the types of proteins encoded by these genes. The cloning and sequencing of disease-causing alleles is expected to play an increasingly important role in the diagnosis and treatment of disease.

Many Genome Sequences Have Been Determined

How do researchers decide which genomes to sequence? Motivation behind genome sequencing projects comes from a variety of sources. For example, basic research scientists can greatly benefit from a genome sequence. It allows a scientist to know which genes a given species has, and it aids in the cloning and characterization of such genes. This has been the impetus for genome projects involving model organisms such as *Escherichia coli*, *Saccharomyces cerevisiae*, *Drosophila melanogaster*, *Caenorhabditis elegans*, *Arabidopsis thaliana*, and the mouse. A second impetus for genome sequencing has involved human disease. As noted previously, researchers expect that the sequencing of the human genome will aid in the identification of genes that, when mutant, play a role in disease. Likewise, the decision to sequence many bacterial, fungal, and protist genomes has been related to the role of these species in infectious diseases. Hundreds of microbial genomes have been sequenced, many of them from species that are pathogenic in humans. The sequencing of such genomes may help to elucidate the genes that play a role in the infectious process.

In addition, the genomes of agriculturally important species have been the subject of genome sequencing projects. An understanding of a species' genome may aid in the development of new strains of livestock and plant species that have improved traits from an agricultural perspective. Finally, genome sequencing projects help us to better understand the evolutionary relationships among living species. This approach, called **comparative genomics,** uses information from genome projects to understand the genetic variation among different populations. We will explore this topic in Chapter 26.

Table 20.2 describes the results of several genome-sequencing projects that have been completed. Newly completed genome sequences are emerging rapidly, particularly those of microbial species. As we obtain more genome sequences, it becomes progressively more interesting to compare them to each other as a way to understand the process of evolution. As we will discover in Chapter 21, the field of functional genomics enables researchers to probe the roles of many genes as they interact to generate the phenotypic traits of the species that contain them.

TABLE **20.2**

Examples of Genomes That Have Been Sequenced

Species	Genome Size (bp)*	Number of Genes**	Description
Prokaryotic genomes			
Bacteria			
Mycoplasma genitalium	580,000	521	An inhabitant of the human genital tract with a very small genome.
Helicobacter pylori	1,668,000	1,590	A common inhabitant of the stomach that may cause gastritis, peptic ulcer, and gastric cancer.
Mycobacterium tuberculosis	4,412,000	4,294	The bacterial species that causes tuberculosis.
Escherichia coli	4,639,000	4,289	A widespread bacterial inhabitant of the gut of animals; also a model research organism.
Archaea			
Thermoplasma volcanium	1,580,000	1,494	An archaea with an optimal growth temperature of 60°C and pH optimum of <2.0.
Pyrococcus abyssi	1,760,000	1,765	Originally isolated from samples taken close to a hot spring situated 3,500 meters deep in the southeast Pacific. Optimal growth conditions are 103°C and 200 atmospheres pressure.
Sulfolobus solfataricus	2,990,000	2,977	Found in terrestrial volcanic hot springs with optimum growth occurring at a temperature of 75–80°C and pH 2–3.
Eukaryotic genomes			
Protists			
Plasmodium falciparum	22,900,000	5,268	A parasitic protist that causes malaria in humans.
Entamoeba histolytica	23,800,000	9,938	An amoeba that causes dysentery in humans.
Fungi			
Saccharomyces cerevisiae (baker's yeast)	12,100,000	6,294	A structurally simple eukaryotic species that has been extensively studied by researchers to understand eukaryotic genetics and cell biology.
Neurospora crassa	40,000,000	10,082	A common bread mold and also a structurally simple eukaryotic species that has been extensively studied by researchers.
Plants			
Arabidopsis thaliana	142,000,000	26,000	A model organism studied by plant biologists.
Oryza sativa (rice)	440,000,000	40,000	A cereal grain with a relatively small genome. It is very important worldwide as a food crop.
Populus trichocarpa (Balsam poplar)	550,000,000	45,555	A tree with a relatively small genome.
Animals			
Caenorhabditis elegans	97,000,000	19,000	A nematode worm that has been used as a model organism to study animal development.
Drosophila melanogaster (fruit fly)	175,000,000	14,000	Used as a model organism to study many genetic phenomena, including development.
Anopheles gambiae	278,000,000	13,683	A mosquito that carries the malaria parasite, *Plasmodium falciparum.*
Canis familiaris (dog)	2,400,000,000	19,300	A common house pet.
Mus musculus (mouse)	2,500,000,000	24,174	A rodent and a model organism studied by researchers.
Pan troglodytes (chimpanzee)	3,100,000,000	25,000	A primate and the closest living relative to humans.
Homo sapiens (human)	3,200,000,000	25,000	A primate.

*In some cases, the values indicate the estimated amount of DNA. For eukaryotic genomes, DNA sequencing is considered completed in the euchromatic regions. The DNA in certain heterochromatic regions is not able to be sequenced, and the total amount is difficult to estimate.

**The numbers of genes were predicted using computer methods described in Chapter 21. These numbers should be considered as estimates of the total gene number.

CONCEPTUAL AND EXPERIMENTAL SUMMARY

The molecular analysis of entire genomes relies on a variety of methods. Cytogenetic mapping uses light microscopy to locate genes or other DNA markers along intact chromosomes. A widely used method to cytogenetically map the locations of genes within large eukaryotic chromosomes is fluorescence *in situ* hybridization (FISH). In this technique, a small fluorescently labeled DNA probe is hybridized to intact chromosomes to detect where a specific DNA sequence is located in the genome. Linkage mapping uses the frequency of genetic recombination between different loci to determine their relative spacing and order along a chromosome. In Chapter 5, we studied linkage mapping that involved crosses between individuals heterozygous for two or more genes. In this chapter, we have considered how a molecular marker, a segment of DNA found at a specific site along a chromosome, can be used as a tool in mapping. It is now common for geneticists to construct maps of entire genomes using molecular markers such as restriction fragment length polymorphisms (RFLPs) and microsatellites. These maps determine the locations of genes by correlating the inheritance of the gene with the inheritance of a particular RFLP or microsatellite. The distance between two linked RFLPs can also be determined, and the likelihood of linkage between them can be evaluated by the lod score method.

Physical mapping relies on the molecular analysis of an organism's genome via DNA cloning methods. Yeast artificial chromosomes (YACs), bacterial artificial chromosomes (BACs), and P1 artificial chromosomes (PACs) have been used as cloning vectors that can accept large DNA inserts, which can be aligned to create a contig, a series of clones that contains contiguous, overlapping pieces of chromosomal DNA. A more refined contig is made by cloning smaller fragments of DNA into cosmid vectors. The methods of constructing a contig are similar in theory to those of positional cloning. A common method used in positional cloning is chromosome walking, in which one gene is first mapped to a region of the chromosome and a clone from that region is used as a starting point in a chromosome walk toward the gene of interest. This is accomplished by sequential hybridization methods. To initiate this type of experiment, a gene's position relative to a marker must be known from mapping studies.

The goal of genome sequencing projects is to determine the DNA sequence of the entire genome of a given species. Shotgun sequencing is the most efficient and inexpensive way to determine the DNA sequence of the entire genome of a given species. Chromosomal DNA is sheared into small fragments that are randomly sequenced, and then computer programs identify overlapping sequences and create one long sequence for each chromosome. In a hierarchical shotgun sequencing approach, large segments of DNA are cloned into BAC contigs, and a physical map is made. In a whole genome shotgun sequencing approach, the entire genome is fragmented into pieces, the ends of which are sequenced. The whole genome shotgun sequencing approach is now the more common method. To date, hundreds of genomes have been sequenced, and the list is increasing steadily. These include the genomes of model organisms, pathogenic microorganisms, and agriculturally important species. By 2006, the Human Genome Project has published completed maps and sequences for all of the human chromosomes. The many benefits of the project, the largest genome project in history, are just beginning to emerge.

PROBLEM SETS & INSIGHTS

Solved Problems

S1. An RFLP marker is located 1 million bp away from a gene of interest. Your goal is to start at this RFLP marker and walk to this gene. The average insert size in the library is 55,000 bp, and the average overlap at each end is 5,000 bp. Approximately how many steps will it take to get there?

Answer: Each step is only 50,000 bp (i.e., 55,000 minus 5,000), because you have to subtract the overlap between adjacent fragments, which is 5,000 bp, from the average inset size. Therefore, it will take about 20 steps to go 1 million bp.

S2. When many RFLPs have been mapped within the genome of a plant, an RFLP analysis can be used to map a herbicide-resistance gene. For example, let's suppose that an agricultural geneticist has two strains, one that is herbicide resistant and one that is herbicide sensitive. The two strains differ with regard to many RFLPs. The sensitive and resistant strains are crossed, and the F_1 offspring are allowed to self-fertilize. The F_2 offspring are then analyzed

with regard to their herbicide sensitivity and RFLP markers. The following results are obtained:

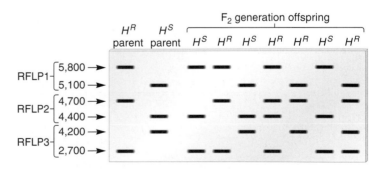

To which RFLP might the herbicide-resistance gene be linked?

Answer: The aim of this experiment is to correlate the presence of a particular RFLP with the herbicide-resistant phenotype. As shown in the data, the herbicide-resistant parent and all the herbicide-resistant offspring have an RFLP that is 4,700 bp in length. In an actual experiment, a thorough lod analysis would be conducted to determine if linkage is considered likely. If so, the 4,700 bp RFLP may either contain the gene that confers herbicide resistance, or, as is more likely, the two may be linked. If the 4,700 bp RFLP has already been mapped to a particular site in the plant's genome, the herbicide-resistance gene also maps to the same site or very close to it. This information may then be used by a plant breeder when making future crosses to produce herbicide-resistant plant strains.

S3. Does a molecular marker have to be polymorphic to be useful in mapping studies? Does a molecular marker have to be polymorphic to be useful in linkage mapping (i.e., involving family pedigree studies or genetic crosses)? Explain why or why not.

Answer: A molecular marker does not have to be polymorphic to be useful in mapping studies. Many sequence-tagged sites (STSs) that are used in physical or cytogenetic studies are monomorphic. Monomorphic markers can provide landmarks in mapping studies.

In linkage mapping studies, a marker must be polymorphic to be useful. Polymorphic molecular markers can be RFLPs, microsatellites, or SNPs. To compute map distances in linkage analysis, individuals must be heterozygous for two or more markers (or genes). For experimental organisms, heterozygotes are testcrossed to homozygotes, and then the number of recombinant offspring and nonrecombinant offspring are determined. For markers that do not assort independently (that is, linked markers), the map distance is computed as the number of recombinant offspring divided by the total number of offspring times 100.

S4. The distance between two molecular markers that are linked along the same chromosome can be determined by analyzing the outcome of crosses. This can be done in humans by analyzing the members of a pedigree. However, the accuracy of linkage mapping in human pedigrees is fairly limited because the number of people in most families is relatively small. As an alternative, researchers can analyze a population of sperm, produced from a single male, and compute linkage distance in this manner. As an example, let's suppose a male is heterozygous for two polymorphic sequence-tagged sites (STSs). STS-1 exists in two sizes: 234 bp and 198 bp. STS-2 also exists in two sizes: 423 bp and 322 bp. A sample of sperm was collected from this man, and individual sperm were placed into 40 separate tubes. In other words, there was one sperm in each tube. Believe it or not, PCR is sensitive enough to allow analysis of DNA in a single sperm! Into each of the 40 tubes were added the primers that amplify STS-1 and STS-2, and then the samples were subjected to PCR. The following results were obtained.

A. What is the arrangement of these two STSs in this individual?

B. What is the map distance between STS-1 and STS-2?

Answer: Keep in mind that mature sperm are haploid, so they will have only one copy of STS-1 and one copy of STS-2.

A. If we look at the 40 lanes, most of the lanes (i.e., 36 of them) have either the 234 bp and 423 bp STSs or the 198 bp and 322 bp. This is the arrangement of STSs in this male. One chromosome has STS-1 that is 234 bp and STS-2 that is 423 bp, and the homologous chromosome has STS-1 that is 198 bp and STS-2 that is 322 bp.

B. There are four recombinant sperm, shown in lanes 15, 22, 25, and 38.

$$\text{Map distance} = \frac{4}{40} \times 100$$
$$= 10.0 \text{ mu}$$

Note: This is a relatively easy experiment compared to a pedigree analysis, which would involve contacting lots of relatives and collecting samples from each of them.

S5. The technique of chromosome walking involves the stepwise cloning of adjacent DNA fragments until the gene of interest is reached. This method was used to clone the gene that (when mutant) causes cystic fibrosis (CF), an autosomal recessive disorder in humans. Linkage mapping studies first indicated that the *CF* gene was located between two markers designated MET and D7S8. Later mapping studies put the *CF* gene between IRP and D7S8. Walking was initiated from the D7S122 marker.

MET   **D7S122**   **IRP**   *CF* **gene**   **D7S8**

(From Kommens, J. M., et al. 1989. Identification of the cystic fibrosis gene: Chromosome walking and jumping. *Science* 245: 1059–65.)

In this study, the researchers restriction mapped this region using two enzymes, *Not*I and *Xho*I. The locations of sites are shown here:

MET	D7S122	IRP	*CF* gene	D7S8
↑A↑	B ↑C↑	D ↑E↑F↑		
N X	X X	X X	N	

N = *Not*I, X = *Xho*I

How would the researchers know they were walking toward the *CF* gene and not away from it?

Answer: The general answer is that they had to determine if they were walking toward the IRP marker or toward the MET marker.

As depicted in the previous diagram, *Not*I would produce a large DNA fragment (including regions A–F) that contained both the D7S122 and IRP markers. Because the D7S122 and IRP markers were available as probes, they could confirm that this region contained both markers by Southern blotting of this *Not*I fragment to both markers. *Xho*I cuts this region into smaller pieces. The piece labeled B carried the D7S122 marker, and the piece labeled D carried the IRP marker. Again, this could be confirmed by hybridization. This map enabled the researchers to determine which direction to walk. They began at fragment B (which contained the D7S122 marker) and walked toward the adjacent fragment labeled C. They then walked to the fragment labeled D, which contains the IRP marker. If they had walked toward the fragment labeled A, they would then have walked to the MET marker. This is the incorrect direction, because they knew from their linkage mapping that IRP is closer to the *CF* gene compared to MET.

Conceptual Questions

C1. A person with a rare genetic disease has a sample of her chromosomes subjected to *in situ* hybridization using a probe that is known to recognize band p11 on chromosome 7. Even though her chromosomes look cytologically normal, the probe does not bind to this person's chromosomes. How would you explain these results? How would you use this information to positionally clone the gene that is related to this disease?

C2. For each of the following, decide if it could be appropriately described as a genome:

A. The *E. coli* chromosome

B. Human chromosome 11

C. A complete set of 10 chromosomes in corn

D. A copy of the single-stranded RNA packaged into human immunodeficiency virus (HIV)

C3. Which of the following statements are true about molecular markers?

A. All molecular markers are segments of DNA that carry specific genes.

B. A molecular marker is a segment of DNA that is found at a specific location in a genome.

C. We can follow the transmission of a molecular marker by analyzing the phenotype (i.e., the individual's bodily characteristics) of offspring.

D. We can follow the transmission of molecular markers using molecular techniques such as gel electrophoresis.

E. An STS is a molecular marker.

Experimental Questions

E1. Would the following methods be described as linkage, cytogenetic, or physical mapping?

A. Fluorescence *in situ* hybridization (FISH)

B. Conducting dihybrid crosses to compute map distances

C. Chromosome walking

D. Examination of polytene chromosomes in *Drosophila*

E. Use of RFLPs in crosses

F. Using BACs and cosmids to construct a contig

E2. In an *in situ* hybridization experiment, what is the relationship between the sequence of the probe DNA and the site on the chromosomal DNA where the probe binds?

E3. Describe the technique of *in situ* hybridization. Explain how it can be used to map genes.

E4. The cells from a malignant tumor were subjected to *in situ* hybridization using a probe that recognizes a unique sequence on chromosome 14. The probe was detected only once in each of these cells. Explain these results and speculate on their significance with regard to the malignant characteristics of these cells.

E5. Figure 20.3 describes the technique of FISH. Why is it necessary to "fix" the cells (and the chromosomes inside of them) to the slides? What does it mean to fix them? Why is it necessary to denature the chromosomal DNA?

E6. Explain how the use of DNA probes with different fluorescence emission wavelengths can be used in a single FISH experiment to map the locations of two or more genes. This method is called chromosome painting. Explain why this is an appropriate term.

E7. A researcher is interested in a gene found on human chromosome 21. Describe the expected results of a FISH experiment using a probe that is complementary to this gene. How many spots would you see if the probe was used on a sample from an individual with 46 chromosomes versus an individual with Down syndrome?

E8. What is a contig? Explain how you would determine that two clones in a contig are overlapping.

E9. Contigs are often made using BAC or cosmid vectors. What are the advantages and disadvantages of these two types of vectors? Which type of contig would you make first, a BAC or cosmid contig? Explain.

E10. Describe the molecular features of a BAC cloning vector. What is the primary advantage of a BAC compared to plasmid or viral vectors?

E11. In general terms, what is a polymorphism? Explain the molecular basis for a restriction fragment length polymorphism (RFLP). How is an RFLP detected experimentally? Why are RFLPs useful in physical mapping studies? How can they be used to clone a particular gene?

E12. Five RFLPs designated 1A and 1B, 2A and 2B, 3A and 3B, 4A and 4B, and 5A and 5B, are known to map along chromosome 4 of corn. A plant breeder has obtained a strain of corn that carries a pesticide-resistance gene that (from previous experiments) is known to map somewhere along chromosome 4. The plant breeder crosses this pesticide-resistance strain that is homozygous for RFLPs 1A, 2B, 3A, 4B, and 5A to a pesticide-sensitive strain that is homozygous for 1B, 2A, 3B, 4A, and 5B. The F_1 generation plants were allowed to self-hybridize to produce the following F_2 plants:

F₂ generation offspring

Based on these results, which RFLP does the pesticide-resistance gene map closest to?

E13. Let's suppose there are two different RFLPs in a species. RFLP 1 can be found in 4,500 bp and 5,200 bp lengths; RFLP 2 can be 2,100 bp and 3,200 bp. A homozygote, 4,500 bp and 2,100 bp, was crossed to a homozygote 5,200 bp and 3,200 bp. The F_1 offspring were then crossed to a homozygote containing the 4,500 bp and 2,100 bp fragments. The following results were obtained:

5,200, 4,500, and 2,100	40 offspring
5,200, 4,500, 3,200, 2,100	98 offspring
4,500 and 2,100	97 offspring
4,500, 2,100, and 3,200	37 offspring

Conduct a chi-square analysis to determine if these two RFLPs are linked. If so, calculate the map distance between them.

E14. Explain how a detailed RFLP map of an organism's genome can be helpful in mapping the location of genes. Describe an experimental strategy you would follow to map a gene near a particular RFLP.

E15. A woman has been married to two different men and produced five children. This group is analyzed with regard to three different STSs: STS-1 is 146 and 122 bp; STS-2 is 102 and 88 bp, and STS-3 is 188 and 204 bp. The mother is homozygous for all three STSs: STS-1 = 122, STS-2 = 88, and STS-3 = 188. Father 1 is homozygous for STS-1 = 122 and STS-2 = 102, and heterozygous for STS-3 = 188/204. Father 2 is heterozygous for STS-1 = 122/146, STS-2 = 88/102, and homozygous for STS-3 = 204. The five children have the following results:

Which children can you definitely assign to father 1 and father 2?

E16. An experimenter used primers to nine different STSs to test their presence along five different BAC clones. The results are shown here.

Alignment of STSs and BACs

	STSs								
	1	2	3	4	5	6	7	8	9
BACs									
1	−	−	−	−	−	+	+	−	+
2	+	−	−	−	+	−	−	+	−
3	−	−	−	+	−	+	−	−	−
4	−	+	+	−	−	−	−	+	−
5	−	−	+	−	−	−	−	−	+

Make a contig map that describes the alignment of the five BACs.

E17. In the Human Genome Project, researchers have collected linkage data from many crosses in which the male was heterozygous for markers and many crosses where the female was heterozygous for markers. The distance between the same two markers, computed in map units or centiMorgans, is different between males and females. In other words, the linkage maps for human males and females are not the same. Propose an explanation for this discrepancy. Do you think the sizes of chromosomes (excluding the Y chromosome) in human males and females are different? How could physical mapping resolve this discrepancy?

E18. Take a look at solved problem S4. Let's suppose a male is heterozygous for two polymorphic sequence-tagged sites. STS-1 exists in two sizes: 211 bp and 289 bp. STS-2 also exists in two sizes: 115 bp and 422 bp. A sample of sperm was collected from this man, and individual sperm were placed into 30 separate tubes.

Into each of the 30 tubes were added the primers that amplify STS-1 and STS-2, and then the samples were subjected to PCR. The following results were obtained:

A. What is the arrangement of these two sequence-tagged sites in this individual?

B. What is the linkage distance between STS-1 and STS-2?

C. Could this approach of analyzing a population of sperm be applied to RFLPs?

E19. A gene affecting flower color in petunias is closely linked to an RFLP. A red allele for this gene is associated with a 4,000 bp RFLP, and a purple allele is linked to a 3,400 bp version of this same RFLP. A second gene in petunias affects flower size. An allele causing big flowers is linked to a 7,200 bp RFLP, and a small-flower allele is linked to the same RFLP that is 1,600 bp. A true-breeding strain with small, red flowers was crossed to a true-breeding strain with big, purple flowers. All the F_1 offspring had big, purple flowers. These F_1 offspring were then crossed to true-breeding petunias with small, red flowers. The following results were obtained:

Red, small   725

Red, big   111

Purple, small 109

Purple, big   729

Are these two genes linked to each other? If so, compute the map distance. What would be the expected outcome regarding the inheritance of the RFLPs among the offspring?

E20. An agricultural geneticist has studied a gene in alfalfa that affects pesticide resistance. It exists in three alleles that confer low, medium, and high levels of resistance. This gene has a significant impact on the yield of alfalfa, depending on seasonal variation in pest problems. This geneticist has followed the basic protocol in Figure 20.6, using *Eco*RI as the enzyme to digest the chromosomal DNA. Unfortunately, after tireless efforts and the analysis of thousands of offspring, it has not been possible to identify an RFLP that is associated with the three alleles of this pesticide-resistance gene. What should the geneticist do next? In other words, discuss ways to vary the RFLP method or propose alternative approaches to identify molecular markers that may be linked to the pesticide-resistance gene.

E21. Figure 20.8 describes the transmission of two RFLPs that were linked and 16 mu apart. If these two RFLPs had not been linked and 100 offspring had been analyzed, what would have been the expected results?

E22. Explain why it is necessary to use the technique of Southern blotting for RFLP mapping.

E23. Compared to a conventional plasmid, what additional sequences are required in a YAC vector so that it can behave like an artificial chromosome? Describe the importance of each required sequence.

E24. In the data of Figure 20.10, individuals who were homozygous for the sickle-cell allele had an 87% likelihood of carrying the 13.0 kbp RFLP. Based on this observation, is it likely that the Hb^S allele originated in an individual with a 13.0 kbp RFLP? How would you explain the observation that the Hb^S allele is found with the 7.6 kbp RFLP 13% of the time?

E25. In the experiment of Figure 20.10, explain the conclusion that the 13.0 kbp fragment is more closely linked to Hb^S compared to the 7.0 and 7.6 fragments. Is it always linked? Why or why not? How is this type of information useful?

E26. Explain the technique of pulsed-field gel electrophoresis (PFGE). What special precautions are needed to prevent the mechanical breakage of the chromosomes? What are the uses of PFGE?

E27. When conducting physical mapping studies, place the following methods in their most logical order:

A. Clone large fragments of DNA to make a BAC library.

B. Determine the DNA sequence of subclones from a cosmid library.

C. Subclone BAC fragments to make a cosmid library.

D. Subclone cosmid fragments for DNA sequencing.

E28. Four cosmid clones, which we will call cosmid A, B, C, and D, were subjected to a Southern blot in pairwise combinations. The insert size of each cosmid was also analyzed. The following results were obtained:

Cosmid	Insert Size (bp)	Hybridized to?
A	6,000	C
B	2,200	C, D
C	11,500	A, B, D
D	7,000	B, C

Draw a map that shows the order of the inserts within these four cosmids.

E29. What is an STS (sequence-tagged site)? How are STSs generated experimentally? What are the uses of STSs? Explain how a microsatellite can produce a polymorphic STS.

E30. A human gene, which we will call gene *X*, is located on chromosome 11 and is found as a normal allele and a recessive disease-causing allele. The location of gene *X* has been approximated on the map shown here that contains four STSs, labeled STS-1, STS-2, STS-3, and STS-4.

STS-1   STS-2   STS-3 Gene *X*   STS-4

A. Explain the general strategy of positional cloning.

B. If you applied the approach of positional cloning to clone gene *X*, where would you begin? As you progressed in your cloning efforts, how would you know if you were walking toward gene *X* or away from gene *X*?

C. How would you know you had reached gene *X*? (Keep in mind that gene *X* exists as a normal allele and a disease-causing allele.)

E31. Describe how you would clone a gene by positional cloning. Explain how a (previously made) contig would make this task much easier.

E32. A bacterium has a genome size of 4.4 Mb. If a researcher carries out shotgun DNA sequencing and sequences a total of 19 Mb, what is the probability that a base will be left unsequenced? What percentage of the total genome will be left unsequenced?

E33. Discuss the general differences between hierarchical shotgun sequencing versus whole genome shotgun sequencing.

Questions for Student Discussion/Collaboration

1. How is it possible to obtain an RFLP linkage map? What kind of experiments would you conduct to correlate the RFLP linkage map with the positions of known genes that had already been cloned? Discuss the uses of RFLPs in genetic analyses.

2. What is a molecular marker? Give two examples. Discuss why it is easier to locate and map many molecular markers rather than functional genes.

3. Which goals of the Human Genome Project do you think are the most important? Why? Discuss the types of ethical problems that might arise as a result of identifying all of our genes.

Note: All answers appear at the website for this textbook; the answers to even-numbered questions are in the back of the textbook.

www.mhhe.com/brookergenetics3e

Visit the Online Learning Center for practice tests, answer keys, and other learning aids for this chapter. Enhance your understanding of genetics with our interactive exercises, quizzes, animations, and much more.

GENOMICS II: FUNCTIONAL GENOMICS, PROTEOMICS, AND BIOINFORMATICS

21

I n Chapter 20, we learned that genomics involves the mapping of an entire genome and, eventually, the determination of a species' complete DNA sequence. The amount of information found within a species' genome is enormous. The goal of **functional genomics** is to elucidate the roles of genetic sequences—DNA, RNA, and amino acid sequences—in a given species. In most cases, functional genomics is aimed at an understanding of gene function. At the genomic level, researchers can study genes as groups. For example, the information gained from a genome-sequencing project can help researchers study entire metabolic pathways. This provides a description of the ways in which gene products interact to carry out cellular processes. In addition, a study of genetic sequences can help to identify regions that play particular functional roles. For example, an analysis of certain species of bacteria helped to identify DNA sequences that promote the uptake of DNA during bacterial transformation.

Because most genes encode proteins, a goal of many molecular biologists is to understand the functional roles of all the proteins a species produces. The entire collection of proteins a given cell or organism can make is called its **proteome,** and the study of the function and interactions of these proteins is termed **proteomics**. An objective of researchers in the field of proteomics is to understand the interplay among many proteins as they function to create cells and, ultimately, the traits of a given species.

From a research perspective, functional genomics and proteomics can be broadly categorized in two ways: experimental and computational. The experimental approach involves the study of groups of genes or proteins using molecular techniques in the laboratory. In the first two sections in this chapter, we will focus on these techniques. By comparison, the computational strategy attempts to analyze genetic sequences using a mathematical approach. This area, which is called **bioinformatics,** has become an important branch of science. The tools of bioinformatics are computers, computer programs, and genetic sequences. In the last section of this chapter, we will consider the field of bioinformatics and discover how this area has provided great insights into the subjects of functional genomics and proteomics.

A DNA microarray. Each spot on the array corresponds to a specific gene. The color of the spots, which occurs via computer imaging techniques, indicates the amount of RNA transcribed from that gene. A DNA microarray allows researchers to simultaneously analyze the expression of many genes.

21.1 FUNCTIONAL GENOMICS

Though the rapid sequencing of genomes, particularly the human genome, has generated great excitement in the field of genetics, many would argue that an understanding of genomic function is fundamentally more interesting. In the past, our ability to study genes involved many of the techniques described in Chapter 18, such as gene cloning, Northern blotting, gel retardation assays, and site-directed mutagenesis. These approaches continue to provide a solid foundation for our understanding of gene function. More recently, genome-sequencing projects have enabled researchers to consider gene function at a more complex level. We now can analyze groups of many genes simultaneously to determine how they work as integrated units that produce the characteristics of cells and the traits of complete organisms. In this section, we will explore how a cDNA library is created and used to identify regions of DNA that contain genes. We also will examine the use of DNA microarrays, which enable researchers to monitor the expression of thousands of genes simultaneously.

Expressed Genes Can Be Identified in a cDNA Library

Chromosomal DNA contains regions where genes are found, as well as intergenic regions that do not contain genes. Therefore, a basic goal of genomic research is to definitively identify the regions of DNA that are actually genes. To do so, one approach is to show that a given region is transcribed into RNA. This can be accomplished by the generation of a **cDNA library.** As described in Chapter 18, cDNA—complementary DNA—is made using RNA as a template, which is reversely transcribed into DNA with reverse transcriptase. A cDNA library used in physical mapping studies is also called an **expressed sequence tag (EST) library,** because the sequences, which are expressed as mRNA, can also be used as molecular markers or tags. The members of an EST library are subjected to DNA sequencing, and EST sequences are then compared to a complete genome sequence. A particular EST sequence will match a genomic sequence, indicating that the genomic region encodes a gene. From the perspective of gene function, a cDNA or EST library provides a reliable identification of a region that is truly transcribed. The mere fact that a region is transcribed indicates that the corresponding genomic region is a gene.

The method of making a cDNA library can be extended to study gene regulation at the genomic level. One strategy is to isolate mRNA under different conditions and then identify particular mRNAs that are expressed only under a specific set of conditions. For example, in the experiment shown in **Figure 21.1,** samples of cells were incubated with and without a particular hormone. The goal is to identify those genes that are turned on by the presence of the hormone. The mRNAs from these cells were isolated and used to make two different groups of cDNAs. The cDNAs derived from cells not exposed to hormone were denatured and tightly bound to a column. (See the Appendix for a description of column chromatography.) The cDNAs derived from the cells exposed to hormone were also denatured and then run over the column. Any of these latter cDNAs will remain bound to the column if they match a cDNA that is already tightly bound to the column. In contrast, cDNAs derived only from the hormone-induced cells will not bind to the column because they do not have a complementary match that is tightly bound to the column. These cDNAs are eluted and cloned into vectors to

+ Hormone   **No hormone**

The red mRNAs are made in the presence and absence of hormone.

Yellow mRNAs are only made in the presence of hormone.

Isolate mRNA and make cDNA as described in Chapter 18. In this case, double-stranded cDNA is made.

Hormone-induced cDNA

Denature cDNA.

Denature cDNA. Covalently attach cDNA strands to column beads. (Note: An excess amount of this cDNA is attached to the column beads.)

Pour cDNA from hormone-induced cells onto column. Allow complementary cDNAs to hybridize. Elute cDNAs that are not complementary.

Clone these cDNAs into vectors.

A subtractive cDNA library

FIGURE 21.1   **Creation of a subtractive cDNA library.**

create a cDNA library termed a **subtractive cDNA library.** This approach is also called **subtractive hybridization.** The members of the subtractive cDNA library can be subjected to DNA sequencing to identify the genes that are turned on by the presence of the hormone.

A Microarray Can Identify Genes That Are Transcribed

Researchers have developed an exciting new technology called a **DNA microarray** (also called a **gene chip**), that makes it possible to monitor the expression of thousands of genes simultaneously. A DNA microarray is a small silica, glass, or plastic slide that is dotted with many different sequences of DNA, each corresponding to a short sequence within a known gene. For example, one spot in a microarray may correspond to a sequence within the β-globin gene, while another spot could correspond to a gene that encodes an iron transporter. A single slide may contain tens of thousands of different spots in an area the size of a postage stamp. The relative location of each spot is known.

How are microarrays made? Some are produced by spotting different samples of DNA onto a slide, much like the way an ink-jet printer works. Different DNA fragments, which are made synthetically (for example, by PCR), are individually spotted onto the slide. The DNA fragments are typically 500 to 5,000 bp in length, and a few thousand to tens of thousands are spotted to make a single array. Alternatively, other microarrays contain shorter DNA segments—oligonucleotides—that are directly synthesized on the surface of the slide. In this case, the DNA sequence at a given spot is produced by selectively controlling the growth of the oligonucleotide using narrow beams of light. Such oligonucleotides are typically 25 to 30 nucleotides in length. Hundreds of thousands of different spots can be found on a single array. Overall, the technology of making DNA microarrays is quite amazing.

Once a DNA microarray has been made, it is used as a hybridization tool, as shown in **Figure 21.2.** In this experiment, RNA was isolated from a sample of cells and then used to make fluorescently labeled cDNA. The labeled cDNAs were then layered onto a DNA microarray. Those cDNAs that are complementary to the DNAs in the microarray will hybridize and thereby remain bound to the microarray. The array is then washed with a buffer to remove any unbound cDNAs and placed in a type of a microscopic device called a laser scanner, which produces higher resolution images than a conventional optical microscope. The device scans each pixel—the smallest element in a visual image—and after correction for local background, the final fluorescence intensity for each spot is obtained by averaging across the pixels in each spot. This results in a group of fluorescent spots at defined locations in the microarray. If the fluorescence intensity in a spot is high, this result means that a large amount of cDNA in the sample hybridized to the DNA at this location. Because the DNA sequence of each spot is already known, a fluorescent spot identifies cDNAs that are complementary to those DNA sequences. Furthermore, because the cDNA was generated from mRNA, this technique identifies RNAs that have been made in a particular cell type under a given set of conditions.

The technology of DNA microarrays has found many important uses (**Table 21.1**). Thus far, its most common use is to study gene expression patterns. This helps us to understand how genes are regulated in a cell-specific manner and how environmental conditions can induce or repress the transcription of

TABLE 21.1

Applications of DNA Microarrays

Application	Description
Cell-specific gene expression	A comparison of microarray data using cDNAs derived from RNA of different cell types can identify genes that are expressed in a cell-specific manner.
Gene regulation	Environmental conditions play an important role in gene regulation. A comparison of microarray data may reveal genes that are induced under one set of conditions and repressed under another set of conditions.
Elucidation of metabolic pathways	Genes that encode proteins that participate in a common metabolic pathway are oftentimes expressed in a parallel manner. This can be revealed from a microarray analysis (as described in Figure 21.3). This application overlaps with the study of gene regulation via microarrays.
Tumor profiling	Different types of cancer cells exhibit striking differences in their profiles of gene expression. This can be revealed by a DNA microarray analysis. This approach is gaining widespread use to subclassify tumors that are sometimes morphologically indistinguishable. Tumor profiling may provide information that can improve a patient's clinical treatment.
Genetic variation	A mutant allele may not hybridize to a spot on a microarray as well as a wild-type allele. Therefore, microarrays are gaining widespread use as a tool to detect genetic variation. They have been used to identify disease-causing alleles in humans and mutations that contribute to quantitative traits in plants and other species. In addition, microarrays are used to detect chromosomal deletions and duplications. As discussed in Chapter 8, a method called comparative genomic hybridization (CGH) is used to determine if cancer cells carry deletions or duplications. In the traditional method, equal amounts of two fluorescently labeled DNA samples are denatured, mixed together, and applied to intact metaphase chromosomes in which the DNA has also been denatured. In a newer approach, called array comparative genomic hybridization (aCGH), the fluorescently labeled DNAs are applied to a microarray that has short segments of DNA from all of the chromosomes. This method is more sensitive in the detection of deletions and duplications.
Microbial strain identification	Microarrays can distinguish between closely related bacterial species and subspecies.
DNA-protein binding	Chromatin immunoprecipitation, which is described in Figure 21.4, can be used with DNA microarrays to determine where in the genome a particular protein binds to the DNA.

FIGURE 21.2 **A DNA microarray.** A mixture of mRNAs is used to create cDNAs that are fluorescently labeled. The cDNAs are applied to the microarray, and any unbound cDNAs are washed away. In this simplified example, three cDNAs specifically hybridize to spots on the microarray. In an actual experiment, there are typically hundreds or thousands of different cDNAs and tens of thousands of different spots on the array. After hybridization, the spots become fluorescent and can be visualized using a laser scanner.

genes. In some cases, microarrays can even help to elucidate the genes encoding proteins that participate in a complicated metabolic pathway. Microarrays can also be used as identification tools. For example, gene expression patterns can aid in the categorization of tumor types. Such identification can be important in the treatment of the disease. Instead of using labeled cDNA, researchers can also hybridize labeled genomic DNA to a micro-

array. This can be used to identify mutant alleles in a population of individuals and to detect deletions and duplications. In addition, this technology is proving useful in the correct identification of closely related bacterial species and subspecies. Finally, microarrays can be used to study DNA–protein interactions as described later in this section.

EXPERIMENT 21A

The Coordinate Regulation of Many Genes Is Revealed by a DNA Microarray Analysis

One way that cells respond to environmental changes is via the coordinate regulation of genes. Under one set of environmental conditions, a particular set of genes may be induced, while under another set of conditions, those same genes may be repressed. In the past, researchers have been able to study this type of gene regulation using tools that can analyze the expression of a few genes at a time. The advent of microarrays, however, has made it possible to study the expression of the whole genome under different sets of environmental conditions.

One of the first studies using this approach involved the analysis of the yeast genome. The genome of baker's yeast, *Saccharomyces cerevisiae*, was the first eukaryotic genome to be sequenced. It encodes approximately 6,300 different genes. An important process in the growth of yeast cells, as well as cells of other species, is the ability to metabolize carbon sources using different metabolic pathways. When yeast cells have glucose available, they metabolize the glucose to smaller products in a process called glycolysis. If oxygen is present, these products can be broken down via the tricarboxylic acid cycle (TCA or citric acid cycle) that occurs in the mitochondria. Therefore, when yeast are first given glucose and then allowed to metabolize it in the presence of oxygen, they first metabolize the carbohydrate via glycolysis, and then, when the glucose is used up, they metabolize the products of glycolysis via the TCA cycle. The process of switching from glycolysis to the TCA cycle, called a diauxic shift, involves major changes in the expression of genes involved with carbohydrate metabolism. The goal of the experiment described in **Figure 21.3** was to identify genes that are induced and repressed as yeast cells shift from glycolysis to the TCA cycle. It was carried out by Joseph DeRisi, Vishwanath Iyer, and Patrick Brown in 1997.

As shown in Figure 21.3, yeast cells were initially given glucose as their carbon source for growth and then allowed to grow for several hours. Over time, the glucose was used up, and the cells shifted from glycolysis to the TCA cycle. At various time points, samples of cells were removed, and the RNA was isolated. The RNA was then exposed to reverse transcriptase, poly-dT primers, and deoxyribonucleotides, one of which was fluorescently labeled. This created fluorescently labeled cDNAs.

To determine the relative changes in RNA synthesis, two different fluorescent dyes were used. The RNA collected at the first time point—when glucose was at a high level—was used to make cDNA that contained a green fluorescent dye. The RNA collected at later time points was used to make cDNAs that contained a red fluorescent dye. A sample of green cDNA (from the first time point) was then mixed with a sample of red cDNA (from later time points), and the mixture was hybridized to a microarray containing about 6,200 yeast genes. A laser scanner was used to measure the amount of red fluorescence and green fluorescence at each spot in the microarray. The fluorescence ratio (red fluorescent units divided by green fluorescent units) provided a way to quantitatively determine how the expression of the genes was changing. For example, if the ratio was high, this means that a gene was being induced as glucose levels fall, because the amount of red cDNA would be higher than the amount of green cDNA. Alternatively, if the red to green ratio was low, this means that a gene was being repressed as glucose is used up.

■ THE GOAL

A diauxic switch from glycolysis to the TCA cycle will involve the induction of certain genes and the repression of other genes. The goal was to identify such groups of genes.

■ ACHIEVING THE GOAL — FIGURE 21.3   The use of a DNA microarray to study carbohydrate metabolism in yeast.

Starting material: A commonly used strain of *Saccharomyces cerevisiae* (baker's yeast) was used in this study. The researchers had made a DNA microarray containing (nearly) all of the known yeast genes.

1. Inoculate yeast into a media containing 2% glucose. Grow for up to 21 hours.

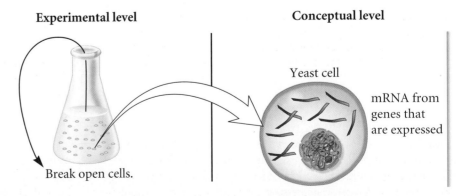

Experimental level

Conceptual level

Break open cells.

Yeast cell

mRNA from genes that are expressed

2. Beginning at 9 hours after inoculation, take out samples of cells and isolate mRNA. This involves breaking open cells and running the cell contents over a poly-dT column under high-salt conditions. Since mRNA has a polyA tail, it will bind to this column while other cell components flow through. The purified mRNA can be then eluted by adding a buffer to the top of the column that contains a low concentration of salt. This breaks the interaction between the poly-dT and polyA tails. In this experiment, mRNA samples were isolated at approximately 2-hour intervals, beginning at 9 hours and ending at 21 hours.

3. Add reverse transcriptase, poly-dT primers, and fluorescently labeled nucleotides to make complementary strands of fluorescently labeled cDNA. Note: The sample at 9 hours was used to make green cDNA, while mRNA samples collected at later time points were used to make red cDNA.

Break open cells.

Load onto column.

Collection tubes

Purified mRNA

High salt   Low salt

mRNA from:

Cells grown 9 hours

Later time points

Incubate with reverse transcriptase and deoxyribonucleotides, one of which is fluorescently labeled.

High salt mRNA binds, while other cellular components do not bind.

Column bead

Low salt mRNA is released.

mRNA from:

Cells grown 9 hours

Later time points

Add reverse transcriptase with a fluorescently labeled nucleotide.

(continued)

4. At each time point, mix together the cDNA derived from that time (i.e., red cDNA) with cDNA derived from the 9-hour time point (i.e., green cDNA).

5. Hybridize the mixture to the yeast DNA microarray as described in Figure 21.2.

6. Examine the DNA microarray with a laser scanner. The data are then analyzed by a computer, which can correlate expression levels among different genes. (This is a cluster analysis.)

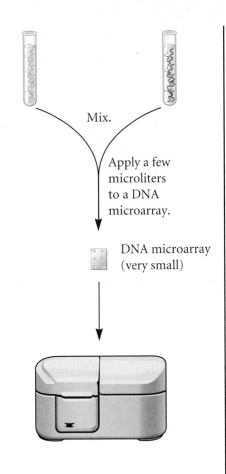

Mix.

Apply a few microliters to a DNA microarray.

DNA microarray (very small)

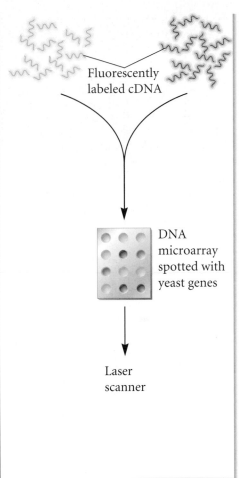

Fluorescently labeled cDNA

DNA microarray spotted with yeast genes

Laser scanner

■ THE DATA*

GLK1

TEF4

* Green spots are genes expressed early in growth, while red spots are expressed later. Yellow spots are expressed more evenly. Spots that are barely visible indicate genes that are not substantially expressed under these growth conditions.

(a) Induced at 15–17 hours

(b) Induced at 19 hours

(c) Repressed at 15–17 hours

(d) Repressed at 19 hours

INTERPRETING THE DATA

A portion of one microarray is shown at the beginning of the data of Figure 21.3. As seen here, the array shows many spots, some of which are relatively green and some of which are relatively red. The green spots indicate genes that are expressed at higher levels during early stages of growth when glucose levels are high. An example is a gene designated *TEF4*, which is involved with protein synthesis. Red spots identify genes that are expressed when the glucose is depleted. The gene *GLK1* encodes an enzyme that phosphorylates glucose but is expressed only when glucose levels are low. In addition, many spots are not fluorescent, indi-

cating there was not much cDNA in the sample to hybridize to the DNA strands in those locations. These spots would correspond to genes that are not greatly expressed under either condition. Judging from the numbers of red and green spots, the shift from glycolysis to the TCA cycle involved a great amount of gene induction and repression. By determining the red:green ratio at each spot, it was found that 710 genes were induced by at least a factor of two, while 1,030 genes were repressed by a factor of two or more. Therefore, it was found that a diauxic shift involves a staggering amount of gene regulation. Of the 6,200 yeast genes, 1,740, or roughly 28%, appeared to be regulated as a result of a diauxic shift.

Because the gene sequence in each spot was known, the next step was to relate the levels of gene expression based on the microarray data to specific genes. A common goal is to identify genes whose pattern of expression seems to strongly correlate with each other. This statistical technique is called a **cluster analysis.** The graphs shown in the data of Figure 21.3 illustrate how microarray data can be used to make this type of comparison. Though the sequences of these genes were known from the yeast genome-sequencing project, the functions of some of these genes were not known. Each time point involved the measurement of the red:green ratio at a particular spot. Figure 21.3 shows the analysis of 28 genes (seven genes in each panel) over the course of 9 to 21 hours following the addition of glucose. The diauxic shift occurred at approximately 15 hours.

Let's examine the data shown in Figure 21.3. In part (a), the genes were induced at 15 to 17 hours of growth. Due to their coordinate regulation, these genes may be controlled by the same transcription factor(s) and may participate in a common metabolic response to the induction of the TCA cycle. In fact, several of the genes in part (a) have already been studied, and they are known to play a role in the ability of the mitochondria to run the TCA cycle. This makes sense because the diauxic shift occurred around 15 hours of growth, which correlates with the time when the genes were induced. By comparison, the genes in part (b) were induced later, after the TCA cycle had operated for a few hours and when the carbon sources were becoming depleted. Therefore, it would seem that these genes were induced as a response to the operation of the TCA cycle or they were induced because the carbon sources in the media were low. Why is this informa-tion useful? Though further work needs to be done to elucidate the functions of some of the genes shown in parts (a) and (b), the data suggest a common regulation and metabolic function for particular groups of genes. It would seem more likely that the transcription factor(s) that regulate the genes in part (a) would be different from those that regulate the genes in part (b). Like-wise, the proteins encoded by these genes may work in different cellular pathways.

The genes shown in parts (c) and (d) illustrate a similar phenomenon, except that the switch to the TCA cycle and the operation of the TCA cycle represses those genes. The genes shown in part (c) were active during glycolysis and then were repressed at the time of the diauxic shift, while those shown in part (d) were repressed after the TCA cycle had operated for a few hours. Most of the genes shown in part (d) were already known to function as ribosomal proteins. Therefore, as the car-bon sources in the media were depleted, these results suggest that one of the cellular responses is to diminish the synthesis of ribo-somes, which, in turn, would slow down the rate of protein syn-thesis. It would seem that the yeast cells were trying to conserve energy at this late stage of growth.

Overall, the data shown in the experiment of Figure 21.3 illustrate how a microarray analysis can shed light on gene func-tion at the genomic level. It provides great insight regarding gene regulation and may help to identify groups of proteins (i.e., clus-ters) that share a common cellular function.

A self-help quiz involving this experiment can be found at the Online Learning Center.

DNA Microarrays Can Be Used to Identify DNA-Protein Binding at the Genome Level

As discussed throughout this textbook, the binding of proteins to specific DNA sites plays a critical role in a variety of molecular processes, including gene transcription and DNA replication. To study these processes, researchers have devised a variety of tech-niques to identify whether or not specific proteins bind to par-ticular sites in the DNA. For example, the techniques of gel retar-dation and DNaseI footprinting, which are described in Chapter 18, are used for this purpose (see Figures 18.12 and 18.13).

More recently, a newer approach called **chromatin immu-noprecipitation (ChIP)** has gained widespread use in the analysis of DNA–protein interactions. This method can deter-mine whether proteins can bind to a particular region of DNA. A distinguishing feature of this method is that it analyzes DNA-protein interactions as they occur in the chromatin of living cells. In contrast, gel retardation and DNaseI footprinting are in vitro techniques, which typically use cloned DNA and purified proteins.

Figure 21.4 describes the steps of the chromatin immu-noprecipitation protocol. Proteins in living cells, which are non-covalently bound to DNA, can be more tightly attached to the chromatin by the addition of formaldehyde or some other chem-ical that covalently cross-links the protein to the DNA. Follow-ing cross-linking, the cells are lysed, and the DNA is broken into pieces of approximately 200 to 1,000 bp by sonication. Next, an antibody is added that is specific for the protein of interest. To conduct a ChIP assay, a researcher must already suspect that a particular protein binds to the DNA and has previously had an antibody made against that protein. As discussed in Chapter 17, antibodies are molecules that specifically bind to antigens. In this case, the antigen is the protein of interest that is thought to be a DNA-binding protein. In Figure 21.4, the binding of antibodies to the DNA–protein complexes causes the complexes to aggre-gate and precipitate out of solution. Because an antibody is made by the immune system of an animal, this step is called immuno-precipitation. The sample is then subjected to centrifugation to collect the DNA–protein complexes in the pellet. (See the Appen-dix for a description of centrifugation.)

The next step is to identify the DNA to which the protein is covalently cross-linked. To do so, the protein is removed by treat-ment with chemicals that break the covalent cross-links. Because the DNA is usually present in very low amounts, researchers must amplify the DNA to analyze it. This is done using PCR, which is described in Chapter 18. If researchers already suspect that a protein binds to a known DNA region, they can use PCR primers that spe-cifically flank the DNA region (see bottom left side of Figure 21.4).

If they get a PCR product, this means that the protein of interest must have been bound (either directly or through other proteins) to this DNA site in living cells.

Alternatively, a researcher may want to determine where the protein of interest binds across the whole genome. In this case, a DNA microarray can be used (see bottom right side of Figure 21.4). Because a DNA microarray is found on a chip, this is called a **ChIP-on-chip assay.** The ends of the precipitated DNA are first ligated to short DNA pieces called linkers. PCR primers are then added that are complementary to the linkers and, therefore, amplify the DNA regions between the linkers. During PCR, the DNA is fluorescently labeled. The labeled DNA is then hybridized to a DNA microarray. Because the DNA was isolated using an antibody against the protein of interest, the fluorescent spots on the microarray identify sites in the genome where the protein binds. In this way, researchers may be able to determine where a protein binds to locations in the genome, even if those site(s) had not been previously determined by other methods.

21.2 PROTEOMICS

Thus far, we have considered ways to characterize the genome of a given species and study its function. Because most genes encode proteins, a logical next step is to examine the functional roles of the proteins that a species can make. As mentioned, this field is called proteomics, and the entire collection of a species' proteins is its proteome.

Genomics represents only the first step in our comprehensive understanding of protein structure and function. Researchers often use genomic information to initiate proteomic studies, but such information must be followed up with research that involves the direct analysis of proteins. For example, as discussed in the preceding section, a DNA microarray may provide insights into the transcription of particular genes under a given set of conditions. However, mRNA levels may not provide an accurate measure of the abundance of a protein that is encoded by a given gene. Protein levels are greatly affected, not only by the level of mRNA, but also by the rate of mRNA translation and by the turnover rate of a given protein. Therefore, DNA microarray data must be corroborated using other methods, such as Western blotting (discussed in Chapter 18), which directly determine the abundance of a protein in a given cell type.

A second way that genomic data can elucidate the workings of the proteome is via homology. As discussed later in this chapter, homology between the genes of different species can be used to predict protein structure and/or function, when the structure or function of the protein in one of the species is already known. However, homology may not provide direct information regarding the regulation of protein structure and function. Also, it may not reveal potential types of protein–protein interactions in which a given protein may participate. Therefore, even though

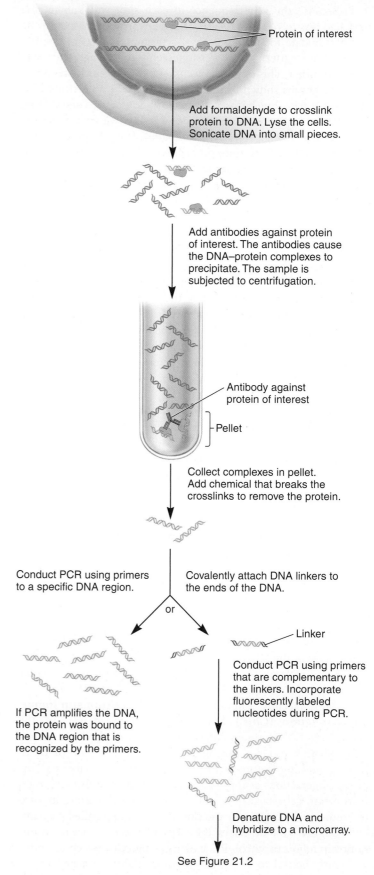

Protein of interest

Add formaldehyde to crosslink protein to DNA. Lyse the cells. Sonicate DNA into small pieces.

Add antibodies against protein of interest. The antibodies cause the DNA–protein complexes to precipitate. The sample is subjected to centrifugation.

Antibody against protein of interest

Pellet

Collect complexes in pellet. Add chemical that breaks the crosslinks to remove the protein.

Conduct PCR using primers to a specific DNA region.

Covalently attach DNA linkers to the ends of the DNA.

or

Linker

Conduct PCR using primers that are complementary to the linkers. Incorporate fluorescently labeled nucleotides during PCR.

If PCR amplifies the DNA, the protein was bound to the DNA region that is recognized by the primers.

Denature DNA and hybridize to a microarray.

See Figure 21.2

FIGURE 21.4   Chromatin immunoprecipitation (ChIP). This method can determine whether proteins bind to a particular region of DNA found within the chromatin of living cells.

homology is profoundly useful as a guiding tool, a full understanding of protein function must involve the direct analysis of proteins as they are found in living cells.

As we move into the twenty-first century, a key challenge facing molecular biologists is the study of proteomes. Much like genomic research, this will require the collective contributions of many research scientists, as well as improvements in technologies that are aimed at unraveling the complexities of the proteome. In this section, we will discuss the phenomena that increase protein diversity beyond genetic diversity. In addition, we will see how the techniques of two-dimensional gel electrophoresis and mass spectrometry can isolate and identify cellular proteins, and explore the use of protein microarrays to study protein expression and function.

The Proteome Is Much Larger Than the Genome

From the sequencing and analysis of an entire genome, researchers can identify all or nearly all of the genes from a given species. The size of the proteome, however, is larger than the genome, and its actual size is somewhat more difficult to determine. What phenomena help to account for the larger size of the proteome? First, changes in pre-mRNAs may ultimately affect the resulting amino acid sequence of a protein. The most important alteration that occurs commonly in eukaryotic species is **alternative splicing.** For many genes, a single pre-mRNA can be spliced into more than one version. The splicing is often cell specific, or it may be related to environmental conditions. As discussed in Chapter 15, alternative splicing is widespread, particularly among complex multicellular organisms. It can lead to the production of several or perhaps dozens of different polypeptide sequences from the same pre-mRNA, which greatly increases the number of potential proteins in the proteome. Similarly, the phenomenon of **RNA editing,** a change in the base sequence of RNA after it has been transcribed (see Chapter 15), can lead to changes in the coding sequence of an mRNA. However, RNA editing is much less common than alternative splicing.

Another process that greatly diversifies the composition of a proteome is the phenomenon of **posttranslational covalent modification.** Certain types of modifications are involved with the assembly and construction of a functional protein. These alterations include proteolytic processing, disulfide bond formation, and the attachment of prosthetic groups, sugars, or lipids. These are typically irreversible changes that are necessary to produce a functional protein. Other types of modifications, such as phosphorylation, acetylation, and methylation, are often reversible modifications that transiently affect the function of a protein. In addition, ubiquitination can promote protein degradation. Because a protein may be subject to several different types of covalent modification, this can greatly increase the forms of a particular protein that are found in a cell at any given time.

Two-Dimensional Gel Electrophoresis Is Used to Separate a Mixture of Cellular Proteins

As we have just learned, the proteome is usually much larger than a species' genome. Even so, any given cell within a complex

multicellular organism will produce only a subset of the proteins found in the proteome of a species. For example, the human genome has approximately 20,000 to 25,000 different genes, yet a muscle cell makes only a subset, perhaps 15,000 types of proteins. The proteins a cell makes depend primarily on the cell type, the stage of development, and the environmental conditions. An objective of researchers in the field of proteomics is the identification and functional characterization of all the proteins a cell type will make. Because cells produce thousands of different proteins, this is a daunting task. Nevertheless, along with genomic research, the past decade has seen important advances in our ability to isolate and identify cellular proteins.

A common technique in the field of proteomics is **two-dimensional gel electrophoresis.** It is a technique that can separate hundreds or even thousands of different proteins within a cell extract. The steps in this procedure are shown in **Figure 21.5.** As its name suggests, the technique involves two different gel electrophoresis procedures. A sample of cells is lysed, and the proteins are loaded onto the top of a tube gel that separates proteins according to their net charge at a given pH. A protein migrates to the point in the gel where its net charge is zero. This is termed **isoelectric focusing.** After the tube gel has run, it is laid horizontally on top of a polyacrylamide slab gel that contains sodium dodecyl sulfate (SDS). The SDS coats each protein with negative charges and denatures them; proteins in the slab gel are separated according to their molecular mass. Smaller proteins move toward the bottom of the gel more quickly than larger ones. After the slab gel has run, the proteins within the gel can be stained with a dye. As seen in Figure 21.5, the end result is a collection of spots, each spot corresponding to a unique cellular protein, with proteins of a larger mass remaining higher in the gel. The resolving power of two-dimensional gel electrophoresis is extraordinary. Proteins that differ by a single charged amino acid can be resolved as two distinct spots using this method.

Various approaches can be followed to identify spots on a gel that may be of interest to researchers. One possibility is that a given cell type may show a few very large spots that are not found when proteins are analyzed from other cell types. The relative abundance of such spots may indicate that a particular protein is important to that cell's structure or function. Secondly, certain spots on a two-dimensional gel may be seen only under a given set of conditions. For example, a researcher may be interested in the effects of a hormone on the function of a particular cell type. Two-dimensional gel electrophoresis could be conducted on a sample in which the cells had not been exposed to the hormone versus a sample in which they had. This may reveal particular spots that are present only when the cells are exposed to a given hormone. Finally, abnormal cells often express proteins that are not found in normal cells. This is particularly true for cancer cells. A researcher may compare normal and cancer cells via two-dimensional gel electrophoresis to identify proteins expressed only in cancer cells.

Mass Spectrometry Is Used to Identify Proteins

Two-dimensional gel electrophoresis may be used as the first step in the separation of cellular proteins. The next goal is to correlate a

Lyse a sample of cells and
load the resulting mixture
of proteins onto an isoelectric
focusing gel.

pH 4.0

Proteins migrate until they
reach the pH where their
net charge is 0. At this
point, a single band could
contain 2 or more
different proteins.

pH 10.0

Lay the tube gel onto an SDS-gel
and separate proteins according to
their molecular mass.

SDS-gel

pH 4.0   pH 10.0

200 kDa

10 kDa

(a) The technique of two-dimensional gel electrophoresis

pH 4.0   pH 10.0

200 kDa

10 kDa

(b) A two-dimensional gel that has been stained to visualize proteins

ONLINE ANIMATION

FIGURE 21.5   **Two-dimensional gel electrophoresis.**
(a) The technique involves two electrophoresis steps. First,
a mixture of proteins is separated on an isoelectric focusing
gel that has the shape of a tube. Proteins migrate to the pH in the gel where
their net charge is zero. This tube gel is placed into a long well on top of an
SDS-polyacrylamide gel. This second gel separates the proteins according
to their mass. In this diagram, only a few spots are seen, but an actual
experiment would involve a mixture of hundreds or thousands of different
proteins. **(b)** A photograph of an SDS-polyacrylamide gel that has been
stained for proteins. Each spot represents a unique cellular protein.

given spot on a two-dimensional gel with a particular protein.
To accomplish this goal, a spot on a two-dimensional gel can be
cut out of the gel to obtain a tiny amount of the protein within
the spot. In essence, the two-dimensional gel electrophoresis procedure purifies a small amount of the cellular protein of interest. The next step is to identify that protein. This can be accomplished via **mass spectrometry,** a technique that measures the
mass of a molecule, such as a peptide fragment.

Figure 21.6 describes how mass spectrometry can determine the amino acid sequence of a protein. As shown here, the
technique actually determines the mass of peptide fragments that
are produced by digesting a purified protein with a protease that
cuts the protein into small peptide fragments. The peptides are
mixed with an organic acid and dried onto a metal slide. The
sample is then struck with a laser. This causes the peptides to
become ejected from the slide in the form of an ionized gas in
which the peptide contains one or more positive charges. The
charged peptides are then accelerated via an electric field and fly
toward a detector. The time of flight is determined by their mass
and net charge. A measurement of the flight time provides an
extremely accurate way to determine the mass of a peptide.

An ultimate goal of mass spectrometry is to determine the
amino acid sequence of a given peptide. This is accomplished
using two mass spectrometers, a method called **tandem mass
spectrometry.** The first mass spectrometer measures the mass
of a given peptide. This same peptide is then analyzed by a second spectrometer after the peptide has been digested into smaller
fragments. The differences in the masses of the peaks in the
spectrum from the second spectrometer reveal the amino acid
sequence of the peptide, because the masses of all 20 amino acids
are known. For example, as shown in Figure 21.6, let's suppose
a peptide had a mass of 1,652 daltons. If one amino acid at the
end was removed and the smaller peptide had a mass that was 87
daltons less (i.e., 1,565 daltons), this would indicate that a serine
is at one end of the peptide, because the mass of serine within a
polypeptide chain is 87 daltons. If two amino acids were removed
at one end and the mass was 224 daltons less, this would correspond to the removal of one serine (87 daltons) and one histidine (137 daltons). If three amino acids were removed and the
mass was decreased by 337 daltons, this would correspond to the
removal of one serine (87 daltons), one histidine (137 daltons),
and one leucine (113 daltons). Therefore, from these measurements, we would conclude that the amino acid sequence from
one end of the peptide was serine–histidine–leucine.

How does this information lead to the identification of a specific protein? After a researcher has obtained a few short peptide sequences from a given protein, genomic information can readily predict the entire amino acid sequence of the protein. For example, if a peptide had the sequence serine–histidine–leucine–asparagine–serine–asparagine, one could determine the possible codon sequences that could encode such a peptide. More than one sequence is possible due to the degeneracy of the genetic code. Using computer software described later, the codon sequences would be used as query sequences to search an entire genomic sequence. This program would locate a match between the predicted codon sequence and a specific gene within the genome. In this way, mass spectrometry makes it possible to identify the gene that encodes the entire protein. The gene sequence, in turn, can be used to predict the remaining amino acid sequence of the entire protein.

Mass spectrometry can also identify protein covalent modifications. For example, if an amino acid within a peptide was phosphorylated, the mass of the peptide would be increased by the mass of a phosphate group. This increase in mass can be determined via mass spectrometry.

Protein Microarrays Can Be Used to Study Protein Expression and Function

Earlier in this chapter, we learned about DNA microarrays, which have gained widespread use to study gene expression at the RNA level. The technology to make DNA microarrays is also being applied to make **protein microarrays.** In this type of technology, proteins, rather than DNA molecules, are spotted onto a glass or silica slide. The development of protein microarrays is more challenging because proteins are much more easily damaged by the manipulations that occur during microarray formation. For example, the three-dimensional structure of a protein may be severely damaged by drying, which usually occurs during the formation of a microarray. This has created additional challenges for researchers who are developing the technology of protein microarrays. In addition, the synthesis and purification of proteins tend to be more time-consuming compared to the production of DNA, which can be amplified by PCR or directly synthesized on the microarray itself. In spite of these technical difficulties, the last few years have seen progress in the production and uses of protein microarrays (**Table 21.2**).

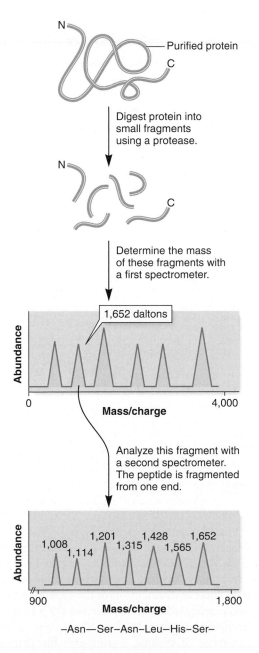

FIGURE 21.6 **The use of tandem mass spectrometry to determine the amino acid sequence of a peptide.**

TABLE 21.2

Some Applications of Protein Microarrays

Application	Description
Protein expression	An antibody microarray can measure protein expression, because each antibody in a given spot recognizes a specific amino acid sequence. This can be used to study the expression of proteins in a cell-specific manner. It can also be used to determine how environmental conditions affect the levels of particular proteins.
Protein function	The substrate specificity and enzymatic activities of groups of proteins can be analyzed by exposing a functional protein microarray to a variety of substrates.
Protein–protein interactions	The ability of two proteins to interact with each other can be determined by exposing a functional protein microarray to fluorescently labeled proteins.
Pharmacology	The ability of drugs to bind to cellular proteins can be determined by exposing a functional protein microarray to different kinds of labeled drugs. This can help to identify the proteins within a cell to which a given drug may bind.

The two common types of protein microarray analyses are antibody microarrays and functional protein microarrays. The purpose of an **antibody microarray** is to study protein expression. As discussed in Chapter 17, antibodies are proteins that recognize antigens. One type of antigen that an antibody can recognize is a short peptide sequence found within another protein. Therefore, an antibody can specifically recognize a cellular protein. Researchers can produce thousands of different antibodies, each one recognizing a different peptide sequence. These can be spotted onto a microarray. Cellular proteins can then be isolated, fluorescently labeled, and exposed to the antibody microarray. When a given protein is recognized by an antibody on the microarray, it will be captured by the antibody and remain bound to that spot. The level of fluorescence at a given spot indicates the amount of a cellular protein that is recognized by a particular antibody.

The other type of array is a **functional protein microarray.** To make this type of array, researchers must purify cellular proteins and then spot them onto a microarray. The microarray can then be analyzed with regard to specific kinds of protein function. In 2000, for example, Heng Zhu, Michael Snyder, and colleagues purified 119 proteins from yeast that were known to function as protein kinases. These kinds of proteins attach phosphate groups onto other cellular proteins. A microarray was made consisting of different possible proteins that may or may not be phosphorylated by these 119 kinases, and then the array was exposed to each of the kinases in the presence of radiolabeled ATP. By following the incorporation of phosphate into the array, they determined the protein specificity of each kinase. On a much larger scale, the same group of researchers purified 5,800 different yeast proteins and spotted them onto a microarray. The array was then exposed to fluorescently labeled calmodulin, which is a regulatory protein that binds calcium ions. Several proteins in the microarray were found to bind calmodulin. While some of these were already known to be regulated by calmodulin, other proteins in the array that had not been previously known to bind calmodulin were identified.

21.3 BIOINFORMATICS

Geneticists use computers to collect, store, manipulate, and analyze data. Molecular genetic data is particularly amenable to computer analysis because it comes in the form of a sequence, such as a DNA, RNA, or amino acid sequence. The ability of computers to analyze data at a rate of millions or even billions of operations per second has made it possible to solve problems concerning genetic information that were thought intractable a few decades ago.

In recent years, the marriage between genetics and computational tools and approaches has yielded an important branch of science known as bioinformatics. Several scientific journals are largely devoted to this topic. Computer analysis of genetic sequences relies on three basic components: a computer, a computer program, and some type of data. In genetic research, the data consist of a particular genetic sequence or several sequences that a researcher or clinician wants to study. For example, this could be a DNA sequence derived from a cloned DNA fragment. A sequence of interest may be relatively short or thousands to millions of nucleotides in length. Experimentally, DNA sequences and related data are obtained using the techniques described in Chapters 18 and 20.

In this section, we will first consider the fundamental concepts that underlie the analysis of genetic sequences. We will then explore how these methods are used to provide insights regarding functional genomics and proteomics. Chapter 26 will describe applications of bioinformatics in the area of evolutionary biology. In addition, you may wish to actually run computer programs, which are widely available at university and government websites (for example, see www.ncbi.nlm.nih.gov/). This type of hands-on learning will help you to see how the computer has become a valuable tool to analyze genetic data.

Sequence Files Are Analyzed by Computer Programs

Most people are familiar with **computer programs,** which consist of a defined series of operations that can manipulate and analyze data in a desired way. For example, a computer program might be designed to take a DNA sequence and translate it into an amino acid sequence. A first step in the computer analysis of genetic data is the creation of a **computer data file** to store the data. This file is simply a collection of information in a form suitable for storage and manipulation on a computer. In genetic studies, a computer data file might contain an experimentally obtained DNA, RNA, or amino acid sequence. For example, a file could contain the DNA sequence of one strand of the *lacY* gene from *Escherichia coli*, as shown here. The numbers to the left represent the base number in the sequence file.

```
    1  ATGTACTATT  TAAAAAACAC  AAACTTTTGG  ATGTTCGGTT  TATTCTTTTT
   51  CTTTTACTTT  TTTATCATGG  GAGCCTACTT  CCCGTTTTTC  CCGATTTGGC
  101  TACATGACAT  CAACCATATC  AGCAAAGTG  ATACGGGTAT  TATTTTGCC
  151  GCTATTTCTC  TGTTCTCGCT  ATTATTCCAA  CCGCTGTTTG  GTCTGCTTC
  201  TGACAAACTC  GGGCTGCGCA  AATACCTGCT  GTGGATTATT  ACGGCATGT
  251  TAGTCATGTT  TGCGCCGTTC  TTTATTTTTA  TCTTCGGGCC  ACTGTTACAA
  301  TACAACATTT  TAGTAGGATC  GATTGTTGGT  GGTATTTATC  TAGGCTTTTG
  351  TTTTAACGCC  GGTGCGCCAG  CAGTAGAGGC  ATTTATTGAG  AAAGTCAGCC
  401  GTCGCAGTAA  TTTCGAATTT  GGTCGCGCGC  GGATGTTTGG  CTGTGTTGGC
  451  TGGGCGCTGT  GTGCCTGAT  TGTCGGCATC  ATGTTCACCA  TCAATAATCA
  501  GTTTGTTTTC  TGGCTGGGCT  CTGGCTGTGC  ACTCATCCTC  GCCGTTTTAC
  551  TCTTTTTCGC  CAAAACGGAT  GCGCCCTCTT  CTGCCACGGT  TGCCAATGCG
  601  GTAGGTGCCA  ACCATTCGGC  ATTTAGCCTT  AAGTGGCAC  TGGAACTGTT
  651  CAGACAGCCA  AAACTGTGGT  TTTGTCTACT  GTATGTTATT  GGCGTTTCCT
  701  GCACCTACGA  TGTTTTTGAC  CAACAGTTTG  CTAATTTCTT  TACTTCGTTC
  751  TTTGCTACCG  GTGAACAGGG  TACGCGGGTA  TTTGGCTACG  TAACGACAAT
  801  GGGCGCAATTA  CTTAACGCCT  CGATTATCTT  CTTGCGCCA  CTGATCATTA
  851  ATCGCATCGG  TGGGAAAAAC  GCCCTGCTGC  TGGCTGGCAC  TATTATCTCT
  901  CTACGTATTA  TTGGCTCATC  GTTCGCCACC  TCAGCGCTGG  AAGTGGTTAT
  951  TCTGAAAACG  CTGCATATGT  TTGAAGTACC  GTTCCTGCTG  GTGGGCTGCT
1,001  TTAAATATAT  TACCAGCCAG  TTTGAAGTGC  GTTTTTCAGC  GACGATTTAT
1,051  CTGGTCTGTT  TCTGCTTCTT  TAAGCAACTG  GCATGATTT  TTATGTCTGT
1,101  ACTGGCGGGC  AATATGTATG  AAAGCATCGG  TTTCCAGGGC  GCTTATCTGG
1,151  TGCTGGGTCT  GGTGGCGCTG  GGCTTCACCT  TAATTTCCGT  GTTCACGCTT
1,201  AGCGGCCCCG  GCCCGCTTTC  CCTGCTGCGT  CGTCAGGTGA  ATGAAGTCGC
1,251  TTAA
```

To store data in a computer data file, a scientist creates the file and enters the data, either by hand or, what is now more common, by laboratory instruments such as densitometers and fluorometers. These instruments have the capability to read data, such as a sequencing ladder, and enter the DNA sequence information directly into a computer file.

The purpose of making a computer file that contains a genetic sequence is to take advantage of the swift speed with which computers can analyze this information. Genetic sequence data in a computer file can be investigated in many different ways, corresponding to the many questions a researcher might ask about the sequence and its functional significance. These include the following:

1. Does a sequence contain a gene?
2. Where are functional sequences such as promoters, regulatory sites, and splice sites located within a particular gene?
3. Does a sequence encode a polypeptide? If so, what is the amino acid sequence of the polypeptide?
4. Does a sequence predict certain structural features for DNA, RNA, or proteins? For example, is a DNA sequence likely to be in a Z-DNA conformation? What is the secondary structure of an RNA sequence or polypeptide sequence?
5. Is a sequence homologous to any other known sequences?
6. What is the evolutionary relationship between two or more genetic sequences?

To answer these and many other questions, different computer programs have been written to analyze genetic sequences in particular ways. These programs have been devised by theoreticians who understand basic genetic principles and can design computational strategies for analyzing genetic sequences. When constructing a computer program, a theoretician has a goal that the program is meant to fulfill. For example, a theoretician may wish to write a program to translate a DNA sequence into an amino acid sequence. Based on knowledge of the genetic code, she/he can devise computational procedures, or algorithms, that relate a DNA sequence to an amino acid sequence. In a computer program, these procedures are executed as a stepwise "plan of operations" that manipulates the data in a sequence file. For a computer to perform the plan of operations, the program instructions must be written in a programming language the computer can decipher. After this has been accomplished, the program is tested to see if it works. Usually, errors (bugs) are found in the program that prevent it from working. After the program is debugged, it is ready for use. An ideal computer program is one that accurately manipulates data and is user-friendly.

As an example, let's consider a computer program aimed at translating a DNA sequence into an amino acid sequence and see how it might work in practice. The operation of the program as it would appear on a computer screen is shown in **Figure 21.7**. The geneticist—the user—has a DNA sequence file that she or he may want to have translated into an amino acid sequence. The user is sitting at a computer that is connected to a program that can translate a DNA sequence into an amino acid sequence. In this example, the program is named TRANSLATION. After the program is launched, it asks a series of questions. These are depicted on the screen in black. The typed answers of the user are shown in red. The first question asks which sequence file the user wants translated. In this case, the user wants the *lacY* gene sequence translated into an amino acid sequence. The name of this file is lacY.SEQ. The program then asks which codon translation table to use. As you may recall from Chapter 13, some spe-

Translation
Selected sequence?
lacY.SEQ
Translation table?
Standard
Translation parameters
  Start translating at: 1
  Stop translating at: 1,254
  Frame(s) to translate: Three forward frames
  Show longest reading frame: Yes
  Name of file to store longest reading frame: lacY.PEP

RUN

FIGURE 21.7   **The operation of a computer program as it would appear on a computer screen.**

cies have slight variations in their usage of the genetic code. In this case, the user selects the standard genetic code. Next, the program requires the parameters of the sequence file that is to be translated. The user decides to begin the translation at the first nucleotide in the sequence file and end the translation at nucleotide number 1,254. The user wants the program to translate the sequence in all three forward reading frames and to show the longest reading frame—the longest amino acid sequence that is uninterrupted by a stop codon. This translated sequence is saved in a file that the user wants to be named lacY.PEP. (Note: The symbol PEP provides a reminder that it is a polypeptide sequence.) The user clicks on the RUN button, and the translation program proceeds to translate the *lacY* sequence and stores the amino acid sequence in a file called lacY.PEP. The contents of the lacY.PEP file are shown here.

```
  1 MYYLKNTNFW  MFGLFFFFYF  FIMGAYFPFF  PIWLHDINHI  SKSDTGIIFA
 51 AISLFSLLFQ  PLFGLLSDKL  GLRKYLLWII  TGMLVMFAPF  FIFIFGPLLQ
101 YNILVGSIVG  GIYLGFCFNA  GAPAVEAFIE  KVSRRSNFEF  GRARMFGCVG
151 WALCASIVGI  MFTINNQFVF  WLGSGCALIL  AVLLFFAKTD  APSSATVANA
201 VGANHSAFSL  KLALELFRQP  KLWFLSLYVI  GVSCTYDVFD  QQFANFFTSF
251 FATGEQGTRV  FGYVTTMGEL  LNASIMFFAP  LIINRIGGKN  ALLLAGTIMS
301 VRIIGSSFAT  SALEVVIKLT  LHMFEVPFLL  VGCFKYITSQ  FEVRFSATIY
351 LVCFCFFKQL  AMIFMSVLAG  NMYESIGFQG  AYLVLGLVAL  GFTLISVFTL
401 SGPGPLSLLR  RQVNEVA
```

In this file, which was created by the computer program TRANSLATION, each of the 20 amino acids is given a single-letter abbreviation (see Figure 13.6).

Why is the TRANSLATION program useful? The advantages of running this program are speed and accuracy. It can translate a relatively long genetic sequence within seconds. By comparison, it would probably take you a few hours to look each codon up in the genetic code table and write the sequence out in the correct order. If you visit a website and actually run a program like TRANSLATION, you will discover that such a program can translate a genetic sequence into six reading frames—three forward and three reverse. This is useful if a researcher does not know where the start codon is located and/or does not know the direction of the coding sequence.

In genetic research, large software packages typically contain many computer programs that can analyze genetic sequences in different ways. For example, one program can translate a DNA sequence into an amino acid sequence, while another program can locate introns within genes. These software packages are found at universities, government facilities, hospitals, and industries. At such locations, a central computer with substantial memory and high-speed computational abilities runs the software, and individuals can connect to this central computer. Many such programs are freely available on the Internet.

The Scientific Community Has Collected Sequence Files and Stored Them in Large Computer Databases

In Chapter 18, we considered how researchers can clone and sequence genes. Likewise, in Chapter 20, we learned how scientists are investigating the genetic sequences of entire genomes from several species, including humans. The amount of genetic information generated by researchers has become enormous. The Human Genome Project, for example, has produced more data than any other undertaking in the history of biology. With these advances, scientists realize that another critical use of computers is to simply store the staggering amount of data produced from genetic research.

A large number of computer data files collected and stored in a single location is called a **database.** In addition to genetic sequences, the files within databases are **annotated,** which means they contain additional information such as a concise description of a sequence, the name of the organism from which this sequence was obtained, and the function of the encoded protein, if it is known. The file may also describe other features of significance and cite a published journal reference that describes the sequence.

The scientific community has collected the genetic information from thousands of research labs and created several large databases. **Table 21.3** describes some of the major genetic databases in use worldwide. These databases enable researchers to access and compare genetic sequences that are obtained by many laboratories. Later, we will learn how researchers can use databases to analyze genetic sequences.

The databases described in Table 21.3 collect genetic information from many different species. Scientists have also created more specialized databases, called **genome databases,** that focus on the genetic sequences and characteristics of a single species. Genome databases have been created for species of bacteria

TABLE 21.3

Examples of Major Computer Databases

Type	Description
Nucleotide sequence	DNA sequence data are collected into three internationally collaborating databases: GenBank (a USA database), EMBL (European Molecular Biology Laboratory Nucleotide Sequence Database), and DDBJ (DNA Databank of Japan). These databases receive sequence and sequence annotation data from genome projects, sequencing centers, individual scientists, and patent offices. These databases are accessed via the Internet and available on CD-ROM.
Amino acid sequence	Amino acid sequence data are collected into a few international databases including Swissprot (Swiss protein database), PIR (Protein Information Resource), Genpept (translated peptide sequences from the GenBank database), and TrEMBL (Translated sequences from the EMBL database).
Three-dimensional structure	PDB (Protein Data Bank) collects the three-dimensional structures of biological macromolecules with an emphasis on protein structure. These are primarily structures that have been determined by X-ray crystallography and nuclear magnetic resonance (NMR), but some models are included in the database. These structures are stored in files that can be viewed on a computer with the appropriate software.
Protein motifs	Prosite is a database containing a collection of amino acid sequence motifs that are characteristic of a protein family, domain structure, or certain posttranslational modifications. Pfam is a database of protein families with multiple amino acid sequence alignments.

(e.g., *E. coli*), yeast (e.g., *Saccharomyces cerevisiae*), worms (e.g., *Caenorhabditis elegans*), fruit flies (e.g., *Drosophila melanogaster*), plants (e.g., *Arabidopsis thaliana*), and mammals (e.g., mice and humans). The primary aim of genome databases is to organize the information from sequencing and mapping projects for a single species. Genome databases identify an organism's known genes and describe their map location within the genome. In addition, a genome database may provide information concerning gene alleles, bibliographic information, a directory of researchers who study the species, and other pertinent information.

Different Computational Strategies Can Identify Functional Genetic Sequences

At the molecular level, the function of the genetic material is based largely on specific genetic sequences that play distinct roles. For example, codons are three-base sequences that specify particular amino acids, and promoters are sequences that provide a binding site for RNA polymerase to initiate transcription.

Computer programs can be designed to scan very long sequences, such as those obtained from genome sequencing projects, and locate meaningful features within them. To illustrate this concept, let's first consider the following sequence file, which contains an alphabetic sequence of 54 letters:

Sequence file:

```
GJTRLLAMAQLHEOGYLTOBWENTMNMTORXXXTGOODNTHEQ
ALLRTLSTORE
```

We will now compare how three different computer programs might analyze this sequence to identify meaningful features. The goal of our first program is to locate all the English words within this sequence. If we ran this program, we would obtain the following result:

```
GJTRLLAMAQLHEOGYLTOBWENTMNMTORXXXTGOODNTHEQ
ALLRTLSTORE
```

In this case, a computer program has identified locations where the sequence of letters forms a word. Several words (which are underlined) have been located within this sequence.

A second computer program could be aimed at locating a series of words that are organized in the correct order to form a grammatically logical English sentence. If we used our sequence file and ran this program, we would obtain the following result:

```
GJTRLLAMAQLHEOGYLTOBWENTMNMTORXXXTGOODNTHEQ
ALLRTLSTORE
```

The second program has identified five words that form a logical sentence.

Finally, a computer program might be used to identify patterns of letters, rather than words or sentences. For example, a computer program could locate a pattern of five letters that occurs in both the forward and reverse directions. If we applied this program to our sequence file, we would obtain the following:

```
GJTRLLAMAQLHEOGYLTOBWENTMNMTORXXXTGOODNTHEQ
ALLRTLSTORE
```

In this case, the program has identified a pattern where five letters are found in both the forward and reverse directions.

In the three previous examples, we can distinguish between **sequence recognition** (as in our first example) and **pattern recognition** (as in our third example). In sequence recognition, the program has the information that a specific sequence of symbols has a specialized meaning. This information must be supplied to the computer program. For example, the first program would have access to the information from a dictionary with all known English words. With this information, the first program can identify sequences of letters that make words. By comparison, the third program does not rely on specialized sequence information. Rather, it is looking for a pattern of symbols that can occur within any group of symbol arrangements.

Overall, the simple programs we have considered illustrate three general types of identification strategies:

1. *Locate specialized sequences within a very long sequence.* A specialized sequence with a particular meaning or function is called a **sequence element** or **motif.** The computer program has a list of predefined sequence elements and can identify such elements within a sequence of interest.

2. *Locate an organization of sequences.* As shown in the second program, this could be an organization of sequence elements. Alternatively, it could be an organization of a pattern of sequences.

3. *Locate a pattern of sequences.* The third program is an example of locating a pattern of sequences.

The great power of computer analysis is that these types of operations can be performed with great speed and accuracy on sequences that may be enormously long.

Now that we understand the general ways that computer programs identify sequences, let's consider specific examples. As we have discussed throughout this textbook, many short nucleotide sequences play specialized roles in the structure or function of genetic material. A geneticist may want to locate a short sequence element within a longer nucleotide sequence in a data file. For example, a sequence of chromosomal DNA might be tens of thousands of nucleotides in length, and a geneticist may want to know whether a sequence element, such as a TATA box, is found at one or more sites within the chromosomal DNA. To do so, a researcher could visually examine the long chromosomal DNA sequence in search of a TATA box. Of course, this process would be tedious and prone to error. By comparison, the appropriate computer program can locate a sequence element within seconds. Therefore, computers are very useful for this type of application. **Table 21.4** lists some examples of sequence elements that can be identified by computer analysis.

TABLE 21.4

Short Sequence Elements That Can Be Identified by Computer Analysis

Type of Sequence	Examples*
Promoter	Many *E. coli* promoters contain TTGACA (−35 site) and TATAAT (−10 site). Eukaryotic core promoters may contain CAAT boxes, GC boxes, TATA boxes, etc.
Response elements	Glucocorticoid response element (AGRACA), cAMP response element (GTGACGTRA)
Start codon	ATG
Stop codons	TAA, TAG, TGA
Splice site	GTRAGT————————YNYTRAC(Y)$_n$AG
Polyadenylation signal	AATAAA
Highly repetitive sequences	Relatively short sequences that are repeated many times throughout a genome
Transposable elements	Usually characterized by a pattern in which direct repeats flank inverted repeats

*The sequences shown in this table would be found in the DNA. For gene sequences, only the coding strand is shown. R = purine (A or G); Y = pyrimidine (T or C); N = A, T, G, or C; U in RNA = T in DNA.

By comparing the amino acid sequences and known functions of proteins in thousands of cases, researchers have also found amino acid sequence motifs that carry out specialized functions within proteins. For example, researchers have determined that the amino acid sequence motif asparagine–X–serine (where X is any amino acid except proline) within eukaryotic proteins is a glycosylation site (i.e., it may have a carbohydrate attached to it). The Prosite database (refer back to Table 21.3) contains a collection of all amino acid sequence motifs known to be functionally important. Researchers can use computer programs to determine whether an amino acid sequence contains any of the motifs found in the Prosite database. This may help them to understand the role of a newly found protein of unknown function.

Several Computer-Based Approaches Can Identify Structural Genes Within a Nucleotide Sequence

A structural gene is composed of nucleotide sequences organized in a particular way. A typical gene contains a promoter, followed by a start codon, a coding sequence, a stop codon, and a transcriptional termination site. In addition, most genes contain regulatory sequences (e.g., eukaryotic response elements or prokaryotic operator sites), and eukaryotic genes are likely to contain introns. After researchers have sequenced a long segment of chromosomal DNA, they frequently want to know if the sequence contains any genes. In an attempt to answer this question, geneticists can use computer programs that are aimed at identifying genes in long genomic DNA sequences.

How do computer programs identify a gene within a long genetic sequence? These programs employ different strategies. A **search by signal** approach relies on known sequences such as promoters, start and stop codons, and splice sites to help predict whether or not a DNA sequence contains a structural gene. The program tries to locate an organization of known sequence elements that normally are found within a gene. It would try to locate a region that contains a promoter sequence, followed by a start codon, a coding sequence, a stop codon, and a transcriptional terminator.

A second strategy is a **search by content** approach. The goal here is to identify sequences with a nucleotide content that differs significantly from a random distribution. Within structural genes, this occurs primarily due to codon usage. Although there are 64 codons, most organisms display a **codon bias** within structural genes. This means that certain codons are used much more frequently than others. For example, UUA, UUG, CUU, CUC, CUA, and CUG all specify leucine. In yeast, however, 80% of the leucine codons are UUG. Codon bias allows organisms to more efficiently rely on a smaller population of tRNA molecules. A search by content strategy, therefore, attempts to locate coding regions by identifying regions where the nucleotide content displays a known codon bias.

A third way to locate coding regions within a DNA sequence is to examine translational reading frames. Recall that the reading frame is a sequence of codons determined by reading bases in groups of three. In a new DNA sequence, researchers must consider that the reading of codons (in groups of three

nucleotides) could begin with the first nucleotide (reading frame 1), the second nucleotide (reading frame 2), or the third nucleotide (reading frame 3). An **open reading frame (ORF)** is a region of a nucleotide sequence that does not contain any stop codons. Because most proteins are several hundred amino acids in length, a relatively long reading frame is required to encode them. In prokaryotic species, long ORFs are contained within the chromosomal gene sequences. In eukaryotic genes, however, the coding sequence may be interrupted by introns. As described earlier, one way to determine eukaryotic ORFs is to clone and sequence cDNA, which is complementary to mRNA. Alternatively, a computer program can translate a genomic DNA sequence in all three reading frames, seeking to identify a long ORF. In **Figure 21.8**, a DNA sequence has been translated in all three reading frames. Only one of the three reading frames (3) contains a very long open reading frame without any stop codons, suggesting that this DNA sequence encodes a protein. In Figure 21.8, it was assumed that the reading frame was from left to right. In an uncharacterized genetic sequence, however, a reading frame could proceed from right to left. Therefore, in a newly discovered genetic sequence, six reading frames are possible—three in the forward direction and three in the reverse direction.

Even though computer programs are a valuable tool, they are not always accurate in their prediction of gene sequences. In particular, it is often difficult for programs to predict the correct start codon and the precise intron/exon boundaries. In some cases, computer programs may even suggest that a region encodes a gene when it does not. Therefore, while a bioinformatic approach is a relatively easy tool to identify potential genes, it should not be viewed as a definitive method. The confirmation that a DNA region encodes an actual gene requires laboratory experimentation to show that it is truly transcribed into RNA.

FIGURE 21.8 **Translation of a DNA sequence in all three reading frames.** The three lines represent the translation of a gene sequence in each of three forward reading frames; the reading frames proceed from left to right. The letter S indicates the location of a stop codon. Reading frame 3 has a very long open reading frame, suggesting that it may be the reading frame for a structural gene. Reading frames 1 and 2 are not likely to be the reading frames for a structural gene, because they contain many stop codons. During the cloning of DNA, the orientation of a gene may become flipped so that the coding sequence is inverted. Therefore, when analyzing many cloned DNA fragments, six reading frames (i.e., three forward and three reverse) are evaluated. Only the three forward frames are shown here.

Computer Programs Can Identify Homologous Sequences

Let's now turn our attention to the uses of computer technology to identify genes that are evolutionarily related. The ability to sequence DNA allows geneticists to examine evolutionary relationships at the molecular level. This has become an extremely powerful tool in the field of genomics. When comparing genetic sequences, researchers frequently find two or more sequences that are similar. For example, the sequence of the *lacY* gene that encodes the lactose permease in *E. coli* is similar to that of the *lacY* gene that encodes the lactose permease in a closely related bacterium, *Klebsiella pneumoniae*. As shown here, when segments of the two *lacY* genes are lined up, approximately 78% of their bases are a perfect match.

```
              151                                                    200
E. coli       TCTTTTTCTT TTACTTTTTT ATCATGGGAG CCTACTTCCC GTTTTTCCCG
K. pneumoniae TCTTTTTCTT TTACTATTTC ATTATGTCAG CCTACTTTCC TTTTTTTCCG

              201                                                    250
              ATTTGGCTAC ATGACATCAA CCATATCAGC AAAAGTGATA CGGGTATTAT
              GTGTGGCTGG CGGAAGTTAA CCATTTAACC AAAACCGAGA CGGGTATTAT
```

In this case, the two sequences are similar because the genes are **homologous** to each other. This means they have been derived from the same ancestral gene. This idea is shown schematically in **Figure 21.9**. An ancestral *lacY* gene was located in a bacterium that preceded the evolutionary divergence between *E. coli* and *K. pneumoniae*. After these two bacteria had diverged from each other, their *lacY* genes accumulated distinct mutations that produced somewhat different base sequences for this gene. Therefore, in these two species of bacteria, the *lacY* genes are similar but not identical. When two homologous genes are found in different species, these genes are termed **orthologs**.

Two or more homologous genes can also be found within a single organism. These are termed paralogous genes, or **paralogs**. As discussed in Chapter 8, abnormal gene duplication may happen several times during evolution, which results in multiple copies of a gene and ultimately leads to the formation of a gene family. A **gene family** consists of two or more paralogs within the genome of a single organism. When a gene family occurs, the concept of orthologs becomes more complex. For example, let's consider the globin gene family found in mammals (see Chapter 8, Figure 8.7). Researchers would say that the β-globin gene in humans is an ortholog to the β-globin gene found in mice. Likewise, α-globin genes found in both species would be considered

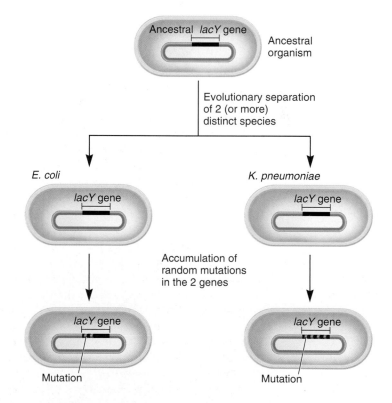

FIGURE 21.9   **The origin of homologous *lacY* genes in *Escherichia coli* and *Klebsiella pneumoniae*.** This figure emphasizes a single gene within an ancestral organism. During evolution, the ancestral organism diverged into two different species, *E. coli* and *K. pneumoniae*. After this divergence, the *lacY* gene in the two separate species accumulated mutations, yielding *lacY* genes with somewhat different sequences.

Genes→Traits   After two species diverge evolutionarily, their genes will accumulate different random mutations. This example concerns the *lacY* gene, which encodes lactose permease. In both species, the function of lactose permease is to transport lactose into the cell. The *lacY* gene in these two species has accumulated different mutations that slightly alter the amino acid sequence of the protein. Researchers have determined that these two species transport lactose at significantly different rates. Therefore, the changes in gene sequences have affected the ability of these two species to transport lactose.

orthologs. However, the α-globin gene in humans would not be called an ortholog of the β-globin gene in mice, though they could be called homologous. The most closely related genes in two different species are considered orthologs.

Homologous genes, whether they are orthologs or paralogs, have similar sequences. What is the distinction between homology and sequence similarity? **Homology** implies a common ancestry. **Similarity** means that two sequences are similar to each other. In many cases, such as the *lacY* example, similarity is due to homology. However, this is not always the case. Short genetic sequences may be similar to each other even though two genes are not related evolutionarily. For example, many nonhomologous bacterial genes contain similar promoter sequences at the −35 and −10 regions.

A Simple Dot Matrix Can Compare the Degree of Similarity Between Two Sequences

To evaluate the similarity between two sequences, a matrix can be constructed. In a general way, **Figure 21.10** illustrates the use of a simple dot matrix. In Figure 21.10a, the sequence GENETICSIS-COOL is compared to itself. Each point in the grid corresponds to one position of each sequence. The matrix allows all such pairs to be compared simultaneously. Dots are placed where the same letter occurs in both sequences. The key observation is that regions of similarity are distinguished by the occurrence of many dots along a diagonal line within the dot matrix. In contrast, Figure 21.10b compares two unrelated sequences: GENETICSIS-COOL and THECOURSEISFUN. In this comparison, no diagonal lines are seen. In some cases, two sequences may be related to each other but differ in length. Figure 21.10c compares the sequences GENETICSISCOOL and GENETICSISVERYCOOL. In this example, the second sequence is four letters longer than the first. In the dot matrix, two diagonal lines occur. To align these two lines, a gap must be created in the first sequence. Figure 21.10d shows the insertion of a gap that aligns the two sequences so that the two diagonal lines fall along the same line.

Overall, Figure 21.10 illustrates two important features of dot matrix methods. First, regions of homology are recognized by a series of dots that lie along a diagonal line. Second, gaps

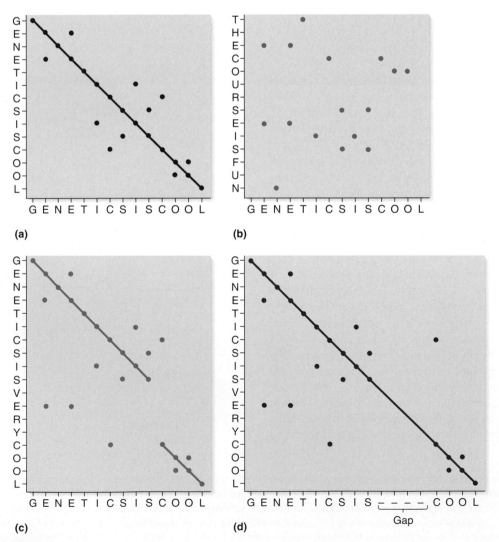

FIGURE 21.10 **The use of dot matrices to evaluate similarity between two sequences.** On a dot matrix, many points that fall on a diagonal line indicate regions of similarity.

can be inserted to align sequences of unequal length but that still have some similarity. These same concepts hold true when genetic sequences are compared to each other. Unfortunately, for anything but the shortest genetic sequences, a simple dot matrix approach is not adequate. Instead, dynamic programming methods are used to identify similarities between genetic sequences. These methods, originally proposed by Saul Needleman and Christian Wunsch, are theoretically similar to a dot matrix, but they involve complex mathematical operations beyond the scope of this textbook.

In their original work of 1970, Needleman and Wunsch demonstrated that whale myoglobin and human β hemoglobin have similar sequences. Since then, this approach has been extended to compare more than two genetic sequences. Newer computer programs can align several genetic sequences and sensibly put in gaps. This produces a **multiple sequence alignment.**

To illustrate the usefulness of a multiple sequence alignment, let's use the general methods of Needleman and Wunsch and apply them to the globin gene family. Hemoglobin, a protein found in red blood cells, is responsible for carrying oxygen through the bloodstream. In humans, nine paralogous globin genes are functionally expressed. (There are also several pseudogenes that are not expressed and one myoglobin gene.) The nine globin genes fall into two categories: the α chains and the β chains. The α-chain genes are α_1, α_2, θ, and ξ; the β-chain genes are β, δ, γ_A, γ_G, and ε. Each hemoglobin protein is composed of two α chains and two β chains.

Because the globin genes are expressed at different stages of human development, the composition of hemoglobin changes during the course of growth. For example, the ξ and ε genes are expressed during early embryonic development, whereas the α and β genes are expressed in the adult.

Insights into the structure and function of the hemoglobin polypeptide chains can be gained by comparing their sequences. In **Figure 21.11**, the sequences of the human globin polypeptides are compared in a multiple sequence alignment using dynamic programming methods. An inspection of a multiple sequence alignment may reveal important features concerning the similarities and differences within a gene family. In this alignment, dots are shown where it is necessary to create gaps to keep the amino acid sequences aligned. As we can see, the sequence similarity is very high between α_1, α_2, θ, and ξ. In fact, the amino acid sequences encoded by the α_1 and α_2 genes are identical. This suggests that the four types of α chains likely carry out very similar functions. Likewise, the β chains encoded by the β, δ, γ_A, γ_G, and ε genes are very similar to each other. In the globin gene family, the α chains are much more similar to each other than they are to the β chains, and vice versa.

In general, amino acids that are highly conserved within a gene family are more likely to be important functionally. The arrows in the multiple sequence alignment point to histidine amino acids that are conserved in all nine members of the hemoglobin gene family. These histidines are involved in the necessary function of binding the heme molecule to the globin polypeptides.

```
                1                                                    50
     β beta     VHLTPEEEKSA VTALWGKV.. NVDEVGGEAL GRLLVVYPWT QRFFESFGDL
     δ delta    VHLTPEEEKSA VNALWGKV.. NVDAVGGEAL GRLLVVYPWT QRFFESFGDL
  γA gamma-A    GHFTEEDKAT ITSLWGKV.. NVEDAGGEAL GRLLVVYPWT QRFFESFGDL
  γG gamma-G    GHFTEEDKAT ITSLWGKV.. NVEDAGGEAL GRLLVVYPWT QRFFESFGDL
     ε epsilon  VHFTAEEKAA VTSLWSKM.. NVEEAGGEAL GRLLVVYPWT QRFFESFGDL
  α1 alpha-1    VLSPADKTN VKAAWGKVGA HAGEGAEAL ERMFLSFPTT KTYFPHF.DL
  α2 alpha-2    VLSPADKTN VKAAWGKVGA HAGEGAEAL ERMFLSFPTT KTYFPHF.DL
     θ theta    ALSAEDRAL VRALWKKLGS NVGVYTTEAL ERTFLAFPAT KTYFSHL.DL
     ξ zeta     SLTKTERTI IVSMWAKIST QADTIGTETL ERLFLSHPQT KTYFPHF.DL

                51                                                   100
     β beta     STPDAVMGNP KVKAHGKKVL GAFSDGLAHL DNLKGTFATL SELHCDKLHV
     δ delta    SSPDAVMGNP KVKAHGKKVL GAFSDGLAHL DNLKGTFSQL SELHCDKLHV
  γA gamma-A    SSASAIMGNP KVKAHGKKVL TSLGDAIKHL DDLKGTFAQL SELHCDKLHV
  γG gamma-G    SSASAIMGNP KVKAHGKKVL TSLGDAIKHL DDLKGTFAQL SELHCDKLHV
     ε epsilon  SSPSAILGNP KVKAHGKKVL TSFGDAIKNM DNLKPAFAKL SELHCDKLHV
  α1 alpha-1    SHGSA..... QVKGHGKKVA DALTNAVAHV DDMPNALSAL SDLHAHKLRV
  α2 alpha-2    SHGSA..... QVKGHGKKVA DALTNAVAHV DDMPNALSAL SDLHAHKLRV
     θ theta    SPGSS..... QVKAHGQKVA DALSLAVERL DDLPHALSAL SHLHACQLRV
     ξ zeta     HPGSA..... QLRAHGSKVV AAVGDAVKSI DDIGGALSKL SELHAYQLRV
                                   ↑                              ↑

                101                                                 148
     β beta     DPENFRLLGN VLVCVLAHHF GKEFTPPVQA AYQKVVAGVA NALAHKYH
     δ delta    DPENFRLLGN VLVCVLARNF GKEFTPQMQA AYQKVVAGVA NALAHKYH
  γA gamma-A    DPENFRLLGN VLVCVLAIHF GKEFTPEVQA SWQKMVTAVA SALSSRYH
  γG gamma-G    DPENFRLLGN VLVCVLAIHF GKEFTPEVQA SWQKMVTAVA SALSSRYH
     ε epsilon  DPENFRLLGN VMVIILATHF GKEFTPEVQA AWQKLVSAVA IALAHKYH
  α1 alpha-1    DPVNFKLLSH CLLVTLAAHL PAEFTPAVHA SLDKFLASVS TVLTSKYR
  α2 alpha-2    DPVNFKLLSH CLLVTLAAHL PAEFTPAVHA SLDKFLASVS TVLTSKYR
     θ theta    DPASFQLLGH CLLVTLARHL PGDFSPALQA SLDKFLSHSVI SALVSEYR
     ξ zeta     DPVNFKLLSH CLLVTLAARF PADFTAEAHA AWDKFLSVVS SVLTEKYR
```

FIGURE 21.11   A multiple sequence alignment among selected members of the globin gene family in humans.

Overall, the alignment shown in Figure 21.11 illustrates the type of information that can be derived from a multiple sequence alignment. In this case, multiple sequence alignment has shown that a group of nine genes falls into two closely related subgroups. The alignment has also identified particular amino acids within the proteins' sequences that are highly conserved. This conservation is consistent with an important role in protein function.

A Database Can Be Searched to Identify Homologous Sequences

Homologous genes usually encode proteins that carry out similar or identical functions. As we have just considered, the members of the globin gene family are all involved with carrying and transporting oxygen. Likewise, the *lacY* genes in *E. coli* and *K. pneumoniae* both encode lactose permease, a protein that transports lactose across the bacterial cell membrane.

A strong correlation is typically found between homology and function. How is this relationship useful with regard to bioinformatics? In many cases, the first indication of the function of a newly determined sequence is through homology to known sequences in a database. An example is the gene that is altered in cystic fibrosis (CF) patients. After the *CF* gene was identified in humans, a database search revealed it is homologous to several genes found in other species. Moreover, a few of the homologous genes were already known to encode proteins that function in the transport of ions and small molecules across the plasma membrane. This observation provided an important clue that cystic fibrosis involves a defect in ion transport.

The ability of computer programs to identify homology between genetic sequences provides a powerful tool for predicting the function of genetic sequences. In 1990, Stephen Altschul, David Lipman, and colleagues developed a program called **BLAST** (for **b**asic **l**ocal **a**lignment **s**earch **t**ool). The BLAST program has been described by many geneticists as the single most important bioinformatic tool. This type of computer program can start with a particular genetic sequence and then locate homologous sequences within a large database. Because there are 20 amino acids but only four bases, homology among protein sequences is easier to identify than DNA sequence homology. Among proteins, sequences that diverged more than 2.5 billion years ago can still be correlated. By comparison, it becomes difficult to identify homologous DNA sequences that diverged more than 100 million years ago.

To see how the BLAST program works, let's consider the human enzyme phenylalanine hydroxylase, which functions in the metabolism of phenylalanine, an amino acid. Recessive mutations in the gene that encodes this enzyme are responsible for the disease called phenylketonuria (PKU). The computational experiment shown in **Table 21.5** started with the amino acid sequence of this protein and used the BLAST program to search the Swissprot database, which contains hundreds of thousands of different protein sequences. The BLAST program can determine which sequences in this database are the closest matches to the amino acid sequence of human phenylalanine hydroxylase. Table 21.5 shows the results—the 10 best matches to human phenylalanine hydroxylase that were identified by the program. Because this enzyme is found in nearly all eukaryotic species, the program identified phenylalanine hydroxylase from many different species. The column to the right of the match number shows the percentage of amino acids that are identical between the species indicated and the human sequence. Because the human phenylalanine hydroxylase sequence is already in the Swissprot database, the closest match of human phenylalanine hydroxylase is to

TABLE 21.5

Results from a BLAST Program Comparing Human Phenylalanine Hydroxylase with Database Sequences

Match	% of Identical Amino Acids*	Species	Function of Sequence†
1	100	Human (*Homo sapiens*)	Phenylalanine hydroxylase
2	99	Orangutan (*Pongo pygmaeus*)	Phenylalanine hydroxylase
3	95	Mouse (*Mus musculus*)	Phenylalanine hydroxylase
4	95	Rat (*Rattus norvegicus*)	Phenylalanine hydroxylase
5	89	Chicken (*Gallus gallus*)	Phenylalanine hydroxylase
6	82	Pipid frog (*Xenopus tropicalis*)	Phenylalanine hydroxylase
7	82	Green pufferfish (*Tetraodon nigroviridis*)	Phenylalanine hydroxylase
8	82	Zebrafish (*Danio rerio*)	Phenylalanine hydroxylase
9	80	Japanese pufferfish (*Takifugu rubripes*)	Phenylalanine hydroxylase
10	75	Fruit fly (*Drosophila melanogaster*)	Phenylalanine hydroxylase

*The number indicates the percentage of amino acids that are identical to the amino acid sequence of human phenylalanine hydroxylase.
†In some cases, the function of the sequence was determined by biochemical assay. In other cases, the function was inferred due to the high degree of sequence similarity with other species.

itself (100%). The next nine sequences are in order of similarity. The next most similar sequence is from the orangutan (99%), a close relative of humans. This is followed by two mammals, the mouse and rat, and then five vertebrates that are not mammals. The tenth best match is from *Drosophila*, an invertebrate.

You can see two trends in Table 21.5. First, the order of the matches follows the evolutionary relatedness of the various species to humans. The similarity between any two sequences is related to the time that has passed since they diverged from a common ancestor. Among the species listed in this table, the human sequence is most similar to the orangutan, a closely related primate. The next most similar sequences are found in other mammals, followed by other vertebrates, and finally invertebrates. A second trend you may have noticed is that several of the matches involve species that are important from a research, medical, or agricultural perspective. Currently, our genetic databases are biased toward organisms that are of interest to humans, particularly model organisms such as mice and *Drosophila*. Over the next several decades, the sequencing of genomes from many different species will tend to lessen this bias.

The results shown in Table 21.5 illustrate the remarkable computational abilities of current computer technology. In minutes, the human phenylalanine hydroxylase sequence can be compared to hundreds of thousands of different sequences.

Genetic Sequences Can Be Used to Predict the Structure of RNA and Proteins

Another topic in which bioinformatics has impacted functional genomics and proteomics is the area of structure prediction. The function of macromolecules such as DNA, RNA, and proteins relies on their three-dimensional structure, which, in turn, depends on the linear sequences of their building blocks. In the case of DNA and RNA, this means a linear sequence of nucleotides; proteins are composed of a linear sequence of amino acids. Currently, the three-dimensional structure of macromolecules is determined primarily through the use of biophysical techniques such as X-ray crystallography and nuclear magnetic resonance (NMR). These methods are technically difficult and very time-consuming. DNA sequencing, by comparison, requires much less effort. Therefore, because the three-dimensional structure of macromolecules depends ultimately on the linear sequence of their building blocks, it would be far easier if we could predict the structure (and function) of DNAs, RNAs, and proteins from their sequence of building blocks.

RNA molecules typically are folded into a secondary structure, which commonly contains double-stranded regions. This secondary structure is further folded and twisted to adopt a tertiary conformation. Such structural features of RNA molecules are functionally important. For example, the folding of RNA into secondary structures, such as stem-loops, affects transcriptional termination and other regulatory events. Therefore, geneticists are interested in the secondary and tertiary structures that RNA molecules can adopt.

Many approaches are available for investigating RNA structure. In addition to biophysical and biochemical techniques,

computer modeling of RNA structure has become an important tool. Modeling programs can consider different types of information. For example, the known characteristics of RNA secondary structure, such as the ability to form double-stranded regions, can provide parameters for use in a modeling program.

A comparative approach can also be used in RNA structure prediction. This method assumes that RNAs of similar function and sequence have a similar structure. For example, the genes that encode certain types of RNAs, such as the 16S rRNA that makes up most of the small ribosomal subunit, have been sequenced from many different species. Among different species, the 16S rRNAs have similar but not identical sequences. Computer programs can compare many different 16S rRNA sequences to aid in the prediction of secondary structure. **Figure 21.12** illustrates a secondary structural model for 16S rRNA from *E. coli* based on a comparative sequence analysis of many bacterial 16S rRNA sequences. This large RNA contains 45 stem-loop regions. As you can imagine, it would be rather difficult to deduce such a model without the aid of a computer! In addition, RNA secondary structure prediction may be aimed at predicting the lowest energy state of a folded molecule. This approach, called free energy minimization, is also related to the base sequence of an RNA molecule.

Structure prediction is also used in the area of proteomics. As described in Chapter 12, proteins contain repeating secondary structural patterns known as α helices and β sheets. Several

FIGURE 21.12 **A secondary structural model for *E. coli* 16S rRNA.**

computer-based approaches attempt to predict secondary structure from the primary amino acid sequence. These programs base their predictions on different types of parameters. Some programs rely on the physical and energetic properties of the amino acids and the polypeptide backbone. More commonly, however, secondary structure predictions are based on the statistical frequency of amino acids within secondary structures that have already been crystallized.

For example, Peter Chou and Gerald Fasman have compiled X-ray crystallographic data to calculate the likelihood that an amino acid will be found in an α helix or a β sheet. Certain amino acids, such as glutamic acid and alanine, are likely to be found in an α helix; others, such as valine and isoleucine, are more likely to be found in a β-sheet structure. Such information can be used to predict whether a sequence of amino acids within a protein is likely to be folded in an α-helix or β-sheet conformation. Secondary structure prediction is correct for approximately 60 to 70% of all sequences. While this degree of accuracy is promising, it is generally not sufficient to predict protein secondary structure reliably. Therefore, one must be cautious in interpreting the results of a secondary structure prediction program.

In recent years, an exciting computer methodology known as a neural network has been applied to protein secondary structure prediction. A computer neural network is a large number of calculation units organized into interconnected layers; this structure is reminiscent of the organization of neurons in the brain. The input layer receives data and may (or may not) transmit that information to the next layer. A neural network can adjust the parameters that define the interconnections among its units in response to data; the network can thus be trained to identify complex patterns coming from the input data. For example, the amino acid sequences of proteins with known crystal structures can be used to train a network (i.e., adjust its parameters) to predict secondary structures for new amino acid sequences. Thus far, neural networks have yielded small improvements in the prediction of protein secondary structure. In the future, a combination of innovative predictive approaches and increased information concerning the biophysical properties of amino acids may make secondary structure prediction a reliable strategy.

While the three-dimensional structure of a protein is extremely difficult to predict solely from its amino acid sequence, researchers have had some success in predicting tertiary structure using a comparative approach. This strategy requires that the protein of interest be homologous to another protein, the tertiary structure of which already has been solved by X-ray crystallography. In this situation, the crystal structure of the known protein can be used as a starting point to model the three-dimensional structure of the protein of interest. This approach is known as homology-based modeling, knowledge-based modeling, or comparative homology. As an example, two research groups independently predicted the three-dimensional structure of a protein, now known as HIV protease, that is encoded by a gene in HIV. The protein is homologous to other proteases with structures that had been solved by X-ray crystallography. Two similar models of the HIV protease were predicted before its actual crystal structure was determined. Both models turned out to be fairly accurate representations of the actual structure, which was later solved by X-ray crystallography. Knowledge of the structure of HIV protease helped scientists synthesize compounds known as HIV protease inhibitors, which are an important part of drug therapy to treat HIV/AIDS.

CONCEPTUAL AND EXPERIMENTAL SUMMARY

The field of functional genomics attempts to elucidate the roles of genetic sequences in a given species. This involves the analysis of groups of genes and even the entire genome. A cDNA or EST library can aid in the identification of genes, because the cDNAs are derived from cellular mRNA (a product of gene transcription). A subtractive cDNA library can identify genes that are expressed only under particular environmental conditions. A DNA microarray is a slide that contains numerous DNA fragments, each corresponding to a short sequence within a known gene. Some arrays contain DNA from every known gene within a species' genome. The arrays are used as hybridization tools to measure the expression levels of thousands of genes simultaneously. Fluorescently labeled cDNA is typically used, although researchers can also hybridize labeled genomic DNA to a microarray. By comparing the results of DNA microarray analyses at different time points or under different sets of environmental conditions, researchers can identify genes whose patterns of expression correlate with each other. This is termed a cluster analysis. These genes often encode proteins that participate in common cellular functions. One of the first studies using a DNA microarray involved the analysis of carbohydrate metabolism in yeast.

The entire collection of proteins a species can produce is called its proteome. The study of the function and interaction of these proteins is termed proteomics. Due to alternative splicing and the posttranslational covalent modification of proteins, the proteome may be much larger than the genome. A common technique to detect proteins is via two-dimensional gel electrophoresis, the end result of which is a collection of spots, each representing a unique cellular protein. Mass spectrometry is used to correlate a given spot with a particular protein. Spots from a two-dimensional gel can be removed, and the amino acid sequences of peptide fragments can be determined via tandem mass spectrometry. Recently, microarray analysis has been applied to proteins. In this type of technology, proteins, rather than DNA molecules, are spotted onto a slide. The two common types of protein microarray analysis are antibody microarrays and functional protein microarrays. An antibody microarray consists of a collection of antibodies spotted onto a microarray. Cellular proteins are isolated and exposed to the antibody array, and the level of protein expression from a given cell type can be assessed. A functional protein microarray consists of many different cellular proteins spotted onto an array. This array can probe the function of proteins, such as their substrate specificity, interactions with drugs, or interactions with other cellular proteins.

The computer has become an invaluable tool in genetic studies and a central instrument in the field of bioinformatics, for both the analysis and the storage of the data produced by genetic research. Genetic sequences contained within computer data files can be analyzed by many different programs. For example, a computer program can scan data and identify meaningful features, such as short sequence elements within very long genetic sequences. Likewise, computer programs can be used to find structural genes within nucleotide sequences and compare several homologous genetic sequences in a multiple sequence alignment. In addition, geneticists may take newly identified sequences and search a database to identify homologous sequences using the BLAST program. The identification of homologous genes within

a database can provide important clues concerning the function of a newly determined sequence.

Computer programs have been developed to predict the structure of RNA and proteins. These programs rely on known biochemical and biophysical techniques as well as computer modeling and may also use a comparative approach that requires the RNA or protein of interest to be homologous to another RNA or protein of known structure. Some predictive programs are more reliable than others. Nevertheless, future research is expected to improve the predictive capabilities of such computer programs.

PROBLEM SETS & INSIGHTS

Solved Problems

S1. When a cell experiences a change in its environment, it may activate one or more genes as a way to respond to the changes. For example, when confronted with a toxic substance, a cell may turn on genes that encode proteins that can degrade or export the toxin. Let's suppose a researcher is interested in the ability of mouse liver cells to protect themselves against toxic heavy metals, such as mercury. To understand the cellular response, the researcher grows mouse liver cells in the laboratory and divides them into two samples. One of the samples is exposed to mercury, while the other is not. Samples of cDNA were made from these cells, and then a subtractive cDNA library was made as described in Figure 21.1. In this case, the cDNAs derived from the cells that were not exposed to mercury were the cDNAs bound to the column. The results showed that the subtractive library contained seven different types of cDNAs.

A. What do these results mean?

B. What would you do next, assuming that the entire mouse genome is contained within a database?

Answer:

A. The results mean that seven different genes are turned on when liver cells are exposed to mercury.

B. Take the seven different cDNAs and subject them to DNA sequencing. Next, the cDNA sequences are used (via computer programs) to find their genomic matches. This will make it possible to determine the entire gene sequences. Then take the entire coding sequence, for each of the seven genes, and look for homology between each mouse gene sequence and sequences within a database. In other words, you would do a BLAST database search, using each of the seven sequences as the query sequence. If you discovered that any of the seven genes were homologous to other genes whose function is already known, this would give you direct information regarding the probable function of particular genes. For example, if one of the genes turned out to be homologous to a gene in bacteria that encodes a protein already known to function in the export of metal ions, this would suggest that one way that mouse liver cells try to avoid the toxic effects of mercury is to try to pump it out.

S2. To answer this question, you will need to look back at the evolution of the globin gene family, which is shown in Chapter 8, Figure 8.7. Throughout the evolution of this gene family, mutations have occurred that have resulted in globin polypeptides

with similar but significantly different amino acid sequences. If we look at the sequence alignment in Figure 21.11, we can make logical guesses regarding the timing of mutations, based on a comparison of the amino acid sequences of family members. What is/are the most probable time(s) that mutations occurred to produce the following amino acid differences? Note: You will have to examine the alignment in Figure 21.11 and the evolutionary timescale in Figure 8.10 to answer this question.

A. Val-111 and Cys-111

B. Met-112 and Leu-112

C. Ser-141, Asn-141, Ile-141, and Thr-141

Answer:

A. We do not know if the original globin gene encoded a cysteine or valine at codon 111. The mutation could have changed cysteine to valine or valine to cysteine. The mutation probably occurred after the duplication that produced the α-globin family and β-globin family (about 300 million years ago) but before the gene duplications that occurred in the last 200 millions years to produce the multiple copies of the globin genes on chromosome 11 and chromosome 16. Therefore, all of the globin genes on chromosome 11 have a valine at codon 111, while all of the globin genes on chromosome 16 have a cysteine.

B. Met-112 occurs only in the ε-globin polypeptide; all of the other globin polypeptides contain a leucine at position 112. Therefore, the primordial globin gene probably contained a leucine codon at position 112. After the gene duplication that produced the ε-globin gene, a mutation occurred that changed this leucine codon into a methionine codon. This would have occurred since the evolution of primates (i.e., within the last 10 or 20 million years).

C. When we look at the possible codons at position 141 (i.e., Ser-141, Asn-141, Ile-141, and Thr-141), we notice that a serine codon is found in θ globin, ξ globin, and γ globin. Because the θ- and ξ-globin genes are found on chromosome 16 and the γ-globin genes are found on chromosome 11, it is probable that serine is the primordial codon and that the other codons (asparagine, isoleucine, and threonine) arose later by mutation of the serine codon. If this is correct, the Thr-141 codon arose after the gene duplications that produced the θ- and ξ-globin genes. The Asn-141 and Ile-141 mutations arose after the gene duplications that produced the γ-globin genes. Therefore, the Thr-141, Asn-141, and Ile-141 arose since the evolution of primates (i.e., within the last 10 or 20 million years).

S3. Using a comparative sequence analysis, the secondary structures of rRNAs have been predicted. Among many homologous rRNAs, one stem-loop usually has the following structure:

You have sequenced a homologous rRNA from a new species and have obtained most of its sequence, but you cannot read the last five bases on your sequencing gel.

5'-GCATTCTACCAGTGCTAG?????-3'

Of course, you will eventually repeat this experiment to determine the last five bases. However, before you get around to doing this, what do you expect will be the sequence of the last five bases?

Answer: AATGC–3'

This will also form a similar stem-loop structure.

S4. How can codon bias be used to search for structural genes within uncharacterized genetic sequences?

Answer: Most species exhibit a bias in the codons they use within the coding sequence of structural genes. This causes the base content within coding sequences to differ significantly from that of noncoding DNA regions. By knowing the codon bias for a particular species, researchers can use a computer to locate regions that display this bias and thereby identify what are likely to be the coding regions of structural genes.

Conceptual Questions

C1. Discuss the meaning of the following terms: structural genomics, functional genomics, and proteomics.

C2. Discuss the reasons why the proteome is larger than the genome of a given species.

C3. What is a database? What types of information are stored within a database? Where does the information come from? Discuss the objectives of a genome database.

C4. Besides the examples listed in Table 21.4, list five types of short sequences that a geneticist might want to locate within a DNA sequence.

C5. Discuss the distinction between sequence recognition and pattern recognition.

C6. A multiple sequence alignment of five homologous proteins is shown here:

C7. What is the difference between similarity and homology?

C8. Which of the following statements uses the term homologous correctly?

A. The two X chromosomes in female mammalian cells are homologous to each other.

B. The α-tubulin gene in *Saccharomyces cerevisiae* is homologous to the α-tubulin gene in *Arabidopsis thaliana*.

C. The promoter of the *lac* operon is homologous to the promoter of the *trp* operon.

D. The *lacY* gene of *E. coli* and *Klebsiella pneumoniae* are approximately 60% homologous to each other.

C9. When comparing (i.e., aligning) two or more genetic sequences, it is sometimes necessary to put in gaps. Explain why. Discuss two changes (i.e., two types of mutations) that could happen during the evolution of homologous genes that would explain the occurrence of gaps in multiple sequence alignments.

```
  1                                                    50
1 MLAFLNQVRK PTLDLPLEVR RKMWFKPFM. QSYLVVFIGY LTMYLIRKNF
2 MLAFLNQVRK PTLDLALDVR RKMWFKPFM. QSYLVVFIGY LTMYLIRKNF
3 MLPFLKAPAD APL.MTDKYE IDARYRYWRR HILLTIWLGY ALFYFTRKSF
4 MLSFLKAPAN APL.ITDKHE VDARYRYWRR HILITIWLGY ALFYFTRKSF
5 MLSIFKPAPH KAR.LPAA.E IDPTYRRLRW QIFLGIFFGY AAYYLVRKNF

  51                                                  100
1 NIAQNDMIST YGLSMTQLGM IGLGFSITYG VGKTLVSYYA DGKNTKQFLP
2 NIAQNDMIST YGLSMTELGM IGLGFSITYG VGKTLVSYYA DGKNTKQFLP
3 NAAVPEILAN GVLSRSDIGL LATLFYITYG VSKFVSGIVS DRSNARYFMG
4 NAAAPEILAS GILTRSDIGL LATLFYITYG VSKFVSGIVS DRSNARYFMG
5 ALAMPYLVEQ .GFSRGDLGF ALSGISIAYG FSKFIMGSVS DRSNPRVFLP
```

Discuss some of the interesting features that this alignment reveals.

Experimental Questions

E1. Explain how a subtractive cDNA library can provide information regarding gene regulation.

E2. Take a look at solved problem S1 regarding a subtractive DNA library. Would you want to load a relatively small amount of cDNA from the cells that had been exposed to mercury, compared to the amount of cDNA that is attached to the column? Explain. What would happen if you ran too much cDNA over the column from the cells that had been exposed to mercury?

E3. With regard to DNA microarrays, answer the following questions:

A. What is attached to the slide? Be specific about the number of spots, the lengths of DNA fragments, and the origin of the DNA fragments.

B. What is hybridized to the microarray?

C. How is hybridization detected?

E4. In the experiment of Figure 21.3, explain how the ratio of red: green fluorescence provides information regarding gene regulation.

E5. What is meant by the term cluster analysis? How is this approach useful?

E6. For two-dimensional gel electrophoresis, what physical properties of proteins promote their separation in the first dimension and the second dimension?

E7. Can two-dimensional gel electrophoresis be used as a purification technique? Explain.

E8. Explain how tandem mass spectroscopy can be used to determine the sequence of a peptide. Once a peptide sequence is known, how is this information used to determine the sequence of the entire protein?

E9. Describe the two general types of protein microarrays. What are their possible applications?

E10. Discuss the strategies that can be used to identify a structural gene using bioinformatics.

E11. What is a motif? Why is it useful for computer programs to identify functional motifs within amino acid sequences?

E12. Discuss why it is useful to search a database to identify sequences that are homologous to a newly determined sequence.

E13. The secondary structure of 16S rRNA has been predicted using a computer-based sequence analysis. In general terms, discuss what type of information is used in a comparative sequence analysis, and explain what assumptions are made concerning the structure of homologous RNAs.

E14. Discuss the basis for secondary structure prediction in proteins. How reliable is it?

E15. To reliably predict the tertiary structure of a protein based on its amino acid sequence, what type of information must be available?

E16. In Figure 21.7, we considered a computer program that can translate a DNA sequence into a polypeptide sequence. A researcher has a sequence file that contains the amino acid sequence of a polypeptide and runs a program that is opposite to the TRANSLATION program. This other program is called BACKTRANSLATE. It can take an amino acid sequence file and determine the sequence of DNA that would encode such a polypeptide. How does this program work? In other words, what

are the genetic principles that underlie this program? What type of sequence file would this program generate: a nucleotide sequence or an amino acid sequence? Would the BACKTRANSLATE program produce only a single sequence file? Explain why or why not.

E17. In experimental question E16, we considered a computer program that can translate a DNA sequence into a polypeptide sequence. Instead of running this program, a researcher could simply look the codons up in a genetic code table and determine the sequence by hand. What are the advantages of running this program compared to the old-fashioned way of doing it by hand?

E18. To identify the following types of genetic occurrences, would a program use sequence recognition, pattern recognition, or both?

A. Whether a segment of *Drosophila* DNA contains a P element (which is a specific type of transposable element)

B. Whether a segment of DNA contains a stop codon

C. In a comparison of two DNA segments, whether there is an inversion in one segment compared to the other segment

D. Whether a long segment of bacterial DNA contains one or more genes

E19. The goal of many computer programs is to identify "sequence elements" within a long segment of DNA. What is a sequence element? Give two examples. How is the specific sequence of a sequence element determined? In other words, is it determined by the computer program or by genetic studies? Explain.

E20. Take a look at the multiple sequence alignment in Figure 21.11 of the globin polypeptides from amino acids 101–148.

A. Which of these amino acids are likely to be most important for globin structure and function? Explain why.

B. Which are likely to be least important?

E21. See solved problem S2 before answering this question. Based on the sequence alignment in Figure 21.11, what is/are the most probable time(s) that mutations occurred in the human globin gene family to produce the following amino acid differences?

A. His-119 and Arg-119

B. Gly-121 and Pro-121

C. Glu-103, Val-103, and Ala-103

E22. Below is short nucleotide sequence within a gene. Via the Internet (e.g., see www.ncbi.nlm.nih.gov/), determine what gene this sequence in found within. Also, determine the species in which this gene sequence is found.

```
5'-GGGCGCAATTACTTAACGCCTCGATT
ATCTTCTTGCGCCACTGATCATTA-3'
```

E23. Take a look at solved problem S2 and the codon table found in Chapter 13 (Table 13.2). Assuming that a mutation involving a single-base change is more likely than a double-base change, propose how the Asn-141, Ile-141, and Thr-141 codons arose. In your answer, describe which of the six possible serine codons is/are likely to be the primordial serine codon of the globin gene family and how that codon changed to produce the Asn-141, Ile-141, and Thr-141 codons.

E24. Membrane proteins often have transmembrane regions that span the membrane in an α-helical conformation. These transmembrane segments are about 20 amino acids long and usually contain amino acids with nonpolar (i.e., hydrophobic) amino acid side chains. Researchers can predict whether a polypeptide sequence has transmembrane segments based on the occurrence of segments that contain 20 nonpolar amino acids. To do so, each amino acid is assigned a hydropathy value, based on the chemistry of its amino acid side chain. Amino acids with very nonpolar side chains are given a high (positive) value, whereas amino acids that are charged and/or polar are given low (negative) values. The hydropathy values usually range from about +4 to −4.

Computer programs have been devised that scan the amino acid sequence of a polypeptide and calculate values based on the hydropathy values of the amino acid side chains. The program usually scans a window of seven amino acids and assigns an average hydropathy value. For example, the program would scan amino acids 1–7 and give an average value, then it would scan 2–8 and give a value, then it would scan 3–9 and give a value, and so on, until it reached the end of the polypeptide sequence (i.e., until it reached the carboxyl terminus).

The program then produces a figure, known as a hydropathy plot, which describes the average hydropathy values throughout the entire polypeptide sequence. An example of a hydropathy plot is shown here.

A. How many transmembrane segments are likely in this polypeptide?

B. Draw the structure of this polypeptide if it were embedded in the plasma membrane. Assume that the amino terminus is found in the cytoplasm of the cell.

E25. Explain how a computer program can predict RNA secondary structure. What is the underlying genetic concept used by the program to predict secondary structure?

E26. Are the following statements about protein structure prediction true or false?

A. The prediction of secondary structure relies on information regarding the known occurrence of amino acid residues in α helices or β sheets from X-ray crystallographic data.

B. The prediction of secondary structure is highly accurate, nearly 100% correct.

C. To predict the tertiary structure of a protein based on its amino acid sequence, it is necessary that the protein of interest is homologous to another protein whose tertiary structure is already known.

Questions for Student Discussion/Collaboration

1. Let's suppose you are in charge of organizing and publicizing a genomic database for the mouse genome. Make a list of innovative strategies you would initiate to make the mouse genome database useful and effective.

2. Let's suppose a 5-year-old told you that she was interested in pursuing a career studying the three-dimensional structure of proteins. (Okay, so she's a bit precocious.) Would you advise her to become a geneticist, a mathematical theoretician, or a biophysicist?

3. If you have access to the necessary computer software, make a sequence file and analyze it in the following ways: What is the translated sequence in all three reading frames? What is the longest open reading frame? Is the sequence homologous to any known sequences? If so, does this provide any clues about the function of the sequence?

Note: All answers appear at the website for this textbook; the answers to even-numbered questions are in the back of the textbook.

Visit the Online Learning Center for practice tests, answer keys, and other learning aids for this chapter. Enhance your understanding of genetics with our interactive exercises, quizzes, animations, and much more.

MEDICAL GENETICS AND CANCER

22

Genetic information is highly personal and unique. Our genes underlie every aspect of human health, both in function and dysfunction. Obtaining a detailed understanding of how genes work together and interact with environmental factors ultimately will allow us to appreciate the differences between the events in normal cellular processes and those that occur in disease pathogenesis. Such knowledge will have a profound impact on the way many diseases are defined, diagnosed, treated, and prevented. Genetic insight is expected to bring about revolutionary changes in medical practices. In fact, changes are already beginning. Currently, several hundred genetic tests are in clinical use, with many more under development. Most of these tests detect mutations associated with rare genetic disorders that follow Mendelian inheritance patterns. These include Duchenne muscular dystrophy, cystic fibrosis, sickle-cell disease, and Huntington disease. In addition, genetic tests are available to detect the predisposition to develop certain forms of cancer.

Approximately 5,000 genetic diseases are known to afflict people, but this is almost certainly an underestimate. Given enough time and effort, scientists can learn how to prevent or treat a great many of them. Most of the genetic disorders discussed in the first part of this chapter are the direct result of a mutation in one gene. However, many diseases have a complex pattern of inheritance involving several genes. These include common medical disorders such as diabetes, asthma, and mental illness. In these cases, a single mutant gene does not determine whether a person has a disease. Instead, a number of genes may each make a subtle contribution to a person's susceptibility to a disease. Unraveling these complexities will be a challenge for some time to come. The availability of the human genome sequence, discussed in Chapter 20, will be of great help.

In this chapter, we will focus our attention on ways that mutant genes contribute to human disease. In the first part of the chapter, we will explore the molecular basis of several genetic disorders and their patterns of inheritance. We will also examine how genetic testing can determine if an individual carries a defective allele. The second part of the chapter will concern cancer, a disease that involves the uncontrolled growth of somatic cells. We will examine the underlying genetic basis for cancer and discuss the roles that many different genes may play in the development of this disease.

CHAPTER OUTLINE

22.1   **Genetic Analysis of Human Diseases**

22.2   **Genetic Basis of Cancer**

Cigarette smoking and lung cancer. Cigarette smoke contains chemicals that are known to mutate genes in the cells of a person's lungs, thereby leading to lung cancer. Lung cancer remains the top cause of cancer death in the United States, with 87% of those deaths linked to smoking.

PART VI
Genetic Analysis of Individuals and Populations

22.1 GENETIC ANALYSIS OF HUMAN DISEASES

Human genetics is a topic hard to resist. Almost everyone who looks at a newborn is tempted to speculate whether the baby resembles the mother, the father, or perhaps a distant relative. In this section, we will focus primarily on the inheritance of human genetic diseases rather than common traits found in the general population. Even so, the study of human genetic diseases provides insights regarding our traits. The disease hemophilia illustrates this point. Hemophilia (also spelled haemophilia) is a condition in which an individual's blood will not clot properly. By analyzing people with this disorder, researchers have identified genes that participate in the process of blood clotting. The study of hemophilia has helped to elucidate a clotting pathway involving several different proteins. Therefore, as with the study of mutants in model organisms such as *Drosophila,* mice, and yeast, when we study the inheritance of genetic diseases, we often learn a great deal about the genetic basis for normal physiological processes as well.

Because thousands of human diseases have an underlying genetic basis, human genetic analysis is of great medical importance. In this section, we will examine the causes and inheritance patterns of human genetic diseases that result from defects in single genes. As you will learn, the mutant genes that cause these diseases often follow simple Mendelian inheritance patterns.

A Genetic Basis for a Human Disease May Be Suggested from a Variety of Observations

When we view the characteristics of people, we usually think that some traits are inherited, whereas others are caused by environmental factors. For example, when the facial features of two related individuals look strikingly similar, we think that this similarity has a genetic basis. The profound resemblance between identical twins is an obvious example. By comparison, other traits are governed by the environment. If we see a person with purple hair, we likely suspect that he or she has used hair dye as opposed to showing an unusual genetic trait.

For human diseases, geneticists would like to know the relative contributions from genetics and the environment. Is a disease caused by a pathogenic microorganism, a toxic agent in the environment, or a faulty gene? Unlike the case with experimental organisms, we cannot conduct human crosses to elucidate the genetic basis for diseases. Instead, we must rely on analyzing the occurrence of a disease in families that already exist. As described in the following list, several observations are consistent with the idea that a disease is caused, at least in part, by the inheritance of mutant genes. When the occurrence of a disease correlates with several of these observations, a geneticist will become increasingly confident that it has a genetic basis.

1. *When an individual exhibits a disease, this disorder is more likely to occur in genetic relatives than in the general population.* For example, if someone has cystic fibrosis, his or her relatives are more likely to have this disease than will randomly chosen members of the general population.

2. *Identical twins share the disease more often than nonidentical twins.* Identical twins, also called **monozygotic (MZ) twins,** are genetically identical to each other, because they were formed from the same sperm and egg. By comparison, nonidentical twins, also called fraternal, or **dizygotic (DZ) twins,** are formed from separate pairs of sperm and egg cells. Fraternal twins share, on average, 50% of their genetic material, the same as any two siblings. When a disorder has a genetic component, a pair of identical twins is more likely to exhibit the disorder than are two fraternal twins.

   Geneticists evaluate a disorder's **concordance,** the degree to which it is inherited, by calculating the percentage of twin pairs in which both twins exhibit the disorder relative to pairs where only one twin shows the disorder. Theoretically, for diseases caused by a single gene, concordance among identical twins should be 100%. For fraternal twins, concordance for dominant disorders is expected to be 50%, assuming only one parent is a heterozygous individual. For recessive diseases, concordance among fraternal twins would be 25% if we assume both parents are heterozygous carriers. However, the actual concordance values observed for most single-gene disorders are usually less than such theoretical values for a variety of reasons. Some disorders are not completely penetrant, meaning that the symptoms associated with the disorder are not always produced. Also, one twin may have a disorder due to a new mutation that occurred after fertilization; it would be very unlikely for the other twin to have the same mutation.

3. *The disease does not spread to individuals sharing similar environmental situations.* Inherited disorders cannot spread from person to person. The only way genetic diseases can be transmitted is from parent to offspring during sexual reproduction.

4. *Different populations tend to have different frequencies of the disease.* Due to evolutionary forces, the frequencies of traits usually vary among different populations of humans. For example, the frequency of the disease sickle-cell anemia is highest among certain African and Asian populations and relatively low in other parts of the world (see Chapter 24, Figure 24.11).

5. *The disease tends to develop at a characteristic age.* Many genetic disorders exhibit a characteristic **age of onset** at which the disease appears. Some mutant genes exert their effects during embryonic and fetal development, so their effects are apparent at birth. Other genetic disorders tend to develop much later in life.

6. *The human disorder may resemble a disorder that is already known to have a genetic basis in an animal.* In animals, where we can conduct experiments, various traits are known to be governed by genes. For example, the albino phenotype is found in humans as well as in many animals (**Figure 22.1**).

FIGURE 22.1 **The albino phenotype in humans and other mammals.**

Genes→Traits Certain enzymes (encoded by genes) are necessary for the production of pigment. A homozygote with two defective alleles in these pigmentation genes exhibits an albino phenotype. This phenotype can occur in humans and other animals.

7. *A correlation is observed between a disease and a mutant human gene or a chromosomal alteration.* A particularly convincing piece of evidence that a disease has a genetic basis is the identification of altered genes or chromosomes that occur only in people exhibiting the disorder. When comparing two individuals, one with a disease and one without, we expect them to have differences in their

genetic material if the disorder has a genetic component. Alterations in gene sequences can be determined by gene cloning and DNA sequencing techniques (see Chapter 18). Also, changes in chromosome structure and number can be detected by the microscopic examination of chromosomes (see Chapter 8).

Inheritance Patterns of Human Diseases May Be Determined via Pedigree Analysis

When a human disorder is caused by a mutation in a single gene, the pattern of inheritance can be deduced by analyzing human pedigrees. How is this accomplished? A geneticist must obtain data from many large pedigrees containing several individuals who exhibit the disorder and then follow its pattern of inheritance from generation to generation. To appreciate the basic features of pedigree analysis, we will examine a few large pedigrees that involve diseases inherited in different ways. You may wish to review Figure 2.10 for the organization and symbols of pedigrees.

The pedigree shown in **Figure 22.2** concerns a genetic disorder called Tay-Sachs disease (TSD), first described by Warren Tay, a British ophthalmologist, and Bernard Sachs, an American neurologist, in the 1880s. Affected individuals appear healthy at birth but then develop neurodegenerative symptoms at 4 to 6 months of age. The primary characteristics are cerebral degeneration, blindness, and loss of motor function. Individuals with TSD typically die in the third or fourth year of life. This disease is particularly prevalent in Ashkenazi (eastern European) Jewish populations, in which it has a frequency of about 1 in 3,600 births, which is over 100 times more frequent than in most other human populations.

At the molecular level, the mutation that causes TSD is in a gene that encodes the enzyme hexosaminidase A (hexA). HexA is responsible for the breakdown of a category of lipids called G_{M2}-gangliosides, which are prevalent in the cells of the central nervous system. A defect in the ability to break down this lipid leads to its excessive accumulation in nerve cells and eventually causes the neurodegenerative symptoms characteristic of TSD. This defect in lipid breakdown was recognized long before the *hexA* gene was identified as being defective in these patients.

As illustrated in Figure 22.2, Tay-Sachs disease is inherited in an autosomal recessive manner. The four common features of autosomal recessive inheritance are as follows:

1. *Frequently, an affected offspring will have two unaffected parents.* For rare recessive traits, the parents are usually unaffected, meaning they do not exhibit the disease. For deleterious alleles that cause early death or infertility, the two parents must be unaffected. This is always the case in TSD.
2. *When two unaffected heterozygotes have children, the percentage of affected children is (on average) 25%.*
3. *Two affected individuals will have 100% affected children.* This observation can be made only when a recessive trait produces fertile, viable individuals. In the case of TSD, the

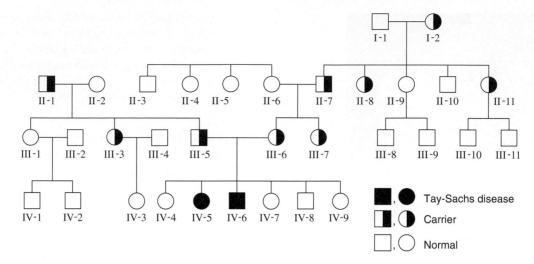

INTERACTIVE
EXERCISE

FIGURE 22.2   **A family pedigree of Tay-Sachs disease, indicating recessive inheritance.** Heterozygotes were determined by genetic testing.

affected individual dies in early childhood, and so it is not possible to observe crosses between two affected people.

4. *The trait occurs with the same frequency in both sexes.*

Autosomal recessive inheritance is a common mode of transmission for genetic disorders, particularly those that involve defective enzymes. Human recessive alleles are often caused by mutations that result in a loss of function in the encoded enzyme. In the case of Tay-Sachs disease, a heterozygous carrier has approximately 50% of the functional enzyme, which is sufficient for a normal phenotype. However, for other genetic diseases, the level of functional protein may vary due to effects of gene regulation. Hundreds of human genetic diseases are inherited in a recessive manner, and in many cases, the mutant genes have been identified. Several of these diseases are described in **Table 22.1**.

Now let's examine a human pedigree involving an autosomal dominant disease (**Figure 22.3**). In this example, the affected individuals have a disorder called Huntington disease (also called Huntington chorea). The major symptoms of this disease, which

usually occurs during middle age, are due to the degeneration of certain types of neurons in the brain, leading to personality changes, dementia, and early death. In 1993, the gene involved in Huntington disease was identified and sequenced. It encodes a protein called huntingtin that is expressed in neurons but is also found in some cells not affected in Huntington disease. In persons with this disorder, a mutation has added a polyglutamine tract—many glutamines in a row—to the amino acid sequence of the huntingtin protein. This causes an aggregation of the protein in neurons. However, additional research will be needed to elucidate the molecular relationship between the abnormality in the huntingtin protein and the disease symptoms. Five common features of autosomal dominant inheritance are as follows:

1. *An affected offspring usually has one or both affected parents.* However, this is not always the case. Some dominant traits show incomplete penetrance (see Chapter 4), so a heterozygote may not exhibit the trait even though it may be passed to offspring who do exhibit the trait. Also, a

TABLE 22.1

Examples of Human Disorders Inherited in an Autosomal Recessive Manner

Disorder	Chromosomal Location of Gene	Gene Product	Effective of Disease-Causing Allele
Adenosine deaminase deficiency	20q	Adenosine deaminase	Defective immune system and skeletal and neurological abnormalities
Albinism (type I)	11q	Tyrosinase	Inability to synthesize melanin, resulting in white skin, hair, etc.
Cystic fibrosis (CF)	7q	CF transmembrane conductance regulator	Water imbalance in tissues of the pancreas, intestine, sweat glands, and lungs due to impaired ion transport; leads to lung damage
Phenylketonuria (PKU)	12q	Phenylalanine hydroxylase	Foul-smelling urine, neurological abnormalities, mental impairment; may be remedied by diet modification at birth
Sickle-cell anemia	11p	β globin	Anemia, blockages in blood circulation
Tay-Sachs disease (TSD)	15q	Hexosaminidase A	Progressive neurodegeneration

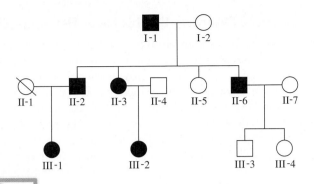

FIGURE 22.3 A family pedigree of Huntington disease, indicating dominant inheritance.

The term **haploinsufficiency** refers to the phenomenon in which a person has only a single functional copy of a gene, and that single functional copy does not produce a normal phenotype. Haploinsufficiency shows a dominant pattern of inheritance because a heterozygote (with one normal allele and one inactive allele) will have the disease. An example is aniridia, which is a rare disorder that results in an absence of the iris of the eye. Aniridia leads to visual impairment and blindness in severe cases.

A second category of dominant disorders involves **gain-of-function mutations.** Such mutations change the gene product so that it "gains" a new or abnormal function. An example is achondroplasia, which is characterized by abnormal bone growth that results in short stature with relatively short arms and legs. This disorder is caused by a point mutation that occurs in the fibroblast growth factor receptor-3 gene. In achondroplasia, the mutant form of the receptor is overactive. This overactivity disrupts the normal signaling pathway and leads to severely shortened bones.

A third category of dominant disorders is characterized by **dominant-negative mutations** in which the altered gene product acts antagonistically to the normal gene product. In humans, Marfan syndrome, which is due to a mutation in the *fibrillin-1* gene, is an example. The *fibrillin-1* gene encodes a glycoprotein that is a structural component of the extracellular matrix that provides structure and elasticity to tissues. The mutant gene encodes a fibrillin-1 protein that antagonizes the effects of the normal protein, thereby weakening the elasticity of certain body parts. For example, the walls of the major arteries such as the aorta, the large artery that leaves the heart, are often affected.

Let's now turn to another inheritance pattern common in humans that is called X-linked recessive inheritance (**Table 22.3**). With regard to recessive genetic diseases, X-linked inheritance poses a special problem for males. Why are males more likely to be affected? Most X-linked genes lack a counterpart on the Y

dominant mutation may occur during gametogenesis, so two unaffected parents may produce an affected offspring.

2. *An affected individual with only one affected parent is expected to produce 50% affected offspring (on average).*

3. *Two affected, heterozygous individuals will have (on average) 25% unaffected offspring.*

4. *The trait occurs with the same frequency in both sexes.*

5. *For most dominant, disease-causing alleles, the homozygote is more severely affected with the disorder. In some cases, a dominant allele may be lethal in the homozygous condition.*

Numerous autosomal dominant diseases have been identified in humans (**Table 22.2**). In these disorders, 50% of the normal protein is not sufficient to produce a normal phenotype. The three common explanations for dominant disorders are haploinsufficiency, a gain-of-function mutation, or a dominant-negative mutation. Let's consider examples of all three types.

TABLE 22.2

Examples of Human Disorders Inherited in an Autosomal Dominant Manner

Disorder	Chromosomal Location of Gene	Gene Product	Effects of Disease-Causing Allele
Aniridia	11p	Pax6 transcription factor	An absence of the iris of the eye, leading to visual impairment and sometimes blindness
Achondroplasia	4p	Fibroblast growth factor receptor-3	A common form of dwarfism associated with a defect in the growth of long bones
Marfan syndrome	15q	Fibrillin-1	Tall and thin individuals with abnormalities in the skeletal, ocular, and cardiovascular systems due to a weakening in the elasticity of certain body parts
Osteoporosis	7q	Collagen (type 1α2)	Brittle, weakened bones
Familial hypercholesterolemia	19p	LDL receptor	Very high serum levels of low-density lipoprotein (LDL), a predisposing factor in heart disease
Huntington disease	4p	Huntingtin	Neurodegeneration that occurs relatively late in life, usually in middle age
Neurofibromatosis I	17q	Neurofibromin	Individuals may exhibit spots of abnormal pigmentation (café-au-lait spots) and growth of noncancerous tumors in the nervous system

TABLE 22.3

Examples of Human Disorders Inherited in an X-linked Recessive Manner

Disorder	Gene Product	Effects of Disease-Causing Allele
Duchenne muscular dystrophy	Dystrophin	Progressive degeneration of muscles that begins in early childhood
Hemophilia A	Clotting factor VIII	Defect in blood clotting
Hemophilia B	Clotting factor IX	Defect in blood clotting
Androgen insensitivity syndrome	Androgen receptor	Missing male steroid hormone receptor; XY individuals have external features that are feminine but internally have undescended testes and no uterus

chromosome. Males are hemizygous—have a single copy—for these genes. Therefore, a female heterozygous for an X-linked recessive gene will pass this trait to 50% of her sons, as shown in the following Punnett square for hemophilia. In this example, X^{h-A} is the chromosome that carries a mutant allele causing hemophilia.

As mentioned previously, hemophilia is a disorder in which the blood cannot clot properly when a wound occurs. For individuals with this trait, a minor cut may bleed for a very long time, and small injuries can lead to large bruises, because internal broken capillaries may leak blood profusely before they are repaired. For hemophiliacs, common injuries pose a threat of severe internal or external bleeding. Hemophilia A, also called classical hemophilia, is caused by a defect in an X-linked gene that encodes the protein clotting Factor VIII. This disease has also been called the "Royal disease," because it affected many members of European royal families. The pedigree shown in **Figure 22.4** illustrates the prevalence of hemophilia A among the descendants of Queen Victoria of England. The pattern of X-linked recessive inheritance is revealed by the following observations:

1. *Males are much more likely to exhibit the trait.*
2. *The mothers of affected males often have brothers or fathers who are affected with the same trait.*
3. *The daughters of affected males will produce, on average, 50% affected sons.*

Many Genetic Disorders Exhibit Locus Heterogeneity

Hemophilia can be used to illustrate another concept in genetics called **locus heterogeneity.** This term refers to the phenomenon in which a particular type of disease may be caused by mutations in two or more different genes. For example, blood clotting involves the participation of several different proteins that take part in a cellular cascade that leads to the formation of a clot. Hemophilia is usually caused by a defect in one of three different clotting factors. In hemophilia A, also called classic hemophilia, a protein called Factor VIII is missing. Hemophilia B is a deficiency in a different clotting factor, called Factor IX. Both Factor VIII and IX are encoded by different genes on the X chromosome. These two types of hemophilia show an X-linked recessive pattern of inheritance. By comparison, hemophilia C is due to a Factor XI deficiency. The gene encoding Factor XI is found on chromosome 4, and this form of hemophilia follows an autosomal recessive pattern of inheritance.

In hemophilia, locus heterogeneity arises from the participation of several proteins in a common cellular process. Another mechanism that may lead to locus heterogeneity occurs when proteins are composed of two or more different subunits, with each subunit being encoded by a different gene. The disease thalassemia is an example of locus heterogeneity caused by a mutation in a protein composed of multiple subunits. This potentially life-threatening disease involves defects in the ability of the red blood cells to transport oxygen. The underlying cause is an alteration in hemoglobin. In adults, hemoglobin is a tetrameric protein composed of two α-globin and two β-globin subunits; α globin and β globin are encoded by separate genes (namely, the α-globin and β-globin genes). Two main types of thalassemia have been discovered in human populations: α thalassemia, in which the α-globin subunit of hemoglobin is defective; and β thalassemia, in which the β-globin subunit is defective.

Unfortunately, locus heterogeneity may greatly confound pedigree analysis. For example, a human pedigree might contain individuals with X-linked hemophilia and other individuals with hemophilia C. A geneticist who assumed all affected individuals had defects in the same gene would be unable to explain the resulting pattern of inheritance. For disorders such as hemophilia and thalassemia, pedigree analysis is not a major problem because the biochemical basis for these diseases is well understood. However, for rare diseases that are poorly understood at the molecular level, locus heterogeneity may profoundly obscure the pattern of inheritance.

Disease-Causing Mutant Genes Are Identified by Mapping and DNA Sequencing

How do geneticists identify genes that cause disease when they are mutant? While a variety of approaches may be followed, the hunt often begins with family pedigrees. As described in Chapter 20, the human genome has been extensively mapped with many molecular markers such as microsatellites (refer back to Figure 20.12). By comparing the transmission patterns of many molecular markers with the occurrence of an inherited disease, researchers can pinpoint particular markers that are closely linked to the disease-causing mutant gene.

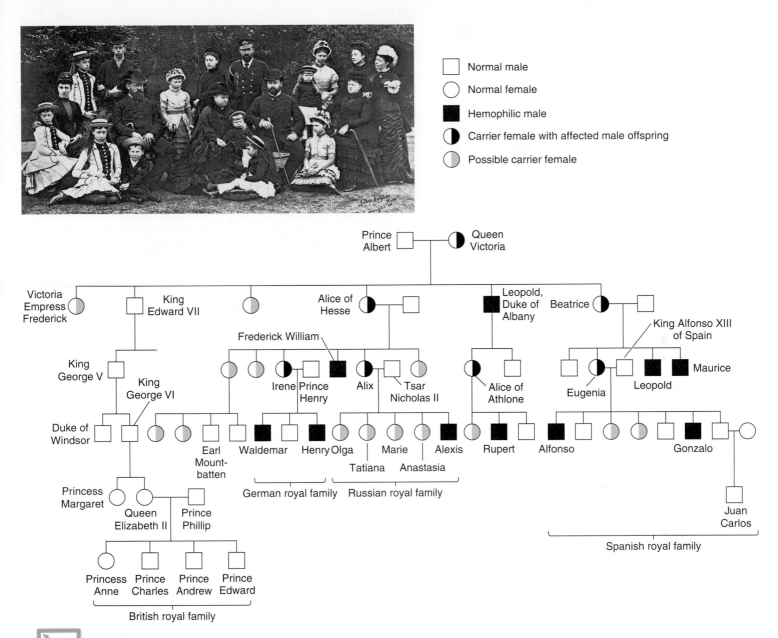

FIGURE 22.4 **A family tree of hemophilia A in the royal families of Europe, indicating an X-linked recessive inheritance.** Pictured are Queen Victoria and Prince Albert of Great Britain with some of their descendants.

As a classic example, Nancy Wexler, whose own mother died of Huntington disease, has studied this disorder among a population of related individuals in Venezuela. To date, over 15,000 individuals have been analyzed. **Figure 22.5** shows a simplified version of a large family pedigree from this Venezuelan population in which many individuals were affected with Huntington disease. A specific polymorphic marker, called G8, is found in four different versions, named A, B, C, or D. In this Venezuelan population, pedigree analysis revealed that the G8-C marker, which is located near the tip of the short arm of chromosome 4, is almost always associated with the mutant gene causing Huntington disease. In other words, the G8-C marker is closely linked to the Huntington allele.

Once a disease-causing mutant gene has been mapped to a chromosome site, the next step is to determine which gene in the region is responsible for the disease. Modern mapping methods can localize a gene to a chromosome region that is typically about 1 Mb (1,000,000 bp) in length. One way to identify a disease-causing allele is chromosome walking (refer back to Figure 20.18). This is how the *Huntingtin* gene was identified. However, the technique of chromosome walking is no longer widely used because the entire human genome has been sequenced and most genes have been identified. Therefore, researchers can analyze the 1 Mb region where a gene has been mapped to determine if a mutant gene in this region is responsible for a disease. In the human genome, a 1 Mb region will usually contain about 5 to 10

FIGURE 22.5   The transmission pattern of a molecular marker for Huntington disease. Affected members are shown with black symbols. The letters under each person indicate the forms of the G8 marker (A, B, C, or D) the individual carries. Affected individuals always carry the C version of this marker. Symbols with slashes indicate deceased individuals. *Data from Gusella et al. (1983) "A polymorphic DNA marker genetically linked to Huntington's disease."* Nature *306, 234–239.*

different genes, though the number can vary greatly. Therefore, mapping does not definitively tell researchers which gene may play a role in human disease, but it usually narrows down the list to a few candidate genes.

How does a researcher determine which of the candidate genes is the correct one? To further narrow down the list, researchers may also consider biological function. As the scientific community explores the functions of genes experimentally, the data are published in the research literature and placed into databases. In some cases, this information may help to narrow down the list of candidate genes. For example, if the disease of interest is neurological, researchers may discover that only certain genes in the mapped region are expressed in nerve cells. Also, researchers may compare data from other organisms. If a mutant gene in a mouse causes neurological problems and a human homolog of the mouse gene is found in the mapped region, this human gene would be a good candidate for being responsible for the human disease symptoms.

After researchers have narrowed down the number of candidate genes to as short a list as possible, the next phase is to sequence the candidate gene(s) from many affected and unaffected individuals, using the dideoxy sequencing method described in Chapter 18. The goal is to identify a gene in which affected individuals always carry a mutation. Once a gene has been identified that has a genetic change found in affected individuals, this is strong evidence that the candidate gene is causing the disease symptoms.

Why is it useful to identify disease-causing mutant genes? In many cases, the identification of these genes will help us to understand how genes contribute to pathogenesis and may even

aid in developing strategies aimed at the treatment of the disease. As described next, the identification of such genes may also result in the development of genetic tests that can enable people to determine if they are carrying disease-causing alleles.

Genetic Testing Can Identify Many Inherited Human Diseases

Because genetic abnormalities occur in the human population at a significant level, people have sought ways to determine whether individuals possess disease-causing alleles. The term **genetic testing** refers to the use of testing methods to determine if an individual carries a genetic abnormality. **Table 22.4** describes several different testing strategies. The term **genetic screening** refers to population-wide genetic testing.

In many cases, single-gene mutations that affect the function of cellular proteins can be examined at the protein level. If a gene encodes an enzyme, biochemical assays to measure that enzyme's activity may be available. As mentioned earlier, Tay-Sachs disease involves a defect in the enzyme hexosaminidase A (hexA). Enzymatic assays for this enzyme involve the use of an artificial substrate in which 4-methylumbelliferone (MU) is covalently linked to N-acetylglucosamine (GlcNAc). HexA cleaves this covalent bond and releases MU, which is fluorescent.

$$\text{MU–GlcNAc} \xrightarrow{\text{hexA}} \text{MU} + \text{GlcNAc}$$
$$\text{(nonfluorescent)} \qquad \text{(fluorescent)}$$

TABLE **22.4**

Testing Methods for Genetic Abnormalities

Method	Description
Protein Level	
Biochemical	As described in this chapter for Tay-Sachs disease, the enzymatic activity of a protein can be assayed in vitro.
Immunological	The presence of a protein can be detected using antibodies that specifically recognize that protein. Western blotting is an example of this type of technique (see Chapter 18).
DNA or Chromosomal Level	
RFLP analysis	As described in the experiment of Figure 20.10, RFLP analysis can also be used to determine the likelihood that a person may carry a disease-causing allele.
Altered restriction site	A gene mutation may alter the presence of a particular restriction enzyme site. A sample of DNA from an individual can be subjected to restriction enzyme digestion and gel electrophoresis (see Chapter 18) to see whether or not the restriction enzyme site is present.
DNA sequencing	If the normal gene has already been identified and sequenced, it is possible to design PCR primers that can amplify the gene from a sample of cells. The amplified DNA segment can then be subjected to DNA sequencing, as described in Chapter 18.
In situ hybridization	A DNA probe that hybridizes to a particular gene or gene segment can be used to determine if the gene is present, absent, or altered in an individual. The technique of fluorescence *in situ* hybridization (FISH) is described in Chapter 20 (see Figures 20.3, 20.4).
Karyotyping	The chromosomes from a sample of cells can be stained and then analyzed microscopically for abnormalities in chromosome structure and number (see Figure 3.2).
DNA microarrays	This method, which is described in Chapter 21, can be used to determine the expression levels of genes, such as those that are mutant in certain forms of cancer. In addition, microarrays are used to detect polymorphisms found in the human population that are associated with diseases.

To perform this assay, a small sample of cells is collected and incubated with MU–GlcNAc, and the fluorescence is measured with a device called a fluorometer. Individuals affected with Tay-Sachs, who do not produce the hexA enzyme, will produce little or no fluorescence, whereas individuals who are homozygous for the normal *hexA* allele produce a high level of fluorescence. Heterozygotes, who have 50% hexA activity, produce intermediate levels of fluorescence.

An alternative and more common approach is to detect single-gene mutations at the DNA level. To apply this testing strategy, researchers must have previously identified the mutant gene using molecular techniques. The identification of many human genes, such as those involved in Duchenne muscular dystrophy, cystic fibrosis, and Huntington disease, has made it possible to test for affected individuals or those who may be carriers of these diseases. Table 22.4 describes several ways to test for gene mutations. These laboratory techniques are described in Chapters 18 and 20.

Many human genetic abnormalities involve changes in chromosome number and/or structure. In fact, changes in chromosome number are a common class of human genetic abnormality. Most of these result in spontaneous abortions. However, approximately 1 in 200 live births are aneuploid—have an abnormal number of chromosomes (see Chapter 8, Table 8.1). About 5% of infant and childhood deaths are related to such genetic abnormalities. With regard to testing, changes in chromosome number and many changes in chromosome structure can be detected by karyotyping the chromosomes with a light microscope.

In the United States, genetic screening for certain disorders has become common medical practice. For example, pregnant women over 35 years old often have tests conducted to see if their fetuses are carrying chromosomal abnormalities. As discussed in Chapter 8, these tests are indicated because the rate of such defects increases with the age of the mother. Another example is the widespread screening for phenylketonuria (PKU). An inexpensive test can determine if newborns have this disease. Those who test positive can then be given a low-phenylalanine diet to avoid PKU's devastating effects.

Genetic screening also has been conducted on specific populations in which a genetic disease is prevalent. For example, in 1971, community-based screening for heterozygous carriers of Tay-Sachs disease was begun among specific Ashkenazi Jewish populations. With the use of this screening, over the course of one generation, the incidence of TSD births was reduced by 90%. For most rare genetic abnormalities, however, genetic screening is not routine practice. Rather, genetic testing is performed only when a family history reveals a strong likelihood that a couple may produce an affected child. This typically involves a couple that already has an affected child or has other relatives with a genetic disease.

Genetic testing can be performed prior to birth, which may affect a woman's decision to terminate a pregnancy. The two common ways of obtaining cellular material from a fetus for the purpose of genetic testing are **amniocentesis** and **chorionic villus sampling.** In amniocentesis, a doctor removes amniotic fluid containing fetal cells, using a needle that is passed through the abdominal wall (**Figure 22.6**). The fetal cells are cultured for several weeks and then karyotyped to determine the number of chromosomes per cell and whether changes in chromosome structure have occurred. In chorionic villus sampling, a small piece of the chorion (the fetal part of the placenta) is removed, and a karyotype is prepared directly from the collected cells. Chorionic villus sampling can be performed earlier

FIGURE 22.6   Techniques to determine genetic abnormalities before birth. In amniocentesis, amniotic fluid is withdrawn, and fetal cells are collected by centrifugation. The cells are then allowed to grow in laboratory culture media for several weeks prior to karyotyping. In chorionic villus sampling, a small piece of the chorion is removed. These cells can be prepared directly for karyotyping.

during pregnancy than amniocentesis, usually around the eighth to tenth week, compared with the fourteenth to sixteenth week for amniocentesis, and results are available sooner. Weighed against these advantages, however, is that chorionic villus sampling may pose a slightly greater risk of causing a miscarriage.

Genetic testing and screening are medical practices with many social and ethical dimensions. For example, people must decide whether or not they want to make use of available tests, particularly when the disease in question has no cure. For example, Huntington disease typically does not affect people until their 50s and can last 20 years. People who learn they are carriers of genetic diseases such as Huntington disease can be devastated by the news. Some argue that people have a right to know about their genetic makeup; others assert that it does more harm than good. Another issue is privacy. Who should have access to personal genetic information, and how could it be used? Could routine genetic testing lead to discrimination by employers or medical insurance companies? In the coming years, we will gain an ever-increasing awareness of our genetic makeup and the underlying causes of genetic diseases. As a society, establishing guidelines for the uses of genetic testing will be a necessary, yet very difficult, task.

Prions Are Infectious Particles That Alter Protein Function Posttranslationally

Before we end this section, let's consider an unusual mechanism in which agents known as prions cause disease. As shown in **Table 22.5**, prions cause several types of neurodegenerative diseases affecting humans and livestock, including mad cow disease. Recent evidence has shown that prions also exist in yeast. In the 1960s, British researchers Tikvah Alper and J. S. Griffith discovered that preparations from animals with certain neurodegenerative diseases remained infectious even after exposure to radiation that would destroy any DNA or RNA. They suggested that the infectious agent was a protein. Furthermore, Alper and Griffith speculated that the protein usually preferred one folding pattern but could sometimes misfold and then catalyze other proteins to do the same. In the early 1970s, Stanley Prusiner, moved by the death of a patient from a neurodegenerative disease, began to search for the causative agent. In 1982, he isolated a disease-causing agent composed entirely of protein, which he called a **prion.** The term emphasizes the prion's unusual character as a proteinaceous infectious agent. Before the discovery of prions, all known infectious agents such as viruses and bacteria contained their own genetic material (either DNA or RNA).

TABLE 22.5

Neurodegenerative Diseases Caused by Prions*

Disease	Description
Infectious Diseases	
Kuru	A human disease that was once common in New Guinea. It begins with a loss of coordination, usually followed by dementia. Infection was spread by cannibalism, a practice that ended in 1958.
Scrapie	A disease of sheep and pigs characterized by intense itching in which the animals tend to scrape themselves against trees, followed by neurodegeneration
Mad cow disease	Begins with changes in posture and temperament, followed by loss of coordination and neurodegeneration
Human Inherited Diseases	
Creutzfeldt-Jakob disease	Characterized by loss of coordination and dementia
Gerstmann-Straussler-Scheinker disease	Characterized by loss of coordination and dementia
Familial fatal insomnia	Begins with sleeping and autonomic nervous system disturbances followed by insomnia and dementia

*All of these diseases are eventually fatal.

Prion-related diseases arise from the ability of the prion protein to exist in two conformational states: a normal form PrP^C, which does not cause disease, and an abnormal form, PrP^{Sc}, which does. (Note: The superscript C refers to the normal conformation, while the superscript Sc refers to the abnormal conformation, such as the one found in the disease called scrapie.) The gene encoding the prion protein (*PrP*) is found in humans and other mammals, and the protein is expressed at low levels in certain types of cells such as nerve cells. The abnormal conformation of the prion protein can come from two sources. An individual can be infected with the abnormal protein by taking the abnormal protein into their bodies. For example, someone may eat products from an animal that had the disease. Alternatively, some people carry alleles of the *PrP* gene that cause their prion protein to convert spontaneously to the abnormal conformation at a very low rate. These individuals have an inherited predisposition to develop a prion-related disease. An example of an inherited prion disease is familial fatal insomnia (Table 22.5).

What is the molecular mechanism through which prions cause disease? As noted, the prion protein can exist in two conformations, PrP^C and PrP^{Sc}. As shown in **Figure 22.7**, the abnormal conformation, PrP^{Sc}, acts as a catalyst to convert normal prion proteins within the cell to the misfolded conformation. As a prion disease progresses, the PrP^{Sc} protein forms dense aggregates in the cells of the brain and peripheral nervous tissues. This deposition is correlated with the disease symptoms affecting the nervous system. Some of the abnormal prion protein is also excreted from infected cells and can travel through the nervous system to infect other cells.

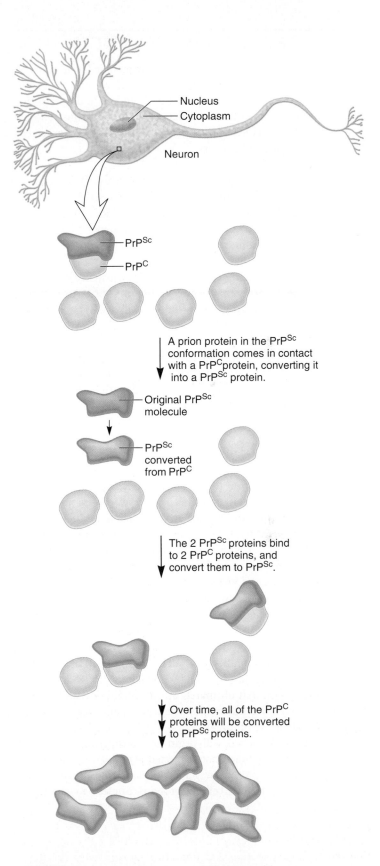

Nucleus
Cytoplasm
Neuron

PrP^{Sc}
PrP^C

A prion protein in the PrP^{Sc} conformation comes in contact with a PrP^C protein, converting it into a PrP^{Sc} protein.

Original PrP^{Sc} molecule

PrP^{Sc} converted from PrP^C

The 2 PrP^{Sc} proteins bind to 2 PrP^C proteins, and convert them to PrP^{Sc}.

Over time, all of the PrP^C proteins will be converted to PrP^{Sc} proteins.

ONLINE ANIMATION

FIGURE 22.7 **A proposed molecular mechanism of prion diseases.** The PrP^{Sc} protein catalyzes the conversion of PrP^C to PrP^{Sc}. Over time, the PrP^{Sc} conformation will accumulate to high levels, leading to symptoms of the prion disease.

22.2 GENETIC BASIS OF CANCER

Cancer is a disease characterized by uncontrolled cell division. It is a genetic disease at the cellular level. More than 100 kinds of human cancers have been identified, and they are classified according to the type of cell that has become cancerous. Though cancer is a diverse collection of many diseases, some characteristics are common to all cancers.

1. Most cancers originate in a single cell. This single cell, and its line of daughter cells, undergoes a series of genetic changes that accumulate during cell division. In this regard, a cancerous growth can be considered to be **clonal** in origin. A hallmark of a cancer cell is that it divides to produce two daughter cancer cells.

2. At the cellular and genetic levels, cancer usually is a multistep process that begins with a precancerous genetic change—a **benign** growth—and is followed by additional genetic changes that lead to cancerous cell growth (**Figure 22.8**).

3. When cells have become cancerous, their growth is described as **malignant.** Cancer cells are **invasive**—they can invade healthy tissues—and **metastatic**—they can migrate to other parts of the body and cause secondary tumors.

In the United States, approximately 1 million people are diagnosed with cancer each year, and about half that number will die from the disease. In 5 to 10% of all cases, a predisposition to develop the cancer is an inherited trait. We will examine some inherited forms of cancer later in this chapter. Most cancers, though, perhaps 90 to 95%, are not passed from parent to offspring. Rather, cancer is usually an acquired condition that typically occurs later in life. While some cancers are caused by spontaneous mutations and viruses, at least 80% of all human cancers are related to exposure to agents that promote genetic changes in somatic cells. These environmental agents, such as UV light and certain chemicals, are mutagens that alter the DNA in a way that affects the function of normal genes. If the DNA is permanently modified in somatic cells, such changes may be transmitted during cell division. These DNA alterations can lead to effects on gene expression that ultimately affect cell division and thereby lead to cancer. An environmental agent that causes cancer in this manner is called a **carcinogen.**

In this section, we will begin by considering some early experimental observations that suggested genes play a role in cancer. We will then explore how genetic abnormalities, which affect the functions of particular cellular proteins, can lead to cancer.

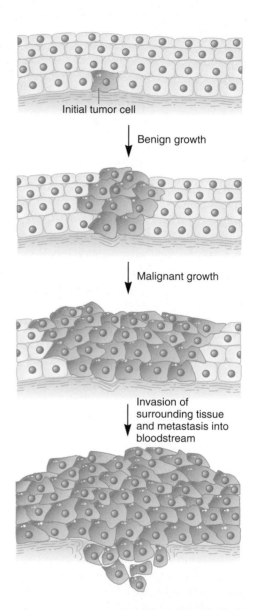

FIGURE 22.8   Progression of cellular growth leading to cancer.
Genes→Traits In a healthy individual, an initial gene mutation converts a normal cell into a tumor cell. This tumor cell divides to produce a benign tumor. Additional genetic changes in the tumor cells may occur, leading to malignant growth. At a later stage in malignancy, the tumor cells will invade surrounding tissues, and some malignant cells may metastasize by traveling through the bloodstream to other parts of the body where they can grow and cause secondary tumors. As a trait, cancer can be viewed as a series of genetic changes that eventually lead to uncontrolled cell growth.

Certain Viruses Can Cause Cancer by Carrying Viral Oncogenes into the Cell

As mentioned, most types of cancers are caused by mutagens that alter the structure and expression of genes. A few viruses, however, are known to cause cancer in plants, animals, and humans. We begin our discussion here, because early studies of cancer-causing viruses identified the first genes that play a role in cancer. Many of these viruses can also infect normal laboratory-grown cells and convert them into malignant cells. In cancer biology, the process of converting a normal cell into a malignant cell is called **transformation.**

Most cancer-causing viruses are not very potent at inducing cancer. An organism must be infected for a long period of time before a tumor actually develops. Furthermore, most viruses are inefficient at transforming or are unable to transform normal cells grown in the laboratory. By comparison, a few types of viruses rapidly induce tumors in animals and efficiently transform cells in culture. These are called **acutely transforming viruses (ACTs).** About 40 such viruses have been isolated from chickens, turkeys, mice, rats, cats, and monkeys. The first virus of this type, the Rous sarcoma virus (RSV), was isolated from chicken sarcomas by Peyton Rous in 1911.

During the 1970s, RSV research led to the identification of the first gene known to cause cancer. Researchers investigated RSV by infecting chicken fibroblast cells in the laboratory. This causes the chicken fibroblasts to grow like cancer cells. During the course of their studies, researchers identified mutant RSV strains that infected and proliferated within chicken cells without transforming them into malignant cells. These RSV strains were determined to contain a defective viral gene. In contrast, in other strains where this gene is functional, cancer occurs. This viral gene was designated *src* for *sarcoma*, the type of cancer it causes. The *src* gene is known as v-*src* because it is found within a viral genome. The v-*src* gene was the first example of an **oncogene,** a gene that promotes cancer.

Because the viral oncogene of RSV is not necessary for viral replication, researchers were curious as to why the virus has this v-*src* oncogene. Harold Varmus and Michael Bishop, in collaboration with Peter Vogt, soon discovered that normal host cells contain a copy of the *src* gene in their chromosomes. This normal copy of the *src* gene found in the chromosomal DNA of the host is termed c-*src*, for *cellular src*. The c-*src* gene does not cause cancer. However, once incorporated into a viral genome, this gene can become a viral oncogene that promotes cancer. Why does cancer occur when the gene is in a viral genome? Three explanations of this phenomenon are possible. First, the many copies of the virus made during viral replication may lead to overexpression of the *src* gene. Second, the incorporation of the *src* gene next to viral regulatory sequences may cause it to be overexpressed. A third possibility is that the v-*src* gene may accumulate additional mutations that convert it to an oncogene.

RSV has acquired the *src* gene by capturing it from a host cell's chromosome. This can occur during the RSV life cycle. RSV is a retrovirus with an RNA genome. During its life cycle, a retrovirus uses reverse transcriptase to make a DNA copy of its genome, which becomes integrated as a provirus into the host cell genome. This integration could occur next to the c-*src* gene. During transcription of the proviral DNA, the *src* gene could be included in the RNA transcript that becomes the retroviral genome.

Since these early studies of RSV, many other retroviruses carrying oncogenes have been investigated. The characterization of such oncogenes has led to the identification of several genes with cancer-causing potential. In addition to retroviruses, several viruses with DNA genomes cause tumors, and some of these are known to cause cancer in humans (**Table 22.6**). Researchers estimate that up to 15% of all human cancers are associated with viruses.

TABLE 22.6

Examples of Viruses That May Cause Cancer

Virus	Description
Retroviruses	
Rous sarcoma virus	Causes sarcomas in chickens.
Simian sarcoma virus	Causes sarcomas in monkeys.
Abelson leukemia virus	Causes leukemia in mice.
Hardy-Zuckerman-4 feline sarcoma virus	Causes sarcomas in cats.
DNA tumor viruses	
Hepatitis B, SV40, polyomavirus	Causes liver cancer in several species including humans. In some cases, these viruses do not cause cancer in their natural hosts but can transform cells in culture.
Papillomavirus	Causes benign tumors and malignant carcinomas in several species, including humans. Causes cervical cancer in humans.
Adenovirus	Does not cause cancer in its natural host but can transform cells in culture.
Herpesvirus	Causes carcinoma in frogs and T-cell lymphoma in chickens. A human herpesvirus, Epstein-Barr virus, is a causative agent in Burkitt's lymphoma, which occurs primarily in immunosuppressed individuals such as AIDS patients.

EXPERIMENT 22A

DNA Isolated from Malignant Mouse Cells Can Transform Normal Mouse Cells into Malignant Cells

The study of retroviruses and other tumor-producing viruses led to the identification of a few dozen viral oncogenes, which were the first genes implicated in causing cancer. However, most cancers are not caused by viruses but instead are due to environmental mutagens, which cause mutations that alter the expression of normal cellular genes. Therefore, researchers also wanted to identify cellular genes that have been altered in a way that leads to malignancy. Methods developed in the study of viral oncogenes proved valuable in this search. Particularly useful were studies conducted in 1971 by Miroslav Hill and Jana Hillova. These experiments demonstrated that the purified DNA of RSV-infected cells could be taken up by chicken fibroblasts and would transform them into malignant cells.

In 1979, Robert Weinberg and colleagues wanted to determine if chromosomal DNA purified from cells that had become malignant due to exposure to mutagens could transform normal cells into malignant cells. If so, this would be the first step in the identification of chromosomally located genes that had been converted to oncogenes. Before we discuss this experiment, let's consider how a researcher can identify malignant cells in a laboratory. A widely used assay relies on the ability to recognize a clump of transformed cells as a distinct **focus,** or pile of raised cells, on a culture dish (**Figure 22.9**). Unlike normal cells, which grow as a single layer, or monolayer, on culture dishes, a malignant focus piles up over the monolayer to form a mass of cells. The focus of cells is a clone derived from a single cell that had become malignant and then proliferated. At the microscopic level, the malignant cells also have altered morphologies.

Now that we understand how malignant cells are identified in a laboratory, let's examine the steps of the experiment of Weinberg and coworkers. They began with several malignant cell lines that had been previously characterized, along with normal cell lines. As shown in **Figure 22.10**, they isolated the DNA from the malignant and normal cells, and then separately mixed this DNA with normal mouse fibroblast cells. Calcium phosphate was added to promote the uptake of the DNA by the mouse fibro-

FIGURE 22.9   The malignant growth of cells. The growth of normal cells leads to the formation of a monolayer, while certain transformed cells grow as a focus (see arrows) or a raised pile of cells. (Reprinted by permission from Macmillan Publishers Ltd. *Nature.* Transforming activity of DNA of chemically transformed and normal cells. Geoffrey M. Cooper, S. Okenquist & L. Silverman. 284:5755, 418-421, 1980.)

blast cells. Once inside the cells, the DNA can be incorporated into the chromosomes of the cells by recombination. The cells were incubated for two to three weeks, and then the researchers examined the plates for the growth of malignant foci of cells. In other words, they looked to see if any of the normal mouse cells had been transformed into malignant cells.

■ THE HYPOTHESIS

Cellular DNA isolated from malignant cells will be taken up by normal mouse fibroblast cells and will transform them into malignant cells.

■ TESTING THE HYPOTHESIS — FIGURE 22.10   Identification of chromosomal oncogenes.

Starting material: Several mouse cell lines. Some of the cell lines were normal, whereas others were malignant due to exposure to chemical or physical mutagens. It was known that none of the cell lines in this experiment were infected with oncogenic viruses.

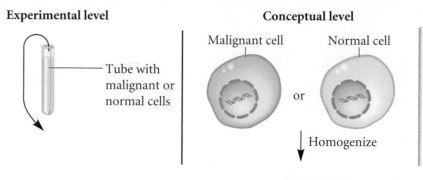

1. Extract the chromosomal DNA from normal or malignant cell lines.

Experimental level

Tube with malignant or normal cells

Conceptual level

Malignant cell   Normal cell

or

Homogenize

Homogenize the cells to break them open.

Isolate the nuclei by differential centrifugation. See the Appendix.

Dissolve the nuclear membrane with detergent, and isolate the chromosomal DNA.

2. Mix the DNA from the normal cells or from the malignant cells with normal mouse fibroblast cells that are growing on a tissue culture plate.

3. Add a buffer containing calcium ions and phosphate ions. This buffer makes the cells permeable to DNA.

4. Incubate for 14–20 days.

5. Examine the plates under a light microscope for cells growing as transformed foci.

   Note: Normal cells in culture grow as a monolayer and very rarely form foci.

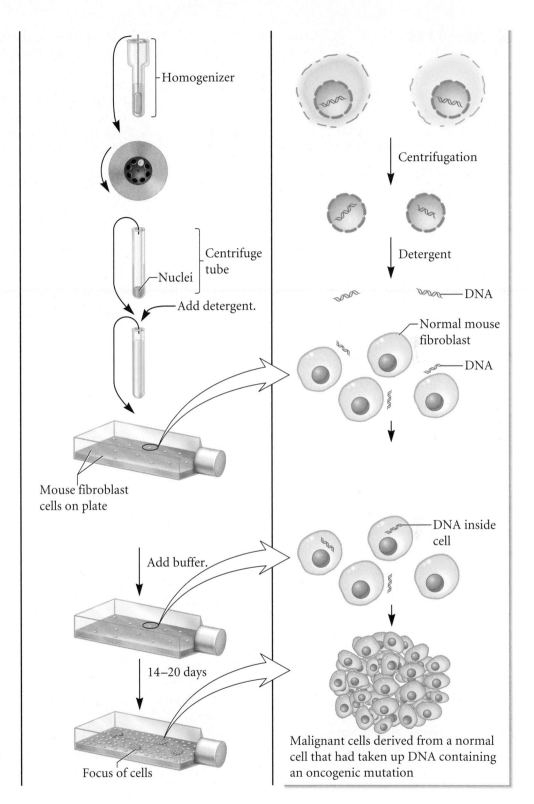

THE DATA

Source of DNA	Recipient Cells	Number of Malignant Foci Found on 12 Plates
Malignant Cell Lines		
MC5-5-0	NIH3T3 (normal fibroblasts)	48*
MCA16	"	5
MB66 MCA ad 36	"	8
MB66 MCA ACL 6	"	0
MB66 MCA ACL 13	"	0
Normal Cell Lines		
NIH3T3	"	1
C3H10T1/2	"	0

*In this experiment, two of the plates were contaminated, so this is 48 foci on 10 plates.

INTERPRETING THE DATA

As shown in the data, the DNA isolated from some (but not all) malignant cell lines could transform normal mouse cells, which then proliferated and produced malignant foci. These results are consistent with the hypothesis that oncogenes had been taken up and expressed in the normal mouse cells, converting them into malignant cells. By comparison, the DNA isolated from normal cells did not cause a significant amount of transformation. From these experiments, it is not clear why the DNA from two of the malignant cell lines, MB66 MCA ACL 6 and MB66 MCA ACL 13, could not transform the normal mouse cells. One possibility is that some malignancies are caused by dominant oncogenes, while others involve genes that act recessively. Recessive genes would be unable to transform normal mouse cells that already contain the normal (nonmalignant) dominant allele. Later, we will learn how another category of genes involved in cancer, called tumor-suppressor genes, act recessively.

Just two years later, in 1981, the laboratories of Weinberg and Geoffrey Cooper identified the first cellular oncogene in humans. They isolated chromosomal DNA from human bladder carcinoma cells and used it to transform mouse cells in vitro. This chromosomal DNA contained a human oncogene that was the result of a mutation. The identity of the gene was eventually determined by transposon tagging, as described in solved problem S2 at the end of this chapter. This observation paved the way for the isolation of many human cellular oncogenes.

A self-help quiz involving this experiment can be found at the Online Learning Center.

Many Oncogenes Have Abnormalities That Affect Proteins Involved in Cell Division Pathways

As described in Chapter 3, eukaryotic cells destined to divide progress through a series of stages known as the cell cycle (**Figure 22.11a**). The phases consist of G_1 (first gap), S (synthesis of DNA, the genetic material), G_2 (second gap), and M phase (mitosis and cytokinesis). The G_1 phase is a period in a cell's life when it may become committed to divide. Depending on the conditions, a cell in the G_1 phase may accumulate molecular changes that cause it to progress through the rest of the cell cycle. When this occurs, cell biologists say that a cell has reached a special control point called the restriction point. The commitment to divide is based on a variety of factors. For example, environmental conditions, such as the presence of sufficient nutrients, are important for cell division. In addition, multicellular organisms rely on signaling molecules to coordinate cell division throughout the body. These signaling molecules are often called **growth factors** because they promote cell division.

As researchers began to identify oncogenes, they wanted to understand how these mutant genes promote abnormal cell division. In parallel with cancer research, cell biologists have studied the roles that normal cellular proteins play in cell division. As mentioned, the cell cycle is regulated in part by growth factors that bind to cell surface receptors and initiate a cascade of cellular events that lead eventually to cell division. **Figure 22.11b** considers a protein called epidermal growth factor (EGF) that is secreted from endocrine cells and stimulates epidermal cells, such as skin cells, to divide. As seen here, EGF binds to its receptor, leading to the activation of an intracellular signaling pathway. This pathway, also known as a signal cascade, leads to a change in gene transcription. In other words, the transcription of specific genes is activated in response to the growth hormone. Once these genes are transcribed and translated, the gene products function to promote the progression through the cell cycle. Figure 22.11b is just one example of a pathway between a growth factor and gene activation. Eukaryotic species produce many different growth factors, and the signaling pathways are often more complex than the one shown here.

What is the relationship between normal genes and oncogenes? A normal, nonmutated gene that has the potential to become an oncogene is termed a **proto-oncogene.** To become an oncogene, a proto-oncogene must incur a mutation that causes its expression to be abnormally active. The mutation typically has one of three possible effects:

1. The amount of the encoded protein is greatly increased.
2. A change occurs in the structure of the encoded protein that causes it to be overly active.
3. The encoded protein is expressed in a cell type where it is not normally expressed.

The mutations that convert proto-oncogenes into oncogenes have been analyzed in many types of cancers. Oncogenes commonly encode proteins that function in cell growth signaling pathways (**Table 22.7**). These include growth factors, growth

FIGURE 22.11   **The eukaryotic cell cycle and activation of a cell signaling pathway by a growth factor.** (a) The cell cycle involves a progression through the G_1, S, G_2, and M phases. Progression through the cell cycle is often stimulated by growth factors. (Note: Chromosomes are not condensed during G_0, G_1, S, and G_2 phases. They are shown that way in this figure so they can be counted.) (b) In this example, epidermal growth factor (EGF) binds to two EGF receptors, causing them to dimerize and phosphorylate each other. An intracellular protein called GRB2 is attracted to the phosphorylated EGF receptor, and it is subsequently bound by another protein called Sos. The binding of Sos to GRB2 enables Sos to activate a protein called Ras. This activation involves the release of GDP and the binding of GTP. The activated Ras/GTP complex then activates Raf-1, which is a protein kinase. Raf-1 phosphorylates MEK, and then MEK phosphorylates MAPK. More than one MAPK may be involved. Finally, the phosphorylated form of MAPK activates transcription factors. such as Myc, Jun, and Fos. This leads to the transcription of genes, which encode proteins that promote cell division.

TABLE 22.7

Examples of Proto-Oncogenes That Can Mutate into Oncogenes

Gene	Cellular Function
Growth Factors*	
sis	Platelet-derived growth factor
int-2	Fibroblast growth factor
Growth Factor Receptors	
erbB	Growth factor receptor for EGF (epidermal growth factor)
trk	Growth factor receptor for CSF-1 (cytostatic factor that inhibits cell division)
fms	Growth factor receptor for NGF (nerve growth factor)
K-sam	Growth factor receptor for FGF (fibroblast growth factor)
Intracellular Signaling Proteins	
ras	GTP/GDP-binding protein
raf	Serine/threonine kinase
src	Tyrosine kinase
abl	Tyrosine kinase
gsp	G-protein α subunit
Transcription Factors	
myc	Transcription factor
jun	Transcription factor
fos	Transcription factor
gli	Transcription factor
erbA	Steroid receptor (which functions as a transcription factor)

*The genes described in this table are found in humans as well as other eukaryotic species. Many of these genes were initially identified in retroviruses. Most of the genes have been given three-letter names that are abbreviations for the type of cancer the oncogene causes or the type of virus in which the gene was first identified.

factor receptors, proteins involved in the intracellular signaling pathways, and transcription factors. Let's consider two examples. Research groups headed by Robert Gallo and Mark Groudine showed that a *myc* gene, designated c-*myc*, was amplified about 10-fold in a human promyelocytic leukemia cell line. Since that time, research has shown that *myc* genes are overexpressed in many forms of cancer, including breast cancer, lung cancer, and colon carcinoma. The overexpression of this transcription factor leads to the transcriptional activation of genes that promote cell division.

FIGURE 22.12 **Functional cycle of the Ras protein.** The binding of GTP to Ras activates the function of Ras and promotes cell division. The hydrolysis of GTP to GDP and P_i converts the active form of Ras to an inactive form.

As a second example, mutations that alter the amino acid sequence of the Ras protein have been shown to cause functional abnormalities. The Ras protein is a GTPase, which hydrolyzes GTP to GDP + P_i (**Figure 22.12**). Therefore, after it has been activated, the Ras protein returns to its inactive state by hydrolyzing GTP. Mutations that convert the normal *ras* gene into an oncogene either decrease the ability of the Ras protein to hydrolyze GTP or increase the rate of exchange of bound GDP for GTP. Both of these functional changes result in a greater amount of the active GTP-bound form of the Ras protein. In this way, these mutations keep the signaling pathway turned on and thereby stimulate the cell to divide.

Genetic Changes in Proto-Oncogenes Convert Them to Oncogenes

How do specific genetic alterations convert proto-oncogenes into oncogenes? By isolating and studying oncogenes at the molecular level, researchers have discovered four main ways this occurs (**Table 22.8**). These changes can be categorized as missense mutations, gene amplifications (i.e., an increase in copy number), chromosomal translocations, and viral integrations.

As mentioned previously, changes in the structure of the Ras protein can cause it to become permanently activated. These changes are caused by a missense mutation in the *ras* gene. The human genome contains four different but evolutionarily related *ras* genes: *ras*H, *ras*N, *ras*K-4a, and *ras*K-4b. All four homologous genes encode proteins with very similar amino acid sequences containing a total of 188 or 189 amino acids. Missense mutants in these normal *ras* genes are associated with particular forms of cancer. For example, a missense mutation in *ras*H that changes a glycine to a valine is responsible for the conversion of *ras*H into an oncogene:

	1	2	3	4	5	6	7	8	9	10	11	12	13		188	189
Normal	Met	Thr	Glu	Tyr	Lys	Leu	Val	Val	Val	Gly	Ala	Gly	Gly		Leu	Ser
Human rasH	ATG	ACG	GAA	TAT	AAG	CTG	GTG	GTG	GTG	GGC	GCC	GGC	GGT	....	CTC	TCC

	1	2	3	4	5	6	7	8	9	10	11	12	13		188	189
Oncogenic rasH	Met	Thr	Glu	Tyr	Lys	Leu	Val	Val	Val	Gly	Ala	**GTC** **Val**	Gly	....	Leu	Ser

TABLE 22.8

Genetic Changes That Convert Proto-Oncogenes into Oncogenes

Type of Change	Description and Examples
Missense mutation	A change in the amino acid sequence of a proto-oncogene protein may cause it to function in an abnormal way. Missense mutations can convert *ras* genes into oncogenes.
Gene amplification	The copy number of a proto-oncogene may be increased by gene duplication. *Myc* genes have been amplified in human leukemias; breast, stomach, lung, and colon carcinomas; and neuroblastomas and glioblastomas. *ErbB* genes have been amplified in glioblastomas, squamous cell carcinomas, and breast, salivary gland, and ovarian carcinomas.
Chromosomal translocations	A piece of chromosome may be translocated to another chromosome and affect the expression of genes at the breakpoint site. In Burkitt's lymphoma, a region of chromosome 8 is translocated to either chromosome 2, 14, or 22. The breakpoint in chromosome 8 causes the overexpression of the c-*myc* gene.
Viral integration	When a virus integrates into the chromosome, it may enhance the expression of nearby proto-oncogenes. In avian lymphomas, the integration of the avian leukosis virus can enhance the transcription of the c-*myc* gene.

Experimentally, chemical carcinogens have been shown to cause these missense mutations and thereby lead to cancer.

Another genetic event that may occur in cancer cells is gene amplification, or an abnormal increase in the copy number of a proto-oncogene. An increase in gene copy number is expected to increase the amount of the encoded protein and thereby contribute to malignancy. Gene amplification does not normally happen in mammalian cells, but it is a common occurrence in cancer cells. As mentioned previously, Gallo and Groudine discovered that c-*myc* was amplified in a human leukemia cell line. Many human cancers are associated with the amplification of particular oncogenes. In such cases, the extent of oncogene amplification may be correlated with the progression of tumors to increasing malignancy. These include the amplification of N-*myc* in neuroblastomas and *erbB-2* in breast carcinomas. In other types of malignancies, gene amplification is more random and may be a secondary event that increases the expression of oncogenes previously activated by other genetic changes.

A third type of genetic alteration that can lead to cancer is a chromosomal translocation. While structural abnormalities are common in cancer cells, very specific types of chromosomal translocations have been identified in certain types of tumors. In 1960, Peter Nowell and David Hungerford discovered that chronic myelogenous leukemia (CML) is correlated with the presence of a shortened version of chromosome 22, which they called the Philadelphia chromosome after the city where it was discovered. Rather than a deletion, this shortened chromosome is the result of a reciprocal translocation between chromosomes 9 and 22. Later studies revealed that this translocation activates a proto-oncogene, *abl*, in an unusual way (**Figure 22.13**). The reciprocal translocation involves breakpoints within the *abl* and *bcr* genes. Following the reciprocal translocation, the coding sequence of the *abl* gene fuses with the promoter and coding sequence of the *bcr* gene. This yields an oncogene that encodes an abnormal fusion protein, which contains the polypeptide sequences encoded from both genes. The *abl* gene encodes a tyrosine kinase enzyme, which uses ATP to attach phosphate groups onto target proteins. This phosphorylation activates certain proteins involved with cell division. Normally, the *abl* gene is highly

FIGURE 22.13   The reciprocal translocation commonly found in people with chronic myelogenous leukemia.

Genes→Traits In healthy individuals, the *abl* gene is located on chromosome 9, and the *bcr* gene is on chromosome 22. In certain forms of myelogenous leukemia, a reciprocal translocation causes the *abl* gene to fuse with the *bcr* gene. This combined gene, under the control of the *bcr* promoter, encodes an abnormal fusion protein that overexpresses the tyrosine kinase function of *abl* and leads to leukemia.

regulated. However, in the Philadelphia chromosome, the fusion gene is controlled by the *bcr* promoter, which is active in white blood cells. This leads to an overexpression of the tyrosine kinase function in such cells and thereby explains why this fusion causes a type of cancer called a leukemia, which involves a proliferation of white blood cells.

Interestingly, the study of the *abl* gene has led to an effective treatment for CML. Until recently, the only successful treatment was to destroy the patient's bone marrow and then restore blood-cell production by infusing stem cells from the bone marrow of a healthy donor. With knowledge about function of the ABL protein, researchers have developed the drug imatinib mesylate (Gleevec®) that appears to dramatically improve survival. This molecule fits into the active site of the ABL protein, preventing ATP from binding there. Without ATP, the ABL protein cannot phosphorylate its target proteins. This prevents the ABL protein from stimulating cell division. In a clinical trial, almost 90% of the CML patients treated with the drug showed no further progression of their disease!

Other forms of cancer also involve chromosomal translocations that cause an overexpression of an oncogene. In Burkitt's lymphoma, for example, a region of chromosome 8 is translocated to either chromosome 2, 14, or 22. The breakpoint in chromosome 8 is near the c-*myc* gene, and the sites on chromosomes 2, 14, and 22 correspond to locations of different immunoglobulin genes that are normally expressed in lymphocytes. The translocation of the c-*myc* gene near the immunoglobulin genes leads to the overexpression of the c-*myc* gene and thereby promotes malignancy in lymphocytes.

A fourth way that oncogenes may occur is via viral integration. As part of their reproductive cycle, certain viruses integrate their genomes into the chromosomal DNA of their host cell. If the integration occurs next to a proto-oncogene, a viral promoter or enhancer sequence may cause the proto-oncogene to be overexpressed. For example, in certain lymphomas that occur in birds, the genome of the avian leukosis virus has been found to be integrated next to the c-*myc* gene and enhances its level of transcription.

Tumor-Suppressor Genes Play a Role in Preventing the Proliferation of Cancer Cells

Thus far, we have considered how oncogenes promote cancer. An oncogene is an abnormally activated gene that leads to uncontrolled cell growth. We will now turn our attention to a second category of genes called **tumor-suppressor genes.** As the name suggests, the role of a tumor-suppressor gene is to prevent cancerous growth. Therefore, when a tumor-suppressor gene becomes inactivated by mutation, it becomes more likely that cancer will occur.

The first identification of a human tumor-suppressor gene involved studies of retinoblastoma, a tumor that occurs in the retina of the eye. Some people have inherited a predisposition to develop this disease within the first few years of life. By comparison, the noninherited form of retinoblastoma, which is caused by environmental agents, tends to occur later in life but only rarely.

Based on these differences, in 1971, Alfred Knudson proposed a "two-hit" model for retinoblastoma. According to this model, retinoblastoma requires two mutations to occur. People with the hereditary form already have received one mutant gene from one of their parents. They need only one additional mutation in the other copy of this tumor-suppressor gene to develop the disease. Because the retina has more than 1 million cells, it is relatively likely that a mutation may occur in one of these cells at an early age, leading to the disease. However, people with the noninherited form of the disease must have two mutations in the same retinal cell to cause the disease. Because two rare events are much less likely to occur than a single such event, the noninher-

ited form of this disease is expected to occur much later in life and only rarely. Therefore, this hypothesis explains the different populations typically affected by the inherited and noninherited forms of retinoblastoma.

Since Knudson's original hypothesis, molecular studies have confirmed the two-hit hypothesis for retinoblastoma. In this case, the gene in which mutations occur is designated *rb* (for <u>r</u>etino<u>b</u>lastoma). This tumor suppressor gene is found on the long arm of chromosome 13. Most people have two normal copies of the *rb* gene. Persons with hereditary retinoblastoma have inherited one normal and one defective copy. In nontumorous cells throughout the body, they have one functional copy and one defective copy of *rb*. However, in retinal tumor cells, the normal *rb* gene has also suffered the second hit (i.e., a mutation), which renders it defective. Without the tumor-suppressor ability, cells are allowed to grow and divide in an unregulated manner, which ultimately leads to cancer. (In contrast, as discussed later, most other forms of cancer involve mutations in several genes.)

More recent studies have revealed how the Rb protein suppresses the proliferation of cancer cells (**Figure 22.14**). The Rb protein regulates a transcription factor called E2F, which activates genes required for cell cycle progression. (The eukaryotic cell cycle is described earlier in Figure 22.11a.) The binding of the Rb protein to E2F inhibits its activity and prevents the cell from progressing through the cell cycle. As discussed later in this chapter, when a normal cell is supposed to divide, cellular proteins called cyclins bind to cyclin-dependent protein kinases. This activates

FIGURE 22.14   Interactions between the Rb and E2F proteins. The binding of the Rb protein to the transcription factor E2F inhibits the ability of E2F to function. This prevents cell division. For cell division to occur, cyclins bind to cyclin-dependent protein kinases, which then phosphorylate the Rb protein. The phosphorylated Rb protein is released from E2F. The free form of E2F can activate target genes needed to progress through the cell cycle.

the kinases, which then leads to the phosphorylation of the Rb protein. The phosphorylated form of the Rb protein is released from E2F, thereby allowing E2F to activate genes needed to progress through the cell cycle. What happens when both copies of the *rb* gene are rendered inactive by mutation? The answer is that the E2F protein is always active. This explains why uncontrolled cell division will occur.

The Vertebrate *p53* Gene Is a Master Tumor-Suppressor Gene That Senses DNA Damage

After the *rb* gene, the second tumor-suppressor gene discovered was the *p53* gene. The *p53* gene is the most commonly altered gene in human cancers. About 50% of all human cancers are associated with defects in *p53*. These include malignant tumors of the lung, breast, esophagus, liver, bladder, and brain as well as sarcomas, lymphomas, and leukemias. For this reason, an enormous amount of research has been aimed at elucidating the functions of the p53 protein.

A primary role of the p53 protein is to determine if a cell has incurred DNA damage. If damage is detected, p53 can promote three types of cellular pathways aimed at preventing the proliferation of cells with damaged DNA. First, when confronted with DNA damage, the cell can try to repair its DNA. This may prevent the accumulation of mutations that activate oncogenes or inactivate tumor-suppressor genes. Second, if a cell is in the process of dividing, it can arrest itself in the cell cycle. By stopping the cell cycle, a cell has more time to repair its DNA and avoid producing two mutant daughter cells. For this to happen, p53 stimulates the expression of another protein termed p21. The p21 protein inhibits cyclin/CDK protein complexes that are needed to progress from the G_1 phase of the cell cycle to the S phase.

The third, and most drastic event, is that a cell can initiate a series of events called **apoptosis,** or programmed cell death. In response to DNA-damaging agents, a cell may self-destruct. Apoptosis is an active process that involves cell shrinkage, chromatin condensation, and DNA degradation. This process is facilitated by proteases known as **caspases.** These types of proteases are sometimes called the "executioners" of the cell. Caspases digest selected cellular proteins such as microfilaments, which are components of the intracellular cytoskeleton. This causes the cell to break down into small vesicles that are eventually phagocytized by cells of the immune system. Apoptosis occurs in some cells as a normal process of embryonic development (see Chapter 23). In addition, it is an important way by which an adult organism can eliminate cells with cancer-causing potential.

Figure 22.15 summarizes how *p53* plays a central role in all three processes. As shown here, the expression of the *p53* gene is caused by the formation of damaged DNA. The inducing signal appears to be double-strand DNA breaks. The p53 protein functions as a transcription factor. It contains a DNA-binding domain and a transcriptional activation domain. Experimental studies have shown that p53 can activate the transcription of several specific target genes. As shown in Figure 22.15, it can activate genes that promote DNA repair, arrest the cell cycle, and

Environmental mutagen causes DNA damage, such as double-strand breaks.

Mutagen

DNA break

p53

Induction of the *p53* gene by double-strand breaks leads to the synthesis of the p53 protein, which functions as a transcription factor. This transcription factor can:

1. Activate genes that promote DNA repair.

2. Activate genes that arrest cell division and may generally repress other genes that are required for cell division.

3. Activate genes that promote apoptosis.

FIGURE 22.15   Central role of *p53* in preventing the proliferation of cancer cells. Expression of the *p53* gene, which encodes a transcription factor, is induced by double-strand DNA breaks. The p53 transcription factor may activate genes that promote DNA repair, activate genes that arrest cell division, repress genes required for cell division, and activate genes that promote apoptosis.

promote apoptosis. In addition, p53 appears to act as a negative regulator by interacting with general transcription factors. This decreases the general expression of many other structural genes. This inhibition may also help prevent the cell from dividing.

Overall, p53 activates the expression of a few specific cellular genes and, at the same time, inhibits the expression of many other genes in the cell. Through the regulation of gene transcription, the p53 protein can prevent the proliferation of cells that have incurred DNA damage.

Tumor-Suppressor Genes Can No Longer Inhibit Cancer When Their Function Is Lost

During the past three decades, researchers have identified many tumor-suppressor genes that, when defective, contribute to the development and progression of cancer (**Table 22.9**). What are

TABLE 22.9

Functions of Selected Tumor-Suppressor Genes

Gene	Function
Genes that negatively regulate cell division	
rb	The Rb protein is a negative regulator of E2F (see Figure 22.14). The inhibition of E2F prevents the transcription of certain genes required for DNA replication and cell division.
p16	A protein kinase that negatively regulates cyclin-dependent kinases. This protein controls the transition from the G1 phase of the cell cycle to the S phase.
NF1	The NF1 protein stimulates Ras to hydrolyze its GTP to GDP. Loss of NF1 function causes the Ras protein to be overactive, which promotes cell division.
APC	A negative regulator of a cell-signaling pathway called the Wnt pathway. The Wnt pathway leads to the activation of genes that promote cell division.
VHL	VHL protein functions in the targeting of certain proteins for (ubiquitin-mediated) protein degradation. When these proteins are not degraded, the cell cycle may proceed in an unregulated manner.
MTS1	The MTS1 protein acts as an inhibitor of cyclin-dependent protein kinases. The activities of these protein kinases promote the progression through the cell cycle.
Genes that maintain genome integrity	
p53	p53 is a transcription factor that is a checkpoint protein that positively regulates a few specific target genes and negatively regulates others in a general manner. It acts as a sensor of DNA damage. It can prevent the progression through the cell cycle and also can promote apoptosis.
BRCA-1, BRCA-2	BRCA1 and BRCA2 proteins are both involved in the cellular defense against DNA damage. These proteins facilitate DNA repair and act as checkpoint proteins.

the general functions of these genes? They tend to fall into two broad categories—genes that negatively regulate cell division or genes that maintain genome integrity.

Some tumor-suppressor genes encode proteins that have direct effects on the regulation of cell division. An example is the *rb* gene. As mentioned earlier, the Rb protein negatively regulates E2F. If both copies of the *rb* gene are inactivated, the growth of cells will be accelerated. Therefore, loss of function of these kinds of negative regulators has a direct impact on the abnormal cell division rates seen in cancer cells. In other words, when the Rb protein is lost, a cell becomes more likely to divide.

Alternatively, other tumor-suppressor genes play a role in the proper maintenance of the integrity of the genome. The term **genome maintenance** refers to cellular mechanisms that either prevent mutations from occurring and/or prevent mutant cells from surviving or dividing. The proteins encoded by such genes help to ensure that gene mutations or changes in chromosome structure and number do not occur and are not transmitted to daughter cells. The proteins that participate in genome maintenance can be subdivided into two classes: checkpoint proteins and those involved directly with DNA repair.

Checkpoint proteins are vital for the detection of abnormalities such as DNA breaks and improperly segregated chromosomes. When such abnormalities are detected, the proteins participate in regulatory pathways that prevent cell division. These proteins are called checkpoint proteins because their role is to check the integrity of the genome and prevent cells from progressing past a certain point in the cell cycle if genetic abnormalities are detected.

Figure 22.16 shows a simplified diagram of cell cycle control. Proteins called cyclins and cyclin-dependent protein kinases (Cdks) are responsible for advancing a cell through the four phases of the cell cycle. For example, an activated G_1-cyclin/Cdk complex is necessary to advance from the G_1 phase to the S phase. Human cells produce several types of cyclins and Cdks. The formation of activated cyclin/Cdk complexes is regulated by a variety of factors. Earlier we considered how growth factor signaling pathways can stimulate cell division (see Figure 22.11b). These pathways operate, in part, by regulating cyclins and Cdks. In addition, checkpoint proteins monitor the state of the cell and stop the progression through the cell cycle if abnormalities are detected. Checkpoint proteins also may exert their effects by regulating the activities of cyclins and Cdks.

Several checkpoint proteins regulate the cell cycle of human cells. Figure 22.16 shows three of the major checkpoints where these proteins exert their effects. Both the G_1 and G_2 checkpoints involve the functions of proteins that can sense if the DNA has incurred damage. If so, these checkpoint proteins, such as p53, can prevent the formation of active cyclin/Cdk complexes. This would stop the progression of the cell cycle. A checkpoint also exists in metaphase. This checkpoint is monitored by proteins that can sense if a chromosome is not correctly attached to the spindle apparatus, making it likely that it will be improperly segregated.

Overall, checkpoint proteins prevent the division of cells that may have incurred DNA damage or harbor abnormalities in chromosome attachment. This provides a mechanism to stop the accumulation of genetic abnormalities that could produce cancer cells within the body. When checkpoint genes are lost, cell division may not be directly accelerated. However, the loss of checkpoint protein function makes it more likely that undesirable genetic changes will occur that could cause cancerous growth.

A second class of proteins involved with genome maintenance consists of DNA repair enzymes, which were discussed in Chapter 16 (refer back to Table 16.6). The genes encoding such enzymes are inactivated in certain forms of cancer. The loss of a DNA repair enzyme makes it more likely for a cell to accumulate mutations that could create an oncogene and/or eliminate the function of a tumor-suppressor gene. In Chapter 16, we considered how defects in the nucleotide excision repair process are responsible for the disease called xeroderma pigmentosum, which results in a predisposition to developing skin cancer (refer back to Figure 16.20). As discussed later in this chapter, defects in DNA mismatch repair enzymes can contribute to colorectal cancer. In these cases, the loss of a DNA repair enzyme contributes to a higher mutation rate, which makes it more likely for other genes to incur cancer-causing mutations.

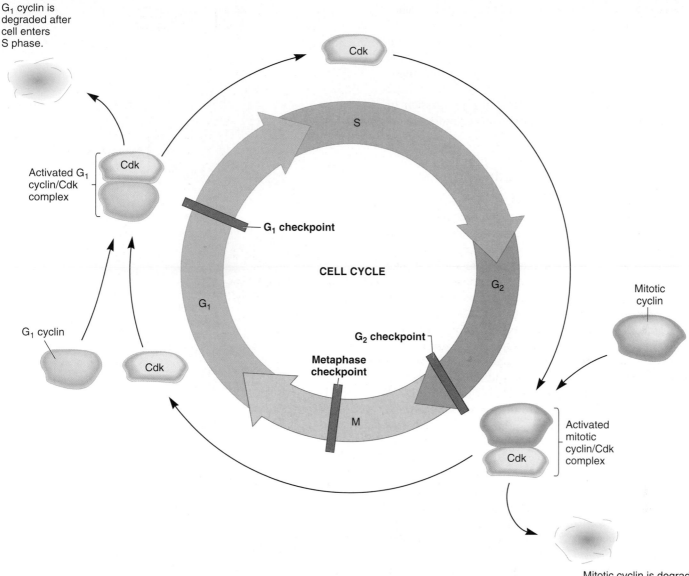

G₁ cyclin is
degraded after
cell enters
S phase.

Activated G₁
cyclin/Cdk
complex

Cdk

CELL CYCLE

S

G₂

G₁

M

G₁ checkpoint

G₂ checkpoint

**Metaphase
checkpoint**

G₁ cyclin

Cdk

Mitotic
cyclin

Activated
mitotic
cyclin/Cdk
complex

Cdk

Mitotic cyclin is degraded as cell
progresses through mitosis.

FIGURE 22.16 **Cell cycle control.** Progression through the cell cycle requires the formation of activated cyclin/Cdk complexes. A cell may produce several different types of cyclin proteins, which are typically degraded after the cell has progressed to the next phase. The formation of activated cyclin/Cdk complexes is regulated by checkpoint proteins. If these proteins detect DNA damage, they will prevent the formation of activated cyclin/Cdk complexes. In addition, other checkpoint proteins can detect chromosomes that are incorrectly attached to the spindle apparatus and prevent further progression through metaphase. Note that this is a simplified diagram of the cell cycle of humans.

Tumor-Suppressor Genes Can Be Silenced in a Variety of Ways

Thus far, we have considered the functions of proteins that are encoded by tumor-suppressor genes. Cancer biologists would also like to understand how tumor-suppressor genes are inactivated, because this knowledge may aid in the prevention of cancer. Researchers have identified three common ways that the function of tumor-suppressor genes can be lost. First, a mutation can occur specifically within a tumor-suppressor gene to inactivate its function. For example, a mutation could inactivate

the promoter of a tumor-suppressor gene or introduce an early stop codon in the coding sequence. Either of these would prevent the expression of a functional protein. A second way that tumor-suppressor genes are inhibited is via DNA methylation. As discussed in Chapter 15, DNA methylation usually inhibits the transcription of eukaryotic genes, particularly when it occurs in the vicinity of the promoter. The methylation of CpG islands near the promoters of tumor-suppressor genes has been found in many types of tumors, suggesting that this form of gene inactivation plays an important role in the formation and/or progression of malignancy. However, further research will be needed to determine why tumor-suppressor genes are aberrantly methylated in cancer cells. Third, many types of cancer are associated with aneuploidy. As discussed in Chapter 8, aneuploidy involves the loss or addition of one or more chromosomes, so the total number of chromosomes is not an even multiple of a set. In some cases, chromosome loss may contribute to the progression of cancer because the lost chromosome carries one or more tumor-suppressor genes.

Most Forms of Cancer Involve Multiple Genetic Changes Leading to Malignancy

The discovery of oncogenes and tumor-suppressor genes, along with molecular techniques that can detect genetic alterations, has enabled researchers to study the progression of certain forms of cancer at the molecular level. Many cancers begin with a benign genetic alteration that, over time and with additional mutations, progresses eventually to malignancy. Furthermore, a malignancy can continue to accumulate genetic changes that make it even more difficult to treat. For example, some tumors may acquire mutations that cause them to be resistant to chemotherapeutic agents.

In 1990, Eric Fearon and Bert Vogelstein proposed a series of genetic changes that lead to colorectal cancer, the second most common cancer in the United States. As shown in **Figure 22.17**, colorectal cancer is derived from cells in the mucosa of the colon. The loss of function of *APC*, a tumor-suppressor gene on chromosome 5, leads to an increased proliferation of mucosal cells and the development of a benign polyp, a noncancerous growth. Additional genetic changes involving the loss of other tumor-suppressor genes and the activation of an oncogene (namely, *ras*) lead eventually to the development of a carcinoma. In Figure 22.17, the genetic changes that lead to colon cancer are portrayed as occurring in an orderly sequence. While the growth of a tumor often begins with mutations in *APC*, it is the total number of genetic changes, not their exact order, that is important. A key issue is that multiple mutations are needed to disable checkpoints.

Among different types of tumors, researchers have identified a large number of genes that are mutated in cancer cells. Though not all of these mutant genes have been directly shown to affect the growth rate of cells, such mutations are likely to be found in tumors because they provide some type of growth advantage for the cell population from which the cancer developed. For example, certain mutations may enable cells to metastasize to neighboring locations. These mutations may not affect growth rate, but they provide the growth advantage that cancer cells are not limited to growing in a particular location, but can migrate to new locations.

Researchers have estimated that about 300 different genes may play a role in the development of human cancer. With an approximate genome size of 20,000 to 25,000 genes, this observation indicates that over 1% of our genes have the potential to promote cancer if their function is altered by a mutation.

In addition to mutations within specific genes, another common genetic change associated with cancer are abnormalities in chromosome structure and number. **Figure 22.18** compares the

FIGURE 22.17   **Multiple genetic changes leading to colorectal cancer.**

FIGURE 22.18 **A comparison between chromosomes found in a normal human cell and a cancer cell from the same person.** The bottom set found in a cancer cell is highly abnormal, with extra copies of some chromosomes and lost copies of others. Chromosomes made of fused pieces of chromosomes (designated mar in this figure) are also common in cancer cells.

chromosome composition of a normal male cell and a tumor cell taken from the same person. The normal composition for this person is 22 pairs of chromosomes plus two sex chromosomes (X and Y). By comparison, the chromosome composition of the tumor cell is quite bizarre, including the fact that the tumor cell has two X chromosomes, which is characteristic of females. There are many cases where chromosomes are missing. If tumor-suppressor genes were on these missing chromosomes, their function is lost as well. Figure 22.18 also shows a few cases of extra chromosomes. If these chromosomes contain proto-oncogenes, the expression of those genes may be overactive. Finally, tumor cells often contain chromosomes that have translocations. Such translocations may create fused genes (as in the case of the Philadelphia chromosome discussed earlier in this chapter), or they may place two genes close together so that the regulatory sequences of one gene affect the expression of the other gene.

DNA Microarrays Are Used to Classify Tumors

As we have just seen, cancer cells are usually the result of multiple genetic alterations that cause the activation of oncogenes and the loss of function of tumor suppressor genes. Therefore, each type of tumor is characterized by a particular set of gene and chromosome alterations. Traditionally, different types of tumors have not been identified on the basis of genetic changes but instead have been classified according to their appearance under a microscope. While this approach is useful, a major drawback is that two tumors may have a very similar microscopic appearance but yet have very different underlying genetic changes and clinical outcomes. For this reason, researchers and clinicians are turning to methods that enable them to understand the molecular changes that occur in diseases such as cancer. This general approach is called **molecular profiling.**

In cancer biology, molecular profiling involves the identification of the genes that play a role in the development of cancer. Why is this useful? First, molecular profiling can distinguish between tumors that look very similar under the microscope. Second, the classification of different tumors via molecular tools may help to predict the clinical outcome. Some tumors respond well to certain treatments, while others do not. Molecular profiling can be very effective in distinguishing such differences. Finally, researchers are optimistic that molecular profiling may lead to improved treatment options. As we gain a better understanding of the genetic changes associated with particular types of cancers, researchers may be able to develop drugs that specifically target the proteins that are encoded by cancer-causing gene mutations. As discussed earlier, the drug imatinib mesylate, which is used to treat chronic myelogenous leukemia, was developed in this way.

DNA microarrays, which are described in Chapter 21, are increasingly used as a tool in the molecular profiling of tumors. The goal is to identify those genes whose pattern of expression correlates with each other—an approach called cluster analysis (see Figure 21.3). In the study of cancer, researchers can compare cancer cells to normal cells and identify groups (clusters) of genes that are turned on in the cancer cells and off in the normal cells, and other groups of genes that are turned off in the cancer cells and on in the normal cells. Likewise, researchers can compare two different types of tumors and identify groups of genes that show different patterns of expression.

As an example, **Figure 22.19a** shows a computer-generated image that illustrates the results of a microarray analysis of 47 samples, most of which came from the tumors of patients with a type of cancer called diffuse large B-cell lymphoma (DLBCL). Each column represents the expression pattern of a set of genes from a particular sample. Genes that are expressed are shown in red; those that are not expressed are shown in green. During the course of these studies, the researchers identified two different patterns of gene expression. The tumor samples on the left side showed a set of genes (next to the orange bar) that tended to be turned on in the tumor and another set of genes (next to the blue bar) that tended to be turned off in the tumor. This pattern of gene expression was similar to the pattern found in a type of B cell called germinal center B cells. In contrast, the tumors

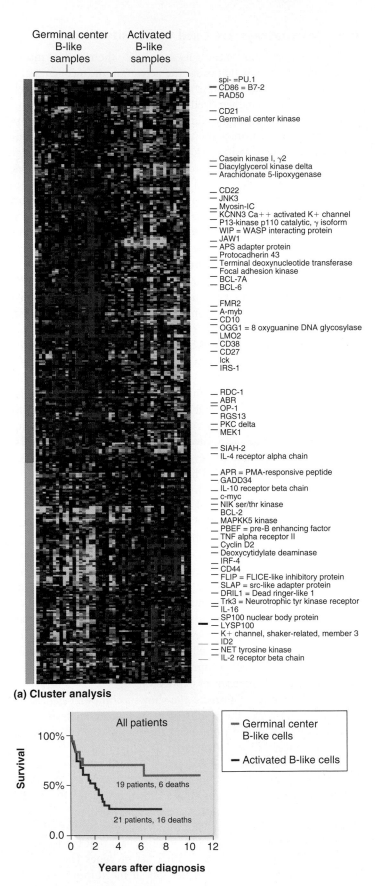

Germinal center B-like samples   Activated B-like samples

spi- =PU.1
CD86 = B7-2
RAD50

CD21
Germinal center kinase

Casein kinase I, γ2
Diacylglycerol kinase delta
Arachidonate 5-lipoxygenase

CD22
JNK3
Myosin-IC
KCNN3 Ca++ activated K+ channel
P13-kinase p110 catalytic, γ isoform
WIP = WASP interacting protein
JAW1
APS adapter protein
Protocadherin 43
Terminal deoxynucleotide transferase
Focal adhesion kinase
BCL-7A
BCL-6

FMR2
A-myb
CD10
OGG1 = 8 oxyguanine DNA glycosylase
LMO2
CD38
CD27
lck
IRS-1

RDC-1
ABR
OP-1
RGS13
PKC delta
MEK1

SIAH-2
IL-4 receptor alpha chain

APR = PMA-responsive peptide
GADD34
IL-10 receptor beta chain
c-myc
NIK ser/thr kinase
BCL-2
MAPKK5 kinase
PBEF = pre-B enhancing factor
TNF alpha receptor II
Cyclin D2
Deoxycytidylate deaminase
IRF-4
CD44
FLIP = FLICE-like inhibitory protein
SLAP = src-like adapter protein
DRIL1 = Dead ringer-like 1
Trk3 = Neurotrophic tyr kinase receptor
IL-16
SP100 nuclear body protein
LYSP100
K+ channel, shaker-related, member 3
ID2
NET tyrosine kinase
IL-2 receptor beta chain

(a) Cluster analysis

All patients

— Germinal center B-like cells
— Activated B-like cells

Survival

100%

50%

0.0

19 patients, 6 deaths

21 patients, 16 deaths

0   2   4   6   8   10   12
Years after diagnosis

(b) Patient outcomes

FIGURE 22.19   The use of DNA microarrays to classify types of tumors. (a) 47 samples, mostly from patients with DLBCL, were subjected to a DNA microarray analysis. The DNA microarray data were then subjected to a cluster analysis to identify genes that are coordinately expressed. The figure shown here is a graphical illustration of a cluster analysis. Each column represents one sample; each row represents the expression of a particular gene. The names of some of the genes are shown along the right side. (Note: The rows and columns are not easily resolved in this illustration). Genes highly expressed are shown in red, while those not expressed are shown in green. One group of samples had an expression pattern similar to that found in germinal center B cells; the other group had an expression pattern typical of activated B cells in the peripheral blood. **(b)** Survival of patients with DLBCL. (a. Reprinted by permission from Macmillan Publishers Ltd. *Nature*. Distinct types of diffuse large B-cell lymphoma identified by gene expression profiling. Ash Alizadeh et al. 403:6769, 503–511, 2000. Image courtesy of Ash Alizadeh.)

on the right side showed the opposite pattern. The upper genes tended to be turned off in these patients, while the lower genes were turned on. These samples showed a gene expression pattern found in normal activated peripheral blood B cells. These results suggest that the two groups of tumors may have originated in B cells at different stages of development—those on the left originated in germinal center B cells, while those on the right originated in activated B cells. Furthermore, the patients from whom these tumors were derived also appeared to have very different clinical outcomes (**Figure 22.19b**). The patients whose tumors had a pattern of gene expression similar to activated B cells had a significantly lower overall survival compared to the other patients.

Inherited Forms of Cancers May Be Caused by Defects in Tumor-Suppressor Genes and DNA Repair Genes

Before we end our discussion of the genetic basis of cancer, let's consider which genes are most likely to be affected in inherited forms of the disease. As mentioned previously, about 5 to 10% of all cases of cancer involve inherited (germ-line) mutations. These familial forms of cancer occur because people have inherited mutations from one or both parents that give them an increased susceptibility to develop cancer. This does not mean they will definitely get cancer, but they are more likely to develop the disease than are individuals in the general population. When individuals have family members who have developed certain forms of cancer, they may be tested to determine if they also carry a mutant gene. For example, von Hippel-Lindau disease and familial adenomatous polyposis are examples of syndromes for which genetic testing to identify at-risk family members is considered the standard of care.

What types of genes are mutant in familial cancers? Most inherited forms of cancer involve a defect in a tumor-suppressor gene (**Table 22.10**). In these cases, the individual is heterozygous, with one normal and one inactive allele. A mutation in the remaining normal allele may occur in a somatic cell, thereby producing a cell with two inactive copies of the

TABLE **22.10**

Inherited Mutant Genes That Confer a Predisposition to Develop Cancer

Gene	Type of Cancer*
Tumor-Suppressor Genes**	
VHL	Causes von Hippel-Lindau disease, which is typically characterized by a clear cell renal carcinoma
APC	Familial adenomatous polyposis and familial colon cancer
rb	Retinoblastoma
p53	Li-Fraumeni syndrome, which is characterized by a wide spectrum of tumors including soft-tissue and bone sarcomas, brain tumors, adenocortical tumors, and premenopausal breast cancers
Pten	Causes Cowden syndrome, which is characterized by the presence of noncancerous, tumor-like growths called hamartomas in the skin, breast, thyroid, gastrointestinal tract, and central nervous system and an increased risk of breast and thyroid carcinomas
BRCA-1	Familial breast cancer
BRCA-2	Familial breast cancer
WT1	Wilm's tumor, which is a nephroblastoma
NF1	Neurofibromatosis
MTS1	Hereditary malignant melanoma
MSH2	Nonpolyposis colorectal cancer
MLH1	Nonpolyposis colorectal cancer
XP-A to XP-G	UV-sensitive forms of cancer such as basal cell carcinoma
Oncogenes	
RET	Multiple endocrine neoplasia type 2

*Many of the genes described in this table are mutated in more than one type of cancer. The cancers listed are those in which it has been firmly established that a predisposition to develop the disease is commonly due to germ-line mutations in the designated gene.

**MSH2, MLH1, and XP-A to XP-G encode proteins that are involved in DNA repair. Some geneticists put these in a separate category from tumor-suppressor genes. However, in this textbook, they are considered tumor-suppressor genes because their loss of function promotes the development of cancer, as shown by the examples in this table.

tumor-suppressor gene. At the phenotypic level, a predisposition for developing cancer is inherited in a dominant fashion because a heterozygote exhibits this predisposition. However, the actual development of cancer is recessive because a second somatic mutation that results in two abnormal alleles (in the same cell) is necessary to promote malignancy. As noted in Table 22.10, not all hereditary forms of cancer are due to defective tumor-suppressor genes. For example, multiple endocrine neoplasia type 2 is due to the activation of an oncogene.

CONCEPTUAL SUMMARY

Throughout this chapter, we have examined many human diseases that have a genetic basis. To discover whether a disease has a genetic basis, researchers rely on pedigree analysis and the identification of altered genes or chromosomes in those who exhibit the disorder. Through pedigree analysis, geneticists have discovered that some diseases are inherited in an autosomal recessive manner, some are autosomal dominant, and some are X linked. Many genetic disorders, such as hemophilia, are characterized by locus heterogeneity, meaning the disease is caused by mutations in two or more different genes. Researchers use mapping and gene sequencing to identify mutant genes and their location on the chromosome. Once a gene is known, genetic testing can determine whether an individual possesses disease-causing alleles.

Cancer is also a human disease with a genetic basis that is characterized by uncontrolled cell division. Cancer can be caused by the inheritance of mutant genes, by spontaneous mutations, and by oncogenic viruses, such as RSV and other acutely transforming viruses (ACVs). However, most forms of cancer are due to environmentally induced mutations in somatic cells.

Cancer-causing mutations may involve two types of genes: oncogenes and tumor-suppressor genes. An oncogene, which is derived from a normal, unmutated proto-oncogene, is an abnormally activated gene that may stimulate cell growth. At the molecular level, a proto-oncogene may be converted to an oncogene by a missense mutation, gene amplification, chromosomal translocation, or viral integration. By comparison, a tumor-suppressor gene normally inhibits cell growth or maintains genome integrity. If tumor-suppressor genes are rendered inactive, cancerous growth may ensue. A master tumor-suppressor gene in vertebrates, called *p53*, plays a critical role in determining if a cell has incurred DNA damage and preventing cell division or eliminating cells that have been damaged. Tumor-suppressor genes can be silenced in many ways, including mutations within the gene, DNA methylation, or chromosomal loss.

EXPERIMENTAL SUMMARY

While researchers cannot experiment with humans as they can with model organisms such as *Drosophila* and mice, they can gain knowledge of human genetic disorders by conducting pedigree analysis and in vitro tests on human cell samples. Often, when a human disease is newly identified, researchers would like to know if it is caused, wholly or in part, by a mutation in a gene. Various observations point to a genetic cause. These include a higher frequency of affected individuals among family members and in identical versus fraternal twins, the inability of the disease to be spread by contact, different rates of the disease among different populations, a characteristic age of onset, a similar disease of known genetic origin in animals, and a correlation between a disease and a mutant gene.

Once genetics has become established as a cause of disease, researchers attempt to determine the pattern of transmission from parents to offspring. Many genetic diseases follow a simple Mendelian pattern of inheritance, and pedigree analysis may indicate whether a disease is transmitted in a recessive, dominant, or X-linked manner. Geneticists also use mapping and DNA sequencing to identify disease-causing genes. Genetic testing and screening involves the use of testing methods to discover if individuals or populations carry genetic abnormalities. Two forms—amniocentesis and chorionic villus sampling—can be done prior to birth.

Research involving the Rous sarcoma virus led to identification of the first oncogene, a gene that promotes cancerous cell growth. Experiments showed that DNA isolated from malignant cells can transform normal cells into cancerous cells. This approach paved the way for the identification of many human cellular oncogenes. The first identification of a human tumor-suppressor gene resulted from studies of retinoblastoma. For this form of cancer, Knudson proposed that cancer required two hits to occur. Research subsequently discovered that tumor-suppressor genes function to suppress the proliferation of cancer cells and can be silenced in a variety of ways. While tumors have traditionally been classified by their appearance under a microscope, newer approaches such as DNA microarrays are helping scientists profile the molecular changes that occur in cancer.

PROBLEM SETS & INSIGHTS

Solved Problems

S1. The pedigree shown here concerns a human disease known as familial hypercholesterolemia.

This disorder is characterized by an elevation of serum cholesterol in the blood. Though relatively rare, this genetic abnormality can

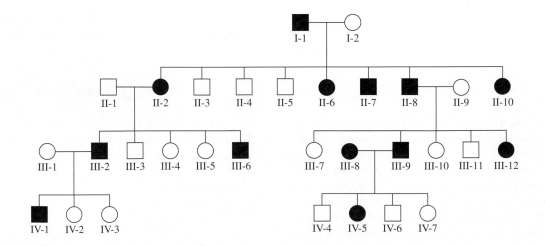

be a contributing factor to heart attacks. At the molecular level, this disease is caused by a defective gene that encodes a protein called the low-density lipoprotein receptor (LDLR). In the bloodstream, serum cholesterol is bound to a carrier protein known as low-density lipoprotein (LDL). LDL binds to LDLR so that cells can absorb cholesterol. When LDLR is defective, it becomes more difficult for the cells to absorb cholesterol. This explains why the levels of LDL blood cholesterol remain high. Based on the pedigree, what is the most likely pattern of inheritance of this disorder?

Answer: The pedigree is consistent with a dominant pattern of inheritance. An affected individual always has an affected parent. Also, individuals III-8 and III-9, who are both affected, produced unaffected offspring. If this trait was recessive, two affected parents should always produce affected offspring. However, because the trait is dominant, two heterozygous parents can produce homozygous unaffected offspring. The ability of two affected parents to have unaffected offspring is a striking characteristic of dominant inheritance. On average, we would expect that two heterozygous parents should produce 25% unaffected offspring. In the family containing IV-4, IV-5, IV-6, and IV-7, three out of four offspring are actually unaffected. This higher-than-expected proportion of unaffected offspring is not too surprising because the family is a very small group and may deviate substantially from the expected value due to random sampling error.

S2. One way to identify a human cellular oncogene is to use human DNA from a malignant cell to transform a mouse cell. A mouse cell that has been transformed by human DNA will have a human oncogene incorporated into its genome; it also is likely to have *Alu* sequences that are closely linked to this human oncogene. (Note: As discussed in Chapter 10, the human genome contains *Alu* sequences interspersed every 5,000 to 6,000 base pairs. *Alu* sequences are not found in the mouse genome.) Discuss how the *Alu* sequence can provide a way to clone human oncogenes.

Answer: One approach is transposon tagging, described in Chapter 17. When a mouse cell is transformed with a human DNA fragment containing an oncogene, that fragment is likely to contain an *Alu* sequence as well. To clone the human oncogene, the chromosomal DNA can be isolated from the transformed mouse cells, digested with a restriction enzyme, and cloned into vectors to create a library of DNA fragments. The members of the library that carry the human oncogene can be identified using a probe complementary to the *Alu* sequence because this sequence is not found in the mouse genome. Using this strategy, researchers have identified several human cellular oncogenes. In human bladder carcinoma, for example, the human cellular oncogene called *ras* was identified this way.

S3. Oncogenes sometimes result from genetic rearrangements (e.g., translocations) that produce gene fusions. An example is the Philadelphia chromosome, in which a reciprocal translocation between chromosomes 9 and 22 leads to fusion of the first part of the *bcr* gene with the *abl* gene. Suggest two different reasons why a gene fusion could create an oncogene.

Answer: An oncogene is derived from a genetic change that abnormally activates the expression of a gene that plays a role in cell division. When a genetic change creates a gene fusion, this can abnormally activate the expression of the gene in two ways.

The first way is at the level of transcription. The promoter and part of the coding sequence of one gene may become fused with the coding sequence of another gene. For example, the promoter and part of the coding sequence of the *bcr* gene may fuse with the coding sequence of the *abl* gene. After this has occurred, the *abl* gene is now under the control of the *bcr* promoter, rather than its own normal promoter. Because the *bcr* promoter is turned on in different cells compared to the *abl* promoter, this leads to the overexpression of the abl protein in certain cell types compared to its normal level of expression.

A second way that a gene fusion can cause abnormal activation is at the level of protein structure. A fusion protein has parts of two different polypeptides. The first portion of a fusion protein may affect the structure of the second portion of the polypeptide in such a way that the second portion becomes abnormally active, or vice versa.

Conceptual Questions

C1. With regard to pedigree analysis, make a list of the patterns that distinguish recessive, dominant, and X-linked genetic diseases from each other.

C2. Explain, at the molecular level, why human genetic diseases often follow a simple Mendelian pattern of inheritance, whereas most normal traits, such as the shape of your nose or the size of your head, are governed by multiple gene interactions.

C3. Many genetic disorders exhibit locus heterogeneity. Define and give two examples of locus heterogeneity. How does locus heterogeneity confound a pedigree analysis?

C4. In general, why do changes in chromosome structure and/or number tend to affect an individual's phenotype? Explain why some changes in chromosome structure, such as reciprocal translocations, do not.

C5. We often speak of diseases such as phenylketonuria (PKU) and achondroplasia as having a "genetic basis." Explain whether the following statements are accurate with regard to the genetic basis of any human disease (not just PKU and achondroplasia).

A. An individual must inherit two copies of a mutant allele to have disease symptoms.

B. A genetic predisposition means that an individual has inherited one or more alleles that make it more likely for them to develop disease symptoms compared to other individuals.

C. A genetic predisposition to develop a disease may be passed from parents to offspring.

D. The genetic basis for a disease is always more important than the environment.

C6. Figure 22.1 illustrates albinism in different species. Describe two other genetic disorders found in both humans and animals.

C7. Discuss why a genetic disease might have a particular age of onset. Would an infectious disease have an age of onset? Explain why or why not.

C8. Gaucher disease (type I) is due to a defect in a gene that encodes a protein called acid β glucosidase. This enzyme plays a role in carbohydrate metabolism within the lysosome. The gene is located on the long arm of chromosome 1. Persons who inherit two defective copies of this gene exhibit Gaucher disease, the major symptoms of which include an enlarged spleen, bone lesions, and changes in skin pigmentation. Let's suppose a phenotypically unaffected woman, whose father had Gaucher disease, has a child with a phenotypically unaffected man, whose mother had Gaucher disease.

A. What is the probability that this child will have the disease?

B. What is the probability that this child will have two normal copies of this gene?

C. If this couple has five children, what is the probability that one of them will have Gaucher disease and four will be phenotypically unaffected?

C9. Ehler-Danlos syndrome is a relatively rare disorder due to a mutation in a gene that encodes a protein called collagen (type 3 A1). Collagen is a protein found in the extracellular matrix that plays an important role in the formation of skin, joints, and other connective tissues. Persons with this syndrome have extraordinarily flexible skin and very loose joints. The pedigree shown here contains several members affected with Ehler-Danlos syndrome, shown with black symbols. Based on this pedigree, does this syndrome appear to be an autosomal recessive, autosomal dominant, X-linked recessive, or X-linked dominant trait? Explain your reasoning.

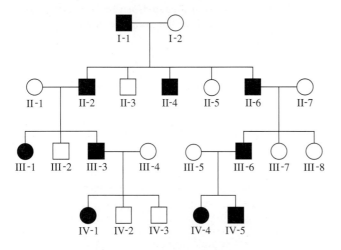

C10. Hurler syndrome is due to a mutation in a gene that encodes a protein called α-L-iduronidase. This protein functions within the lysosome as an enzyme that breaks down mucopolysaccharides (a type of polysaccharide that has many acidic groups attached). When this enzyme is defective, excessive amounts of the mucopolysaccharides dermatan sulfate and heparin sulfate accumulate within the lysosomes, especially in liver cells and connective tissue cells. This leads to symptoms such as an enlarged liver and spleen, bone abnormalities, corneal clouding, heart problems, and severe neurological problems. The pedigree shown here contains three members affected with Hurler syndrome, indicated with black symbols. Based on this pedigree, does this syndrome appear to be an autosomal recessive, autosomal

dominant, X-linked recessive, or X-linked dominant trait? Explain your reasoning.

C11. Like Hurler syndrome, Fabry disease involves an abnormal accumulation of substances within lysosomes. However, the lysosomes of individuals with Fabry disease show an abnormal accumulation of lipids. The defective enzyme is α-galactosidase A, which is a lysosomal enzyme that functions in lipid metabolism. This defect causes cell damage, especially to the kidneys, heart, and eyes. The gene that encodes α-galactosidase A is found on the X chromosome. Let's suppose a phenotypically unaffected couple produces two sons with Fabry disease and one phenotypically unaffected daughter. What is the probability that the daughter will have an affected son?

C12. Achondroplasia is a rare form of dwarfism caused by an autosomal dominant mutation that affects the gene that encodes a fibroblast growth factor receptor. Among 1,422,000 live births, the number of babies born with achondroplasia was 31. Among those 31 babies, 18 of them had one parent with achondroplasia. The remaining babies had two unaffected parents. How do you explain these 13 babies, assuming that the mutant allele has 100% penetrance? What are the odds that these 13 individuals will pass this mutant gene to their offspring?

C13. Lesch-Nyhan syndrome is due to a mutation in a gene that encodes a protein called hypoxanthine-guanine phosphoribosyltransferase (HPRT). HPRT is an enzyme that functions in purine metabolism. Persons afflicted with this syndrome have severe neurodegeneration and loss of motor control. The pedigree shown here contains several members with Lesch-Nyhan syndrome. Affected members are shown with black symbols. Based on this pedigree, does this syndrome appear to be an autosomal recessive, autosomal dominant, X-linked recessive, or X-linked dominant trait? Explain your reasoning.

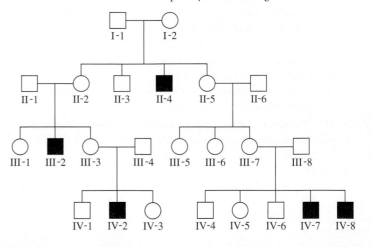

C14. Marfan syndrome is due to a mutation in a gene that encodes a protein called fibrillin-1. It is inherited as a dominant trait. The fibrillin-1 protein is the main constituent of extracellular microfibrils. These microfibrils can exist as individual fibers or associate with a protein called elastin to form elastic fibers. The gene that encodes fibrillin-1 is located on the long arm of chromosome 15. It has been suggested that Abraham Lincoln may have been afflicted with this disorder. Let's suppose a phenotypically unaffected woman has a child with a man who has Marfan syndrome.

A. What is the probability this child will have the disease?

B. If this couple has three children, what is the probability none of them will have Marfan syndrome?

C15. Sandhoff disease is due to a mutation in a gene that encodes a protein called hexosaminidase B. This disease has symptoms that are similar to Tay-Sachs disease. Weakness begins in the first six months of life. Individuals exhibit early blindness and progressive mental and motor deterioration. The pedigree shown here contains several members with Sandhoff disease. Affected members are shown with black symbols.

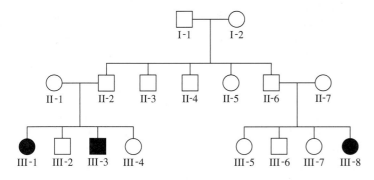

A. Based on this pedigree, does this syndrome appear to be an autosomal recessive, autosomal dominant, X-linked recessive, or X-linked dominant trait? Explain your reasoning.

B. What is the likelihood that II-1, II-2, II-3, II-4, II-5, II-6, and II-7 carry a mutant allele for the hexosaminidase B gene?

C16. What is a prion? Explain how a prion relies on normal cellular proteins to cause a disease such as mad cow disease.

C17. Some people have a genetic predisposition for developing prion diseases. Examples are described in Table 22.5. In the case of Gerstmann-Straussler-Scheinker disease, the age of onset is typically at 30 to 50 years, and the duration of the disease (which leads to death) is about five years. Suggest a possible explanation why someone can live for a relatively long time without symptoms and then succumb to the disease in a relatively short time.

C18. Familial fatal insomnia (described in Table 22.5) is a prion disease inherited as an autosomal dominant trait. Researchers have identified the *PrP* gene, located on the short arm of chromosome 20, and tried to understand the relationship between the *PrP* gene sequence and the molecular basis of familial fatal insomnia.

A. It has been found that a rare mutation at codon 178, changing an aspartic acid to an asparagine, can cause this disease. In addition, codon 129 seems to play a role. The human population is polymorphic at codon 129, which may encode valine or methionine. If codon 178 is the normal aspartic acid (Asp) codon, it does not seem to matter if valine or methionine

is found at position 129. In other words, if codon 178 is aspartic acid, Met-129 and Val-129 codons are not associated with disease symptoms. However, if codon 178 specifies asparagine (Asn), then it does matter. Familial fatal insomnia seems to require an asparagine at codon 178 and a methionine at codon 129 in the *PrP* gene sequence. Although the actual biochemical mechanism is not known, suggest a possible reason why this is the case.

B. Also, researchers have compared the sequences of the *PrP* gene in many people with familial fatal insomnia. Keep in mind that this is a dominant autosomal trait, so people with this disorder have one mutant copy of the *PrP* gene and one normal copy. People with familial fatal insomnia must have one abnormal copy of the gene that contains Asn-178 and Met-129. The second copy of the gene has Asp-178, and it may have Met-129 or Val-129. Some results suggest that people having a Met-129 codon in this second copy of the *PrP* gene causes the disease to develop more rapidly than people who have Val-129 in the second copy. Propose an explanation why disease symptoms may occur more rapidly when Met-129 is found in the second copy of the *PrP* gene.

C19. What is the difference between an oncogene and a tumor-suppressor gene? Give two examples.

C20. What is a proto-oncogene? What are the typical functions of proteins encoded by proto-oncogenes? At the level of protein function, what are the general ways that proto-oncogenes can be converted into oncogenes?

C21. What is a retroviral oncogene? Is it necessary for viral infection and proliferation? How have retroviruses acquired oncogenes?

C22. A genetic predisposition to develop cancer is usually inherited as a dominant trait. At the level of cellular function, are the alleles involved actually dominant? Explain why some individuals who have inherited these dominant alleles do not develop cancer during their lifetimes.

C23. Describe the types of genetic changes that commonly convert a proto-oncogene into an oncogene. Give three examples. Explain how the genetic changes are expected to alter the activity of the gene product.

C24. Relatively few inherited forms of cancer involve the inheritance of mutant oncogenes. Instead, most inherited forms of cancer are defects in tumor-suppressor genes or DNA repair genes. Give two or more reasons why we seldom see inherited forms of cancer involving activated oncogenes.

C25. The *rb* gene encodes a protein that inhibits E2F, a transcription factor that activates several genes involved in cell division. Mutations in *rb* are associated with certain forms of cancer, such as retinoblastoma. Under each of the following conditions, would you expect cancer to occur?

A. One copy of *rb* is defective; both copies of *E2F* are functional.

B. Both copies of *rb* are defective; both copies of *E2F* are functional.

C. Both copies of *rb* are defective; one copy of *E2F* is defective.

D. Both copies of *rb* and E2F are defective.

C26. A *p53* knockout mouse in which both copies of *p53* are defective has been produced by researchers. This type of mouse appears normal at birth. However, it is highly sensitive to UV light. Based on your knowledge of *p53*, explain the normal appearance at birth and the high sensitivity to UV light.

C27. With regard to cancer cells, which of the following statements are true?

A. Cancer cells are clonal, which means they are derived from a single mutant cell.

B. To become cancerous, cells usually accumulate multiple genetic changes that eventually result in uncontrolled growth.

C. Most cancers are caused by oncogenic viruses.

D. Cancer cells have lost the ability to properly regulate cell division.

C28. When the DNA of a human cell becomes damaged, this leads to the activation of the *p53* gene. What is the general function of the p53 protein? Is it an enzyme, transcription factor, cell cycle protein, or something else? Describe three ways that the synthesis of the p53 protein affects cellular function. Why is it beneficial for these three things to happen when a cell's DNA has been damaged?

Experimental Questions

E1. Which of the following experimental observations would suggest that a disease has a genetic basis?

A. The frequency of the disease is less likely in relatives that live apart compared to relatives that live together.

B. The frequency of the disease is unusually high in a small group of genetically related individuals who live in southern Spain.

C. The disease symptoms usually begin around the age of 40.

D. It is more likely that both monozygotic twins will be affected by the disease compared to dizygotic twins.

E2. At the beginning of Chapter 22, we discussed the types of experimental observations that suggest a disease is inherited. Which of these observations do you find the least convincing? Which do you find the most convincing? Explain your answer.

E3. What is meant by the term genetic testing? What is different between testing at the protein level versus testing at the DNA level? Describe five different techniques used in genetic testing.

E4. A particular disease is found in a group of South American Indians. During the 1920s, many of these people migrated to Central America. In the Central American group, the disease is never found. Discuss whether or not you think the disease has a genetic component. What types of further observations would you make?

E5. Chapter 18 describes a blotting method known as Western blotting that can be used to detect a polypeptide that is translated from a particular mRNA. In this method, a particular polypeptide or protein is detected by an antibody that specifically recognizes a segment of its amino acid sequence. After the antibody binds to the polypeptide within a gel, a secondary antibody (which is labeled) is used to visualize the polypeptide as a dark band. For example, an antibody that recognizes α-galactosidase A could be used to specifically detect the amount of α-galactosidase A protein on a gel. The enzyme α-galactosidase A is defective in individuals with Fabry disease, which shows an X-linked recessive pattern of inheritance. Amy, Nan, and Pete are siblings, and Pete has Fabry disease. Aileen, Jason, and Jerry are brothers and sister, and Jerry has Fabry disease. Amy, Nan, and Pete are not related to Aileen, Jason, and Jerry. Amy, Nan, and Aileen are concerned that they could be carriers of a defective α-galactosidase A gene. A sample of cells was obtained from each of these six individuals and subjected to Western blotting, using an antibody against α-galactosidase A.

Samples were also obtained from two unrelated normal females (lanes 7 and 8). The results are shown here.

Samples from:

Lane 1. Amy
Lane 2. Nan
Lane 3. Pete
Lane 4. Aileen
Lane 5. Jason
Lane 6. Jerry
Lane 7. Normal male
Lane 8. Normal female

Note: Due to X inactivation in females, the amount of expression of genes on the single X chromosome in males is equal to the amount of expression from genes on both X chromosomes in females.

A. Explain the type of mutation (i.e., missense, nonsense, promoter, etc.) that causes Fabry disease in Pete and Jerry.

B. What would you tell Amy, Nan, and Aileen regarding the likelihood that they are carriers of the mutant allele and the probability of having affected offspring?

E6. An experimental assay for the blood clotting protein called Factor IX is available. A blood sample was obtained from each member of the pedigree shown here. The amount of Factor IX protein is shown within the symbol of each member and is expressed as a percent of the amount observed in normal individuals who do not carry a mutant copy of the gene.

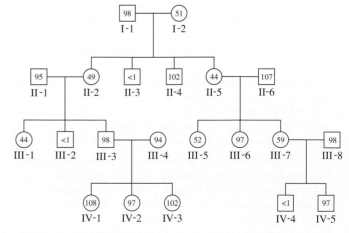

What are the likely genotypes of each member of this pedigree?

E7. Let's suppose a cell line has become malignant because it has accumulated mutations that inactivate two different tumor-suppressor genes. A researcher followed the protocol described in the experiment of Figure 22.10 and isolated DNA from this mutant cell line. The DNA was used to transform normal fibroblast (NIH3T3) cells. What results would you expect? Would you expect a high number of malignant foci or not? Explain your answer.

E8. What is a transformed cell? Describe three different methods to transform cells in a laboratory.

E9. In the experiment of Figure 22.10, what would be the results if the DNA sample had been treated with either DNase, RNase, or protease prior to the treatment with calcium and phosphate ions?

E10. Explain how the experimental study of cancer-causing viruses has increased our understanding of cancer.

E11. Discuss ways to distinguish whether a particular form of cancer involves an inherited predisposition or is due strictly to (postzygotic) somatic mutations. In your answer, consider that only one mutation may be inherited, but the cancer might develop only after several somatic mutations.

E12. The codon change (Gly-12 to Val-12) in human *ras*H that converts it to oncogenic *ras*H has been associated with many types of cancers. For this reason, researchers would like to develop drugs to inhibit oncogenic *ras*H. Based on your molecular understanding of the Ras protein, what types of drugs might you develop? In other words, what would be the structure of the drugs, and how would they inhibit Ras protein? How would you test the efficacy of the drugs? What might be some side effects?

E13. Describe how normal mammalian cells grow on solid media in a laboratory and how cancer cells grow. Is a malignant focus derived from a single cancer cell or from many independent cancer cells that happen to be in the same region of a tissue culture plate?

Questions for Student Discussion/Collaboration

1. Make a list of the benefits that may arise from genetic testing as well as possible negative consequences. Discuss the items on your list.

2. Our government has finite funds to devote to cancer research. Discuss which aspects of cancer biology you would spend the most money pursuing.

   A. Identifying and characterizing oncogenes and tumor-suppressor genes

   B. Identifying agents in our environment that cause cancer

   C. Identifying viruses that cause cancer

   D. Devising methods aimed at killing cancer cells in the body

   E. Informing the public of the risks involved in exposure to carcinogens

   In the long run, which of these areas would you expect to be the most effective in decreasing mortality due to human cancer?

   Note: All answers appear at the website for this textbook; the answers to even-numbered questions are in the back of the textbook.

Visit the Online Learning Center for practice tests, answer keys, and other learning aids for this chapter. Enhance your understanding of genetics with our interactive exercises, quizzes, animations, and much more.

23

DEVELOPMENTAL GENETICS

Multicellular organisms, such as animals and plants, begin their lives with a fairly simple organization, namely, a fertilized egg, and then proceed step by step to a much more complex arrangement. As this occurs, cells divide, migrate, and change their characteristics as they become highly specialized units within a multicellular individual. Each cell in an organism has its own particular role. In animals, for example, muscle cells allow an organism to move, while intestinal cells facilitate the absorption of nutrients. This division of labor among the various cells and organs of the body promotes the survival of the individual.

Developmental genetics is concerned with the roles that genes play in orchestrating the changes that occur during development. In this chapter, we will examine how the sequential actions of genes provide a program for the development of an organism from a fertilized egg to an adult. The last couple of decades have seen staggering advances in our understanding of developmental genetics at the molecular level. Scientists have chosen a few experimental organisms, such as the fruit fly (*Drosophila melanogaster*), nematode (*Caenorhabditis elegans*), mouse (*Mus musculus*), and a flowering plant (*Arabidopsis thaliana*), and worked toward the identification and characterization of the genes required for running their developmental programs. In certain organisms, notably *Drosophila*, most of the genes that play a critical role in the early stages of development have been identified. Researchers are now studying how the proteins encoded by these genes control the course of development. In this chapter, we will explore the body plans of invertebrates, vertebrates, and plants, and consider several examples in which geneticists understand how the actions of genes govern the developmental process.

The fruit fly (*Drosophila melanogaster*), a model organism in the study of developmental genetics.

23.1 INVERTEBRATE DEVELOPMENT

We will begin our discussion of multicellular development by considering two model organisms, *Drosophila melanogaster* and *Caenorhabditis elegans*, that have been pivotal in our understanding of developmental genetics. As we first saw in Chapter 3, *Drosophila* has been a favorite subject of geneticists since 1910, when Thomas Hunt Morgan isolated his first white-eyed mutant. The fruit fly has been used to determine many of the fundamental principles of genetics, including the chromosome theory of inheritance, the random mutation theory, and linkage mapping, to mention only a few.

In developmental genetics, *Drosophila* is also useful for a variety of reasons. First, the techniques for generating and analyzing mutants in this organism are highly advanced, and researchers have identified many mutant strains with altered developmental pathways. Second, at the embryonic and larval stages, *Drosophila* is large enough to conduct transplantation experiments, yet small enough to determine where particular genes are expressed at critical stages of development.

By comparison, *C. elegans* is used by developmental geneticists because of its simplicity. The adult organism is a small transparent worm composed of only about 1,000 somatic cells. Starting with the fertilized egg, the pattern of cell division and the fate of each cell are completely known.

In this section, we will begin by examining the general features of *Drosophila* development. We will then focus our attention on embryonic development (embryogenesis), because it is during this stage that the overall body plan is determined. We will see how the expression of particular genes and the localization of gene products within the embryo influence the developmental process. We will then briefly consider development in *C. elegans*. In this organism, we will examine how the timing of gene expression plays a key role in determining the developmental fate of particular cells.

The Generation of a Body Pattern Depends on the Positional Information That Each Cell Receives During Development

Multicellular development in animals and plants follows a body plan or pattern. The term **pattern** refers to the spatial arrangement of different regions of the body. At the cellular level, the body pattern is due to the arrangement of cells and their specialization.

The progressive growth of a fertilized egg into an adult organism involves four types of cellular events: cell division, cell migration, cell differentiation, and cell death (**Figure 23.1**). The coordination of these four events leads to the formation of a body with a particular pattern. As we will see, the temporal expression of genes and the localization of gene products at precise regions in the fertilized egg and early embryo are the critical phenomena that underlie this coordination.

Before we consider how genes affect development, let's consider a central concept in developmental biology known as **positional information.** For an organism to develop a body pattern with unique morphological and cellular features, each cell of the body must receive signals—positional information—that cause it to become the appropriate cell type based on its position relative to the other cells.

How may a cell respond to positional information? Various responses are possible. For example, positional information may cause a cell or group of cells to divide into two daughter cells or migrate in a particular direction from one region of the embryo to another. In addition, positional information may stimulate a cell to differentiate into a particular cell type. Finally, positional information or other stimuli may cause a cell to die. This process, known as **apoptosis,** is a necessary event during normal devel-

(a) Cell division

(b) Cell migration

Undifferentiated cell — Nerve cell
(c) Cell differentiation

(d) Cell death (apoptosis)

FIGURE 23.1   Four events that may happen to cells during the course of development.

opment to sculpt tissues and organs. For example, in the later stages of development of the retina in *Drosophila*, excess cells are eliminated by apoptosis. During mammalian development, individual fingers and toes are created when embryonic cells that lie between each future finger and toe undergo this type of cell death.

A **morphogen** is a molecule that conveys positional information and promotes developmental changes. It is generally a diffusible molecule that acts in a concentration-dependent manner to influence the developmental fate of a cell. Morphogens provide the positional information that stimulate a cell to divide, migrate, differentiate, or die. For example, when regions of a *Drosophila* embryo are exposed to a high concentration of the morphogen known as Bicoid, they differentiate into structures characteristic of the anterior region of the body. Many such morphogens function as transcription factors that regulate the expression of many genes. This topic will be described in greater detail later. How do morphogens control pattern development? Within an oocyte and during embryonic development, morphogens typically are distributed along a concentration gradient.

In other words, the concentration of a morphogen varies from low to high in different regions of the developing organism. A key feature of morphogens is that they act in a concentration-dependent manner. In a particular concentration range, a morphogen or a combination of two or more different morphogens will restrict a cell into a specific developmental pathway.

During the earliest stages of development, morphogenic gradients are preestablished within the oocyte (**Figure 23.2a**). Following fertilization, the zygote subdivides into many smaller cells. Due to the preestablished gradient of morphogens within the oocyte, these smaller cells have higher or lower concentrations of morphogens, depending on their location in the embryo. In this way, the morphogen gradients in the oocyte can provide positional information that is important in establishing the general polarity of an embryo along two main axes: the antero-posterior axis and the dorso-ventral axis, which are described later.

(a) Asymmetric distribution of morphogens in an oocyte

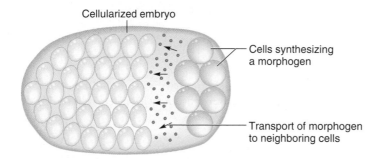

(b) Asymmetric synthesis and extracellular distribution of a morphogen in an embryo

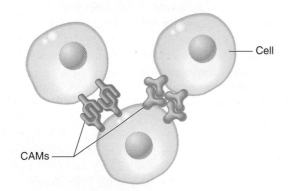

(c) Cell-to-cell contact conveys positional information

FIGURE 23.2   **Three molecular mechanisms of positional information.**

A morphogen gradient can also be established in the embryo by cell secretion and transport. A certain cell or group of cells may synthesize and secrete a morphogen at a specific stage of development. After secretion, the morphogen may be transported to neighboring cells. The concentration of the morphogen is usually highest near the cells that secrete it (**Figure 23.2b**). The morphogen may then influence the developmental fate of the cells exposed to it. The process by which a cell or group of cells governs the developmental fate of neighboring cells is known as **induction.** An example of a secreted morphogen in *Drosophila* is a protein known as Hedgehog. This protein is secreted from cells on the posterior side of body segments and forms a gradient from the posterior to anterior region of each segment. Hedgehog plays a role in creating the morphological differences between the anterior and posterior parts of individual body segments and also influences the formation of certain adult structures. Hedgehog mutants result in larvae that are short and stubby and have a higher density of denticles, which are cuticle spikes projecting from the body surface. The appearance of the stubby and spikey larvae inspired the name hedgehog.

In addition to morphogens, positional information is conveyed by **cell adhesion** (**Figure 23.2c**). Each cell makes its own collection of surface receptors that enable it to adhere to other cells and/or to the extracellular matrix (ECM), which consists primarily of carbohydrates and fibrous proteins. Such receptors are known as **cell adhesion molecules (CAMs).** A cell may gain positional information via the combination of contacts it makes with other cells or with the ECM.

The phenomenon of cell adhesion, and its role in multicellular development, was first recognized by Henry Wilson in 1907. He took multicellular sponges and passed them through a sieve, which disaggregated them into individual cells. Remarkably, the cells actively migrated until they adhered to one another to form a new sponge, complete with the chambers and canals that characterize a sponge's internal structure! When sponge cells from different species were mixed, they sorted themselves properly, adhering only to cells of the same species. Overall, these results indicate that cell adhesion plays an important role in governing the position that a cell will adopt during development.

The Study of Mutants with Disrupted Developmental Patterns Has Identified Genes That Control Development

Mutations that alter the course of development in experimental organisms, such as *Drosophila*, have greatly contributed to our understanding of the normal process of development. For example, **Figure 23.3** shows a photograph of a fly that carries multiple mutations in a complex of genes called *bithorax*. This mutant fly has four wings instead of two; the halteres (balancing organs that resemble miniature wings), which are found on the third thoracic segment, are changed into wings, normally found on the second thoracic segment. (The term bithorax refers to the observation that the characteristics of the second thoracic segment are duplicated.)

FIGURE 23.3   The bithorax mutation in *Drosophila*.

Genes→Traits   A normal fly contains two wings on the second thoracic segment and two halteres on the third thoracic segment. However, this mutant fly contains multiple mutations in a complex of genes called the *bithorax* complex. In this fly, the third thoracic segment has the same characteristics as the second thoracic segment, thereby producing a fly with four wings instead of the normal number of two.

TABLE **23.1**

Examples of *Drosophila* Genes That Play a Role in Pattern Development

Description	Examples
Some genes play a role in determining the axes of development. Also certain genes govern the formation of the extreme terminal (anterior and posterior) regions.	Anterior: *bicoid, exuperantia, swallow, staufen* Posterior: *nanos, cappuccino, oskar pumilio, spire, staufen, tudor, vasa* Terminal: *torso, torsolike, Trunk, NTF-1* Dorso-ventral: *Toll, cactus, dorsal, easter, gurken, nudel, pelle, pipe, snake, spatzle*
Some genes play a role in promoting the subdivision of the embryo into segments. These are called segmentation genes. The three types are gap genes, pair-rule genes, and segment-polarity genes.	Gap genes: *empty spiracles, giant, huckebein, hunchback, knirps, Krüppel, tailless, orthodenticle* Pair-rule genes: *even-skipped, hairy, runt, fushi tarazu, paired* Segment-polarity genes: *frizzled, frizzled-2, engrailed, patched, smoothened, hedgehog, wingless, gooseberry*
Some genes play a role in determining the fate of particular segments. These are known as homeotic genes. *Drosophila* has two clusters of homeotic genes known as the *Antennapedia* complex and the *bithorax* complex	*Antennapedia* complex: *labial, proboscipedia, Deformed, Sex combs reduced, Antennapedia P* *Bithorax* complex: *Ultrabithorax, abdominal A, Abdominal B*

Edward Lewis, a pioneer in the genetic study of development, became interested in the bithorax phenotype and began investigating it in 1946. Researchers later discovered that the mutant chromosomal region actually contained a complex of three genes involved in specifying developmental pathways in the fly. A gene that plays a central role in specifying the final identity of a body region is called a **homeotic gene.** We will discuss particular examples of homeotic genes later in this chapter.

During the 1960s and 1970s, interest in the relationship between genetics and embryology blossomed as biologists began to appreciate the role of genetics at the molecular and cellular levels. It soon became clear that the genomes of multicellular organisms contain groups of genes that initiate a program of development involving networks of gene regulation. By identifying mutant alleles that disrupt development, geneticists have begun to unravel the pattern of gene expression that underlies the normal pattern of multicellular development.

In *Drosophila*, the establishment of the body axes and division of the body into segments involves the participation of a few dozen genes. Table 23.1 lists many of the important genes governing pattern formation during embryonic development. These genes were identified by characterizing mutants that had altered development patterns. They are often given interesting names based on the observed phenotype when they are mutant. It is beyond the scope of this textbook to describe how all of these genes exert their effects during embryonic development. Instead, we will consider a few examples that illustrate how the expression of a particular gene and the localization of its gene product have a defined effect on the pattern of development.

The Early Stages of Embryonic Development Determine the Pattern of Structures in the Adult Organism

Figure 23.4 illustrates the general sequence of events in *Drosophila* development. The oocyte is the most critical cell in determining the pattern of development in the adult organism. It is an elongated cell with preestablished axes (Figure 23.4a). After fertilization takes place, the zygote goes through a series of nuclear divisions that are not accompanied by cytoplasmic division. Initially, the resulting nuclei are scattered throughout the yolk, but eventually they migrate to the periphery of the cytoplasm. This is the syncytial blastoderm stage (Figure 23.4b).

After the nuclei have lined up along the cell membrane, individual cells are formed as portions of the cell membrane surround each nucleus, creating a cellular blastoderm (Figure 23.4c). This structure is composed of a sheet of cells on the outside with yolk in the center. In this arrangement, the cells are distributed asymmetrically. At the posterior end are a group of cells called the pole cells, the primordial germ cells that eventually give rise to gametes in the adult organism. After blastoderm formation is complete, dramatic changes occur during gastrulation (Figure 23.4d). This stage produces three cell layers known as the ectoderm, mesoderm, and endoderm.

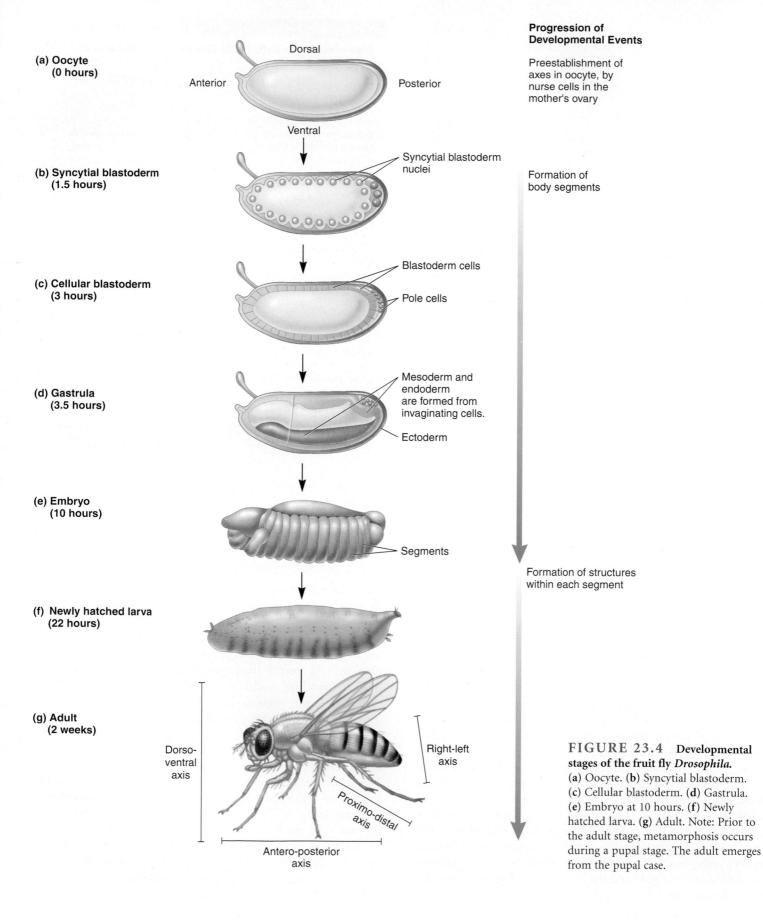

(a) Oocyte
(0 hours)

Dorsal

Anterior   Posterior

Ventral

(b) Syncytial blastoderm
(1.5 hours)

Syncytial blastoderm nuclei

(c) Cellular blastoderm
(3 hours)

Blastoderm cells

Pole cells

(d) Gastrula
(3.5 hours)

Mesoderm and endoderm are formed from invaginating cells.

Ectoderm

(e) Embryo
(10 hours)

Segments

(f) Newly hatched larva
(22 hours)

(g) Adult
(2 weeks)

Dorso-ventral axis

Right-left axis

Proximo-distal axis

Antero-posterior axis

Progression of Developmental Events

Preestablishment of axes in oocyte, by nurse cells in the mother's ovary

Formation of body segments

Formation of structures within each segment

FIGURE 23.4 **Developmental stages of the fruit fly *Drosophila*.** (**a**) Oocyte. (**b**) Syncytial blastoderm. (**c**) Cellular blastoderm. (**d**) Gastrula. (**e**) Embryo at 10 hours. (**f**) Newly hatched larva. (**g**) Adult. Note: Prior to the adult stage, metamorphosis occurs during a pupal stage. The adult emerges from the pupal case.

As this process occurs, the embryo begins to be subdivided into detectable units. Initially, shallow grooves divide the embryo into 14 **parasegments,** which are only transient subdivisions. A short time later, these grooves disappear, and new boundaries are formed that divide the embryo into morphologically discrete **segments.** Figure 23.4e shows the segmented pattern of a *Drosophila* embryo about 10 hours after fertilization. Later in this section, we will explore how the coordination of gene expression underlies the formation of these parasegments and segments.

At the end of **embryogenesis,** which lasts about 18 to 22 hours, a larva hatches from the egg (Figure 23.4f) and begins feeding on its own. *Drosophila* has three larval stages, separated by molts. During molting, the larva sheds its cuticle, a hardened extracellular shell that is secreted by the epidermis.

After the third larval stage, *Drosophila* forms a pupa that undergoes metamorphosis. During metamorphosis, groups of cells called imaginal disks that were produced earlier in development differentiate into the structures found in the adult fly, such as the head, wings, legs, and abdomen. The adult fly then emerges from the pupal case. The adult is produced about two weeks after fertilization. In *Drosophila*, as in all bilateral animals, the final result of development is an adult body organized along three axes: the **antero-posterior axis,** the **dorso-ventral axis,** and the **right-left axis** (Figure 23.4g). An additional axis, used mostly for designating limb parts, is the **proximo-distal axis.**

Although many interesting developmental events occur during the three larval stages and the pupal stage, we will focus most of our attention on the events that occur during embryonic development. Even before hatching, the embryo develops the basic body plan that will be found in the adult organism. In other words, during the early stages of development, the embryo is divided into segments that correspond to the segments of the larva and adult. Therefore, an understanding of how these segments form in the embryo is critical to our understanding of pattern formation.

The Gene Products of Maternal Effect Genes Are Deposited Asymmetrically into the Oocyte and Establish the Antero-posterior and Dorso-ventral Axes at a Very Early Stage of Development

The first stage in *Drosophila* embryonic pattern development is the establishment of the body axes. This occurs before the embryo becomes segmented. In fact, the morphogens necessary to establish these axes are distributed prior to fertilization. During oogenesis, gene products such as mRNA, which are important in early developmental stages, are deposited asymmetrically within the egg. Later, after the egg has been fertilized and development begins, these gene products will establish independent developmental programs that govern the formation of the body axes of the embryo. Proper development is dependent on the products of several different genes.

As shown in **Figure 23.5,** a few gene products act as key morphogens, or receptors for morphogens, that initiate changes in embryonic development. As shown here, these gene products are deposited asymmetrically in the egg. For example, the product of the *bicoid* gene is necessary to initiate development of the anterior structures of the organism. During oogenesis, the mRNA for *bicoid* accumulates in the anterior region of the oocyte. In contrast, the mRNA from the *nanos* gene accumulates in the posterior end. Later in development, the *nanos* mRNA will be translated into protein, which functions to influence posterior development. *Nanos* is required for the formation of the abdomen.

In addition to morphogens such as Bicoid and Nanos, the development of the structures at the extreme anterior and posterior ends of the embryo are regulated in part by a receptor protein called Torso. This receptor, which is activated by ligand binding only at the anterior and posterior ends of the egg, is necessary

bicoid mRNA

(a) Anterior distribution of *bicoid* mRNA

nanos mRNA

(b) Posterior distribution of *nanos* mRNA

Ligand

Activated Torso
receptor protein

(c) Terminal distribution of Torso receptor protein

Ligand

Activated
Toll receptor
protein

(d) Ventral distribution of Toll receptor protein

FIGURE 23.5   The establishment of the axes of polarity in the *Drosophila* embryo. This figure shows some of the maternal gene products deposited in the oocyte that are critical in the establishment of the antero-posterior, terminal, and dorso-ventral axes. (**a**) *Bicoid* mRNA is distributed in the anterior end of the oocyte and promotes the formation of anterior structures. (**b**) *Nanos* mRNA is localized to the posterior end and promotes the formation of posterior structures. (**c**) The Torso receptor protein is found in the membrane and is activated by ligand binding at either end of the oocyte. It causes the formation of structures that are found only at the ends of the organism. (**d**) The Toll receptor protein is activated by ligand binding at the ventral side of the embryo and establishes the dorso-ventral axis. Note: The Torso and Toll receptor proteins are distributed throughout the plasma membranes of the oocyte and embryo, but they are activated by ligand binding only in the regions shown in this figure.

for the formation of the terminal ends of the embryo. The dorso-ventral axis is governed by the actions of several proteins, including a receptor protein known as Toll. This receptor is found in plasma membranes throughout the embryo but is activated by ligand binding only along the ventral midline of the embryo.

Let's now take a closer look at the molecular mechanism of one morphogen, Bicoid. The *bicoid* gene got its name because a larva defective in this gene develops with two posterior ends (**Figure 23.6**). This allele exhibits a maternal effect pattern of inheritance, in which the genotype of the mother determines the phenotypic traits of the offspring (see Chapter 7). A female fly that is phenotypically normal (because its mother was heterozygous for the normal *bicoid* allele), but genotypically homozygous for an inactive *bicoid* allele (because it inherited an inactive allele from both its mother and father), will produce 100% abnormal offspring even when mated to a male that is homozygous for the normal *bicoid* allele. In other words, the genotype of the mother determines the phenotype of the offspring. This occurs because the *bicoid* gene product is provided to the oocyte via the nurse cells.

In the ovaries of female flies, the nurse cells are localized asymmetrically toward the anterior end of the oocyte. During oogenesis, gene products are transferred from nurse cells into the oocyte via cell-to-cell connections called cytoplasmic bridges. Maternally encoded gene products enter one side of the oocyte, which will eventually become the anterior end of the embryo (**Figure 23.7a**). The *bicoid* gene is actively transcribed in the nurse cells, and *bicoid* mRNA is transported into the anterior end of the oocyte. The 3' end of *bicoid* mRNA contains a signal that is recognized by RNA-binding proteins thought necessary for the transport of this mRNA into the oocyte. After it enters the oocyte, the *bicoid* mRNA is trapped at the anterior end.

How can researchers determine the location of *bicoid* mRNA in the oocyte and resulting zygote? **Figure 23.7b** shows an *in situ* hybridization experiment in which a *Drosophila* egg was examined via a probe complementary to the *bicoid* mRNA. (The technique of *in situ* hybridization is described in Chapter 20.) As seen here, the *bicoid* mRNA is highly concentrated near the anterior end of

the egg cell. When the *bicoid* mRNA subsequently is translated, a gradient of Bicoid protein is established as shown in **Figure 23.7c.**

After fertilization occurs, the Bicoid protein functions as a transcription factor. A remarkable feature of this protein is that its ability to influence gene expression is tuned exquisitely to its concentration. Depending on the distribution of the Bicoid protein, this transcription factor will activate genes only in certain regions of the embryo. For example, Bicoid stimulates the transcription of a gene called *hunchback* in the anterior half of the embryo, but its concentration is too low in the posterior half to activate the *hunchback* gene there.

Gap, Pair-Rule, and Segment-Polarity Genes Act Sequentially to Divide the *Drosophila* Embryo into Segments

After the antero-posterior and dorso-ventral regions of the embryo have been established by maternal effect genes, the next develop-

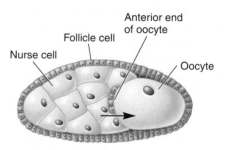

(a) Transport of maternal effect gene products into the oocyte

(b) *In situ* hybridization of *bicoid* mRNA

(c) Immunostaining of Bicoid protein

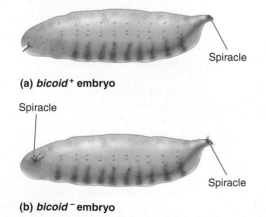

(a) *bicoid⁺* embryo

(b) *bicoid⁻* embryo

FIGURE 23.6   The bicoid mutation in *Drosophila*. (a) A normal *bicoid⁺* embryo. **(b)** A *bicoid⁻* embryo in which both ends of the larva develop posterior structures. For example, both ends develop a spiracle, which normally is found only at the posterior end. This is a lethal condition.

FIGURE 23.7   Asymmetrical localization of gene products during oogenesis in *Drosophila*. (a) The nurse cells transport gene products into the anterior (left) end of the developing oocyte. **(b)** An *in situ* hybridization experiment showing that the *bicoid* mRNA is trapped near the anterior end. **(c)** The *bicoid* mRNA is translated into protein soon after fertilization. The location of the Bicoid protein is revealed by immunostaining using an antibody that specifically recognizes this protein.

mental process organizes the embryo transiently into parasegments and then permanently into segments. The segmentation pattern of the embryo is shown in **Figure 23.8**. This pattern of positional information will be maintained, or "remembered," throughout the rest of development. In other words, each segment of the embryo gives rise to unique morphological features in the adult. For example, T2 becomes a thoracic segment with a pair of legs and a pair of wings, and A8 becomes a segment of the abdomen.

Figure 23.8 shows the overlapping relationship between parasegments and segments. As seen here, the boundaries of the segments are out of register with the boundaries of the parasegments. An appreciation of this feature is critical to our understanding of segmentation. From the viewpoint of genes, we will see that the parasegments are the locations where gene expression is controlled spatially. The anterior compartment of each segment coincides with the posterior region of a parasegment; the posterior compartment of a segment coincides with the anterior region of the next parasegment. The pattern of gene expression that occurs in the posterior region of one parasegment and the anterior region of an adjacent parasegment results in the formation of a segment.

Now that we have a general understanding of the way the *Drosophila* embryo is subdivided, we can examine how particular genes cause it to become segmented into this pattern. Three classes of genes, collectively called **segmentation genes,** play a role in the formation of body segments: gap genes, pair-rule genes, and segment-polarity genes. The expression and activation patterns of these genes in specific regions of the embryo cause it to become segmented.

How were the three classes of segmentation genes discovered? In the 1970s, segmentation genes were identified by Christiane Nüsslein-Volhard and Eric Wieschaus, who undertook a systematic search for mutations affecting embryonic development. Their pioneering effort identified most of the genes required for the embryo to develop a segmented pattern.

Figure 23.9 represents a few of the phenotypic effects that occur when a particular segmentation gene is defective. The gray boxes indicate the regions missing in the resulting larvae. In flies

Krüppel

(a) Gap

Even-skipped

(b) Pair-rule

Gooseberry

(c) Segment-polarity

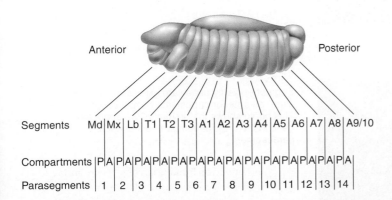

Segments	Md	Mx	Lb	T1	T2	T3	A1	A2	A3	A4	A5	A6	A7	A8	A9/10
Compartments	P A	P A	P A	P A	P A	P A	P A	P A	P A	P A	P A	P A	P A	P A	
Parasegments	1	2	3	4	5	6	7	8	9	10	11	12	13	14	

FIGURE 23.8   A comparison of segments and parasegments in the *Drosophila* embryo. Note that the parasegments and segments are out of register. The posterior (P) and anterior (A) regions are shown for each segment.

FIGURE 23.9   Phenotypic effects in *Drosophila* larvae with mutations in segmentation genes. The effects shown here are caused by mutations in (**a**) a gap gene, (**b**) a pair-rule gene, and (**c**) a segment-polarity gene.

with a mutation in a gap gene known as *Krüppel*, a contiguous section of the larva is missing (Figure 23.9a). In other words, a gap of several segments has occurred. By comparison, a defect in a pair-rule gene causes alternating regions to be deleted (Figure 23.9b). For example, when the *even-skipped* gene is defective, portions of alternating segments in the resulting larva are missing. Finally, segment-polarity mutations cause individual segments to be missing either an anterior or a posterior region. Figure 23.9c shows a mutation in a segmentation gene known as

gooseberry. When this gene is defective, the anterior portion of each segment is missing from the larva. In the case of segment-polarity mutants, the segments adjacent to the deleted regions exhibit a mirror-image duplication. Overall, the phenotypic effects of mutant segmentation genes provided geneticists with important clues regarding the roles of these genes in the developmental process of segmentation.

Figure 23.10 presents a partial, simplified scheme of the genetic hierarchy that leads to a segmented pattern in the *Dro-*

1. **Maternal effect genes** establish the antero-posterior and dorso-ventral axes. They activate gap genes.

Asymmetric localization of maternal effect gene products

Examples

Asymmetric localization of Bicoid protein. Other maternal effect gene products (not shown) are also asymmetrically localized.

2. **Gap gene products** act as genetic regulators of pair-rule genes. They bind to stripe-specific enhancers that are located adjacent to pair-rule genes.

Gap gene expression

Gap gene expression occurs as broad bands in the embryo. In this photo, Krüppel protein is shown in red and Hunchback in green. Their region of overlap is yellow. Other gap genes (not shown) are also expressed.

3. The expression of a **pair-rule gene** in a stripe defines the boundary of a parasegment. Pair-rule gene products regulate the expression of segment-polarity genes.

Pair-rule gene expression

Pair-rule genes are expressed in alternating stripes. Each stripe corresponds to a parasegment. In this photo, the *even-skipped* gene product is expressed in the light bands that correspond to odd-numbered parasegments.

4. **Segment-polarity genes** define the anterior or posterior compartment of each parasegment.

Segment-polarity gene expression

Segment-polarity genes are expressed in either an anterior or posterior compartment. This photo shows the product of the *engrailed* gene in the anterior compartment of each parasegment.

FIGURE 23.10   Overview of the genetic hierarchy leading to segmentation in *Drosophila*. Note: When comparing steps 3 and 4, the embryo has undergone a 180° turn, folding back on itself.

sophila embryo. Keep in mind that while this figure presents the general sequence of events that occurs during the early stages of embryonic development, many more genes are actually involved in this process (refer back to Table 23.1). As described in this figure, the following steps occur:

1. Maternal effect gene products, such as *bicoid* mRNA, are deposited asymmetrically into the oocyte. These gene products form a gradient that will later influence the formation of axes, such as the antero-posterior axis.

2. In contrast to maternal effect genes, which are expressed during oogenesis, **zygotic genes** are expressed after fertilization. The first zygotic genes to be activated are the gap genes. Maternal effect genes are responsible for activating the gap genes. As shown in step 2 of Figure 23.10, gap genes divide the embryo into a series of broad bands or regions. These bands do not correspond to parasegments or segments within the embryo.

3. The gap genes and maternal effect genes then activate the pair-rule genes. The photograph shown in step 3 of Figure 23.10 illustrates the alternating pattern of expression of the *even-skipped* gene. The expression of pair-rule genes in stripes defines the boundaries of parasegments. *Even-skipped* is expressed in the odd-numbered parasegments. Solved problem S4 at the end of the chapter examines how the *even-skipped* gene can be expressed in this pattern of alternating stripes. The pair-rule genes divide the broad regions established by gap genes into seven bands or stripes.

4. Once the pair-rule genes are activated in an alternating banding arrangement, their gene products then regulate the expression of segment-polarity genes. The segment-polarity genes divide the embryo into 14 stripes, one within each parasegment. As shown in step 4 of Figure 23.10, the segment-polarity gene *engrailed* is expressed in the anterior region of each parasegment. Another segment-polarity gene, *wingless*, is expressed in the posterior region. Later in development, the anterior end of one parasegment and the posterior end of another parasegment will develop into a segment with particular morphological characteristics.

The Expression of Homeotic Genes Controls the Phenotypic Characteristics of Segments

Thus far, we have considered how the *Drosophila* embryo becomes organized along axes and then into a segmented body pattern. Now let's examine how each segment develops its unique morphological features. Geneticists often use the term **cell fate** to describe the ultimate morphological features that a cell or group of cells will adopt. For example, the fate of the cells in segment T2 in the *Drosophila* embryo is to develop into a thoracic segment containing two legs and two wings. In *Drosophila*, the fate of the cells in each segment of the body is determined at a very early stage of embryonic development, long before the morphological features become apparent.

Our understanding of developmental fate has been greatly aided by the identification of mutant genes that alter cell fates. In animals, the first mutant of this type was described by the German entomologist Ernst Gustav Kraatz in 1876. He observed a sawfly (*Cimbex axillaris*) in which part of an antenna was replaced with a leg. During the late nineteenth century, the English zoologist William Bateson collected many of these types of observations and published them in 1894 in a book entitled *Materials for the Study of Variation Treated with Especial Regard to Discontinuity in the Origin of Species*. In this book, Bateson coined the term **homeotic** to describe mutants in which one body part is replaced by another, such as the transformation of the antennae of insects into legs.

As mentioned earlier, Edward Lewis began to study strains of *Drosophila* having homeotic mutations. This work, which began in 1946, was the first systematic study of homeotic genes. Each homeotic gene controls the fate of a particular region of the body. *Drosophila* contains two clusters of homeotic genes called the *Antennapedia* complex and the *bithorax* complex. **Figure 23.11** shows the organization of genes within these complexes.

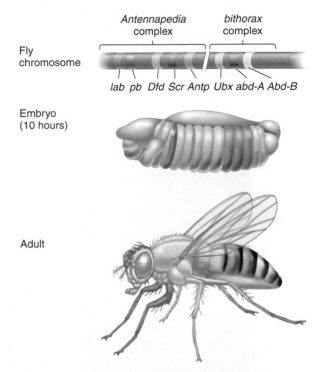

FIGURE 23.11 **Expression pattern of homeotic genes in *Drosophila*.** The order of homeotic genes, *labial* (*lab*), *proboscipedia* (*pb*), *Deformed* (*Dfd*), *Sex combs reduced* (*Scr*), *Antennapedia* (*Antp*), *Ultrabithorax* (*Ubx*), *abdominal A* (*abd-A*), and *Abdominal B* (*Abd-B*), correlates with the order of expression in the embryo. The expression pattern of seven of these genes is shown. *Lab* (purple) is expressed in the region that will eventually give rise to mouth structures. *Dfd* (light green) is expressed in the region that will form much of the head. *Scr* (forest green) and *Antp* (blue) are expressed in embryonic segments that give rise to thoracic segments. *Ubx* (gray), *abd-A* (red), and *Abd-B* (yellow) are expressed in posterior segments that will form the abdomen. The order of gene expression, from anterior to posterior, parallels the order of genes along the chromosome. The expression pattern of the *pb* gene is not well understood.

The *Antennapedia* complex contains five genes, designated *lab*, *pb*, *Dfd*, *Scr*, and *Antp*. The *bithorax* complex has three genes, *Ubx*, *abd-A*, and *Abd-B*. Both of these complexes are located on chromosome 3, but a large segment of DNA separates them.

As noted in Figure 23.11, the order of these genes along chromosome 3 correlates with their pattern of gene expression along the antero-posterior axis of the body. For example, *lab* is expressed in an anterior segment and governs the formation of mouth structures. The *Antp* gene is expressed strongly in the thoracic region during embryonic development and controls the formation of thoracic structures. Transcription of the *Abd-B* gene occurs in the posterior region of the embryo. This gene controls the formation of the posterior-most abdominal segments.

The role of homeotic genes in determining the identity of particular segments has been revealed by mutations that alter their function. For example, the *Antp* gene is normally expressed in the thoracic region. The *Antennapedia* mutation causes the *Antp* gene to also be expressed in the region where the antennae are made. These mutant flies have legs in the place of antennae (**Figure 23.12**)! This is an example of a **gain-of-function muta-tion.** In this case, the *Antp* gene is also expressed abnormally in the anterior segment that normally gives rise to the antennae.

Investigators have also studied many loss-of-function mutations in homeotic genes. When a particular homeotic gene is defective, the region it normally governs will usually be controlled by the homeotic gene that acts in the adjacent anterior

region. For example, the *Ubx* gene functions within parasegments 5 and 6. If this gene is missing, this section of the fly becomes converted to the structures found in parasegment 4.

The homeotic genes are part of the genetic hierarchy that produces the morphological characteristics of the fly. How are they regulated? The homeotic genes are controlled by gap genes and pair-rule genes, and they are also regulated by interactions among themselves. Other genes maintain the patterns of homeotic gene expression after the initial patterns are established by segmentation genes. A group of genes known as *Polycomb* genes represses the expression of homeotic genes in regions of the embryo where they should not act. This is accomplished by the remodeling of chromatin structure to convert it to a closed conformation in which transcription is difficult or impossible. Alternatively, in regions where the homeotic genes are active, the products of *Trithorax* genes maintain an open conformation in which transcription can take place. Overall, the concerted actions of many gene products cause the homeotic genes to be expressed only in the appropriate region of the embryo, as shown in Figure 23.11.

Because they are part of a genetic hierarchy in which genes activate other genes, perhaps it is not surprising that homeotic genes encode transcription factors. The coding sequence of homeotic genes contains a 180 base pair consensus sequence known as a **homeobox** (**Figure 23.13a**). This sequence was first discovered in the *Antp* and *Ubx* genes, and it has since

(a) Normal fly

(b) *Antennapedia* mutant

FIGURE 23.12   The *Antennapedia* mutation in *Drosophila.*

Genes→Traits **(a)** A normal fly with antennae. **(b)** This mutant fly has a gain-of-function mutation in which the *Antp* gene is expressed in the embryonic segment that normally gives rise to antennae. The expression of *Antp* causes this region to have legs rather than antennae.

(a) Homeotic gene

(b) Homeotic protein bound to DNA

FIGURE 23.13   Molecular features of homeotic proteins. (a) A homeotic gene (shown in tan and orange) contains a 180-bp sequence called the homeobox (shown in orange). **(b)** When a homeotic gene is expressed, it produces a protein that functions as a transcription factor. The homeobox encodes a region of the protein called a homeodomain, which binds to the major groove of DNA. These DNA-binding sites are found within genetic regulatory elements (i.e., enhancers). The enhancers are found in the vicinity of promoters that are turned on by homeotic proteins. For this to occur, the homeotic protein also contains a transcriptional activation domain, which will activate the transcription of a gene after the homeodomain has bound to the DNA.

been found in all *Drosophila* homeotic genes and in some other genes affecting pattern development, such as *bicoid*. The protein domain encoded by the homeobox is called a **homeodomain.** The arrangement of α helices within the homeodomain promotes the binding of the protein to the major groove of DNA (**Figure 23.13b**). In this way, homeotic proteins can bind to DNA in a sequence-specific manner. In addition to DNA-binding ability, homeotic proteins also contain a transcriptional activation domain that functions to activate the genes to which the homeodomain can bind.

The transcription factors encoded by homeotic genes activate genes encoding proteins that produce the morphological characteristics of each segment. Much current research attempts to identify these genes and determine how their expression in particular regions of the embryo leads to morphological changes in the embryo, larva, and adult. In some cases, these genes also encode transcription factors that control the expression of other sets of genes that will alter the morphological characteristics of cells. In other cases, genes that are activated by homeotic proteins play a role in cell-to-cell signaling pathways. The activation of these signaling pathways enables cells and groups of cells to adopt their correct morphologies. It is expected that research during the next few decades will shed considerable light on the pathways by which genes control morphological changes in the fruit fly.

The Developmental Fate of Each Cell in the Nematode *Caenorhabditis elegans* Is Known

We now turn our attention to another invertebrate, *C. elegans,* a nematode that has been the subject of numerous studies in developmental genetics. The embryo develops within the eggshell and hatches when it reaches a size of 550 cells. After hatching, it continues to grow and mature as it passes through four successive molts. It takes about three days for a fertilized egg to develop into an adult worm about 1 mm in length. With regard to sex, *C. elegans* can be a male (and only produce sperm) or a hermaphrodite (capable of producing sperm and egg cells). An adult male is composed of 1,031 somatic cells and produces about 1,000 sperm. A hermaphrodite consists of 959 somatic cells and produces about 2,000 gametes (both sperm and eggs).

A remarkable feature of this organism is that the pattern of cellular development is extremely constant from worm to worm. In the early 1960s, Sydney Brenner pioneered the effort to study the pattern of cell division in *C. elegans* and establish it as a model organism. Because *C. elegans* is transparent and composed of relatively few cells, researchers can follow cell division step by step under the microscope, beginning with a fertilized egg and ending with an adult worm. Researchers can identify a particular cell at an embryonic stage, follow that cell as it divides, and observe where its descendant cells will be located in the adult. An illustration that depicts how cell division proceeds is called a **lineage diagram.** It describes the cell division patterns and fates of any cell's descendants.

Figure 23.14 shows a partial lineage diagram for a *C. elegans* hermaphrodite. At the first cell division, the egg divides to produce two cells, called AB and P_1. AB then divides into two cells, ABa and ABp; and P_1 divides into two cells, EMS and P_2. The EMS cell then divides into two cells, called MS and E. The cellular descendants of the E cell give rise to the worm's intestine.

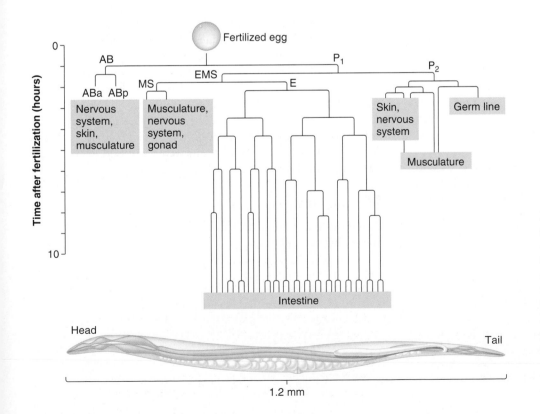

FIGURE 23.14   A cell lineage diagram of the nematode *Caenorhabditis elegans.* This partial lineage diagram illustrates how the cells divide to produce different regions of the adult. The fate of the intestinal cell lineage is shown in greater detail than that of other cell lineages. A complete lineage diagram is known for this organism, although its level of detail is beyond the scope of this textbook. Note: The lowercase letters "a" and "p" refer to anterior and posterior, respectively, and thereby indicate the relative positions of the daughter cells.

In other words, the fate of the E cell's descendants is to develop into intestinal cells. This diagram also illustrates the concept of a **cell lineage**—a series of cells that are derived from each other by cell division. For example, the EMS cell, E cell, and the intestinal cells are all part of the same cell lineage.

Why is a lineage diagram for an organism an important experimental advantage? It allows researchers to investigate how gene expression in any cell, at any stage of development, may affect the outcome of a cell's fate. In the experiment described next, we will see how the timing of gene expression is an important factor in the fate of a cell's descendants.

EXPERIMENT 23A

Heterochronic Mutations Disrupt the Timing of Developmental Changes in *C. elegans*

Our discussion of *Drosophila* development focused on how the spatial expression and localization of gene products can lead to a particular pattern of embryonic development. Another important issue in development is timing. The cells of a multicellular organism must know when to divide and when to differentiate into a particular cell type. If the timing of these processes is not coordinated, certain tissues will develop too early or too late, disrupting the developmental process.

In *C. elegans*, the timing of developmental events can be examined carefully at the cellular level. As mentioned, the fate of each cell in *C. elegans* has been determined. Using a microscope, a researcher can focus on a particular cell within this transparent worm and watch it divide into two cells, then four cells, and so forth. Therefore, a scientist can judge whether a cell is behaving as it should during the developmental process.

To identify genes that play a role in the timing of cell fates, researchers have searched for mutant alleles that disrupt the normal timing process. In a collaboration in the late 1970s, H. Robert Horvitz and John Sulston set out to identify mutant alleles in *C. elegans* that disrupt cell fates or the timing of cell fates. Using a microscope, they screened thousands of worms for altered morphologies that might indicate an abnormality in development. During this screening process, one of the phenotypic abnormalities they found was a defective egg-laying phenotype. They reasoned that because the egg-laying system depends on a large number of cell types (vulval cells, muscle cells, and nerve cells), an abnormality in any of the cell lineages leading to these cell types might cause an inability to lay eggs.

In *C. elegans*, a defective egg-laying phenotype is easy to identify, because the hermaphrodite will be able to fertilize its own eggs but will be unable to lay them. When this occurs, the eggs actually hatch within the hermaphrodite's body. This leads to the death of the hermaphrodite as it becomes filled with hatching worms. This defective egg-laying phenotype, in which the her-

Normal *C. elegans*

C. elegans with the defective egg-laying phenotype, the "bag of worms"

maphrodite becomes filled with its own offspring, is called a "bag of worms." Eventually, the newly hatched larvae eat their way out and can be saved for further study.

In their initial study, published in 1980, the defective egg-laying phenotype yielded several mutant strains that were defective in particular cell lineages. A few years later, in the experiment described in **Figure 23.15**, Victor Ambros and Horvitz took this same approach and were able to identify genes that play a key role in the timing of cell fate. They began with wild-type *C. elegans* and three mutant lines designated *n536*, *n355*, and *n540*. All three of the mutant lines showed an egg-laying defect. Right after hatching, they observed the fates of particular cells via microscopy. This involved spending hours looking at the nuclei of specific cells (which are relatively easy to see in this transparent worm) and timing the patterns of cell division. The patterns in the mutant and wild-type strains were then used to create lineage diagrams for particular cells.

■ THE HYPOTHESIS

Mutations that cause a defective egg-laying phenotype may affect the timing of cell lineages.

TESTING THE HYPOTHESIS — FIGURE 23.15   **Identification of heterochronic mutations in *C. elegans*.**

Starting material: Prior to this work, many laboratories had screened thousand of *C. elegans* worms and identified many different mutant strains that were defective egg-laying. (Note: When mutated, many different genes may cause a defective egg-laying phenotype. Only some of them are expected to be genes that alter the timing of cell fate within a particular cell lineage.)

Experimental level	**Conceptual level**

1. Obtain a large number of *C. elegans* strains that have a defective egg-laying phenotype. The wild-type strain was also studied as a control.

2. Right after hatching, observe the fate of particular cells via microscopy. In this example, a researcher began watching a cell called the T cell and monitored its division pattern, and the pattern of subsequent daughter cells, during the larval stages. These patterns were examined in both wild-type and defective egg-laying worms.

■ THE DATA

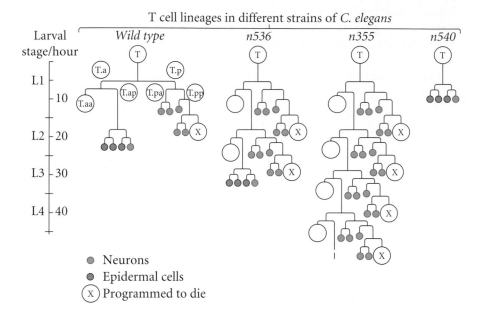

T cell lineages in different strains of *C. elegans*

- ● Neurons
- ● Epidermal cells
- Ⓧ Programmed to die

■ INTERPRETING THE DATA

Lineage diagrams involving cells derived from a cell called a T cell are shown in the data of Figure 23.15. As seen here, the wild-type strain follows a particular pattern of cell division for the T cell lineage. Each division event occurs at specific times during the L1 and L2 larval stages. In the normal strain, the T cell divides during the L1 larval stage to produce a T.a and T.p cell. The T.a cell also divides during L1 to produce a T.aa and T.ap cell. The T.p cell divides during L1 to produce T.pa and T.pp. These cells also divide during L1, eventually producing five neurons (labeled in blue) and one cell that is programmed to die (designated with an X). During the L2 larval stage, the T.ap cell resumes division to produce four cells: three epidermal cells (labeled in red) and one neuron.

The other T cell lineages in the data are from worms that carry mutations causing an egg-laying defect. Later research revealed that these three mutations are located in a gene called *lin-14*. The allele designated *n536* has caused the reiteration of the normal events of L1 during the L2 larval stage. In L2, the only cell of this lineage that is supposed to divide is T.ap. In worms carrying the *n536* allele, however, this cell behaves as if it were a T cell, rather than a T.ap cell, and produces a group of cells identical to what a T cell produces during the L1 stage. In the L3 stage, the cells in the *n536* strain behave as if they were in L2. Besides the egg-laying defect, the phenotypic outcome of this irregularity in the timing of cell fates is a worm that has several additional cells derived from the T cell lineage.

An allele that causes multiple reiterations is the *n355* allele. This strain continues to reiterate the normal events of L1 during the L2, L3, and L4 stages. In contrast, the *n540* allele has an opposite effect on the T cell lineage. During the L1 larval stage,

the T cell behaves as if it were a T.ap cell in the L2 stage. In this case, it skips the divisions and cell fates of the L1 and proceeds directly to cell fates that occur during the L2 stage.

The types of mutations described here are called **heterochronic mutations.** The term heterochrony refers to a change in the relative timing of developmental events. In heterochronic mutations, the timing of the fate of particular cell lineages is not synchronized with the development of the rest of the organism. More recent molecular data have shown that this is due to an irregular pattern of gene expression. In wild-type worms, the LIN-14 protein accumulates during the L1 stage and promotes the T cell division pattern shown for the wild type. During L2, the LIN-14 protein diminishes to negligible levels. The *n536* and *n355* alleles are examples of gain-of-function mutations. In strains with these alleles, the LIN-14 protein persists during later larval stages. For the *n536* allele, it is made for one additional cell division, while the *n355* allele continues to express *lin-14* for several cell divisions. By comparison, the *n540* allele is a loss-of-function mutation. This allele causes LIN-14 to be inactive during L1, so it cannot promote the normal L1 pattern of cell division and cell fate.

Overall, the results described in this experiment are consistent with the idea that the precise timing of *lin-14* expression during development is necessary to correctly control the fate of particular cells in *C. elegans*. Mutations that alter the expression of *lin-14* lead to phenotypic abnormality, including the inability to lay eggs. This detrimental phenotypic consequence illustrates the importance of the correct timing for cell division and differentiation during development.

A self-help quiz involving this experiment can be found at the Online Learning Center.

23.2 VERTEBRATE DEVELOPMENT

Biologists have studied the morphological features of development in many vertebrate species. Historically, amphibians and birds have been studied extensively, because their eggs are rather large and easy to manipulate. For example, certain developmental stages of the frog and chicken have been described in great detail. In more recent times, the successes obtained in *Drosophila* have shown the great power of genetic analyses in elucidating the underlying molecular mechanisms that govern biological development. With this knowledge, many researchers are attempting to understand the genetic pathways that govern the development of the more complex body structure found in vertebrate organisms.

Several vertebrate species have been the subject of developmental studies. These include the mouse (*Mus musculus*), the frog (*Xenopus laevis*), and the small aquarium zebrafish (*Danio rerio*). In this section, we will primarily discuss the genes that are important in mammalian development, particularly those that have been characterized in the mouse, one of the best-studied mammals. As we will see, several genes affecting its developmental pathways have been cloned and characterized. In this section, we will examine how these genes affect the course of vertebrate development.

Researchers Have Identified Homeotic Genes in Vertebrates

Vertebrates typically have long generation times and produce relatively few offspring. Therefore, it is usually not practical to screen large numbers of embryos or offspring in search of mutant phenotypes with developmental defects. As an alternative, a successful way of identifying genes that affect vertebrate development has been the use of molecular techniques to identify vertebrate genes similar to those that control development in simpler organisms such as *Drosophila*.

As discussed in Chapters 21 and 26, species that are evolutionarily related to each other often contain genes with similar DNA sequences. When two or more genes have similar sequences because they are derived from the same ancestral gene, they are called **homologous genes.** Homologous genes found in different species are termed **orthologs.** Experimentally, a DNA strand from one gene will hybridize to a complementary strand of a homologue, because they have similar sequences.

Researchers initially followed a strategy of using cloned *Drosophila* genes as probes to identify homologous vertebrate genes, which has been quite successful. With this method, researchers have found complexes of homeotic genes in many vertebrate species that are homologous to those in the fruit fly. These groups of adjacent homeotic genes are called ***Hox* complexes.** As shown in **Figure 23.16**, the mouse has four *Hox* complexes, designated *HoxA* (on chromosome 6), *HoxB* (on chromosome 11), *HoxC* (on chromosome 15), and *HoxD* (on chromosome 2). A total of 38 genes are found in the four complexes. Thirteen different types of homeotic genes occur within the four *Hox* complexes, although none of the four complexes contains representatives of all 13 types of genes. The addition of *Hox* genes into the genomes

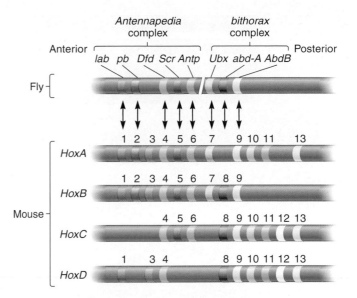

FIGURE 23.16   A comparison of homeotic genes in *Drosophila* and the mouse. The mouse contains four gene clusters, *HoxA–D*, that correspond to certain homeotic genes found in *Drosophila*. Thirteen different types of homeotic genes are found in the mouse, although each *Hox* gene cluster does not contain all 13 genes. In this drawing, orthologous genes are aligned in columns. For example, *lab* is the ortholog to *HoxA-1*, *HoxB-1*, and *HoxD-1*; *Ubx* is orthologous to *HoxA-7* and *HoxB-7*.

of animals has allowed animals to develop more complex body plans. We will consider the evolutionary origin of the *Hox* genes and their importance in the evolution of animal species in Chapter 26.

Remarkably, several of the homeotic genes in fruit flies and mammals are strikingly similar. Among the first six types of genes, five of them are homologous to genes found in the *Antennapedia* complex of *Drosophila*. Among the last seven, three are homologous to the genes of the *bithorax* complex. These results indicate there are fundamental similarities in the ways that animals as different as fruit flies and mammals undergo embryonic development. This suggests that a "universal body plan" underlies animal development.

Like the *Antennapedia* and *bithorax* complexes in *Drosophila*, the arrangement of *Hox* genes along the mouse chromosomes reflects their pattern of expression from the anterior to the posterior end (**Figure 23.17a**). This phenomenon is seen in more detail in **Figure 23.17b**, which shows the expression pattern for a group of *HoxB* genes in a mouse embryo. Overall, these results are consistent with the idea that the *Hox* genes play a role in determining the fates of segments along the antero-posterior axis.

Currently, researchers are trying to understand the functional roles of the genes within the *Hox* complexes in vertebrate development. In *Drosophila*, great advances in developmental genetics have been made by studying mutant alleles in genes that control development. In mice, however, few natural mutations have been identified that affect development. This has made it

(a) Correlation between *Hox* gene arrangement and expression

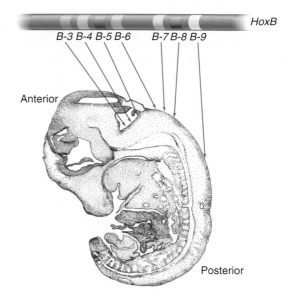

(b) Anterior expression boundaries for a series of *HoxB* genes

FIGURE 23.17 **Expression pattern of *Hox* genes in the mouse.** (a) A schematic illustration of the *Hox* gene expression in the embryo and the corresponding regions in the adult. (b) A more-detailed description of *HoxB* expression in a mouse embryo. The arrows indicate the anterior-most boundaries for the expression of *HoxB-3* to *HoxB-9*. The order of *Hox* gene expression, from the anterior end to posterior of the embryo, is in the same order as the genes are found along the chromosome.

difficult to understand the role that genetics plays in the development of the mouse and other vertebrate organisms.

How are researchers attempting to overcome this problem? Geneticists are taking an approach known as **reverse genetics.** In this strategy, researchers first identify the wild-type gene using cloning methods. In this case, the *Hox* genes in vertebrates have been cloned using *Drosophila* genes as probes. The next step is to create a mutant version of a *Hox* gene in vitro. This mutant allele is then reintroduced into a mouse using gene replacement techniques described in Chapter 19. When the function of the wild-type gene is eliminated, it is called a **gene knockout.** In this way, researchers can determine how the mutant allele affects the phenotype of the mouse.

The term reverse genetics reflects the idea that the experimental steps occur in an order opposite to that in the conventional approach used in *Drosophila* and other organisms. In the fly, mutant alleles were identified by their phenotype first, and then they were cloned. In the mouse, the genes were cloned first, the mutations were made in vitro, and then the mutated genes were introduced into the mouse to observe their phenotypic effects.

In recent years, several laboratories have used a reverse genetic approach to understand how many different genes, including *Hox* genes, affect vertebrate development. In *Drosophila*, loss-of-function alleles for homeotic genes usually show an

anterior transformation. This means that the segment where the defective homeotic gene is expressed now exhibits characteristics that resemble the adjacent anterior segment. Similarly, certain gene knockouts (e.g., *HoxA-2*, *B-4*, and *C-8*) also show anterior transformations within particular regions of the mouse. However, knockouts of other *Hox* genes (e.g., *A-11*) have posterior transformations, and knockouts of *A-3* and *A-1* exhibit abnormalities in morphology but no clear homeotic transformations. Interestingly, a *HoxA-5* knockout in mice shows evidence of both anterior and posterior transformations, which are also seen in *Drosophila* when its ortholog, *Scr*, is knocked out.

Overall, the current picture indicates that the *Hox* genes in vertebrates play a key role in patterning the antero-posterior axis. During the evolution of animals, increases in the number of *Hox* genes and changes in their patterns of expression have had an important impact on their morphologies. As an example, let's consider the expression of the *HoxC-6* gene. Changes in its pattern of expression among vertebrate species are associated with changes in the boundary between cervical (neck) vertebrae and thoracic (chest) vertebrae. The *HoxC-6* gene is expressed during embryonic development prior to vertebrae formation. Differences in the relative position of its expression correlate with the number of neck vertebrae produced (**Figure 23.18**). In the mouse, which has a relatively short neck, *HoxC-6* expression begins in the region of the early embryo that will later develop into vertebrae 7 and 8.

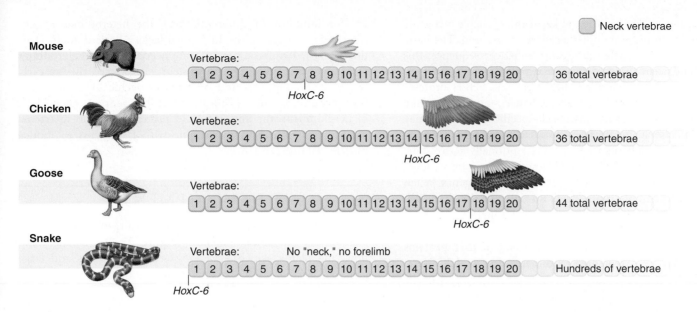

FIGURE 23.18 **Expression of the *HoxC-6* gene in different species of vertebrates.** This figure indicates the anterior boundary where the *HoxC-6* gene is expressed in the mouse, chicken, goose, and snake.

In contrast, *HoxC-6* expression in the chicken and goose begins much farther back, between vertebrae 14 and 15, or 17 and 18, respectively. The forelimbs also arise at this boundary in all vertebrates. However, snakes, which have no neck or forelimbs, do not have such a boundary because *HoxC-6* expression begins toward their heads.

Genes That Encode Transcription Factors Also Play a Key Role in Cell Differentiation

Thus far, we have focused our attention on patterns of gene expression that occur during the very early stages of development. These genes control the basic body plan of the organism. As this process occurs, cells become **determined.** This term refers to the phenomenon that a cell will be destined to become a particular cell type. In other words, its fate has been predetermined to eventually become a particular type of cell such as a nerve cell. This occurs long before a cell becomes **differentiated.** This later term means that a cell's morphology and function have changed, usually permanently, into a highly specialized cell type. For example, an undifferentiated mesodermal cell may differentiate into a specialized muscle cell, or an ectodermal cell may differentiate into a nerve cell.

At the molecular level, the profound morphological differences between muscle cells and nerve cells arise from gene regulation. Though muscle and nerve cells contain the same set of genes, they regulate the expression of those genes in very different ways. Certain genes that are transcriptionally active in muscle cells are completely inactive in nerve cells, and vice versa. Therefore, muscle and nerve cells express different proteins, which affect the morphological and physiological characteristics of the respective cells in distinct ways. In this manner, differential gene regulation underlies cell differentiation.

We learned earlier that a hierarchy of gene regulation is responsible for establishing the body pattern in *Drosophila*. Maternal effect genes control the expression of gap genes, which control the expression of pair-rule genes, and so forth. A similar type of hierarchy is thought to underlie cell differentiation in vertebrates. Researchers have identified specific genes that cause cells to differentiate into particular cell types. These genes trigger undifferentiated cells to differentiate and follow their proper cell fates.

In 1987, Harold Weintraub and colleagues identified a gene, which they called *MyoD*. This gene plays a key role in skeletal muscle cell differentiation. Experimentally, when the cloned *MyoD* gene was introduced into fibroblast cells in a laboratory, the fibroblasts differentiated into skeletal muscle cells. This result was particularly remarkable because fibroblasts normally differentiate into osteoblasts (bone cells), chondrocytes (cartilage cells), adipocytes (fat cells), and smooth muscle cells, but in vivo, they never differentiate into skeletal muscle cells.

Since this initial discovery, researchers have found that *MyoD* belongs to a small group of genes that initiate muscle development. Besides *MyoD*, these include *Myogenin*, *Myf5*, and *Mrf4*. All four of these genes encode transcription factors that contain a **basic domain** and a **helix-loop-helix domain (bHLH).** The basic domain is responsible for DNA binding and the activation of skeletal muscle-cell-specific genes. The helix-loop-helix domain is necessary for dimer formation between transcription factor proteins. Because of their common structural features and their role in muscle differentiation, MyoD, Myogenin, Myf5, and Mrf4 constitute a family of proteins called **myogenic bHLH proteins.** They are found in all vertebrates and have been identified in several invertebrates, such as *Drosophila* and *C. elegans*. In all cases, the myogenic bHLH genes are activated during skeletal muscle cell development.

At the molecular level, certain key features enable myogenic bHLH proteins to promote muscle cell differentiation. The basic domain binds to a muscle-cell-specific enhancer sequence; this sequence is adjacent to genes that are expressed only in muscle cells (**Figure 23.19**). Therefore, when myogenic bHLH proteins are activated, they can bind to these enhancers and activate the expression of many different muscle-cell-specific genes. They may exert their effects via alterations in chromatin structure or via the activation of RNA polymerase to a transcriptionally active state. In this way, myogenic bHLH proteins function as master switches that activate the expression of many muscle-specific genes. When the encoded proteins are synthesized, they change the characteristics of an undifferentiated cell into those of a highly specialized skeletal muscle cell.

Another important aspect of myogenic bHLH proteins is that their activity is regulated by dimerization, which occurs via the helix-loop-helix domain. As shown in Figure 23.19, heterodimers—dimers formed from two different proteins—may be activating or inhibitory. When a heterodimer forms between a myogenic bHLH protein and an E protein, which also contains a basic domain, the heterodimer binds to the DNA and activates gene expression (Figure 23.19a). However, when a heterodimer forms between a myogenic bHLH protein and a protein called

Id (for <u>i</u>nhibitor of <u>d</u>ifferentiation), the heterodimer cannot bind to DNA because the Id protein lacks a basic domain (Figure 23.19b). Two basic domains are needed for the heterodimer to bind to the DNA. The Id protein is produced during early stages of development and prevents myogenic bHLH proteins from promoting muscle differentiation too soon. At later stages of development, the amount of Id protein falls, and myogenic bHLH proteins can then combine with E proteins to induce muscle-cell differentiation.

23.3 PLANT DEVELOPMENT

In developmental plant biology, the model organism for genetic analysis is *Arabidopsis thaliana* (**Figure 23.20**). Unlike most flowering plants, which have long generation times and large genomes, *Arabidopsis* has a generation time of about two months

(a) Action of bHLH–E heterodimer

(b) Action of bHLH–Id heterodimer

FIGURE 23.19 **Regulation of muscle-cell-specific genes by myogenic bHLH proteins. (a)** A heterodimer formed from a myogenic bHLH protein and an E protein can bind to a muscle-cell-specific enhancer sequence and activate gene expression. **(b)** When a myogenic bHLH protein forms a heterodimer with an Id protein, it cannot bind to the DNA and therefore does not activate gene transcription.

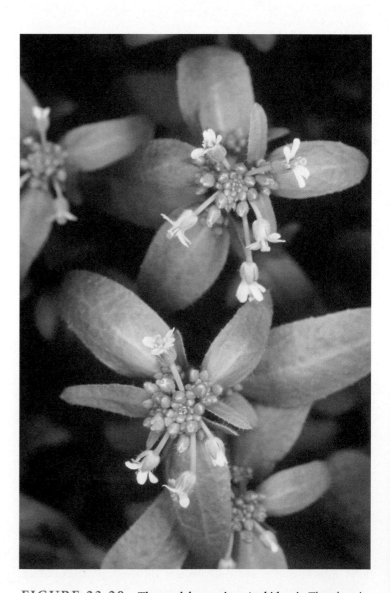

FIGURE 23.20 **The model organism *Arabidopsis.*** The plant is relatively small, making it easy to grow many of them in the laboratory.

and a genome size of 14×10^7 bp, which is similar to *Drosophila* and *C. elegans*. A flowering *Arabidopsis* plant produces a large number of seeds and is small enough to be grown in the laboratory. Like *Drosophila*, *Arabidopsis* can be subjected to mutagens to generate mutations that alter developmental processes. The small genome size of this organism makes it relatively easy to map these mutant alleles and eventually clone the relevant genes (as described in Chapters 18 and 20).

The morphological patterns of growth are markedly different between animals and plants. As described previously, animal embryos become organized along antero-posterior, dorso-ventral, and right-left axes, and then they subdivide into segments. By comparison, the form of plants has two key features. The first is the root-shoot axis. Most plant growth occurs via cell division near the tips of the shoots and the bottoms of the roots.

Second, this growth occurs in a well-defined radial pattern. For example, early in *Arabidopsis* growth, a rosette of leaves or flowers is produced from buds that emanate in a spiral pattern directly from the main shoot (see Figure 23.20). Later, the shoot generates branches that will also produce leaf buds as they grow. Overall, the radial pattern in which a plant shoot gives off the buds that produce leaves, flowers, and branches is an important mechanism that determines much of the general morphology of the plant.

At the cellular level, too, plant development differs markedly from animal development. For example, cell migration does not occur during plant development. In addition, the development of a plant does not rely on morphogens that are deposited asymmetrically in the oocyte. In plants, an entirely new individual can be regenerated from many types of somatic cells. In other words, many plant cells are **totipotent,** meaning that they have the ability to differentiate into every cell type and to produce an entire individual. By comparison, animal development typically relies on the organization within an oocyte as a starting point for development.

In spite of these apparent differences, the underlying molecular mechanisms of pattern development in plants still share some similarities with those in animals. In this section, we will consider a few examples in which the genes encoding transcription factors play a key role in plant development.

Plant Growth Occurs from Meristems Formed During Embryonic Development

Figure 23.21 illustrates a common sequence of events that takes place in the development of seed plants such as *Arabidopsis*. After fertilization, the first cellular division is asymmetrical and produces a smaller cell, called the apical cell, and a larger basal cell (Figure 23.21a). The apical cell will give rise to most of the embryo, and it will later develop into the shoot of the plant. The basal cell will give rise to the root, along with the suspensor that produces extraembryonic tissue required for seed formation. At the heart stage, which is composed of only about 100 cells, the basic organization of the plant has been established (Figure 23.21d). The **shoot meristem** will arise from a group of cells located between the cotyledons. These cells are the precursors that will produce the shoot of the plant, along with lateral structures such as leaves and flowers. The **root meristem** is located at the opposite side and will create the root.

A meristem contains an organized group of actively dividing stem cells. As discussed in Chapter 19, stem cells retain the ability to divide and differentiate into multiple cell types. As they grow, meristems produce offshoots of proliferating cells. On a shoot meristem, for example, these offshoots or buds give rise to structures such as leaves and flowers. The organization of a shoot meristem is shown in **Figure 23.22**. It is organized into three areas called the **organizing center,** the **central zone,** and the **peripheral zone.** The role of the organizing center is to ensure the proper organization of the meristem and preserve the correct number

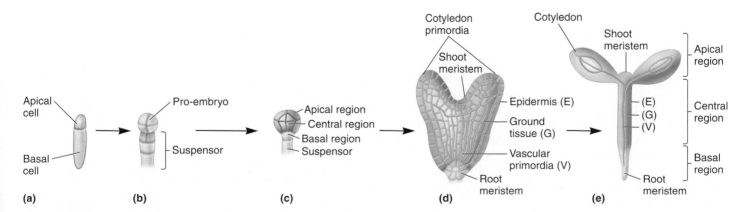

FIGURE 23.21 **Developmental steps in the formation of a plant embryo.** (**a**) The two-cell stage consists of the apical cell and basal cell. (**b**) The eight-cell stage consists of a pro-embryo and a suspensor. The suspensor gives rise to extraembryonic tissue, which is needed for seed formation. (**c**) At this stage of embryonic development, the three main regions of the embryo (i.e., apical, central, and basal) have been determined. (**d**) At the heart stage, all of the plant tissues have begun to form. Note that the shoot meristem is located between the future cotyledons, and the root meristem is on the opposite side. (**e**) A seedling.

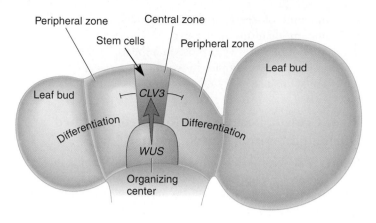

FIGURE 23.22   **Organization of a shoot meristem.** The organization of a shoot meristem is controlled by the *WUS* and *CLV3* genes, which are acronyms for *Wuschel* and *clavata*, respectively. The *WUS* gene is expressed in the organizing center and induces the cells in the central zone to become undifferentiated stem cells. The red arrow indicates that the WUS protein induces these stem cells to turn on the *CLV3* gene, which encodes a secreted protein that binds to receptors in the cells of the peripheral zone. The black lines with a vertical slash indicate that the CLV3 protein prevents the cells in the peripheral zone from expressing the *WUS* gene. This limits the area of *WUS* gene expression to the underlying organizing center and thereby maintains a small population of stem cells at the growing tip. The cells in the peripheral zone are allowed to grow and differentiate into lateral structures such as leaves.

of actively dividing stem cells. The central zone is an area where undifferentiated stem cells are always maintained. The peripheral zone contains dividing cells that will eventually differentiate into plant structures. For example, the peripheral zone may form a bud, which will produce a leaf or flower.

In *Arabidopsis*, the organization of a shoot meristem is controlled by two critical genes termed *WUS* and *CLV3*. The *WUS* gene encodes a transcription factor that is expressed in the organizing center (Figure 23.22). The expression of the *WUS* gene induces the adjacent cells in the central zone to become undifferentiated stem cells. These stem cells then turn on the *CLV3* gene, which encodes a secreted protein. The CLV3 protein binds to receptors in the cells of the peripheral zone, preventing them from expressing the *WUS* gene. This limits the area of *WUS* gene expression to the underlying organizing center and thereby maintains a small population of stem cells at the growing tip. A shoot meristem in *Arabidopsis* contains only about 100 cells. The inhibition of *WUS* expression in the peripheral cells also allows them to embark on a path of cell differentiation so they can produce structures such as leaves and flowers.

In the seedling shown earlier in Figure 23.21e, three main regions are observed. The **apical region** produces the leaves and flowers of the plant. The **central region** (not to be confused with the central zone) creates the stem. The radial pattern of cells in the central region causes the radial growth observed in plants. Finally, the **basal region** produces the roots. Each of these three

regions develops differently as indicated by their unique cell division patterns and distinct morphologies. In addition, by analyzing mutants that disrupt the developmental process, researchers have discovered that these three regions express different sets of genes. Plant biologists have identified a category of genes, known as **apical-basal-patterning genes,** that are important in the early stages of development. As described in **Table 23.2,** defects in apical-basal-patterning genes cause dramatic effects in each of the three main regions. For example, the *gurke* gene is necessary for apical development. When it is defective, the embryo lacks apical structures. Currently, a great amount of effort is directed toward the identification of genes that govern pattern formation in the three regions of *Arabidopsis* and other plants.

TABLE 23.2

Examples of *Arabidopsis* Apical-Basal-Patterning Genes That Affect the Development of the Apical, Central, or Basal Region

Gene	Description
Apical	
Gurke	Minor loss-of-function alleles in the *Gurke* gene produce seedlings with highly reduced or no cotyledons. Complete loss-of-function eliminates the entire shoot.
Pin1	Encodes a putative auxin exporter that is expressed in the peripheral zone. It plays a role in the location of bud sites, which give rise to leaves and flowers.
Aintegumenta	Encodes a transcription factor that is also expressed in the peripheral zone. Its expression maintains the proliferative cell state during the growth of lateral buds.
Central	
Fackel	Encodes an enzyme involved in plant sterol synthesis. Loss-of-function alleles show a severe defect in stem growth.
Scarecrow	Encodes a transcription factor protein that plays a role in the asymmetric division that produces the radial pattern of growth in the stem. Note: The scarecrow protein also affects cell division patterns in roots and plays a role in sensing gravity.
Basal	
Monopterous	Encodes a transcription factor. When this gene is defective, the plant embryo cannot initiate the formation of root structures, but root structures can be formed postembryonically under the correct growth conditions. This gene seems to be required for organizing root formation in the embryo but is not required for root formation per se.
Hobbit	Encodes a subunit of a protein that functions as a cell cycle checkpoint during anaphase. *Hobbit* function may be required to couple cell division to cell differentiation in the root meristem or to restrict the response to growth hormones such as auxin. Loss-of-function alleles result in plants that are incapable of forming a root.
Overall Organization	
Gnom	Plays a role in the stable fixation of the apical-basal axis of the *Arabidopsis* embryo.

Plant Homeotic Genes Control Flower Development

Although the term homeotic was coined by William Bateson to describe homeotic mutations in animals, the first known mutations in homeotic genes were observed in plants. In ancient Greece and Rome, for example, double flowers in which stamens were replaced by petals were noted. In current research, geneticists have been studying these types of mutations to better understand developmental pathways in plants. Many homeotic mutations affecting flower development have been identified in *Arabidopsis* and also in the snapdragon (*Antirrhinum majus*).

Examples of homeotic mutants in *Arabidopsis* are shown in **Figure 23.23**. A normal *Arabidopsis* flower is composed of four concentric whorls of structures (Figure 23.23a). The outer whorl contains four sepals, which protect the flower bud before it opens. The second whorl is composed of four petals, and the third whorl contains six stamens, the structures that make the male gametophyte, pollen. Finally, the innermost whorl contains two carpels, which are fused together. The carpel produces the female gametophyte. The homeotic mutants shown in Figure 23.23 have undergone transformations of particular whorls. For example, in Figure 23.23b, the sepals have been transformed into carpels, and the petals into stamens.

By analyzing the effects of many different homeotic mutations in *Arabidopsis*, Elliot Meyerowitz and colleagues proposed the **ABC model** for flower development. In this model, three classes of genes, called *A*, *B*, and *C*, govern the formation of sepals, petals, stamens, and carpels. More recently, a fourth category of genes called the *Sepallata* genes (*SEP* genes) have been

found to be required for this process. **Figure 23.24** illustrates how these genes affect normal flower development in *Arabidopsis*. Gene *A* products are made in the outermost whorl (whorl 1), promoting sepal formation. In whorl 2, gene *A*, gene *B*, and *SEP* gene products are made, which promotes petal formation. The expression of gene *B*, gene *C*, and *SEP* genes in whorl 3 causes stamens to be made. Finally, in whorl 4, gene *C* and *SEP* genes promote carpel formation.

What are the phenotypic effects of homeotic mutants affecting the *A*, *B*, *C*, or *SEP* genes? In the original ABC model, it was proposed that genes *A* and *C* repress each other's expression, and gene *B* functions independently. In a mutant defective in gene *A* expression, gene *C* will also be expressed in whorls 1 and 2. This produces a carpel-stamen-stamen-carpel arrangement. When gene *B* is defective, a flower cannot make petals or stamens. Therefore, a gene *B* defect yields a flower with a sepal-sepal-carpel-carpel arrangement. When gene *C* is defective, gene *A* is expressed in all four whorls. This results in a sepal-petal-petal-sepal pattern. If the expression of *SEP* genes is defective, the flower consists entirely of sepals, which is the origin of the gene's name.

Overall, the types of genes described in Figure 23.24 promote cell differentiation that leads to sepal, petal, stamen, or carpel structures. But what happens if genes *A*, *B*, and *C* are all defective? As shown in Figure 23.23c, this produces a "flower" that is composed entirely of leaves! These results indicate that the leaf structure is the default pathway and that the *A*, *B*, *C*, and *SEP* genes cause development to deviate from a leaf structure in order to make something else. In this regard, the sepals, petals,

(a) **Normal flower**

(b) **Single homeotic mutant**

(c) **Triple mutant**

FIGURE 23.23   **Examples of homeotic mutations in *Arabidopsis*.**

Genes→Traits   **(a)** A normal flower. It is composed of four concentric whorls of structures: sepals, petals, stamens, and carpel. **(b)** A homeotic mutant in which the sepals have been transformed into carpels and the petals have been transformed into stamens. **(c)** A triple mutant in which all of the whorls have been changed into leaves.

FIGURE 23.24 **The ABC model of homeotic gene action in *Arabidopsis*.** Note: This is a revised model based on the recent identification of *SEP* genes. The black lines with a vertical slash indicate that the *A* gene product represses the *C* gene product, and vice versa.

stamens, and carpels can be viewed as modified leaves. Interestingly, the German philosopher and poet Johann Goethe originally proposed this idea—that flower formation comes from modifications of the leaf—over 200 years ago.

Arabidopsis has two types of gene *A* (*apetala1* and *apetala2*), two types of gene *B* (*apetala3* and *pistillata*), one type of gene *C* (*agamous*), and three *SEP* genes (*SEP1*, *SEP2*, and *SEP3*). All of these plant homeotic genes encode transcription factor proteins that contain a DNA-binding domain and a dimerization domain. However, the *Arabidopsis* homeotic genes do not contain a sequence similar to the homeobox found in animal homeotic genes.

Like the *Drosophila* homeotic genes, plant homeotic genes are part of a hierarchy of gene regulation. Genes that are expressed within a flower bud produce proteins that activate the expression of these homeotic genes. Once they are transcriptionally activated, the homeotic genes then regulate the expression of other genes, the products of which promote the formation of sepals, petals, stamens, or carpels.

23.4 SEX DETERMINATION IN ANIMALS AND PLANTS

To end our discussion of development, let's consider how genetics plays a role in the development of male and female individuals. In the animal kingdom, the existence of two sexes is nearly universal. In plants, most species produce male and female haploid reproductive cells on the same individual. However, some plant species are sexually dimorphic. Such species produce two distinct types of individuals that have different abilities to produce male and female gametophytes. We will consider such types of plants in this section.

The underlying factors that distinguish female versus male development vary widely. In animals, sex determination is often caused by differences in chromosomal composition (Table 23.3). In flies and worms, the ratio of the number of X chromosomes to the number of sets of autosomes determines sex. By comparison, in mammals, the presence of the *Sry* gene on the Y chromosome causes maleness. Similarly, plants that are sexually dimorphic

TABLE 23.3

Mechanisms of Sex Determination in Selected Species

Species	Mechanism
Drosophila, *C. elegans*	Ratio of the number of X chromosomes to the number of sets of autosomes determines sex. An X:A ratio of 1.0 results in females (or hermaphrodites in *C. elegans*), and an X:A ratio of 0.5 produces males.
Mammals	The presence of the *Sry* gene on the Y chromosome causes maleness.
Birds	Females are ZW, and males are ZZ.
Bees	Males are haploid, and females (workers and queen bees) are diploid.
Certain species of reptiles, fish, and turtles	Environmental conditions, usually temperature, influence the ratio of male and female offspring.
Plants	The male plant produces pollen. In some sexually dimorphic species, the males are XY. In other species, there are not distinct sex chromosomes, but the male plant is heterozygous for a dominant allele that suppresses the female pathway.

typically have sex chromosomes in which the plant that produces the male gametophyte has two types of sex chromosomes, similar to the XY system in mammals. In birds, the female has heteromorphic sex chromosomes: ZW birds are female, and ZZ birds are male. Finally, sex determination is not always caused by two types of sex chromosomes. In bees, for example, males are haploid, and females are diploid.

In many animal species, variation in chromosome composition is not the underlying factor that distinguishes female from male development. Sex determination in many species of reptiles and fish is controlled by environmental factors such as temperature. For example, in the American alligator (*Alligator mississippiensis*), temperature controls sex determination. Eggs incubated at 33°C typically result in ~100% male individuals, whereas eggs incubated at lower or higher temperatures such as 30°C or 34.5°C result in ~100% or 95% females, respectively.

The adoption of one of two sexual fates is an event that has been studied in great detail in several species. Researchers have discovered that sex determination is a process controlled genetically by a hierarchy of genes that exert their effects in early embryonic development. In this section, we will consider features of these hierarchies in *Drosophila*, *C. elegans*, mammals, and plants.

In *Drosophila*, Sex Determination Involves a Regulatory Cascade That Includes Alternative Splicing

In diploid fruit flies, XX flies develop into females, and X0 flies become males. The ratio of the number of X chromosomes to

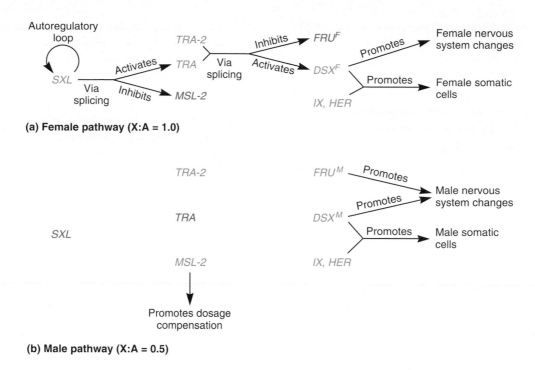

(a) Female pathway (X:A = 1.0)

(b) Male pathway (X:A = 0.5)

FIGURE 23.25 **Sex determination pathway for *Drosophila melanogaster*.** Genes or gene products that are functionally expressed are shown in light orange, while those that are not expressed are shown in blue. The gene names are acronyms for the phenotypes that result from mutations that cause loss-of-function or aberrant expression. These are as follows: *SXL* (<u>s</u>e<u>x</u> <u>l</u>ethal), *MSL* (<u>m</u>ale <u>s</u>ex <u>l</u>ethal), *TRA* (<u>tra</u>nsformer), *DSX* (<u>d</u>ouble <u>s</u>e<u>x</u>), *FRU* (<u>fru</u>itless), *IX* (<u>i</u>nterse<u>x</u>), and *HER* (<u>her</u>maphrodite). Note: This is a simplified pathway. More gene products are involved than shown here.

the number of sets of autosomes is the determining factor. In a diploid fly that carries two sets of chromosomes, this ratio is 1.0 in females versus 0.5 in males. Although male fruit flies usually carry a Y chromosome, it is not necessary for male development. The mechanism of sex determination begins in early embryonic development and involves a regulatory cascade composed of several genes. Females and males follow one of two alternative pathways. Simplified versions of these pathways are depicted in **Figure 23.25**.

Let's begin with the pathway that produces female flies. In females, the higher ratio of X chromosomes results in the embryonic expression of a gene designated *SXL* (Figure 23.25a). The *SXL* gene product is a protein that functions in the splicing of pre-mRNA. In female embryos, the Sxl protein enhances its own expression by splicing its own pre-mRNA, an event termed an **autoregulatory loop**. In addition, it splices the pre-mRNA from two other genes called *MSL-2* and *TRA*. The Sxl protein promotes the splicing of the *MSL-2* pre-mRNA in a way that introduces an early stop codon in the coding sequence and thereby produces a shortened version of the Msl-2 protein that is functionally inactive. By comparison, the Sxl protein promotes the splicing of *TRA* pre-mRNA to produce an mRNA that is translated into a functional protein. Therefore, *SXL* activates *TRA*.

The *TRA* gene product and a constitutively expressed product from a gene called *TRA-2* are also splicing factors. In the female, they cause the alternative splicing of the pre-mRNAs that are expressed from the *FRU* and *DSX* genes.

The tra and tra-2 proteins cause these pre-mRNAs to be spliced into mRNAs designated *FRU^F* and *DSX^F*, respectively. The female-specific *FRU^F* mRNA is not translated into a sex-specific gene product. However, the *DSX^F* mRNA, together with two other gene products from the *IX* and *HER* genes, promotes female sexual development and controls some aspects of female-specific behavior via the central nervous system. The dsx^F protein is known to be a transcription factor that regulates certain genes that promote these changes.

How are males produced? In X0 or XY flies, the *SXL* gene is transcriptionally activated, but it is spliced in a way that places an early stop codon in the coding sequence. Therefore, a functional Sxl protein is not made (Figure 23.25b). This permits the expression of *MSL-2*, which promotes dosage compensation. In fruit flies, dosage compensation is accomplished by turning up the expression of X-linked genes in the male to a level that is twofold higher. Therefore, even though the male has only one X chromosome, the expression of X-linked genes in the male and female are approximately equal. The absence of *SXL* expression in male embryos also promotes the development of maleness. Without *SXL*, the *TRA* mRNA is not properly spliced, so *TRA* is not expressed. Without the tra protein, the *FRU* and *DSX* mRNAs are spliced in a different way to produce mRNAs designated *FRU^M* and *DSX^M*. Along with *IX* and *HER* gene products, the dsx^M protein promotes male development as well as male-specific behavior. Like the dsx^F protein, the dsx^M protein is a transcription factor that regulates certain genes. Because *DSX^M* is spliced differently from DSX^F, the dsx^M protein's structure is different from the dsx^F protein, and this difference alters the regulation pattern of dsx^M. In addition, the FRU^M gene product is necessary for the regulation of genes involved in male-specific behaviors.

In *C. elegans* the Ratio of X Chromosomes to Sets of Autosomes Initiates a Regulatory Cascade That Determines Sex

C. elegans has two sexes: hermaphrodites and males. During larval development, the hermaphrodites produce sperm that are stored in a structure called the spermathecae. In adulthood, hermaphrodites are anatomically female. They produce oocytes that are fertilized when they are forced through the spermathecae by muscular contractions. Fertilization can take place in either of two ways. One possibility is that a sperm can fertilize an oocyte from the same worm. In other words, the hermaphrodite is capable of self-fertilization. Alternatively, a hermaphrodite can mate with a male worm, which produces only sperm.

The sexual identity of *C. elegans*, hermaphrodite versus male, is a trait determined very early in embryonic development. As in *Drosophila*, sexual identity is controlled by the activities of many genes that interact in a regulatory cascade that determines the sex of the worm and controls dosage compensation. In *C. elegans*, the expression of genes on the X chromosome in the hermaphrodite, which carries two X chromosomes, is decreased to 50% compared to males, which carry one X chromosome.

As in *Drosophila*, the ratio of the number of X chromosomes to the number of sets of autosomes (the X:A ratio) is the factor that causes the regulatory cascade to follow one of two alternative pathways (**Figure 23.26**). Let's first consider the events that occur during early embryonic development in the hermaphrodite. Because hermaphrodites have two X chromosomes, this results in an X:A ratio of 1.0, compared to the ratio of 0.5 for males. This higher X:A ratio results in an enhanced expression of *fox-1* and *sex-1*, which are located on the X chromosome. In the XX hermaphrodite, the expression of *fox-1* and *sex-1* inhibits the expression of *xol-1*. The protein product of the *xol-1* gene inhibits the expression of three genes called *sdc-1*, *sdc-2*, and *sdc-3*. Because *xol-1* activity is inhibited in hermaphrodites, this permits the expression of the *sdc* genes. The expression of *sdc-1*, *sdc-2*, and *sdc-3* has two effects. First, these three genes are necessary for dosage compensation, and second, they are necessary to permit the downstream expression of *tra-1*, which is needed to promote hermaphrodite development. This second effect occurs via a chain of events that are also shown in Figure 23.26a. The *sdc* gene products inhibit the expression of *her-1*. When *her-1* is inhibited, this permits the expression of *tra-2*. When *tra-2* is active, this inhibits the activities of *fem-1*, *fem-2*, and *fem-3*. When the *fem* gene products are inhibited, this permits the expression of *tra-1*. It is the *tra-1* gene product that promotes hermaphrodite development and inhibits male development.

Figure 23.26b describes the pathway for male development. Because males contain a single X chromosome, the expression of *fox-1* and *sex-1* is insufficient to inhibit the expression of *xol-1*. When *xol-1* is expressed, its protein product inhibits the expression of the *sdc* genes. This prevents dosage compensation and permits the expression of *her-1*. The *her-1* gene product

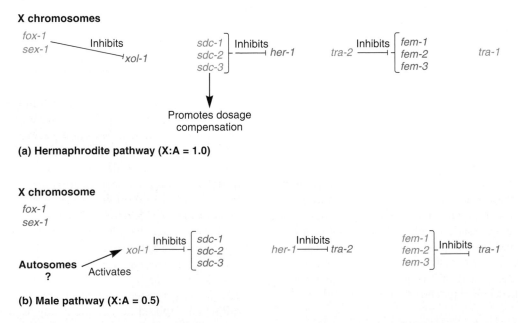

(a) Hermaphrodite pathway (X:A = 1.0)

(b) Male pathway (X:A = 0.5)

FIGURE 23.26 **Sex determination pathway for *C. elegans*.** Genes or gene products that are expressed at relatively high levels are shown in light orange, while those that are not expressed or are expressed at low levels are shown in blue. The gene names are acronyms that usually refer to the effects of mutations. These are as follows: *fox* (feminizing on X), *sex* (sex determination), *xol* (XO lethal), *sdc* (sex determination and dosage compensation), *her* (hermaphrodite), *tra* (transformer), and *fem* (feminization). Note: This is a simplified pathway.

then inhibits *tra-2*. This, in turn, permits the expression of the *fem* genes, which inhibit the expression of *tra-1*. Without *tra-1* expression, the worm develops into a male instead of a hermaphrodite. Experimentally, it is has been shown that XX worms lacking a functional *tra-1* gene develop into males.

In Mammals, the *Sry* Gene on the Y Chromosome Determines Maleness

In most mammals, such as humans, mice, and marsupials, the presence of the *Sry* gene on the Y chromosome determines maleness. In cases of abnormal sex chromosome composition such as XXY, an individual develops into a male. The *Sry* gene, which is located on the Y chromosome, causes the sex determination pathway to follow a male developmental scheme. The *Sry* gene encodes a protein that contains a DNA-binding domain called an HMG box, which is found in a broad category of DNA-binding proteins known as the high mobility group. In several of these proteins, the HMG box is known to cause DNA bends. Thus far, the ability of the Sry protein to promote male sex determination is not well understood, but it may act as a transcription factor and/or promote changes in DNA architecture, much like the DNA remodeling proteins discussed in Chapters 10 and 15.

Sex determination in mammals, like that of fruit flies and worms, involves a cascade that is initiated in early embryonic development. However, the details of the pathway are not completely elucidated. **Figure 23.27** illustrates a simplified pathway.

(a) Male pathway (XY)

(b) Female pathway (XX)

*The *Sry* gene is located on the Y chromosome, so it is found only in males.

FIGURE 23.27   **Sex determination pathway for mammals.** Genes or gene products that are expressed at relatively high levels are shown in light orange, while those that are not expressed or are inactive are shown in blue. The regulation of the *DAX1* gene is not well understood, but it may involve activation via *SF1* and/or *WT1* gene products. The gene names are acronyms that usually refer to the effects of mutations and/or their relatedness to other genes or chromosomal regions. These are as follows: *SF1* (steroidogenic factor-1), *WT1* (Wilm's tumor gene), *Sry* (sex determining region of the Y chromosome), *DMRT1* (doublesex- and mab-3-related transcription factor), *SOX9* (Sry-like HMG box), and *DAX1* (dosage-sensitive sex-reversal, adrenal hypoplasia congenital, X chromosome). Note: This is a simplified pathway.

Several genes in mammals have been identified that are expressed very early in embryonic development and may be directly or indirectly involved in turning on the *Sry* gene. For example, two genes, designated *SFI* and *WTI*, are expressed in the early embryo prior to sexual differentiation and may regulate *Sry* expression. Once activated, researchers postulate that *Sry* expression causes the expression of other genes that promote male development. However, the target genes that are turned on by the Sry protein have not been definitively identified. One likely candidate is a gene called *SOX9*. Like the *Sry* gene, *SOX9* encodes a protein with an HMG domain. In addition, the SOX9 protein is larger than the Sry protein because it also contains two transcriptional activation domains, indicating that it is a transcriptional activator. *SOX9* expression is necessary for male development, because a loss-of-function allele of *SOX9* results in female development in XY animals. Furthermore, when researchers produced XX mice with three copies of the *SOX9* gene, such individuals developed into males.

Two other genes, designated *DAX1* and *DMRT1*, also play a role in sex determination. The *DAX1* gene is X linked, and its gene product is thought to prevent male development. In XX animals, its expression remains high, in contrast to XY males. XY animals with two copies of the *DAX1* gene develop into females when *Sry* gene expression is low due to a weak allele in the *Sry* gene. This suggests that *DAX1* can inhibit the effects of the *Sry* gene. However, the *DAX1* gene is not needed for female development because XX mice lacking the *DAX1* gene develop as normal females. *DAX1* encodes a hormone-receptor protein. Further research is needed to understand the pathways that are activated via this receptor. Finally, the *DMRT1* gene in mammals is evolutionarily related to the *mab-3* gene in *C. elegans* and the *DSX* gene in *Drosophila*. These genes encode transcription factors involved in the differentiation of the gonads. In mammals, *DMRT1* expression is gonad specific, and it is expressed at higher levels in the testes. XY mice that are lacking the *DMRT1* gene develop as males, but they are infertile due to severe defects in testis structure and the inability to produce sperm cells.

The mechanism of dosage compensation in mammals involves the process of X inactivation. In normal females, one of the two X chromosomes in females is compacted into a Barr body. This topic is described in Chapter 7.

In Sexually Dimorphic Plants, the Male Plant Is Usually Heteromorphic

As mentioned earlier, most species of plants, about 95%, are sexually monomorphic. Only a single type of individual is produced that can make both male and female gametophytes. By comparison, sexually dimorphic species have two separate types of individuals. The most common types are **dioecious,** in which one kind of individual produces male gametophytes, while the other produces female gametophytes. A few plants species are gynodioecious, which produce hermaphrodite plants and female plants. Alternatively, a few species are androdioecious and produce hermaphrodite plants and male plants.

The genetics of sex determination in dioecious plant species is beginning to emerge. Because *Arabidopsis* is sexually monomorphic, researchers have had to turn to other species to study sex determination in plants. For example, the white campion, *Silene latifolia*, has been the subject of numerous investigations. In this species, sex chromosomes, designated X and Y, are responsible for sex determination. The male plant has heteromorphic sex chromosomes, XY; the female plant is XX. In other dioecious species, cytological examination of the chromosomes does not always reveal distinct types of sex chromosomes. Nevertheless, the male plants appear to be the heterozygote. This has been determined because male plants are often "inconstant," which means that they produce an occasional fruit. When a male plant undergoes such self-fertilization, the seeds from the fruit produce a 3:1 ratio of male to female plants. This is the expected result if the male parent was heterozygous for a dominant gene that promotes maleness.

Researchers are beginning to identify genes that are important for sex determination in plants, though further studies are needed to unravel the genetic hierarchy that promotes sexual development. A chromosomal site designated Su^F plays a role in this process. The Su^F locus is believed to contain a gene that acts as a dominant suppressor of the female pathway and prevents carpel development. It is responsible for producing the 3:1 ratio when male plants undergo self-fertilization. In addition, other loci that appear to control early and late anther development have been discovered. Along with Su^F, these loci are linked on the Y chromosome in *S. latifolia*. Eventually, the discovery of more genes that control sexual development will allow researchers to propose alternative pathways that promote male and female development in dioecious plant species.

CONCEPTUAL SUMMARY

In this chapter, we have examined the role that genetics plays in the development of multicellular organisms. Each organism has its own developmental program, driven by a genetic hierarchy that encodes transcription factors and other types of regulatory proteins. In *Drosophila*, the developmental program begins in the oocyte. It is here that a few key gene products act as morphogens, molecules that convey positional information and promote developmental changes. Morphogens provide positional information that initiates body plan development. Maternal effect gene products are deposited asymmetrically in the oocyte and serve to organize the antero-posterior and dorso-ventral axes. Once these regions have been established, the next step of the developmental process organizes the embryo into segments.

Three categories of genes, known as gap genes, pair-rule genes, and segment-polarity genes, act sequentially in the formation of body segments. Gap genes divide the embryo into broad regions; the pair-rule genes control the identity of each parasegment; and the segment-polarity genes divide each parasegment into anterior and posterior regions. A segment, which is a morphological feature, is formed from the posterior region of one parasegment and the anterior region of an adjacent parasegment. The expression of homeotic genes dictates the phenotypic characteristics of each segment. These genes encode transcription factors that activate other genes responsible for determining the morphological characteristics of cells. A protein domain called the homeodomain promotes the binding of homeotic proteins to DNA.

In *C. elegans*, the availability of a complete lineage diagram has allowed geneticists to identify heterochronic mutations that alter the timing of cell fate and lead to abnormalities in morphology. This phenomenon illustrates the importance of coordinated cell division in determining cell fate during development.

In vertebrates, much less is known about the programs that promote development. Nevertheless, they appear to bear many similarities to the programs in simpler invertebrates. For example, *Hox* complexes in mice contain groups of genes that are homologous to homeotic genes in *Drosophila*. Furthermore, experiments with gene knockouts of particular *Hox* genes indicate that they play a key role in patterning the antero-posterior axis. Researchers have identified genes involved in the differentiation of cells into highly specialized cell types. As an example, the myogenic bHLH proteins play a critical role in the differentiation of skeletal muscle cells.

Morphologically, plant development differs markedly from animal development. Plants are organized along a root-shoot axis and growth occurs in a well-defined radial pattern. Plant geneticists have identified some of the genes that control the development of the meristems, organized groups of actively dividing cells. Geneticists have also discovered many homeotic genes that dictate the morphology of particular structures later in development. In the ABC model, four classes of homeotic genes, *A*, *B*, *C*, and *SEP*, are involved in the formation of the four whorls that make up a flower. These results show that plant development is dictated by a genetic program that bears some similarities to the programs found in animals.

Sex determination in animals and plants relies on a regulatory cascade that begins in early embryonic development. Different factors provide the basis for choosing between the female- and male-specific pathways. The hierarchy in any given species may involve a variety of control mechanisms including regulation of transcription factors, alternative splicing, and dosage compensation. The outcome of such regulation is to promote dosage compensation and the development of female or male structures and behaviors.

EXPERIMENTAL SUMMARY

We have seen that a genetic approach, namely, the identification of mutant alleles, has been essential to our understanding of development. Researchers, particularly those studying invertebrates and *Arabidopsis*, have identified many different mutant alleles that alter key steps in the developmental process. In *Drosophila*, for example, these include mutations in maternal effect genes, segmentation genes, and homeotic genes. The molecular characterization of these genes has shown that many of them encode transcription factors that influence the expression of other genes. The study of developmental mutations has enabled scientists to piece together a genetic hierarchy that underlies many developmental programs.

The characterization of developmental mutants has also been correlated with the spatial localization of gene products using techniques such as *in situ* hybridization. This enables researchers to determine when and where a gene product is made during oogenesis or embryogenesis. This approach has shown that some mutations that alter development are gain-of-function alleles, which cause a gene to be expressed in the wrong place or at the wrong time. Other mutations are loss-of-function alleles, which cause a defect in gene expression. In *C. elegans*, the availability of a lineage diagram has also allowed researchers to examine how heterochronic mutations, such as those that occur in the *lin-14* gene, can affect the timing of developmental steps.

The key experimental advantage of studying invertebrates is the ability to identify mutations that alter developmental steps. This approach is not as easy in vertebrates, but a reverse genetics

approach is proving to be successful. Using invertebrate genes as probes, researchers have identified vertebrate genes, such as the *Hox* genes in mice, that are homologous to the homeotic genes of invertebrates. To understand their role, the wild-type *Hox* genes have been mutated in vitro and reintroduced into mice, replacing the normal gene. This is called a gene knockout. The results of this method are consistent with the idea that the *Hox* genes function as the mouse homologues of the *Drosophila* homeotic genes. However, further research will be needed to understand the individual roles of these 38 genes.

In plants, a genetic approach is also proving effective in elucidating the development process. Early in development, a category of genes known as the apical-basal-patterning genes appear necessary for the formation of the apical and basal regions of the embryo. These genes were identified by investigating the effects of mutations on the developing plant embryo. Similarly, mutations in plant homeotic genes have been discovered by their ability to abnormally transform one plant structure into another (e.g., sepals into petals). A comparison of single, double, and triple homeotic mutations provided the framework for the ABC model of flower development.

PROBLEM SETS & INSIGHTS

Solved Problems

S1. Discuss and distinguish the functional roles of the maternal effect genes, gap genes, pair-rule genes, and segment-polarity genes in *Drosophila*.

Answer: These genes are involved in pattern formation of the *Drosophila* embryo. The asymmetric distribution of maternal effect gene products in the oocyte establishes the antero-posterior and dorso-ventral axes. These gene products also control the expression of the gap genes, which are expressed as broad bands in certain regions of the embryo. The overlapping expression of maternal effect genes and gap genes controls the pair-rule genes, which are expressed in alternating stripes. A stripe corresponds to a parasegment. Within each parasegment, the expression of segment-polarity genes defines an anterior and posterior compartment. With regard to morphology, an anterior compartment of one parasegment and the posterior compartment of an adjacent parasegment will form a segment of the fly.

S2. With regard to genes affecting development, what are the phenotypic effects of gain-of-function mutations versus loss-of-function mutations?

Answer: Gain-of-function mutations cause a gene to be expressed in the wrong place, at the wrong time, or at an abnormal level. When the gene is expressed in the wrong place, that region may develop into an inappropriate structure. For example, when *Antp* is abnormally expressed in an anterior segment, this segment develops legs in place of antennae. When gain-of-function mutations cause a gene to be expressed at the wrong time, this can also disrupt the development process. Gain-of-function heterochronic mutations cause cell lineages to be reiterated and thereby alter the course of development. By comparison, loss-of-function mutations result in a defect in the expression of a gene with the result that protein function is reduced or abolished. This usually will disrupt the developmental process, because the cells in the region where the gene is normally expressed will not be directed to develop along the correct pathway.

S3. Mutations in genes that control the early stages of development are often lethal (e.g., see Figure 23.6b). To circumvent this problem, developmental geneticists may try to isolate temperature-sensitive developmental mutants or *ts* alleles. If an embryo carries a *ts* allele, it will develop correctly at the permissive temperature (e.g., 25°C) but will fail to develop if incubated at the nonpermissive temperature (e.g., 30°C). In most cases, *ts* alleles have missense mutations that slightly alter the amino acid sequence of a protein, causing a change in its structure that prevents it from working properly at the nonpermissive temperature. *Ts* alleles are particularly useful because they can provide insight regarding the stage of development when the protein is necessary. Researchers can take groups of embryos that carry a *ts* allele and expose them to the permissive and nonpermissive temperature at different stages of development. In the experiment described next, embryos were divided into five groups and exposed to the permissive or nonpermissive temperature at different times after fertilization.

Time After Fertilization (hours) Group:	1	2	3	4	5
0–1	25°C	25°C	25°C	25°C	25°C
1–2	25°C	30°C	25°C	25°C	25°C
2–3	25°C	25°C	30°C	25°C	25°C
3–4	25°C	25°C	25°C	30°C	25°C
4–5	25°C	25°C	25°C	25°C	30°C
5–6	25°C	25°C	25°C	25°C	25°C
SURVIVAL:	Yes	Yes	Yes	No	Yes

Explain these results.

Answer: By varying the temperature during different stages of development, researchers can pinpoint the stage when the function of the protein encoded by this *ts* allele is critical. As shown, embryos fail to survive if they are subjected to the nonpermissive temperature between

3 and 4 hours after fertilization, but they do survive if subjected to the nonpermissive temperature at other times of development. These results indicate that this protein plays a crucial role at the 3–4-hour stage of development.

S4. An intriguing question in developmental genetics is, How can a particular gene, such as *even-skipped*, be expressed in a multiple banding pattern as seen in Figure 23.10? Another way of asking this question is, How is the positional information within the broad bands of the gap genes able to be deciphered in a way that causes the pair-rule genes to be expressed in this alternating banding pattern? The answer lies in a complex mechanism of genetic regulation. Certain pair-rule genes have several stripe-specific enhancers that are controlled by multiple transcription factors. A stripe-specific enhancer is typically a short segment of DNA, 300 to 500 bp in length, that contains binding sequences recognized by several different transcription factors. This term is a bit misleading because a stripe-specific enhancer is a regulatory region that contains both enhancer and silencer elements.

In 1992, Michael Levine and his colleagues investigated stripe-specific enhancers that are located near the promoter of the *even-skipped* gene. A segment of DNA, termed the stripe 2 enhancer, controls the expression of the *even-skipped* gene; this enhancer is responsible for the expression of the *even-skipped* gene in stripe 2, which corresponds to parasegment 3 of the embryo. The stripe 2 enhancer is a segment of DNA that contains binding sites for four transcription factors that are the products of the *Krüppel, bicoid, hunchback,* and *giant* genes. The Hunchback and Bicoid transcription factors bind to this enhancer and activate the transcription of the *even-skipped* gene. In contrast, the transcription factors encoded by the *Krüppel* and *giant* genes bind to the stripe 2 enhancer and repress transcription. The figure shown below describes the concentrations of these four transcription factor proteins in the region of parasegment 3 (i.e., stripe 2) in the *Drosophila* embryo.

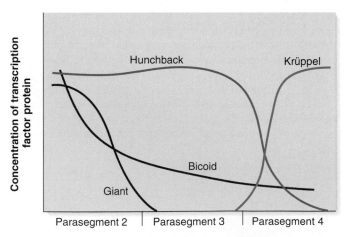

Region of *Drosophila* embryo

To study stripe-specific enhancers, researchers have constructed artificial genes in which the enhancer is linked to a reporter gene, the expression of which is easy to detect. The next figure shows the results of an experiment in which an artificial gene was made by putting the stripe 2 enhancer next to the β-galactosidase gene. This artificial gene was introduced into *Drosophila*, and then embryos containing this gene were analyzed for β-galactosidase activity. If a region of the embryo is expressing β galactosidase, the region will stain darkly because β galactosidase converts a colorless compound into a dark blue compound.

Explain these results.

Answer: As shown in the first figure to this problem, the concentrations of the Hunchback and Bicoid transcription factors are relatively high in the region of the embryo corresponding to stripe 2 (which is parasegment 3). The levels of Krüppel and Giant are very low in this region. Therefore, the high levels of activators and low levels of repressors cause the *even-skipped* gene to be transcribed. In this experiment, β galactosidase was made only in stripe 2 (i.e., parasegment 3). These results show that the stripe-2-specific enhancer controls gene expression only in parasegment 3. Because we know that the *even-skipped* gene is expressed as several alternating stripes (as seen in Figure 23.10), the *even-skipped* gene must contain other stripe-specific enhancers that allow it to be expressed in these other alternating parasegments.

Conceptual Questions

C1. In the case of multicellular animals, what are the four types of cellular processes that occur so a fertilized egg can develop into an adult organism? Briefly discuss the role of each process.

C2. The arrangement of body axes of the fruit fly are shown in Figure 23.4g. Are the following statements true or false with regard to body axes in the mouse?

A. Along the antero-posterior axis, the head is posterior to the tail.

B. Along the dorso-ventral axis, the vertebrae of the back are dorsal to the stomach.

C. Along the dorso-ventral axis, the feet are dorsal to the hips.

D. Along the proximo-distal axis, the toes on the hind legs are distal to the hips of the hind legs.

C3. If you observed fruit flies with the following developmental abnormalities, would you guess that a mutation has occurred in a segmentation gene or a homeotic gene? Explain your guess.

A. Three abdominal segments were missing.

B. One abdominal segment had legs.

C. A fly with the correct number of segments had two additional thoracic segments and two fewer abdominal segments.

C4. Which of the following statements are true with regard to positional information in *Drosophila*?

A. Morphogens are a type of molecule that conveys positional information.

B. Morphogenetic gradients are established only in the oocyte, prior to fertilization.

C. Cell adhesion molecules also provide a way for a cell to obtain positional information.

C5. Discuss the morphological differences between the parasegments and segments of *Drosophila*. Discuss the evidence, providing specific examples, that suggests the parasegments of the embryo are the subdivisions for the organization of gene expression.

C6. Here are schematic diagrams of mutant larvae.

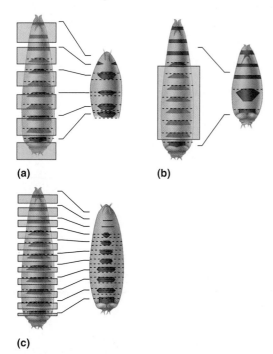

(a)   (b)

(c)

The left side of each pair shows a wild-type larva, with gray boxes showing the sections that are missing in the mutant larva. Which type of gene is defective in each larva: a gap gene, a pair-rule gene, or a segment-polarity gene?

C7. Describe what a morphogen is and how it exerts its effects. What do you expect will happen when a morphogen is expressed in the wrong place in an embryo? List five examples of morphogens that function in *Drosophila*.

C8. What is the meaning of positional information? Discuss three different ways that cells can obtain positional information. Which of these three ways do you think is the most important for the formation of a segmented body pattern in *Drosophila*?

C9. Gradients of morphogens can be preestablished in the oocyte. Also, later in development, morphogens can be secreted from cells. How are these two processes similar and different?

C10. Discuss how the anterior portion of the antero-posterior axis is established. What aspects of oogenesis are critical in establishing this axis? What do you think would happen if the *bicoid* mRNA was not trapped at the anterior end but instead diffused freely throughout the oocyte?

C11. Describe the function of the Bicoid protein. Explain how its ability to exert its effects in a concentration-dependent manner is a critical feature of its function.

C12. With regard to development, what are the roles of the maternal effect genes versus the zygotic genes? Which types of genes are needed earlier in the development process?

C13. Discuss the role of homeotic genes in development. Explain what happens to the phenotype of a fly when a gain-of-function homeotic gene mutation occurs in an abnormal region of the embryo. What are the consequences of a loss-of-function mutation in such a gene?

C14. Describe the molecular features of the homeobox and homeodomain. Explain how these features are important in the function of homeotic genes.

C15. What would you predict to be the phenotype of a *Drosophila* larva whose mother was homozygous for a loss-of-function allele in the *nanos* gene?

C16. Based on the photographs in Figure 23.12, in which segments would the *Antp* gene normally be expressed?

C17. If a mutation in a homeotic gene produced the following phenotypes, would you expect the mutation to be a loss-of-function or a gain-of-function allele? Explain your answer.

A. An abdominal segment has antennae attached to it.

B. The most anterior abdominal segment resembles the most posterior thoracic segment.

C. The most anterior thoracic segment resembles the most posterior abdominal segment.

C18. Explain how loss-of-function mutations in the following categories of genes would affect the morphologies of *Drosophila* larvae:

A. Gap genes

B. Pair-rule genes

C. Segment-polarity genes

C19. What is the difference between a maternal effect gene and a zygotic gene? Of the following genes that play a role in *Drosophila* development, which would be maternal effect genes and which would be zygotic? Explain your answer.

A. *nanos*

B. *Antp*

C. *bicoid*

D. *lab*

C20. Cloning of mammals (such as Dolly) is described in Chapter 19. Based on your understanding of animal development, explain why an enucleated egg is needed to clone mammals. In other words, what features of the oocyte are essential for animal development?

C21. A hypothetical cell lineage is shown here.

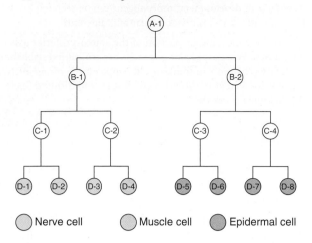

Nerve cell   Muscle cell   Epidermal cell

A gene, which we will call gene *X*, is activated in the B-1 cell, so the B-1 cell will progress through the proper developmental stages to produce three nerve cells (D-1, D-2, and D-3) and one muscle cell (D-4). Gene *X* is normally inactivated in A-1, C-1, and C-2 cells, as well as the four D cells. Draw the expected cell lineages if a heterochronic mutation had the following effects:

A. Gene *X* is turned on one cell division too early.

B. Gene *X* is turned on one cell division too late.

C22. What is a heterochronic mutation? How does it affect the phenotypic outcome of an organism? What phenotypic effects would you expect if a heterochronic mutation affected the cell lineage that determines the fates of intestinal cells?

C23. Discuss the similarities and differences between the *bithorax* and *Antennapedia* complexes in *Drosophila* and the *Hox* gene complexes in mice.

C24. What is cell differentiation? Discuss the role of bHLH proteins in the differentiation of muscle cells. Explain how they work at the molecular level. In your answer, indicate how protein dimerization is a key feature in gene regulation.

C25. The *MyoD* gene in mammals plays a role in muscle cell differentiation, whereas the *Hox* genes are homeotic genes that play a role in the differentiation of particular regions of the body. Explain how the functions of these genes are similar and different.

C26. What is a totipotent cell? In the following examples, which cells would be totipotent? In the case of multicellular organisms, such as humans and corn, explain which cell types would be totipotent.

A. Humans

B. Corn

C. Yeast

D. Bacteria

C27. What is a meristem? Explain the role of meristems in plant development.

C28. Discuss the morphological differences between animals and plants. How are they different at the cellular level? How are they similar at the genetic level?

C29. Predict the phenotypic consequences of each of the following mutations:

A. *apetala1* defective

B. *pistillata* defective

C. *apetala1* and *pistillata* defective

Experimental Questions

E1. Researchers have used the cloning methods described in Chapter 18 to clone the *bicoid* gene and express large amounts of the Bicoid protein. The Bicoid protein was then injected into the posterior end of a zygote immediately after fertilization. What phenotypic results would you expect to occur? What do you think would happen if the Bicoid protein was injected into a segment of a larva?

E2. Compare and contrast the experimental advantages of *Drosophila* and *C. elegans* in the study of developmental genetics.

E3. What is meant by the term cell fate? What is a lineage diagram? Discuss the experimental advantage of having a lineage diagram. What is a cell lineage?

E4. Explain why a lineage diagram is necessary to determine if a mutation is heterochronic.

E5. Explain the rationale behind the use of the "bag of worms" phenotype as a way to identify heterochronic mutations.

E6. Here are the results of cell lineage analyses of hypodermal cells in wild-type and mutant strains of *C. elegans*.

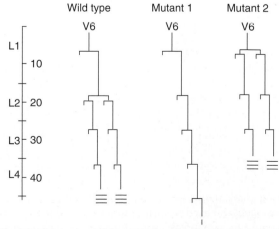

Explain the nature of the mutations in the altered strains.

E7. Take a look at solved problem S3 before answering this question. *Drosophila* embryos carrying a *ts* mutation were exposed to the permissive (25°C) or nonpermissive (30°C) temperature at different stages of development. Explain these results.

Time After Fertilization (hours) Group:	1	2	3	4	5
0–1	25°C	25°C	25°C	25°C	25°C
1–2	25°C	30°C	25°C	25°C	25°C
2–3	25°C	25°C	30°C	25°C	25°C
3–4	25°C	25°C	25°C	30°C	25°C
4–5	25°C	25°C	25°C	25°C	30°C
5–6	25°C	25°C	25°C	25°C	25°C
SURVIVAL:	Yes	No	No	Yes	Yes

E8. All of the homeotic genes in *Drosophila* have been cloned. As discussed in Chapter 18, cloned genes can be manipulated in vitro. They can be subjected to cutting and pasting, site-directed mutagenesis, etc. After *Drosophila* genes have been altered in vitro, they can be inserted into a *Drosophila* transposon vector (i.e., a P element vector), and then the genetic construct containing the altered gene within a P element can be injected into *Drosophila* embryos. The P element will then transpose into the chromosomes and thereby introduce one or more copies of the altered gene into the *Drosophila* genome. This method is termed P element transformation.

With these ideas in mind, how would you make a mutant gene with a "gain-of-function" in which the *Antp* gene would be expressed where the *abd-A* gene is normally expressed? What phenotype would you expect for flies that carried this altered gene?

E9. You will need to understand solved problem S4 before answering this question. If the artificial gene containing the stripe 2 enhancer and the β-galactosidase gene was found within an embryo that also contained the following loss-of-function mutations, what results would you expect? In other words, would there be a stripe or not? Explain why.

A. *Krüppel*

B. *bicoid*

C. *hunchback*

D. *giant*

E10. Two techniques commonly used to study the expression patterns of genes that play a role in development are Northern blotting and *in situ* hybridization. As described in Chapter 18, Northern blotting can be used to detect RNA that is transcribed from a particular gene. In this method, a specific RNA is detected by using a short segment of cloned DNA as a probe. The DNA probe, which is radioactive, is complementary to the RNA that the researcher wishes to detect. After the radioactive probe DNA binds to the RNA within a blot of a gel, the RNA is visualized as a dark (radioactive) band on an X-ray film. For example, a DNA probe that is complementary to the *bicoid* mRNA could be used to specifically detect the amount and size of the mRNA in a blot.

A second technique, termed fluorescence *in situ* hybridization (FISH), can be used to identify the locations of genes on chromosomes. This technique can also be used to locate gene products within oocytes, embryos, and larvae. For this reason,

it has been commonly used by developmental geneticists to understand the expression patterns of genes during development. The photograph in Figure 23.7b is derived from the application of the FISH technique. In this case, the probe was complementary to bicoid mRNA.

Now here is the question. Suppose a researcher has three different *Drosophila* strains that have loss-of-function mutations in the *bicoid* gene. We will call them *bicoid-A*, *bicoid-B*, and *bicoid-C*; the wild type is designated *bicoid⁺*. To study these mutations, phenotypically normal female flies that are homozygous for the bicoid mutation were obtained, and their oocytes were analyzed using these two techniques. A wild-type strain was also analyzed as a control. In other words, RNA was isolated from some of the oocytes and analyzed by Northern blotting, and some oocytes were subjected to *in situ* hybridization. In both cases, the probe was complementary to the bicoid mRNA. The results are shown here.

Northern blot

Lane 1. Wild type (*bicoid⁺*)
Lane 2. *bicoid-A*
Lane 3. *bicoid-B*
Lane 4. *bicoid-C*

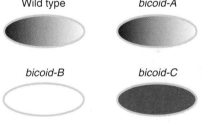

In situ hybridization

Wild type   bicoid-A

bicoid-B   bicoid-C

A. How can phenotypically normal female flies be homozygous for a loss-of-function allele in the *bicoid* gene?

B. Explain the type of mutation (e.g., deletion, point mutation, etc.) in each of the three strains. Explain how the mutation may cause a loss of normal function for the *bicoid* gene product.

C. Discuss how the use of both techniques provides more definitive information compared to the application of just one of the two techniques.

E11. Explain one experimental strategy you could follow to determine the functional role of the mouse *HoxD-3* gene.

E12. In the experiment of Figure 23.15, suggest reasons why the *n536*, *n355*, and *n540* strains have an egg-laying defect.

E13. Another way to study the role of proteins (e.g., transcription factors) that function in development is to microinject the mRNA that encodes a protein, or the purified protein itself, into an oocyte or embryo, and then determine how this affects the subsequent development of the embryo, larva, and adult. For example, if Bicoid protein is injected into the posterior region of an oocyte, the resulting embryo will develop into a larva that has anterior structures at both ends. Based on your understanding of the function of these developmental genes, what would be the predicted phenotype if the following proteins or mRNAs were injected into normal oocytes?

A. *Nanos* mRNA injected into the anterior end of an oocyte

B. Antp protein injected into the posterior end of an embryo

C. *Toll* mRNA injected the dorsal side of an early embryo

E14. Why have geneticists been forced to use reverse genetics to study the genes involved in vertebrate development? Explain how this strategy differs from traditional genetic analyses like those done by Mendel.

Questions for Student Discussion/Collaboration

1. Compare and contrast the experimental advantages and disadvantages of *Drosophila*, *C. elegans*, mammals, and *Arabidopsis*.

2. It seems that developmental genetics boils down to a complex network of gene regulation. Try to draw how this network is structured for *Drosophila*. How many genes do you think are necessary to describe a complete developmental network for the fruit fly? How many genes do you think are needed for a network to specify one segment? Do you think it is more difficult to identify genes that are involved in the beginning, middle, or end of this network? Knowing what you know about *Drosophila* development, suppose you were trying to identify all the genes needed for development in a chicken. Would you first try to identify genes necessary for early development, or would you begin by identifying genes involved in cell differentiation?

3. At the molecular level, how do you think a gain-of-function mutation in a developmental gene might cause it to be expressed in the wrong place or at the wrong time? Explain what type of DNA sequence would be altered.

Note: All answers appear at the website for this textbook; the answers to even-numbered questions are in the back of the textbook.

www.mhhe.com/brookergenetics3e

Visit the Online Learning Center for practice tests, answer keys, and other learning aids for this chapter. Enhance your understanding of genetics with our interactive exercises, quizzes, animations, and much more.

POPULATION GENETICS

Until now, we have primarily focused our attention on genes within individuals and their related family members. In this chapter and Chapters 25 and 26, we will turn to the study of genes in a population or species. The field of **population genetics** is concerned with changes in genetic variation within a group of individuals over time. Population geneticists want to know the extent of genetic variation within populations, why it exists, and how it changes over the course of many generations. The field of population genetics emerged as a branch of genetics in the 1920s and 1930s. Its mathematical foundations were developed by theoreticians who extended the principles of Gregor Mendel and Charles Darwin by deriving formulas to explain the occurrence of genotypes within populations. These foundations can be largely attributed to three scientists: Sir Ronald Fisher, Sewall Wright, and J. B. S. Haldane. As we will see, support for their mathematical theories was provided by several researchers who analyzed the genetic composition of natural and experimental populations. More recently, population geneticists have used techniques to probe genetic variation at the molecular level. In addition, staggering advances in computer technology have aided population geneticists in the analysis of their genetic theories and data. In this chapter, we will explore the genetic variation that occurs in populations and consider the reasons why the genetic composition of populations may change over the course of several generations.

The African cheetah. This species has a relatively low level of genetic variation because the population was reduced to a small size approximately 10,000 to 12,000 years ago.

24.1 GENES IN POPULATIONS, AND THE HARDY-WEINBERG EQUATION

Population genetics may seem like a significant departure from other topics in this textbook, but it is a direct extension of our understanding of Mendel's laws of inheritance, molecular genetics, and the ideas of Darwin. In the field of population genetics, the focus shifts away from the individual and toward the population of which the individual is a member. Conceptually, all of the alleles of every gene in a population make up the **gene pool.** In this regard, each member of the population is viewed as receiving its genes from its parents, which, in turn, are members of the gene pool. Furthermore, individuals that reproduce contribute to the gene pool of the next generation. Population geneticists study the genetic variation within the gene pool and how such variation changes from one generation to the next. The emphasis is often on allelic variation. In this introductory section, we will examine some of the general features of populations and gene pools.

A Population Is a Group of Interbreeding Individuals That Share a Gene Pool

In genetics, the term population has a very specific meaning. With regard to sexually reproducing species, a **population** is a group of individuals of the same species that occupy the same region and can interbreed with one another. Many species occupy a wide geographic range and are divided into discrete populations. For example, distinct populations of a given species may be located on different continents, or populations on the same continent could be divided by a geographical feature such as a large mountain range.

A large population usually is composed of smaller groups called **local populations** or **demes.** The members of a local population are far more likely to breed among themselves than with other members of the general population. Local populations are often separated from each other by moderate geographic barriers. As shown in **Figure 24.1,** two groups of Douglas fir (*Pseudotsuga menziesii*) are separated by a wide river, where the fir trees cannot grow. Trees on opposite sides of this river constitute local populations. Breeding is much more apt to occur among members of each local population than between members of neighboring populations. On relatively rare occasions, however, pollen can be blown across the river, which allows breeding between these two local populations.

Populations typically are dynamic units that change from one generation to the next. A population may change its size, geographic location, and genetic composition. With regard to size, natural populations commonly go through cycles of "feast or famine," during which the population swells or shrinks. In addition, natural predators or disease may periodically decrease the size of a population to significantly lower levels; the population later may rebound to its original size. Populations or individuals within populations may migrate to a new site and establish a distinct population in this location. The environment at this new geographic location may differ from the original site. What are the consequences of such changes? As population sizes and locations change, their genetic composition generally changes as well. As described later, population geneticists have developed mathematical theories that predict how the gene pool will change in response to fluctuations in size, migration, and new environments.

At the Population Level, Some Genes May Be Monomorphic, But Most Are Polymorphic

In population genetics, the term **polymorphism** (meaning many forms) refers to the observation that many traits display variation within a population. Historically, polymorphism first referred to the variation in traits that are observable with the naked eye. Polymorphisms in color and pattern have long attracted the attention of population geneticists. These include studies involving yellow and red varieties of the elder-flowered orchid, and brown, pink, and yellow land snails, which are discussed later in this chapter. **Figure 24.2** illustrates a striking example of polymorphism in the Hawaiian happy-face spider (*Theridion grallator*). The three individuals shown in this figure are from the same species, but they differ in alleles that affect color and pattern.

What is the underlying cause of polymorphism? At the DNA level, polymorphism may be due to two or more alleles that influence the phenotype of the individual that inherits them. In other words, it is due to genetic variation. Geneticists also use the term **polymorphic** to describe a gene that commonly exists as two or more alleles in a population. By comparison, a **monomorphic** gene exists predominantly as a single allele in a population. By convention, when a single allele is found in at least 99% of all cases, the gene is considered monomorphic. (Some geneticists view an allele frequency of 95% or greater to be monomorphic.)

At the level of a particular gene, a polymorphism may involve various types of changes such as a deletion of a significant region of the gene, a duplication of a region, or a change in a single nucleotide. This last phenomenon is called a **single-nucleotide polymorphism (SNP).** SNPs are the smallest type of genetic change that can occur within a given gene and are also the most common. In humans, for example, SNPs represent 90% of all the variation in DNA sequences that occurs among different people. SNPs are found very frequently in genes. In the human population, a gene that is 2,000 to 3,000 bp in length will, on average, contain 10 different sites that are polymorphic. The high frequency of SNPs indicates that polymorphism is the norm for most human genes. Likewise, relatively large, healthy populations of nearly all species exhibit a high level of genetic variation as evidenced by the occurrence of SNPs within most genes.

Within a population, the alleles of a given gene may arise by different types of genetic changes. **Figure 24.3** considers a gene that exists in multiple forms in humans. This example is a short segment of DNA found within the human β-globin gene. The top sequence is an allele designated *Hb^A*, whereas the middle sequence is called *Hb^S*. These alleles differ from each other

FIGURE 24.1 **Two local populations of Douglas fir.** A wide river separates the two local populations. A much higher frequency of breeding occurs among members of the same local population than between the two different local populations.

FIGURE 24.2   Polymorphism in the Hawaiian happy-face spider.

Genes→Traits These three spiders are members of the same species and carry the same genes. However, several genes that affect pigmentation patterns are polymorphic, meaning there is more than one allele for each gene within the population. This polymorphism within the Hawaiian happy-face spider population produces members that look quite different from each other.

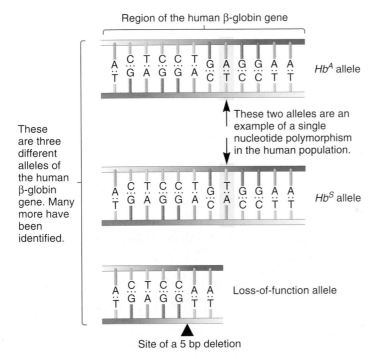

FIGURE 24.3   The relationship between alleles and various types of mutations. The DNA sequence shown here is a small portion of the β-globin gene in humans. Mutations have altered the gene to create the three different alleles in this figure. The top two alleles differ by a single base pair and are referred to as a single nucleotide polymorphism (SNP). The bottom allele has a 5 bp deletion that results in a nonfunctional polypeptide. It is a loss-of-function allele.

by a single nucleotide, so they are an example of a single nucleotide polymorphism. As discussed in Chapter 4, the Hb^S allele causes sickle-cell disease in a homozygote. The bottom sequence contains a short, 5 bp deletion compared to the other two alleles. This deletion results in a nonfunctional β-globin polypeptide. Therefore, the bottom sequence is an example of a loss-of-function allele.

Population Genetics Is Concerned with Allele and Genotype Frequencies

As we have seen, population geneticists want to understand the prevalence of polymorphic genes within populations. Much of their work evaluates the frequency of alleles in a quantitative way. Two fundamental calculations are central to population genetics: **allele frequencies** and **genotype frequencies.** The allele and genotype frequencies are defined as

$$\text{Allele frequency} = \frac{\text{Number of copies of an allele in a population}}{\text{Total number of all alleles for that gene in a population}}$$

$$\text{Genotype frequency} = \frac{\text{Number of individuals with a particular genotype in a population}}{\text{Total number of individuals in a population}}$$

Though these two frequencies are related, a clear distinction between them must be kept in mind. As an example, let's consider a population of 100 pea plants with the following genotypes:

64 tall plants with the genotype *TT*

32 tall plants with the genotype *Tt*

4 dwarf plants with the genotype *tt*

When calculating an allele frequency, homozygous individuals have two copies of an allele, whereas heterozygotes have only one. For example, in tallying the *t* allele, each of the 32 heterozygotes has one copy of the *t* allele, and each dwarf plant has two copies. The allele frequency for *t* equals

$$t = \frac{32 + 2(4)}{2(64) + 2(32) + 2(4)}$$

$$= \frac{40}{200} = 0.2, \text{ or } 20\%$$

This result tells us that the allele frequency of *t* is 20%. In other words, 20% of the alleles for this gene in the population are the *t* allele.

Let's now calculate the genotype frequency of *tt* (dwarf) plants:

$$tt = \frac{4}{64 + 32 + 4}$$

$$= \frac{4}{100} = 0.04, \text{ or } 4\%$$

We see that 4% of the individuals in this population are dwarf plants.

Allele and genotype frequencies are always less than or equal to 1 (i.e., less than or equal to 100%). If a gene is monomorphic, the allele frequency for the single allele will equal or be close to a value of 1.0. For polymorphic genes, if we add up the frequencies for all the alleles in the population, we should obtain a value of 1.0. In our pea plant example, the allele frequency of *t* equals 0.2. The frequency of the other allele, *T*, equals 0.8. If we add the two together, we obtain a value of 0.2 + 0.8 = 1.0.

The Hardy-Weinberg Equation Can Be Used to Calculate Genotype Frequencies Based on Allele Frequencies

Now that we have a general understanding of genes in populations, we can begin to relate these concepts to mathematical expressions as a way to examine whether allele and genotype frequencies will change over the course of many generations. In 1908, a British mathematician, Godfrey Harold Hardy, and a German physician, Wilhelm Weinberg, independently derived a simple mathematical expression that predicted stability of allele and genotype frequencies from one generation to the next. It is called the **Hardy-Weinberg equilibrium,** because (under a given set of conditions, described later) the allele and genotype frequencies do not change over the course of many generations.

This relationship establishes a framework on which to understand changes in allele frequencies within a population when such an equilibrium is violated.

Let's begin by considering a situation in which a gene is polymorphic and exists as two different alleles, *A* and *a*. If the allele frequency of *A* is denoted by the variable *p*, and the allele frequency of *a* by *q*, then

$$p + q = 1$$

For example, if $p = 0.8$, then q must be 0.2. In other words, if the allele frequency of *A* equals 80%, the remaining 20% of alleles must be *a*, because together they equal 100%.

The Hardy-Weinberg equation is used to relate allele frequencies and genotype frequencies. For a gene that exists in two alleles:

$$p^2 + 2pq + q^2 = 1 \text{ (\textbf{Hardy-Weinberg equation})}$$

If this equation is applied to a gene that exists in alleles designated *A* and *a*, then

p^2	equals the genotype frequency of *AA*
$2pq$	equals the genotype frequency of *Aa*
q^2	equals the genotype frequency of *aa*

If $p = 0.8$ and $q = 0.2$ and if the population is in Hardy-Weinberg equilibrium, then

$$AA = p^2 = (0.8)^2 = 0.64$$

$$Aa = 2pq = 2(0.8)(0.2) = 0.32$$

$$aa = q^2 = (0.2)^2 = 0.04$$

In other words, if the allele frequency of *A* is 80% and the allele frequency of *a* is 20%, the genotype frequency of *AA* is 64%, *Aa* is 32%, and *aa* is 4%.

To illustrate the relationship between allele frequencies and genotypes, **Figure 24.4** compares the Hardy-Weinberg equation with the way that gametes combine randomly with each other to produce offspring. In a population, the frequency of a gamete carrying a particular allele is equal to the allele frequency in that population. In this example, the frequency of a gamete carrying the *A* allele equals 0.8.

We can use the product rule to determine the frequency of genotypes. For example, the frequency of producing an *AA* homozygote is 0.8 × 0.8 = 0.64, or 64%. Likewise, the probability of inheriting both *a* alleles is 0.2 × 0.2 = 0.04, or 4%. As seen in Figure 24.4, heterozygotes can be produced in two different ways. An offspring could inherit the *A* allele from its father and *a* from its mother, or *A* from its mother and *a* from its father. Therefore, the frequency of heterozygotes is *pq* + *pq*, which equals 2*pq*; in our example, this is 2(0.8)(0.2) = 0.32, or 32%.

One useful feature of the Hardy-Weinberg equation is that it allows us to determine the frequency of heterozygotes for recessive genetic diseases. As an example, let's consider cystic fibrosis, which involves a gene that encodes a chloride transporter. Persons afflicted with this disorder have an irregularity in salt and water

	♂ A 0.8	a 0.2
♀ A 0.8	AA (0.8)(0.8) = 0.64	Aa (0.8)(0.2) = 0.16
a 0.2	Aa (0.8)(0.2) = 0.16	aa (0.2)(0.2) = 0.04

AA genotype = 0.64 = 64%
Aa genotype = 0.16 + 0.16 = 0.32 = 32%
aa genotype = 0.04 = 4%

FIGURE 24.4 **A comparison between allele frequencies and the union of alleles in a Punnett square.** In a population in Hardy-Weinberg equilibrium, the frequency of gametes carrying a particular allele is equal to the allele frequency in the population.

balance. One of the symptoms is thick mucus in the lungs that can contribute to repeated lung infections. In Caucasian populations, the frequency of affected individuals is approximately 1 in 2,500. Because this is a recessive disorder, affected individuals are homozygotes. If q represents the allele frequency of the disease-causing allele, then

$$q^2 = 1/2,500$$

$$q^2 = 0.0004$$

We take the square root to determine q:

$$q = \sqrt{0.0004}$$

$$q = 0.02$$

If p represents the normal allele,

$$p = 1 - q$$

$$p = 1 - 0.02 = 0.98$$

The frequency of heterozygous carriers is

$$2pq = 2(0.98)(0.02) = 0.0392, \text{ or } 3.92\%$$

The Hardy-Weinberg equation predicts an equilibrium—unchanging allele and genotype frequencies from generation to generation—if certain conditions are met in a population. With regard to the gene of interest, these conditions are as follows:

1. No new mutations: The gene of interest does not incur any new mutations.
2. No genetic drift: The population is so large that allele frequencies do not change due to random sampling effects.
3. No migration: Individuals do not travel between different populations.
4. No natural selection: All of the genotypes have the same reproductive success.

5. Random mating: With respect to the gene of interest, the members of the population mate with each other without regard to their phenotypes and genotypes.

The Hardy-Weinberg equation provides a quantitative relationship between allele and genotype frequencies in a population. **Figure 24.5** describes this relationship for different allele frequencies of a and A. As expected, when the allele frequency of a is very low, the AA genotype predominates; when the a allele frequency is high, the aa homozygote is most prevalent in the population. When the allele frequencies of a and A are intermediate in value, the heterozygote predominates.

In reality, no population satisfies the Hardy-Weinberg equilibrium completely. Nevertheless, in large natural populations with little migration and negligible natural selection, the Hardy-Weinberg equilibrium may be nearly approximated for certain genes. In addition, the Hardy-Weinberg equilibrium can be extended to situations in which a single gene exists in three or more alleles, as described in solved problem S4 at the end of the chapter.

As discussed in Chapter 2, the chi square test can determine if observed and expected data are in agreement. Therefore, we can use a chi square test to determine whether a population exhibits Hardy-Weinberg equilibrium for a particular gene. To do so, it is necessary to distinguish between the homozygotes and heterozygotes, either phenotypically or at the molecular level. This is necessary so that we can determine both the allele and genotype frequencies. As an example, let's consider a human blood type called the MN type. In this case, the blood type is determined by two codominant alleles, M and N. In an Inuit population in East

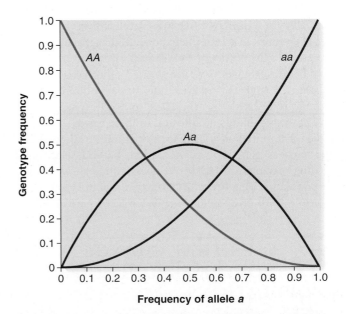

FIGURE 24.5 **The relationship between allele frequencies and genotype frequencies according to the Hardy-Weinberg equilibrium.** This graph assumes that a and A are the only two alleles for this gene.

Greenland, it was found that among 200 people, 168 were *MM*, 30 were *MN*, and 2 were *NN*. We can use these observed data to calculate the expected number of each genotype based on the Hardy-Weinberg equation.

$$\text{Allele frequency of } M = \frac{2(168) + 30}{400} = 0.915$$

$$\text{Allele frequency of } N = \frac{2(2) + 30}{400} = 0.085$$

Expected frequency of $MM = p^2 = (0.915)^2 = 0.837$

Expected number of *MM* individuals $= 0.837 \times 200 = 167.4$ (or 167 if rounded to the nearest individual)

Expected frequency of $NN = q^2 = (0.085)^2 = 0.007$

Expected number of *NN* individuals $= 0.007 \times 200 = 1.4$ (or 1 if rounded to the nearest individual)

Expected frequency of $MN = 2pq = 2(0.915)(0.085) = 0.156$

Expected number of *MN* individuals $= 0.156 \times 200 = 31.2$ (or 31 if rounded to the nearest individual)

$$\chi^2 = \frac{(O_1 - E_1)^2}{E_1} + \frac{(O_2 - E_2)^2}{E_2} + \frac{(O_3 - E_3)^2}{E_3}$$

$$\chi^2 = \frac{(168 - 167)^2}{167} + \frac{(30 - 31)^2}{31} + \frac{(2 - 1)^2}{1}$$

$$= 1.04$$

If we refer back to Table 2.1 in Chapter 2, the calculated chi square value for 1 degree of freedom is well within the acceptable range. Therefore, we accept the hypothesis that the alleles for this gene are in Hardy-Weinberg equilibrium.

When researchers have investigated other genes in various populations, a high chi square value may be obtained, and the hypothesis is rejected that the allele and genotype frequencies are in Hardy-Weinberg equilibrium. In these cases, we would say that such a population is in **disequilibrium.** As discussed next, factors such as genetic drift, migration, natural selection, and inbreeding may disrupt Hardy-Weinberg equilibrium. Therefore, when population geneticists discover that a population is not in equilibrium, they try to determine which factors are at work.

24.2 FACTORS THAT CHANGE ALLELE AND GENOTYPE FREQUENCIES IN POPULATIONS

The genetic variation in natural populations typically changes over the course of many generations. The term **microevolution** describes changes in a population's gene pool from generation to generation. Such change is rooted in two related phenomena

(Table 24.1). First, the introduction of new genetic variation into a population is one essential aspect of microevolution. As discussed later in this chapter, gene variation can originate by a variety of molecular mechanisms. For example, new alleles of preexisting genes can arise by random mutations. Such events provide a continuous source of new variation to populations. However, due to their low rate of occurrence, mutations do not act as a major force in promoting widespread changes in a population. If mutations were the only type of change occurring in a population, that population would not evolve at a significant rate because mutations are so rare.

Microevolution also involves the action of evolutionary mechanisms that alter the prevalence of a given allele or genotype in a population. These mechanisms are random genetic drift, migration, natural selection, and nonrandom mating (see Table 24.1). The collective contributions of these evolutionary mechanisms over the course of many generations have the potential to promote widespread genetic changes in a population. In this section, we will examine how these mechanisms may affect the

TABLE **24.1**	
Factors that Govern Microevolution	
Source of New Allelic Variation*	
Mutation	In this section, we will consider allelic variation. Random mutations within pre-existing genes introduce new alleles into populations, but at a very low rate. New mutations may be beneficial, neutral, or deleterious. For new alleles to rise to a significant percentage in a population, evolutionary mechanisms (i.e., random genetic drift, migration, natural selection) must operate on them.
Mechanisms That Alter Existing Genetic Variation	
Random genetic drift	This is a change in genetic variation from generation to generation due to random sampling error. Allele frequencies may change as a matter of chance from one generation to the next. This tends to have a greater impact in a small population.
Migration	Migration can occur between two different populations that have different allele frequencies. The introduction of migrants into a recipient population may change the allele frequencies of that population.
Natural selection	This is the phenomenon in which the environment selects for individuals that possess certain traits. Natural selection can favor the survival of members with beneficial traits or disfavor the survival of individuals with unfavorable traits.
Nonrandom mating	This is the phenomenon in which individuals select mates based on their phenotypes or genetic lineage. This can alter the relative proportion of homozygotes and heterozygotes predicted by the Hardy-Weinberg equation but will not change allele frequencies.

*Allelic variation is just one source of new genetic variation. Section 24.3 considers a variety of mechanisms through which new genetic variation can occur.

type of genetic variation that occurs when a gene exists in two or more alleles in a population. As you will learn, these mechanisms may cause a particular allele to be favored, or they may create a balance where two or more alleles are maintained in a population.

Mutations Provide the Source of Genetic Variation

As discussed in Chapters 8 and 16, mutations involve changes in gene sequences, chromosome structure, and/or chromosome number. Mutations are random events that occur spontaneously at a low rate or are caused by mutagens at a higher rate. In 1926, the Russian geneticist Sergei Chetverikov was the first to suggest that mutational variability provides the raw material for evolution but does not constitute evolution itself. In other words, mutation can provide new alleles to a population but does not substantially alter allele frequencies. Chetverikov proposed that populations in nature absorb mutations like a sponge and retain them in a heterozygous condition, thereby providing a source of variability for future change.

Population geneticists often consider how new mutations affect the survival and reproductive potential of the individual that inherits them. A new mutation may be deleterious, neutral, or beneficial. For genes that encode proteins, the effects of new mutations will depend on their impact on protein function. Deleterious and neutral mutations are far likelier to occur than beneficial mutations. For example, alleles can be altered in many different ways that render an encoded protein defective. As discussed in Chapter 16, deletions and point mutations such as frameshift mutations, missense mutations, and nonsense mutations all may cause a gene to express a protein that is nonfunctional or less functional than the wild-type protein. Also, mutations in noncoding regions can alter gene expression (refer back to Table 16.3). Neutral mutations can also occur in several different ways. For example, a neutral mutation can change the wobble base without affecting the amino acid sequence of the encoded protein, or it can be a missense mutation that has no effect on protein function. Such point mutations occur at specific sites within the coding sequence. Neutral mutations can also occur within introns, the noncoding sequences of genes. By comparison, beneficial mutations are relatively uncommon. To be advantageous, a new mutation could alter the amino acid sequence of a protein to yield a better-functioning product. While such mutations do occur, they are expected to be very rare for a population in a stable environment.

The **mutation rate** is defined as the probability that a gene will be altered by a new mutation. The rate is typically expressed as the number of new mutations in a given gene per generation. A common value for the mutation rate is in the range of 1 in 100,000 to 1 in 1,000,000, or 10^{-5} to 10^{-6} per generation. However, mutation rates vary depending on species, cell types, chromosomal location, and gene size. Furthermore, in experimental studies, the mutation rate is usually measured by following the change of a normal (functional) gene to a deleterious (nonfunctional) allele. The mutation rate producing beneficial alleles is expected to be substantially less.

It is clear that new mutations provide genetic variability, but population geneticists also want to know how much the mutation rate affects the allele frequencies in a population. Can random mutations have a large impact on allele frequencies over time? To answer this question, let's take the simple case where a gene exists in a functional allele, A; the allele frequency of A is denoted by the variable p. A deleterious mutation can convert the A allele into a nonfunctional allele, a. The allele frequency of a is designated by q. The conversion of the A allele into the a allele by mutation will occur at a rate that is designated μ. If we assume that the rate of the reverse mutation (a to A) is negligible, the increase in the frequency of the a allele after one generation will be

$$\Delta q = \mu p$$

For example, let's consider the following conditions:

$p = 0.8$ (i.e., frequency of A is 80%)

$q = 0.2$ (i.e., frequency of a is 20%)

$\mu = 10^{-5}$ (i.e., the mutation rate of converting A to a)

$\Delta q = (10^{-5})(0.8) = (0.00001)(0.8) = 0.000008$

Therefore, in the next generation (designated $n + 1$),

$q_{n+1} = 0.2 + 0.000008 = 0.200008$

$p_{n+1} = 0.8 - 0.000008 = 0.799992$

As we can see from this calculation, new mutations do not significantly alter the allele frequencies in a single generation.

We can use the following equation to calculate the change in allele frequency after any number of generations:

$$(1 - \mu)^t = \frac{p_t}{p_0}$$

where

μ is the mutation rate of the conversion of A to a

t is the number of generations

p_0 is the allele frequency of A in the starting generation

p_t is the allele frequency of A after t generations

As an example, let's suppose that the allele frequency of A is 0.8, $\mu = 10^{-5}$, and we want to know what the allele frequency will be after 1,000 generations ($t = 1,000$). Plugging these values into the preceding equation and solving for p_t,

$$(1 - 0.00001)^{1,000} = \frac{p_t}{0.8}$$

$$p_t = 0.792$$

Therefore, after 1,000 generations, the frequency of A has dropped only from 0.8 to 0.792. Again, these results point to how slowly the occurrence of new mutations will change allele frequencies. In natural populations, the rate of new mutation is rarely a significant catalyst in shaping allele frequencies. Instead, other processes such as genetic drift, migration, and natural selection have far greater effects on allele frequencies. Next, we will examine how these other factors work.

In Small Populations, Allele Frequencies Can Be Altered by Random Genetic Drift

In the 1930s, geneticist Sewall Wright played a key role in developing the concept of **random genetic drift,** which refers to changes in allele frequencies in a population due to chance fluctuations. As a matter of random chance, the frequencies of alleles found in gametes that unite to form zygotes will vary from generation to generation. Over the long run, genetic drift favors either the loss of an allele or its fixation at 100% in the population. The rate at which this occurs depends on the population size and on the initial allele frequencies. **Figure 24.6** illustrates the potential consequences of genetic drift in one large ($N = 1,000$) and five small ($N = 20$) populations. At the beginning of this hypothetical simulation, all of these populations have identical allele frequencies: $A = 0.5$ and $a = 0.5$. In the five small populations, this allele frequency fluctuates substantially from generation to generation. Eventually, one of the alleles is eliminated and the other is fixed at 100%. At this point, the allele has become monomorphic and cannot fluctuate any further. By comparison, the allele frequencies in the large population fluctuate much less, because random sampling error is expected to have a smaller impact. Nevertheless, genetic drift will lead to homozygosity even in large populations, but this will take many more generations to occur.

Now let's ask two questions:

1. How many new mutations do we expect in a natural population?
2. How likely is it that any new mutation will be either fixed in, or eliminated from, a population due to random genetic drift?

With regard to the first question, the average number of new mutations depends on the mutation rate (μ) and the number of individuals in a population (N). If each individual has two copies of the gene of interest, the expected number of new mutations in this gene is

$$\text{Expected number of new mutations} = 2N\mu$$

From this, we see that a new mutation is more likely to occur in a large population than in a small one. This makes sense, because the larger population has more copies of the gene to be mutated. With regard to the second question, the probability of fixation of a newly arising allele due to genetic drift is

$$\text{Probability of fixation} = \frac{1}{2N} \quad \begin{array}{l}\text{(assuming equal numbers of}\\ \text{males and females contribute}\\ \text{to the next generation)}\end{array}$$

In other words, the probability of fixation is the same as the initial allele frequency in the population. For example, if $N = 20$, the probability of fixation of a new allele equals $1/(2 \times 20)$, or 2.5%. Conversely, a new allele may be lost from the population.

$$\text{Probability of elimination} = 1 - \text{probability of fixation}$$
$$= 1 - \frac{1}{2N}$$

If $N = 20$, the probability of elimination equals $1 - 1/(2 \times 20)$, or 97.5%. As you may have noticed, the value of N has opposing effects with regard to new mutations and their eventual fixation in a population. When N is very large, new mutations are much more likely to occur. Each new mutation, however, has a greater chance of being eliminated from the population due to random genetic drift. On the other hand, when N is small, the probability of new mutations is also small, but if they occur, the likelihood of fixation is relatively large.

Now that we have an appreciation for the phenomenon of genetic drift, we can ask a third question:

3. If fixation of a new allele does occur, how many generations is it likely to take?

FIGURE 24.6 **A hypothetical simulation of random genetic drift.** In all cases, the starting allele frequencies are $A = 0.5$ and $a = 0.5$. The colored lines illustrate five populations in which $N = 20$; the black line shows a population in which $N = 1,000$.

The formula for calculating this also depends on the number of individuals in the population:

$$\bar{t} = 4N$$

where

$\bar{t}$ equals the average number of generations to achieve fixation

N equals the number of individuals in the population, assuming that males and females contribute equally to each succeeding generation

As you may have expected, allele fixation will take much longer in large populations. If a population has 1 million breeding members, it will take, on average, 4 million generations, perhaps an insurmountable period of time, to reach fixation. In a small group of 100 individuals, however, fixation will take only 400 generations, on average. As discussed in Chapter 26, the drifting of neutral alleles among different populations and species provides a way to measure the rate of evolution and can be used to determine evolutionary relationships.

The preceding discussion of random genetic drift has emphasized two important points. First, genetic drift ultimately operates in a random manner with regard to allele frequency and, over the long run, leads to either allele fixation or elimination. The process is random with regard to particular alleles. Genetic drift can lead to the fixation of deleterious, neutral, or beneficial alleles. A second important feature of genetic drift is that its impact is greatly affected by population size. Genetic drift may lead more quickly to allele loss or fixation in a small population.

In nature, there are different ways that geography and population size influence how genetic drift affects the genetic composition of a species. Some species occupy wide ranges in which small, local populations become geographically isolated from the rest of the species. The allele frequencies within these small populations are more susceptible to genetic drift. Because this is a random process, small isolated populations tend to be more genetically disparate in relation to other populations.

Changes in population size may influence genetic drift via the **bottleneck effect.** In nature, a population can be reduced dramatically in size by events such as earthquakes, floods, drought, or human destruction of habitat. Such events may randomly eliminate most of the members of the population without regard to genetic composition. The time period of the bottleneck, when the population size is very small, may be greatly influenced by genetic drift because the surviving members may have allele frequencies that differ from those of the original population. In addition, allele frequencies are expected to drift substantially during the generations when the population size is small. In extreme cases, alleles may even be eliminated. Eventually, the bottlenecked population may regain its original size (**Figure 24.7**). However, the new population will have less genetic variation than the original large population. As an example, the African cheetah population lost a substantial amount of its genetic variation due to a bottleneck effect. DNA analysis by population geneticists has suggested that a severe bottleneck occurred approximately 10,000 to 12,000 years ago when the population size was dramatically reduced. The population eventually rebounded, but the bottleneck significantly decreased the genetic variation.

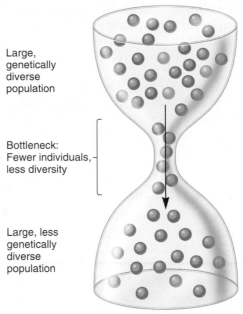

Large, genetically diverse population

Bottleneck: Fewer individuals, less diversity

Large, less genetically diverse population

(a) Bottleneck effect

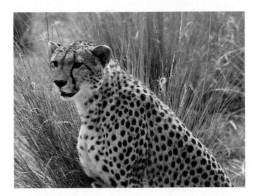

(b) An African cheetah

FIGURE 24.7 **The bottleneck effect, an example of genetic drift.** (a) A representation of the bottleneck effect. Note that the genetic variation denoted by the green balls has been lost. (b) The African cheetah. The modern species has low genetic variation due to a genetic bottleneck that is thought to have occurred about 10,000 to 12,000 years ago.

Geography and population size may also influence genetic drift via the **founder effect.** Compared to the bottleneck effect, the key difference is that the founder effect involves migration; a small group of individuals separates from a larger population and establishes a colony in a new location. For example, a few individuals may migrate from a large continental population and become the founders of an island population. The founder effect has two important consequences. First, the founding population is expected to have less genetic variation than the original population from which it was derived. Second, as a matter of chance, the allele frequencies in the founding population may differ markedly from those of the original population.

Population geneticists have studied many examples where isolated populations have been started from a few members of another population. In the 1960s, Victor McKusick studied allele frequencies in the Old Order Amish of Lancaster County, Pennsylvania. At that time, this was a group of about 8,000 people, descended from just three couples that immigrated to the United States in 1770. Among this population of 8,000, a genetic disease known as the Ellis-van Creveld syndrome (a recessive form of dwarfism) was found at a frequency of 0.07, or 7%. By comparison, this disorder is extremely rare in other human populations, even the population from which the founding members had originated. The high frequency of dwarfism in the Lancaster County population is a chance occurrence due to the founder effect.

Migrations Between Two Populations Can Alter Allele Frequencies

We have just seen how migration to a new location by a relatively small group can result in a population with an altered genetic composition. In addition, migration between two different established populations can alter allele frequencies. For example, a species of birds may occupy two geographic regions that are separated by a large body of water. On rare occasions, the prevailing winds may allow birds from the western population to fly over this body of water and become members of the eastern population. If the two populations have different allele frequencies and if migration occurs in sufficient numbers, this may alter the allele frequencies in the eastern population.

After migration has occurred, the new (eastern) population is called a conglomerate. To calculate the allele frequencies in the conglomerate, we need two kinds of information. First, we must know the original allele frequencies in the donor and recipient populations. Second, we must know the proportion of the conglomerate population that is due to migrants. With these data, we can calculate the change in allele frequency in the conglomerate population using the following equation:

$$\Delta p_C = m(p_D - p_R)$$

where

Δp_C is the change in allele frequency in the conglomerate population

p_D is the allele frequency in the donor population

p_R is the allele frequency in the original recipient population

m is the proportion of migrants that make up the conglomerate population

$$m = \frac{\text{number of migrants in the conglomerate population}}{\text{total number of individuals in the conglomerate population}}$$

As an example, let's suppose the allele frequency of A is 0.7 in the donor population and 0.3 in the recipient population. A group of 20 individuals migrates and joins the recipient population, which originally had 80 members. Thus,

$$m = \frac{20}{20 + 80}$$
$$= 0.2$$

$$\Delta p_C = m(p_D - p_R)$$
$$= 0.2(0.7 - 0.3)$$
$$= 0.08$$

We can now calculate the allele frequency in the conglomerate:

$$p_C = p_R + \Delta p_C$$
$$= 0.3 + 0.08 = 0.38$$

Therefore, in the conglomerate population, the allele frequency of A has changed from 0.3 (its value before migration) to 0.38. This increase in allele frequency arises from the higher allele frequency of A in the donor population. **Gene flow** occurs whenever individuals migrate between populations having different allele frequencies, and the migrants are able to breed successfully with the members of the recipient population. Gene flow depends not only on migration, but also on the ability of the migrants' alleles to be passed to subsequent generations.

In our previous example, we considered the consequences of a unidirectional migration from a donor to a recipient population. In nature, it is common for individuals to migrate in both directions. What are the main consequences of bidirectional migration? Depending on its rate, migration tends to reduce differences in allele frequencies between neighboring populations. In fact, population geneticists can analyze allele frequencies in two different populations to evaluate the rate of migration between them. Populations that frequently mix their gene pools via migration tend to have similar allele frequencies, whereas isolated populations are expected to be more disparate. In addition, migration can enhance genetic diversity within a population. As discussed earlier, new mutations are relatively rare events. Therefore, a particular mutation may arise only in one population. Migration may then introduce this new allele into neighboring populations.

Natural Selection Is Based on the Relative Reproductive Success of Genotypes

In the 1850s, Charles Darwin and Alfred Russel Wallace independently proposed the theory of **natural selection.** We will discuss the phenotypic consequences of natural selection in greater detail in Chapter 26. According to this theory, the conditions found in nature result in the selective survival and reproduction of individuals whose characteristics make them well adapted to their environment. These surviving individuals are more likely to reproduce and contribute offspring to the next generation. Natural selection can be related not only to differential survival but also to mating efficiency and fertility.

A modern restatement of the principles of natural selection can relate our knowledge of molecular genetics to the phenotypes of individuals.

1. Within a population, allelic variation arises in various ways, such as through random mutations that cause differences in DNA sequences. A mutation that creates a new allele may alter the amino acid sequence of the

encoded protein, which, in turn, may alter the function of the protein.

2. Some alleles may encode proteins that enhance an individual's survival or reproductive capability compared to that of other members of the population. For example, an allele may produce a protein that is more efficient at a higher temperature, conferring on the individual a greater probability of survival in a hot climate.

3. Individuals with beneficial alleles are more likely to survive and contribute to the gene pool of the next generation.

4. Over the course of many generations, allele frequencies of many different genes may change through this process, thereby significantly altering the characteristics of a species. The net result of natural selection is a population that is better adapted to its environment and more successful at reproduction.

As mentioned at the beginning of the chapter, Fisher, Wright, and Haldane developed mathematical relationships to explain the theory of natural selection. As our knowledge of the process of natural selection has increased, it has become apparent that it operates in many different ways. In this chapter, we will consider a few examples of natural selection involving a single trait or a single gene that exists in two alleles. In reality, however, natural selection acts on populations of individuals in which many genes are polymorphic and each individual contains thousands or tens of thousands of different genes.

To begin our quantitative discussion of natural selection, we must examine the concept of **Darwinian fitness**—the relative likelihood that a genotype will contribute to the gene pool of the next generation as compared to other genotypes. Natural selection acts on phenotypes that are derived from individuals' genotypes. Although Darwinian fitness often correlates with physical fitness, the two ideas should not be confused. Darwinian fitness is a measure of reproductive superiority. An extremely fertile genotype may have a higher Darwinian fitness than a less fertile genotype that appears more physically fit.

To consider Darwinian fitness, let's use our example of a gene existing in the A and a alleles. If the three genotypes have the same level of mating success and fertility, we can assign fitness values to each of the three genotype classes based on their likelihood of surviving to reproductive age. For example, let's suppose that the relative survival to adulthood of each of the three genotype classes is as follows: For every five AA individuals that survive, four Aa individuals survive, and one aa individual survives. By convention, the genotype with the highest reproductive ability is given a fitness value of 1.0. Relative fitness values are denoted by the variable W. The fitness values of the other genotypes are assigned values relative to this 1.0 value:

Fitness of AA: $W_{AA} = 1.0$

Fitness of Aa: $W_{Aa} = 4/5 = 0.8$

Fitness of aa: $W_{aa} = 1/5 = 0.2$

Keep in mind that differences in reproductive achievement among genotypes may stem from various reasons. In this case,

the fittest genotype is more likely to survive to reproductive age. In other situations, the most fit genotype is more likely to mate. For example, a bird with brightly colored feathers may have an easier time attracting a mate than a bird with duller plumage. Finally, a third possibility is that the fittest genotype may be more fertile. It may produce a higher number of gametes or gametes that are more successful at fertilization.

By studying species in their native environments, population geneticists have discovered that natural selection can occur in several ways. The patterns of natural selection depend on the relative fitness values of the different genotypes and on the variation of environmental effects. The four patterns of natural selection that we will consider are called directional, stabilizing, disruptive, and balancing selection. In most of the examples described next, natural selection leads to adaptation so that a species is better able to survive to reproductive age.

Directional Selection Favors the Extreme Phenotype

Directional selection favors individuals at one extreme of a phenotypic distribution that are more likely to survive and reproduce in a particular environment. Different phenomena may initiate the process of directional selection. One way that directional selection may arise is that a new allele may be introduced into a population by mutation, and the new allele may promote a higher fitness in individuals that carry it (**Figure 24.8**). If the homozygote carrying the favored allele has the highest fitness value, directional selection may cause this favored allele to eventually become the predominant allele in the population, perhaps even becoming a monomorphic allele.

Another possibility is that a population may be exposed to a prolonged change in its living environment. Under the new environmental conditions, the relative fitness values may change to favor one genotype, and this will promote the elimination of other genotypes. As an example, let's suppose a population of finches on the mainland already has genetic variation in beak size. A small number of birds migrate to an island where the seeds are generally larger than they are on the mainland. In this new environment, birds with larger beaks would have a higher fitness because they would be better able to crack open the larger seeds and thereby survive to reproductive age. Over the course of many generations, directional selection would produce a population of birds carrying alleles that promote larger beak size.

In the case of directional selection, allele frequencies may change in a step-by-step, generation-per-generation way. To appreciate how this occurs, let's take a look at how fitness affects the Hardy-Weinberg equilibrium and allele frequencies. Again, let's suppose a gene exists in two alleles, A and a. The three fitness values, which are based on relative survival levels, are

$$W_{AA} = 1.0$$
$$W_{Aa} = 0.8$$
$$W_{aa} = 0.2$$

In the next generation, we expect that the Hardy-Weinberg equilibrium will be modified in the following way due to directional selection:

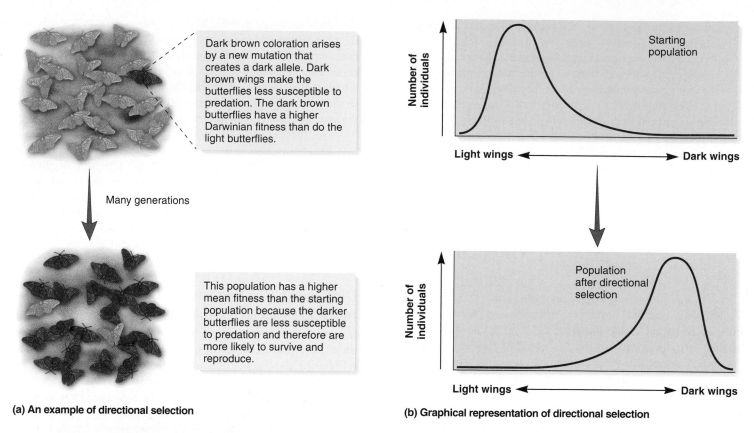

(a) An example of directional selection

Dark brown coloration arises by a new mutation that creates a dark allele. Dark brown wings make the butterflies less susceptible to predation. The dark brown butterflies have a higher Darwinian fitness than do the light butterflies.

Many generations

This population has a higher mean fitness than the starting population because the darker butterflies are less susceptible to predation and therefore are more likely to survive and reproduce.

(b) Graphical representation of directional selection

Starting population

Light wings ⟵ ⟶ Dark wings

Population after directional selection

Light wings ⟵ ⟶ Dark wings

FIGURE 24.8   Directional selection. (a) A new mutation arises in a population that confers higher Darwinian fitness. In this example, butterflies with dark wings are more likely to survive and reproduce. Over many generations, directional selection will favor the prevalence of darker individuals. (b) A graphical representation of directional selection.

Frequency of AA: p^2W_{AA}

Frequency of Aa: $2pqW_{Aa}$

Frequency of aa: q^2W_{aa}

In a population that is changing due to natural selection, these three terms may not add up to 1.0, as they would in the Hardy-Weinberg equilibrium. Instead, the three terms sum to a value known as the **mean fitness of the population ($\overline{W}$):**

$$p^2W_{AA} + 2pqW_{Aa} + q^2W_{aa} = \overline{W}$$

Dividing both sides of the equation by the mean fitness of the population,

$$\frac{p^2W_{AA}}{\overline{W}} + \frac{2pqW_{Aa}}{\overline{W}} + \frac{q^2W_{aa}}{\overline{W}} = 1$$

Using this equation, we can calculate the expected genotype and allele frequencies after one generation of directional selection:

Frequency of AA genotype: $\dfrac{p^2W_{AA}}{\overline{W}}$

Frequency of Aa genotype: $\dfrac{2pqW_{Aa}}{\overline{W}}$

Frequency of aa genotype: $\dfrac{q^2W_{aa}}{\overline{W}}$

Allele frequency of A: $p_A = \dfrac{p^2W_{AA}}{\overline{W}} + \dfrac{pqW_{Aa}}{\overline{W}}$

Allele frequency of a: $q_a = \dfrac{q^2W_{aa}}{\overline{W}} + \dfrac{pqW_{Aa}}{\overline{W}}$

As an example, let's suppose that the starting allele frequencies are $A = 0.5$ and $a = 0.5$, and use fitness values of 1.0, 0.8, and 0.2 for the three genotypes, AA, Aa, and aa, respectively. We begin by calculating the mean fitness of the population:

$$p^2W_{AA} + 2pqW_{Aa} + q^2W_{aa} = \overline{W}$$

$$\overline{W} = (0.5)^2(1) + 2(0.5)(0.5)(0.8) + (0.5)^2(0.2)$$

$$\overline{W} = 0.25 + 0.4 + 0.05 = 0.7$$

After one generation of directional selection,

Frequency of AA genotype: $\dfrac{p^2 W_{AA}}{\overline{W}} = \dfrac{(0.5)^2(1)}{0.7} = 0.36$

Frequency of Aa genotype: $\dfrac{2pq W_{Aa}}{\overline{W}} = \dfrac{2(0.5)(0.5)(0.8)}{0.7} = 0.57$

Frequency of aa genotype: $\dfrac{q^2 W_{aa}}{\overline{W}} = \dfrac{(0.5)^2(0.2)}{0.7} = 0.07$

Allele frequency of A: $p_A = \dfrac{p^2 W_{AA}}{\overline{W}} + \dfrac{pq W_{Aa}}{\overline{W}}$

$$= \dfrac{(0.5)^2(1)}{0.7} + \dfrac{(0.5)(0.5)(0.8)}{0.7} = 0.64$$

Allele frequency of a: $q_a = \dfrac{q^2 W_{aa}}{\overline{W}} + \dfrac{pq W_{Aa}}{\overline{W}}$

$$= \dfrac{(0.5)^2(0.2)}{0.7} + \dfrac{(0.5)(0.5)(0.8)}{0.7} = 0.36$$

After one generation, the allele frequency of A has increased from 0.5 to 0.64, while the frequency of a has decreased from 0.5 to 0.36. This has occurred because the AA genotype has the highest fitness, while the Aa and aa genotypes have lower fitness values. Another interesting feature of natural selection is that it raises the mean fitness of the population. If we assume the individual fitness values are constant, the mean fitness of this next generation is

$$\overline{W} = p^2 W_{AA} + 2pq W_{Aa} + q^2 W_{aa}$$
$$= (0.64)^2(1) + 2(0.64)(0.36)(0.8) + (0.36)^2(0.2)$$
$$= 0.80$$

The mean fitness of the population has increased from 0.7 to 0.8.

What are the consequences of natural selection at the population level? This population is better adapted to its environment than the previous one. Another way of viewing this calculation is that the subsequent population has a greater reproductive potential than the previous one. We could perform the same types of calculations to find the allele frequencies and mean fitness value in the next generation. If we assume the individual fitness values remain constant, the frequencies of A and a in the next generation are 0.85 and 0.15, respectively, and the mean fitness increases to 0.931. As we can see, the general trend is to increase A, decrease a, and increase the mean fitness of the population.

In the previous example, we considered the effects of natural selection by beginning with allele frequencies at intermediate levels (namely, $A = 0.5$ and $a = 0.5$). **Figure 24.9** illustrates what would happen if a new mutation introduced the A allele into a population that was originally monomorphic for the a allele. As before, the AA homozygote has a fitness of 1.0, the Aa heterozygote 0.8, and the recessive aa homozygote 0.2. Initially, the A allele is at a very low frequency in the population. If it is not lost initially due to genetic drift, its frequency slowly begins to rise and then, at intermediate values, rises much more rapidly.

Eventually, this type of natural selection may lead to fixation of a beneficial allele. However, a new beneficial allele is in a

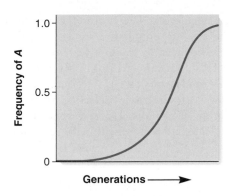

FIGURE 24.9 **The fate of a beneficial allele that is introduced as a new mutation into a population.** A new allele (A) is beneficial in the homozygous condition: $W_{AA} = 1.0$. The heterozygote, Aa ($W_{Aa} = 0.8$), and homozygote, aa ($W_{aa} = 0.2$), have lower fitness values.

precarious situation when its frequency is very low. As we have seen, random genetic drift is likely to eliminate new mutations, even beneficial ones, due to chance fluctuations.

Researchers have identified many examples of directional selection in nature. As mentioned in Chapter 6, resistance to antibiotics is a growing concern among members of the medical profession. The selection of bacterial strains resistant to one or more antibiotics typically occurs in a directional manner. Similarly, the resistance of insects to pesticides, such as DDT (dichlorodiphenyltrichloroethane), occurs in a directional manner. DDT usage began in the 1940s as a way to decrease the populations of mosquitoes and other insects. However, certain insect species can become resistant to DDT by a dominant mutation in a single enzyme-encoding gene. The mutant enzyme detoxifies DDT, making it harmless to the insect. **Figure 24.10** shows the results of an experiment in which mosquito larva (*Aedes*

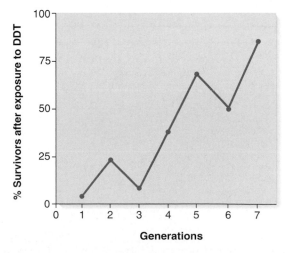

FIGURE 24.10 **Directional selection for DDT-resistance in a mosquito population.** In this experiment, mosquito larvae (*A. aegypti*) were exposed to 10 mg/liter of DDT. The percentage of survivors was recorded, and then the survivors of each generation were used as parents for the next generation.

aegypti) were exposed to DDT over the course of seven generations. The starting population showed a low level of DDT resistance, as evidenced by the low percentage of survivors after exposure to DDT. By comparison, in seven generations, nearly 100% of the population was DDT resistant. These results illustrate the power of directional selection in promoting change in a population. Since the 1950s, resistance to nearly every known insecticide has evolved within 10 years of its commercial introduction!

Stabilizing Selection Favors Individuals with Intermediate Phenotypes

In **stabilizing selection,** the extreme phenotypes for a trait are selected against, and those individuals with the intermediate phenotypes have the highest fitness values. Stabilizing selection tends to decrease genetic diversity for a particular gene because it eliminates alleles that cause a greater variation in phenotypes. An example of stabilizing selection involves clutch size in birds, which was first proposed by British biologist David Lack in 1947. Under stabilizing selection, birds that lay too many or too few eggs will have lower fitness values compared to those that lay an intermediate value (**Figure 24.11**). Laying too many eggs means that many offspring will die due to inadequate parental care and food. In addition, the strain on the parents themselves may decrease their likelihood of survival and therefore their ability to produce more offspring. Having too few offspring, on the

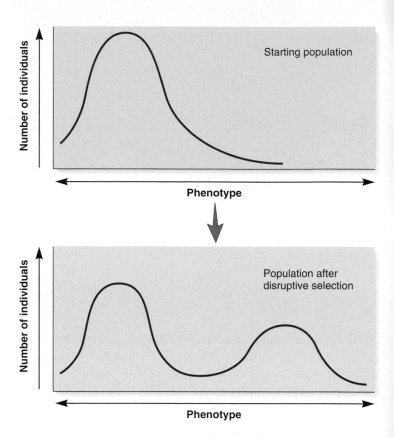

FIGURE 24.12   **Disruptive selection.** Over time, this form of selection favors two or more phenotypes due to heterogeneous environments.

other hand, does not contribute many individuals to the next generation. Therefore, the most successful parents are those that produce an intermediate clutch size. In the 1980s, Swedish evolutionary biologist Lars Gustafsson and colleagues examined the phenomenon of stabilizing selection in the collared flycatcher, *Ficedula albicollis*, on the island of Gotland, which is southeast of the mainland of Sweden. They discovered that Lack's hypothesis that clutch size is subject to the action of stabilizing selection also appears to be true for this species.

Disruptive Selection Favors Multiple Phenotypes

Disruptive selection, also known as diversifying selection, favors the survival of two or more different genotypes that produce different phenotypes (**Figure 24.12**). In disruptive selection, the fitness values of a particular genotype are higher in one environment and lower in a different environment. Disruptive selection is likely to occur in populations that occupy diverse environments so that some members of the species will survive in each type of environmental condition.

As an example, **Figure 24.13a** shows a photograph of land snails, *Cepaea nemoralis*, that live in woods and open fields. This snail is polymorphic in color and banding patterns. In 1954, Arthur Cain and Phillip Sheppard found that snail color was correlated with the environment. As shown in **Figure 24.13b**, the highest frequency of brown shell color was found in snails in the

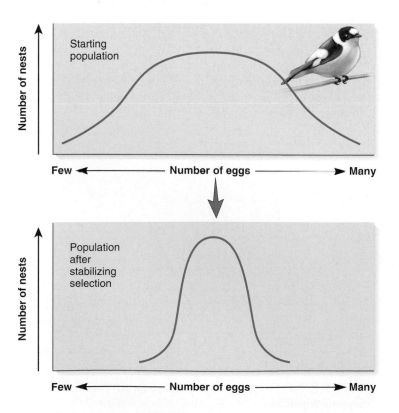

FIGURE 24.11   **Stabilizing selection.** In this pattern of natural selection, the extremes of a phenotypic distribution are selected against. Those individuals with intermediate traits have the highest fitness. This results in a population with less diversity and more uniform traits.

(a) Land snails

Habitat	Brown	Pink	Yellow
Beechwoods	0.23	0.61	0.16
Deciduous woods	0.05	0.68	0.27
Hedgerows	0.05	0.31	0.64
Rough herbage	0.004	0.22	0.78

(b) Frequency of snail color

FIGURE 24.13 **Polymorphism in the land snail,** *Cepaea nemoralis.* (a) This species of snail can exist in several different colors and banding patterns. (b) Coloration of the snails is correlated with the specific environments where they are located.

Genes→Traits Snail coloration is an example of genetic polymorphism due to heterogeneous environments; the genes governing shell coloration are polymorphic. The predation of snails is correlated with their ability to be camouflaged in their natural environment. Snails with brown shells are most prevalent in beechwoods, where the soil is dark. Pink snails are most abundant in the leaf litter of beechwoods and deciduous woods. Yellow snails are most prevalent in more sunny locations, such as hedgerows and rough herbage.

beechwoods, where there are wide expanses of dark soil. Their frequency was substantially less in other environments. By comparison, pink snails are most common in the leaf litter of forest floors, and the yellow snails are most abundant in the sunny, grassy areas of hedgerows and rough herbage. Researchers have suggested that this disruptive selection can be explained by different levels of predation by thrushes. Depending on the environment, certain snail phenotypes may be more easily seen by their predators compared to others. Migration can occasionally occur between the snail populations, which keeps the polymorphism in balance among these different environments.

Balanced Polymorphisms May Occur Due to Heterozygote Advantage or Negative Frequency-Dependent Selection

As we have just seen, polymorphisms may occur when a species occupies a diverse environment. Researchers have discovered other patterns of natural selection that favor the maintenance of two or more alleles in a more homogeneous environment. This pattern, called **balancing selection,** results in a genetic polymorphism in a population.

For genetic variation involving a single gene, balancing selection may arise when the heterozygote has a higher fitness than either corresponding homozygote, a situation called **heterozygote advantage.** In this case, an equilibrium is reached in which both alleles are maintained in the population. If the fitness values are known for each of the genotypes, the allele frequencies at equilibrium can be calculated. To do so, we must consider the **selection coefficient** (*s*), which measures the degree to which a genotype is selected against.

$$s = 1 - W$$

By convention, the genotype with the highest fitness has an *s* value of zero. Genotypes at a selective disadvantage have *s* values that are greater than zero but less than or equal to 1.0. An extreme case is a recessive lethal allele. It would have an *s* value of 1.0 in the homozygote, while the *s* value in the heterozygote could be zero.

Let's consider the following case of relative fitness, where

$$W_{AA} = 0.7$$
$$W_{Aa} = 1.0$$
$$W_{aa} = 0.4$$

The selection coefficients are

$$s_{AA} = 1 - 0.7 = 0.3$$
$$s_{Aa} = 1 - 1.0 = 0$$
$$s_{aa} = 1 - 0.4 = 0.6$$

The population will reach an equilibrium when

$$s_{AA}p = s_{aa}q$$

If we take this equation, let $q = 1 - p$, and then solve for *p*:

$$p = \text{Allele frequency of } A = \frac{s_{aa}}{s_{AA} + s_{aa}}$$

$$= \frac{0.6}{0.3 + 0.6} = 0.67$$

If we let $p = 1 - q$ and then solve for *q*:

$$q = \text{Allele frequency of } a = \frac{s_{AA}}{s_{AA} + s_{aa}}$$

$$= \frac{0.3}{0.3 + 0.6} = 0.33$$

In this example, balancing selection maintains the two alleles in the population at frequencies in which *A* equals 0.67 and *a* equals 0.33.

Heterozygote advantage can sometimes explain the high frequency of alleles that are deleterious in a homozygous condition. A classic example is the Hb^S allele of the human β-globin gene. A homozygous $Hb^S Hb^S$ individual displays sickle-cell disease, a disorder that leads to the sickling of the red blood cells. The $Hb^S Hb^S$ homozygote has a lower fitness than a homozygote with two normal copies of the β-globin gene, $Hb^A Hb^A$. However, the heterozygote, $Hb^A Hb^S$, has a higher level of fitness than either

(a) Malaria prevalence

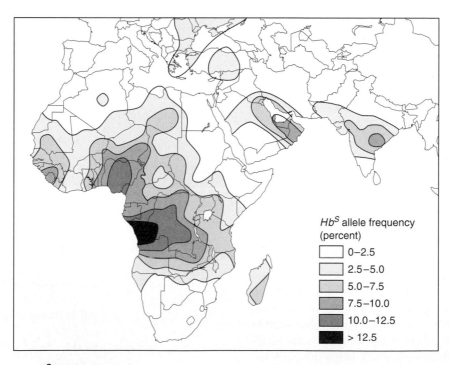

Hb^S allele frequency
(percent)

	0–2.5
	2.5–5.0
	5.0–7.5
	7.5–10.0
	10.0–12.5
	> 12.5

(b) *Hb^S* allele frequency

FIGURE 24.14   **The geographic relationship between malaria and the frequency of the sickle-cell allele in human populations.**
(**a**) The geographic prevalence of malaria in Africa and surrounding areas. (**b**) The frequency of the *Hb^S* allele in the same areas.

Genes→Traits The sickle-cell allele of the β-globin gene is maintained in human populations as a balanced polymorphism. In areas where malaria is prevalent, the heterozygote carrying one copy of the *Hb^S* allele has a greater fitness than either of the corresponding homozygotes (*Hb^A Hb^A* and *Hb^S Hb^S*). Therefore, even though the *Hb^S Hb^S* homozygotes suffer the detrimental consequences of sickle-cell disease, this negative aspect is balanced by the beneficial effects of malarial resistance in the heterozygotes.

homozygote in areas where malaria is endemic (**Figure 24.14**). Compared to $Hb^A Hb^A$ homozygotes, heterozygotes have a 10 to 15% better chance of survival if infected by the malarial parasite, *Plasmodium falciparum*. Therefore, the Hb^S allele is maintained in populations where malaria is prevalent, even though the allele is detrimental in the homozygous state.

In addition to sickle-cell disease, other gene mutations that cause human disease in the homozygous state are thought to be prevalent because of heterozygote advantage. For example, the high prevalence of the allele causing cystic fibrosis may be related to this phenomenon, but the advantage that a heterozygote may possess is not understood.

Negative frequency-dependent selection is a second mechanism of balancing selection. In this pattern of natural selection, the fitness of a genotype decreases when its frequency becomes higher. In other words, rare individuals have a higher fitness than more common individuals. Therefore, rare individuals are more likely to reproduce, while common individuals are less likely, thereby producing a balanced polymorphism in which no genotype becomes too rare or too common.

An interesting example of negative frequency-dependent selection involves the elder-flowered orchid, *Dactylorhiza sambucina* (**Figure 24.15**). Throughout its range, both yellow- and red-flowered individuals are prevalent. The explanation for this polymorphism is related to its pollinators, which are mainly bumblebees such as *Bombus lapidarius* and *B. terrestris*. The pollinators increase their visits to the flower color of *D. sambucina* as it becomes less common in a given area. One reason why this may occur is because *D. sambucina* is a rewardless flower; that is, it does not provide its pollinators with any reward for visiting, such as sweet nectar. Pollinators learn that the more common color of *D. sambucina* in a given area does not offer a reward, and they increase their visits to the less-common flower. Thus, the relative fitness of the less-common flower increases.

FIGURE 24.15   **The two color variations found in the elder-flowered orchid, *D. sambucina.*** The two colors are maintained in the population due to negative frequency-dependent selection.

EXPERIMENT 24A

The Grants Have Observed Natural Selection in Galápagos Finches

Let's now turn to a study that demonstrates natural selection in action. Since 1973, Peter Grant, Rosemary Grant, and their colleagues have studied the process of natural selection in finches found on the Galápagos Islands. For over 30 years, the Grants have focused much of their research on one of the Galápagos Islands known as Daphne Major (**Figure 24.16**). This small island (0.34 square kilometer) has a moderate degree of isolation (8 km from the nearest island), an undisturbed habitat, and a resident population of finches, including the medium ground finch, *Geospiza fortis*.

To study natural selection, the Grants observed various traits in the medium ground finch, including beak size, over the course of many years. The medium ground finch has a relatively small crushing beak, suitable for breaking open small, tender seeds. The Grants quantified beak size among the medium ground finches of Daphne Major by carefully measuring beak

(a)   (b)

FIGURE 24.16   **Location and species used in the Grants' investigation of natural selection in finches.** (a) Daphne Major, one of the Galápagos Islands. (b) A medium ground finch (*Geospiza fortis*) that populates this island.

depth (a measurement of the beak from top to bottom, at its base) on individual birds. During the course of their studies, they compared the beak sizes of parents and offspring by examining many broods over several years. The depth of the beak was transmitted from parents to offspring, regardless of environmental conditions, indicating that differences in beak sizes are due to genetic differences in the population. In other words, they found that beak depth is a heritable trait.

By measuring many birds every year, the Grants were able to assemble a detailed portrait of natural selection from gen-eration to generation. In the study shown in **Figure 24.17**, they measured beak depth in 1976 and 1978.

■ THE HYPOTHESIS

Beak size will be influenced by natural selection. Environments that produce larger seeds will select for birds with large beaks.

■ TESTING THE HYPOTHESIS — FIGURE 24.17   Natural selection in medium ground finches of Daphne Major.

	Experimental level	**Conceptual level**
1. In 1976, measure beak depth in parents and offspring of the species *G. fortis*.	Capture birds and measure beak depth.	This is a way to measure a trait that may be subject to natural selection.
2. Repeat the procedure on offspring that were born in 1978 and had reached mature size. A drought had occurred in 1977 that caused plants on the island to produce mostly larger seeds and relatively few small seeds.	Capture birds and measure beak depth.	This is a way to measure a trait that may be subject to natural selection.

■ THE DATA

Beak depth (mm)

■ INTERPRETING THE DATA

In the wet year of 1976, the plants of Daphne Major produced the small seeds that these finches were able to eat in abundance. However, a drought occurred in 1977. During this year, the plants on Daphne Major tended to produce few of the smaller seeds, which the finches rapidly consumed. To survive, the finches resorted to eating larger, drier seeds, which are harder to crush. As a result, the birds that survived tended to have larger beaks, because they were better able to break open these large seeds. In the year after the drought, the average beak depth of offspring in the population increased to approximately 9.8 mm because the surviving birds with larger beaks passed this trait on to their offspring. This is likely to be due to directional selection (see Figure 24.8), although genetic drift could also contribute to these data. Overall, these results illustrate the power of natural selection to alter the nature of a trait, in this case, beak depth, in a given population.

A self-help quiz involving this experiment can be found at the Online Learning Center.

Nonrandom Mating May Occur in Populations

As mentioned earlier, one of the conditions required to establish the Hardy-Weinberg equilibrium is random mating. This means that individuals choose their mates irrespective of their genotypes and phenotypes. In many cases, particularly in human populations, this condition is violated frequently.

When mating is nonrandom in a population, the process is called **assortative mating.** Positive assortative mating occurs when individuals with similar phenotypes choose each other as mates. The opposite situation, where dissimilar phenotypes mate preferentially, is called negative assortative mating. In addition, individuals may choose a mate that is part of the same genetic lineage. The mating of two genetically related individuals, such as cousins, is called **inbreeding.** This is also termed consanguinity. Inbreeding sometimes occurs in human societies and is more likely to take place in nature when population size becomes very limited. In Chapter 25, we will examine how inbreeding is a useful strategy for developing agricultural breeds or strains with desirable characteristics. Conversely, **outbreeding,** which involves mating between unrelated individuals, can create hybrids that are heterozygous for many genes.

In the absence of other evolutionary processes, inbreeding and outbreeding do not affect allele frequencies in a population. However, these patterns of mating do disrupt the balance of genotypes that is predicted by the Hardy-Weinberg equation. Let's first consider inbreeding in a family pedigree. **Figure 24.18** illustrates a human pedigree involving a mating between cousins. Individuals III-2 and III-3 are cousins and have produced the daughter labeled IV-1. She is said to be inbred, because her parents are genetically related to each other.

During inbreeding, the gene pool is smaller, because the parents are related genetically. In the 1940s, Gustave Malécot developed methods to quantify the degree of inbreeding. The **inbreeding coefficient** is the probability that two alleles in a particular individual will be identical for a given gene because both copies are due to descent from a common ancestor. An inbreed-

ing coefficient (F) can be computed by analyzing the degree of relatedness within a pedigree.

As an example, let's determine the inbreeding coefficient for individual IV-1. To begin this problem, we must first identify all of this individual's common ancestors. A common ancestor is anyone who is an ancestor to both of an individual's parents. In Figure 24.18, IV-1 has one common ancestor, I-2, her great-grandfather. I-2 is the grandfather of III-2 and III-3.

Our next step is to determine the inbreeding paths. An inbreeding path for an individual is the shortest path through the pedigree that includes both parents and the common ancestor. In a pedigree, there is an inbreeding path for each common ancestor. The length of each inbreeding path is calculated by adding together all of the individuals in the path except the individual of interest. In this case, there is only one path because IV-1 has only one common ancestor. To add the members of the path, we begin with individual IV-1, but we do not count her. We then move to her father (III-2); to her grandfather (II-2); to I-2, her great-grandfather (the common ancestor); back down to her other grandmother (II-3); and finally to her mother (III-3). This path has five members. Finally, to calculate the inbreeding coefficient, we use the following formula:

$$F = \Sigma (1/2)^n (1 + F_A)$$

where

F is the inbreeding coefficient of the individual of interest

n is the number of individuals in the inbreeding path, excluding the inbred offspring

F_A is the inbreeding coefficient of the common ancestor

Σ indicates that we add together $(1/2)^n (1 + F_A)$ for each inbreeding path

In this case, there is only one common ancestor and, therefore, only one inbreeding path. Also, we do not know anything about the heritage of the common ancestor, so we assume that F_A is zero. Thus, in our example of Figure 24.18,

$$F = \Sigma (1/2)^n (1 + 0)$$

$$= (1/2)^5 = 1/32 = 3.125\%$$

What does this value mean? Our inbreeding coefficient, 3.125%, tells us the probability that a gene in the inbred individual (IV-1) is homozygous due to its inheritance from a common ancestor (I-2). In this case, therefore, each gene in individual IV-1 has a 3.125% chance of being homozygous because she has inherited the same allele twice from her great-grandfather (I-2), once through each parent.

As an example, let's suppose that the common ancestor (I-2) is heterozygous for the gene involved with cystic fibrosis. His genotype would be Cc, where c is the recessive allele that causes cystic fibrosis. There is a 3.125% probability that the inbred individual (IV-1) is homozygous (CC or cc) for this gene because

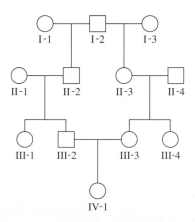

FIGURE 24.18   A human pedigree containing inbreeding.
Individual IV-1 is the result of inbreeding because her parents are related.

she has inherited both copies from her great-grandfather. She has a 1.56% probability of inheriting both normal alleles (*CC*) and a 1.56% probability of inheriting both mutant alleles (*cc*). The inbreeding coefficient is denoted by the letter *F* (for Fixation) because it is the probability that an allele will be fixed in the homozygous condition. The term fixation signifies that the homozygous individual can pass only one type of allele to their offspring.

In other pedigrees, an individual may have two or more common ancestors. In this case, the inbreeding coefficient *F* is calculated as the sum of the inbreeding paths. Such an example is described in solved problem S2 at the end of the chapter.

The effects of inbreeding and outbreeding can also be considered within a population as well as within individuals. For example, let's consider the situation in which the frequency of $A = p$ and the frequency of $a = q$. In a given population, the genotype frequencies are determined in the following way:

$p^2 + fpq$ equals the frequency of *AA*

$2pq(1 - f)$ equals the frequency of *Aa*

$q^2 + fpq$ equals the frequency of *aa*

where *f* is a measure of how much the genotype frequencies deviate from Hardy-Weinberg equilibrium due to nonrandom mating. The value of *f* ranges from −1 to +1. When inbreeding occurs, the value is greater than zero. When outbreeding occurs, the value is less than zero.

As an example, let's suppose that $p = 0.8$, $q = 0.2$, and $f = 0.25$. We can calculate the frequencies of the *AA*, *Aa*, and *aa* genotypes under these conditions as follows:

$$AA = p^2 + fpq = (0.8)^2 + (0.25)(0.8)(0.2) = 0.68$$

$$Aa = 2pq(1 - f) = 2(0.8)(0.2)(1 - 0.25) = 0.24$$

$$aa = q^2 + fpq = (0.2)^2 + (0.25)(0.8)(0.2) = 0.08$$

There will be 68% *AA* homozygotes, 24% heterozygotes, and 8% *aa* homozygotes. If mating had been random (i.e., $f = 0$), the genotype frequencies of *AA* would be p^2, which equals 64%, and *aa* would be q^2, which equals 4%. The frequency of heterozygotes would be $2pq$, which equals 32%. When comparing these numbers, we see that inbreeding raises the proportions of homozygotes and decreases the proportion of heterozygotes. In natural populations, the value of *f* tends to become larger as a population becomes smaller, because each individual has a more limited choice in mate selection.

What are the consequences of inbreeding in a population? From an agricultural viewpoint, it results in a higher proportion of homozygotes, which may exhibit a desirable trait. For example, an animal breeder may use inbreeding to produce animals that are larger because they have become homozygous for alleles promoting larger size. On the negative side, many genetic diseases are inherited in a recessive manner (see Chapter 22). For these disorders, inbreeding increases the likelihood that an individual will be homozygous and therefore afflicted with the disease. Also, in natural populations, inbreeding will lower the mean

fitness of the population if homozygous offspring have a lower fitness value. This can be a serious problem as natural populations become smaller due to human habitat destruction. As the population shrinks, inbreeding becomes more likely because individuals have fewer potential mates from which to choose. The inbreeding, in turn, produces homozygotes that are less fit, thereby decreasing the reproductive success of the population. This phenomenon is called **inbreeding depression.** Conservation biologists sometimes try to circumvent this problem by introducing individuals from one population into another. For example, the endangered Florida panther (*Felis concolor coryi*) suffers from inbreeding-related defects, which include poor sperm quality and quantity, and morphological abnormalities. To help alleviate these effects, panthers of the same species from Texas have been introduced into the Florida population.

24.3 SOURCES OF NEW GENETIC VARIATION

In the previous section, we primarily focused on genetic variation in which a single gene exists in two or more alleles. This simplified scenario allows us to appreciate the general principles behind evolutionary mechanisms. As researchers have analyzed genetic variation at the molecular, cellular, and population level, they have come to understand that new genetic variation occurs in many ways (**Table 24.2**). Among eukaryotic species, sexual reproduction is an important way that new genetic variation occurs among offspring. In Chapters 3 and 5, we considered how independent assortment and crossing over during sexual reproduction may produce new combinations of alleles among different genes and thereby produce new genetic variation in the resulting offspring. Similarly, in Chapter 26, we will consider how breeding between members of different species may produce hybrid offspring that harbor new combinations of genetic material. Such hybridization events have been important in the evolution of new species, particularly those in the plant kingdom. Though prokaryotic species reproduce asexually, they also possess mechanisms for gene transfer, such as conjugation, transduction, and transformation (see Chapter 6). These mechanisms are important for fostering genetic variation among bacterial and archaeal populations.

Rare mutations in DNA may also give rise to new types of variation (see Table 24.2). As discussed earlier in this chapter (see Figure 24.3) and in Chapter 16, mutations may occur within a particular gene to create new alleles of that gene. Such allelic variation is common in natural populations. Also, as described in Chapter 8, gene duplications may create a gene family; each family member will acquire independent mutations and often times will evolve more specialized functions. An example is the globin gene family (see Figure 8.7). Changes in chromosome structure and number are also important in the evolution of new species (see Chapter 26). In this section, we will examine some additional mechanisms through which an organism can acquire new genetic variation. These include exon shuffling, horizontal gene transfer, and changes in repetitive sequences. The myriad of mechanisms that foster genetic variation underscores its profound importance

TABLE 24.2

Sources of New Genetic Variation That Occur in Populations

Type	Description
Independent assortment	The independent segregation of different homologous chromosomes may give rise to new combinations of alleles in offspring (see Chapter 3).
Crossing over	Recombination (crossing over) between homologous chromosomes can also produce new combinations of alleles that are located on the same chromosome (see Chapter 5).
Interspecies crosses	On occasion, members of different species may breed with each other to produce hybrid offspring. This topic is discussed in Chapter 26.
Prokaryotic gene transfer	Prokaryotic species possess mechanisms of genetic transfer such as conjugation, transduction, and transformation (see Chapter 6).
New alleles	Point mutations can occur within a gene to create single nucleotide polymorphisms (SNPs). In addition, genes can be altered by small deletions and additions. Gene mutations are also discussed in Chapter 16.
Gene duplications	Events, such as misaligned crossovers, can add additional copies of a gene into a genome and lead to the formation of gene families. This topic is discussed in Chapter 8.
Chromosome structure and number	Chromosome structure may be changed by deletions, duplications, inversions, and translocations. Changes in chromosome number result in aneuploid, polyploid, and alloploid offspring. These mechanisms are discussed in Chapters 8 and 26.
Exon shuffling	New genes can be created when exons of preexisting genes are rearranged to make a gene that encodes a protein with a new combination of protein domains.
Horizontal gene transfer	Genes from one species can be introduced into another species and become incorporated into that species' genome.
Changes in repetitive sequences	Short repetitive sequences are common in genomes due to the occurrence of transposable elements and due to tandem arrays. The number and lengths of repetitive sequences tend to show considerable variation in natural populations.

in the evolution of species that are well adapted to their native environments and successful at reproduction.

New Genes Are Created in Eukaryotes Via Exon Shuffling

Sources of new genetic variation are revealed when the parts of genes that encode protein domains are compared within a single species. Many proteins, particularly those found in eukary-

otic species, have a modular structure composed of two or more domains with different functions. For example, certain transcription factors have discrete domains involved with hormone binding, dimerization, and DNA binding. As described in Chapter 15, the glucocorticoid receptor has a domain that binds the hormone, a second domain that facilitates protein dimerization, and a third domain that allows the glucocorticoid receptor to bind to glucocorticoid response elements (GREs) next to genes (see Figure 15.6). By comparing the modular structure of eukaryotic proteins with the genes that encode them, geneticists have discovered that each domain tends to be encoded by one coding sequence, or exon, or by a series of two or more adjacent exons.

During the evolution of eukaryotic species, many new genes have been created by a process known as **exon shuffling,** in which an exon and its flanking introns are inserted into a gene, thereby producing a new gene that encodes a protein with an additional domain (**Figure 24.19**). This process may also involve the duplication and rearrangement of exons. Exon shuffling results in novel genes that express proteins with diverse functional modules. Such proteins can then alter traits in the organism and may be acted upon by evolutionary forces such as genetic drift and natural selection.

Exon shuffling may occur by more than one mechanism. As described in Chapter 17, transposable elements may promote the insertion of exons into the coding sequences of other genes. Alternatively, a double crossover event could promote the insertion of an exon into another gene (this is the case in Figure 24.19). This is called nonhomologous recombination because the two regions involved in the crossover are not homologous to each other.

New Genes Are Acquired Via Horizontal Gene Transfer

Species also accumulate genetic changes by a process called **horizontal gene transfer,** which involves the exchange of genetic material among different species. **Figure 24.20** illustrates one possible mechanism for horizontal gene transfer. In this example, a eukaryotic cell has engulfed a bacterium by endocytosis. During the degradation of the bacterium, a bacterial gene escapes to the nucleus of the cell, where it is inserted into one of the chromosomes. In this way, a gene has been transferred from a bacterial species to a eukaryotic species. By analyzing gene sequences among many different species, researchers have discovered that horizontal gene transfer is a common phenomenon. This process can occur from prokaryotes to eukaryotes, from eukaryotes to prokaryotes, between different species of prokaryotes, and between different species of eukaryotes.

Gene transfer among bacterial species is relatively widespread. As discussed in Chapter 6, bacterial species may carry out three natural mechanisms of gene transfer known as conjugation, transduction, and transformation. By analyzing the genomes of bacterial species, scientists have determined that many genes within a given bacterial genome are derived from horizontal gene transfer. Genome studies have suggested that as much as 20 to 30% of the variation in the genetic composition of modern prokaryotic species can be attributed to this process. For example, in

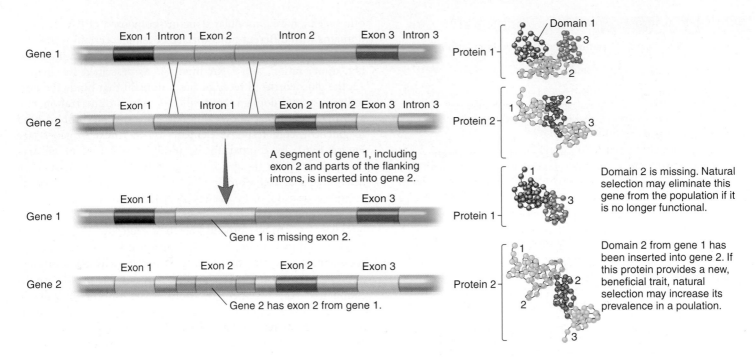

FIGURE 24.19   The process of exon shuffling. In this example, a segment of one gene containing an exon and its flanking introns has been inserted into another gene. A rare, abnormal crossing over event called nonhomologous recombination may cause this to happen. This results in proteins that have new combinations of domains and possibly new functions.

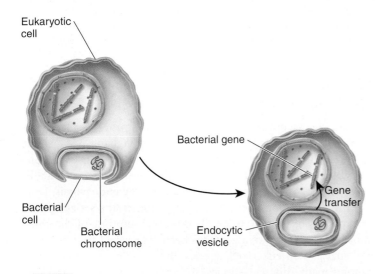

FIGURE 24.20   Horizontal gene transfer from a bacterium to a eukaryote. In this example, a bacterium is engulfed by a eukaryotic cell, and a bacterial gene is transferred to one of the eukaryotic chromosomes.

E. coli and *Salmonella typhimurium*, roughly 17% of their genes have been acquired via horizontal gene transfer during the past 100 million years. The roles of these acquired genes are quite varied, though they commonly involve functions that are readily acted upon by natural selection. These include genes that confer antibiotic resistance, the ability to degrade toxic compounds, and pathogenicity (the ability to cause disease).

Genetic Variation Is Produced Via Changes in Repetitive Sequences

Another source of genetic variation comes from changes in **repetitive sequences**—short sequences typically a few base pairs to a few thousand base pairs that are repeated many times within a species' genome. Repetitive sequences usually come from two types of sources. First, transposable elements are genetic sequences that can move from place to place in a species' genome (see Chapter 17). The prevalence and movement of transposable elements provides a great deal of genetic variation between species and within a single species. In certain eukaryotic species, transposable elements have become fairly abundant (see Table 17.3).

A second type of repetitive sequence is nonmobile and involves short sequences that are tandemly repeated. In a **microsatellite** (also called short tandem repeats, STRs), the repeat unit is usually 1 to 6 bp, and the whole tandem repeat is less than a couple hundred bp in length. For example, the most common microsatellite encountered in humans is a sequence $(CA)_N$, where N may range from 5 to more than 50. In other words, this dinucleotide sequence can be tandemly repeated 5 to 50 or more times. The $(CA)_N$ microsatellite is found, on average, about every 10,000 bases in the human genome. In a **minisatellite,** the repeat unit is typically 6 to 80 bp in length, and the size of the minisatellite ranges from 1 kbp to 20 kbp. An example of a minisatellite in humans is telomeric DNA. In a human sperm cell, for example, the repeat unit is 6 bp and the size of a telomere is about 15 kbp. (Note: Tandem repeat sequences are called satellites because they sediment away from the rest of the chromosomal DNA during equilibrium gradient centrifugation.)

Tandem repetitive sequences such as microsatellites and minisatellites tend to undergo mutation in which the number of tandem repeats changes. For example, a microsatellite with a 4 bp repeat unit and a length of 64 bp may undergo a mutation that adds three more repeat units and become 76 bp long. The mechanism whereby micro- and minisatellites change in length is not well understood, but it may involve errors in DNA replication and homologous recombination.

Because repetitive sequences tend to vary within a population, they have become a common tool that geneticists use in a variety of ways. For example, as described in Chapters 20 and 22, microsatellites can be used as molecular markers to map the locations of genes (see Figure 22.5). Likewise, population geneticists analyze microsatellites or minisatellites to study variation at the population level and to determine the relationships among individuals and neighboring populations. The sizes of microsatellites and minisatellites found in closely related individuals tend to be more similar compared to unrelated individuals. As described next, this phenomenon is the basis for DNA fingerprinting.

DNA Fingerprinting Is Used for Identification and Relationship Testing

The technique of **DNA fingerprinting,** also known as **DNA profiling,** analyzes individuals based on the occurrence of repetitive sequences in their genome. When subjected to traditional DNA fingerprinting, the chromosomal DNA gives rise to a series of bands on a gel (**Figure 24.21**). The sizes and order of bands is an individual's DNA fingerprint. Like the human fingerprint, the DNA of each individual has a distinctive pattern. It is the unique pattern of these bands that makes it possible to distinguish individuals.

A comparison of the DNA fingerprints among different individuals has found two applications. First, DNA fingerprinting can be used as a method of identification. In forensics, DNA fingerprinting can identify a crime suspect. In medicine, the technique can identify the type of bacterium that is causing an infection in a particular patient. A second use of DNA fingerprinting is relationship testing. Closely related individuals have more similar fingerprints compared to distantly related ones (see solved problem S6). In humans, this can be used in paternity testing. In population genetics, DNA fingerprinting can provide evidence regarding the degree of relatedness among members of a population. Such information may help geneticists determine if a population is likely to be suffering from inbreeding depression.

The development of DNA fingerprinting has relied on the identification of DNA fragments that are quite variable among members of a population. This naturally occurring variation causes each individual to have a unique DNA fingerprint. In the 1980s, Alec Jeffries and his colleagues found that certain locations within human chromosomes are particularly variable in their lengths. These minisatellites tend to vary within the human population due to changes in the number of tandem repeats at each minisatellite. For this reason, minisatellites are also referred to as loci with a <u>v</u>ariable <u>n</u>umber of <u>t</u>andem <u>r</u>epeats (VNTRs).

The occurrence of minisatellites is shown schematically in **Figure 24.22**, where the chromosomal DNA from two individ-

FIGURE 24.21 **A comparison of two DNA fingerprints.** The chromosomal DNA from two different individuals (Suspect 1—S1, and Suspect 2—S2) was subjected to DNA fingerprinting. The DNA evidence at a crime scene, E(vs), was also subjected to DNA fingerprinting. Following the hybridization of a radiolabeled probe, the DNA appears as a series of bands on a gel. The dissimilarity in the pattern of these bands distinguishes different individuals, much as the differences in physical fingerprint patterns can be used for identification. As seen here, S2 matches the DNA found at the crime scene.

uals is compared. The diagram emphasizes two features of the chromosomal DNA. First, it shows the sites recognized by a particular restriction enzyme (designated with an arrow). As a matter of chance, these sites are interspersed throughout the genome of both individuals. Second, minisatellites (designated by their orange color with a series of repeat units labeled as R's) are also found. For the minisatellite shown on the left, the two individuals have the same number of repeats. For the other two minisatellites, the number of repeats differs substantially. The variation in the sizes of minisatellites affects the sizes of fragments produced when the chromosomal DNAs are digested with the restriction

FIGURE 24.22   A comparison of minisatellites between two individuals. The restriction sites found in both individuals are represented by arrows. Each individual carries three minisatellites, shown in orange. The repeat units within each minisatellite are depicted with the letter R. The variation in the number of repeats affects the sizes of the DNA fragments produced when the DNA is digested with the restriction enzyme that cleaves at the designated sites. (Note: Minisatellites actually have more repeat units than shown here.)

enzyme; these differences yield a distinct pattern of DNA fragments when analyzed via gel electrophoresis.

Figure 24.22 depicts only a short segment of DNA. When the chromosomal DNA from a sample of actual cells is digested with a restriction enzyme, this would yield too many DNA fragments to analyze. In traditional DNA fingerprinting, DNA probes were used that hybridized specifically to the repeat sequence located within selected minisatellites. DNA fingerprinting probes were made that recognized a selected minisatellite sequence found at a relatively small number of sites (say, 5 to 30 sites) in the human genome. Using such probes, a DNA fingerprinting analysis examined the length of the DNA fragments at the sites of the chosen minisatellite sequence. This involved the analysis of 5 to 30 bands that correspond to minisatellite sites, as in Figure 24.21.

Figure 24.23a outlines the steps in a traditional DNA fingerprinting experiment. This procedure is a Southern blot using a radiolabeled probe complementary to a selected minisatellite sequence. The chromosomal DNA is isolated from a sample and digested with a restriction enzyme. The resulting DNA fragments are then separated by gel electrophoresis. The fragments in the gel are blotted onto a nylon membrane, the DNA is denatured, and the membrane is exposed to the radiolabeled probe. Because the probe is complementary to a selected minisatellite sequence, it hybridizes to approximately 5 to 30 fragments of DNA that

contain this sequence and thereby labels 5 to 30 bands. In Figure 24.23a, the results of a DNA fingerprinting experiment on two samples are compared. Because the pattern of band sizes does not match between the two samples, it is concluded that the samples came from different individuals.

In the past decade, the technique of DNA fingerprinting has become automated, much like the automation that changed the procedure of DNA sequencing described in Chapter 18. DNA fingerprinting is now done using the technique of polymerase chain reaction (PCR), which amplifies microsatellites. Like minisatellites, microsatellites are found in multiple sites in the genome of humans and other species and are variable among different individuals. In this procedure, the microsatellites from a sample of DNA are amplified by PCR using primers that flank the repetitive region and then separated by gel electrophoresis according to their molecular masses. As in automated DNA sequencing, the amplified microsatellite fragments are fluorescently labeled. A laser excites the fluorescent molecule within a microsatellite, and a detector records the amount of fluorescence emission for each microsatellite. As shown in **Figure 24.23b**, this type of DNA fingerprint yields a series of peaks, each peak having a characteristic molecular mass. In this automated approach, the pattern of peaks rather than bands constitutes an individual's DNA fingerprint.

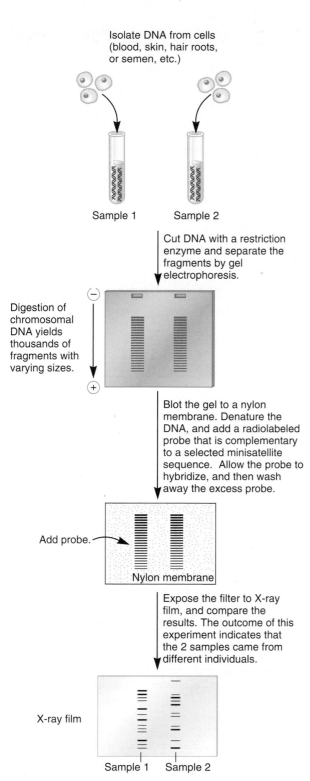

Isolate DNA from cells (blood, skin, hair roots, or semen, etc.)

Sample 1   Sample 2

Cut DNA with a restriction enzyme and separate the fragments by gel electrophoresis.

Digestion of chromosomal DNA yields thousands of fragments with varying sizes.

Blot the gel to a nylon membrane. Denature the DNA, and add a radiolabeled probe that is complementary to a selected minisatellite sequence. Allow the probe to hybridize, and then wash away the excess probe.

Add probe.

Nylon membrane

Expose the filter to X-ray film, and compare the results. The outcome of this experiment indicates that the 2 samples came from different individuals.

X-ray film

Sample 1   Sample 2

(a) Traditional DNA fingerprinting

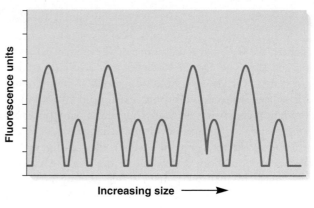

(b) Automated DNA fingerprinting

FIGURE 24.23 Protocol for DNA fingerprinting.
(a) In traditional DNA fingerprinting, chromosomal DNA is isolated from a sample of cells (often a blood sample, hair roots, or semen) and cut with a restriction enzyme, and then the DNA fragments are separated by gel electrophoresis. The DNA fragments within the gel are then transferred to a nylon membrane. The DNA is denatured, and a radiolabeled probe is added that is complementary to a selected minisatellite sequence. The probe is given sufficient time to hybridize to any complementary DNA fragments, and any unbound probe is washed away. Finally, the membrane is exposed to X-ray film. The outcome of this experiment indicates that the two samples came from different individuals. **(b)** In automated DNA fingerprinting, a sample of DNA is amplified, using primers that recognize the ends of microsatellites. The microsatellite fragments are fluorescently labeled and then separated by gel electrophoresis. The fluorescent molecules within each microsatellite are excited with a laser, and the amount of fluorescence is measured via a fluorescence detector. A printout from the detector is shown here.

In this chapter, we have surveyed some of the critical issues in population genetics. The primary focus of this field is an understanding of genetic variation within populations. In natural populations, most genes are polymorphic, existing in two or more alleles. Population geneticists want to understand why this variation in the gene pool exists and how it may change.

The Hardy-Weinberg equation relates allele and genotype frequencies in a population. According to the Hardy-Weinberg equilibrium, allele frequencies remain stable as long as several conditions are met. These include no new mutations, no genetic drift, no migration, no natural selection, and random mating. While these conditions are never truly met, they can be approximated in certain large populations.

However, we know that many processes in nature promote genetic change. The term microevolution describes changes in a population's gene pool over generations. The initial source of genetic variation is mutation. Evolutionary forces may act to increase the frequency of some new alleles. Genetic drift can alter allele frequencies due to random sampling error; a new mutation may be eliminated from a population or it may become fixed in it. Genetic drift can occur via a bottleneck effect, when a population is reduced in size due to negative environmental events, or via a founder effect, when a small population moves to a new site. Allele frequencies can also be altered by migration between two populations that differ in their allele frequencies, a process known as gene flow.

Natural selection favors individuals with the greatest reproductive potential (i.e., the greatest Darwinian fitness). These may be individuals who are more likely to survive, to mate, or to be more fertile. Depending on the relative fitness values of genotypes and the complexities of the environmental effects, natural selection may follow different patterns, including directional, stabilizing, disruptive, and balancing,. Finally, the process of nonrandom mating, including inbreeding, can alter genotype frequencies in a population.

A central approach in population genetics is to measure variation experimentally and then examine the mathematical relationships between genetic variation and the occurrence of genotypes and phenotypes within populations. In this chapter, we have focused on how population geneticists explain phenotypic variation via mathematical equations. Two fundamental calculations are allele frequencies and genotype frequencies. Experimentally, phenotypic variation can be measured by tallying the organisms within a population that display particular phenotypes. For example, we can count the number of dark snails and light snails that live in a particular location. This provides a measure of the phenotypic variation within this population. In Chapter 26, we will see that many other molecular approaches, particularly DNA sequencing, allow population geneticists to measure the genetic variation within populations.

After variation is measured, population geneticists attempt to develop and apply mathematical relationships that provide insight into the causes and dynamics of the variation. Although geneticists understand that mutation is the source of new variation, their mathematical formulas indicate that the rate of new mutations is too low to explain the experimentally observed variation in natural populations. Instead, other processes, such as genetic drift, migration, and natural selection, must alter the frequency of new mutations after they have occurred. For example, the mathematical theories of population geneticists allow us to predict how natural selection may be directional, leading to the eventual fixation or elimination of an allele, or promote a balanced polymorphism within a population. As we have seen, Peter and Rosemary Grant observed natural selection in action by tracking changes in beak size in a population of medium ground finches over the course of many years. Their results indicated that beak size is subject to directional selection.

DNA fingerprinting relies on variation in the length of microsatellites and minisatellites that are found at multiple sites in a given species' genome. This method is used to identify individuals based on their pattern of bands and also to study genetic relationships between individuals (e.g., paternity testing) and among members of a population.

PROBLEM SETS & INSIGHTS

Solved Problems

S1. The phenotypic frequency of people who cannot taste phenylthiocarbamide (PTC) is approximately 0.3. The inability to taste this bitter substance is due to a recessive allele. If we assume there are only two alleles in the population (namely, tasters, T, and nontasters, t) and that the population is in Hardy-Weinberg equilibrium, calculate the frequencies of these two alleles.

Answer: Let p = allele frequency of the taster allele and q = the allele frequency of the nontaster allele. The frequency of nontasters is 0.3.

This is the frequency of the genotype tt, which in this case is equal to q^2:

$$q^2 = 0.3$$

To determine the frequency q of the nontaster allele, we take the square root of both sides of this equation:

$$q = 0.55$$

With this value, we can calculate the frequency p of the taster allele:

$$p = 1 - q$$
$$= 1 - 0.55 = 0.45$$

S2. In the pedigree shown here, answer the following questions with regard to individual VII-1:

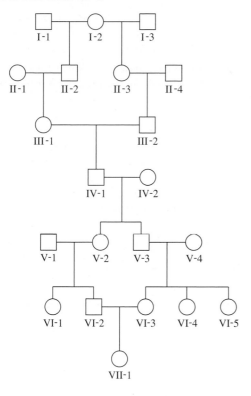

A. Who are the common ancestors of her parents?

B. What is the inbreeding coefficient?

Answer:

A. The common ancestors are IV-1 and IV-2. They are the grandparents of VI-2 and VI-3, who are the parents of VII-1.

B. The inbreeding coefficient is calculated using the formula

$$F = \Sigma(1/2)^n(1 + F_A)$$

In this case, there are two common ancestors, IV-1 and IV-2. Also, IV-1 is inbred, because I-2 is a common ancestor to both of IV-1's parents. The first step is to calculate F_A, the inbreeding coefficient for this common ancestor. The inbreeding path for IV-1 contains five people: III-1, II-2, I-2, II-3, and III-2. Therefore,

$$n = 5$$

$$F_A = (1/2)^5 = 0.03$$

Now we can calculate the inbreeding coefficient for VII-1. Each inbreeding path contains five people: VI-2, V-2, IV-1, V-3, and VI-3; and VI-2, V-2, IV-2, V-3, and VI-3. Thus,

$$F = (1/2)^5 (1 + 0.03) + (1/2)^5 (1 + 0)$$

$$= 0.032 + 0.031 = 0.063$$

S3. The Hardy-Weinberg equation provides a way to predict genotype frequency based on allele frequency. In the case of mammals, males are hemizygous for X-linked genes, whereas females have two copies. Among males, the frequency of any X-linked trait will equal the frequency of males with the trait. For example, if an allele frequency for an X-linked disease-causing allele was 5%, then 5% of all males would be affected with the disorder. Female

genotype frequencies are computed using the Hardy-Weinberg equation.

As a specific example, let's consider the human X-linked trait known as hemophilia A (see Chapter 22 for a description of this disorder). In human populations, the allele frequency of the hemophilia A allele is approximately 1 in 10,000, or 0.0001. The other allele for this gene is the normal allele. Males can be affected or unaffected, whereas females can be affected, unaffected carriers, or unaffected noncarriers.

A. What are the allele frequencies for the mutant and normal allele in the human population?

B. Among males, what is the frequency of affected individuals?

C. Among females, what is the frequency of affected individuals and heterozygous carriers?

D. Within a population of 100,000 people, what is the expected number of affected males? In this same population, what is the expected number of carrier females?

Answer: Let p represent the normal allele and q represent the allele that causes hemophilia.

A. X^H normal allele, frequency = 0.9999 = p

  X^h hemophilia allele, frequency = 0.0001 = q

B. X^hY genotype frequency of affected males = q = 0.0001

C. X^hX^h genotype frequency of affected females = q^2 = $(0.0001)^2$ = 0.00000001

  X^HX^h genotype frequency of carrier females = $2pq$ = 2(0.9999)(0.0001) = 0.0002

D. We will assume this population is composed of 50% males and 50% females.

  Number of affected males = 50,000 × 0.0001 = 5

  Number of carrier females = 50,000 × 0.0002 = 10

S4. The Hardy-Weinberg equation can be modified to include situations of three or more alleles. In its standard (two-allele) form, the Hardy-Weinberg equation reflects the Mendelian notion that each individual inherits two copies of each allele, one from both parents. For a two-allele situation, it can also be written as

  $(p + q)^2 = 1$ (Note: The number 2 in this equation reflects the idea that the genotype is due to the inheritance of two alleles, one from each parent.)

  This equation can be expanded to include three or more alleles. For example, let's consider a situation in which a gene exists as three alleles: *A1*, *A2*, and *A3*. The allele frequency of *A1* is designated by the letter p, *A2* by the letter q, and *A3* by the letter r. Under these circumstances, the Hardy-Weinberg equation becomes

  $(p + q + r)^2 = 1$

  $p^2 + q^2 + r^2 + 2pq + 2pr + 2qr = 1$

where

  p^2 is the genotype frequency of *A1A1*

  q^2 is the genotype frequency of *A2A2*

  r^2 is the genotype frequency of *A3A3*

  $2pq$ is the genotype frequency of *A1A2*

  $2pr$ is the genotype frequency of *A1A3*

  $2qr$ is the genotype frequency of *A2A3*

Now here is the question. As discussed in Chapter 4, the gene that affects human blood type can exist in three alleles. In a Japanese population, the allele frequencies are

$$I^A = 0.28$$

$$I^B = 0.17$$

$$i = 0.55$$

Based on these allele frequencies, calculate the different possible genotype frequencies and blood type frequencies.

Answer: If we let p represent I^A, q represent I^B, and r represent i, then

p^2 is the genotype frequency of $I^A I^A$, which is type A blood $= (0.28)^2 = 0.08$

q^2 is the genotype frequency of $I^B I^B$, which is type B blood $= (0.17)^2 = 0.03$

r^2 is the genotype frequency of ii, which is type O blood $= (0.55)^2 = 0.30$

$2pq$ is the genotype frequency of $I^A I^B$, which is type AB blood $= 2(0.28)(0.17) = 0.09$

$2pr$ is the genotype frequency of $I^A i$, which is type A blood $= 2(0.28)(0.55) = 0.31$

$2qr$ is the genotype frequency of $I^B i$, which is type B blood $= 2(0.17)(0.55) = 0.19$

Type A $= 0.08 + 0.31 = 0.39$, or 39%

Type B $= 0.03 + 0.19 = 0.22$, or 22%

Type O $= 0.30$, or 30%

Type AB $= 0.09$, or 9%

S5. Let's suppose that pigmentation in a species of insect is controlled by a single gene existing in two alleles, D for dark and d for light. The heterozygote Dd is intermediate in color. In a heterogeneous environment, the allele frequencies are $D = 0.7$ and $d = 0.3$. This polymorphism is maintained because the environment contains some dimly lit forested areas and some sunny fields. During a hurricane, a group of 1,000 insects is blown to a completely sunny area. In this environment, the fitness values are $DD = 0.3$, $Dd = 0.7$, and $dd = 1.0$. Calculate the allele frequencies in the next generation.

Answer: The first step is to calculate the mean fitness of the population:

$$p^2 W_{DD} + 2pq W_{Dd} + q^2 W_{dd} = \overline{W}$$

$$\overline{W} = (0.7)^2(0.3) + 2(0.7)(0.3)(0.7) + (0.3)^2(1.0)$$

$$= 0.15 + 0.29 + 0.09 = 0.53$$

After one generation of selection, we get

Allele frequency of D: $p_D = \dfrac{p^2 W_{DD}}{\overline{W}} + \dfrac{pq W_{Dd}}{\overline{W}}$

$$= \frac{(0.7)^2(0.3)}{0.53} + \frac{(0.7)(0.3)(0.7)}{0.53}$$

$$= 0.55$$

Allele frequency of d: $q_d = \dfrac{q^2 W_{dd}}{\overline{W}} + \dfrac{pq W_{Dd}}{\overline{W}}$

$$= \frac{(0.3)^2(1.0)}{0.53} + \frac{(0.7)(0.3)(0.7)}{0.53}$$

$$= 0.45$$

After one generation, the allele frequency of D has decreased from 0.7 to 0.55, while the frequency of d has increased from 0.3 to 0.45.

S6. An important application of DNA fingerprinting is relationship testing. Persons who are related genetically will have some bands or peaks in common. The number they share depends on the closeness of their genetic relationship. For example, an offspring is expected to receive half of his or her minisatellites from one parent and the rest from the other. The diagram below schematically shows a traditional DNA fingerprint of an offspring, mother, and two potential fathers.

| Mother | Child | Male 1 | Male 2 |

In paternity testing, the offspring's DNA fingerprint is first compared to that of the mother. The bands that the offspring has in common with the mother are depicted in purple. The bands that are not similar between the offspring and the mother must have been inherited from the father. These bands are depicted in red. Which male could be the father?

Answer: Male 2 does not have many of the paternal bands. Therefore, he can be excluded as being the father of this child. However, male 1 has all of the paternal bands. He is very likely to be the father.

Geneticists can calculate the likelihood that the matching bands between the offspring and a prospective father could occur just as a matter of random chance. To do so, they must analyze the frequency of each band in a reference population (e.g., Caucasians living in the United States). For example, let's suppose that DNA fingerprinting analyzed 40 bands. Of these, 20 bands matched with the mother and 20 bands matched with a prospective father. If the probability of each of these bands in a reference population was 1/4, the likelihood of such a match occurring by random chance would be $(1/4)^{20}$, or roughly 1 in 1 trillion. Therefore, a match between two samples is rarely a matter of random chance.

Conceptual Questions

C1. What is the gene pool? How is a gene pool described in a quantitative way?

C2. In genetics, what does the term population mean? Pick any species you like and describe how its population might change over the course of many generations.

C3. What is a genetic polymorphism? What is the source of genetic variation?

C4. For each of the following, state whether it is an example of an allele, genotype, and/or phenotype frequency:

   A. Approximately 1 in 2,500 Caucasians is born with cystic fibrosis.

   B. The percentage of carriers of the sickle-cell allele in West Africa is approximately 13%.

   C. The number of new mutations for achondroplasia, a genetic disorder, is approximately 5×10^{-5}.

C5. The term polymorphism can refer to both genes and traits. Explain the meaning of a polymorphic gene and a polymorphic trait. If a gene is polymorphic, does the trait that the gene affects also have to be polymorphic? Explain why or why not.

C6. Cystic fibrosis is a recessive autosomal trait. In certain Caucasian populations, the number of people born with this disorder is about 1 in 2,500. Assuming Hardy-Weinberg equilibrium for this trait,

   A. What are the frequencies for the normal and CF alleles?

   B. What are the genotype frequencies of homozygous normal, heterozygous, and homozygous affected individuals?

   C. Assuming random mating, what is the probability that two phenotypically unaffected heterozygous carriers will choose each other as mates?

C7. Does inbreeding affect allele frequencies? Why or why not? How does it affect genotype frequencies? With regard to rare recessive diseases, what are the consequences of inbreeding in human populations?

C8. For a gene existing in two alleles, what are the allele frequencies when the heterozygote frequency is at its maximum value? What if there are three alleles?

C9. In a population, the frequencies of two alleles are $B = 0.67$ and $b = 0.33$. The genotype frequencies are $BB = 0.50$, $Bb = 0.37$, and $bb = 0.13$. Do these numbers suggest inbreeding? Explain why or why not.

C10. The ability to roll your tongue is inherited as a recessive trait. The frequency of the rolling allele is approximately 0.6, and the dominant (nonrolling) allele is 0.4. What is the frequency of individuals who can roll their tongues?

C11. In the pedigree shown here, answer the following questions for individual VI-1:

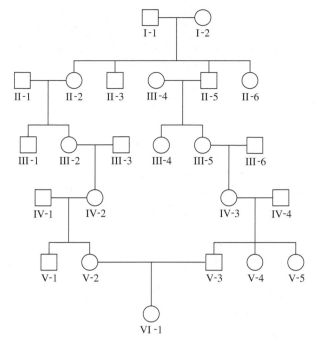

   A. Is this individual inbred?

   B. If so, who are her common ancestor(s)?

   C. Calculate the inbreeding coefficient for VI-1.

   D. Are the parents of VI-1 inbred?

C12. A family pedigree is shown here.

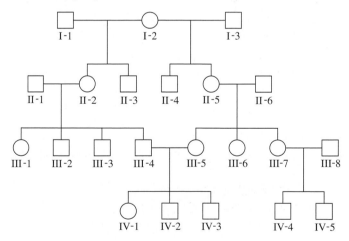

   A. What is the inbreeding coefficient for individual IV-3?

   B. Based on the data shown in this pedigree, is individual IV-4 inbred?

C13. A family pedigree is shown here.

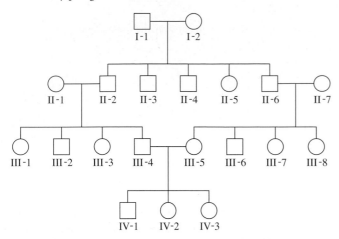

A. What is the inbreeding coefficient for individual IV-2? Who is/are her common ancestors?

B. Based on the data shown in this pedigree, is individual III-4 inbred?

C14. What evolutionary forces can cause allele frequencies to change and possibly lead to a genetic polymorphism? Discuss the relative importance of each type of process.

C15. In the term genetic drift, what is drifting? Why is this an appropriate term to describe this phenomenon?

C16. What is the difference between a random and an adaptive evolutionary process? Describe two or more examples of each. At the molecular level, explain how mutations can be random or adaptive.

C17. Let's suppose the mutation rate for converting a B allele into a b allele is 10^{-4}. The current allele frequencies are $B = 0.6$ and $b = 0.4$. How long will it take for the allele frequencies to equal each other, assuming that no genetic drift is taking place?

C18. Why is genetic drift more significant in small populations? Why does it take longer for genetic drift to cause allele fixation in large populations than in small ones?

C19. A group of four birds flies to a new location and initiates the formation of a new colony. Three of the birds are homozygous AA, and one bird is heterozygous Aa.

A. What is the probability that the a allele will become fixed in the population?

B. If fixation occurs, how long will it take?

C. How will the growth of the population, from generation to generation, affect the answers to parts A and B? Explain.

C20. Describe what happens to allele frequencies during the bottleneck effect. Discuss the relevance of this effect with regard to species that are approaching extinction.

C21. When two populations frequently intermix due to migration, what are the long-term consequences with regard to allele frequencies and genetic variation?

C22. Discuss the similarities and differences among directional, disruptive, balancing, and stabilizing selection.

C23. What is Darwinian fitness? What types of characteristics can promote high fitness values? Give several examples.

C24. What is the intuitive meaning of the mean fitness of a population? How does its value change in response to natural selection?

C25. Antibiotics are commonly used to combat bacterial and fungal infections. During the past several decades, however, antibiotic-resistant strains of microorganisms have become alarmingly prevalent. This has undermined the ability of physicians to treat many types of infectious disease. Discuss how the following processes that alter allele frequencies may have contributed to the emergence of antibiotic-resistant strains:

A. Random mutation

B. Genetic drift

C. Natural selection

C26. With regard to genetic drift, are the following statements true or false? If a statement is false, explain why.

A. Over the long run, genetic drift will lead to allele fixation or loss.

B. When a new mutation occurs within a population, genetic drift is more likely to cause the loss of the new allele rather than the fixation of the new allele.

C. Genetic drift promotes genetic diversity in large populations.

D. Genetic drift is more significant in small populations.

C27. Two populations of antelope are separated by a mountain range. The antelope are known to occasionally migrate from one population to the other. Migration can occur in either direction. Explain how migration will affect the following phenomena:

A. Genetic diversity in the two populations

B. Allele frequencies in the two populations

C. Genetic drift in the two populations

C28. Do the following examples describe directional, disruptive, balancing, or stabilizing selection?

A. Polymorphisms in snail color and banding pattern as described in Figure 24.13

B. Thick fur among mammals exposed to cold climates

C. Birth weight in humans

D. Sturdy stems and leaves among plants exposed to windy climates

Experimental Questions

E1. You will need to be familiar with the techniques described in Chapter 18 to answer this question. Gene polymorphisms can be detected using a variety of cellular and molecular techniques. Which techniques would you use to detect gene polymorphisms at the following levels?

A. DNA level

B. RNA level

C. Polypeptide level

E2. You will need to understand solved problem S4 to answer this question. As described in Chapter 4, the gene for coat color in rabbits can exist in four alleles termed C (full coat color), c^{ch} (chinchilla), c^h (Himalayan), and c (albino). In a population of rabbits in Hardy-Weinberg equilibrium, the allele frequencies are

$C = 0.34$

$c^{ch} = 0.17$

$c^h = 0.44$

$c = 0.05$

Assume that C is dominant to the other three alleles. c^{ch} is dominant to c^h and c, and c^h is dominant to c.

A. What is the frequency of albino rabbits?

B. Among 1,000 rabbits, how many would you expect to have a Himalayan coat color?

C. Among 1,000 rabbits, how many would be heterozygotes with a chinchilla coat color?

E3. In a large herd of 5,468 sheep, 76 animals have yellow fat, compared to the rest of the members of the herd, which have white fat. Yellow fat is inherited as a recessive trait. This herd is assumed to be in Hardy-Weinberg equilibrium.

A. What are the frequencies of the white and yellow fat alleles in this population?

B. Approximately how many sheep with white fat are heterozygous carriers of the yellow allele?

E4. The human MN blood group is determined by two codominant alleles, M and N. The following data were obtained from various human populations:

		Percentages		
Population	Place	MM	MN	NN
Inuit	East Greenland	83.5	15.6	0.9
Navajo Indians	New Mexico	84.5	14.4	1.1
Finns	Karajala	45.7	43.1	11.2
Russians	Moscow	39.9	44.0	16.1
Aborigines	Queensland	2.4	30.4	67.2

(Data from Speiss, E. B. (1990). *Genes in Populations*, 2d ed. NY: Wiley-Liss.)

A. Calculate the allele frequencies in these five populations.

B. Which populations appear to be in Hardy-Weinberg equilibrium?

C. Which populations do you think have had significant intermixing due to migration?

E5. You will need to understand solved problem S4 before answering this question. In an island population, the following data were obtained regarding the numbers of people with each of the four blood types:

Type O	721
Type A	932
Type B	235
Type AB	112

Is this population in Hardy-Weinberg equilibrium? Explain your answer.

E6. In a donor population, the allele frequencies for the normal (Hb^A) and sickle-cell alleles (Hb^S) are 0.9 and 0.1, respectively. A group of 550 individuals migrates to a new population containing 10,000 individuals; in the recipient population, the allele frequencies are $Hb^A = 0.99$ and $Hb^S = 0.01$.

A. Calculate the allele frequencies in the conglomerate population.

B. Assuming that the donor and recipient populations are each in Hardy-Weinberg equilibrium, calculate the genotype frequencies in the conglomerate population prior to further mating between the donor and recipient populations.

C. What will be the genotype frequencies of the conglomerate population in the next generation, assuming that it achieves Hardy-Weinberg equilibrium in one generation?

E7. A recessive lethal allele has achieved a frequency of 0.22 due to genetic drift in a very small population. Based on natural selection, how would you expect the allele frequencies to change in the next three generations? Note: Your calculation can assume that genetic drift is not altering allele frequencies in either direction.

E8. Among a large population of 2 million gray mosquitoes, one mosquito is heterozygous for a body color gene; this mosquito has one gray allele and one blue allele. There is no selective advantage or disadvantage between gray and blue body color. All the other mosquitoes carry the gray allele.

A. What is the probability of fixation of the blue allele?

B. If fixation happens to occur, how many generations is it likely to take?

C. Qualitatively, how would the answers to parts A and B be affected if the blue allele conferred a slight survival advantage?

E9. Resistance to the poison warfarin is a genetically determined trait in rats. Homozygotes carrying the resistance allele (WW) have a lower fitness because they suffer from vitamin K deficiency, but heterozygotes (Ww) do not. However, the heterozygotes are still resistant to warfarin. In an area where warfarin is applied, the heterozygote has a survival advantage. Due to warfarin resistance, the heterozygote is also more fit than the normal homozygote (ww). If the relative fitness values for Ww, WW, and ww individuals are 1.0, 0.37, and 0.19, respectively, in areas where warfarin is applied, calculate the allele frequencies at equilibrium. How would this equilibrium be affected if the rats were no longer exposed to warfarin?

E10. Describe, in as much experimental detail as possible, how you would test the hypothesis that snail color distribution is due to predation.

E11. In the Grants' study of the medium ground finch, do you think the pattern of natural selection was directional, stabilizing, disruptive, or balancing? Explain your answer. If the environment remained dry indefinitely (for many years), what do you think would be the long-term outcome?

E12. Here are traditional DNA fingerprints of five people: a child, mother, and three potential fathers:

Mother   Offspring   Male 1   Male 2   Male 3

Which males can be ruled out as being the father? Explain your answer. If one of the males could be the father, explain the general strategy for calculating the likelihood that he could match the offspring's DNA fingerprint by chance alone. (See solved problem S6 before answering this question.)

E13. What is DNA fingerprinting? How can it be used in human identification?

E14. What roles do PCR and Southern blotting play in the analysis of DNA fingerprints?

E15. What is a minisatellite (also called a VNTR)? Discuss the relationship between minisatellites and DNA fingerprinting.

E16. When analyzing the DNA fingerprints of a father and his biological daughter, a technician examined 50 bands and found that 30 of them were a perfect match. In other words, 30 out of 50 bands, or 60%, were a perfect match. Is this percentage too high, or would you expect a value of only 50%? Explain why or why not.

E17. What would you expect to be the minimum percentage of matching bands in a DNA fingerprint for the following pairs of individuals?

A. Mother and son

B. Sister and brother

C. Uncle and niece

D. Grandfather and grandson

Questions for Student Discussion/Collaboration

1. Discuss examples of positive and negative assortative mating in natural populations, human populations, and agriculturally important species.

2. Discuss the role of mutation in the origin of genetic polymorphisms. Suppose that a genetic polymorphism has two alleles at frequencies of 0.45 and 0.55. Describe three different scenarios to explain these observed allele frequencies. You can propose that the alleles are neutral, beneficial, or deleterious.

3. Most new mutations are detrimental, yet rare beneficial mutations can be adaptive. With regard to the fate of new mutations, discuss whether you think it is more important for natural selection to select against detrimental alleles or to select in favor of beneficial alleles. Which do you think is more significant in human populations?

Note: All answers appear at the website for this textbook; the answers to even-numbered questions are in the back of the textbook.

www.mhhe.com/brookergenetics3e

Visit the Online Learning Center for practice tests, answer keys, and other learning aids for this chapter. Enhance your understanding of genetics with our interactive exercises, quizzes, animations, and much more.

QUANTITATIVE GENETICS

In this chapter, we will examine **complex traits**—characteristics that are determined by several genes and are significantly influenced by environmental factors. Many complex traits are viewed as **quantitative traits** because they can be described numerically. In humans, quantitative traits include height, the shape of our noses, and the rate at which we metabolize food, to name a few examples. The field of genetics that studies the mode of inheritance of complex or quantitative traits is called **quantitative genetics.** Quantitative genetics is an important branch of genetics for several reasons. In agriculture, most of the key characteristics of interest to plant and animal breeders are quantitative traits. These include traits such as weight, fruit size, resistance to disease, and the ability to withstand harsh environmental conditions. As we will see later in this chapter, genetic techniques have improved our ability to develop strains of agriculturally important species with desirable quantitative traits. In addition, many human diseases are viewed as complex traits that are influenced by several genes.

Quantitative genetics is also important in the study of evolution. Many of the traits that allow a species to adapt to its environment are quantitative. Examples include the long neck of a giraffe, the swift speed of the cheetah, and the sturdy branches of trees in windy climates. The importance of quantitative traits in the evolution of species will be discussed in Chapter 26. In this chapter, we will examine how genes and the environment contribute to the phenotypic expression of complex or quantitative traits. We will begin with an examination of quantitative traits and how to analyze them using statistical techniques. We will then look at the inheritance of polygenic traits and at quantitative trait loci, locations on chromosomes containing genes that affect the outcome of quantitative traits. Advances in genetic mapping strategies have enabled researchers to identify these genes. Last, we look at heritability and consider various ways of calculating and modifying the genetic variation that affects phenotype.

Domesticated wheat. The color of wheat ranges from a dark red to white, which is an example of a complex or quantitative trait.

25.1 QUANTITATIVE TRAITS

When we compare characteristics among members of the same species, the differences are often quantitative rather than qualitative. Humans, for example, all have the same basic anatomical features (two eyes, two ears, and so on), but they differ in quantitative ways. People vary with regard to height, weight, the shape of facial features, pigmentation, and many other characteristics. As shown in **Table 25.1**, quantitative traits can be categorized as anatomical, physiological, and behavioral. In addition, many human diseases exhibit characteristics and inheritance patterns analogous to those of quantitative traits. Three of the leading causes of death worldwide—heart disease, cancer, and diabetes—are considered complex traits.

TABLE 25.1

Types of Quantitative Traits

Trait	Examples
Anatomical traits	Height, weight, number of bristles in *Drosophila*, ear length in corn, and the degree of pigmentation in flowers and skin
Physiological traits	Metabolic traits, speed of running and flight, ability to withstand harsh temperatures, and milk production in mammals
Behavioral traits	Mating calls, courtship rituals, ability to learn a maze, and the ability to grow or move toward light
Diseases	Atherosclerosis, hypertension, cancer, diabetes, arthritis, and obesity

In many cases, quantitative traits are easily measured and described numerically. Height and weight can be measured in centimeters (or inches) and kilograms (or pounds). Speed can be measured in kilometers per hour, and metabolic rate can be assessed as the grams of glucose burned per minute. Behavioral traits can also be quantified. A mating call can be evaluated with regard to its duration, sound level, and pattern. The ability to learn a maze can be described as the time and/or repetitions it takes to learn the skill. Finally, complex diseases such as diabetes can also be studied and described via numerical parameters. For example, the severity of the disease can be assessed by the age of onset or by the amount of insulin needed to prevent adverse symptoms.

From a scientific viewpoint, the measurement of quantitative traits is essential when comparing individuals or evaluating groups of individuals. It is not very informative to say that two people are tall. Instead, we are better informed if we know that one person is 6 feet 4 inches and the other is 6 feet 7 inches. In this branch of genetics, the measurement of a quantitative trait is how we describe the phenotype.

In the early 1900s, Francis Galton in England and his student Karl Pearson showed that many traits in humans and domesticated animals are quantitative in nature. To understand the underlying genetic basis of these traits, they founded what became known as the **biometric field** of genetics, which involved the statistical study of biological traits. During this period, Galton and Pearson developed various statistical tools for studying the variation of quantitative traits within groups of individuals; many of these tools are still in use today. In this section, we will examine how quantitative traits are measured and how statistical tools are used to analyze their variation within groups.

Quantitative Traits Exhibit a Continuum of Phenotypic Variation That May Follow a Normal Distribution

In Part II of this textbook, we discussed many traits that fall into discrete categories. For example, fruit flies might have white eyes or red eyes, and pea plants might have wrinkled seeds or smooth seeds. The alleles that govern these traits affect the phenotype in a qualitative way. In analyzing crosses involving these types of traits, each offspring can be put into a particular phenotypic category. Such attributes are called **discontinuous traits.**

In contrast, quantitative traits show a continuum of phenotypic variation within a group of individuals. For such traits, it is often impossible to place organisms into a discrete phenotypic class. For example, **Figure 25.1a** is a classic photograph from 1914 showing the range of heights of 175 students at the Connecticut Agricultural College. Though height is found at minimum and maximum values, the range of heights between these values is fairly continuous.

How do geneticists describe traits that show a continuum of phenotypes? Because quantitative traits do not naturally fall into a small number of discrete categories, an alternative way to describe them is a **frequency distribution.** To construct a frequency distribution, the trait is divided arbitrarily into a number of convenient, discrete phenotypic categories. For example, in Figure 25.1, the range of heights is partitioned into 1-inch intervals. Then a graph is made that shows the numbers of individuals found in each of the categories.

Figure 25.1b shows a frequency distribution for the heights of students pictured in Figure 25.1a. The measurement of height is plotted along the x-axis, and the number of individuals who exhibit that phenotype is plotted on the y-axis. The values along the x-axis are divided into the discrete 1-inch intervals that define the phenotypic categories, even though height is essentially continuous within a group of individuals. For example, in Figure 25.1a, 22 students were between 64.5 and 65.5 inches in height, which is plotted as the point (65 inches, 22 students) on the graph in Figure 25.1b. This type of analysis can be conducted on any group of individuals who vary with regard to a quantitative trait.

The line in the frequency distribution depicts a **normal distribution,** a distribution for a large sample in which the trait of interest varies in a symmetric way around an average value. The distribution of measurements of many biological characteristics is approximated by a symmetrical bell curve like that in Figure 25.1b. Normal distributions are common when the phenotype is determined by the cumulative effect of many small independent factors. We will consider the significance of this type of distribution next.

Statistical Methods Are Used to Evaluate a Frequency Distribution Quantitatively

Statistical tools are used to analyze a normal distribution in a number of ways. One measure that you are probably familiar with is a parameter called the **mean,** which is the sum of all the values in the group divided by the number of individuals in the group. The mean is computed using the following formula:

$$\overline{X} = \frac{\Sigma X}{N}$$

where

$\overline{X}$	is the mean
ΣX	is the sum of all the values in the group
N	is the number of individuals in the group

Number of students	1	0	0	1	5	7	7	22	25	26	27	17	11	17	4	4	1
Height (inches)	58	59	60	61	62	63	64	65	66	67	68	69	70	71	72	73	74

(a)

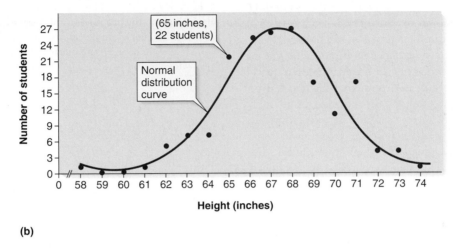

(b)

FIGURE 25.1 **Normal distribution of a quantitative trait.** (a) The distribution of heights in 175 students at the Connecticut Agricultural College in 1914. (b) A frequency distribution for the heights of students shown in (a).

A more generalized form of this equation can be used:

$$\overline{X} = \frac{\Sigma f_i X_i}{N}$$

where

$\overline{X}$ is the mean

$\Sigma f_i X_i$ is the sum of all the values in the group; each value in the group is multiplied by its frequency (f_i) in the group

N is the number of individuals in the group

For example, suppose a bushel of corn had ears with the following lengths (rounded to the nearest centimeter): 15, 14, 13, 14, 15, 16, 16, 17, 15, and 15. Then

$$\overline{X} = \frac{4(15) + 2(14) + 13 + 2(16) + 17}{10}$$

$$\overline{X} = 15 \text{ cm}$$

In genetics, we are often interested in the amount of phenotypic variation that exists in a group. As we will see later in this chapter and in Chapter 26, variation lies at the heart of breeding experiments and of evolution. Without variation, selective breeding is not possible, and natural selection cannot favor one phenotype over another. A common way to evaluate variation within a population is with a statistic called the **variance,** which is a measure of the variation around the mean. The variance is the sum of the squared deviations from the mean divided by the degrees of freedom (*df* equals $N - 1$; see Chapter 2 for a review of degrees of freedom).

$$V_X = \frac{\Sigma f_i (X_i - \overline{X})^2}{N - 1}$$

where

V_X is the variance

$X_i - \overline{X}$ is the difference between each value and the mean

N equals the number of observations

For example, if we use the values given previously for the lengths of ears of corn, the variance in this group is calculated as follows:

$$\Sigma f_i (X_i - \overline{X})^2 = 4(15-15)^2 + 2(14-15)^2 +$$
$$(13-15)^2 + 2(16-15)^2 + (17-15)^2$$

$$\Sigma f_i (X_i - \overline{X})^2 = 0 + 2 + 4 + 2 + 4$$

$$\Sigma f_i (X_i - \overline{X})^2 = 12 \text{ cm}^2$$

$$V_X = \frac{\Sigma f_i (X_i - \overline{X})^2}{N - 1}$$

$$V_X = \frac{12 \text{ cm}^2}{9}$$

$$V_X = 1.33 \text{ cm}^2$$

Because the variance is computed from squared deviations, it is a statistic that may be difficult to understand intuitively. For example, weight can be measured in grams; the corresponding variance would be measured in square grams. Even so, variances are centrally important in the analysis of quantitative traits because they are additive under certain conditions. This means that the variances for different factors that contribute to a quantitative trait, such as genetic and environmental factors, can be added together to predict the total variance for that trait. Later, we will examine how this property is useful in predicting the outcome of genetic crosses.

To gain a more intuitive grasp of variation, we can take the square root of the variance. This statistic is called the **standard deviation (SD)**. Again, using the same values for length, the standard deviation is

$$SD = \sqrt{V_X} = \sqrt{1.33}$$

$$SD = 1.15 \text{ cm}$$

If the values in a population follow a normal distribution, it is easier to appreciate the amount of variation by considering the standard deviation. **Figure 25.2** illustrates the relationship between the standard deviation and the percentages of individuals that deviate from the mean. Approximately 68% of all individuals have values within one standard deviation from the mean, either in the positive or negative direction. About 95% are within two standard deviations, and 99.7% are within three standard deviations. When a quantitative characteristic follows a normal distribution, less than 0.3% of the individuals will have values that are more or less than three standard deviations from the mean of the population. In our corn example, three standard deviations equal 3.45 cm. Therefore, we would expect that approximately 0.3% of the ears of corn would be less than 11.55 cm or greater than 18.45 cm, assuming that length follows a normal distribution.

Some Statistical Methods Compare Two Variables to Each Other

In many biological problems, it is useful to compare two different variables. For example, we may wish to compare the occur-

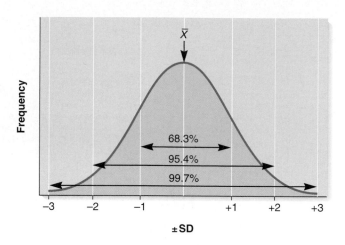

FIGURE 25.2   **The relationship between the standard deviation and the proportions of individuals in a normal distribution.** For example, approximately 68% of the individuals in a population are between the mean and one standard deviation above or below the mean.

rence of two different phenotypic traits. Do obese animals have larger hearts? Are brown eyes more likely to occur in people with dark skin pigmentation? A second type of comparison is between traits and environmental factors. Does insecticide resistance occur more frequently in areas that have been exposed to insecticides? Is heavy body weight more prevalent in colder climates? Finally, a third type of comparison is between traits and genetic relationships. Do tall parents tend to produce tall offspring? Do women with diabetes tend to have brothers with diabetes?

To gain insight into such questions, a statistic known as the correlation coefficient is often applied. To calculate this statistic, we first need to determine the **covariance,** which describes the relationship between two variables within a group. The covariance is similar to the variance, except that we multiply together the deviations of two different variables rather than squaring the deviations from a single factor.

$$CoV_{(X,Y)} = \frac{\Sigma[(X - \overline{X})(Y - \overline{Y})]}{N - 1}$$

where

  $CoV_{(X,Y)}$ is the covariance between X and Y values

  X represents the values for one variable, and $\overline{X}$ is the mean value in the group

  Y represents the values for another variable, and $\overline{Y}$ is the mean value in that group

  N is the total number of pairs of observations

As an example, let's compare the weight of cows and that of their adult female offspring. A farmer might be interested in this relationship to determine if genetic variation plays a role in the weight of cattle. The data here describe the weights at 5 years of age for 10 different cows and their female offspring.

Mother's Weight (kg)	Offspring's Weight (kg)	$X - \overline{X}$	$Y - \overline{Y}$	$(X - \overline{X})(Y - \overline{Y})$
570	568	−26	−30	780
572	560	−24	−38	912
599	642	3	44	132
602	580	6	−18	−108
631	586	35	−12	−420
603	642	7	44	308
599	632	3	34	102
625	580	29	−18	−522
584	605	−12	7	−84
575	585	−21	−13	273
$\overline{X}$ = 596	$\overline{Y}$ = 598			Σ = 1,373

$SD_X = 21.1$ $SD_Y = 30.5$

$$CoV_{(X,Y)} = \frac{\Sigma[(X - \overline{X})(Y - \overline{Y})]}{N - 1}$$

$$CoV_{(X,Y)} = \frac{1,373}{10 - 1}$$

$$CoV_{(X,Y)} = 152.6$$

After we have calculated the covariance, we can evaluate the strength of the association between the two variables by calculating a **correlation coefficient (r)**. This value, which ranges between −1 and +1, indicates how two factors vary in relation to each other. The correlation coefficient is calculated as

$$r_{(X,Y)} = \frac{CoV_{(X,Y)}}{SD_X SD_Y}$$

A positive r value means that two factors tend to vary in the same way relative to each other; as one factor increases, the other will increase with it. A value of zero indicates that the two factors do not vary in a consistent way relative to each other; the values of the two factors are not related. Finally, an inverse correlation, in which the correlation coefficient is negative, indicates that the two factors tend to vary in opposite ways to each other; as one factor increases, the other will decrease.

Let's use the data of 5-year weights for mother and offspring to calculate a correlation coefficient.

$$r_{(X,Y)} = \frac{152.6}{(21.1)(30.5)}$$

$$r_{(X,Y)} = 0.237$$

The result is a positive correlation between the 5-year weights of mother and offspring. In other words, the positive correlation value suggests that heavy mothers tend to have heavy offspring and that lighter mothers have lighter offspring.

How do we evaluate the value of r? After a correlation coefficient has been calculated, one must consider whether the r value represents a true association between the two variables or whether it could be simply due to chance. To accomplish this, we can test the hypothesis that there is no real correlation (i.e., the null hypothesis, $r = 0$). The null hypothesis is that the

r value differs from zero due only to random sampling error. We followed a similar approach in the chi square analysis described in Chapter 2. Like the chi square value, the significance of the correlation coefficient is directly related to sample size and the degrees of freedom (df). In testing the significance of correlation coefficients, df equals $N - 2$, which is one less than the degrees of freedom of variance (i.e., df for variance equals $N - 1$). **Table 25.2** shows the relationship between the r values and degrees of freedom at the 5% and 1% confidence intervals.

The use of Table 25.2 is valid only if several assumptions are met. First, the values of X and Y in the study must have been obtained by an unbiased sampling of the entire population. In addition, this approach assumes that the scores of X and Y follow a normal distribution, like that of Figure 25.1, and that the relationship between X and Y is linear.

TABLE **25.2**

Values of r at the 5% and 1% Confidence Intervals

Degrees of Freedom (*df*)	5%	1%	Degrees of Freedom (*df*)	5%	1%
1	.997	1.000	24	.388	.496
2	.950	.990	25	.381	.487
3	.878	.959	26	.374	.478
4	.811	.917	27	.367	.470
5	.754	.874	28	.361	.463
6	.707	.834	29	.355	.456
7	.666	.798	30	.349	.449
8	.632	.765	35	.325	.418
9	.602	.735	40	.304	.393
10	.576	.708	45	.288	.372
11	.553	.684	50	.273	.354
12	.532	.661	60	.250	.325
13	.514	.641	70	.232	.302
14	.497	.623	80	.217	.283
15	.482	.606	90	.205	.267
16	.468	.590	100	.195	.254
17	.456	.575	125	.174	.228
18	.444	.561	150	.159	.208
19	.433	.549	200	.138	.181
20	.423	.537	300	.113	.148
21	.413	.526	400	.098	.128
22	.404	.515	500	.088	.115
23	.396	.505	1,000	.062	.081

Note: *df* equals $N - 2$.
From Spence, J. T. et al. (1976). *Elementary Statistics*. Prentice-Hall, Englewood Cliffs, New Jersey.

To illustrate the use of Table 25.2, let's consider the correlation we have just calculated for 5-year weights of cows and their female offspring. In this case, we obtained a value of 0.237 for r, and the value of N was 10. Under these conditions, df equals 8. To be valid at a 5% confidence interval, the value of r would have to be 0.632 or higher. Because the value that we obtained is much less than this, it is fairly likely that this value could have occurred as a matter of random sampling error. In this case, we cannot reject the null hypothesis, and, therefore, we cannot conclude the positive correlation is due to a true association between the weights of mothers and offspring.

In an actual experiment, however, a researcher would examine many more pairs of cows and offspring, perhaps 500 to 1,000. If a correlation of 0.237 was observed for $N = 1,000$, the value would be significant at the 1% level. We would therefore reject the null hypothesis that weights are not associated with each other. Instead, we would conclude that a real association occurs between the weights of mothers and their offspring. In fact, these kinds of experiments have been done for cattle weights, and the correlations between mothers and offspring have often been found to be significant.

If a statistically significant correlation is obtained, how do we interpret its meaning? An r value that is statistically significant need not imply a cause-and-effect relationship. When parents and offspring display a significant correlation for a trait, we should not jump to the conclusion that genetics is the underlying cause of the positive association. In many cases, parents and offspring share similar environments, so the positive association might be rooted in environmental factors. In general, correlations are quite useful in identifying positive or negative associations between two variables. We should use caution, however, because this statistic, by itself, cannot prove that the association is due to cause and effect. When it has been established that the variables are related due to cause and effect—that one variable (the independent variable) affects the outcome of another (the dependent variable)—researchers may use a **regression analysis** to predict how much the dependent variable will change in response to the independent variable. This approach is described in solved problem S4 at the end of the chapter.

25.2 POLYGENIC INHERITANCE

In Section 25.1, we saw that quantitative traits tend to show a continuum of variation and can be analyzed with various statistical tools. At the beginning of the 1900s, a great debate focused on the inheritance of quantitative traits. The biometric school, founded by Francis Galton and Karl Pearson, argued that these types of traits are not controlled by discrete genes that affect phenotypes in a predictable way. To some extent, the biometric school favored the idea of blending inheritance, which had been proposed many years earlier (see Chapter 2).

Alternatively, the followers of Mendel, led by William Bateson in England and William Castle in the United States, held firmly to the idea that traits are governed by genes, which are inherited as discrete units. As we know now, Bateson and Castle were correct. However, as we will see in this section, the difficulty of studying quantitative traits lies in the fact that these traits

are controlled by multiple genes and substantially influenced by environmental factors.

Most quantitative traits are polygenic and exhibit a continuum of phenotypic variation. The term **polygenic inheritance** refers to the transmission of a trait governed by two or more different genes. The location on a chromosome that harbors one or more genes that affect the outcome of a quantitative trait is called a **quantitative trait locus (QTL)**. As discussed later, QTLs are chromosomal regions that are identified by genetic mapping. Because such mapping usually locates the QTL to a relatively large chromosomal region, a QTL may contain a single gene or two or more closely linked genes that affect a quantitative trait.

Just a few years ago, it was extremely difficult for geneticists to determine the inheritance patterns for genes underlying polygenic traits, particularly those determined by three or more genes having multiple alleles for each gene. Recently, however, molecular genetic tools (described in Chapters 19 and 20) have greatly enhanced our ability to find regions in the genome where QTLs are likely to reside. This has been a particularly exciting advance in the field of quantitative genetics. In some cases, the identification of QTLs may allow the improvement of quantitative traits in agriculturally important species.

Polygenic Inheritance and Environmental Factors Create Overlaps Between Genotypes and Phenotypes

The first experiment demonstrating that continuous variation is related to polygenic inheritance was conducted by the Swedish geneticist Herman Nilsson-Ehle in 1909. He studied the inheritance of red pigment in the hull of bread wheat, *Triticum aestivum* (**Figure 25.3a**). When true-breeding plants with white hulls were crossed to a variety with red hulls, the F_1 generation had an intermediate color. When the F_1 generation was allowed to self-fertilize, great variation in redness was observed in the F_2 generation, ranging from white, light red, intermediate red, medium red, and dark red. An unsuspecting observer might conclude that this F_2 generation displayed a continuous variation in hull color. However, as shown in **Figure 25.3b**, Nilsson-Ehle carefully categorized the colors of the hulls and discovered that they followed a 1:4:6:4:1 ratio. He concluded that this species is diploid for two different genes that control hull color, each gene existing in a red or white allelic form. He hypothesized that these two loci must contribute additively to the color of the hull; the contribution of each red allele to the color of the hull is additive.

Later, researchers discovered a third gene that also affects hull color. The strains that Nilsson-Ehle had used in his original experiments must have been homozygous for the white allele of this third gene. It makes sense that wheat would be diploid for three genes that affect hull color because we now know that *T. aestivum* is a hexaploid derived from three closely related diploid species, as discussed in Chapter 8. Therefore, *T. aestivum* has six copies of many genes.

As we have just seen, Nilsson-Ehle categorized wheat hull colors into several discrete genotypic categories. However, for many polygenic traits, this is difficult or impossible. In general, as the number of genes controlling a trait increases and the influ-

(a) Red and white hulls of wheat

♂	R1R2	R1r2	r1R2	r1r2
♀				
R1R2	R1R1R2R2 Dark red	R1R1R2r2 Medium red	R1r1R2R2 Medium red	R1r1R2r2 Intermediate red
R1r2	R1R1R2r2 Medium red	R1R1r2r2 Intermediate red	R1r1R2r2 Intermediate red	R1r1r2r2 Light red
r1R2	R1r1R2R2 Medium red	R1r1R2r2 Intermediate red	r1r1R2R2 Intermediate red	r1r1R2r2 Light red
r1r2	R1r1R2r2 Intermediate red	R1r1r2r2 Light red	r1r1R2r2 Light red	r1r1r2r2 White

(b) R1r1R2r2 x R1r1R2r2

FIGURE 25.3 **The Nilsson-Ehle experiment studying how continuous variation is related to polygenic inheritance in wheat. (a)** Red (top) and white (bottom) varieties of wheat, *Triticum aestivum.* **(b)** Nilsson-Ehle carefully categorized the colors of the hulls in the F$_2$ generation and discovered that they followed a 1:4:6:4:1 ratio. This occurs because the contributions of the red alleles are additive.

Genes→Traits In this example, two genes, with two alleles each (red and white), govern hull color. Offspring can display a range of colors, depending on how many copies of the red allele they inherit. If an offspring is homozygous for the red allele of both genes, it will have very dark red hulls. By comparison, if it carries three red alleles and one white allele, it will be medium red (which is not quite as deep in color). In this way, this polygenic trait can exhibit a range of phenotypes from dark red to white.

ence of the environment increases, the categorization of phenotypes into discrete genotypic classes becomes increasingly difficult, if not impossible. Therefore, a Punnett square cannot be used to analyze most quantitative traits. Instead, statistical methods, which are described later, must be employed.

Figure 25.4 illustrates how genotypes and phenotypes may overlap for polygenic traits. In this example, the environment (sunlight, soil conditions, and so forth) may affect the phenotypic outcome of a trait in plants (namely, seed weight). Figure 25.4a considers a situation in which seed weight is controlled by one gene with light (*w*) and heavy (*W*) alleles. A heterozygous plant (*Ww*) is allowed to self-fertilize. When the weight is only slightly influenced by variation in the environment, as seen on the left, the light, intermediate, and heavy seeds fall into separate, well-defined categories. When the environmental variation has a greater impact on seed weight, as shown on the right, more phenotypic variation is found in seed weight within each genotypic class. The variance

in the frequency distribution on the right is much higher. Even so, most individuals can be classified into the three main categories.

By comparison, Figure 25.4b illustrates a situation in which seed weight is governed by three genes instead of one, each existing in light and heavy alleles. When the environmental variation is low and/or plays a minor role in the outcome of this trait, the expected 1:6:15:20:15:6:1 ratio is observed. As shown in the upper illustration in Figure 25.4b, nearly all individuals fall within a phenotypic category that corresponds to their genotype. When the environment has a more variable effect on phenotype, as shown in the lower illustration, the situation becomes more ambiguous. For example, individuals with one *w* allele and five *W* alleles have a phenotype that overlaps with that of individuals having six *W* alleles or two *w* alleles and four *W* alleles. Therefore, it becomes difficult to categorize each phenotype into a unique genotypic class. Instead, the trait displays a continuum ranging from light to heavy seed weight.

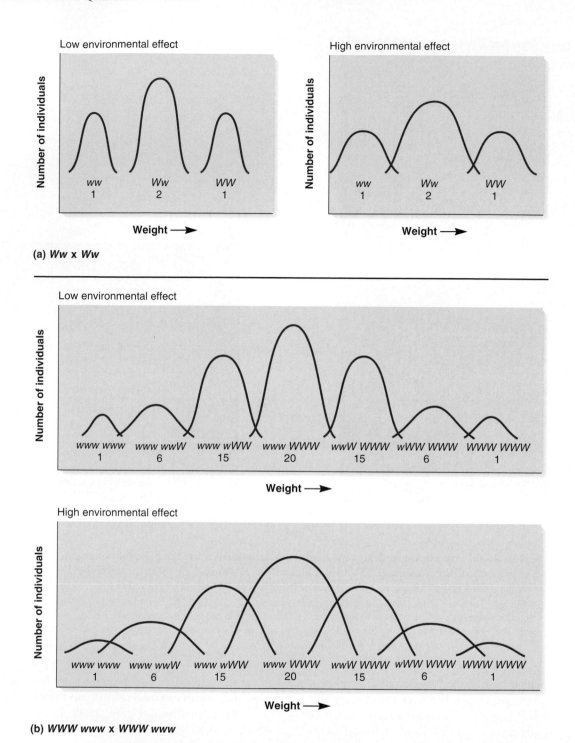

FIGURE 25.4   **How genotypes and phenotypes may overlap for polygenic traits.** **(a)** Situations in which seed weight is controlled by one gene, existing in light (*w*) and heavy (*W*) alleles. **(b)** Situations in which seed weight is governed by three genes instead of one, each existing in light and heavy alleles. Note: The 1:2:1 and 1:6:15:20:15:6:1 ratios were derived by analyzing these crosses using a Punnett square.

Genes→Traits The ability of geneticists to correlate genotype and phenotype depends on how many genes are involved and how much the environment causes the phenotype to vary. In **(a)**, a single gene influences weight. In the graph on the left side, the environment does not cause much variation in weight. This makes it easy to distinguish the three genotypes. There is no overlap in the weights of *ww*, *Ww*, and *WW* individuals. In the graph on the right side, the environment causes more variation in weight. In this case, a few individuals with *ww* genotypes will have the same weight as a few individuals with *Ww* genotypes; and a few *Ww* genotypes will have the same weight as *WW* genotypes. As shown in **(b)**, it becomes even more difficult to distinguish genotype based on phenotype when three genes are involved. The overlaps are minor when the environment does not cause much weight variation. However, when the environment causes substantial phenotypic variation, the overlaps between genotypes and phenotypes are very pronounced and greatly confound genetic analysis.

EXPERIMENT 25A

Polygenic Inheritance Explains DDT Resistance in *Drosophila*

As we have just learned, the phenotypic overlap for a quantitative trait may be so great that it may not be possible to establish discrete phenotypic classes. This is particularly true if many genes contribute to the trait. One way to identify the genes affecting polygenic inheritance is to look for linkage between genes affecting quantitative traits and genes affecting discontinuous traits. This approach was first studied in *Drosophila melanogaster* because many alleles had been identified and mapped to particular chromosomes.

In 1957, James Crow conducted one of the earliest studies to show linkage between genes affecting quantitative traits and genes affecting discontinuous traits. Crow, who was interested in evolution, spent time studying insecticide resistance in *Drosophila*. He noted, "Insecticide resistance is an example of evolutionary change, the insecticide acting as a powerful selective sieve for concentrating resistant mutants that were present in low frequencies in the population." His aim was to determine the genetic basis for insecticide resistance in *Drosophila melanogaster*. Many alleles were already known in this species, and these could serve as **genetic markers** for each of the four different chromosomes. Dominant alleles are particularly useful because they allow the experimenter to determine which chromosomes are inherited from either parent. The general strategy in identifying QTLs is to cross two strains that are homozygous for different genetic markers and also differ with regard to the quantitative trait of interest. This produces an F_1 generation that is heterozygous for the markers and usually exhibits an intermediate phenotype for the quantitative trait. The next step is to backcross the F_1 offspring to the parental strains. This backcross produces a population of F_2 offspring that differ with regard to their combinations of parental chromosomes. A few offspring may contain all of their chromosomes from one parental strain or the other, but most offspring will contain a few chromosomes from one parental strain and the rest from the other strain. The genetic markers on the chromosomes provide a way to determine whether particular chromosomes were inherited from one parental strain or the other.

To illustrate how genetic markers work, **Figure 25.5** considers a situation in which two strains differ with regard to a quantitative trait, resistance to DDT, and also differ with regard to dominant alleles on chromosome 3. The dominant alleles serve as markers for this chromosome. One strain is resistant to DDT and carries a dominant allele that causes minute bristles (M), while another strain is sensitive to DDT and carries a dominant allele that causes a rough eye (R). The wild-type alleles, which are recessive, produce long bristles (m) and smooth eyes (r). At the start of this experiment, it is not known if alleles affecting DDT resistance are located on this chromosome. If offspring from a backcross inherit both copies of chromosome 3 from the DDT-resistant strain, they will have smooth eyes and minute bristles. If they inherit both copies from the DDT-sensitive strain, they will have rough eyes and long bristles. By comparison, a fly with

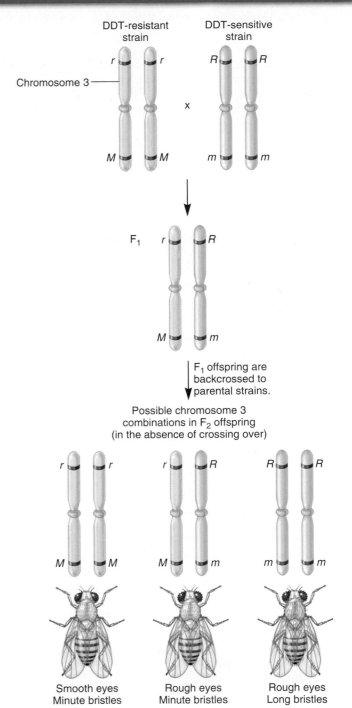

FIGURE 25.5 **The use of genetic markers to map a QTL affecting DDT resistance.** One strain is DDT resistant. On chromosome 3, it also carries a dominant allele that causes minute bristles (M). The other strain is DDT sensitive and carries a dominant allele that causes a roughness to the eye (R). The wild-type alleles, which are recessive, produce long bristles (m) and smooth eyes (r). F_2 offspring can have either both copies of chromosome 3 from the DDT-resistant strain, both from the sensitive strain, or one of each. This can be discerned by the phenotypes of the F_2 offspring.

rough eyes (*R*) and minute bristles (*M*) inherited one copy of chromosome 3 from the DDT-resistant strain and one copy from the DDT-sensitive strain. The transmission of the other *Drosophila* chromosomes can also be followed in a similar way. Therefore, the phenotypes of the offspring from the backcross provide a way to discern whether particular chromosomes were inherited from the DDT-resistant or DDT-sensitive strain.

Figure 25.6 shows the protocol followed by James Crow. He began with a DDT-resistant strain that had been produced by exposing flies to DDT for many generations. This DDT-resistant strain was crossed to a sensitive strain. As described previously in Figure 25.5, the two strains had allelic markers that made it possible to determine the origins of the different *Drosophila* chromosomes. Recall that *Drosophila* has four chromosomes. In this study, only chromosomes X, 2, and 3 were marked with alleles. Chromosome 4 was neglected due to its very small size. The F_1 flies were backcrossed to both parental strains, and then the F_2 female progeny were examined in two ways. First, their phenotypes were examined to determine whether particular chromosomes were inherited from the DDT-resistant or DDT-sensitive strain. Next, the female flies were exposed to filter paper impregnated with DDT. It was then determined if the flies survived this exposure for 18 to 24 hours.

■ THE HYPOTHESIS

DDT resistance is a polygenic trait.

■ TESTING THE HYPOTHESIS — FIGURE 25.6   Polygenic inheritance of DDT-resistance alleles in *Drosophila melanogaster*.

Starting material: DDT-resistant and DDT-sensitive strains of fruit flies.

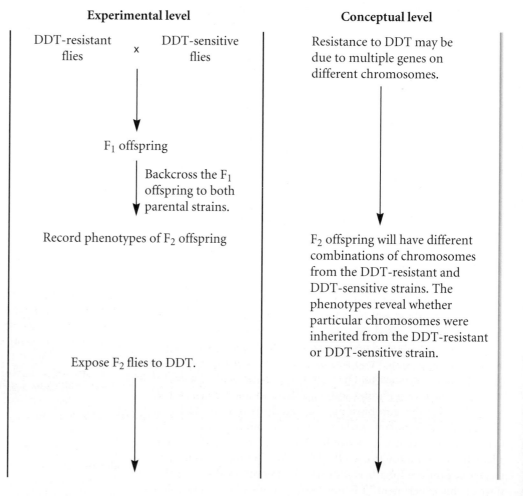

	Experimental level	**Conceptual level**
1. Cross the DDT-resistant strain to the sensitive strain. In each strain, chromosomes X, 2, and 3 were marked with alleles that provided easily discernible phenotypes.	DDT-resistant flies x DDT-sensitive flies	Resistance to DDT may be due to multiple genes on different chromosomes.
2. Take the F_1 flies and backcross to both parental strains.	F_1 offspring Backcross the F_1 offspring to both parental strains.	
3. Identify the origin of the chromosomes in the F_2 flies according to their phenotypes.	Record phenotypes of F_2 offspring	F_2 offspring will have different combinations of chromosomes from the DDT-resistant and DDT-sensitive strains. The phenotypes reveal whether particular chromosomes were inherited from the DDT-resistant or DDT-sensitive strain.
4. Expose the F_2 female flies to DDT on a filter paper for 18–24 hours.	Expose F_2 flies to DDT.	

5. Record the number of survivors.

Filter paper with DDT

The goal is to determine if survival is correlated with the inheritance of specific chromosomes from the DDT-resistant strain.

■ THE DATA

Chromosomes in the F$_2$ offspring

Percent survival of the F$_2$ offspring when exposed to DDT

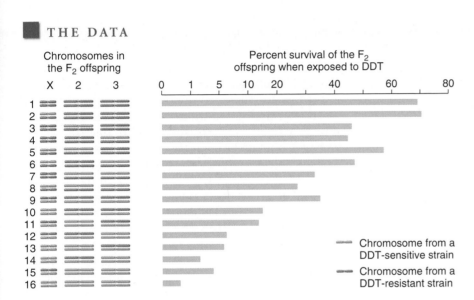

— Chromosome from a DDT-sensitive strain

— Chromosome from a DDT-resistant strain

■ INTERPRETING THE DATA

The results of this analysis are shown in the data of Figure 25.6. As a matter of chance, some offspring contained all of their chromosomes from one parental strain or the other, but most offspring contained a few chromosomes from one parental strain and the rest from the other. The data in Figure 25.6 suggest that each copy of the X chromosome and chromosomes 2 and 3 from the DDT-resistant strain confer a significant amount of insecticide resistance. A general trend was observed in which flies inheriting more chromosomes from the DDT-resistant strain had greater levels of resistance. (However, exceptions to the trend did occur; compare examples 1 and 2.) Overall, the results are consistent with the hypothesis that insecticide resistance is a polygenic trait involving multiple genes that reside on the X chromosome and on chromosomes 2 and 3.

A self-help quiz involving this experiment can be found at the Online Learning Center.

Quantitative Trait Loci (QTLs) Are Now Mapped by Linkage to Molecular Markers

In the previous experiment, we have seen how the minimum number of genes affecting a quantitative trait, such as DDT resistance, was determined by the linkage of such unknown genes to known genes on the *Drosophila* chromosomes. In the past few years, newer research techniques have identified molecular markers, such as RFLPs and microsatellites, that serve as reference points along chromosomes. This topic is discussed in Chapter 20. These markers have been used to construct detailed genetic maps of several species' genomes. Once a genome map is obtained, it becomes much easier to determine the number of genes that affect a quantitative trait. In addition to model organisms such as *Drosophila, Arabidopsis, Caenorhabditis elegans*, and mice, detailed molecular maps have been obtained for many species of agricultural importance. These include crops such as corn, rice, and tomatoes, as well as livestock such as cattle, pigs, sheep, and chickens.

To map the genes in a eukaryotic species, researchers now determine their locations by identifying molecular markers that are close to such genes. In 1989, Eric Lander and David Botstein extended this technique to identify QTLs that govern a quantitative trait. The basis of **QTL mapping** is the association between genetically determined phenotypes for quantitative traits and molecular markers such as RFLPs, microsatellites, and single nucleotide polymorphisms (SNPs). The general strategy for QTL mapping is shown in **Figure 25.7**. This figure depicts two different strains of a diploid species with four chromosomes per set. The strains are highly inbred, which means they are homozygous for most molecular markers and genes. They differ in two important ways. First, the two strains differ with regard to many molecular markers. These markers are designated 1A and 1B, 2A and 2B, and so forth. The markers 1A and 1B would mark the same chromosomal location in this species, namely, the upper tip of chromosome 1. However, the two markers would be distinguishable in the two strains at the molecular level. For example, 1A might be a microsatellite that is 148 bp, while 1B might be 212 bp. Second, the two strains differ with regard to a quantitative trait of interest. In this example, the strain on the left produces large fruit, while the strain on the right produces small fruit. The unknown genes affecting this trait are designated with the letter X. A black X indicates a QTL that harbors alleles that promote large fruit, and a blue X is the same site that carries alleles that promote small fruit. Prior to conducting their crosses, researchers would not know the chromosomal locations of the QTLs shown in this figure. The purpose of the experiment is to determine their locations.

With these ideas in mind, the protocol shown in Figure 25.7 begins by mating the two inbred strains to each other and then backcrossing the F_1 offspring to both parental strains. This would produce an F_2 generation with a great degree of variation. The F_2 offspring would then be characterized in two ways. First, they would be examined for their fruit size, and second, a sample of cells would be analyzed to determine which molecular markers were found in their chromosomes. The goal is to find an association between particular molecular markers and fruit size. For example, 2A would be strongly associated with large size, while 2B would be strongly associated with small size. By comparison, 9A and 9B would not be associated with large or small size, because a QTL affecting this trait is not found on this chromosome. Also, markers such as 14A and 14B, which are fairly far away from a QTL, would not be strongly associated with a particular QTL. Markers that are on the same chromosome but far away from a QTL would often be separated from the QTL in the F_1 heterozygote due to crossing over. Only closely linked markers will be strongly associated with a particular QTL.

Overall, QTL mapping involves the analysis of a large number of markers and offspring. The data are analyzed by computer programs that can statistically associate the phenotype (e.g., fruit size) with particular markers. Markers found throughout the genome of a species provide a way to identify the locations of several different genes that possess allelic differences that may affect the outcome of a quantitative trait.

As an actual example of QTL mapping, in 1988, Andrew Paterson and his colleagues examined quantitative trait inheritance in the tomato. They studied a domestic strain of tomato and a South American green-fruited variety. These two strains differed in their RFLPs, and they also exhibited dramatic differences in three agriculturally important characteristics: fruit mass, soluble solids content, and fruit pH. The researchers crossed the two strains and then backcrossed the offspring to the domestic tomato. A total of 237 plants were then examined with regard to 70 known RFLP markers. In addition, between 5 and 20 tomatoes from each plant were analyzed with regard to fruit mass, soluble solids content, and fruit pH. Using this approach, they were able to map genes contributing much of the variation in these traits to particular intervals along the tomato chromosomes. They identified six loci causing variation in fruit mass, four affecting soluble solids content, and five with effects on fruit pH.

25.3 HERITABILITY

As we have just seen, recent approaches in molecular mapping have enabled researchers to identify the genes that contribute to a quantitative trait. The other key factor that affects the phenotypic outcomes of quantitative traits is the environment. All traits of biological organisms are influenced by genetics and the environment, and this is particularly pertinent in the study of quantitative traits. Researchers want to understand how variation, both genetic and environmental, will affect the phenotypic results.

The term **heritability** refers to the amount of phenotypic variation within a group of individuals that is due to genetic variation. Genes play a role in the development of essentially all of an organism's traits. Even so, variation of a trait in a population may be due entirely to environmental variation, entirely to genetic variation, or to a combination of the two. If all the phenotypic variation in a group is due to genetic variation, the heritability would have a value of 1. If all the variation is due to environmental effects, the heritability would equal 0. For most groups of organisms, the heritability for a given trait lies between

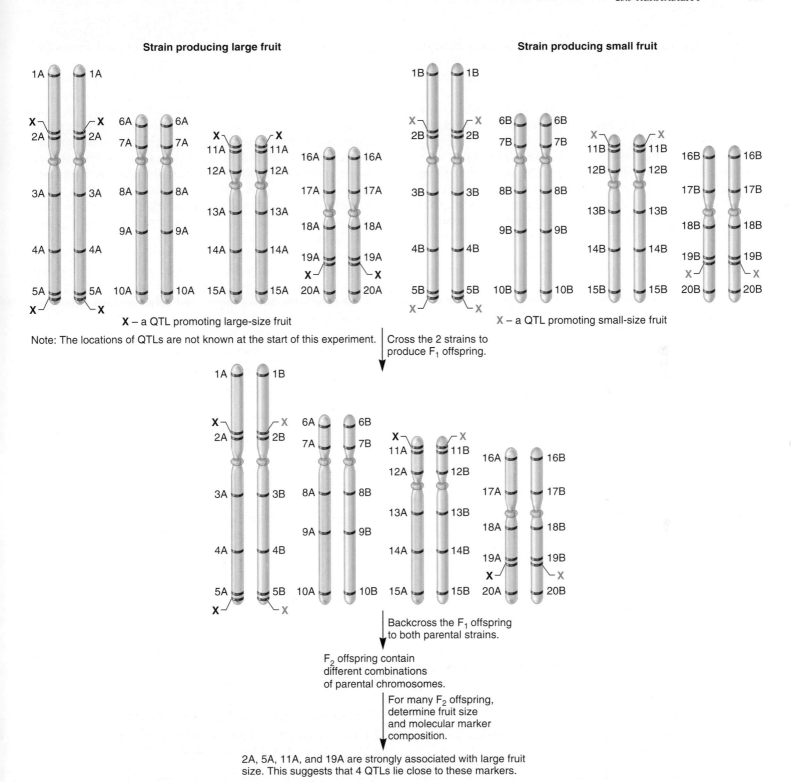

Strain producing large fruit

X – a QTL promoting large-size fruit

Note: The locations of QTLs are not known at the start of this experiment.

Strain producing small fruit

X – a QTL promoting small-size fruit

Cross the 2 strains to produce F₁ offspring.

Backcross the F₁ offspring to both parental strains.

F₂ offspring contain different combinations of parental chromosomes.

For many F₂ offspring, determine fruit size and molecular marker composition.

2A, 5A, 11A, and 19A are strongly associated with large fruit size. This suggests that 4 QTLs lie close to these markers.

FIGURE 25.7   **The general strategy for QTL mapping via molecular markers.** Two different inbred strains have four chromosomes per set. The strain on the left produces large fruit, and the strain on the right produces small fruit. The goal of this mapping strategy is to locate the unknown genes affecting this trait, which are designated with the letter X. A black X indicates a site promoting large fruit, and a blue X is a site promoting small fruit. The two strains differ with regard to many molecular markers designated 1A and 1B, 2A and 2B, and so forth. The two strains are mated, and then the F₁ offspring are backcrossed to the parental strains. Many F₂ offspring are then examined for their fruit size and to determine which molecular markers are found in their chromosomes. The data are analyzed by computer programs that can statistically associate the phenotype (e.g., fruit size) with particular markers. Markers found throughout the genome of this species provide a way to locate many different genes that may affect the outcome of a single quantitative trait. In this case, the analysis would predict four QTLs promoting heavier fruit weight that would be linked to regions of the chromosomes containing the following markers: 2A, 5A, 11A, and 19A.

these two extremes. For example, both genes and diet affect the size that an individual will attain. Some individuals will inherit alleles that tend to make them large, and a proper diet will also promote larger size. Other individuals will inherit alleles that make them small, and an inadequate diet may contribute to small size. Taken together, both genetics and the environment influence the phenotypic results.

In the study of quantitative traits, a primary goal is to determine how much of the phenotypic variation arises from genetic variation and how much comes from environmental variation. In this section, we will examine how geneticists analyze the genetic and environmental components that affect quantitative traits. As we will see, this approach has been applied with great success in breeding strategies to produce domesticated species with desirable and commercially valuable characteristics.

Genetic Variance and Environmental Variance Both May Contribute to Phenotypic Variance

Earlier, we examined the amount of phenotypic variation within a group by calculating the variance. In studying quantitative trait variation, geneticists partition this variation into components that are attributable to different causes. These include

Genetic variation (V_G)

Environmental variation (V_E)

Variation due to interactions between genetic and environmental factors ($V_{G \times E}$)

Variation due to associations between genetic and environmental factors ($V_{G \leftrightarrow E}$)

Let's begin by considering a simple situation in which V_G and V_E are the only factors that determine phenotypic variance and they are independent of each other. If so, then the total variance for a trait in a group of individuals is

$$V_T = V_G + V_E$$

where

V_T is the total variance. It reflects the amount of variation that is measured at the phenotypic level.

V_G is the relative amount of variance due to genetic variation.

V_E is the relative amount of variance due to environmental variation.

Why is this equation useful? The partitioning of variance into genetic and environmental components allows us to estimate their relative importance in influencing the variation within a group. If V_G is very high and V_E is very low, genetics plays a greater role in promoting variation within a group. Alternatively, if V_G is low and V_E is high, environmental factors underlie much of the phenotypic variation. As described later in this chapter, a livestock breeder might want to apply selective breeding if V_G for an important (quantitative) trait is high. In this way, the characteristics of the herd may be improved. Alternatively, if V_G is neg-

ligible, it would make more sense to investigate (and manipulate) the environmental causes of phenotypic variation.

With experimental and domesticated species, one possible way to determine V_G and V_E is by comparing the variation in traits between genetically identical and genetically disparate groups. For example, researchers have followed the practice of **inbreeding** to develop genetically homogeneous strains of mice. Inbreeding in mice involves many generations of brother-sister matings, which eventually produces strains that are **monomorphic** for all of their genes. The term monomorphic means that all the members of a population are homozygous for the same allele of a given gene. Within such an inbred strain of mice, V_G equals zero. Therefore, all phenotypic variation is due to V_E. When studying quantitative traits such as weight, an experimenter might want to know the genetic and environmental variance for a different, genetically heterogeneous group of mice. To do so, the genetically homogeneous and heterogeneous mice could be raised under the same environmental conditions and their weights measured. The phenotypic variance for weight could then be calculated as described earlier. Let's suppose we obtained the following results:

$V_T = 0.30$ sq oz for the group of genetically homogeneous mice

$V_T = 0.52$ sq oz for the group of genetically heterogeneous mice

In the case of the homogeneous mice, $V_T = V_E$, because V_G equals zero. Therefore, V_E equals 0.30 sq oz. To estimate V_G for the heterogeneous group of mice, we could assume that V_E (i.e., the environmentally produced variance) is the same for them as it is for the homogeneous mice, because the two groups were raised in identical environments. This assumption allows us to calculate the genetic variance for the heterogeneous mice.

$$V_T = V_G + V_E$$
$$0.52 = V_G + 0.30$$
$$V_G = 0.22 \text{ sq oz}$$

This result tells us that some of the phenotypic variance in the genetically heterogeneous group is due to the environment (namely, 0.30 sq oz) and some (0.22 sq oz) is due to genetic variation in alleles that affect weight.

Phenotypic Variation May Also Be Influenced by Interactions and Associations Between Genotype and the Environment

Thus far, we have considered the simple situation in which genetic variation and environmental variation are independent of each other and affect the phenotypic variation in an additive way. As another example, let's suppose that three genotypes, *TT*, *Tt*, and *tt*, affect height, producing tall, medium, and dwarf plants, respectively. Greater sunlight makes the plants grow taller regardless of their genotypes. In this case, our assumption that $V_T = V_G + V_E$ would be reasonably valid.

However, let's consider a different environmental factor such as minerals in the soil. Perhaps the *Tt* and *tt* plants are shorter because they cannot take up a sufficient supply of certain minerals to support maximal growth, while the *TT* plants may not be limited by mineral uptake. According to this hypothetical scenario, adding minerals to the soil would enhance the growth rate of *tt* plants by a large amount and the *Tt* plants by a smaller amount (**Figure 25.8**). The height of *TT* plants would not be affected by mineral supplementation. When the environmental effects on phenotype differ according to genotype, this phenomenon is called a **genotype-environment interaction.** Variation due to interactions between genetic and environmental factors is termed $V_{G \times E}$ as noted earlier.

Interactions of genetic and environmental factors are common. As an actual example, **Table 25.3** shows results from a study conducted in 2000 by Cristina Vieira, Trudy Mackay, and colleagues in which they investigated genotype-environment interaction for quantitative trait loci affecting life span in *Drosophila*

TABLE 25.3

Longevity of Two Strains of *Drosophila melanogaster**

Temperature	Strain A Male	Strain A Female	Strain B Male	Strain B Female
Standard	33.6	39.5	37.5	28.9
High	36.3	33.9	23.2	28.6
Low	77.5	48.3	45.8	77.0

*Longevity was measured in the mean number of days of survival. Strains A and B were inbred strains of *D. melanogaster* called Oregon and 2b, respectively. The standard, high, and low temperature conditions were 25°C, 29°C, and 14°C, respectively.

Genotype	Standard soil	Soil supplemented with minerals	
TT			No environmental effect
Tt			Small environmental effect
tt			Large environmental effect

FIGURE 25.8 **A schematic example of genotype-environment interaction.** When grown in standard soil, the three genotypes *TT*, *Tt*, and *tt* show large, medium, and small heights, respectively. When the soil is supplemented with minerals, a great effect is seen on the *tt* genotype and a smaller effect on the *Tt* genotype. The *TT* genotype is unaffected by the environmental change.

melanogaster. The data seen here compare the life span in days of male and female flies from two different strains of *D. melanogaster* raised at different temperatures. Because males and females differ in their sex chromosomes and gene expression patterns, they can be viewed as having different genotypes. The effects of environmental changes depended greatly on the strain and the sex of the flies. Under standard culture conditions, the females of strain A had the longest life span, while females of strain B had the shortest. In strain A, high temperature increased the longevity of males and decreased the longevity of females. In contrast, under hotter conditions, the longevity of males of strain B was dramatically reduced, while females of this same strain were not significantly affected. Lower growth temperature also had different effects in these two strains. While low temperature increased the longevity of both strains, the effects were most dramatic in the males of strain A and the females of strain B. Taken together, these results illustrate the potential complexity of genotype-environmental interaction when measuring a quantitative trait such as life span.

Another issue confronting geneticists is that genotypes may not be randomly distributed in all possible environments. When certain genotypes are preferentially found in particular environments, this phenomenon is called a **genotype-environment association** ($V_{G \leftrightarrow E}$). When such an association occurs, the effects of genotype and environment are not independent of each other, and the association needs to be considered when determining the effects of genetic and environmental variation on the total phenotypic variation. Genotype-environment associations are very common in the study of human genetics, in which large families tend to have more similar environments compared to the population as a whole. One way to evaluate this effect is to compare different genetic relationships, such as identical versus fraternal twins. We will examine this approach later in the chapter. Another strategy that geneticists might follow is to analyze siblings that have been adopted by different parents at birth. Their environment conditions tend to be more disparate, and this may help to minimize the effects of genotype-environment association.

Heritability Is the Relative Amount of Phenotypic Variation That Is Due to Genetic Variation

Another way to view variance is to focus our attention on the genetic contribution to phenotypic variation. Heritability is the proportion of the phenotypic variance that is attributable to genetic variation. If we assume again that environment and genetics are independent and the only two factors affecting phenotype, then

$$h_B{}^2 = V_G/V_T$$

where

$h_B{}^2$ is the heritability in the broad sense

V_G is the variance due to genetics

V_T is the total phenotypic variance, which
    equals $V_G + V_E$

The heritability defined here, $h_B{}^2$, called the **broad-sense heritability,** takes into account different types of genetic variation that may affect the phenotype. As we have seen throughout this textbook, genes can affect phenotypes in various ways. As described earlier, the Nilsson-Ehle experiment showed that the alleles determining hull color in wheat affect the phenotype in an additive way. Alternatively, alleles affecting other traits may show a dominant/recessive relationship. In this case, the alleles are not strictly additive, because the heterozygote has a phenotype closer to, or perhaps the same as, the homozygote containing two copies of the dominant allele. For example, both *TT* and *Tt* pea plants show a tall phenotype. In addition, another complicating factor is epistasis (described in Chapter 4), in which the alleles for one gene can mask the phenotypic expression of the alleles of another gene. To account for these differences, geneticists usually subdivide V_G into these three different genetic categories:

$$V_G = V_A + V_D + V_I$$

where

V_A is the variance due to additive alleles. A
    heterozygote shows a phenotype that is
    intermediate between the respective homozygotes.

V_D is the variance due to alleles that follow a
    dominant/recessive pattern of inheritance.

V_I is the variance due to genes that interact in an
    epistatic manner.

In analyzing quantitative traits, geneticists may focus on V_A and neglect the contributions of V_D and V_I. They do this for scientific as well as practical reasons. For some quantitative traits, the additive effects of alleles may play a primary role in the phenotypic outcome. In addition, when the alleles behave additively, we can predict the outcomes of crosses based on the quantitative characteristics of the parents. The heritability of a trait due to the additive effects of alleles is called the **narrow-sense heritability:**

$$h_N{}^2 = V_A/V_T$$

The narrow-sense heritability ($h_N{}^2$) may be an inaccurate measure of the actual heritability if V_D and V_I are not small

values. However, for certain quantitative traits, the value of V_A may be relatively large compared to V_D and V_I. In such cases, the broad-sense and the narrow-sense heritabilities are similar to each other.

How can the narrow-sense heritability be determined? In this chapter, we will consider two common ways. As discussed later, one way to estimate the narrow-sense heritability involves selective breeding practices, which are done with agricultural species. A second common strategy to determine $h_N{}^2$ involves the measurement of a quantitative trait among groups of genetically related individuals. For example, agriculturally important traits, such as egg weight in poultry, can be analyzed in this way. To calculate the heritability, a researcher would determine the observed egg weights between individuals whose genetic relationships are known, such as a mother and her female offspring. These data could then be used to compute a correlation between the parent and offspring, using the methods described earlier. The narrow-sense heritability is then calculated as

$$h_N{}^2 = r_{obs}/r_{exp}$$

where

r_{obs} is the observed phenotypic correlation between
    related individuals

r_{exp} is the expected correlation based on the known
    genetic relationship

In our example, r_{obs} is the observed phenotypic correlation between parent and offspring. In particular research studies, the observed phenotypic correlation for egg weights between mothers and daughters has been found to be about 0.25 (although this varies among strains). The expected correlation, r_{exp}, is based on the known genetic relationship. A parent and child share 50% of their genetic material, so r_{exp} equals 0.50. So,

$$h_N{}^2 = r_{obs}/r_{exp}$$

$$= 0.25/0.50$$

$$= 0.50$$

(Note: For siblings, $r_{exp} = 0.50$; for identical twins, $r_{exp} = 1.0$; and for an uncle-niece relationship, $r_{exp} = 0.25$.)

According to this calculation, about 50% of the phenotypic variation in egg weight is due to additive genetic variation; the other half is due to the environment.

When calculating heritabilities from correlation coefficients, keep in mind that this computation assumes that genetics and the environment are independent variables. However, this is not always the case. The environments of parents and offspring are often more similar to each other than they are to those of unrelated individuals. As mentioned earlier, there are several ways to minimize this confounding factor. First, in human studies, one may analyze the heritabilities from correlations between adopted children and their biological parents. Alternatively, one can examine a variety of relationships (uncle-niece, identical twins versus fraternal twins, and so on) and see if the heritability values are roughly the same in all cases. This approach was applied in the study that is described next.

Heritability of Dermal Ridge Count in Human Fingerprints Is Very High

Fingerprints are inherited as a quantitative trait. It has long been known that identical twins have fingerprints that are very similar, whereas fraternal twins show considerably less agreement. Galton was the first researcher to study fingerprint patterns, but this trait became more amenable to genetic studies in the 1920s, when Kristine Bonnevie, a Norwegian geneticist, developed a method for counting the number of ridges within a human fingerprint.

As shown in **Figure 25.9**, human fingerprints can be categorized as having an arch, loop, or whorl, or a combination of these patterns. The primary difference among these patterns is the number of triple junctions, each known as a triradius (Figure 25.9b and c). At a triradius, a ridge emanates in three different directions. An arch has zero triradii, a loop has one, and a whorl has two. In Bonnevie's method of counting, a line is drawn from a triradius to the center of the fingerprint. The ridges that touch this line are then counted. (Note: The triradius ridge itself is not counted, and the last ridge is not counted if it forms the center of the fingerprint.) With this method, one can obtain a ridge count for all 10 fingers. Bonnevie conducted a study on a small population and found that ridge count correlations were relatively high in genetically related individuals.

Sarah Holt, who was also interested in the inheritance of this quantitative trait, carried out a more exhaustive study of ridge counts in a British population. As shown in the two graphs below, in groups of 825 males and 825 females, the ridge count on all 10 fingers varied from 0 to 300, with mean values of approximately 145 for males ($SD = 51.1$) and 127 for females ($SD = 52.5$).

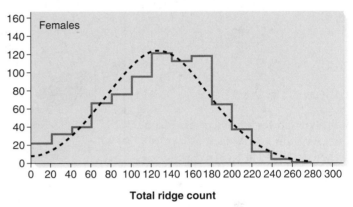

Based on these results, Holt decided to conduct a more detailed analysis of ridge counts by examining the fingerprint patterns of a large group of people and their close relatives. In the experiment of **Figure 25.10**, the ridge counts for pairs of related individuals were determined by the method described in Figure 25.9. The correlation coefficients for ridge counts were then calculated among the pairs of related or unrelated individuals. To estimate the narrow-sense heritability, the observed correlations were then divided by the expected correlations based on the known genetic relationships.

■ THE HYPOTHESIS

Dermal ridge count has a genetic component. The goal of this experiment is to determine the contribution of genetics in the variation of dermal ridge counts.

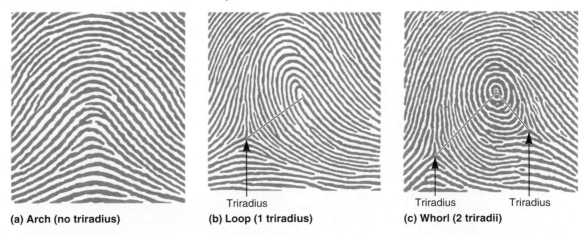

(a) Arch (no triradius)   (b) Loop (1 triradius)   (c) Whorl (2 triradii)

FIGURE 25.9   Human fingerprints and the ridge count method of Bonnevie. (a) This print has an arch rather than a triradius. The ridge count is zero. **(b)** This print has one triradius. A straight line is drawn from the triradius to the center of the print. The number of ridges dissecting this straight line is 13. **(c)** This print has two triradii. Straight lines are drawn from both triradii to the center. There are 16 ridges touching the left line and 7 touching the right line, giving a total ridge count of 23.

■ **TESTING THE HYPOTHESIS — FIGURE 25.10**   **Heritability of human fingerprint patterns.**

Starting material: A group of human subjects from Great Britain.

| | **Experimental level** | **Conceptual level** |

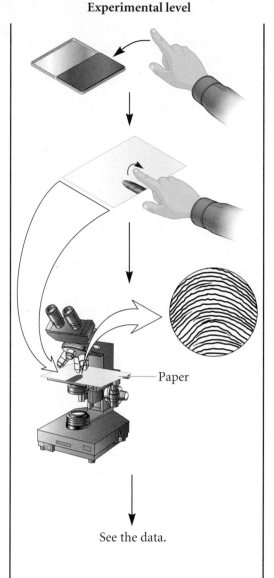

1. Take a person's finger and blot it onto an ink pad.

2. Roll the person's finger onto a recording surface to obtain a print.

This is a method to measure a quantitative trait.

3. With a low-power binocular microscope, count the number of ridges, using the method described in Figure 25.9.

Paper

4. Calculate the correlation coefficients between different pairs of individuals as described earlier in this chapter.

See the data.

The correlation coefficient provides a way to determine the heritability for the quantitative trait.

■ **THE DATA**

Type of Relationship	Number of Pairs Examined	Correlation Coefficient(r_{obs})	Heritability r_{obs}/r_{exp}
Parent-child	810	0.48 ± 0.04*	0.96
Parent-parent	200	0.05 ± 0.07	—†
Sibling-sibling	642	0.50 ± 0.04	1.00
Identical twins	80	0.95 ± 0.01	0.95
Fraternal twins	92	0.49 ± 0.08	0.98
			0.97 average heritability

*± = Standard error of the mean.
†Note: We cannot calculate a heritability value because the value for r_{exp} is not known. Nevertheless, the value for r_{obs} is very low, suggesting that there is a negligible correlation between unrelated individuals.
Adapted from Holt, S. B. (1961). Quantitative Genetics of Finger-Print Patterns. *British Med. Bull.* 17, 247–250.

INTERPRETING THE DATA

As seen in the data table, the results indicate that genetics plays the major role in explaining the variation in this trait. Genetically unrelated individuals (namely, parent-parent relationships) have a negligible correlation for this trait. By comparison, individuals who are genetically related have a substantially higher correlation. When the observed correlation coefficient is divided by the expected correlation coefficient based on the known genetic relationships, the average heritability value is 0.97, which is very close to 1.0.

What do these high heritability values mean? They indicate that nearly all of the variation in fingerprint pattern is due to genetic variation. Significantly, fraternal and identical twins have substantially different observed correlation coefficients, even though we expect that they have been raised in very similar environments. These results support the idea that genetics is playing the major role in promoting variation and that the results are not biased heavily by environmental similarities that may be associated with genetically related individuals. From an experimental viewpoint, the results show us how the determination of correlation coefficients between related and unrelated individuals can provide insight regarding the relative contributions of genetics and environment to the variation of a quantitative trait.

A self-help quiz involving this experiment can be found at the Online Learning Center.

Heritability Values Are Relevant Only to Particular Groups Raised in a Particular Environment

Table 25.4 describes heritability values that have been calculated for traits in particular populations. Unfortunately, heritability is a widely misunderstood concept. Heritability describes the amount of phenotypic variation due to genetic variation for a particular population raised in a particular environment. The words *variation*, *particular population*, and *particular environment* cannot be overemphasized. For example, in one population of cattle, the heritability for milk production may be 0.35, while in another group (with less genetic variation), the heritability may be 0.1. Second, if a group displays a heritability of 1.0 for a particular trait, this does not mean that the environment is unimportant in affecting the outcome of the trait. A heritability value of 1.0 only means that the amount of variation within this group is due to genetics. Perhaps the group has been raised in a relatively homogeneous environment, so the environment has not caused a significant amount of variation. Nevertheless, the environment may be quite important. It just is not causing much variation within this particular group.

As a hypothetical example, let's suppose that we take a species of rodent and raise a group on a poor diet; we find their weights range from 1.5 to 2.5 pounds, with a mean weight of 2 pounds. We allow them to mate and then raise their offspring on a healthy diet of rodent chow. The weights of the offspring range from 2.5 to 3.5 pounds, with a mean weight of 3 pounds. In this hypothetical experiment, we might find a positive correlation in which the small parents tended to produce small offspring, and the large parents produce large offspring. The correlation of weights between parent and offspring might be, say, 0.5. In this case, the heritability for weight would be calculated as r_{obs}/r_{exp}, which equals 0.5/0.5, or 1.0. The value of 1.0 means that all the variation within the groups is due to genetic factors. The offspring vary from 2.5 to 3.5 pounds because of genetic variation, and also the parents range from 1.5 to 2.5 pounds because of genetics. However, as we see here, environment has played an important role. Presumably, the mean weight of the offspring is higher because of their better diet. This example is meant to emphasize the point that heritability tells us only the relative contributions of genetic variation and environment in influencing phenotypic *variation* in a *particular population* in a *particular environment*. Heritability does not describe the relative importance of these two factors in determining the outcomes of traits. When a heritability value is high, it does not mean that a change in the environment cannot have a major impact on the outcome of the trait.

With regard to the roles of genetics and environment (sometimes referred to as nature versus nurture), the topic of human intelligence has been hotly debated. As a trait, intelligence

TABLE 25.4
Examples of Heritabilities for Quantitative Traits

Trait	Heritability Value*
Humans	
Stature	0.65
IQ testing ability	0.60
Cattle	
Body weight	0.65
Butterfat, %	0.40
Milk yield	0.35
Mice	
Tail length	0.40
Body weight	0.35
Litter size	0.20
Poultry	
Body weight	0.55
Egg weight	0.50
Egg production	0.10

*As emphasized in this chapter, these values apply to particular populations raised in particular environments. The value for IQ testing ability is an average value from many independent studies. The other values were taken from Falconer, D. S. (1989). *Introduction to Quantitative Genetics*, 3d ed. Longman, Essex, England.

is difficult to define or to measure. Nevertheless, performance on an IQ test has been taken by some people as a reflection of intelligence ever since 1916 when Alfred Binet's test was used in the United States. Even though such tests may have inherent bias and consider only a limited subset of human cognitive abilities, IQ tests still remain a method of assessing intelligence. By comparing IQ scores among related and unrelated individuals, various studies have attempted to estimate heritability values in selected human populations. These values have ranged from 0.3 to 0.8. A heritability value of around 0.6 is fairly common among many studies. Such a value indicates that over half of the heritability for IQ testing ability is due to genetic factors.

Let's consider what a value of 0.6 means, and what it does not mean. It means that 60% of the variation in IQ testing ability is due to genetic variation in a selected population raised in a particular environment. It does not mean that 60% of an individual's IQ testing ability is due to genetics and 40% is due to the environment. Heritability is meaningless at the level of a single individual. Furthermore, even at the population level, a heritability value of 0.6 does not mean that 60% of the IQ testing ability is due to genetics and 40% is due to the environment. Rather, it means that in the selected population that was examined, 60% of the variation in IQ testing ability is due to genetics, while 40% of the variation is due to the environment. Heritability is strictly a population value that pertains to variation.

Selective Breeding of Species Can Alter Quantitative Traits Dramatically

The term **selective breeding** refers to programs and procedures designed to modify phenotypes in species of economically important plants and animals. This phenomenon, also called **artificial selection,** is related to natural selection, discussed in Chapter 24. In forming his theory of natural selection, Charles Darwin was influenced by his observations of selective breeding by pigeon fanciers and other breeders. The primary difference between artificial and natural selection is how the parents are chosen. Natural selection is due to natural variation in reproductive success. In artificial selection, the breeder chooses individuals that possess traits that are desirable from a human perspective.

For centuries, humans have been practicing selective breeding to obtain domestic species with interesting or agriculturally useful characteristics. The common breeds of dogs and cats have been obtained by selective breeding strategies (**Figure 25.11**). As shown here, it is very striking how selective breeding can modify the quantitative traits in a species. When comparing a greyhound with a bulldog, the magnitude of the differences is fairly amazing. They hardly look like members of the same species.

Likewise, most of the food we eat is obtained from species that have been modified profoundly by selective breeding strategies. This includes products such as grains, fruits, vegetables, meat, milk, and juices. **Figure 25.12** illustrates how certain characteristics in the wild mustard plant (*Brassica oleracea*) have been modified by selective breeding to create several varieties of important domesticated crops. This plant is native to Europe and Asia, and plant breeders began to modify its traits approximately

Greyhound

German shepherd

Bulldog

Cocker spaniel

FIGURE 25.11   **Some common breeds of dogs that have been obtained by selective breeding.**

Genes→Traits By selecting parents carrying the alleles that influence certain quantitative traits in a desired way, dog breeders have produced breeds with distinctive sets of traits. For example, the bulldog has alleles that give it short legs and a flat face. By comparison, the corresponding genes in a German shepherd are found in alleles that produce longer legs and a more pointy snout. All the dogs shown in this figure carry the same kinds of genes (e.g., many genes that affect their sizes, shapes, and fur color). However, the alleles for many of these genes are different among these dogs, thereby producing breeds with strikingly different phenotypes.

4,000 years ago. As seen here, certain quantitative traits in the domestic strains, such as stems and lateral buds, differ considerably from those of the original wild species.

The phenomenon that underlies selective breeding is variation. Within a group of individuals, allelic variation may affect the outcome of quantitative traits. The fundamental strategy of the selective breeder is to choose parents that will pass on alleles to their offspring that produce desirable phenotypic characteristics. For example, if a breeder wants large cattle, the largest members of the herd will be chosen as parents for the next generation. These large cattle will transmit an array of alleles to their offspring that confer large size. The breeder will often choose genetically related individuals (e.g., brothers and sisters) as the parental stock. As mentioned previously, the practice of mating between genetically related individuals is known as inbreeding. Some of the consequences of inbreeding are also described in Chapter 24.

What is the outcome when selective breeding is conducted for a quantitative trait? **Figure 25.13a** shows the results of a program begun at the Illinois Agricultural Experiment Station in 1896, even before the rediscovery of Mendel's laws. This experiment began with 163 ears of corn with an oil content ranging from 4 to 6%. In each of 80 succeeding generations, corn plants

Wild mustard plant

FIGURE 25.12   **Crop plants developed by selective breeding of the wild mustard plant (*Brassica oleracea*).**

Genes→Traits The wild mustard plant carries a large amount of genetic (i.e., allelic) variation, which was used by plant breeders to produce modern strains that are agriculturally desirable and economically important. For example, by selecting for alleles that promote the formation of large lateral leaf buds, the strain of Brussels sprouts was created. By selecting for alleles that alter the leaf morphology, kale was developed. Although these six agricultural plants look quite different from each other, they carry many of the same alleles as the wild mustard. However, they differ in alleles affecting the formation of stems, leaves, flower buds, and leaf buds.

Strain	Modified trait
Kohlrabi	Stem
Kale	Leaves
Broccoli	Flower buds and stem
Brussels sprouts	Lateral leaf buds
Cabbage	Terminal leaf bud
Cauliflower	Flower buds

were divided into two separate groups. In one group, members with the highest oil content were chosen as parents of the next generation. In the other group, members with the lowest oil content were chosen. After 80 generations, the oil content in the first group rose to over 18%; in the other group, it dropped to less than 1%. These results show that selective breeding can modify quantitative traits in a very directed manner.

Similar results have been obtained for many other quantitative traits. **Figure 25.13b** shows an experiment by Kenneth Mather conducted in the 1940s, in which flies were selected on the basis of their bristle number. The starting group had an average of 40 bristles for females and 35 bristles for males. After eight generations, the group selected for high bristle number had an average of 46 bristles for females and 40 for males, while the group selected for low bristle number had an average of 36 bristles for females and 30 for males.

When comparing the curves in Figure 25.13, keep in mind that quantitative traits are often at an intermediate value in unselected populations. Therefore, artificial selection can increase or decrease the magnitude of the trait. Oil content can go up or down, and bristle number can increase or decrease. Artificial selection tends to be the most rapid and effective in changing the frequency of alleles that are at intermediate range in a starting population, such as 0.2 to 0.8.

Figure 25.13 also shows the phenomenon known as a **selection limit**—after several generations a plateau is reached where artificial selection is no longer effective. A selection limit may occur for two reasons. Presumably, the starting population possesses a large amount of genetic variation, which contributes to the diversity in phenotypes. By carefully choosing the parents, each succeeding generation has a higher proportion of the desirable alleles. However, after many generations, the population may be nearly monomorphic for all or most of the desirable alleles that affect the trait of interest. At this point, additional selective breeding will have no effect. When this occurs, the heritability for the trait is near zero, because nearly all genetic variation for the trait of interest has been eliminated from the population. Without the introduction of new mutations into the population, further selection is not possible. A second reason for a selection limit is related to fitness. Some alleles that accumulate in a population due to artificial selection have a negative impact on the population's overall fitness. A selection limit is reached in which the desired effects of artificial selection are balanced by the negative effects on fitness.

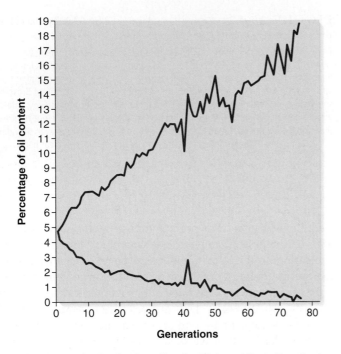

(a) Results of selective breeding for high and low oil content in corn

(b) Results of selective breeding for high and low bristle number in flies

FIGURE 25.13   **Common results of selective breeding for a quantitative trait.** (**a**) Selection for high and low oil content in corn. (**b**) Selection for high and low bristle number in flies.

Using artificial selection experiments, the response to selection is a common way to estimate the narrow-sense heritability in a starting population. The narrow-sense heritability measured in this way is also called the **realized heritability.** It is calculated as

$$h_N^2 = \frac{R}{S}$$

where

R is the response in the offspring to selection, or the difference between the mean of the offspring and the mean of the population of the parents' generation.

S is the selection differential in the parents, or the difference between the mean of the parents and the mean of the population.

Here,

$$R = \overline{X}_O - \overline{X}$$
$$S = \overline{X}_P - \overline{X}$$

where

$\overline{X}$ is the mean of the starting population
$\overline{X}_O$ is the mean of the offspring
$\overline{X}_P$ is the mean of the parents

So,

$$h_N^2 = \frac{\overline{X}_O - \overline{X}}{\overline{X}_P - \overline{X}}$$

The narrow-sense heritability is the proportion of the variance in phenotype that can be used to predict changes in the population mean when selection is practiced.

As an example, let's suppose we began with a population of fruit flies in which the average bristle number for both sexes was 37.5. The parents chosen from this population had an average bristle number of 40. The offspring of the next generation had an average bristle number of 38.7. With these values, the realized heritability is

$$h_N^2 = \frac{38.7 - 37.5}{40 - 37.5}$$

$$h_N^2 = \frac{1.2}{2.5}$$

$$h_N^2 = 0.48$$

This result tells us that about 48% of the phenotypic variation is due to the additive effects of alleles.

An important aspect of narrow-sense heritabilities is their ability to predict the outcome of selective breeding. Solved problem S1 at the end of the chapter illustrates this idea.

Heterosis May Be Explained by Dominance or Overdominance

As we have just seen, selective breeding can alter the phenotypes of domesticated species in a highly directed way. An unfortunate consequence of inbreeding, however, is that it may inadvertently promote homozygosity for deleterious alleles. This phenomenon is called **inbreeding depression.** In addition, genetic drift, described in Chapter 24, may contribute to the loss of beneficial alleles. In agriculture, it is widely observed that when two different inbred strains are crossed to each other, the resulting offspring are more vigorous (e.g., larger or longer-lived) than either of the inbred parental strains. This phenomenon is called **heterosis, or hybrid vigor.**

In modern agricultural breeding practices, many strains of plants and animals consist of hybrids produced by crossing two different inbred lines. In fact, much of the success of agricultural breeding programs is founded in heterosis. In rice, for example, hybrid strains have a 15 to 20% yield advantage over the best conventional inbred varieties under similar cultivation conditions.

As shown in **Figure 25.14**, two different phenomena may contribute to heterosis. In 1908, Charles Davenport developed the dominance hypothesis, in which the effects of dominant

alleles explain the favorable outcome in a heterozygote. He suggested that highly inbred strains have become homozygous for one or more recessive genes that are somewhat deleterious (but not lethal). Because the homozygosity occurs by chance, two different inbred strains are likely to be homozygously recessive for different genes. Therefore, when they are crossed to each other, the resulting hybrids are heterozygous and do not suffer the consequences of homozygosity for deleterious recessive alleles. In other words, the benefit of the dominant alleles explains the observed heterosis. Steven Tanksley, working with colleagues in China, found that heterosis in rice seems to be due to the phenomenon of dominance. This is a common explanation for heterosis.

In 1908, George Shull and Edward East proposed a second hypothesis, known as the overdominance hypothesis (Figure 25.14). As described in Chapter 4, overdominance occurs when the heterozygote is more vigorous than either corresponding homozygote. According to this idea, heterosis can occur because the resulting hybrids are heterozygous for one or more genes that display overdominance. The heterozygote is more vigorous than either homozygote. In corn, Charles Stuber and his colleagues have found that several QTLs for grain yield support the overdominance hypothesis.

Finally, it should be pointed out that overdominance is very difficult to distinguish from **pseudo-overdominance,** a phenomenon initially suggested by James Crow. Pseudo-overdominance is really the same as dominance, except that the chromosomal region contains two or more genes that are very closely linked. For example, a QTL may be identified in a mapping experiment to be close to a particular molecular marker. However, a QTL could contain two genes, both affecting the same quantitative trait. For example, at a single QTL, two genes, *a* and *B*, may be closely linked in one strain, while *A* and *b* would be closely linked in another strain. The hybrid is really heterozygous (*AaBb*) for two different genes, but this may be difficult to discern in mapping experiments because the genes are so close together. If a researcher assumed there was only one gene at a QTL, the overdominance hypothesis would be favored, whereas if two genes were actually present, the dominance hypothesis would be correct. Therefore, without very fine mapping, which is rarely done for QTLs, it is hard to distinguish between overdominance and pseudo-overdominance.

The recessive alleles (*a* and *b*) are slightly harmful in the homozygous condition.

The hybrid offspring is more vigorous, because the harmful effects of the recessive alleles are masked by the dominant alleles.

The Dominance Hypothesis

Neither the *A1* nor *A2* allele is recessive.

The hybrid offspring is more vigorous, because the heterozygous combination of alleles exhibits overdominance. This means that the *A1A2* heterozygote is more vigorous than either the *A1A1* or *A2A2* homozygote.

The Overdominance Hypothesis

FIGURE 25.14 **Mechanisms to explain heterosis.** The two common explanations are the dominance hypothesis and the overdominance hypothesis.

CONCEPTUAL SUMMARY

Quantitative genetics is the study of traits that vary in a continuous way in populations. Such complex or quantitative traits can be anatomical, physiological, or behavioral, and some diseases exhibit characteristics and inheritance patterns analogous to those of quantitative traits. Typically, such traits are determined by two or more genes and are said to be polygenic. Within populations, quantitative traits commonly exhibit a continuum of phenotypic variation that follows a normal distribution. Statistical methods can be used to analyze such a distribution. The mean ($\overline{X}$) describes the average value among the population; the variance (V_x) measures the spread of values from the mean. The standard deviation (SD) is the square root of the vari-

ance. About 68% of all values will be within one SD above or below the mean, and about 95% are within two SDs of the mean. The correlation coefficient (r) evaluates the strength of association between two variables. In genetics, it is often used to see if there are phenotypic correlations between genetically related individuals.

The chromosomal location of one or more genes that influence quantitative traits is called a quantitative trait locus (QTL). The reasons most quantitative traits show a continuum are that each trait is governed by several genes existing in two or more alleles, and the environment contributes a substantial amount of variation to the phenotypic outcome.

In the simplest situation, phenotypic variance is due to the additive effects of genetic variance and environmental variance. More commonly, however, interactions and associations may occur between genetic and environmental factors. Heritability is the fraction of the phenotypic variance that is attributable to genetic variation; it describes the amount of phenotypic variation that is due to genetic variation for a particular population raised in a particular environment. Broad-sense heritability takes into account all of the genetic categories that could affect the phenotype. By comparison, the heritability of a trait due to the additive effects of alleles is called the narrow-sense heritability. For many quantitative traits, the broad-sense and narrow-sense heritabilities are similar.

The term selective breeding refers to programs and procedures designed to modify phenotypes in species of economically important plants and animals in a highly directed way. The fundamental strategy is to choose parents that will pass on alleles to their offspring that produce desirable phenotypic characteristics. Selective breeding typically involves the repeated breeding of related individuals to produce strains that are highly inbred. Much of the success of agricultural breeding programs is founded in heterosis, or hybrid vigor, the production of offspring that are more vigorous than either of the inbred parental strains. Heterosis may be due to the dominant effects of masking harmful recessive alleles or to overdominance of alleles in the heterozygous state.

EXPERIMENTAL SUMMARY

Geneticists measure the quantitative traits of an individual and describe them numerically. In a population, experimentally obtained numerical values from each individual are used to calculate statistics such as the mean, standard deviation, variance, and correlation coefficient.

Several methods can be used to determine the number of QTLs that influence a quantitative trait. Crow used genetic markers to explore the linkage between genes affecting quantitative traits and genes affecting discontinuous traits. His results showed that insecticide resistance is a polygenic trait involving multiple genes on different chromosomes. Newer research techniques have used molecular markers, such as RFLPs, microsatellites, and single nucleotide polymorphisms (SNPs), to determine the regions in the genome where QTLs for a given quantitative trait are likely to reside.

A central issue for quantitative geneticists is the relative contributions of genetics and the environment that underlie the phenotypic variation seen in quantitative traits. In this chapter, we have considered two experimental methods to measure the heritability of quantitative traits. One method is to compare the correlation coefficients among related and unrelated individuals. Holt's experiments with fingerprint patterns showed how the determination of correlation coefficients between related and unrelated individuals can provide insight regarding the relative contributions of genetics and environment to the variation of a quantitative trait. A second method is selective breeding, which is frequently used to improve quantitative traits in agriculturally important species. Quantitative traits are often at an intermediate value in unselected populations. Therefore, artificial selection can increase or decrease the magnitude of a trait. The calculation of realized heritability can be used to predict changes in the population mean when selection is practiced.

PROBLEM SETS & INSIGHTS

Solved Problems

S1. The narrow-sense heritability for potato weight in a starting population of potatoes is 0.42, and the mean weight is 1.4 lb. If a breeder crosses two strains with average potato weights of 1.9 and 2.1 lb, respectively, what is the predicted average weight of potatoes in the offspring?

Answer: The mean weight of the parental strains is 2.0 lb. To solve for the mean weight of the offspring:

$$h_N^2 = \frac{\overline{X}_O - \overline{X}}{\overline{X}_P - \overline{X}}$$

$$0.42 = \frac{\overline{X}_O - 1.4}{2.0 - 1.4}$$

$$\overline{X}_O = 1.65 \text{ lb}$$

S2. A farmer wants to increase the average body weight in a herd of cattle. She begins with a herd having a mean weight of 595 kg and chooses individuals to breed that have a mean weight of 625 kg. Twenty offspring were obtained, having the following weights in kilograms: 612, 587, 604, 589, 615, 641, 575, 611, 610,

598, 589, 620, 617, 577, 609, 633, 588, 599, 601, and 611. Calculate the realized heritability in this herd with regard to body weight.

Answer:

$$h_N^2 = \frac{R}{S}$$

$$= \frac{\overline{X}_O - \overline{X}}{\overline{X}_P - \overline{X}}$$

We already know the mean weight of the starting herd (595 kg) and the mean weight of the parents (625 kg). The only calculation missing is the mean weight of the offspring, $\overline{X}_O$.

$$\overline{X}_O = \frac{\text{Sum of the offspring's weights}}{\text{Number of offspring}}$$

$$\overline{X}_O = 604 \text{ kg}$$

$$h_N^2 = \frac{604 - 595}{625 - 595}$$

$$= 0.3$$

S3. The following data describe the 6-week weights of mice and their offspring of the same sex:

Parent (g)	Offspring (g)
24	26
21	24
24	22
27	25
23	21
25	26
22	24
25	24
22	24
27	24

Calculate the correlation coefficient.

Answer: To calculate the correlation coefficient, we first need to calculate the means and standard deviations for each group:

$$\overline{X}_{parents} = \frac{24 + 21 + 24 + 27 + 23 + 25 + 22 + 25 + 22 + 27}{10} = 24$$

$$\overline{X}_{offspring} = \frac{26 + 24 + 22 + 25 + 21 + 26 + 24 + 24 + 24 + 24}{10} = 24$$

$$SD_{parents} = \sqrt{\frac{0 + 9 + 0 + 9 + 1 + 1 + 4 + 1 + 4 + 9}{9}} = 2.1$$

$$SD_{offspring} = \sqrt{\frac{4 + 0 + 4 + 1 + 9 + 4 + 0 + 0 + 0 + 0}{9}} = 1.6$$

Next, we need to calculate the covariance.

$$CoV_{(parents,\ offspring)} = \frac{\Sigma[(X_P - \overline{X}_P)(X_O - \overline{X}_O)]}{N - 1}$$

$$= \frac{0 + 0 + 0 + 3 + 3 + 2 + 0 + 0 + 0 + 0}{9}$$

$$= 0.9$$

Finally, we calculate the correlation coefficient:

$$r_{(parent,\ offspring)} = \frac{CoV_{(P,O)}}{SD_P SD_O}$$

$$r_{(parent,\ offspring)} = \frac{0.9}{(2.1)(1.6)}$$

$$r_{(parent,\ offspring)} = 0.27$$

S4. As described in this chapter, the correlation coefficient provides a way to determine the strength of association between two variables. When the variables are related due to cause and effect (i.e., one variable affects the outcome of another variable), researchers may use a regression analysis to predict how much one variable will change in response to the other variable. This is easier to understand if we plot the data for two variables. The graph shown here compares mothers' and offspring's body weights in cattle. The line running through the data points is called a regression line. It is the straight line that is closest to all the data points. It is the minimal distance away from the squared vertical distance of all the points.

For many types of data, particularly those involving quantitative traits, the regression line can be represented by the equation

$$Y = bX + a$$

where

b is the regression coefficient

a is a constant

In this example, X is the value of an offspring's weight, and Y is the value of its mother's weight. The value of b, known as the **regression coefficient,** represents the slope of the regression line. The value of a is the y-intercept (i.e., the value of Y when X equals zero). The equation shown is very useful because it allows us to predict the value of X at any given value of Y, and vice versa. To do so, we first need to determine the values of b and a. This can be accomplished in the following manner:

$$b = \frac{CoV_{(X,Y)}}{V_X}$$

$$a = \overline{Y} - b\overline{X}$$

Once the values of a and b have been computed, we can use them to predict the values of X or Y by using the equation

$$Y = bX + a$$

For example, if $b = 0.5$, $a = 2$, and $Y = 31$, this equation can be used to compute a value of X that equals 58. It is important to keep in mind that this equation predicts the average value of X. As we see in the preceding figure, the data points tend to be scattered around the regression line. Deviations may occur between the data points and the line. The equation predicts the values that are the most likely to occur. In an actual experiment, however, some deviation will occur between the predicted values and the experimental values due to random sampling error.

Now here is the question. Using the data found in this chapter regarding weight in cattle, what is the predicted weight of an offspring if its mother weighed 660 lb?

Answer: We first need to calculate a and b.

$$b = \frac{CoV_{(X,Y)}}{V_X}$$

We need to use the data on page 701 to calculate V_X, which is the variance for the mothers' weights. The variance equals 445.1. The covariance is already calculated on page 701; it equals 152.6.

$$b = \frac{152.6}{445.1}$$

$$b = 0.34$$

$$a = \overline{Y} - b\overline{X}$$

$$a = 598 - (0.34)(596) = 395.4$$

Now we are ready to calculate the predicted weight of the offspring using the equation

$$Y = bX + a$$

In this problem, X = 660 pounds

$$Y = 0.34(660) + 395.4$$

$$Y = 619.8 \text{ lb}$$

The average weight of the offspring is predicted to be 619.8 lb.

S5. Genetic variance can be used to estimate the number of genes affecting a quantitative trait by using the following equation:

$$n = \frac{D^2}{8V_G}$$

where

n is the number of genes affecting the trait

D is the difference between the mean values of the trait in two strains that have allelic differences at every gene that influences the trait

V_G is the genetic variance for the trait; it is calculated using data from both strains

For this method to be valid, several assumptions must be met. In particular, the alleles of each gene must be additive, each gene must contribute equally to the trait, all the genes must assort independently, and the two strains must be homozygous for alternative alleles of each gene. For example, if three genes affecting a quantitative trait exist in two alleles each, one strain could be *AA bb CC* and the other would be *aa BB cc*. In addition, the strains must be raised under the same environmental conditions. Unfortunately, these assumptions are not typically met with regard to most quantitative traits. Even so, when one or more assumptions are invalid, the calculated value of n is smaller than the actual number. Therefore, this calculation can be used to estimate the minimum number of genes that affect a quantitative trait.

Now here is the question. The average bristle number in two strains of flies was 35 and 42. The genetic variance for bristle number calculated for both strains was 0.8. What is the minimum number of genes that affect bristle number?

Answer: We apply the equation described previously.

$$n = \frac{D^2}{8V_G}$$

$$n = \frac{(35 - 42)^2}{8(0.8)}$$

$$n = 7.7 \text{ genes}$$

Because genes must come in whole numbers and because this calculation is a minimum estimate, one would conclude there must be at least eight genes that affect bristle number.

S6. Are the following statements regarding heritability true or false?

A. Heritability applies to a specific population raised in a particular environment.

B. Heritability in the narrow sense takes into account all types of genetic variance.

C. Heritability is a measure of the amount that genetics contributes to the outcome of a trait.

Answer:

A. True

B. False. Narrow-sense heritability considers only the effects of additive alleles.

C. False. Heritability is a measure of the amount of phenotypic variation that is due to genetic variation; it applies to the variation of a specific population raised in a particular environment.

Conceptual Questions

C1. Give several examples of quantitative traits. How are these quantitative traits described within groups of individuals?

C2. At the molecular level, explain why quantitative traits often exhibit a continuum of phenotypes within a population. How does the environment help produce this continuum?

C3. What is a normal distribution? Discuss this curve with regard to quantitative traits within a population. What is the relationship between the standard deviation and the normal distribution?

C4. Explain the difference between a continuous trait and a discontinuous trait. Give two examples of each. Are quantitative traits likely to be continuous or discontinuous? Explain why.

C5. What is a frequency distribution? Explain how the graph is made for a quantitative trait that is continuous.

C6. The variance for weight in a particular herd of cattle is 484 lb^2. The mean weight is 562 lb. How heavy would an animal have to be if it was in the top 2.5% of the herd? The bottom 0.13%?

C7. Two different varieties of potatoes both have the same mean weight of 1.5 lb. One group has a very low variance, and the other has a much higher variance.

A. Discuss the possible reasons for the differences in variance.

B. If you were a potato farmer, would you rather raise a variety with a low or high variance? Explain your answer from a practical point of view.

C. If you were a potato breeder and you wanted to develop potatoes with a heavier weight, would you choose the variety with a low or high variance? Explain your answer.

C8. If an r value equals 0.5 and $N = 4$, would you conclude a positive correlation is found between the two variables? Explain your answer. What if $N = 500$?

C9. What does it mean when a correlation coefficient is negative? Can you think of examples?

C10. When a correlation coefficient is statistically significant, what do you conclude about the two variables? What do the results mean with regard to cause and effect?

C11. What is polygenic inheritance? Discuss the issues that make polygenic inheritance difficult to study.

C12. What is a quantitative trait locus (QTL)? Does a QTL contain one gene or multiple genes? What technique is commonly used to identify QTLs?

C13. Let's suppose that weight in a species of mammals is polygenic, and each gene exists as a heavy and light allele. If the allele frequencies in the population were equal for both types of allele (i.e., 50% heavy alleles and 50% light alleles), what percentage of individuals would be homozygous for the light alleles at all of the genes affecting this trait, if the trait was determined by the following number of genes?

A. Two

B. Three

C. Four

C14. The broad-sense heritability for a trait equals 1.0. In your own words, explain what this value means. Would you conclude that the environment is unimportant in the outcome of this trait? Explain your answer.

C15. Compare and contrast the dominance and overdominance hypotheses. Based on your knowledge of mutations and genetics, which do you think tends to be the more common explanation for heterosis?

C16. What is hybrid vigor (also known as heterosis)? Give examples that you might find in a vegetable garden.

C17. From an agricultural point of view, discuss the advantages and disadvantages of selective breeding. It is common for plant breeders to take two different, highly inbred strains, which are the product of many generations of selective breeding, and cross them to make hybrids. How does this approach overcome some of the disadvantages of selective breeding?

C18. Many beautiful varieties of roses have been produced, particularly in the last few decades. These newer varieties often have very striking and showy flowers, making them desirable as horticultural specimens. However, breeders and novices alike have noticed that some of these newer varieties do not have very fragrant flowers compared to the older, more traditional varieties. From a genetic point of view, suggest an explanation why some of these newer varieties with superb flowers are not as fragrant.

C19. In your own words, explain the meaning of the term heritability. Why is a heritability value valid only for a particular population of individuals raised in a particular environment?

C20. What is the difference between broad-sense heritability and narrow-sense heritability? Why is narrow-sense heritability such a useful concept in the field of agricultural genetics?

C21. The heritability for egg weight in a group of chickens on a farm in Maine is 0.95. Are the following statements regarding heritability true or false? If a statement is false, explain why.

A. The environment in Maine has very little impact on the outcome of this trait.

B. Nearly all of the phenotypic variation for this trait in this group of chickens is due to genetic variation.

C. The trait is polygenic and likely to involve a large number of genes.

D. Based on the observation of the heritability in the Maine chickens, it is reasonable to conclude that the heritability for egg weight in a group of chickens on a farm in Montana is also very high.

C22. In a fairly large population of people living in a commune in the southern United States, everyone cares about good nutrition. All the members of this population eat very nutritional foods, and their diets are very similar to each other. With regard to height, how do you think this commune population would compare to the general population in the following categories?

A. Mean height

B. Heritability for height

C. Genetic variation for alleles that affect height

C23. When artificial selection is practiced over many generations, it is common for the trait to reach a plateau in which further selection has little effect on the outcome of the trait. This phenomenon is illustrated in Figure 25.13. Explain why.

C24. Discuss whether a natural population of wolves or a domesticated population of German shepherds is more likely to have a higher heritability for the trait of size.

C25. With regard to heterosis, would the following statements be consistent with the dominance hypothesis, the overdominance hypothesis, or both?

A. Strains that have been highly inbred have become monomorphic for one or more recessive alleles that are somewhat detrimental to the organism.

B. Hybrid vigor occurs because highly inbred strains are monomorphic for many genes, while hybrids are more likely to be heterozygous for those same genes.

C. If a gene exists in two alleles, hybrids are more vigorous because heterozygosity for the gene is more beneficial than homozygosity of either allele.

Experimental Questions

E1. Here are data for height and weight among 10 male college students.

Height (cm)	Weight (kg)
159	48
162	50
161	52
175	60
174	64
198	81
172	58
180	74
161	50
173	54

A. Calculate the correlation coefficients for this group.

B. Is the correlation coefficient statistically significant? Explain.

E2. The abdomen length (in millimeters) was measured in 15 male *Drosophila,* and the following data were obtained: 1.9, 2.4, 2.1, 2.0, 2.2, 2.4, 1.7, 1.8, 2.0, 2.0, 2.3, 2.1, 1.6, 2.3, and 2.2. Calculate the mean, standard deviation, and variance for this population of male fruit flies.

E3. You will need to understand solved problem S5 before answering this question. The average weights for two varieties of cattle were 514 kg and 621 kg. The genetic variance for weight calculated for both strains was 382 kg². What is the minimum number of genes that affect weight variation in these two varieties of cattle?

E4. Using the same strategy as the experiment of Figure 25.6, the following data are the survival of F_2 offspring obtained from backcrosses to insecticide-resistant and control strains:

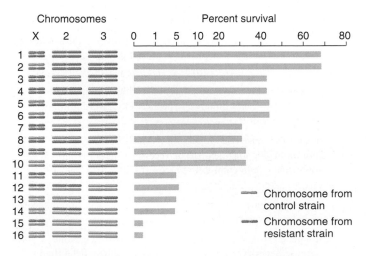

Interpret these results with regard to the locations of QTLs.

E5. In one strain of cabbage, you conduct an RFLP analysis with regard to head weight; you determine that seven QTLs affect this trait. In another strain of cabbage, you find that only four QTLs affect this trait. Note that both strains of cabbage are from the same species, although they may have been subjected to different degrees of inbreeding. Explain how one strain can have seven

QTLs and another strain four QTLs for exactly the same trait. Is the second strain missing three genes?

E6. From an experimental viewpoint, what does it mean to say that an RFLP is associated with a trait? Let's suppose that two strains of pea plants differ in two RFLPs that are linked to two genes governing pea size. RFLP-1 is found in 2,000 bp and 2,700 bp bands, and RFLP-2 is found in 3,000 bp and 4,000 bp bands. The plants producing large peas have RFLP-1 (2,000 bp) and RFLP-2 (3,000 bp); those producing small peas have RFLP-1 (2,700 bp) and RFLP-2 (4,000 bp). A cross is made between these two strains, and the F_1 offspring are allowed to self-fertilize. Five phenotypic classes are observed: small peas, small-medium peas, medium peas, medium-large peas, and large peas. We assume that each of the two genes makes an equal contribution to pea size and that the genetic variance is additive. Draw a gel and explain what RFLP banding patterns you would expect to observe for these five phenotypic categories. Note: Certain phenotypic categories may have more than one possible banding pattern.

E7. Let's suppose that two strains of pigs differ in 500 RFLPs. One strain is much larger than the other. The pigs are crossed to each other, and the members of the F_1 generation are also crossed among themselves to produce an F_2 generation. Three distinct RFLPs are associated with F_2 pigs that are larger. How would you interpret these results?

E8. Outline the steps you would follow to determine the number of genes that influence the yield of rice. Describe the results you might get if rice yield is governed by six different genes.

E9. A researcher has two highly inbred strains of mice. One strain is susceptible to infection by a mouse leukemia virus, while the other strain is resistant. Susceptibility/resistance is a polygenic trait. The two strains were crossed together, and all the F_1 mice were resistant. The F_1 mice were then allowed to interbreed, and 120 F_2 mice were obtained. Among these 120 mice, 118 were resistant to the viral pathogen, and 2 were sensitive. Discuss how many different genes may be involved in this trait. How would your answer differ if none of the F_2 mice had been susceptible to the leukemia virus? Hint: You should assume that the inheritance of one viral-resistance allele is sufficient to confer resistance.

E10. In a wild strain of tomato plants, the phenotypic variance for tomato weight is 3.2 g². In another strain of highly inbred tomatoes raised under the same environmental conditions, the phenotypic variance is 2.2 g². With regard to the wild strain,

A. Estimate V_G.

B. What is h_B^2?

C. Assuming all the genetic variance is additive, what is h_N^2?

E11. The average thorax length in a *Drosophila* population is 1.01 mm. You want to practice selective breeding to make larger *Drosophila*. To do so, you choose 10 parents (five males and five females) of the following sizes: 0.97, 0.99, 1.05, 1.06, 1.03, 1.21, 1.22, 1.17, 1.19, and 1.20. You mate them and then analyze the thorax sizes of 30 offspring (half male and half female):

0.99, 1.15, 1.20, 1.33, 1.07, 1.11, 1.21, 0.94, 1.07, 1.11, 1.20, 1.01, 1.02, 1.05, 1.21, 1.22, 1.03, 0.99, 1.20, 1.10, 0.91, 0.94, 1.13, 1.14, 1.20, 0.89, 1.10, 1.04, 1.01, 1.26

Calculate the realized heritability in this group of flies.

E12. In a strain of mice, the average 6-week body weight is 25 g and the narrow-sense heritability for this trait is 0.21.

A. What would be the average weight of the offspring if parents with a mean weight of 27 g were chosen?

B. What weight of parents would you have to choose to obtain offspring with an average weight of 26.5 g?

E13. Two tomato strains, A and B, both produce fruit that weighs, on average, 1 lb each. All of the variance is due to V_G. When these two strains are crossed to each other, the F_1 offspring display heterosis with regard to fruit weight, with an average weight of 2 lb. You take these F_1 offspring and backcross them to strain A. You then grow several plants from this cross and measure the weights of their fruit. What would be the expected results for each of the following scenarios?

A. Heterosis is due to a single overdominant gene.

B. Heterosis is due to two dominant genes, one in each strain.

C. Heterosis is due to two overdominant genes.

D. Heterosis is due to dominance of several genes each from strains A and B.

E14. You will need to understand solved problem S4 before answering this question. With regard to height, the variance for fathers (in square inches) was 112, the variance for sons was 122, and the covariance was 144. The mean height for fathers was 68 in., and the mean height for sons was 69 in. If a father had a height of 70 in., what is the most probable height of his son?

E15. A danger in computing heritability values from studies involving genetically related individuals is the possibility that these individuals share more similar environments than do unrelated individuals. In the experiment of Figure 25.10, which data are the most compelling evidence that ridge count is not caused by genetically related individuals sharing common environments? Explain.

E16. A large, genetically heterogeneous group of tomato plants was used as the original breeding stock by two different breeders, named Mary and Hector. Each breeder was given 50 seeds and began an artificial selection strategy, much like the one described in Figure 25.13. The seeds were planted, and the breeders selected the 10 plants with the highest mean tomato weights as the breeding stock for the next generation. This process was repeated over the course of 12 growing seasons, and the following data were obtained:

	Mean Weight of Tomatoes (lb)	
Year	Mary's Tomatoes	Hector's Tomatoes
1	0.7	0.8
2	0.9	0.9
3	1.1	1.2
4	1.2	1.3
5	1.3	1.3
6	1.4	1.4
7	1.4	1.5
8	1.5	1.5
9	1.5	1.5
10	1.5	1.5
11	1.5	1.5
12	1.5	1.5

A. Explain these results.

B. Another tomato breeder, named Martin, got some seeds from Mary's and Hector's tomato strains (after 12 generations), grew the plants, and then crossed them to each other. The mean weight of the tomatoes in these hybrids was about 1.7 lb. For a period of five years, Martin subjected these hybrids to the same experimental strategy that Mary and Hector had followed, and he obtained the following results:

	Mean Weight of Tomatoes (lb)
Year	Martin's Tomatoes
1	1.7
2	1.8
3	1.9
4	2.0
5	2.0

Explain Martin's data. Is heterosis occurring? Why was Martin able to obtain tomatoes heavier than 1.5 lb, while Mary's and Hector's strains appeared to plateau at this weight?

E17. The correlations for height were determined for 15 pairs of individuals with the following genetic relationships:

Mother/daughter: 0.36

Mother/granddaughter: 0.17

Sister/sister: 0.39

Sister/sister (fraternal twins): 0.40

Sister/sister (identical twins): 0.77

What is the average heritability for height in this group of females?

E18. An animal breeder had a herd of sheep with a mean weight of 254 lb at 3 years of age. He chose animals with mean weights of 281 lb as parents for the next generation. When these offspring reached 3 years of age, their mean weights were 269 lb.

A. Calculate the narrow-sense heritability for weight in this herd.

B. Using the heritability value that you calculated in part A, what weight of animals would you have to choose to get offspring that weigh 275 lb (at 3 years of age)?

E19. The trait of blood pressure in humans has a frequency distribution that is similar to a normal distribution. The following graph shows the ranges of blood pressures for a selected population of people. The red line depicts the frequency distribution of the systolic pressures for the entire population. Several individuals with high blood pressure were identified, and the blood pressures of their relatives were determined. This frequency distribution is depicted

(continued)

with a blue line. (Note: The blue line does not include the people who were identified with high blood pressure; it includes only their relatives.)

What do these data suggest with regard to a genetic basis for high blood pressure? What statistical approach could you use to determine the heritability for this trait?

Questions for Student Discussion/Collaboration

1. Discuss why heritability is an important phenomenon in agriculture. Discuss how it is misunderstood.

2. From a biological viewpoint, speculate as to why many traits seem to fit a normal distribution. Students with a strong background in math and statistics may want to explain how a normal distribution is generated, and what it means. Can you think of biological examples that do not fit a normal distribution?

3. What is heterosis? Discuss whether it is caused by a single gene or several genes. Discuss the two major hypotheses proposed to explain heterosis. Which do you think is more likely to be correct?

Note: All answers appear at the website for this textbook; the answers to even-numbered questions are in the back of the textbook.

www.mhhe.com/brookergenetics3e

Visit the Online Learning Center for practice tests, answer keys, and other learning aids for this chapter. Enhance your understanding of genetics with our interactive exercises, quizzes, animations, and much more.

EVOLUTIONARY GENETICS

26

Biological evolution is a heritable change in one or more characteristics of a population or species across many generations. Evolution can be viewed on a small scale as it relates to a single gene or it can be viewed on a larger scale as it relates to the formation of new species. In Chapter 24, we examined several factors that cause allele frequencies to change in populations. This process, also known as **microevolution,** concerns the changing composition of gene pools with regard to particular alleles over measurable periods of time. As we have seen, several evolutionary mechanisms, such as mutation, genetic drift, migration, natural selection, and inbreeding, affect the allele and genotypic frequencies within natural populations. On a microevolutionary scale, evolution can be viewed as a change in allele frequency over time.

The goal of this chapter is to relate phenotypic changes that occur during evolution to the underlying genetic changes that cause them to happen. In the first part of the chapter, we will be concerned with evolution on a large scale, which leads to the origin of new species. The question of how species form has been central to the development of evolutionary theory. The term **macroevolution** refers to large-scale evolutionary changes that create new species and higher taxa. It concerns the establishment of the diversity of organisms over long periods of time through the accumulated evolution and extinction of many species.

In the second part of this chapter, we will link molecular genetics to the evolution of species. The development of techniques for analyzing chromosomes and DNA sequences has greatly enhanced our understanding of evolutionary processes at the molecular level. The term **molecular evolution** refers to the molecular changes in genetic material that underlie the phenotypic changes associated with evolution. In this section, we will see how molecular data can provide information about the phylogenetic relationships among different organisms. Finally, as discussed in the third part of the chapter, we will see how genes that affect embryonic development can have a dramatic impact on the phenotypes of organisms. The field of evolutionary developmental biology (evo-devo) focuses on the role of developmental genes in the formation of traits that are important in the evolution of new species.

The topics of molecular evolution and evo-devo are a fitting way to end our discussion of genetics because they integrate the ongoing theme of this textbook—the relationship between molecular genetics and traits—in the broadest and most profound ways. Theodosius Dobzhansky, an influential evolutionary scientist, once said, "Nothing in biology makes sense except in the light of evolution." The

The evolution of eyes. Developmental biologists have recently discovered that the eyes of many diverse species, including fruit flies, frogs, mice, and people, are under the control of the homologous gene called *Pax6*, suggesting that animal eyes may have evolved from the same ancestral species.

extraordinarily diverse and seemingly bizarre array of species on our planet can be explained naturally within the context of evolution. An examination of molecular evolution allows us to make sense of the existence of these species at both the population and the molecular levels.

26.1 ORIGIN OF SPECIES

Charles Darwin, a British naturalist born in 1809, proposed the theory of evolution and provided evidence that existing species have evolved from preexisting species. Like many great scientists, Darwin had a broad background in science, which enabled him to see connections among different disciplines. His thinking was influenced by theories of geology indicating that the Earth is very old and that slow geological processes can lead eventually to substantial changes in the Earth's characteristics.

Darwin's own experimental observations also had a great impact on his thinking. His famous voyage on the *HMS Beagle*, which lasted from 1832 to 1836, involved a careful examination of many different species. He observed the similarities among many discrete species, yet noted the differences that enabled them to be adapted to their environmental conditions. He was particularly struck by the distinctive adaptations of island species. For example, the finches found on the Galápagos Islands had unique phenotypic characteristics compared to those of similar finches found on the mainland.

A third important influence on Darwin was a paper published in 1798, "Essay on the Principle of Population," by Thomas Malthus, an English economist. Malthus asserted that the population size of humans can, at best, increase arithmetically due to increased land usage and improvements in agriculture, whereas the reproductive potential of humans can increase geometrically. He argued that famine, war, and disease work to limit population growth, especially among the poor.

With these three ideas in mind, Darwin had largely formulated his theory of evolution by the mid-1840s. He then spent several years studying barnacles without having published his ideas. The geologist Charles Lyell, who had greatly influenced Darwin's thinking, strongly encouraged Darwin to publish his theory of evolution. In 1856, Darwin began to write a long book to explain his ideas. In 1858, however, Alfred Wallace, a naturalist working in the East Indies, sent Darwin an unpublished manuscript to read prior to its publication. In it, Wallace proposed the same ideas concerning evolution. Darwin therefore quickly excerpted some of his own writings on this subject, and two papers, one by Darwin and one by Wallace, were published in the *Proceedings of the Linnaean Society of London*. These papers were not widely recognized. A short time later, however, Darwin finished his book, *On the Origin of Species by Means of Natural Selection*, which expounded his ideas in greater detail and with experimental support. This book, which received high praise from many scientists and scorn from others, started a great debate concerning evolution. Although some of his ideas were incomplete because the genetic basis of traits was not understood at that time, Darwin's work represents one of the most important contributions to our understanding of biology.

Darwin called evolution "the theory of descent with modification through variation and natural selection." As its name suggests, evolution is based on two fundamental principles: genetic variation and natural selection. A modern interpretation of evolution can view these two principles at the species level (macroevolution) and at the level of genes in populations (microevolution).

1. Genetic variation at the species level: As we have seen in Chapter 24, genetic variation is a consistent feature of natural populations. Darwin observed that many species exhibit a great amount of phenotypic variation. Although the theory of evolution preceded Mendel's pioneering work in genetics, Darwin (as well as many other people before him) observed that offspring resemble their parents more than they do unrelated individuals. Therefore, he assumed that traits are passed from parent to offspring. However, the genetic basis for the inheritance of traits was not understood at that time.

   At the gene level: Genetic variation can involve allelic differences in genes, changes in chromosome structure, and alterations in chromosome number. These differences are caused by random mutations. Alternative alleles may affect the functions of the proteins they encode and thereby affect the phenotype of the organism. Likewise, changes in chromosome structure and number may affect gene expression and thereby influence the phenotype of the individual.

2. Natural selection at the species level: Darwin agreed with Malthus that most species produce many more offspring than will survive and reproduce, creating an ever-present struggle for existence. Over the course of many generations, those individuals who happen to possess the most favorable traits will dominate the composition of the population. The result of natural selection is to make a species better adapted to its environment and more successful at reproduction.

   At the gene level: Some alleles encode proteins that provide the individual with a selective advantage. Over time, natural selection may change the allele frequencies of genes and thereby lead to the fixation of beneficial alleles and the elimination of detrimental alleles.

In this section, we will examine the features of evolution as it occurs in natural populations over time.

Species Concepts Are Used in the Identification of Species

Before we begin to consider how biologists study the evolution of new species, we need to consider how species are defined and identified. A **species** refers to a group of organisms that maintains a distinctive set of attributes in nature. The difficulty of identifying whether certain groups constitute unique species is often rooted in the phenomenon that a single species may exist in two or more distinct populations that are in the slow process of evolving into different species. The amount of time that populations are separated from each other will have an important impact. If the time is short, the populations are likely to be very similar, so they would be considered the same species. If the time is long, significant changes may have occurred, so the different populations would show unequivocal differences that allow them to maintain their distinctive set of features in nature. When studying natural populations, evolutionary biologists are often confronted with situations where some differences between two populations are apparent, but it is difficult to decide whether they truly represent separate species. When two or more groups within the same species display one or more traits that are somewhat different but not enough to warrant their placement into different species, biologists sometimes classify such groups as subspecies.

Biologists adopt methods of species identification that are based on their own experience with the organisms they wish to study. The characteristics that a biologist uses to identify a species depend, in large part, on the species in question. For example, the traits used to distinguish insect species would be quite different from those used to identify bacterial species. The most commonly used characteristics are physical or morphological traits, the ability to interbreed, evolutionary lineages, and ecological factors. These various ways of considering species have led to the use of different approaches for distinguishing species called **species concepts** (Table 26.1). A comparison of these concepts will help you to appreciate the various factors involved in categorizing the species on our planet.

Phylogenetic Species Concept According to the **phylogenetic species concept,** as advocated by Quentin Wheeler and Norman Platnick, the members of a single species are identified by having a unique combination of character states.[1] Historically, the first way to categorize species was based on their physical characteristics. Organisms are classified as the same species if their anatomical traits appear to be very similar. Likewise, microorganisms can be classified according to morphological traits at the cellular level. In addition, molecular features, such as DNA sequences, can now be used to compare organisms. An advantage of the phylogenetic species concept is that it can be applied to all types of organisms.

Biological Species Concept In the late 1920s, Theodosius Dobzhansky proposed that each species is reproductively isolated from other species. Such **reproductive isolation** prevents one species from successfully interbreeding with other species. In 1942, Ernst Mayr expanded on the ideas of Dobzhansky to provide a biological definition of a species. According to Mayr's **biological species concept,** a species is a group of individuals whose members have the potential to interbreed with one another in nature to produce viable, fertile offspring, but cannot successfully interbreed with members of other species. How does reproductive isolation occur? **Table 26.2** describes several ways. These are classified as **prezygotic mechanisms,** which prevent the formation

TABLE 26.1

Species Concepts

Concept	Description
Phylogenetic species concept	Various physical characteristics can be analyzed to distinguish between species. These often include morphological (anatomical) traits. In the case of unicellular organisms, characteristics such as cell wall structure and other cellular traits may be examined. Molecular characteristics can also be compared.
Biological species concept	Two species are often judged to be separate species if they are unable to interbreed in nature to produce viable, fertile offspring.
Evolutionary species concept	An analysis of ancestry may help biologists determine if two groups are members of the same species or represent evolutionarily distinct species.
Ecological species concept	The ability of organisms to successfully occupy their own ecological niche or habitat, including their use of resources and impact on the environment, may be used to distinguish species.

[1] The phylogenetic species concept is related to other species concepts called the morphological species concept and typological species concept.

TABLE 26.2

Types of Reproductive Isolation Among Different Species

Prezygotic Mechanisms	
Habitat isolation	Species may occupy different habitats so that they never come in contact with each other.
Temporal isolation	Species have different mating or flowering seasons, mate at different times of day, or become sexually active at different times of the year.
Sexual isolation	Sexual attraction between males and females of different animal species is limited due to differences in behavior, physiology, or morphology.
Mechanical isolation	The anatomical structures of genitalia prevent mating between different species.
Gametic isolation	Gametic transfer takes place, but the gametes fail to unite with each other. This can occur because the male and female gametes fail to attract, because they are unable to fuse, or because the male gametes are inviable in the female reproductive tract of another species.
Postzygotic Mechanisms	
Hybrid inviability	The egg of one species is fertilized by the sperm from another species, but the fertilized egg fails to develop past early embryonic stages.
Hybrid sterility	The interspecies hybrid survives, but it is sterile. For example, the mule, which is sterile, is a cross between a female horse (*Equus caballus*) and a male donkey (*Equus asinus*).
Hybrid breakdown	The F_1 interspecies hybrid is viable and fertile, but succeeding generations (i.e., F_2, etc.) become increasingly inviable. This is usually due to the formation of less fit genotypes by genetic recombination.

of a zygote, and **postzygotic mechanisms,** which prevent the development of a viable and fertile individual after fertilization has taken place. Reproductive isolation in nature may be circumvented when species are kept in captivity. For example, different species of the genus *Drosophila* rarely mate with each other in nature. In the laboratory, however, it is fairly easy to produce interspecies hybrids.

Evolutionary Species Concept   In 1961, American paleontologist George Gaylord Simpson proposed a species concept based on ancestry. According to the **evolutionary species concept,** a species is derived from a single lineage that is distinct from other lineages and has its own evolutionary tendencies and historical fate. A lineage is the genetic relationship between an individual or group of individuals and their ancestors. The evolutionary species concept is a theoretical viewpoint that is focused on the pathway that has led to the formation of each distinct species. In this regard, it can be applied to the formation of all species. One drawback of the evolutionary species concept is that it does not provide an easy way to identify a unique species. In most cases, lineages are difficult to examine and evaluate quantitatively. The interpretation of lineages involving fossils may be particularly difficult to evaluate due to incomplete fossil remains.

Ecological Species Concept   The **ecological species concept,** described by American evolutionary biologist Leigh Van Valen in 1976, is a viewpoint that considers a species within its native environment. Each species occupies an ecological niche, which is the unique set of habitat resources that a species requires and the influence a species has on the environment and other species. Within their own niche, members of a given species compete with each other for survival. If two organisms are very similar, their needs will overlap, which results in competition. Such competing individuals are likely to be of the same species. According to this concept, species are formed because evolutionary mechanisms control how each type of species uses resources. This species concept is particularly useful in distinguishing bacterial species that do not reproduce sexually. Bacterial cells of the same species are likely to use the same types of resources (such as sugars and vitamins) and grow under the same types of conditions (temperature, pH, and so on).

Speciation Often Occurs Via a Branching Process Called Cladogenesis

Speciation is the process by which new species are formed via evolution. By examining the fossil record, evolutionary biologists have found two different patterns of speciation. During **anagenesis** (from the Greek *ana*, up, and *genesis*, origin), a single species evolves into a different species over the course of many generations (**Figure 26.1a**). During this process, the characteristics of the species change due to both random genetic drift and natural selection. As a result of natural selection, the new species may be better adapted to survive in its original environment, or the environment may have changed so that the new species is better adapted to the new surroundings.

By comparison, **cladogenesis** (from the Greek *clados*, branch) involves the division of a species into two or more species (**Figure 26.1b**). This form of speciation increases species diversity. Although cladogenesis is usually thought of as a splitting process, it commonly occurs as a budding process, which results in the original species plus one or more new species with different characteristics. If we view evolution as a tree, the new species bud from the original species and develop characteristics that prevent them from breeding with the original one.

(a) Anagenesis   **(b) Cladogenesis**

FIGURE 26.1   **A comparison between anagenesis and cladogenesis, two patterns of speciation.**
(**a**) Anagenesis is the change of one species into another.
(**b**) Cladogenesis involves a process in which one original species is separated into two or more different species.

Cladogenesis Can Be Allopatric, Parapatric, or Sympatric

Depending on the geographic locations of the evolving population(s), cladogenesis is categorized as allopatric, parapatric, or sympatric (**Table 26.3**). **Allopatric speciation** (from the Greek *allos*, other, and Latin *patria*, homeland) is thought to be the most prevalent way for a species to diverge. It occurs when members of a species become geographically separated from the other members. This form of speciation can occur by the geographic subdivision of large populations via geological processes. For example, a mountain range may emerge and split a species that occupies the lowland regions, or a creeping glacier may divide a population. **Figure 26.2** shows an interesting example in which geological separation promoted speciation. Two species of antelope squirrels occupy opposite rims of the Grand Canyon. On the south rim is Harris's antelope squirrel (*Ammospermophilus harrisi*), while a closely related white-tailed antelope squirrel (*Ammospermophilus leucurus*) is found on the north rim. Presumably, these two species evolved from a common species that existed before the canyon was formed. Over time, the accumulation of genetic changes in the two separated populations led to the formation of two morphologically distinct species. Interestingly, birds that can easily fly across the canyon have not diverged into different species on the opposite rims.

Allopatric speciation can also occur via a second mechanism, known as the **founder effect,** which is thought to be more rapid and frequent than allopatric speciation caused by geological events. The founder effect, which was discussed in Chapter 24, occurs when a small group migrates to a new location that is geographically separated from the main population. For example, a storm may force a small group of birds from the mainland to a distant island. In this case, the migration of individuals between the island and the mainland is a very infrequent event. In a relatively short period of time, the founding population on the island may evolve into a new species. Two evolutionary mechanisms may contribute to this rapid evolution. First, genetic drift

A. harrisi   *A. leucurus*

FIGURE 26.2   **An example of allopatric speciation: two closely related species of antelope squirrels that occupy opposite rims of the Grand Canyon.**

Genes→Traits Harris's antelope squirrel (*Ammospermophilus harrisi*) is found on the south rim of the Grand Canyon, while the white-tailed antelope squirrel (*Ammospermophilus leucurus*) is found on the north rim. These two species evolved from a common species that existed before the canyon was formed. After the canyon was formed, the two separated populations accumulated genetic changes due to mutation, genetic drift, and natural selection that eventually led to the formation of two distinct species.

may quickly lead to the random fixation of certain alleles and the elimination of other alleles from the population. Another factor is natural selection. The environment on an island may differ significantly from the mainland environment. For this reason, natural selection on the island may favor different types of alleles.

Parapatric speciation (from the Greek *para*, beside) occurs when members of a species are separated partially or when a species is very sedentary. In these cases, the geographic separation is not complete. For example, a mountain range may divide a species into two populations but have breaks in the range where the two groups are connected physically. In these zones of contact, the members of two populations can interbreed, although this tends to occur infrequently. Likewise, parapatric speciation may occur among very sedentary species even though no geographic isolation exists. Certain organisms are so sedentary that 100 to 1,000 meters may be sufficient to limit the interbreeding between neighboring groups. Plants, terrestrial snails, rodents, grasshoppers, lizards, and many flightless insects may speciate in a parapatric manner.

TABLE **26.3**

Common Genetic Mechanisms That Underlie Allopatric, Parapatric, and Sympatric Speciation

Type of Speciation	Common Genetic Mechanisms Responsible for Speciation
Allopatric—two large populations are separated by geographic barriers	Many small genetic differences may accumulate over a long period of time, leading to reproductive isolation. Some of these genetic differences may be adaptive, while others will be neutral.
Allopatric—a small founding population separates from the main population	Genetic drift may lead to the rapid formation of a new species. If the group has moved to an environment that is different from its previous environment, natural selection is expected to favor beneficial alleles and eliminate harmful alleles.
Parapatric—two populations occupy overlapping ranges, so a limited amount of interbreeding occurs	A new combination of alleles or chromosomal rearrangement may rapidly limit the amount of gene flow between neighboring populations because hybrid offspring have a very low fitness.
Sympatric—within a population occupying a single habitat in a continuous range, a small group evolves into a reproductively isolated species	An abrupt genetic change leads to reproductive isolation. For example, a mutation may affect gamete recognition. In plants, the formation of a tetraploid often leads to the formation of a new species because the interspecies hybrid (e.g., diploid × tetraploid) is triploid and sterile.

During parapatric speciation, **hybrid zones** may exist where two populations can interbreed. For speciation to occur, the amount of gene flow within the hybrid zones must become very limited. In other words, there must be selection against the offspring produced in the hybrid zone. One way that this can happen is that each of the two parapatric populations may accumulate different chromosomal rearrangements, such as inversions and balanced translocations. How do chromosomal rearrangements, such as inversions, prevent interbreeding? As discussed in Chapter 8, if a hybrid individual has one chromosome with a large inversion and one that does not carry the inversion, crossing over during meiosis can lead to the production of grossly abnormal chromosomes. Therefore, such an individual is substantially less fertile. By comparison, an individual homozygous for two normal chromosomes or for two chromosomes carrying the same inversion will be fertile, because crossing over can proceed normally.

Sympatric speciation (from the Greek *sym*, together) occurs when a new species arises in the same geographic area as the species from which it was derived. In plants, a common way for sympatric speciation to occur is the formation of polyploids. As discussed in Chapter 8, complete nondisjunction of chromosomes during gamete formation can increase the number of chromosome sets within a single species (autopolyploidy) or between different species (allopolyploidy). Polyploidy is so frequent in plants that it is a major form of speciation. In ferns and flowering plants, at least 30% of the species are polyploid. By comparison, polyploidy is much less common in animals, but it can occur. For example, roughly 30 species of reptiles and amphibians have been identified that are polyploids derived from diploid relatives.

The formation of a polyploid can abruptly lead to reproductive isolation. As an example, let's consider the probable events that led to the formation of a natural species of common hemp nettle known as *Galeopsis tetrahit*. This species is thought to be an allotetraploid derived from two diploid species, *Galeopsis pubescens* and *Galeopsis speciosa*. As shown in **Figure 26.3a**, *G. tetrahit* has 32 chromosomes, whereas the two diploid species contain 16 chromosomes each (2*n* = 16). **Figure 26.3b** illustrates what would happen in crosses between the allotetraploid

(a) Chromosomal composition of 3 *Galeopsis* species

(b) Outcome of intraspecies and interspecies crosses

FIGURE 26.3 **A comparison of crosses between three natural species of hemp nettle with different ploidy levels.** *Galeopsis tetrahit* is an allotetraploid that is thought to be derived from *Galeopsis pubescens* and *Galeopsis speciosa*. If *G. tetrahit* is mated with the other two species, the F₁ hybrid offspring will be monoploid for one chromosome set and diploid for the other set. The F₁ offspring are likely to be sterile, because they will produce highly aneuploid gametes.

and the diploid species. The allotetraploid crossed to another allotetraploid produces an allotetraploid. The allotetraploid is fertile, because all of its chromosomes occur in homologous pairs that can segregate evenly during meiosis. However, a cross between an allotetraploid and a diploid produces an offspring that is monoploid for one chromosome set and diploid for the other set. These offspring are expected to be sterile, because they will produce highly aneuploid gametes that have incomplete sets of chromosomes. This hybrid sterility renders the allotetraploid reproductively isolated from the diploid species.

In the 1930s, Arne Müntzing first proposed that *Galeopsis tetrahit* is an allotetraploid that arose from an interspecies cross between *G. pubescens* and *G. speciosa*. To test this hypothesis, he performed a series of crosses between *G. pubescens* and *G. speciosa* and succeeded in producing an allotetraploid that had two sets of chromosomes from both species. This artificial *G. tetrahit* had traits similar to the natural *G. tetrahit* species but different from *G. pubescens* and *G. speciosa*. Furthermore, the artificial and natural *G. tetrahit* strains could be mated to each other to produce fertile offspring. Overall, Müntzing came to the conclusion that "not only the artificial tetraploid but probably also natural *G. tetrahit* represents a synthesis of *pubescens*- and *speciosa*-genomes."

Speciation Can Be Gradual or Punctuated by Periods of Rapid Change

As we have seen, many different genetic mechanisms give rise to new species. For this reason, the rates of evolutionary change are not constant, although the degree of inconstancy has been debated since the time of Darwin. Even Darwin himself suggested that evolution can occur at fast and slow paces. **Figure 26.4** illustrates contrasting views concerning the rates of evolutionary change. These views are not mutually exclusive but represent two different ways to consider the tempo of evolution. The concept of **gradualism** suggests that each new species evolves gradually over long spans of time (Figure 26.4a). The principal idea is that large phenotypic differences that produce the emergence of new species are due to the slow accumulation of many small genetic changes. By comparison, the concept of **punctuated equilibrium,** proposed by Niles Eldredge and Stephen Jay Gould, suggests that the tempo of evolution is more sporadic (Figure 26.4b). According to this model, species exist relatively unchanged for many generations. During this period, the species is in equilibrium with its environment. These long periods of equilibrium are punctuated by relatively short periods (on a geological timescale) during which evolution occurs at a far more rapid rate.

In reality, neither of the views presented in Figure 26.4 fully account for evolutionary change. The phenomenon of punctuated equilibrium is often supported by the fossil record. Paleontologists

(a) Gradualism

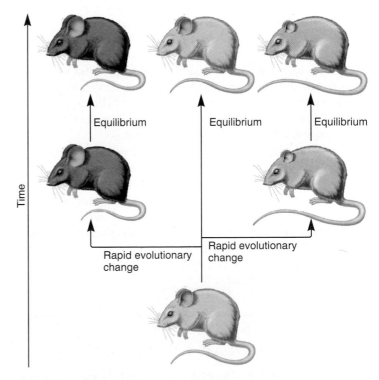

(b) Punctuated equilibrium

FIGURE 26.4   **A comparison of gradualism and punctuated equilibrium. (a)** During gradualism, a species gradually changes due to the accumulation of small genetic changes. **(b)** During punctuated equilibrium, a species exists essentially unchanged for long periods of time, during which it is in equilibrium with its environment. These equilibrium periods are punctuated by short periods of evolutionary change, during which its features may change rapidly.

rarely find a gradual transition of fossil forms. Instead, it is much more common to observe species appearing as new forms rather suddenly in a layer of rocks, persisting relatively unchanged for a very long period of time, and then suddenly becoming extinct. It is presumed that the transition period during which a previous species evolved into a new, morphologically distinct species was so short that few, if any, of the transitional members were preserved as fossils. Even so, these rapid periods of change were likely followed by long periods of equilibrium that involved the additional accumulation of many small genetic changes, consistent with gradualism.

As discussed earlier, rapid evolutionary change can be explained by genetic phenomena. As we have seen throughout this textbook, single-gene mutations can have dramatic effects on phenotypic characteristics. Therefore, only a small number of new mutations may be required to alter phenotypic characteristics, eventually producing a group of individuals that make up a new species. Likewise, genetic events such as changes in chromosome structure (e.g., inversions and translocations) or chromosome number may abruptly create individuals with new phenotypic traits. On an evolutionary timescale, these types of events can be rather rapid because one or only a few genetic changes can have a major impact on the phenotype of the organism.

In conjunction with genetic changes, species may also be subjected to sudden environmental shifts that quickly drive the gene pool in a particular direction via natural selection. For example, a small group may migrate to a new environment in which certain alleles provide better adaptation to the new surroundings. Alternatively, a species may be subjected to a relatively sudden environmental event that has a major impact on survival. There may be a change in climate, or a new predator may infiltrate the geographic range of the species. Natural selection may lead to rapid evolution of the gene pool by favoring those genetic changes that allow members of the population to survive the climatic change or more easily avoid the predator.

Overall, the fossil record and known genetic phenomena tend to support the idea that the tempo of evolution can be quite variable. In some cases, rapid evolutionary change has taken place and led to the formation of new species. During other periods, smaller phenotypic changes may occur over a longer timescale. In conjunction with phenotypic changes, the gradual accumulation of variations in gene sequences has been revealed by molecular analyses of DNA. Next, we will explore the use of molecular data to analyze evolution.

26.2 PHYLOGENETIC TREES AND MOLECULAR EVOLUTION

Thus far, we have considered the various factors that play a role in the formation of new species. In this section, we will examine **phylogeny**—the sequence of events involved in the evolutionary development of a species or group of species. A systematic approach is followed to produce a **phylogenetic tree,** which is a diagram that describes a phylogeny. Such a tree is a hypothesis of the evolutionary relationships among various species, based on

the information available to and gathered by biologists termed **systematists.**

Historically, morphological differences have been used to construct evolutionary trees. In this approach, species that are more similar in appearance tend to be placed closer together on the tree. In addition, species have been categorized based on differences in physiology, biochemistry, and even behavior. While these approaches continue to be used, systematists are increasingly using molecular data to infer evolutionary relationships. With this approach, species are identified and placed into related groups based on the properties of their genetic material (i.e., their genotypes). Those species that are closely related evolutionarily are expected to have greater similarities in their genetic material than are distantly related species.

The advent of molecular approaches for analyzing DNA and gene products has revolutionized the field of evolution. Differences in nucleotide sequences are quantitative and can be analyzed using mathematical principles in conjunction with computer programs. Evolutionary changes at the DNA level can be objectively compared among different species to establish evolutionary relationships. Furthermore, this approach can be used to compare any two existing organisms, no matter how greatly they differ in their morphological traits. For example, we can compare DNA sequences between humans and bacteria, or between plants and fruit flies. Such comparisons would be very difficult at a morphological level. In this section, we will examine how phylogenetic trees are constructed, with an emphasis on the use of molecular data.

A Phylogenetic Tree Depicts the Evolutionary Relationships Among Different Species

Let's first take a look at what information is found within a phylogenetic tree and the form in which it is presented. **Figure 26.5** shows a hypothetical phylogenetic tree of the relationships between various insect species in which the species (butterflies) are labeled A through K. The vertical axis represents time, with the oldest species at the bottom.

As discussed previously, new species can form through anagenesis, in which a single species evolves into a different species, or through cladogenesis, in which a species diverges into two or more species. In cladogenesis, the original species may remain in existence when a second new species is formed. The nodes or branch points in a phylogenetic tree illustrate times when cladogenesis has occurred. For example, approximately 12 million years ago, species A diverged into species A and species B by cladogenesis. Figure 26.5 also shows anagenesis. After species B split into species B and D, species D then evolved into species G by anagenesis. The tips of branches represent either species that became extinct in the past, such as species C, B, and E, or modern species, such as F, I, G, J, H, and K, that are at the top of the tree. Species A and D are also extinct, but gave rise to modern species.

By studying the branch points of a phylogenetic tree, researchers can group species according to common ancestry. A **monophyletic group,** also known as a **clade,** is a group of species

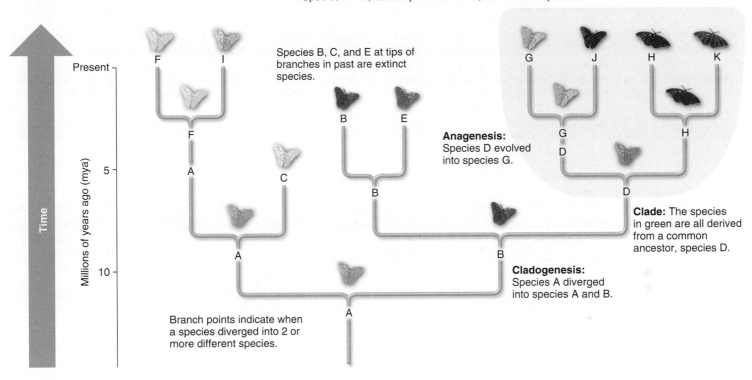

Species F – K, at the tips of branches, are modern species.

Species B, C, and E at tips of branches in past are extinct species.

Anagenesis: Species D evolved into species G.

Clade: The species in green are all derived from a common ancestor, species D.

Cladogenesis: Species A diverged into species A and B.

Branch points indicate when a species diverged into 2 or more different species.

Present

Millions of years ago (mya)

5

10

Time

FIGURE 26.5 **How to read a phylogenetic tree.** This hypothetical tree shows the proposed relationships among various butterfly species.

consisting of all descendents of the group's most common ancestor. For example, the group highlighted in light green is a clade derived from the common ancestor labeled D. The present-day descendents of a common ancestor can also be called a clade. In this case, species G, J, H, and K would form a modern clade. Likewise, the entire tree shown in Figure 26.5 forms a clade, with species A as a common ancestor. As we see in this figure, smaller and more recent clades are subsets of larger clades.

Homologous Genes Are Derived from a Common Ancestral Gene

A phylogenetic tree is based on **homology,** which refers to similarities among various species that occur because the species are derived from a common ancestor. Attributes that are the result of homology are said to be homologous. For example, the wing of a bat, the arm of a human, and the front leg of a cat are homologous structures. When constructing phylogenetic trees, researchers identify homologous features that are shared by some species but not by others. This allows them to group species based on their similarities.

Researchers typically study homology at the level of morphological traits or at the level of genes. We will focus on genetic homology. Two genes are said to be **homologous** if they are derived from the same ancestral gene. During evolution, a single species may become divided into two or more different spe-

cies. When two homologous genes are found in different species, these genes are termed **orthologs.** In Chapter 23, we considered orthologs of *Hox* genes. In that case, several homologous genes were identified in the fruit fly and the mouse. In addition, two or more homologous genes can be found within a single species. These are termed paralogous genes, or **paralogs.** As discussed in Chapter 8, this can occur because abnormal gene duplication events can produce multiple copies of a gene and ultimately lead to the formation of a **gene family.** A gene family consists of two or more paralogs within the genome of a particular species.

Figure 26.6 shows examples of both orthologs and paralogs in the globin gene family. Hemoglobin is an oxygen-carrying protein found in all vertebrate species. It is composed of two different subunits, encoded by the α-globin and β-globin genes. Figure 26.6 shows the deduced amino acid sequences encoded by these genes. The sequences are homologous between humans and horses because of their evolutionary relationship. We would say that the human and horse α-globin genes are orthologs of each other, as are the human and horse β-globin genes. The α-globin and β-globin genes in humans are paralogs of each other.

If you scan the sequences shown in Figure 26.6, you may notice that the sequences of the orthologs are more similar to each other than the paralogs. For example, the sequences of human and horse β globins are much more similar than the sequences of the human α globin and human β globin. What do these results mean? They indicate that the gene duplications that

Hemoglobin

FIGURE 26.6 **A comparison of the α- and β-globin polypeptides from humans and horses.** This figure shows the deduced amino acid sequences obtained by sequencing the exon portions of the corresponding genes. The gaps indicate where additional amino acids are found in the sequence of myoglobin, another member of this gene family.

created the α-globin and β-globin genes occurred long before the evolutionary divergence that produced different species of mammals. For this reason, there was a greater amount of time for the α- and β-globin genes to accumulate changes compared to the amount of time that has elapsed since the evolutionary divergence of mammalian species. This idea is schematically shown in **Figure 26.7.**

Based on the analysis of genetic sequences, evolutionary biologists have estimated that the gene duplication that created the α-globin and β-globin gene lineages occurred approximately 400 million years ago, while the speciation events that created different species of mammals occurred less than 200 million years ago. Therefore, the α-globin and β-globin genes have had much more time to accumulate changes relative to each other.

Genetic Variation at the Molecular Level Is Associated with Neutral Changes in Gene Sequences

As we have seen, the globin genes exhibit variation in their sequences. Researchers have asked the question, Is such variation due primarily to mutations that are favored by natural selection or due to random genetic drift?

A **nonneutral mutation** is one that affects the phenotype of the organism and can be acted on by natural selection. Such a mutation may only subtly alter the phenotype of an organism, or it may have a major impact. According to Darwin, natural selection is the agent that leads to evolutionary change in populations. It selects for individuals with the highest Darwinian fitness and often promotes the establishment of beneficial alleles and the elimination of deleterious ones. Therefore, many geneticists have assumed that natural selection is the dominant factor in changing the genetic composition of natural populations, thereby leading to variation.

In opposition to this viewpoint, in 1968, Motoo Kimura proposed the **neutral theory of evolution.** According to this theory, most genetic variation observed in natural populations is due to the accumulation of neutral mutations that do not affect the phenotype of the organism and are not acted on by natural selection. For example, a mutation within a structural gene that changes a glycine codon from GGG to GGC would not affect the amino acid sequence of the encoded protein. Because neutral mutations do not affect phenotype, they spread throughout a population according to their frequency of appearance and to random genetic drift. This theory has been called the "survival of the luckiest" and also **non-Darwinian evolution** to contrast

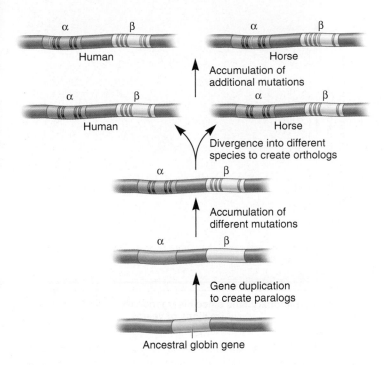

FIGURE 26.7 **Evolution of paralogs and orthologs.** In this schematic example, the ancestral globin gene duplicated to create the α- and β-globin genes, which are paralogous genes. (Note: Actually, gene duplications occurred several times to create a large globin gene family, but this is a simplified example that shows only one gene duplication event. Also, chromosomal rearrangements have placed the α- and β-globin genes on different chromosomes.) Over time, these two paralogs accumulated different mutations, designated with red and blue lines. At a later point in evolution, a species divergence occurred to create different mammalian species including humans and horses. Over time, the orthologs also accumulated different mutations, designated with green lines. As noted here, the orthologs have fewer differences compared to the paralogs because the gene duplication occurred prior to the species divergence.

it with Darwin's theory, which focuses on fitness. Kimura agreed with Darwin that natural selection is responsible for adaptive changes in a species during evolution. His main argument is that most modern variation in gene sequences is neutral with respect to natural selection.

In support of the theory, Kimura and his colleague Tomoko Ohta outlined five principles that govern the evolution of genes at the molecular level:

1. For each protein, the rate of evolution, in terms of amino acid substitutions, is approximately constant with regard to neutral substitutions that do not affect protein structure or function.

   Evidence: As an example, the amount of genetic variation between the coding sequence of the human α-

globin and β-globin genes is approximately the same as the difference between the α-globin and β-globin genes in the horse (shown previously in Figure 26.6). This type of comparison holds true in many different genes compared among many different species.

2. Proteins that are functionally less important for the survival of an organism, or parts of a protein that are less important for its function, tend to evolve faster than more important proteins or regions of a protein. In other words, during evolution, less important proteins accumulate amino acid substitutions more rapidly than important proteins.

   Evidence: Certain proteins are critical for survival, and their structure is precisely suited to their function. Examples are the histone proteins necessary for nucleosome formation in eukaryotes. Histone genes tolerate very few mutations and have evolved extremely slowly. By comparison, fibrinopeptides, which bind to fibrinogen to form a blood clot, evolve very rapidly. Presumably, the sequence of amino acids in this polypeptide is not very important for allowing it to aggregate and form a clot. Another example concerns the amino acid sequences of enzymes. Amino acid substitutions are very rare within the active site, which is critical for function, but are more frequent in other parts of the protein.

3. Amino acid substitutions that do not significantly alter the existing structure and function of a protein are found more commonly than disruptive amino acid changes.

   Evidence: When comparing the coding sequences within homologous genes of modern species, nucleotide differences are more likely to be observed in the wobble base than in the first or second base within a codon. Mutations in the wobble base are often silent because they do not change the amino acid sequence of the protein. In addition, conservative substitutions (i.e., a substitution with a similar amino acid, such as a nonpolar amino acid for another nonpolar amino acid) are fairly common. By comparison, nonconservative substitutions—those that significantly alter the structure and function of a protein—are less frequent. Nonsense and frameshift mutations are very rare within the coding sequences of genes. Also, intron sequences evolve more rapidly than exon sequences.

4. Gene duplication often precedes the emergence of a gene having a new function.

   Evidence: When a single copy of a gene exists in a species, it usually plays a functional role similar to that of the homologous gene found in another species. Gene duplications have created gene families in which each family member can evolve somewhat different functional roles. Examples include the globin family described in Chapter 8 and the *Hox* genes described in Chapter 23.

5. Selective elimination of definitely deleterious mutations and the random fixation of selectively neutral or very

slightly deleterious alleles occur far more frequently in evolution than selection of advantageous mutants.

Evidence: As mentioned in principle 3, silent and conservative mutations are much more common than nonconservative substitutions. Presumably these nonconservative mutations usually have a negative effect on the phenotype of the organism, so they are effectively eliminated from the population by natural selection. On rare occasions, however, an amino acid substitution due to a mutation may have a beneficial effect on the phenotype. For example, a nonconservative mutation in the β-globin gene produces the *Hb^S* allele, which gives an individual resistance to malaria in the heterozygous condition.

In general, the DNA sequencing of hundreds of thousands of different genes from thousands of species has provided compelling support for these five principles of gene evolution at the molecular level. When it was first proposed, the neutral theory sparked a great debate. Some geneticists, called selectionists, strongly opposed the neutralist theory of evolution. However, the debate largely cooled after Ohta incorporated the concept of nearly neutral mutations into the theory. Nearly neutral mutations have a minimal impact on phenotype—they may be slightly beneficial or slightly detrimental. Ohta suggested that the prevalence of such alleles can depend mostly on natural selection or mostly on genetic drift, depending on the population size.

Why do evolutionary biologists care about neutral or nearly neutral mutations? One reason is that their prevalence is used as a tool to construct phylogenetic trees. This topic is discussed next.

Molecular Clocks Can Be Used to Date the Divergence of Species

In 1963, Linus Pauling and Emile Zuckerkandl were the first to suggest the use of molecular data to establish evolutionary relationships. When comparing homologous genes in different species, the DNA sequences from closely related species are more similar to each other than are the sequences from distantly related species. According to the neutral theory of evolution, most of the observed variation is due to neutral mutations. In a sense, the relatively constant rate of neutral mutations acts as a **molecular clock** on which to measure evolutionary time. According to this idea, neutral mutations will become fixed in a population at a rate that is proportional to the rate of mutation per generation. On this basis, the genetic divergence between species that is due to neutral mutations reflects the time elapsed since their last common ancestor.

Figure 26.8 illustrates the concept of a molecular clock. The y-axis is a measure of the number of nucleotide sequence differences between pairs of species. The x-axis plots the amount of time that has elapsed since a pair of species shared a common ancestor. As seen in this diagram, the number of sequence differences is higher when two species shared a common ancestor in the very distant past compared to pairs that shared a more recent common ancestor. The explanation for this phenomenon is that species accumulate independent mutations after they have

FIGURE 26.8   A molecular clock. According to the concept of a molecular clock, neutral mutations accumulate over evolutionary time. When comparing homologous genes between species, those species that diverged more recently tend to have fewer differences compared to those whose common ancestor occurred in the very distant past.

diverged from each other; a longer period of time since their divergence allows for a greater accumulation of mutations that makes their sequences different.

Figure 26.8 suggests a linear relationship between the number of sequence changes and the time of divergence. Such a relationship indicates that the observed rate of neutral mutations remains constant over millions of years. For example, a linear relationship predicts that a pair of species showing 20 nucleotide differences in a given gene sequence would have a (most recent) common ancestor that existed twice as long ago compared to another pair showing 10 nucleotide differences. While actual data sometimes show a relatively linear relationship over a defined time period, evolutionary biologists do not believe that molecular clocks are perfectly linear over very long periods of time. Several factors can contribute to nonlinearity of molecular clocks. These include differences in the generation times of the species being analyzed, the presence of mutations that are acted upon by natural selection, and variation in mutation rates between different species.

To obtain reliable data, researchers must calibrate their molecular clocks. How much time does it take to accumulate a certain percentage of nucleotide changes? To perform such a calibration, researchers must have information regarding the date when two species shared a common ancestor. Such information could come from the fossil record, for instance. The genetic differences between those species are then divided by the time elapsed since their last common ancestor to calculate a rate of change. For example, research suggests that humans and chimpanzees diverged from a common ancestor approximately 6 million years ago. The percentage of nucleotide differences in mitochondrial DNA between humans and chimpanzees is 12%. From these data, the molecular clock for changes in mitochondrial DNA sequences of primates is calibrated at roughly 2% nucleotide changes per million years.

To understand the concept of a molecular clock, let's consider the evolution of some species of primates. **Figure 26.9** illustrates a phylogenetic tree for several species that was derived by comparing DNA sequences in a mitochondrial gene that encodes a protein called cytochrome oxidase subunit II, which is involved with cellular respiration. The vertical scale represents time, and the branch points that are labeled with letters represent common ancestors. Let's take a look at three branch points (labeled A, D, and E) and relate them to the concept of a molecular clock. The common ancestor labeled A diverged into two species that ultimately gave rise to siamangs and the other five species. Since this divergence, a long time has elapsed (approximately 23 million years), allowing the siamang genome to accumulate random neutral changes that would be different from the random changes that have occurred in the genomes of the other five species (see yellow bar in Figure 26.9). Therefore, the cytochrome oxidase gene in the siamangs is more different compared with the genes in the other five species. Now let's compare humans and chimpanzees.

The common ancestor that gave rise to these species is labeled D. This species diverged into two species that eventually gave rise to humans and chimpanzees. This divergence occurred a moderate time ago, approximately 6 million years ago, as illustrated by the red bar. Compared to humans and chimpanzees, humans and siamangs have more differences in their gene sequences because there has been more time for them to accumulate random neutral mutations. Finally, let's consider the two species of chimpanzees, whose common ancestor is labeled E. The time for them to accumulate random mutations has been relatively short, as depicted by the green bar. Therefore, the two modern species of chimpanzees have very similar gene sequences.

A Phylogenetic Tree Can Be Constructed Using a Cladistic Approach

Now that we appreciate the concept of a molecular clock, let's now turn our attention to how evolutionary biologists actually compare gene sequences to construct phylogenetic trees. A **cladistic approach** reconstructs a phylogenetic tree, also called a **cladogram,** by considering the various possible pathways of evolution and then choosing the most plausible tree. Cladistics is now the most commonly used method for the construction of phylogenetic trees.

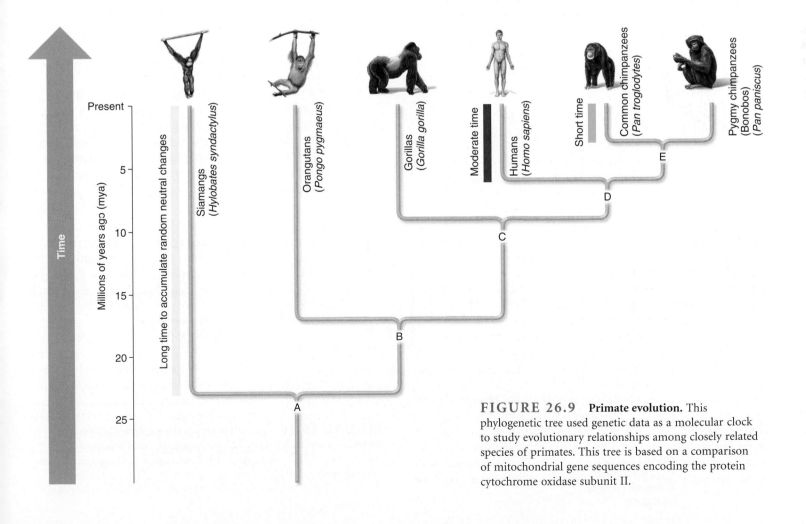

FIGURE 26.9 Primate evolution. This phylogenetic tree used genetic data as a molecular clock to study evolutionary relationships among closely related species of primates. This tree is based on a comparison of mitochondrial gene sequences encoding the protein cytochrome oxidase subunit II.

A cladistic approach compares traits, also called characters, that are either shared or not shared by different species. These can be morphological traits, such as the shapes of birds' beaks, or they can be molecular traits, such as sequences of homologous genes. Such characters may come in different versions called character states. Those that are shared with a distant ancestor are called **ancestral,** or **primitive characters.** Such characters are viewed as being older—ones that arose earlier in evolution. In contrast, a **shared derived character,** or **synapomorphy,** is a trait that is shared by a group of organisms but not by a distant common ancestor. Compared to primitive characters, shared derived characters are more recent traits on an evolutionary timescale. For example, among mammals, only some species have flippers, such as whales and dolphins. In this case, flippers were derived from the two front limbs of an ancestral species. The word "derived" refers to the observation that evolution involves the modification of traits in preexisting species. In other words, the features of newer populations of organisms are derived from changes in preexisting populations. The basis of the cladistic approach is to analyze many shared derived characters among groups of species to deduce the pathway that gave rise to the species.

To understand the concept of primitive and shared derived characters, let's consider how a cladogram can be constructed based on molecular data such as a sequence of a gene. Our example uses molecular data obtained from seven different hypothetical species called A through G. In these species, a homologous gene was sequenced; a portion of the gene sequence is shown in the following.

A: GATAGTACCC	E: GGTATAACCC
B: GATAGTTCCC	F: GGTAGTACCA
C: GATAGTTCCG	G: GGTAGTACCC
D: GGTATTACCC	

In a cladogram, an **ingroup** is a monophyletic group in which we are interested. By comparison, an **outgroup** is a species or group of species that is most closely related to an ingroup. The root of a cladogram is placed between the outgroup and the ingroup. In the cladogram of **Figure 26.10,** the outgroup is species E. This

may have been inferred because the other species may share traits that are not found in species E. The other species (A, B, C, D, F, and G) form the ingroup. For these data, a mutation that changes the DNA sequence is analogous to a modification of a characteristic. Species that share such genetic changes possess shared derived characters because the new genetic sequence was derived from a more primitive sequence.

Now that we have an understanding of some of the general principles of cladistics, let's consider the steps a researcher would follow to construct a cladogram using this approach.

1. **Choose the species in whose evolutionary relationships you are interested.** In a simple cladogram, individual species are compared to each other. In more complex cladograms, species may be grouped into larger taxa (e.g., families) and compared with each other. If such grouping is done, the groups must be clades for the results to be reliable.
2. **Choose characters for comparing different species.** As mentioned, a character is a general feature of an organism. Characters may come in different versions called character states. For example, a base at a particular location in a gene can be considered a character, and this character could exist in different character states such as A, T, G, or C, due to mutations.
3. **Determine the polarity of character states.** In other words, determine if a character state is primitive or derived. In the case of morphological traits, this information may be available by examining the fossil record. If possible, identify an outgroup.
4. **Group species (or higher taxa) based on shared derived characters.**
5. **Build a cladogram based on the following principles:**
   - All species (or higher taxa) are placed on tips in the phylogenetic tree, not at branch points.
   - Each cladogram branch point should have a list of one or more shared derived characters that are common to all species above the branch point unless the character is later modified.

FIGURE 26.10 **Primitive versus shared derived characters involving a molecular trait.** This phylogenetic tree illustrates a cladogram of relationships involving homologous gene sequences found in seven species. Mutations that alter a primitive DNA sequence are shared among certain species and thereby allow the construction of a cladogram.

- All shared derived characters appear together only once in a cladogram unless they independently arose during evolution more than once in the ancestors of different clades.

6. **Choose the best cladogram among possible options.** When grouping species (or higher taxa), more than one cladogram may be possible. Therefore, analyzing the data and producing the best possible cladogram is a key aspect of this process. As described next, different theoretical approaches can be followed to achieve this goal.

The greatest challenge in a cladistic approach is to determine the correct order of events. It may not always be obvious which traits are primitive and came earlier and which are derived and came later in evolution. Different approaches can be used to deduce the correct order. We will discuss three of them here. First, for morphological traits, a common way to deduce the order of events is to analyze fossils and determine the relative dates that certain traits arose. A second strategy that can be used to deduce the correct order is to assume that the best hypothesis is the one that requires the fewest number of evolutionary changes. This concept, called the **principle of parsimony,** states that the preferred hypothesis is the one that is the simplest. For example, if two species possess a tail, we would initially assume that a tail arose once during evolution and that both species have descended from a common ancestor with a tail. Such a hypothesis is simpler than assuming that tails arose twice during evolution and that the tails in the two species are not due to descent from a common ancestor. A third strategy is to incorporate known features of genetic changes into the analysis. This method, known as **maximum likelihood,** allows researchers to choose the best trees based on inferences about the most likely ways that DNA incurs mutations. For example, a researcher might assume that the rate of neutral genetic changes is constant over a given time period or that neutral mutations are more likely to occur in the third base of a codon because of the redundancy of the genetic code. This approach can analyze different phylogenetic trees and identify which tree has the maximum likelihood of occurring, based on assumptions about how DNA changes over the course of evolution. When constructing their trees, researchers often have preferences about the kinds of strategies they employ. More than one strategy may be used to construct a cladogram.

Let's consider a simple example to illustrate the process. Our example uses molecular data obtained from four different hypothetical species called A through D. In these species, a homologous region of DNA was sequenced as shown in the following.

```
      1 2 3 4 5
A: GTACA
B: GACAG
C: GTCAA
D: GACCG
```

Given this information, three different trees are shown in **Figure 26.11,** although more are possible. In these examples, tree 1 requires seven mutations, tree 2 requires six, and tree 3 requires only five. Because tree 3 requires the fewest number of mutations, it is considered the most parsimonious. Based on the principle of parsimony, tree 3 would be the more likely choice.

FIGURE 26.11   The cladistic approach: choosing a cladogram from molecular genetic data. This figure shows three different phylogenetic trees for the evolution of a short DNA sequence, but many more are possible. According to the principle of parsimony, the cladogram shown in tree 3 is the more plausible choice because it requires only five mutations. When constructing cladograms based on long genetic sequences, researchers use computers to generate trees with the fewest possible genetic changes.

Phylogenetic Trees Refine Our Understanding of Evolutionary Relationships

For molecular evolutionary studies, the DNA sequences of many genes have been obtained from a wide range of sources. Several different types of gene sequences have been used to construct phylogenetic trees. One very commonly analyzed gene is that encoding 16S rRNA, an rRNA found in the small ribosomal subunit. This gene has been sequenced from thousands of different species. It is as reliable a molecular measure of phylogenetic relationships among organisms as is now available. Because rRNA is universal in all living organisms, its function was established at an early stage in the evolution of life on this planet, and its sequence has changed fairly slowly. Presumably, most mutations in this gene are deleterious, so few neutral or beneficial alleles can occur. This limitation causes this gene sequence to change very slowly during evolution. Furthermore, 16S rRNA is a rather large molecule, and therefore it contains a large amount of sequence information.

In 1977, Carl Woese analyzed 16S rRNA sequences and identified a new domain of life called Archaea. **Figure 26.12** illustrates a phylogenetic tree of all life based on Woese's work. It proposes three main evolutionary branches: the **Bacteria,** the **Archaea,** and the **Eukaryotes** (also called Eukarya). From these types of genetic analyses, it has become apparent that all living organisms are connected through a complex evolutionary tree.

Although the work of Woese was a breakthrough in our appreciation of evolution, more recent molecular genetic data have shed new light regarding the classification of species. With regard to eukaryotic species, biologists once categorized them into four kingdoms: protists (Protista), fungi (Fungi), animals (Animalia), and plants (Plantae). However, recent models propose several major groups, called **supergroups,** as a way to organize eukaryotes into monophyletic groups. **Figure 26.13** shows a diagram that hypothesizes several supergroups. Of the four traditional kingdoms, both Fungi and Animalia are within the Opisthokonta supergroup, while Plantae is found within Archaeplastida. The remaining branches and supergroups used to be classified within the single kingdom Protista. As seen in this figure, molecular data and newer ways of building trees reveal that protists played a key role in the evolution of many diverse groups of eukaryotic species, producing several large monophyletic supergroups.

Horizontal Gene Transfer Also Contributes to the Evolution of Species

The types of phylogenetic trees considered thus far are examples of **vertical evolution,** in which species evolve from preexisting species by the accumulation of gene mutations and by changes in

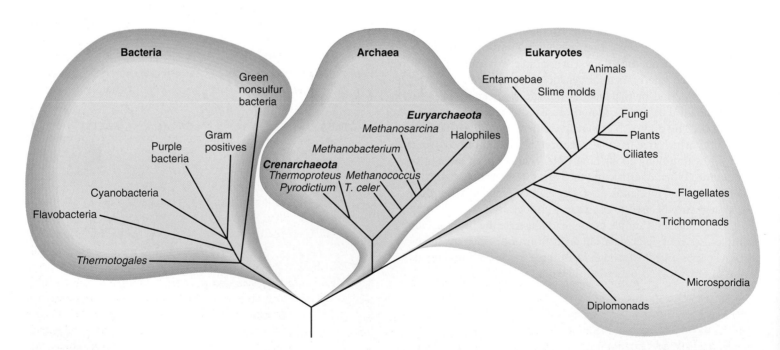

FIGURE 26.12   A phylogenetic tree of all life on Earth based on 16S rRNA sequences. The relationships in this figure among eukaryotic species have been substantially revised based on newer molecular data (see Figure 26.13).

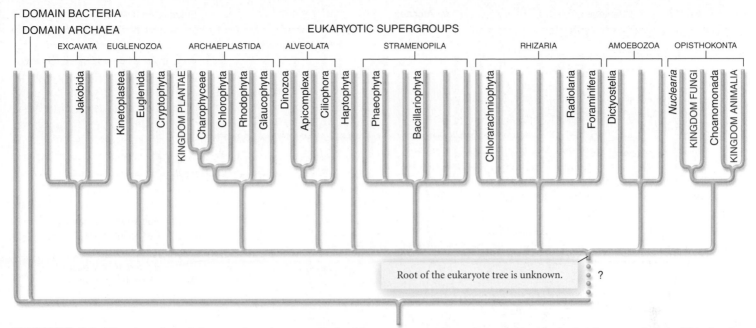

FIGURE 26.13 **A modern cladogram for eukaryotes.** Each of the supergroups shown here are hypothesized to be monophyletic. This drawing should be considered a working hypothesis. The arrangement of these supergroups relative to each other is not entirely certain.

chromosome structure and number. Vertical evolution involves genetic changes in a series of ancestors that form a lineage. In addition to vertical evolution, however, species accumulate genetic changes by another process called **horizontal gene transfer.** As discussed in Chapter 24, this involves the exchange of genetic material among different species.

An analysis of many genomes suggests that horizontal gene transfer was prevalent during the early stages of evolution, when all organisms were unicellular, but continued even after the divergence of the three major domains of life. With regard to modern organisms, horizontal gene transfer remains prevalent among prokaryotic species. By comparison, this process is less common among eukaryotes, though it does occur. Researchers have speculated that multicellularity and sexual reproduction have presented barriers to horizontal gene transfer in many eukaryotes. For a gene to be transmitted to eukaryotic offspring, it would have to be transferred into a eukaryotic cell that is a gamete or a cell that gives rise to gametes.

How has horizontal gene transfer impacted our understanding of evolution? In the past few decades, scientists have debated the role of horizontal gene transfer in the earliest stages of evolution, prior to the emergence of the two prokaryotic domains. The traditional viewpoint was that the three domains

of life arose from a single type of prokaryotic cell (or preprokaryotic cell) called the universal ancestor. However, genomic research has suggested that horizontal gene transfer may have been particularly common during the early stages of evolution on Earth, when all species were unicellular. Rather than proposing that all life arose from a single type of prokaryotic cell, horizontal gene transfer may have been so prevalent that the universal ancestor may have actually been an ancestral community of cell lineages that evolved as a whole. If that were the case, the tree of life cannot be traced back to a single universal ancestor.

Figure 26.14 illustrates a schematic scenario for the evolution of life on Earth that includes the roles of both vertical evolution and horizontal gene transfer. This has been described as a "web of life" rather than a "tree of life." Instead of a universal ancestor, a web of life began with a community of primitive cells that transferred genetic material in a horizontal fashion. Horizontal gene transfer was also prevalent during the early evolution of bacteria and archaea and when eukaryotes first emerged as unicellular species. While horizontal gene transfer remains a prominent way to foster evolutionary change in modern bacteria and archaea, the region of the diagram that contains eukaryotic species has a more treelike structure, because horizontal gene transfer has become much less common in these species.

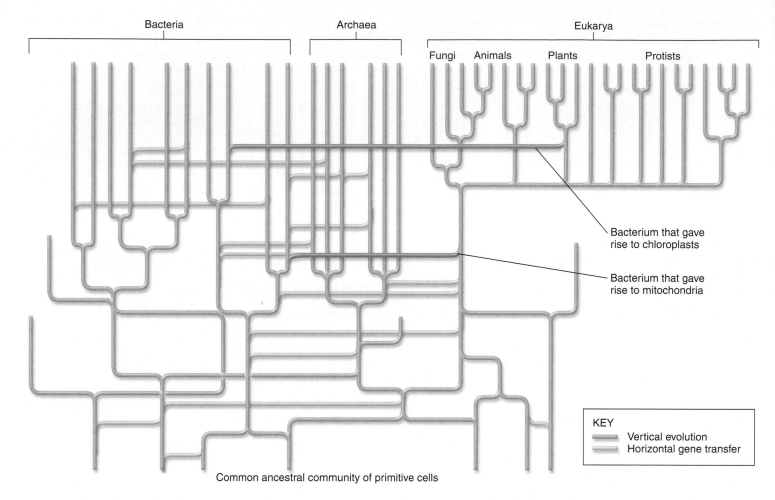

FIGURE 26.14   A revised view regarding the evolution of life, incorporating the concept of horizontal gene transfer. This phylogenetic tree shows a classification of life on Earth that includes the contribution of horizontal gene transfer in the evolution of species on our planet. This phenomenon was prevalent during the early stages of evolution when all organisms were unicellular. Horizontal gene transfer continues to be a prominent factor in the speciation of Bacteria and Archaea. Note: This tree is meant to be schematic. Figure 26.13 is a more realistic representation of the evolutionary relationships among modern species.

EXPERIMENT 26A

Scientists Can Analyze Ancient DNA to Examine the Relationships Between Living and Extinct Flightless Birds

The majority of phylogenetic trees have been constructed from molecular data using DNA samples collected from living species. With this approach, we can infer the prehistoric changes that gave rise to present-day DNA sequences. As an alternative, scientists have discovered that it is occasionally possible to obtain DNA sequence information from species that have lived in the past. In 1984, the first successful attempt at determining DNA sequences from an extinct species was accomplished by groups at the University of California at Berkeley and the San Diego Zoo, including Russell Higuchi, Barbara Bowman, Mary Freiberger, Oliver Ryder, and Allan Wilson. They obtained a sample of dried muscle from a museum specimen of the quagga (*Equus quagga*), a zebralike species that became extinct in 1883. This piece of

muscle tissue was obtained from an animal that had died 140 years ago. A sample of its skin and muscle had been preserved in salt in the Museum of Natural History at Mainz, Germany. The researchers extracted DNA from the sample, cloned pieces of it into vectors, and then sequenced hybrid vectors containing the quagga DNA. This pioneering study opened the field of **ancient DNA analysis,** also known as **molecular paleontology.**

Since the mid-1980s, many researchers have become excited about the information that might be derived from sequencing DNA obtained from older specimens. Currently there is debate concerning how long DNA can remain significantly intact after an organism has died. Over time, the structure of DNA is degraded by hydrolysis and the loss of purines. Nevertheless, under certain conditions (e.g., cold temperature, low oxygen), DNA samples may remain stable for as long as 50,000 to 100,000 years and perhaps longer.

In most studies involving prehistoric specimens (in particular, those that are much older than the salt-preserved quagga sample), the ancient DNA is extracted from bone, dried muscle, or preserved skin. These samples are often obtained from museum specimens that have been gathered by archaeologists. However, it is unlikely that enough DNA can be extracted to enable a researcher to directly clone the DNA into a vector. Since 1985, however, the advent of PCR technology, described in Chapter 18, has made it possible to amplify very small amounts of DNA using PCR primers that flank a region within the 12S rRNA gene, a slowly changing gene. In recent years, this approach has been used to elucidate the phylogenetic relationships between modern and extinct species.

In the experiment described in **Figure 26.15**, Alan Cooper, Cécile Mourer-Chauviré, Geoffrey Chambers, Arndt von Haeseler, Allan Wilson, and Svante Pääbo investigated the evolutionary relationships among some extinct and modern species of flightless birds. Two groups of flightless birds, the moas and the kiwis, existed in New Zealand during the Pleistocene era. The moas are now extinct, although 11 species were formerly present. In this study, the researchers investigated the phylogenetic relationships among four extinct species of moas that were available as museum samples, three kiwis of New Zealand, and several other (nonextinct) species of flightless birds. These included the emu and the cassowary (found in Australia and New Guinea), the ostrich (found in Africa and formerly Asia), and two rheas (found in South America).

The samples from the various species were subjected to PCR to amplify the 12S rRNA gene. This provided enough DNA to subject the gene to DNA sequencing. The sequences of the genes were aligned using computer programs described in Chapter 21.

■ THE GOAL

Because DNA is a relatively stable molecule, it can be amplified by PCR from a preserved sample of a deceased organism and subjected to DNA sequencing. A comparison of these DNA sequences with modern species may help elucidate the phylogenetic relationships between extinct and modern species.

■ ACHIEVING THE GOAL — FIGURE 26.15   DNA analysis reveals phylogenetic relationships among extinct and modern flightless birds.

Starting material: Tissue samples from four extinct species of moas were obtained from museum specimens. Tissue samples were also obtained from three species of kiwis, one emu, one cassowary, one ostrich, and two species of rhea.

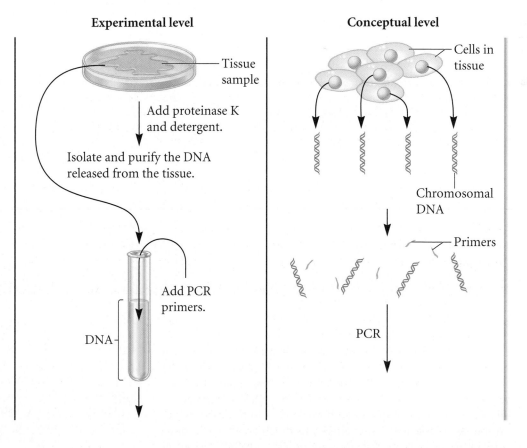

Experimental level

1. For soft tissue samples, treat with proteinase K (which digests protein) and a detergent that dissolves cell membranes. This releases the DNA from the cells.

Tissue sample

Add proteinase K and detergent.

Isolate and purify the DNA released from the tissue.

2. Individually, mix the DNA samples with a pair of PCR primers that are complementary to the 12S rRNA gene.

Add PCR primers.

DNA

Conceptual level

Cells in tissue

Chromosomal DNA

Primers

PCR

3. Subject the samples to PCR. See chapter 18 (Figure 18.6) for a description of PCR.

PCR technique

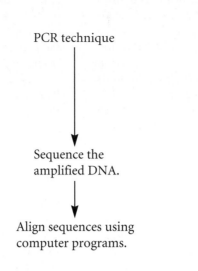

Many copies of the 12S rRNA gene are made.

4. Subject the amplified DNA fragments to DNA sequencing. See chapter 18 for a description of DNA sequencing.

Sequence the amplified DNA.

The amplification of the 12S rRNA gene allows it to be subjected to DNA sequencing.

5. Align the DNA sequences to each other. Methods of DNA sequence alignment are described in chapter 21.

Align sequences using computer programs.

Align sequences to compare the degree of similarity.

■ THE DATA

```
MOA 1       GCTTAGCCCTAAATCCAGATACTTACCCTACACAAGTATCCGCCCGAGAACTACGAGCACAAACGCTTAAAACTCTAAGGACTTGGCGGTGCCCCAAACCCACCTAGAGGAGCCTGTTCTATAATCGATAATCCACGATA
MOA 2       · · · · · · · · · · · · · · · · · · · · · · · · · · · · · · · · · · · · · · · · · · · · · · · · · · · · · · · · · · · · · · · · · · · · · · · · · · · · · · · · · · · · · · · · · · · · · · · · · · · · · · · · · · · · · · · · · · · · ·
MOA 3       · · · · · G · · · · · · · · · · · · T · · · · · · · · · · · · · · · · · · · · · · · T · · · · · · · · · · · · · · · · · · · · · · · · · · · · · · · · · · · · · · · · · · · · · · · · · · · · · · · · · · · · · · · · · · · · · · · ·
MOA 4       · · · · · · · · · · · · · · · · · · · · · · · · · · · · C · · · · · · · · · · · · · · · · · · · · · · · · · · · · · · · · · · · · · · · · · · · · · · · · · · · · · · · · · · · · · · · · · · · · · · · · · · · · · · C · · · · T · · ·
KIWI 1      · · · · · · · · · · · · T · G · · · · · GT · · CT · · · C · · · · · · · · · · · · · · · · · · · · · · · · · · · · · · · · · · · · · · T · · · · · · · · · · · · · · · · · C · · · · · · ·
KIWI 2      · · · · · · · · · · · · T · G · G · · AT · · CT · · · C · · · · · · · · · · · · · · · · · · · · · · · · · · · · · · · · · · · T · · · · · · · · · · · · · · · · · C · · · · · · ·
KIWI 3      · · · · · · · · · · · · T · G · G · · G · AT · · C · · · · C · · · · · · · · · · · · · · · · · · · · · · · · · · · · · · · · · · T · · · · · · · · · · · · · · · · · C · · · · · · ·
EMU         · · · · · · · · · · · · TT · · · C · · T · · CAG · C · · · · · T · · · · · · · · · · · · · · · · · · · · · · · · · · · · · · T · · · · · · · · · · · · · · · · · C · · · · · · ·
CASSOWARY   · · · · · · · · · · · · TT · · · · CG · TA · · CTG · · · · · · · · · · · · · · · · · · · · · · · · · · · · · · · · · · · T · · · · · · · · · · · · · · · · · C · · · · · · ·
OSTRICH     · · · · · · · T · · · · AT · · · · · · · · C · · CT · · · · · · · · · · · · · · · · · · · · · · · · · · · · · · · · · · · T · · · · · · · · · · · · · · · · · · · · · · · · T
RHEA 1      · · · · · · · · · · · · T · · · · C · · CT · · · · · · · · · · · · · · · · · · · · · · · · · · · · · · · · · · · · · · · · · T · · · · · · · · · · · · · · · · · C · · · · · · ·
RHEA 2      · · · · · · · · · · · · C · · · · · · C · C · · · · · · · · · · · · · · · · · · · · · · · · · · · · · · · · · · · · · · · · · T · · · · · · · · · · · · · · · · · C · · · · · · ·
```

```
MOA 1       CACCCGACCATCCCTCGCCCGT-GCAGCCTACATACCGCCGTCCCCAGCCCGCCT--AATGAAAG-AACAAATAGCGAGCACAACAGCCCTCCCCGCTAACAAGACAGGTCAAGGTATAGCATATGAGATGGAAGAAATG
MOA 2       · · · · · · · · · · · · · · · · A · - · · · · · · · · · · · · · · · · · · · -- · · · · - · · · · - · · · · · · · · · · · · · · · · · · · TCA- · · · · · · · · · · · · · · · · · · · · · · · · · · · · · · · · · ·
MOA 3       · · · · · · · · T · T · · · A · - · · · · · · · · · · · · · · · · · -- · · · · - · · · · · · · · TA · - - · T · · · · · · · · · · · · · · · · · · · · · · · · · · · · · · · · · · · · · · · · ·
MOA 4       · · · · · · · · T · T · · · A · - · · · · · · · · · · · · · · · · · -- · · · · - · · · · · · · T · · AC-- · · · · · · · · · · · · · · · · · · · · · · · · · · · · · · · · · · · · · · · · ·
KIWI 1      · · · · A · · · · T · T · · AAC-A · · · · T · · · · · · · · G · · T · · AA · · · G · - · · · C · · A · · · · TA · - · A · · · · · · · · · C · · · · · · · · · · · · · · · · · · · · · · ·
KIWI 2      · · · · A · · · · T · T · · AAC-A · · · · T · · · · · · · · G · · T · · AA · · · G · - · · · C · · A · · · · TA · - · A · · · · · · · · · C · · · · · · · · · · · · · · · · · · · · · · ·
KIWI 3      · · · · A · · · · T · T · · AAC-A · · · · · · · · · · · · · · · GC · · · AA · · · · · · · · TACA-A · · · · · · · · · · · · · CC · C · · · · G · · · · · · · · ·
EMU         · · · · AG · · · T · T · · AA-A · · · · · G · · · · · · · · · · · -- · · · · · · · · · T · · AC--TT · · · · · · · · · · · · · · G · · · · · · · · · · · · · · · ·
CASSOWARY   · · · · A · · · · T · T · · AA · TA · · · · · · G · · · · · -- · G · · G · - · · · · · · · · · T · · AC--T · · · · · · · · · · · · · · G · · · · · · · · · · · · · · · ·
OSTRICH     · · · · A · · · C · · T · · A-- T · · · · · · · · · · · · C-- · · · G · · · · · · · · T · · A · · · · · · · · · · · · · · · · · · · · · · · · GAG · · ·
RHEA 1      · · · · · · · T · T · · · A--A · · · · · · · · · · · · TA · G · · · · · · C · · AG · · T · T · · TA--- · · · · · · · · · · · G · · · · · · · · ·
RHEA 2      · · · · · · · · · · · · · · · · · TA · · · · · · · C · · A · · T · T · · TA---A · · · G · · · · · · · · · · · · · · · · · · · · · · · · · · · ·
```

```
MOA 1       GGCTACATTTTCTAACATAGAACACCC-------------ACGAAAGAGAAGGTGAAACCCTCCTCAAAAGGCGGATTTAGCAGTAAAATAGAACAAGAATGCCTATTTTAAGCCCGGCCCTGGGGC
MOA 2       · · · · · · · · · · · · · · · · · · · · · · · · · A · · · · · T · · · · G · · · · · · · · · · · · · · · · · · · · · · · · · · · · · · · · · · · · · T · · · · ·
MOA 3       · · · · · · · · · · · · T------------- · · · · · · · · G · · · · · · · · · · · · · · · · G · · · · · · C · · · C · · · · T · · · · · · · · ·
MOA 4       · · · · · · · · · · · · · A------------- · · · · · · · · G · · · · · · · · · · · · G · · · · · · C · · · C · · · · T · · · · · · · · ·
KIWI 1      · · · · · · A · · · · T · T------------- · · · A · GGT · · · T · - C · · T · G · · · · · · · C · · · T · · GA · T · · · · · - · T · · · A · · ·
KIWI 2      · · · · · · A · · · · T · T------------- · · · A · GGT · · · T · - C · · T · G · · · · · · · C · · · T · · GA · T · · · · · - · T · · · A · · ·
KIWI 3      · · · · · · A · · · · T · T------------- · · · A · GGTA · · · T · · C · · T · G · · · · · A · · · · C · · · T · · A · T · · · · · · · · A · · ·
EMU         · · · · · · · · T · T------------- · · · · · AG · T · · T · AC · T · G · · · · · · · C · · · T · · GA · T · · A- · T · · T · A · · ·
CASSOWARY   · · · · · · · · · A · G · T------------- · · · A · G · T · · T · A · · T · G · · · · · · · · C · · · · GA · T · · · A- · · · · A · · ·
OSTRICH     · · · · · · · · T · A------------- · · · · · G · TA · · · T · A · · T · G · · · · · · · · T · · GA · T · · · · · - T · · · T · A · · ·
RHEA 1      · · · · · · TC · · · · A------------- · G · · · GGCA · · · AC- · · CG · · · · · · G · · G · TC · · A · · C · C · · - · · · · · A · · ·
RHEA 2      · · · · · GTC · · · G------------- · · · GGCA · · · AC · · CG · · · · · · G · G · G · TC · · A · · C · C · · - · · · · · A · · ·
```

■ INTERPRETING THE DATA

The data of Figure 26.15 illustrate a multiple sequence alignment of the amplified DNA sequences. The first line shows the DNA sequence of one extinct moa species, and underneath it are the sequences of the other species. When the other sequences are identical to the first sequence, a dot is placed in the corresponding position. When the sequences are different, the nucleotide base (A, T, G, or C) is placed there. In a few regions, the genes are different lengths. In these cases, a dash is placed at the corresponding position.

As you can see from the large number of dots, the sequences among these flightless birds are very similar. To establish evolutionary relationships, researchers focus on the few differences that occur. Some surprising results were obtained. The sequences from the kiwis (a New Zealand species) are actually more similar to the sequence from the ostrich (an African species) than they are to those of the moas, which were once found in New Zealand. Likewise, the kiwis are more similar to the emu and cassowary (found in Australia and New Guinea) than they are to the moas. Contrary to their original expectations, the authors concluded that the kiwis

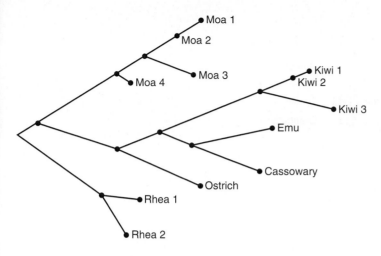

FIGURE 26.16   **A revised phylogenetic tree of moas, kiwis, emus, cassowaries, ostriches, and rheas.**

FIGURE 26.17   **Extinct organisms from which DNA sequences have been obtained.** Right panel, from bottom right to top right: quagga, marsupial wolf, sabre-toothed cat, moa, mammoth, cave bear, blue antelope, giant ground sloth, and Aurochs. Left panel, from bottom left to top right: mastodon, New Zealand coot, South Island piopio, Steller's sea cow, Neanderthal man, South Island adzebill (*Aptornis defossor*), Shasta ground sloth, pig-footed bandicoot, moa-nalo, and Balearic Islands cave goat (*Myotragus balearicus*). (Adapted from Hofreiter et al. [2001], "Ancient DNA," *Nature Reviews Genetics*, vol. 2, p. 357.)

are more closely related to Australian and African flightless birds than they are to the moas. They proposed that New Zealand was colonized twice by ancestors of flightless birds. As shown in **Figure 26.16**, the researchers constructed a new evolutionary tree to illustrate the relationships among these modern and extinct species.

Since these early studies, sequences of ancient DNA have been derived from a variety of species. **Figure 26.17** shows some extinct organisms from which DNA sequences have been determined. Many of these samples were tens of thousands of years old. For example, the sample of a Neanderthal man was approximately 30,000 years old. The oldest samples are likely to be in the range of 50,000 to 100,000 years old.

A self-help quiz involving this experiment can be found at the Online Learning Center.

Speciation Is Associated with Changes in Chromosome Structure and Number

In this section, we have focused on mutations that alter the DNA sequences within genes. In addition to gene mutations, other types of changes, such as gene duplications, transpositions, inversions, translocations, and changes in chromosome number, are important features of evolution.

As discussed earlier, changes in chromosome structure and/or number may not always be adaptive, but they can lead to reproductive isolation and the origin of new species. As an example of variation in chromosome structure among closely related species, **Figure 26.18** compares the banding pattern of the three largest chromosomes in humans and the corresponding chromosomes in chimpanzees, gorillas, and orangutans. The banding patterns are strikingly similar because these species are closely related evolutionarily. However, some interesting differences are observed. Humans have one large chromosome 2, but this chromosome is divided into two separate chromosomes in the other three species. This explains why humans have 23 types

of chromosomes and the other species have 24. This may have occurred by a fusion of the two smaller chromosomes during the development of the human lineage. Another interesting change in chromosome structure is seen in chromosome 3. The banding patterns among humans, chimpanzees, and gorillas are very similar, but the orangutan has a large inversion that flips the arrangement of bands in the centromeric region.

With the advent of molecular techniques, researchers can analyze the chromosomes of two or more different species and identify regions that contain the same groups of linked genes, which are called **synteny groups.** Within a particular synteny group, the same types of genes are found in the same order. In 1995, Graham Moore and colleagues analyzed the locations of molecular markers along the chromosomes of several cereal grasses including rice (*Oryza sativa*), wheat (*Triticum aestivum*), maize (*Zea mays*), foxtail millet (*Setaria italica*), sugarcane (*Saccharum officinarum*), and sorghum (*Sorghum vulgare*). From this analysis, they were able to identify several large synteny groups that are common to most of these species (**Figure 26.19**). As an example, let's compare rice (12 chromosomes per set) to wheat

FIGURE 26.18   **A comparison of banding patterns among the three largest human chromosomes and the corresponding chromosomes in apes.** This is a schematic drawing of late prophase chromosomes. The conventional numbering system of the banding patterns is shown next to the human chromosomes.

FIGURE 26.19   **Synteny groups in cereal grasses.** The synteny groups are named according to their relative arrangement in rice, which is shown on the left. In some cases, a synteny group or portion of a group may have incurred an inversion, but these are not shown in this figure. Also, the genome of maize contains two copies of most synteny groups, suggesting that it is derived from an ancestral polyploid species.

(7 chromosomes per set). In rice, chromosome 6 contains two synteny groups designated R6a and R6b, while chromosome 8 contains a single synteny group, R8. In wheat, chromosome 7 consists of R6a and R6b at either end, while R8 is sandwiched in the middle. One possible explanation for this difference could be a chromosomal rearrangement during the evolution of wheat in which R8 became inserted into the middle of a chromosome containing R6a and R6b. Overall, the evolution of cereal grass species has maintained most of the same types of genes, but many chromosomal rearrangements have occurred. These types of rearrangements would promote reproductive isolation.

26.3 EVO-DEVO: EVOLUTIONARY DEVELOPMENTAL BIOLOGY

The genetic changes that result in morphological and physiological differences distinguish one species from another. In recent years, researchers have begun to investigate how genetic variation produces species and groups of species with novel shapes and forms. The underlying basis for these changes is often rooted in the developmental pathways that control an organism's morphology. **Evolutionary developmental biology** (informally referred to as **evo-devo**) is a relatively new field of biology that compares the development of species in an attempt to understand ancestral relationships between different species and the developmental mechanisms that bring about evolutionary change. During the past few decades, developmental geneticists have gained a better understanding of biological development at the molecular level. Much of this work has involved the discovery of genes that control development in experimental organisms. As more and more organisms have been analyzed, researchers have become interested in the similarities and differences that occur between closely related and distantly related species. Evolutionary developmental biology has arisen in response to this trend.

How do new morphological forms come into being? For example, how does a nonwebbed foot evolve into a webbed foot? Or how does a new organ, such as an eye, come into existence? As we will learn, such novelty arises through genetic changes, also called genetic innovations. Certain types of genetic innovation have been so advantageous that they have resulted in groups of many new species. For example, the innovation of feathered wings resulted in the evolution of many different species of birds. In this section, we will learn that proteins that control developmental changes, such as cell-signaling proteins and transcription factors, often play a key role in promoting the morphological changes that occur during evolution.

FIGURE 26.20 **The role of cell-signaling proteins in the morphology of birds' feet.** (a) Expression of the *BMP4* gene in the developing limbs. BMP4 protein is stained blue. (b) Expression of the *gremlin* gene in the developing limbs. Gremlin protein is stained brown and is expressed in the interdigit region only in the duck. BMP4 causes programmed cell death; Gremlin inhibits BMP4. (c) Because BMP4 is not inhibited in the interdigit regions in the chicken, the cells in this region die, and the foot is not webbed. By comparison, inhibition of BMP4 in the interdigit regions in the duck results in a webbed foot.

Variation in the Expression Patterns of Developmental Genes Can Dramatically Affect Morphology

As discussed in Chapter 23, genes that play a role in development may influence cell division, cell differentiation, cell migration, and cell death. These four processes create an organism with a specific body form and function. Developmental genes affect traits such as the size of a flower or the shape of your nose. In recent years, researchers have determined that developmental genes are key players in the evolution of many types of traits. Changes in such genes affect traits that can be acted on by natural selection. Variation in the expression of development genes may be commonly involved in the acquisition of new traits that promote the formation of new species.

To appreciate this concept, let's compare the formation of a chicken's foot with that of a duck. Developmental biologists have discovered that the morphological differences between a webbed and a nonwebbed foot are due to the differential expression of two cell-signaling proteins called bone morphogenetic protein 4 (BMP4) and gremlin. As shown in **Figure 26.20a**, the *BMP4* gene is expressed throughout the developing limb of both chickens

Chicken   Duck

(a) BMP4 protein levels

Future interdigit regions

(b) Gremlin protein levels

(c) Comparison of a chicken and a duck foot

and ducks. The BMP4 protein, which is stained in blue, causes cells to undergo apoptosis and die. The gremlin protein, which is stained brown in **Figure 26.20b**, inhibits the function of BMP4 and thereby allows cells to survive. In the developing chicken limb, the gremlin gene is expressed throughout the limb, except in the regions between each digit. Therefore, these cells die, and a chicken develops a nonwebbed foot (**Figure 26.20c**). In contrast, *gremlin* expression in the duck occurs throughout the entire limb, including the interdigit regions, resulting in a webbed foot. Interestingly, researchers have been able to introduce gremlin protein into the interdigit regions of developing chicken limbs. This produces a chicken with webbed feet!

How do these observations relate to bird evolution? As we have seen, variation in the expression of these genes determines whether or not a bird's feet are webbed. At some point in the evolution of birds, mutations occurred that provided variation in the expression of the *BMP4* and *gremlin* genes, which resulted in nonwebbed or webbed feet. In terrestrial settings, having nonwebbed feet is an advantage because these are more effective at holding on to perches, running along the ground, and snatching prey. Therefore, natural selection would maintain nonwebbed feet in terrestrial environments. This process explains the occurrence of nonwebbed feet in chickens, hawks, crows, and many other terrestrial birds. In aquatic environments, webbed feet are an advantage because they act as paddles for swimming, so genetic variation that produced webbed feet would have been promoted by natural selection. Over the course of many generations, this gave rise to webbed feet that are now found in ducks, geese, penguins, and other aquatic birds.

How does having webbed or nonwebbed feet influence speciation? This trait may not directly affect the ability of two individuals to mate. However, due to natural selection, birds with webbed feet would become more prevalent in aquatic environments, while birds with nonwebbed feet would be found in terrestrial locations. Therefore, reproductive isolation would occur because the populations would occupy different environments.

The Evolution of Animal Body Plans Is Related to Changes in *Hox* Gene Number and Expression

Hox genes, which are discussed in Chapter 23, are found in all animals. Developmental biologists have speculated that genetic variation in the *Hox* genes may have spawned the formation of many new body types, yielding many different animal species. As shown in **Figure 26.21**, the number and arrangement of *Hox* genes varies considerably among different types of animals.

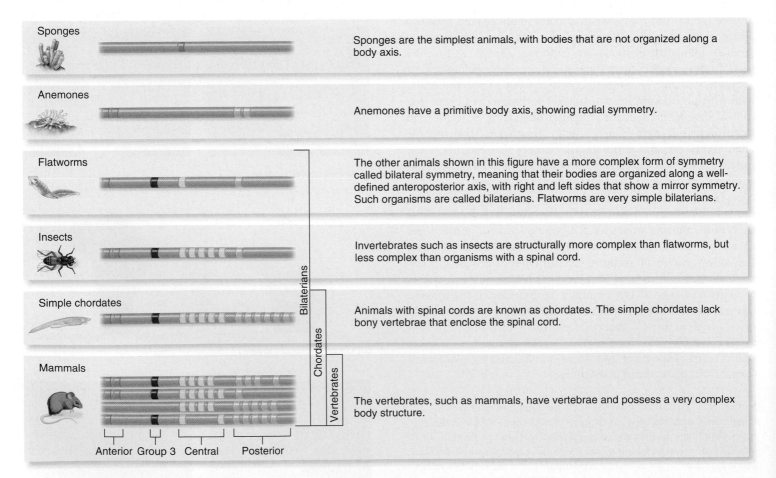

Sponges Sponges are the simplest animals, with bodies that are not organized along a body axis.

Anemones Anemones have a primitive body axis, showing radial symmetry.

Flatworms The other animals shown in this figure have a more complex form of symmetry called bilateral symmetry, meaning that their bodies are organized along a well-defined anteroposterior axis, with right and left sides that show a mirror symmetry. Such organisms are called bilaterians. Flatworms are very simple bilaterians.

Insects Invertebrates such as insects are structurally more complex than flatworms, but less complex than organisms with a spinal cord.

Simple chordates Animals with spinal cords are known as chordates. The simple chordates lack bony vertebrae that enclose the spinal cord.

Mammals The vertebrates, such as mammals, have vertebrae and possess a very complex body structure.

Anterior Group 3 Central Posterior

Bilaterians / Chordates / Vertebrates

FIGURE 26.21 *Hox* **gene composition in different types of animals.** Researchers speculate that the duplication of *Hox* genes and *Hox* gene clusters played a key role in the evolution of more complex body plans in animals. The *Hox* genes are divided into four groups, called anterior, group 3, central, and posterior, based on their relative similarities. Each group is represented by a different color in this figure.

Sponges, the simplest of animals, have at least one *Hox* gene, whereas insects typically have nine or more. In most cases, multiple *Hox* genes occur in a cluster in which the genes are close to each other along a chromosome. In mammals, such *Hox* gene clusters have been duplicated twice during the course of evolution to form four clusters with a total of 38 genes.

How would an increase in *Hox* genes enable more complex body forms to evolve? Part of the answer lies in the spatial expression of the *Hox* genes. In fruit flies, for example, different *Hox* genes are expressed in different segments of the body along the anteroposterior axis (see Figure 23.16). Therefore, an increase in the number of *Hox* genes allows each of these master control genes to become more specialized in the region that it controls. One segment in the middle of the fruit fly body can be controlled by a particular *Hox* gene and form wings and legs, while a segment in the head region can be controlled by a different *Hox* gene and develop antennae. Therefore, research suggests that one way that new, more complex body forms evolved was through an increase in the number of *Hox* genes, thereby making it possible to form many specialized parts of the body that are organized along a body axis.

Three lines of evidence support the idea that *Hox* gene complexity has been instrumental in the evolution and speciation of animals with different body patterns. First, *Hox* genes are known to control body development. Second, as described in Figure 26.21, a general trend is observed in which animals with a more complex body structure tend to have more *Hox* genes and more *Hox* clusters in their genomes than do simpler animals. Finally, a comparison of *Hox* gene evolution and animal evolution bears striking parallels. Researchers can analyze *Hox* gene sequences among modern species and make estimates regarding the timing of past events via molecular clock data. Geneticists can estimate when the first *Hox* gene arose by gene innovation; the date is difficult to pinpoint but is well over 600 million years ago. The single *Hox* gene found in the sponge has descended from this primordial *Hox* gene. In addition, gene duplications of this primordial gene produced clusters of *Hox* genes in other species. Clusters, such as those found in modern insects, were likely to have arisen approximately 600 million years ago. A duplication of that cluster is estimated to have occurred around 520 million years ago. Remarkably, these estimates of *Hox* gene origins correlate with major speciation events in the history of animals. The Cambrian explosion, which occurred from 533 to 525 million years ago, saw a phenomenal diversification in the body plan of invertebrate species. This diversification occurred after the *Hox* cluster was formed and was possibly undergoing its first duplication to create two *Hox* clusters. Also, approximately 420 million years ago, a second duplication produced species with four *Hox* clusters. This event precedes the proliferation of tetrapods—vertebrates with four limbs—that occurred approximately 417 to 354 million years ago. Modern tetrapods, such as mammals, have four *Hox* clusters. This second duplication may have been a critical event that led to the evolution of complex terrestrial vertebrates with four limbs, such as lizards, bears, and humans.

The Study of the *Pax6* Gene Indicates That Different Types of Eyes May Have Evolved from a Simpler Form

Explaining how a complex organ comes into existence is another major challenge for evolutionary biologists. While it is relatively easy to understand how a limb could undergo evolutionary modifications to become a wing, flipper, or arm, it is more difficult to understand how a body structure comes into being in the first place. In his book *The Origin of Species*, Charles Darwin addressed this question and pointed out that the existence of complex organs, such as the eye, was difficult to understand. As Darwin noted, the eye of vertebrate species is exceedingly complex, being able to adjust focus, let in different amounts of light, and detect a spectrum of colors. He suggested that such a complex eye must have evolved from a simpler structure through the process of descent with modification. With amazing insight, he speculated that a very simple eye could be composed of just two cells, a photoreceptor cell and an adjacent pigment cell. The photoreceptor cell, which is a type of nerve cell, is able to absorb light and respond to it. The function of the pigment cell is to stop the light from reaching one side of the photoreceptor cell. This primitive, two-cell arrangement would allow an organism to sense both light and the direction from which the light comes.

How would natural selection play a role in the evolution of eyes? A primitive eye would provide an additional way for a mobile organism to sense its environment, possibly allowing it to avoid predators or locate food. Vision is nearly universal among animals, which indicates that there must be a strong selective advantage to better eyesight. Over time, eyes could become more complex by enhancing the ability to absorb different amounts and wavelengths of light and also by refinements in structures such as the addition of lenses that detect the direction of light.

Since the time of Darwin, many evolutionary biologists have wrestled with the question of eye evolution. From an anatomical point of view, researchers have discovered many different types of eyes. For example, the eyes of fruit flies, squid, and humans are quite different from each other. Furthermore, species that are closely related evolutionarily sometimes have different types of eyes. This observation led evolutionary biologists Luitfried von Salvini-Plawen and Ernst Mayr to propose that eyes may have independently arisen many different times during evolution. Based solely on morphology, such a hypothesis seemed reasonable, and for many years was accepted by the scientific community.

The situation took a dramatic turn when geneticists began to study eye development. Researchers identified a gene, *Pax6*,[2] that influences eye development in both rodents and humans. In mice and rats, a mutation in this gene results in small eyes. In humans, a mutation in the *Pax6* gene results in an eye disorder called aniridia. In heterozygotes carrying one defective copy of the gene, the iris does not form. Researchers subsequently discovered a gene in *Drosophila*, named *eyeless*, that also causes a defect in eye development when mutant. DNA sequencing revealed the *eyeless* and *Pax6* genes are homologous; they are derived from the same ancestral gene.

---

[2] *Pax* is an acronym for p̲a̲i̲r̲e̲d̲ b̲o̲x̲. The protein encoded by this gene contains a domain called a paired box.

In 1995, Walter Gehring and his colleagues were able to show experimentally that the abnormal expression of the *eyeless* gene in other parts of the fruit fly body could promote the formation of additional eyes. For example, using genetic engineering techniques, they were able to express the *eyeless* gene in the region where antennae should form. As seen in **Figure 26.22a**, this resulted in the formation of fruit fly eyes where antennae are normally found! Remarkably, the expression of the mouse *Pax6* gene in *Drosophila* can also cause the formation of eyes in unusual places. **Figure 26.22b** shows the formation of an eye on the leg of a fruit fly. This eye was caused by the expression of the mouse *Pax6* gene in this region.

The mouse *Pax6* gene switches on eye formation in *Drosophila*, but the eye produced is a *Drosophila* eye, not a mouse eye. This happens because the genes activated by the *Pax6* gene are all from the *Drosophila* genome. In *Drosophila*, the *eyeless* gene switches on a cascade involving 2,500 different genes required for eye morphogenesis. The *Pax6* gene and its *Drosophila* homolog, *eyeless*, are master control genes that promote the formation of an eye. In addition, Gehring has suggested that the eyes of *Drosophila* and mammals are evolutionarily derived from the modification of an eye that arose once during evolution. If *Drosophila* and mammalian eyes had arisen independently, the *Pax6* gene from mice would not be expected to induce the formation of eyes in *Drosophila*.

Since the initial discovery of the *Pax6* and *eyeless* genes, homologs of this gene have been discovered in many different species. In all cases where it has been tested, this gene directs eye development. The *Pax6* gene and its homologs encode a transcription factor protein that controls the expression of many different genes. Gehring and colleagues have hypothesized that the eyes from many different species all evolved from a common ancestral form consisting of, as proposed by Darwin, one photoreceptor cell and one pigment cell (**Figure 26.23**). As mentioned previously, such a very simple eye can accomplish some rudimentary form of vision by detecting light and its direction. Eyes such as these are still found in modern species such as the larvae of certain types of mollusks. Over the course of evolution, simple eyes were transformed into more complex types of eyes by modifications that resulted in the addition of more types of cells such as lens and muscle cells.

(a) Abnormal expression of *Drosophila eyeless* gene

(b) Abnormal expression of mouse *Pax6* gene in a fruit fly leg

FIGURE 26.22 **Formation of additional eyes in *Drosophila* due to the abnormal expression of a master control gene for eye morphogenesis. (a)** When the *Drosophila eyeless* gene is expressed in the antenna region, eyes are formed where antennae should be located. **(b)** When the mouse *Pax6* gene is expressed in the leg region of *Drosophila*, a small eye is formed there.

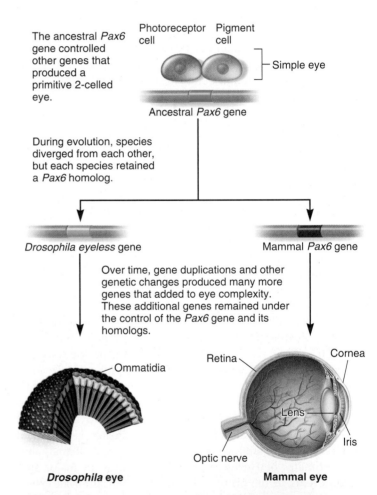

FIGURE 26.23 **Eye evolution.** In this diagram, genetic changes, under control of the ancestral *Pax6* gene, led to the evolution of different types of eyes. Ommatidia are units of the insect compound eye.

CONCEPTUAL SUMMARY

Biological evolution involves heritable changes in one or more characteristics in a population or species over the course of many generations. Charles Darwin's theory of evolution is based on two fundamental principles: genetic variation and natural selection. In the first part of this chapter, we were concerned with how these features of evolution result in the formation of new species. Biolo-gists use four different species concepts—the phylogenetic, biological, evolutionary, and ecological species concepts—to define and identify species. Speciation is often a branching process (clado-genesis), although anagenesis (transformation of a single species) occasionally occurs. Speciation may be allopatric, parapatric, or sympatric, depending on whether or not geographic barriers are present. Allopatric speciation, which involves geographic barriers, is thought to be the most widespread form of speciation. It can occur slowly, due to the gradual formation of a geographic barrier, or rapidly, due to the founder effect. Parapatric speciation is similar to allopatric speciation, except that the geographic barriers are less complete, with ever-decreasing levels of interbreeding taking place in hybrid zones. Sympatric speciation does not require geographic barriers. Instead, an abrupt genetic event may lead to reproductive isolation. In plants, the formation of polyploids is a common form of sympatric speciation. The fossil record suggests that speciation occurs through punctuated equilibrium, a relatively rapid process that punctuates long periods during which a species is in equilibrium with its environment. In contrast, gradualism proposes a slower but steady rate of evolution due to the accumulation of many small genetic changes.

Molecular evolution is the study of the molecular changes in the genetic material during evolution. Molecular data are increasingly used to construct phylogenetic trees, which are diagrams that describe the evolutionary relationships among different species. Homologous genes are derived from a common ancestral gene. The analysis of DNA sequences and deduced amino acid sequences of organisms are the most commonly used methods to make phylogenetic trees. Those species with greater similarity in genetic material generally are assumed to be closely related evolutionarily. The cladistic approach constructs a phylogenetic tree, called a cladogram, by considering the various possible pathways of evolution and then choosing the most plausible route. According to the neutral theory of evolution, most variation within and between species is due to the accumulation of neutral or nearly neutral mutations. Kimura and Ohta proposed five principles of molecular evolution that are consistent with our present knowledge of gene sequences. The accumulation of neutral mutations can be used as a molecular clock on which to measure evolutionary time.

In addition to random changes in gene sequences, other mechanisms contribute to the evolution of new species. For example, horizontal gene transfer is particularly important in the evolution of prokaryotic species and was probably very prevalent during the earliest stages of evolution, when all organisms were unicellular. Also, changes in chromosome structure and number may promote reproductive isolation.

A relatively new area of evolutionary studies involves the study of how developmental genes play a role in evolutionary processes, a field called evolutionary developmental biology (evo-devo). Researchers have identified several genes that control developmental pathways. For example, *BMP4* and *gremlin* are key genes that control webbed versus nonwebbed feet in birds. In addition, an increase in the number of *Hox* genes appears to be a critical factor that led to the development of animals with more complex body plans. Finally, we considered how the study of the *Pax6* gene, a master control gene, has revealed that all types of eyes most likely evolved from a single type of eye, perhaps a simple two-celled eye composed of a photoreceptor cell and a pigment cell.

EXPERIMENTAL SUMMARY

In a broad sense, evolutionary biologists would like to know how genetic changes have led to the phenotypic characteristics of present-day species. At the heart of this question is speciation: How do we explain the existence of so many different species? To answer this question, several species concepts can be applied. Biologists also try to understand how two different, but closely related, species have become reproductively isolated. By observing species in nature, they have identified several prezygotic and postzygotic mechanisms that prevent the production of viable, interspecies offspring. Researchers have also correlated reproductive isolation with experimentally observable genetic changes. For example, parapatric speciation may be associated with the accumulation of chromosomal inversions that prevent the production of fertile offspring within hybrid zones of contact. Sympatric speciation can occur via the formation of allotetraploids. This speciation mechanism can be confirmed by producing "artificial" polyploids.

Evolutionary biologists are also interested in the rates of evolutionary change. Experimentally, the fossil record suggests that it is more common for evolution to follow a pattern of punctuated equilibrium. According to this idea, a species is well adapted to its environment for a long period of time (the equilibrium). This equilibrium is punctuated by short periods in which the species undergoes rapid evolutionary change. These short periods involve abrupt genetic changes (e.g., allopolyploidism or new alleles that have a dramatic effect on phenotype), or there may be a short period of strong selective pressure (e.g., the founder effect, a new predator in the region, or a significant environmental change).

At the molecular level, evolution is due to alterations in DNA sequences and changes in chromosome structure and number. Geneticists have found a great amount of variation within most species. The analysis of DNA sequences is now the most common way to study molecular evolution. To construct a phylogenetic tree, researchers usually follow a cladistic approach in which species are compared with regard to shared derived characters and primitive characters. Often times, the analysis involves homologous gene sequences among different species. It has even been possible to analyze DNA sequences from extinct species.

PROBLEM SETS & INSIGHTS

Solved Problems

S1. A codon for leucine is UUA. A mutation causing a single-base substitution in a gene can change this codon in the transcribed mRNA into GUA (valine), AUA (isoleucine), CUA (leucine), UGA (stop), UAA (stop), UCA (serine), UUG (leucine), UUC (phenylalanine), or UUU (phenylalanine). According to the neutral theory, which of these mutations would you expect to see within the genetic variation of a natural population? Explain.

Answer: The neutral theory proposes that neutral mutations will accumulate to the greatest extent in a population. Leucine is a nonpolar amino acid. For a UUA codon, single-base changes of CUA and UUG are silent, so they would be the most likely to occur in a natural population. Likewise, conservative substitutions to other nonpolar amino acids such as isoleucine (AUA), valine (GUA), and phenylalanine (UUC and UUU) may not affect protein structure and function, so they may also occur and not be eliminated rapidly by natural selection. The polar amino acid serine (UCA) is a nonconservative substitution; one would predict that it is more likely to disrupt protein function. Therefore, it may be less likely to be found. Finally, the stop codons, UGA and UAA, would be expected to diminish or eliminate protein function, particularly if they occur early in the coding sequence. These types of mutations are selected against and, therefore, are not usually found in natural populations.

S2.   Explain why homologous genes have sequences that are similar but not identical.

Answer: Homologous genes are derived from the same ancestral gene. Therefore, as a starting point, they had identical sequences. Over time, however, each gene accumulates random mutations that the other homologous genes did not acquire. These random mutations change the gene from its original sequence. Therefore, much of the sequence between homologous genes remains identical, but some of the sequence will be altered due to the accumulation of independent random mutations.

S3.   Explain why plants are more likely to evolve by sympatric speciation compared to animals.

Answer: A common way for sympatric speciation to occur is by the formation of polyploids. For example, if one species is diploid ($2n$), nondisjunction could produce an individual that is tetraploid ($4n$). If the tetraploid individual was a plant and if it was monoecious (i.e., produces both pollen and egg cells), the plant could multiply to produce many tetraploid offspring. These offspring would be reproductively isolated from the diploid plants that are found in the same geographic area. This isolation occurs because the offspring of a cross between a diploid and tetraploid plant would be infertile. For example, if the pollen from a diploid plant fertilized the egg from a tetraploid plant, this would produce triploid offspring. The triploid offspring might be viable, but it is very likely that triploids would be sterile because they would produce highly aneuploid gametes. The gametes would be aneuploid because there are an odd number of homologous chromosomes. In this case, there would be three copies of each homologous chromosome, and these could not be equally distributed into gametes. Therefore, when a tetraploid is produced, it

can immediately create its own unique species that is reproductively isolated from other species.

In contrast, polyploidy rarely occurs in animals. Perhaps the main reason is because animals usually cannot tolerate polyploidy. Tetraploid animals typically die during early stages of development. In addition, most species of animals have male and female sexes. To develop a tetraploid species from a diploid species, nondisjunction would have to produce both a male and female offspring that could reproduce with each other. The chance of producing one tetraploid offspring is relatively rare. The chances of producing two tetraploid offspring that happen to be male and female and happen to mate with each other to produce many offspring would be extremely rare.

S4.   As described in Figure 26.18, evolution is associated with changes in chromosome structure and number. As seen here, chromosome 2 in humans is divided into two distinct chromosomes in chimpanzees, gorillas, and orangutans. In addition, chromosome 3 in the orangutan has a large inversion not found in the other three primates. Discuss the potential role of these types of changes in the evolution of these primate species.

Answer: As discussed in Chapter 8, changes in chromosome structure, such as inversions and balanced translocations, may not have any phenotypic effects. Likewise, the division of a single chromosome into two distinct chromosomes may not have any phenotypic effect as long as the total amount of genetic material remains the same. Overall, the types of changes in chromosome structure and number shown in Figure 26.18 may not have caused any changes in the phenotypes of primates. However, the changes would be expected to promote reproductive isolation. For example, if a gorilla mated with an orangutan, the offspring would be an inversion heterozygote for chromosome 3. As shown in Figure 8.10, crossing over during gamete formation in an inversion heterozygote may produce chromosomes that have too much or too little genetic material. This is particularly likely if the inversion is fairly large (such as the one shown in Figure 26.18). The inheritance of too much or too little genetic material is likely to be detrimental or even lethal. For this reason, the hybrid offspring of a gorilla and orangutan would probably not be fertile. (Note: In reality, there are several other reasons why interspecies matings between gorillas and orangutans do not produce viable offspring.)

Overall, the primary effect of changes in chromosome structure and number, like the ones shown in Figure 26.18, is to promote reproductive isolation. Once two populations become reproductively isolated, they will accumulate different mutations, and over the course of many generations, this will lead to two different species that have distinct characteristics.

It should be noted that changes in chromosome number in plants are more likely to have effects that abruptly lead to the formation of new species. This idea is discussed in solved problem S3.

Conceptual Questions

C1.   Discuss the two principles on which evolution is based.

C2.   Evolution, which involves genetic change in a population of organisms over time, is often described as the unifying theme in biology. Discuss how evolution is unifying at the molecular and cellular levels.

C3.   What is meant by the term reproductive isolation? Give several examples.

C4.   Briefly describe four different species concepts.

C5.   Would the following examples of reproductive isolation be considered a prezygotic or postzygotic mechanism?

A. Horses and donkeys can interbreed to produce mules, but the mules are infertile.

B. Three species of the orchid genus *Dendrobium* produce flowers 8 days, 9 days, and 11 days after a rainstorm. The flowers remain open for one day.

C. Two species of fish release sperm and eggs into seawater at the same time, but the sperm of one species do not fertilize the eggs of the other species.

D. Two tree frogs, *Hyla chrysoscelis* (diploid) and *H. versicolor* (tetraploid), can produce viable offspring, but the offspring are sterile.

C6. Distinguish between anagenesis and cladogenesis. Which type of speciation is more prevalent? Why?

C7. Describe three or more genetic mechanisms that may lead to the rapid evolution of a new species. Which of these genetic mechanisms are influenced by natural selection, and which are not?

C8. Explain the type of speciation (allopatric, parapatric, or sympatric) most likely to occur under each of the following conditions:

A. A pregnant female rat is transported by an ocean liner to a new continent.

B. A meadow containing several species of grasses is exposed to a pesticide that promotes nondisjunction.

C. In a very large lake containing several species of fish, the water level gradually falls over the course of several years. Eventually, the large lake becomes subdivided into smaller lakes, some of which are connected by narrow streams.

C9. Alloploids are created by crosses involving two different species. Explain why alloploids are reproductively isolated from the two original species from which they were derived. Explain why alloploids are usually sterile, whereas allotetraploids (containing a diploid set from each species) are commonly fertile.

C10. Discuss the evidence in favor of the punctuated equilibrium model of evolution. What mechanisms could account for this pattern of evolution? In contrast, what type of genetic changes are consistent with gradualism?

C11. Discuss whether the phenomenon of reproductive isolation applies to bacteria, which reproduce asexually. How would a geneticist divide bacteria into separate species?

C12. Discuss the major differences between allopatric, parapatric, and sympatric speciation.

C13. The following are two DNA sequences from homologous genes:

TTGCATAGGCATACCGTATGATATCGAAAACTAGAAAAATAGGGCGATAGCTA

GTATGTTATCGAAAAGTAGCAAAATAGGGCGATAGCTACCCAGACTACCGGAT

The two sequences, however, do not begin and end at the same location. Try to line them up according to their homologous regions.

C14. What is meant by the term molecular clock? How is this concept related to the neutral theory of evolution?

C15. Would the rate of deleterious or beneficial mutations be a good molecular clock? Why or why not?

C16. Which would you expect to exhibit a faster rate of evolutionary change, the nucleotide sequence of a gene or the amino acid sequence of the encoded polypeptide of the same gene? Explain your answer.

C17. When comparing the coding region of structural genes among closely related species, it is commonly found that certain regions of the gene have evolved more rapidly (i.e., have tolerated more changes in sequence) compared to other regions of the gene. Explain why different regions of a structural gene evolve at different rates.

C18. Plant seeds contain storage proteins that are encoded by plant genes. When the seed germinates, these proteins are rapidly hydrolyzed (i.e., the covalent bonds between amino acids within the polypeptides are broken), which releases amino acids for the developing seedling. Would you expect the genes that encode plant storage proteins to evolve slowly or rapidly compared to genes that encode enzymes? Explain your answer.

C19. Figure 26.12 shows a phylogenetic tree of all life on Earth based on 16S rRNA data. Based on your understanding of molecular genetics (in Chapter 26 and other chapters), describe three or more observations that suggest that all life-forms on Earth evolved from a common ancestor or group of ancestors.

C20. Take a look at the α-globin and β-globin sequences in Figure 26.6. Which sequences are more similar, the α-globin in humans and the α-globin in horses, or the α-globin in humans and the β-globin in humans? Based on your answer, would you conclude that the gene duplication that gave rise to the α-globin and β-globin genes occurred before or after the divergence of humans and horses? Explain your reasoning.

C21. Compare and contrast the neutral theory of evolution versus the Darwinian (i.e., selectionist) theory of evolution. Explain why the neutral theory of evolution is sometimes called non-Darwinian evolution.

C22. For each of the following examples, discuss whether it would be the result of neutral mutation or mutation that has been acted upon by natural selection, or both:

A. When comparing sequences of homologous genes, differences in the coding sequence are most common at the wobble base (i.e., the third base in each codon).

B. For a structural gene, the regions that encode portions of the polypeptide that are vital for structure and function are less likely to incur mutations compared to other regions of the gene.

C. When comparing the sequences of homologous genes, introns usually have more sequence differences compared to exons.

C23. As discussed in Chapter 24, genetic variation is prevalent in natural populations. This variation is revealed in the DNA sequencing of genes. According to the neutral theory of evolution, discuss the relative importance of natural selection against detrimental mutations, natural selection in favor of beneficial mutations, and neutral mutations in accounting for the genetic variation we see in natural populations.

C24. If you were comparing the karyotypes of species that are closely related evolutionarily, what types of similarities and differences would you expect to find?

C25. In the developing bud that gives rise to a hand in a human embryo, where would you expect the *gremlin* gene to be expressed?

C26. Discuss how *Hox* gene number is related to body complexity.

Experimental Questions

E1. Two populations of snakes are separated by a river. The snakes cross the river only on rare occasions. The snakes in the two populations look very similar to each other except that the members of the population on the eastern bank of the river have a yellow spot on the top of their head, while the members of the western population have an orange spot on the top of their head. Discuss two experimental methods that you might follow to determine whether the two populations are members of the same species or members of different species.

E2. Sympatric speciation by allotetraploidy has been proposed as a common mechanism for speciation. Let's suppose you were interested in the origin of certain grass species in Southern California. Experimentally, how would you go about determining if some of the grass species are the result of allotetraploidy?

E3. Two diploid species of closely related frogs, which we will call species A and species B, were analyzed with regard to genes that encode an enzyme called hexokinase. Species A has two distinct copies of this gene: *A1* and *A2*. In other words, this diploid species is *A1A1 A2A2*. The other species has three copies of the hexokinase gene, which we will call *B1*, *B2*, and *B3*. A diploid individual of species B would be *B1B1 B2B2 B3B3*. These hexokinase genes from the two species were subjected to DNA sequencing, and the percentage of sequence identity was compared among these genes. The results are shown here:

Percentage of DNA Sequence Identity

	A1	*A2*	*B1*	*B2*	*B3*
A1	100	62	54	94	53
A2	62	100	91	49	92
B1	54	91	100	67	90
B2	94	49	67	100	64
B3	53	92	90	64	100

If we assume that hexokinase genes were never lost in the evolution of these frog species, how many distinct hexokinase genes do you think there were in the most recent ancestor that preceded the divergence of these two species? Explain your answer. Also explain why species B has three distinct copies of this gene, whereas species A has only two.

E4. A researcher sequenced a portion of a bacterial gene and obtained the following sequence, beginning with the start codon, which is underlined:

<u>ATG</u> CCG GAT TAC CCG GTC CCA AAC AAA ATG
ATC GGC CGC CGA ATC TAT CCC

The bacterial strain that contained this gene has been maintained in the laboratory and grown serially for many generations. Recently, another person working in the laboratory isolated DNA from the bacterial strain and sequenced the same region. The following results were obtained.

<u>ATG</u> CCG GAT TAT CCG GTC CCA AAT AAA ATG
ATC GGC CGC CGA ATC TAC CCC

Explain why these sequencing differences may have occurred.

E5. F_1 hybrids between two species of cotton, *Gossypium barbadense* and *G. hirsutum*, are very vigorous plants. However, F_1 crosses produce many seeds that do not germinate and a low percentage of very weak F_2 offspring. Suggest two reasons for these observations.

E6. A species of antelope contains 20 chromosomes per set. The species is divided by a mountain range into two separate populations, which we will call the eastern and western population. When comparing the karyotypes between these two populations, it was discovered that the members of the eastern population are homozygous for a large inversion within chromosome 14. How would this inversion affect the interbreeding between the two populations? Could such an inversion play an important role in speciation?

E7. Explain why molecular techniques were needed as a way to provide evidence for the neutral theory of evolution.

E8. Prehistoric specimens often contain minute amounts of ancient DNA. What technique can be used to increase the amount of DNA in an older sample? Explain how this technique is performed and how it increases the amount of a specific region of DNA.

E9. In the experiment of Figure 26.15, explain how we know that the kiwis are more closely related to the emu and cassowary than to the moas. Cite particular regions in the sequences that support your answer.

E10. In Chapter 20, we learned about a technique called fluorescence *in situ* hybridization (FISH), during which a labeled piece of DNA is hybridized to a set of chromosomes. Let's suppose that we cloned a piece of DNA from *G. pubescens* (see Figure 26.3) and used it as a labeled probe for *in situ* hybridization. What would you expect to happen if we hybridized it to the *G. speciosa*, the natural *G. tetrahit*, or the artificial *G. tetrahit* strains? Describe your expected results.

E11. A team of researchers has obtained a dinosaur bone (*Tyrannosaurus rex*) and has attempted to extract ancient DNA from it. Using primers to the 12S rRNA gene, they have used PCR and obtained a DNA segment that yields a sequence homologous to crocodile DNA. Other scientists are skeptical that this sequence is really from the dinosaur. Instead, they believe that it may be due to contamination from more recent DNA, such as the remains of a reptile that lived much more recently. What criteria might you use to establish the credibility of the dinosaur sequence?

E12. Discuss how the principle of parsimony can be used in a cladistics approach of constructing a phylogenetic tree.

E13. As discussed in this chapter and Chapter 24, genes are sometimes transferred between different species via horizontal gene transfer. Discuss how horizontal gene transfer might give misleading results when constructing a phylogenetic tree. How could you overcome this problem?

E14. If a researcher used genetic engineering techniques to express the *Drosophila eyeless* gene in the embryo at the region that will become the tip of the mouse's tail, what results would you expect in the resulting offspring?

Questions for Student Discussion/Collaboration

1. The raw material for evolution is random mutation. Discuss whether or not you view evolution as a random process.

2. Compare the forms of speciation that are slow with those that occur more rapidly. Make a list of the slow and fast forms. With regard to mechanisms of genetic change, what features do slow and rapid speciation have in common? What features are different?

3. Do you think that Darwin would object to the neutral theory of evolution?

Note: All answers appear at the website for this textbook; the answers to even-numbered questions are in the back of the textbook.

www.mhhe.com/brookergenetics3e

Visit the Online Learning Center for practice tests, answer keys, and other learning aids for this chapter. Enhance your understanding of genetics with our interactive exercises, quizzes, animations, and much more.

A

A.1 METHODS OF CELL GROWTH

Researchers often grow cells in a laboratory as a way to study their properties. This is known as a **cell culture.** Cell culturing offers several technical advantages. The primary advantage is that the growth medium is defined and can be controlled. Minimal growth medium contains the bare essentials for cell growth: salts, a carbon source, an energy source, essential vitamins, amino acids, and trace elements. In their experiments, geneticists often compare strains that can grow in minimal media and mutant strains that cannot grow unless the medium is supplemented with additional components. A rich growth medium contains many more components than are required for growth.

Researchers also add substances to the culture medium for other experimental reasons. For example, radioactive isotopes can be added to the culture medium to radiolabel cellular macromolecules. In addition, an experimenter could add a hormone to the growth medium and then monitor the cells' response to the hormone. In all of these cases, cell culturing is advantageous because the experimenter can control and vary the composition of the growth medium.

The first step in creating a cell culture is the isolation of a cell population that the researchers wish to study. For bacteria, such as *Escherichia coli,* and eukaryotic microorganisms, such as yeast and *Neurospora,* the researchers simply obtain a sample of cells from a colleague or a stock center. For animal or plant tissues, the procedure is a bit more complicated. When cells are contained within a complex tissue, they must first be dispersed by treating the tissue with agents that separate it into individual cells to create a cell suspension.

Once a desired population of cells has been obtained, researchers can grow them in a laboratory (i.e., in vitro) either suspended in a liquid growth medium or attached to a solid surface such as agar. Both methods have been used commonly in the experiments considered throughout this textbook. Liquid culture is often used when researchers want to obtain a large quantity of cells and isolate individual cellular components, such as nuclei or DNA. By comparison, **Figure A.1** shows animal cells and bacteria

(a) Fibroblast (animal cell) culture

(b) Bacterial colonies

FIGURE A.1   Growth of cells on solid growth media. (**a**) This micrograph shows fibroblasts growing as a monolayer on a solid growth medium. (**b**) Bacterial cells form colonies that are a clonal population of cells derived from a single cell.

cells that are grown on solid growth media. As shown in Figure 22.9, solid media are used to study cancer cells, because such cells can be distinguished by the formation of foci in which malignant cells pile up on top of each other. In gene cloning experiments with bacteria and yeast, solid media are also used. Each colony of cells is a clone of cells that is derived from a single cell that divided to produce many cells (**Figure A.1b**). As discussed in Chapter 18, a solid medium is used in the isolation of individual clones that contain a desired gene.

A.2 MICROSCOPY

Microscopy is a technique to observe things that are not visible (or are hardly visible) with the naked eye. A key concept in microscopy is **resolution,** which is defined as the minimum distance between two objects that can be seen as separate from each other. The ability to resolve two points as being separate depends on several factors, including the wavelength of the illumination source (light or electron beam), the medium in which the sample is immersed, and the structural features of the microscope (which are beyond the scope of this textbook).

As shown in **Figure A.2**, two widely used kinds of microscopes are the optical (light) microscope and the transmission electron microscope (TEM). The light microscope is used to resolve cellular structures to a limit of approximately 0.3 μm. (For comparison, a typical bacterium is about 1 μm long.) At this resolution, the individual cell organelles in eukaryotic cells can be discerned easily, and chromosomes are also visible. Karyotyping is accomplished via light microscopy after the chromosomes have been treated with stains. A variation of light microscopy known as fluorescence microscopy is often used to highlight a particular feature of a chromosome or cellular structure. The technique of fluorescence *in situ* hybridization (FISH; see Chapter 20) makes

use of this type of microscope. Also, optical modifications in certain light microscopes (e.g., phase contrast and differential interference) can be used to exaggerate the differences in densities between neighboring cells or cell structures. These kinds of light microscopes are useful in monitoring cell division in living (unstained) cells or in transparent worms (as in Figure 23.15).

The structural details of large macromolecules such as DNA and ribosomes are not observable by light microscopy. The coarse topology of these macromolecules can be determined by electron microscopy. Electron microscopes have a limit of resolution of about 2 nm, which is about 100 times finer than the best light microscopes. The primary advantage of electron microscopy over light microscopy is its better resolution. Disadvantages include a much higher expense and more extensive sample preparation. In transmission electron microscopy, the sample is bombarded with an electron beam. This requires that the sample be dried, fixed, and usually coated with a heavy metal that absorbs electrons.

A.3 SEPARATION METHODS

Biologists often wish to take complex systems and separate them into less complex components. For example, the cells within a complex tissue can be separated into individual cells, or the macromolecules within cells can be separated from the other cellular components. In this section, we will focus primarily on methods aimed at separating and purifying macromolecules.

Disruption of Cellular Components

In many experiments described in this textbook, researchers have obtained a sample of cells and then wish to isolate particular components from the cells. For example, a researcher may want to purify a protein that functions as a transcription factor. To do so,

(a) Light microscope

- Ocular lens
- Beam of light
- Objective lens
- Specimen
- Condenser lens
- Light source

(b) Transmission electron microscope

- Tungsten filament (electron source)
- Beam of electrons
- Condenser lens
- Specimen
- Objective lens
- Projector lens
- Binoculars
- Final image on photographic plate or screen

FIGURE A.2   Design of (a) optical (light) and (b) transmission electron microscopes.

the researcher would begin with a sample of cells that synthesize this protein and then break open the cells using one of the methods described in **Table A.1**. In eukaryotes, the breakage of cells releases the soluble proteins from the cell; it also dissociates the cell organelles that are bounded by membranes. This mixture of proteins and cell organelles can then be isolated and purified by centrifugation and chromatographic methods, which are described next.

Centrifugation

Centrifugation is a method commonly used to separate cell organelles and macromolecules. A **centrifuge** contains a motor, which causes a rotor holding centrifuge tubes to spin very rap-

idly. As the rotor spins, particles move toward the bottom of the centrifuge tube; the rate at which they move depends on several factors, including their densities, sizes, and shapes and the viscosity of the medium. The rate at which a macromolecule or cell organelle sediments to the bottom of a centrifuge tube is called its **sedimentation coefficient,** which is normally expressed in Svedberg units (S). A sedimentation coefficient has the units of seconds: $1\,S = 1 \times 10^{-13}$ seconds.

When a sample contains a mixture of macromolecules or cell organelles, it is likely that different components will sediment at different rates. This phenomenon, known as **differential centrifugation,** is shown in **Figure A.3**. As seen here, particles with large sedimentation coefficients reach the bottom of the tube more quickly than those with smaller coefficients. Researchers can follow two different strategies that use differential centrifugation as a separation technique. One way is to separate the **supernatant** from the **pellet** following centrifugation. The pellet is a collection of particles found at the bottom of the tube, and the supernatant is the liquid found above the pellet. In Figure A.3, when the experimenter had subjected the sample to a low-speed spin, most of the particles with large sedimentation coefficients would be found in the pellet while most of the particles with small and intermediate coefficients would be found in the supernatant. A high-speed spin of the supernatant would then separate the small and intermediate particles. Therefore, differential centrifugation provides a way to segregate these three types of particles.

A second way to separate particles using centrifugation is to collect fractions. A **fraction** is a portion of the liquid contained

TABLE **A.1**	
Common Methods of Cell Disruption	
Method	**Description**
Sonication	The exposure of cells to intense sound waves, which breaks the cell membranes.
French press	The passage of cells through a small aperture under high pressure, which breaks the cell membranes and cell wall.
Homogenization	Cells are placed in a tube that contains a pestle. When the pestle is spun, the cells are squeezed through the small space between the pestle and the glass wall of the tube, thereby breaking them.
Osmotic shock	The transfer of cells into a hypo-osmotic medium. The cells take up water and eventually burst.

- Particles with large sedimentation coefficients
- Particles with intermediate sedimentation coefficients
- Particles with small sedimentation coefficients

FIGURE A.3　The method of differential centrifugation. A sample containing a mixture of particles with different sedimentation coefficients is placed in a centrifuge tube. The tube is subjected to a low-speed spin that pellets the particles with large sedimentation coefficients. After a high-speed spin of the supernatant, the particles with an intermediate sedimentation coefficient are found in the pellet, while those with a small sedimentation coefficient are in the liquid supernatant.

within a centrifuge tube. The collection of fractions is done when the solution within the centrifuge tube contains a gradient. For example, as shown in **Figure A.4**, the solution at the top of the tube has a lower concentration of cesium chloride (CsCl) than that at the bottom. In this experiment, a sample is layered on the top of the gradient and then centrifuged. In this example, the DNA and RNA separate from each other, because they have different sedimentation coefficients. The experimenter then punctures the bottom of the tube and collects fractions. The DNA fragments, which are heavier, come out of the tube in the earlier fractions; the RNA molecules will be collected in later fractions.

A type of gradient centrifugation that may also be used to separate macromolecules and organelles is **equilibrium density centrifugation.** In this method, the particles will sediment through the gradient, reaching a position where the density of the particle matches the density of the solution. At this point, the particle is at equilibrium and does not move any farther toward the bottom of the tube.

Chromatography and Gel Electrophoresis

Chromatography is a method to separate different macromolecules and small molecules based on their chemical and physical properties. In this method, a sample is dissolved in a liquid solvent and exposed to some type of matrix, such as a column containing beads or a thin strip of paper. The degree to which the molecules interact with the matrix depends on their chemical and physical characteristics. For example, a positively charged molecule will bind tightly to a negatively charged matrix, whereas a neutral molecule will not.

FIGURE A.4　Gradient centrifugation and the collection of fractions.

Figure A.5 illustrates how column chromatography can be used to separate molecules that differ with respect to charge. Prior to this experiment, a column is packed with beads that are positively charged. There is plenty of space between the beads for molecules to flow from the top of the column to the bottom. However, if the molecules are negatively charged, they will spend some of their time binding to the positive charges on the surface of the beads. In the example shown in Figure A.5, the red proteins are positively charged and, therefore, flow rapidly from the top of the column to the bottom. They emerge in the fractions that are collected early in this experiment. The blue proteins, however, are negatively charged and tend to bind to the beads. The binding of the blue proteins to the beads can be disrupted by changing the ionic strength or pH of the solution that is added to the column. Eventually, the blue proteins will be eluted (i.e., leave the column) in later fractions.

Load sample onto top of column.

— Negatively charged protein

— Positively charged protein

— Collect fractions.

1 2 3 4 5 6 7 8

Add more buffer solution.

Positively charged gel bead

1 2 3 4 5 6 7 8

└ Positively charged protein

Add buffer at high pH or low ionic strength.

1 2 3 4 5 6 7 8

Negatively charged protein

FIGURE A.5   Ion-exchange chromatography.

Researchers use many variations of chromatography to separate molecules and macromolecules. The type shown in Figure A.5 is called ion-exchange chromatography, because its basis for separation depends on the charge of the molecules. In another type of column chromatography, known as gel filtration chromatography, the beads are porous. Small molecules are temporarily trapped within the beads, while large molecules flow between the beads. In this way, gel filtration separates molecules on the basis of size. To separate different types of macromolecules, such as proteins, researchers may use another type of bead; this bead has a preattached molecule that binds specifically to the protein they want to purify. For example, if a transcription factor binds a particular DNA sequence as part of its function, the beads within a column may have this DNA sequence preattached to them. Therefore, the transcription factor will bind tightly to the DNA attached to these beads, while all other proteins will be eluted rapidly from the column. This form of chromatography is called affinity chromatography, because the beads have a special affinity for the macromolecule of interest.

Besides column chromatography, in which beads are packed into a column, a matrix can be made in other ways. In paper chromatography, molecules pass through a matrix composed of paper. The rate of movement of molecules through the paper depends on their degree of interaction with the solvent and paper. In thin-layer chromatography, a matrix is spread out as a very thin layer on a rigid support such as a glass plate. In general, paper and thin-layer chromatography are effective at separating small molecules, whereas column chromatography is used to separate macromolecules such as DNA fragments or proteins.

Gel electrophoresis combines chromatography and electrophoresis to separate molecules and macromolecules. As its name

suggests, the matrix used in gel electrophoresis is composed of a gel. As shown in **Figure A.6**, samples are loaded into wells at one end of the gel, and an electric field is applied across the gel. This electric field causes charged molecules to migrate from one side of the gel to the other. The migration of molecules in response to an electric field is called **electrophoresis.** In the examples of gel electrophoresis found in this textbook, the macromolecules within the sample migrate toward the positive end of the gel. In most forms of gel electrophoresis, a mixture of macromolecules is separated according to their molecular masses. Small proteins or DNA fragments move to the bottom of the gel more quickly than larger ones. Because the samples are loaded in rectangular wells at the top of the gel, the molecules within the sample are separated into bands within the gel. These bands of separated macromolecules can be visualized with stains. For example, ethidium bromide is a stain that binds to DNA and RNA and can be seen under ultraviolet light.

The two most commonly used gels are polymers made from acrylamide or agarose. Proteins typically are separated on polyacrylamide gels, whereas DNA fragments are separated on agarose gels. Occasionally, researchers use polyacrylamide gels to separate DNA fragments that are relatively small (namely, less than 1,000 bp in length).

A.4 METHODS TO MEASURE CONCENTRATIONS OF MOLECULES AND TO DETECT RADIOISOTOPES AND ANTIGENS

To understand the structure and function of cells, researchers often need to detect the presence of molecules and macromolecules and to measure their concentrations. In this section, we will consider a variety of methods to detect and measure the concentrations of biological molecules and macromolecules.

Spectroscopy

Macromolecules found in living cells, such as proteins, DNA, and RNA, are fairly complex molecules that can absorb radiation (i.e., light). Likewise, small molecules such as amino acids and nucleotides can also absorb light. A device known as a **spectrophotometer** is used by researchers to determine how much radiation at various wavelengths a sample can absorb. The amount of absorption can be used to determine the concentration of particular molecules within a sample, because each type of molecule or macromolecule has its own characteristic wavelength(s) of absorption, called its absorption spectrum.

A spectrophotometer typically has two light sources, which can emit ultraviolet or visible light. As shown in **Figure A.7**, the light source is passed through a monochromator, which emits the light at a desired wavelength. This incident light then strikes a sample contained within a cuvette. Some of the incident light is absorbed, and some is not. The amount and wavelengths of light that are absorbed depend on the concentration and structures of the molecules and macromolecules in the cuvette. The

Load samples into wells.

Apply an electric field.

Longer time

Higher-mass particles

Lower-mass particles

(a) Separation of a mixture of particles by gel electrophoresis

Upper buffer solution

Gel is sandwiched between 2 glass plates.

Lower buffer solution

Power supply

Electrode

Gel

Electrode

(b) Apparatus used in gel electrophoresis

FIGURE A.6 Acrylamide gel electrophoresis of DNA fragments.

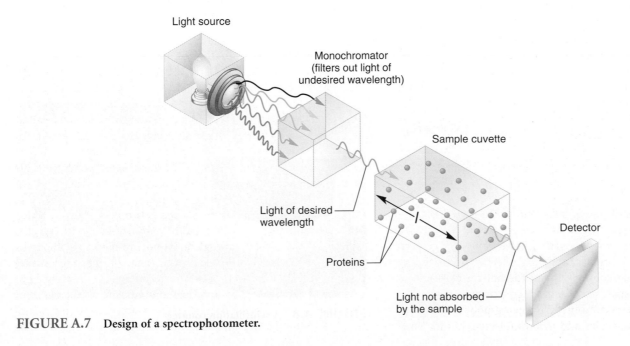

Light source

Monochromator (filters out light of undesired wavelength)

Sample cuvette

Light of desired wavelength

Detector

Proteins

Light not absorbed by the sample

FIGURE A.7 Design of a spectrophotometer.

unabsorbed light passes through the sample and is detected by the spectrophotometer. The amount of light that strikes the detector is subtracted from the amount of incident light, yielding the measure of absorption. In this way, the spectrophotometer provides an absorption reading for the sample. This reading can be used to calculate the concentration of particular molecules or macromolecules in a sample.

Detection of Radioisotopes

A radioisotope is an unstable form of an atom that decays to a more stable form by emitting α, β, or γ rays, which are types of ionizing radiation. In research, radioisotopes that are β and/or γ emitters are commonly used. A β ray is an emitted electron, and a γ ray is an emitted photon. Some radioisotopes commonly used in biological experiments are shown in **Table A.2**.

Experimentally, radioisotopes are used because they are easy to detect. Therefore, if a particular compound is radiolabeled, its presence can be detected specifically throughout the course of the experiment. For example, if a nucleotide is radiolabeled with ^{32}P, a researcher can determine whether the isotope becomes incorporated into newly made DNA or whether it remains as the free nucleotide. Researchers commonly use two different methods to detect radioisotopes: scintillation counting and autoradiography.

The technique of **scintillation counting** permits a researcher to count the number of radioactive emissions from a sample containing a population of radioisotopes. In this approach, the sample is dissolved in a solution (called the scintillant) that contains organic solvents and one or more compounds known as fluors. When radioisotopes emit ionizing radiation, the energy is absorbed by the fluors in the solvent. This excites the fluor molecules, causing their electrons to be boosted to higher energy levels. The excited electrons return to lower, more stable

energy levels by releasing photons of light. When a fluor is struck by ionizing radiation, it also absorbs the energy and then releases a photon of light within a particular wavelength range. The role of a device known as a scintillation counter is to count the photons of light emitted by the fluor. **Figure A.8** shows a scintillation counter. To use this device, a researcher dissolves her or his sample in a scintillant and then places the sample in a scintillation vial. The vial is then placed in the scintillation counter, which detects the amount of radioactivity. The scintillation counter has a digital meter that displays the amount of radioactivity in the sample, and it also provides a printout of the amount of radioactivity in counts per minute. A scintillation counter contains several rows for the loading and analysis of many scintillation vials. After they have been loaded, the scintillation counter will count the amount of radioactivity in each vial and provide the researcher with a printout of the amount of radioactivity in each vial.

A second way to detect radioisotopes is via **autoradiography**. This technique is not as quantitative as scintillation counting, because it does not provide the experimenter with a precise measure of the amount of radioactivity in counts per minute. However, autoradiography has the great advantage that it can detect the location of radioisotopes as they are found in macromolecules or cells. For example, autoradiography is used to detect a particular band on a gel or to map the location of a gene within an intact chromosome.

To conduct autoradiography, a sample containing a radioisotope is fixed and usually dried. If it is a cellular sample, it also may be thin sectioned. The sample is then pressed next to X-ray film (in the dark) and placed in a lightproof cassette. When a radioisotope decays, it will emit a β or γ ray, which may strike a thin layer of photoemulsion next to the film. The photoemulsion contains silver salts such as AgBr. When a radioactive particle is emitted and strikes the photoemulsion, a silver grain is deposited on the film. This produces a dark spot on the film, which correlates with the original location of the radioisotope in the sample. In this way, the dark image on the film reveals the location(s) of the radioisotopes in the sample. Figure 11.4b shows how autora-

---

TABLE A.2

Some Useful Isotopes in Genetics

Isotope	Stable or Radioactive	Emission	Half-life
^2H	Stable		
^3H	Radioactive	β	12.3 years
^{13}C	Stable		
^{14}C	Radioactive	β	5,730 years
^{15}N	Stable		
^{18}O	Stable		
^{24}Na	Radioactive	β (and γ)	15 hours
^{32}P	Radioactive	β	14.3 days
^{35}S	Radioactive	β	87.4 days
^{45}Ca	Radioactive	β	164 days
^{59}Fe	Radioactive	β (and γ)	45 days
^{131}I	Radioactive	β (and γ)	8.1 days

FIGURE A.8   A scintillation counter.

diography can be used to visualize the process of bacterial chromosome replication. In this case, radiolabeled nucleotides were incorporated into the DNA, making it possible to picture the topology of the chromosome as two replication forks pass around the circular chromosome.

Detection of Antigens by Radioimmunoassay

Antibodies, also known as **immunoglobulins,** are proteins that are used to ward off infection by foreign substances; they are produced by cells of the immune system. Antibodies bind to structures on the surface of foreign substances known as **epitopes;** the foreign substance is called an **antigen.** A particular antibody binds to a particular antigen with a very high degree of specificity. For this reason, antibodies have been used extensively by researchers to detect particular antigens. For example, a human protein such as hemoglobin can be injected into a rabbit. Human hemoglobin is a foreign substance in the rabbit's bloodstream. Therefore, the rabbit will make antibodies that specifically recognize human hemoglobin and are designed to destroy it. Researchers can isolate and purify these antibodies from a sample of the rabbit's blood and then use them to detect human hemoglobin in their experiments.

A **radioimmunoassay** is a method to measure the amount of an antigen in a biological sample. The steps in this method are shown in **Figure A.9a.** The researcher begins with two tubes that have a known amount of radiolabeled antigen (shown in blue). An unknown amount of the same antigen, which is not radiolabeled (shown in orange), is added to the tube on the right. The nonradiolabeled antigen comes from a biological sample; the goal of this experiment is to determine how much of this antigen is contained within the sample. Next, a known amount of antibody is added to each of the two tubes. The amount of the antibody is less than the amount of the antigen, so the nonlabeled and radiolabeled antigens compete with each other for binding to the antibody. After binding, a precipitating agent such as an anti-immunoglobulin antibody is added, and the precipitate is centrifuged to the bottom of the tube. The radioactivity in the precipitate is then determined by scintillation counting.

To calculate the amount of antigen in the sample being assayed, the researcher must determine the percentage of antibody that has bound to nonlabeled antigen. To do so, a second component of the experiment is to develop a standard curve in which a fixed amount of radiolabeled antigen is mixed with varying amounts of unlabeled antigen (**Figure. A.9b**). Using this standard curve, a researcher can determine how much antigen is found in the unknown sample. For example, as shown in the dashed line, if the unknown sample had about 45% of the antibody bound, then the concentration of antigen in the sample would be between 50 and 75 nanomolar.

Radioimmunoassays are used to determine the concentrations of many different kinds of antigens. This includes small molecules such as hormones (as in Figure 19.1) or macromolecules such as proteins.

(a) A radioimmunoassay

(b) Standard curve

FIGURE A.9 **The method of radioimmunoassay (a) and the construction of a standard curve (b).** In (**b**), the dashed line corresponds to the amount of antigen (Ag) bound by an unknown sample. This amounts to a concentration between 50 and 75 nanomolar of antigen.

SOLUTIONS TO EVEN-NUMBERED PROBLEMS

::

CHAPTER 1

Conceptual Questions

C2. A chromosome is a very long polymer of DNA. A gene is a specific sequence of DNA within that polymer; the sequence of bases creates a gene and distinguishes it from other genes. Genes are located in chromosomes, which are found within living cells.

C4. At the molecular level, a gene (a sequence of DNA) is first transcribed into RNA. The genetic code within the RNA is used to synthesize a protein with a particular amino acid sequence. This second process is called translation.

C6. Genetic variation involves the occurrence of genetic differences within members of the same species or different species. Within any population, variation may occur in the genetic material. Variation may occur in particular genes so that some individuals carry one allele and other individuals carry a different allele. An example would be differences in coat color among mammals. There also may be variation in chromosome structure and number. In plants, differences in chromosome number can affect disease resistance.

C8. You could pick almost any trait. For example, flower color in petunias would be an interesting choice. Some petunias are red and others are purple. There must be different alleles in a flower color gene that affect this trait in petunias. In addition, the amount of sunlight, fertilizer, and water also affects the intensity of flower color.

C10. A DNA sequence is a sequence of nucleotides. Each nucleotide may have one of four different bases (i.e., A, T, G, or C). When we speak of a DNA sequence, we focus on the sequence of bases.

C12. A. A gene is a segment of DNA. For most genes, the expression of the gene results in the production of a functional protein. The functioning of proteins within living cells affects the traits of an organism.

B. A gene is a segment of DNA that usually encodes the information for the production of a specific protein. Genes are found within chromosomes. Many genes are found within a single chromosome.

C. An allele is an alternative version of a particular gene. For example, suppose a plant has a flower color gene. One allele could produce a white flower, while a different allele could produce an orange flower. The white allele and orange allele are alleles of the flower color gene.

D. A DNA sequence is a sequence of nucleotides. The information within a DNA sequence (which is transcribed into an RNA sequence) specifies the amino acid sequence within a protein.

C14. A. How genes and traits are transmitted from parents to offspring

B. How the genetic material functions at the molecular and cellular levels

C. Why genetic variation exists in populations, and how it changes over the course of many generations

Experimental Questions

E2. This would be used primarily by molecular geneticists. The sequence of DNA is a molecular characteristic of DNA. In addition, as we will learn throughout this textbook, the sequence of DNA is interesting to transmission and population geneticists as well.

E4. A. Transmission geneticists. Dog breeders are interested in how genetic crosses affect the traits of dogs.

B. Molecular geneticists. This is a good model organism to study genetics at the molecular level.

C. Both transmission geneticists and molecular geneticists. Fruit flies are easy to cross and study the transmission of genes and traits from parents to offspring. Molecular geneticists have also studied many genes in fruit flies to see how they function at the molecular level.

D. Population geneticists. Most wild animals and plants would be the subject of population geneticists. In the wild, you cannot make controlled crosses. But you can study genetic variation within populations and try to understand its relationship to the environment.

E. Transmission geneticists. Agricultural breeders are interested in how genetic crosses affect the outcome of traits.

CHAPTER 2

Conceptual Questions

C2. In the case of plants, cross-fertilization occurs when the pollen and eggs come from different plants, whereas in self-fertilization, they come from the same plant.

C4. A homozygote that has two copies of the same allele

C6. Diploid organisms contain two copies of each type of gene. When they make gametes, only one copy of each gene is found in a gamete. Two alleles cannot stay together within the same gamete.

C8. Genotypes: 1:1 Tt and tt

Phenotypes: 1:1 Tall and dwarf

C10. c is the recessive allele for constricted pods; Y is the dominant allele for yellow color. The cross is $ccYy \times CcYy$. Follow the directions for setting up a Punnett square, as described in Chapter 2. The genotypic ratio is 2 $CcYY$: 4 $CcYy$: 2 $Ccyy$: 2 $ccYY$: 4 $ccYy$: 2 $ccyy$. This 2:4:2:2:4:2 ratio could be reduced to a 1:2:1:1:2:1 ratio.

The phenotypic ratio is 6 smooth pods, yellow seeds : 2 smooth pods, green seeds : 6 constricted pods, yellow seeds : 2 constricted pods, green seeds. This 6:2:6:2 ratio could be reduced to a 3:1:3:1 ratio.

C12. Offspring with a nonparental phenotype are consistent with the idea of independent assortment. If two different traits were always transmitted together as unit, it would not be possible to get nonparental phenotypic combinations. For example, if a true-breeding parent had two dominant traits and was crossed to a true-breeding parent having the two recessive traits, the F_2 generation could not have offspring with one recessive and one dominant phenotype. However, because independent assortment can occur, it is possible for F_2 offspring to have one dominant and one recessive trait.

C14. A. Barring a new mutation during gamete formation, the chance is 100% because they must be heterozygotes in order to produce a child with a recessive disorder.

    B. Construct a Punnett square. There is a 50% chance of heterozygous children.

    C. Use the product rule. The chance of being phenotypically normal is 0.75 (i.e., 75%), so the answer is $0.75 \times 0.75 \times 0.75 = 0.422$, which is 42.2%.

    D. Use the binomial expansion equation, where $n = 3$, $x = 2$, $p = 0.75$, $q = 0.25$. The answer is 0.422, or 42.2%.

C16. First construct a Punnett square. The chances are 75% of producing a solid pup and 25% of producing a spotted pup.

    A. Use the binomial expansion equation, where $n = 5$, $x = 4$, $p = 0.75$, $q = 0.25$. The answer is 0.396 = 39.6% of the time.

    B. You can use the binomial expansion equation for each litter. For the first litter, $n = 6$, $x = 4$, $p = 0.75$, $q = 0.25$; for the second litter, $n = 5$, $x = 5$, $p = 0.75$, $q = 0.25$. Because the litters are in a specified order, we use the product rule and multiply the probability of the first litter times the probability of the second litter. The answer is 0.070, or 7.0%.

    C. To calculate the probability of the first litter, we use the product rule and multiply the probability of the first pup (0.75) times the probability of the remaining four. We use the binomial expansion equation to calculate the probability of the remaining four, where $n = 4$, $x = 3$, $p = 0.75$, $q = 0.25$. The probability of the first litter is 0.316. To calculate the probability of the second litter, we use the product rule and multiply the probability of the first pup (0.25) times the probability of the second pup (0.25) times the probability of the remaining five. To calculate the probability of the remaining five, we use the binomial expansion equation, where $n = 5$, $x = 4$, $p = 0.75$, $q = 0.25$. The probability of the second litter is 0.025. To get the probability of these two litters occurring in this order, we use the product rule and multiply the probability of the first litter (0.316) times the probability of the second litter (0.025). The answer is 0.008, or 0.8%.

    D. Because this is a specified order, we use the product rule and multiply the probability of the firstborn (0.75) times the probability of the second born (0.25) times the probability of the remaining four. We use the binomial expansion equation to calculate the probability of the remaining four pups, where $n = 4$, $x = 2$, $p = 0.75$, $q = 0.25$. The answer is 0.040, or 4.0%.

C18. A. Use the product rule:

      $(^1/_4)(^1/_4) = ^1/_{16}$

    B. Use the binomial expansion equation:

      $n = 4$, $p = ^1/_4$, $q = ^3/_4$, $x = 2$

      $P = 0.21 = 21\%$

    C. Use the product rule:

      $(^1/_4)(^3/_4)(^3/_4) = 0.14$, or 14%

C20. A. $^1/_4$

    B. 1, or 100%

    C. $(^3/_4)(^3/_4)(^3/_4) = ^{27}/_{64} = 0.42$, or 42%

    D. Use the binomial expansion equation, where

      $n = 7$, $p = ^3/_4$, $q = ^1/_4$, $x = 3$

      $P = 0.058$, or 5.8%

    E. The probability that the first plant is tall is $^3/_4$. To calculate the probability that among the next four, any two will be tall, we use the binomial expansion equation, where $n = 4$, $p = ^3/_4$, $q = ^1/_4$, and $x = 2$.

      The probability P equals 0.21.

      To calculate the overall probability of these two events:

      $(^3/_4)(0.21) = 0.16$, or 16%

C22. It violates the law of segregation because two copies of one gene are in the gamete. The two alleles for the A gene did not segregate from each other.

C24. Based on this pedigree, it is likely to be dominant inheritance because an affected child always has an affected parent. In fact, it is a dominant disorder.

C26. It is impossible for the F_1 individuals to be true-breeding because they are all heterozygotes.

C28. 2 TY, tY, 2 Ty, ty, TTY, TTy, 2 TtY, 2 Tty

    It may be tricky to think about, but you get 2 TY and 2 Ty because either of the two T alleles could combine with Y or y. Also, you get 2 TtY and 2 Tty because either of the two T alleles could combine with t and then combine with Y or y.

C30. The genotype of the F_1 plants is $Tt\ Yy\ Rr$. According to the laws of segregation and independent assortment, the alleles of each gene will segregate from each other, and the alleles of different genes will randomly assort into gametes. A $Tt\ Yy\ Rr$ individual could make eight types of gametes: TYR, TyR, Tyr, TYr, tYR, tyR, tYr, and tyr, in equal proportions (i.e., $^1/_8$ of each type of gamete). To determine genotypes and phenotypes, you could make a large Punnett square that would contain 64 boxes. You would need to line up the eight possible gametes across the top and along the side and then fill in the 64 boxes. Alternatively, you could use one of the two approaches described in solved problem S3. The genotypes and phenotypes would be

1 *TT YY RR*

2 *TT Yy RR*

2 *TT YY Rr*

2 *Tt YY RR*

4 *TT Yy Rr*

4 *Tt Yy RR*

4 *Tt YY Rr*

8 *Tt Yy Rr*   = 27 tall, yellow, round

1 *TT yy RR*

2 *Tt yy RR*

2 *TT yy Rr*

4 *Tt yy Rr*   = 9 tall, green, round

1 *TT YY rr*

2 *TT Yy rr*

2 *Tt YY rr*

4 *Tt Yy rr*   = 9 tall, yellow, wrinkled

1 *tt YY RR*

2 *tt Yy RR*

2 *tt YY Rr*

4 *tt Yy Rr*   = 9 dwarf, yellow, round

1 *TT yy rr*

2 *Tt yy rr*   = 3 tall, green, wrinkled

1 *tt yy RR*

2 *tt yy Rr*   = 3 dwarf, green, round

1 *tt YY rr*

2 *tt Yy rr*   = 3 dwarf, yellow, wrinkled

1 *tt yy rr*   = 1 dwarf, green, wrinkled

C32. The wooly-haired male is a heterozygote, because he has the trait and his mother did not. (He must have inherited the normal allele from his mother.) Therefore, he has a 50% chance of passing the wooly allele to his offspring; his offspring have a 50% of passing the allele to their offspring; and these grandchildren have a 50% chance of passing the allele to their offspring (the wooly-haired man's great-grandchildren). Because this is an ordered sequence of independent events, we use the product rule: $0.5 \times 0.5 \times 0.5 = 0.125$, or 12.5%. Because no other Scandinavians are on the island, the chance is 87.5% for the offspring being normal (because they could not inherit the wooly hair allele from anyone else). We use the binomial expansion equation to determine the likelihood that one out of eight great-grandchildren will have wooly hair, where $n = 8$, $x = 1$, $p = 0.125$, $q = 0.875$. The answer is 0.393, or 39.3%, of the time.

C34. Use the product rule. If the woman is heterozygous, there is a 50% chance of having an affected offspring: $(0.5)^7 = 0.0078$, or 0.78%, of the time. This is a pretty small probability. If the woman has an eighth child who is unaffected, however, she has to be a heterozygote, because it is a dominant trait. She would have to pass a normal allele to an unaffected offspring. The answer is 100%.

Experimental Questions

E2. The experimental difference depends on where the pollen comes from. In self-fertilization, the pollen and eggs come from the same plant. In cross-fertilization, they come from different plants.

E4. According to Mendel's law of segregation, the genotypic ratio should be 1 homozygote dominant : 2 heterozygotes : 1 homozygote recessive. This data table considers only the plants with a dominant phenotype. The genotypic ratio should be 1 homozygote dominant : 2 heterozygotes. The homozygote dominants would be true-breeding, while the heterozygotes would not be true-breeding. This 1:2 ratio is very close to what Mendel observed.

E6. All three offspring had black fur. The ovaries from the albino female could produce eggs with only the dominant black allele (because they were obtained from a true-breeding black female). The actual phenotype of the albino mother does not matter. Therefore, all offspring would be heterozygotes (*Bb*) and have black fur.

E8. If we construct a Punnett square according to Mendel's laws, we expect a 9:3:3:1 ratio. Because a total of 556 offspring were observed, the expected number of offspring are

$556 \times \ ^9/_{16} = 313$ round, yellow

$556 \times \ ^3/_{16} = 104$ wrinkled, yellow

$556 \times \ ^3/_{16} = 104$ round, green

$556 \times \ ^1/_{16} = 35$ wrinkled, green

If we plug the observed and expected values into the chi square equation, we get a value of 0.51. With four categories, our degrees of freedom equal $n - 1$, or 3. If we look up the value of 0.51 in the chi square table (see Table 2.1), we see that it falls between the P values of 0.80 and 0.95. This means that the probability is 80% to 95% that any deviation between observed results and expected results was caused by random sampling error. Therefore, we accept the hypothesis. In other words, the results are consistent with the law of independent assortment.

E10. A. If we let c^+ represent normal wings and c represent curved wings, and e^+ represents gray body and e represents ebony body,

Parental Cross: $cce^+e^+ \times c^+c^+ee$.

F_1 generation is heterozygous c^+ce^+e

An F_1 offspring crossed to flies with curved wings and ebony bodies is

$c^+ce^+e \times ccee$

The F_2 offspring would be a 1:1:1:1 ratio of flies:

$c^+ce^+e : c^+cee : cce^+e : ccee$

B. The phenotypic ratio of the F_2 flies would be a 1:1:1:1 ratio of flies:

normal wings, gray body : normal wings, ebony bodies : curved wings, gray bodies : curved wings, ebony bodies

C. From part B, we expect $^1/_4$ of each category. There are a total of 444 offspring. The expected number of each category is $^1/_4 \times 444$, which equals 111.

$$\chi^2 = \frac{(114 - 111)^2}{111} + \frac{(105 - 111)^2}{111} + \frac{(111 - 111)^2}{111} + \frac{(114 - 111)^2}{111}$$
$$\chi^2 = 0.49$$

With 3 degrees of freedom, a value of 0.49 or greater is likely to occur between 80% and 95% of the time. Therefore, we accept our hypothesis.

E12. Follow through the same basic chi square strategy as before. We expect a 3:1 ratio, or $^3/_4$ of the dominant phenotype and $^1/_4$ of the recessive phenotype.

The observed and expected values are as follows (rounded to the nearest whole number):

Observed*	Expected	$\frac{(O-E)^2}{E}$
5,474	5,493	0.066
1,850	1,831	0.197
6,022	6,017	0.004
2,001	2,006	0.012
705	697	0.092
224	232	0.276
882	886	0.018
299	295	0.054
428	435	0.113
152	145	0.338
651	644	0.076
207	215	0.298
787	798	0.152
277	266	0.455
		$\chi^2 = 2.15$

*Due to rounding, the observed and expected values may not add up to precisely the same number.

Because $n = 14$, there are 13 degrees of freedom. If we look up this value in the chi square table, we have to look between 10 and 15 degrees of freedom. In either case, we would expect the value of 2.15 or greater to occur more than 99% of the time. Therefore, we accept the hypothesis.

E14. The dwarf parent with terminal flowers must be homozygous for both genes, because it is expressing these two recessive traits: *ttaa*, where *t* is the recessive dwarf allele, and *a* is the recessive allele for terminal flowers. The phenotype of the other parent is dominant for both traits. However, because this parent was able to produce dwarf offspring with axial flowers, it must have been heterozygous for both genes: *TtAa*.

E16. Our hypothesis is that blue flowers and purple seeds are dominant traits and they are governed by two genes that assort independently. According to this hypothesis, the F_2 generation should yield a ratio of 9 blue flowers, purple seeds : 3 blue flowers, green seeds : 3 white flowers, purple seeds : 1 white flower, green seeds. Because a total of 300 offspring are produced, the expected numbers would be

$^9/_{16} \times 300 = 169$ blue flowers, purple seeds

$^3/_{16} \times 300 = 56$ blue flowers, green seeds

$^3/_{16} \times 300 = 56$ white flowers, purple seeds

$^1/_{16} \times 300 = 19$ white flowers, green seeds

$$\chi^2 = \frac{(103 - 169)^2}{169} + \frac{(49 - 56)^2}{56} + \frac{(44 - 56)^2}{56} + \frac{(104 - 19)^2}{19}$$

$$\chi^2 = 409.5$$

If we look up this value in the chi square table under 3 degrees of freedom, the value is much higher than would be expected 1% of the time by chance alone. Therefore, we reject the hypothesis. The idea that the two genes are assorting independently seems to be incorrect. The F_1 generation supports the idea that blue flowers and purple seeds are dominant traits.

Questions for Student Discussion/Collaboration

2.   If we construct a Punnett square, the following probabilities will be obtained:

tall with axial flowers, $^3/_8$

dwarf with terminal flowers, $^1/_8$

To calculate the probability of being tall with axial flowers or dwarf with terminal flowers, we use the sum rule:

$^3/_8 + ^1/_8 = ^4/_8 = ^1/_2$

We use the product rule to calculate the ordered events of the first three offspring being tall/axial or dwarf/terminal, and the fourth offspring being tall axial:

$(^1/_2)(^1/_2)(^1/_2)(^3/_8) = ^3/_{64} = 0.047 = 4.7\%$

CHAPTER 3

Conceptual Questions

C2.   The term homologue refers to the members of a chromosome pair. Homologues are usually the same size and carry the same types and order of genes. They may differ in that the genes they carry may be different alleles.

C4.   Metaphase is the organization phase, and anaphase is the separation phase.

C6.   In metaphase I of meiosis, each pair of chromatids is attached to only one pole via the kinetochore microtubules. In metaphase of mitosis, there are two attachments (i.e., to both poles). If the attachment was lost, a chromosome would not migrate to a pole and may not become enclosed in a nuclear membrane after telophase. If left out in the cytoplasm, it would eventually be degraded.

C8.   The reduction occurs because there is a single DNA replication event but two cell divisions. Because of the nature of separation during anaphase I, each cell receives one copy of each type of chromosome.

C10. It means that the maternally derived and paternally derived chromosomes are randomly aligned along the metaphase plate during metaphase I. Refer to Figure 3.17.

C12. There are three pairs of chromosomes. The number of different, random alignments equals 2^n, where n equals the number of chromosomes per set. So the possible number of arrangements equals 2^3, which is 8.

C14. The probability would be much lower because pieces of maternal chromosomes would be mixed with the paternal chromosomes. Therefore, it is unlikely to inherit a chromosome that was completely paternally derived.

C16. During interphase, the chromosomes are greatly extended. In this conformation, they might get tangled up with each other and not sort properly during meiosis and mitosis. The condensation process probably occurs so that the chromosomes easily align along the equatorial plate during metaphase without getting tangled up.

C18. During prophase II, your drawing should show four replicated chromosomes (i.e., four structures that look like Xs). Each chromosome is one homologue. During prophase of mitosis, there should be eight replicated chromosomes (i.e., eight Xs). During prophase of mitosis, there are pairs of homologues. The main difference is that prophase II has a single copy of each of the four chromosomes, whereas prophase of mitosis has four pairs of homologues. At the end of meiosis I, each daughter cell has received only one copy of a homologous pair, not both. This is due to the alignment of homologues during metaphase I and their separation during anaphase I.

C20. DNA replication does not take place during interphase II. The chromosomes at the end of telophase I have already replicated (i.e., they are found in pairs of sister chromatids). During meiosis II, the sister chromatids separate from each other, yielding individual chromosomes.

C22. A. 20

      B. 10

      C. 30

      D. 20

C24. A. Dark males and light females; reciprocal: all dark offspring

      B. All dark offspring; reciprocal: dark females and light males

      C. All dark offspring; reciprocal: dark females and light males

      D. All dark offspring; reciprocal: dark females and light males

C26. To produce sperm, a spermatogonial cell first goes through mitosis to produce two cells. One of these remains a spermatogonial cell and the other progresses through meiosis. In this way, the testes continue to maintain a population of spermatogonial cells.

C28. A. *A B C, A B c, A b C, A b c, a B C, a b C, a B c, a b c*

      B. *A B C, A b C*

      C. *A B C, A B c, a B C, a B c*

      D. *A b c, a b c*

C30. A. The fly is a male because the ratio of X chromosomes to sets of autosomes is $^1/_2$, or 0.5.

      B. The fly is female because the ratio of X chromosomes to sets of autosomes is 1.0.

      C. The fly is male because the ratio of X chromosomes to sets of autosomes is 0.5.

      D. The fly is female because the ratio of X chromosomes to sets of autosomes is 1.0.

Experimental Questions

E2.   Perhaps the most convincing observation was that all of the white-eyed flies of the F_2 generation were males. This suggests a link between sex determination and the inheritance of this trait. Because sex determination in fruit flies is determined by the number of X chromosomes, this suggests a relationship between the inheritance of the X chromosome and the inheritance of this trait.

E4.   The basic strategy is to set up a pair of reciprocal crosses. The phenotype of sons is usually the easiest way to discern the two patterns. If it is Y linked, the trait will be passed only from father to son. If it is X linked, the trait will be passed from mother to son.

E6.   The 3:1 sex ratio occurs because the female produces 50% gametes that are XX (and must produce female offspring) and 50% that are X (and produce half male and half female offspring). The original female had one X chromosome carrying the red allele and two other

X chromosomes carrying the eosin allele. Set up a Punnett square assuming that this female produces the following six types of gametes: $X^{w+}X^{w-e}$, $X^{w+}X^{w-e}$, $X^{w-e}X^{w-e}$, X^{w+}, X^{w-e}, X^{w-e}. The male of this cross is X^wY.

Male gametes

♀ \ ♂	X^w	Y
$X^{w+}X^{w-e}$	$X^{w+}X^{w-e}X^w$ **Red, female**	$X^{w+}X^{w-e}Y$ **Red, female**
$X^{w+}X^{w-e}$	$X^{w+}X^{w-e}X^w$ **Red, female**	$X^{w+}X^{w-e}Y$ **Red, female**
$X^{w-e}X^{w-e}$	$X^{w-e}X^{w-e}X^w$ **Eosin female**	$X^{w-e}X^{w-e}Y$ **Eosin female**
X^{w+}	$X^{w+}X^w$ **Red, female**	$X^{w+}Y$ **Red, male**
X^{w-e}	$X^{w-e}X^w$ **Light-eosin female**	$X^{w-e}Y$ **Light-eosin male**
X^{w-e}	$X^{w-e}X^w$ **Light-eosin female**	$X^{w-e}Y$ **Light-eosin male**

Female gametes (left axis)

E8. If we use the data from the F_1 mating (i.e., F_2 results), there were 3,470 red-eyed flies. We would expect a 3:1 ratio between red- and white-eyed flies. Therefore, assuming that all red-eyed offspring survived, there should have been about 1,157 (i.e., $^{3,470}/_3$) white-eyed flies. However, there were only 782. If we divide 782 by 1,157, we get a value of 0.676, or a 67.6% survival rate.

E10. You need to make crosses to understand the pattern of inheritance of traits (determined by genes) from parents to offspring. And you need to microscopically examine cells to understand the pattern of transmission of chromosomes. The correlation between the pattern of transmission of chromosomes during meiosis, and Mendel's laws of segregation and independent assortment is what led to the chromosome theory of inheritance.

E12. Originally, individuals who had abnormalities in their composition of sex chromosomes provided important information. In mammals, X0 individuals are females, whereas in flies, X0 individuals are males. In mammals, XXY individuals are males, while in flies, XXY individuals are females. These results indicate that the presence of the Y chromosome causes maleness in mammals, but it does not in flies. A further analysis of flies with abnormalities in the number of sets of autosomes revealed that it is the ratio between the number of X chromosomes and the number of sets of autosomes that determines sex in flies.

Questions for Student Discussion/Collaboration

2. It's not possible to give a direct answer, but the point is for students to be able to draw chromosomes in different configurations and understand the various phases. The chromosomes may or may not be

    1. In homologous pairs

    2. Connected as sister chromatids

    3. Associated in bivalents

    4. Lined up in metaphase

    5. Moving toward the poles.

    And so on.

CHAPTER 4

Conceptual Questions

C2. Sex-influenced traits are influenced by the sex of the individual even though the gene that governs the trait may be autosomally inherited. Pattern baldness in people is an example. Sex-limited traits are an extreme example of sex influence. The expression of a sex-limited trait is limited to one sex. For example, colorful plumage in certain species of birds is limited to the male sex. Sex-linked traits involve traits whose genes are found on the sex chromosomes. Examples in humans include hemophilia and color blindness.

C4. If the normal allele is dominant, it tells you that one copy of the gene produces a sufficient amount of the protein encoded by the gene. Having twice as much of this protein, as in the normal homozygote, does not alter the phenotype. If the allele is incompletely dominant, this means that one copy of the normal allele does not produce the same trait as the homozygote.

C6. The ratio would be 1 normal : 2 star-eyed individuals.

C8. If individual 1 is ii, individual 2 could be I^Ai, I^AI^A, I^Bi, I^BI^B, or I^AI^B.

    If individual 1 is I^Ai or I^AI^A, individual 2 could be I^Bi, I^BI^B, or I^AI^B.

    If individual 1 is I^Bi or I^BI^B, individual 2 could be I^Ai, I^AI^A, or I^AI^B.

    Assuming individual 1 is the parent of individual 2:

    If individual 1 is ii, individual 2 could be I^Ai or I^Bi.

    If individual 1 is I^Ai, individual 2 could be I^Bi or I^AI^B.

    If individual 1 is I^AI^A, individual 2 could be I^AI^B.

    If individual 1 is I^Bi, individual 2 could be I^Ai or I^AI^B.

    If individual 1 is I^BI^B, individual 2 could be I^AI^B.

C10. The father could not be I^AI^B, I^BI^B, or I^AI^A. He is contributing the O allele to his offspring. Genotypically, he could be I^Ai, I^Bi, or ii and have type A, B, or O blood, respectively.

C12. Perhaps it should be called codominant at the "hair level" because one or the other allele is dominant with regard to a single hair. However, this is not the same as codominance in blood types, in which every cell can express both alleles.

C14. A. X-linked recessive (unaffected mothers transmit the trait to sons)

    B. Autosomal recessive (affected daughters and sons are produced from unaffected parents)

C16. First set up the following Punnett square:

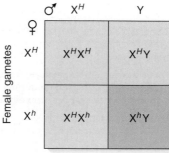

Male gametes

♀ \ ♂	X^H	Y
X^H	X^HX^H	X^HY
X^h	X^HX^h	X^hY

There is a $^1/_4$ probability of each type of offspring.

    A. $^1/_4$

    B. $(^3/_4)(^3/_4)(^3/_4)(^3/_4) = {}^{81}/_{256}$

    C. $^3/_4$

    D. The probability of an affected offspring is $^1/_4$, and the probability of an unaffected offspring is $^3/_4$. For this problem, you use the binomial expansion equation where $x = 2$, $n = 5$, $p = {}^1/_4$, and $q = {}^3/_4$. The answer is 0.26, or 26%, of the time.

C18. 1 affected daughter : 1 unaffected daughter : 1 affected son : 1 unaffected son

C20. On a standard vegetarian diet, the results would be 50% white and 50% yellow.

On a xanthophyll-free diet, the results would be 100% white.

C22. Set up a Punnett square, but keep in mind that the eosin gene is X linked.

	♂ CX^{w-e}	CY
♀		
CX^w	$CCX^{w-e}X^w$	CCX^wY
c^aX^w	$Cc^aX^{w-e}X^w$	Cc^aX^wY

Because C is dominant, the phenotypic ratios are two light-eosin females and two white males. Note: Females that have one eosin allele and one white allele have light-eosin eyes, whereas females with two eosin alleles or males with one eosin allele have eosin eyes, which are a bit darker than light-eosin eyes.

C24. A. Could be.

B. No, because an unaffected father has an affected daughter.

C. No, because two unaffected parents have affected children.

D. No, because an unaffected father has an affected daughter.

E. No, because both sexes exhibit the trait.

F. Could be.

C26. You would look at the pattern within families over the course of many generations. For a recessive trait, 25% of the offspring within a family are expected to be affected if both parents are unaffected carriers, and 50% of the offspring would be affected if one parent was affected. You could look at many families and see if these 25% and 50% values are approximately true. Incomplete penetrance would not necessarily predict such numbers. Also, for very rare alleles, incomplete penetrance would probably have a much higher frequency of affected parents producing affected offspring. For rare recessive disorders, it is most likely that both parents are heterozygous carriers. Finally, the most informative pedigrees would be situations in which two affected parents produce children. If they can produce an unaffected offspring, this would indicate incomplete penetrance. If all of their offspring were affected, this would be consistent with recessive inheritance.

C28. The probability of a heterozygote passing the allele to his/her offspring is 50%. The probability of an affected offspring expressing the trait is 80%. We use the product rule to determine the likelihood of these two independent events.

$(0.5)(0.8) = 0.4$, or 40% of the time

C30. This is an example of incomplete dominance. The heterozygous horses are palominos. For example, if C represents chestnut and c represents cremello, the chestnut horses are CC, the cremello horses are cc, and the palominos are Cc.

Experimental Questions

E2. Two redundant genes are involved in feathering. The unfeathered Buff Rocks are homozygous recessive for the two genes. The Black Langhans are homozygous dominant for both genes. In the F_2 generation (which is a double heterozygote crossed to another double heterozygote), 1 out of 16 offspring will be doubly homozygous for both recessive genes. All the others will have at least one dominant allele for one of the two (redundant) genes.

E4. The reason why all the puppies have black hair is because albino alleles are found in two different genes. If we let the letters A and B represent the two different pigmentation genes, then one of the dogs is $AAbb$, and the other is $aaBB$. Their offspring are $AaBb$ and therefore are not albinos because they have one dominant copy of each gene.

E6. In general, you cannot distinguish between autosomal and pseudoautosomal inheritance from a pedigree analysis. Mothers and fathers have an equal probability of passing the alleles to sons and daughters. However, if an offspring had a chromosomal abnormality, you might be able to tell. For example, in a family tree involving the *Mic2* allele, an offspring that was X0 would have less of the gene product, and an offspring that was XXX or XYY or XXY would have extra amounts of the gene products. This may lead you to suspect that the gene is located on the sex chromosomes.

E8. One parent must be $RRPp$. The other parent could be $RRPp$ or $RrPp$. All the offspring would inherit (at least) one dominant R allele. With regard to the other gene, $3/4$ would inherit at least one copy of the dominant P allele. These offspring would have a walnut comb. The other $1/4$ would be homozygous pp and have a rose comb (because they would also have a dominant R allele).

E10. Let's use the letters A and B for these two genes. Gene A exists in two alleles, which we will call A and a. Gene B exists in two alleles, B and b. The uppercase alleles are dominant to the lowercase alleles. The true-breeding long-shaped squash is $aabb$, and the true-breeding disk-shaped squash is $AABB$. The F_1 offspring are $AaBb$. You can construct a Punnett square, with 16 boxes, to determine the outcome of self-fertilization of the F_1 plants.

To get the disk-shaped phenotype, an offspring must inherit at least one dominant allele from both genes.

1 $AABB$ + 2 $AaBB$ + 2 $AABb$ + 4 $AaBb$ = 9 disk-shaped offspring

To get the round phenotype, an offspring must inherit at least one dominant allele for one of the two genes but must be homozygous recessive for only one of the two genes.

1 $aaBB$ + 1 $AAbb$ + 2 $aaBb$ + 2 $Aabb$ = 6 round-shaped offspring

To get the long phenotype, an offspring must inherit all recessive alleles:

1 $aabb$ = 1 long-shaped offspring

E12. The results obtained when crossing two F_1 offspring appear to yield a 9:3:3:1 ratio, which would be expected if eye color is affected by two different genes that exist in dominant and recessive alleles. Neither gene is X linked. Let pr^+ represent the red allele of the first gene and pr the purple allele. Let sep^+ represent the red allele of the second gene and sep the sepia allele.

The first cross is $prpr\ sep^+sep^+ \times pr^+pr^+\ sep\ sep$

All the F_1 offspring would be $pr^+pr\ sep^+sep$. They have red eyes because they have a dominant red allele for each gene. When the F_1 offspring are crossed to each other, the following results would be obtained:

♀ \ ♂	pr^+sep^+	pr^+sep	$pr\ sep^+$	$pr\ sep$
pr^+sep^+	pr^+pr^+ sep^+sep^+ Red	pr^+pr^+ sep^+sep Red	pr^+pr sep^+sep^+ Red	pr^+pr sep^+sep Red
pr^+sep	pr^+pr^+ sep^+sep Red	pr^+pr^+ $sep\ sep$ Sepia	pr^+pr sep^+sep Red	pr^+pr $sep\ sep$ Sepia
$pr\ sep^+$	pr^+pr sep^+sep^+ Red	pr^+pr sep^+sep Red	$pr\ pr$ sep^+sep^+ Purple	$pr\ pr$ sep^+sep Purple
$pr\ sep$	pr^+pr sep^+sep Red	pr^+pr $sep\ sep$ Sepia	$pr\ pr$ sep^+sep Purple	$pr\ pr$ $sep\ sep$ Pur/Sepia

In this case, one gene exists as the red (dominant) or purple (recessive) allele, and the second gene exists as the red (dominant) or sepia (recessive) allele. If an offspring is homozygous for the purple allele, it will have purple eyes. Similarly, if an offspring is homozygous for the sepia allele, it will have sepia eyes. An offspring homozygous for both recessive alleles has purplish sepia eyes. To have red eyes, it must have at least one copy of the dominant red allele for both genes. Based on an expected 9 red : 3 purple : 3 sepia : 1 purplish sepia, the observed and expected numbers of offspring are as follows:

Observed	*Expected*
146 purple eyes	148 purple eyes ($791 \times {}^{3}/_{16}$)
151 sepia eyes	148 sepia eyes ($791 \times {}^{3}/_{16}$)
50 purplish sepia eyes 4	49 purplish sepia eyes ($791 \times {}^{1}/_{16}$)
<u>444 red eyes</u>	445 red eyes ($791 \times {}^{9}/_{16}$)
791 total offspring	

If we plug the observed and expected values into our chi square formula, we obtain a chi square value of about 0.11. With 3 degrees of freedom, this is well within our expected range of values, so we cannot reject our hypothesis that purple and sepia alleles are in two different genes and that these recessive alleles are epistatic to each other.

E14. To see if the allele is X linked, the pink-eyed male could be crossed to a red-eyed female. All the offspring would have red eyes, assuming that the pink allele is recessive. When crossed to red-eyed males, the F_1 females will produce $^1/_2$ red-eyed daughters, $^1/_4$ red-eyed sons, and $^1/_4$ pink-eyed sons if the pink allele is X linked.

If the pink allele is X linked, then one could determine if it is in the same X-linked gene as the white and eosin alleles by crossing pink-eyed males to white-eyed females. (Note: We already know that white and eosin are alleles of the same gene.) If the pink and white alleles are in the same gene, the F_1 female offspring should have pink eyes (assuming that the pink allele is dominant over white). However, if the pink and white alleles are in different genes, the F_1 females will have red eyes (assuming that pink is recessive to red). This is because the F_1 females will be heterozygous for two genes, $X^{w+p} X^{wp+}$, in which the X^{w+} and X^{p+} alleles are the dominant wild-type alleles that produce red eyes, and the X^w and X^p alleles are recessive alleles for these two different genes, which produce white eyes and pink eyes, respectively.

Questions for Student Discussion/Collaboration

2.   Perhaps the easiest way to solve this problem is to take one trait at a time. With regard to combs, all the F_1 generation would be *RrPp* or walnut comb. With regard to shanks, they would all be feathered, because they would inherit one dominant copy of a feathered allele. With regard to hen- or cock-feathering: 1 male cock-feathered : 1 male hen-feathered : 2 females hen-feathered. Overall then, we would have a 1 : 1 : 2 ratio of

walnut comb/feathered shanks/cock-feathered males

walnut comb/feathered shanks/hen-feathered males

walnut comb/feathered shanks/hen-feathered females

CHAPTER 5

C2.   An independent assortment hypothesis is used because it enables us to calculate the expected values based on Mendel's ratios. Using the observed and expected values, we can calculate whether or not the deviations between the observed and expected values are too large to occur as a matter of chance. If the deviations are very large, we reject the hypothesis of independent assortment.

C4.   If the chromosomes (on the right side) labeled 2 and 4 move into one daughter cell, that will lead to a patch that is albino and has long fur.

The other cell will receive chromosomes 1 and 3, which will produce a patch that has dark, short fur.

C6.   A single crossover produces *A B C*, *A b c*, *a B C*, and *a b c*.

A.   Between 2 and 3, between genes *B* and *C*

B.   Between 1 and 4, between genes *A* and *B*

C.   Between 1 and 4, between genes *B* and *C*

D.   Between 2 and 3, between genes *A* and *B*

C8.   The likelihood of scoring a basket would be greater if the basket was larger. Similarly, the chances of a crossover initiating in a region between two genes is proportional to the size of the region between the two genes. There are a finite number (usually a few) of crossovers that occur between homologous chromosomes during meiosis, and the likelihood that a crossover will occur in a region between two genes depends on how big that region is.

C10. The pedigree suggests a linkage between the dominant allele causing nail-patella syndrome and the I^B allele of the ABO blood type gene. In every case, the individual who inherits the I^B allele also inherits this disorder.

C12. <u>*Ass-1*   43   *Sdh-1*   5   *Hdc* 9   *Hao-1*   6   *Odc-2*   8   *Ada-1*</u>

C14. The inability to detect double crossovers causes the map distance to be underestimated. In other words, more crossovers occur in the region than we realize. When we have a double crossover, we do not get a recombinant offspring (in a dihybrid cross). Therefore, the second crossover cancels out the effects of the first crossover.

C16. The key feature is that all the products of a single meiosis are contained within a single sac. The spores in this sac can be dissected, and then their genetic traits can be analyzed individually.

C18. In an unordered ascus, the products of meiosis are free to move around. In an ordered octad (or tetrad), they are lined up according to their relationship to each other during meiosis and mitosis. An ordered octad can be used to map the distance between a single gene and its centromere.

C20. The percentage would be higher with respect to gene *A*. First-division segregation patterns occur when there is not a crossover between the centromere and the gene of interest. Because gene *A* is closer to the centromere compared to gene *B*, it would be less likely to have a crossover between gene *A* and the centromere. This would make it more likely to observe first-division segregation.

Experimental Questions (Includes Most Mapping Questions)

E2.   They could have used a strain with two abnormal chromosomes. In this case, the recombinant chromosomes would either look normal or have abnormalities at both ends.

E4.   A gene on the Y chromosome in mammals would be transmitted only from father to son. It would be difficult to genetically map Y-linked genes because a normal male has only one copy of the Y chromosome, so you do not get any crossing over between two Y chromosomes. Occasionally, abnormal males (XYY) are born with two Y chromosomes. If such males were heterozygous for alleles of Y-linked genes, one could examine the normal male offspring of XYY fathers and determine if crossing over has occurred.

E6. The answer is explained in solved problem S5. We cannot get more than 50% recombinant offspring because the pattern of multiple crossovers can yield an average maximum value of only 50%. When a testcross does yield a value of 50% recombinant offspring, it can mean two different things. Either the two genes are on different chromosomes or the two genes are on the same chromosome but at least 50 mu apart.

E8. If two genes are at least 50 mu apart, you would need to map genes between them to show that the two genes were actually in the same linkage group. For example, if gene *A* was 55 mu from gene *B*, there might be a third gene (e.g., gene *C*) that is 20 mu from *A* and 35 mu from *B*. These results would indicate that *A* and *B* are 55 mu apart, assuming dihybrid testcrosses between genes *A* and *B* yielded 50% recombinant offspring.

E10. Sturtevant used the data involving the following pairs: *y* and *w*, *w* and *v*, *v* and *r*, and *v* and *m*.

E12. A. Because they are 12 mu apart, we expect 12% (or 120) recombinant offspring. This would be approximately 60 *Aabb* and 60 *aaBb* plus 440 *AaBb* and 440 *aabb*.

  B. We would expect 60 *AaBb*, 60 *aabb*, 440 *Aabb*, and 440 *aaBb*.

E14. Due to the large distance between the two genes, they will assort independently even though they are actually on the same chromosome. According to independent assortment, we expect 50% parental and 50% recombinant offspring. Therefore, this cross will produce 150 offspring in each of the four phenotypic categories.

E16. A. If we hypothesize two genes independently assorting, then the predicted ratio is 1:1:1:1. There are a total of 390 offspring. The expected number of offspring in each category is about 98. Plugging the figures into our chi square formula,

$$\chi^2 = \frac{(117-98)^2}{98} + \frac{(115-98)^2}{98} + \frac{(78-98)^2}{98} + \frac{(80-98)^2}{98}$$

$$\chi^2 = 3.68 + 2.95 + 4.08 + 3.31$$

$$\chi^2 = 14.02$$

Looking up this value in the chi square table under 1 degree of freedom, we reject our hypothesis, because the chi square value is above 7.815.

  B. Map distance:

$$\text{Map Distance} = \frac{78 + 80}{117 + 115 + 78 + 80}$$
$$= 40.5 \text{ mu}$$

Because the value is relatively close to 50 mu, it is probably a significant underestimate of the true distance between these two genes.

E18. The percentage of recombinants for the green, yellow and wide, narrow is 7%, or 0.07; there will be 3.5% of the green, narrow and 3.5% of the yellow, wide. The remaining 93% parentals will be 46.5% green, wide and 46.5% yellow, narrow. The third gene assorts independently. There will be 50% long and 50% short with respect to each of the other two genes. To calculate the number of offspring out of a total of 800, we multiply 800 by the percentages in each category.

(0.465 green, wide)(0.5 long)(800) = 186 green, wide, long

(0.465 yellow, narrow)(0.5 long)(800) = 186 yellow, narrow, long

(0.465 green, wide)(0.5 short)(800) = 186 green, wide, short

(0.465 yellow, narrow)(0.5 short)(800) = 186 yellow, narrow, short

(0.035 green, narrow)(0.5 long)(800) = 14 green, narrow, long

(0.035 yellow, wide)(0.5 long)(800) = 14 yellow, wide, long

(0.035 green, narrow)(0.5 short)(800) = 14 green, narrow, short

(0.035 yellow, wide)(0.5 short)(800) = 14 yellow, wide, short

E20. Let's use the following symbols: *G* for green pods, *g* for yellow pods, *S* for green seedlings, *s* for bluish green seedlings, *C* for normal plants, *c* for creepers. The parental cross is *GG SS CC* crossed to *gg ss cc*.

The F$_1$ plants would all be *Gg Ss Cc*. If the genes are linked, the alleles *G*, *S*, and *C* would be linked on one chromosome, and the alleles *g*, *s*, and *c* would be linked on the homologous chromosome.

The testcross is F$_1$ plants, which are *Gg Ss Cc* crossed to *gg ss cc*.

To measure the distances between the genes, we can separate the data into gene pairs.

Pod color, seedling color

  2,210 green pods, green seedlings—nonrecombinant

  296 green pods, bluish green seedlings—recombinant

  2,198 yellow pods, bluish green seedlings—nonrecombinant

  293 yellow pods, green seedlings—recombinant

$$\text{Map Distance} = \frac{296 + 293}{2{,}210 + 296 + 2{,}198 + 293} \times 100 = 11.8 \text{ mu}$$

Pod color, plant stature

  2,340 green pods, normal—nonrecombinant

  166 green pods, creeper—recombinant

  2,323 yellow pods, creeper—nonrecombinant

  168 yellow pods, normal—recombinant

$$\text{Map Distance} = \frac{166 + 168}{2{,}340 + 166 + 2{,}323 + 168} \times 100 = 6.7 \text{ mu}$$

Seedling color, plant stature

  2,070 green seedlings, normal—nonrecombinant

  433 green seedlings, creeper—recombinant

  2,056 bluish green seedlings, creeper—nonrecombinant

  438 bluish green seedlings, normal—recombinant

$$\text{Map Distance} = \frac{433 + 438}{2{,}070 + 433 + 2{,}056 + 438} \times 100 = 17.4 \text{ mu}$$

The order of the genes is seedling color, pod color, and plant stature (or you could say the opposite order). Pod color is in the middle. If we use the two shortest distances to construct our map:

<u>S_____11.8_____G_____6.7_____C</u>

E22. To answer this question, we can consider genes in pairs. Let's consider the two gene pairs that are closest together. The distance between the wing length and eye color genes is 12.5 mu. From this cross, we expect 87.5% to have long wings and red eyes or short wings and purple eyes, and 12.5% to have long wings and purple eyes or short wings and red eyes. Therefore, we expect 43.75% to have long wings and red eyes, 43.75% to have short wings and purple eyes, 6.25% to have long wings and purple eyes, and 6.25% to have short wings and red eyes. If we have 1,000 flies, we expect 438 to have long wings and red eyes, 438 to have short wings and purple eyes, 62 to have long wings and purple eyes, and 62 to have short wings and red eyes (rounding to the nearest whole number).

The distance between the eye color and body color genes is 6 mu. From this cross, we expect 94% to have a parental combination (red eyes and gray body or purple eyes and black body) and 6% to have a nonparental combination (red eyes and black body or purple eyes and gray body). Therefore, of our 438 flies with long wings and red eyes, we expect 94% of them (or about 412) to have long wings, red eyes, and gray body and 6% of them (or about 26) to have long wings, red eyes, and black bodies. Of our 438 flies with short wings and purple eyes, we expect about 412 to have short wings, purple eyes, and black bodies and 26 to have short wings, purple eyes, and gray bodies.

Of the 62 flies with long wings and purple eyes, we expect 94% of them (or about 58) to have long wings, purple eyes, and black bodies and 6% of them (or about 4) to have long wings, purple eyes, and gray bodies. Of the 62 flies with short wings and red eyes, we expect 94% (or about 58) to have short wings, red eyes, and gray bodies and 6% (or about 4) to have short wings, red eyes, and black bodies.

In summary,

Long wings, red eyes, gray body	412
Long wings, purple eyes, gray body	4
Long wings, red eyes, black body	26
Long wings, purple eyes, black body	58
Short wings, red eyes, gray body	58
Short wings, purple eyes, gray body	26
Short wings, red eyes, black body	4
Short wings, purple eyes, black body	412

The flies with long wings, purple eyes, and gray bodies, or short wings, red eyes, and black bodies, are produced by a double-crossover event.

E24. Yes. Begin with females that have one X chromosome that is X^{Nl} and the other X chromosome that is X^{nL}. These females have to be mated to $X^{NL}Y$ males because a living male cannot carry the n or l allele. In the absence of crossing over, a mating between $X^{Nl}X^{nL}$ females to $X^{NL}Y$ males should not produce any surviving male offspring. However, during oogenesis in these heterozygous female mice, there could be a crossover in the region between the two genes, which would produce an X^{NL} chromosome and an X^{nl} chromosome. Male offspring inheriting these recombinant chromosomes will be either $X^{NL}Y$ or $X^{nl}Y$ (whereas nonrecombinant males will be $X^{nL}Y$ or $X^{Nl}Y$). Only the male mice that inherit $X^{NL}Y$ will live. The living males represent only half of the recombinant offspring. (The other half are $X^{nl}Y$, which are born dead.)

To compute map distance:

$$\text{Map Distance} = \frac{2(\text{Number of male living offspring})}{\text{Number of males born dead} + \text{Number of males born alive}} \times 100$$

E26.
$$\text{Map Distance} = \frac{(^1/_2)(\text{SDS})}{\text{Total}} \times 100$$

$$= \frac{(^1/_2)(22 + 21 + 21 + 23)}{22 + 21 + 21 + 21 + 451 + 23 + 455} \times 100$$

$$= 4.4 \text{ mu}$$

Questions for Student Discussion/Collaboration

2. The X and Y chromosomes are not completely distinct linkage groups. One might describe them as overlapping linkage groups having some genes in common, but most genes are not common to both.

CHAPTER 6

Conceptual Questions

C2. It is not a form of sexual reproduction, in which two distinct parents produce gametes that unite to form a new individual. However, conjugation is similar to sexual reproduction in the sense that the genetic material from two cells are somewhat mixed. In conjugation, there is not the mixing of two genomes, one from each gamete. Instead, there is a transfer of genetic material from one cell to another. This transfer can alter the combination of genetic traits in the recipient cell.

C4. An F^+ strain contains a separate, circular piece of DNA that has its own origin of transfer. An Hfr strain has its origin of transfer integrated into the bacterial chromosome. An F^+ strain can transfer only the DNA contained on the F factor. If given enough time, an Hfr strain can actually transfer the entire bacterial chromosome to the recipient cell.

C6. Sex pili promote the binding of donor and recipient cells and provide a passageway for the transfer of genetic material from the donor to the recipient cell.

C8. Though exceptions are common, interspecies genetic transfer via conjugation is not as likely because the cell surfaces do not interact correctly. Interspecies genetic transfer via transduction is also not very likely because each species of bacteria is sensitive to particular bacteriophages. The correct answer is transformation. A consequence of interspecies genetic transfer is that new genes can be introduced into a bacterial species from another species. For example, interspecies genetic transfer could provide the recipient bacterium with a new trait, such as resistance to an antibiotic. Evolutionary biologists call this horizontal gene transfer, while the passage of genes from parents to offspring is termed vertical gene transfer.

C10. Cotransduction is the transduction of two or more genes. The distance between the genes determines the frequency of cotransduction. When two genes are close together, the cotransduction frequency would be higher compared to two genes that are relatively farther apart.

C12. If a site that frequently incurred a breakpoint was between two genes, the cotransduction frequency of these two genes would be much less than expected. This is because the site where the breakage occurred would separate the two genes from each other.

C14. The transfer of conjugative plasmids such as F factor DNA

C16. A. If it occurred in a single step, transformation is the most likely mechanism because conjugation does not usually occur between different species, particularly distantly related species, and different species are not usually infected by the same bacteriophages.

B. It could occur in a single step, but it may be more likely to have involved multiple steps.

C. The use of antibiotics selects for the survival of bacteria that have resistance genes. If a population of bacteria is exposed to an antibiotic, those carrying resistance genes will survive, and their relative numbers will increase in subsequent generations.

C18. The term allele means alternative forms of the same gene. Therefore, mutations in the same gene among different phages are alleles of each other; the mutations may be at different positions within the same gene. When we map the distance between mutations in the same gene, we are mapping the distance between the mutations that create different alleles of the same gene. An intragenic map describes the locations of mutations within the same gene.

Experimental Questions

E2. Mix the two strains together and then put some of them on plates containing streptomycin and some of them on plates without streptomycin. If mated colonies are present on both types of plates, then the thr^+, leu^+, and thi^+ genes were transferred to the met^+ bio^+ thr^- leu^- thi^- strain. If colonies are found only on the plates that lack streptomycin, then the met^+ and bio^+ genes are being transferred to the met^- bio^- thr^+ leu^+ thi^+ strain. This answer assumes a one-way transfer of genes from a donor to a recipient strain.

E4. An interrupted mating experiment is a procedure in which two bacterial strains are allowed to mate, and then the mating is interrupted at various time points. The interruption occurs by agitation of the solution in which the bacteria are found. This type of study is used to map the locations of genes. It is necessary to interrupt mating so that you can vary the time and obtain information about the order of transfer; which gene transferred first, second, and so on.

E6. Mate unknown strains A and B to the F^- strain in your lab that is resistant to streptomycin and cannot use lactose. This is done in two separate tubes (i.e., strain A plus your F^- strain in one tube, and strain B plus your F^- strain in the other tube). Plate the mated cells on growth media containing lactose plus streptomycin. If you get growth of colonies, the unknown strain had to be strain A, the F^+ strain that had lactose utilization genes on its F factor.

E8. A. If we extrapolate these lines back to the x-axis, the *hisE* intersects at about 3 minutes, and the *pheA* intersects at about 24 minutes. These are the values for the times of entry. Therefore, the distance between these two genes is 21 minutes (i.e., 24 minus 3).

B.

↑	4	↑	17	↑
hisE		*pabB*		*pheA*

E10. One possibility is that you could treat the P1 lysate with DNase I, an enzyme that digests DNA. (Note: If DNA were digested with DNase I, the function of any genes within the DNA would be destroyed.) If the DNA were within a P1 phage, it would be protected from DNase I digestion. This would allow you to distinguish between transformation (which would be inhibited by DNase I) versus transduction (which would not be inhibited by DNase I). Another possibility is that you could try to fractionate the P1 lysate. Naked DNA would be smaller than a P1 phage carrying DNA. You could try to filter the lysate to remove naked DNA, or you could subject the lysate to centrifugation and remove the lighter fractions that contain naked DNA.

E12. Cotransduction frequency = $(1 - d/L)^3$

For the normal strain,

Cotransduction frequency = $(1 - {}^{0.7}/_2)^3 = 0.275$, or 27.5%

For the new strain,

Cotransduction frequency = $(1 - {}^{0.7}/_5)^3 = 0.64$, or 64%

The experimental advantage is that you could map genes that are farther than 2 minutes apart. You could map genes that are up to 2 minutes apart.

E14. Cotransduction frequency = $(1 - {}^d/_L)^3$

$$0.53 = (1 - {}^d/_2 \text{ minutes})^3$$

$$(1 - {}^d/_2 \text{ minutes}) = \sqrt[3]{0.53}$$

$$(1 - {}^d/_2 \text{ minutes}) = 0.81$$

$$d = 0.38 \text{ minutes}$$

E16. A. We first need to calculate the cotransformation frequency, which equals 2/70, or 0.029.

Contransformation frequency = $(1 - {}^d/_L)^3$

$$0.029 = (1 - {}^d/_2 \text{ minutes})^3$$

$$d = 1.4 \text{ minutes}$$

B. Contransformation frequency = $(1 - {}^d/_L)^3$

$$= (1 - {}^{1.4}/_4)^3$$

$$= 0.27$$

As you may have expected, the cotransformation frequency is much higher when the transformation involves larger pieces of DNA.

E18. Benzer could use this observation as a way to evaluate if intragenic recombination had occurred. If two *rII* mutations recombined to make a wild-type gene, the phage would produce plaques in this *E. coli* K12(λ) strain.

E20. *rIIA*: L47 and L92

rIIB: L33, L40, L51, L62, L65, and L91

E22. Benzer first determined the individual nature of each gene by showing that mutations within the same gene did not complement each other. He then could map the distance between two mutations within the same gene. The map distances defined each gene as a linear, divisible unit. In this regard, the gene is divisible due to crossing over.

Questions for Student Discussion/Collaboration

2. Flower color in sweet peas is another example. When two different genes affect flower color, the white allele of one gene may be epistatic to the purple allele of another gene. A heterozygote for both genes would have purple flowers. In other words, the two purple alleles are complementing the two white alleles. Other similar examples are discussed in Chapter 4 in the section on Gene Interactions.

CHAPTER 7

Conceptual Questions

C2. A maternal effect gene is one in which the genotype of the mother determines the phenotype of the offspring. At the cellular level, this happens because maternal effect genes are expressed in diploid nurse cells and then the gene products are transported into the egg. These gene products play key roles in the early steps of embryonic development.

C4. The genotype of the mother must be *bic⁻ bic⁻*. That is why it produces abnormal offspring. Because the mother is alive and able to produce offspring, its mother (the maternal grandmother) must have been *bic⁺ bic⁻* and passed the *bic⁻* allele to its daughter (the mother in this problem). The maternal grandfather also must have passed the *bic⁻* allele to its daughter. The maternal grandfather could be either *bic⁺ bic⁻* or *bic⁻ bic⁻*.

C6. The mother must be heterozygous. She is phenotypically abnormal because her mother must have been homozygous for the abnormal recessive allele. However, because she produces all normal offspring, she must have inherited the normal dominant allele from her father. She produces all normal offspring because this is a maternal effect gene, and the gene product of the normal dominant allele is transferred to the egg.

C8. Maternal effect genes exert their effects because the gene products are transferred from nurse cells to eggs. The gene products, mRNA and proteins, do not last a very long time before they are eventually degraded. Therefore, they can exert their effects only during early stages of embryonic development.

C10. Dosage compensation refers to the phenomenon that the level of expression of genes on the sex chromosomes is the same in males and females, even though they have different numbers of sex chromosomes. In many species it seems necessary so that the balance of gene expression between the autosomes and sex chromosomes is similar between the two sexes.

C12. In mammals, one of the X chromosomes is inactivated in females; in *Drosophila*, the level of transcription on the X chromosome in males is doubled; in *C. elegans*, the level of transcription of the X chromosome in hermaphrodites is decreased by 50% of that of males.

C14. X inactivation begins with the counting of Xics. If there are two X chromosomes, in the process of initiation, one is targeted for inactivation. During embryogenesis, this inactivation begins at the Xic locus and spreads to both ends of the X chromosome until it becomes a highly condensed Barr body. The *Tsix* gene may play a role in the choice of the X chromosome that remains active. The *Xist* gene, which is located in the Xic region, remains transcriptionally active on the inactivated X chromosome. It is thought to play an important role in X inactivation by coating the inactive X chromosome. After X inactivation is established, it is maintained in the same X chromosome in somatic cells during subsequent cell divisions. In germ cells, however, the X chromosomes are not inactivated, so an egg can transmit either copy of an active (noncondensed) X chromosome.

C16. A. One

B. Zero

C. Two

D. Zero

C18. The offspring inherited X^B from its mother and X^O and Y from its father. It is an XXY animal, which is male (but somewhat feminized).

C20. Erasure and reestablishment of the imprint occurs during gametogenesis. It is necessary to erase the imprint because each sex will transmit either inactive or active alleles of a gene. In somatic cells, the two alleles for a gene are imprinted according to the sex of the parent from which the allele was inherited.

C22. A person born with paternal uniparental disomy 15 would have Angelman syndrome, because this individual would not have an active copy of the *AS* gene; the paternally inherited copies of the *AS* gene are silenced. This individual would have normal offspring, because she does not have a deletion in either copy of chromosome 15.

C24. In some species, such as marsupials, X inactivation depends on the sex. This is similar to imprinting. Also, once X inactivation occurs during embryonic development, it is remembered throughout the rest of the life of the organism, which is also similar to imprinting. X inactivation in mammals is different from genomic imprinting, in that it is not sex dependent. The X chromosome that is inactivated could be inherited from the mother or the father. There was no marking process on the X chromosome that occurred during gametogenesis. In contrast, genomic imprinting always involves a marking process during gametogenesis.

C26. The term reciprocal cross refers to two parallel crosses that involve the same genotypes of the two parents, but their sexes are opposite in the two crosses. For example, the reciprocal cross of female $BB \times$ male bb is the cross female $bb \times$ male BB. Autosomal inheritance gives the same result because the autosomes are transmitted from parent to offspring in the same way for both sexes. However, for extranuclear inheritance, the mitochondria and plastids are not transmitted via the gametes in the same way for both sexes. For maternal inheritance, the reciprocal crosses would show that the gene is always inherited from the mother.

C28. The phenotype of a petite mutant is that it forms small colonies on growth media, as opposed to wild-type strains that formed larger colonies. These mutants are unable to grow when the cells only have an energy source that requires mitochondrial function. Because nuclear and mitochondrial genes are necessary for mitochondrial function, it is possible for a petite mutation to involve a gene in the nucleus or in the mitochondrial genome. The difference between neutral and suppressive petites is that neutral petites lack most of their mitochondrial DNA, whereas suppressive petites usually lack small segments of the mitochondrial genetic material.

C30. The mitochondrial and chloroplast genomes are composed of a circular chromosome found in one or more copies in a region of the organelle known as the nucleoid. The number of genes per chromosome varies from species to species. Chloroplasts genomes tend to be larger than mitochondria genomes. See Table 7.3 for examples of the variation among mitochondrial and chloroplast genomes.

C32. A. Yes

B. Yes

C. No, it is determined by a gene in the chloroplast genome.

D. No, it is determined by a gene in the mitochondria.

C34. Biparental extranuclear inheritance would resemble Mendelian inheritance in that offspring could inherit alleles of a given gene from both parents. It differs, however, when you think about it from the perspective of heterozygotes. For a Mendelian trait, the law of segregation tells us that a heterozygote passes one allele for a given gene to an offspring, but not both. In contrast, if a parent has a mixed population of mitochondria (e.g., some carrying a mutant gene and some carrying a normal gene), that parent could pass both types of genes (mutant and normal) to a single offspring, because more than one mitochondrion could be contained within a sperm or egg cell.

Experimental Questions

E2. The first type of observation was based on cytological studies. The presence of the Barr body in female cells was consistent with the idea that one of the X chromosomes was highly condensed. The second type of observation was based on genetic mutations. A variegated phenotype that is found only in females is consistent with the idea that certain patches express one allele and other patches express the other allele. This variegated phenotype would occur only if the inactivation of one X chromosome happened at an early stage of embryonic development and was inherited permanently thereafter.

E4. The pattern of inheritance is consistent with imprinting. In every cross, the allele that is inherited from the father is expressed in the offspring, but the allele inherited from the mother is not.

E6. We assume that the snails in the large colony on the second island are true-breeding, DD. Let the male snail from the deserted island mate with a female snail from the large colony. Then let the F_1 snails mate with each other to produce an F_2 generation. Then let the F_2 generation mate with each other to produce an F_3 generation. Here are the expected results:

Female $DD \times$ Male DD

All F_1 snails coil to the right.

All F_2 snails coil to the right.

All F_3 snails coil to the right.

Female $DD \times$ Male Dd

All F_1 snails coil to the right.

All F_2 snails coil to the right because all of the F_1 females are DD or Dd.

$^{15}/_{16}$ of F_3 snails coil to the right; $^1/_{16}$ of F_3 snails coil to the left (because $^1/_{16}$ of the F_2 females are dd).

Female $DD \times$ Male dd.

All F_1 snails coil to the right.

All F_2 snails coil to the right because all of the F_1 females are Dd.

$^3/_4$ of F_3 snails coil to the right; $^1/_4$ of F_3 snails coil to the left (because $^1/_4$ of the F_2 females are dd).

E8. Let's first consider the genotypes of male A and male B. Male A must have two normal copies of the *Igf2* gene. We know this because male A's mother was *Igf2 Igf2*; the father of male A must have been a heterozygote *Igf2 Igf2⁻* because half of the litter that contained male A also contained dwarf offspring. But because male A was not dwarf, it must have inherited the normal allele from its father. Therefore, male A must be *Igf2 Igf2*. We cannot be completely sure of the genotype of male B. It must have inherited the normal *Igf2* allele from its father because male B is phenotypically normal. We do not know the genotype of male B's mother, but she could be either *Igf2⁻ Igf2⁻* or *Igf2 Igf2⁻*. In either case, the mother of male B could pass the *Igf2* allele to an offspring, but we do not know for sure if she did. So, male B could be either *Igf2 Igf2⁻* or *Igf2 Igf2*.

For the *Igf2* gene, we know that the maternal allele is inactivated. Therefore, the genotypes and phenotypes of females A and B are irrelevant. The phenotype of the offspring is determined only by the allele that is inherited from the father. Because we know that male A has to be *Igf2 Igf2*, we know that it can produce only normal offspring. Because both females A and B both produced dwarf offspring, male A cannot be the father. In contrast, male B could be either *Igf2 Igf2* or *Igf2 Igf2⁻*. Because both females gave birth to dwarf babies (and because male A and male B were the only two male mice in the cage), we conclude that male B must be *Igf2 Igf2⁻* and is the father of both litters.

E10. In fruit flies, the expression of a male's X-linked genes is turned up twofold. In mice, one of the two X chromosomes is inactivated; that is why females and males produce the same total amount of mRNA for most X-linked genes. In *C. elegans*, the expression of hermaphrodite X-linked genes is turned down twofold. Overall, the total amount of expression of X-linked genes is the same in males and females (or hermaphrodites) of these three species. In fruit flies and *C. elegans*, heterozygous females and hermaphrodites express 50% of each allele compared to a homozygous male, so that heterozygous females and hermaphrodites produce the same total amount of mRNA from X-linked genes compared to males. Note: In heterozygous females of fruit flies, mice, and *C. elegans*, there is 50% of each gene product (compared to hemizygous males and homozygous females).

A.

B.

C.

E12. In the absence of UV light, we would expect all *sm^r* offspring. With UV light, we would expect a greater percentage of *sm^s* offspring.

Questions for Student Discussion/Collaboration

2. An infective particle is something in the cytoplasm that contains its own genetic material and isn't an organelle. Some symbiotic infective particles, such as those found in killer paramecia, are similar to mitochondria and chloroplasts in that they contain their own genomes and are known to be bacterial in origin. The observation that these endosymbiotic relationships can initiate in modern species tells us that endosymbiosis can spontaneously happen. Therefore, it is reasonable that it happened a long time ago and led to the evolution of mitochondria and chloroplasts.

CHAPTER 8

Conceptual Questions

C2. Small deletions and duplications are less likely to affect phenotype simply because they usually involve fewer genes. If a small deletion did have a phenotypic effect, you would conclude that a gene or genes in this region is required to have a normal phenotype.

C4. A gene family is a group of genes that are derived from the process of gene duplications. They have similar sequences, but the sequences have some differences due to the accumulation of mutations over many generations. The members of a gene family usually encode proteins with similar but specialized functions. The specialization may occur in different cells or at different stages of development.

C6. It has a pericentric inversion.

C8. There are four products from meiosis. One would be a normal chromosome, and one would contain the inversion shown in the drawing to conceptual question C6. The other two chromosomes would be dicentric or acentric with the following order of genes:

centromere		centromere	
A ↓	*B C D E F G H I J*	*D C B* ↓	*A* Dicentric
	M L K J I H G F E	*K L M*	Acentric

C10. In the absence of crossing over, alternate segregation would yield half the cells with two normal chromosomes and half with a balanced translocation. For adjacent-1 segregation, all cells will be unbalanced. Two cells would be

$A\ B\ C\ D\ E$ + $A\ I\ J\ K\ L\ M$

And the other two cells would be

$H\ B\ C\ D\ E$ + $H\ I\ J\ K\ L\ M$

C12. One of the parents may carry a balanced translocation between chromosomes 5 and 7. The phenotypically abnormal offspring has inherited an unbalanced translocation due to the segregation of translocated chromosomes during meiosis.

C14. A deficiency and an unbalanced translocation are more likely to have phenotypic effects because they create genetic imbalances. For a deficiency, there are too few copies of several genes, and for an unbalanced translocation, there are too many.

C16. It is due to a crossover within the inverted region. You should draw the inversion loop (as is done in Figure 8.12a). The crossover occurred between *P* and *U*.

C18. This person has a total of 46 chromosomes. However, this person would be considered aneuploid rather than euploid, because one of the sets is missing a sex chromosome and one set has an extra copy of chromosome 21.

C20. It may be related to genetic balance. In aneuploidy, there is an imbalance in gene expression between the chromosomes found in their normal copy number versus those that are either too many or too few. In polyploidy, the balance in gene expression is still maintained.

C22. The male offspring is the result of nondisjunction during oogenesis. The female produced an egg without any sex chromosomes. The male parent transmitted a single X chromosome carrying the red allele. This produces an X0 male offspring with red eyes.

C24. Trisomies 13, 18, and 21 survive because the chromosomes are small and probably contain fewer genes compared to the larger chromosomes. Individuals with abnormal numbers of X chromosomes can survive because the extra copies are converted to transcriptionally inactive Barr bodies. The other trisomies are lethal because they cause a great amount of imbalance between the level of gene expression on the normal diploid chromosomes relative to the chromosomes that are trisomic.

C26. Endopolyploidy means that a particular somatic tissue is polyploid even though the rest of the organism is not. The biological significance is not entirely understood, although it has been speculated that an increase in chromosome number in certain cells may enhance their ability to produce specific gene products that are needed in great abundance.

C28. In certain types of cells, such as salivary cells, the homologous chromosomes pair with each other and then replicate approximately nine times to produce a polytene chromosome. The centromeres from each chromosome aggregate with each other at the chromocenter. This structure has six arms that arise from the single arm of two telocentric chromosomes (the X and 4) and two arms each from the metacentric chromosomes 2 and 3.

C30. The turtles are two distinct species that appear phenotypically identical. The turtles with 48 chromosomes are polyploid relatives (i.e., tetraploids) of the species with 24 chromosomes. In animals, it is somewhat hard to imagine how this could occur because animals cannot self-fertilize, so there had to be two animals (i.e., one male and one female) that became tetraploids. It is easy to imagine how one animal could become a tetraploid; complete nondisjunction could occur during the first cell division of a fertilized egg, thereby creating a tetraploid cell that continued to develop into a tetraploid animal. This would have to happen independently (i.e., in two individuals of opposite sex) to create a tetraploid species. If you mated a tetraploid turtle with a diploid turtle, the offspring would be triploid and probably phenotypically normal. However, the triploid offspring would be sterile because they would make highly aneuploid gametes.

C32. Polyploid, triploid, and euploid should not be used.

C34. The boy carries a translocation involving chromosome 21: probably a translocation in which nearly all of chromosome 21 is translocated to chromosome 14. He would have one normal copy of chromosome 14, one normal copy of chromosome 21, and the translocated chromosome that contains both chromosome 14 and chromosome 21. This boy is phenotypically normal because the total amount of genetic material is normal, although the total number of chromosomes is 45 (because chromosome 14 and chromosome 21 are fused into a single chromosome). His sister has familial Down syndrome because she has inherited the translocated chromosome, but she also must have one copy of chromosome 14 and two copies of chromosome 21. She has the equivalent of three copies of chromosome 21 (i.e., two normal copies and one copy fused with chromosome 14). This is why she has familial Down syndrome. One of the parents of these two children is probably normal with regard to karyotype (i.e., the parent has 46 normal chromosomes). The other parent would have a karyotype that would be like the phenotypically normal boy.

C36. Nondisjunction is a mechanism whereby the chromosomes do not segregate equally into the two daughter cells. This can occur during meiosis to produce cells with altered numbers of chromosomes, or it can occur during mitosis to produce a genetic mosaic individual. A third way to alter chromosome number is by interspecies crosses that produce an alloploid.

C38. A mutation occurred during early embryonic development to create the blue patch of tissue. One possibility is a mitotic nondisjunction in which the two chromosomes carrying the b allele went to one cell and the two chromosomes carrying the B allele went to the other daughter cell. A second possibility is that the chromosome carrying the B allele was lost. A third possibility is that the B allele was deleted. This would cause the recessive b allele to exhibit pseudodominance.

C40. Homeologous chromosomes are chromosomes from two species that are evolutionarily related to each other. For example, chromosome 1 in the sable antelope and the roan antelope are homeologous; they carry many of the same genes.

C42. In general, Turner syndrome could be due to nondisjunction during either oogenesis or spermatogenesis. However, the Turner individual with color blindness is due to nondisjunction during spermatogenesis. The sperm lacked a sex chromosome, due to nondisjunction, and the egg carried an X chromosome with the recessive color blindness allele. This X chromosome had to be inherited from the mother, because the father was not color-blind. The mother must be heterozygous for the recessive color blindness allele, and the father is hemizygous for the normal allele. Therefore, the mother must have transmitted a single X chromosome carrying the color blindness allele to her offspring, indicating that nondisjunction did not occur during oogenesis.

Experimental Questions

E2. Due to the mix up, the ratio of green/red fluorescence would be 0.5 in regions where the cancer cells had a normal amount of DNA. If a duplication occurred on both chromosomes of cancer cells, the ratio would be 1.0. If a deletion occurred on a single chromosome, the ratio would be 0.25.

E4. Colchicine interferes with the spindle apparatus and thereby causes nondisjunction. At high concentrations, it can cause complete nondisjunction and produce polyploid cells.

E6. The primary purpose is to generate diploid strains that are homozygous for all of their genes. Because the pollen is haploid, it has

only one copy of each gene. The strain can later be made diploid (and homozygous) by treatment with colchicine.

E8. First, you would cross the two strains together. It is difficult to predict the phenotype of the offspring. Nevertheless, you would keep crossing offspring to each other and backcrossing them to the parental strains until you obtained a great-tasting tomato strain that was resistant to heat and the viral pathogen. You could then make this strain tetraploid by treatment with colchicine. If you crossed the tetraploid strain with your great-tasting diploid strain that was resistant to heat and the viral pathogen, you may get a triploid that had these characteristics. This triploid would probably be seedless.

E10. A polytene chromosome is formed when a chromosome replicates many times, and the chromatids lie side by side, as shown in Figure 8.21. The homologous chromosomes also lie side by side. Therefore, if there is a deletion, there will be a loop. The loop is the segment that is not deleted from one of the two homologues.

E12. Because the plant giving the pollen is heterozygous for many genes, some of the pollen grains may be haploid for recessive alleles, which are nonbeneficial or even lethal. However, some pollen grains may inherit only the dominant (beneficial) alleles and grow quite well.

Questions for Student Discussion/Collaboration

2. There are many possibilities. The students could look in agriculture and botany books to find many examples. In the insect world, there are interesting examples of euploidy affecting gender determination. Among amphibians and reptiles, there are also several examples of closely related species that have euploid variation.

4. 1. Polyploids are often more robust and disease resistant.

   2. Allopolyploids may have useful combinations of traits.

   3. Hybrids are often more vigorous; they can be generated from monoploids.

   4. Strains with an odd number of chromosome sets (e.g., triploids) are usually seedless.

CHAPTER 9

Conceptual Questions

C2. The transformation process is described in Chapter 6.

   1. A fragment of DNA binds to the cell surface.

   2. It penetrates the cell wall/cell membrane.

   3. It enters the cytoplasm.

   4. It recombines with the chromosome.

   5. The genes within the DNA are expressed (i.e., transcription and translation).

   6. The gene products create a capsule. That is, they are enzymes that synthesize a capsule using cellular molecules as building blocks.

C4. The building blocks of a nucleotide are a sugar (ribose or deoxyribose), a nitrogenous base, and a phosphate group. In a nucleotide, the phosphate is already linked to the 5' position on the sugar. When two nucleotides are hooked together, a phosphate on one nucleotide forms a covalent bond with the 3' hyrdroxyl group on another nucleotide.

C6. The structure is a phosphate group connecting two sugars at the 3' and 5' positions, as shown in Figure 9.11.

C8. 3'–CCGTAATGTGATCCGGA–5'

C10. A drawing of a DNA helix with 10 bp per turn would look like Figure 9.17 in the textbook. To make 15 bp per turn, you would have to add 5 more base pairs, but the helix should still make only one complete turn.

C12. The nucleotide bases occupy the major and minor grooves. Phosphate and sugar are found in the backbone. If a DNA-binding protein does not recognize a nucleotide sequence, it probably is not binding in the grooves but instead is binding to the DNA backbone (i.e., sugar–phosphate sequence). DNA-binding proteins that recognize a base sequence must bind into a major or minor groove of the DNA, which is where the bases would be accessible to a DNA-binding protein. Most DNA-binding proteins that recognize a base sequence fit into the major groove. By comparison, other DNA-binding proteins, such as histones, which do not recognize a base sequence, bind to the DNA backbone.

C14. The structure is shown in Figure 9.8. You begin numbering at the carbon that is to the right of the ring oxygen and continue to number the carbon atoms in a clockwise direction. Antiparallel means that the backbones are running in the opposite direction. In one strand, the sugar carbons are oriented in a 3' to 5' direction, while in the other strand they are oriented in a 5' to 3' direction.

C16. Double-stranded RNA is more like A DNA than B DNA. See the text for a discussion of A-DNA structure.

C18. Its nucleotide base sequence

C20. G = 32%, C = 32%, A = 18%, T = 18%.

C22. One possibility is a sequential mechanism. First, the double helix could unwind and replicate itself as described in Chapter 11. This would produce two double helices. Next, the third strand (bound in the major groove) could replicate itself via a semiconservative mechanism. This new strand could be copied to make a copy that is identical to the strand that lies in the major groove. At this point, you would have two double helices and two strands that could lie in the major groove. These could assemble to make two triple helices.

C24. Lysines and arginines, and also polar amino acids

C26. This DNA molecule contains 280 bp. There are 10 base pairs per turn, so there are 28 complete turns.

C28. A hydroxyl group is at the 3' end, and a phosphate group is at the 5' end.

C30. Not necessarily. The AT/GC rule is required only of double-stranded DNA molecules.

C32. The first thing we need to do is to determine how many base pairs are in this DNA molecule. The linear length of 1 base pair is 0.34 nm, which equals 0.34×10^{-9} m. One centimeter equals 10^{-2} meters.

$$\frac{10^{-2}}{0.34 \times 10^{-9}} = 2.9 \times 10^7 \, bp$$

There are approximately 2.9×10^7 bp in this DNA molecule, which equals 5.8×10^7 nucleotides. If 15% are adenine, then 15% must also be thymine. This leaves 70% for cytosine and guanine. Because cytosine and guanine bind to each other, there must be 35% cytosine and 35% guanine. If we multiply 5.8×10^7 times 0.35, we get

$(5.8 \times 10^7)(0.35) = 2.0 \times 10^7$ cytosines, or about 20 million cytosines

C34. The methyl group is not attached to one of the atoms that hydrogen bonds with guanine, so methylation would not directly affect hydrogen bonding. It could indirectly affect hydrogen bonding if it perturbed the structure of DNA. Methylation may affect gene expression because

it could alter the ability of proteins to recognize DNA sequences. For example, a protein might bind into the major groove by interacting with a sequence of bases that includes one or more cytosines. If the cytosines are methylated, this may prevent a protein from binding into the major groove properly. Alternatively, methylation could enhance protein binding. In Chapter 7, we considered DNA-binding proteins that were influenced by the methylation of DMRs (differentially methylated regions) that occur during genomic imprinting.

Experimental Questions

E2. A. There are different possible reasons why most of the cells were not transformed.

   1. Most of the cells did not take up any of the type IIIS DNA.

   2. The type IIIS DNA was usually degraded after it entered the type IIR bacteria.

   3. The type IIIS DNA was usually not expressed in the type IIR bacteria.

   B. The antibody/centrifugation steps were used to remove the bacteria that had not been transformed. It enabled the researchers to determine the phenotype of the bacteria that had been transformed. If this step was omitted, there would have been so many colonies on the plate it would have been difficult to identify any transformed bacterial colonies, because they would have represented a very small proportion of the total number of bacterial colonies.

   C. They were trying to demonstrate that it was really the DNA in their DNA extract that was the genetic material. It was possible that the extract was not entirely pure and could contain contaminating RNA or protein. However, treatment with RNase and protease did not prevent transformation, indicating that RNA and protein were not the genetic material. In contrast, treatment with DNase blocked transformation, confirming that DNA is the genetic material.

E4. A. There are several possible explanations why about 35% of the DNA is in the supernatant. One possibility is that not all of the DNA was injected into the bacterial cells. Alternatively, some of the cells may have been broken during the shearing procedure, thereby releasing the DNA.

   B. If the radioactivity in the pellet had been counted instead of the supernatant, the following figure would be produced:

   C. ^{32}P and ^{35}S were chosen as radioisotopes to label the phages because phosphorous is found in nucleic acids, while sulfur is found only in proteins.

   D. There are multiple reasons why less than 100% of the phage protein is removed from the bacterial cells during the shearing process. For example, perhaps the shearing just is not strong enough to remove all of the phages, or perhaps the tail fibers remain embedded in the bacterium and only the head region is sheared off.

E6.   This is really a matter of opinion. The Avery, MacLeod, and McCarty experiment seems to indicate directly that DNA is the genetic material, because DNase prevented transformation and RNase and protease did not. However, one could argue that the DNA is required for the rough bacteria to take up some other contaminant in the DNA preparation. It would seem that the other contaminant would not be RNA or protein. The Hershey and Chase experiments indicate that DNA is being injected into bacteria, although quantitatively the results are not entirely convincing. Some ^{35}S-labeled protein was not sheared off, so the results do not definitely rule out the possibility that protein could be the genetic material. But the results do indicate that DNA is the more likely candidate.

E8.   A. The purpose of chromatography was to separate the different types of bases.

B. It was necessary to separate the bases and determine the total amount of each type of base. In a DNA strand, all the bases are found within a single molecule, so it is difficult to measure the total amount of each type of base. When the bases are removed from the strand, each type can be purified, and then the total amount of each type of base can be measured by spectroscopy.

C. Chargaff's results would probably not be very convincing if done on a single species. The strength of his data was that all species appeared to conform to the AT/GC rule, suggesting that this is a consistent feature of DNA structure. In a single species, the observation is that A = T and G = C could occur as a matter of chance.

Questions for Student Discussion/Collaboration

2.   There are many possibilities. You could use a DNA-specific chemical and show that it causes heritable mutations. Perhaps you could inject an oocyte with a piece of DNA and produce a mouse with a new trait.

CHAPTER 10

Conceptual Questions

C2.   Viruses also need sequences that enable them to be replicated. These sequences are equivalent to the origins of replication found in bacterial and eukaryotic chromosomes.

C4.   A bacterium with two nucleoids is similar to a diploid eukaryotic cell because it would have two copies of each gene. The bacterium is different, however, with regard to alleles. A eukaryotic cell can have two different alleles for the same gene. For example, a cell from a pea plant could be heterozygous, *Tt*, for the gene that affects height. By comparison, a bacterium with two nucleoids has two identical chromosomes. Therefore, a bacterium with two nucleoids is homozygous for its chromosomal genes. Note: As discussed in Chapter 6, a bacterium can contain another piece of DNA, called an F' factor, that can carry a few genes. The alleles on an F' factor can be different from the alleles on the bacterial chromosome.

C6.   A. One loop is 40,000 bp. One base pair is 0.34 nm, which equals 0.34×10^{-3} μm. If we multiply the two together:

$$(40,000)(0.34 \times 10^{-3}) = 13.6 \ \mu m$$

B. Circumference = πD

$$13.6 \ \mu m = \pi D$$

$$D = 4.3 \ \mu m$$

C. No, it is too big to fit inside of *E. coli*. Supercoiling is needed to make the loops more compact.

C8.   These drugs would diminish the amount of negative supercoiling in DNA. Negative supercoiling is needed to compact the chromosomal DNA, and it also aids in strand separation. Bacteria might not be able to survive and/or transmit their chromosomes to daughter cells if their DNA was not compacted properly. Also, because negative supercoiling aids in strand separation, these drugs would make it more difficult for the DNA strands to separate. Therefore, the bacteria would have

a difficult time transcribing their genes and replicating their DNA, because both processes require strand separation. As discussed in Chapter 11, DNA replication is needed to make new copies of the genetic material to transmit from mother to daughter cells. If DNA replication was inhibited, the bacteria could not grow and divide into new daughter cells. As discussed in Chapters 12–14, gene transcription is necessary for bacterial cells to make proteins. If gene transcription was inhibited, the bacteria could not make many proteins that are necessary for survival.

C10.

C12.   The centromere is the attachment site for the kinetochore, which attaches to the spindle. If a chromosome is not attached to the spindle, it is free to "float around" within the cell, and it may not be near a pole when the nuclear membrane re-forms during telophase. If a chromosome is left outside of the nucleus, it is degraded during interphase. That is why the chromosome without a centromere may not be found in daughter cells.

C14.   Highly repetitive DNA, as its name suggests, is a DNA sequence that is repeated many times, from tens of thousands to millions of times throughout the genome. It can be interspersed in the genome or found clustered in a tandem array, in which a short nucleotide sequence is repeated many times in a row. In DNA renaturation studies, highly repetitive DNA renatures at a much faster rate because there are many copies of the complementary sequences.

C16.   During interphase (i.e., G_1, S, and G_2), the euchromatin is found primarily as a 30 nm fiber in a radial loop configuration. Most interphase chromosomes also have some heterochromatic regions where the radial loops are more highly compacted. During M phase, each chromosome becomes entirely heterochromatic, which is needed for the proper sorting of the chromosomes during nuclear division.

C18.

C20.   During interphase, the chromosomes are found within the cell nucleus. They are less tightly packed and are transcriptionally active. Segments of chromosomes are anchored to the nuclear matrix. During M phase, the chromosomes become highly condensed, and the nuclear membrane is fragmented into vesicles. The chromosomal DNA remains anchored to a scaffold, formed from the nuclear matrix. The chromosomes eventually become attached to the spindle apparatus via microtubules that attach to the kinetochore, which is attached to the centromere.

C22.   There are 146 bp around the core histones. If the linker region is 54 bp, we expect 200 bp of DNA (i.e., 146 + 54) for each nucleosome and linker region. If we divide 46,000 bp by 200 bp, we get 230. Because there are two molecules of H2A for each nucleosome, there would be 460 molecules of H2A in a 46,000-bp sample of DNA.

C24.   The role of the core histones is to form the nucleosomes. In a nucleosome, the DNA is wrapped 1.65 times around the core histones. Histone H1 binds to the linker region. It may play a role in compacting the DNA into a 30 nm fiber.

C26.   The answer is B and E. A Barr body is composed of a type of highly compacted chromatin called heterochromatin. Euchromatin is not so compacted. A Barr body is not composed of euchromatin. A Barr body is one chromosome, the X chromosome. The term genome refers to all the types of chromosomes that make up the genetic composition of an individual.

C28. First, they may directly influence interactions between nucleosomes. Second, histone modifications provide binding sites that are recognized by proteins. The pattern of covalent modifications of amino terminal tails acts much like a code in specifying alterations in chromatin structure. One pattern of histone modification may attract proteins that cause the chromatin to become even more compact, which would silence the transcription of genes in the region. Alternatively, a different combination of histone modifications may attract proteins that serve to loosen the chromatin and thereby promote gene transcription.

Experimental Questions

E2. This type of experiment gives the relative proportions of highly repetitive, moderately repetitive, and unique DNA sequences within the genome. The highly repetitive sequences renature at a fast rate, the moderately repetitive sequences renature at an intermediate rate, and the unique sequences renature at a slow rate.

E4. Supercoiled DNA would look curled up into a relatively compact structure. You could add different purified topoisomerases and see how they affect the structure via microscopy. For example, DNA gyrase relaxes positive supercoils, and topoisomerase I relaxes negative supercoils. If you added topoisomerase I to a DNA preparation and it became less compacted, then the DNA was negatively supercoiled.

E6.

E8. You would get DNA fragments of about 446 to 496 bp (i.e., 146 bp plus 300 to 350 bp).

E10. Histones are positively charged, and DNA is negatively charged. They bind to each other by these ionic interactions. Salt is composed of positively charged ions and negatively charged ions. For example, when dissolved in water, NaCl becomes individual ions of Na^+ and Cl^-. When chromatin is exposed to a salt such as NaCl, the positively charged Na^+ could bind to the DNA, and the negatively charged Cl^- could bind to the histones. This would prevent the histones and DNA from binding to each other.

Questions for Student Discussion/Collaboration

2. This is a matter of opinion. It seems strange to have so much DNA that seems to have no obvious function. It's a waste of energy. Perhaps it has a function that we don't know about yet. On the other hand, evolution does allow bad things to accumulate within genomes, such as genes that cause diseases, etc. Perhaps this is just another example of the negative consequences of evolution.

CHAPTER 11

Conceptual Questions

C2. Bidirectional replication refers to DNA replication in both directions starting from one origin.

C4. A.
```
      TTGGHTGUTGG
      HHUUTHUGHUU
```
B.
```
      TTGGHTGUTGG
      HHUUTHUGHUU
            ↓
   TTGGHTGUTGG CCAAACACCAA
   AACCCACAACC HHUUTHUGHUU
            ↓
TTGGHTGUTGG TTGGGTGTTGG CCAAACACCAA CCAAACACCAA
AACCCACAACC AACCCACAACC GGTTTGTGGTT HHUUTHUGHUU
```

C6. Let's assume there are 4,600,000 bp of DNA and that DNA replication is bidirectional at a rate of 750 nucleotides per second.

If there were just a single replication fork,

4,600,000/750 = 6,133 seconds, or 102.2 minutes.

Because replication is bidirectional, $^{102.2}/_2$ = 51.1 minutes.

Actually, this is an average value based on a variety of growth conditions. Under optimal growth conditions, replication can occur substantially faster.

With regard to errors, if we assume an error rate of one mistake per 100,000,000 nucleotides,

4,600,000 × 1,000 bacteria = 4,600,000,000 nucleotides of replicated DNA.

4,600,000,000/100,000,000 = 46 mistakes.

When you think about it, this is pretty amazing. In this population, DNA polymerase would cause only 46 single mistakes in a total of 1,000 bacteria, each containing 4.6 million bp of DNA.

C8. DNA polymerase would slide from right to left. The new strand would be 3'–CTAGGGCTAGGCGTATGTAAATGGTCTAGTGGTGG–5'

C10. A. When looking at Figure 11.5, the first, second, and fourth DnaA boxes are running in the same direction, and the third and fifth are running in the opposite direction. Once you realize that, you can see the sequences are very similar to each other.

B. According to the direction of the first DnaA box, the consensus sequence is

```
TGTGGATAA
ACACCTATT
```

C. This sequence is nine nucleotides long. Because there are four kinds of nucleotides (i.e., A, T, G, and C), the chance of this sequence occurring by random chance is 4^{-9}, which equals once every 262,144 nucleotides. Because the *E. coli* chromosome is more than 10 times longer than this, it is fairly likely that this consensus sequence occurs elsewhere. The reason why there are not multiple origins, however, is because the origin has five copies of the consensus sequence very close together. The chance of having five copies of this consensus sequence occurring close together (as a matter of random chance) is very small.

C12. 1. According to the AT/GC rule, a pyrimidine always hydrogen bonds with a purine. A transition still involves a pyrimidine hydrogen bonding to a purine, but a transversion causes a purine to hydrogen bond with a purine or a pyrimidine to hydrogen bond with a pyrimidine. The structure of the double helix makes it much more difficult for this latter type of hydrogen bonding to occur.

2. The induced-fit phenomenon of the active site of DNA polymerase makes it unlikely for DNA polymerase to catalyze covalent bond formation if the wrong nucleotide is bound to the template strand. A transition mutation creates a somewhat bad interaction between the bases in opposite strands, but it is not as bad as the fit caused

by a transversion mutation. In a transversion, a purine is opposite another purine, or a pyrimidine is opposite a pyrimidine. This is a very bad fit.

3. The proofreading function of DNA polymerase is able to detect and remove an incorrect nucleotide that has been incorporated into the growing strand. A transversion is going to cause a larger distortion in the structure of the double helix and make it more likely to be detected by the proofreading function of DNA polymerase.

C14. Primase and DNA polymerase are able to knock the single-strand binding proteins off the template DNA.

C16. A. The right Okazaki fragment was made first. It is farthest away from the replication fork. The fork (not seen in this diagram) would be to the left of the three Okazaki fragments, and moving from right to left.

B. The RNA primer in the right Okazaki fragment would be removed first. DNA polymerase would begin by elongating the DNA strand of the middle Okazaki fragment and removing the right RNA primer with its 5' to 3' exonuclease activity. DNA polymerase I would use the 3' end of the DNA of the middle Okazaki fragment as a primer to synthesize DNA in the region where the right RNA primer is removed. If the middle fragment was not present, DNA polymerase could not fill in this DNA (because it needs a primer).

C. You need DNA ligase only at the right arrow. DNA polymerase I begins at the end of the left Okazaki fragment and synthesizes DNA to fill in the region as it removes the middle RNA primer. At the left arrow, DNA polymerase I is simply extending the length of the left Okazaki fragment. No ligase is needed here. When DNA polymerase I has extended the left Okazaki fragment through the entire region where the RNA primer has been removed, it hits the DNA of the middle Okazaki fragment. This occurs at the right arrow. At this point, the DNA of the middle Okazaki fragment has a 5' end that is a monophosphate. DNA ligase is needed to connect this monophosphate with the 3' end of the region where the middle RNA primer has been removed.

D. As mentioned in the answer to part C, the 5' end of the DNA in the middle Okazaki fragment is a monophosphate. It is a monophosphate because it was previously connected to the RNA primer by a phosphoester bond. At the location of the right arrow, there was only one phosphate connecting this deoxyribonucleotide to the last ribonucleotide in the RNA primer. For DNA polymerase to function, the energy to connect two nucleotides comes from the hydrolysis of the incoming triphosphate. In this location shown at the right arrow, however, the nucleotide is already present at the 5' end of the DNA, and it is a monophosphate. DNA ligase needs energy to connect this nucleotide with the left Okazaki fragment. It obtains energy from the hydrolysis of ATP or NAD$^+$.

C18. 1. It recognizes the origin of replication.

2. It initiates the formation of a replication bubble.

3. It recruits helicase to the region.

C20. The picture would depict a ring of helicase proteins traveling along a DNA strand and separating the two helices, as shown in Figure 11.6.

C22. The leading strand is primed once, at the origin, and then DNA polymerase III synthesizes DNA continuously in the direction of the replication fork. In the lagging strand, many short pieces of DNA (Okazaki fragments) are made. This requires many RNA primers. The primers are removed by DNA polymerase I, which then fills in the gaps with DNA. DNA ligase then covalently connects the Okazaki fragments. Having the enzymes within a complex such as a primosome or replisome provides coordination among the different steps in the replication process and thereby allows it to proceed faster and more efficiently.

C24. A processive enzyme is one that remains clamped to one of its substrates. In the case of DNA polymerase, it remains clamped to the template strand as it makes a new daughter strand. This is important to ensure a fast rate of DNA synthesis.

C26. The reason is the inability to synthesize DNA in the 3' to 5' direction and the need for a primer prevent replication at the 3' end of the DNA strands. Telomerase is different than DNA polymerase in that it uses a short RNA sequence, which is part of its structure, as a template for DNA synthesis. Because it uses this sequence many times in row, it produces a tandemly repeated sequence in the telomere at the 3' ends of linear chromosomes.

C28. Fifty, because two replication forks emanate from each origin of replication. DNA replication is bidirectional.

C30. A. Both reverse transcriptase and telomerase use an RNA template to make a complementary strand of DNA.

B. Because reverse transcriptase does not have a proofreading function, it is more likely for mistakes to occur. This creates many mutant strains of the virus. Some mutations might prevent the virus from proliferating. However, other mutations might prevent the immune system from battling the virus. These kinds of mutations would enhance the proliferation of the virus.

Experimental Questions

E2. A. You would probably still see a band of DNA, but you would only see a heavy band.

B. You would probably not see a band because the DNA would not be released from the bacteria. The bacteria would sediment to the bottom of the tube.

C. You would not see a band. UV light is needed to see the DNA, which absorbs light in the UV region.

E4. If you started with single-stranded DNA, you would need to add a primer (or primase), dNTPs, and DNA polymerase. If you started with double-stranded DNA, you would also need helicase. Adding single-strand binding protein and topoisomerase may also help.

E6. This is a critical step because you need to separate the radioactivity in the free nucleotides from the radioactivity in the newly made DNA strands. If you used an acid that precipitated free nucleotides and DNA strands, all of the radioactivity would be in the pellet. You would get the same amount of radioactivity in the pellet no matter how much DNA was synthesized into newly made strands. For this reason, the perchloric acid step is very critical. It separates radioactivity in the free nucleotides from radioactivity in the newly made strands.

E8. A. The left end is the 5' end. If you flip the sequence of the first primer around, you will notice it is complementary to the right end of the template DNA. The 5' end of the first primer binds to the 3' end of the template DNA.

B. The sequence would be 3'–CGGGGCCATG–5'. It could not be used because the 3' end of the primer is at the end of the template DNA. There wouldn't be any place for nucleotides to be added to the 3' end of the primer and bind to the template DNA strand.

E10. A. Heat is used to separate the DNA strands, so you do not need helicase.

B. Each primer must be a sequence that is complementary to one of the DNA strands. There are two types of primers, and each type binds to one of the two complementary strands.

C. A thermophilic DNA polymerase is used because DNA polymerases isolated from nonthermophilic species would be permanently inactivated during the heating phase of the PCR cycle. Remember that DNA polymerase is a protein, and most proteins are denatured by heating. However, proteins from thermophilic organisms have evolved to withstand heat, which is how thermophilic organisms survive at high temperatures.

D. With each cycle, the amount of DNA is doubled. Because there are initially 10 copies of the DNA, there will be 10×2^{27} copies after 27 cycles. $10 \times 2^{27} = 1.34 \times 10^9 = 1.34$ billion copies of DNA. As you can see, PCR can amplify the amount of DNA by a staggering amount!

Questions for Student Discussion/Collaboration

2. Basically, the idea is to add certain combinations of enzymes and see what happens. You could add helicase and primase to double-stranded DNA, along with radiolabeled ribonucleotides. Under these conditions, you would make short (radiolabeled) primers. If you also added DNA polymerase and deoxyribonucleotides, you would make DNA strands. Alternatively, if you added DNA polymerase plus an RNA primer, you wouldn't need to add primase. Or, if you used single-stranded DNA as a template rather than double-stranded DNA, you wouldn't need to add helicase.

CHAPTER 12

Conceptual Questions

C2. The release of sigma factor marks the transition to the elongation stage of transcription.

C4. GGCATTGTCA

C6. The most highly conserved positions are the first, second, and sixth. In general, when promoter sequences are conserved, they are more likely to be important for binding. That explains why changes are not found at these positions; if a mutation altered a conserved position, the promoter would probably not work very well. By comparison, changes are occasionally tolerated at the fourth position and frequently at the third and fifth positions. The positions that tolerate changes are less important for binding by sigma factor.

C8. This will not affect transcription. However, it will affect translation by preventing the initiation of polypeptide synthesis.

C10. Sigma factor can slide along the major groove of the DNA. In this way, it is able to recognize base sequences that are exposed in the groove. When it encounters a promoter sequence, hydrogen bonding between the bases and the sigma factor protein can promote a tight and specific interaction.

C12. DNA-G/RNA-C

DNA-C/RNA-G

DNA-A/RNA-U

DNA-T/RNA-A

The template strand is 3'–CCGTACGTAATGCCGTAGTGTGATCCCTAG–5' and the coding strand is 5'–GGCATGCATTACGGCATCACACTAGGGATC–3'. The promoter would be to the left (in the 3' direction) of the template strand.

C14. Transcriptional termination occurs when the hydrogen bonding is broken between the DNA and the part of the newly made RNA transcript that is located in the open complex.

C16. DNA helicase and ρ protein bind to a nucleic acid strand and travel in the 5' to 3' direction. When they encounter a double-stranded region, they break the hydrogen bonds between complementary strands. ρ protein is different from DNA helicase in that it moves along an RNA strand, while DNA helicase moves along a DNA strand. The purpose of DNA helicase function is to promote DNA replication; the purpose of ρ protein function is to promote transcriptional termination.

C18. A. Mutations that alter the uracil-rich region by introducing guanines and cytosines, and mutations that prevent the formation of the stem-loop structure.

B. Mutations that alter the termination sequence, and mutations that alter the ρ recognition site

C. Eventually, somewhere downstream from the gene, another transcriptional termination sequence would be found, and transcription would terminate there. This second termination sequence might be found randomly, or it might be at the end of an adjacent gene.

C20. Eukaryotic promoters are somewhat variable with regard to the pattern of sequence elements that may be found. In the case of structural genes that are transcribed by RNA polymerase II, it is common to have a TATA box, which is about 25 bp upstream from a transcriptional start site. The TATA box is important in the identification of the transcriptional start site and the assembly of RNA polymerase and various transcription factors. The transcriptional start site defines where transcription actually begins.

C22. It is primarily an accessibility problem. When the DNA is tightly wound around histones, it becomes difficult for large proteins, like RNA polymerase and transcription factors, to recognize the correct base sequence in the DNA and to catalyze the movement of the open complex. It is thought that a loosening or complete disruption of the nucleosome structure is necessary for transcription to occur.

C24. Hydrogen bonding is usually the predominant type of interaction when proteins and DNA follow an assembly and disassembly process. In addition, ionic bonding and hydrophobic interactions could occur. Covalent interactions would not occur. High temperature and high salt concentrations tend to break hydrogen bonds. Therefore, high temperature and high salt would inhibit assembly and stimulate disassembly.

C26. In bacteria, the 5' end of the tRNA is cleaved by RNase P. The 3' end is cleaved by a different endonuclease, and then a few nucleotides are digested away by an exonuclease that removes nucleotides until it reaches a CCA sequence.

C28. A ribozyme is an enzyme whose catalytic part is composed of RNA. Examples are RNase P and self-splicing group I and II introns. It is thought that the spliceosome may contain catalytic RNAs as well.

C30. Self-splicing means that an RNA molecule can splice itself without the aid of a protein. Group I and II introns can be self-splicing, although proteins can also enhance the rate of processing.

C32. In alternative splicing, variation occurs in the pattern of splicing, so the resulting mRNAs contain alternative combinations of exons. The biological significance is that two or more different proteins can be produced from a single gene. This is a more efficient use of the genetic material. In multicellular organisms, alternative splicing is often used in a cell-specific manner.

C34. As shown at the left side of Figure 12.20, the guanosine, which binds to the guanosine-binding site, does not have a phosphate group attached to it. This guanosine is the nucleoside that winds up at the 5' end of the intron. Therefore, the intron does not have a phosphate group at its 5' end.

C36. U5

Experimental Questions

E2. An R loop is a loop of DNA that occurs when RNA is hybridized to double-stranded DNA. While the RNA is hydrogen bonding to one of the DNA strands, the other strand does not have a partner to hydrogen bond with, so it bubbles out as a loop. RNA is complementary to the template strand, so that is the strand it binds to.

E4. The 1,100-nucleotide band would be observed from a normal individual (lane 1). A deletion that removed the −50 to −100 region would greatly diminish transcription, so the homozygote would produce hardly any of the transcript (just a faint amount, as shown in lane 2), and the heterozygote would produce roughly half as much of the 1,100-nucleotide transcript (lane 3) compared to a normal individual. A nonsense codon would not have an effect on transcription; it affects only translation. So the individual with this mutation would produce a normal amount of the 1,100-nucleotide transcript (lane 4). A mutation that removed the splice acceptor site would prevent splicing. Therefore, this individual would produce a 1,550-nucleotide transcript (actually, 1,547 to be precise, 1,550 minus 3). The Northern blot is shown here:

E6. A. It would not be retarded because ρ protein would not bind to the mRNA encoded by a gene that is terminated in a ρ-independent manner. The mRNA from such genes does not contain the sequence near the 3' end that acts as a recognition site for the binding of ρ protein.

B. It would be retarded because ρ protein would bind to the mRNA.

C. It would be retarded because U1 would bind to the pre-mRNA.

D. It would not be retarded because U1 would not bind to mRNA that has already had its introns removed. U1 binds only to pre-mRNA.

E8. A. mRNA molecules would bind to this column because they have a polyA tail. The string of adenine nucleotides in the polyA tail is complementary to stretch of thymine in the poly-dT column, so the two would hydrogen bond to each other. To purify mRNAs, one begins with a sample of cells; the cells need to be broken open by some technique such as homogenization or sonication. This would release the RNAs and other cellular macromolecules. The large cellular structures (organelles, membranes, etc.) could be removed from the cell extract by a centrifugation step. The large cellular structures would be found in the pellet, while soluble molecules such as RNA and proteins would stay in the supernatant. At this point, you would want the supernatant to contain a high salt concentration and neutral pH. The supernatant would then be poured over the poly-dT column. The mRNAs would bind to the poly-dT column and other molecules (i.e., other types of RNAs and proteins) would flow through the column. Because the mRNAs would bind to the poly-dT column via hydrogen bonds, to break the bonds, you could add a solution that contains a low salt concentration and/or a high pH. This would release the mRNAs, which would then be collected in a low salt/high pH solution as it dripped from the column.

B. The basic strategy is to attach a short stretch of DNA nucleotides to the column matrix that is complementary to the type of RNA that you want to purify. For example, if an rRNA contained a sequence 5'–AUUCCUCCA–3', a researcher could chemically synthesize an oligonucleotide with the sequence 3'–TAAGGAGGT–5' and attach it to the column matrix. To purify rRNA, one would use this 3'–TAAGGAGGT–5' column and follow the general strategy described in part A.

Questions for Student Discussion/Collaboration

2. RNA transcripts come in two basic types: those that function as RNA (e.g., tRNA, rRNA, etc.) versus those that are translated (i.e., mRNA). As described in this chapter, they play a myriad of functional roles. RNAs that form complexes with proteins carry out some interesting roles. In some cases, the role is to bind other types of RNA molecules. For example, rRNA in bacteria plays a role in binding mRNA. In other cases, the RNA plays a catalytic role. An example is RNaseP. The structure and function of RNA molecules may be enhanced by forming a complex with proteins.

CHAPTER 13

Conceptual Questions

C2. When we say the genetic code is degenerate, it means that more than one codon can specify the same amino acid. For example, GGG, GGC, GGA, and GGU all specify glycine. In general, the genetic code is nearly universal, because it is used in the same way by viruses, prokaryotes, fungi, plants, and animals. As discussed in Table 13.3, there are a few exceptions, which occur primarily in protists and yeast and mammalian mitochondria.

C4. A. This mutant tRNA would recognize glycine codons in the mRNA but would put in tryptophan amino acids where glycine amino acids are supposed to be in the polypeptide chain.

B. This mutation tells us that the aminoacyl-tRNA synthetase is primarily recognizing other regions of the tRNA molecule besides the anticodon region. In other words, tryptophanyl-tRNA synthetase (the aminoacyl-tRNA synthetase that attaches tryptophan) primarily recognizes other regions of the tRNA^{trp} sequence (that is, other than the anticodon region), such as the T- and D-loops. If aminoacyl-tRNA synthetases recognized only the anticodon region, we would expect glycyl-tRNA synthetase to recognize this mutant tRNA and attach glycine. That is not what happens.

C6. A. The answer is three. There are six leucine codons: UUA, UUG, CUU, CUC, CUA, and CUG. The anticodon AAU would recognize UUA and UUG. You would need two other tRNAs to efficiently recognize the other four leucine codons. These could be GAG and GAU or GAA and GAU.

B. The answer is one. There is only one codon, AUG, so you need only one tRNA with the anticodon UAC.

C. The answer is three. There are six serine codons: AGU, AGC, UCU, UCC, UCA, and UCG. You would need only one tRNA to recognize AGU and AGC. This tRNA could have the anticodon UCG or UCA. You would need two tRNAs to efficiently recognize the other four tRNAs. These could be AGG and AGU or AGA and AGU.

C8. 3'–CUU–5' or 3'–CUC–5'

C10. It can recognize 5'–GGU–3', 5'–GGC–3', and 5'–GGA–3'. All of these specify glycine.

C12. All tRNA molecules have some basic features in common. They all have a cloverleaf structure with three stem-loop structures. The second stem-loop contains the anticodon sequence that recognizes the codon sequence in mRNA. At the 3' end, there is an acceptor stem, with the sequence CCA, that serves as an attachment site for an amino acid. Most tRNAs also have base modifications that occur within their nucleotide sequences.

C14. The role of aminoacyl-tRNA synthetase is to specifically recognize tRNA molecules and attach the correct amino acid to them. This ability is sometimes described as the second genetic code because the specificity of the attachment is a critical step in deciphering the genetic code. For example, if a tRNA has a 3'–GGG–5' anticodon, it will recognize a 5'–CCC–3' codon, which should specify proline. It is essential that the aminoacyl-tRNA synthetase known as prolyl-tRNA-synthetase recognizes this tRNA and attaches proline to the 3' end. The other aminoacyl-tRNA synthetases should not recognize this tRNA.

C16. Bases that have been chemically modified can occur at various locations throughout the tRNA molecule. The significance of all of these modifications is not entirely known. However, within the anticodon region, base modification alters base pairing to allow the anticodon to recognize two or more different bases within the codon.

C18. No, it is not. Due to the wobble rules, the 5' base in the anticodon of a tRNA can recognize two or more bases in the third (3') position of the mRNA. Therefore, any given cell type synthesizes far fewer than 61 types of tRNAs.

C20. The assembly process is very complex at the molecular level. In eukaryotes, 33 proteins and one rRNA assemble to form a 40S subunit, and 49 proteins and three rRNAs assemble to form a 60S subunit. This assembly occurs within the nucleolus.

C22. A. On the surface of the 30S subunit and at the interface between the two subunits

B. Within the 50S subunit

C. From the 50S subunit

D. To the 30S subunit

C24. Most bacterial mRNAs contain a Shine-Dalgarno sequence, which is necessary for the binding of the mRNA to the small ribosomal subunit. This sequence, UUAGGAGGU, is complementary to a sequence in the 16S rRNA. Due to this complementarity, these sequences will hydrogen bond to each other during the initiation stage of translation.

C26. The ribosome binds at the 5' end of the mRNA and then scans in the 3' direction in search of an AUG start codon. If it finds one that reasonably obeys Kozak's rules, it will begin translation at that site. Aside from an AUG start codon, two other important features are a guanosine at the +4 position and a purine at the −3 position.

C28. The A (aminoacyl) site is the location where a tRNA carrying a single amino acid initially binds. The only exception is the initiator tRNA, which binds to the P (peptidyl) site. The growing polypeptide chain is removed from the tRNA in the P site and transferred to the amino acid attached to the tRNA in the A site. The ribosome translocates in the 3' direction, with the result that the two tRNAs in the P and A sites are moved to the E (exit) and P sites, and the uncharged tRNA in the E site is released.

C30. Sorting signals direct a protein to its correct location (i.e., compartment) within the cell. Proteins destined for the ER, Golgi complex, lysosomes, plasma membrane, or secretory vesicles have an SRP sorting signal at their amino terminal end. Nuclear proteins have a nuclear-localization sequence (NLS), and so on. These sorting signals are recognized by specific cellular components that sort the proteins to their correct destination.

C32. The initiation phase involves the binding of the Shine-Dalgarno sequence to the rRNA in the 30S subunit. The elongation phase involves the binding of anticodons in tRNA to codons in mRNA.

C34. A. The E site and P sites. (Note: A tRNA without an amino acid attached is only briefly found in the P site, just before translocation occurs.)

B. P site and A site. (Note: A tRNA with a polypeptide chain attached is only briefly found in the A site, just before translocation occurs.)

C. Usually the A site, except the initiator tRNA, which can be found in the P site.

C36. The tRNAs bind to the mRNA because their anticodon and codon sequences are complementary. When the ribosome translocates in the 5' to 3' direction, the tRNAs remain bound to their complementary codons, and the two tRNAs shift from the A site and P site to the P site and E site. If the ribosome moved in the 3' direction, it would have to dislodge the tRNAs and drag them to a new position where they would not (necessarily) be complementary to the mRNA.

C38. 52

Experimental Questions

E2. A. There could have been other choices, but this template would be predicted to contain a cysteine codon, UGU, but would not contain any alanine codons.

B. You do not want to use ^{35}S because the radiolabel would be removed during treatment with Raney nickel.

C. There would not be a significant amount of radioactivity incorporated into newly made polypeptides with or without Raney nickel treatment. The only radiolabeled amino acid in this experiment was cysteine, which became attached to tRNA^{cys}. When exposed to Raney nickel, these cysteines were converted to alanine but only after they were already attached to tRNA^{cys}. If there were not any cysteine codons in the mRNA template, the tRNA^{cys} would not recognize this mRNA. Therefore, we would not expect to see much radioactivity in the newly made polypeptides.

E4. The initiation phase of translation is very different in bacteria and in eukaryotes, so they would not be translated very efficiently. A bacterial mRNA would not be translated very efficiently in a eukaryotic translation system, because it lacks a cap structure attached to its 5' end. A eukaryotic mRNA would not have a Shine-Dalgarno sequence near its 5' end, so it would not be translated very efficiently in a bacterial translation system.

E6. Looking at the figure, the 5' end of the template DNA strand is toward the right side. The 5' ends of the mRNAs are farthest from the DNA, and the 3' ends of the mRNAs are closest to the DNA. The start codons are slightly downstream from the 5' ends of the RNAs.

E8.

E10. A. If codon usage were significantly different between kangaroo and yeast cells, this would inhibit the translation process. For example, if the preferred leucine codon in kangaroos was CUU, translation would probably be slow in a yeast translation system. We would expect the cell-free translation system from yeast cells to primarily contain leucine tRNAs with an anticodon sequence that is AAC, because this tRNA^{leu} would match the preferred yeast leucine codon, which is UUG. In a yeast translation system, there probably would not be a large amount of tRNA with an anticodon of GAA, which would match the preferred leucine codon, CUU, of kangaroos. For this reason, kangaroo mRNA would not be translated very well in a yeast translation system, but it probably would be translated to some degree.

B. The advantage of codon bias is that a cell can rely on a smaller population of tRNA molecules to efficiently translate its proteins. A disadvantage is that mutations, which do not change the amino acid sequence but do change a codon (e.g., UUG to UUA), may inhibit the production of a polypeptide if a preferred codon is changed to a nonpreferred codon.

Questions for Student Discussion/Collaboration

2. This could be a very long list. There are similarities along several lines:

1. There is a lot of molecular recognition going on, either between two nucleic acid molecules or between proteins and nucleic acid molecules. Students may see these as similarities or differences, depending on their point of view.

2. There is biosynthesis going on in both processes. Small building blocks are being connected together. This requires an input of energy.

3. There are genetic signals that determine the beginning and ending of these processes.

There are also many differences:

1. Transcription produces an RNA molecule with a similar structure to the DNA, whereas translation produces a polypeptide with a structure that is very different from RNA.

2. Depending on your point of view, it seems that translation is more biochemically complex, requiring more proteins and RNA molecules to accomplish the task.

CHAPTER 14

Conceptual Questions

C2. In bacteria, gene regulation greatly enhances the efficiency of cell growth. It takes a lot of energy to transcribe and translate genes. Therefore, a cell is much more efficient and better at competing in its environment if it expresses genes only when the gene product is needed. For example, a bacterium will express only the genes that are necessary for lactose metabolism when a bacterium is exposed to lactose. When the environment is missing lactose, these genes are turned off. Similarly, when tryptophan levels are high within the cytoplasm, the genes required for tryptophan biosynthesis are repressed.

C4. A. Regulatory protein

   B. Effector molecule

   C. DNA segment

   D. Effector molecule

   E. Regulatory protein

   F. DNA segment

   G. Effector molecule

C6. A mutation that has a *cis*-effect is within a genetic regulatory sequence, such as an operator site, that affects the binding of a genetic regulatory protein. A *cis*-effect mutation affects only the adjacent genes that the genetic regulatory sequence controls. A mutation having a *trans*-effect is usually in a gene that encodes a genetic regulatory protein. A *trans*-effect mutation can be complemented in a merozygote experiment by the introduction of a normal gene that encodes the regulatory protein.

C8. A. No transcription would take place. The *lac* operon could not be expressed.

   B. No regulation would take place. The operon would be continuously turned on.

   C. The rest of the operon would function normally, but none of the transacetylase would be made.

C10. Diauxic growth refers to the phenomenon in which a cell first uses up one type of sugar (such as glucose) before it begins to metabolize a second sugar (such as lactose). In this case, it is caused by gene regulation. When a bacterial cell is exposed to both sugars, the uptake of glucose causes the cAMP levels in the cell to fall. When this occurs, the catabolite activator protein (CAP) is removed from the *lac* operon, so it is not able to be activated by CAP.

C12. A mutation that prevented the lac repressor from binding to the operator would make the *lac* operon constitutive only in the absence of glucose. However, this mutation would not be entirely constitutive because transcription would be inhibited in the presence of glucose. The disadvantage of constitutive expression of the *lac* operon is that the bacterial cell would waste a lot of energy transcribing the genes and translating the mRNA when lactose was not present.

C14. A. Without *araO₂*, the repression of the *ara* operon could not occur. The operon would be constitutively expressed at high levels because AraC protein could still activate transcription of the *ara* operon by binding to *araI*. The presence of arabinose would have no effect. Note: The binding of arabinose to AraC is not needed to form an AraC dimer at *araI*. The dimer is able to form because the loop has been broken. This point may be figured out if you notice that an AraC dimer is bound to *araO₁* in the presence and absence of arabinose (see Figure 14.12).

   B. Without *araO₁*, the AraC protein would be overexpressed. It would probably require more arabinose to alleviate repression. In addition, activation might be higher because there would be more AraC protein available.

   C. Without *araI*, transcription of the *ara* operon cannot be activated. You might get a very low level of constitutive transcription.

D. Without *araO₂*, the repression of the *ara* operon could not occur. However, without *araI*, transcription of the *ara* operon cannot be activated. You might get a very low level of constitutive transcription.

C16. A. Attenuation will not occur because loop 2–3 will form.

   B. Attenuation will occur because 2–3 cannot form, so 3–4 will form.

   C. Attenuation will not occur because 3–4 cannot form.

   D. Attenuation will not occur because 3–4 cannot form.

C18. The addition of Gs and Cs into the U-rich sequence would prevent attenuation. The U-rich sequence promotes the dissociation of the mRNA from the DNA, when the terminator stem-loop forms. This causes RNA polymerase to dissociate from the DNA and thereby causes transcriptional termination. The UGGUUGUC sequence would probably not dissociate because of the Gs and Cs. Remember that GC base pairs have three hydrogen bonds and are more stable than AU base pairs, which have only two hydrogen bonds.

C20. It takes a lot of cellular energy to translate mRNA into a protein. A cell wastes less energy if it prevents the initiation of translation rather than a later stage such as elongation or termination.

C22. One mechanism is that histidine could act as corepressor that shuts down the transcription of the histidine synthetase gene. A second mechanism would be that histidine could act as an inhibitor via feedback inhibition. A third possibility is that histidine inhibits the ability of the mRNA encoding histidine synthetase to be translated. Perhaps it induces a gene that encodes an antisense RNA. If the amount of histidine synthetase protein was identical in the presence and absence of extracellular histidine, a feedback inhibition mechanism is favored, because this affects only the activity of the histidine synthetase enzyme, not the amount of the enzyme. The other two mechanisms would diminish the amount of this protein.

C24. The two proteins are similar in that both bind to a segment of DNA and repress transcription. They are different in three ways. (1) They recognize different effector molecules (i.e., the lac repressor recognizes allolactose, and the trp repressor recognizes tryptophan). (2) Allolactose causes the lac repressor to release from the operator, while tryptophan causes the trp repressor to bind to its operator. (3) The sequences of the operator sites that these two proteins recognize are different from each other. Otherwise, the lac repressor could bind to the *trp* operator, and the trp repressor could bind to the *lac* operator.

C26. In the lytic cycle, the virus directs the bacterial cell to make more virus particles until eventually the cell lyses and releases them. In the lysogenic cycle, the viral genome is incorporated into the host cell's genome as a prophage. It remains there in a dormant state until some stimulus causes it to excise itself from the bacterial chromosome and enter the lytic cycle.

C28. The O_R region contains three operator sites and two promoters. P_{RM} and P_R transcribe in opposite directions. The λ repressor will first bind to O_{R1} and then O_{R2}. The binding of the λ repressor to O_{R1} and O_{R2} inhibits transcription from P_R and thereby switches off the lytic cycle. Early in the lysogenic cycle, the λ repressor protein concentration may become so high that it will occupy O_{R3}. Later, when the λ repressor concentration begins to drop, it will first be removed from O_{R3}. This allows transcription from P_{RM} and maintains the lysogenic cycle. By comparison, the cro protein has its highest affinity for O_{R3}, and so it binds there first. This blocks transcription from P_{RM} and thereby switches off the lysogenic cycle. The cro protein has a similar affinity for O_{R2} and O_{R1}, and so it may occupy either of these sites next. It will bind to both O_{R2} and O_{R1}. This turns down the expression from P_R, which is not needed in the later stages of the lytic cycle.

C30. It would first increase the amount of cro protein, so the lytic cycle would be favored.

C32. Neither cycle could be followed. As shown in Figure 14.19, N protein is needed to make a longer transcript from P_L for the lysogenic cycle and also to make a longer transcript from P_R for the lytic cycle.

C34. If the F^- strain is lysogenic for phage λ, the λ repressor is already being made in that cell. If the F^- strain receives genetic material from an *Hfr* strain, you would not expect it to have an effect on the lysogenic cycle, which is already established in the F^- cell. However, if the *Hfr* strain is lysogenic for λ and the F^- strain is not, the *Hfr* strain could transfer the integrated λ DNA (i.e., the prophage) to the F^- strain. The cytoplasm of the F^- strain would not contain any λ repressor. Therefore, this λ DNA could choose between the lytic and lysogenic cycle. If it follows the lytic cycle, the F^- recipient bacterium will lyse.

Experimental Questions

E2. In samples loaded in lanes 1 and 4, we expect the repressor to bind to the operator because no lactose is present. In the sample loaded into lane 4, the CAP protein could still bind cAMP because there is no glucose. However, there really is no difference between lanes 1 and 4, so it does not look like the CAP can activate transcription when the lac repressor is bound. If we compare samples loaded into lanes 2 and 3, the lac repressor would not be bound in either case, and the CAP would not be bound in the sample loaded into lane 3. There is less transcription in lane 3 compared to lane 2, but because there is some transcription seen in lane 3, we can conclude that the removal of the CAP (because cAMP levels are low) is not entirely effective at preventing transcription. Overall, the results indicate that the binding of the lac repressor is more effective at preventing transcription of the *lac* operon compared to the removal of the CAP.

E4. A. Yes, if you do not sonicate, then β-galactosidase will not be released from the cell, and not much yellow color will be observed. (Note: You may observe a little yellow color because some β-ONPG may be taken into the cell.)

   B. No, you should still get yellow color in the first two tubes even if you forgot to add lactose because the unmated strain does not have a functional lac repressor.

   C. Yes, if you forgot to add β-ONPG, you could not get yellow color because the cleavage of β-ONPG by β-galactosidase is what produces the yellow color.

E6. You could mate a strain that has an F' factor carrying a normal *lac* operon and a normal *lacI* gene to this mutant strain. Because the mutation is in the operator site, you would still continue to get expression of β-galactosidase, even in the absence of lactose.

E8. In this case, things are more complex because AraC acts as a repressor and an activator protein. If AraC were missing due to mutation, there would not be repression or activation of the *ara* operon in the presence or absence of arabinose. It would be expressed constitutively at low levels. The introduction of a normal *araC* gene into the bacterium on an F' factor would restore normal regulation (i.e., a *trans*-effect).

Questions for Student Discussion/Collaboration

2. A DNA loop may inhibit transcription by preventing RNA polymerase from recognizing the promoter. Or, it may inhibit transcription by preventing the formation of the open complex. Alternatively, a bend may enhance transcription by exposing the base sequence that the sigma factor of RNA polymerase recognizes. The bend may expose the major groove in such a way that this base sequence is more accessible to binding by sigma factor and RNA polymerase.

CHAPTER 15

Conceptual Questions

C2. Regulatory elements are relatively short genetic sequences that are recognized by regulatory transcription factors. After the regulatory transcription factor has bound to the regulatory element, it will affect the rate of transcription, either activating it or repressing it, depending on the action of the regulatory protein. Regulatory elements are typically located in the upstream region near the promoter, but they can be located almost anywhere (i.e., upstream and downstream) and even quite far from the promoter.

C4. Transcriptional activation occurs when a regulatory transcription factor binds to a regulatory element and activates transcription. Such proteins, also called transactivators, may interact with TFIID and/or mediator to promote the assembly of RNA polymerase and general transcription factors at the promoter region. They also could alter the structure of chromatin so that RNA polymerase and transcription factors are able to gain access to the promoter. Transcriptional inhibition occurs when a regulatory transcription factor inhibits transcription. Such repressors also may interact with TFIID and/or mediator to inhibit RNA polymerase.

C6. A. DNA binding

   B. DNA binding

   C. Protein dimerization

C8. For the glucocorticoid receptor to bind to a GRE, a steroid hormone must first enter the cell. The hormone then binds to the glucocorticoid receptor, which releases HSP90. The release of HSP90 exposes a nuclear localization signal (NLS) within the receptor, which enables it to dimerize and then enter the nucleus. Once inside the nucleus, the dimer binds to a pair of GREs, which activates transcription of the adjacent genes.

C10. Phosphorylation of the CREB protein causes it to act as a transcriptional activator. The unphosphorylated CREB protein can still bind to CREs, but it does not stimulate transcription.

C12. A. Eventually, the glucocorticoid hormone will be degraded by the cell. The glucocorticoid receptor binds the hormone with a certain affinity. The binding is a reversible process. Once the concentration of the hormone falls below the affinity of the hormone for the receptor, the receptor will no longer have the glucocorticoid hormone bound to it. When the hormone is released, the glucocorticoid receptor will change its conformation, and it will no longer bind to the DNA.

   B. An enzyme known as a phosphatase will eventually cleave the phosphate groups from the CREB protein. When the phosphates are removed, the CREB protein will stop activating transcription.

C14. The enhancer found in A would work, but the ones found in B and C would not. The sequence that is recognized by the transcriptional activator is 5'–GTAG–3' in one strand and 3'–CATC–5' in the opposite strand. This is the same arrangement found in A. In B and C, however, the arrangement is 5'–GATG–3' and 3'–CATC–5'. In the arrangement found in B and C, the two middle bases (i.e., A and T) are not in the correct order.

C16.

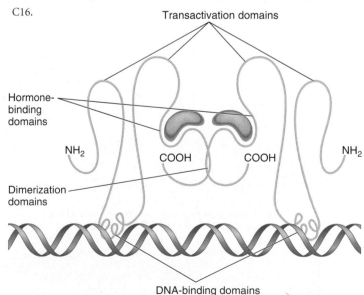

This is one hypothetical drawing of the glucocorticoid receptor. As discussed in your textbook, the glucocorticoid receptor forms a homodimer. The dimer shown here has rotational symmetry. If you flip the left side around, it has the same shape as the right side. The hormone-binding, DNA-binding, dimerization, and transactivation domains are labeled. The glucocorticoid hormone is shown in orange.

C18. The 30 nm fiber is the predominant form of chromatin during interphase. The chromatin must be converted from a closed conformation to an open conformation for transcription to take place. This "opening" process involves less packing of the chromatin and may involve changes in the locations of histone proteins. Transcriptional activators recruit histone acetyltransferase and ATP-dependent remodeling enzymes to the region, which leads to a conversion to the open conformation.

C20. The binding of the DNA to the core histones could

1. Prevent transcriptional activators from recognizing enhancer elements that are required for transcriptional activation.

2. Prevent RNA polymerase from binding to the core promoter.

3. Prevent RNA polymerase from forming an open complex in which the DNA strands separate.

C22. Perhaps the methyltransferase is responsible for methylating and inhibiting a gene that causes a cell to become a muscle cell. The methyltransferase is inactivated by the mutation.

C24. The function of splicing factors is to influence the selection of splice sites in RNA. In certain cell types, the concentration of particular splicing factors is higher than in other tissues. The high concentration of particular splicing factors, and the regulation of their activities, may promote the selection of particular splice sites and thereby lead to tissue-specific splicing.

C26. This person would be unable to make ferritin, because the IRP would always be bound to the IRE. The amount of transferrin receptor mRNA would be high, even in the presence of high amounts of iron, because the IRP would always remain bound to the IRE and stabilize the transferrin receptor mRNA. Such a person would not have any problem taking up iron into his/her cells. In fact, this person would take up a lot of iron via the transferrin receptor, even when the iron concentrations were high. Therefore, this person would not need more iron in the diet. However, excess iron in the diet would be very toxic for two reasons. First, the person cannot make ferritin, which prevents the toxic buildup of iron in the cytosol. Second, when iron levels are high, the person would continue to synthesize the transferrin receptor, which functions in the uptake of iron.

C28. A disadvantage of mRNAs with a short half-life is that the cells probably waste a lot of energy making them. If a cell needs the protein encoded by a short-lived mRNA, the cell has to keep transcribing the gene that encodes the mRNA because the mRNAs are quickly degraded. An advantage of short-lived mRNAs is that the cell can rapidly turn off protein synthesis. If a cell no longer needs the polypeptide encoded by a short-lived mRNA, it can stop transcribing the gene, and the mRNA will be quickly degraded. This will shut off the synthesis of more proteins rather quickly. With most long-lived mRNAs, it will take much longer to shut off protein synthesis after transcription has been terminated.

C30. First, the double-stranded miRNA could come from the transcription of a gene, as pre-miRNA. Second, it could come from a virus. Third, multiple insertions of the same gene in the genome may lead to both DNA strands being transcribed, thereby forming double-stranded RNA.

C32. If mRNA stability is low, this means it is degraded more rapidly. Therefore, low stability results in a low mRNA concentration. The length of the polyA tail is one factor that affects stability. A longer tail makes mRNA more stable. Certain mRNAs have sequences that affect their half-lives. For example, AU-rich elements (AREs) are found in many short-lived mRNAs. The AREs are recognized by cellular proteins that cause the mRNAs to be rapidly degraded.

Experimental Questions

E2. S1 nuclease cuts single-stranded DNA but not double-stranded DNA. If a gene is in an open conformation and cut by DNase I, it would not hybridize to the probe that is complementary to it. In this case, the single-stranded probe would be digested by the S1 nuclease. In contrast, if the gene is in a closed conformation and resistant to DNase I, the probe will hybridize to the gene and will not be digested by the S1 nuclease. Therefore, the ability of S1 nuclease to digest the probe tells you whether or not the gene is being cleaved by DNase I.

The precipitation step is needed to separate the DNA fragments from free nucleotides. If the radioactive probe is bound to a complementary DNA strand, it will be protected from degradation by S1 nuclease, and it will be found in the pellet. This occurs when the globin gene is in a closed conformation. In contrast, if the radioactive probe does not bind to a complementary DNA strand, it will be digested into free nucleotides, and the free nucleotides will remain in the supernatant. This occurs when the globin gene is in an open conformation.

E4. If the DNA band is 3,800 bp, it means the site is unmethylated, because it has been cut by NotI (NotI cannot cut methylated DNA). If the DNA fragment is 5,300 bp, the DNA is not cut by NotI, so we assume it is methylated. Now let's begin our interpretation of these data with lane 4. When the gene is isolated from root tissue, the DNA is not methylated because it runs at 3,800 bp. This suggests that gene T is expressed in root tissue. In the other samples, the DNA runs at 5,300 bp, indicating that the DNA is methylated. The pattern of methylation seen here is consistent with the known function of gene T. We would expect it to be expressed in root cells, because it functions in the uptake of phosphate from the soil. It would be silenced in the other parts of the plant via methylation.

E6. A. Based on the transformation of kidney cells, a silencer is in region B. Based on the transformation of pancreatic cells, an enhancer is in region A.

B. The pancreatic cells must not express the repressor protein that binds to the silencer in region B.

C. The kidney cells must express a repressor protein that binds to region B and represses transcription. Kidney cells express the downstream gene only if the silencer is removed. As mentioned in part B, this repressor is not expressed in pancreatic cells.

E8. When they injected just antisense mex-3 RNA, they observed lower but detectable levels of mex-3 mRNA. However, the injection of both sense and antisense RNA, which would form a double-stranded structure, resulted in a complete loss of mex-3 mRNA in the cells. In this way, double-stranded RNA would silence the expression of the mex-3 gene.

Questions for Student Discussion/Collaboration

2. Probably the most efficient method would be to systematically make deletions of progressively smaller sizes. For example, you could begin by deleting 20,000 bp on either side of the gene and see if that affects transcription. If you found that only the deletion on the 5' end of the gene had an effect, you could then start making deletions from the 5' end, perhaps in 10,000 bp or 5,000 bp increments until you localized response elements. You would then make smaller deletions in the putative region until it was down to a hundred or a few dozen nucleotides. At this point, you might conduct site-directed mutagenesis, as described in Chapter 18, as a way to specifically identify the regulatory element sequence.

CHAPTER 16

Conceptual Questions

C2. It is a gene mutation, a point mutation, a base substitution, a transition mutation, a deleterious mutation, a mutant allele, a nonsense mutation, a conditional mutation, and a temperature-sensitive lethal mutation.

C4. A. It would probably inhibit protein function, particularly if it was not near the end of the coding sequence.

   B. It may or may not affect protein function, depending on the nature of the amino acid substitution and whether the substitution is in a critical region of the protein.

   C. It would increase the amount of functional protein.

   D. It may affect protein function if the alteration in splicing changes an exon in the mRNA that results in a protein with a perturbed structure.

C6. A. Not appropriate, because the second mutation is at a different codon

   B. Appropriate

   C. Not appropriate, because the second mutation is in the same gene as the first mutation

   D. Appropriate

C8. A. Silent, because the same amino acid (glycine) is encoded by GGA and GGT

   B. Missense, because a different amino acid is encoded by CGA compared to GGA

   C. Missense, because a different amino acid is encoded by GTT compared to GAT

   D. Frameshift, because an extra base is inserted into the sequence

C10. One possibility is that a translocation may move a gene next to a heterochromatic region of another chromosome and thereby diminish its expression, or it could be moved next to a euchromatic region and increase its expression. Another possibility is that the translocation breakpoint may move the gene next to a new promoter or regulatory sequences that may now influence the gene's expression.

C12. A. No; the position (i.e., chromosomal location) of a gene has not been altered.

   B. Yes; the expression of a gene has been altered because it has been moved to a new chromosomal location.

   C. Yes; the expression of a gene has been altered because it has been moved to a new chromosomal location.

C14. If a mutation within the germ line is passed to an offspring, all the cells of the offspring's body will carry the mutation. A somatic mutation affects only the somatic cell in which it originated and all of the daughter cells that the somatic cell produces. If a somatic mutation occurs early during embryonic development, it may affect a fairly large region of the organism. Because germ-line mutations affect the entire organism, they are potentially more harmful (or beneficial), but this is not always the case. Somatic mutations can cause quite harmful effects, such as cancer.

C16. A thymine dimer can interfere with DNA replication because DNA polymerase cannot slide past the dimer and add bases to the newly growing strand. Alkylating mutagens such as nitrous acid will cause DNA replication to make mistakes in the base pairing. For example, an alkylated cytosine will base-pair with adenine during DNA replication, thereby creating a mutation in the newly made strand. A third example is 5-bromouracil, which is a thymine analogue. It may base-pair with guanine instead of adenine during DNA replication.

C18. During TNRE, a trinucleotide repeat sequence gets longer. If someone was mildly affected with a TNRE disorder, he or she might be concerned that an expansion of the repeat might occur during gamete formation, yielding offspring more severely affected with the disorder, a phenomenon called anticipation. This phenomenon may depend on the sex of the parent with the TNRE.

C20. According to the random mutation theory, spontaneous mutations can occur in any gene and do not involve exposure of the organism to a particular environment that selects for specific types of mutation. However, the structure of chromatin may cause certain regions of the DNA to be more susceptible to random mutations. For example, DNA

in an open conformation may be more accessible to mutagens and more likely to incur mutations. Similarly, hot spots—certain regions of a gene that are more likely to mutate than other regions—can occur within a single gene. Also, another reason that some genes mutate at a higher rate is that some genes are larger than others, which provides a greater chance for mutation.

C22. Excision repair systems could fix this damage. Also, homologous recombination repair could fix the damage.

C24. Anticipation means that the TNRE expands even further in future generations. Anticipation may depend on the sex of the parent with the TNRE.

C26. The mutation frequency is the total number of mutant alleles divided by the total number of alleles in the population. If there are 1,422,000 babies, there are 2,844,000 copies of this gene (because each baby has two copies). The mutation frequency is $^{31}/_{2,844,000}$, which equals 1.09×10^{-5}. The mutation rate is the number of new mutations per generation. There are 13 babies who did not have a parent with achondroplasia; thus, 13 is the number of new mutations. If we calculate the mutation rate as the number of new mutations in a given gene per generation, then we should divide 13 by 2,844,000. In this case, the mutation rate would be 4.6×10^{-6}.

C28. The effects of mutations are cumulative. If one mutation occurs in a cell, this mutation will be passed to the daughter cells. If a mutation occurs in the daughter cell, now there will be two mutations. These two mutations will be passed to the next generation of daughter cells, and so forth. The accumulation of many mutations eventually kills the cells. That is why mutagens are more effective at killing dividing cells compared to nondividing cells. It is because the number of mutations accumulates to a lethal level.

There are two main side effects to this treatment. First, some normal (noncancerous) cells of the body, particularly skin cells and intestinal cells, are actively dividing. These cells are also killed by chemotherapy and radiation therapy. Secondly, it is possible that the therapy may produce mutations that will cause noncancerous cells to become cancerous. For these reasons, there is a maximal dose of chemotherapy or radiation therapy that is recommended.

C30. A. Yes

   B. No; the albino trait affects the entire individual.

   C. No; the early apple-producing trait affects the entire tree.

   D. Yes

C32. Mismatch repair is aimed at eliminating mismatches that may have occurred during DNA replication. In this case, the wrong base is in the newly made strand. The binding of MutH, which occurs on a hemimethylated sequence, provides a sensing mechanism to distinguish between the unmethylated and methylated strands. In other words, MutH binds to the hemimethylated DNA in a way that allows the mismatch repair system to distinguish which strand is methylated and which is not.

C34. Because sister chromatids are genetically identical, an advantage of homologous recombination is that it can be an error-free mechanism to repair a DSB. A disadvantage, however, is that it occurs only during the S and G_2 phase of the cell cycle in eukaryotes or following DNA replication in bacteria. An advantage of NHEJ is that it doesn't involve the participation of a sister chromatid, so it can occur at any stage of the cell cycle. However, a disadvantage is that NHEJ can result in small deletions in the region that has been repaired. Overall, NHEJ is a quick but error-prone repair mechanism, while HR is a more accurate method of repair that is limited to certain stages of the cell cycle.

C36. In E. coli, the TRCF recognizes when RNA polymerase is stalled on the DNA. This stalling may be due to DNA damage such as a thymine dimer. The TRCF removes RNA polymerase and recruits the excision DNA repair system to the region, thereby promoting the repair of the template strand of DNA. It is beneficial to preferentially repair actively transcribed DNA because it is functionally important. It is a DNA region that encodes a gene.

C38. The underlying genetic defect that causes xeroderma pigmentosum is a defect in one of the genes that encode a polypeptide involved with nucleotide excision repair. These individuals are defective in repairing DNA abnormalities such as thymine dimers and abnormal bases. Therefore, they are very sensitive to environmental agents such as UV light. Because they are defective at repair, UV light is more likely to cause mutations in these people compared to unaffected individuals. For this reason, people with XP develop pigmentation abnormalities and premalignant lesions and have a high predisposition to skin cancer.

C40. Both types of repair systems recognize an abnormality in the DNA and excise the abnormal strand. The normal strand is then used as a template to synthesize a complementary strand of DNA. The systems differ in the types of abnormalities they detect. The mismatch repair system detects base pair mismatches, while the excision repair system recognizes thymine dimers, chemically modified bases, missing bases, and certain types of cross-links. The mismatch repair system operates immediately after DNA replication, allowing it to distinguish between the daughter strand (which contains the wrong base) and the parental strand. The excision repair system can operate at any time in the cell cycle.

Experimental Questions

E2. When cells from a master plate were replica plated onto two plates containing selective media with the T1 phage, T1-resistant colonies were observed at the same locations on both plates. These results indicate that the mutations occurred randomly while on the master plate (in the absence of T1) rather than occurring as a result of exposure to T1. In other words, mutations are random events, and selective conditions may promote the survival of mutant strains that occur randomly.

To show that antibiotic resistance is due to random mutation, one could follow the same basic strategy except the secondary plates would contain the antibiotic instead of T1 phage. If the antibiotic resistance arose as a result of random mutation on the master plate, one would expect the antibiotic-resistant colonies to appear at the same locations on two different secondary plates.

E4. Perhaps the X-rays also produce mutations that make the *ClB* daughters infertile. Many different types of mutations could occur in the irradiated males and be passed to the *ClB* daughters. Some of these mutations could prevent the *ClB* daughters from being fertile. These mutations could interfere with oogenesis, etc. Such *ClB* daughters would be unable to have any offspring.

E6. You would conclude that chemical A is not a mutagen. The percentage of *ClB* daughters (whose fathers had been exposed to chemical A) that did not produce sons was similar to the control (compare 3 out of 2,108 with 2 out of 1,402). In contrast, chemical B appears to be a mutagen. The percentage of *ClB* daughters (whose fathers had been exposed to chemical B) that did not produce sons was much higher than the control (compare 3 out of 2,108 with 77 out of 4,203).

E8. You would expose the bacteria to the physical agent. You could also expose the bacteria to the rat liver extract, but it is probably not necessary for two reasons. First, a physical mutagen is not something that a person would eat. Therefore, the actions of digestion via the liver are probably irrelevant if you are concerned that the agent might be a mutagen. Second, the rat liver extract would not be expected to alter the properties of a physical mutagen.

E10. The results suggest that the strain is defective in excision repair. If we compare the normal and mutant strains that have been incubated for 2 hours at 37°C, much of the radioactivity in the normal strain has been transferred to the soluble fraction because it has been excised. In the mutant strain, however, less of the radioactivity has been transferred to the soluble fraction, suggesting that it is not as efficient at removing thymine dimers.

Questions for Student Discussion/Collaboration

2. The worst time to be exposed to mutagens would be at very early stages of embryonic development. An early embryo is most sensitive to mutation because it will affect a large region of the body. Adults must also worry about mutagens for several reasons. Mutations in somatic cells can cause cancer, a topic discussed in Chapter 22. Also, adults should be careful to avoid mutagens that may affect the ovaries or testes because these mutations could be passed to offspring.

CHAPTER 17

Conceptual Questions

C2. Branch migration will not create a heteroduplex during SCE because the sister chromatids are genetically identical. There should not be any mismatches between the complementary strands. Gene conversion cannot take place because the sister chromatids carry alleles that are already identical to each other.

C4. The two molecular mechanisms that can explain the phenomenon of gene conversion are mismatch DNA repair and gap repair synthesis. Both mechanisms could occur in the double-strand break model.

C6. A recombinant chromosome is one that has been derived from a crossover and contains a combination of alleles that is different from the parental chromosomes. A recombinant chromosome is a hybrid of the parental chromosomes.

C8. Gene conversion occurs when a pair of different alleles is converted to a pair of identical alleles. For example, a pair of *Bb* alleles could be converted to *BB* or *bb*.

C10. Gene conversion is likely to take place near the breakpoint. According to the double-strand break model, a gap may be created by the digestion of one DNA strand in the double helix. Gap repair synthesis may result in gene conversion. A second way that gene conversion can occur is by mismatch repair. A heteroduplex may be created after DNA strand migration and may be repaired in such a way as to cause gene conversion.

C12. No, it does not necessarily involve DNA mismatch repair; it could also involve gap repair synthesis. The double-strand break model involves a migration of DNA strands and the digestion of a gap. Therefore, in this gap region, only one chromatid is providing the DNA strands. As seen in Figure 17.7, the top chromosome is using the top DNA strand from the bottom chromosome in the gap region. The bottom chromosome is using the bottom DNA strand from the bottom chromosome. After DNA synthesis, both chromosomes may have the same allele.

C14. First, gene rearrangement of V, D, and J domains occurs within the light- and heavy-chain genes. Second, within a given B cell, different combinations of light and heavy chains are possible. And third, imprecise fusion may occur between the V, D, and J domains.

C16. One segment, which includes some variable (V) domains and perhaps one or more joining (J) domain, of DNA is removed from the κ light-chain gene. One segment, which may include one or more J domains and the region between the J domain and C domain, is removed during pre-mRNA splicing.

C18. The ends of a short region would be flanked by direct repeats. This is a universal characteristic of all transposable elements. In addition, many elements contain IRs or LTRs that are involved in the transposition process. One might also look for the presence of a transposase or reverse transcriptase gene, although this is not an absolute requirement, because nonautonomous transposable elements typically lack transposase or reverse transcriptase.

C20. Direct repeats occur because transposase or integrase produces staggered cuts in the two strands of chromosomal DNA. The transposable element is then inserted into this site, which temporarily leaves two gaps. The gaps are filled in by DNA polymerase. Because this gap filling is due to complementarity of the base sequences, the two gaps end up with the exact same sequence.

C22. Transposable elements are mutagens, because they alter (disrupt) the sequences of chromosomes and genes within chromosomes. They do this by inserting themselves into genes.

C24. A. Viral-like retroelements and nonviral-like retroelements

   B. Insertion sequences, composite transposons, and replicative transposons

   C. All five types have direct repeats

   D. Insertion sequences, composite transposons, and replicative transposons

C26. A. As shown in solved problem S2a, a crossover between direct repeats (DRs) will excise the region between the repeats. In this case, it will excise a large chromosomal region, which does not contain a centromere, and will be lost. The remaining portion of the chromosome is shown here. It has one direct repeat that is formed partly from DR-1 and partly from DR-4. It is designated DR-$\frac{1}{4}$.

DR
$\frac{1}{4}$

   B. A crossover between IR-1 and IR-4 will cause an inversion between the intervening region. This is a similar effect to the crossover within a single transposable element, as described in solved problem S2b. The difference is that the crossover causes a large chromosomal region to be inverted.

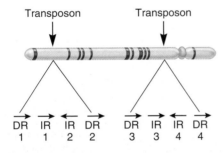

C28. At a frequency of about 1 in 10,000, recombination occurs between the inverted repeats within this transposon. As shown in solved problem S2b, recombination between inverted repeats causes the sequence within the transposon to be reversed. If this occurred in a strain that expressed the *H2/rH1* operon, the promoter would be flipped in the opposite direction, so the *H2* gene and the *rH1* gene would not be expressed. The *H2* flagellar protein would not be made, and the *H1* repressor protein would not be made. The *H1* flagellar protein would be made because the expression of the *H1* gene would not be repressed. A strain expressing the *H1* gene could also "switch" at a frequency of about 1 in 10,000 by the same mechanism. If a crossover occurred between the inverted repeats within the transposable element, the promoter would be flipped again, and the *H2* and *rH1* genes would be expressed again.

Experimental Questions

E2. You would conclude that the substance is a mutagen. Substances that damage DNA tend to increase the level of genetic exchange such as SCE.

E4. The drawing here shows the progression through three rounds of BrdU exposure. After one round, all of the chromosomes would be dark. After two rounds, all of the chromosomes would be harlequin. After three rounds, the number of light sister chromatids would be twice as much as the number of light sister chromatids found after two rounds of replication.

E6. When McClintock started with a colorless strain containing *Ds*, she identified 20 cases where *Ds* had moved to a new location to produce red and shrunken kernels. This identification was possible because the 20 strains had a higher frequency of chromosomal breaks at a specific site and because of the mutability of particular genes. She also had found a strain where *Ds* had inserted into the red-color-producing gene, resulting in the colorless phenotype, so its transposition out of the gene would produce a red patch. Overall, her analysis of the data showed that the sectoring (i.e., mutability) phenotype was consistent with the transposition of *Ds*.

E8. Transposon tagging is an experimental method aimed at cloning genes. In this approach, a transposon is introduced into a strain, and the researcher identifies individuals in which the function of a gene of interest has been inactivated. In many cases, the loss of function has occurred because the transposon has been inserted into the gene. If so, DNA can be isolated, digested into fragments, and cloned into viral vectors, creating a DNA library. The library is then probed using a radiolabeled fragment of DNA that is complementary to the transposon in order to identify clones in which the transposon has been inserted.

E10. A transposon creates a mutable site because the excision of a transposon causes chromosomal breakage if the two ends are not reconnected. After it has moved out of its original locus (causing chromosomal breakage), it may be inserted into a new locus somewhere else. You could experimentally determine this by examining a strain that has incurred a chromosomal breakage at the first locus. You could microscopically examine many cells that had such a broken chromosome and see if a new locus had been formed. This new locus would be the site into which the transposon had moved. On occasion, there would be chromosomal breakage at this new locus, which could be observed microscopically.

Questions for Student Discussion/Collaboration

2. Beneficial consequences: You wouldn't get (as many) translocations, inversions, and the accumulation of selfish DNA.

   Harmful consequences: The level of genetic diversity would be decreased, because linked combinations of alleles would not be able to recombine. You wouldn't be able to produce antibody diversity in the same way. Gene duplication could not occur, so the evolution of new genes would be greatly inhibited.

CHAPTER 18

Conceptual Questions

C2. A restriction enzyme recognizes and binds to a specific DNA sequence and then cleaves a (covalent) ester bond in each of two DNA strands.

C4. The term cDNA refers to DNA that is made using RNA as the starting material. Compared to genomic DNA, it lacks introns.

Experimental Questions

E2. Remember that AT base pairs form two hydrogen bonds, while GC base pairs form three hydrogen bonds. The order (from stickiest to least sticky) would be

   *Bam*HI = *Pst*I = *Sac*I > *Eco*RI > *Cla*I

E4. In conventional gene cloning, many copies are made because the vector replicates to a high copy number within the cell, and the cells divide to produce many more cells. In PCR, the replication of the DNA to produce many copies is facilitated by primers, deoxyribonucleoside triphosphates (dNTPs), and *Taq* polymerase.

E6. A recombinant vector is a vector that has a piece of "foreign" DNA inserted into it. The foreign DNA came from somewhere else, such as the chromosomal DNA of some organism. To construct a recombinant vector, the vector and source of foreign DNA are digested with the same restriction enzyme. The complementary ends of the fragments are allowed to hydrogen bond to each other (i.e., sticky ends are allowed to bind), and then DNA ligase is added to create covalent bonds. In some cases, a piece of the foreign DNA will become ligated to the vector, thereby creating a recombinant vector. In other cases, the two ends of the vector ligate back together, restoring the vector to its original structure.

   As described in Figure 18.2, the insertion of foreign DNA can be detected using X-Gal. As seen here, the insertion of the foreign DNA causes the inactivation of the *lacZ* gene. The *lacZ* gene encodes the enzyme β-galactosidase, which converts the colorless compound X-Gal to a blue compound. If the *lacZ* gene is inactivated by the insertion of foreign DNA, the enzyme will not be produced, and the bacterial colonies will be white. If the vector has simply recircularized and the *lacZ* gene remains intact, the enzyme will be produced, and the colonies will be blue.

E8. If the *Eco*RI fragment containing the *kan^R* gene also had an origin of replication, it is possible that this fragment could circularize and

become a very small plasmid. The electrophoresis results would be consistent with the idea that the same bacterial cells could contain two different plasmids: pSC101 and a small plasmid corresponding to the segment of DNA (now circularized) that carried the *kan^R* gene. However, the results shown in step 9 rule out this possibility. The density gradient centrifugation showed a single peak, corresponding to a plasmid that had an intermediate size between pSC101 and pSC102. In contrast, if our alternative explanation had been correct (i.e., that bacterial cells contain two plasmids), there would be two peaks from the density gradient centrifugation. One peak would correspond to pSC101, and the other peak would indicate a very small plasmid (i.e., smaller than pSC101 and pSC102).

E10. 3×2^{27}, which equals 4.0×10^8, or about 400 million copies.

E12. Initially, the mRNA would be mixed with reverse transcriptase and nucleotides to create a complementary strand of DNA. Reverse transcriptase also needs a primer. This could be a primer that is known to be complementary to the β-globin mRNA. Alternatively, mature mRNAs have a polyA tail, so one could add a primer that consists of many Ts, called a poly-dT primer. After the complementary DNA strand has been made, the sample would then be mixed with primers, *Taq* polymerase, and nucleotides and subjected to the standard PCR protocol. Note: The PCR reaction would have two kinds of primers. One primer would be complementary to the 5' end of the mRNA and would be unique to the β-globin sequence. The other primer would be complementary to the 3' end. This second primer could be a poly-dT primer, or it could be a unique primer that would bind slightly upstream from the polyA-tail region.

E14. A DNA library is a collection of recombinant vectors that contain different pieces of DNA from a source of chromosomal DNA. Because it is a diverse collection of many different DNA pieces, the name library seems appropriate.

E16. Hybridization occurs due to the hydrogen bonding of complementary sequences. Due to the chemical properties of DNA and RNA strands, they form double-stranded regions when the base sequences are complementary. In a Southern and Northern experiment, the cloned DNA is labeled and used as a probe.

E18. The purpose of a Northern blotting experiment is to identify a specific RNA within a mixture of many RNA molecules, using a fragment of cloned DNA as a probe. It can tell you if a gene is transcribed in a particular cell or at a particular stage of development. It can also tell you if a pre-mRNA is alternatively spliced into two or more mRNAs of different sizes.

E20. It appears that this mRNA is alternatively spliced to create a high molecular mass and a lower molecular mass product. Nerve cells produce a very large amount of the larger mRNA, whereas spleen cells produce a moderate amount of the smaller mRNA. Both types are produced in small amounts by the muscle cells. It appears that kidney cells do not transcribe this gene.

E22. 1. β (detected at the highest stringency)

   2. δ

   3. γ_A and ε

   4. α_1

   5. *Mb* (detected only at the lowest stringency)

E24. The Western blot is shown here. The sample in lane 2 came from a plant that was homozygous for a mutation that prevented the expression of this polypeptide. Therefore, no protein was observed in this lane. The sample in lane 4 came from a plant that is homozygous for a mutation that introduces an early stop codon into the coding sequence; thus, the polypeptide is shorter than normal (13.3 kDa). The sample in lane 3 was from a heterozygote that expresses about 50% of each type of polypeptide. Finally, the sample in lane 5 came from a plant homozygous for a mutation that changed one amino acid to another amino acid. This type of mutation, termed a missense mutation, may not be detectable on a gel. However, a single amino acid substitution could affect polypeptide function.

E26. The products of structural genes are proteins with a particular amino acid sequence. Antibodies can specifically recognize proteins due to their amino acid sequence. Therefore, an antibody can detect whether or not a cell is making a particular type of protein.

E28. In this case, the transcription factor binds to the response element when the hormone is present. Therefore, the hormone promotes the binding of the transcription factor to the DNA and thereby promotes transcriptional activation.

E30. The levels of cAMP affect the phosphorylation of CREB, and this affects whether or not it can activate transcription. However, CREB can bind to CREs whether or not it is phosphorylated. Therefore, in a gel retardation assay, we would expect CREB to bind to CREs and retard their mobility whether or not the cell extracts were pretreated with epinephrine.

E32. The glucocorticoid receptor will bind to GREs if glucocorticoid hormone is also present (lane 2). The glucocorticoid receptor does not bind without hormone (lane 1), and it does not bind to CREs (lane 6). The CREB protein will bind to CREs with or without hormone (lanes 4 and 5), but it will not bind to GREs (lane 3). The expected results are shown here. In this drawing, the binding of CREB protein to the 700 bp fragment results in a complex with a higher mass compared to the glucocorticoid receptor binding to the 600 bp fragment.

E34. The rationale behind a footprinting experiment has to do with accessibility. If a protein is bound to the DNA, it will cover up the part of the DNA where it is bound. This region of the DNA will be inaccessible to the actions of chemicals or enzymes that cleave the DNA, such as DNase I.

E36. A. AGGTCGGTTGCCATCGCAATAATTTCTGCCTGAACCCAATA

B. Automated sequencing has several advantages. First, the reactions are done in a single tube as opposed to four tubes. Second, the detector can "read" the sequence and provide the researcher with a printout of the sequence. This is much easier than looking at an X-ray film and writing the sequence out by hand. It also avoids human error.

E38. There are lots of different strategies one could follow. For example, you could mutate every other base and see what happens. It would be best to make very nonconservative mutations, such as a purine for a pyrimidine or a pyrimidine for a purine. If the mutation prevents protein binding in a gel retardation assay, then the mutation is probably within the response element. If the mutation has no effect on protein binding, it probably is outside the response element.

Questions for Student Discussion/Collaboration

2. A. Does a particular amino acid within a protein sequence play a critical role in the protein's structure or function?

B. Does a DNA sequence function as a promoter?

C. Does a DNA sequence function as a regulatory site?

D. Does a DNA sequence function as a splicing junction?

E. Is a sequence important for correct translation?

F. Is a sequence important for RNA stability?

And many others. . .

CHAPTER 19

Conceptual Questions

C2. *A. radiobacter* synthesizes an antibiotic that kills *A. tumefaciens*. The genes, which are necessary for antibiotic biosynthesis and resistance, are plasmid encoded and can be transferred during interspecies conjugation. If *A. tumefaciens* received this plasmid during conjugation, it would be resistant to killing. Therefore, the conjugation-deficient strain prevents the occurrence of *A. tumefaciens*–resistant strains.

C4. A biological control agent is an organism that prevents the harmful effects of some other agent in the environment. Examples include *Bacillus thuringiensis*, a bacterium that synthesizes compounds that act as toxins to kill insects, *Ice⁻* bacteria that inhibit the proliferation of *Ice⁺* bacteria, and the use of *Agrobacterium radiobacter* to prevent crown gall disease caused by *Agrobacterium tumefaciens*.

C6. A mouse model is a strain of mice that carries a mutation in a mouse gene that is analogous to a mutation in a human gene that causes disease. These mice can be used to study the disease and to test potential therapeutic agents.

C8. The T DNA gets transferred to the plant cell; it then is incorporated into the plant cell's genome.

C10. A. With regard to maternal effect genes, the phenotype would depend on the animal that donated the egg. It is the cytoplasm of the egg that accumulates the gene products of maternal effect genes.

B. The extranuclear traits depend on the mitochondrial genome. Mitochondria are found in the egg and in the somatic cell. So, theoretically, both cells could contribute extranuclear traits. In reality, however, researchers have found that the mitochondria in Dolly were from the animal that donated the egg. It is not clear why she had no mitochondria from the mammary cell.

C. The cloned animal would be genetically identical to the animal that donated the nucleus with regard to traits that are determined by nuclear genes, which are expressed during the lifetime of the organism. The cloned animal would/could differ from the animal that donated the nucleus with regard to traits that are determined by maternal effect genes and mitochondrial genes. Such an animal is not a true clone, but it is likely that it would greatly resemble the animal that donated the nucleus, because the vast majority of genes are found in the cell nucleus.

C12. Some people are concerned with the release of genetically engineered microorganisms into the environment. The fear is that such organisms may continue to proliferate and it may not be possible to "stop them." A second concern involves the use of genetically engineered organisms in the food we eat. Some people are worried that genetically engineered organisms may pose an unknown health risk. A third issue is ethics. Some people feel that it is morally wrong to tamper with the genetics of organisms. This opinion may also apply to genetic techniques such as cloning, stem cell research, and gene therapy.

Experimental Questions

E2.   The plasmid with the incorrect orientation would not work because the coding sequence would be in the wrong direction relative to the promoter sequence. Therefore, the region containing the somatostatin sequence would not be translated into the somatostatin polypeptide.

E4.   To construct the coding sequence for somatostatin, the researchers synthesized eight oligonucleotides, labeled A–H. When these oligonucleotides were mixed together, they would hydrogen bond to each other due to their complementary sequences. The addition of ligase would covalently link the DNA backbones. The left side of the coding sequence had an *Eco*RI site, and the right side had a *Bam*HI site; these sites made it possible to insert this sequence at the end of the β-galactosidase gene. You may also notice that an ATG codon (AUG in the mRNA) precedes the first alanine codon in somatostatin. This AUG codon specifies methionine, which allows the somatostatin to be cleaved from β-galactosidase using cyanogen bromide. This was necessary because somatostatin, by itself, is rapidly degraded in *E. coli*, whereas the fusion protein is not.

E6.   A kanamycin-resistance gene is contained within the T DNA. Exposure to kanamycin selects for the growth of plant cells that have incorporated the T DNA into their genome. The carbenicillin kills the *A. tumefaciens*. The plant growth hormones promote the regeneration of an entire plant from somatic cells. If kanamycin were left out, it would not be possible to select for the growth of cells that had taken up the T DNA.

E8.   The term gene knockout refers to an organism in which the function of a particular gene has been eliminated. For autosomal genes in animals and plants, a gene knockout is a homozygote for a defect in both copies of the gene. If a gene knockout has no phenotypic effect, the gene may be redundant. In other words, there may be multiple genes within the genome that can carry out the same function. Another reason why a gene knockout may not have a phenotypic effect is because of the environment. As an example, let's say a mouse gene is required for the synthesis of a vitamin. If the researchers were providing food that contained the vitamin, the knockout mouse that was lacking this gene would have a normal phenotype; it would survive just fine. Sometimes, researchers have trouble knowing the effects of a gene knockout until they modify the environmental conditions in which the animals are raised.

E10. Gene replacement occurs by homologous recombination. For homologous recombination to take place, two crossovers must occur, one at each end of the target gene. After homologous recombination, only the *Neo^R* gene, which is inserted into the target gene, can be incorporated into the chromosomal DNA of the embryonic stem cell. In contrast, nonhomologous recombination can involve two crossovers anywhere in the cloned DNA. Because the *TK* gene and target gene are adjacent to each other, nonhomologous recombination usually transfers both the *TK* gene and the *Neo^R* gene. If both genes are transferred to an embryonic stem cell, it will die because the cells are grown in the presence of gancyclovir. The product of the *TK* gene kills the cell under these conditions. In contrast, cells that have acquired the *Neo^R* gene due to homologous recombination but not the *TK* gene will survive. Stem cells, which have not taken up any of the cloned DNA, will die because they will be killed by the neomycin. In this way, the presence of gancyclovir and neomycin selects for the growth of stem cells that have acquired the target gene by homologous recombination.

E12. A. Dolly's chromosomes may seem so old because they were already old when they were in the nucleus that was incorporated into the enucleated egg. They had already become significantly shortened in the mammary cells. This shortening was not repaired by the egg.

      B. Dolly's age does not matter. Remember that shortening does not occur in germ cells. However, Dolly's eggs are older than they seem by about six or seven years, because Dolly's germ-line cells received their chromosomes from a sheep that was 6 years old, and the cells were grown in culture for a few doublings before a mammary

cell was fused with an enucleated egg. Therefore, the calculation would be 6 or 7 years (the age of the mammary cells that produced Dolly's germ-line cells) plus 8 years (the age of Molly), which equals 14 or 15 years. However, only half of Molly's chromosomes would appear to be 14 or 15 years old. The other half of her chromosomes, which she inherited from her father, would appear to be 8 years old.

      C. Chromosome shortening is a bit disturbing, because it suggests that aging has occurred in the somatic cell, and this aging is passed to the cloned organism. If cloning was done over the course of many generations, this may eventually have a major impact on the life span of the cloned organism. It may die much earlier than a noncloned organism. However, chromosome shortening may not always occur. It did not seem to occur in mice that were cloned for six consecutive generations.

E14. The term molecular pharming refers to the practice of making transgenic animals that will synthesize human products in their milk. It also refers to the manufacture of medical products by agricultural plants. It can be advantageous when bacterial cells are unable to make a functional protein product from a human gene. For example, some proteins are posttranslationally modified by the attachment of carbohydrate molecules. This type of modification does not occur in bacteria, but it may occur correctly in transgenic animals and plants. Also, dairy cows produce large amounts of milk, which may improve the yield of the human product. Likewise, plants can produce a large amount of recombinant proteins.

E16. You would first need to clone the normal mouse gene. Cloning methods are described in Chapter 18. After the normal gene was cloned, you would then follow the protocol shown in Figure 19.7. The normal gene would be inactivated by the insertion of the *Neo^R* gene, and the *TK* gene would be cloned next to it. This DNA segment would be introduced into mouse embryonic stem cells and grown in the presence of neomycin and gancyclovir. This selects for homologous recombinants. The surviving embryonic stem cells would be injected into early embryos, which would then develop into chimeras. The chimeric mice would be identified by their patches of light and dark fur. At this point, if all has gone well, a portion of the mouse is heterozygous for the normal gene and the gene that has the *Neo^R* insert. This mouse would be bred, and then brothers and sisters from the litter would be bred to each other. Southern blots would be conducted to determine if the offspring carried the gene with the *Neo^R* insert. At first, one would identify heterozygotes that had one copy of the inserted gene. These heterozygotes would be crossed to each other to obtain homozygotes. The homozygotes are gene knockouts because the function of the gene has been "knocked out" by the insertion of the *Neo^R* gene. Perhaps, these mice would be dwarf and exhibit signs of mental retardation. At this point, the researcher would have a mouse model to study the disease.

E18. In cystic fibrosis gene therapy, an aerosol spray, containing the normal *CF* gene in a retrovirus or liposome, is used to get the normal *CF* gene into the lung cells. The epithelial cells on the surface of the lung will take up the gene. However, these surface epithelial cells have a finite life span, so it is necessary to have repeated applications of the aerosol spray. Ideally, scientists hope to devise methods whereby the normal *CF* gene will be able to penetrate more deeply into the lung tissue.

Questions for Student Discussion/Collaboration

2.   From a genetic viewpoint, the recombinant and nonrecombinant strains are very similar. The main difference is their history. The recombinant strain has been subjected to molecular techniques to eliminate a particular gene. The nonrecombinant strain has had the same gene eliminated by a spontaneous mutation or via mutagens. The nonrecombinant strain has the advantage of a better public perception. People are less worried about releasing nonrecombinant strains into the environment.

CHAPTER 20

Conceptual Questions

C2. A. Yes

B. No; this is only one chromosome in the genome.

C. Yes

D. Yes

Experimental Questions

E2. They are complementary to each other.

E4. Because normal cells contain two copies of chromosome 14, one would expect that a probe would bind to complementary DNA sequences on both of these chromosomes. If a probe recognized only one of two chromosomes, this means that one of the copies of chromosome 14 has been lost, or it has suffered a deletion in the region where the probe binds. With regard to cancer, the loss of this genetic material may be related to the uncontrollable cell growth.

E6. After the cells and chromosomes have been fixed to the slide, it is possible to add two or more different probes that recognize different sequences (i.e., different sites) within the genome. Each probe has a different fluorescence emission wavelength. Usually, a researcher will use computer imagery that recognizes the wavelength of each probe and then assigns that probe a bright color. The color seen by the researcher is not the actual color emitted by the probe; it is a secondary color assigned by the computer. In a sense, the probes, with the aid of a computer, are "painting" the regions of the chromosomes that are recognized by a probe. An example of chromosome painting is shown in Figure 20.4. In this example, human chromosome 5 is painted with six different colors.

E8. A contig is a collection of clones that contain overlapping segments of DNA that span a particular region of a chromosome. To determine if two clones are overlapping, one could conduct a Southern blotting experiment. In this approach, one of the clones is used as a probe. If it is overlapping with the second clone, it will bind to it in a Southern blot. Therefore, the second clone is run on a gel, and the first clone is used as a probe. If the band corresponding to the second clone is labeled, this means that the two clones are overlapping.

E10. BAC cloning vectors have the replication properties of a bacterial chromosome and the cloning properties of a plasmid. To replicate like a chromosome, the BAC vector contains an origin of replication from an F factor. Therefore, in a bacterial cell, a BAC can behave as a chromosome. Like a plasmid, BACs also contain selectable markers and convenient cloning sites for the insertion of large segments of DNA. The primary advantage is the ability to clone very large pieces of DNA.

E12. The resistance gene appears to be linked to RFLP 4B.

E14. For most organisms, it is usually easy to locate many RFLPs throughout the genome. The RFLPs can be used as molecular markers to make a map of the genome. This is done using the strategy described in Figure 20.8. To map a functional gene, one would also follow the same general strategy described in Figure 20.8, except that the two strains would also have an allelic difference in the gene of interest. The experimenter would make crosses, such as dihybrid crosses, and determine the number of parental and recombinant offspring based on the alleles and RFLPs they had inherited. If an allele and RFLP are linked, there will be a much lower percentage (i.e., less than 50%) of the recombinant offspring.

E16.
Deduced Outcome

STSs

```
      4  6        7  9      3  2     8  5  1
YAC
  3   <------->
  1        <-------->
  5               <------->
  4            <-------->
  2                     <------->
```

E18. A. One homologue contains the STS-1 that is 289 bp and STS-2 that is 422 bp, while the other homologue contains STS-1 that is 211 bp and STS-2 that is 115 bp. This is based on the observation that 28 of the sperm have either the 289 bp and 422 bp bands or the 211 bp and 115 bp bands.

B. There are two recombinant sperm; see lanes 12 and 18. Because there are two recombinant sperm out of a total of thirty,

$$\text{Map Distance} = \frac{2}{30} \times 100$$

$$= 6.7 \text{ mu}$$

C. In theory, this method could be used. However, there is not enough DNA in one sperm to carry out an RFLP analysis unless the DNA is amplified by PCR.

E20. One possibility is that the geneticist could try a different restriction enzyme. Perhaps there is sequence variation in the vicinity of the pesticide-resistance gene that affects the digestion pattern of a restriction enzyme other than *Eco*RI. There are hundreds of different restriction enzymes that recognize a myriad of different sequences.

Alternatively, the geneticist could give up on the RFLP approach and try to identify one or more sequence-tagged sites that are in the vicinity of the pesticide-resistance gene. In this case, the geneticist would want to identify STSs that are also microsatellites. As shown in Figure 20.12, the transmission of microsatellites can be followed in genetic crosses. Therefore, if the geneticist could identify microsatellites in the vicinity of the pesticide-resistance gene, this would make it possible to predict the outcome of crosses. For example, let's suppose a microsatellite linked to the pesticide-resistance gene existed in three forms: 234 bp, 255 bp, and 311 bp. And let's also suppose that the 234 bp form was linked to the high-resistance allele, the 255 bp form was linked to the moderate-resistance allele, and the 311 bp form was linked to the low-resistance allele. According to this hypothetical example, the geneticist could predict the level of resistance in an alfalfa plant by analyzing the inheritance of these microsatellites.

E22. When chromosomal DNA is isolated and digested with a restriction enzyme, this produces thousands of DNA fragments of different sizes. This makes it impossible to see any particular band on a gel if you simply stained the gel for DNA. Southern blotting allows you to detect one or more RFLPs that are complementary to the radioactive probe that is used.

E24. Based on these results, it is likely that the sickle-cell allele originated in an individual with the 13.0 kbp RFLP. This would explain why the Hb^S allele is usually transmitted with the 13.0 kbp RFLP. On occasion, however, a crossover could occur in the region between the β-globin gene and the distal *Hpa*I site.

After crossing over, the Hb^S allele is now linked to a 7.6 kbp RFLP.

E26. PFGE is a method of electrophoresis that is used to separate small chromosomes and large DNA fragments. The electrophoresis devices used in PFGE have two sets of electrodes. The two sets of electrodes produce alternating pulses of current, and this facilitates the separation of large DNA fragments.

It is important to handle the sample gently to prevent the breakage of the DNA due to mechanical forces. Cells are first embedded in agarose blocks, and then the blocks are loaded into the wells of the gel. The agarose keeps the sample very stable and prevents shear forces that might mechanically break the DNA. After the blocks are in the gel, the cells within the blocks are lysed, and, if desired, restriction enzymes can be added to digest the DNA. For PFGE, a restriction enzyme that cuts very infrequently might be used.

PFGE can be used as a preparative technique to isolate and purify individual chromosomes or large DNA fragments. PFGE can also be used, in conjunction with Southern blotting, as a mapping technique.

E28. Note that the insert of cosmid B is contained completely within the insert of cosmid C.

E30. A. The general strategy is shown in Figure 20.18. The researcher begins at a certain location and then walks toward the gene of interest. You begin with a clone that has a marker that is known to map relatively close to the gene of interest. A piece of DNA at the end of the insert is subcloned and then used in a Southern blot to identify an adjacent clone in a cosmid DNA library. This is the first "step." The end of this clone is subcloned to make the next step. And so on. Eventually, after many steps, you will arrive at your gene of interest.

B. In this example, you would begin at STS-3. If you walked a few steps and happened upon STS-2, you would know that you were walking in the wrong direction.

C. This is a difficult aspect of chromosome walking. Basically, you would walk toward gene X using DNA from a normal individual and DNA from an individual with a mutant gene X. When you have found a site where the sequences are different between the normal and mutant individual, you may have found gene X. You would eventually have to confirm this by analyzing the DNA sequence of this region and determining that it encodes a functional gene.

E32. We can calculate the probability that a base will not be sequenced using this approach with the following equation:

$$P = e^{-m}$$

where

P is the probability that a base will be left unsequenced

e is the base of the natural logarithm; $e = 2.72$

m is the number of bases sequenced divided by the total genome size

In this case, $m = 19$ divided by 4.4, which equals 4.3

$$P = e^{-m} = e^{-4.3} = 0.0136 = 1.36\%$$

This means that if we randomly sequence 19.0 Mb, we are likely to miss only 1.36% of the genome. With a genome size of 4.4 Mb, we would miss about 57,840 base pairs out of approximately 4,400,000.

Questions for Student Discussion/Collaboration

2. A molecular marker is a segment of DNA, not usually encoding a gene, that has a known location within a particular chromosome. It marks the location of a site along a chromosome. RFLPs and STSs are examples. It is easier to use these types of markers because they can be readily identified by molecular techniques such as restriction digestion analysis and PCR. The locations of functional genes is usually more difficult because this relies on conventional genetic mapping approaches whereby allelic differences in the gene are mapped by making crosses or following a pedigree. For monomorphic genes, this approach doesn't work.

CHAPTER 21

Conceptual Questions

C2. There are two main reasons why the proteome is larger than the genome. The first reason involves the processing of pre-mRNA, a phenomenon that occurs primarily in eukaryotic species. RNA splicing and editing can alter the codon sequence of mRNA and thereby produce alternative forms of proteins that have different amino acid sequences. The second reason for protein diversity is posttranslational modifications. There are many ways that a given protein's structure can be covalently modified by cellular enzymes. These include proteolytic processing, disulfide bond formation, glycosylation, attachment of lipids, phosphorylation, methylation, and acetylation, to name a few.

C4. Centromeric sequences, origins of replication, telomeric sequences, repetitive sequences, and enhancers. Other examples are possible.

C6. There are a few interesting trends. Sequences 1 and 2 are similar to each other, as are sequences 3 and 4. There are a few places where amino acid residues are conserved among all five sequences. These amino acids may be particularly important with regard to function.

C8. A. Correct

B. Correct

C. This is not correct. These are short genetic sequences that happen to be similar to each other. The *lac* operon and *trp* operon are not derived from the same primordial operon.

D. This is not correct. The two genes are homologous to each other. It is correct to say that their sequences are 60% identical.

Experimental Questions

E2. As described in solved problem S1, one reason for making a subtractive DNA library is to determine which RNAs are produced when environmental conditions change. You want to load a small amount of cDNA on the column that was derived from cells that had been exposed to mercury. Remember that the cDNA, which is derived from mRNA that is made in the absence of mercury, is already bound to the column. You want the cDNA that is made in the presence of mercury to bind to this cDNA if it is complementary. If too much of this cDNA is loaded, all of the cDNAs will have complementary cDNA bound to them, and some of them will not bind to the column, even though they may be complementary to the cDNAs. In other words, if you load too much cDNA (derived from the mercury-exposed cells), you will have saturated the binding sites for cDNAs that are made in the absence of mercury. You do not want this to happen, because you want only the cDNAs that are derived from mRNAs that are specifically expressed in the presence of mercury to flow through the column. These cDNAs are not complementary to any cDNAs attached to the column.

E4. The cDNA labeled with a green dye is derived from mRNA obtained from cells at an early time point, when glucose levels were high. The other samples of cDNA were derived from cells collected at later time points, when glucose levels were falling and when the diauxic shift was occurring. These were labeled with a red dye. The green fluorescence provides a baseline for gene expression when glucose is high. At later time points, if the red : green ratio is high (i.e., greater than one), this means that a gene is induced as glucose levels fall, because there is more red cDNA compared to green cDNA. If the ratio is low (i.e., less than one), this means that a gene is being repressed.

E6. In the first dimension (i.e., in the tube gel), proteins migrate to the point in the gel where their net charge is zero. In the second dimension (i.e., the slab gel), proteins are coated with SDS and separated according to their molecular mass.

E8. In tandem mass spectroscopy, the first spectrometer determines the mass of a peptide fragment from a protein of interest. The second spectrometer determines the masses of progressively smaller fragments derived from that peptide. Because the masses of each amino acid are known, the molecular masses of these smaller fragments reveal the amino acid sequence of the peptide. With peptide sequence information, it is possible to use the genetic code and produce DNA sequences that could encode such a peptide. More than one sequence is possible due to the degeneracy of the genetic code. These sequences are used as query sequences to search a genomic database. This program will (hopefully) locate a match. The genomic sequence can then be analyzed to determine the entire coding sequence for the protein of interest.

E10. One strategy is a search by signal approach, which relies on known sequences such as promoters, start and stop codons, and splice sites to help predict whether or not a DNA sequence contains a structural gene. It attempts to identify a region that contains a promoter sequence, then a start codon, a coding sequence, and a stop codon. A second strategy is a search-by-content approach. The goal is to identify sequences whose nucleotide content differs significantly from a random distribution, which is usually due to codon bias. A search-by-content approach attempts to locate coding regions by identifying regions where the nucleotide content displays a bias. A third method to locate structural genes is to search for long open reading frames within a DNA sequence. An open reading frame is a sequence that does not contain any stop codons.

E12. By searching a database, one can identify genetic sequences that are homologous to a newly determined sequence. In most cases, homologous sequences carry out identical or very similar functions. Therefore, if one identifies a homologous member of a database whose function is already understood, this provides an important clue regarding the function of the newly determined sequence.

E14. The basis for secondary structure prediction is that certain amino acids tend to be found more frequently in α helices or β sheets. This information is derived from the statistical frequency of amino acids within secondary structures that have already been crystallized. Predictive methods are perhaps 60 to 70% accurate, which is not very good.

E16. The BACKTRANSLATE program works by knowing the genetic code. Each amino acid has one or more codons (i.e., three-base sequences) that are specified by the genetic code. This program would produce a sequence file that was a nucleotide base sequence. The BACKTRANSLATE program would produce a degenerate base sequence because the genetic code is degenerate. For example, lysine can be specified by AAA or AAG. The program would probably store a single file that had degeneracy at particular positions. For example, if the amino acid sequence was lysine–methionine–glycine–glutamine, the program would produce the following sequence:

```
5'-AA(A/G)ATGGG(T/C/A/G)CA(A/G)
```

The bases found in parentheses are the possible bases due to the degeneracy of the genetic code.

E18. A. To identify a specific transposable element, a program would use sequence recognition. The sequence of P elements is already known. The program would be supplied with this information and scan a sequence file looking for a match.

B. To identify a stop codon, a program would use sequence recognition. There are three stop codons that are specific three-base sequences. The program would be supplied with these three sequences and scan a sequence file to identify a perfect match.

C. To identify an inversion of any kind, a program would use pattern recognition. In this case, the program would be looking for a pattern in which the same sequence was running in opposite directions in a comparison of the two sequence files.

D. A search by signal approach uses both sequence recognition and pattern recognition as a means to identify genes. It looks for an organization of sequence elements that would form a functional gene. A search-by-content approach identifies genes based on patterns, not on specific sequence elements. This approach looks for a pattern in which the nucleotide content is different from a random distribution. The third approach to identify a gene is to scan a genetic sequence for long open reading frames. This approach is a combination of sequence recognition and pattern recognition. The program is looking for specific sequence elements (i.e., stop codons), but it is also looking for a pattern in which the stop codons are far apart.

E20. A. The amino acids that are most conserved (i.e., the same in all of the family members) are most likely to be important for structure and/or function. This is because a mutation that changed the amino acid might disrupt structure and function, and these kinds of mutations would be selected against. Completely conserved amino acids are found at the following positions: 101, 102, 105, 107, 108, 116, 117, 124, 130, 134, 139, 143, and 147.

B. The amino acids that are least conserved are probably not very important because changes in the amino acid does not seem to inhibit function. (If it did inhibit function, natural selection would eliminate such a mutation.) At one location, position 118, there are five different amino acids.

E22. This sequence is within the *lacY* gene of the *lac* operon of *E. coli*. It is found on page 589, nucleotides 801–850.

E24. A. This sequence has two regions that are about 20 amino acids long and very hydrophobic. Therefore, it is probable that this polypeptide has two transmembrane segments.

B.

E26. A. True

B. False. The programs are only about 60 to 70% accurate.

C. True

Questions for Student Discussion/Collaboration

2. This is a very difficult question. In 20 years, we may have enough predictive information so that the structure of macromolecules can be predicted from their genetic sequences. If so, it would be better to be a mathematical theoretician with some genetics background. If not, it is probably better to be a biophysicist with some genetics background.

CHAPTER 22

Conceptual Questions

C2. When a disease-causing allele affects a trait, it is causing a deviation from normality, but the gene involved is not usually the only gene that governs the trait. For example, an allele causing hemophilia prevents the normal blood clotting pathway from operating correctly. It follows a simple Mendelian pattern because a single gene affects the phenotype. Even so, it is known that normal blood clotting is due to the actions of many genes.

C4. Changes in chromosome number and unbalanced changes in chromosome structure tend to affect phenotype because they create an imbalance of gene expression. For example, in Down syndrome, there are three copies of chromosome 21 and, therefore, three copies of all the genes on chromosome 21. This leads to a relative overexpression of genes that are located on chromosome 21 compared to the other chromosomes. Balanced translocations and inversions often are without phenotypic consequences because the total amount of genetic material is not altered, and the level of gene expression is not significantly changed.

C6. There are lots of possible answers; here are a few. Dwarfism occurs in people and dogs. Breeds like the dachshund and basset hound are types of dwarfism in dogs. There are diabetic people and mice. There are forms of inherited obesity in people and mice. Hip dysplasia is found in people and dogs.

C8. A. Because a person must inherit two defective copies of this gene and it is known to be on chromosome 1, the mode of transmission is autosomal recessive. Both members of this couple must be heterozygous, because they have one affected parent (who had to transmit the mutant allele to them) and their phenotypes are unaffected (so they must have received the normal allele from their other parent). Because both parents are heterozygotes, there is a $1/4$ chance of producing an affected child (a homozygote) with Gaucher disease. If we let G represent the nonmutant allele and g the mutant allele:

	♂ G	g
♀		
G	GG Normal	Gg Normal
g	Gg Normal	gg Gaucher

B. From this Punnett square, we can also see that there is a $1/4$ chance of producing a homozygote with both normal copies of the gene.

C. We need to apply the binomial expansion equation to solve this problem (see Chapter 2 for a description of this equation). In this problem, $n = 5$, $x = 1$, $p = 0.25$, $q = 0.75$. The answer is 0.396, or 39.6%.

C10. The mode of transmission is autosomal recessive. All of the affected individuals do not have affected parents. Also, the disorder is found in both males and females. If it were X-linked recessive, individual III-1 would have to have an affected father, which she does not.

C12. The 13 babies have acquired a new mutation. In other words, during spermatogenesis or oogenesis, or after the egg was fertilized, a new mutation occurred in the fibroblast growth factor gene. These 13 individuals have the same chances of passing the mutant allele to their offspring as the 18 individuals who inherited the mutant allele from a parent. The chance is 50%.

C14. Because this is a dominant trait, the mother must have two normal copies of the gene, and the father (who is affected) is most likely to be a heterozygote. (Note: The father could be a homozygote, but this is extremely unlikely because the dominant allele is very rare.) If we let M represent the mutant Marfan allele and m the normal allele, the following Punnett square can be constructed:

	♂ M	m
♀		
m	Mm Marfan	mm Normal
m	Mm Marfan	mm Normal

A. There is a 50% chance that this couple will have an affected child.

B. We use the product rule. The odds of having an unaffected child are 50%. So if we multiply $0.5 \times 0.5 \times 0.5$, this equals 0.125, or a 12.5% chance of having three unaffected offspring.

C16. A prion is a protein that behaves like an infectious agent. The infectious form of the prion protein has an abnormal conformation. This abnormal conformation is termed PrP^{Sc}, and the normal conformation of the protein is termed PrP^C. An individual can be "infected" with the abnormal conformation of the protein by eating products from another animal that had the disease, or the prion protein may convert spontaneously to the normal conformation. A prion protein in the PrP^{Sc} conformation can bind to a prion protein in the PrP^C conformation and convert it to the PrP^{Sc} form. An accumulation of prions in the PrP^{Sc} form is what causes the disease symptoms.

C18. A. Keep in mind that a conformational change is a stepwise process. It begins in one region of a protein and involves a series of small changes in protein structure. The entire conformational change from PrP^C to PrP^{Sc} probably involves many small changes in protein structure that occur in a stepwise manner. Perhaps the conformational change (from PrP^C to PrP^{Sc}) begins in the vicinity of position 178 and then proceeds throughout the rest of the protein. If there is a methionine at position 129, the complete conformational change (from PrP^C to PrP^{Sc}) can take place. However, perhaps the valine at position 129 somehow blocks one of the steps needed to complete the conformational change.

B. Once the PrP^{Sc} conformational change is completed, a PrP^{Sc} protein can bind to another prion protein in the PrP^C conformation. Perhaps it begins to convert it (to the PrP^{Sc} conformation) by initiating a small change in protein structure in the vicinity of position 178. The conversion would then proceed in a stepwise manner until the PrP^{Sc} conformation has been achieved. If a valine is at position 129, this could somehow inhibit one of the steps that are needed to complete the conformational change. If an individual had Val-129 in the polypeptide encoded by the second PrP gene, half of their prion proteins would be less sensitive to conversion by PrP^{Sc}, compared to individuals who had Met-129. This would explain why individuals with Val-129 in half of the prion proteins would have disease symptoms that would progress more slowly.

C20. A proto-oncogene is a normal cellular gene that typically plays a role in cell division. It can be altered by mutation to become an oncogene and thereby cause cancer. At the level of protein function, a proto-oncogene can become an oncogene by synthesizing too much of a protein or synthesizing the same amount of a protein that is abnormally active.

C22. The predisposition to develop cancer is inherited in a dominant fashion because the heterozygote has the higher predisposition. The mutant allele is actually recessive at the cellular level. But because we have so many cells in our bodies, it becomes relatively likely that a defective mutation will occur in the single normal gene and lead to a cancerous cell. Some heterozygous individuals may not develop the disease as a matter of chance. They may be lucky and not get a defective mutation in the normal gene, or perhaps their immune system is better at destroying cancerous cells once they arise.

C24. If an oncogene was inherited, it may cause uncontrollable cell growth at an early stage of development and thereby cause an embryo to develop improperly. This could lead to an early spontaneous abortion and thereby explain why we do not observe individuals with inherited oncogenes. Another possibility is that inherited oncogenes may adversely affect gamete survival, which would make it difficult for them to be passed from parent to offspring. A third possibility would be that oncogenes could affect the fertilized zygote in a way that would prevent the correct implantation in the uterus.

C26. The role of *p53* is to sense DNA damage and prevent damaged cells from proliferating. Perhaps, prior to birth, the fetus is in a protected environment, so DNA damage may be minimal. In other words, the fetus may not really need *p53*. After birth, agents such as UV light may cause DNA damage. At this point, *p53* is important. A *p53* knockout is more sensitive to UV light because it cannot repair its DNA properly in response to this DNA-damaging agent, and it cannot kill cells that have become irreversibly damaged.

C28. The *p53* protein is a regulatory transcription factor; it binds to DNA and influences the transcription rate of nearby genes. This transcription factor (1) activates genes that promote DNA repair; (2) activates a protein that inhibits cyclin/CDK protein complexes that are required for cell division; and (3) activates genes that promote apoptosis. If a cell is exposed to DNA damage, it has a greater potential to become malignant. Therefore, an organism wants to avoid the proliferation of such a cell. When exposed to an agent that causes DNA damage, a cell will try to repair the damage. However, if the damage is too extensive, the *p53* protein will stop the cell from dividing and program it to die. This helps to prevent the proliferation of cancer cells in the body.

Experimental Questions

E2. Perhaps the least convincing is the higher incidence of the disease in particular populations. Because populations living in specific geographic locations are exposed to their own unique environment, it is difficult to distinguish genetic versus environmental causes for a particular disease. The most convincing evidence might be the higher incidence of a disease in related individuals and/or the ability to correlate a disease with the presence of a mutant gene. Overall, however, the reliability that a disease has a genetic component should be based on as many observations as possible.

E4. You would probably conclude that it is less likely to have a genetic component. If it were rooted primarily in genetics, it would be likely to be found in the Central American population. Of course, there is a chance that very few or none of the people who migrated to Central America were carriers of the mutant gene, which is somewhat unlikely for a large migrating population. By comparison, one might suspect that an environmental agent present in South America but not present in Central America may underlie the disease. Researchers could try to search for this environmental agent (e.g., a pathogenic organism).

E6. Males I-1, II-1, II-4, II-6, III-3, III-8, and IV-5 have a normal copy of the gene. Males II-3, III-2, and IV-4 are hemizygous for an inactive mutant allele. Females III-4, III-6, IV-1, IV-2, and IV-3 have two normal copies of the gene, whereas females I-2, II-2, II-5, III-1, III-5, and III-7 are heterozygous carriers of a mutant allele.

E8. A transformed cell is one that has become malignant. In a laboratory, this can be done in three ways. First, the cells could be treated with

a mutagen that would convert a proto-oncogene into an oncogene. Second, cells could be exposed to the DNA from a malignant cell line. Under the appropriate conditions, this DNA can be taken up by the cells and integrated into their genome so that they become malignant. A third way to transform cells is by exposure to an oncogenic virus.

E10. By comparing oncogenic viruses with strains that have lost their oncogenicity, researchers have been able to identify particular genes that cause cancer. This has led to the identification of many oncogenes. From this work, researchers have also learned that normal cells contain proto-oncogenes that usually play a role in cell division. This suggests that oncogenes exert their effects by upsetting the cell division process. In particular, it appears that oncogenes are abnormally active and keep the cell division cycle in a permanent "on" position.

E12. One possible category of drugs would be GDP analogues (i.e., compounds that resemble the structure of GDP). Perhaps one could find a GDP analogue that binds to the Ras protein and locks it in the inactive conformation.

One way to test the efficacy of such a drug would be to incubate the drug with a type of cancer cell that is known to have an overactive Ras protein and then plate the cells on solid media. If the drug locked the Ras protein in the inactive conformation, it should inhibit the formation of malignant growth or malignant foci.

There are possible side effects of such drugs. First, they might block the growth of normal cells, because Ras protein plays a role in normal cell proliferation. Second (if you have taken a cell biology course), there are many GTP/GDP-binding proteins in cells, and the drugs could somehow inhibit cell growth and function by interacting with these proteins.

Questions for Student Discussion/Collaboration

2. There isn't a clearly correct answer to this question, but it should stimulate a large amount of discussion.

CHAPTER 23

Conceptual Questions

C2. A. False; the head is anterior to the tail.

B. True

C. False; the feet are ventral to the hips.

D. True

C4. A. True

B. False; because gradients are also established after fertilization during embryonic development.

C. True

C6. A. This is a mutation in a pair-rule gene (*runt*).

B. This is a mutation in a gap gene (*knirps*).

C. This is a mutation in a segment-polarity gene (*patched*).

C8. Positional information refers to the phenomenon whereby the spatial locations of morphogens and cell adhesion molecules (CAMs) provide a cell with information regarding its position relative to other cells. In *Drosophila*, the formation of a segmented body pattern relies initially on the spatial location of maternal gene products. These gene products lead to the sequential activation of the segmentation genes. Positional information can come from morphogens that are found within the oocyte, from morphogens secreted from cells during development, and from cell-to-cell contact. Although all three are important, morphogens in the oocyte have the greatest impact on the overall body structure.

C10. The anterior portion of the antero-posterior axis is established by the action of Bicoid. During oogenesis, the mRNA for Bicoid enters the anterior end of the oocyte and is sequestered there to establish an anterior (high) to posterior (low) gradient. Later, when the mRNA is translated, the Bicoid protein in the anterior region establishes a genetic hierarchy that leads to the formation of anterior structures. If Bicoid was not trapped in the anterior end, it is likely that anterior structures would not form.

C12. Maternal effect gene products influence the formation of the main body axes, including the antero-posterior, dorso-ventral, and terminal regions. They are expressed during oogenesis and needed very early in development. Zygotic genes, particularly the three classes of the segmentation genes, are necessary after the axes have been established. The segmentation genes are expressed after fertilization.

C14. The coding sequence of homeotic genes contains a 180-bp consensus sequence known as a homeobox. The protein domain encoded by the homeobox is called a homeodomain. The homeodomain contains three conserved sequences that are folded into α-helical conformations. The arrangement of these α helices promotes the binding of the protein to the major groove of the DNA. Helix III is called the recognition helix because it recognizes a particular nucleotide sequence within the major groove. In this way, homeotic proteins are able to bind to DNA in a sequence-specific manner and thereby activate particular genes.

C16. It would normally be expressed in the three thoracic segments that have legs (T1, T2, and T3).

C18. A. When a mutation inactivates a gap gene, a contiguous section of the larva is missing.

B. When a mutation inactivates a pair-rule gene, some regions that are derived from alternating parasegments are missing.

C. When a mutation inactivates a segment-polarity gene, portions are missing at either the anterior or posterior end of the segments.

C20. Proper development in mammals is likely to require the products of maternal effect genes that play a key role in initiating embryonic development. The adult body plan is merely an expansion of the embryonic body plan, which is established in the oocyte. Because the starting point for the development of an embryo is the oocyte, this explains why an enucleated oocyte is needed to clone mammals.

C22. A heterochronic mutation is one that alters the timing when a gene involved in development is normally expressed. The gene may be expressed too early or too late, which causes certain cell lineages to be out of sync with the rest of the animal. If a heterochronic mutation affected the intestine, the animal may end up with too many intestinal cells if it is a gain-of-function mutation or too few if it is a loss-of-function mutation. In either case, the effects might be detrimental because the growth of the intestine must be coordinated with the growth of the rest of the animal.

C24. Cell differentiation is the specialization of a cell into a particular cell type. In the case of skeletal muscle cells, the bHLH proteins play a key role in the initiation of cell differentiation. When bHLH proteins are activated, they are able to bind to enhancers and activate the expression of many different muscle-specific genes. In this way, myogenic bHLH proteins turn on the expression of many muscle-specific proteins. When these proteins are synthesized, they change the characteristics of the cell into those of a muscle cell. Myogenic bHLH proteins are regulated by dimerization. When a heterodimer forms between a myogenic bHLH protein and an E protein, it activates gene expression. However, when a heterodimer forms between myogenic bHLH proteins and a protein called Id, the heterodimer is unable to bind to DNA. The Id protein is produced during early stages of development and prevents myogenic bHLH proteins from promoting muscle differentiation too soon. At later stages of development, the amount of Id protein falls, and myogenic bHLH proteins can combine with E proteins to induce muscle differentiation.

C26. A totipotent cell is a cell that has the potential to create a complete organism.

A. In humans, a fertilized egg is totipotent, and the cells during the first few embryonic divisions are totipotent. However, after several divisions, embryonic cells lose their totipotency and, instead, are determined to become particular tissues within the body.

B. In plants, many living cells are totipotent.

C. Because yeast are unicellular, one cell is a complete individual. Therefore, yeast cells are totipotent; they can produce new individuals by cell division.

D. Because bacteria are unicellular, one cell is a complete individual. Therefore, bacteria are totipotent; they can produce new individuals by cell division.

C28. Animals begin their development from an egg and then form antero-posterior and dorso-ventral axes. The formation of an adult organism is an expansion of the embryonic body plan. Plants grow primarily from two meristems: shoot and root meristems. At the cellular level, plant development is different in that it does not involve cell migration, and most plant cells are totipotent. Animals require the organization within an oocyte to begin development. At the genetic level, however, animal and plant development are similar in that a genetic hierarchy of transcription factors governs pattern formation and cell specialization.

Experimental Questions

E2. *Drosophila* has an advantage in that researchers have identified many more mutant alleles that alter development in specific ways. The hierarchy of gene regulation is particularly well understood in the fruit fly. *C. elegans* has the advantage of simplicity and a complete knowledge of cell fate. This enables researchers to explore how the timing of gene expression is critical to the developmental process.

E4. To determine that a mutation is affecting the timing of developmental decisions, a researcher needs to know the normal time or stage of development when cells are supposed to divide and what types of cells will be produced. A lineage diagram provides this. With this information, one can then determine if particular mutations alter the timing when cell division occurs.

E6. Mutant 1 is a gain-of-function allele; it keeps reiterating the L1 pattern of division. Mutant 2 is a loss-of-function allele; it skips the L1 pattern and immediately follows an L2 pattern.

E8. As discussed in Chapter 15, most eukaryotic genes have a core promoter that is adjacent to the coding sequence; regulatory elements that control the transcription rate at the promoter are typically upstream from the core promoter. Therefore, to get the *Antp* gene product expressed where the *abd-A* gene product is normally expressed, you would link the upstream genetic regulatory region of the *abd-A* gene to the coding sequence of the *Antp* gene. This construct would be inserted into the middle of a P element (see next). The construct shown here would then be introduced into an embryo by P element transformation.

	abd-A regulatory		*Antp* coding	
P element	region	/	sequence	P element

The *Antp* gene product is normally expressed in the thoracic region and produces segments with legs, as illustrated in Figure 23.11. Therefore, because the *abd-A* gene product is normally expressed in the anterior abdominal segments, one might predict that the genetic construct shown above would produce a fly with legs attached to the segments that are supposed to be the anterior abdominal segments. In other words, the anterior abdominal segments might resemble thoracic segments with legs.

E10. A. The female flies must have had mothers that were heterozygous for a (dominant) normal allele and the mutant allele. Their fathers were either homozygous for the mutant allele or heterozygous. The female flies inherited a mutant allele from both their father and mother. Nevertheless, because their mother was heterozygous for the normal (dominant) allele and mutant allele, and because this is a maternal effect gene, their phenotype is based on the genotype of their mother. The normal allele is dominant, so they have a normal phenotype.

B. *Bicoid-A* appears to have a deletion that removes part of the sequence of the gene and thereby results in a shorter mRNA. *Bicoid-B* could also have a deletion that removes all of the sequence of the *bicoid* gene, or it could have a promoter mutation that prevents the expression of the bicoid gene. *Bicoid-C* seems to have a point mutation that does not affect the amount of the bicoid mRNA.

With regard to function, all three mutations are known to be loss-of-function mutations. *Bicoid-A* probably eliminates function by truncating the Bicoid protein. The Bicoid protein is a transcription factor. The *bicoid-A* mutation probably shortens this protein and thereby inhibits its function. The *bicoid-B* mutation prevents expression of the bicoid mRNA. Therefore, none of the Bicoid protein would be made, and this would explain the loss of function. The *bicoid-C* mutation seems to prevent the proper localization of the *bicoid* mRNA in the oocyte. There must be proteins within the oocyte that recognize specific sequences in the *bicoid* mRNA and trap it in the anterior end of the oocyte. This mutation must change these sequences and prevent these proteins from recognizing the bicoid mRNA.

E12. An egg-laying defect is somehow related to an abnormal anatomy. The *n540* strain has fewer neurons compared to a normal worm. Perhaps the *n540* strain is unable to lay eggs because it is missing neurons that are needed for egg laying. The *n536* and *n355* strains have an abnormal abundance of neurons. Perhaps this overabundance also interferes with the proper neural signals needed for egg laying.

E14. Geneticists interested in mammalian development have used reverse genetics because it has been difficult for them to identify mutations in developmental genes based on phenotypic effects in the embryo. This is because it is difficult to screen a large number of mammalian embryos in search of abnormal ones that carry mutant genes. It is easy to have thousands of flies in a laboratory, but it is not easy to have thousands of mice. Instead, it is easier to clone the normal gene based on its homology to invertebrate genes and then make mutations in vitro. These mutations can be introduced into a mouse to create a gene knockout. This strategy is opposite to that of Mendel, who characterized genes by first identifying phenotypic variants (e.g., tall versus dwarf, green seeds versus yellow seeds, etc.).

Questions for Student Discussion/Collaboration

2. In this problem, the students should try to make a flow diagram that begins with maternal effect genes, then gap genes, pair-rule genes, and segment polarity genes. These genes then lead to homeotic genes and finally genes that promote cell differentiation. It's almost impossible to make an accurate flow diagram because there are so many gene interactions, but it is instructive to think about developmental genetics in this way. It is probably easier to identify mutant phenotypes that affect later stages of development because they are less likely to be lethal. However, modern methods can screen for conditional mutants as described in solved problem S3. To identify all of the genes necessary for chicken development, you may begin with early genes, but this assumes you have some way to identify them. If they had been identified, you would then try to identify the genes that they stimulate or repress. This could be done using molecular methods described in Chapters 14, 15, and 18.

CHAPTER 24

Conceptual Questions

C2. A population is a group of interbreeding individuals. Let's consider a squirrel population in a forested area. Over the course of many generations, several things could happen to this population. A forest fire, for example, could dramatically decrease the number of individuals and thereby cause a bottleneck. This would decrease the genetic diversity of the population. A new predator may enter the region and natural selection may select for the survival of squirrels that are best able to evade the predator. Another possibility is that a group of squirrels within the population may migrate to a new region and found a new squirrel population.

C4. A. Phenotype frequency and genotype frequency

B. Genotype frequency

C. Allele frequency

C6. A. The genotype frequency for the *CF* homozygote is $1/2{,}500$, or 0.004. This would equal q^2. The allele frequency is the square root of this value, which equals 0.02. The frequency of the corresponding normal allele equals $1 - 0.02 = 0.98$.

B. The frequency for the *CF* homozygote is 0.004; for the unaffected homozygote, $(098)^2 = 0.96$; and for the heterozygote, $2(0.98)(0.02)$, which equals 0.039.

C. If a person is known to be a heterozygous carrier, the chances that this particular person will happen to choose another as a mate is equal to the frequency of heterozygous carriers in the population, which equals 0.039, or 3.9%. The chances that two randomly chosen individuals will choose each other as mates equals $0.039 \times 0.039 = 0.0015$, or 0.15%.

C8. For two alleles, the heterozygote is at a maximum when they are 0.5 each. For three alleles, the two heterozygotes are at a maximum when each allele is 0.33.

C10. Because this is a recessive trait, only the homozygotes for the rolling allele will be able to roll their tongues. If p equals the rolling allele and q equals the nonrolling allele, the Hardy-Weinberg equation predicts that the frequency of homozygotes who can roll their tongues would be p^2. In this case, $p^2 = (0.6)^2 = 0.36$, or 36%.

C12. A. The inbreeding coefficient is calculated using the formula

$$F = \Sigma \, (1/2)^n \, (1 + F_A)$$

In this case, there is one common ancestor, I-2. Because we have no prior history on I-2, we assume she is not inbred, which makes $F_A = 0$. The inbreeding loop for IV-3 contains five people, III-4, II-2, I-2, II-5, and III-5. Therefore, $n = 5$.

$$F = (1/2)^5(1 + 0) = 1/32 = 0.031$$

B. Based on the data shown in this pedigree, individual IV-4 is not inbred.

C14. Migration, genetic drift, and natural selection are the driving forces that alter allele frequencies within a population. Natural selection acts to eliminate harmful alleles and promote beneficial alleles. Genetic drift involves random changes in allele frequencies that may eventually lead to elimination or fixation of alleles. It is thought to be important in the establishment of neutral alleles in a population. Migration is important because it introduces new alleles into neighboring populations. According to the neutral theory, genetic drift is largely responsible for the variation seen in natural populations.

C16. A random force is one that alters allele frequencies without any regard to whether the changes are beneficial or not. Genetic drift and migration are the two main ways this can occur. An example is the founder effect, when a group of individuals migrate to a new location such as an island. Adaptive forces increase the reproductive success of a species. Natural selection is the adaptive force that tends to eliminate harmful alleles from a population and increase the frequency of beneficial alleles. An example would be the long neck of the giraffe, which enables it to feed in tall trees. At the molecular level, beneficial mutations may alter the coding sequence of a gene and change the structure and function of the protein in a way that is beneficial. For example, the sickle-cell anemia allele alters the structure of hemoglobin, and in the heterozygous condition, this inhibits the sensitivity of red blood cells to the malaria pathogen.

C18. Genetic drift is due to sampling error, and the degree of sampling error depends on the population size. In small populations, the relative proportion of sampling error is much larger. If genetic drift is moving an allele toward fixation, it will take longer in a large population because the degree of sampling error is much smaller.

C20. During the bottleneck effect, allele frequencies are dramatically altered due to genetic drift. In extreme cases, some alleles are lost, while others may become fixed at 100%. The overall effect is to decrease genetic diversity within the population. This may make it more difficult for the species to respond in a positive way to changes in the environment. Species that are approaching extinction also face a bottleneck as their numbers decrease. The loss of genetic diversity may make it even more difficult for the species to rebound.

C22. In all cases, these forms of natural selection favor one or more phenotypes because such phenotypes have a reproductive advantage. However, the patterns differ with regard to whether a single phenotype or multiple phenotypes are favored and whether the favored phenotype is in the middle of the phenotypic range or at one or both extremes. Directional selection favors one phenotype at a phenotypic extreme. Over time, natural selection is expected to favor the fixation of alleles that cause these phenotypic characteristics. Disruptive selection favors two or more phenotypic categories. It will lead to a population with a balanced polymorphism for the trait. Examples of balancing selection are heterozygote advantage and negative-frequency dependent selection. These promote a stable polymorphism in a population. Stabilizing selection favors individuals with intermediate phenotypes. It tends to decrease genetic diversity because alleles that favor extreme phenotypes are eliminated.

C24. The intuitive meaning of the mean fitness of a population is the relative likelihood that members of a population will reproduce. If the mean fitness is high, it is likely that an average member will survive and produce offspring. Natural selection increases the mean fitness of a population.

C26. A. True

B. True

C. False; it causes allele loss or fixation, which results in less diversity.

D. True

C28. A. Disruptive. There are multiple environments that favor different phenotypes.

B. Directional. The thicker the fur, the more likely that survival will occur.

C. Stabilizing. Low birth weight is selected against because it results in low survival. Also, very high birth weight is selected against because it could cause problems in delivery, which also could decrease the survival rate.

D. Directional. Sturdy stems and leaves will promote survival in windy climates.

Experimental Questions

E2. Solved problem S4 shows how the Hardy-Weinberg equation can be modified to include situations of three or more alleles. In this case

$(p + q + r + s)^2 = 1$

$p^2 + q^2 + r^2 + s^2 + 2pq + 2qr + 2qs + 2rp + 2rs + 2sp = 1$

Let $p = C$, $q = c^{ch}$, $r = c^h$, and $s = c$.

A. The frequency of albino rabbits is s^2:

$s^2 = (0.05) = 0.0025 = 0.25\%$

B. Himalayan is dominant to albino but recessive to full and chinchilla. Therefore, Himalayan rabbits would be represented by r^2 and by $2rs$:

$r^2 + 2rs = (0.44)^2 + 2(0.44)(0.05) = 0.24 = 24\%$

Among 1,000 rabbits, about 240 would have a Himalayan coat color.

C. Chinchilla is dominant to Himalayan and albino but recessive to full coat color. Therefore, heterozygotes with chinchilla coat color would be represented by $2qr$ and by $2qs$.

$2qr + 2qs = 2(0.17)(0.44) + 2(0.17)(0.05) = 0.17$, or 17%

Among 1,000 rabbits, about 170 would be heterozygotes with chinchilla fur.

E4. A.

Inuit	$M = 0.913$	$N = 0.087$
Navajo	$M = 0.917$	$N = 0.083$
Finns	$M = 0.673$	$N = 0.327$
Russians	$M = 0.619$	$N = 0.381$
Aborigines	$M = 0.176$	$N = 0.824$

B. To determine if these populations are in equilibrium, we can use the Hardy-Weinberg equation and calculate the expected number of individuals with each genotype. For example

Inuit $MM = (0.913)^2 = 0.833 = 83.3\%$

$MN = 2(0.913)(0.087) = 0.159 = 15.9\%$

$NN = (0.087)^2 = 0.0076 = 0.76\%$

In general, the values agree pretty well with an equilibrium. The same is true for the other four populations.

C. Based on similar allele frequencies, the Inuit and Navajo Indians seem to have interbred, as well as the Finns and Russians.

E6. A. $\Delta p_C = m(p_D - p_R)$

With regard to the sickle-cell allele,

$\Delta p_C = (550/10,550)(0.1 - 0.01) = 0.0047$

$p_C = p_R + \Delta p_C = 0.01 + 0.0047 = 0.0147$

B. We need to calculate the genotypes separately:

For the 550 migrating individuals,

$Hb^A Hb^A = (0.9)2 = 0.81$, or 81%
We expect $(0.81)550 = 445.5$ individuals to have this genotype.

$Hb^A Hb^S = 2(0.9)(0.1) = 0.18$
We expect $(0.18)550 = 99$ heterozygotes.

$Hb^S Hb^S = (0.1)^2 = 0.01$
We expect $(0.01)550 = 5.5$ $Hb^S Hb^S$.

For the original recipient population,

$Hb^A Hb^A = (0.99)^2 = 0.98$
We expect 9,801 individuals to have this genotype.

$Hb^A Hb^S = 2(0.99)(0.01) = 0.0198$
We expect 198 with this genotype.

$Hb^S Hb^S = (0.01)^2 = 0.0001$
We expect 1 with this genotype.

To calculate the overall population,

$(445.5 + 9801)/10,550 = 0.971$ $Hb^A Hb^A$ homozygotes

$(99 + 198)/10,550 = 0.028$ heterozygotes

$(5.5 + 1)/10,550 = 0.00062$ $Hb^S Hb^S$ homozygotes

C. After one round of mating, the allele frequencies in the conglomerate (calculated in part A), should yield the expected genotype frequencies according to the Hardy-Weinberg equilibrium.

Allele frequency of $Hb^S = 0.0147$, so $Hb^A = 0.985$.

$Hb^A Hb^A = (0.985)^2 = 0.97$

$Hb^A Hb^S = 2(0.985)(0.0147) = 0.029$

$Hb^S Hb^S = (0.0147)^2 = 0.0002$

E8. A. Probability of fixation $= 1/2N$ (Assuming equal numbers of males and females contributing to the next generation)

Probability of fixation $= 1/2(2,000,000)$

$= 1$ in 4,000,000 chance

B. $\bar{t} = 4N$

Where $\bar{t} =$ the average number of generations to achieve fixation

$N =$ the number of individuals in population, assuming that males and females contribute equally to each succeeding generation

$\bar{t} = 4(2 \text{ million}) = 8$ million generations

C. If the blue allele had a selective advantage, the value calculated in part A would be slightly larger; there would be a higher chance of allele fixation. The value calculated in part B would be smaller; it would take a shorter period of time to reach fixation.

E10. You could mark snails with a dye and release equal numbers of dark and light snails into dimly lit forested regions and sunny fields. At a later time, recapture the snails and count them. It would be important to have a method of unbiased recapture because the experimenter would have an easier time locating the light snails in a forest and the dark snails in a field. Perhaps one could bait the region with something that the snails like to eat and only collect snails that are at the bait. In addition to this type of experiment, one could also observe predation as it occurs.

E12. Male 2 is the potential father, because he contains the bands found in the offspring but not found in the mother. To calculate the probability, one would have to know the probability of having each of the types of bands that match. In this case, for example, male 2 and the offspring have four bands in common. As a simple calculation, we could eliminate the four bands the offspring shares with the mother. If the probability of having each paternal band is $1/4$, the chances that this person is not the father are $(1/4)^4$.

E14. PCR is used to amplify DNA if there is only a small amount of it (e.g., a small sample at a crime scene). It is also used to amplify microsatellites. Southern blotting, using a probe complementary to a minisatellite, is needed to specifically identify a limited number of bands (20 or so) that are variable within human populations.

E16. This percentage is not too high. Based on their genetic relationship, we expect that a father and daughter must share at least 50% of the same bands in a DNA fingerprint. However, the value can be higher than that because the mother and father may have some bands in common, even though they are not genetically related. For example, at one site in the genome, the father may be heterozygous for a 4,100 bp and 5,200 bp microsatellite, and the mother may also be heterozygous in this same region and have 4,100 bp and 4,700 bp microsatellites. The father could pass the 5,200 bp band to his daughter, and the mother could pass the 4,100 bp band. Thus, the daughter would inherit the 4,100 bp and 5,200 bp bands. This would be a perfect match to both of the father's bands, even though the father transmitted only the 5,200 bp band to his daughter. The 4,100 bp band matches because the father and mother happened to have a microsatellite in common. Therefore, the 50% estimate of matching bands in a DNA fingerprint based on genetic relationships is a minimum estimate. The value can be higher than that.

Questions for Student Discussion/Collaboration

2. Mutation is responsible for creating new alleles, but the rate of new mutations is so low that it cannot explain allele frequencies in this range. Let's call the two alleles B and b and assume that B was the original allele and b is more recent allele that arose as a result of mutation. Three scenarios to explain the allele frequencies:

   1. The b allele is neutral and reached its present frequency by genetic drift. It hasn't reached elimination or fixation yet.

   2. The b allele is beneficial, and its frequency is increasing due to natural selection. However, there hasn't been enough time to reach fixation.

   3. The Bb heterozygote is at a selective advantage, leading to a balanced polymorphism.

CHAPTER 25

Conceptual Questions

C2. At the molecular level, quantitative traits often exhibit a continuum of phenotypic variation because they are usually influenced by multiple genes that exist as multiple alleles. A large amount of environmental variation will also increase the overlap between genotypes and phenotypes for polygenic traits.

C4. A discontinuous trait is one that falls into discrete categories. Examples include brown eyes versus blue eyes in humans and purple versus white flowers in pea plants. A continuous trait is one that does not fall into discrete categories. Examples include height in humans and fruit weight in tomatoes. Most quantitative traits are continuous; the trait falls within a range of values. The reason why quantitative traits are continuous is because they are usually polygenic and greatly influenced by the environment. As shown in Figure 25.4b, this tends to create ambiguities between genotypes and a continuum of phenotypes.

C6. To be in the top 2.5% is about two standard deviation units. If we take the square root of the variance, the standard deviation would be 22 lb. To be in the top 2.5%, an animal would have to weigh at least 44 lb heavier than the mean, which equals 606 lb. To be in the bottom 0.13%, an animal would have to be three standard deviations lighter, which would be at least 66 lb lighter than the mean, or 496 lb.

C8. There is a positive correlation, but it could have occurred as a matter of chance alone. According to Table 25.2, this value could have occurred through random sampling error. You would need to conduct more experimentation to determine if there is a significant correlation, such as examining a greater number of pairs of individuals. If $N = 500$, the correlation would be statistically significant, and you would conclude that the correlation did not occur as a matter of random chance. However, you could not conclude cause and effect.

C10. When a correlation coefficient is statistically significant, it means that the association is likely to have occurred for reasons other than random sampling error. It may indicate cause and effect but not necessarily. For example, large parents may have large offspring due to genetics (cause and effect). However, the correlation may be related to the sharing of similar environments rather than cause and effect.

C12. Quantitative trait loci are sites within chromosomes that contain genes that affect a quantitative trait. It is possible for a QTL to contain one gene, or it may contain two or more closely linked genes. QTL mapping, which involves linkage to known molecular markers, is commonly used to determine the locations of QTLs.

C14. If the broad sense heritability equals 1.0, it means that all of the variation in the population is due to genetic variation rather than environmental variation. It does not mean that the environment is unimportant in the outcome of the trait. Under another set of environmental conditions, the trait may have turned out quite differently.

C16. Hybrid vigor is the phenomenon in which an offspring produced from two inbred strains is more vigorous than the corresponding parents. Tomatoes and corn are often the products of hybrids.

C18. When a species is subjected to selective breeding, the breeder is focusing his/her attention on improving one particular trait. In this case, the rose breeder is focused on the size and quality of the flowers. Because the breeder usually selects a small number of individuals (e.g., the ones with best flowers) as the breeding stock for the next generation, this may lead to a decrease in the allelic diversity at other genes. For example, several genes affect flower fragrance. In an unselected population, these genes may exist as "fragrant alleles" and "nonfragrant alleles." After many generations of breeding for large flowers, the fragrant alleles may be lost from the population, just as a matter of random chance. This is a common problem of selective breeding. As you select for an improvement in one trait, you may inadvertently diminish the quality of an unselected trait.

Others have suggested that the lack of fragrance may be related to flower structure and function. Perhaps the amount of energy that a flower uses to make beautiful petals somehow diminishes its capacity to make fragrance.

C20. Broad-sense heritability takes into account all genetic factors that affect the phenotypic variation in a trait. Narrow-sense heritability considers only alleles that behave in an additive fashion. In many cases, the alleles affecting quantitative traits appear to behave additively. More importantly, if a breeder assumes that the heritability of a trait is due to the additive effects of alleles, it is possible to predict the outcome of selective breeding. This is also termed the *realized heritability*.

C22. A. Because of their good nutrition, you may speculate that they would grow to be taller.

B. If the environment is rather homogeneous, then heritability values tend to be higher because the environment contributes less to the amount of variation in the trait. Therefore, in the commune, the heritability might be higher, because they uniformly practice good nutrition. On the other hand, because the commune is a smaller size than the general population, the amount of genetic variation might be less, so this would make the heritability lower. However, because the problem states that the commune population is large, we would probably assume that the amount of genetic variation is similar to that in the general population. Overall, the best guess would be that the heritability in the commune population is higher because of the uniform nutrition standards.

C. As stated in part B, the amount of variation would probably be similar, because the commune population is large. As a general answer, larger populations tend to have more genetic variation. Therefore, the general population probably has a bit more variation.

C24. A natural population of animals is more likely to have a higher genetic diversity compared to a domesticated population. This is because domesticated populations have been subjected to many generations of selective breeding, which decreases the genetic diversity. Therefore, V_G is likely to be higher for the natural population. The other issue is the environment. It is difficult to say which group would have a more homogeneous environment. In general, natural populations tend to have a more heterogeneous environment, but not always. If the environment is more heterogeneous, this tends to cause more phenotypic variation, which makes V_E higher.

$$\text{Heritability} = V_G/V_T$$
$$= V_G/(V_G + V_E)$$

When V_G is high, heritability increases. When V_E is high, heritability decreases. In the natural wolf population, we would expect that V_G would be high. In addition, we would guess that V_E might be high as well (but that is less certain). Nevertheless, if this were the case, the heritability of the wolf population might be similar to the domestic population. This is because the high V_G in the wolf population is balanced by its high V_E. On the other hand, if V_E is not that high in the wolf population, or if it is fairly high in the domestic population, then the wolf population would have a higher heritability for this trait.

Experimental Questions

E2. To calculate the mean, we add the values together and divide by the total number.

$$\text{Mean} = \frac{1.9 + 2(2.4) + 2(2.1) + 3(2.0) + 2(2.2) + 1.7 + 1.8 + 2(2.3) + 1.6}{15}$$

Mean = 2.1

The variance is the sum of the squared deviations from the mean divided by $N - 1$. The mean value of 2.1 must be subtracted from each value, and then the square is taken. These 15 values are added together and then divided by 14 (which is $N - 1$).

$$\text{Variance} = \frac{0.85}{14}$$
$$= 0.061$$

The standard deviation is the square root of the variance.

Standard deviation = 0.25

E4. The results are consistent with the idea that there are QTLs for this trait on chromosomes 2 and 3 but not on the X chromosome.

E6. When we say an RFLP is associated with a trait, we mean that a gene that influences a trait is closely linked to an RFLP. At the chromosomal level, the gene of interest is so closely linked to the RFLP that a crossover almost never occurs between them.

Note: Each plant inherits four RFLPs, but it may be homozygous for one or two of them.

Small: 2,700 and 4,000 (homozygous for both)

Small-medium: 2,700 (homozygous), 3,000, and 4,000; or 2,000, 2,700, and 4,000 (homozygous)

Medium: 2,000 and 4,000 (homozygous for both); or 2,700 and 3,000 (homozygous for both); or 2,000, 2,700, 3,000, and 4,000

Medium-large: 2,000 (homozygous), 3,000, and 4,000; or 2,000, 2,700, and 3,000 (homozygous)

Large: 2,000 and 3,000 (homozygous for both)

E8. Let's assume there is an extensive molecular marker map for the rice genome. We would begin with two strains of rice, one with a high yield and one with a low yield, that greatly differ with regard to the molecular markers they carry. We would make a cross between these two strains to get F_1 hybrids. We would then backcross the F_1 hybrids to either of the parental strains and then examine hundreds of offspring with regard to their rice yields and molecular markers. In this case, our expected results would be that six different markers in the high-producing strain would be correlated with offspring that produce higher yields. We might get fewer than six bands if some of these genes are closely linked and associate with the same marker. We also might get fewer than six if the two parental strains have the same marker that is associated with one or more of the genes that affect yield.

E10. A. If we assume that the highly inbred strain has no genetic variance,

V_G (for the wild strain) $= 3.2 \text{ g}^2 - 2.2 \text{ g}^2 = 1.0 \text{ g}^2$

B. $h_B^2 = 1.0 \text{ g}^2/3.2 \text{ g}^2 = 0.31$

C. It is the same as h_B^2, so it also equals 0.31.

E12. A. $h_N^2 = \dfrac{\overline{X}_O - \overline{X}}{\overline{X}_P - \overline{X}}$

$0.21 = (26.5 \text{ g} - 25 \text{ g})/(27 \text{ g} - 25 \text{ g})$

$\overline{X}_O - 25 \text{ g} = 2 \text{ g} (0.21)$

$\overline{X}_O = 25.42 \text{ g}$

B. $0.21 = (26.5 \text{ g} - 25 \text{ g})/(\overline{X}_P - 25 \text{ g})$

$(\overline{X}_P - 25 \text{ g})(0.21) = 1.5 \text{ g}$

$\overline{X}_P = 32.14 \text{ g parents}$

However, because this value is so far from the mean, there may not be 32.14 g parents in the population of mice that you have available.

E14. We first need to calculate a and b. In this calculation, X represents the height of fathers, and Y represents the height of sons.

$b = \dfrac{144}{112} = 1.29$

$a = 69 - (1.29)(68) = -18.7$

For a father who is 70 in. tall,

$Y = (1.29)(70) + (-18.7) = 71.6$

The most likely height of the son would be 71.6 in.

E16. A. After six or seven generations, the selective breeding seems to have reached a plateau. This suggests that the tomato plants have become monomorphic for the alleles that affect tomato weight.

B. There does seem to be heterosis, because the first generation has a weight of 1.7 lb, which is heavier than either Mary's or Hector's tomatoes. This partially explains why Martin has obtained tomatoes heavier than 1.5 lb. However, heterosis is not the whole story; it does not explain why Martin obtained tomatoes that weigh 2 lb. Even though Mary's and Hector's tomatoes were selected for heavier weight, they may not have all of the "heavy alleles" for each gene that controls weight. For example, let's suppose there are 20 genes that affect weight, with each gene existing in a light and heavy allele. During the early stages of selective breeding, when Mary and Hector picked their 10 plants as seed producers for the next generation, as a matter of random chance, some of these plants may have been homozygous for the light alleles at a few of the 20 genes that control weight. Therefore, just as a matter of chance, they probably "lost" a few of the heavy alleles that affect weight. So, after 12 generations of breeding, they have predominantly heavy alleles but also have light alleles for some of the genes. If we represent heavy alleles with a capital letter and light alleles with a lowercase letter, Mary's and Hector's strains could be the following:

Mary's strain: *AA BB cc DD EE FF gg hh II JJ KK LL mm NN OO PP QQ RR ss TT*

Hector's strain: *AA bb CC DD EE ff GG HH II jj kk LL MM NN oo PP QQ RR SS TT*

As we see here, Mary's strain is homozygous for the heavy allele at 15 of the genes but carries the light allele at the other 5. Similarly, Hector's strain is homozygous for the heavy allele at 15 genes and carries the light allele at the other 5. It is important to note, however, that the light alleles in Mary's and Hector's strains are not in the same genes. Therefore, when Martin crosses them together, he will initially get the following:

Martin's F_1 offspring: *AA Bb Cc DD EE Ff Gg Hh II Jj Kk LL Mm NN Oo PP QQ RR Ss TT*

If the alleles are additive and contribute equally to the trait, we would expect about the same weight (1.5 lb), because this hybrid has a total of 10 light alleles. However, if heterosis is occurring, genes (which were homozygous recessive in Mary's and Hector's strains) will become heterozygous in the F_1 offspring, and this may make the plants healthier and contribute to a higher weight. If Martin's F_1 strain is subjected to selective breeding, the 10 genes that are heterozygous in the F_1 offspring may eventually become homozygous for the heavy allele. This would explain why Martin's tomatoes achieved a weight of 2.0 pounds after five generations of selective breeding.

E18. A. $h_N^2 = \dfrac{\overline{X}_O - \overline{X}}{\overline{X}_P - \overline{X}}$

$h_N^2 = \dfrac{269 - 254}{281 - 254} = 0.56$

B. $0.56 = \dfrac{275 - 254}{\overline{X}_P - 254}$

$\overline{X}_P = 291.5 \text{ lb}$

Questions for Student Discussion/Collaboration

2. Most traits depend on the influence of many genes. Also, genetic variation is a common phenomenon in most populations. Therefore, most individuals have a variety of alleles that contribute to a given trait. For quantitative traits, some alleles may make the trait bigger, and other alleles may make the trait turn out smaller. If a population contains many different genes and alleles that govern a quantitative trait, most individuals will have an intermediate phenotype because they will have inherited some large and some small alleles. Fewer individuals will inherit a predominance of large alleles or a predominance of small alleles. An example of a quantitative trait that does not fit a normal distribution is snail pigmentation. The dark snails and light snails are favored rather than the intermediate colors because they are less susceptible to predation.

CHAPTER 26

Conceptual Questions

C2. Evolution is unifying because all living organisms on this planet evolved from the same primordial organism. At the molecular level, all organisms have a great deal in common. With the exception of a few viruses, they all use DNA as their genetic material. This DNA is found within chromosomes, and the sequence of the DNA is organized into units called genes. Most genes are structural genes that encode the amino acid sequence of polypeptides. Polypeptides fold to form functional units called proteins. At the cellular level, all living organisms also share many similarities. For example, living cells share many of the same basic features including a plasma membrane, ribosomes, enzymatic pathways, and so on. In addition, as discussed in Chapter 7, the mitochondria and chloroplasts of eukaryotic cells are evolutionarily derived from bacterial cells.

C4. 1. Phylogenetic—Each species has distinct morphological, cellular, and/or molecular traits.

2. Biological—Each species is considered distinct if it is reproductively isolated from other species.

3. Evolutionary—Each species is derived from a single lineage that is distinct from other lineages and has its own evolutionary tendencies and historical fate.

4. Ecological—Each species occupies its own ecological niche.

C6. Anagenesis is the evolution of one species into another, whereas cladogenesis is the divergence of one species into two or more species. Of the two, cladogenesis is more prevalent. There may be many reasons why. It is common for an abrupt genetic change such as alloploidy to produce a new species from a preexisting one. Also, migrations of a few members of species into a new region may lead to the formation of a new species in the region (i.e., allopatric speciation).

C8. A. Allopatric

B. Sympatric

C. At first, it may involve parapatric speciation with a low level of intermixing. Eventually, when smaller lakes are formed, allopatric speciation will occur.

C10. The main evidence in favor of punctuated equilibrium is the fossil record. Paleontologists rarely find a gradual transition of fossil forms. The transition period in which environment pressure and genetic changes cause a previous species to evolve into a new species is thought to be so short that few, if any, of the transitional members would be preserved as fossils. Therefore, the fossil record primarily contains representatives from the long equilibrium periods. Also, rapid evolutionary change is consistent with known genetic phenomena, including single-gene mutations that have dramatic effects on phenotypic characteristics, the founder effect, and genetic events such as changes in chromosome structure (e.g., inversions and translocations) or chromosome number, which may abruptly create individuals with new phenotypic traits. In some cases, however, gradual changes are observed in certain species over long periods of time. In addition, the gradual accumulation of mutations is known to occur from the molecular analyses of DNA.

C12. Allopatric speciation involves a physical separation of a species into two or more separate populations. Over time, each population accumulates mutations that alter the characteristics of each population. Because the populations are separated, each will evolve different characteristics and eventually become distinct species. In parapatric speciation, there is some physical separation of two or more populations, but the separation is not absolute. On occasion, members of different populations can interbreed. Even so, the (somewhat) separated populations will tend to accumulate different genetic changes (e.g., inversions) that will ultimately lead to reproductive isolation among the different populations. In sympatric speciation, members of a population are not physically separated, but something happens (e.g., polyploidy) that abruptly results in reproductive isolation between members of the population. For example, a species could be diploid and a member of the population could become tetraploid. The tetraploid member would be reproductively isolated from the diploid members because hybrid offspring would be triploid and sterile. Therefore, the tetraploid individual has become a separate species.

C14. The relatively constant rate of neutral mutations acts as a molecular clock on which to measure evolutionary time. Neutral mutations occur at a rate proportional to the rate of mutation per generation. Therefore, the genetic divergence between species that is due to neutral mutations reflects the time elapsed since their last common ancestor. The concept is related to the neutral theory of evolution because it assumes that most genetic variation is due to the accumulation of neutral or nearly neutral mutations.

C16. A gene sequence can evolve more rapidly. The purpose of structural genes is to encode a polypeptide with a defined amino acid sequence. Many nucleotide changes will have no effect on the amino acid sequence of the polypeptide. For example, mutations in intron sequences and mutations at the wobble base may not affect the amino acid sequence of the encoded polypeptide. These neutral mutations will happen rather rapidly on an evolutionary timescale because natural selection will not remove them from the population. In contrast, changes in the amino acid sequence may alter the structure and function of the polypeptide. Most random mutations that affect the polypeptide sequence are more likely to be detrimental than beneficial, and detrimental mutations will be eliminated by natural selection. This makes it more difficult for the amino acid sequence of the polypeptide to evolve. Only neutral changes and beneficial changes will happen rapidly, and these are less likely to occur in the amino acid sequence compared to the gene sequence.

C18. You would expect the sequences of plant storage proteins to evolve rapidly. The polypeptide sequence is not particularly important for the structure or function of the protein. The purpose of the protein is to provide nutrients to the developing embryo. Changing the sequence would likely be tolerated. However, major changes in the amino acid composition (not the sequence) may be selected against. For example, the storage protein would have to contain some cysteine in its amino acid sequence because the embryo would need some cysteine to grow. However, the location of cysteine codons within the amino acid sequence would not be important; it would only be important that the gene sequence have some cysteine codons.

C20. The α-globin sequences in humans and horses are more similar to each other, compared to the α-globin in humans and the β-globin in humans. This suggests that the gene duplication that produced the α-globin and β-globin genes occurred first. After this gene duplication occurred, each gene accumulated several different mutations that caused the sequences of the two genes to diverge. At a much later time, during the evolution of mammals, a split occurred that produced different branches in the evolutionary tree of mammals. One branch eventually led to the formation of horses and a different branch led to the formation of humans. During the formation of these mammalian branches (which has been more recent), some additional mutations could occur in the α- and β-globin genes. This would explain why the α-globin gene in humans and horses is not exactly the same. However, it is more similar than the α- and β-globin genes within humans because the divergence of humans and horses occurred much more recently than the gene duplication that produced the α- and β-globin genes. In other words, there has been much less time for the α-globin gene in humans to diverge from the α-globin gene in horses.

C22. A. This is an example of neutral mutation. Mutations in the wobble base are neutral when they do not affect the amino acid sequence.

B. This is an example of natural selection. Random mutations that occur in vital regions of a polypeptide sequence are likely to inhibit function. Therefore, these types of mutations are eliminated by natural selection. That is why they are relatively rare.

C. This is a combination of neutral mutation and natural selection. The prevalence of mutations in introns is due to the accumulation of neutral mutations. Most mutations within introns do not have any effect on the expression of the exons, which contain the polypeptide sequence. In contrast, mutations within the exons are more likely to be affected by natural selection. As mentioned in the answer to part B, mutations in vital regions are likely to inhibit function. Natural selection tends to eliminate these mutations. Therefore, mutations within exons are less likely than mutations within introns.

C24. Generally, one would expect a similar number of chromosomes with very similar banding patterns. However, there may be a few notable differences. An occasional translocation could change the size or chromosomal number between two different species. Also, an occasional inversion may alter the banding pattern between two species.

C26. Animals with very simple body plans, such as the sponge, have relatively few *Hox* genes. In contrast, animals with more complicated bodies, such as mammals, have many. Researchers speculate that each *Hox* gene can govern the morphological features of a particular region of the body. By having multiple *Hox* genes, different regions of the body can become more specialized, and therefore, more complex.

Experimental Questions

E2. Perhaps the easiest way to determine allotetraploidy is by the chromosomal examination of closely related species. A researcher could karyotype the chromosomes from many different species and look for homologous chromosomes that have similar banding patterns. This may enable them to identify allotetraploids that contain a diploid set of chromosomes from two different species.

E4. The mutations that have occurred in this sequence are neutral mutations. In all cases, the wobble base has changed, and this change would not affect the amino acid sequence of the encoded polypeptide. Therefore, a reasonable explanation is that the gene has accumulated random neutral mutations over the course of many generations. This observation would be consistent with the neutral theory of evolution. A second explanation would be that one of these two researchers made a few experimental mistakes when determining the sequence of this region.

E6. Inversions do not affect the total amount of genetic material. Usually, inversions do not affect the phenotype of the organism. Therefore, if members of the two populations were to interbreed, the offspring would probably be viable because they would have inherited a normal amount of genetic material from each parent. However, such offspring would be inversion heterozygotes. As described in Chapter 8 (see Figure 8.12), crossing over during meiosis may create chromosomes that have too much or too little genetic material. If these unbalanced chromosomes are passed to the next generation of offspring, the offspring may not survive. For this reason, inversion heterozygotes (that are phenotypically normal) may not be very fertile because many of their offspring will die. Because inversion heterozygotes are less fertile, this would tend to keep the eastern and western populations reproductively isolated. Over time, this would aid in the independent evolution of the two populations and would ultimately promote the evolution of the two populations into separate species.

E8.   The technique of PCR is used to amplify the amount of DNA in a sample. To accomplish this, one must use oligonucleotide primers that are complementary to the region that is to be amplified. For example, as described in the experiment of Figure 26.15, PCR primers that were complementary to and flank the 12S rRNA gene can be used to amplify the 12S rRNA gene. The technique of PCR is described in Chapter 18.

E10. We would expect the probe to hybridize to the natural *G. tetrahit* and also the artificial *G. tetrahit*, because both of these strains contain two sets of chromosomes from *G. pubescens*. We would expect two bright spots in the *in situ* experiment. Depending on how closely related *G. pubescens* and *G. speciosa* are, the probe may also hybridize to two sites in the *G. speciosa* genome, but this is difficult to predict *a priori*. If so, the *G. tetrahit* species would show four spots.

E12. The principle of parsimony chooses a phylogenetic tree that requires the fewest number of evolutionary changes. When using molecular data, researchers can use computer programs that compare DNA sequences from homologous genes of different species and construct a tree that requires the fewest numbers of mutations. Such a tree is the most likely pathway for the evolution of such species.

E14. Possibly, the mouse would have an eye at the tip of its tail!

Questions for Student Discussion/Collaboration

2.   The founder effect and allotetraploidy are examples of rapid forms of evolution. In addition, some single gene mutations may have a great impact on phenotype and lead to the rapid evolution of new species by cladogenesis. Geological processes may promote the slower accumulation of alleles and alter a species' characteristics more gradually. In this case, it is the accumulation of many phenotypically minor genetic changes that ultimately leads to reproductive isolation. Slow and fast mechanisms of evolution have the common theme that they result in reproductive isolation. This is a prerequisite for the evolution of new species. Fast mechanisms tend to involve small populations and a few number of genetic changes. Slower mechanisms may involve larger populations and involve the accumulation of a large number of genetic changes that each contributes in a small way.

GLOSSARY

::

A

A an abbreviation for adenine.

ABC model a model for flower development.

acentric describes a chromosome without a centromere.

acquired antibiotic resistance the acquisition of antibiotic resistance because a bacterium has taken up a gene or plasmid from another bacterial strain.

acridine dye a type of chemical mutagen that causes frameshift mutations.

acrocentric describes a chromosome with the centromere significantly off center, but not at the very end.

activator a transcriptional regulatory protein that increases the rate of transcription.

acutely transforming virus (ACT) a virus that readily transforms normal cells into malignant cells, when grown in a laboratory.

adaptor hypothesis a hypothesis that proposes a tRNA has two functions: recognizing a three-base codon sequence in mRNA and carrying an amino acid that is specific for that codon.

adenine a purine base found in DNA and RNA. It base-pairs with thymine in DNA.

A DNA a right-handed DNA double helix with 11 base pairs per turn. Does not occur in living cells.

age of onset for alleles that cause genetic diseases, the time of life at which disease symptoms appear.

alkaptonuria a human genetic disorder involving the accumulation of homogentisic acid due to a defect in homogentisic acid oxidase.

alkyltransferase an enzyme that can remove methyl or ethyl groups from guanine bases.

allele an alternative form of a specific gene.

allele frequency the number of copies of a particular allele in a population divided by the total number of all alleles for that gene in the population.

allelic variation genetic variation in a population that involves the occurrence of two or more different alleles for a particular gene.

allodiploid an organism that contains one set of chromosomes from two different species.

allopatric speciation (Greek, *allos*, "other"; Latin, *patria*, "homeland") an evolutionary phenomenon in which speciation occurs when members of a species become geographically separated from the other members.

alloploid an organism that contains chromosomes from two (or more) different species.

allopolyploid an organism that contains two (or more) sets of chromosomes from two (or more) species.

allosteric enzyme an enzyme that contains two binding sites—a catalytic site and a regulatory site.

allosteric regulation the phenomenon that an effector molecule binds to a noncatalytic site on a protein and causes a conformational change that regulates its function.

allosteric site the site on a protein where a small effector molecule binds to regulate the function of the protein.

allotetraploid an organism that contains two sets of chromosomes from two different species.

allozymes two or more enzymes (encoded by the same type of gene) with alterations in their amino acid sequences, which may affect their gel mobilities.

α helix a type of secondary structure found in proteins.

alternative exon an exon that is not always found in mRNA. It is only found in certain types of alternatively spliced mRNAs.

alternative splicing refers to the phenomenon in which a pre-mRNA can be spliced in more than one way.

amber a stop codon with the sequence UAG.

Ames test a test using strains of a bacterium, *Salmonella typhimurium*, to determine if a substance is a mutagen.

amino acid a building block of polypeptides and proteins. It contains an amino group, a carboxyl group, and a side chain.

aminoacyl site (A site) a site on the ribosome where a charged tRNA initially binds.

aminoacyl tRNA a tRNA molecule that has an amino acid covalently attached to its 3' end.

aminoacyl-tRNA synthetase an enzyme that catalyzes the attachment of a specific amino acid to the correct tRNA.

2-aminopurine a base analogue that acts as a chemical mutagen.

amino terminus the location of the first amino acid in a polypeptide chain. The amino acid at the amino terminus still retains a free amino group that is not covalently attached to the second amino acid.

amniocentesis a method of obtaining cellular material from a fetus for the purpose of genetic testing.

amplified restriction fragment length polymorphism (AFLP) a RFLP that is amplified via PCR.

anabolic enzyme an enzyme involved in connecting organic molecules to create larger molecules.

anagenesis (Greek, *ana*, "up," and *genesis*, "origin") the evolutionary phenomenon in which a single species is transformed into a different species over the course of many generations.

anaphase the fourth stage of M phase. As anaphase proceeds, half of the chromosomes move to one pole, and the other half move to the other pole.

ancestral character see *primitive character*.

ancient DNA analysis analysis of DNA that is extracted from the remains of extinct species.

aneuploid not euploid. Refers to a variation in chromosome number such that the total number of chromosomes is not an exact multiple of a set or *n* number.

annealing the process in which two complementary segments of DNA bind to each other.

annotated in files involving genetic sequences, annotation is a description of the known function and features of the sequence, as well as other pertinent information.

antero-posterior axis in animals, the axis that runs from the head (anterior) to the tail or base of the spine (posterior).

anther the structure in flowering plants that gives rise to pollen grains.

anther culture the generation of monoploid plants by cold-shock treatment of anthers.

antibodies proteins produced by the B cells of the immune system that recognize foreign substances (namely, viruses, bacteria, and so forth) and target them for destruction.

antibody microarray a small silica, glass, or plastic slide that is dotted with many different antibodies, which recognize particular amino acid sequences within proteins.

anticipation the phenomenon in which the severity of an inherited disease tends to get worse in future generations.

anticodon a three-nucleotide sequence in tRNA that is complementary to a codon in mRNA.

antigens foreign substances that are recognized by antibodies.

antiparallel an arrangement in a double helix in which one strand is running in the 5' to 3' direction, while the other strand is 3' to 5'.

antisense RNA an RNA strand that is complementary to a strand of mRNA.

antisense strand also called the template strand. It is the strand of DNA that is used as a template for RNA synthesis.

antitermination the function of certain proteins, such as N protein in bacteria, that prevents transcriptional termination.

AP endonuclease a DNA repair enzyme that recognizes a DNA region that is missing a base, and makes a cut in the DNA backbone near that site.

apical–basal-patterning gene one of several plant genes that play a role in embryonic development.

apical region in plants, the region that produces the leaves and flowers.

apoptosis programmed cell death.

apurinic site a site in DNA that is missing a purine base.

archaea also called archaebacteria; one of the three domains of life. Archaea are prokaryotic species. They tend to live in extreme environments and are less common than bacteria (also called eubacteria).

ARS elements DNA sequences found in yeast that function as origins of replication.

artificial chromosomes cloning vectors that can accommodate large DNA inserts and behave like chromosomes when inside of living cells.

artificial selection see *selective breeding*.

artificial transformation transformation of bacteria that occurs via experimental treatments.

ascus (pl. **asci**) a sac that contains haploid spores of fungi (i.e., yeast or molds).

asexual reproduction a form of reproduction that does not involve the union of gametes; at the cellular level, a preexisting cell divides to produce two new cells.

A site see *aminoacyl site*.

assortative mating breeding in which individuals preferentially mate with each other based on their phenotypes.

AT/GC rule in DNA, the phenomenon in which an adenine base in one strand always hydrogen bonds with a thymine base in the opposite strand, and a guanine base always hydrogen bonds with a cytosine.

ATP-dependent chromatin remodeling see *chromatin remodeling*.

attachment site a site in a host cell chromosome where a virus will integrate during site-specific recombination.

attenuation a mechanism of genetic regulation, seen in the *trp* operon, in which a short RNA is made but its synthesis is terminated before RNA polymerase can transcribe the rest of the operon.

attenuator sequence a sequence found in certain operons (e.g., *trp* operon) in bacteria that stops transcription soon after it has begun.

AU-rich element (ARE) a sequence found in many short-lived mRNAs that contains the consensus sequence AUUUA.

automated sequencing the use of fluorescently labeled dideoxyribonucleotides and a fluorescence detector to sequence DNA.

autonomous transposable element a transposable element that contains all of the information necessary for transposition or retroposition to take place.

autopolyploid a polyploid produced within a single species due to nondisjunction.

autoradiography a technique, which involves the use of X-ray film, to detect the location of radioisotopes as they are found in macromolecules or cells. It is used to detect a particular band on a gel or to map the location of a gene within an intact chromosome.

autoregulatory loop a form of gene regulation in which a protein, such as a splicing factor or a transcription factor, regulates its own expression.

autosomes chromosomes that are not sex chromosomes.

auxotroph a strain that cannot synthesize a particular nutrient and needs that nutrient supplemented in its growth medium or diet.

B

BAC see *bacterial artificial chromosome*.

backbone the portion of a DNA or RNA strand that is composed of the repeated covalent linkage of the phosphates and sugar molecules.

backcross in genetics, this usually refers to a cross of F_1 hybrids to individuals that have genotypes of the parental generation.

bacterial artificial chromosome (BAC) a cloning vector that propagates in bacteria and is used to clone large fragments of DNA.

bacteriophages (or **phages**) viruses that infect bacteria.

balanced polymorphism when natural selection favors the maintenance of two or more alleles in a population.

balanced translocation a translocation, such as a reciprocal translocation, in which the total amount of genetic material is normal or nearly normal.

balancing selection a pattern of natural selection that favors the maintenance of two or more alleles. Examples include heterozygote advantage and negative frequency-dependent selection.

band shift assay see *gel retardation assay*.

Barr body a structure in the interphase nuclei of somatic cells of female mammals that is a highly condensed X chromosome.

basal region in plants, the region that produces the roots.

basal transcription in eukaryotes, a low level of transcription via the core promoter. The binding of transcription factors to enhancer elements may increase transcription above the basal level.

basal transcription apparatus the minimum number of proteins needed to transcribe a gene.

base a nitrogen-containing molecule that is a portion of a nucleotide in DNA or RNA. Examples of bases are adenine, thymine, guanine, cytosine, and uracil.

base excision repair a type of DNA repair in which a modified base is removed from a DNA strand. Following base removal, a short region of the DNA strand is removed, which is then resynthesized using the complementary strand as a template.

base mismatch when two bases opposite each other in a double helix do not conform to the AT/GC rule. For example, if A is opposite C, that would be a base mismatch.

base pair the structure in which two nucleotides in opposite strands of DNA hydrogen bond with each other. For example, an AT base pair is a structure in which an adenine-containing nucleotide in one DNA strand hydrogen bonds with a thymine-containing nucleotide in the complementary strand.

base substitution a point mutation in which one base is substituted for another base.

basic domain a protein domain containing several basic amino acids, which is often involved in binding to DNA.

B DNA the predominant form of DNA in living cells. It is a right-handed DNA helix with 10 base pairs per turn.

behavioral trait a trait that involves behavior. An example would be the ability to learn a maze.

beneficial mutation a mutation that is beneficial with regard to its effect on phenotype.

benign refers to a noncancerous tumor that is not invasive and cannot metastasize.

β sheet a type of secondary structure found in proteins.

bHLH refers to a structure found in transcription factor proteins with a basic domain involved in DNA binding and a helix-loop-helix domain involved in dimerization.

bidirectionally means that two replication forks move in opposite directions outward from the origin.

bidirectional replication the phenomenon in which two DNA replication forks emanate in both directions from an origin of replication.

bilateral gynandromorph an animal in which one side is phenotypically male and the other side is female.

binary fission the physical process whereby a bacterial cell divides into two daughter cells. During this event, the two daughter cells become divided by the formation of a septum.

binomial expansion equation an equation to solve genetic problems involving two types of unordered events.

biodegradation the breakdown of a larger molecule into a smaller molecule via cellular enzymes.

bioinformatics literally, this term means the study of biological information. Recently, this term has been associated with the analysis of genetic sequences, using computers and computer programs.

biolistic gene transfer the use of microprojectiles to introduce DNA into plant cells.

biological control the use of microorganisms or products from microorganisms to alleviate plant diseases or damage from undesirable environmental conditions (e.g., frost damage).

biological evolution the accumulation of genetic changes in a species or population over the course of many generations.

biological species concept definition of a species as a group of individuals whose members have the potential to interbreed with one another in nature to produce viable, fertile offspring, but who cannot interbreed successfully with members of other species.

biometric field a field of genetics that involves the statistical study of biological traits.

bioremediation the use of microorganisms to decrease pollutants in the environment.

biotechnology technologies that involve the use of living organisms, or products from living organisms, as a way to benefit humans.

biotransformation the conversion of one molecule into another via cellular enzymes. This term is often used to describe the conversion of a toxic molecule into a nontoxic molecule.

bivalent a structure in which two pairs of homologous sister chromatids have synapsed (i.e., aligned) with each other.

BLAST (basic local alignment search tool) a computer program that can start with a particular genetic sequence and then locate homologous sequences within a large database.

blending hypothesis of inheritance an early, incorrect hypothesis of heredity. According to this view, the seeds that dictate hereditary traits are able to blend together from generation to generation. The blended traits would then be passed to the next generation.

bottleneck effect a type of genetic drift that occurs when most members of a population are eliminated without any regard to their genetic composition.

box in genetics, a term used to describe a sequence with a specialized function.

branch migration the lateral movement of a Holliday junction.

breakpoint the region where two chromosome pieces break and rejoin with other chromosome pieces.

broad sense heritability heritability that takes into account all genetic factors.

5-bromodeoxyuridine a base analogue that can be incorporated into chromosomes during DNA replication. The presence of this analogue can affect the ability of the chromosomes to absorb certain dyes. This is the basis for the staining of harlequin chromosomes.

5-bromouracil a base analogue that acts as a chemical mutagen.

C

C an abbreviation for cytosine.

cAMP see *cyclic AMP*.

cAMP response element (CRE) a short DNA sequence found next to certain eukaryotic genes that is recognized by the cAMP response element–binding (CREB) protein.

cancer a disease characterized by uncontrolled cell division.

cancer cell a cell that has lost its normal growth control. Cancer cells are invasive (i.e., they can invade normal tissues) and metastatic (i.e., they can migrate to other parts of the body).

CAP an abbreviation for the catabolite activator protein, a genetic regulatory protein found in bacteria.

capping the covalent attachment of a 7-methylguanosine nucleotide to the 5' end of mRNA in eukaryotes.

CAP site the sequence of DNA that is recognized by CAP.

carbohydrate organic molecules with the general formula $C(H_2O)$. An example of a simple carbohydrate would be the sugar glucose. Large carbohydrates are composed of multiple sugar units.

carboxyl terminus the location of the last amino acid in a polypeptide chain. The amino acid at the carboxyl terminus still retains a free carboxyl group that is not covalently attached to another amino acid.

carcinogen an agent that can cause cancer.

caspases proteolytic enzymes that play a role in apoptosis.

catabolic enzyme an enzyme that is involved in the breakdown of organic molecules into smaller units.

catabolite activator protein see *CAP*.

catabolite repression the phenomenon in which a catabolite (such as glucose) represses the expression of certain genes (such as the *lac* operon).

catenane interlocked circular molecules.

cDNA see *complementary DNA*.

cDNA library a DNA library made from a collection of cDNAs.

cell adhesion when the surfaces of cells bind to each other or to the extracellular matrix.

cell adhesion molecule (CAM) a molecule (e.g., surface protein or carbohydrate) that plays a role in cell adhesion.

cell culture refers to the growth of cells in a laboratory.

cell cycle in eukaryotic cells, a series of stages through which a cell progresses in order to divide. The phases are G for growth, S for synthesis (of the genetic material), and M for mitosis. There are two G phases, G_1 and G_2.

cell fate the final morphological features that a cell or group of cells will adopt.

cell-free translation system an experimental mixture that can synthesis polypeptides.

cell fusion describes the process in which individual cells are mixed together and made to fuse with each other.

cell lineage a series of cells that are descended from a cell or group of cells by cell division.

cell plate the structure that forms between two daughter plant cells that leads to the separation of the cells by the formation of an intervening cell wall.

cellular trait a trait that is observed at the cellular level. An example would be the shape of a cell.

centiMorgans (cM) (same as a map unit) a unit of map distance obtained from genetic crosses. Named in honor of Thomas Hunt Morgan.

central dogma of molecular biology the idea that the usual flow of genetic information is from DNA to RNA to polypeptide (protein). In addition, DNA replication serves to copy the information so that it can be transmitted from cell to cell and from parent to offspring.

central region in plants, it is the region that creates the stem. It is the radial pattern of cells in the central region that causes the radial growth observed in plants.

central zone in plants, an area in the meristem where undifferentiated stem cells are always maintained.

centrifugation a method to separate cell organelles and macromolecules in which samples are placed in tubes and spun very rapidly. The rate at which particles move toward the bottom of the tube depends on their densities, sizes, shapes, and the viscosity of the medium.

centrifuge a machine that contains a motor, which causes a rotor holding centrifuge tubes to spin very rapidly.

centromere a segment of eukaryotic chromosomal DNA that provides an attachment site for the kinetochore.

centrosome a cellular structure from which microtubules emanate.

chain termination refers to the stoppage of growth of a DNA strand, RNA strand, or polypeptide sequence.

chaperone a protein that aids in the folding of polypeptides.

character in genetics, this word has the same meaning as trait.

Chargaff's rule see *AT/GC rule.*

charged tRNA a tRNA that has an amino acid attached to its 3' end by an ester bond.

checkpoint protein a protein that monitors the conditions of DNA and chromosomes and may prevent a cell from progressing through the cell cycle if an abnormality is detected.

chiasma (pl. **chiasmata**) the site where crossing over occurs between two chromosomes. It resembles the Greek letter chi, χ.

chimera an organism composed of cells that are embryonically derived from two different individuals.

ChIP-on-chip assay a form of chromatin immunoprecipitation that utilizes a microarray to determine where in the genome a particular protein binds.

chi square (χ^2) test a commonly used statistical method to determine the goodness of fit. This method can be used to analyze population data in which the members of the population fall into different categories. It is particularly useful for evaluating the outcome of genetic crosses, because these usually produce a population of offspring that differ with regard to phenotypes.

chloroplast DNA (cpDNA) the genetic material found within a chloroplast.

chorionic villus sampling a method to obtain cellular material from a fetus for the purpose of genetic testing.

chromatid following chromosomal replication in eukaryotes, the two copies remain attached to each other in the form of sister chromatids.

chromatin the association between DNA and proteins that is found within chromosomes.

chromatin immunoprecipitation (ChIP) a method to determine whether proteins bind to a particular region of DNA. This method analyzes DNA–protein interactions as they occur in the chromatin of living cells.

chromatin remodeling a change in chromatin structure that alters the degree of compaction and/or the spacing of nucleosomes.

chromatography a method to separate different macromolecules and small molecules based on their chemical and physical properties. A sample is dissolved in a liquid solvent and exposed to some type of matrix, such as a gel, a column containing beads, or a thin strip of paper.

chromocenter the central point where polytene chromosomes aggregate.

chromomere a dark band within a polytene chromosome.

chromosome the structures within living cells that contain the genetic material. Genes are physically located within the structure of chromosomes. Biochemically, chromosomes contain a very long segment of DNA, which is the genetic material, and proteins, which are bound to the DNA and provide it with an organized structure.

chromosome territory in the cell nucleus, each chromosome occupies a nonoverlapping region called a chromosome territory.

chromosome theory of inheritance a theory of Sutton and Boveri, which indicated that the inheritance patterns of traits can be explained by the transmission patterns of chromosomes during gametogenesis and fertilization.

chromosome walking a common method used in positional cloning in which a mapped gene or RFLP marker provides a starting point to molecularly walk toward a gene of interest via overlapping clones.

***cis*-acting element** a sequence of DNA, such as a regulatory element, that exerts a *cis*-effect.

***cis*-effect** an effect on gene expression due to genetic sequences that are within the same chromosome and often are immediately adjacent to the gene of interest.

cistron refers to the smallest genetic unit that produces a positive result in a complementation experiment. A cistron is equivalent to a gene.

clade a group of species consisting of all descendents of the group's most common ancestor.

cladistic approach a way to construct a phylogenetic tree, also called a cladogram, by considering the various possible pathways of evolution and then choosing the most plausible tree.

cladogenesis (Greek, *clados*, "branch") during evolution, a form of speciation that involves the division of a species into two or more species.

cladogram a phylogenetic tree that has been constructed using a cladistic approach.

cleavage furrow a constriction that causes the division of two animal cells during cytokinesis.

clonal an adjective to describe a clone. For example, a clonal population of cells is a group of cells that are derived from the same cell.

clone the general meaning of this term is to make many copies of something. In genetics, this term has several meanings: (1) a single cell that has divided to produce a colony of genetically identical cells; (2) an individual that has been produced from a somatic cell of another individual, such as the sheep Dolly; (3) many copies of a DNA fragment that are propagated within a vector or produced by PCR.

closed complex the complex between transcription factors, RNA polymerase, and a promoter before the DNA has denatured to form an open complex.

closed conformation a tightly packed conformation of chromatin that cannot be transcribed.

cluster analysis the analysis of microarray data to determine if certain groups (i.e., clusters) of genes are expressed under the same conditions.

cM an abbreviation for centiMorgans; also see *map unit.*

coding strand the strand in DNA that is not used as a template for mRNA synthesis.

codominance a pattern of inheritance in which two alleles are both expressed in the heterozygous condition. For example, a person with the genotype $I^A I^B$ will have the blood type AB and will express both surface antigens A and B.

codon a sequence of three nucleotides in mRNA that functions in translation. A start codon, which usually specifies methionine, initiates translation, and a stop codon terminates translation. The other codons specify the amino acids within a polypeptide sequence according to the genetic code.

codon bias the phenomenon that, in a given species, certain codons are used more frequently than other codons.

coefficient of inbreeding (F) see *inbreeding coefficient.*

cohesin a protein complex that facilitates the alignment of sister chromatids.

colinearity the correspondence between the sequence of codons in the DNA coding strand and the amino acid sequence of a polypeptide.

colony hybridization a technique in which a probe is used to identify bacterial colonies that contain a hybrid vector with a gene of interest.

combinatorial control refers to the phenomenon that is common in eukaryotes in which the combination of many factors determines the expression of any given gene.

common ancestor someone who is an ancestor to both of an individual's parents.

comparative genomic hybridization (CGH) a hybridization technique to determine if cells (e.g., cancer cells) have changes in chromosome structure, such as deletions or duplications.

comparative genomics uses information from genome projects to understand the genetic variation between different populations and evolutionary relationships among different species.

competence factors proteins that are needed for bacterial cells to become naturally transformed by extracellular DNA.

competence-stimulating peptide (CSP) a peptide secreted by certain species of bacteria that allow them to become competent for transformation.

competent cells cells that can be transformed by extracellular DNA.

complementary describes sequences in two DNA strands that match each other according to the AT/GC rule. For example, if one strand has the sequence of ATGGCGGATTT, then the complementary strand must be TACCGCCTAAA.

complementary DNA (cDNA) DNA that is made from an RNA template by the action of reverse transcriptase.

complementation a phenomenon in which the presence of two different mutant alleles in the same organism produces a wild-type phenotype. It usually happens because the two mutations are in different genes, so

the organism carries one copy of each mutant allele and one copy of each wild-type allele.

complementation test an experimental procedure in which the goal is to determine if two different mutations that affect the same trait are in the same gene or in two different genes.

complete nondisjunction during meiosis or mitosis, when all of the chromosomes fail to disjoin and remain in one of the two daughter cells.

complete transposable element see *autonomous transposable element.*

complex traits characteristics that are determined by several genes and are significantly influenced by environmental factors.

composite transposon a transposon that contains additional genes, such as antibiotic resistance genes, that are not necessary for transposition *per se.*

computer data file a file (a collection of information) stored by a computer.

computer program a series of operations that can analyze data in a defined way.

concordance in genetics, the degree to which pairs of individuals (e.g., identical twins or fraternal twins) exhibit the same trait.

condensation a change in chromatin structure to become more compact.

condense refers to chromosomes forming a more compact structure

condensin a protein complex that plays a role in the condensation of interphase chromosomes to become metaphase chromosomes.

conditional alleles alleles in which the phenotypic expression depends on the environmental conditions. An example is temperature-sensitive alleles, which affect the phenotype only at a particular temperature.

conditional lethal allele an allele that is lethal, but only under certain environmental conditions.

conditional mutant a mutant whose phenotype depends on the environmental conditions, such as a temperature-sensitive mutant.

conglomerate a population composed of members of an original population plus new members that have migrated from another population.

conjugation a form of genetic transfer between bacteria that involves direct physical interaction between two bacterial cells. One bacterium acts as donor and transfers genetic material to a recipient cell.

conjugation bridge a connection between two bacterial cells that provides a passageway for DNA during conjugation.

conjugative plasmid a plasmid that can be transferred to a recipient cell during conjugation.

consensus sequence the most commonly occurring bases within a sequence element.

conservative model an incorrect model for DNA in which both strands of parental DNA remain together following DNA replication.

conservative transposition see *simple transposition.*

constitutive exon an exon that is always found in mRNA following splicing.

constitutive gene a gene that is not regulated and has essentially constant levels of expression over time.

constitutive heterochromatin regions of chromosomes that are always heterochromatic and are permanently transcriptionally inactive.

contig a series of clones that contain overlapping pieces of chromosomal DNA.

control element see *regulatory sequence or element.*

core enzyme the subunits of an enzyme that are needed for catalytic activity, as in the core enzyme of RNA polymerase.

corepressor a small effector molecule that binds to a repressor protein, thereby causing the repressor protein to bind to DNA and inhibit transcription.

core promoter a DNA sequence that is absolutely necessary for transcription to take place. It provides the binding site for general transcription factors and RNA polymerase.

correlation coefficient (r) a statistic with a value that ranges between -1 and 1. It describes how two factors vary with regard to each other.

cosmid a vector that is a hybrid between a plasmid vector and phage λ. Cosmid DNA can replicate in a cell like a plasmid or be packaged into a protein coat like a phage. Cosmid vectors can accept fragments of DNA that are typically tens of thousands of base pairs in length.

C_0t **curve** a plot of C/C_0 versus C_0t.

cotransduction the phenomenon in which bacterial transduction transfers a piece of DNA carrying two closely linked genes.

cotransformation the phenomenon in which bacterial transformation transfers a piece of DNA carrying two closely linked genes.

cotranslational events that occur during translation.

cotranslational import during the synthesis of certain eukaryotic proteins, translation begins in the cytosol and then is temporarily halted by the signal recognition particle (SRP). Translation resumes when the ribosome has become bound to the ER membrane and the polypeptide is synthesized into the ER lumen or ER membrane.

cotranslational sorting refers to the sorting of proteins into the ER. The protein is actually translated into the ER lumen or ER membrane.

covariance a statistic that describes the degree of variation between two variables within a group.

cpDNA an abbreviation for chloroplast DNA.

CpG island a group of CG sequences that may be clustered near a promoter region of a gene. The methylation of the cytosine bases usually inhibits transcription.

CREB protein (cAMP response element-binding protein) a regulatory transcription factor that becomes activated in response to specific cell-signaling molecules that cause the synthesis of cAMP.

cross a mating between two distinct individuals. An analysis of their offspring may be conducted to understand how traits are passed from parent to offspring.

cross-fertilization same meaning as *cross*. It requires that the male and female gametes come from separate individuals.

crossing over a physical exchange of chromosome pieces that most commonly occurs during prophase of meiosis I.

C-terminus see *carboxyl terminus*.

cyclic AMP (cAMP) in bacteria, a small effector molecule that binds to CAP (catabolite activator protein). In eukaryotes, cAMP functions as a second messenger in a variety of intracellular signaling pathways; in some cases, it binds to transcription factors such as the CREB protein.

cyclin a type of protein that plays a role in the regulation of the eukaryotic cell cycle.

cyclin-dependent protein kinases (CDKs) enzymes that are regulated by cyclins and can phosphorylate other cellular proteins by covalently attaching a phosphate group.

cytogeneticist a scientist who studies chromosomes under the microscope.

cytogenetic mapping the mapping of genes or genetic sequences using microscopy.

cytogenetics the field of genetics that involves the microscopic examination of chromosomes.

cytokinesis the division of a single cell into two cells. The two nuclei produced in M phase are segregated into separate daughter cells during cytokinesis.

cytological mapping see *cytogenetic mapping*.

cytoplasmic inheritance (also known as *extranuclear inheritance*) refers to the inheritance of genetic material that is not found within the cell nucleus.

cytosine a pyrimidine base found in DNA and RNA. It base-pairs with guanine in DNA.

D

Dam see *DNA adenine methyltransferase*.

Darwinian fitness the relative likelihood that a phenotype will survive and contribute to the gene pool of the next generation as compared with other phenotypes.

database a computer storage facility that stores many data files such as those containing genetic sequences.

daughter strand in DNA replication, the newly made strand of DNA.

deamination the removal of an amino group from a molecule. For example, the removal of an amino group from cytosine produces uracil.

decoding function the ability of the 16S rRNA to detect when an incorrect tRNA is bound at the A site and prevent elongation until the mispaired tRNA is released from the A site.

decondensed refers to chromosomes forming a less compact structure.

deficiency condition in which a segment of chromosomal material is missing.

degeneracy in genetics, this term means that more that one codon specifies the same amino acid. For example, the codons GGU, GGC, GGA, and GGG all specify the amino acid glycine.

degrees of freedom in a statistical analysis, the number of categories that are independent of each other.

deleterious mutation a mutation that is detrimental with regard to its effect on phenotype.

deletion condition in which a segment of DNA is missing.

deletion mapping the use of strains carrying deletions within a defined region to map a mutation of unknown location.

deme see *local population*.

de novo methylation the methylation of DNA that has not been previously methylated. This is usually a highly regulated event.

deoxyribonucleic acid (DNA) the genetic material. It is a double-stranded structure, with each strand composed of repeating units of deoxyribonucleotides.

deoxyribose the sugar found in DNA.

depurination the removal of a purine base from DNA.

determined cell a cell that is destined to differentiate into a specific cell type.

developmental genetics the area of genetics concerned with the roles of genes in orchestrating the changes that occur during development.

diakinesis the fifth stage of prophase of meiosis I.

diauxic growth the sequential use of two sugars by a bacterium.

dicentric describes a chromosome with two centromeres.

dicentric bridge the region between the two centromeres in a dicentric chromosome.

dicer an endonuclease that makes a cut in double-stranded RNA.

dideoxyribonucleotide a nucleotide used in DNA sequencing that is missing the 3'—OH group. If a dideoxyribonucleotide is incorporated into a DNA strand, it stops any further growth of the strand.

dideoxy sequencing a method of DNA sequencing that uses dideoxyribonucleotides to terminate the growth of DNA strands.

differential centrifugation a form of centrifugation involving a series of centrifugation steps in which the supernatant or pellet is used in each subsequent centrifugation step.

differentially methylated region (DMR) in the case of imprinting, a site that is methylated during spermatogenesis or oogenesis, but not both.

differentiated cell a cell that has become a specialized type of cell within a multicellular organism.

dihybrid cross a cross in which an experimenter follows the outcome of two different traits.

dihybrid testcross a cross in which an experimenter crosses an individual that is heterozygous for two genes to an individual that is homozygous recessive for the same two genes.

dimeric DNA polymerase a complex of two DNA polymerase proteins that move as a unit during DNA replication.

dioecious in plants, a species in which male and female gametophytes are produced on a single (sporophyte) individual.

diploid an organism or cell that contains two copies of each type of chromosome.

diplotene the fourth stage of prophase of meiosis I.

directionality in DNA and RNA, refers to the 5' to 3' arrangement of nucleotides in a strand; in proteins, refers to the linear arrangement of amino acids from the N-terminal to C-terminal ends.

directional selection natural selection that favors an extreme phenotype. This usually leads to the fixation of the favored allele.

direct repeat (DR) short DNA sequences that flank transposable elements in which the DNA sequence is repeated in the same direction.

discontinuous trait a trait in which each offspring can be put into a particular phenotypic category.

discovery-based science experimentation that does not require a preconceived hypothesis. In some cases, the goal is to collect data to be able to formulate a hypothesis.

disequilibrium in population genetics, refers to a population that is not in Hardy-Weinberg equilibrium.

dispersive model an incorrect model for DNA replication in which segments of parental DNA and newly made DNA are interspersed in both strands following the replication process.

disruptive selection natural selection that favors both extremes of a phenotypic category. This results in a balanced polymorphism.

dizygotic twins also known as fraternal twins; twins formed from separate pairs of sperm and egg cells.

DNA the abbreviation for deoxyribonucleic acid.

DNA adenine methyltransferase an enzyme in bacteria that attaches methyl groups to the adenine base in DNA that is found within the sequence GATC.

DnaA box sequence serves as a recognition site for the binding of the DnaA protein, which is involved in the initiation of bacterial DNA replication.

DnaA protein a protein that binds to the dnaA box sequence at the origin of replication in bacteria and initiates DNA replication.

DNA fingerprinting a technology to identify a particular individual based on the properties of their DNA.

DNA gap repair synthesis the synthesis of DNA in a region where a DNA strand has been previously removed, usually by a DNA repair enzyme or by an enzyme involved in homologous recombination.

DNA gyrase also known as topoisomerase II; an enzyme that introduces negative supercoils into DNA using energy from ATP. Gyrase can also relax positive supercoils when they occur.

DNA helicase an enzyme that separates the two strands of DNA.

DNA library a collection of many hybrid vectors, each vector carrying a particular fragment of DNA from a larger source. For example, each hybrid vector in a DNA library might carry a small segment of chromosomal DNA from a particular species.

DNA ligase an enzyme that catalyzes a covalent bond between two DNA fragments.

DNA methylation the phenomenon in which an enzyme covalently attaches a methyl group (—CH_3) to a base (usually adenine or cytosine) in DNA.

DNA methyltransferase the enzyme that attaches methyl groups to adenine or cytosine bases.

DNA microarray a small silica, glass, or plastic slide that is dotted with many different sequences of DNA, corresponding to short sequences within known genes.

DNA-N-glycosylase an enzyme that can recognize an abnormal base and cleave the bond between it and the sugar in the DNA backbone.

DNA polymerase an enzyme that catalyzes the covalent attachment of nucleotides together to form a strand of DNA.

DNA primase an enzyme that synthesizes a short RNA primer for DNA replication.

DNA probe in a hybridization experiment, a single-stranded DNA fragment with a base sequence that is complementary to a gene of interest. The DNA probe is labeled to detect a gene of interest.

DNA profiling see *DNA fingerprinting*.

DNA replication the process in which original DNA strands are used as templates for the synthesis of new DNA strands.

DNA replication licensing in eukaryotes, occurs when MCM helicase is bound at an origin, enabling the formation of two replication forks.

DNase an enzyme that cuts the sugar–phosphate backbone in DNA.

DNase I an endonuclease that cleaves DNA.

DNase I footprinting a method to study protein–DNA interactions in which the binding of a protein to DNA protects the DNA from digestion by DNase I.

DNase I sensitivity the phenomenon in which DNA is in an open conformation and able to be digested by DNase I.

DNA sequencing a method to determine the base sequence in a segment of DNA.

DNA supercoiling the formation of additional coils in DNA due to twisting forces.

DNA uptake signal sequences DNA sequences found in certain species of bacteria that are needed for a DNA fragment to be taken up during transformation.

domain a segment of a protein that has a specific function.

dominant describes an allele that determines the phenotype in the heterozygous condition. For example, if a plant is Tt and has a tall phenotype, the T (tall) allele is dominant over the t (dwarf) allele.

dominant negative mutation a mutation that produces an altered gene product that acts antagonistically to the normal gene product. Shows a dominant pattern of inheritance.

dorso-ventral axis in animals, the axis from the spine (dorsal) to the stomach (ventral).

dosage compensation refers to the phenomenon that in species with sex chromosomes, one of the sex chromosomes is altered so that males and females will have similar levels of gene expression, even though they do not contain the same complement of sex chromosomes.

double-barrel shotgun sequencing a type of shotgun sequencing in which DNA fragments are randomly sequenced from both ends.

double helix the arrangement in which two strands of DNA (and sometimes RNA) interact with each other to form a double-stranded helical structure.

double-strand break model a model for homologous recombination in which the event that initiates recombination is a double-strand break in one of the double helices.

down promoter mutation a mutation in a promoter that inhibits the rate of transcription.

down regulation genetic regulation that leads to a decrease in gene expression.

duplication the copying of a segment of DNA.

E

ecological species concept a species concept in which each species occupies an ecological niche, which is the unique set of habitat resources that a species requires, as well as the species' influence on the environment and other species.

editosome a complex that catalyzes RNA editing.

egg cell also known as an ovum; it is a female gamete that is usually very large and nonmotile.

electrophoresis the migration of ions or molecules in response to an electric field.

electroporation the use of electric current that creates transient pores in the plasma membrane of a cell to allow entry of DNA.

elongation (1) in transcription, the synthesis of RNA using DNA as a template; (2) in translation, the synthesis of a polypeptide using the information within mRNA.

embryogenesis an early stage of animal and plant development leading to the production of an embryo with organized tissue layers and a body plan organization.

embryonic carcinoma cell (EC cell) a type of pluripotent stem cell found in a specific type of human tumor.

embryonic germ cell (EG cell) a type of pluripotent stem cell found in the gonads of the fetus.

embryonic stem cell (ES cell) a type of pluripotent stem cell found in the early blastocyst.

embryo sac in flowering plants, the female gametophyte that contains an egg cell.

empirical approach a strategy in which experiments are designed to determine quantitative relationships as a way to derive laws that govern biological, chemical, or physical phenomena.

empirical laws laws that are discovered using an empirical (observational) approach.

endonuclease an enzyme that can cut in the middle of a DNA strand.

endopolyploidy in a diploid individual, the phenomenon in which certain cells of the body may be polyploid.

endosperm in flowering plants, the material in the seed, which is $3n$, that nourishes the developing embryo.

endosymbiosis a symbiotic relationship in which the symbiont actually lives inside (*endo*) the larger of the two species.

endosymbiosis theory the theory that the ancient origin of plastids and mitochondria was the result of certain species of bacteria taking up residence within a primordial eukaryotic cell.

enhancer a DNA sequence that functions as a regulatory element. The binding of a regulatory transcription factor to the enhancer increases the level of transcription.

environment the surroundings an organism experiences.

enzyme a protein that functions to accelerate chemical reactions within the cell.

enzyme adaptation the phenomenon in which a particular enzyme appears within a living cell after the cell has been exposed to the substrate for that enzyme.

epigenetic inheritance an inheritance pattern in which a modification to a nuclear gene or chromosome alters gene expression in an organism, but the expression is not changed permanently over the course of many generations.

episome a segment of bacterial DNA that can exist as an F factor and also integrate into the chromosome.

epistasis an inheritance pattern where one gene can mask the phenotypic effects of a different gene.

epitope the structure on the surface of an antigen that is recognized by an antibody.

equilibrium density centrifugation a form of centrifugation in which the particles will sediment through the gradient, reaching a position where the density of the particle matches the density of the solution.

E site see *exit site*.

essential gene a gene that is essential for survival.

EST library see *expressed sequence tagged library*.

ethyl methanesulfonate (EMS) a type of chemical mutagen that alkylates bases (i.e., attaches methyl or ethyl groups).

eubacteria one of the three domains of life. Eubacteria, more commonly known as bacteria, are prokaryotic species.

euchromatin DNA that is not highly compacted and may be transcriptionally active.

eukaryotes (Greek, "true nucleus") one of the three domains of life. A defining feature of these organisms is that their cells contain nuclei bounded by cell membranes. Some simple eukaryotic species are single-celled protists and yeast; more complex multicellular species include fungi, plants, and animals.

euploid describes an organism in which the chromosome number is an exact multiple of a chromosome set.

evo-devo see *evolutionary developmental biology*.

evolution see *biological evolution*.

evolutionary developmental biology (evo-devo) a field of biology that focuses on the role of developmental genes in the formation of traits that are important in the evolution of new species.

evolutionary species concept a species concept in which a species is derived from a single lineage that is distinct from other lineages and has its own evolutionary tendencies and historical fate.

excisionase an enzyme that excises a prophage from a host cell's chromosome.

exit site (E site) a site on the ribosome from which an uncharged tRNA exits.

exon a segment of RNA that is contained within the RNA after splicing has occurred. In mRNA, the coding sequence of a polypeptide is contained within the exons.

exon shuffling the phenomenon that exons have been transferred between different genes during evolution. Some researchers believe that transposable elements have played a role in this phenomenon.

exon skipping when an exon is spliced out of a pre-mRNA.

exonuclease an enzyme that digests an RNA or DNA strand from the end.

expressed sequence tagged (EST) library a DNA library containing many clones that have different cDNA inserts.

expression vector a cloning vector that contains a promoter so that the gene of interest will be transcribed into RNA when the vector is introduced into a host cell.

expressivity the degree to which a trait is expressed. For example, flowers with deep red color would have a high expressivity of the red allele.

extragenic suppressor a mutation in a gene that suppresses the phenotypic effect of a mutation in a different gene.

extranuclear inheritance (also known as *cytoplasmic inheritance*) refers to the inheritance of genetic material that is not found within the nucleus.

ex vivo approach in the case of gene therapy, refers to genetic manipulations that occur outside the body.

F

F see *inbreeding coefficient*.

facultative heterochromatin heterochromatin that is derived from the conversion of euchromatin to heterochromatin.

fate map a diagram that depicts how cell division proceeds in an organism.

feedback inhibition the phenomenon in which the final product of a metabolic pathway inhibits an enzyme that acts early in the pathway.

fertilization the union of gametes (e.g., sperm and egg) to begin the life of a new organism.

F factor a fertility factor found in certain strains of bacteria in addition to their circular chromosome. Strains of bacteria that contain an F factor are designated F^+; strains without F factors are F^-.

F' factors an F factor that also carries genes derived from the bacterial chromosome.

F_1 generation the offspring produced from a cross of the parental generation.

F_2 generation the offspring produced from a cross of the F_1 generation.

fidelity a term used to describe the accuracy of a process. If there are few mistakes, a process has a high fidelity.

fine structure mapping also known as intragenic mapping; the aim of fine structure mapping is to ascertain the distances between two (or more) different mutations within the same gene.

first-division segregation (FDS) in an ordered octad, a 4:4 arrangement of spores that occurs because the two alleles have segregated from each other after the first meiotic division.

fitness see *Darwinian fitness*.

fluctuation test the experimental procedure of Luria and Delbrück in which there was a greater fluctuation in the number of random mutations in small bacterial cultures compared to one large culture.

fluorescence *in situ* hybridization (FISH) a form of *in situ* hybridization in which the probe is fluorescently labeled.

focus with regard to cancer cells, it means a clump of raised cells that grow without regard to contact inhibition.

footprinting see *DNase I footprinting*.

fork see *replication fork*.

forked-line method a method to solve independent assortment problems in which lines are drawn to connect particular genotypes.

forward mutation a mutation that changes the wild-type genotype into some new variation.

founder with regard to genetic diseases, an individual who lived many generations ago and was the person in which a genetic disease originated.

founder effect changes in allele frequencies that occur when a small group of individuals separates from a larger population and establishes a colony in a new location.

fraction (1) following centrifugation, a portion of the liquid contained within a centrifuge tube; (2) following column chromatography, a portion of the liquid that has been eluted from a column.

frameshift mutation a mutation that involves the addition or deletion of nucleotides not in a multiple of three and thereby shifts the reading frame of the codon sequence downstream from the mutation.

frequency distribution a graph that describes the numbers of individuals that are found in each of several phenotypic categories.

functional genomics the study of gene function at the genome level. It involves the study of many genes simultaneously.

G

G an abbreviation for guanine.

gain-of-function mutation a mutation that causes a gene to be expressed in an additional place where it is not normally expressed or during a stage of development when it is not normally expressed.

gamete a reproductive cell (usually haploid) that can unite with another reproductive cell to create a zygote. Sperm and egg cells are types of gametes.

gametogenesis the production of gametes (e.g., sperm or egg cells).

gametophyte the haploid generation of plants.

gap gene one category of segmentation genes.

G bands the chromosomal banding pattern that is observed when the chromosomes have been treated with the chemical dye Giemsa.

gel electrophoresis a method that combines chromatography and electrophoresis to separate molecules and macromolecules. Samples are loaded into wells at one end of the gel, and an electric field is applied across the gel that causes charged molecules to migrate from one side of the gel to the other.

gel retardation assay a technique to study protein–DNA interactions in which the binding of protein to a DNA fragment retards it mobility during gel electrophoresis.

gene a unit of heredity that may influence the outcome of an organism's traits. At the molecular level, a gene contains the information to make a functional product, either RNA or protein.

gene addition the addition of a cloned gene into a site in a chromosome of a living cell.

gene amplification an increase in the copy number of a gene.

gene chip see *DNA microarray.*

gene cloning the production of many copies of a gene using molecular methods such as PCR or the introduction of a gene into a vector that replicates in a host cell.

gene conversion the phenomenon in which one allele is converted to another allele due to genetic recombination and DNA repair.

gene dosage effect when the number of copies of a gene affects the phenotypic expression of a trait.

gene duplication an increase in the copy number of a gene. Can lead to the evolution of gene families.

gene expression the process in which the information within a gene is accessed, first to synthesize RNA (and proteins), and eventually to affect the phenotype of the organism.

gene family two or more different genes within a single species that are homologous to each other because they were derived from the same ancestral gene.

gene flow changes in allele frequencies due to migration.

gene interaction when two or more different genes influence the outcome of a single trait.

gene knockin a type of gene addition in which a gene of interest has been added to a particular site in the mouse genome.

gene knockout when both copies of a normal gene have been replaced by an inactive mutant gene.

gene modifer effect when the allele of one gene modifies the phenotypic effect of the allele of a different gene.

gene mutation a relatively small mutation that affects only a single gene.

gene pool the totality of all genes within a particular population.

generalized transduction a form of transduction in which any piece of the bacterial chromosomal DNA can be incorporated into a phage.

general transcription factor one of several proteins that are necessary for basal transcription at the core promoter.

gene rearrangement a rearrangement in segments of a gene, as occurs in antibody precursor genes.

gene redundancy the phenomenon in which an inactive gene is compensated for by another gene with a similar function.

gene regulation the phenomenon in which the level of gene expression can vary under different conditions.

gene replacement the swapping of a cloned gene made experimentally with a normal chromosomal gene found in a living cell.

gene therapy the introduction of cloned genes into living cells in an attempt to cure or alleviate disease.

genetically modified organism (GMO) an organism that has received genetic material via recombinant DNA technology.

genetic approach in research, refers to the study of mutant genes that have abnormal function. By studying mutant genes, researchers may better understand normal genes and normal biological processes.

genetic code the correspondence between a codon (i.e., a sequence of three bases in an mRNA molecule) and the functional role that the codon plays during translation. Each codon specifies a particular amino acid or the end of translation.

genetic cross a mating between two individuals and the analysis of their offspring in an attempt to understand how traits are passed from parent to offspring.

genetic drift random changes in allele frequencies due to sampling error.

genetic map a chart that describes the relative locations of genes or other DNA segments along a chromosome.

genetic mapping any method used to determine the linear order of genes as they are linked to each other along the same chromosome. This term is also used to describe the use of genetic crosses to determine the linear order of genes.

genetic marker any genetic sequence that is used to mark a specific location on a chromosome.

genetic mosaic see *mosaicism.*

genetic polymorphism when two or more alleles occur in population; each allele is found at a frequency of 1% or higher.

genetic recombination (1) the process in which chromosomes are broken and then rejoined to form a novel genetic combination. (2) the process in which alleles are assorted and passed to offspring in combinations that are different from the parents.

genetics the study of heredity.

genetic screening the use of testing methods to determine if an individual is a heterozygous carrier for or has a genetic disease.

genetic testing the analysis of individuals with regard to their genes or gene products. In many cases, the goal is to determine if an individual carries a mutant gene.

genetic transfer describes the physical transfer of genetic material from one bacterial cell to another.

genetic variation genetic differences among members of the same species or among different species.

genome all of the chromosomes and DNA sequences that an organism or species can possess.

genome database a database that focuses on the genetic sequences and characteristics of a single species.

genome maintenance refers to cellular mechanisms that either prevent mutations from occurring and/or prevent mutant cells from surviving or dividing.

genome sequencing projects research endeavors that have the ultimate goal of determining the sequence of DNA bases of the entire genome of a given species.

genomic clone a clone made from the digestion and cloning of chromosomal DNA.

genomic imprinting a pattern of inheritance that involves a change in a single gene or chromosome during gamete formation. Depending on whether the modification occurs during spermatogenesis or oogenesis, imprinting governs whether an offspring will express a gene that has been inherited from its mother or father.

genomic library a DNA library made from chromosomal DNA fragments.

genomics the molecular analysis of the entire genome of a species.

genotype the genetic composition of an individual, especially in terms of the alleles for particular genes.

genotype-environment association when certain genotypes are preferentially found in particular environments.

genotype-environment interaction when the environmental effects on phenotype differ according to genotype.

genotype frequency the number of individuals with a particular genotype in a population divided by the total number of individuals in the population.

germ cells the gametes (i.e., sperm and egg cells).

germ line a lineage of cells that gives rise to gametes.

germ-line mutation a mutation in a cell of the germ line.

GloFish genetically modified aquarium fish that glow due to the introduction of genes that encode fluorescent proteins.

glucocorticoid receptor a type of steroid receptor that functions as a regulatory transcription factor.

goodness of fit the degree to which the observed data and expected data are similar to each other. If the observed and predicted data are very similar, the goodness of fit is high.

gradualism an evolutionary hypothesis suggesting that each new species evolves continuously over long spans of time. The principal idea is that large phenotypic differences that cause the divergence of species are due to the accumulation of many small genetic changes.

grande normal (large-sized) yeast colonies.

grooves in DNA, the indentations where the atoms of the bases are in contact with the surrounding water. In B DNA, there is a smaller minor groove and a larger major groove.

group I intron a type of intron found in self-splicing RNA that uses free guanosine in its splicing mechanism.

group II intron a type of intron found in self-splicing RNA that uses an adenine nucleotide within the intron itself in its splicing mechanism.

growth factors protein factors that influence cell division.

guanine a purine base found in DNA and RNA. It base-pairs with cytosine in DNA.

guide RNA in trypanosome RNA editing, an RNA molecule that directs the addition of uracil residues into the mRNA.

gyrase see *DNA gyrase.*

H

haplodiploid a species, such as certain bees, in which one sex is haploid (e.g., male) and the other sex is diploid (e.g., female).

haploid describes the phenomenon that gametes contain half the genetic material found in somatic cells. For a species that is diploid, a haploid gamete contains a single set of chromosomes.

haploinsufficiency the phenomenon in which a person has only a single functional copy of a gene, and that single functional copy does not produce a normal phenotype. Shows a dominant pattern of inheritance.

Hardy-Weinberg equation $p^2 + 2pq + q^2 = 1$.

Hardy-Weinberg equilibrium the phenomenon that, under certain conditions, allele frequencies will be maintained in a stable condition and genotypes can be predicted according to the Hardy-Weinberg equation.

helicase see *DNA helicase.*

helix–loop–helix domain a domain found in transcription factors that enables them to dimerize.

helix-turn-helix motif a structure found in transcription factor proteins that promotes binding to the major groove of DNA.

hemizygous describes the single copy of an X-linked gene in the male. A male mammal is said to be hemizygous for X-linked genes.

heritability the amount of phenotypic variation within a particular group of individuals that is due to genetic factors.

heterochromatin highly compacted DNA. It is usually transcriptionally inactive.

heterochronic mutation a mutation that alters the timing of expression of a gene and thereby alters the outcome of cell fates.

heterodimer when two polypeptides encoded by different genes bind to each other to form a dimer.

heteroduplex a double-stranded region of DNA that contains one or more base mismatches.

heterogametic sex in species with two types of sex chromosomes, the heterogametic sex produces two types of gametes. For example, in mammals, the male is the heterogametic sex, because a sperm can contain either an X or a Y chromosome.

heterogamous describes a species that produces two morphologically different types of gametes (i.e., sperm and eggs).

heterogeneity see *locus heterogeneity.*

heterogeneous nuclear RNA (hnRNA) same as pre-mRNA.

heterokaryon a cell produced from cell fusion that contains two separate nuclei.

heteroplasmy when a cell contains variation in a particular type of organelle. For example, a plant cell could contain some chloroplasts that make chlorophyll and other choroplasts that do not.

heterosis the phenomenon in which hybrids display traits superior to either corresponding parental strain. Heterosis is usually different from overdominance, because the hybrid may be heterozygous for many genes, not just a single gene, and because the superior phenotype may be due to the masking of deleterious recessive alleles.

heterozygote an individual who is heterozygous.

heterozygote advantage a pattern of inheritance in which a heterozygote is more vigorous than either of the corresponding homozygotes.

heterozygous describes a diploid individual who has different copies (i.e., two different alleles) of the same gene.

***Hfr* strain (for high frequency of recombination)** a bacterial strain in which an F factor has become integrated into the bacterial chromosome. During conjugation, an *Hfr* strain can transfer segments of the bacterial chromosome.

hierarchical shotgun sequencing a genome sequencing strategy in which small DNA fragments are mapped prior to DNA sequencing.

highly repetitive sequences sequences that are found tens of thousands or even millions of times throughout the genome.

high stringency refers to highly selective hybridization conditions that promote the binding of DNA or RNA fragments that are perfect or almost perfect matches.

histone acetyltransferase an enzyme that attaches acetyl groups to the amino terminal tails of histone proteins.

histone code hypothesis the hypothesis that the pattern of histone modification acts much like a language or code in specifying alterations in chromatin structure.

histone deacetylase an enzyme that removes acetyl groups from the amino terminal tails of histone proteins.

histones a group of proteins involved in forming the nucleosome structure of eukaryotic chromatin.

hnRNA an abbreviation for heterogeneous nuclear RNA.

holandric gene a gene on the Y chromosome.

Holliday junction a site where an unresolved crossover has occurred between two homologous chromosomes.

Holliday model a model to explain the molecular mechanism of homologous recombination.

holoenzyme an enzyme containing all of its subunits, as in the holoenzyme of RNA polymerase that has σ factor along with the core enzyme.

homeobox a 180 base pair consensus sequence found in homeotic genes.

homeodomain the protein domain encoded by the homeobox. The homeodomain promotes the binding of the protein to the DNA.

homeologous describes the homologous chromosomes from closely related species.

homeotic an adjective that was originally used to describe mutants in which one body part is replaced by another.

homeotic gene a gene that functions in governing the developmental fate of a particular region of the body.

homoallelic describes two or more alleles in different organisms that are due to mutations at exactly the same base within a gene.

homodimer when two polypeptides encoded by the same gene bind to each other to form a dimer.

homogametic sex in species with two types of sex chromosomes, the homogametic sex produces only one type of gamete. For example, in mammals, the female is the homogametic sex, because an egg can only contain an X chromosome.

homologous in the case of genes, this term describes two genes that are derived from the same ancestral gene. Homologous genes have similar DNA sequences. In the case of chromosomes, the two homologues of a chromosome pair are said to be homologous to each other.

homologous recombination the exchange of DNA segments between homologous chromosomes.

homologous recombination repair also called homology-directed repair, occurs when the DNA strands from a sister chromatid are used to repair a lesion in the other sister chromatid.

homologue one of the chromosomes in a pair of homologous chromosomes.

homology structures that are similar to each other because they evolved from a common ancestor.

homozygous describes a diploid individual who has two identical alleles of a particular gene.

horizontal gene transfer the transfer of genes between different species.

host cell a cell that is infected with a virus or bacterium.

host range the spectrum of host species that a virus or other pathogen can infect.

hot spots sites within a gene that are more likely to be mutated than other locations.

housekeeping gene a gene that encodes a protein required in most cells of a multicellular organism.

***Hox* complexes** a group of several *Hox* genes located in a particular chromosomal region.

***Hox* genes** mammalian genes that play a role in development. They are homologous to homeotic genes in *Drosophila.*

Human Genome Project a worldwide collaborative project that provided a detailed map of the human genome and obtained a complete DNA sequence of the human genome.

hybrid (1) an offspring obtained from a hybridization experiment; (2) a cell produced from a cell fusion experiment in which the two separate nuclei have fused to make a single nucleus.

hybrid cell a cell produced from the fusion of two different cells, usually from two different species.

hybrid dysgenesis a syndrome involving defective *Drosophila* offspring, due to the phenomenon that P elements can transpose freely.

hybridization (1) the mating of two organisms of the same species with different characteristics; (2) the phenomenon in which two single-stranded molecules renature together to form a hybrid molecule.

hybrid vector a cloning vector that contains an insert of foreign DNA.

hybrid vigor see *heterosis.*

hybrid zones during parapatric speciation, places where two populations can interbreed.

hypothesis testing using statistical tests to determine if the data from experimentation are consistent with a hypothesis.

I

illegitimate recombination see *nonhomologous recombination.*

immunoglobulin (IgG) see *antibodies.*

imprinting see *genomic imprinting.*

inborn error of metabolism a genetic disease that involves a defect in a metabolic enzyme.

inbreeding the practice of mating between genetically related individuals.

inbreeding coefficient (F) the probability that two alleles in a particular individual will be identical for a given gene because both copies are due to descent from a common ancestor.

inbreeding depression the phenomenon in which inbreeding produces homozygotes that are less fit, thereby decreasing the reproductive success of a population.

incomplete dominance a pattern of inheritance in which a heterozygote that carries two different alleles exhibits a phenotype that is intermediate to the corresponding homozygous individuals. For example, an *Rr* heterozygote may be pink, while the *RR* and *rr* homozygotes are red and white, respectively.

incomplete penetrance a pattern of inheritance in which a dominant allele does not always control the phenotype of the individual.

incomplete transposable element see *nonautonomous transposable element.*

induced refers to a gene that has been transcriptionally activated by an inducer.

induced mutation a mutation caused by environmental agents.

inducer a small effector molecule that binds to a genetic regulatory protein and thereby increases the rate of transcription.

inducible gene a gene that is regulated by an inducer, which is a small effector molecule that causes transcription to increase.

induction (1) the effects of an inducer on increasing the transcription of a gene; (2) the process by which a cell or group of cells governs the developmental fate of neighboring cells.

infective particle genetic material found within the cytoplasm of eukaryotic cells that differs from the genetic material normally found in cell organelles.

ingroup in cladistics, a group of species in which a researcher is interested.

inhibitor a small effector molecule that binds to an activator protein, causing the protein to be released from the DNA and thereby inhibiting transcription.

initiation (1) in transcription, the stage that involves the initial binding of RNA polymerase to the promoter in order to begin RNA synthesis; (2) in translation, the formation of a complex between mRNA, the initiator tRNA, and the ribosomal subunits.

initiator tRNA during translation, the tRNA that binds to the start codon.

insertion sequences the simplest transposable elements. They are commonly found in bacteria.

***in situ* hybridization** a technique used to cytologically map the locations of genes or other DNA sequences within large eukaryotic chromosomes. In this method, a complementary probe is used to detect the location of a gene within a set of chromosomes.

integrase an enzyme that functions in the integration of viral DNA or retroelements into a chromosome.

interference see *positive interference.*

intergenic region in a chromosome, a region of DNA that lies between two different genes.

intergenic suppressor a suppressor mutation that is in a different gene from the gene that contains the first mutation.

internal nuclear matrix a network of irregular protein fibers and other proteins that is connected to the nuclear lamina and fills the interior of the nucleus.

interphase the series of phases G_1, S, and G_2, during which a cell spends most of its life.

interrupted mating a method used in conjugation experiments in which the length of time that the bacteria spend conjugating is stopped by a blender treatment or other type of harsh agitation.

interstitial deficiency see *interstitial deletion.*

interstitial deletion when an internal segment is lost from a linear chromosome.

intervening sequence also known as an *intron.* A segment of RNA that is removed during RNA splicing.

intragenic mapping see *fine structure mapping.*

intragenic suppressor a suppressor mutation that is within the same gene as the first mutation that it suppresses.

intrinsic termination transcriptional termination that does not require the function of the rho protein.

intron intervening sequences that are found between exons. Introns are spliced out of the RNA prior to translation.

invasive refers to a tumor that can invade surrounding tissue.

inversion a change in the orientation of genetic material along a chromosome such that a segment is flipped in the reverse direction.

inversion heterozygote a diploid individual that carries one normal chromosome and a homologous chromosome with an inversion.

inversion loop the loop structure that is formed when the homologous chromosomes of an inversion heterozygote attempt to align themselves (i.e., synapse) during meiosis.

inverted repeats DNA sequences found in transposable elements that are identical (or very similar) but run in the opposite directions.

iron regulatory protein a translational regulatory protein that recognizes iron response elements that are found in specific mRNAs. It may inhibit translation or stabilize the mRNA.

iron response element an RNA sequence that is recognized by the iron regulatory protein.

isoacceptor tRNAs two different tRNAs that can recognize the same codon.

isoelectric focusing a form of gel electrophoresis in which a protein migrates to the point in the gel where its net charge is zero.

isogamous describes a species that makes morphologically similar gametes.

isolating mechanism a mechanism that favors reproductive isolation. These can be prezygotic or postzygotic.

K

karyotype a photographic representation of all the chromosomes within a cell. It reveals how many chromosomes are found within an actively dividing somatic cell.

kinetochore a group of cellular proteins that attach to the centromere during meiosis and mitosis.

knockin see *gene knockin.*

knockout see *gene knockout.*

Kozac's rules a set of rules that describes the most favorable types of bases that flank a eukaryotic start codon.

L

lac repressor a protein that binds to the operator site of the *lac* operon and inhibits transcription.

lagging strand a strand during DNA replication that is synthesized as short Okazaki fragments in the direction away from the replication fork.

lariat an excised intron structure composed of a circle and a tail.

law of independent assortment see *Mendel's law of independent assortment.*

law of segregation see *Mendel's law of segregation.*

leading strand a strand during DNA replication that is synthesized continuously toward the replication fork.

leptotene the first stage of prophase of meiosis I.

lesion-replicating polymerase a type of DNA polymerase that can replicate over a DNA region that contains an abnormal structure (i.e., a lesion).

lethal allele an allele that may cause the death of an organism.

lethal mutation a mutation that produces a lethal allele that causes the death of an organism.

library see *DNA library.*

ligase see *DNA ligase.*

lineage diagram in developmental biology, a depiction of a cell and all of its descendants that are produced by cell division.

LINEs in mammals, long interspersed elements that are usually 1 to 5 kbp in length and found in 20,000 to 100,000 copies per genome.

linkage refers to the occurrence of two or more genes along the same chromosome.

linkage group a group of genes that are linked together because they are found on the same chromosome.

linkage mapping the mapping of genes or other genetic sequences along a chromosome by analyzing the outcome of crosses.

lipid a general name given to an organic molecule that is insoluble in water. Cell membranes contain a large amount of lipid.

liposome a vesicle that is surrounded by a phospholipid bilayer.

local population a segment of a population that is slightly isolated. Members of a local population are more likely to breed with each other than with members that are outside of the local population.

locus (pl. **loci**) the physical location of a gene within a chromosome.

locus control region (LCR) a segment of DNA that is involved in the regulation of chromatin opening and closing.

locus heterogeneity the phenomenon in which a particular type of disease or trait may be caused by mutations in two or more different genes.

lod score method a method that analyzes pooled data from a large number of pedigrees or crosses to determine the probability that two genetic markers exhibit a certain degree of linkage. A lod score value of >3 or higher is usually accepted as strong evidence that two markers are linked.

long terminal repeats (LTRs) sequences containing many short segments that are tandemly repeated. They are found in retroviruses and viral-like retroelements.

loop domain a segment of chromosomal DNA that is anchored by proteins, so it forms a loop.

loss-of-function allele an allele of a gene that encodes an RNA or protein that is nonfunctional or compromised in function.

loss-of-function mutation a change in a genetic sequence that creates a loss-of-function allele.

low stringency refers to hybridization conditions in which DNA or RNA fragments that have some mismatches are still able to recognize and bind to each other.

LTRs see *long terminal repeats.*

Lyon hypothesis a hypothesis to explain the pattern of X inactivation seen in mammals. Initially, both X chromosomes are active. However, at an early stage of embryonic development, one of the two X chromosomes is randomly inactivated in each somatic cell.

lysis cell breakage.

lysogenic cycle a type of growth cycle for a phage in which the phage integrates its genetic material into the chromosome of the bacterium. This integrated phage DNA can exist in a dormant state for a long time, during which no new bacteriophages are made.

lytic cycle a type of growth cycle for a phage in which the phage directs the synthesis of many copies of the phage genetic material and coat proteins. These components then assemble to make new phages. When synthesis and assembly is completed, the bacterial host cell is lysed, and the newly made phages are released into the environment.

M

macroevolution evolutionary changes at or above the species level involving relatively large changes in form and function that are sufficient to produce new species and higher taxa.

macromolecule a large organic molecule composed of smaller building blocks. Examples include DNA, RNA, proteins, and large carbohydrates.

MADS box a domain found in several transcription factors that play a role in plant development.

maintenance methylation the methylation of hemimethylated DNA following DNA replication.

major groove a wide indentation in the DNA double helix in which the bases have access to water.

malignant describes a tumor composed of cancerous cells.

mammalian cloning experimentally, this refers to the use of somatic cell nuclei and enucleated eggs to create a clone of a mammal.

map distance the relative distance between sites (e.g., genes) along a single chromosome.

mapping the experimental process of determining the relative locations of genes or other segments of DNA along individual chromosomes.

map unit (mu) a unit of map distance obtained from genetic crosses. One map unit is equivalent to 1% recombinant offspring in a testcross.

mass spectrometry a technique to accurately measure the mass of molecule, such as a peptide fragment.

maternal effect an inheritance pattern for certain nuclear genes in which the genotype of the mother directly determines the phenotypic traits of her offspring.

maternal inheritance inheritance of DNA that occurs through the cytoplasm of the egg.

matrix-attachment region (MAR) a site in the chromosomal DNA that is anchored to the nuclear matrix or scaffold.

maturase a protein that enhances the rate of splicing of Group I and II introns.

maximum likelihood a strategy for choosing the best phylogenetic tree based on assumptions about the most likely ways that DNA incurs mutations.

MCM helicase a group of eukaryotic proteins needed to complete a process called DNA replication licensing, which is necessary for the formation of two replication forks at an origin of replication.

mean the sum of all the values in a group divided by the number of individuals in the group.

mean fitness of the population the average fitness of a population that is calculated by considering the frequencies and fitness values for all genotypes.

mediator a protein complex that interacts with RNA polymerase II and various regulatory transcription factors. Depending on its interactions with regulatory transcription factors, mediator may stimulate or inhibit RNA polymerase II.

meiosis a form of nuclear division in which the sorting process results in the production of haploid cells from a diploid cell.

meiotic nondisjunction the event in which chromosomes do not segregate equally during meiosis.

Mendelian inheritance the common pattern of inheritance observed by Mendel, which involves the transmission of eukaryotic genes that are located on the chromosomes found within the cell nucleus.

Mendel's law of independent assortment two different genes will randomly assort their alleles during gamete formation (if they are not linked).

Mendel's law of segregation the two copies of a gene segregate from each other during transmission from parent to offspring.

meristem in plants, an organized group of actively dividing cells.

merozygote a partial diploid strain of bacteria containing F' factor genes.

messenger RNA (mRNA) a type of RNA that contains the information for the synthesis of a polypeptide.

metacentric describes a chromosome with the centromere in the middle.

metaphase the third stage of M phase. The chromosomes align along the center of the spindle apparatus, and the formation of the spindle apparatus is complete.

metaphase plate the plane at which chromosomes align during metaphase.

metastatic describes cancer cells that migrate to other parts of the body.

methylation see *DNA methylation.*

methyl-CpG-binding protein a protein that binds to a CpG island when it is methylated.

methyl-directed mismatch repair a DNA repair system that detects a mismatch and specifically removes the segment from the newly made daughter strand.

methyltransferase see *DNA methyltransferase.*

microarray see *DNA microarray, protein microarray,* or *antibody microarray.*

microevolution changes in the gene pool with regard to particular alleles that occur over the course of many generations.

microinjection the use of microscopic-sized needles or pipettes to inject a substance into cells, such as DNA.

microsatellite short simple sequences (typically a couple hundred base pairs in length) that are interspersed throughout a genome and are quite variable in length among different individuals. They can be amplified by PCR.

microscopy the use of a microscope to view cells or subcellular structures.

minichromosomes a structure formed from many copies of circular DNA molecules.

minisatellite a repetitive sequence that was formerly used in DNA fingerprinting. Its use has been largely superceded by smaller repetitive sequences called microsatellites.

minor groove a narrow indentation in the DNA double helix in which the bases have access to water.

minute a unit of measure in bacterial conjugation experiments. This unit refers to the relative time it takes for genes to first enter a recipient strain during conjugation.

mismatch repair system see *methyl-directed mismatch repair.*

missense mutation a base substitution that leads to a change in the amino acid sequence of the encoded polypeptide.

mitochondrial DNA the DNA found within mitochondria.

mitosis a type of nuclear division into two nuclei, such that each daughter cell will receive the same complement of chromosomes.

mitotic nondisjunction an event in which chromosomes do not segregate equally during mitosis.

mitotic recombination crossing over that occurs during mitosis.

mitotic spindle apparatus (also known as the *mitotic spindle*) the structure that organizes and separates the chromosomes during M phase of the eukaryotic cell cycle.

model organism an organism studied by many scientists so that researchers can more easily compare their results and begin to unravel the properties of a given species.

moderately repetitive sequences sequences that are found a few hundred to several thousand times in the genome.

molecular clock refers to the phenomenon that the rate of neutral mutations can be used as a tool to measure evolutionary time.

molecular evolution the molecular changes in the genetic material that underlie the phenotypic changes associated with evolution.

molecular genetics an examination of DNA structure and function at the molecular level.

molecular marker a segment of DNA that is found at a specific site in the genome and has properties that enable it to be uniquely recognized using molecular tools such as gel electrophoresis.

molecular paleontology the analysis of DNA sequences from extinct species.

molecular pharming a recombinant technology that involves the production of medically important proteins in the mammary glands of livestock.

molecular profiling methods that enable researchers to understand the molecular changes that occur in diseases such as cancer.

molecular trait a trait that is observed at the molecular level. An example would be the amount of a given protein in a cell.

monoallelic expression in the case of imprinting, refers to the phenomenon that only one of the two alleles of a given gene is transcriptionally expressed.

monohybrid an individual produced from a monohybrid cross.

monohybrid cross a cross in which an experimenter is following the outcome of only a single trait.

monomorphic a term used to describe a gene that is found as only one allele in a population.

monophyletic group see *clade.*

monoploid an organism with a single set of chromosomes within its somatic cells.

monosomic a diploid cell that is missing a chromosome (i.e., $2n - 1$).

monozygotic twins twins that are genetically identical to each other because they were formed from the same sperm and egg.

morph a form or phenotype in a population. For example, red eyes and white eyes are different eye color morphs.

morphogen a molecule that conveys positional information and promotes developmental changes.

morphological trait a trait that affects the morphology (physical form) of an organism. An example would be eye color.

mosaicism when the cells of part of an organism differ genetically from the rest of the organism.

motif the name given to a domain or amino acid sequence that functions in a similar manner in many different proteins.

M phase a general name given to nuclear division that can apply to mitosis or meiosis. It is divided into prophase, prometaphase, metaphase, anaphase, and telophase.

mRNA see *messenger RNA.*

mtDNA an abbreviation for mitochondrial DNA.

multinomial expansion equation an equation to solve genetic problems involving three or more types of unordered events.

multiple alleles when the same gene exists in two or more alleles within a population.

multiple sequence alignment an alignment of two or more genetic sequences based on their homology to each other.

multiplication method a method to solve independent assortment problems in which the probabilities of the outcome for each gene are multiplied together.

multipotent a type of stem cell that can differentiate into several different types of cells.

mutable site a site in a chromosome that tends to break at a fairly high rate due to the presence of a transposable element.

mutagen an agent that causes alterations in the structure of DNA.

mutant alleles alleles that have been created by altering a wild-type allele by mutation.

mutation a permanent change in the genetic material that can be passed from cell to cell or from parent to offspring.

mutation frequency the number of mutant genes divided by the total number of genes within the population.

mutation rate the likelihood that a gene will be altered by a new mutation.

myogenic bHLH protein a type of transcription factor involved in muscle cell differentiation.

N

n an abbreviation that designates the number of chromosomes in a set. In humans, $n = 23$, and a diploid cell has $2n = 46$ chromosomes.

narrow sense heritability heritability that takes into account only those genetic factors that are additive.

natural selection refers to the process whereby differential fitness acts on the gene pool. When a mutation creates a new allele that is beneficial, the allele may become prevalent within future generations because the individuals possessing this allele are more likely to reproduce and pass the beneficial allele to their offspring.

natural transformation a natural process of transformation that occurs in certain strains of bacteria.

negative control transcriptional regulation by repressor proteins.

negative frequency-dependent selection a pattern of natural selection in which the fitness of a genotype decreases when its frequency becomes higher.

neutral mutation a mutation that has no detectable effect on protein function and/or no detectable effect on the survival of the organism.

neutral theory of evolution the theory that most genetic variation observed in natural populations is due to the accumulation of neutral mutations.

nick translation the phenomenon in which DNA polymerase uses its 5' to 3' exonuclease activity to remove a region of DNA and, at the same time, replaces it with new DNA.

nitrogen mustard an alkylating agent that can cause mutations in DNA.

nitrous acid a type of chemical mutagen that deaminates bases, thereby changing amino groups to keto groups.

nonautonomous element a transposable element that lacks a gene such as transposase or reverse transcriptase that is necessary for transposition.

noncoding strand the strand of DNA within a structural gene that is complementary to the mRNA. The noncoding strand is used as a template to make mRNA.

noncomplementation the phenomenon in which two mutant alleles in the same organism do not produce a wild-type phenotype.

non-Darwinian evolution see *neutral theory of evolution.*

nondisjunction event in which chromosomes do not segregate properly during mitosis or meiosis.

nonessential genes genes that are not absolutely required for survival, although they are likely to be beneficial to the organism.

nonhomologous end-joining protein (NHEJ) a protein that joins the ends of DNA fragments that are not homologous. This occurs during site-specific recombination of immunoglobulin genes.

nonhomologous recombination the exchange of DNA between nonhomologous segments of chromosomes or plasmids.

nonneutral mutation a mutation that affects the phenotype of the organism and can be acted on by natural selection.

nonparental see *nonrecombinant.*

nonparental ditype (NPD) an ascus that contain cells that all have a nonparental combination of alleles.

nonrecombinant in a testcross, refers to a phenotype or arrangement of alleles on a chromosome that is not found in the parental generation.

nonsense codon a stop codon.

nonsense mutation a mutation that involves a change from a sense codon to a stop codon.

nonviral-like retroelement a type of retroelement in which the sequence does not resemble a modern virus.

normal distribution a distribution for a large sample in which the trait of interest varies in a symmetrical way around an average value.

norm of reaction the effects of environmental variation on an individual's traits.

Northern blotting a technique used to detect a specific RNA within a mixture of many RNA molecules.

N-terminus see *amino terminus.*

nuclear genes genes that are located on chromosomes found in the cell nucleus of eukaryotic cells.

nuclear lamina a collection of fibers that line the inner nuclear membrane.

nuclear matrix (or **nuclear scaffold**) a group of proteins that anchor the loops found in eukaryotic chromosomes.

nucleic acid RNA or DNA. A macromolecule that is composed of repeating nucleotide units.

nucleoid a darkly staining region that contains the genetic material of mitochondria, chloroplasts, or bacteria.

nucleolus a region within the nucleus of eukaryotic cells where the assembly of ribosomal subunits occurs.

nucleoprotein a complex of DNA (or RNA) and protein.

nucleoside structure in which a base is attached to a sugar, but no phosphate is attached to the sugar.

nucleosome the repeating structural unit within eukaryotic chromatin. It is composed of double-stranded DNA wrapped around an octamer of histone proteins.

nucleotide the repeating structural unit of nucleic acids, composed of a sugar, phosphate, and base.

nucleotide excision repair (NER) a DNA repair system in which several nucleotides in the damaged strand are removed from the DNA and the undamaged strand is used as a template to resynthesize a normal strand.

nucleus a membrane-bound organelle in eukaryotic cells where the linear sets of chromosomes are found.

null hypothesis a hypothesis that assumes there is no real difference between the observed and expected values.

O

ocher a stop codon with the sequence UGA.

octad a group of eight fungal spores contained within an ascus.

Okazaki fragments short segments of DNA that are synthesized in the lagging strand during DNA replication.

oligonucleotide a short strand of DNA, typically a few or a few dozen nucleotides in length.

oncogene a mutant gene that promotes cancer.

one gene–one enzyme hypothesis the idea, which later needed to be expanded, that one gene encodes one enzyme.

oogenesis the production of egg cells.

opal a stop codon with the sequence UAA.

open complex the region of separation of two DNA strands produced by RNA polymerase during transcription.

open conformation a loosely packed chromatin structure that is capable of transcription.

open reading frame a genetic sequence that does not contain stop codons.

operator (or **operator site**) a sequence of nucleotides in bacterial DNA that provides a binding site for a genetic regulatory protein.

operon an arrangement in DNA where two or more structural genes are found within a regulatory unit that is under the transcriptional control of a single promoter.

ordered octad an ascus composed of eight cells whose order depends on crossing over during meiosis.

ordered tetrad an ascus composed of four cells whose order depends on crossing over during meiosis.

ORF see *open reading frame.*

organelle a large specialized structure within a cell, which is often surrounded by a single or double membrane.

organismal cloning the cloning of whole organisms. An example is the cloning of the sheep named Dolly.

organism level when the level of observation or experimentation involves a whole organism.

organizing center in plants, a region of the meristem that ensures the proper organization of the meristem and preserves the correct number of actively dividing stem cells.

orientation independent refers to certain types of genetic regulatory elements that can function in the forward or reverse direction. Certain enhancers are orientation independent.

origin of replication a nucleotide sequence that functions as an initiation site for the assembly of several proteins required for DNA replication.

origin of transfer the location on an F factor or within the chromosome of an Hfr strain that is the initiation site for the transfer of DNA from one bacterium to another during conjugation.

origin recognition complex (ORC) a complex of six proteins found in eukaryotes that is necessary to initiate DNA replication.

ortholog homologous genes in different species that were derived from the same ancestral gene.

outbreeding mating between genetically unrelated individuals.

outgroup in cladistics, a species or group of species that is most closely related to the ingroup.

ovary (1) in plants, the structure in which the ovules develop; (2) in animals, the structure that produces egg cells and female hormones.

overdominance an inheritance pattern in which a heterozygote is more vigorous than either of the corresponding homozygotes.

ovule the structure in higher plants where the female gametophyte (i.e., embryo sac) is produced.

ovum a female gamete, also known as an egg cell.

P

p an abbreviation for the short arm of a chromosome.

pachytene the third stage of prophase of meiosis I.

pair-rule gene one category of segmentation genes.

palindromic when a sequence is the same in the forward and reverse direction.

pangenesis an incorrect hypothesis of heredity. It suggested that hereditary traits could be modified depending on the lifestyle of the individual. For example, it was believed that a person who practiced a particular skill would produce offspring that would be better at that skill.

paracentric inversion an inversion in which the centromere is found outside of the inverted region.

paralogs homologous genes within a single species that constitute a gene family.

parapatric speciation (Greek, *para,* "beside"; Latin, *patria,* "homeland") a form of speciation that occurs when members of a species are only partially separated or when a species is very sedentary.

parasegments transient subdivisions that occur in the *Drosophila* embryo prior to the formation of segments.

parental in a testcross, refers to a phenotype or arrangement of alleles on a chromosome that is the same as one or both members of the parental generation.

parental ditype (PD) an ascus that contains four spores with the parental combinations of alleles.

parental generation in a genetic cross, the first generation in the experiment. In Mendel's studies, the parental generation was true-breeding with regard to particular traits.

parental strand in DNA replication, the DNA strand that is used as a template.

parthenogenesis the formation of an individual from an unfertilized egg.

particulate theory of inheritance a theory proposed by Mendel. It states that traits are inherited as discrete units that remain unchanged as they are passed from parent to offspring.

P1 artificial chromosome (PAC) an artificial chromosome developed from P1 bacteriophage chromosomes.

paternal leakage the phenomenon that in species where maternal inheritance is generally observed, the male parent may, on rare occasions, provide mitochondria or chloroplasts to the zygote.

pattern the spatial arrangement of different regions of the body. At the cellular level, the body pattern is due to the arrangement of cells and their specialization.

pattern recognition in bioinformatics, this term refers to a program that recognizes a pattern of symbols.

PCR see *polymerase chain reaction.*

pedigree analysis a genetic analysis using information contained within family trees. In this approach, the aim is to determine the type of inheritance pattern that a gene follows.

pellet a collection of particles found at the bottom of a centrifuge tube.

peptide bond a covalent bond formed between the carboxyl group in one amino acid in a polypeptide chain and the amino group in the next amino acid.

peptidyl site (P site) a site on the ribosome that carries a tRNA along with a polypeptide chain.

peptidyl transfer the step during the elongation stage of translation in which the polypeptide is removed from the tRNA in the P site and transferred to the amino acid at the A site.

peptidyltransferase a complex that functions during translation to catalyze the formation of a peptide bond between the amino acid in the A site of the ribosome and the growing polypeptide chain.

pericentric inversion an inversion in which the centromere is located within the inverted region of the chromosome.

peripheral zone in plants, an area in the meristem that contains dividing cells that will eventually differentiate into plant structures.

petites mutant strains of yeast that form small colonies due to defects in mitochondrial function.

PFGE see *pulsed-field gel electrophoresis.*

P generation the parental generation in a genetic cross.

phage see *bacteriophage.*

pharming see *molecular pharming.*

phenotype the observable traits of an organism.

phenylketonuria (PKU) a human genetic disorder arising from a defect in phenylalanine hydroxylase.

phosphodiester linkage in a DNA or RNA strand, a linkage in which a phosphate group connects two sugar molecules together.

photolyase an enzyme found in yeast and plants that can repair thymine dimers by splitting the dimers, which returns the DNA to its original condition.

phylogenetic species concept a species concept in which the members of a single species are identified by having a unique combination of character states.

phylogenetic tree a diagram that describes the evolutionary relationships among different species.

phylogeny the sequence of events involved in the evolutionary development of a species or group of species.

physical mapping the mapping of genes or other genetic sequences using DNA cloning methods.

physiological trait a trait that affects a cellular or body function. An example would be the rate of glucose metabolism.

PKU see *phenylketonuria.*

plaque a clear zone within a bacterial lawn on a petri plate. It is due to repeated cycles of viral infection and bacterial lysis.

plasmid a general name used to describe circular pieces of DNA that exist independently of the chromosomal DNA. Some plasmids are used as vectors in cloning experiments.

pleiotrophy refers to the multiple effects of a single gene on the phenotype of an organism.

pluripotent a type of stem cell that can differentiate into all or nearly all the types of cells of the adult organism.

point mutation a change in a single base pair within DNA.

polarity in genetics, the phenomenon in which a nonsense mutation in one gene will affect the translation of a downstream gene in an operon.

pollen also, pollen grain; the male gametophyte of flowering plants.

polyA-binding protein a protein that binds to the 3' polyA tail of mRNAs and protects the mRNA from degradation.

polyadenylation the process of attachment of a string of adenine nucleotides to the 3' end of eukaryotic mRNAs.

polyA tail the string of adenine nucleotides at the 3' end of eukaryotic mRNAs.

polycistronic mRNA an mRNA transcribed from an operon that encodes two or more proteins.

polygenic inheritance refers to the transmission of traits that are governed by two or more different genes.

polymerase chain reaction (PCR) the method to amplify a DNA region involving the sequential use of oligonucleotide primers and *Taq* polymerase.

polymerase switch during DNA replication, when one type of DNA polymerase (such as α) is switched for another type (such as β).

polymorphic a term used to describe a trait or gene that is found in two or more forms in a population.

polymorphism (1) the prevalence of two or more phenotypic forms in a population; (2) the phenomenon in which a gene exists in two or more alleles in a population.

polypeptide a sequence of amino acids that is the product of mRNA translation. One or more polypeptides will fold and associate with each other to form a functional protein.

polyploid an organism or cell with three or more sets of chromosomes.

polyribosome an mRNA transcript that has many bound ribosomes in the act of translation.

polysome see *polyribosome.*

polytene chromosome chromosomes that are found in certain cells, such as *Drosophila* salivary cells, in which the chromosomes have replicated many times and the copies lie side by side.

population a group of individuals of the same species that are capable of interbreeding with one another.

population genetics the field of genetics that is primarily concerned with the prevalence of genetic variation within populations.

population level when the level of observation or experimentation involves a population of organisms.

positional cloning a cloning strategy in which a gene is cloned based on its mapped position along a chromosome.

positional information chemical substances and other environmental cues that enable a cell to deduce its position relative to other cells.

position effect a change in phenotype that occurs when the position of a gene is changed from one chromosomal site to a different location.

positive control genetic regulation by activator proteins.

positive interference the phenomenon in which a crossover that occurs in one region of a chromosome decreases the probability that another crossover will occur nearby.

posttranslational describes events that occur after translation is completed.

posttranslational covalent modification the covalent attachment of a molecule to a protein after it has been synthesized via ribosomes.

posttranslational sorting refers to protein sorting to an organelle that occurs after the protein has been completely synthesized in the cytosol.

postzygotic isolating mechanism a mechanism of reproductive isolation that prevents an offspring from being viable or fertile.

preinitiation complex the stage of transcription in which the assembly of RNA polymerase and general transcription factors occurs at the core promoter, but the DNA has not yet started to unwind.

pre-mRNA in eukaryotes, the transcription of structural genes produces a long transcript known as pre-mRNA, which is located within the nucleus. This pre-mRNA is usually altered by splicing and other modifications before it exits the nucleus.

prereplication complex (preRC) in eukaryotes, an assembly of at least 14 different proteins, including a group of six proteins called the origin recognition complex (ORC), that acts as the initiator of eukaryotic DNA replication.

prezygotic isolating mechanism a mechanism for reproductive isolation that prevents the formation of a zygote.

Pribnow box the TATAAT sequence that is often found at the −10 region of bacterial promoters.

primary structure with regard to proteins, the linear sequence of amino acids.

primase an enzyme that synthesizes a short RNA primer during DNA replication.

primitive character a trait that is shared with a distant ancestor.

primosome a multiprotein complex composed of DNA helicase, primase, and several accessory proteins.

principle of parsimony the principle that the preferred hypothesis is the one that is the simplest.

prion an infectious particle that causes several types of neurodegenerative diseases affecting humans and livestock. It is composed entirely of protein.

probability the chance that an event will occur in the future.

processive enzyme an enzyme, such as RNA and DNA polymerase, which glides along the DNA and does not dissociate from the template strand as it catalyzes the covalent attachment of nucleotides.

product rule the probability that two or more independent events will occur is equal to the products of their individual probabilities.

proflavin a type of chemical mutagen that causes frameshift mutations.

prokaryotes (Greek, "prenucleus") another name for bacteria and archaea. The term refers to the fact that their chromosomes are not contained within a separate nucleus of the cell.

prometaphase the second phase of M phase. During this phase, the nuclear membrane vesiculates, and the mitotic spindle is completely formed.

promoter a sequence within a gene that initiates (i.e., promotes) transcription.

promoter bashing the approach of making deletions in the vicinity of a promoter as a way to identify the core promoter and response elements.

proofreading function the ability of DNA polymerase to remove mismatched bases from a newly made strand.

prophage phage DNA that has been integrated into the bacterial chromosome.

prophase the first stage of M phase. The chromosomes have already replicated and begin to condense. The mitotic spindle starts to form.

protease an enzyme that digests the polypeptide backbone found in proteins.

protein a functional unit composed of one or more polypeptides.

protein microarray a small silica, glass, or plastic slide that is dotted with many different proteins.

proteome the collection of all proteins that a given cell or species can make.

proteomics the study of protein function at the genome level. It involves the study of many proteins simultaneously.

proto-oncogene a normal cellular gene that does not cause cancer but which may incur a gain-of-function mutation or become incorporated into a viral genome and thereby lead to cancer.

protoplast a plant cell without a cell wall.

prototroph a strain that does not need a particular nutrient supplemented in its growth medium or diet.

provirus viral DNA that has been incorporated into the chromosome of a host cell.

proximo-distal axis in animals, an axis involving limbs in which the part of the limb attached to the trunk is located proximal, while the end of the limb is located distal.

pseudoautosomal inheritance the inheritance pattern of genes that are found on both the X and Y chromosomes. Even though such genes are located physically on the sex chromosomes, their pattern of inheritance is identical to that of autosomal genes.

pseudodominance a pattern of inheritance that occurs when a single copy of a recessive allele is phenotypically expressed because the second copy of the gene has been deleted from the homologous chromosome.

pseudo-overdominance occurs when two closely linked genes exhibit heterosis. The heterosis is really due to the masking of recessive alleles.

P site see *peptidyl site.*

pulse/chase experiment an experimental strategy in which cells are given a radiolabeled compound for a brief period of time—a pulse—followed by an excess amount of unlabeled compound—a chase.

pulsed-field gel electrophoresis (PFGE) a method of gel electrophoresis used to separate small chromosomes or very large pieces of chromosomes.

punctuated equilibrium an evolutionary theory proposing that species exist relatively unchanged for many generations. These long periods of equilibrium are punctuated by relatively short periods during which evolution occurs at a relatively rapid rate.

Punnett square a diagrammatic method in which the gametes that two parents can produce are aligned next to a square grid as a way to predict the types of offspring the parents will produce and in what proportions.

purine a type of nitrogenous base that has a double-ring structure. Examples are adenine and guanine.

P value in a chi square table, the probability that the deviations between observed and expected values are due to random chance.

pyrimidine a type of nitrogenous base that has a single-ring structure. Examples are cytosine, thymine, and uracil.

Q

q an abbreviation for the long arm of a chromosome.

QTL see *quantitative trait loci.*

QTL mapping the determination of the location of QTLs using mapping methods such as genetic crosses coupled with the analysis of molecular markers.

quantitative genetics the area of genetics concerned with traits that can be described in a quantitative way.

quantitative trait a trait, usually polygenic in nature, that can be described with numbers.

quantitative trait loci (QTLs) the locations on chromosomes where the genes that influence quantitative traits reside.

quaternary structure in proteins, refers to the binding of two or more polypeptides to each other to form a protein.

R

radial loop domains organization of chromatin during interphase into loops, often 25,000 to 200,000 bp in size, which are anchored to the nuclear matrix at matrix-attachment regions.

radioimmunoassay a method to measure the amount of an antigen in a biological sample.

RAG1/RAG2 enzymes that recognize recombination signal sequences and make double-stranded cuts. In the case of V/J recombination in immunoglobulin genes, a cut is made at the end of one V region and the beginning of one J region.

random genetic drift see *genetic drift.*

random mutation theory according to this theory, mutations are a random process—they can occur in any gene and do not involve exposure of an organism to a particular condition that selects for specific types of mutations.

random sampling error the deviation between the observed and expected outcomes due to chance.

reading frame a series of codons beginning with the start codon as a frame of reference.

realized heritability a form of narrow sense heritability that is observed when selective breeding is practiced.

recessive describes a trait or gene that is masked by the presence of a dominant trait or gene.

recessive epistasis a form of epistasis in which an individual must be homozygous for either recessive allele to mask a particular phenotype.

reciprocal crosses a pair of crosses in which the traits of the two parents differ with regard to sex. For example, one cross could be a red-eyed female fly and a white-eyed male fly, and the reciprocal cross would be a red-eyed male fly and a white-eyed female fly.

reciprocal translocation when two different chromosomes exchange pieces.

recombinant (1) refers to combinations of alleles or traits that are not found in the parental generation; (2) describes DNA molecules that are produced by molecular techniques in which segments of DNA are joined to each other in ways that differ from their original arrangement in their native chromosomal sites. The cloning of DNA into vectors is an example.

recombinant DNA molecules molecules that are produced in a test tube by covalently linking DNA fragments from two different sources.

recombinant DNA technology the use of in vitro molecular techniques to isolate and manipulate different pieces of DNA.

recombinant vector a vector that contains an inserted fragment of DNA, such as a gene from a chromosome.

recombination see *genetic recombination.*

recombinational repair DNA repair via homologous recombination.

recombination signal sequence a specific DNA sequence that is involved in site-specific recombination. Such sequences are found in immunoglobulin genes.

redundancy see *gene redundancy.*

regression analysis the approach of using correlated data to predict the outcome of one variable when given the value of a second variable.

regulatory sequence or element a sequence of DNA (or possibly RNA) that binds a regulatory protein and thereby influences gene expression. Bacterial operator sites and eukaryotic enhancers and silencers are examples.

regulatory transcription factor a protein or protein complex that binds to a regulatory element and influences the rate of transcription via RNA polymerase.

relaxosome a protein complex that recognizes the origin of transfer in F factors and other conjugative plasmids, cuts one DNA strand, and aids in the transfer of the T DNA.

release factor a protein that recognizes a stop codon and promotes translational termination and the release of the completed polypeptide.

repetitive sequences DNA sequences that are present in many copies in the genome.

replica plating a technique in which a replica of bacterial colonies is transferred to a new petri plate.

replication see *DNA replication.*

replication fork the region in which two DNA strands have separated and new strands are being synthesized.

replicative transposition a form of transposition in which the end result is that the transposable element remains at its original site and also is found at a new site.

replicative transposon a transposable element that increases in number during the transposition process.

replisome a complex that contains a primosome and dimeric DNA polymerase.

repressible gene a gene that is regulated by a corepressor or inhibitor, which are small effector molecules that cause transcription to decrease.

repressor a regulatory protein that binds to DNA and inhibits transcription.

reproductive isolation the inability of a species to successfully interbreed with other species.

resolution the last stage of homologous recombination, in which the entangled DNA strands become resolved into two separate structures.

resolvase an endonuclease that makes the final cuts in DNA during the resolution phase of recombination.

response element see *regulatory sequence or element.*

restriction endonuclease (or **restriction enzyme**) an endonuclease that cleaves DNA. The restriction enzymes used in cloning experiments bind to specific base sequences and then cleave the DNA backbone at two defined locations, one in each strand.

restriction fragment length polymorphism (RFLP) genetic variation within a population in the lengths of DNA fragments that are produced when chromosomes are digested with particular restriction enzymes.

restriction mapping a technique to determine the locations of restriction endonuclease sites within a segment of DNA.

restriction point a point in the G$_1$ phase of the cell cycle that causes a cell to progress to cell division.

retroelement a type of transposable element that moves via an RNA intermediate.

retroposon see *retroelement*.

retrospective testing a procedure in which a researcher formulates a hypothesis and then collects observations as a way to confirm or refute the hypothesis. This approach relies on the observation and testing of already existing materials, which themselves are the result of past events.

retrotransposition a form of transposition in which the element is transcribed into RNA. The RNA is then used as a template via reverse transcriptase to synthesize a DNA molecule that is integrated into a new region of the genome via integrase.

retrotransposon see *retroelement*.

reverse genetics an experimental strategy in which researchers first identify the wild-type gene using cloning methods. The next step is to make a mutant version of the wild-type gene, introduce it into an organism, and see how the mutant gene affects the phenotype of the organism.

reverse mutation see *reversion*.

reverse transcriptase an enzyme that uses an RNA template to make a complementary strand of DNA.

reverse transcriptase PCR (RT-PCR) a modification of PCR in which the first round of replication involves the use of RNA and reverse transcriptase to make a complementary strand of DNA.

reversion a mutation that returns a mutant allele back to the wild-type allele.

R factor a type of plasmid found commonly in bacteria that confers resistance to a toxic substance such as an antibiotic.

RFLP see *restriction fragment length polymorphism*.

RFLP map a genetic map composed of many RFLP markers.

RFLP mapping the mapping of a gene or other genetic sequence relative to the known locations of RFLPs within a genome.

R group the side chain of an amino acid.

rho (ρ) a protein that is involved in transcriptional termination for certain bacterial genes.

rho-dependent termination transcriptional termination that requires the function of the rho protein.

rho-independent termination transcription termination that does not require the rho protein. It is also known as intrinsic termination.

ribonucleic acid (RNA) a nucleic acid that is composed of ribonucleotides. In living cells, RNA is synthesized via the transcription of DNA.

ribose the sugar found in RNA.

ribosomal-binding site a sequence in bacterial mRNA that is needed to bind to the ribosome and initiate translation.

ribosome a large macromolecular structure that acts as the catalytic site for polypeptide synthesis. The ribosome allows the mRNA and tRNAs to be positioned correctly as the polypeptide is made.

ribozyme an RNA molecule with enzymatic activity.

right-left axis in bilateral animals, this axis determines the two sides of the body relative to the anteroposterior axis.

R loop experimentally, a DNA loop that is formed because RNA is displacing it from its complementary DNA strand.

RNA see *ribonucleic acid*.

RNA editing the process in which a change occurs in the nucleotide sequence of an RNA molecule that involves additions or deletions of particular bases, or a conversion of one type of base to a different type.

RNA-induced silencing complex (RISC) the complex that mediates RNA interference.

RNA interference the phenomenon that double-stranded RNA targets complementary RNAs within the cell for degradation.

RNA polymerase an enzyme that synthesizes a strand of RNA using a DNA strand as a template.

RNA primer a short strand of RNA, made by DNA primase, that is used to elongate a strand of DNA during DNA replication.

RNase an enzyme that cuts the sugar–phosphate backbone in RNA.

RNaseP a bacterial enzyme that is an endonuclease and cuts precursor tRNA molecules. RNaseP is a ribozyme, which means that its catalytic ability is due to the action of RNA.

RNA splicing the process in which pieces of RNA are removed and the remaining pieces are covalently attached to each other.

Robertsonian translocation the structure produced when two telocentric chromosomes fuse at their short arms.

root meristem an actively dividing group of cells that gives rise to root structures.

RT-PCR see *reverse transcriptase PCR*.

S

satellite DNA in a density centrifugation experiment, a peak of DNA that is separated from the majority of the chromosomal DNA. It is usually composed of highly repetitive sequences.

scaffold a collection of proteins that holds the DNA in place and gives chromosomes their characteristic shapes.

scaffold-attachment region (SAR) a site in the chromosomal DNA that is anchored to the nuclear matrix or scaffold.

science a way of knowing about our natural world. The science of genetics allows us to understand how the expression of genes produces the traits of an organism.

scientific method a basis for conducting science. It is a process that scientists typically follow so that they may reach verifiable conclusions about the world in which they live.

scintillation counting a technique that permits a researcher to count the number of radioactive emissions from a sample containing a population of radioisotopes.

search by content in bioinformatics, an approach to predict the location of a gene because the nucleotide content of a particular region differs significantly (due to codon bias) from a random distribution.

search by signal in bioinformatics, an approach that relies on known sequences such as promoters, start and stop codons, and splice sites to help predict whether or not a DNA sequence contains a gene.

secondary structure a regular repeating pattern of molecular structure, such as the DNA double helix or the α helix and β sheet found in proteins.

second-division segregation (SDS) in an ordered octad, a 2:2:2:2 or 2:4:2 arrangement of spores that occurs because two alleles do not segregate until the second meiotic division is completed.

sedimentation coefficient a measure of centrifugation that is normally expressed in Svedberg units (S). A sedimentation coefficient has the units of seconds: $1\ S = 1 \times 10^{-13}$ seconds.

segmentation gene in animals, a gene, whose encoded product is involved in the development of body segments.

segment polarity gene one category of segmentation genes.

segments anatomical subdivisions that occur during the development of species such as *Drosophila*.

segregate when two things are kept in separate locations. For example, homologous chromosomes segregate into different gametes.

selectable marker a gene that provides a selectable phenotype in a cloning experiment. Many selectable markers are genes that confer antibiotic resistance.

selection coefficient one minus the fitness value.

selectionists scientists who oppose the neutral theory of evolution.

selection limit the phenomenon in which several generations of artificial selection results in a plateau where artificial selection is no longer effective.

selective breeding refers to programs and procedures designed to modify the phenotypes in economically important species of plants and animals.

self-fertilization fertilization that involves the union of male and female gametes derived from the same parent.

selfish DNA hypothesis the idea that transposable elements exist because they possess characteristics that allow them to multiply within the host cell DNA and inhabit the host without offering any selective advantage.

self-splicing refers to RNA molecules that can remove their own introns without the aid of other proteins or RNA.

semiconservative model the correct model for DNA replication in which the newly made double-stranded DNA contains one parental strand and one daughter strand.

semiconservative replication refers to the net result of DNA replication in which the DNA contains one original strand and one newly made strand.

semilethal alleles lethal alleles that kill some individuals but not all.

semisterility when an individual has a lowered fertility.

sense codon a codon that encodes a specific amino acid.

sense strand the strand of DNA within a structural gene that has the same sequence as mRNA except that T is found in the DNA instead of U.

sequence complexity the number of times a particular base sequence appears throughout the genome of a given species.

sequence element in genetics, a sequence with a specialized function.

sequence recognition in bioinformatics, this term refers to a program that recognizes particular sequence elements.

sequence-tagged site (STS) a short segment of DNA, usually between 100 and 400 base pairs long, the base sequence of which is found to be unique within the entire genome. Sequence-tagged sites are identified by PCR.

sequencing see *DNA sequencing*.

sequencing ladder a series of bands on a gel that can be followed in order (e.g., from the bottom of the gel to the top of the gel) to determine the base sequence of DNA.

sex chromosomes a pair of chromosomes (e.g., X and Y in mammals) that determines sex in a species.

sex-influenced inheritance an inheritance pattern in which an allele is dominant in one sex but recessive in the opposite sex. In humans, pattern baldness is an example of a sex-influenced trait.

sex-limited inheritance an inheritance pattern in which a trait is found in only one of the two sexes. An example would be beard development in men.

sex-limited traits traits that occur in only one of the two sexes.

sex linkage the phenomenon that certain genes are found on one of the two types of sex chromosomes but not both.

sex-linked gene a gene that is located on one of the sex chromosomes.

sex pilus (pl. **pili**) a structure on the surface of bacterial cells that acts as an attachment site to promote the binding of bacteria to each other. The sex pilus provides a passageway for the movement of DNA during conjugation.

sexual reproduction the process whereby parents make gametes (e.g., sperm and egg) that fuse with each other in the process of fertilization to begin the life of a new organism.

sexual selection natural selection that acts to promote characteristics that give individuals a greater chance of reproducing.

shared derived character a trait shared by a group of organisms but not by a distant common ancestor.

Shine–Dalgarno sequence a sequence in bacterial mRNAs that functions as a ribosomal binding site.

shoot meristem an actively dividing group of cells that gives rise to shoot structures.

short tandem repeat sequences (STRs) short DNA sequences that are repeated many times in a row. Often found in centromeric and telomeric regions.

shotgun sequencing a genome sequencing strategy in which DNA fragments to be sequenced are randomly generated from larger DNA fragments.

shuttle vector a cloning vector that can propagate in two or more different species, such as *E. coli* and yeast.

side chain in an amino acid, the chemical structure that is attached to the carbon atom (i.e., the α carbon) that is located between the amino group and carboxyl group.

sigma factor a transcription factor that recognizes bacterial promoter sequences and facilitates the binding of RNA polymerase to the promoter.

silencer a DNA sequence that functions as a regulatory element. The binding of a regulatory transcription factor to the silencer decreases the level of transcription.

silent mutation a mutation that does not alter the amino acid sequence of the encoded polypeptide even though the nucleotide sequence has changed.

similarity with regard to DNA, refers to a comparison of DNA sequences that have regions where the bases match up. Similarity may be due to homology.

simple Mendelian inheritance an inheritance pattern involving a simple, dominant/recessive relationship that produces observed ratios in the offspring that readily obey Mendel's laws.

simple translocation when one piece of a chromosome becomes attached to a different chromosome.

simple transposition a cut-and-paste mechanism for transposition in which a transposable element is removed from one site and then inserted into another site.

SINEs in mammals, short interspersed elements that are less than 500 base pairs in length.

single-factor cross see *monohybrid cross*.

single-nucleotide polymorphism a genetic polymorphism within a population in which two alleles of the gene differ by a single nucleotide.

single-strand binding protein a protein that binds to both of the single strands of DNA during DNA replication and prevents them from re-forming a double helix.

sister chromatid exchange (SCE) the phenomenon in which crossing over occurs between sister chromatids, thereby exchanging identical genetic material.

sister chromatids pairs of replicated chromosomes that are attached to each other at the centromere. Sister chromatids are genetically identical to each other.

site-directed mutagenesis a technique that enables scientists to change the sequence of cloned DNA segments.

site-specific recombination when two different DNA segments break and rejoin with each other at a specific site. This occurs during the integration of certain viruses into the host chromosome and during the rearrangement of immunoglobulin genes.

SMC proteins proteins that use energy from ATP to catalyze changes in chromosome structure.

snRNP refers to a complex containing small nuclear RNA and a set of proteins, which are components of the spliceosome.

somatic cell refers to any cell of the body except for germ-line cells that give rise to gametes.

somatic mutation a mutation in a somatic cell.

sorting signal an amino acid sequence or posttranslational modification that directs a protein to the correct region of the cell.

SOS response a response to extreme environmental stress in which bacteria replicate their DNA using DNA polymerases that are likely to make mistakes.

Southern blotting a technique used to detect the presence of a particular genetic sequence within a mixture of many chromosomal DNA fragments.

speciation the process by which new species are formed via evolution.

species a group of organisms that maintains a distinctive set of attributes in nature.

species concepts different approaches for distinguishing species.

spectrophotometer a device used by researchers to determine how much radiation at various wavelengths a sample can absorb.

spermatids immature sperm cells produced from spermatocytes.

spermatogenesis the production of sperm cells.

sperm cell a male gamete. Sperm are small and usually travel relatively far distances to reach the female gamete.

spindle see *mitotic spindle apparatus*.

spliceosome a multisubunit complex that functions in the splicing of eukaryotic pre-mRNA.

splicing see *RNA splicing*.

splicing factor a protein that regulates the process of RNA splicing.

spontaneous mutation a change in DNA structure that results from random abnormalities in biological processes.

spores haploid cells that are produced by certain species such as fungi (i.e., yeast and molds).

sporophyte the diploid generation of plants.

SR protein a type of splicing factor.

stabilizing selection natural selection that favors individuals with an intermediate phenotype.

stamen the structure found in the flower of higher plants that produces the male gametophyte (i.e., pollen).

standard deviation a statistic that is computed as the square root of the variance.

start codon a three-base sequence in mRNA that initiates translation. It is usually 5'-AUG-3' and encodes methionine.

stem cell a cell that has the capacity to divide and to differentiate into one or more specific cell types.

steroid receptor a category of transcription factors that respond to steroid hormones. An example is the glucocorticoid receptor.

stigma the structure in flowering plants on which the pollen land and the pollen tube starts to grows so that sperm cells can reach the egg cells.

stop codon a three-base sequence in mRNA that signals the end of translation of a polypeptide. The three stop codons are 5'–UAA–3', 5'–UAG–3', and 5'–UGA–3'.

strain a variety that continues to produce the same characteristic after several generations.

strand in DNA or RNA, nucleotides covalently linked together to form a long, linear polymer.

stripe-specific enhancer in *Drosophila*, a regulatory region that controls the expression of a gene so that it occurs only in a particular parasegment during early embryonic development.

STRs see *short tandem repeat sequences*.

structural gene a gene that encodes the amino acid sequence within a particular polypeptide or protein.

STS see *sequence-tagged site*.

subcloning the procedure of making smaller DNA clones from a larger one.

submetacentric describes a chromosome in which the centromere is slightly off center.

subtractive cDNA library a cDNA library that contains cDNA inserts derived from mRNA that is expressed only under certain conditions.

subtractive hybridization a method used to create a subtractive cDNA library.

subunit this term may have multiple meanings. In a protein, each subunit is a single polypeptide.

sum rule the probability that one of two or more mutually exclusive events will occur is equal to the sum of their individual probabilities.

supercoiling see *DNA supercoiling*.

supergroups a relatively recent way for evolutionary biologists to subdivide the eukaryotic domain.

supernatant following centrifugation, the fluid that is found above the pellet.

suppressor (or **suppressor mutation**) a mutation at a second site that suppresses the phenotypic effects of another mutation.

SWI/SNF family a group of related proteins that catalyze chromatin remodeling.

sympatric speciation (Greek, *sym*, "together"; Latin, *patria*, "homeland") a form of speciation that occurs when members of a species diverge while occupying the same habitat within the same range.

synapomorphy see *shared derived character*.

synapsis the event in which homologous chromosomes recognize each other and then align themselves along their entire lengths.

synaptonemal complex a complex of proteins that promote the interconnection between homologous chromosomes during meiosis.

synonymous codons two different codons that specify the same amino acid.

synteny group a group of genes that are found in the same order on the chromosomes of different species.

systematists biologists who study the evolutionary relationships among different species.

T

T an abbreviation for thymine.

tandem array (or **tandem repeat**) a short nucleotide sequence that is repeated many times in a row.

tandem mass spectrometry the sequential use of two mass spectrometers. It can be used to determine the sequence of amino acids in a polypeptide.

***Taq* polymerase** a thermostable form of DNA polymerase used in PCR experiments.

TATA box a sequence found within eukaryotic core promoters that determines the starting site for transcription. The TATA box is recognized by a TATA-binding protein, which is a component of TFIID.

tautomer refers to the forms of certain small molecules, such as bases, which can spontaneously interconvert between chemically similar forms.

tautomeric shift a change in chemical structure such as an alternation between the keto- and enol-forms of the bases that are found in DNA.

T DNA a segment of DNA found within a Ti plasmid that is transferred from a bacterium to infected plant cells. The T DNA from the Ti plasmid becomes integrated into the chromosomal DNA of the plant cell by recombination.

TE see *transposable element*.

telocentric describes a chromosome with its centromere at one end.

telomerase the enzyme that recognizes telomeric sequences at the ends of eukaryotic chromosomes and synthesizes additional numbers of telomeric repeat sequences.

telomeres specialized DNA sequences found at the ends of linear eukaryotic chromosomes.

telophase the fifth stage of M phase. The chromosomes have reached their respective poles and decondense.

temperate phage a bacteriophage that usually exists in the lysogenic cycle.

temperature-sensitive allele an allele in which the resulting phenotype depends on the environmental temperature.

temperature-sensitive (ts) lethal allele an allele that is lethal at a certain environmental temperature.

template DNA a strand of DNA that is used to synthesize a complementary strand of DNA or RNA.

template strand see *template DNA*.

terminal deficiency see *terminal deletion*.

terminal deletion when a segment is lost from the end of a linear chromosome.

termination (1) in transcription, the release of the newly made RNA transcript and RNA polymerase from the DNA; (2) in translation, the release of the polypeptide and the last tRNA and the disassembly of the ribosomal subunits and mRNA.

termination codon see *stop codon*.

termination sequences (ter sequences) in *E. coli*, a pair of sequences in the chromosome that bind a protein known as the termination utilization substance (Tus), which stops the movement of the replication forks.

terminator a sequence within a gene that signals the end of transcription.

tertiary structure the three-dimensional structure of a macromolecule, such as the tertiary structure of a polypeptide.

testcross an experimental cross between a recessive individual and an individual whose genotype the experimenter wishes to determine.

tetrad (1) the association among four sister chromatids during meiosis; (2) a group of four fungal spores contained within an ascus.

tetraploid having four sets of chromosomes (i.e., 4*n*).

tetratype (T) an ascus that has two parental cells and two nonparental cells.

TFIID a type of general transcription factor in eukaryotes that is needed for RNA polymerase II function. It binds to the TATA box and recruits RNA polymerase II to the core promoter.

thermocycler a device that automates the timing of temperature changes in each cycle of a PCR experiment.

30 nm fiber the association of nucleosomes to form a more compact structure that is 30 nm in diameter.

thymine a pyrimidine base found in DNA. It base pairs with adenine in DNA.

thymine dimer a DNA lesion involving a covalent linkage between two adjacent thymine bases in a DNA strand.

Ti plasmid a tumor-inducing plasmid found in *Agrobacterium tumefaciens*. It is responsible for inducing tumor formation after a plant has been infected.

tissue-specific gene a gene that is highly regulated and is expressed in a particular cell type.

topoisomerase an enzyme that alters the degree of supercoiling in DNA.

topoisomers DNA conformations that differ with regard to supercoiling.

totipotent a cell that possesses the genetic potential to produce an entire individual. A somatic plant cell or a fertilized egg is totipotent.

traffic signal see *sorting signal*.

trait any characteristic that an organism displays. Morphological traits affect the appearance of an organism. Physiological traits affect the ability of an organism to function. A third category of traits are those that affect an organism's behavior (behavioral traits).

***trans*-acting factor** a regulatory protein that binds to a regulatory element in the DNA and exerts a *trans* effect.

transcription the process of synthesizing RNA from a DNA template.

transcriptional start site the site in a gene where transcription begins.

transcription factors a broad category of proteins that influence the ability of RNA polymerase to transcribe DNA into RNA.

transcription-repair coupling factor (TRCF) a protein that recognizes when RNA polymerase is stalled over a damaged region of DNA and recruits DNA repair enzymes to fix the damaged site.

transduction a form of genetic transfer between bacterial cells in which a bacteriophage transfers bacterial DNA from one bacterium to another.

***trans*-effect** an effect on gene expression that occurs even though two DNA segments are not physically adjacent to each other. *Trans*-effects are mediated through diffusible genetic regulatory proteins.

transfection (1) when a viral vector is introduced into a bacterial cell; (2) the introduction of any type of recombinant DNA into a eukaryotic cell.

transfer RNA (tRNA) a type of RNA used in translation that carries an amino acid. The anticodon in tRNA is complementary to a codon in the mRNA.

transformation (1) when a plasmid vector or segment of chromosomal DNA is introduced into a bacterial cell; (2) when a normal cell is converted into a malignant cell.

transgene a gene from one species that is introduced into another species.

transgenic organism an organism that has DNA from another organism incorporated into its genome via recombinant DNA techniques.

transition a point mutation involving a change of a pyrimidine to another pyrimidine (e.g., C to T) or a purine to another purine (e.g., A to G).

translation the synthesis of a polypeptide using the codon information within mRNA.

translational regulatory protein a protein that regulates translation.

translational repressor a protein that binds to mRNA and inhibits its ability to be translated.

translesion synthesis (TLS) the synthesis of DNA over a template strand that harbors some type of DNA damage. This occurs via lesion-replicating polymerases.

translocation (1) when one segment of a chromosome breaks off and becomes attached to a different chromosome; (2) when a ribosome moves from one codon in an mRNA to the next codon.

translocation cross the structure that is formed when the chromosomes of a reciprocal translocation attempt to synapse during meiosis. This structure contains two normal (nontranslocated chromosomes) and two translocated chromosomes. A total of eight chromatids are found within the cross.

transposable element (TE) a small genetic element that can move to multiple locations within the chromosomal DNA.

transposase the enzyme that catalyzes the movement of transposable elements.

transposition the phenomenon of transposon movement.

transposon see *transposable element*.

transposon tagging a technique for cloning genes in which a transposon inserts into a gene and inactivates it. The transposon-tagged gene is then cloned using a complementary transposon as a probe to identify the gene.

transversion a point mutation in which a purine is interchanged with a pyrimidine, or vice versa.

trihybrid cross a cross in which an experimenter follows the outcome of three different traits.

trinucleotide repeat expansion (TNRE) a type of mutation that involves an increase in the number of tandemly repeated trinucleotide sequences.

triplex DNA a double-stranded DNA that has a third strand wound around it to form a triple-stranded structure.

triploid an organism or cell that contains three sets of chromosomes.

trisomic a diploid cell with one extra chromosome (i.e., $2n + 1$).

tRNA see *transfer RNA*.

trp repressor a protein that binds to the operator site of the *trp* operon and inhibits transcription.

true-breeding line a strain of a particular species that continues to produce the same trait after several generations of self-fertilization (in plants) or inbreeding.

tumor-suppressor gene a gene that functions to inhibit cancerous growth.

two-dimensional gel electrophoresis a technique to separate proteins that involves isoelectric focusing in the first dimension and SDS-gel electrophoresis in the second dimension.

two-factor cross see *dihybrid cross*.

U

U an abbreviation for uracil.

unbalanced translocation a translocation in which a cell has too much genetic material compared to a normal cell.

unipotent a type of stem cell that can differentiate into only a single type of cell.

universal in genetics, this term refers to the phenomenon that nearly all organisms use the same genetic code with just a few exceptions.

unordered octad an ascus composed of eight unordered cells.

unordered tetrad an ascus composed of four unordered cells.

up promoter mutation a mutation in a promoter that increases the rate of transcription.

up regulation genetic regulation that leads to an increase in gene expression.

uracil a pyrimidine base found in RNA.

UTR an abbreviation for the untranslated region of mRNA.

U-tube a U-shaped tube that has a filter at bottom of the U. The pore size of the filter allows the passage of small molecules (e.g., DNA molecules) from one side of the tube to the other, but restricts the passage of bacterial cells.

V

variance the sum of the squared deviations from the mean divided by the degrees of freedom.

variants individuals of the same species that exhibit different traits. An example is tall and dwarf pea plants.

V(D)J recombination site-specific recombination that occurs within immunoglobulin genes.

vector a small segment of DNA that is used as a carrier of another segment of DNA. Vectors are used in DNA cloning experiments.

vertical evolution refers to the phenomenon that species evolve from preexisting species by the accumulation of gene mutations and by changes in chromosome structure and number. Vertical evolution involves genetic changes in a series of ancestors that form a lineage.

vertical gene transfer the transfer of genetic material from parents to offspring or from mother cell to daughter cell.

viral genome the genetic material of a virus.

viral-like retroelements retroelements that are evolutionarily related to known retroviruses. These transposable elements have retained the ability to move around the genome, though, in most cases, mature viral particles are not produced.

viral vector a vector used in gene cloning that is derived from a naturally occurring virus.

virulent phage a phage that follows the lytic cycle.

virus a small infectious particle that contains nucleic acid as its genetic material, surrounded by a capsid of proteins. Some viruses also have an envelope consisting of a membrane embedded with spike proteins.

VNTRs also called minisatellites; segments of DNA that are located in several places in a genome and have a variable number of tandem repeats. The pattern of VNTRs was originally used in DNA fingerprinting.

W

Western blotting a technique used to detect a specific protein among a mixture of proteins.

whole genome shotgun sequencing a genome sequencing strategy that bypasses the mapping step. The whole genome is subjected to shotgun sequencing.

wild-type allele an allele that is fairly prevalent in a natural population, generally greater than 1% of the population. For polymorphic genes, there may be more than one wild-type allele.

wobble base the first base (from the 5' end) in an anticodon. This term suggests that the first base in the anticodon can wobble a bit to recognize the third base in the mRNA.

wobble rules rules that govern the binding specificity between the third base in a codon and the first base in an anticodon.

X

X chromosomal controlling element (Xce) a region adjacent to Xic that influences the choice of the active X chromosome during the process of X inactivation.

X inactivation a process in which mammals equalize the expression of X-linked genes by randomly turning off one X chromosome in the somatic cells of females.

X-inactivation center (Xic) a site on the X chromosome that appears to play a critical role in X inactivation.

X-linked genes (alleles) genes (or alleles of genes) that are physically located within the X chromosome.

X-linked inheritance an inheritance pattern in certain species that involves genes that are located only on the X chromosome.

X-linked recessive an allele or trait in which the gene is found on the X chromosome and the allele is recessive relative to a corresponding dominant allele.

Y

YAC see *yeast artificial chromosome*.

yeast artificial chromosome (YAC) a cloning vector propagated in yeast that can reliably contain very large insert fragments of DNA.

Y-linked genes (alleles) genes (or alleles of genes) that are located only on the Y chromosome.

Z

Z DNA a left-handed DNA double helix that is found occasionally in the DNA of living cells.

zygote a cell formed from the union of a sperm and egg.

zygotene the second stage of prophase of meiosis I.

zygotic gene a gene that is expressed after fertilization.

CREDITS

∷

Photographs

Chapter 1
Opener: © Photo courtesy of the College of Veterinary Medicine, Texas A&M University/CORBIS; **1.2(left):** © Roslin Institute; **1.3a(right):** Advanced Cell Technology, Inc., Worcester, Massachusetts; **1.3b:** Photo taken by Flaminia Catteruccia, Jason Benton, and Andrea Crisanti, and assembled by www.luciariccidesign.com; **1.4:** © Biophoto Associates/Photo Researchers; **1.5:** © CNRI/SCIENCE PHOTO LIBRARY/Photo Researchers; **1.8(top left):** © Edmund D. Brodie III; **1.9a(top right):** © Joseph Sohm/ChromoSohm Inc./CORBIS; **1.9b:** © Paul Edmondson/Photodisc Red/Getty Images; **1.10:** © March of Dimes Birth Defects Foundation; **1.13a:** © George Musil/Visuals Unlimited; **1.13b:** © SciMAT/Photo Researchers, Inc.; **1.13c:** © Steve Hopkin/ARDEA LONDON; **1.13d:** © Brad Mogen/Visuals Unlimited; **1.13e:** © Mark Smith/Photo Researchers, Inc.; **1.13f:** © J-M. Labat/Photo Researchers, Inc.; **1.13g:** © Wally Eberhart/Visuals Unlimited.

Chapter 2
Opener: © CHRIS MARTIN BAHR/SCIENCE PHOTO LIBRARY; **2.1:** © SPL/Photo Researchers, Inc.; **2.2:** © Nigel Cattlin/Photo Researchers, Inc.

Chapter 3
Opener: © Photomicrographs by Dr. Conly L. Rieder, Wadsworth Center, Albany, New York 12201-0509; **3.2b(right top):** © Burger/Photo Researchers; **3.2c(right bottom):** © Leonard Lessin/Peter Arnold; **3.6a(left):** © Leonard Lessin/Peter Arnold; **3.6a(right):** © Biophoto Associates/Photo Researchers, Inc.; **3.8a-f:** © Photomicrographs by Dr. Conly L. Rieder, Wadsworth Center, Albany, New York 12201-0509; **3.9a(1):** © Dr. David M. Phillips/Visuals Unlimited; **3.9a(2):** © Carolina Biological Supply Company/Phototake.com; **3.11:** © Diter von Wettstein.

Chapter 4
Opener: © Robert Calentine/Visuals Unlimited; **4.1:** © Blickwinkel/Alamy; **4.5b:** © Bob Shanley/The Palm Beach Post; **4.6a(1):** © PETER WEIMANN/Animals Animals-Earth Scenes; **4.6a(2):** © Tom Walker/Visuals Unlimited; **4.6b:** © Sally Haugen/Virginia Schuett, www.pkunews.org; **4.6c:** © PHOTOTAKE Inc./Alamy; **4.7a(left), 4.7b(right):** © Stan Flegler/Visuals Unlimited; **4.9b:** © John T. Fowler/Alamy; **4.9c:** © Wegner, P./Peter Arnold, Inc.; **4.9d:** © Gary Randall/Visuals Unlimited; **4.9a(top left):** © Zig Leszczynski/AnimalsAnimals-Earth Scenes; **4.10(top right):** © Alan & Sandy Carey/Photo Researchers; **4.13a:** © AP Images; **4.15a(top left), 4.15b(top right), 4.15c(bottom left):** National Parks Service, Adams National Historical Park; **4.15d(bottom right):** © Bettmann/CORBIS; **4.17a(left), 4.17b(right):** © Robert Maier/AnimalsAnimals-Earth Scenes; **4.18a:** © Jane Burton/naturepl.com.

Chapter 5
Opener: © Carolina Biological/Visuals Unlimited.

Chapter 6
Opener: © David Scharf/Peter Arnold; **6.4b:** © Dr. L. Caro/Science Photo Library/Photo Researchers; **6.15a(left), 6.15b(right):** © Carolina Biological Supply/Phototake.

Chapter 7
Opener: © John Mendenhall, Institute for Cellular and Molecular Biology, University of Texas at Austin; **7.3a(1), 7.3a(2):** Courtesy of I. Solovei, University of Munich (LMU); **7.6:** © G.W. Willis, MD/Visuals Unlimited; **TA 7.1:** Ronald G. Davidson, Harold M. Nitowsky, and Barton Childs. "Demonstration of Two Populations of Cells in the Human Female Heterozygous for Glucose-6-Phosphate Dehydrogenase Variants." *PNAS* 50 (1963) f. 2, p. 484. Courtesy Harold M. Nitowsky; **7.9:** Courtesy of Dr. Argiris Efstratiadis; **7.13a(left):** Reproduced with permission from *The Journal of Cell Biology*, 1977, 72:687-694. Copyright 1977 The Rockefeller University Press; **7.13b(right):** Gibbs, SP., Mak, R., Ng, R., & Slankis, T. "The chloroplast nucleoid in Ochromonas danica, II. Evidence for an increase in plastid DNA during greening." *J Cell Sci.* 1974 Dec;16(3):579-91. Fig. 1. By permission of the Company of Biologists Limited.

Chapter 8
Opener: © BSIP/Phototake, Inc.; **8.1a(top left):** © Scott Camazine/Photo Researchers; **8.1a(center):** © Michael Abbey/Photo Researchers; **8.1a(top right):** © Carlos R Carvalho/Universidade Federal de Viçosa; **8.1c(bottom left):** © C.N.R.I./Phototake; **8.4a(left):** © Biophoto Associates/Science Source/Photo Researchers; **8.4b(right):** © Jeff Noneley; **8.12b(left):** © Paul Benke/University of Miami School of Medicine; **8.12c(right):** © Will Hart/PhotoEdit; **8.18a(top), 8.18b(bottom):** © A. B. Sheldon; **8.19b:** © David M. Phillips/Visuals Unlimited; **8.20a(left):** © James Steinberg/Photo Researchers; **8.20a(top right), 8.20b(bottom right):** © Biophoto Associates/Science Source/Photo Researchers; **8.26:** Robinson, T.J. & Harley, E.H. "Absence of geographic chromosomal variation in the roan and sable antelope and the cytogenetics of a naturally occurring hybrid." *Cytogenet Cell Genet.* 1995; 71(4):363-9. Permission granted by S. Karger AG, Basel.

Chapter 9
Opener: © Ken Eward/Photo Researchers; **9.4(left):** © Omikron/Photo Researchers; **9.12b(left):** © Pictorial Parade/Getty Images; **9.16a(left):** © Barrington Brown/Photo Researchers; **9.16b(right):** © Hulton Archive by Getty Images; **9.18b:** © Michael Freeman/Phototake; **9.24b:** © Alfred Pasieka/Photo Researchers, Inc.

Chapter 10
Opener: © Dr. Gopal Murti/Visuals Unlimited; **10.3:** ©American Society for Microbiology; **10.10b(top):** © Simpson's Nature Photography; **10.10c(bottom):** © William Leonard; **10.13c:** Image courtesy of Timothy J. Richmond. Figure 1 from" Crystal structure of the nucleosome core particle at 2.8 Å resolution." *Nature* 1997 Sept 18; 389(6648): 251-60. Reprinted by permission from Macmillan Publishers Ltd.; **TA 10.14:** Reprinted by permission from Macmillan Publishers Ltd. *Nature* Subunit structure of chromatin. Markus Noll. 251: 5472, 249-251. 1974.; **10.17b, 10.17c:** Nickerson et al. "The nuclear matrix revealed by eluting chromatin from a cross-linked nucleus." *PNAS* 94: 4446-4450. Figure 2a &b. © 1997 National Academy of Sciences, U.S.A.; **10.18a(left), 10.18b(right):** Reprinted by permission from Macmillan Publishers Ltd. *Nature Reviews/Genetics.* Chromosome territories, nuclear architecture and gene regulation in mammalian cells. Cremer, T. & Cremer, C., 2: 4, 292-301, 2001; **10.21a(2nm):** © Dr. Gopal Murit/Visuals Unlimited; **10.21a(11nm):** © Olins and Olins/Biological Photo Service; **10.21b(30nm):** Jerome Rattner/University of Calgary; **10.21c(300nm):** This article was published in *Cell.* Nov. 12(3). Paulson, JR. & Laemmli, UK. "The structure of histone-depleted metaphase chromosomes." 817-2 8. f. 5. Copyright Elsevier, 1977; **10.21d(700nm), 10.21d(1400nm), 10.22a:** © Peter Engelhardt/Department of Virology, Haartman Institute.; **10.22b:** © Dr. Donald Fawcett/Visuals Unlimited.

Chapter 11
Opener: © Clive Freeman, The Royal Institution/Photo Researchers; **TA 11.1:** Meselson M, Stahl, F. "The Replication of DNA in Escherichia Coli." *PNAS* Vol. 44, 1958. f. 4a, p. 673. Courtesy of M. Meselson; **11.4b:** From *Cold Spring Harbor Symposia of Quantitative Biology*, 28, p. 43 (1963). Copyright holder is Cold Spring Habour Laboratory Press; **11.8b:** Reprinted by permission from Macmillan Publishers Ltd. *The Embo Journal.* "Crystal structures of open and closed forms of binary and ternary complexes of the large fragment of Thermus aquaticus DNA polymerase I: structural basis for nucleotide incorporation." Ying Li et al. 17: 24, 7514-7525, 1998.; **11.20:** This article was published in *Journal of Molecular Biology.* Mar 14; 32(2). Huberman JA. Riggs AD. "Links On the mechanism of DNA replication in mammalian chromosomes":327-41. Copyright Elsevier, 1968; **11.21b:** Henry J. Kriegstein and David S. Hogness "Mechanism of DNA Replication in Drosophila Chromosomes: Structure of Replication Forks and Evidence for Bidirectionality." *PNAS.* January 1, 1974. vol. 71, no. 1. 135-139. Permission courtesy D. Hogness.

Chapter 12
Opener: From Patrick Cramer, David A. Bushnell, Roger D. Kornberg. "Structural Basis of Transcription: RNA Polymerase II at 2.8 Angstrom Resolution." *Science,*

Vol. 292:5523, 1863-1876, June 8, 2001. Reprinted with permission of AAAS; **TA 12.1:** Tilghman et al. "Intervening sequence of DNA identified in the structural protein of a mouse ß-glovin gene." *PNAS* 1978, Vol 75. f. 2, p. 727. This image is in the public domain; **12.12a(left):** From Seth Darst, Bacterial RNA polymerase. *Current Opinion* in Structural Biology; **12.12b(right):** Image and permission courtesy of David Bushnell.

Chapter 13
Opener: © Tom Pantages; **13.16a:** © E. Kiseleva and Donald Fawcett/Visuals Unlimited; **13.16b:** © Tom Pantages; **13.22:** Miller, O. L. *Scientific American.* Vol 228:3, 1973. p.35. Image courtesy O.L. Miller.

Chapter 14
14.10b: © Mitchell Lewis, University of Pennsylvania Medical Center/Science Photo Library/Photo Researchers.

Chapter 15
Opener: Zif268 zinc fingers bound to DNA (Pavletich and Pabo, 1991). Courtesy Song Tan, Penn State University; **15.8(left):** Courtesy of Dr. Joseph G. Gall; **15.8(right):** Courtesy Dr. Oscar L. Miller; **15.22(a), 15.22(b), 15.22(c):** Reprinted by permission from Macmillan Publishers Ltd. *Nature.* Andrew Fire et al. Potent and specific genetic interference by double-stranded RNA in Caenorhabditis elegans, 391: 6669. 806-811. 1998.

Chapter 16
Opener: © Robert Brooker; **16.1(left), 16.1(right):** © PHOTOTAKE Inc./Alamy; **16.3a, 16.3b:** Aulner et al. "The AT-Hook Protein D1 Is Essential for Drosophila melanogaster Development and Is Implicated in Position-Effect Variegation." Molecular and Cellular Biology. 22: 4. 1218-1232, February 2002. Figure. 7. © 2002 American Society for Microbiology and E. Kas; **16.5:** ©Scott Aitken, www.scottpix.com; **16.20:** © Dr. Kenneth Greer/Visuals Unlimited.

Chapter 17
Opener: © Matt Meadows/Peter Arnold Inc.; **17.3(the data):** Reprinted by permission from Macmillan Publishers Ltd. *Nature.* New Giemsa method for the differential staining of sister chromatids. Perry, P. & Wolff, S. 251:5471, 156-158, 1974; **17.4b:** © Dr. John D. Cunningham/Visuals Unlimited.

Chapter 18
Opener: © Argus Fotoarchive/Peter Arnold Inc.; **TA18.4:** Cohen, Stanely et al."Construction of Biologically Functional Bacterial Plasmids In Vitro." *PNAS* 1973. Vol. 40. Figure 4. p. 3242. © Stanley Cohen.

Chapter 19
Opener: © Najlah Feanny/SABA/CORBIS; **19.3:** © M. Greenlar/Image Works; **19.4:** R. L. Brinster and R. E. Hammer, School of Veterinary Medicine, University of Pennsylvania; **19.6:** Glofish © www.glofish.com; **19.7:** © Yoav Levy/Phototake; **19.11:** Photo courtesy of the College of Veterinary Medicine, Texas A&M University/CORBIS; **19.15:** © Jack Bostrack/Visuals Unlimited; **19.17:** Courtesy Monsanto; **19.18:** © Richard T. Nowitz/Phototake; **19.19a:** © Bill Barksdale/AGStockUSA.

Chapter 20
Opener: © Dr. Peter Lansdorp/Visuals Unlimited; **20.4:** From Ried, T., Baldini, A., Rand, T.C., and Ward, D.C. "Simultaneous visualization of seven different DNA probes by in situ hybridization using combinatorial fluorescence and digital imaging microscopy. PNAS 89: 4.1388-92. 1992. Courtesy Thomas Ried; **20.19a:** This image was published in *Pulsed Field Gel* Electrophoresis, A Practical Guide. Bruce Birren and Eric Lai. Figure .1, p. 109. Copyright Elsevier 1993; **20.19b:** From Peter D. Butler and E. Richard Moxon. "A physical map of the genome of Haemophilus influenzae type b." *J Gen Microbiol.* 1990 Dec; 136 (Pt 12):2333-42. Reprinted by permission of the Society for General Microbiology.

Chapter 21
Opener: © Alfred Pasieka/Photo Researchers; **TA21.1:** Courtesy Joseph DeRisi; **21.5b:** © Medical School, University of Newcastle upon Tyne/Simon Fraser/Photo Researchers.

Chapter 22
Opener: © Royalty-Free/CORBIS; **22.1(top):** © Dr. P. Marazzi/Photo Researchers; **22.1(middle):** © Hiroya Minakuchi/Seapics.com; **22.1(bottom):** © Mitch Reardon/Photo Researchers; **22.4:** © Hulton-Deutsch Collection/Corbis; **22.6:** © CNRI/SCIENCE PHOTO LIBRARY/Photo Researchers; **22.9:** Reprinted by permission from Macmillan Publishers Ltd. *Nature.* Transforming activity of DNA of chemically transformed and normal cells. Authors: Geoffrey M. Cooper, S. Okenquist & L. Silverman. 284: 5755, 418-421, 1980; **22.18(top), 22.18(bottom):** Dr. Ruhong Li, Molecular and Cell Biology, University of California at Berkeley. Permission granted by The Duesberg Laboratory. Appeared in *Science*, Vol. 297, Number 5581, Issue 26, Jul 2002, p. 544. Author: Jean Marx, Title: "Debate Surges Over the Origins of Genomic Defects in Cancer"; **22.19a:** Reprinted by

permission from Macmillan Publishers Ltd. *Nature.* Distinct types of Diffuse large B-cell lymphoma Identified by Gene Expression Profiling. Ash Alizadeh et al. 403:6769, 503-511, 2000. Image courtesy Ash Alizadeh.

Chapter 23
Opener: © Herman Eisenbeiss/Photo Researchers, Inc.; **TA 23.1, TA 23.2:** Horvitz, H.R. & Sulston, J. (1980) Isolation and genetic characterization of cell lineage mutants of the nematode caenorhabditis elegans. *Genetics* 96, 435-454. fig. 1a&b (1980). Courtesy Dr. Horvitz; **23.3:** Courtesy of E. B. Lewis, California Institute of Technology; **TA 23.3:** Courtesy Stephen Small/New York University; **23.7b, 23.7c, 23.10(1):** Christiane Nusslein-Volhard, *Development,* Supplement 1, 1991. © The Company of Biologists Limited; **23.10(2). 23.10(3), 23.10(4):** Jim Langeland, Steve Paddock and Sean Carroll/University of Wisconsin-Madison; **23.12a:** © Juergen Berger/Photo Researchers; **23.12b:** © F. R. Turner, Indiana University/Visuals Unlimited; **23.17:** © Photodisc/Object Series/Vol. 50; **23.20:** © Jeremy Burgess/Photo Researchers; **23.23a,c:** Elliott Meyerowitz and John Bowman. Development. Vol. 112:1-20, 1991. Courtesy of Elliott Meyerowitz; **23.23b:** Courtesy of Elliott Meyerowitz.

Chapter 24
Opener: © PhotoDisc website; **24.1:** © Color-Pic/Animals Animals-Earth Scenes; **24.2(1), 24.2(2), 24.2(3):** © Geoff Oxford; **24.7b:** © Corbis/RF website; **24.13a:** © OSF/Animals Animals-Earth Scenes; **24.15:** © Paul Harcourt Davies/SPL/Photo Researchers, Inc; **24.16(left):** © D. Parer & E .Parer-Cook/ardea.com; **24.21:** Leonard Lessin/Peter Arnold Inc.; **24.26(right):** © Gerald & Buff Corsi/Visuals Unlimited.

Chapter 25
Opener: © PhotoDisc/Vol. 74; **25.1a:** Image originally appeared in: Albert and Blakeslee, "Corn and Man," *Journal of Heredity.* 1914, Vol. 5, pg. 51. Oxford University Press. This image is in the public domain; **25.3a(1):** © GRANT HEILMAN /Grant Heilman Photography; **25.3a(2):** © Nigel Cattlin/Visuals Unlimited; **25.11(top left):** © Henry Ausloos/Animals Animals-Earth Scenes; **25.11(top right):** © Roger Tidman/Corbis; **25.11(bottom left):** © Philip Gould/Corbis; **25.11(bottom right):** © Juniors Bildarchiv/Alamy; **25.12(1):** © Inga Spence/Visuals Unlimited; **25.12(2):** © Nigel Cattlin/Photo Researchers Inc.; **25.12(3):** © Valerie Giles/ Photo Researchers Inc.; **25.12(4-7):** © Michael P. Gadomski/Photo Researchers Inc.

Chapter 26
Opener: © LWA/Stephen Welstead/Blend Images/Corbis Royalty-Free; **26.2(left):** © Paul & Joyce Berquist/Animals Animals-Earth Scenes; **26.2(right):** © Gerald and Buff Corsi/Visuals Unlimited; **26.20a(left), 26.20a(right):** Courtesy Ed Laufer; **26.20b(left), 26.20b(right), 26.20c(left), 26.20c(right):** Courtesy of Dr. J.M. Hurle (Originally published in *Development.* 1999 Dec; 126 (23):5515-22.); **26.22a, 26.22b:** © Prof. Walter J. Gehring, University of Basel.

Appendixes
1a: © Michael Gabridge/Visuals Unlimited; **1b:** © Fred Hossler/Visuals Unlimited; **8:** © Richard Wehr/Custom Medical Stock Photo.

Line Art

Chapter 1
1.1a, b: Courtesy of the U.S. Department of Energy Genome Programs. http://genomics.energy.gov.

Chapter 9
9.19: Illustration, Irving Geis. Rights owned by Howard Hughes Medical Institute. Not to be reproduced without permission.

Chapter 12
12.12: Reprinted from Structural Biology, vol. 11, Darst, "Current Opinion," page 157, Copyright © 2001, with permission from Elsevier.

Chapter 14
14.10: The binding of the lac repressor to two operator sites, from M. Lewis et al., *Science,* vol. 271, pp. 1247-1254, March 1, 1996. Reprinted by permission.

Chapter 20
20.1: A complete map of the bacterium Hemophilus influenzae, from R. D. Fleishman, *Science,* vol. 269, p.28, 1995. Reprinted by permission.

Chapter 22
22.19a,b: Reprinted by permission from Macmillan Publishers Ltd: Nature, Alizadeh et al., February 2000, vol. 403, pp. 503-511, Copyright 2000.

Chapter 26
26.12: FASEB Journal by Olsen and Woesl. Copyright 1993 by Fedn of Am Societies for Experimental Bio (FASEB). Reproduced with permission of Fedn of Am Societies for Experimental Bio (FASEB) in the format Textbook via Copyright Clearance Center.

INDEX

::